CHILTON'S GUIDE TO
EMISSION DIAGNOSIS TUNE-UP and VACUUM DIAGRAMS

Vice President & General Manager John P. Kushnerick
Executive Editor Kerry A. Freeman, S.A.E.
Senior Editor Richard J. Rivele, S.A.E.

CHILTON BOOK COMPANY
Chilton Way, Radnor, PA 19089

Manufactured in USA

ISBN 0-8019-7649-9

1234567890 4321098765

Index

Page

SAFETY NOTICE

Proper service and repair procedures are vital to the safe, reliable operation of all motor vehicles, as well as the personal safety of those performing repairs. This manual outlines procedures for servicing and repairing vehicles using safe effective methods. The procedures contain many NOTES, CAUTIONS and WARNINGS which should be followed along with standard safety procedures to eliminate the possibility of personal injury or improper service which could damage the vehicle or compromise its safety.

It is important to note that repair procedures and techniques, tools and parts for servicing motor vehicles, as well as the skill and experience of the individual performing the work vary widely. It is not possible to anticipate all of the conceivable ways or conditions under which vehicles may be serviced, or to provide cautions as to all of the possible hazards that may result. Standard and accepted safety precautions and equipment should be used when handling toxic or flammable fluids, and safety goggles or other protection should be used during cutting, grinding, chiseling, prying, or any other process that can cause material removal or projectiles.

Some procedures require the use of tools specially designed for a specific purpose. Before substituting another tool or procedure, you must be completely satisfied that neither your personal safety, nor the performance of the vehicle will be endangered.

Although information in this manual is based on industry sources and is as complete as possible at the time of publication, the possibility exists that some car manufacturers made later changes which could not be included here. While striving for total accuracy, Chilton Book Company cannot assume responsibility for any errors, changes, or omissions that may occur in the compilation of this data.

CAUSES OF EMISSION INSPECTION FAILURES

When a customer drives into your shop and hands you a piece of paper that shows his car has failed an emission inspection, it doesn't mean you are in for a lot of trouble. If your infra-red exhaust emission tester is calibrated frequently, and checked often against a known gas, then you have nothing to fear. The first thing to do is hook up your infra-red and do a normal idle CO and HC check to see if the car has high emissions. Don't forget to disconnect the air pump and plug the opening in the exhaust manifold to keep air from being drawn in.

If your infra-red does not show high emissions, or in other words does not agree with the inspection report the customer has brought in, the best thing you can do is send the car back to the inspection lane for a retest. If you attempt to reduce the emissions from the car, when your equipment already shows there is nothing wrong, then you are asking for trouble.

But in most cases (unless somebody has worked on the car before you see it) a car that has failed an emission inspection will also fail, or show high emissions on your shop equipment. If it does, here are some of the reasons for the high emission readings.

High CO

Imagine that the engine is divided into two areas or sections. One area is the intake system, from the fresh air tube, through the air cleaner, the carburetor, and ending at the carburetor flange. This area is what is responsible for high carbon monoxide (CO). There is only one way that an engine can put out too much CO. It has to burn too much fuel for the amount of air. This is the same as saying that the mixture is too rich. Anything that makes the mixture too rich will cause high CO. Here is a list of some specific causes in this area.

1. Restricted air cleaner or intake tube .
2. Dirty air filter.
3. Rich choke setting.
4. Choke stuck closed.
5. High fuel level in the carburetor bowl.
6. Dirty air bleeds in the carburetor.
7. Drilled out jets or improperly adjusted metering rods (tampering).
8. Excessive blowby which feeds a rich mixture into the engine through the PCV valve.
9. Idle mixture screws adjusted for performance instead of emissions.

As you can see, anything that richens the mixture will cause high CO. The biggest cause of high CO is incorrect adjustment of the idle mixture screws. Chrysler Corporation cars must be adjusted with an infra-red analyzer. Other cars are adjusted with an analyzer, or by the speed drop method. Almost any engine will have high CO and fail a state inspection if you adjust by the sound of the engine.

High HC

The second area that we divide the engine into consists of everything from the base of the carburetor (which overlaps the first area) into the engine, including the intake manifold, combustion chamber, valves, rings, and camshaft. Hydrocarbons come only from unburned fuel, so when we say that an engine has high HC, it means simply that there is unburned fuel coming out the exhaust, usually in the form of fuel vapor. It's a little harder to understand the reasons for high HC, because some of them involve engine design. It may be easier if you remember that an engine was designed to burn up all its fuel in the combustion chamber. If anything in the engine wears out, it can easily result in some of the fuel passing out of the combustion chamber without being burned. Here are some specific causes.

1. Leaking exhaust valves. Unburned fuel is forced out through the leaking valve into the exhaust.
2. Idle mixture screws adjusted lean. This causes lean misfire at idle.
3. Misfiring plugs. Fuel passes into the exhaust without being burned.
4. Incorrect spark timing. Too much spark advance lets the exhaust system cool off, and the HC does not burn up in the exhaust. This can be caused not only by incorrect initial timing, but also by faulty advance mechanisms.
5. Lean carburetion. If the carburetor is too lean at part throttle, it will cause lean misfire.
6. Worn camshaft lobes. If an intake cam lobe is worn, it does not allow as much charge into the cylinder, which reduces the pressure in the cylinder and the fuel doesn't burn up. The worn lobe won't necessarily show up in the way the engine runs, but can be easily spotted by removing the rocker cover and checking the lift at the pushrod.
7. Vacuum leaks. Too much air leans the mixture and causes lean misfire.
8. Worn piston rings. If the rings are worn enough, the pressure in the cylinder will be low, and the fuel doesn't burn up.

High NOx

Very few states are now testing for nitrogen oxides (NOx), but as more states begin inspection programs, they will probably include NOx sniffing in their inspection

NOx is controlled mainly by the exhaust gas recirculation system. Too much spark advance can increase NOx, but usually the cause of an NOx failure is that the EGR system is not working at all. Normal testing and checking of the EGR system will usually show that an EGR passageway is blocked, or that the EGR valve is not getting vacuum.

NOx sniffing is not possible in the field, because there are no exhaust analyzers that will detect it. Until such analyzers are made, most of the field testing will have to be with an infra-red, concentrating on HC and CO.

With a quality infra-red analyzer, frequently calibrated and checked often against a known gas, there is no reason why you can't tackle emission inspection failures with the same confidence you go after any other automotive problem.

HOW TO IDENTIFY A CALIFORNIA CAR

California cars are usually made with different emission control equipment, and adjusted to different specifications, such as idle speed and timing. There is only one positive way of identifying a California car. If the engine decal definitely states that the car conforms to California standards or regulations, then it is a "California car." If the decal says nothing about California, or states that the car conforms, except for California, then it is a "49-State car."

Some people think that the location where a car was assembled indicates whether it is a California or 49-State car. This is not true in all cases. It is also not true that a car first sold outside of California has to be a 49-State car. Cars built to California emission standards can be sold anywhere in the United States, and frequently are. The California emission hardware is usually listed on dealer order forms as an option group. All the dealer has to do is tell the factory, and they will build the car to California emission standard's, no matter where the dealer happens to be. Sometimes a dealer has no control over the car he gets. It may be easier for the factory to send an out-of-California dealer California cars, even though he didn't order them that way. Also, dealers frequently send cars to other dealers, and there is nothing to prevent a California dealer from sending surplus cars to another state. It does not work the other way, however. 49-State cars can not be sold in California, although they are welcome as used cars, after they have been first sold in another state.

VEHICLE EMISSION CONTROL INFORMATION

ENGINE = Cu. In. FAMILY

This vehicle conforms to U.S. Environmental Protection Agency regulations applicable to model year new motor vehicles, with the exception of the carbon monoxide standard for California.

Chrysler Corporation label on 49-State car

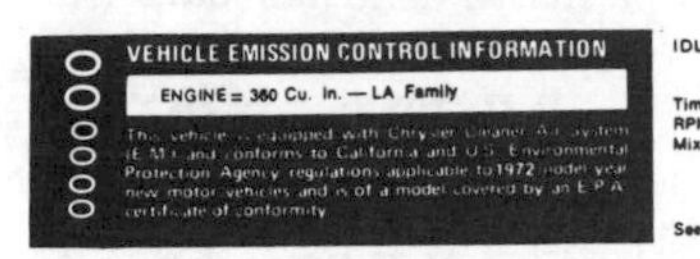

VEHICLE EMISSION CONTROL INFORMATION

ENGINE = 360 Cu. In. — LA Family

This vehicle is equipped with Chrysler Cleaner Air System (E.M.) and conforms to California and U.S. Environmental Protection Agency regulations applicable to 1972 model year new motor vehicles and is of a model covered by an E.P.A. certificate of conformity.

IDLE SETTINGS

Timing = 00
RPM = 000 *
Mixture = See Procedure

Make all adjustments with Engine fully warmed up, Transmission in neutral, Headlights off and Air Conditioning compressor not operating.

When checking dwell or timing remove hose at distributor and plug hose. Timing may be set within + 00 – 0° to compensate for detonation or altitude.

Dwell values are not applicable when electronic ignition is used.

*NOTE: On a new vehicle (under 00 miles) idle speed setting should be reduced 0 rpm.

See Service Manual for detailed instructions.

MIXTURE SETTING–PREFERRED PROCEDURE Use exhaust analyzer. Adjust idle speed screw and mixture screws to give 0.0% carbon monoxide (14.3 A/F ratio) with lowest hydrocarbon level or smoothest idle at 0 RPM.* *See note

MIXTURE SETTING–ALTERNATE PROCEDURE (Speed Drop Method).
1. Remove idle limiter caps.
2. Adjust idle speed screw and mixture screws to lean best idle at 000 RPM.*
3. Without touching speed screw, turn mixture screws clockwise (lean) equally to reduce speed to 000 RPM.*
4. Reinstall limiter caps with stop ¼ turn from rich limit. *See note

CHRYSLER CORPORATION

The complete Chrysler Corporation label has tune-up information

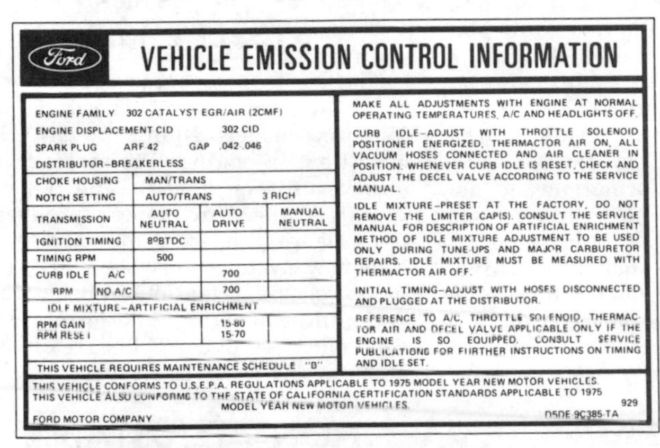

VEHICLE EMISSION CONTROL INFORMATION

ENGINE FAMILY 302 CATALYST EGR/AIR (2CMF)
ENGINE DISPLACEMENT CID 302 CID
SPARK PLUG ARF 42 GAP .042-.046
DISTRIBUTOR–BREAKERLESS

CHOKE HOUSING		MAN/TRANS		
NOTCH SETTING		AUTO/TRANS		3 RICH
TRANSMISSION		AUTO NEUTRAL	AUTO DRIVE	MANUAL NEUTRAL
IGNITION TIMING		8°BTDC		
TIMING RPM		500		
CURB IDLE	A/C		700	
RPM	NO A/C		700	
IDLE MIXTURE–ARTIFICIAL ENRICHMENT				
RPM GAIN RPM RESET			15-80 15-70	

THIS VEHICLE REQUIRES MAINTENANCE SCHEDULE "B"

MAKE ALL ADJUSTMENTS WITH ENGINE AT NORMAL OPERATING TEMPERATURES, A/C AND HEADLIGHTS OFF.

CURB IDLE–ADJUST WITH THROTTLE SOLENOID POSITIONER ENERGIZED, THERMACTOR AIR ON, ALL VACUUM HOSES CONNECTED AND AIR CLEANER IN POSITION. WHENEVER CURB IDLE IS RESET, CHECK AND ADJUST THE DECEL VALVE ACCORDING TO THE SERVICE MANUAL.

IDLE MIXTURE–PRESET AT THE FACTORY, DO NOT REMOVE THE LIMITER CAP(S). CONSULT THE SERVICE MANUAL FOR DESCRIPTION OF ARTIFICIAL ENRICHMENT METHOD OF IDLE MIXTURE ADJUSTMENT TO BE USED ONLY DURING TUNE-UPS AND MAJOR CARBURETOR REPAIRS. IDLE MIXTURE MUST BE MEASURED WITH THERMACTOR AIR OFF.

INITIAL TIMING–ADJUST WITH HOSES DISCONNECTED AND PLUGGED AT THE DISTRIBUTOR.

REFERENCE TO A/C, THROTTLE SOLENOID, THERMACTOR AIR AND DECEL VALVE APPLICABLE ONLY IF THE ENGINE IS SO EQUIPPED. CONSULT SERVICE PUBLICATIONS FOR FURTHER INSTRUCTIONS ON TIMING AND IDLE SET.

THIS VEHICLE CONFORMS TO U.S.E.P.A. REGULATIONS APPLICABLE TO 1975 MODEL YEAR NEW MOTOR VEHICLES. THIS VEHICLE ALSO CONFORMS TO THE STATE OF CALIFORNIA CERTIFICATION STANDARDS APPLICABLE TO 1975 MODEL YEAR NEW MOTOR VEHICLES. 929

FORD MOTOR COMPANY D5DE-9C385-TA

Ford Motor Co. California car label

VEHICLE EMISSION CONTROL INFORMATION

ENGINE FAMILY 302 "A" CATALYST EGR/AIR (1 CEF)
ENGINE DISPLACEMENT CID 302 CID
SPARK PLUG ARF 42 GAP .042-.046
DISTRIBUTOR–BREAKERLESS

CHOKE HOUSING		MAN/TRANS		INDEX
NOTCH SETTING		AUTO/TRANS		3 RICH
TRANSMISSION		AUTO NEUTRAL	AUTO DRIVE	MANUAL NEUTRAL
IGNITION TIMING				6°BTDC
TIMING RPM				550
CURB IDLE	A/C			900
RPM	NO A/C			900
IDLE MIXTURE–ARTIFICIAL ENRICHMENT				
RPM GAIN RPM RESET				10-100 10-60

THIS VEHICLE REQUIRES MAINTENANCE SCHEDULE "A"

MAKE ALL ADJUSTMENTS WITH ENGINE AT NORMAL OPERATING TEMPERATURES, A/C AND HEADLIGHTS OFF.

CURB IDLE–ADJUST WITH THROTTLE SOLENOID POSITIONER ENERGIZED, THERMACTOR AIR ON, ALL VACUUM HOSES CONNECTED AND AIR CLEANER IN POSITION. WHENEVER CURB IDLE IS RESET, CHECK AND ADJUST THE DECEL VALVE ACCORDING TO THE SERVICE MANUAL.

IDLE MIXTURE–PRESET AT THE FACTORY, DO NOT REMOVE THE LIMITER CAP(S). CONSULT THE SERVICE MANUAL FOR DESCRIPTION OF ARTIFICIAL ENRICHMENT METHOD OF IDLE MIXTURE ADJUSTMENT TO BE USED ONLY DURING TUNE-UPS AND MAJOR CARBURETOR REPAIRS. IDLE MIXTURE MUST BE MEASURED WITH THERMACTOR AIR OFF.

INITIAL TIMING–ADJUST WITH HOSES DISCONNECTED AND PLUGGED AT THE DISTRIBUTOR.

REFERENCE TO A/C, THROTTLE SOLENOID, THERMACTOR AIR AND DECEL VALVE APPLICABLE ONLY IF THE ENGINE IS SO EQUIPPED. CONSULT SERVICE PUBLICATIONS FOR FURTHER INSTRUCTIONS ON TIMING AND IDLE SET.

THIS VEHICLE CONFORMS TO U.S.E.P.A. REGULATIONS APPLICABLE TO 1975 MODEL YEAR NEW MOTOR VEHICLES D5TE 9C485-AEA 948

FORD MOTOR COMPANY

Ford Motor Co. 49-State label

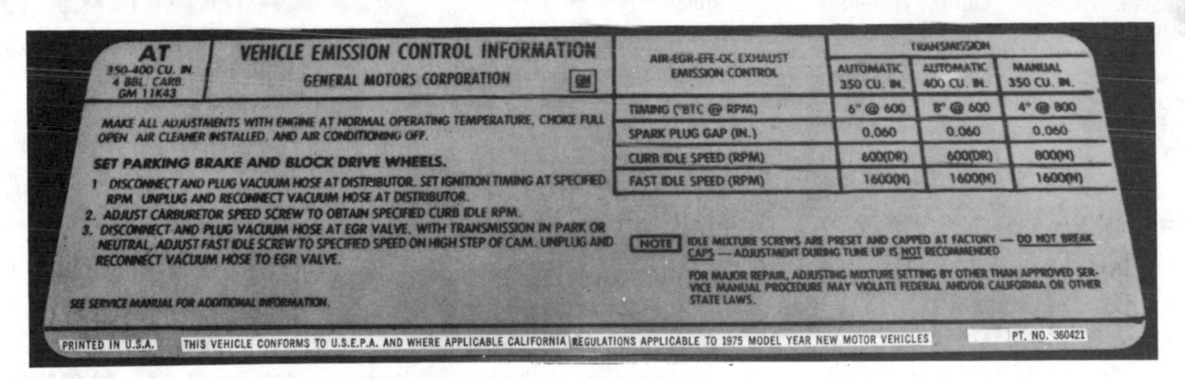

AT
350-400 CU. IN.
4 BBL. CARB.
GM 11K43

VEHICLE EMISSION CONTROL INFORMATION
GENERAL MOTORS CORPORATION

AIR-EGR-EFE-OC EXHAUST EMISSION CONTROL	TRANSMISSION		
	AUTOMATIC 350 CU. IN.	AUTOMATIC 400 CU. IN.	MANUAL 350 CU. IN.
TIMING (°BTC @ RPM)	6° @ 600	8° @ 600	4° @ 800
SPARK PLUG GAP (IN.)	0.060	0.060	0.060
CURB IDLE SPEED (RPM)	600(DR)	600(DR)	800(N)
FAST IDLE SPEED (RPM)	1600(N)	1600(N)	1600(N)

MAKE ALL ADJUSTMENTS WITH ENGINE AT NORMAL OPERATING TEMPERATURE, CHOKE FULL OPEN, AIR CLEANER INSTALLED, AND AIR CONDITIONING OFF.

SET PARKING BRAKE AND BLOCK DRIVE WHEELS.
1. DISCONNECT AND PLUG VACUUM HOSE AT DISTRIBUTOR. SET IGNITION TIMING AT SPECIFIED RPM. UNPLUG AND RECONNECT VACUUM HOSE AT DISTRIBUTOR.
2. ADJUST CARBURETOR SPEED SCREW TO OBTAIN SPECIFIED CURB IDLE RPM.
3. DISCONNECT AND PLUG VACUUM HOSE AT EGR VALVE. WITH TRANSMISSION IN PARK OR NEUTRAL, ADJUST FAST IDLE SCREW TO SPECIFIED SPEED ON HIGH STEP OF CAM. UNPLUG AND RECONNECT VACUUM HOSE TO EGR VALVE.

SEE SERVICE MANUAL FOR ADDITIONAL INFORMATION.

NOTE: IDLE MIXTURE SCREWS ARE PRESET AND CAPPED AT FACTORY — DO NOT BREAK CAPS — ADJUSTMENT DURING TUNE UP IS NOT RECOMMENDED

FOR MAJOR REPAIR, ADJUSTING MIXTURE SETTING BY OTHER THAN APPROVED SERVICE MANUAL PROCEDURE MAY VIOLATE FEDERAL AND/OR CALIFORNIA OR OTHER STATE LAWS.

PRINTED IN U.S.A. THIS VEHICLE CONFORMS TO U.S.E.P.A. AND WHERE APPLICABLE CALIFORNIA REGULATIONS APPLICABLE TO 1975 MODEL YEAR NEW MOTOR VEHICLES PT. NO. 360421

General Motors 1975 label for California cars

LB
1.4-1.6 LITRE
85-97.6 CU. IN.
1 BBL. CARB.
11W1V

VEHICLE EMISSION CONTROL INFORMATION
GENERAL MOTORS CORPORATION

AIR-EGR-OC EXHAUST EMISSION CONTROL	ENGINE				
	1.4 LITRE (85 CU. IN.)		1.6 LITRE (97.6 CU. IN.)		
	TRANSMISSION		TRANSMISSION		
	AUTO. W/O A/C	MANUAL ALL	AUTO. W/O A/C	AUTO. W A/C	MANUAL ALL
TIMING (°BTC @ RPM)	10° @ 850	10°@1000	10° @ 850	10° @ 850	8°@1000
SPARK PLUG GAP (IN.)	0.035	0.035	0.035	0.035	0.035
SOLENOID ASM. ADJ. (RPM)	850 (DR)	1000 (N)	850 (DR)	950 (DR)	1000 (N)
SOLENOID ⅛ HEX. ADJ. (RPM)	600 (DR)	600 (N)	600 (DR)	850 (DR)	600 (N)
FAST IDLE SPEED (RPM)	2000 (P) OR (N)	2000 (N)	2000 (P) OR (N)	2000 (P) OR (N)	2000 (N)

MAKE ALL ADJUSTMENTS WITH ENGINE AT NORMAL OPERATING TEMPERATURE, CHOKE FULL OPEN, AIR CLEANER INSTALLED, AND AIR CONDITIONING OFF.

SET PARKING BRAKE AND BLOCK DRIVE WHEELS
1. DISCONNECT AND PLUG PCV HOSE AT VAPOR CANISTER.
2. DISCONNECT AND PLUG VACUUM HOSE AT DISTRIBUTOR.
3. VEHICLES, EXCEPT THOSE EQUIPPED WITH AUTO. TRANS. AND AIR COND. — ADJUST THE SOLENOID ASSEMBLY SPEED BY TURNING SOLENOID ASM. TO SPECIFIED RPM. DISCONNECT ELECTRICAL LEAD AT SOLENOID. ADJUST BASE IDLE SPEED TO SPECIFICATION BY TURNING 1/8 HEX SCREW LOCATED IN END OF SOLENOID BODY. RECONNECT ELECTRICAL LEAD.
4. VEHICLES, WITH AUTO. TRANS. AND AIR COND. — TURN THE 1/8 HEX SCREW LOCATED IN END OF SOLENOID BODY UNTIL SCREW IS FULLY BOTTOMED. ADJUST SOLENOID ASM. SPEED BY TURNING SOLENOID ASM. TO SPECIFIED RPM. ADJUST BASE IDLE SPEED TO SPECIFICATION BY TURNING 1/8 HEX SCREW OUT UNTIL THE BASE SPEED IS OBTAINED.
5. SET IGNITION TIMING AT SPECIFIED RPM. RESET SOLENOID ASM. AND BASE IDLE SPEEDS IF NECESSARY.
6. CONNECT VACUUM HOSE AT DISTRIBUTOR. DISCONNECT AND PLUG VACUUM HOSE AT EGR VALVE. WITH TRANSMISSION IN PARK OR NEUTRAL, SET FAST IDLE SPEED ON HIGH STEP OF CAM. UNPLUG AND RECONNECT VACUUM HOSE TO EGR VALVE.
7. RECONNECT PCV HOSE AT CANISTER.

NOTE: IDLE MIXTURE SCREWS ARE PRESET AND CAPPED AT FACTORY — DO NOT BREAK CAPS — ADJUSTMENT DURING TUNE UP IS NOT RECOMMENDED.

FOR MAJOR REPAIR, ADJUSTING MIXTURE SETTING BY OTHER THAN APPROVED SERVICE MANUAL PROCEDURE MAY VIOLATE FEDERAL AND/OR CALIFORNIA OR OTHER STATE LAWS.

SEE SERVICE MANUAL FOR ADDITIONAL INFORMATION.

PRINTED IN U.S.A. THIS VEHICLE CONFORMS TO U.S.E.P.A. AND WHERE APPLICABLE CALIFORNIA REGULATIONS APPLICABLE TO 1976 MODEL YEAR NEW MOTOR VEHICLES PT. NO. 372554

General Motors 1976 label for California cars

THE "GRAMS PER MILE" LABEL

Every new car is required to have a label on the outside of the car (usually on a side window) giving emission test information about the car. Most of these labels list the grams per mile of hydrocarbons or carbon monoxide that were emitted in the federal test. This grams per mile figure has nothing to do with testing in the field using an infra-red unit. There is no way that this grams per mile figure can be translated into percentage of CO or parts per million of HC.

If the car maker recommends measuring the HC or CO in the field (some don't) the figures you need will be on the underhood label. The label can usually be found on the radiator brace, the underside of the hood, or on the rocker cover.

HC and CO READINGS

Some carmakers want you to measure HC and CO at the tailpipe with the air pump disconnected. Whenever you disconnect the pump prior to taking an HC-CO reading, you must plug the opening to the exhaust so that air will not be sucked in. Exhaust pulsations can draw in enough air through the check valve to change the CO readings if a hose connection is left open. When plugging the hose or connection, be sure you are plugging the opening that leads to the exhaust, not the hose that goes back to the pump.

GENERAL MOTORS DIAGNOSTIC CONNECTOR

Full size General Motors cars now have the same underhood diagnostic connectors that originally appeared factory with considerably more horsepower than before. The fuel mixtures have been richened and the spark has been advanced, which makes the engine much more responsive and easier to drive. On 49-State cars, the increased emissions from the engine can be cleaned up by the converter alone, but California engines usually need an air pump.

Because the converter is doing more to clean up the exhaust, an infra-red tester has a better chance of detecting a failed converter. Before the '77 models came out, you could be sure that high tailpipe readings were caused by the engine, because an engine in good shape was clean. Now, a bad converter can cause high readings.

The right way to correct high emissions is to check the engine first. If you are sure the engine is okay, then the problem is in the converter.

ENGINES 1979

A number of engines have been dropped by the various automobile manufacturers for the model year 1979.

The engines can be identified by referring to the vehicle identification number, visible through the left side of the windshield.

The fifth character of the vehicle identification number represents the engine that is being used in that particular vehicle for General Motors, Chrysler Corporation, and Ford Motor Co. The seventh character in the vehicle identification number represents the engine that is being used in vehicles that are being made by the American Motors Corporation.

For complete information and description about the particular vehicle identification number codes used by each individual automobile manufacturer, refer to the beginning of the various car sections in this book.

CATALYTIC CONVERTER (PHASE II) 1979

The Phase II catalytic converter helps to reduce all three exhaust emission pollutants, Hydrocarbons, Carbon Monoxide, and Oxides of Nitrogen, but is particularly selective in the reduction of oxides of nitrogen. This converter is designed to maintain high efficiency, thereby making it necessary to closely control the air fual ratio. The Phase II catalytic converter beads are coated with Platinum and Rhodium rather than Platinum and Palladium, which are used in the other types of catalytic converters.

TURBO CHARGER 1979

Buick Motor Division uses a turbo charged V6 engine in some of it's 1979 models. The turbo charger is used to increase power on a demand basis, thus allowing a smaller engine to do the work of a larger and more powerful engine. Buick's turbo V6 engine is more complicated than it's non-turbo V6 engine, but it does incorporate many of the design features utilized by the non-turbo engine.

New for Ford Motor Co. this year is the turbo charged 2.3 Liter 4 cyl. engine. This engine is available in the Ford Mustang and Mercury Capri.

When the turbo charger is in operation a green light will glow on the instrument panel, If, during turbo operation, a problem should occur, a red warning light and buzzer will warn the driver of a malfunction within the system. Also, should the engine oil temperature become critically high, the red light will flash on the instrument panel.

FORD MOTOR COMPANY ELECTRONIC ENGINE CONTROL SYSTEM 1979

The electronic engine control system, called EEC-II, is used on all Mercury models equipped with the 5.8 Liter V8 engine. The system will also be used on all 5.8 Liter V8 engines for the Ford models produced for sale in the state of California. The new system controls spark, exhaust gas recirculation, and air/fuel ratio, in incorpating an electronically controlled "feedback" carburetor and a three way catalyst system.

NON-ADJUSTABLE CARBURETORS 1979

General Motors Corporation for 1979, has made it impossible to adjust the air/fuel mixtures on it's carburetors. The only way adjustment can be accomplished is at the time of major carbureton overhaul. On American Motors Corporation cars, no adjustments can be made, with regards to the air/fuel mixture on the 4 cyl. engine, but the 6 and 8 cyl. engines are adjustable, within a narrow range.

GENERAL MOTORS FRONT WHEEL DRIVE (X BODY CARS) 1979

General Motors Corporation is introducing a mid year production automobile in it's Buick, Pontiac, Oldsmobile, and Chevrolet Divisions. This new vehicle will be equipped with front wheel drive and powered by a 60°, V6 engine, which is a performance option, or the standard 151 4 cyl. engine, which is now being made by Pontiac Motor Division. The terminology that is being used to identify these new cars is "X-body". The name is derived from the construction design of the automobile's chassis.

TUNE-UP SPECIFICATIONS

ENGINE CODE, 7th CHARACTER OF THE VIN NUMBER

Year	ENGINE V.I.N. Code	ENGINE No. Cyl. Disp. (cu in)	SPARK PLUGS Orig. Type	SPARK PLUGS Gap (in)	IGNITION Distributor	IGNITION Timing① (deg. B.T.D.C.) Man. Trans.	IGNITION Timing① (deg. B.T.D.C.) Auto Trans.	Valves Intake Opens (deg. B.T.D.C.)	CARBURETION Idle Speed (r.p.m.) Manual Fed.	CARBURETION Idle Speed (r.p.m.) Manual Cal.	CARBURETION Idle Speed (r.p.m.) Auto (in drive) Fed.	CARBURETION Idle Speed (r.p.m.) Auto (in drive) Cal.
1979	G	4-121	N-8L	.035	②	12	12(8)	25③	900	900	800	800
	E	6-232	N-13L	.035	E.I.	8	10	12	600	600	550	550
	A	6-258	N-13L	.035	E.I.	—	8	12	—	—	700	700
	C	6-258	N-13L	.035	E.I.	4	8	14½	700	700	600	600
	H	8-304	N-12Y	.035	E.I.	5	8	12	800	800	600	600
1980	B	4-151	R44TSX	.060	E.I.	④		33	④	④	④	④
	C	6-258	N-14LY	.035	E.I.	④		14½	700	700	600	600
	C	6-258	N-13L⑤	.035	E.I.	④		④	700	700	600	600

NOTE: Should the information provided in this manual deviate from the specifications on the underhood tune-up label, the label specifications should be used, as they may reflect production changes.

① Figure in parentheses is for California applications
② Set point gap at .018 in. and dwell at 47°
③ Set valve cearance with engine warm to: intake .006-.009 in., exhaust .016-.019 in.
④ See underhood specifications decal
⑤ 1980 Eagle

DISTRIBUTOR SPECIFICATIONS

Year	DISTRIBUTOR IDENTIFICATION	CENTRIFUGAL ADVANCE Start Dist. Deg. @ Dist. RPM	CENTRIFUGAL ADVANCE Maximum Dist. Deg. @ Dist. RPM	VACUUM ADVANCE Start @ In. Hg.	VACUUM ADVANCE Maximum Dist. Deg. @ In. Hg.
1979	3231340	—1-1.5 @ 500	13-16 @ 2200	2-7	6.5-8.5 @ 12.5
	3231915	—1-0.5 @ 500	8-10.5 @ 2200	1.5-3.5	11.5-13.5 @ 11.5
	3232434	—1-0.5 @ 500	8-10 @ 2200	2-7	7-8.5 @13.5
	3233174	—1-0.5 @ 450	7-9 @ 2200	1.5-4	11-13 @ 12
	3233959	—1-0.5 @ 500	12-14.5 @ 2200	1.5-2	14.5-17.5 @ 12.5
	3234693	—1-0.5 @ 500	12-14 @ 2200	2-3.5	11.5-12 @ 12.5
	3250163	0-2 @ 750	15-17 @ 2200	2-4	7-8.5 @ 10
	3250497	—1-0.5 @ 500	8-11 @ 1700	2-4	15-17.5 @ 14
1980	110560①	—2-0 @ 1000	12-16 @ 4400	3.5-4.5	19-22 @ 10
	110561②	—2-0 @ 800	5-9 @ 2000	2.5-3.5	14-16 @ 7
	110650③	—2-0 @ 1200	12-16 @ 4400	3.5-4.5	19-21 @ 10
	3238428④	—2-1 @ 700	8-13 @ 4400	—1.5-1.5	19-25 @ 19
	3235141⑤	—2-1 @ 900	8-14 @ 4400	—1.5-1.5	18-25 @ 18
	3235141⑥	—2-1 @ 1000	8-13 @ 4400	—1-1	22-24 @ 16
	3235141⑦	—1-1 @ 1000	8-13 @ 4400	—1-1	21-25 @ 15

① 151-49 state, manual
② 151-49 state, automatic
③ 151 California
④ 258—All, manual w/8132820 vacuum unit
⑤ 258—All, manual w/8131971 vacuum unit
⑥ 258—Eagle, auto. w/8131971 vacuum unit
⑦ 258—All auto. w/8131971 vacuum unit

2SE, E2SE CARBURETOR ADJUSTMENTS

Year	Carburetor Identification	Float Level (in.)	Pump Rod (in.)	Fast Idle (rpm)	Choke Coil Lever (in.)	Fast Idle Cam (deg./in.)	Air Valve Rod (in.)	Primary Vacuum Break (deg./in.)	Choke Setting (notches)	Secondary Vacuum Break (deg./in.)	Choke Unloader (deg./in.)	Secondary Lockout (in.)
1980	17080681	3/16	17/32	2400	.142	18/0.096	.018	20/.110	Fixed	—	32/.195	N.A.
	17080683	3/16	½	2400	.142	18/0.096	.018	20/.110	Fixed	—	32/.195	N.A.
	17080686	3/16	½	2600	.142	18/0.096	.018	20/.110	Fixed	—	32/.195	N.A.
	17080688	3/16	½	2600	.142	18/0.096	.018	20/.110	Fixed	—	32/.195	N.A.

CARTER BBD SPECIFICATIONS

AMERICAN MOTORS

Year	Model ④	Float Level (in.)	Accelerator Pump Travel (in.)	Bowl Vent (in.)	Choke Unloader (in.)	Choke Vacuum Kick	Fast Idle Cam Position	Fast Idle Speed (rpm)	Automatic Choke Adjustment
1979	8185	¼	0.470	—	0.280	0.140	0.110	1600	1 Rich
	8186	¼	0.520	—	0.280	0.150	0.110	1500	1 Rich
	8187	¼	0.470	—	0.280	0.140	0.110	1600	1 Rich
	8221	¼	0.530	—	0.280	0.150	0.110	1600	1 Rich
1980	8216	¼	0.520	—	0.280	0.140	0.090	1850	2 Rich
	8246	¼	0.520	—	0.280	0.140	0.095	1850	2 Rich
	8247	¼	0.520	—	0.280	0.150	0.095	1700	1 Rich
	8248	¼	0.520	—	0.280	0.150	0.095	1700	1 Rich
	8253	¼	0.470	—	0.280	0.128	0.095	1850	2 Rich
	8256	¼	0.470	—	0.280	0.128	0.093	1850	2 Rich
	8278	¼	0.542	—	0.280	0.140	0.093	1850	Index

① Indicates the drill bit number.
③ At idle
④ Model numbers located on the tag or casting

CARTER YF, YFA SPECIFICATIONS

AMERICAN MOTORS

Year	Model ①	Float Level (in.)	Fast Idle Cam (in.)	Unloader (in.)	Choke
1978-79	7201	0.476	0.195	0.275	Index
	7228	0.476	0.195	0.275	1 Rich
	7229	0.476	0.195	0.275	1 Rich
	7235	0.476	0.195	0.275	Index
	7267	0.476	0.195	0.275	1 Rich
	7232	0.476	0.201	0.275	2 Rich
	7233	0.476	0.201	0.275	1 Rich

FORD, AUTOLITE, MOTORCRAFT MODELS 2100, 2150 SPECIFICATIONS

American Motors

Year	(9510)* Carburetor Identification	Dry Float Level (in.)	Wet Float Level (in.)	Pump Setting Hole #①	Choke Plate Pulldown (in.)	Fast Idle Cam Linkage Clearance (in.)	Fast Idle (rpm)	Dechoke (in)	Choke Setting	Dashpot (in.)
1979	9DA2	0.313	0.780	3	0.125	0.113	1600⑦	0.300	1 Rich	—

Model 5210-C

AMC OHC 4 Cylinder

Year	Carb. Part No. ① ②	Float Level (Dry) (in.)	Float Drop (in.)	Pump Position	Fast Idle Cam (in.)	Choke Plate Pulldown* (in.)	Secondary Vacuum Break (in.)	Fast Idle Setting (rpm)	Choke Unloader (in.)	Choke Setting
1979	8548	0.420	—	—	0.204	0.191	—	1800	0.300	1 Rich
	8549	0.420	—	—	0.191	0.266	—	1800	0.300	1 Rich
	7846	0.420	—	—	0.193	0.191	—	1800	0.300	1 Rich
	8675	0.420	—	—	0.173	0.177	—	1800	0.300	Index

CAR SERIAL NUMBER AND ENGINE IDENTIFICATION

1979

Mounted behind the windshield on the driver's side is a plate with the vehicle identification number. The second character is the model year, with 9 for 1979. The seventh character is the engine code, as follows:

A ....258 1-bbl. 6-cyl.
C ....258 2-bbl. 6-cyl.
E ....232 1-bbl. 6-cyl.
G ....2-liter 2-bbl. 4-cyl.
H 304 2-bbl. V-8

NOTE: *The 2-liter 4-cylinder engine is actually 1984 cubic centimeters, or 121 cubic inches.*

Engines can be identified by a build date code attached to the right bank rocker cover on the V-8, or to the right side of the block between numbers 2 and 3 cylinder on the 6-cylinder. The 4-cylinder engine has the code on the left rear of the block near the dipstick. The fourth character in the code is a letter identifying the engine. The letters are the same as in the vehicle identification number listed above.

1980

Mounted behind the windshield on the driver's side is a plate with the vehicle identification number. The second character is the model year, with 0 for 1980. The seventh character is the engine code, as follows:

B .... 151 2-bbl. 4-cyl.
C ....258 2-bbl. 6-cyl.

NOTE: *For 1980, the 121 (2 liter) 4-cylinder, the 232 6-cylinder the V-8 engine have been cancelled. The new 151 4-cylinder engine for 1980 is from General Motors' Pontiac division.*

The 151 (2.5 liter) 4-cylinder engine can be identified by the two-character engine identification code letters stamped into the front top left hand corner of the engine block.

The 258 (4.2 liter) 6-cylinder engine can be identified by the engine Build Date Code located on the right side of the block between the No. 2 and the No. 3 cylinders. The numbers identify the year, month and day that the engine was built. The letters identify the engine size, carburetor type and compression ratio.

EMISSION EQUIPMENT

1979

Air pump
Closed positive crankcase ventilation
Emission calibrated carburetor
Emission calibrated distributor
Single diaphragm vacuum advance
Exhaust gas recirculation
Vapor control, canister storage
Heated air cleaner
Thermostatically Controlled Air Cleaner (Vacuum Controlled)
Transmission controlled spark
  49-States
    Not used
  Calif.
    All models,
Catalytic converter, warm-up
  49-States
    Not used
  Calif.
    All 6-cyl. & V-8 models (with pellet type.)
    Not used on 4-cyl.
Catalytic converter, pellet type
  49-States
    All models
  Calif.
    All 6-cyl. & V-8 models (with warm-up type.)
    All 4-cyl.
Electric choke
  49-States
    All 4-cyl.
    Not used on 6-cyl. & V-8
  Calif.
    Not used
Throttle Solenoid
  49-States
    4-cyl. models
    6-cyl. 258 CID w/Auto
    V-8 All
  Calif.
    All models

1980

Pulsair Air Guard System (151)
Catalytic Converter
EGR system
EGR CTO switch
Fuel tank vapor control
PCV valve
TAC system
TAC TVS switch
Spark CTO
Non-linear valve
Carb. vent to canister
Electric choke
Throttle solenoid
EGR TVS switch
Forward delay valve
Air control valve (258)
AGE CTO switch (258)
Decelerator valve
Reverse delay valve
Spark TVS switch
Spark control solenoid
WOT TVS switch (258)
EGR/TVS switch (258)
Diverter valve (258)
Vacuum source dump valve
C4 System   Calif. (151)
CEC System (258)
Pre-converter   Calif.
Delay 2-way valve   Calif. (258)

NOTE: *Not all vehicles use all of the above equipment, depending on application.*

IDLE SPEED AND MIXTURE ADJUSTMENTS

1979

Air cleaner ....In place
Air cond. ....Off
Auto. trans. ....Drive
Mix. adj. ......See rpm drop below
Idle CO ....Not used
Idle Spec
121 2-bbl. 4-Cyl.
  All auto. trans.
    Solenoid connected 800
    Solenoid disconnected 500
    Mixture adj. ......45 rpm drop (845-800)
  49-States man. trans.
    Solenoid connected 900
    Solenoid disconnected 500
    Mixture adj. ....120 rpm drop (1020-900)
  Code "EH" only ①
    Solenoid connected 1000
    Solenoid disconnected 500
    Mixture adj. ....120 rpm drop (1120-1000)
  Calif. man. trans.
    Solenoid connected 1000
    Solenoid disconnected 500
    Mixture adj. ....120 rpm drop (1120-1000)
232 1-bbl. 6-Cyl.
  49-States auto. trans. 550
    Mixture adj. ......25 rpm drop (575-550)
  49-States man. trans. 600
    Mixture adj. ......50 rpm drop (650-600)

IDLE SPEED AND MIXTURE ADJUSTMENTS

258 1-bbl. 6-Cyl.
Calif. auto. trans.
Solenoid connected 700
Solenoid disconnected 500
Mixture adj. ......25 rpm drop (725-700)

258 2-bbl. 6-Cyl.
49-States auto. trans.
Solenoid connected 600
Solenoid disconnected 500
Mixture adj. ......25 rpm drop (625-600)
49-States man. trans. 600
Mixture adj. ......50 rpm drop (650-600)

304 V-8
49-States all trans.
Solenoid connected 600
Solenoid disconnected 500
Mixture adj. ......40 rpm drop (640-600)

①—"EH" noted on upper right corner of vehicle emission control information label

1980

Air cleaner ....In place
Air cond. ....See below
Auto. trans. ....Drive
Mix. adj. ......See rpm drop below
Idle CO ....Not used
Idle Spec

151 2-bbl. 4-Cyl. w/2SE carb.
49-States auto. trans.
Solenoid connected, A.C. off 700
Solenoid disconnected, A.C. off 500
Mixture adj. ......20 rpm drop
49-States man. trans.
Solenoid connected, A.C. on 900
Solenoid disconnected, A.C. off 500
Mixture adj. ......70 rpm drop

151 2-bbl. 4-Cyl. w/E2SE carb.
California auto. trans. w/A.C.
Solenoid connected, A.C. on 950
Solenoid disconnected, A.C. off 700
Mixture adj. ...Non-adjustable
California auto. trans. w/o A.C.
Solenoid connected 700
Solenoid disconnected 500
Mixture adj. ...Non-adjustable
California man. trans. w/A.C.
Solenoid connected, A.C. on 1250
Solenoid disconnected, A.C. off 900
Mixture adj. ...non-adjustable
California man. trans. w/o A.C.
Solenoid connected 900
Solenoid disconnected 500
Mixture adj. ...non-adjustable

258 2-bbl. 6-Cyl. w/BBD carb.
Solenoid connected
All, auto. trans. 600
All, man. trans. 700
Mixture adj. man. trans. ...40 rpm drop
Mixture adj. auto. trans. ...20 rpm drop

CAUTION: *Mixture adjustment should only be performed if the screws have been removed during carburetor overhaul. An accurate, expanded scale tachometer will be required. Idle speed is set in Drive if automatic, and Neutral if manual transmission. Set parking brake firmly. Do not accelerate engine.*

INITIAL TIMING

1979

NOTE: *Distributor vacuum hose must be disconnected and plugged. Set timing at idle speed. Variation allowed: Plus or minus 2°. On auto. trans. cars, idle speed in Neutral may be fast enough to bring in mechanical advance. To prevent this, make timing setting with transmission in Drive, and wheels safely blocked.*

4-Cyl.
49-States ....12° BTDC
Code "EH" only ① ..16° BTDC
Calif. .... 8° BTDC

232 6-Cyl.
49-States
Auto. trans. ....10° BTDC
Man. trans. .... 8° BTDC
Code "EH" only ① ..12 BTDC

258 1-bbl. 6-Cyl.
Calif.
Auto. trans. .... 8° BTDC

258 2-bbl. 6-Cyl.
49-States
Auto. trans. .... 8° BTDC
Man. trans. .... 4° BTDC

304 2-bbl. V-8
49-States
Auto. trans. .... 8° BTDC
Man. trans. .... 5° BTDC

①—"EH" noted on upper right corner of vehicle emission control information label

1980

NOTE: *Distributor vacuum hose must be disconnected and plugged. Set timing at idle speed. Variation allowed: Plus or minus 2°. On auto. trans. cars, idle speed in Neutral may be fast enough to bring in mechanical advance. To prevent this, make timing setting with transmission in Drive, and wheels safely blocked.*

4-Cyl. 151
49-States
Auto. trans ....12° BTDC
Man. trans. ....10° BTDC
California
Auto. trans ....10° BTDC
Man. trans. ....12° BTDC

6-Cyl. 258
Auto. trans. (Pacer) .... 8° BTDC
Auto. trans. (Spirit, Concord) ....10° BTDC
Man. trans. .... 6° BTDC
Auto. trans. (Eagle, 49 States) ....10° BTDC
Auto. trans. (Eagle, Calif.) .... 8° BTDC

VACUUM ADVANCE 1979-80

All engines ....Ported

SPARK PLUGS

1979

All 4-Cyl. ...... CH-N8L ... .035
All 6-Cyl. ...... CH-N13L .. .035
All 8-Cyl. ...... CH-N12Y .. .035

NOTE: *Resistor plugs may be used.*

1980

All 4-Cyl. ...... R44TSX ... .060
All 6-Cyl. ...... N14LY ..... .035
6-Cyl. Eagle .... N13L ...... .035

NOTE: *Resistor plugs may be used.*

EMISSION CONTROL SYSTEMS

IGNITION SYSTEM 1979-80

4-cylinder—O.H.C.

The four cylinder engines use the Bosch mechanical, point-type ignition system. The operation of the ignition system is in the conventional manner, which uses the three groups of components working together, to deliver high voltage to the spark plugs at the correct time.

The three groups consist of the distributor assembly, the ignition coil and the secondary distributor cap and wiring assembly.

The primary side of the distributor contains the ignition points and condensor, with mechanical and vacuum advances to control the opening and closing of the ignition points. The current to the ignition points is controlled by a resistance wire in the body harness. The coil is a self-contained unit which changes the primary voltage to secondary voltage through the induction process. The third group consists of the distributor rotor, distributor cap, secondary wiring and spark plugs.

NOTE: *This engine was discontinued at the end of the 1979 model year.*

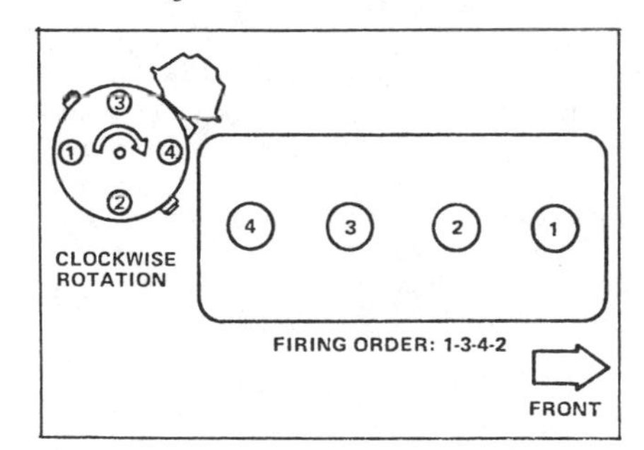

Distributor wiring sequence and firing order. 1979 4-cylinder only (© American Motors)

6-Cylinder and V-8 Engines

All 6-Cylinder and 8-Cylinder 1979 AMC cars use the Motorcraft Solid State Ignition, manufactured by Ford Motor Co. This ignition uses a magnetic trigger wheel in place of the distributor cam, and a sensor coil to detect the position of the trigger wheel. There are no distributor points, and periodic service of the distributor is not required, except for inspecting the cap, rotor, and wires.

The control box and coil are part of the system purchased from Motorcraft, so they are of Ford design also. The primary connector at the coil is the typical Ford slip on design. Tachometers should be connected to the coil negative terminal, which has enough metal exposed so that an alligator clip can be connected without removing the coil connector.

The spark from this ignition can jump comparatively great distances. And like any ignition, the spark will jump to the closest metallic object. If the plug wires to cylinders No. 3 or 5 on a 6-cylinder, or No. 3 or 4 on an 8-cylinder are removed with the engine running, the spark will probably jump inside the cap to the sensor bracket. This sudden surge of high voltage can damage the sensor. For this reason, those plug wires must never be removed while the engine is running, not even for an instant. When removing a plug wire to check available secondary voltage, do it with any plug wire other than those mentioned above.

NOTE: *V-8 engines were discontinued at the end of the 1979 model year.*

4-cylinder—O.H.V.

The 1980 151 CID four cylinder engine uses the GM type High Energy Ignition system. The HEI distributor combines all of the ignition components into one unit. The unit has external connections for the ignition switch wire, the tachometer pick-up and the spark plug leads. The magnetic pick-up assembly is shifted by the vacuum control unit to provide vacuum advance. Conventional advance weights provide centrifugal advance. The module automatically controls the dwell period and features a longer spark duration made possible by the higher amount of energy stored in the coil primary. This helps to fire the lean mixtures, especially at higher speeds.

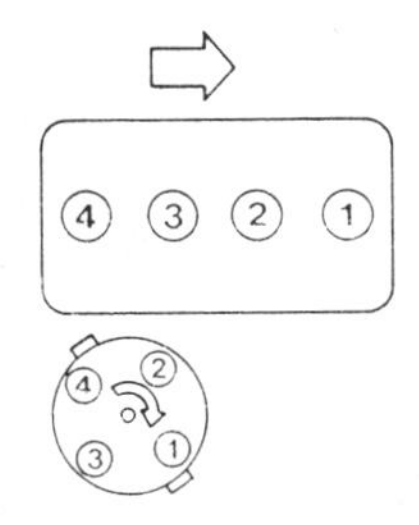

Distributor wiring sequence and firing order. 1980 4-cylinder only (© American Motors)

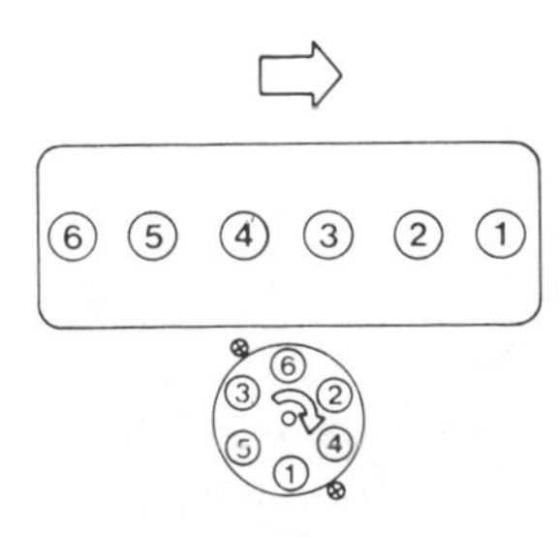

Distributor wiring sequence and firing order—6-cylinder (© American Motors)

Timing

On both the 4-cylinder HEI equipped engine and the 6-cylinder SSI equipped engine, a magnetic timing probe hole is built into the timing case cover. This hole accepts the special timing probe now being used with diagnostic equipment. The probe senses the notch in the vibration damper. The probe is inserted through the probe hole until it touches the vibration damper. It is automatically spaced away from the damper by the damper eccentricity. Ignition timing can then be read from a meter or computer printout, depending on the type of equipment being used.

The probe hole is located at 9.5° ATDC, and the equipment is calibrated for this reading. Do not use the probe location to check timing using a conventional timing light.

VAPOR CONTROL SYSTEM 1979-80

Carbon canisters have four hose connections, and a purge valve. The hose to the purge valve is connected to ported vacuum. When the engine is at idle, the port in the carburetor is above the throttle blade, and the purge valve is closed. Above idle, the port is exposed to vacuum, which acts on the purge valve in the canister and allows manifold vacuum to pull the fumes from the canister to the engine. As the throttle is opened farther, ported vacuum increases. When it gets to 12 in. Hg. or more, the purge valve opens into a second stage and allows a higher rate of purging.

Gasoline fumes are fed to the canister through two hoses, one from the tank and one from the carburetor. The carburetor hose connects to an external vent on the carburetor bowl. This vent is operated either by manifold vacuum or mechanical linkage, depending on the model of carburetor.

On the Model 5210 Holley-Weber carburetor, used on the 4-cylinder engine, up to 1979 a small diaphragm and valve are built into the bowl cover. The operating side of the diaphragm is connected to manifold vacuum, which closes the valve whenever the engine is running. If the engine is not running, a spring behind the diaphragm opens the valve and vents the bowl to the carbon canister.

Carter's YF 1-bbl. carburetor, used until the end of 1979 on the 6-cylinder engine, has an external bowl vent operated by mechanical linkage from the throttle. The vent opens only

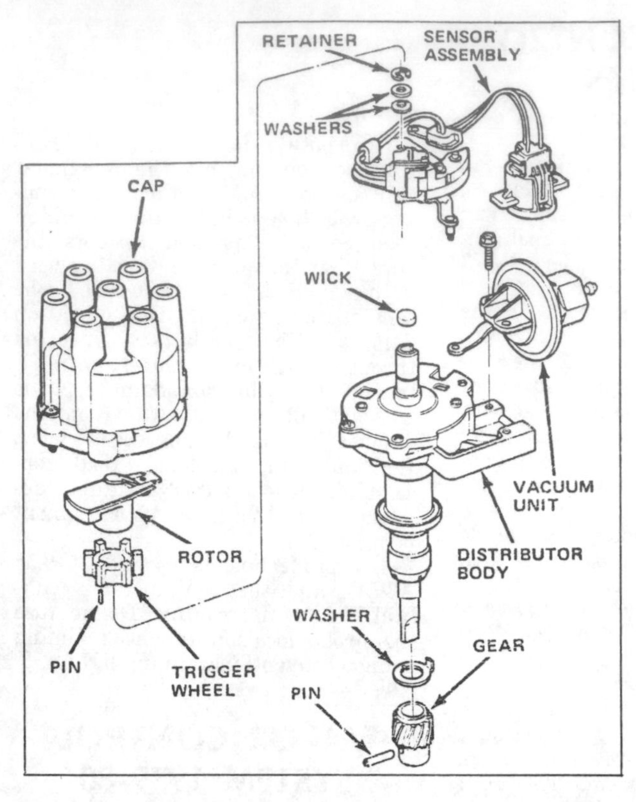

SSI distributor components—six cylinder shown (© American Motors)

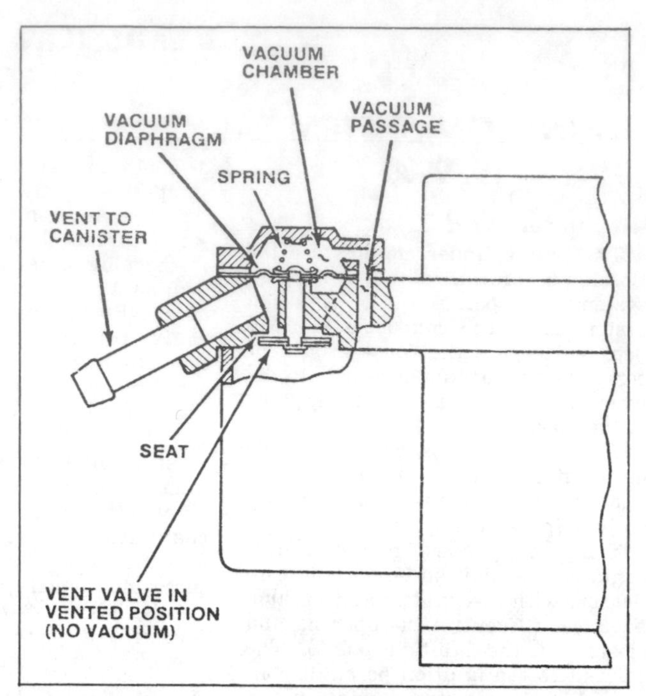

External bowl vent—Holley Weber 5210 (© American Motors)

when the throttle is in the curb idle position. The vent is closed at all other throttle positions.

Because the YF bowl vent has no spring, high vapor pressure in the carburetor bowl can also open the vent. This will occur, when the throttle is above idle, if the pressure in the bowl goes above 0.14 in. of water.

Carter's 2-bbl. BBD model, used on the 6-cylinder, has two external bowl vents built into the bowl cover. The upper vent, with the plastic hose fitting, is pressure operated, and opens only if there is high pressure in the bowl. The lower vent, with the metal hose fitting, is operated by the throttle linkage, and opens only with the throttle in the curb idle position. Both of these vents are connected by short hoses to a "T", and then a single hose goes to the canister.

The Motorcraft 2100 2-bbl. used until the end of 1979 on the V-8 engines has a mechanical external bowl vent operated by the accelerating pump lever. The vent is a plunger type, with a small spring that holds it in the open position. Whenever the throttle is above idle, a lever with a stronger spring overrides the plunger spring and closes the vent. At curb idle, the lever lifts off the plunger, which opens the vent.

External vents not only help prevent air pollution, but also keep gasoline vapors from entering the intake manifold through the internal vent that most carburetors have. This helps avoid hard starting after a car has been parked in the hot sun.

The E2SE model is a new design for the 1980 4 cylinder engine. The float chamber is externally vented

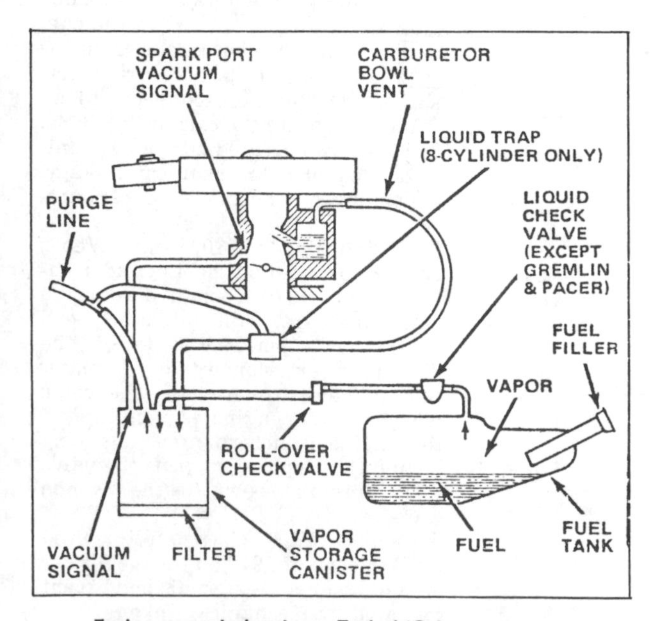

Fuel vapor control system—Typical (© American Motors)

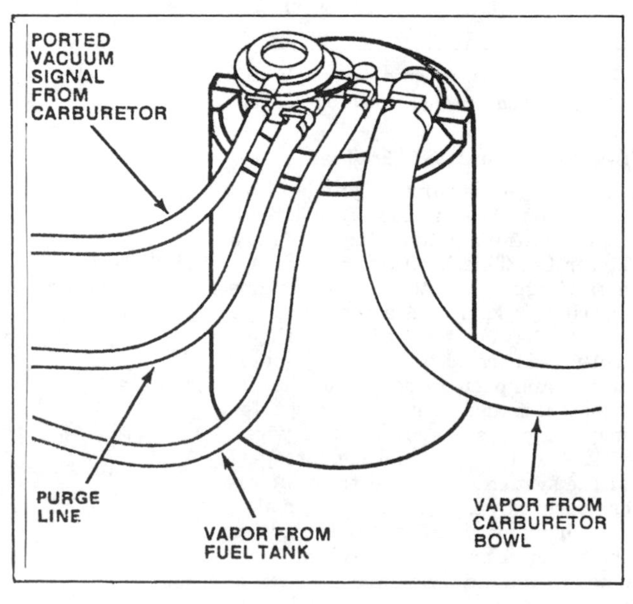

Charcoal canister and hoses (© American Motors)

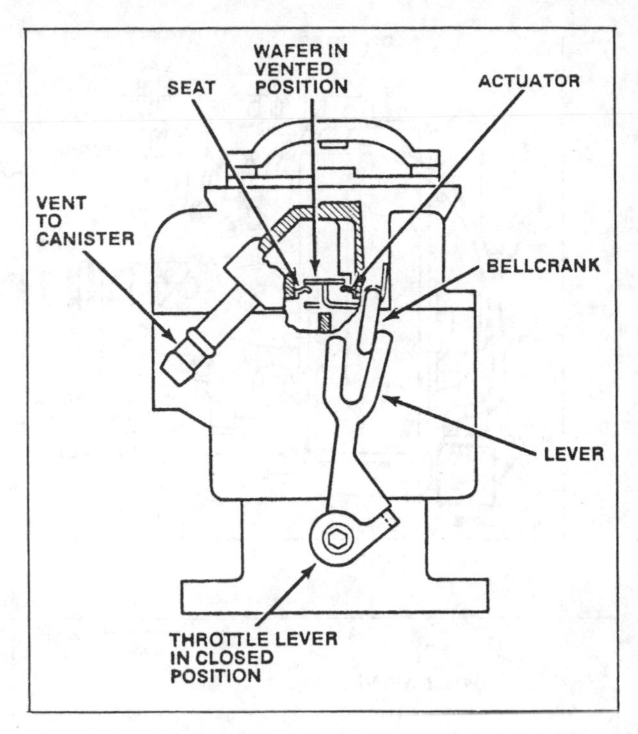

External fuel bowl vent—Carter YF (© American Motors)

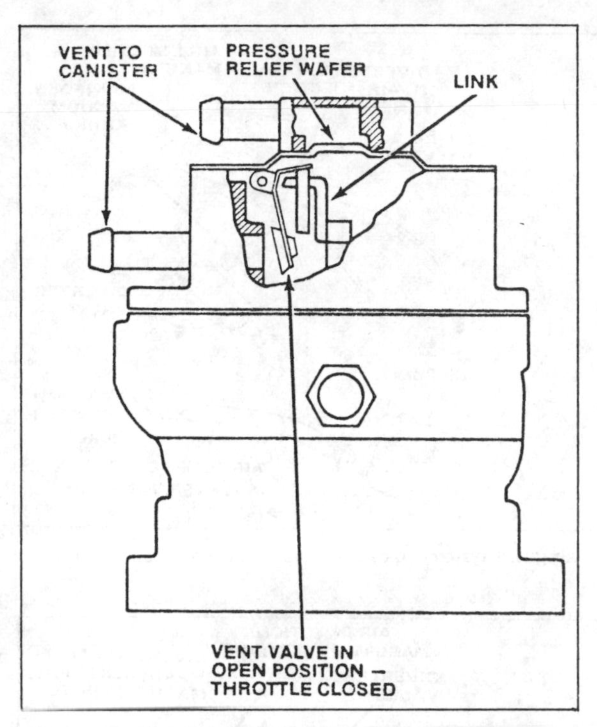

External bowl vent—Carter BBD (© American Motors)

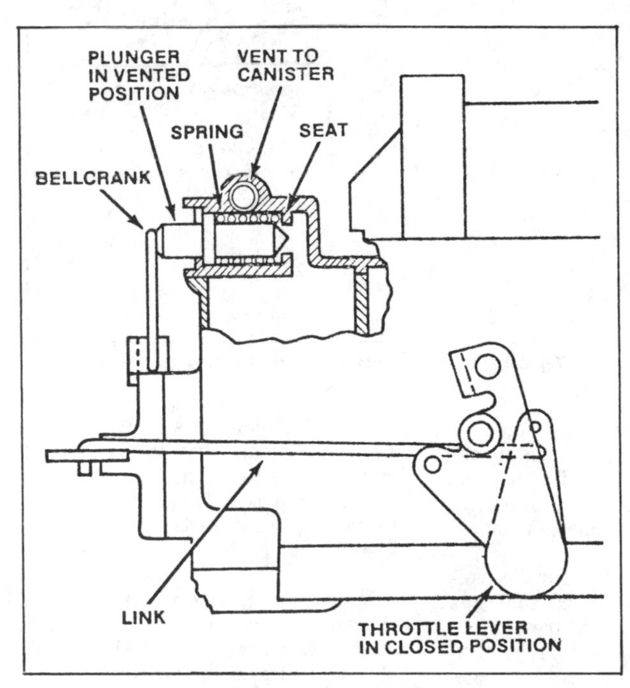

External bowl vent—Motorcraft 2100 (© American Motors)

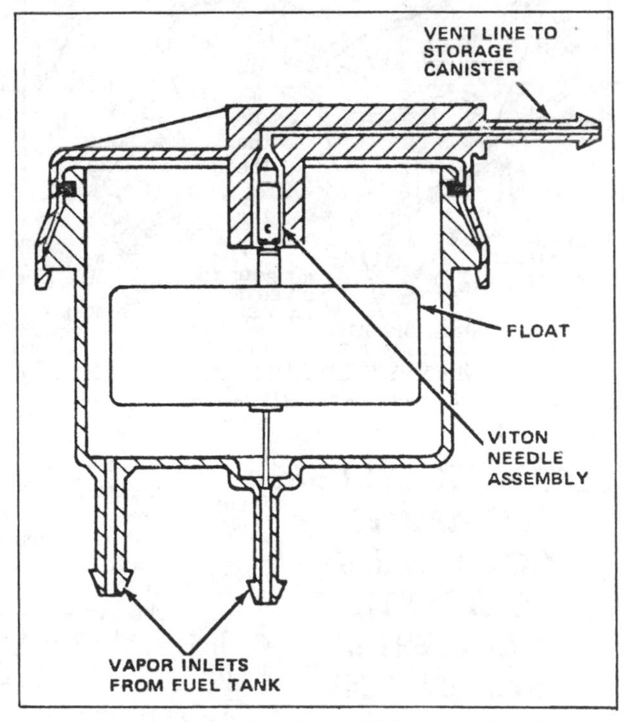

Liquid check valve—Typical (© American Motors)

through a tube in the airhorn. A hose connects this tube directly to a vacuum operated vapor vent in the vapor canister. When the engine is not running, the canister vapor vent is open, and fuel vapor is allowed to flow from the float chamber to the canister to be stored until purged. The venting to the canister conforms to evaporative emission requirements and improves hot engine starting.

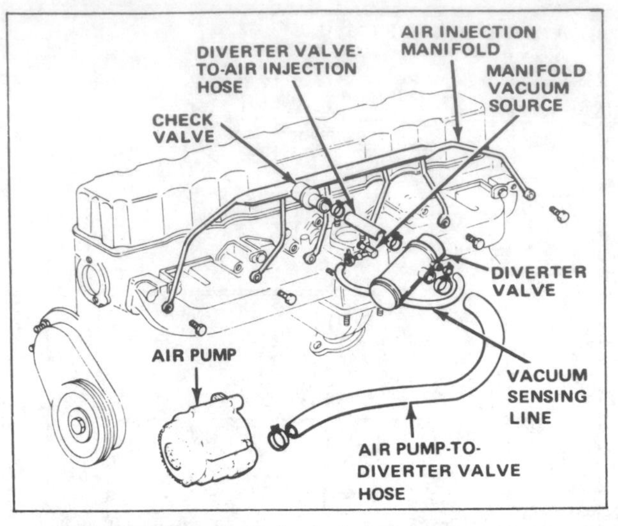

Air guard system—6-cylinder, typical (© American Motors)

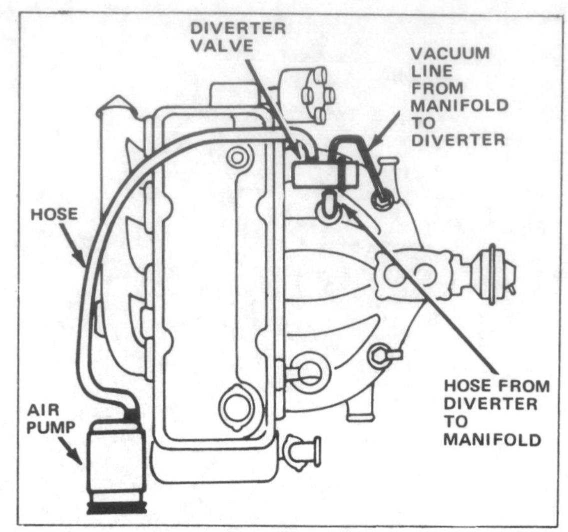

Air guard system—1979 4 cylinder engine

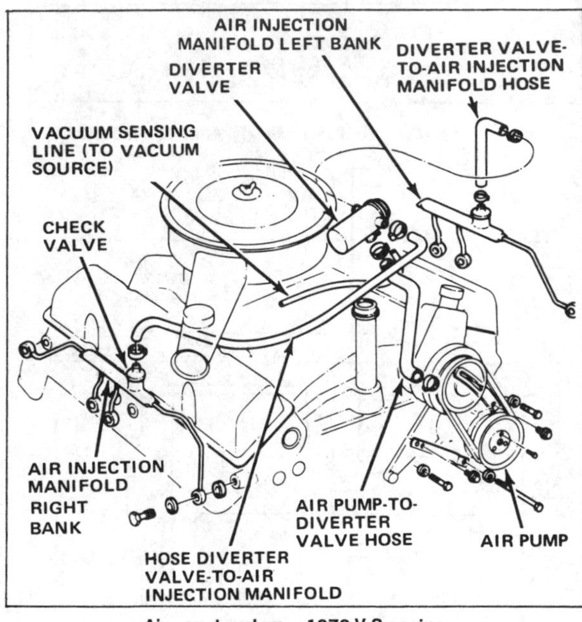

Air guard system—1979 V-8 engine

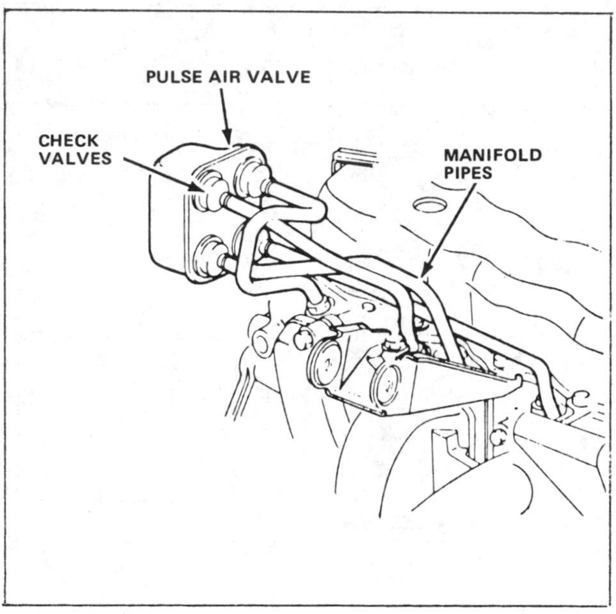

Typical Pulsair air injection system (© American Motors)

FEEDBACK SYSTEMS COMPUTER CONTROLLED CATALYTIC CONVERTER SYSTEM (C4)

4-Cylinder Engine

The purpose of the C-4 System is to maintain the ideal air/fuel ratio at which the catalytic converter is most effective.

Major components of the system include an Electronic Control Module (ECM), an oxygen sensor, an electronically controlled carburetor, and a three-way oxidation-reduction catalytic converter. The system also includes a maintenance reminder flag connected to the odometer which becomes visible in the instrument cluster at regular intervals, signaling the need for oxygen sensor replacement.

The oxygen sensor generates a voltage which varies with exhaust gas oxygen content. Lean mixtures (more oxygen) reduce voltage; rich mixtures (less oxygen) increase voltage. Voltage output is sent to the ECM.

An engine temperature sensor installed in the engine coolant outlet monitors engine coolant temperatures. Vacuum control switches and throttle position sensors also monitor engine conditions and supply signals to the ECM.

The Electronic Control Module receives input signals from all sensors. It processes these signals and generates a control signal sent to the carburetor. The control signal cycles between on (lean command) and off (rich command). The amount of on and off time is a function of the input voltage sent to the ECM by the oxygen sensor.

Basically, an electrically operated mixture control solenoid is installed in the carburetor float bowl. The solenoid controls the air/fuel mixture metered to the idle and main metering systems. Air metering to the idle system is controlled by an idle air bleed valve. It follows the movement of the mixture solenoid to control the idle system, enrichening or leaning out the mixture as appropriate. Air/fuel mixture enrichment occurs when the fuel valve is open and the air bleed valve is closed. All cycling of this system, which occurs ten times per second, is controlled by the ECM.

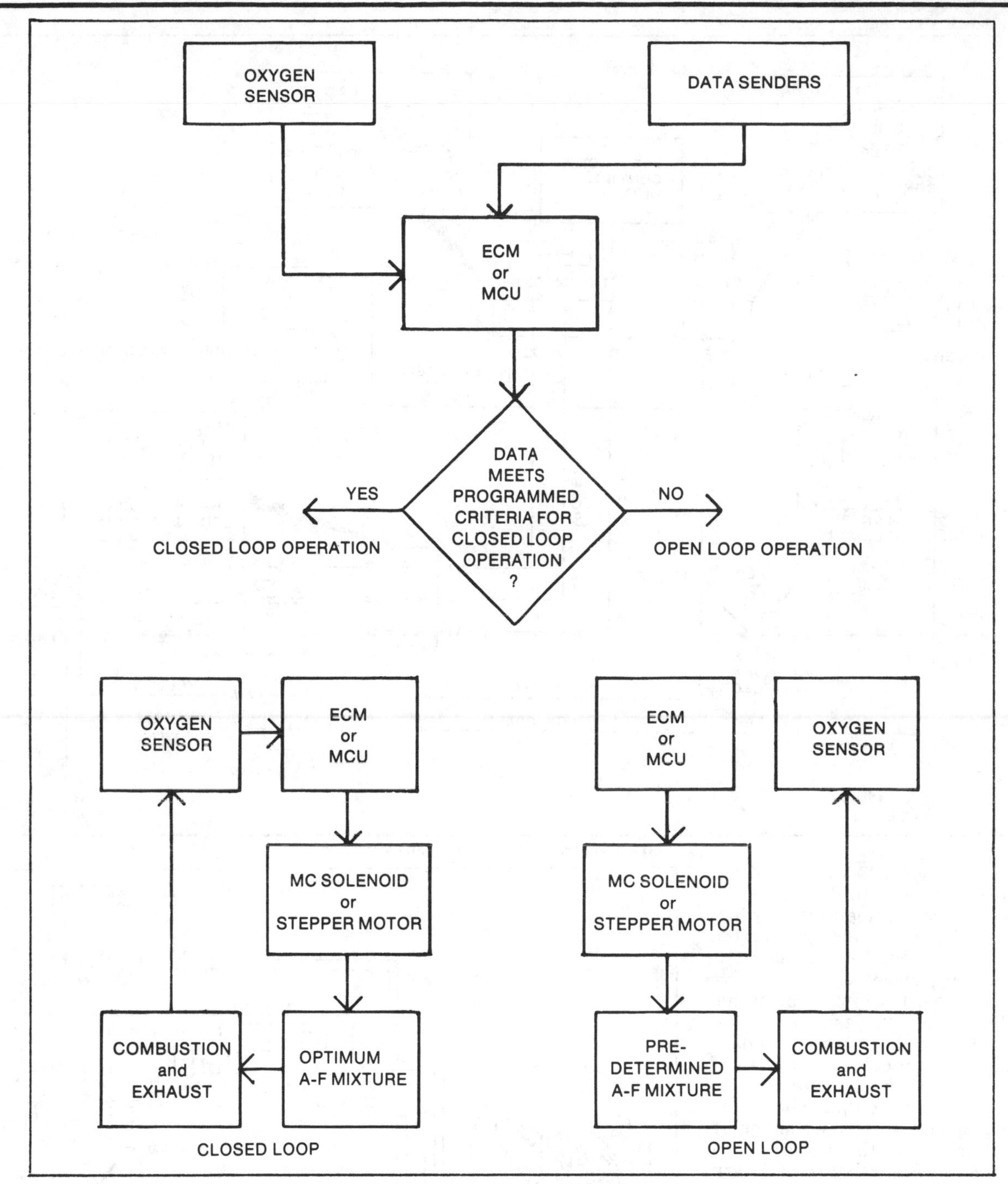

Open and closed loop modes (© American Motors)

A throttle position switch informs the ECM of open or closed throttle operation. A number of different switches are used, varying with application.

When the ECM receives a signal from the throttle switch, indicating a change of position, it immediately searches its memory for the last set of operating conditions that resulted in an ideal air/fuel ratio, and shifts to that set of conditions. The memory is continually updated during normal operation.

A "Check Engine" light is included in the C-4 System installation. When a fault develops, the light comes on, and a trouble code is set into the ECM memory. However, if the fault is intermittent, the light will go out, but the trouble code will remain in the ECM memory as long as the engine is running. The trouble codes are used as a diagnostic aid, and are pre-programmed.

Unless the required tools are available, troubleshooting the C-4 System should be confined to mechanical checks of electrical connectors, vacuum hoses and the like.

COMPUTERIZED EMISSION CONTROL SYSTEM (CEC)

The CEC system is used on the six cylinder 258 engine equipped vehicles with the exception of the Eagle.

Major components of the system include a Micro Computer Unit (MCU), an oxygen sensor, and a stepper motor in the carburetor to vary the size of the air bleed orifices and thus the mixture. In addition,

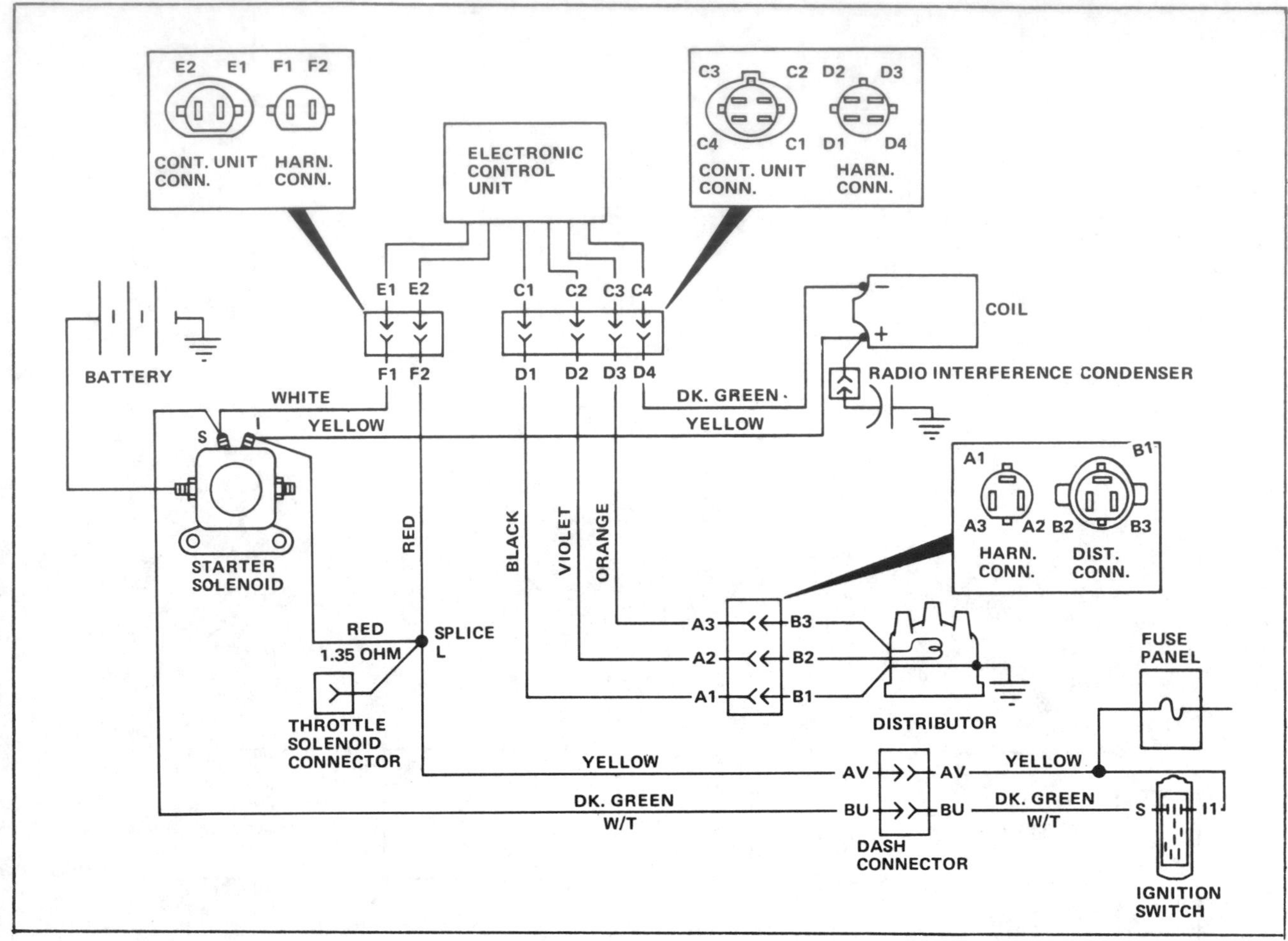

SSI system schematic (© American Motors)

vacuum switches detect and send the throttle position data to the MCU and a coolant temperature switch supplies data to the MCU to keep the system in the Open Loop Mode until sufficiently warmed up.

In operation, the system performs basically in the same manner as the C4 system described above. The data senders send signals to the MCU which monitors the voltage signals. This generates an output signal for the BBD carburetor stepper motor. If the system is in the Closed Loop Mode, the air/fuel mixture will vary according to the oxygen content in the exhaust gas. If the system is in the Open Loop Mode, the mixture will be based on a predetermined ratio dependant on the engine rpm. The stepper motor moves the metering pins in and out of the orifices in steps according to the signal from the MCU. The motor has a range of 100 steps although normally only the middle range is used. Both the ECM of the C4 system and the MCU of the CEC system are mounted inside the passenger compartment beneath the dash panel on the right-hand side.

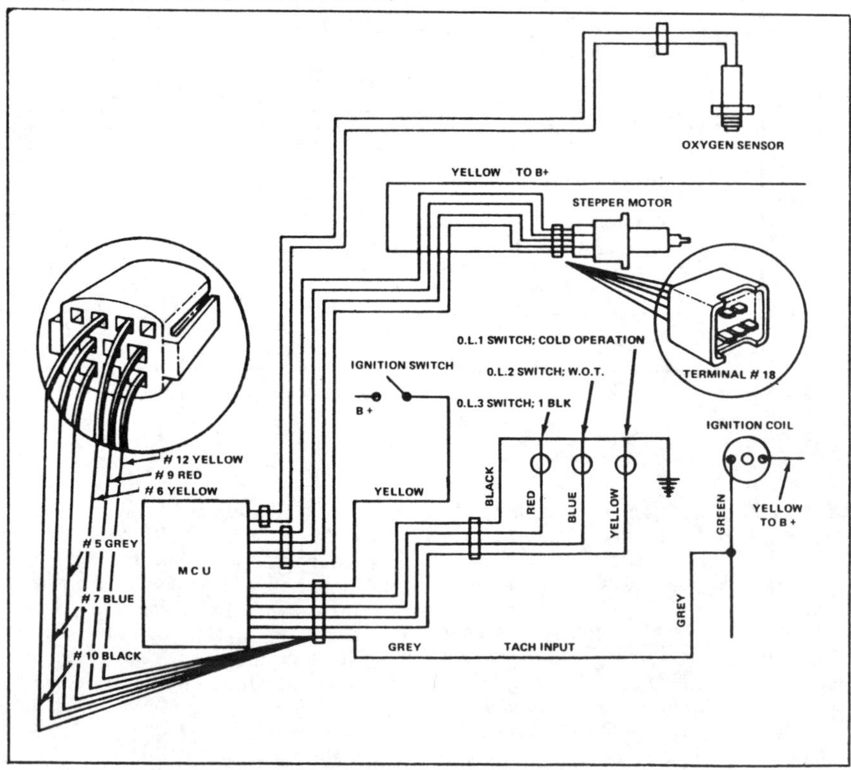

CEC system schematic (© American Motors)

VACUUM CIRCUITS

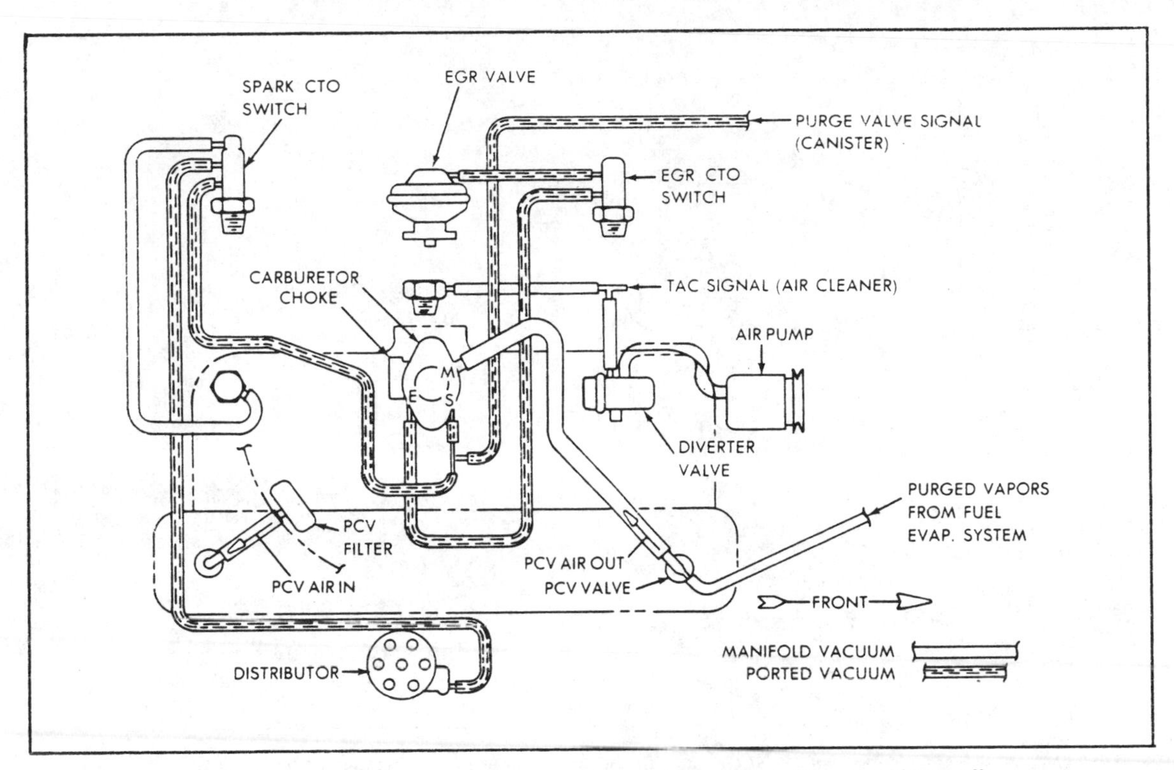

1978-79 Vacuum hose schematic—49 States—232 CID 6 cylinder engine with manual/automatic transmission

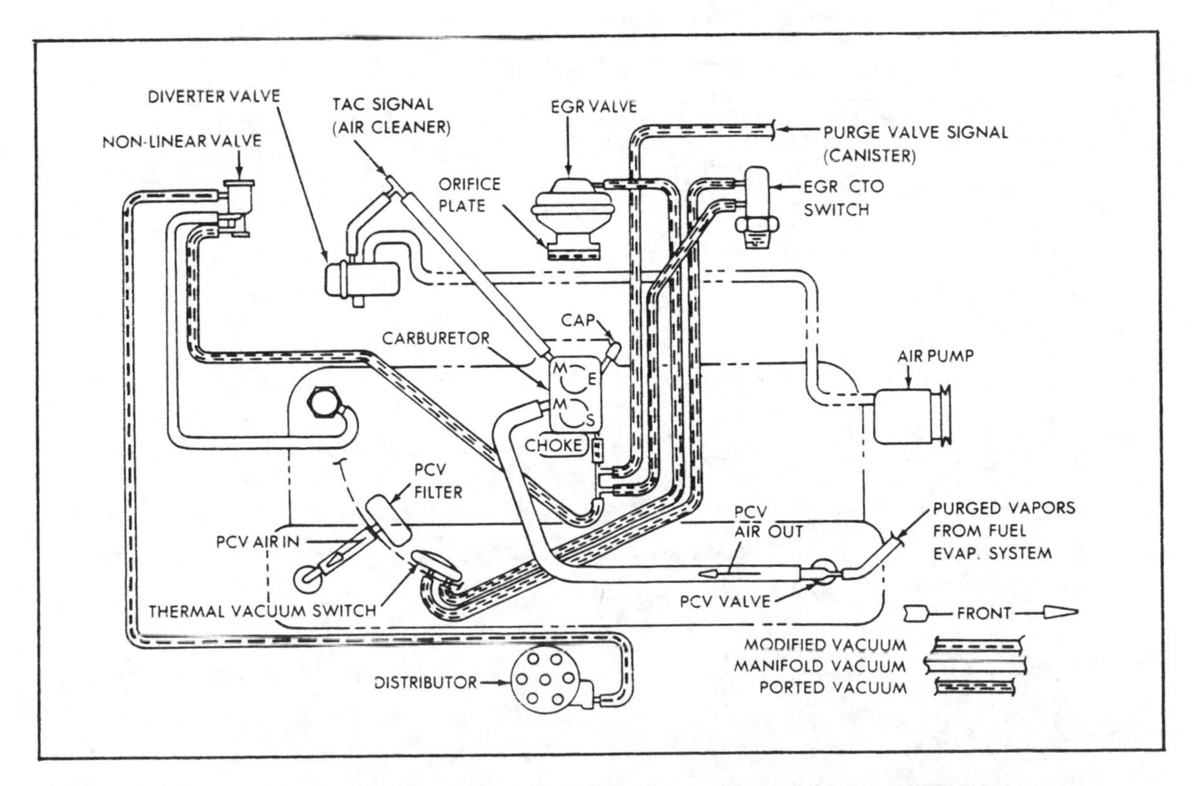

1978-79 Vacuum hose schematic—49 States—258 CID 6 cylinder engine with manual transmission (EGR valve orifice plate is deleted on the 258 CID engine with automatic transmission)

VACUUM CIRCUITS

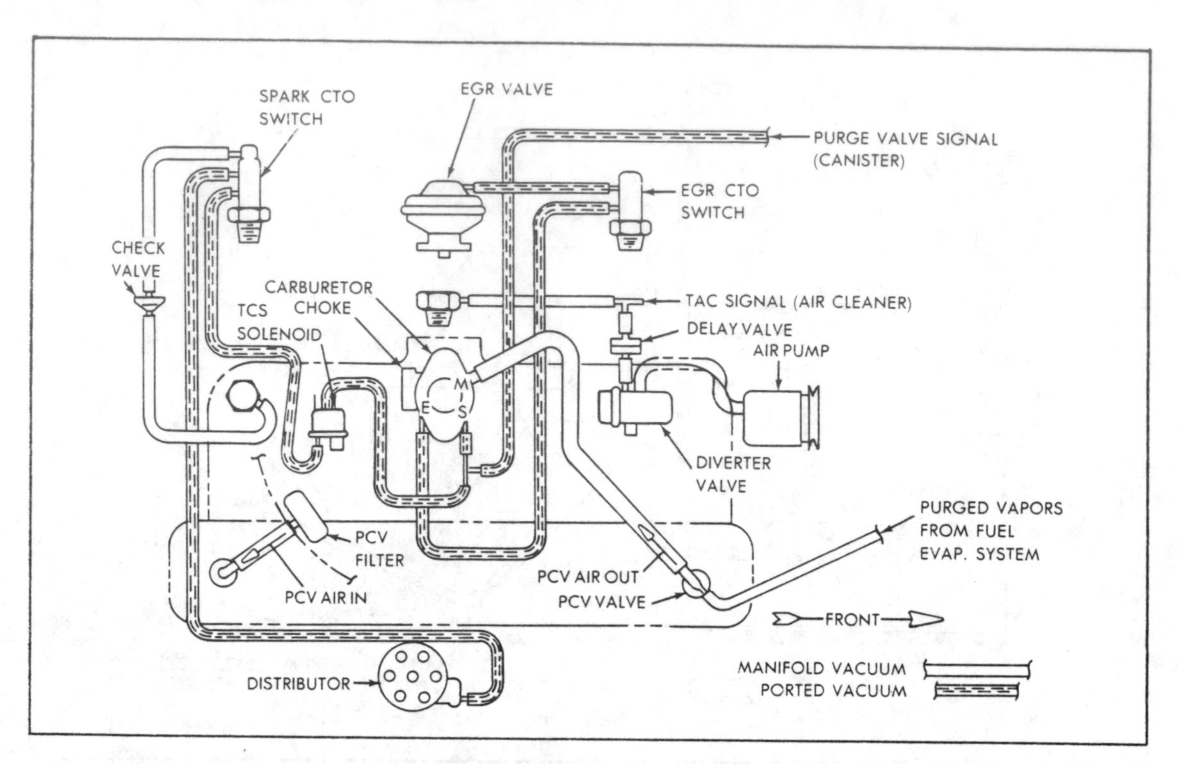

1978-79 Vacuum hose schematic—California 258 CID 6 cylinder engine with automatic transmission

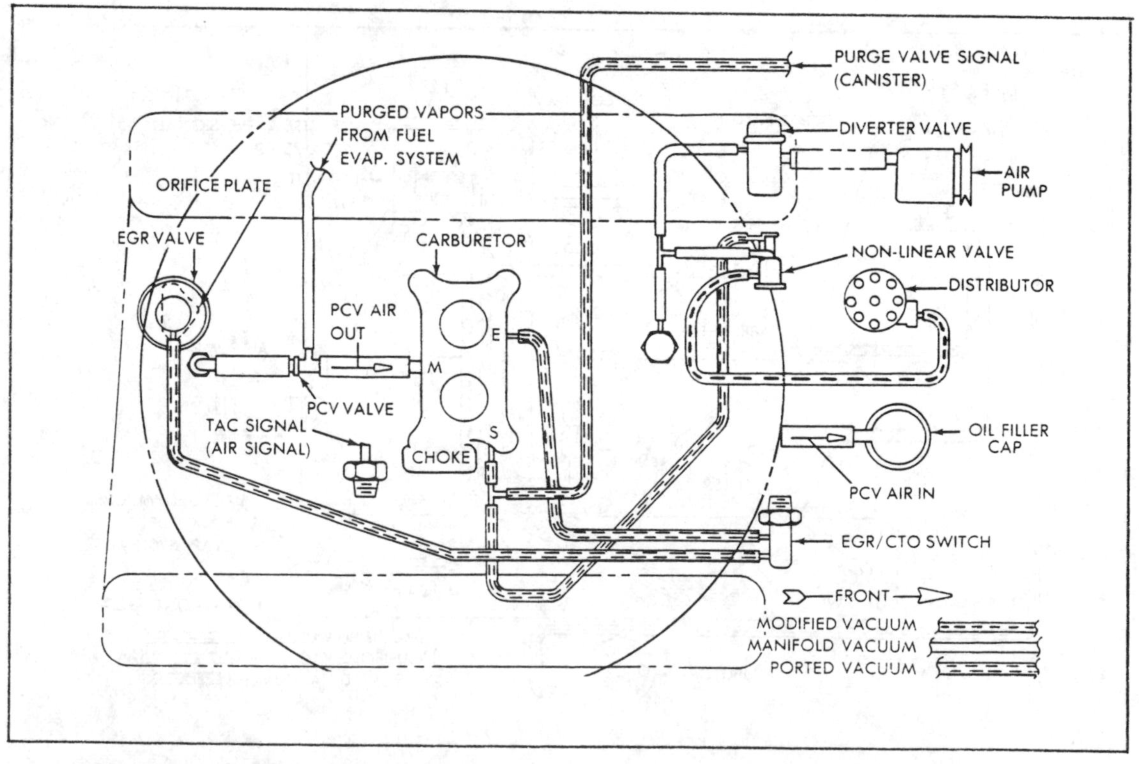

1978-79 Vacuum hose schematic—49 States—304 CID V-8 engine with automatic transmission (EGR valve, orifice plate and the non-linear valve are deleted on the 304 CID V-8 engine with manual transmission)

VACUUM CIRCUITS

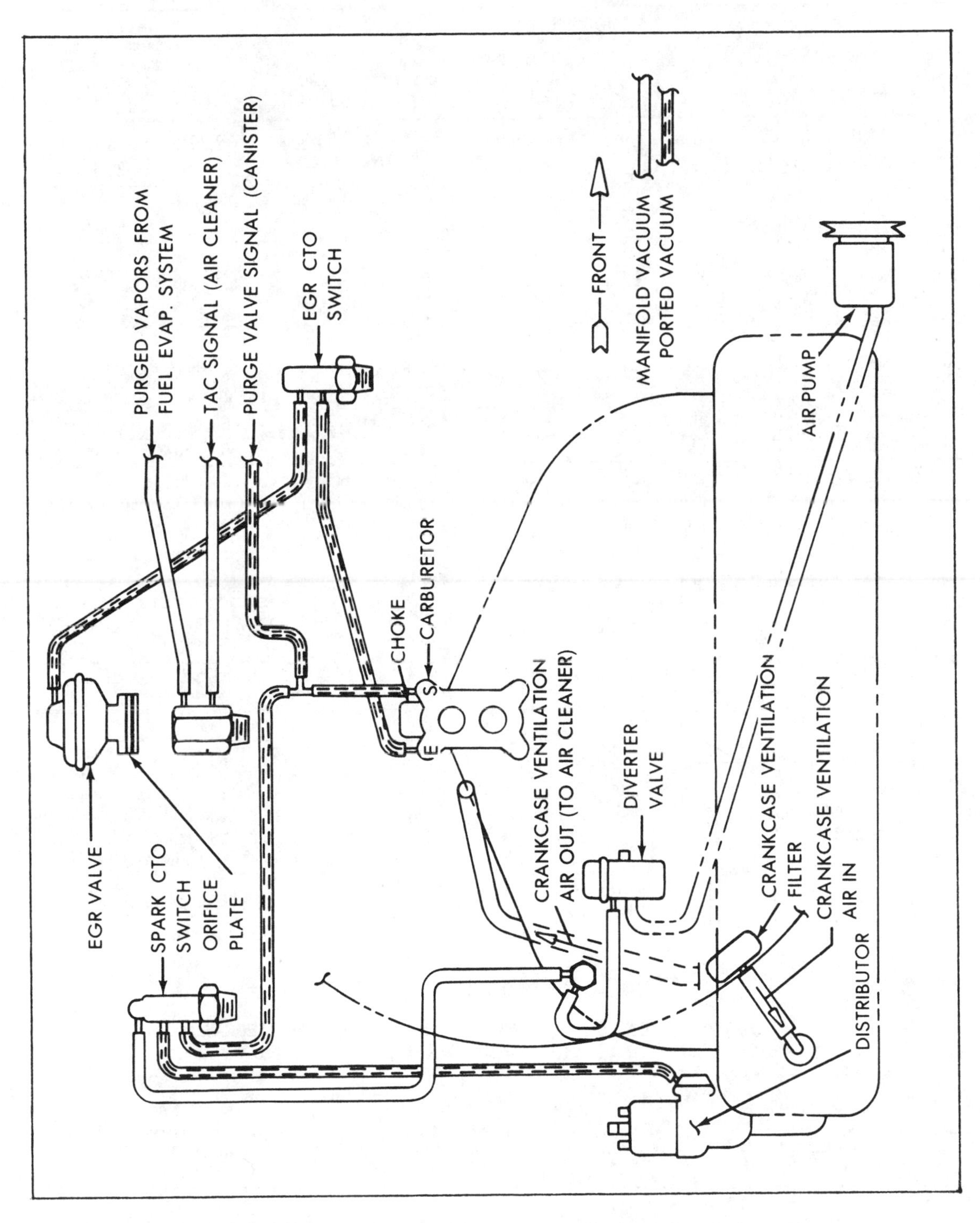

1978-79 Vacuum hose schematic—49 States—121 CID 4 cylinder engine with manual/automatic transmission

BUICK Except Skyhawk & '80 Skyark

TUNE-UP SPECIFICATIONS

ENGINE CODE, 5th CHARACTER OF THE VIN NUMBER
MODEL YEAR CODE, 6th CHARACTER OF THE VIN NUMBER

	ENGINE		SPARK PLUGS		IGNITION			CARBURETION Idle Speed② (rpm)			
						Timing① (deg B.T.D.C. @ rpm)			Auto Trans.		
Year	V.I.N. Code	No. Cyl Disp. (cu in)	Orig. Type	Gap (in)	Dist.	Man. Trans.	Auto Trans.	Man. Trans.	Fed.	Cal.	Hi. Alt.⑤
1979	A	6-231	R46TSX	.060	E.I.	15 @ 800	15 @ 600	600/800	550/670	600	600
	2	6-231	R46TSX	.060	E.I.	—	15	—	—	580/670	—
	C	6-196	R46TSX	.060	E.I.	15 @ 800	15 @ 600	600/800	550/670	—	—
	3	6-231	R46TSX	.060	E.I.	—	15 @ 650	—	650④	650④	650
	H	8-305	R45TS	.045	E.I.	—	③	—	—	500/600	600/650
	K	8-403	R46SZ	.060	E.I.	—	20 @ 1100	—	550/650	500/600	600/700
	L	8-350	R45TS	.045	E.I.	—	8	—	—	500/600	600/650
	X	8-350	R46TSX	.060	E.I.	—	15	—	550	—	—
	G	8-305	R45TS	.045	E.I.	—	4 @ 600	—	525/600	—	—
	R	8-350	R46SZ	.060	E.I.	—	20 @ 1100	—	550/650	500/600	600/700
	Y	8-301	R46TSX	.060	E.I.	—	12 @ 650	—	500/650	—	—
	W	8-301	R46TSX	.060	E.I.	—	12 @ 650	—	500/650	—	—
1980	A	6-231	R45TSX	.060	E.I.	15 @ 550	15 @ 550	600/800	550/670	550/620	—
	3	6-231	R45TS	.040	E.I.	—	15 @ 650	—	650⑥	650⑥	—
	4	6-252	R45TSX	.060	E.I.	—	15 @ 550	—	680/550	—	—
	S	8-256	R45TSX	.060	E.I.	—	10 @ 700	—	650/550	—	—
	W	8-301	R45TSX	.060	E.I.	—	12 @ 500	—	650/550	—	—
	H	8-305	R45TS	.035	E.I.	—	4 @ 550	—	—	650/500	—
	X	8-350	R45TSX	.060	E.I.	—	15 @ 550	—	650/550	—	—
	R	8-350	R46TSX R47TSX	.080 .080	E.I. —	—	18 @ 1100	—	600/500	650/550	—
	N	8-350	Diesel	—	—	—	7 @ 800	—	600	—	—

NOTE: The underhood certification/specification decal is the authority for performance specifications affecting exhaust emissions. Use this manual's information only when that decal is not available.

FED Federal

CAL California

HI. ALT. High Altitude

— Not Applicable

① Time at curb idle speed unless otherwise indicated

② Set idle speed with automatic transmission in Drive, manual transmission in Neutral. Where two figures appear, the lower figure indicates idle speed with A/C solenoid deenergized

③ On California applications, set timing at 4° at 500 rpm. On high altitude applications, set timing to 8° at 600 rpm

④ 650 on Riviera

⑤ See underhood certification/specification decal

⑥ 650/600 on Riviera

BUICK SKYHAWK

TUNE-UP SPECIFICATIONS

ENGINE CODE, 5th CHARACTER OF THE VIN NUMBER
MODEL YEAR CODE, 6th CHARACTER OF THE VIN NUMBER

						SPARK PLUGS		DIST		IGNITION① TIMING (deg BTDC)		VALVES	FUEL	IDLE SPEED① (rpm)	
Year	Eng. V.I.N. Code	Engine No. Cyl. Disp. (cu in)	Eng.⑦ Mfg.	Carb Bbl	H.P.	Orig. Type	Gap (in)	Point Dwell (deg)	Point Gap (in)	Man	Auto	Intake Opens (deg BTDC)	Pump Pres. (psi)	Man Trans	Auto Trans In Drive
1979	1	4-151	P	2v	85	R43TSX	.060	Electronic		12	12	33	4-5.5	②	②
	V	4-151	P	2v	85	R43TSX	.060	Electronic		14	14	33	4-5.5	1000⑤	650⑤
	9	4-151	P	2v	85	R43TSX	.060	Electronic		14	14	33	4-5.5	②	②
	C	6-196	B	2v	90	R46TSX	.060	Electronic		②	②	18	3-4.5③	②	②
	A	6-231	B	2v	115	R46TSX	.060	Electronic		15	15	17	3-4.5③	800	600
	2	6-231	B	2v	115	R46TSX	.060	Electronic		15	15	17	3-4.5③	800	600
	G	8-305	C	2v	145	R45TS	.045	Electronic		4	4(4)	28	7.5-9.0	600	500④
1980	A	6-231	B	2v	115	R45TSX	.060	Electronic		15⑥	15⑥	—	3-4.5③	600	550

① Figures in parentheses are California specs.
② See underhood tune-up decal.
③ At 12.6 volts.
④ High alt. base idle is 600 rpm.
⑤ Base idle 500 rpm.
⑥ At 550 rpm.
⑦ B—Buick
C—Chevrolet
O—Oldsmobile
P—Pontiac

BUICK (Except Skyhawk & '80 Skylark)

DISTRIBUTOR SPECIFICATIONS

Year	Distributor Identification	Centrifugal Advance① Start Crank. Deg. @ Eng. RPM	Centrifugal Advance① Finish Crank. Deg. @ Eng. RPM	Vacuum Advance Start @ In. Hg.	Vacuum Advance Finish Crank. Deg. @ In. Hg.
1979	1103266	0 @ 500	10.5 @ 2000	5	8 @ 11
	1103281	—	18-22 @ 3800	3-6	18 @ 11-13
	1103314	—	18-24 @ 3400	3-5	26 @ 11-13
	1103322	—	26-32 @ 4000	4-6	24 @ 12-14
	1103323	—	16-22 @ 4000	4-5	16 @ 10-12
	1103324	—	20-26 @ 3600	4-6	24 @ 12-14
	1103325	—	12-16 @ 3600	4-5	16 @ 10-12
	1103342	—	14-18 @ 4400	5-7	24 @ 11-13
	1103346	—	16-22 @ 4000	4-6	24 @ 12-14
	1103347	—	12-16 @ 3600	4-6	24 @ 12-14
	1103368	—	18-22 @ 3800	3-6	10 @ 7-9
	1103379	—	18-22 @ 3800	2-4	20 @ 6-9
	1103399	—	18-22 @ 4400	3-5	26 @ 11-13
	1103400	—	14-28 @ 4600	3-5	26 @ 10-13
	1110765	—	12-18 @ 4400	2-4	20 @ 11-13
	1110766	—	12-18 @ 3600	4-5	24 @ 10-12
	1110767	—	12-18 @ 3600	2-4	20 @ 11-13
	1110768	—	12-18 @ 3600	2-4	20 @ 11-13
	1110769	—	12-18 @ 3600	4-5	24 @ 10-12
	1110770	—	12-18 @ 3600	2-4	20 @ 8-10
	1110772	—	12-18 @ 3600	2-4	24 @ 9-11
	1110774	—	8-12 @ 4400	2-4	20 @ 11-13
	1110775	—	12-18 @ 3600	2-4	20 @ 8-10
	1110779	—	12-18 @ 3600	2-4	24 @ 8-11
1980	1110550	0-6 @ 1000	12-18 @ 4400	2-4	20
	1110571	0-6 @ 1000	12-18 @ 4400	2-4	20
	1110551	0-8 @ 1000	8-12 @ 4400	—	—
	1110572	0-8 @ 1000	8-12 @ 4400	—	—
	1110573	Controlled by ECM②			
	1103417	0-4 @ 2200	14-18 @ 4400	5-6	24
	1103447	0-4 @ 2200	14-18 @ 4400	—	—
	1103384	0-8 @ 1000	18-22 @ 4000	3-6	24
	1103386	0-6 @ 1200	18-22 @ 3800	3-5	16
	1103379	0-6 @ 1200	18-22 @ 3800	2-4	20
	1103412	EMR③			
	1103398	0-4 @ 1100	24-28 @ 4400	—	—
	1103413	EMR			
	1103414	EMR			
	1103450	0-6 @ 1300	16-22 @ 3200	2-4	14
	1103444	0-8 @ 1400	12-18 @ 2000	7-8	20
	1103407	0-4 @ 1200	20-26 @ 4400	4-5	20

① RPM and advance given for distributor on engine when using distributor machine divide rpm and degrees by 2.
② Electronic control module (not the HEI module).
③ Electronic module retard which is part of the HEI module on 5.7 Liter V8 engines.

BUICK SKYHAWK

DISTRIBUTOR SPECIFICATIONS

Year	Distributor Identification	Centrifugal Advance: Start Dist. Deg. @ Dist. RPM	Centrifugal Advance: Finish Dist. Deg. @ Dist. RPM	Vacuum Advance: Start @ In. Hg.	Vacuum Advance: Finish Dist. Deg. @ In. Hg.
1979	1103229	0 @ 600	10 @ 2200	3.5	10 @ 12
	1103231	0 @ 600	10 @ 2200	3.5	10 @ 12
	1103239	0 @ 600	10 @ 2100	4	5 @ 8
	1103244	0 @ 500	10 @ 1900	4	10 @ 10
	1103282	0 @ 500	10 @ 1900	4	10 @ 10
	1103285	0 @ 600	11 @ 2100	4	5 @ 8
	1103365	0 @ 850	10 @ 2325	5	8 @ 11
	1110726	0 @ 500	9 @ 2000	4	10 @ 10
	1110757	0 @ 600	9 @ 2000	4	10 @ 10
	1110766	0 @ 810	7.5 @ 1800	3	10 @ 9
	1110767	0 @ 840	7.5 @ 1800	4	12 @ 11
1980	1110555	2-8 @ 1600	12-24 @ 3600	3-5	24
	1110554	0-4 @ 2000	12-18 @ 3600	2-4	24
	1110784	Controlled by ECM			

2MC, M2MC, M2ME, E2ME CARBURETOR SPECIFICATIONS

BUICK

Year	Carburetor Identification①	Float Level (in.)	Choke Rod (in.)	Choke Unloader (in.)	Vacuum Break Lean or Front (in.)	Vacuum Break Rich or Rear (in.)	Pump Rod (in.)	Choke Coil Lever (in.)	Automatic Choke (notches)
1979	17059134	15/32	0.243	0.243	0.157	—	1/4	0.120	1 Lean
	17059136	15/32	0.243	0.243	0.157	—	1/4	0.120	1 Lean
	17059193	13/32	0.139	0.220	0.103	0.090	1/4②	0.120	2 Rich
	17059194	11/32	0.139	0.220	0.103	0.090	1/4②	0.120	2 Rich
	17059190	11/32	0.139	0.243	0.103	0.090	1/4②	0.120	2 Rich
	17059191	11/32	0.139	0.243	0.103	0.090	9/32②	0.120	2 Rich
	17059491	11/32	0.139	0.277	0.129	0.117	9/32②	0.120	1 Rich
	17059492	11/32	0.139	0.277	0.129	0.117	9/32②	0.120	1 Rich
	17059196	11/32	0.139	0.277	0.129	0.117	1/4②	0.120	1 Rich
	17059498	11/32	0.139	0.277	0.129	0.117	9/32②	0.120	2 Rich
	17059180	11/32	0.139	0.243	0.103	0.090	1/4②	0.120	2 Rich
	17059184	11/32	0.139	0.220	0.103	0.090	1/4②	0.120	2 Rich
	17059496	5/16	0.139	0.243	0.117	0.179	3/8②	0.120	2 Rich
1980	17080496	5/16	0.139	0.243	0.117	0.203	3/8	0.120	Fixed
	17080498	5/16	0.139	0.243	0.117	0.203	3/8	0.120	Fixed
	17080490	5/16	0.139	0.243	0.117	0.203	3/8	0.120	Fixed
	17080492	5/16	0.139	0.243	0.117	0.203	3/8	0.120	Fixed
	17080491	5/16	0.139	0.243	0.117	0.220	3/8	0.120	Fixed
	17080190	9/32	0.139	0.243	0.123	0.110	1/4②	0.120	Fixed
	17080191	11/32	0.139	0.243	0.096	0.096	1/4②	0.120	Fixed
	17080195	9/32	0.139	0.243	0.103	0.071	1/4②	0.120	Fixed
	17080197	9/32	0.139	0.243	0.103	0.071	1/4②	0.120	Fixed
	17080192	9/32	0.139	0.243	0.123	0.110	1/4②	0.120	Fixed
	17080160	5/16	0.074	0.239	0.168	0.207	1/4②	0.120	Fixed

QUADRAJET CARBURETOR SPECIFICATIONS

BUICK

Year	Carburetor Identification①	Float Level (in.)	Air Valve Spring (turn)	Pump Rod (in.)	Primary Vacuum Break (in.)	Secondary Vacuum Break (in.)	Secondary Opening (in.)	Choke Rod (in.)	Choke Unloader (in.)	Fast Idle Speed ④ (rpm)
1979	17059240	7/32	3/4	9/32	0.117	0.117	②	0.074	0.179	⑥
	17059243	7/32	3/4	9/32	0.117	0.117	②	0.074	0.179	⑥
	17059540	7/32	3/4	9/32	0.117	0.129	②	0.074	0.243	⑥
	17059543	7/32	3/4	9/32	0.117	0.129	②	0.074	0.243	⑥
	17059242	7/32	3/4	9/32	0.066	0.066	②	0.074	0.179	⑥
	17059553	13/32	1/2	9/32	0.136	0.230	②	0.103	0.220	⑥
	17059555	13/32	1/2	9/32	0.149	0.230	②	0.103	0.220	⑥
	17059250	13/32	1/2	9/32	0.129	0.182	②	0.096	0.220	⑥
	17059253	13/32	1/2	9/32	0.129	0.182	②	0.096	0.220	⑥
	17059208	15/32	7/8	9/32	—	0.129	②	0.314	0.277	⑥
	17059209	15/32	7/8	9/32	—	0.129	②	0.314	0.277	⑥
	17059210	15/32	1	9/32	0.157	—	②	0.243	0.243	⑥
	17059211	15/32	1	9/32	0.157	—	②	0.243	0.243	⑥
	17059228	15/32	1	9/32	0.157	—	②	0.243	0.243	⑥
	17059241	5/16	3/4	3/8	0.120	0.113	②	0.096	0.243	⑥
	17059247	5/16	3/4	3/8	0.110	0.103	②	0.096	0.243	⑥
	17059272	15/32	5/8	3/8	0.136	0.195	②	0.074	0.220	⑥
1980	17080240	3/16	9/16	9/32③	0.083	0.083	②	0.074	0.179	⑥
	17080241	7/16	3/4	9/32③	0.129	0.114	②	0.096	0.243	⑥
	17080242	13/32	9/16	9/32③	0.077	0.096	②	0.074	0.220	⑥
	17080243	3/16	9/16	9/32③	0.083	0.083	②	0.074	0.179	⑥
	17080244	5/16	5/8	9/32③	0.096	0.071	②	0.139	0.243	⑥
	17080249	7/16	3/4	9/32③	0.129	0.114	②	0.096	0.243	⑥
	17080253	13/32	1/2	9/32③	0.149	0.211	②	0.090	0.220	⑥
	17080259	13/32	1/2	9/32③	0.149	0.211	②	0.090	0.220	⑥
	17080270	15/32	5/8	3/8⑦	0.149	0.211	②	0.074	0.220	⑥
	17080271	15/32	5/8	3/8⑦	0.142	0.211	②	0.110	0.203	⑥
	17080272	15/32	5/8	3/8⑦	0.129	0.175	②	0.074	0.203	⑥
	17080502	1/2	7/8	Fixed	0.136	0.179	②	0.110	0.243	⑥
	17080504	1/2	7/8	Fixed	0.136	0.179	②	0.110	0.243	⑥
	17080540	3/8	9/16	Fixed	0.103	0.129	②	0.074	0.243	⑥
	17080542	3/8	9/16	Fixed	0.103	0.066	②	0.074	0.243	⑥
	17080543	3/8	9/16	Fixed	0.103	0.129	②	0.074	0.243	⑥
	17080553	15/32	1/2	Fixed	0.142	0.220	②	0.090	0.220	⑥
	17080554	15/32	1/2	Fixed	0.142	0.211	②	0.090	0.220	⑥

① The carburetor identification number is stamped on the float bowl, near the secondary throttle lever.
② No measurement necessary on two point linkage; see text
③ Inner hole
④ On low step of cam, automatic in Drive through 1974; on high step of cam, automatic in Park starting 1975.
⑤ 3 turns after contacting lever for preliminary setting
⑥ 2 turns after contacting lever for preliminary setting
⑦ Outer hole

CAR SERIAL NUMBER AND ENGINE IDENTIFICATION

1979

The vehicle identification number is stamped on a plate which is attached to the top left side of the instrument panel. It is visible through the lower left hand corner of the windshield. The sixth character designates the model year; 9 represents 1979. The fifth character designates the engine code, as follows:

A .... 3.8L (231 CID) 2 bbl V-6 Buick
2 .... 3.8L (231 CID) 2 bbl V-6 Buick
3 (turbocharged) .. 3.8L (231 CID) 4 bbl V-6 Buick
C .... 3.2L (196 CID) 2 bbl V-6 Buick
G (turbocharged) .. 5.0L (305 CID) 2 bbl V-8 Buick
H .... 5.0L (305 CID) 4 bbl V-8 Chevrolet
K .... 6.5L (403 CID) 4 bbl V-8 Oldsmobile
L .... 5.7L (350 CID) 4 bbl V-8 Chevrolet
R .... 5.7L (350 CID) 4 bbl V-8 Oldsmobile
W .... 4.9L (301 CID) 4 bbl V-8 Pontiac
X .... 5.7L (350 CID) 4 bbl V-8 Buick
Y .... 4.9L (301 CID) 2 bbl V-8 Pontiac

NOTE: *Codes "3" and "G" are the turbocharged engines. Code "A" is the standard 2 bbl/V-6 engine. Code "2" is the 2 bbl/V-6 engine equipped with the electronic fuel control system.*

1980

The vehicle identification plate is attached to the top left side of the instrument panel. It is visible through the lower left hand corner of the windshield. The sixth character designates the model year: A represents 1980. The fifth character represents the engine code.

A .... 3.8L (231 CID) 2bbl V-6 Buick
3 (turbocharged) .. 3.8L (231 CID) 2bbl V-6 Buick
4 .... 4.1L (252 CID) 4bbl V-6 Buick
W .... 4.9L (301 CID) 4bbl V-8 Pontiac
H .... 5.0L (305 CID) 4bbl V-8 Chevrolet
N (diesel) .... 5.7L (350 CID) Diesel V-8 Oldsmobile
R .... 5.7L (350 CID) 4bbl V-8 Oldsmobile
X .... 5.7L (350 CID) 4bbl V-8 Buick

EMISSION EQUIPMENT

1979

The emission control systems for most engines consist of the following:

Calibrated carburetion
Calibrated spark distribution
Catalytic converter
Early fuel evaporation (EFE)
Exhaust gas recirculation (EGR)
Positive crankcase ventilation (PCV)
Calibrated carburetor choke
Thermostatic air cleaner (TAC)
Evaporative emission control system (EECS)

Included in some systems is an additional emission control device called an air injection reaction (AIR).

EFE thermal vacuum switch (EFE-TVS)
- 305 CID (G)
- 231 CID (3)
- 301 CID (Y)
- 305 CID (H)
- 350 CID (L)
- 301 CID (W)

EFE-EGR thermal vacuum switch (EFE-EGR-TVS)
- 196 CID (C)
- 231 CID (3)
- 231 CID (A)
- 231 CID (2)
- 350 CID (X)

EFE check valve (EFE-CV)
- 196 CID (C)
- 305 CID (G)
- 231 CID (3)
- 231 CID (A)
- 305 CID (H)
- 350 CID (L)

EGR thermal control valve (EGR-TCV)
- 350 CID (R)
- 403 CID (K)

Canister purge thermal vacuum switch (CP-TVS)
- 305 CID (G)
- 305 CID (H)
- 350 CID (X)
- 350 CID (L)

Spark delay valve (SDV)
- 305 CID (G)
- 305 CID (H)

Early fuel evaporation distributor thermal vacuum switch (EFE-DTVS)
- 305 CID (G) California only

Distributor vacuum regulating valve
- 301 CID (Y)

Spark retard delay valve
- 301 CID (Y)

Distributor thermal vacuum switch (DTVS)
- 196 CID (C)
- 305 CID (G)
- 231 CID (A)
- 301 CID (Y)
- 305 CID (H)
- 350 CID (R)
- 403 CID (K)
- 301 CID (W)

EGR distributor thermal vacuum switch (EGR-DTVS)
- 350 CID (R) California only
- 403 CID (K)

Spark advance vacuum modulator system (SAVM)
- 196 CID (C)
- 231 CID (A)

Distributor spark vacuum modulator valve (DS-VMV)
- 301 CID (W)

Distributor vacuum delay valve (DVDV)
- 301 CID (W)

Choke thermal vacuum switch (CTVS)
- 301 CID (Y)
- 350 CID (X)
- 350 CID (R)
- 403 CID (K)

Secondary vacuum break thermal vacuum switch (SUB-TVS)
- 231 CID (3) 49 states only
- 301 CID (W)

Idle speed up solenoid
- 196 CID (C) with automatic transmission and air conditioning
- 231 CID (3) E series vehicles
- 231 CID (A) with automatic transmission and air conditioning

Anti-dieseling solenoid
- 196 CID (C) A, H, and X series vehicles with manual transmission
- 231 CID (A) A, H, and X series vehicles with manual transmission

Air injection reaction system (AIR)
- 305 CID (G) California only
- 231 CID (3)
- 231 CID (A) California only
- 305 CID (H) California and high altitude only
- 350 CID (L) California and high altitude only
- 350 CID (R) California only
- 403 CID (K) California only

Thermostatic air cleaner thermal check valve (TAC-TCV)
- 231 CID (3) E series vehicles

1980

The emission control systems for most engines consist of the following:

EMISSION EQUIPMENT

Calibrated carburetion
Catalytic converter
Early fuel evaporation (EFE)
Exhaust gas recirculation (EGR)
Positive crankcase ventilation (PCV)
Calibrated carburetor choke
Thermostatic air cleaner (TAC)
Evaporative emission control system (EECS)

Included in some systems is an additional emission control device called and air injection reaction (AIR).

EFE thermal vacuum switch (EFE-TVS)
- 305 CID (G)
- 231 CID (3)
- 305 CID (H)
- 301 CID (W)

EFE-EGR thermal vacuum switch (EFE-EGR-TVS)
- 231 CID (3)
- 231 CID (A)
- 350 CID (X)

EFE check valve (EFE-CV)
- 231 CID (3)
- 231 CID (A)
- 305 CID (H)

EGR thermal control valve (EGR-TCV)
- 350 CID (R)

Canister purge thermal vacuum switch (CP-TVS)
- 305 CID (H)
- 350 CID (X)

Spark delay valve (SDV)
- 305 CID (H)

Early fuel evaporation distributor thermal vacuum switch (EFE-DTVS)
- 305 CID (H)

Distributor spark vacuum regulator valve (DS-VRV)
- 256 CID (S)

Distributor thermal vacuum switch (DTVS)
- 231 CID (A)
- 256 CID (S)
- 301 CID (W)
- 305 CID (H)
- 350 CID (R)

Engine temperature Sensor (ETS)
- 350 Diesel (N)

EGR distributor thermal vacuum switch (EGR-DTVS)
- 350 CID (R) California only

EGR vacuum switch (EGR-VS)
- 350 Diesel (N)

EGR canister purge thermal vacuum switch (EGR/CP-TVS)
- 305 CID (H)

EGR early fuel evaporation thermal vacuum switch (EGR/EFE-TVS)
- 231 CID (A)
- 350 CID (R)
- 350 CID (X)

Spark advance vacuum modulator (SAVM)
- 231 CID (A)

Distributor spark vacuum modulator valve (DS-VMV)
- 301 CID (W)

Distributor vacuum delay valve (DVDV)
- 301 CID (W)
- 256 CID (S)

Choke thermal vacuum switch (CTVS)
- 350 CID (R)
- 350 CID (X)

Secondary vacuum break thermal vacuum switch (SUB-TVS)
- 231 CID (3) 49 states only
- 301 CID (W)

Idle speed up solenoid
- 231 CID (3) E series vehicles
- 231 CID (A) Automatic transmission and air conditioning

Anti-dieseling solenoid
- 231 CID (A) A, H and X series vehicles with manual trans.

Air injection reaction (AIR)
- 231 CID (3)
- 231 CID (A) California only
- 305 CID (H) California only
- 350 CID (R) California only

Thermostatic air cleaner thermal check valve (TAC-TCV)
- 231 CID (3) E series vehicles

Vacuum regulator valve (VRV)
- 350 Diesel (N)

CAUTION: *Always check the emission label under the hood of the vehicle being serviced for complete emission equipment listings.*

IDLE SPEED AND MIXTURE ADJUSTMENT

1979

Air cleaner .... In place
A/C compressor lead .. Disconnected
Auto. trans. .... Drive

NOTE: *Refer to the emission label under the hood of the vehicle being serviced*

RPM

196 V-6 (6) 49 states
- Auto. Trans.
  - Propane enriched 575
  - OFF idle 550
  - Solenoid screw (Curb idle) . 670
  - Fast idle (Park) 2200
- Man. Trans.
  - Propane enriched 1000
  - OFF idle 600
  - Solenoid screw (Curb idle) . 800
  - Fast idle 2200

231 V-6 (A) 49 states
- Auto. Trans.
  - Propane enriched 575
  - OFF idle 550
  - Solenoid screw (Curb idle) . 670
  - Fast idle (Park) 2200
- Man. Trans.
  - Propane enriched 1000
  - OFF idle 600
  - Solenoid screw (Curb idle) . 800
  - Fast idle 2200

231 V-6 (A) Calfornia and high altitude
- Auto. Trans.
  - Propane enriched 615
  - OFF idle 600
  - Solenoid screw (Curb idle) . NA
  - Fast idle (Park) 2200
- Man. Trans.
  - Propane enriched 840
  - OFF idle 600
  - Solenoid screw (Curb idle) 800
  - Fast idle 2200

231 V-6 (2) California only
- Auto. Trans.
  - OFF idle 580
  - Solenoid screw (Curb idle) . 670
  - Fast idle (Park) 2200

231 V-6 (3) 49 states
- Auto. Trans.
  - OFF idle 650
  - Fast idle (Park 2500

231 V-6 (3) California and high altitude
- Auto. Trans.
  - OFF idle 650
  - Fast idle (Park) 2500

231 V-6 (3) 49 states and California "E" series only
- Auto. Trans.
  - OFF idle 600
  - Solenoid screw (Curb idle) . 650
  - Fast idle (Park) 2500

301 V-8 (Y) 49 states
- Auto. Trans.
  - Propane enriched 530
  - OFF idle 500
  - Solenoid screw (Curb idle) . 650
  - Fast idle (Park) 2000

301 V-8 (W) 49 states
- Auto. Trans.
  - Propane enriched 540
  - OFF idle 500
  - Solenoid screw (Curb idle) . 650
  - Fast idle (Park) 2200

305 V-8 (G) 49 states
- Auto. Trans.
  - Propane enriched 520-540
  - OFF idle .... 550 A/C —500 without A/C
  - Solenoid screw (Curb idle) . 600
  - Fast idle (Park) 1600

305 V-8 (H) California only
- Auto. Trans.
  - Propane enriched 520-560
  - OFF idle 500
  - Solenoid screw (Curb idle) . 600
  - Fast idle (Park) 1600

305 V-8 (H) High altitude
- Auto. Trans.
  - Propane enriched 630-670
  - OFF idle 600
  - Solenoid screw (Curb idle) . 650

IDLE SPEED AND MIXTURE AJDUSTMENT

Fast idle (Park) 1750

350 V-8 (L) California only
Auto. Trans.
Propane enriched 520-560
OFF idle 500
Solenoid screw (Curb idle) . 600
Fast idle (Park) 1600

350 V-8 (L) High altitude
Auto. Trans.
Propane enriched 630-670
OFF idle 600
Solenoid screw (Curb idle) . 650
Fast idle (Park) 1750

350 V-8 (R) 49 states
Auto. Trans.
Propane enriched 625-640
OFF idle 550
Solenoid screw (Curb idle) . 650
Fast idle (Park) 900

350 V-8 (R) California only
Auto. Trans.
Propane enriched 565-585
OFF idle 500-550
Solenoid screw (Curb idle) . 600
Fast idle (Park) 1000

350 V-8 (R) High altitude
Auto. Trans.
Propane enriched 590
OFF idle 550-600
Solenoid screw Curb idle) . 700
Fast idle (Park) 900

350 V-8 (X) 49 states
Auto. Trans.
Propane enriched 590
OFF idle 550
Fast idle (Park) 1500

403 V-8 (K) 49 states
Auto. Trans.
Propane enriched 625-645
OFF idle 550
Solenoid screw (Curb idle) . 650
Fast idle (Park) 900

403 V-8 (K) California only
Auto. Trans.
Propane enriched 565-585
OFF idle 500-550
Solenoid screw (Curb idle) . 600
Fast idle (Park) 1000

403 V-8 (K) High altitude
Auto. Trans.
Propane enriched 590
OFF idle 600
Solenoid screw (Curb idle) . 700
Fast idle (Park) 1000

1980

Air cleaner ....In place
A/C compressor lead ..Disconnected
Automatic trans. lever .....In drive

NOTE: *Refer to the emission label under the hood of the vehicle being serviced.*

231 V-6 (A) 49 States
Auto Trans.
Propane enriched 600
Curb idle 550
Fast idle ....2000 in park
A/C on (solenoid) 670
Manual Trans. 49 States
Propane enriched 830
Curb idle ....800 in neutral
Fast idle ....2200 in park
A/C on (solenoid) 800 in neutral

231 V-6 (A) California
Auto. Trans.
Propane enriched 600
Curb idle 550
Fast idle ....2200 in park
A/C on (solenoid) 620

231 V-6 (3) 49 States
Auto. Trans.
Propane enriched 715
Curb idle 650
Fast idle ....2200 in park
A/C on solenoid) 650

231 V-6 (3) Riviera 49 States
Auto. Trans.
Propane enriched 700
Curb idle 600
Fast idle ....2200 in park
A/C on (solenoid) 650

231 V-6 (3) Riviera California
Auto. Trans.
Propane enriched 700
Curb idle 600
Fast idle ....2500 in park
A/C on solenoid) 650

256 V-8 (S) 49 States
Auto. Trans.
Propane enriched 590
Curb idle 550
Fast idle ....2200 in neutral
A/C in (solenoid) 650

301 V-8 (W) 49 States
Auto. Trans.
Propane enriched 550
Curb idle 500
Fast idle ....2500 in neutral
A/C on (solenoid) 650

305 V-8 (H) California
Auto. Trans.
Propane enriched 560
Curb idle 550
Fast idle ....220 in neutral
A/C on (solenoid) 650

252 V-6 (4) 49 States
Auto. Trans.
Propane enriched 590
Curb idle 550
Fast idle ....2000 in park
A/C on (solenoid) 650

350 V-8 (X) 49 States
Auto. Trans.
Propane enriched 590
Curb idle 550
Fast idle 1850
A/C on (solenoid) 670

350 V-8 (R) 49 States
Auto. Trans.
Propane enriched 575
Curb idle 500
Fast idle ....700 in drive
A/C on (solenoid) 600

350 V-8 (R) California
Auto. Trans.
Propane enriched 575
Curb idle 550
Fast idle ....700 in drive
A/C on (solenoid) 650

INITIAL TIMING

1979

NOTE: *Distributor vacuum hose must be disconnected and plugged. Set timing at idle speed unless shown otherwise.*

196 V-6 ....15° BTDC
231 V-6 ....15° BTDC
231 V-6 Turbo ....15° BTDC
301 V-8 ....12° BTDC
305 2-bbl. V-8
49-States .... 4° BTDC
Calif. .... 4° BTDC
Altitude .... 8° BTDC
305 4-bbl. V-8 .... 4° BTDC
350 V-8 "X" ....15° BTDC
350 V-8 "L" .... 8° BTDC
350 V-8 "R" (at 1100 rpm)
All ....20° BTDC
403 V-8 "K" (at 1100 rpm)
All ....20° BTDC

1980

NOTE: *Distributor vacuum hose must be disconnected and plugged. Set timing at idle speed unless shown otherwise.*

231 V-6 ....15° BTDC
231 V-6 Turbo ....15° BTDC
252 V-6 ....15° BTDC
256 V-8 ....10° BTDC
305 V-8 California ...... 4° BTDC
350 V-8 "X" ....15° BTDC
350 V-8 "R" 49 States
(at 1100 rpm) ....18° BTDC
350 V-8 "R" California C-4
(at 1100 rpm) ....18° BTDC
350 V-8 "R" California Riviera
(at 1100 rpm) ....16° BTDC
350 V-8 Diesel
(at 800 rpm) .... 7° BTDC

SPARK PLUGS

1979

196	V-6	AC-R46TSX—	.060
231	V-6	AC-R46TSX	.060
231	V-6 Turbo	AC-R44TSX	.060
301	V-8	AC-R46TSX	.060
305	V-8	AC-R45TS	.045
350	V-8 "X"	AC-R46TSX	.060
350	V-8 "L"	AC-R45TS	.045
350	V-8 "R"	AC-R46SZ	.060
403	V-8	AC-R46SZ	.060

1980

231	V-6	R45TSX	.060
231	V-6 Turbo	R45TS	.040
252	V-6	R45TSX	.060
256	V-8	R45TSX	.060
305	V-8 California	R45TS	.035
350	V-8 "X"	R45TSX	.060
350	V-8 "R" All	R46SX	.080
Optional heat range		R47SX	.080

VACUUM ADVANCE

1979

Diaphragm type ....Single
Vacuum source
All V-6 ....Ported
301 V-8
Auto. trans. no A.C. ...Manifold
Auto. trans. A.C. ....Modulated
305 V-8
49-States ....Manifold
Calif. ....Ported
Altitude ....Manifold
350 V-8 "X" ....Ported
350 V-8 "L" ....Manifold
350 V-8 "R"
Calif. ....Ported
Altitude ....Ported
403 V-8
49-States ....Manifold
Calif. ....Ported
Altitude ....Ported

1980

Use the vehicle emission control information label for the most up to date information on the vacuum hose routings.

EMISSION CONTROL SYSTEMS

V-6 VACUUM 1979-80

49-State manual transmission V-6 engines (no turbocharger) have the distributor vacuum advance connected to the EFE-EGR Thermal Vacuum Switch. The vacuum advance hose connects, through a "T" to the EGR part of the switch. The switch is closed below approximately 120°F. engine coolant temperature, to eliminate exhaust gas recirculation when the engine is cold. Because the vacuum advance is connected to the EGR vacuum source, there is no vacuum advance until the engine warms up above 120°F. Above that temperature, the vacuum advance operates off of ported vacuum which is only available above idle.

With the vacuum advance and EGR diaphragms both connected to the same vacuum source, a leak in either diaphragm will affect the other. For example, a leaking EGR diaphragm might make the EGR valve stay closed, but it might also bleed off so much vacuum that the distributor vacuum unit does not advance the spark.

TURBOCHARGER 1979-80

A turbocharger is a compressor, driven by the engine exhaust. It compresses the air-fuel mixture and forces it into the combustion chamber for increased power and acceleration.

A turbocharger consists of two turbine wheels, each in its own separate housing, and connected by a shaft. One of the turbine wheels is driven by the pressure of the exhaust gas leaving the engine. Buick takes the exhaust gas from the rear of the left bank exhaust manifold and routes it into the rear of the right bank manifold. All

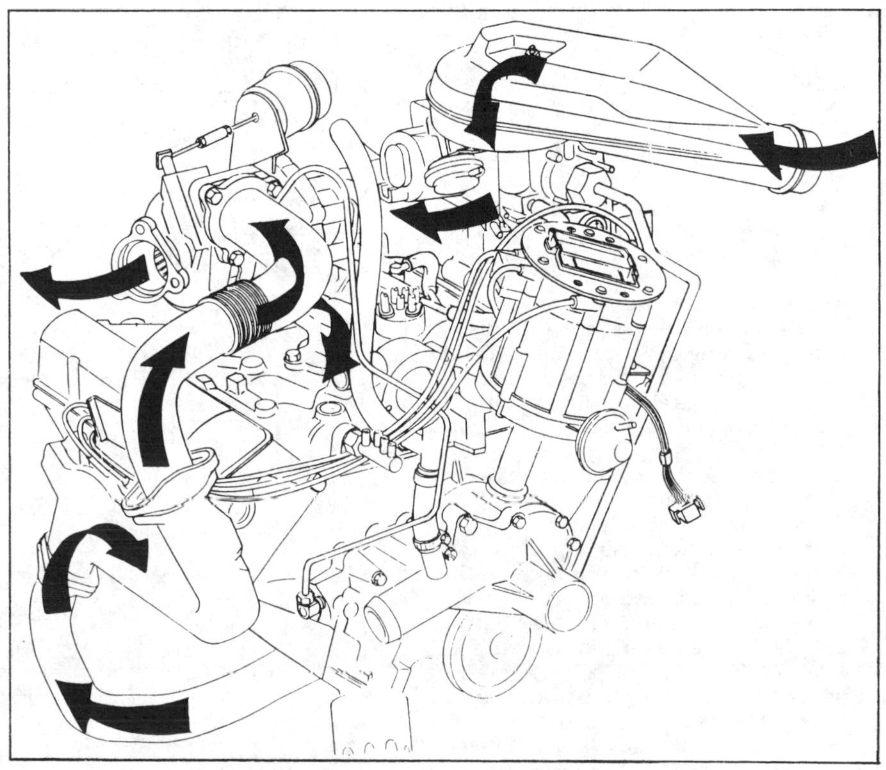

Air flow of 1979 Buick Turbocharged V-6 engine

the exhaust from the engine exits the front of the right exhaust manifold, and goes to the turbine housing.

The pressure and flow of the exhaust makes the turbine wheel turn, just like air blowing on a windmill. The turbine wheel is connected by a shaft to the compressor wheel. The compressor wheel accepts the air-fuel mixture from the carburetor at its inlet, and pushes this mixture into the engine through its outlet. The outlet of the compressor is bolted to the intake manifold to the engine cylinders the same as any intake manifold.

The two turbine wheels, connected by the shaft, are completely free to turn at any time. There is no clutch or brake of any kind that prevents them from turning. When the engine is idling, exhaust pressure is low, so the compressor wheel turns from the force of the incoming air-fuel mixture going into the engine. The compressor contributes a pressure increase, known as "boost" only when the exhaust pressure is high. Usually, the only time boost occurs is during wide open throttle.

When the driver steps on the gas pedal to accelerate, he will get a normal amount of acceleration immediately. One or more seconds later, the exhaust presure will build up enough so that the compressor starts supplying boost. Then he will get additional acceleration. This delay is always present, and does not indicate that anything is wrong. It takes time for the exhaust pressure to build up and start the boost.

To keep the boost from getting too high, and possibly causing engine damaging detonation, a pressure-operated wastegate is used. A vacuum diaphram is connected to the pressure side of the compressor housing. When the pressure reaches 8 pounds, the diaphragm moves a rod attached to the wastegate. The wastegate then opens and lets some of the exhaust pressure bypass the exhaust on the turbine and the compressor wheel slows slightly, which lowers the pressure.

At 8 pounds of pressure, the wastegate actuating rod only moves about .008 in. This is such a small movement that you can barely see it move, but it is enough to open the wastegate slightly and lower the pressure on the turbine wheel. A threaded end on the rod allows adjustment.

The vacuum diaphragm actuator that moves the wastegate has two hoses attached to it. The hose on the end is from the compressor housing and senses pressure. The hose on the side is connected to intake manifold vacuum. The amount of vacuum that acts on the diaphragm has very little effect. The main reason for the hose is to provide the escape path for gas vapors in case the diaphragm should rupture. If vapors should pass through the diaphragm they will immediately be drawn into the engine.

The wastegate opening can be checked by applying 8 lbs. pressure with a hand pump to the acuator, or by temporarily installing a pressure gauge and driving the car. A full throttle acceleration from zero to 50 should produce 7-8 psi for the 2-bbl. engine, and 8-9 psi on the 4-bbl. engine. A maximum of 10 psi is allowed on the 4-bbl. engine.

Vacuum and pressure switches are mounted in the engine compartment and connected to the intake manifold with hoses. When the intake manifold has low vacuum, the vacuum switch illuminates a yellow light on the instrument panel, indicating moderate acceleration. When the intake manifold is under pressure, the pressure switch lights an orange light on the instrument panel, indicating full acceleration.

Because turbochargers spin at extremely high speeds, their shaft bearings need full pressure lubrication. An oil pressure line comes from the engine block to the turbocharger shaft. Oil runoff from the shaft bearings runs back into the engine through a rubber tube under the turbocharger shaft.

The turbocharger is downstream from the carburetor. With this design, the turbocharger not only pumps air-fuel mixture into the engine, but also creates a tremendous suction that pulls air through the carburetor. This suction is highest at wide open throttle, when the turbocharger is providing full boost. This suction creates high vacuum between the carburetor and the turbocharger. A high vacuum under the carburetor throttle blades at wide open throttle is exactly the opposite of what you get on an unblown engine, where the vacuum drops to zero at wide open throttle.

If the high vacuum at wide open throttle, is allowed to operate the distributor vacuum advance, the EGR valve, and the carburetor power piston, those units will act as if the engine is decelerating or cruising. The distributor would go to full vacuum advance, while the EGR valve opened wide, and the carburetor power piston would close the power valve. Of course, this is exactly the opposite of what is needed at wide open throttle. We need no vacuum advance, no EGR and want the power valve wide open.

To get the action needed, an extra valve is added to the intake manifold. It senses intake manifold pressure, and shuts off the vacuum when the turbo is operating. On the 2-bbl. engine, the valve is called the Turbocharger Vacuum Bleed Valve (TVBV).

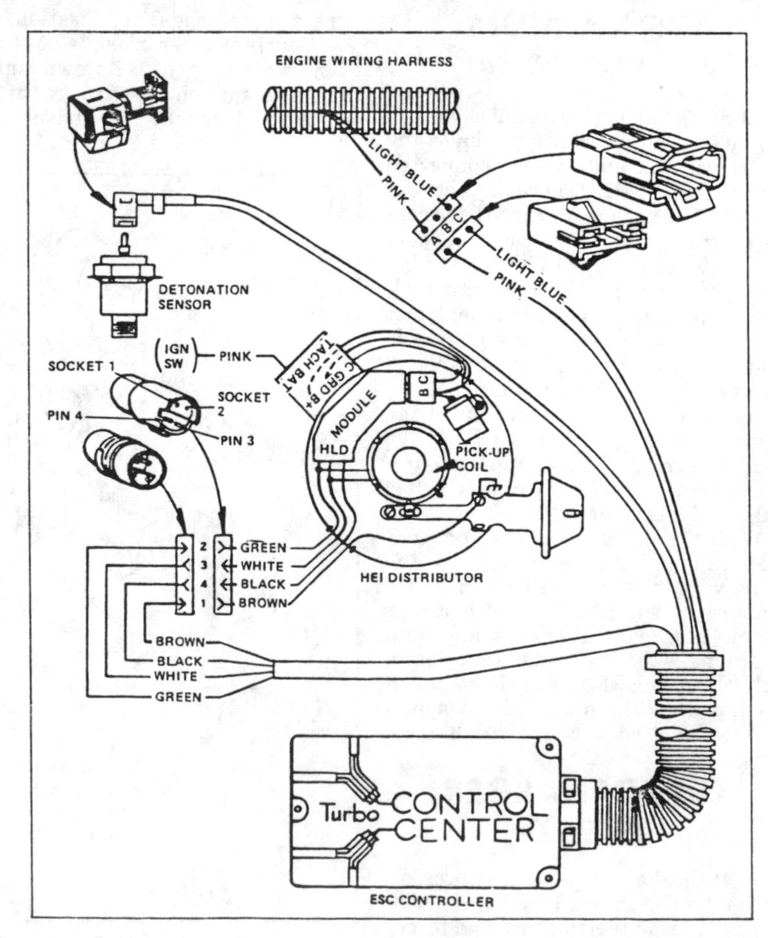

1979 Buick Turbo V-6 electronic spark control system

When pressure in the intake manifold gets above 3 psi. the TVBV shuts off and bleeds down the vacuum to the distributor vacuum advance, EGR valve, and carburetor power piston.

On the 4-bbl. engine, the valve is called the Power Enrichment Control Valve (PECV). Because the 4-bbl. carburetor has a large area at wide open throttle, the vacuum does not go as high. It is only necessary to shut off vacuum to the carburetor power piston. Any time pressure in the intake manifold goes above 3 psi, the PECV shuts off the vacuum to the carburetor power piston.

The TVBV used on the 2-bbl. engine has seven hoses connected to it. Each pair of hoses brings vacuum to the valve, and then out to the EGR valve, distributor, or power piston. The seventh hose, in the middle of the valve, is the bleed. It is connected to the carburetor so that only clean air will enter the system.

The PECV used on the 4-bbl. engine has only three hoses connected to it. Two of them supply vacuum to the power valve. and the third, on the right side, is the bleed.

ELECTRONIC SPARK CONTROL 1979-80

The boost pressure that a turbocharger provides is similar to increasing the compression ratio on an engine. Higher compression and added boost both require less spark advance to avoid detonation. Even though the spark advance on the turbocharged engine has been tailored to the engine, there is stil a chance of detonation, mainly because of the low octane, nonleaded gas used.

To prevent detonation that might cause engine damage, Buick uses an Electronic Spark Control that is a retarder. Mounted on the intake manifold above the engine thermostat is a detonation sensor. If the engine detonates, the metal of the intake manifold will vibrate. Then sensor picks up this vibration and sends a signal to an electronic black box on the fan shroud. The black box then retards the timing as much as 18-22°.

Testing the retarder is done by tapping on the intake manifold next to the sensor. Do not tap on the sensor itself. With the engine running at 2000 rpm in Neutral, tapping on the manifold will make the spark retard. If you have a timing light hooked up you can actually see it retard while you are tapping. Within 20 seconds after the tapping stops, the timing will return to its normal setting.

COMPUTER CONTROLLED CATALYTIC CONVERTER SYSTEM 1979-80

C-4 System

This new, electronically regulated, emission control system is available on 1980 vehicles with engine codes "A", "3", "H", and "R". This special exhaust system, also known as the closed loop system, is comprised of an exhaust gas oxygen sensor, an electronic control module (ECM), a controlled air-fuel ratio carburetor, a Phase II catalytic converter, and an oxygen sensor maintenance reminder (located in the instrument panel) which signals when it is time to replace the oxygen sensor.

The oxygen sensor (zirconia sensor), located in the left side exhaust manifold, develops a voltage which varies correspondingly with the volume of oxygen present in the exhaust gas. A rise in the voltage indicates a decrease in the oxygen content, and a drop or decrease in voltage indicates an increase in the oxygn content.

Sensor voltage is monitored by the ECM, which, in turn, signals the mixture control solenoid in the carburetor. As input voltage to the ECM increases (indicating a rich mixture), the output signal to the mixture control solenoid increases resulting in a leaner mixture at the carburetor. As input voltage to the ECM decreases (indicating a lean mixture), the output signal to the mixture control solenoid decreases resulting in a richer mixture at the carburetor.

Three conditions must be met before the ECM can begin regulating the carburetor air/fuel ratio: (1) a minimum of ten seconds must have elapsed since the engine was started before C-4 system operation can occur, (2) the coolant temperature must be higher than 90 degrees F., and (3) the ECM performs an oxygen sensor output-voltage check to determine when it has warmed up sufficiently to provide good information for C-4 system operation.

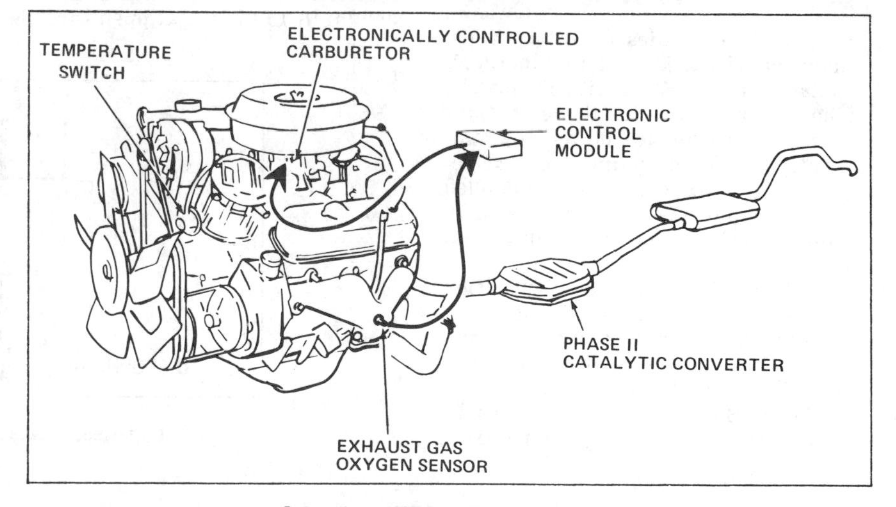

C-4 system 1979 V-6 vin code 2

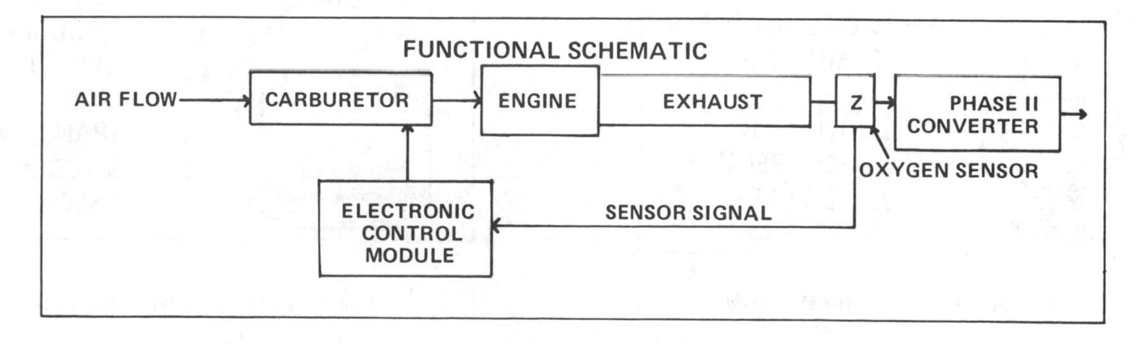

C-4 system schematic

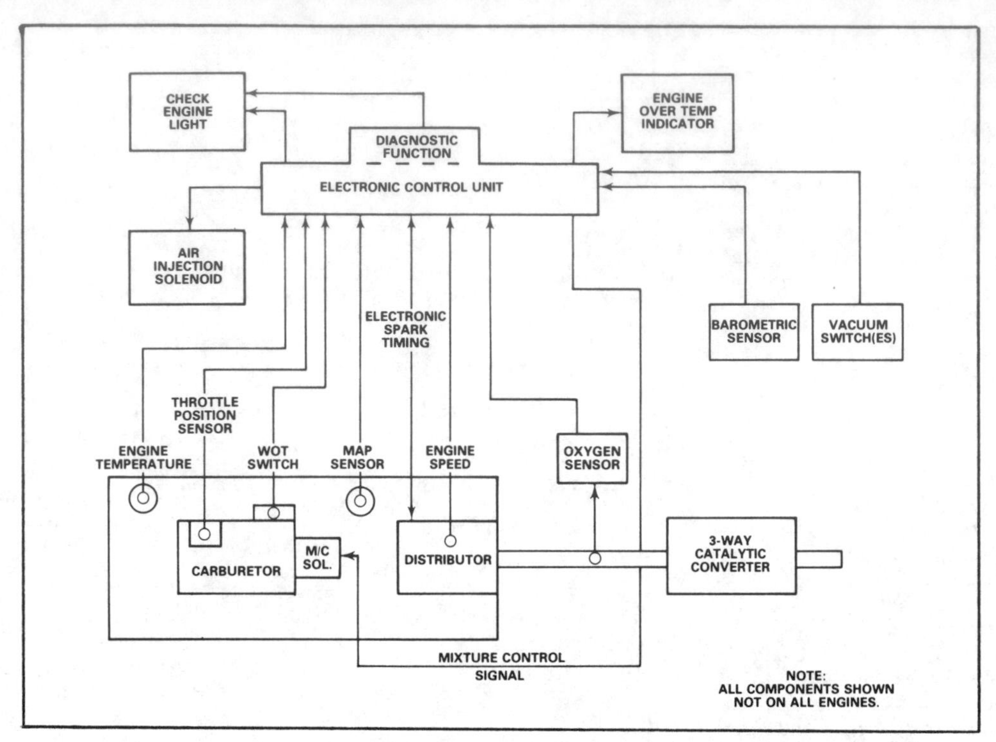

C-4 System Schematic

DIESEL ENGINE 1980

The 350 CID diesel built by Oldsmobile Division is available in Buick LeSabre, Electra and Riviera models. These engines are available in California with standard emission equipment. The diesel engine does not require the C-4 computer controlled catalytic converter system. The emission control devices on the diesel include:

1. PCV positive crankcase ventilation
2. EGR exhaust gas recirculation
3. ETS engine temperature sensor
4. VRV vacuum regulator valve
5. EGR-VS EGR vacuum switch

The ETS turns off the fast idle solenoid when engine temperature reaches 120°F. The vacuum regulator valve controls vacuum in proportion to throttle position. At closed throttle vacuum in 15 in. and at open throttle vacuum is 0 in. The VRV supplies vacuum to the EGR-VS, the transmission vacuum modulator and the transmission converter clutch when used.

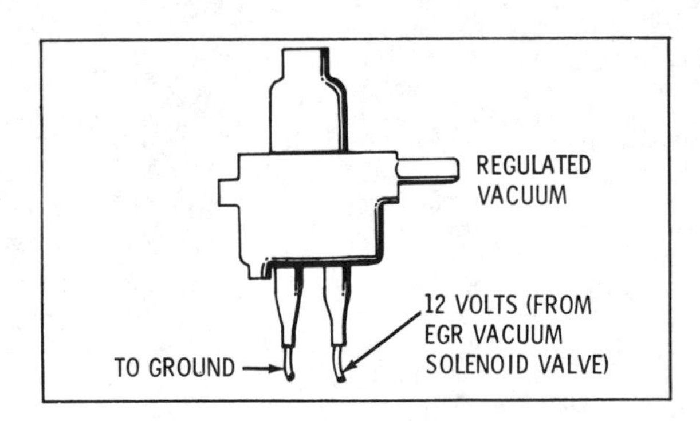

EGR-vacuum switch (EGR-VS) 350 Diesel

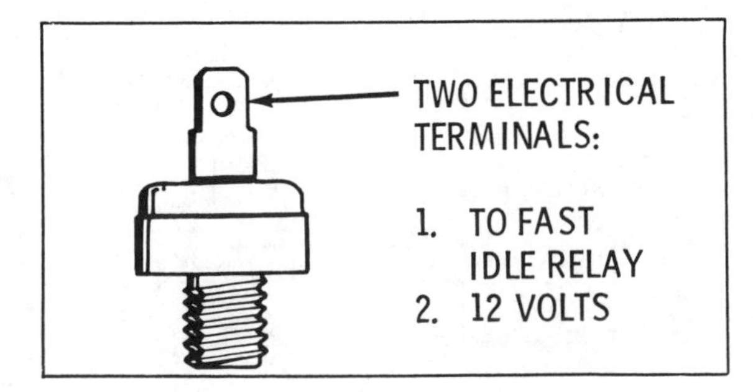

Engine temperature sensor (ETS) 350 Diesel

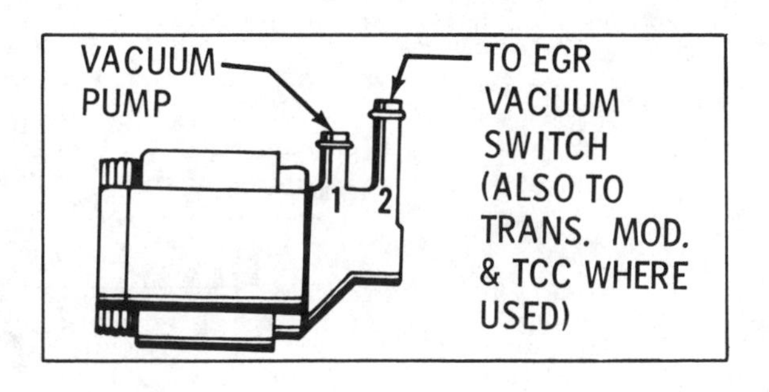

Vacuum regulator valve (VRV) 350 Diesel

VACUUM CIRCUITS

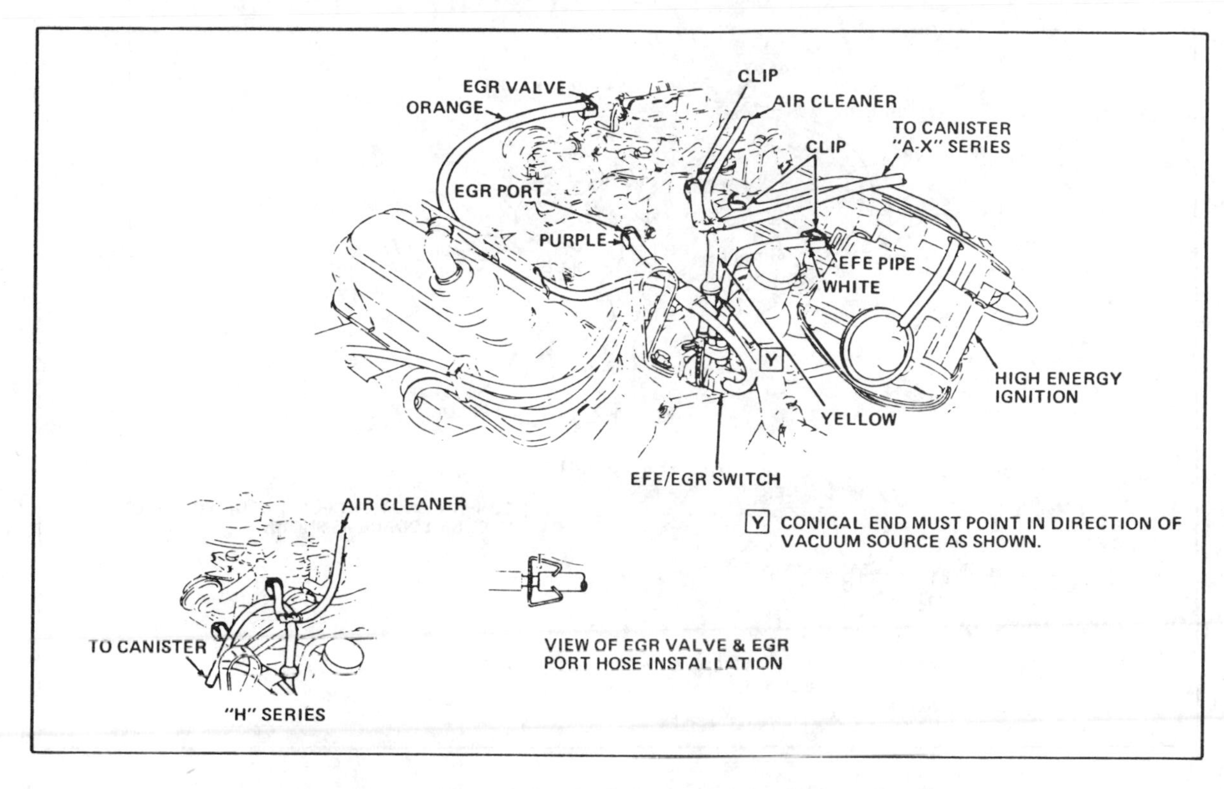

1979 Buick Vacuum hose routing—Code A, C—Series A, H, X (49 states/man. trans.)

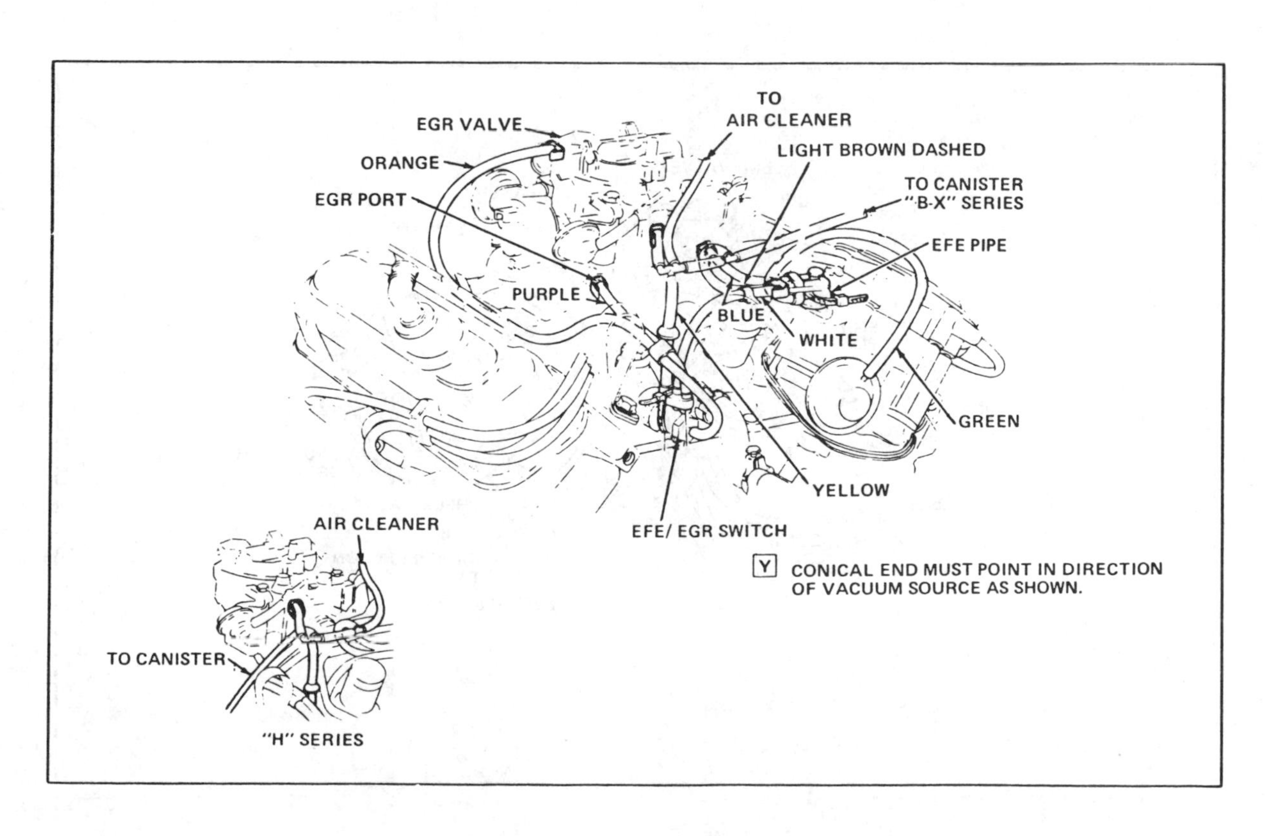

1979 Buick Vacuum hose routing—Code A—Series B, H, X (Calif. & high alt./auto. trans.)

VACUUM CIRCUITS

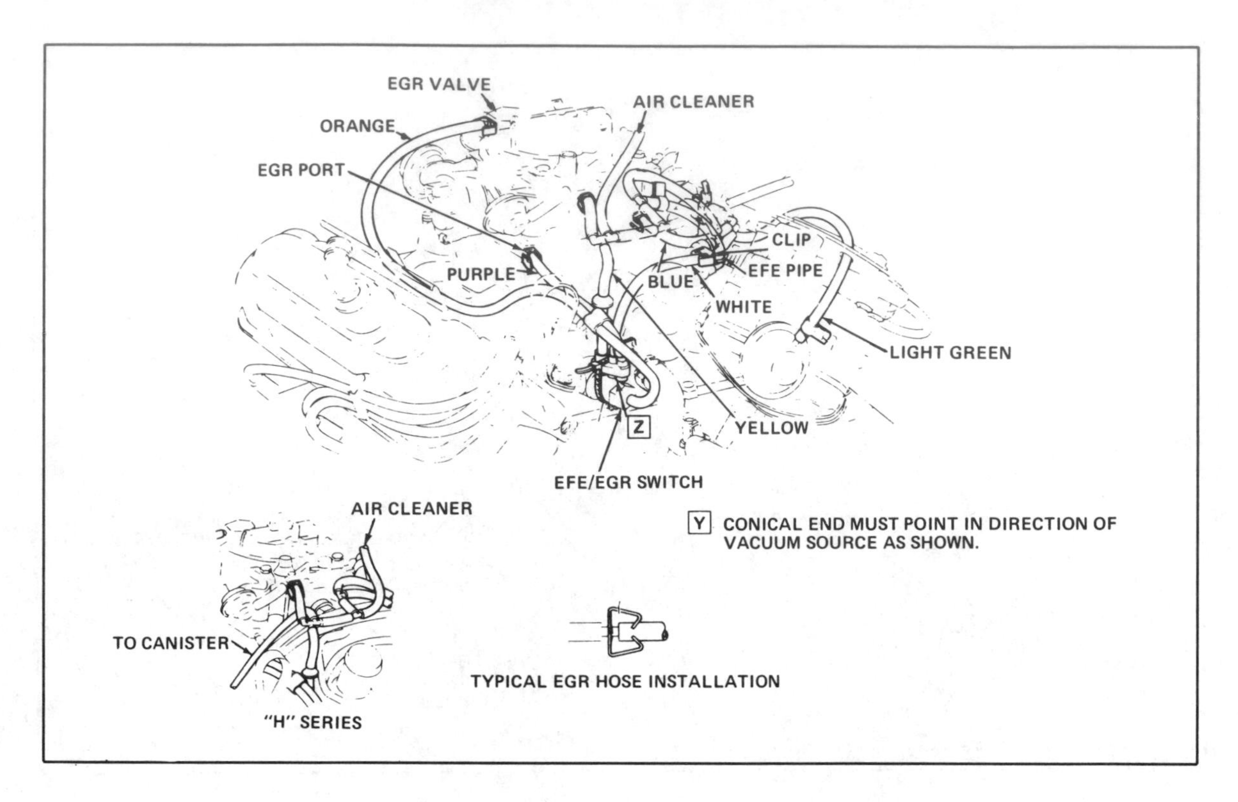

1979 Buick Vacuum hose routing—Code A, C—Series A, B, H, X (49 states/auto. trans.)

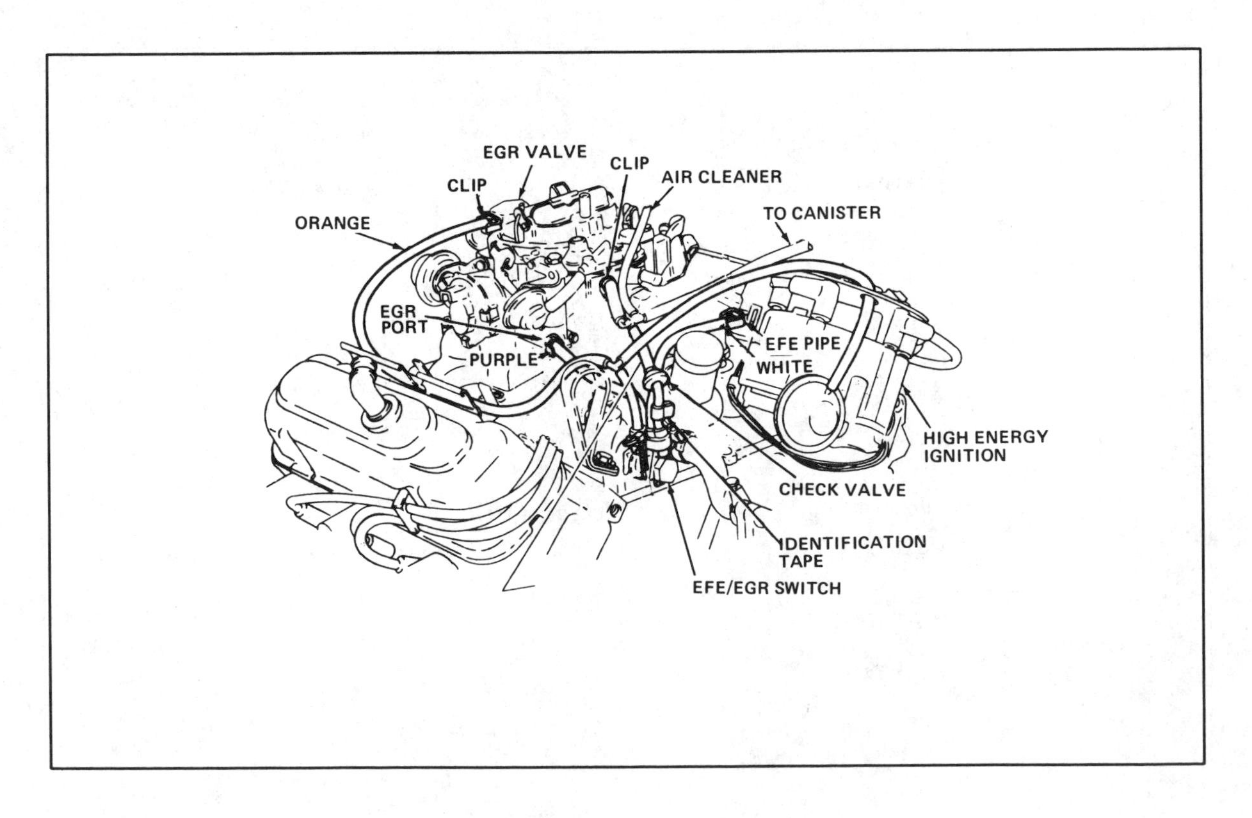

1979 Buick Vacuum hose routing—Code A—Series A (Calif./auto. trans.)

VACUUM CIRCUITS

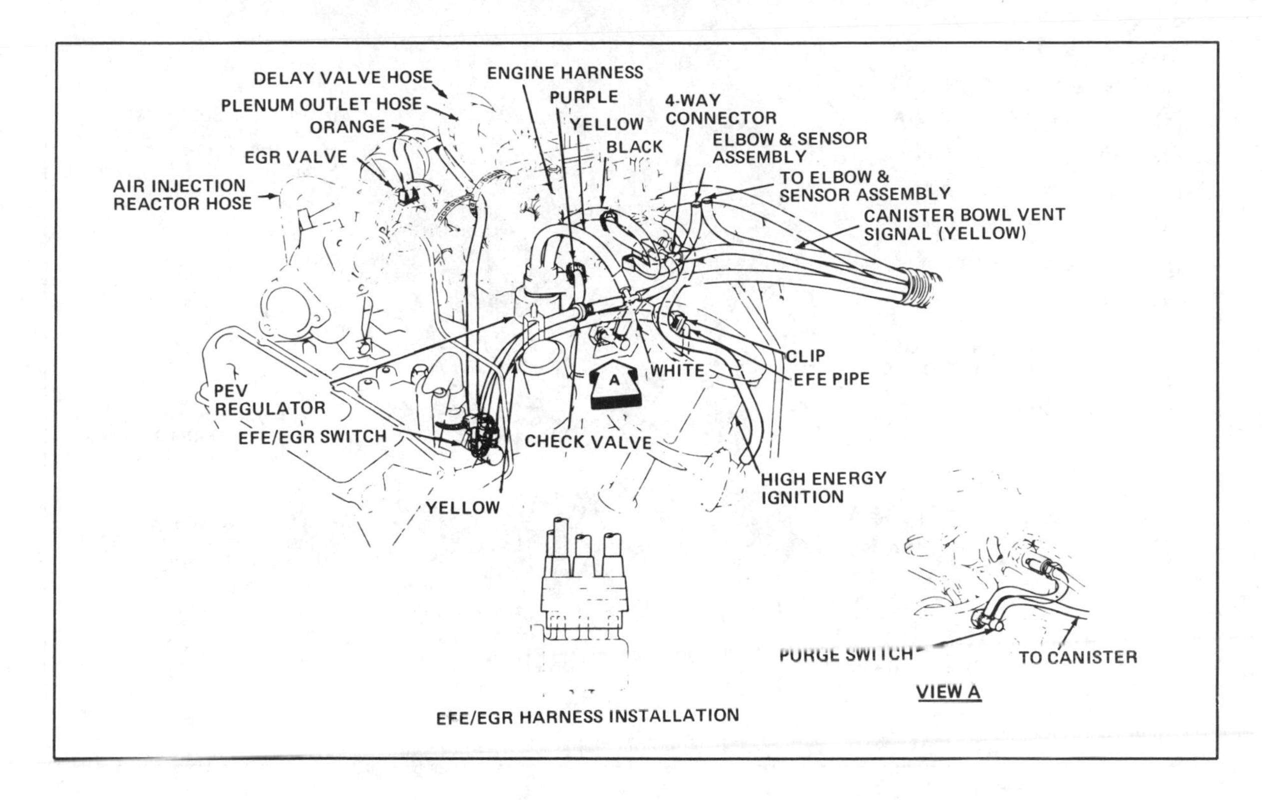

1979 Buick Vacuum hose routing—Code 3—Series A, B (49 states)

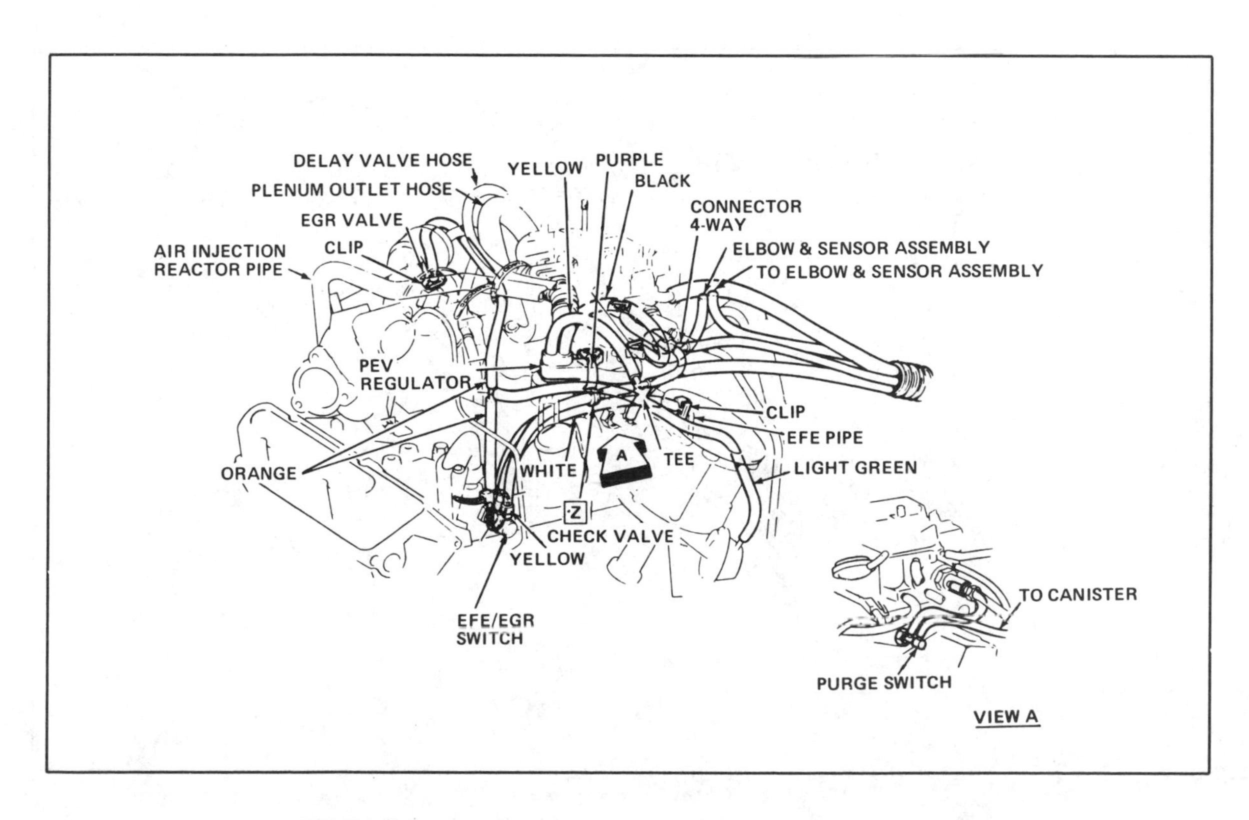

1979 Buick Vacuum hose routing—Code 3—Series A, B (Calif.)

VACUUM CIRCUITS

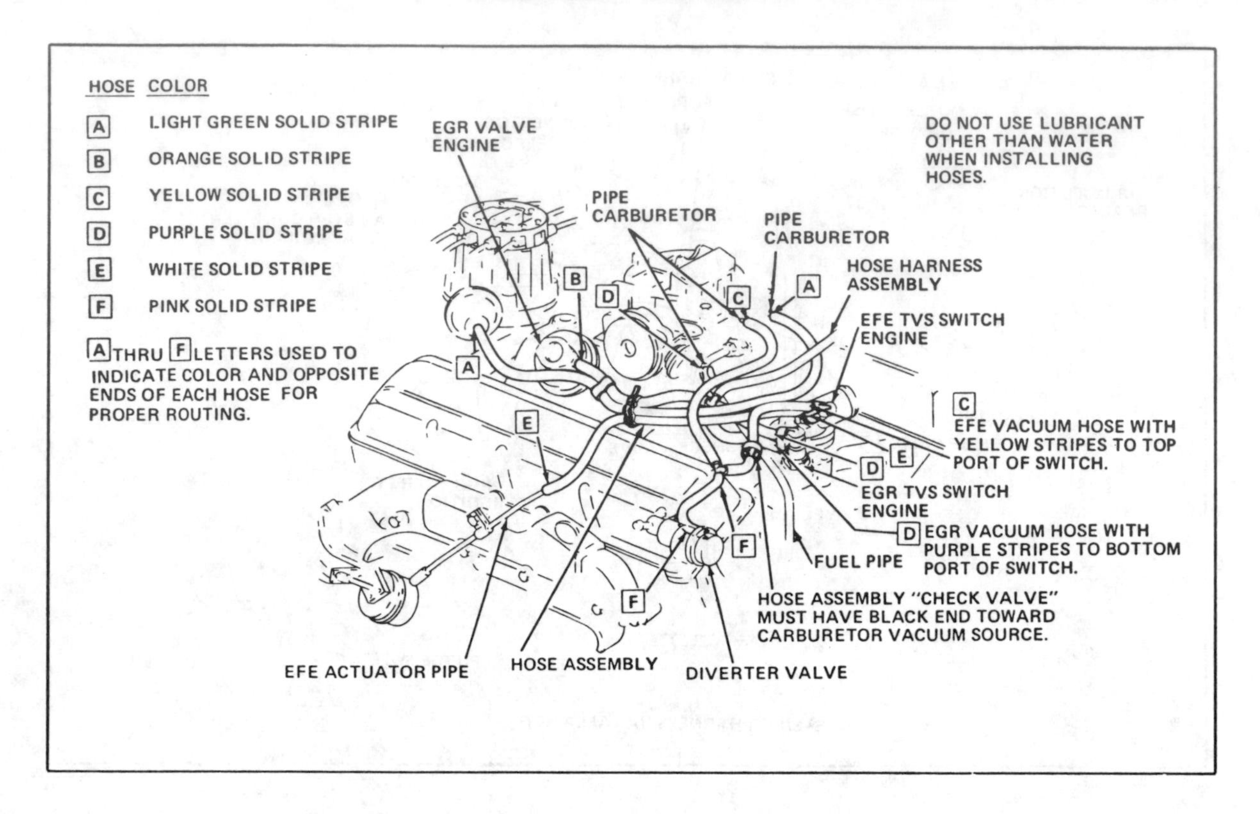

1979 Buick Vacuum hose routing—Code H, L—Series A

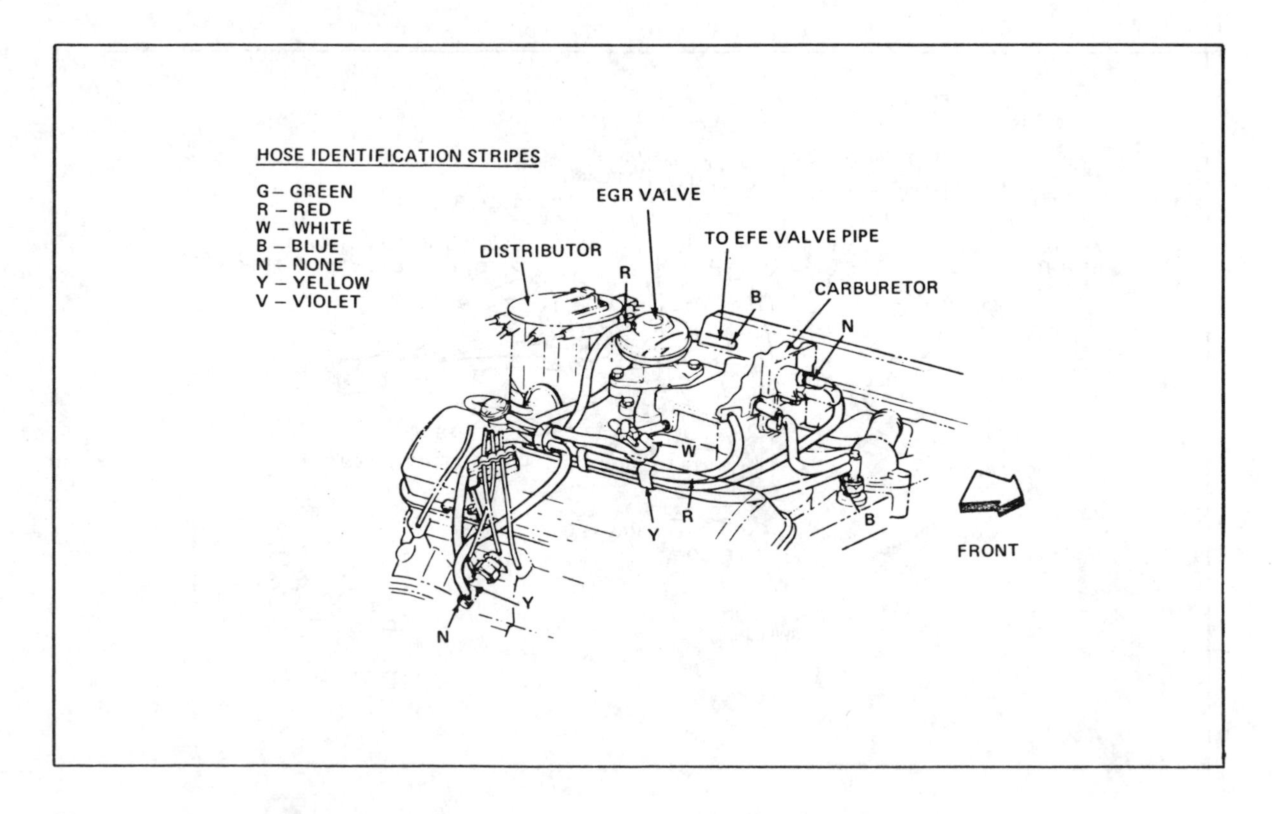

1979 Buick Vacuum hose routing—Code Y—Series A, B (49 states with A/C)

VACUUM CIRCUITS

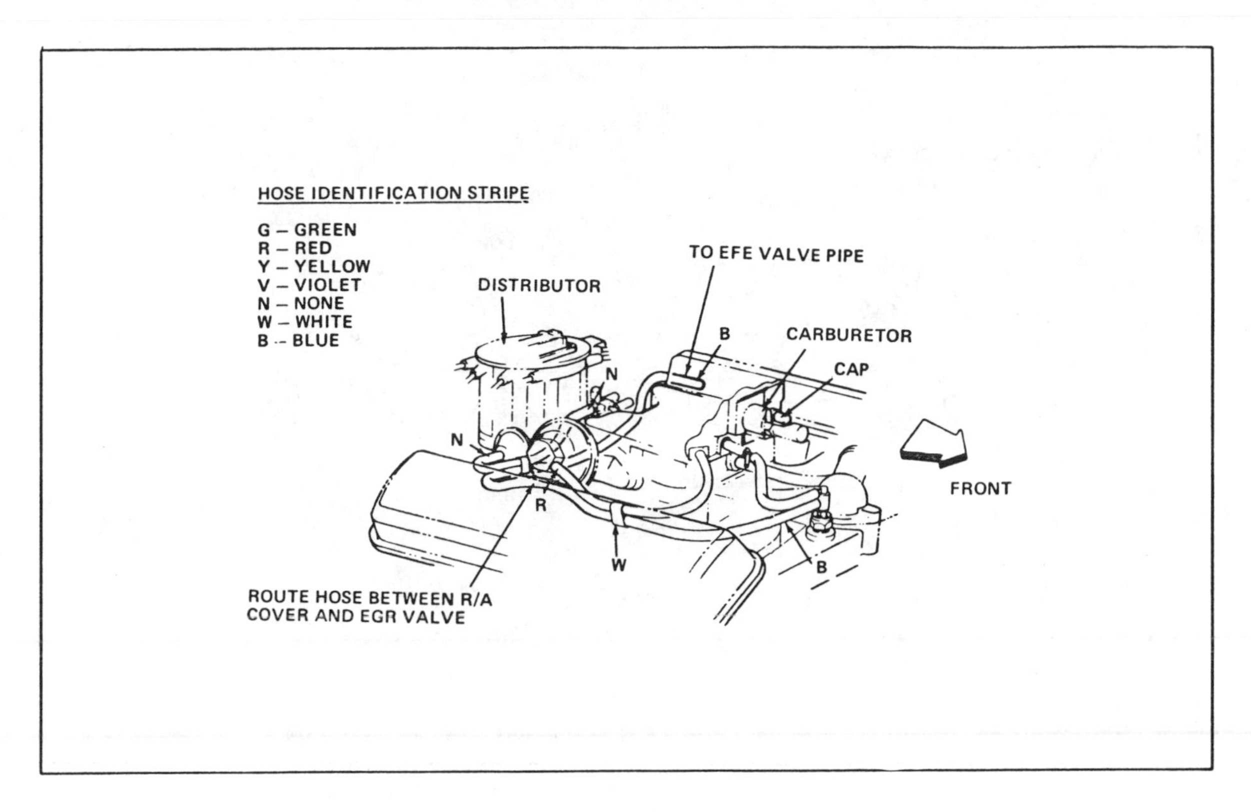

1979 Buick Vacuum hose routing—Code W—Series A (49 states)

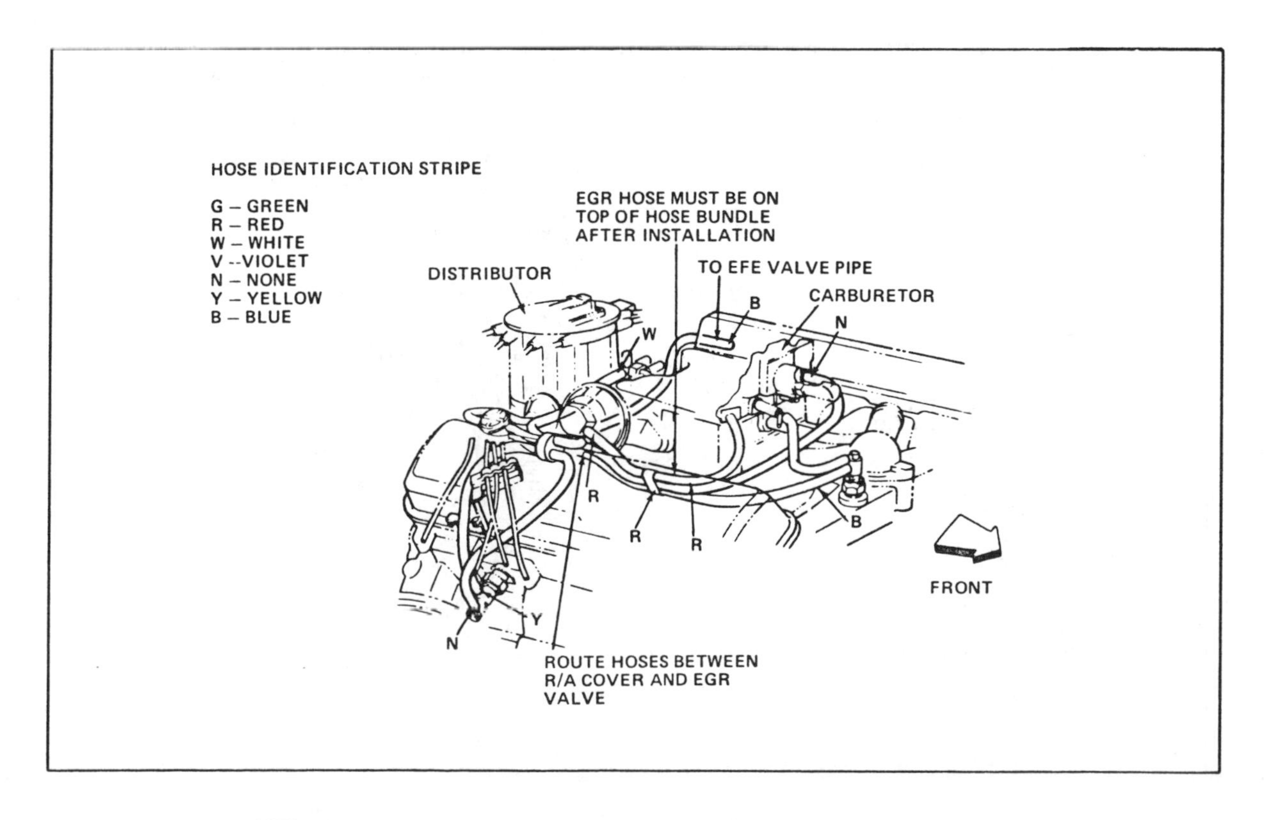

1979 Buick Vacuum hose routing—Code W—Series A (49 states with A/C)

VACUUM CIRCUITS

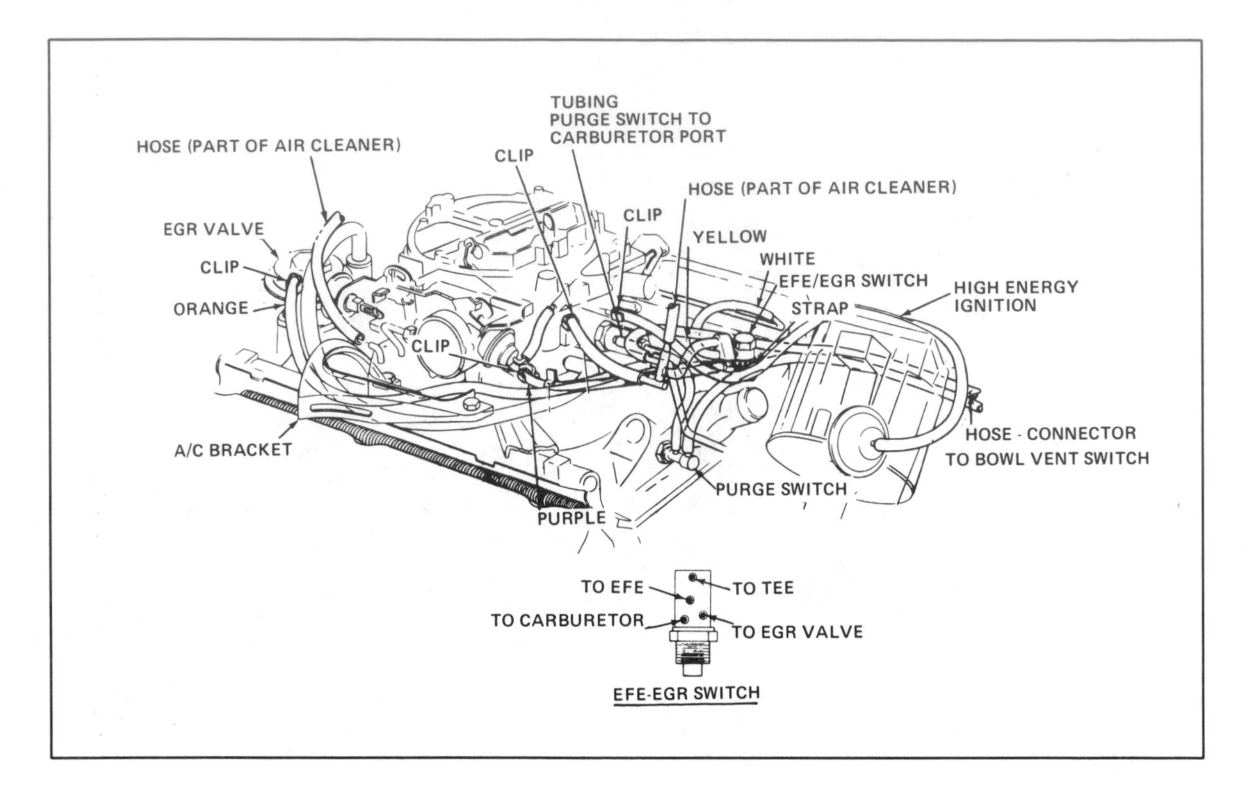

1979 Buick Vacuum hose routing—Code X—Series B, C

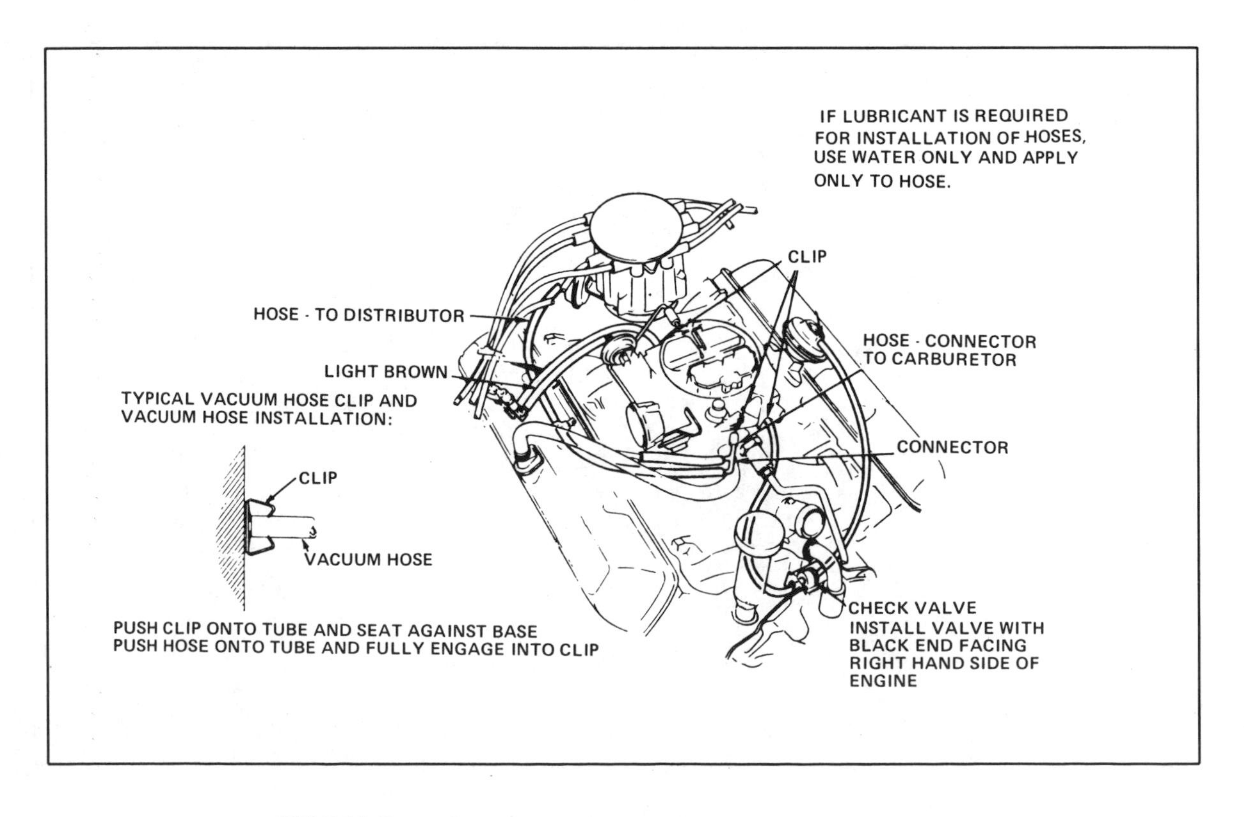

1979 Buick Vacuum hose routing—Code K—Series B, C (49 states)

VACUUM CIRCUITS

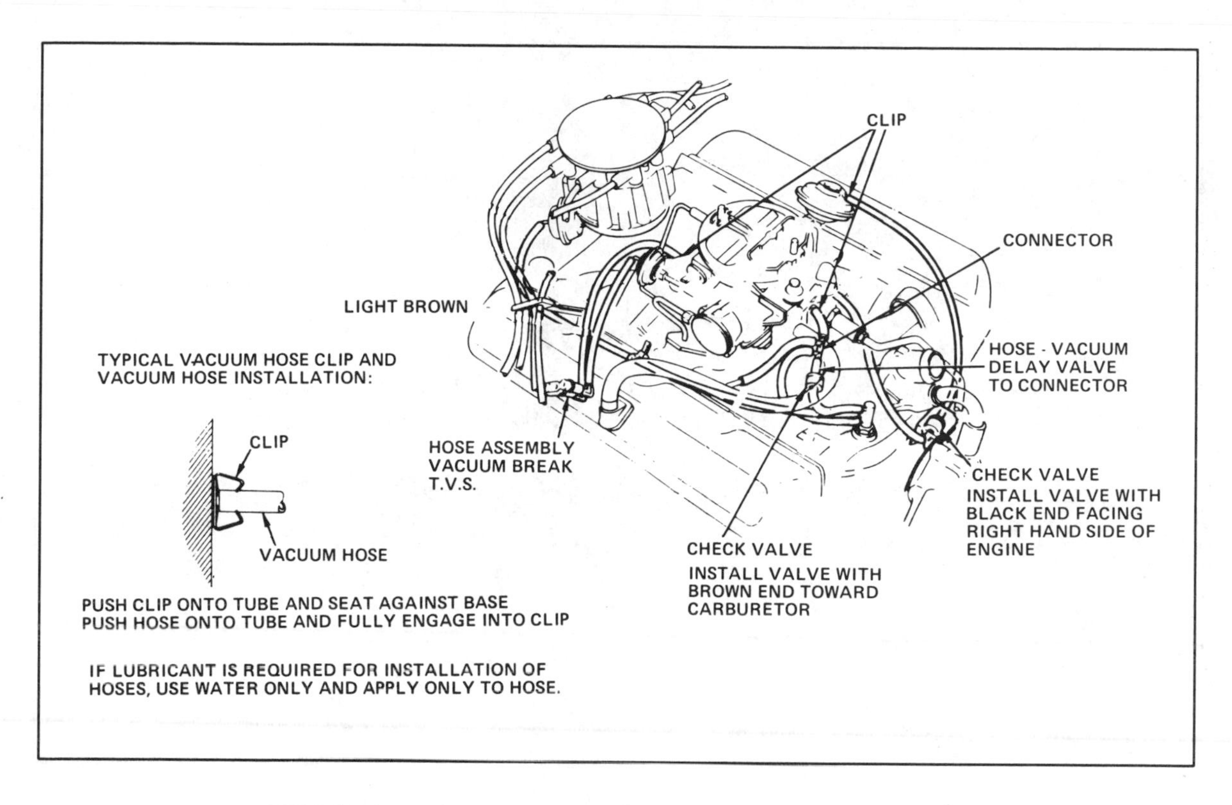

1979 Buick Vacuum hose routing—Code R, K—Series B, C (High alt.)

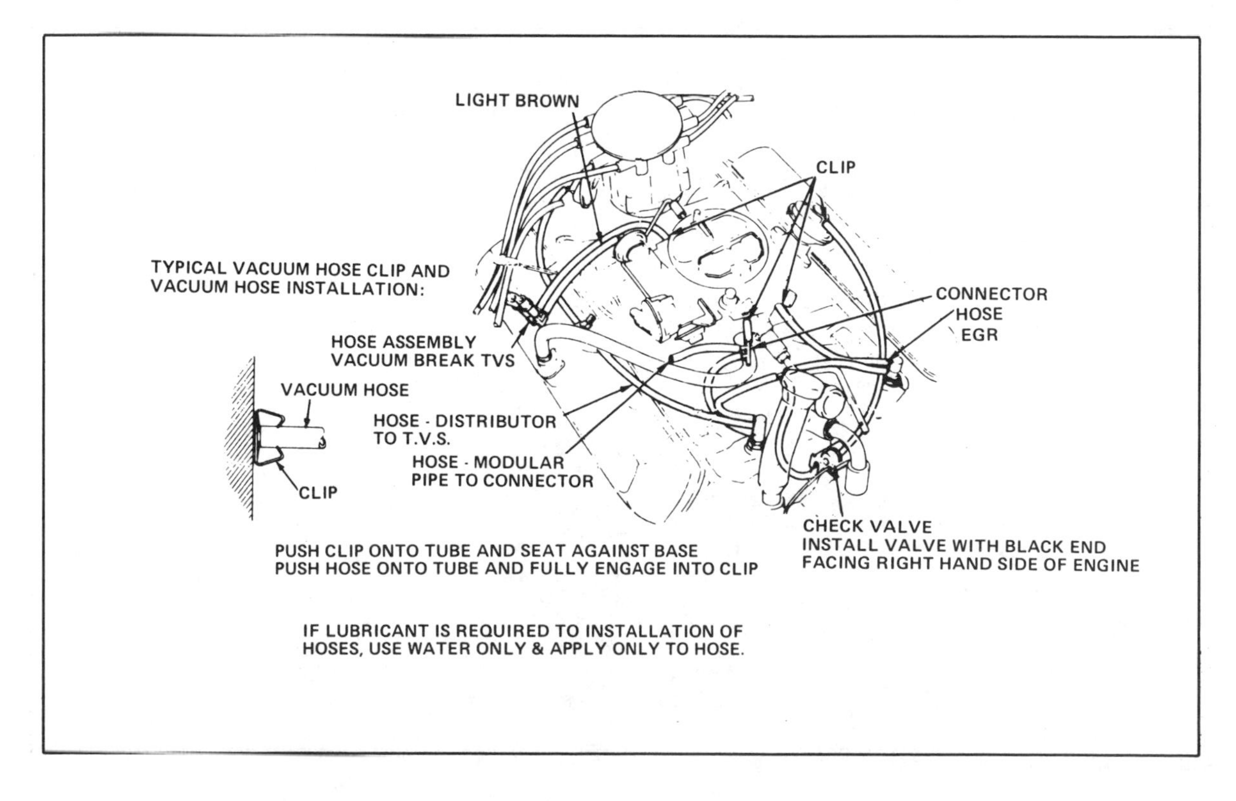

1979 Buick Vacuum hose routing—Code R, K—Series B, C (Calif.)

VACUUM CIRCUITS

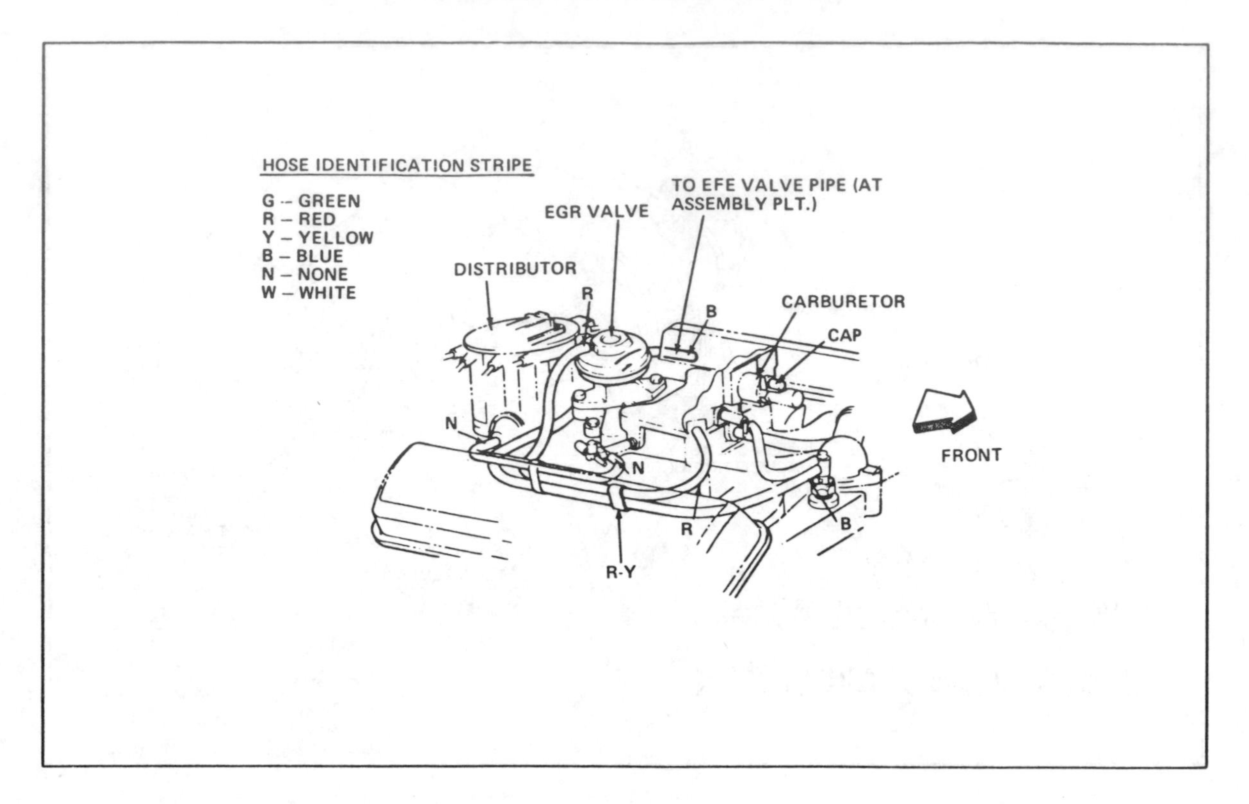

1979 Buick Vacuum hose routing—**Code Y—Series A, B (49 states)**

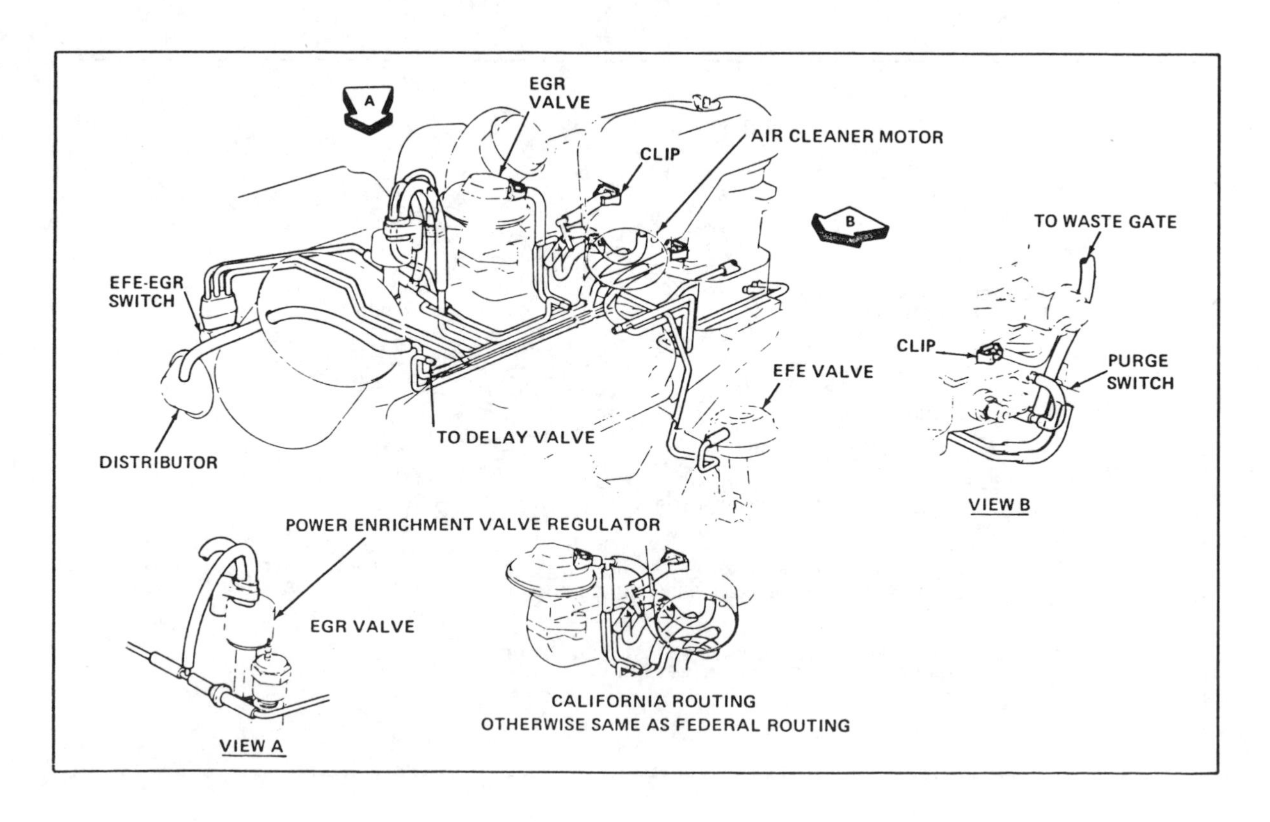

1979 Buick Vacuum hose routing—**Code 3—Series E**

VACUUM CIRCUITS

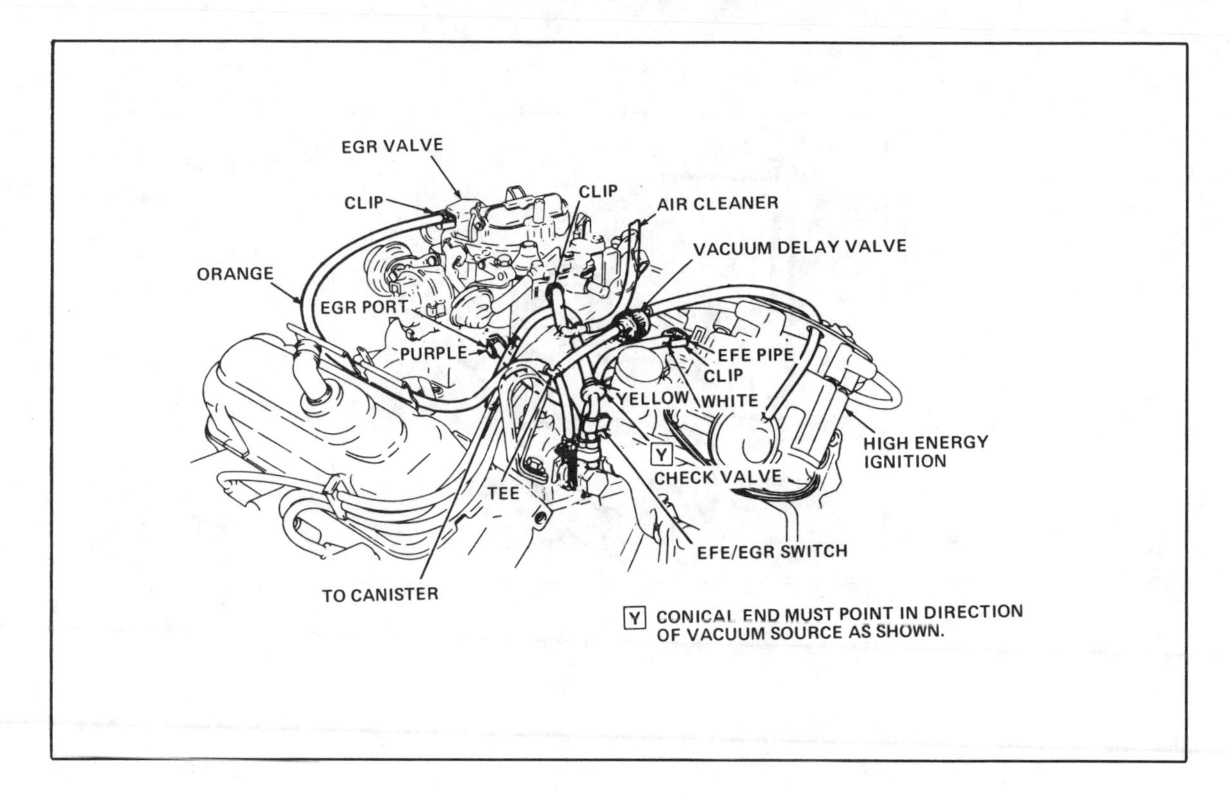

1979 Buick Vacuum hose routing—Code A—Series H (Calif./man. trans.)

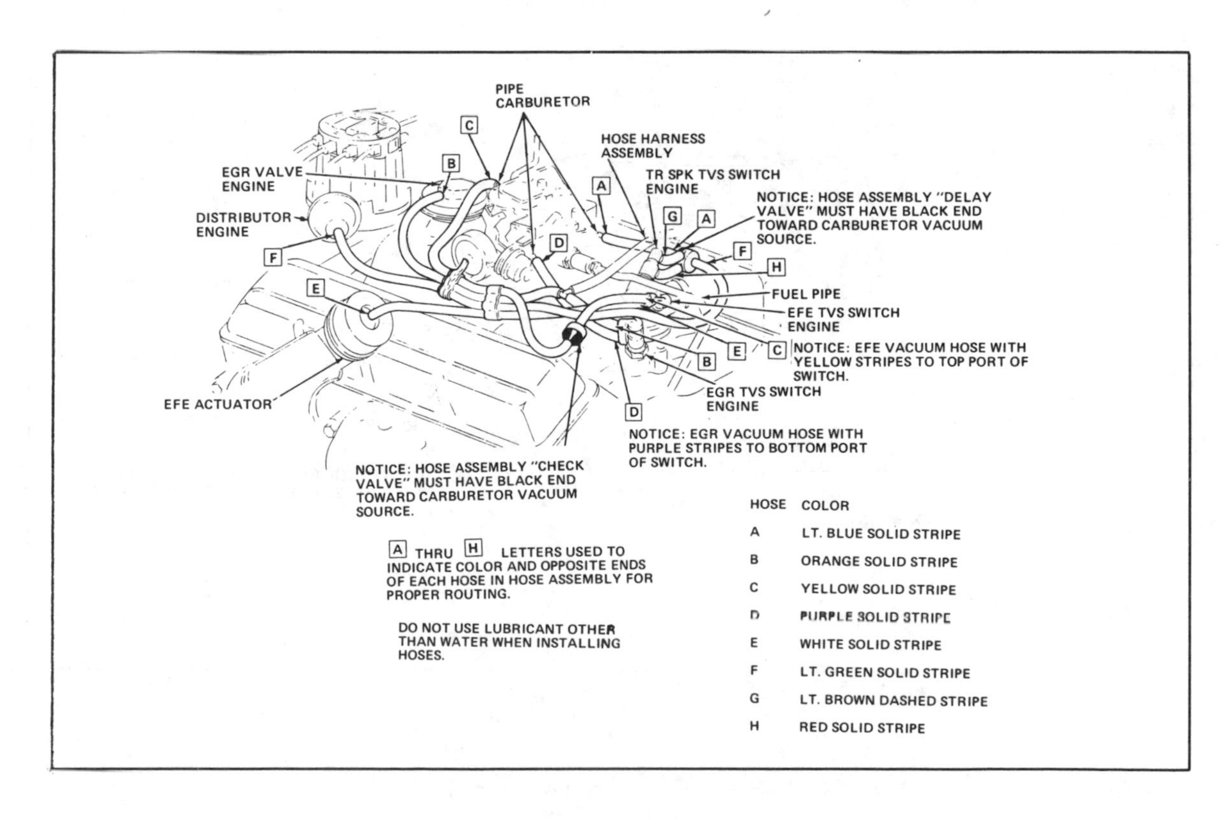

1979 Buick Vacuum hose routing—Code G—Series X

VACUUM CIRCUITS

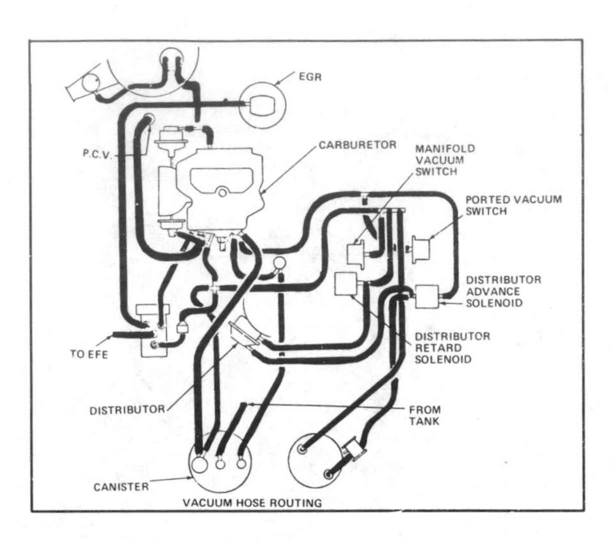

1979 Buick Vacuum hose schematic—C-4 System

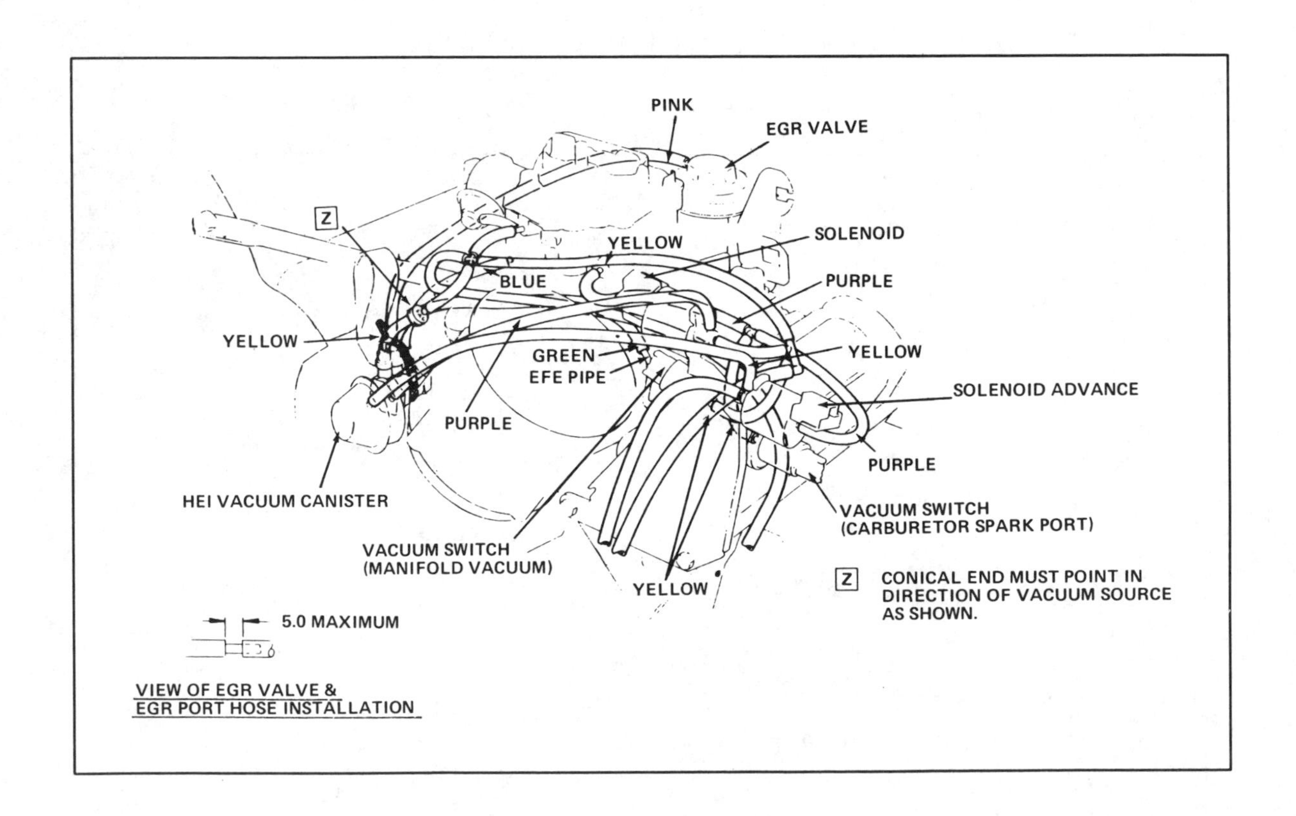

1979 Buick Vacuum hose routing—C-4 System

Note: For 1980—Use the Vehicle Emission Control Information Label for the most up to date information concerning hose routing and tune up data.

TUNE-UP SPECIFICATIONS

ENGINE CODE, 5th CHARACTER OF THE VIN NUMBER
MODEL YEAR CODE, 6th CHARACTER OF THE VIN NUMBER

Year	ENGINE V.I.N. Code	No. Cyl. Disp. (cu in)	SPARK PLUGS* Orig. Type	Gap (in)	DIST.	IGNITION TIMING (deg BTDC) Lo-Alt.	Calif.	CARBURETOR HOT IDLE In Drive③ Lo-Alt.	Calif.	Hi-Alt.
1979	B	8-350	R-47SX	.060	E.I.	10	10	600	600	600
	S	8-425	R-45NSX	.060	E.I.	23 @ 1600	23 @ 1600	600	600	600
	T	8-425	R-45NSX	.060	E.I.	18 @ 1400	18 @ 1400	600	600	600
	N	8-350	DIESEL	—	—	—	—	650	②	②
1980	N	350	DIESEL	—	—	—	—	650	②	②
	8	350	R-47SX	.060	E.I.	10	10	600	600	600
	6	368	R-45NSX	.060	E.I.	②	—	600	600	600
	9	368	R-45NSX	.060	E.I.	②	—	600	600	600

NOTE: Information listed on the emission control tune-up decal supersedes all published information as it may reflect production changes.

② Information not available at time of publication, see emission oontrol docal under hood.

③ Disconnect air leveling compressor hose at air cleaner and plug hose. Disconnect parking brake hose and plug. Air conditioner turned off.

DISTRIBUTOR SPECIFICATIONS

Year	DISTRIBUTOR IDENTIFICATION	CENTRIFUGAL ADVANCE Start Crankshaft Deg @ Eng. RPM	Finish Crankshaft Deg @ Eng. RPM	VACUUM ADVANCE Start @ In Hg	Finish Crankshaft Deg @ In Hg
1979	1103392	.25-0 @ 550	13-19 @ 4800	3.0	19-21 @ 14
	1103307	.047 @ 800	19.5-22 @ 6000	4.0	15-19 @ 17
	1103389	.25-.5 @ 600	14-19 @ 4000	4.0	27-29 @ 16
	1103393	.5-0 @ 800	19.5-22 @ 6000	5.625	22-25 @ 13
	1103394	.5-0 @ 800	19.5-22 @ 6000	5.625	27-29 @ 16
	1103395	.5-0 @ 550	11-19 @ 4800	8.4375	9-11 @ 12
	1103332	.125-0 @ 575	13-19 @ 4800	4.0	19-21 @ 12
	1103334	.25-0 @ 575	14-19 @ 5000	5.625	27-29 @ 16
	1103335	.25-0 @ 600	11-16 @ 5800	5.0	19-21 @ 13

QUADRAJET CARBURETOR SPECIFICATIONS

CADILLAC

Year	Carburetor Identification①	Float Level (in.)	Air Valve Spring (turn)	Pump Rod (in.)	Primary Vacuum Break (in.)	Secondary Vacuum Break (in.)	Secondary Opening (in.)	Choke Rod (in.)	Choke Unloader (in.)	Fast Idle Speed ④ (rpm)
1979	17059230	13/32	1/2	9/32②	0.142	0.234	0.015	0.083	0.142	1000
	17059232	13/32	1/2	9/32②	0.142	0.234	0.015	0.083	0.142	1500
	17059530	13/32	1/2	9/32②	0.149	0.164	0.015	0.083	0.142	1500
	17059532	13/32	1/2	9/32②	0.149	0.164	0.015	0.083	0.142	1500
1980	17080230	7/16	1/2	9/32②	0.149	0.136	③	0.083	0.220	1450
	17080530	17/32	1/2	Fixed	0.142	0.400	③	0.083	0.260	1350

① The carburetor identification number is stamped on the float bowl, near the secondary throttle lever.

② Inner hole

③ No measurement necessary on two point linkage; see text.

CAR SERIAL NUMBER AND ENGINE IDENTIFICATION

1979

The vehicle identification number is mounted on a plate behind the windshield on the driver's side. The sixth character is the model year, with 9 for 1979. The fifth character is the engine code, as follows:

B....................350 F.I. V-8
N...................350 diesel V-8
S...................425 4-bbl. V-8
T...................425 F.I. V-8

Part of the car serial number is stamped on the rear of the block behind the intake manifold on all except Seville, Eldorado and Diesel models. On these models the engine identification number is located on the left hand side of the cylinder block at the front below the cylinder head.

The 425-inch V-8's engines are made by Cadillac. The 350 V-8 engine is based on a block assembly produced by Oldsmobile, which is similar to other Olds 350 V-8's.

1980

The vehicle identification number is mounted on a plate behind the windshield on the driver's side. The sixth character is the model year, with O for 1980. The fifth character is the engine code, as follows:

8 .350 Electronic Fuel Enjection, V8
N ....350 Diesel V8
6 ....368 V8
9 ....368 Digital Electronic Fuel Enjection V8

On the "C" car, part of the engine serial number is stamped on the rear of the block behind the intake manifold. The engine identification number of Seville, Eldorado and Diesel models is located on the left hand side of the cylinder block at the front below the cylinder head.

EMISSION EQUIPMENT

1979-80

Carburetor Models

Closed positive crankcase ventilation
Emission calibrated carburetor
Sealed idle mixture screws
Emission calibrated distributor
Heated air cleaner
Vapor control, canister storage
Exhaust gas recirculation
Catalytic converter
Electric choke
Early fuel evaporation
Intake manifold with riser tubes
Air pump
  49-States
    Limousines and commercial chassis only
  Calif.
    All models
  Altitude
    All models

Temperature controlled choke vacuum break
  49-States
    All models
  Calif.
    Not used
  Altitude
    All models

Air management system (1980)

Fuel Injection Models

Closed positive crankcase ventilation
Air pump, except Seville for Calif.
Emission calibrated distributor
Vapor control, canister storage
Exhaust gas recirculaion
Catalytic converter
"Closed Loop" Fuel Injection Sensor Calif.
Air management system (1980)
Digital electronic fuel injection (1980)

Diesel Models

Positive crankcase ventilation
EGR valve (1980)

IDLE SPEED AND MIXTURE ADJUSTMENTS

1979-80

NOTE: *The underhood specifications sticker, which is located on the radiator support, often reflects changes that have been made in production. Sticker figures must be used if they disagree with those listed here.*

Air cleaner ....In place
Auto. trans. ....Drive
Air cond. ....Off

Parking brake vac. hose .Disconnect at cylinder and plug

Air leveling comp. hose ..Disconnect at air cleaner and plug

Other hoses ....See underhood label

Idle speed
  Carburetor engines 600
    AC idle speedup 675
  Fuel inj. engines 600
  Diesel engines 575

NOTE: *Idle Air/Fuel Adjustment should only be made at times of major carburetor overhaul, throttle base replacement or high idle CO as determined by state or local emission inspections.*

INITIAL TIMING

1979

CAUTION: *Emission control adjustment changes are noted on the Vehicle Emission Information Label by the manufacturer. Refer to the label before any adjustments are made.*

NOTE: *The distributor vacuum hose must be disconnected and plugged. Also disconnect and plug the EGR vacuum hose on Cadillac engines, both 4-bbl. and F.I. The hole in the timing bracket on the engine front cover is for magnetic timing equipment only. Set timing with engine running at rpm shown below.*

425 carburetor except Eldorado
  49-States (1600 rpm) ..21° BTDC
  Calif. (1600 rpm) .....21° BTDC
  Altitude (1600 rpm) ..23° BTDC
425 fuel inj. engine
  All (1400) rpm) ......18° BTDC
Eldorado
  49-States (1600 rpm) ..21° BTDC
  Calif. 1600 rpm) ......18° BTDC
  Altitude (1600 rpm) ..23° BTDC

INITIAL TIMING

Com. chassis and limousine
49-States (1600 rpm) . . 18° BTDC
Calif. (1600 rpm) . . . . . 18° BTDC
Altitude (1600 rpm) . . 23° BTDC
Seville 350
49-States (600 rpm) . . 10° BTDC
Calif. (600 rpm) . . . . . . 8° BTDC
Altitude (600 rpm) . . . . 10° BTDC

1980

With the vehicle engine at the proper operating temperature refer to the engine specification sticker which is located on the radiator support and set the engine timing to its proper specification.

SPARK PLUGS

1979

All 425 . . . . . . AC-R45NSX 060
Seville and Eldorado w/EFI
AC-R47SX 060

1980

6.0 liter R45NSX 060
5.7 liter w/EFI R47SX 060

VACUUM ADVANCE 1979-80

Diaphragm . . . . Single
Vacuum source
All with carburetor . . . . Manifold
All fuel inj. . . . . Ported

EMISSION CONTROL SYSTEMS

ELECTRONIC SPARK SELECTION 1979-80

The ESS system is now standard on the Eldorado, Seville, Limousine, Commercial Chassis and carbureted California "C" cars.

The maximum advance RPM to put the ESS system into operation is as follows;

Eldorado 1,200
Seville 1,450
Seville w/EFI (1980) 1,200
All other models 1,350

Because the EGR solenoids are deleted on the carbureted cars, a new three-way coolant temperature switch is used to send a signal to the ESS decoder, to control the spark retard during cold engine operation and to prevent over advance during hot engine operation.

EFE-EGR THERMAL VACUUM SWITCH 1979-80

Carburetor engines use a four or five-nozzle thermal vacuum switch to control vacuum to the EGR and EFE systems. On California cars, the same switch also controlled vacuum to the distributor. In 1979, the switch is the same, and it still controls vacuum to the EGR system and to the distributor. The EFE system is controlled by a separate switch, mounted on the right side of the thermostat housing.

The EFE switch applies vacuum to the EFE actuator below 165°F. This allows the EFE system stay on longer for better cold driveability.

The EGR switch keeps the EGR off below 120°F. which blocks vacuum to the distributor so there is no vacuum advance until the engine coolant temperature goes over 120°F.

TEMPERATURE CONTROLLED CHOKE VACUUM BREAK SYSTEM 1979-80

All carbureted 49-State cars use a thermal vacuum switch (TVS) mounted in the air cleaner to control vacuum to the secondary vacuum break at the rear of the carburetor. Below approximately 62°F. air cleaner temperature, the TVS is closed, preventing the secondary vacuum break from opening the choke. This delays the full opening of the choke for better cold operation.

The air cleaner thermal sensor also connects to the TVS. This connection is for convenience only. The vacuum path through the TVS to the thermal sensor is open at all times, and is not affected by the part of the TVS that shuts off vacuum to the vacuum break unit.

EGR VALVES 1979-80

The exhaust gas recirculation system is used on all engines to reduce oxides of nitrogen in the engine exhaust.

On all engines equipped with a carburetor the E.G.R. valve is back pressure operated. This design includes

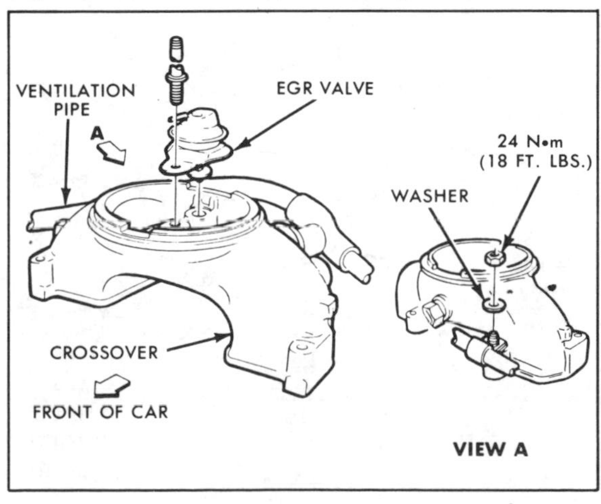

Diesel engine EGR valve location (© General Motors Corp.)

the transducer incorporated into the valve. An orifice gasket is not used with this type of valve. The E.G.R. valve is mounted at the rear of the intake manifold.

On all electronic fuel injection equipped vehicles the E.G.R. valve is controlled by the exhaust pressure transducer and the ECU. This valve uses four different orifice gaskets. The E.G.R. valve is mounted at the side of the intake manifold.

On diesel engine equipped vehicles for 1980 the E.G.R. valve is located in the crossover with exhaust gas being directed from an adapter on the exhaust manifold through a pipe leading to the air crossover.

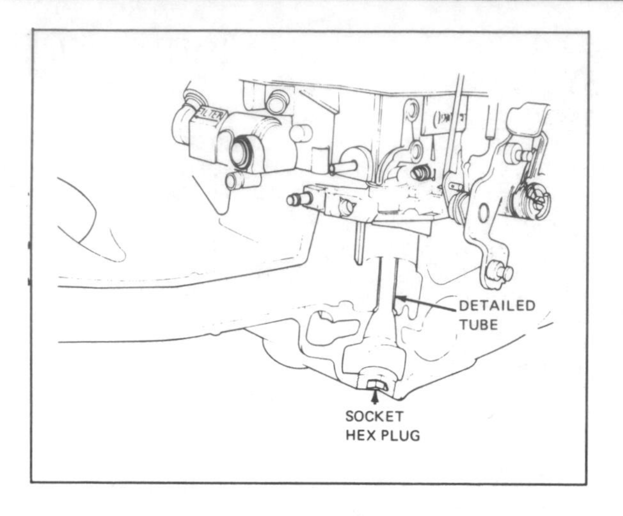

Riser tube location in the floor of the intake manifold

EGR GASKET REVISION 1979

The EGR gaskets have been revised on the Electronic Fuel Injection engines, to meter the exhaust gases more evenly. The applications are as follows;

Application	*Orifice Size*	*Part Number*
All "C" cars (EFI)	0.600 in.	1608508
All Eldorados	0.500 in.	417174
Seville Federal	0.484 in.	1608505
Seville Calif.	0.375 in.	551516

CATALYTIC CONVERTER 1979-80

Rodium pellets are used in the converter, in addition to the platium and paladium pellets, to meet the stricter Oxides of Nitrogen standards in the state of California, on 1979-80 Seville and 1980 Eldorado.

1980

The dual bed catalytic converter used on vehicles with the C-4 system contains a three-way reducing/oxidizing bed in its upstream section to reduce oxides of nitrogen while at the same time oxidizing hydrocarbons and carbon monoxide. An air supply pipe from the AIR pump introduces an extra amount of air between the dual beds, so that the second bed oxidizes any remaining hydrocarbons and carbon monoxide with high conversion efficiency, to minimize overall emissions.

CHANGE CATALYST 1980

Eldorado and Seville equipped with Digital Electronic Fuel Injection (DEFI) will require a catalyst change every 30,000 miles.

INTAKE MANIFOLDS 1979-80

(Carbureted "C" cars)

Riser tubes have been added to the floor of the intake manifold to more evenly distribute the exhaust gases, thereby making the EGR system more efficient.

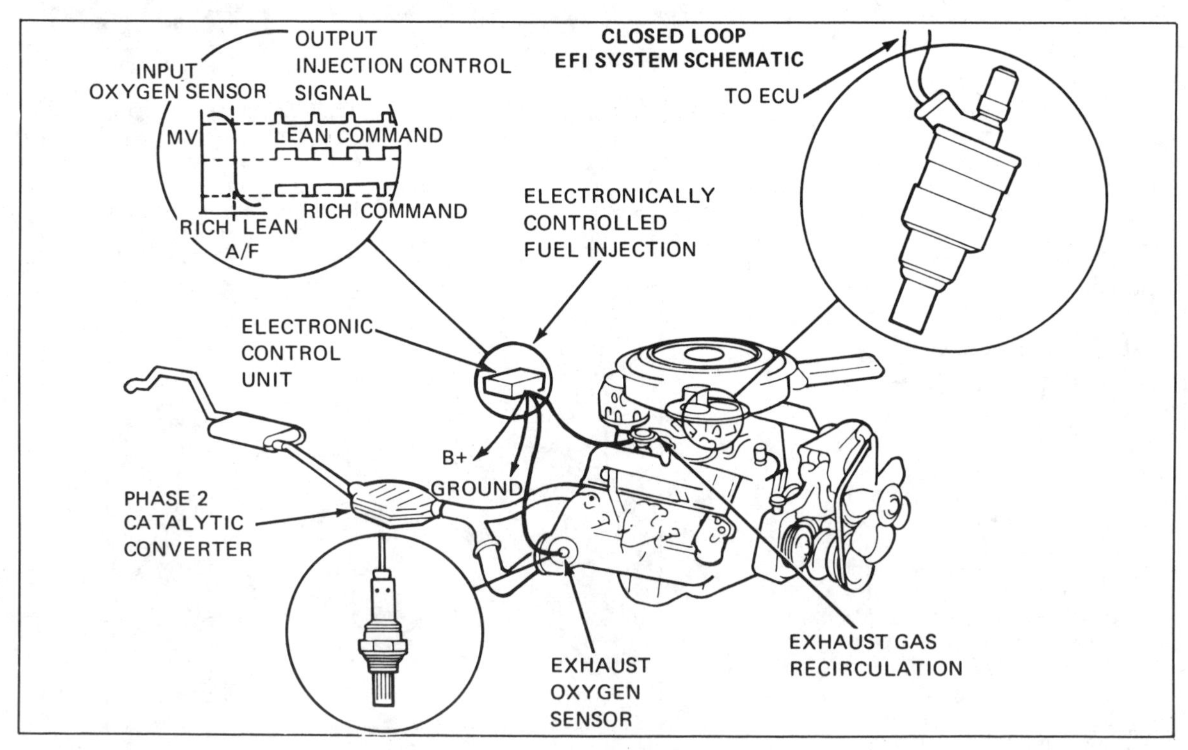

Closed loop EFI system

CLOSED LOOP EPI SYSTEM 1979-80

(All 1979-80 Sevilles and 1980 Eldorados sold in California)

The term "Closed Loop" is derived from the systems ability to sense the air/fuel ratio and correct it as necessary through the use of a feedback signal to the Electronic Control Unit (ECU). This process is accomplished by the use of an exhaust oxygen sensor, located in the top of the right exhaust manifold.

The sensor is an electro-chemical device, using Zirconium Dioxide to produce a variable voltage in relation to the oxygen content in the exhaust gases.

The sensor is exposed at one end to the exhaust gases as they leave the engine, while the sensor center is vented to the atmosphere, thereby comparing the oxygen content of both the exhaust gases and the atmosphere air. The variable voltage produced is dependent upon the richness or leanness of the exhaust gases of the oxygen content. The electrical signal is sent to the ECU, which adjusts the air/fuel mixture entering the engine to the ideal ratio of 14.7:1.

A "SENSOR" indicator dash light will operate at an interval of 15,000 miles. This indicates the Exhaust Oxygen sensor must be replaced. The dash "SENSOR" light is reset from behind the instrument cluster by pulling the sensor reset cable.

AIR MANAGEMENT SYSTEM 1980

The Air Management System of controlling hydrocarbons and carbon monoxide is used on all vehicles. The Air Management System, reduces the amount of hydrocarbons and carbon monoxide in the exhaust gas by injecting air directly into the exhaust port of each cylinder (except 4 and 5 on EFI cars). The air added to the hot exhaust gases causes further oxidation of the gases before they enter the exhaust manifold.

This system consists of a belt-driven pump, three formed rubber air hoses (five on California vehicles), a metal tubing manifold between the cylinder heads, air management valves providing additional functions along with the standard divert air function, an air supply pipe from the A.I.R. pump to the catalytic converter and a check valve to protect the hoses and the pump from hot gases. A single valve system is used on all vehicles except those sold in California which use a dual valve system.

The single valve system diverts air from the A.I.R. pump to either the exhaust ports or to the air cleaner. Air injected into the exhaust ports reacts with the hot exhaust gases to help complete combustion and decrease carbon monoxide and hydrocarbon levels. Under deceleration, air is diverted to the air cleaner for backfire protection, fuel economy and catalytic converter protection.

The dual valve system is used with the dual bed catalytic converter emission control system (C-4). The air switching function will direct the air flow either to the exhaust ports, or between the catalytic converter beds. When air is directed to the exhaust ports, it reacts with the exhaust gases to complete combustion. The burning of exhaust gases in the exhaust ports increases the exhaust gas temperature which aids in bringing the catalytic converter up to operating temperature. When the catalytic converter *is* up to operating temperature and the emission system switches to "closed-loop", the air is directed between the beds of the converter. This holds the amount of oxygen in the first (reducing) converter bed to a minimum to aid in decreasing (NOx) levels, while making sufficient oxygen available to the second (oxidizing) converter bed to aid in oxidizing carbon monoxide and hydrocarbons.

DIGITAL ELECTRONIC FUEL INJECTION 1980

Eldorados sold in all states except California, are equipped with a Digital Electronic Fuel Injection system (DEFI) as standard equipment. The system is available as an option on the Seville. This system (DEFI) provides a means of fuel distribution for controlling exhaust emissions within the legislated limits by precisely controlling the air/fuel mixture under all

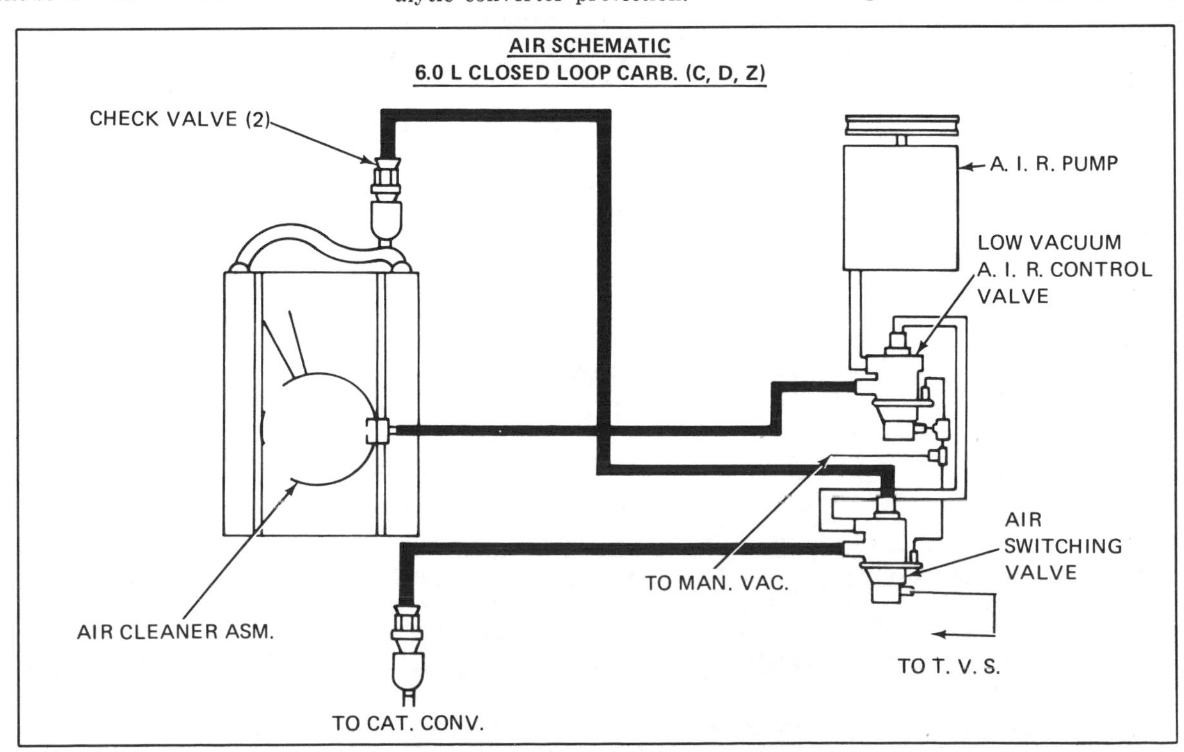

Air management system—typical (© General Motors Corp.)

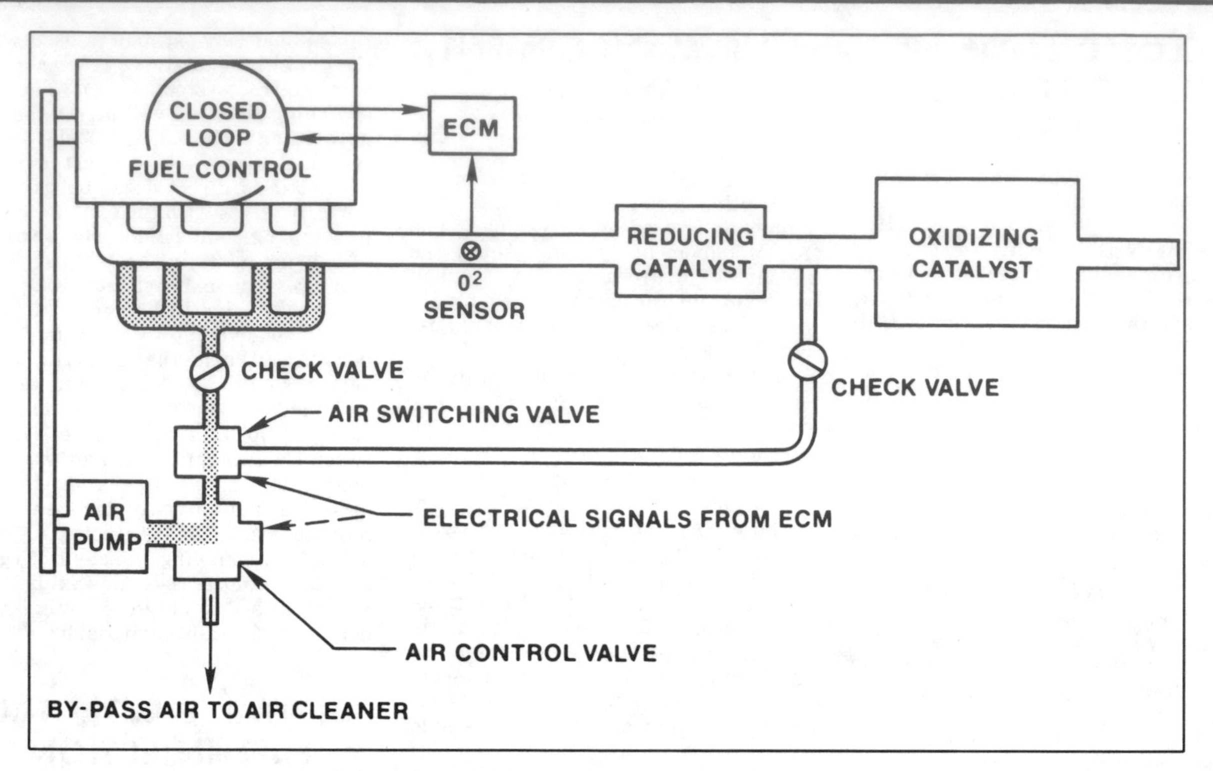

A.I.R. schematic—all except Calif. (© General Motors Corp.)

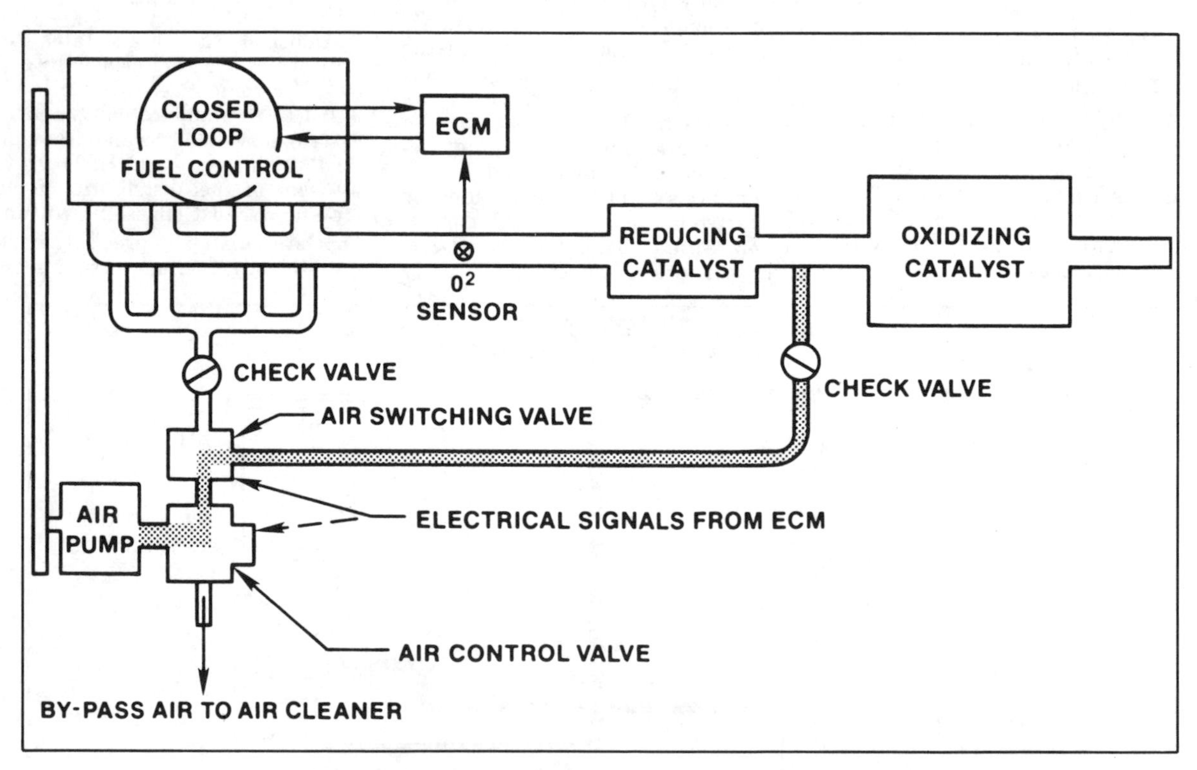

A.I.R. schematic—California (© General Motors Corp.)

operating conditions for complete combustion. This is accomplished by establishing a program for the digital Electronic Control Module (ECM), which will provide the correct quantity of fuel for a wide range of operating conditions. Several sensors are used to determine which operating conditions exist and the ECM can then signal the injectors to provide the precise amount of fuel required.

The Digital Electronic Fuel Injection system consists of a pair of electrically actuated fuel metering valves which, when actuated, spray a calculated quantity of fuel into the engine intake manifold. These valves or injectors are mounted on the throttle body above the throttle blades with the metering tip pointed into the throttle throats. The injectors are normally actuated alternately. Fuel is supplied to the inlet of the injectors through the fuel lines and is maintained at a constant pressure across the injector inlets. When the solenoid operated valves are energized, the injector ball valve moves to the full open position and since the pressure differential across the valve is con-

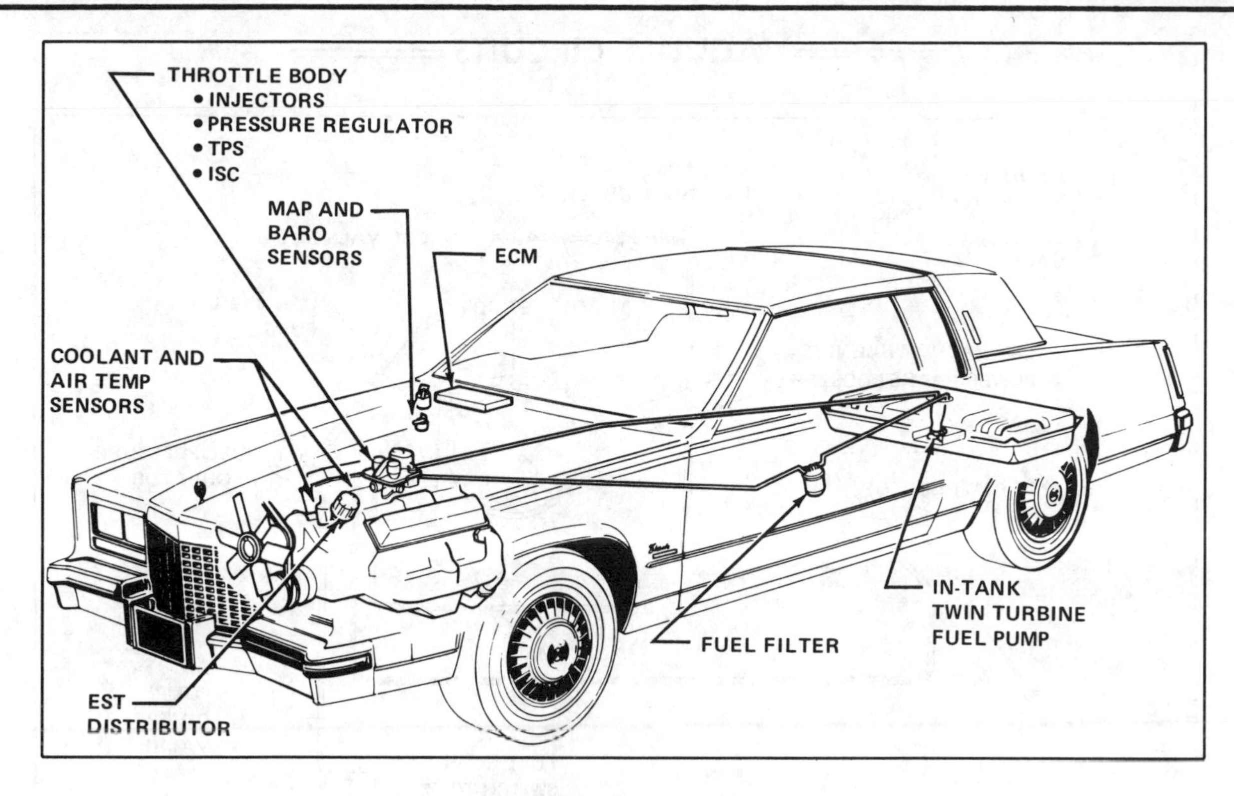

Major components of the DEFI system (© General Motors Corp.)

stant, the fuel quantity is changed by varying the time that the injector is held open. The amount of air that enters the engine is measured by monitoring the intake manifold absolute pressure (MAP), the intake manifold air temperature (MAT) and the engine speed (in RPM). This information allows the ECM to compute the flow rate of air being inducted into the engine and, consequently, the flow rate of fuel required to achieve the desired air/fuel mixture for the particular engine operating condition.

DIESEL ENGINE 1979-80

The idle speed on the diesel is 575 rpm, set with an adjustment on the side of the injection pump. Because the diesel has no electric ignition system, special equipment must be used to measure the idle rpm. A bracket for holding a magnetic tachometer pickup is on the front of the engine.

Diesel engines for 1980 are now using EGR valves. The EGR valve is located in the intake manifold crossover.

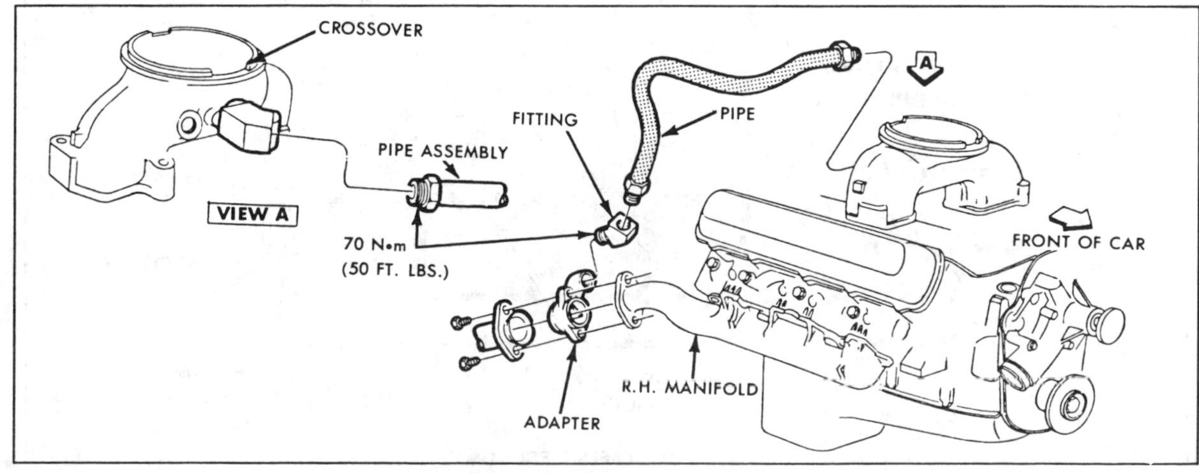

E.G.R. Pipe and Adapter - Diesel

VACUUM CIRCUITS

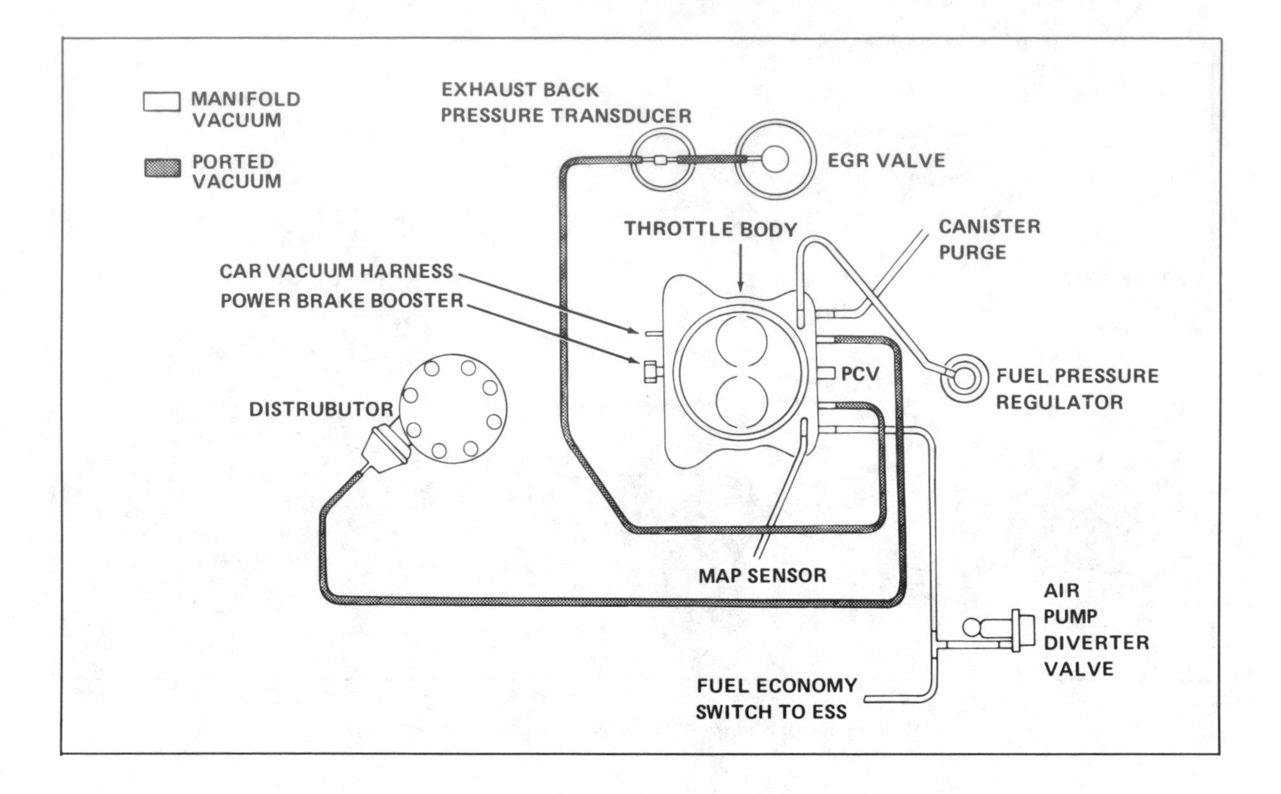

1979 Vacuum hose schematic—Eldorado for 49 States

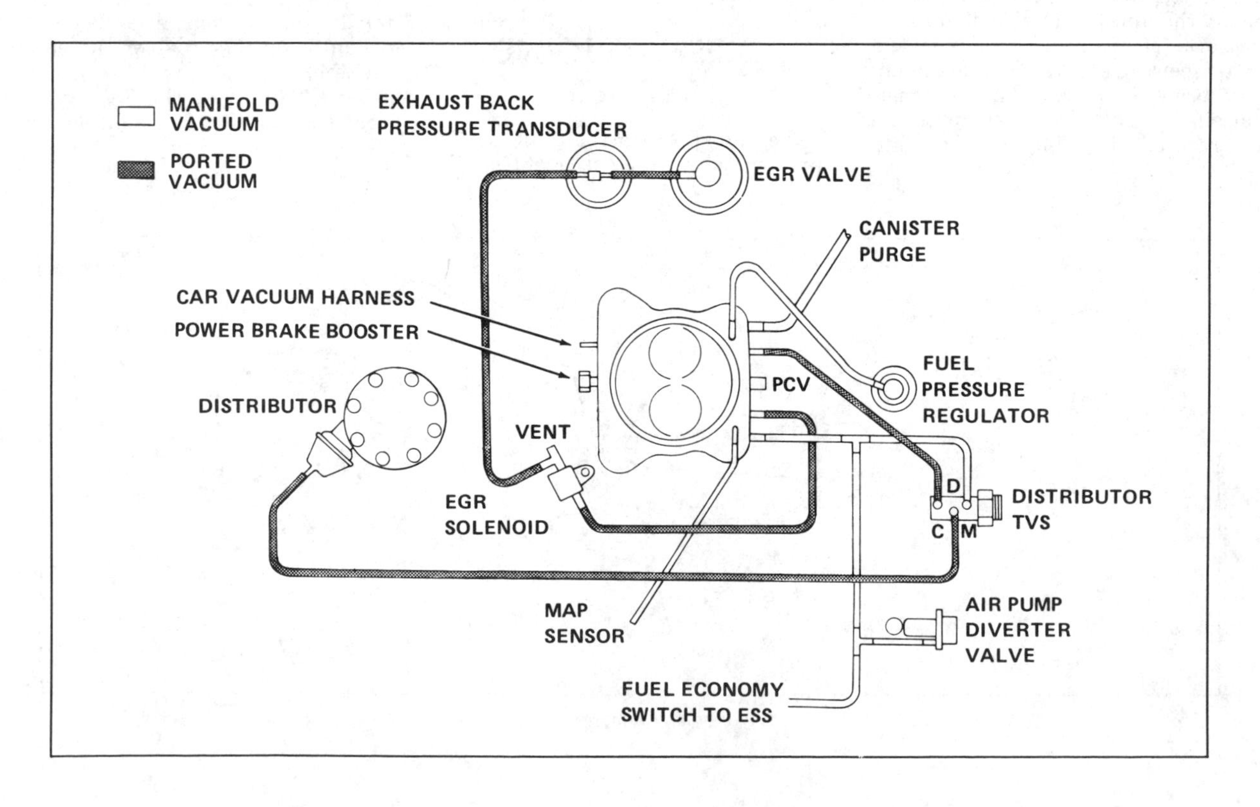

1979 Vacuum hose schematic—Eldorado for California

VACUUM CIRCUITS

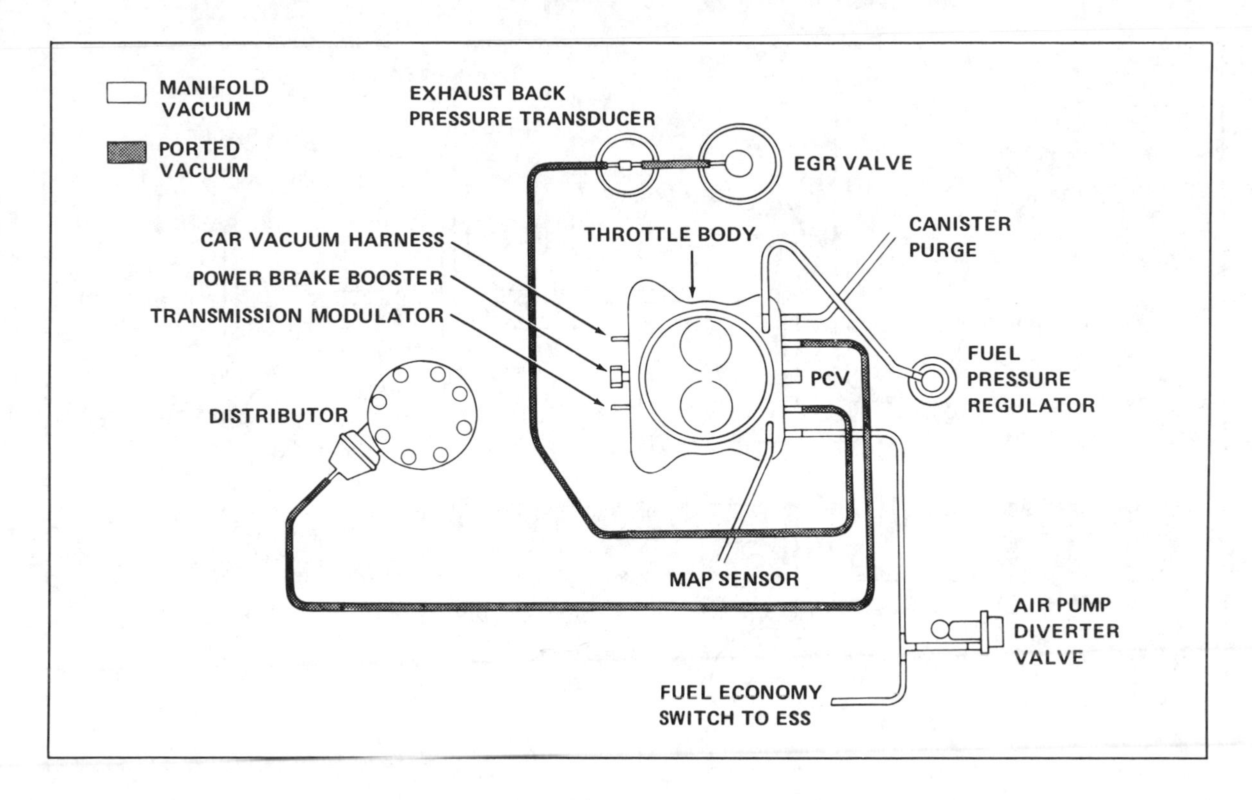

1979 Vacuum hose schematic—Seville for 49 States

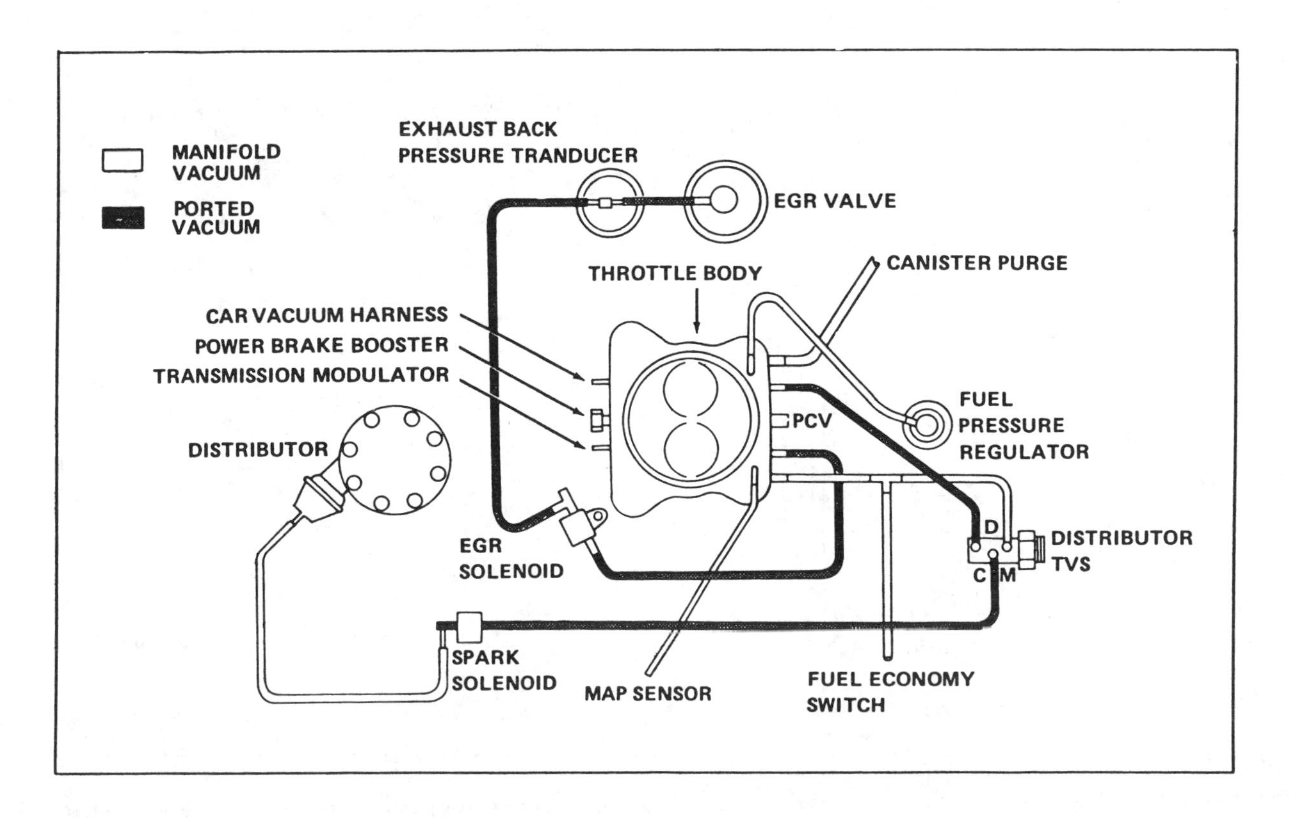

1979 Vacuum hose schematic—Seville for California

VACUUM CIRCUITS

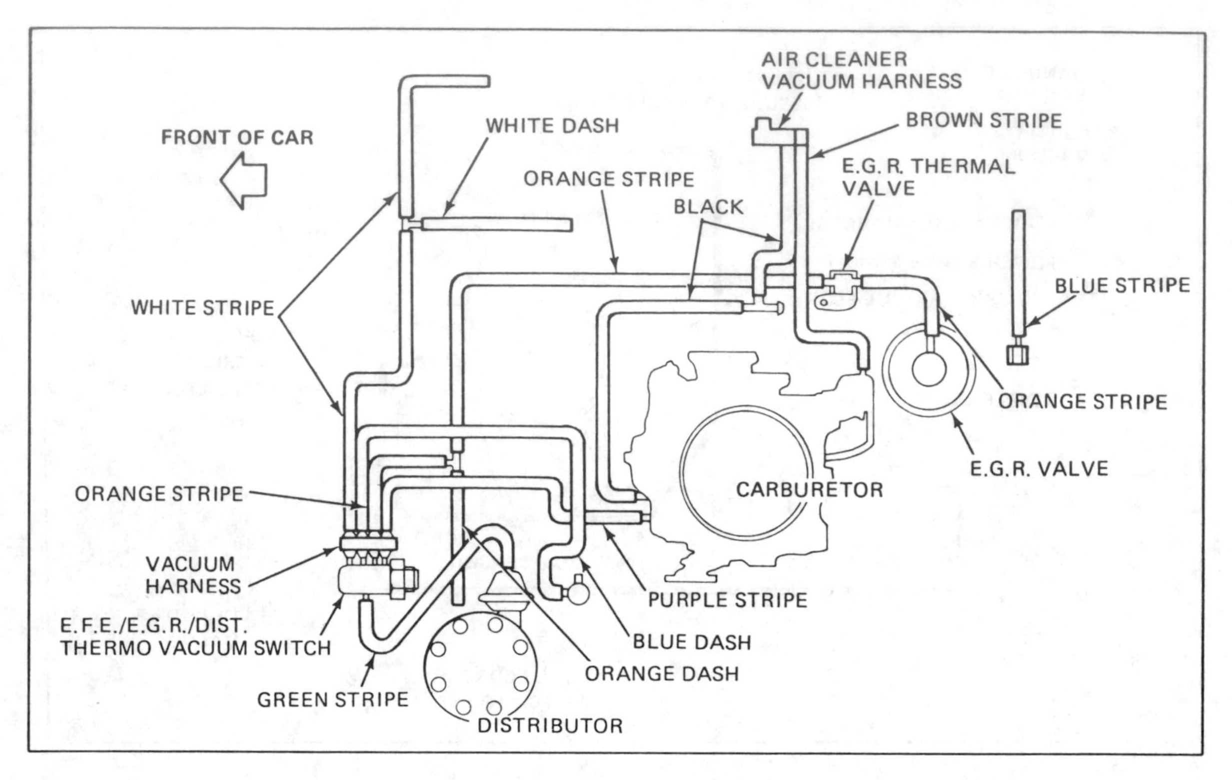

1980 Vacuum hose schematic Fleetwood & DeVille, For California (© General Motors Corp.)

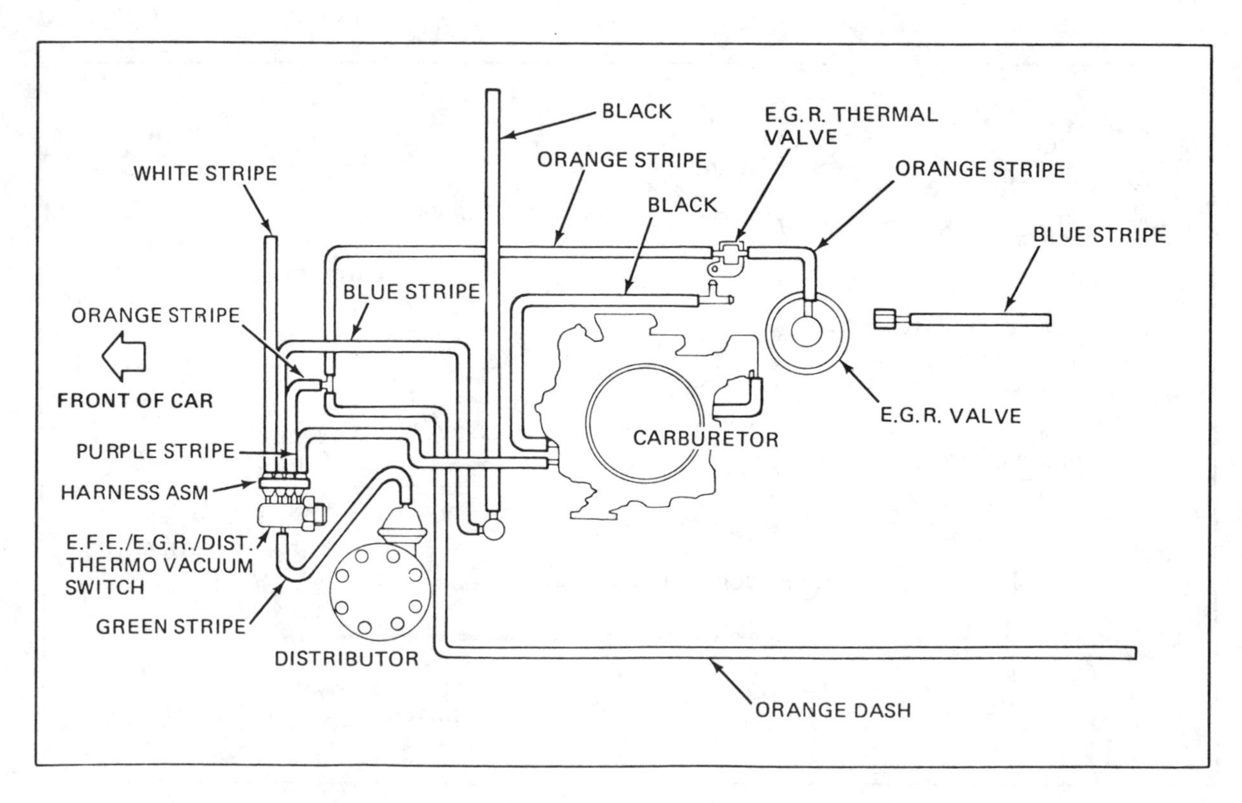

1980 Vacuum hose schematic Fleetwood & DeVille, exc. California (© General Motors Corp.)

VACUUM CIRCUITS

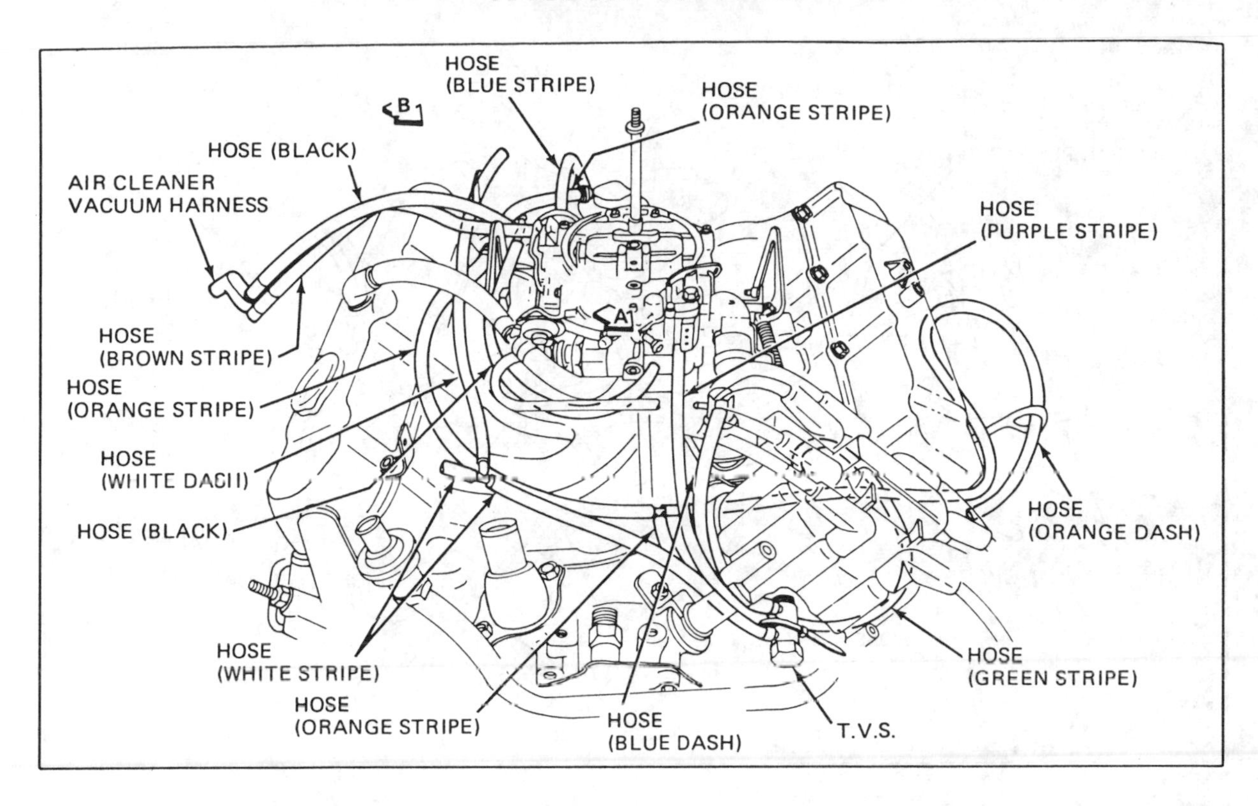

1980 Vacuum hose routing Fleetwood & DeVille, For California (© General Motors Corp.)

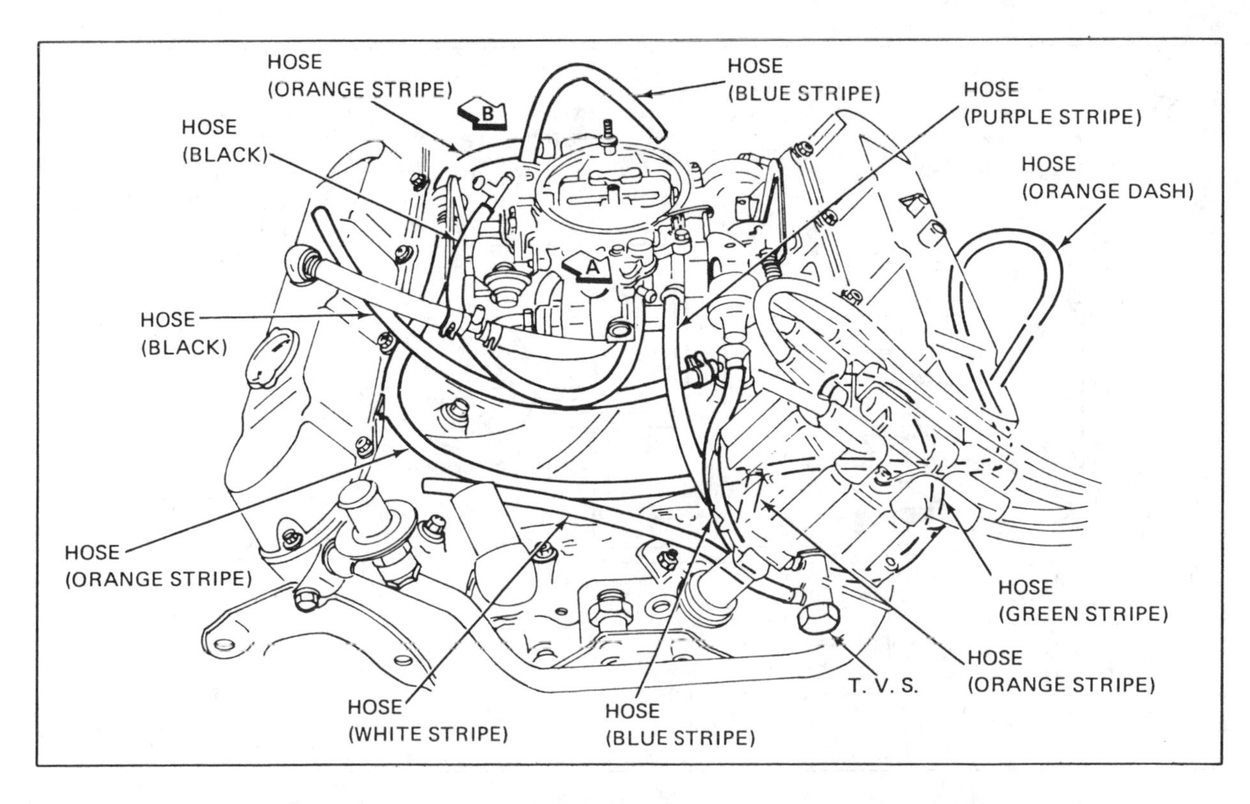

1980 Vacuum hose routing Fleetwood & DeVille, exc. California (© General Motors Corp.)

VACUUM CIRCUITS

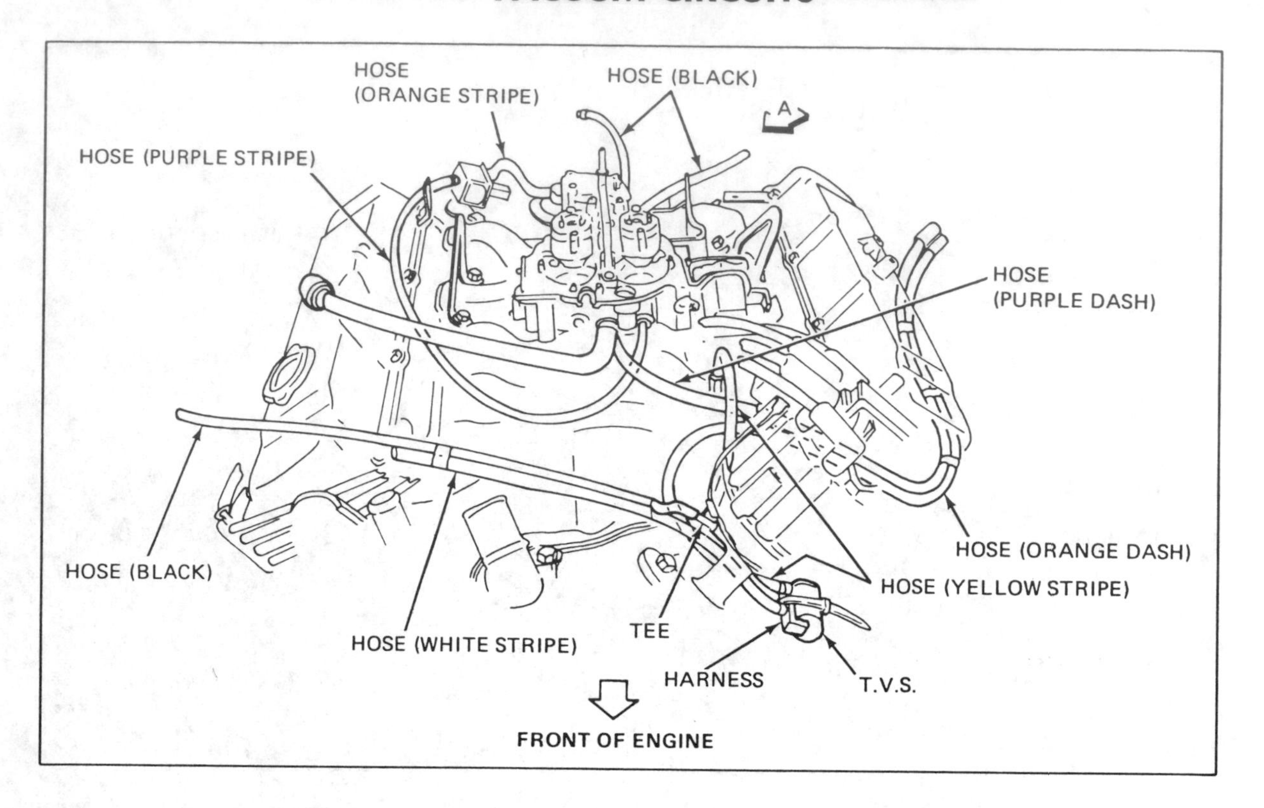

1980 Vacuum hose routing Eldorado & Seville (DEFI) (© General Motors Corp.)

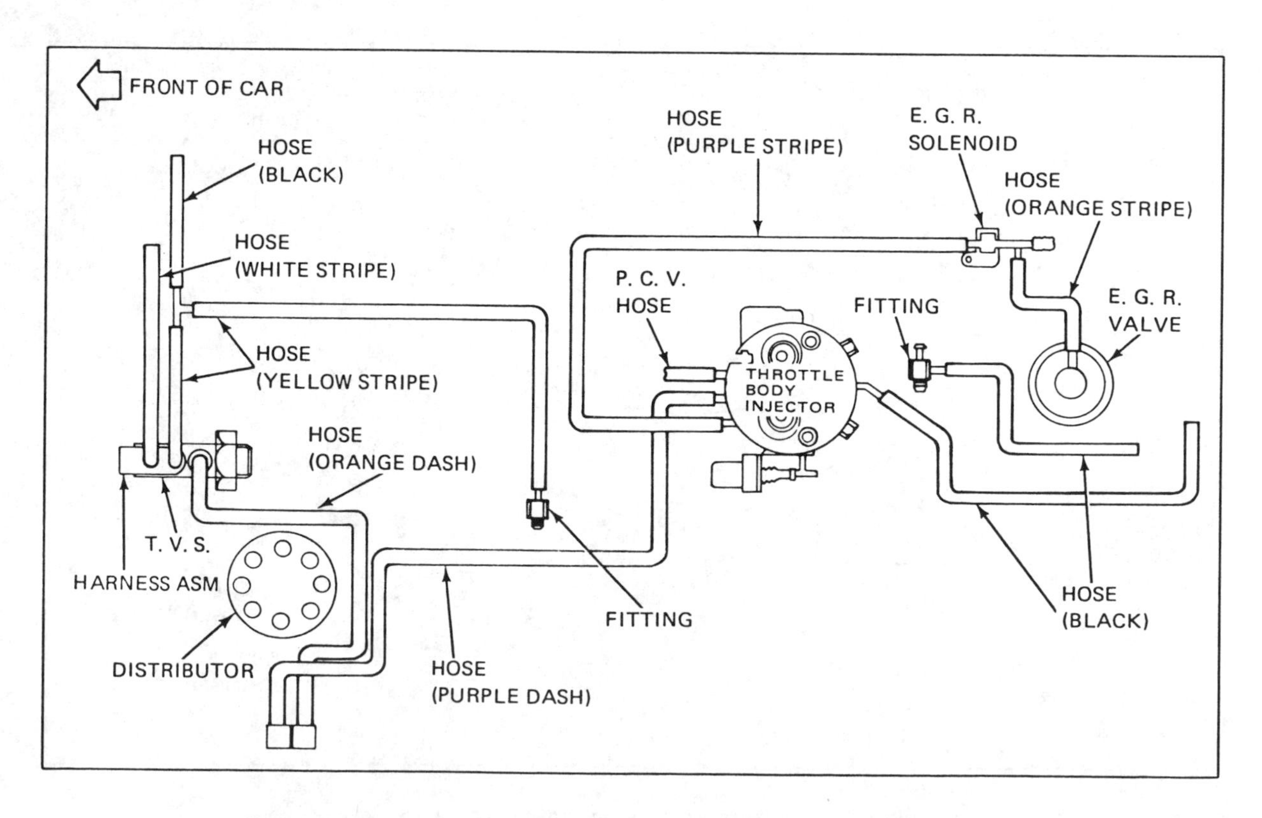

1980 Vacuum hose schematic Eldorado & Seville (DEFI) (© General Motors Corp.)

VACUUM CIRCUITS

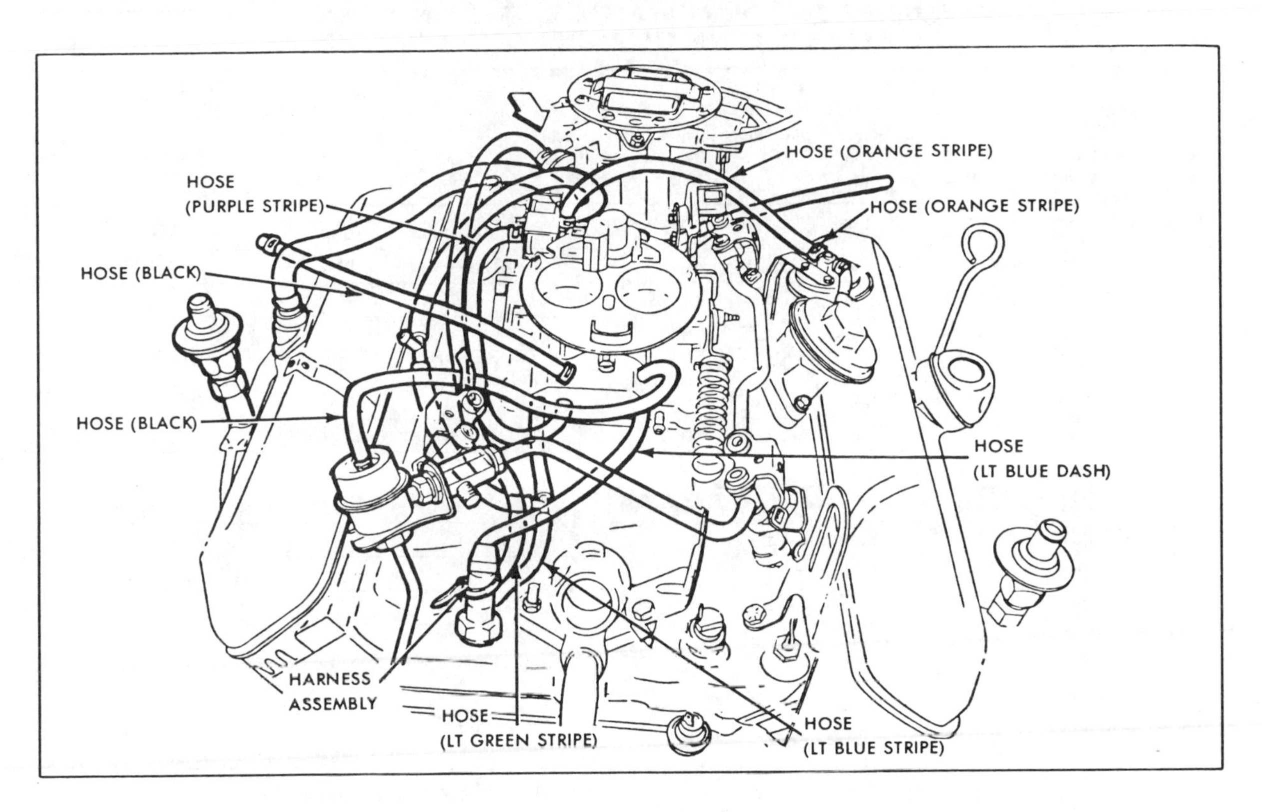

1980 Vacuum hose routing Eldorado & Seville (EFI) (© General Motors Corp.)

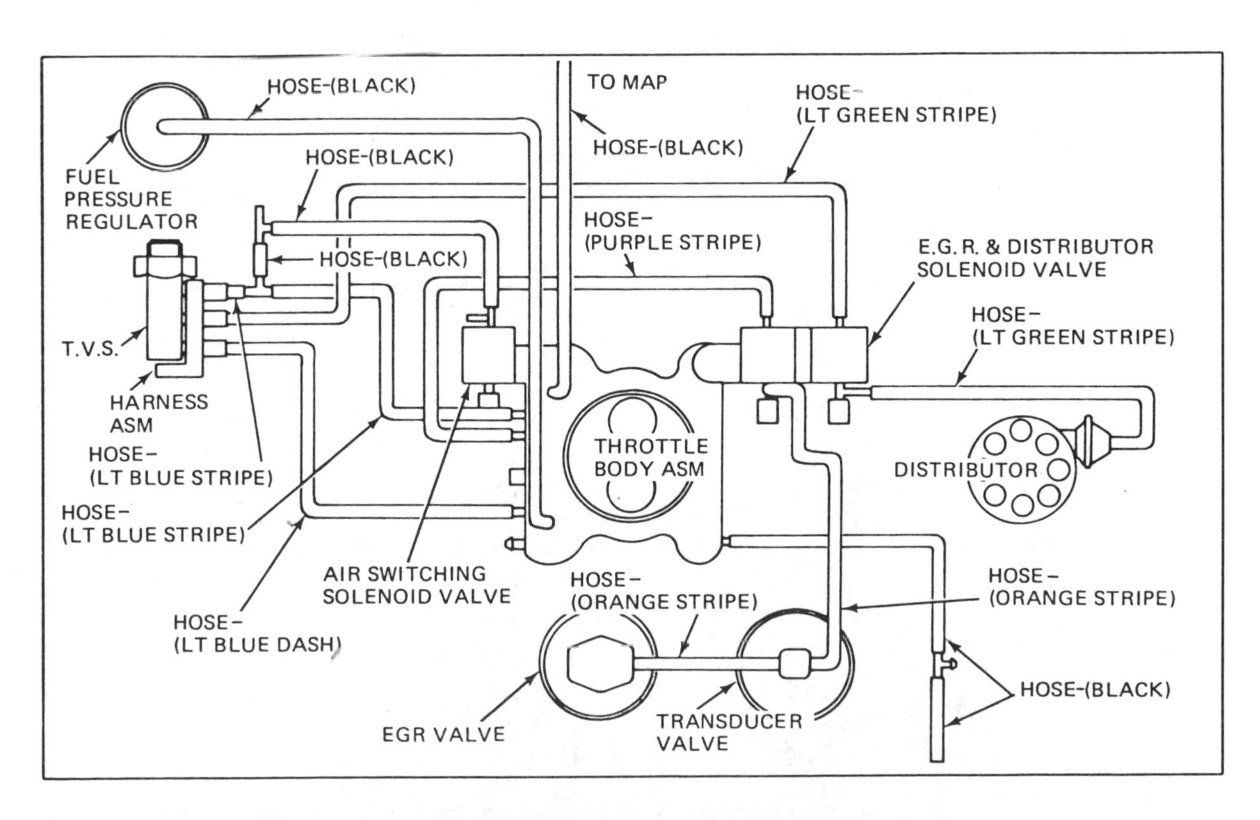

1980 Vacuum hose schematic Eldorado & Seville (EFI) (© General Motors Corp.)

TUNE-UP SPECIFICATIONS

ENGINE CODE, 5th CHARACTER OF THE VIN NUMBER
MODEL YEAR CODE, 6th CHARACTER OF THE VIN NUMBER

	ENGINE		SPARK PLUGS*		IGNITION			CARBURETION Idle Speed② (r.p.m.)			
						Timing① (deg B.T.D.C. @ rpm)				Auto. Trans.	
Year	V.I.N. Code	No. Cyl. Disp. (cu in)	Orig. Type	Gap (in)	Distributor	Man. Trans.	Auto Trans.	Man. Trans.	Fed.	Cal.	Hi. Alt.
1979	M	V6-200	R45TS	.045	E.I.	8	12	700/800	600/700	—	—
	A	V6-231	R45TSX	.060	E.I.	—	15	—	550/670	600	600
	D	6-250	R46TS	.035	E.I.	12	8 (6)	800	675	600	—
	J	8-267	R45TS	.045	E.I.	4	8	600/700	500/600	—	—
	G	8-305	R45TS	.045	E.I.	4	4	600/700	500/600	600/650	—
	H	8-305	R45TSX	.045	E.I.	4	4③	700	500/600	500/600	600/650
	L	8-350	R45TS	.045	E.I.	6	6 (8)③	700	500/600	500/600	600/650
1980	K	6-229	R45TS	.045	E.I.	8	12	700/800	600/675	—	—
	A	6-231	R45TSX	.060	E.I.	—	15	—	560/670	600	—
	3	6-231	R46TSX	.060	E.I.	④			④		
	S	8-265	R45TSX	.060	E.I.	④			④		
	J	8-267	R45TS	.045	E.I.	—	6	—	500/600	—	—
	W	8-301	R45TSX	.060	E.I.	—	8	—	500/600	—	—
	H	8-305	R45TSX	.045	E.I.	4	4	700	500/600	550/650	—
	N	8-350	—	—	Diesel	④		—	650	650	650
	L	8-350	R45TS	.045	E.I.	6	6	700	500/600	—	—

NOTE: The underhood certification/specification decal is the authority for performance specifications affecting vehicle emissions. Use this manual's information when that decal is not available.

— Not applicable

E.I. Electronic Ignition

① The curb idle speed unless otherwise indicated. Figures in parentheses are California applications.

② Set idle speed with automatic transmission in drive and manual transmission in neutral. Where two figures appear, the lower figure indicates idle speed with A/C solenoid deenergized.

③ High altitude—8.

④ See underhood certification/specification decal.

DISTRIBUTOR SPECIFICATIONS

		CENTRIFUGAL ADVANCE		VACUUM ADVANCE	
Year	DISTRIBUTOR IDENTIFICATION	Start Crank. Deg. @ Eng. RPM	Finish Crank. Deg. @ Eng. RPM	Start @ In. Hg.	Finish Crank. Deg. @ In. Hg.
1979	1103281	0 @ 1000	20 @ 3800	4	18 @ 12
	1103282	0 @ 1000	20 @ 3800	4	20 @ 10
	1103285	0 @ 1200	22 @ 4200	4	10 @ 8
	1103337	0 @ 1100	22 @ 4600	4	24 @ 11
	1103353	0 @ 1100	22 @ 4600	4	20 @ 10
	1103368	0 @ 1000	20 @ 3800	4	10 @ 8
	1103370	0 @ 1300	16 @ 4200	3	24 @ 10
	1103371	0 @ 1000	22 @ 4400	3	24 @ 10
	1103379	0 @ 1000	20 @ 3800	3	20 @ 8.5
	1110695	0 @ 1680	15 @ 3600	3.9	24 @ 10.9
	1110696	0 @ 1000	20 @ 3800	2	16 @ 7.5
	1110716	0 @ 1000	20 @ 4200	4	15 @ 12
	1110731	0 @ 1680	15 @ 3600	4.9	16 @ 8.3
	1110737	0 @ 1000	20 @ 3800	3	30 @ 9.5

DISTRIBUTOR SPECIFICATIONS

Year	DISTRIBUTOR IDENTIFICATION	CENTRIFUGAL ADVANCE Start Crank. Deg. @ Eng. RPM	CENTRIFUGAL ADVANCE Finish Crank. Deg. @ Eng. RPM	VACUUM ADVANCE Start @ In. Hg.	VACUUM ADVANCE Finish Crank. Deg. @ In. Hg.
1979	1110748	0 @ 1000	20 @ 4200	4	20 @ 11
	1110756	0 @ 1400	14 @ 3800	2	24 @ 10
	1110766	0-4 @ 2000	14 @ 3600	4	24 @ 12
	1110767	0-4 @ 2000	14-18 @ 3600	3	20 @ 12
1980	1110696	0 @ 1000	10 @ 3800	2	16 @ 7.5
	1110756	0 @ 1400	14 @ 3800	2	24 @ 10
	1110766	0-4 @ 2000	13-17 @ 3600	4	24 @ 12
	1110767	0-4 @ 2000	13-17 @ 3600	4	20 @ 12
	1110716	0 @ 1000	20 @ 4200	4	15 @ 12
	1103371	0 @ 1000	22 @ 4400	3	24 @ 10
	1103370	0 @ 1300	16 @ 4200	3	24 @ 10
	1103282	0 @ 1000	20 @ 3800	4	20 @ 11
	1103379	0 @ 1000	20 @ 3800	3	20 @ 8.5
	1103368	0 @ 1000	20 @ 3800	4	10 @ 8
	1103353	0 @ 1100	22 @ 4600	4	20 @ 10
	1103337	0 @ 1100	22 @ 4600	4	24 @ 11

Model 6510-C

General Motors Corp.

Year	Part Number	Vacuum Break Adjustment (in.)	Fast Idle Cam Adjustment (in.)	Unloader Adjustment (in.)	Fast Idle Adjustment (rpm)	Float Level Adjustment (in.)	Choke Setting
1979	10008489, 10008490	.250	.150	.350	2400	.520	1 Rich
	10008491, 10008492	.250	.150	.350	2200	.520	2 Rich
	10009973, 10009974	.275	.150	.350	2400	.520	2 Rich
1980	All w/manual	.275	.130	.350	2600	.500	Fixed
	All w/automatic	.300	.130	.350	2500	.500	Fixed

1ME CARBURETOR SPECIFICATIONS

CHEVROLET PRODUCTS, CHEVETTE

Year	Carburetor Identification① Number	Float Level (in.)	Metering Rod (in.)	Fast Idle Speed (rpm)	Fast Idle Cam (in.)	Vacuum Break (in.)	Choke Unloader (in.)	Choke Setting (notches)
1979	17059014	3/8	0.095	2000	0.180	0.200	0.400	Index
	17059020	3/8	0.095	2000	0.180	0.200	0.400	Index
	17059013	3/8	0.095	1800	0.180	0.200	0.400	Index
	17059314	3/8	0.100	2000	0.190	0.245	0.400	Index

① Stamped on float bowl, next to fuel inlet nut
② 2200 rpm for the first two numbers
③ 2200 rpm for the last two numbers

2MC, M2MC, M2ME, E2ME CARBURETOR SPECIFICATIONS

CHEVROLET MONZA

Year	Carburetor Identification①	Float Level (in.)	Choke Rod (in.)	Choke Unloader (in.)	Vacuum Break Lean or Front (in.)	Vacuum Break Rich or Rear (in.)	Pump Rod (in.)	Choke Coil Lever (in.)	Automatic Choke (notches)
1980	17080191	11/32	0.139	0.243	0.096	0.096	1/4②	0.120	Fixed
	17080195	9/32	0.139	0.243	0.103	0.090	1/4②	0.120	Fixed
	17080197	9/32	0.139	0.243	0.103	0.090	1/4②	0.120	Fixed
	17080491	5/16	0.139	0.243	0.117	—	3/8	0.120	Fixed
	17080496	5/16	0.139	0.243	0.117	0.203	3/8	0.120	Fixed
	17080498	5/16	0.139	0.243	0.117	0.203	3/8	0.120	Fixed

2MC, M2MC, M2ME, E2ME CARBURETOR SPECIFICATIONS

CHEVROLET (EXCEPT MONZA)

Year	Carburetor Identification①	Float Level (in.)	Choke Rod (in.)	Choke Unloader (in.)	Vacuum Break Lean or Front (in.)	Vacuum Break Rich or Rear (in.)	Pump Rod (in.)	Choke Coil Lever (in.)	Automatic Choke (notches)
1979	17059180	11/32	0.139	0.243	0.103	0.090	1/4②	0.120	1 Lean
	17059190	11/32	0.139	0.243	0.103	0.090	1/4②	0.120	1 Lean
	17059196	11/32	0.139	0.277	0.129	0.117	1/4②	0.120	1 Lean
	17059134	13/32	0.243	0.243	0.157	—	1/4②	0.120	1 Lean
	17059135	13/32	0.243	0.243	0.157	—	1/4②	0.120	1 Lean
	17059136	13/32	0.243	0.243	0.157	—	1/4②	0.120	1 Lean
	17059137	13/32	0.243	0.243	0.157	—	1/4②	0.120	1 Lean
	17059434	13/32	0.243	0.243	0.171	—	1/4②	0.120	1 Lean
	17059436	13/32	0.243	0.243	0.171	—	1/4②	0.120	1 Lean
	17059130	9/32	0.243	0.243	0.157	—	1/4②	0.120	Index
	17059131	9/32	0.243	0.243	0.157	—	1/4②	0.120	Index
	17059132	9/32	0.243	0.243	0.157	—	1/4②	0.120	1 Lean
	17059133	9/32	0.243	0.243	0.157	—	1/4②	0.120	1 Lean
	17059138	9/32	0.243	0.243	0.164	—	1/4②	0.120	1 Lean
	17059139	9/32	0.243	0.243	0.164	—	1/4②	0.120	1 Lean
	17059140	9/32	0.243	0.243	0.164	—	1/4②	0.120	1 Lean
	17059141	9/32	0.243	0.243	0.164	—	1/4②	0.120	1 Lean
	17059430	9/32	0.243	0.243	0.157	—	1/4②	0.120	1 Lean
	17059432	9/32	0.243	0.243	0.157	—	1/4②	0.120	1 Lean
	17059496	5/16	0.139	0.243	0.117	0.179	3/8②	0.120	2 Rich
1980	17080108	3/8	0.243	0.243	0.142	—	5/16②	0.120	Fixed
	17080110	3/8	0.243	0.243	0.142	—	5/16②	0.120	Fixed
	17080130	5/16	0.243	0.243	0.142	—	5/16②	0.120	Fixed
	17080131	5/16	0.243	0.243	0.142	—	5/16②	0.120	Fixed
	17080132	5/16	0.243	0.243	0.142	—	5/16②	0.120	Fixed
	17080133	5/16	0.243	0.243	0.142	—	5/16②	0.120	Fixed
	17080138	3/8	0.243	0.243	0.142	—	5/16②	0.120	Fixed
	17080140	3/8	0.243	0.243	0.142	—	5/16②	0.120	Fixed
	17080493	5/16	0.139	0.243	0.117	0.179	Fixed	0.120	Fixed
	17080495	5/16	0.139	0.243	0.117	0.179	Fixed	0.120	Fixed
	17080496	5/16	0.139	0.243	0.117	0.203	Fixed	0.120	Fixed
	17080498	5/16	0.139	0.243	0.117	0.203	Fixed	0.120	Fixed

QUADRAJET CARBURETOR SPECIFICATIONS

CHEVROLET

Year	Carburetor Identification①	Float Level (in.)	Air Valve Spring (turn)	Pump Rod (in.)	Primary Vacuum Break (in.)	Secondary Vacuum Break (in.)	Secondary Opening (in.)	Choke Rod (in.)	Choke Unloader (in.)	Fast Idle Speed ④ (rpm)
1979	17059203	15/32	7/8	1/4	0.157	—	⑤	0.243	0.243	⑦
	17059207	15/32	7/8	1/4	0.157	—	⑤	0.243	0.243	⑦
	17059216	15/32	7/8	1/4	0.157	—	⑤	0.243	0.243	⑦
	17059217	15/32	7/8	1/4	0.157	—	⑤	0.243	0.243	⑦
	17059218	15/32	7/8	1/4	0.164	—	⑤	0.243	0.243	⑦
	17059222	15/32	7/8	1/4	0.164	—	⑤	0.243	0.243	⑦
	17059502	15/32	7/8	1/4	0.164	—	⑤	0.243	0.243	⑦
	17059504	15/32	7/8	1/4	0.164	—	⑤	0.243	0.243	⑦
	17059582	15/32	7/8	11/32	0.203	—	⑤	0.243	0.314	⑦
	17059584	15/32	7/8	11/32	0.203	—	⑤	0.243	0.314	⑦
	17059210	15/32	1	9/32	0.157	—	⑤	0.243	0.243	⑦
	17059211	15/32	1	9/32	0.157	—	⑤	0.243	0.243	⑦
	17029228	15/32	1	9/32	0.157	—	⑤	0.243	0.243	⑦
1980	17080202	7/16	7/8	1/4⑧	0.157	—	⑤	0.110	0.243	⑩
	17080204	7/16	7/8	1/4⑧	0.157	—	⑤	0.110	0.243	⑩
	17080207	7/16	7/8	1/4⑧	0.157	—	⑤	0.110	0.243	⑩
	17080228	7/16	7/8	9/32⑧	0.179	—	⑤	0.110	0.243	⑩
	17080243	3/16	9/16	9/32⑧	0.016	0.083	⑤	0.074	0.179	⑩
	17080274	15/32	5/8	5/16⑨	0.110	0.164	⑤	0.083	0.203	⑩
	17080282	7/16	7/8	11/32⑨	0.142	—	⑤	0.110	0.243	⑩
	17080284	7/16	7/8	11/32⑨	0.142	—	⑤	0.110	0.243	⑩
	17080502	1/2	7/8	Fixed	0.136	0.179	⑤	0.110	0.243	⑩
	17080504	1/2	7/8	Fixed	0.136	0.179	⑤	0.110	0.243	⑩
	17080542	3/8	9/16	Fixed	0.103	0.066	⑤	0.074	0.243	⑩
	17080543	3/8	9/16	Fixed	0.103	0.129	⑤	0.074	0.243	⑩

① The carburetor identification number is stamped on the float bowl, near the secondary throttle lever.
② Without vacuum advance.
③ With automatic transmission; vacuum advance connected and EGR disconnected and the throttle positioned on the high step of cam.
④ With manual transmission; without vacuum advance and the throttle positioned on the high step of cam.
⑤ No measurement necessary on two point linkage; see text.
⑥ 3 turns after contacting lever for preliminary setting.
⑦ 2 turns after contacting lever for preliminary setting.
⑧ Inner hole
⑨ Outer hole
⑩ 4 turns after contacting lever for preliminary setting.

2SE, E2SE CARBURETOR ADJUSTMENTS

Chevrolet Monza

Year	Carburetor Identification	Float Level (in.)	Pump Rod (in.)	Fast Idle (rpm)	Choke Coil Lever (in.)	Fast Idle Cam (deg./in.)	Air Valve Rod (in.)	Primary Vacuum Break (deg./in.)	Choke Setting (notches)	Secondary Vacuum Break (deg./in.)	Choke Unloader (deg./in.)	Secondary Lockout (in.)
1979	17059674	13/64	1/2	2400	.120	18/0.096	.025	19/.103	2 Rich	—	32/.195	.030
	17059675	13/64	17/32	2200	.120	18/0.096	.025	21/.117	1 Rich	—	32/.195	.030
	17059676	13/64	1/2	2400	.120	18/0.096	.025	19/.103	2 Rich	—	32/.195	.030
	17059677	13/64	17/32	2200	.120	18/0.096	.025	21/.117	1 Rich	—	32/.195	.030
1980	All	3/16	1/2	①	.085	18/0.096	0.18	—	Fixed	—	32/.195	.120

① See Underhood Decal

CAR SERIAL NUMBER AND ENGINE IDENTIFICATION

1979

The vehicle identification number is mounted on a plate behind the windshield on the driver's side. The sixth character is the model year, with 9 for 1979. The fifth character is the engine code, as follows:

A . . . . .231 2-bbl. V-6 LD-5 Buick
2 . . . . .231 2-bbl. V-6 LC-6 Buick
D . . . . . .250 1-bbl. 6-cyl. L-22 Chev.
H . . . . . . .305 4-bbl. V-8 LG-4 Chev.
L . .350 4-bbl. V-8 LM-1 & L-48 Chev.
M . . . . .200 2-bbl. V-6 L-26 Chev.
G . . . . . . .305 2-bbl. V-8 LG-3 Chev.

NOTE: *The LM-1 and the L-48 are the same engine. When used as an optional engine in cars other than Corvette, it is known as an LM-1. When used as the standard engine in the Corvette, it is known as the L-48. The L-82 is the high performance V-8, used only in the Corvette.*

1980

The vehicle identification number is mounted on a plate behind the windshield on the drivers side. The sixth character is the model year with A designating 1980. The fifth character is the engine code as follows:

A . . . . . . .3.8L (231 CID) V-6 Buick
H . . . . .5.0L (305 CID) V-8 Chev
J . . . . .4.4L (267 CID) V-8 Chev
K . . . . . . .3.8L (229 CID) V-6 Chev
L . . . . .5.7L (350 CID) V-8 Chev
N . . . . .5.7L (350 CID) V-8 Diesel Olds
T . . . . .4.9L (301 CID) V-8 Turbo Pont
3 . . . . .3.8L (231 CID) V-6 Turbo Buick

EMISSION EQUIPMENT 1979-80

All Models

Closed positive crankcase ventilation
Emission calibrated carburetor
Emission calibrated distributor
Heated air cleaner
Vapor control, canister storage
Exhaust gas recirculation
Catalytic converter, underfloor
Catalytic converter, manifold
  6-cyl. Calif. only
Early fuel evaporation
  All models, except
  Not used on Camaro 350 4-bbl. V-8
Vacuum delay valve, 4-nozzle
  49-States
    200 2-bbl. V-6
    250 6-cyl.
  Calif.
    250 6-cyl.
  Altitude
    Not used
Spark delay valve
  49-States
    Not used
  Calif.
    305 2-bbl. V-8
  Altitude
    Not used
Air pump
  49-States
    Corvette 350 4-bbl. V-8 L-82
  Calif.
    All models
  Altitude
    All models
Pulse Air System
  All 250 6-cyl.
Electric choke
  All 6-cyl.

Various thermal vacuum switches (TVS) are used to control the timing and extent of application of some emission systems. The switches usually measure engine temperature. A TVS might cause an EGR valve to be inoperative on a cold engine but allow it to work at normal engine temperature. If an engine is overheating a TVS could switch to higher vacuum and give more vacuum advance.
  EFE-TVS
  EGR-TVS
  Distributor vacuum-TVS

NOTE: *Refer to emission label for particular applications of these switches.*

IDLE SPEED AND MIXTURE ADJUSTMENTS

1979

CAUTION: *Emission control adjustment changes are noted on the Vehicle Emission Information Label by the manufacturer. Refer to the label before any adjustments are made.*

Air cleaner . . . . .In place
Air cond. .Off
Auto. trans. . . . . .Drive
Hoses . . . . .See underhood label
Idle CO . . . . .Not used
Mixture adj. . See propane rpm below
200 V-6 49-States
  Auto. trans. 600
    Propane enriched 620-640
    AC idle speedup 700
  Man. trans. 700
    Propane enriched 800-840
    AC idle speedup 800
231 V-6 49-States
  Auto. trans. 600
    Propane enriched . . . .
  Man. trans.
    Solenoid connected 800
    Solenoid disconnected 600
    Propane enriched . . . .
231 V-6 Calif.
  Auto. trans. 600
    Propane enriched . . . .
  Man. trans.
    Solenoid connected 800
    Solenoid disconnected 600
    Propane enriched . . . .
231 V-6 Altitude
  Auto. trans. 600
    Propane enriched . . . .
  Man. trans.
    Solenoid connected 800
    Solenoid disconnected 600
    Propane enriched . . . .
250 6-cyl. 49-States
  Auto. trans. 550
    Propane enriched 600-630
    AC idle speedup 600
  Man. trans.
    Solenoid connected 800
    Solenoid disconnected 425
    Propane enriched 800-1000
250 6-cyl. Calif.
  Auto. trans.
    Solenoid connected 600
    Solenoid disconnected 400
    Propane enriched 600-605
305 2-bbl. V-8 49-States
  Auto. trans. 500
    Propane enriched 520-540
    AC idle speedup 600
  Man. trans. 600
    Propane enriched 700-740
    AC idle speedup 700
305 2-bbl. V-8 Calif.
  Auto. trans. 500
    Propane enriched 520-540
    AC idle speedup 650
305 2-bbl. V-8 Altitude
  Auto. trans. 600
    Propane enriched 620-640
    AC idle speedup 700
350 4-bbl. V-8 "L" 49-States
  Auto. trans. 500
    Propane enriched 530-570
    AC idle speedup 600
  Man. trans. 700
    Propane enriched 850-900
350 4-bbl. V-8 "L" Calif.
  Auto. trans. 500
    Propane enriched 530-570
    AC idle speedup 600

IDLE SPEED AND MIXTURE ADJUSTMENTS

350 4-bbl. V-8 "L" Altitude
Auto. trans. 600
Propane enriched 630-670
AC idle speedup 650
350 4-bbl. V-8 L82 Engine
49-States
Auto. trans. 700
Propane enriched 760-800
AC idle speedup 750
Man. trans. 900
Propane enriched 1050-1100

1980

CAUTION: *Emission control specifications are listed on the vehicle emission label. Refer to the label when making adjustments.*

Air cleaner ....In place
Air cond. ....Off
Auto. Trans. ....Drive
Hoses ....See label
229 V-6 49 States
Auto trans
Curb 600
Propane enriched 630-650
A/C Solenoid 675

Manual trans
Curb 700
Propane enriched 850-900
Solenoid 705
231 V-6 49 States
Auto trans
Curb 650
Propane enriched 610
Solenoid ....
231 V-6 California
Auto trans
Curb 550
Propane enriched ....
Solenoid ....
231 V-6 Turbo California
Auto trans
Curb 650
Propane enriched ....
Solenoid ....
267 V-8 49 States
Auto trans
Curb 500
Propane enriched ....
Solenoid 600
301 V-8 49 States
Auto trans
Curb 600
Propane enriched ....
Solenoid 650

301 V-8 Turbo 49 States
Auto trans
Curb 500
Propane enriched 540-550
Solenoid 650
305 V-8 49 States
Auto trans
Curb 500
Propane enriched ....
Solenoid 600
Manual trans
Curb 700
Propane enriched ....
Solenoid ....
305 V-8 California
Auto trans
Curb 550
Propane enriched ....
Solenoid 650
350 V-8 49 States
Auto trans
Curb 500
Propane enriched ....
Solenoid 600
Manual trans
Curb 700
Propane enriched ....
Solenoid 500

INITIAL TIMING

1979

200 2-bbl. V-6 .... 8° BTDC
231 2-bbl. V-6 ....15° BTDC
250 6-cyl.
49-States
Auto. trans. ...10° or 8° BTDC (See label)
Calif.
Auto trans. .... 6° BTDC
305 4-bbl. V-8 .... 4° BTDC
305 2-bbl. V-8
49-States .... 4° BTDC
Calif. .... 6° BTDC
Altitude .... 8° BTDC
350 4-bbl. V-8 "L" engine
49-States .... 6° BTDC
Calif. .... 8° BTDC
Altitude ....6° or 8° BTDC (See label)
350 4-bbl. V-8 "H" engine (Corvette)
All ....12° BTDC
267 2-bbl. V-8 ....10° BTDC

1980

229 V-6 (K) Man/trans. . 8° BTDC
229 V-6 (K) Auto trans. .12° BTDC

231 V-6 All ....15° BTDC
267 V-8 (J) ....6° or 4° BTDC (see label)

301 V-8 (W) Turbo ..... 8° BTDC
305 V-8 (H) 49 States .. 4° BTDC
350 V-8 (L) 49 States ... 6° BTDC

SPARK PLUGS

1979

200 V-6	AC-R45TS	.045
231 V-6	AC-R46TSX	.060
250 6-cyl.	AC-R46TS	.035
305 V-8	AC-R45TS	.045
(4-bbl.)	AC-R43TS	.045
350 V-8	AC-R45TS	.045
267 V-8	AC-B45TS	.045

1980

229 V-6 (K)	AC-R45TS	.045
231 V-6 (All)	AC-R45TSX	.060
267 V-8 (J) A/T	AC-R46TS	.035
267 V-8 (J) /MT	AC-R45TS	.045
301 V-8 (T) Turbo	AC-R45TS	.045
305 V-8 (H) 49 States	AC-R45TS	.045
350 V-8 (L) 49 States	AC-R45TS	.045

VACUUM ADVANCE

1979

Diaphragm type ....Single
Vacuum source
200 V-6 ....Manifold
231 V-6
49-States ....Manifold
Calif. ....Ported
Altitude ....Ported
250 6-cyl. ....Manifold
305 V-8
49-States ....Manifold
Calif. ....Ported
Altitude ....Manifold
350 V-8 ....Manifold

1980

229 V-6 (K) .... ①
231 V-6 ....Manifold
231 V-6 California ....Ported
267 V-8 (J) .... ①
301 V-8 (T) .... ①
305 V-8 (H) ....Manifold
305 V-8 (H) Cal ....Ported
350 V-8 (L) ....Manifold

① See Emissions Label

EMISSION CONTROL SYSTEMS

POSITIVE CRANKCASE VENTILATION (PCV) 1979-80

All engines are equipped with a closed Positive Crankcase Ventilation System (PCV). This system is designed to provide a more complete dispersion of crankcase vapors.

An engine that is operated without any crankcase ventilation can be seriously damaged, therefore it is important to check and replace the PCV valve periodically.

EMISSION CALIBRATED CARBURETOR 1979-80

While the carburetor's main function is to provide the engine with a combustible air/fuel mixture, it's calibration is also very important to maintaining proper emission levels. The carburetor's idle, off-idle, main metering, power enrichment, and accelerating pump systems are calibrated in order to provide a combination of engine performance, fuel economy, and exhaust emission control. Adjustments and service must be performed according to recommended procedures.

EMISSION CALIBRATED DISTRIBUTOR 1979-80

Distributor calibration is a critical part of engine exhaust emission control. Therefore the initial timing centrifugal advance and vacuum advance are calibrated to provide the best engine performance and fuel economy at varying speeds and loads while remaining within exhaust emission limits.

THERMOSTATIC AIR CLEANER (TAC) 1979-80

The thermostatic air cleaner system (TAC) used on all 1979 engines, which controls the amount of heated air that enters the carburetor to maintain a controlled air temperature, is exactly the same as the system that was used on the 1978 engines.

EVAPORATIVE CONTROL SYSTEM (ECS) 1979-80

All vehicles are equipped with this system, which is designed to prevent the escape of fuel vapor into the atmosphere. Before this system was introduced the vapor generated by the evaporation of fuel in the gas tank was expelled into the atmosphere. With the inception of the evaporative control system (ECS) the vapor is transfered by an emissions line to the engine compartment. When the vehicle is operating vapors are fed directly into the engine for consumption. During periods of inoperation, an activated charcoal canister stores any vapor generated for consumption during the next period of vehicle operation.

EXHAUST GAS RECIRCULATION (EGR) 1979-80

All engines are equipped with exhaust gas recirculation (EGR). This system consists of a metering valve, a vacuum line to the carburetor, and cast-in exhaust gas passages in the intake manifold. The EGR valve is controlled by carburetor vacuum, and accordingly opens and closes to admit exhaust gases into the fuel/air mixture. The exhaust gases lower the combustion temperature, and reduce

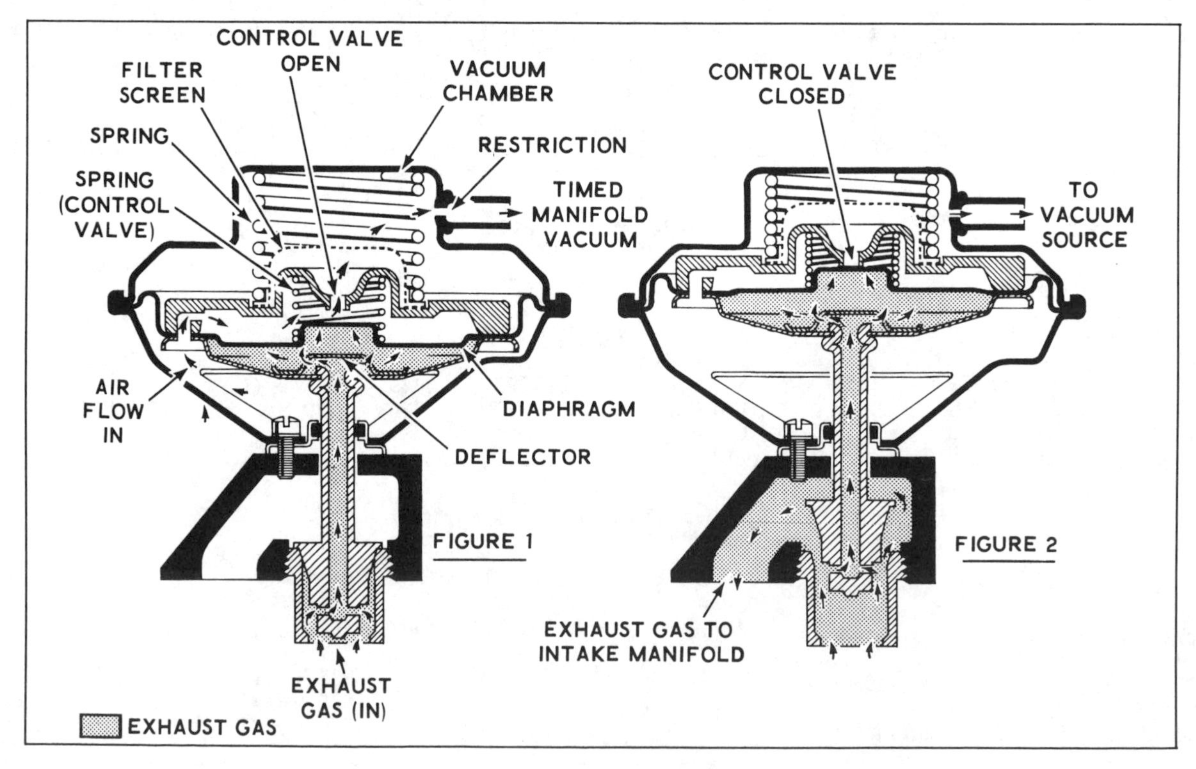

Positive back pressure EGR valve

the amount of oxides of nitrogen (NO_x) produced. The valve is closed at idle between the two extreme throttle positions.

All California models and cars delivered in areas above 4000 ft. are equipped with back pressure EGR valves. The EGR valve receives exhaust back pressure through its hollow shaft. This exerts a force on the bottom of the control valve diaphragm, opposed by a light spring. Under low exhaust pressure (low engine load and partial throttle), the EGR signal is reduced by an air bleed. Under conditions of high exhaust pressure (high engine load and large throttle opening), the air bleed is closed and the EGR valve responds to an unmodified vacuum signal. At wide open throttle, the EGR flow is reduced in proportion to the amount of vacuum signal available.

The negative transducer backpressure EGR valve assembly has the same function as the positive transducer backpressure EGR valve except the transducer is designed to allow the valve to open with a negative exhaust backpressure. The flow of the valve is controlled by manifold vacuum, negative exhaust backpressure and the carburetor ported vacuum signal.

1979-80 CATALYTIC CONVERTER

The catalytic converter, which is an emission control device, is located midway in the exhaust system. It is designed to reduce hydrocarbons (HC) and carbon monoxide (CO) pollutants from the exhaust gas stream. The catalytic converter contains beads that are coated with a material containing platinum and palladium. All six cylinder engines have an additional converter located directly under the exhaust manifold.

EARLY FUEL EVAPORATION (EFE) 1979-80

The early fuel evaporation system (EFE) is used to provide a fast heat source to the engine induction system during cold driveway. This heat provides quick fuel evaporation and more uniform fuel distribution to aid cold vehicle operation. It also reduces the length of time carburetor choking is required making reductions in exhaust emission levels possible.

VACUUM DELAY VALVE 1979-80

The vacuum delay valve is nothing more than a retard delay valve. It allows the manifold vacuum to act on the distributor diaphragm without any restriction. But when the manifold vacuum drops to zero, as during heavy acceleration or if the engine should die at idle, the valve "traps" the vacuum and lets it bleed down slowly. This keeps the spark advanced for better performance, or for easier starting.

Above 120° F. coolant temperature, a thermal vacuum switch, that is connected in the system by hoses, opens and bypasses the vacuum delay valve so that the vacuum advance works normally.

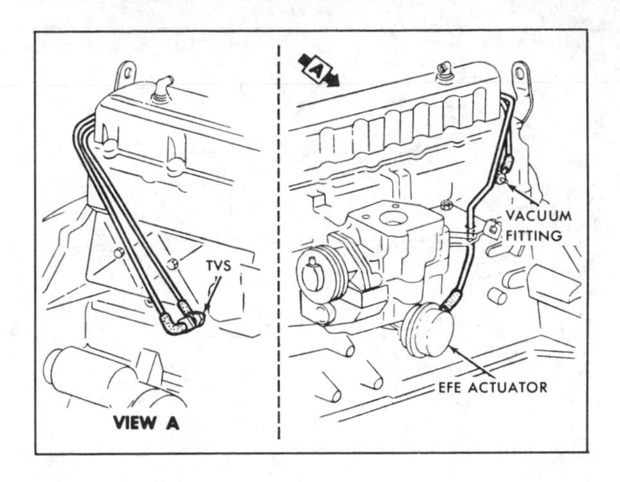

Early fuel evaporation—6-cyl. engine

SPARK DELAY VALVE 1979-80

The spark delay valve assembly is used on all California 305 V8 engines. This device controls vehicle exhaust emissions by delaying vacuum advance during vehicle acceleration modes. The valve is connected between the vaccum advance can of the distributor and the carburetor.

AIR PUMP 1979-80

The air injection reactor (AIR) injects compressed air into the exhaust system, close enough to the exhaust valves to continue the burning of the normally unburned segment of the exhaust gases. To do this it employs an air injection pump and a system of hoses, valves, tubes necessary to carry the compressed air from the pump to the exhaust manifolds.

A diverter valve is used to prevent backfiring. The valve senses sudden increases in manifold vacuum and ceases the injection of air during fuel-rich periods. During coasting, this valve diverts the entire air flow through the muffler and during high engine speeds, expels it through a relief valve. Check valves in the system prevent exhaust gases from entering the pump.

PULSE AIR SYSTEM 1979-80

The Pulse Air system consists of four pipes of equal length. Each pipe is to be secured to a check valve which is inserted into a grommet in the "pod" which is a plenum welded to the top of the rocker arm cover. The other end of the pipe routes to and is secured into an exhaust manifold passage in the head assembly using a tube nut. A separate pipe is

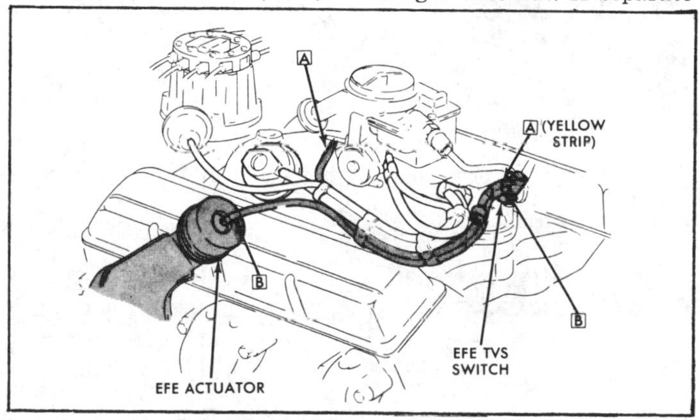

Early fuel evaporation system—V-8

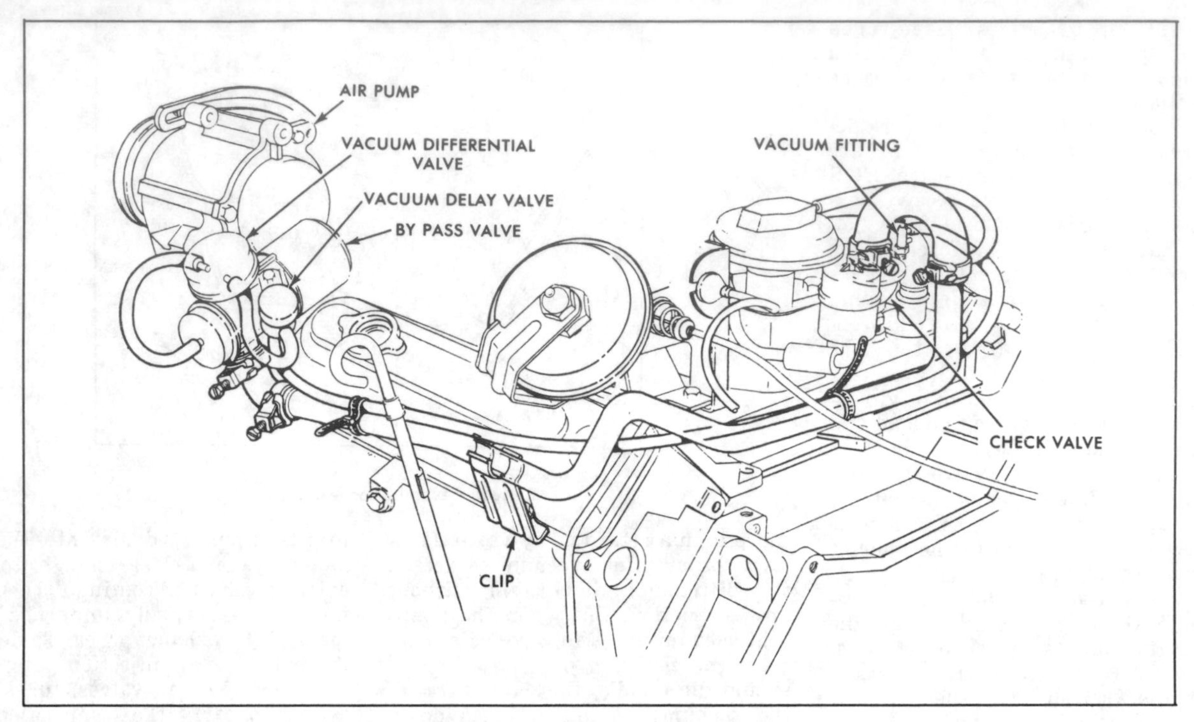

Air injector system—V-6 (231 CID) Calif.

installed between the pods as a fresh air supply source. A tee is part of this pipe onto which is to be slipped a hose which leads to a nipple on the clean side of the air cleaner.

A deceleration valve is used on California engines. This valve mounts on the forward leg of the inlet manifold and dumps air into the engine during deceleration.

ELECTRIC CHOKE 1979-80

The Electric Choke is used on select models of 1979, and operates in the same manner as the 1978 models. At temperatures below 50° F., the electrical current is directed to the small segment resistor of the choke to allow gradual opening of the choke valve. At temperatures of 70° F., and above, the electrical current is directed to both the small and the large segment resistors, to allow quicker opening of the choke valve.

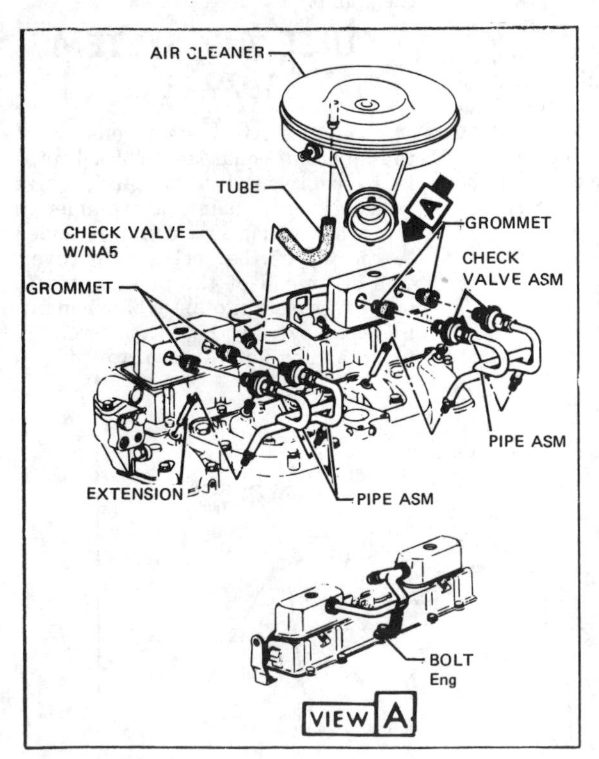

Pulse air system, 6-cyl. engines (49-States)

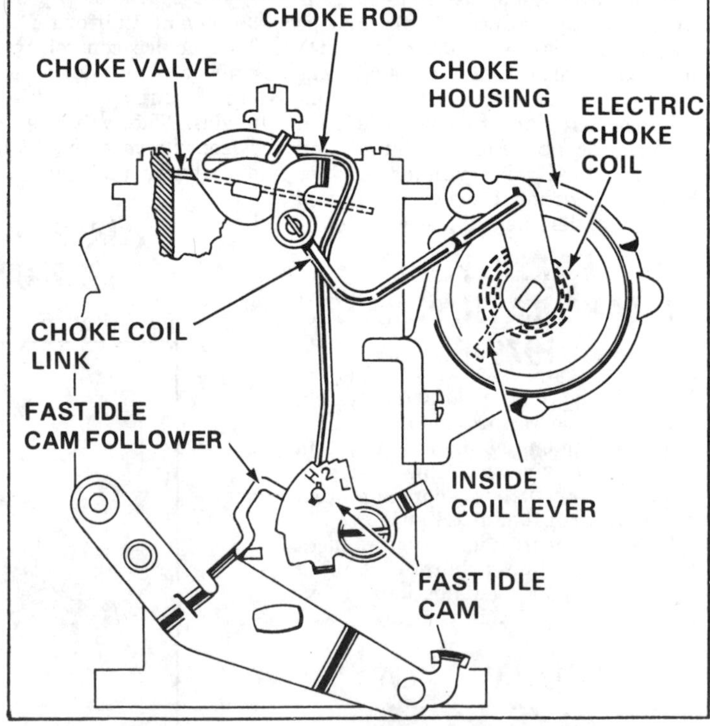

Electric choke assembly and carburetor components

TURBOCHARGER 1979-80

A turbocharger is a compressor, driven by the engine exhaust. It compresses the air-fuel mixture and forces it into the combustion chamber for increased power and acceleration.

A turbocharger consists of two turbine wheels, each in its own separate housing, and connected by a shaft. One of the turbine wheels is driven by the pressure of the exhaust gas leaving the engine. Buick takes the exhaust gas from the rear of the left bank exhaust manifold and routes it into the rear of the right bank manifold. All the exhaust from the engine exits the front of the right exhaust manifold, and goes to the turbine housing.

The pressure and flow of the exhaust makes the turbine wheel turn, just like air blowing on a windmill. The turbine wheel is connected by a shaft to the compressor wheel. The compressor wheel accepts the air-fuel mixture from the carburetor at its inlet, and pushes this mixture into the engine through its outlet. The outlet of the compressor is bolted to the intake manifold to the engine cylinders the same as any intake manifold.

The two turbine wheels, connected by the shaft are completely free to turn at any time. There is no clutch or brake of any kind that prevents them from turning. When the engine is idling, exhaust pressure is low, so the compressor wheel turns from the force of the incoming air-fuel mixture going into the engine. The compressor contributes a pressure increase, known as "boost" only when the exhaust presure is high. Usually, the only time boost occurs is during wide open throttle.

When the driver steps on the gas pedal to accelerate, he will get a normal amount of acceleration immediately. One or more seconds later the exhaust pressure will build up enough so that the compressor starts supplying boost. Then he will get additional acceleration. This delay is always present, and does not indicate that anything is wrong. It takes time for the exhaust pressure to build up and start the boost.

To keep the boost from getting too high, and possibly causing engine damaging detonation, a pressure-operated wastegate is used. A vacuum diaphram is connected to the pressure side of the compressor housing. When the pressure reaches 8 pounds, the diaphragm moves a rod attached to the wastegate. The wastegate then opens and lets some of the exhaust pressure bypass the exhaust on the turbine and the compressor wheel slows slightly, which lowers the pressure.

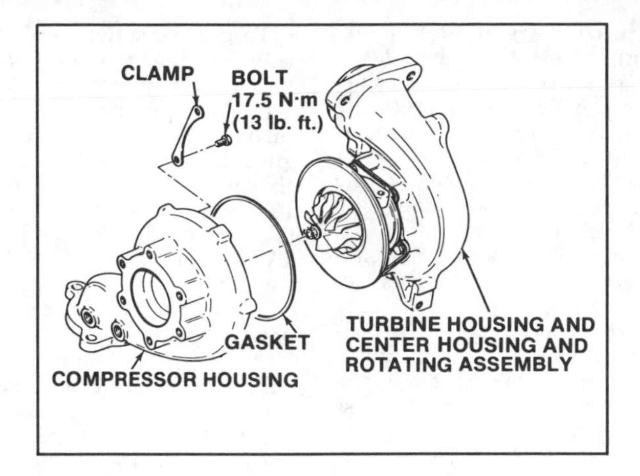

Turbo Charger Compressor Housing (© General Motors Corp.)

At 8 pounds of pressure, the wastegate actuating rod only moves about .008 in. This is such a small movement that you can barely see it move, but it is enough to open the wastegate slightly and lower the pressure on the turbine wheel. A threaded end on the rod allows adjustment.

The vacuum diaphragm actuator that moves the wastegate has two hoses attached to it. The hose on the end is from the compressor housing and senses pressure. The hose on the side is connected to intake manifold vacuum. The amount of vacuum that acts on the diaphragm has very little effect. The main reason for the hose is to provide the escape path for gas vapors in case the diaphragm should rupture. If vapors should pass through the diaphragm they will immediately be drawn into the engine.

The wastegate opening can be checked by applying 8 lbs. pressure with a hand pump to the acuator, or by temporarily installing a pressure gauge and driving the car. A full throttle acceleration from zero to 50 should produce 7-8 psi for the 2-bbl. engine, and 8-9 psi on the 4-bbl. engine. A maximum of 10 psi is allowed on the 4-bbl. engine.

Vacuum and pressure switches are mounted in the engine compartment and connected to the intake manifold with hoses. When the intake manifold has low vacuum, the vacuum switch illuminates a yellow light on the instrument panel, indicating moderate acceleration. When the intake manifold is under pressure, the pressure switch lights an orange light on the instrument panel, indicating full acceleration.

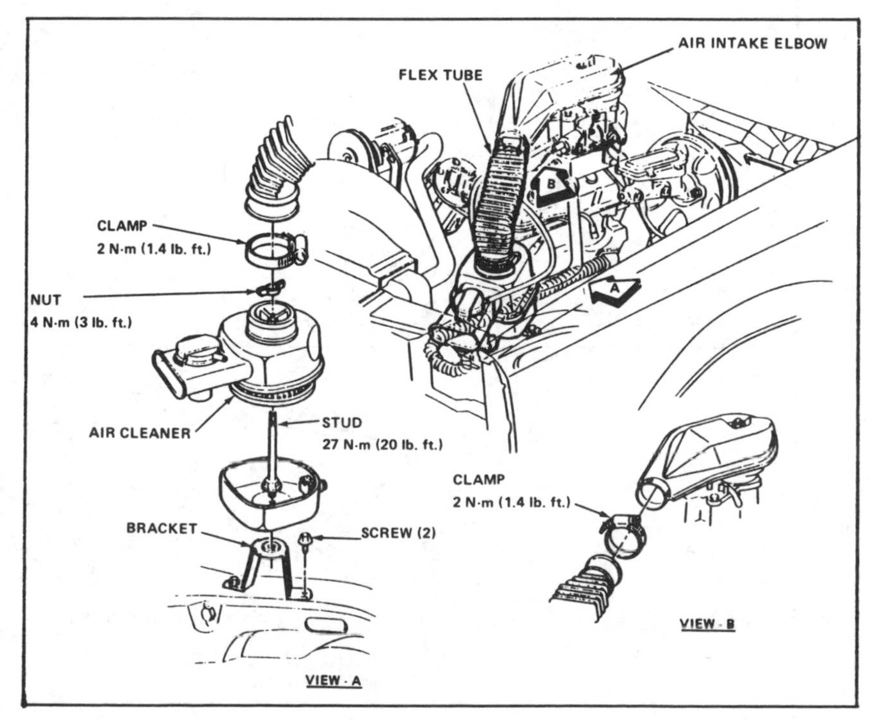

Turbo Charger Air Cleaner (© General Motors Corp.)

Because turbochargers spin at extremely high speeds, their shaft bearings need full pressure lubrication. An oil pressure line comes from the engine block to the turbocharger shaft. Oil runoff from the shaft bearings runs back into the engine through a rubber tube under the turbocharger shaft.

The turbocharger is downstream from the carburetor. With this design, the turbocharger not only pumps air-fuel mixture into the engine, but also creates a tremendous suction that pulls air through the carburetor. This suction is highest at wide open throttle, when the turbocharger is providing full boost. This suction creates high vacuum between the carburetor and the turbocharger. A high vacuum under the carburetor throttle blades at wide open throttle is exactly the opposite of what you get on an unblown engine, where the vacuum drops to zero at wide open throttle.

If the high vacuum at wide open throttle, is allowed to operate the distributor vacuum advance, the EGR valve, and the carburetor power piston, those units will act as if the engine is decelerating or cruising. The distributor would go to full vacuum advance, while the EGR valve opened wide, and the carburetor power piston would close the power valve. Of course, this is exactly the opposite of what is needed at wide open throttle. We need no vacuum advance, no EGR and want the power valve wide open.

To get the action needed, an extra valve is added to the intake manifold. It senses intake manifold pressure, and shuts off the vacuum when the turbo is operating. On the 2-bbl. engine, the valve is called the Turbocharger Vacuum Bleed Valve (TVBV). When pressure in the intake manifold gets above 3 psi. the TVBV shuts off and bleeds down the vacuum to the distributor vacuum advance, EGR valve, and carburetor power piston.

On the 4-bbl. engine, the valve is called the Power Enrichment Control Valve (PECV). Because the 4-bbl. carburetor has a large area at wide open throttle, the vacuum does not go as high. It is only necessary to shut off vacuum to the carburetor power piston. Any time pressure in the intake manifold goes above 3 psi, the PECV shuts off the vacuum to the carburetor power piston.

The TVBV used on the 2-bbl. engine has seven hoses connected to it. Each pair of hoses brings vacuum to the valve, and then out to the EGR valve, distributor, or power piston. The seventh hose, in the middle of the valve, is the bleed. It is connected to the carburetor so that only clean air will enter the system.

The PECV used on the 4-bbl. engine has only three hoses connected to it. Two of them supply vacuum to the power valve, and the third, on the right side, is the bleed.

ELECTRONIC SPARK CONTROL 1979-80

The boost pressure that a turbocharger provides is similar to increasing the compression ratio on an engine. Higher compression and added boost both require less spark advance to avoid detonation. Even though the spark advance on the turbocharged engine has been tailored to the engine, there is still a chance of detonation, mainly because of the low octane, non-leaded gas used.

To prevent detonation that might cause engine damage, Buick uses an Electronic Spark Control that is a retarder. Mounted on the intake manifold above the engine thermostat is a detonation sensor. If the engine detonates, the metal of the intake manifold will vibrate. Then sensor picks up this vibration and sends a signal to an electronic black box on the fan shroud. The black box then retards the timing as much as 18-22°.

Testing the retarder is done by tapping on the intake manifold next to the sensor. Do not tap on the sensor itself. With the engine running at 2000 rpm in Neutral, tapping on the manifold will make the spark retard. If you have a timing light hooked up you can actually see it retard while you are tapping. Within 20 seconds after the tapping stops, the timing will return to its normal setting.

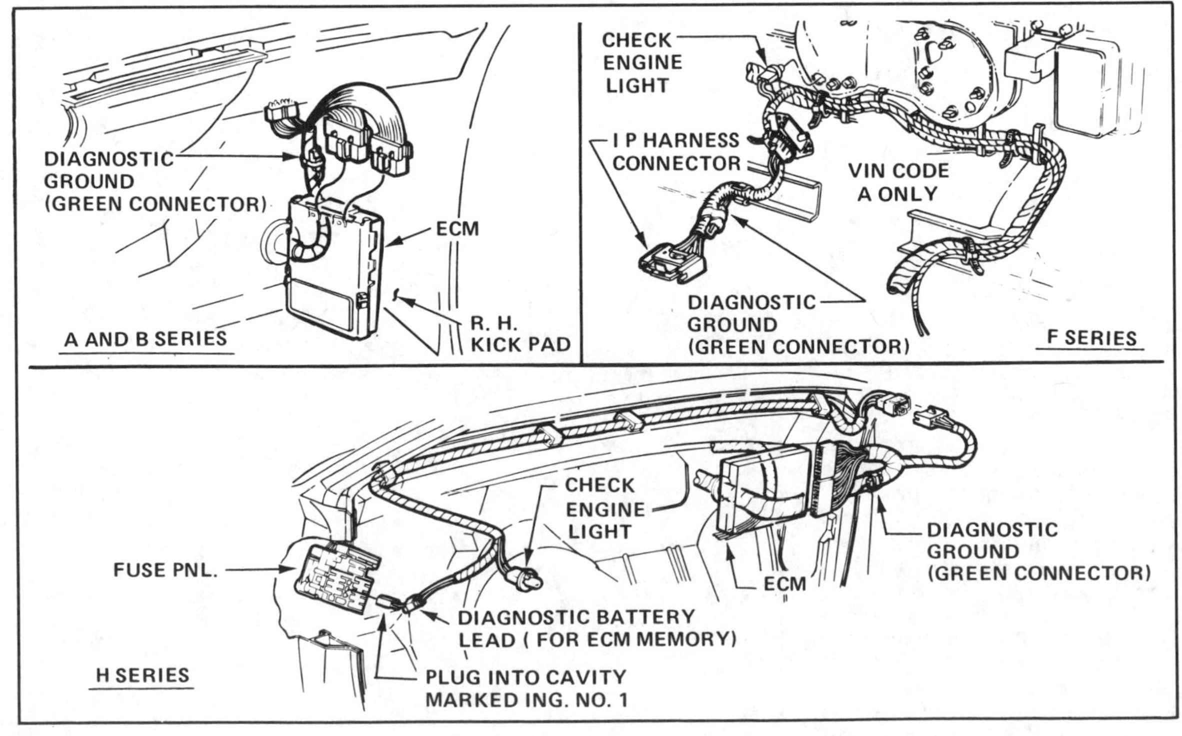

C-4 System Diagnosis Test lead (© General Motors Corp.)

COMPUTER CONTROLLED CATALYTIC CONVERTER C-4 SYSTEM 1979-80

The C-4 system is basically the same as the 1979 Electronic Fuel Control. The new C-4 system has an oxygen sensor in the catalytic converter and the following additional controls:

1. Barometric pressure sensor BARO
2. Engine temperature sensor ETS
3. Manifold absolute pressure sensor MAP
4. Lean authority limit switch
5. Throttle position sensor TPS
6. Vacuum control switch VCS
7. Electronic spark timing EST

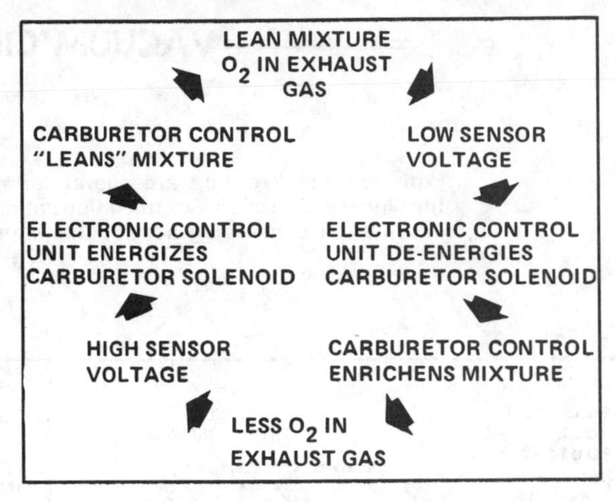

Basic Cycle of Operation (© General Motors Corp.)

The electronic control unit is called the electronic control module ECM.

Another difference in the C-4 is its self diagnostic feature. A check engine light in the dash warns the driver of malfunctions in the system. The same lamp, with the diagnostic system activated, flashes a trouble code which assists in locating malfunctions.

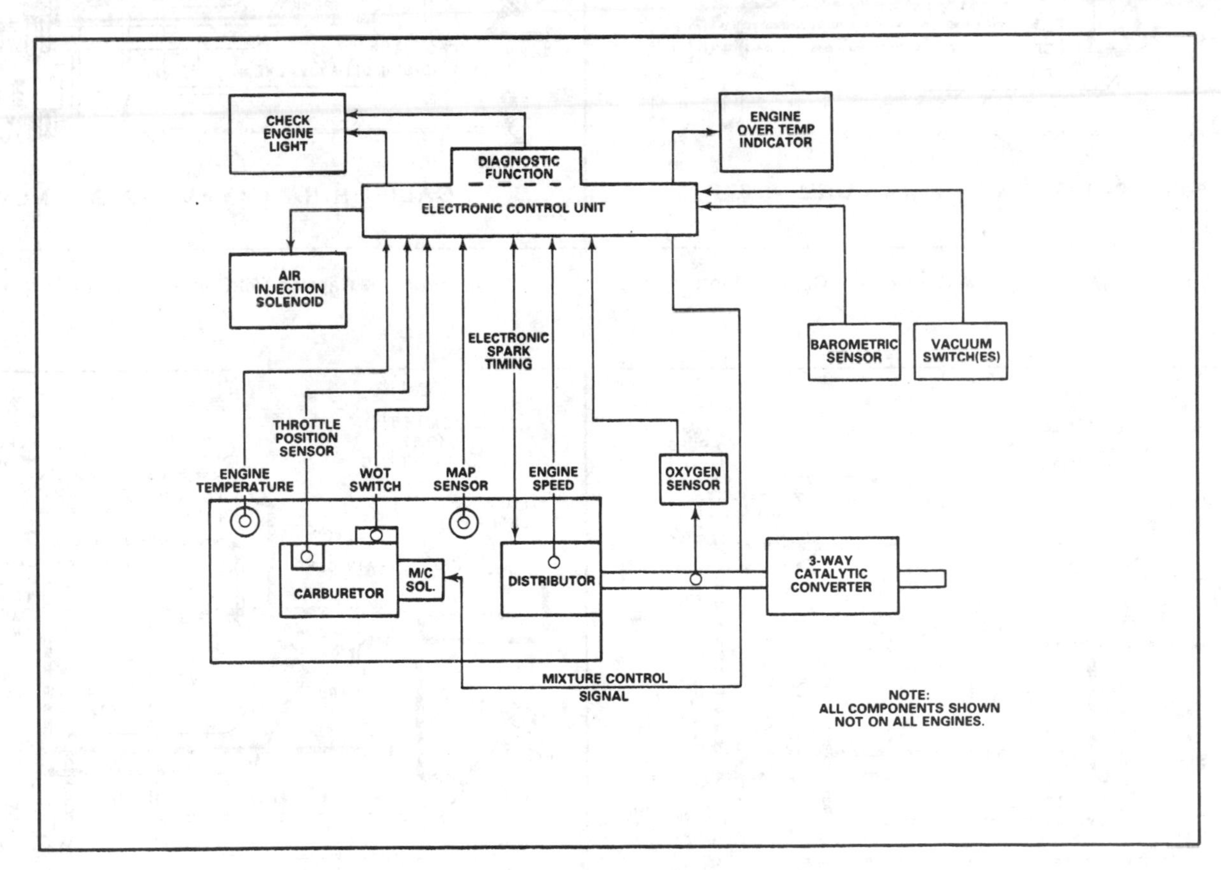

Computer Controlled Catalytic Converter System (© General Motors Corp.)

VACUUM CIRCUITS

Emission hose routing are shown for reference information only. Use the Vehicle Emission Control Information Label for the most up to date information concerning hose routings and tune up data.

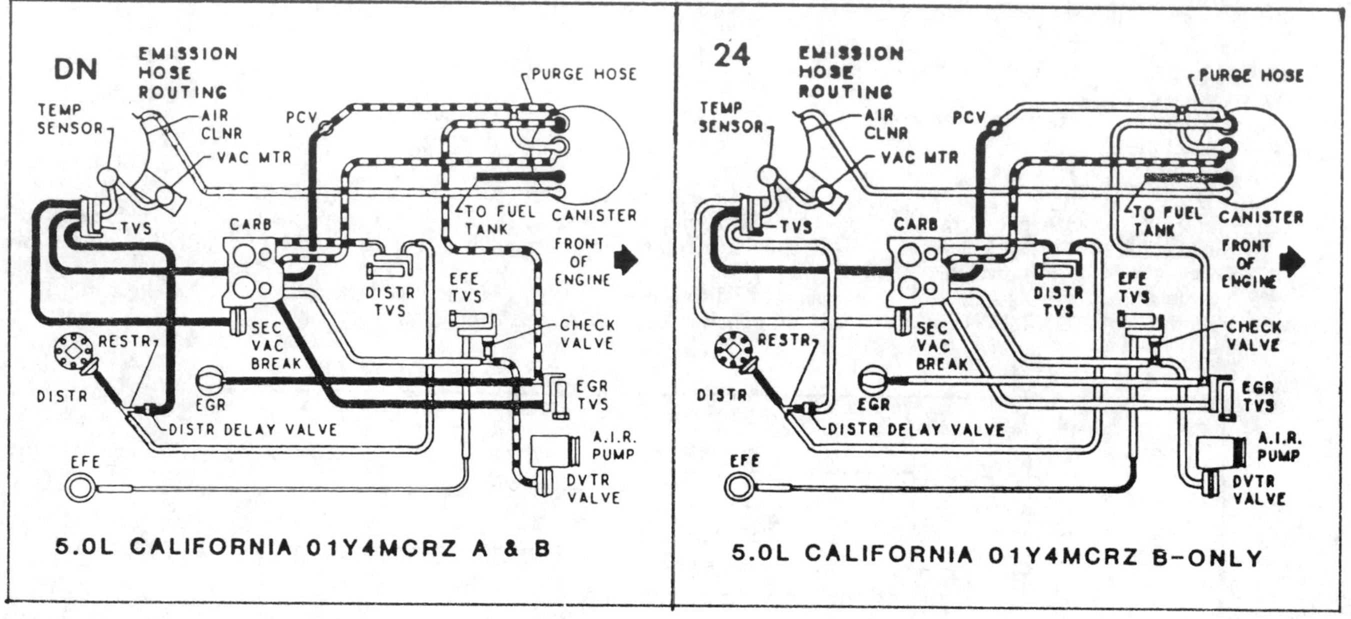

1980 Malibu, Monte Carlo & Caprice (© General Motors Corp.)

1980 Impala & Caprice (© General Motors Corp.)

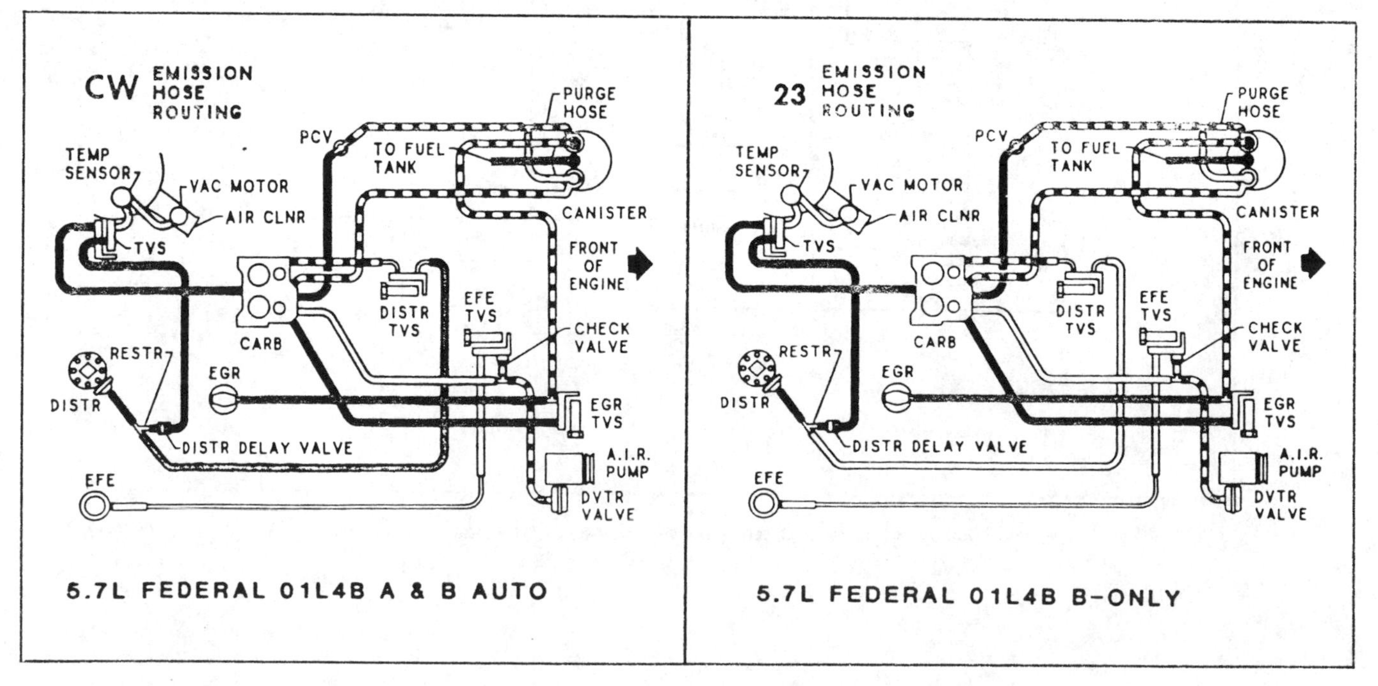

1980 Malibu, Monte Carlo & Caprice (© General Motors Corp.)

1980 Impala & Caprice (© General Motors Corp.)

VACUUM CIRCUITS

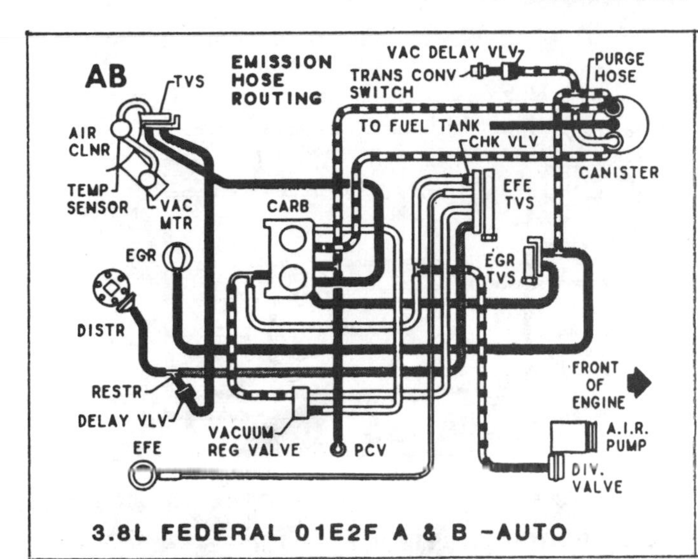

1980 Malibu, Monte Carlo & Caprice (© General Motors Corp.)

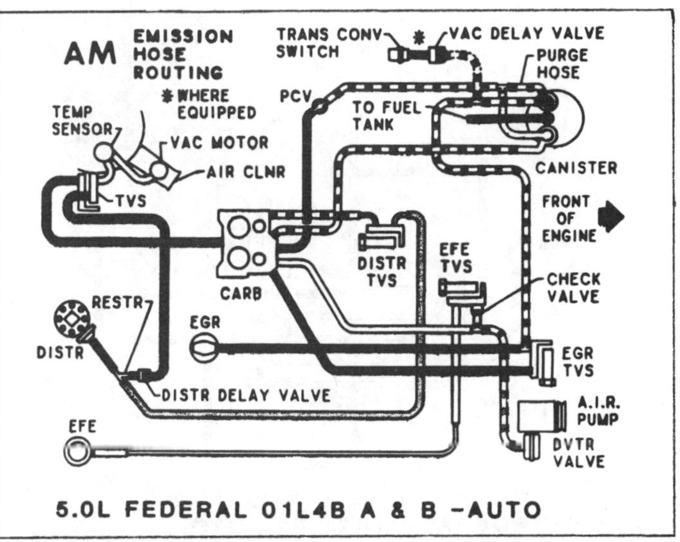

1980 Malibu, Monte Carlo & Caprice (© General Motors Corp.)

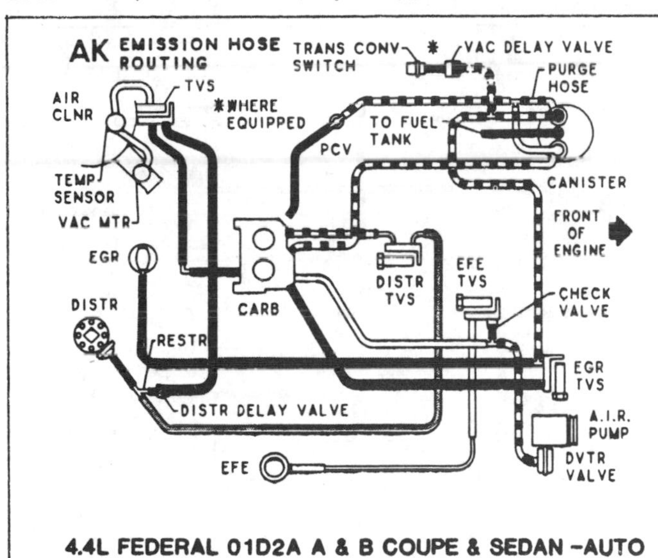

1980 Malibu, Monte Carlo & Caprice (© General Motors Corp.)

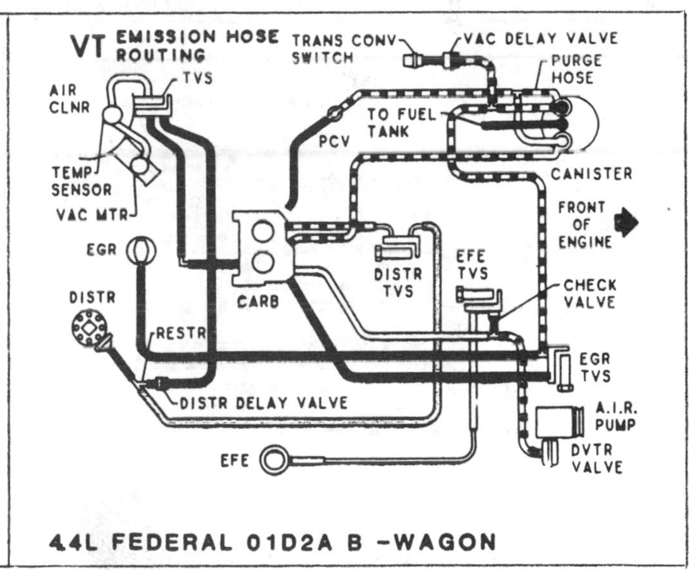

1980 Impala & Caprice Station Wagons (© General Motors Corp.

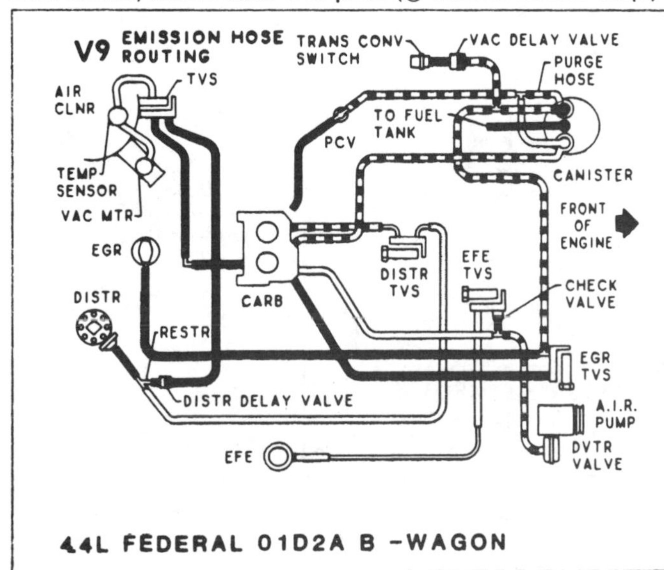

1980 Impala & Caprice Station Wagons (© General Motors Corp.

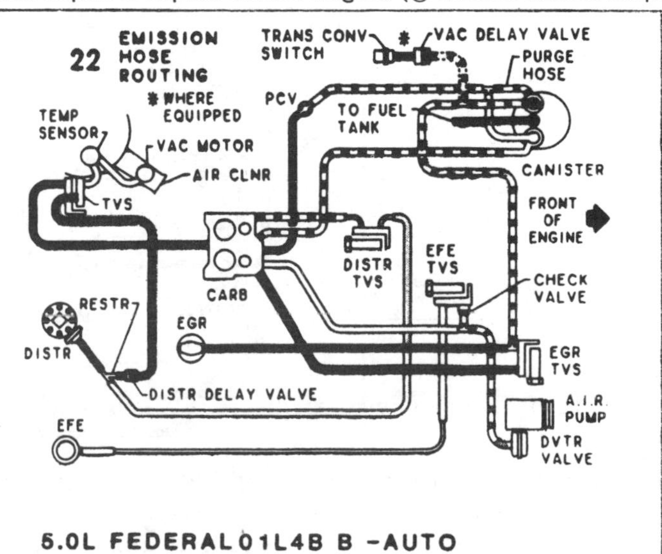

1980 Impala & Caprice (© General Motors Corp.)

VACUUM CIRCUITS

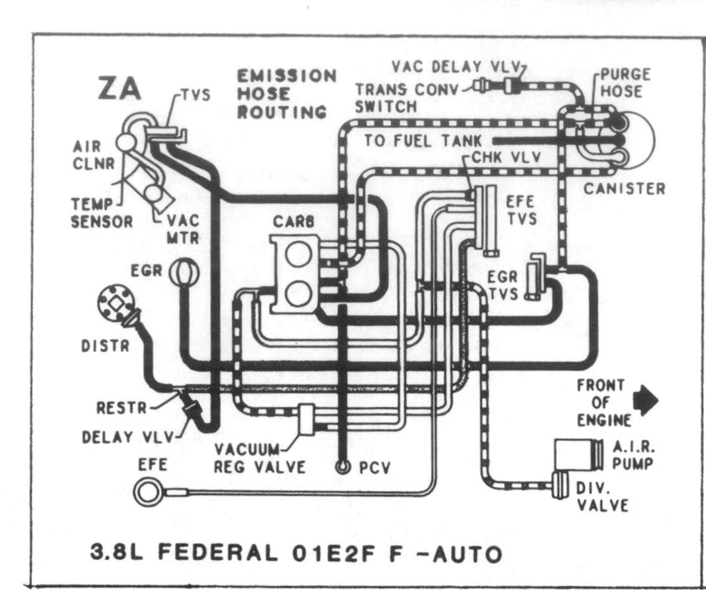

3.8L FEDERAL 01E2F F -AUTO

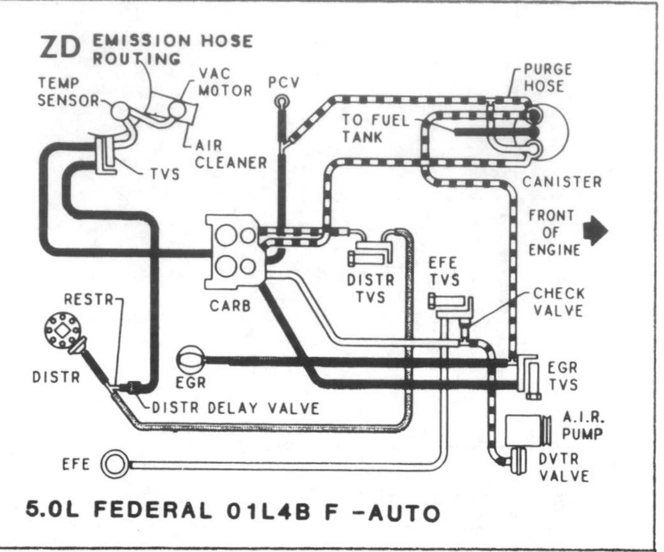

5.0L FEDERAL 01L4B F -AUTO

1980 Camaro (© General Motors Corp.)

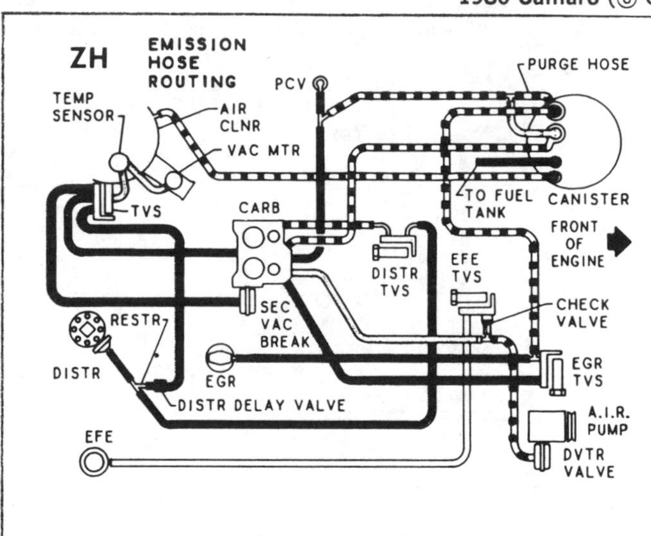

5.0L CALIFORNIA 01Y4MCRZ F EXCEPT Z28

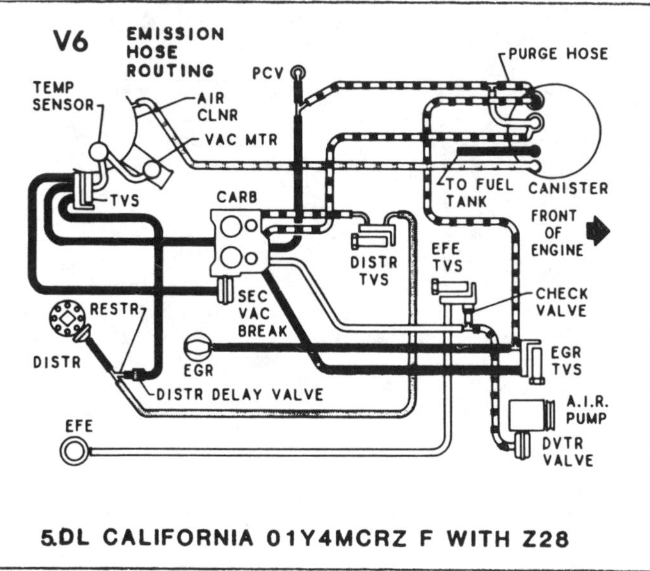

5.0L CALIFORNIA 01Y4MCRZ F WITH Z28

1980 Camaro (© General Motors Corp.)

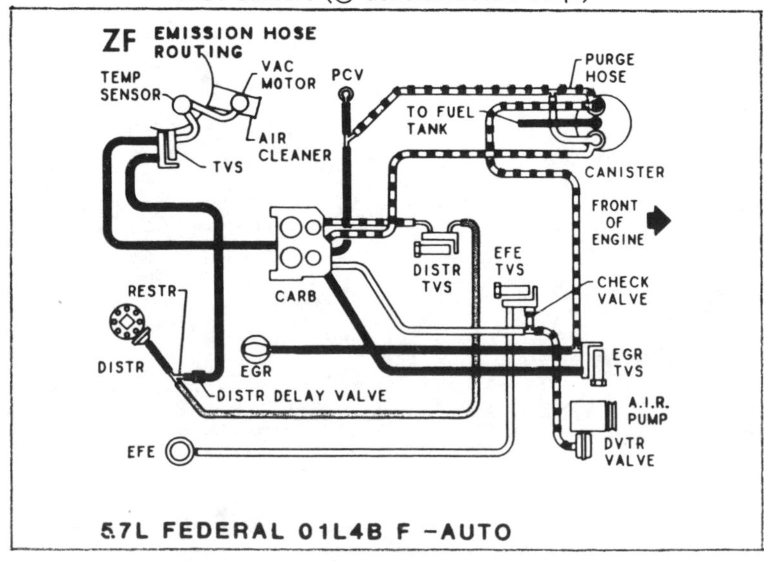

5.7L FEDERAL 01L4B F -AUTO

1980 Camaro (© General Motors Corp.)

TUNE-UP SPECIFICATIONS

ENGINE CODE, 5th CHARACTER OF THE VIN NUMBER
MODEL YEAR CODE, 6th CHARACTER OF THE VIN NUMBER

Year	Eng. V.I.N. Code	Engine No. Cyl. Disp. (cu in)	Eng.① Mfg.	Carb Bbl	H.P.	Spark Plugs Orig. Type	Spark Plugs Gap (in)	Dist. Point Dwell (deg)	Dist. Point Gap (in)	Ignition② Timing (deg BTDC) Man	Ignition② Timing (deg BTDC) Auto	Valves Intake Opens (deg BTDC)	Fuel Pump Pres. (psi)	Idle Speed (rpm) Man Trans	Idle Speed (rpm) Auto Trans In Drive
1979	1	4-151	P	2v	85	R43TSX	.060	Electronic		12	12	33	4-5.5	③	③
	V	4-151	P	2v	85	R43TSX	.060	Electronic		14	14	33	4-5.5	1000⑤	650⑤
	9	4-151	P	2v	85	R43TSX	.060	Electronic		14	14	33	4-5.5	③	③
	C	6-196	B	2v	90	R46TSX	.060	Electronic		③	③	18	3-4.5④	③	③
	A	6-231	B	2v	115	R46TSX	.060	Electronic		15	15	17	3-4.5④	800	600
	2	6-231	B	2v	115	R46TSX	.060	Electronic		15	15	17	3-4.5④	800	600
	G	8-305	C	2v	145	R45TS	.045	Electronic		4	4(4)	28	7.5-9.0	600	500⑥
1980	V	4-151	P	2v	85	R43TSX	.060	E.I.		14	14	33	4-5.5	650/800	650/800
	A	6-231	B	2v	115	R46TSX	.060	E.I.		15	15	17	3-4.5	600	—

Should the information provided in this manual deviate from the specifications on the underhood tune-up label, the label specifications should be used, as they may reflect production changes.

① B—Buick
C—Chevrolet
O—Oldsmobile
P—Pontiac

② Figures in parentheses are California specs.
③ See underhood tune-up decal.
④ At 12.6 volts
⑤ Base idle 500 rpm
⑥ High alt base idle is 600 rpm.

DISTRIBUTOR SPECIFICATIONS

Year	Distributor Identification	Centrifugal Advance Start Dist. Deg. @ Dist. RPM	Centrifugal Advance Finish Dist. Deg. @ Dist. RPM	Vacuum Advance Start @ In. Hg.	Vacuum Advance Finish Dist. Deg. @ In. Hg.
1979	1103229	0 @ 600	10 @ 2200	3.5	10 @ 12
	1103231	0 @ 600	10 @ 2200	3.5	10 @ 12
	1103239	0 @ 600	10 @ 2100	4	5 @ 8
	1103244	0 @ 500	10 @ 1900	4	10 @ 10
	1103282	0 @ 500	10 @ 1900	4	10 @ 10
	1103285	0 @ 600	11 @ 2100	4	5 @ 8
	1103365	0 @ 850	10 @ 2325	5	8 @ 11
	1110726	0 @ 500	9 @ 2000	4	10 @ 10
	1110757	0 @ 600	9 @ 2000	4	10 @ 10
	1110766	0 @ 810	7.5 @ 1800	3	10 @ 9
	1110767	0 @ 840	7.5 @ 1800	4	12 @ 11
1980	1110558	0 @ 1000	14 @ 4000	3	15 @ 5
	1110559	0 @ 1200	14 @ 4000	3	15 @ 5
	1110560	0 @ 1000	14 @ 4400	4	20 @ 10

2SE, E2SE CARBURETOR ADJUSTMENTS

Chevrolet Monza

Year	Carburetor Identification	Float Level (in.)	Pump Rod (in.)	Fast Idle (rpm)	Choke Coil Lever (in.)	Fast Idle Cam (deg./in.)	Air Valve Rod (in.)	Primary Vacuum Break (deg./in.)	Choke Setting (notches)	Secondary Vacuum Break (deg./in.)	Choke Unloader (deg./in.)	Secondary Lockout (in.)
1979	17059674	13/64	1/2	2400	.120	18/0.096	.025	19/.103	2 Rich	—	32/.195	.030
	17059675	13/64	17/32	2200	.120	18/0.096	.025	21/.117	1 Rich	—	32/.195	.030
	17059676	13/64	1/2	2400	.120	18/0.096	.025	19/.103	2 Rich	—	32/.195	.030
	17059677	13/64	17/32	2200	.120	18/0.096	.025	21/.117	1 Rich	—	32/.195	.030
1980	All	3/16	1/2	①	.085	18/0.096	0.18	—	Fixed	—	32/.195	.120

① See Underhood Decal

2MC, M2MC, M2ME, E2ME CARBURETOR SPECIFICATIONS

CHEVROLET MONZA

Year	Carburetor Identification①	Float Level (in.)	Choke Rod (in.)	Choke Unloader (in.)	Vacuum Break Lean or Front (in.)	Vacuum Break Rich or Rear (in.)	Pump Rod (in.)	Choke Coil Lever (in.)	Automatic Choke (notches)
1980	17080191	11/32	0.139	0.243	0.096	0.096	1/4②	0.120	Fixed
	17080195	9/32	0.139	0.243	0.103	0.090	1/4②	0.120	Fixed
	17080197	9/32	0.139	0.243	0.103	0.090	1/4②	0.120	Fixed
	17080491	5/16	0.139	0.243	0.117	—	3/8	0.120	Fixed
	17080496	5/16	0.139	0.243	0.117	0.203	3/8	0.120	Fixed
	17080498	5/16	0.139	0.243	0.117	0.203	3/8	0.120	Fixed

2SE, E2SE CARBURETOR ADJUSTMENTS

Pontiac (except Phoenix)

Year	Carburetor Identification	Float Level (in.)	Pump Rod (in.)	Fast Idle (rpm)	Choke Coil Lever (in.)	Fast Idle Cam (deg./in.)	Air Valve Rod (in.)	Primary Vacuum Break (deg./in.)	Choke Setting (notches)	Secondary Vacuum Break (deg./in.)	Choke Unloader (deg./in.)	Secondary Lockout (in.)
1979	17059674	3/16	1/2	2400	.120	18/0.096	.025	19/.103	2 Rich	—	32/.195	.01-.04
	17059675	3/16	17/32	2200	.120	18/0.096	.025	21/.117	1 Rich	—	32/.195	.01-.04
	17059676	3/16	1/2	2400	.120	18/0.096	.025	19/.103	2 Rich	—	32/.195	.01-.04
	17059677	3/16	17/32	2200	.120	18/0.096	.025	21/.117	1 Rich	—	32/.195	.01-.04
1980	17080674	3/16	1/2	①	.085	18/0.096	.018	19/.103	Fixed	—	32/.195	.012
	17080675	3/16	1/2	①	.085	18/0.096	.018	21/.117	Fixed	—	32/.195	.012
	17080676	3/16	1/2	①	.085	18/0.096	.018	19/.103	Fixed	—	32/.195	.012
	17080677	3/16	1/2	①	.085	18/0.096	.018	21/.117	Fixed	—	32/.195	.012
	17059774	5/32	1/2	①	.085	18/0.096	.018	19/.103	Fixed	—	32/.195	.012
	17059775	5/32	17/32	①	.085	18/0.096	.018	21/.117	Fixed	—	32/.195	.012
	17059776	5/32	1/2	①	.085	18/0.096	.018	19/.103	Fixed	—	32/.195	.012
	17059777	5/32	17/32	①	.085	18/0.096	.018	21/.117	Fixed	—	32/.195	.012

① See Underhood Decal

Note: —Use the Vehicle Emission Control Information Label for the most up to date information concerning hose routing and tune up data.

NOTE: *California C-4 system does not use the propane enrichment method of checking and setting emissions.*

CAR SERIAL NUMBER AND ENGINE IDENTIFICATION

1979

The vehicle identification number is on a plate behind the windshield on the driver's side. The sixth character is the model year, with 9 for 1979.

The fifth character is the engine code, as follows.

V ....151 2-bbl. 4-cyl. LX8, LS6/LS8 Pont.
C ....196 2-bbl. V-6 LC9 Buick
A ....231 2-bbl. V-6 LD5 Buick
G ....305 2-bbl. V-8 LG3 Chev.

NOTE: *LS8 to replace LS6 in January 1979.*

1980

The vehicle identification number is on a plate behind the windshield on the drivers side. The sixth character is the model year, with A designating 1980.

V ....151 2 bbl. 4 cyl.
A ....231 2 bbl. V-6

EMISSION EQUIPMENT

1979-80

All Models

Closed positive crankcase ventilation
Emission calibration carburetor
Emission calibrated distributor
Heated air cleaner
Vapor control, canister storage
Exhaust gas recirculation
Electric choke
Catalytic converter
Choke vacuum delay valve
Early fuel evaporation
All engines, except
Not used with electronic fuel control
Vacuum delay valve, 4-nozzle
All engines, except
Not used with electronic fuel control
Electronic fuel control
Cold engine air bleed TVS

IDLE SPEED AND MIXTURE ADJUSTMENTS

1979

CAUTION: *Emission control adjustment changes are noted on the Vehicle Emission Information Label by the manufacturer. Refer to the label before any adjustments are made.*

Air cleaner ....In place
Air cond. ....Off
Hoses ....See label
Auto. trans. ....Drive
Idle speed
151 4-cyl. no air cond.
Auto. trans.
Solenoid connected 650
Solenoid disconnected 500
Propane enriched 690
Man. trans.
Solenoid connected 1000
Solenoid disconnected 500
Propane enriched 1150
151 4-cyl. with air cond.
Auto. trans 650
Propane enriched 690
A.C. idle speedup 850
Man. trans. 1000
Propane enriched 1150
A.C. idle speedup 1200

NOTE: *All 151 4-cyl. engines have a carburetor solenoid. On engines without air conditioning, the solenoid is for anti-dieseling, and the idle speed is set with the solenoid adjustment. On air conditioned cars, the solenoid is for idle speedup, and is only energized when the air conditioning is on. The curb idle speed adjustment on air conditioned engines is made with the throttle screw.*

196 V6
Auto. trans. 600
Propane enriched ....
Man. trans.
Solenoid connected 800
Solenoid disconnected 600
231 V-6 49-States
Auto. trans. 600
Propane enriched ....
Man. trans.
Solenoid connected 800
Solenoid disconnected 600
Propane enriched ....
231 V-6 Calif.
Auto. trans. 600
Propane enriched ....
Man. trans.
Solenoid connected 800
Solenoid disconnected 600
Propane enriched ....
231 V-6 Altitude
Auto. trans. 600
Propane enriched ....
Man. trans.
Solenoid connected 800
Solenoid disconnected 600
Propane enriched ....
305 2-bbl. V-8 49-States
Auto. trans. 500
Propane enriched 520-540
AC idle speedup 600
Man. trans. 600
Propane enriched 700-740
AC idle speedup 700
305 2-bbl. V-8 Calif.
Auto. trans. 500
Propane enriched 520-540
AC idle speedup 650
305 2-bbl. V-8 Altitude
Auto. trans. 600
Propane enriched 520-640
AC idle speedup 700

1980

CAUTION: *Emission control adjustment changes are noted on the vehicle underhood emission label. Refer to the label before making any adjustments.*

Air cleaner ....In place
Air conditioner ....Off
Vacuum hoses ....See label
Automatic Trans. ....In Drive
Idle Speed
151 4 Cyl no air cond.
Auto. Trans.
Solenoid 650
Curb 550①
Propane enriched 700
① California 500
Manual Trans.
Solenoid 1000
Curb 550
Propane enriched 1150
151 4 Cyl with air cond.
Auto. Trans.
Solenoid 850
Curb 650
Propane enriched 700
Manual Trans.
Solenoid 1250②
Curb 1000
Propane enriched 1150
② California 1200
231 V-6 49 States
Auto Trans.
Solenoid 670
Curb 550
Propane enriched 610
Manual Trans.
Solenoid 800
Curb 600
Propane enriched 830
231 V-6 California
Auto Trans.
Solenoid ③
Curb W/air 620
Curb Without Air 550
Propane enriched ③
③ see C-4 adjustments
Manual Trans
Solenoid ④
Curb ④
Propane enriched ④
④ see C-4 adjustments

INITIAL TIMING
1979-80

NOTE: *Vehicle timing information is noted on the Vehicle Emission Information Label by the manufacturer. Refer to this information before any adjustments are made.*

151 4-cyl. ....12° BTDC
196 V-6 .... 8° BTDC
231 V-6 ....15° BTDC
305 V-8 .... 4° BTDC

SPARK PLUGS
1979

151 4-cyl. ......AC-R43TSX.. .060
196 & 231 V-6 ..AC-R46TSX.. .060
305 V-8 ....AC-R45TS.. .045

1980

151 4 Cyl .... AC-R43TSX .. .060
151 4 Cyl
California .. AC-R44TSX .. .060
231 V-6 All .. AC-R45TSX .. .060

VACUUM ADVANCE

1979

Diaphragm type ....Single
Vacuum source
4-cyl.
With air cond..........Ported
No air cond. ....Manifold
196 V-6 ....Manifold
231 V-6
49-States ....Manifold
Calif. ....Ported
Altitude ....Ported
305 V-8
49-States ....Manifold
Calif. ....Ported
Altitude ....Manifold

1980

Diaphragm type ....Single
Vacuum source
California C-4
151 4 Cyl
Manual trans ....Manifold
151 4 Cyl
Auto Trans ....Mixed①
231 V-6 Manual trans .Ported
231 V-6 Auto trans ...Ported
49 States
151 4 Cyl
All except A/T ..Manifold with air cond
151 4 Cyl
A/T with A/C ....Mixed①
231 V-6
Manual Trans ......Mixed①
231 V-6
Auto Trans ....Mixed①

① Ported and manifold vacuum mixed.

EMISSION CONTROL SYSTEMS

ELECTRONIC FUEL CONTROL 1979

This system controls both the idle and main metering fuel mixture according to how much oxygen there is in the exhaust. Most tune-up men have used an infra-red exhaust gas analyzer to measure the carbon monoxide (CO) and hydrocarbon (HC) in the exhaust. The infra-red measures the actual pollutants, except nitrogen oxides (NOx). Since these pollutants are what we want to control, it would seem that the best way to control them would be to install an analyzer on every car. The electronic fuel control system does just that, but instead of using an infra-red analyzer, it uses an oxygen sensor.

The easiest thing to analyze in the exhaust is the amount of oxygen. Because repair shops are not used to analyzing oxygen with their exhaust gas analyzers, most emission control technicians do not realize that the exhaust contains any oxygen. But oxygen is always present in the exhaust to some degree. If the mixture is rich, the extra fuel will combine with the oxygen in the air and the exhaust will have very little oxygen. If the mixture is lean, there won't be enough fuel to use up all the oxygen and more of it will pass through the engine.

As the oxygen content of the exhaust goes up, (lean mixture) the amount of CO goes down, because there is an excess of oxygen to combine with the CO and turn it into harmless $CO2$ (carbon dioxide). However, if the mixture gets too lean there is a tendency to form more NOx. The electronic fuel control solves all these problems by controlling the mixture within very narrow limits.

The oxygen sensor has a zirconia element, and looks very much like a spark plug. It screws into the exhaust pipe close to the exhaust manifold. Zirconia, when combined with heat and oxygen, has the peculiar property of being able to generate current. The amount of currents is very small, but it is enough that it will pass through a wire and can be amplified by electronics.

A small black box under the hood is the electronic control unit. It receives the electric signals from the oxygen sensor and amplifies them enough to operate a vacuum modulator. The vacuum modulator is a vibrating electrical device. Engine manifold vacuum connects to the modulator, and then goes to the carburetor. The amount of vacuum that the carburetor receives is controlled by the modulator.

The carburetor is specially constructed with an idle air bleed valve and a main metering fuel valve. Both of the valves are operated by vacuum diaphragms. The idle air bleed gives a richer mixture at low vacuum because a spring pushes the regulating needle into the bleed. At high vacuum, the diaphragm compresses the spring and pulls the needle out of the bleed, letting more air into the idle system, which leans the mixture.

The main metering fuel valve actually takes the place of the power valve. It is separate from the main metering jet. It works the same as a power valve, in that it can add fuel to the system or shut off, but it cannot change the amount of fuel going through the main metering jet.

The spring tension on both the diaphragms is adjustable with small screws. However, the main metering diaphragm is factory adjusted, and there are no specifications for adjusting it in the field. Both the idle diaphragm screw and the idle mixture screw on the carburetor are covered with press-in plugs. The idle mixture needle is covered by a cup plug that can be removed with a screw extractor. The screw extractor should fit the cup plug without drilling a hole. If a hole is drilled in the plug, the mixture needle will be damaged. The idle diaphragm plug is soft lead, and can be carefully pried out. Idle mixture settings are normally made only when the carburetor is overhauled. The necessary plugs are in the overhaul kit.

The Electronic Fuel Control is not a complicated system. Except for the

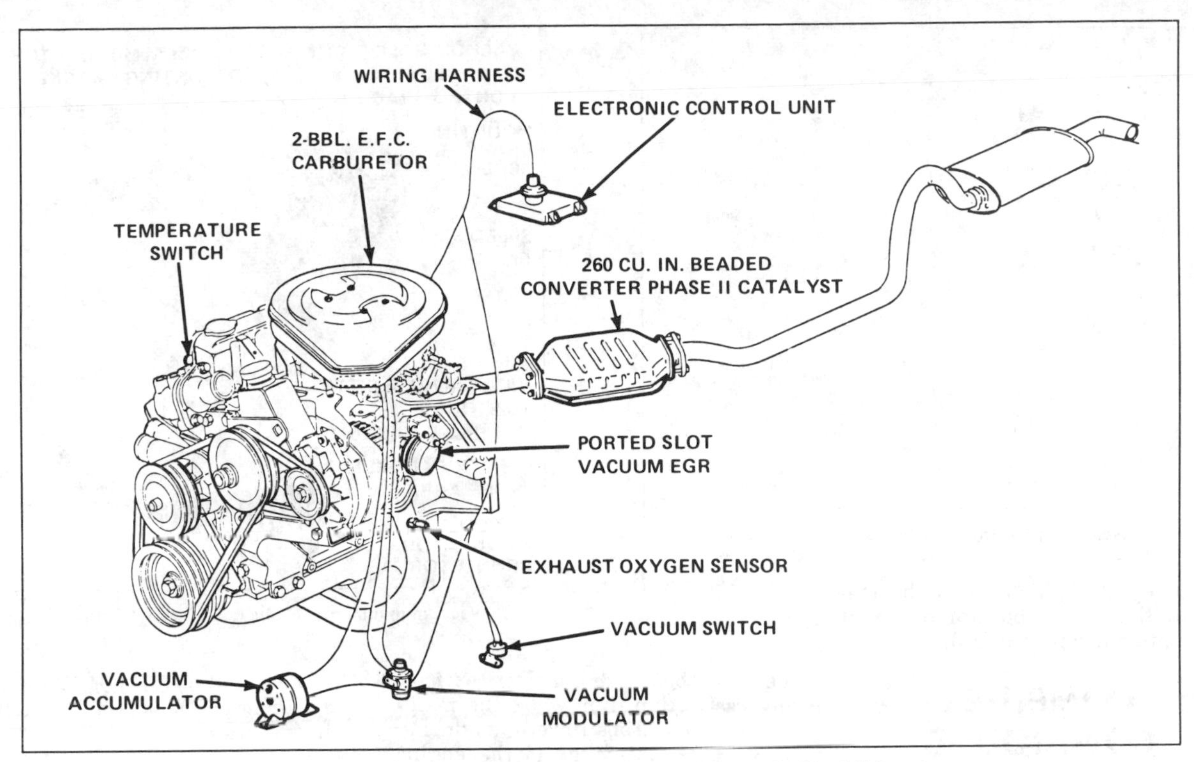

electronic fuel control system—L-4 (Vin) (© Chevrolet Div. GM Corp.)

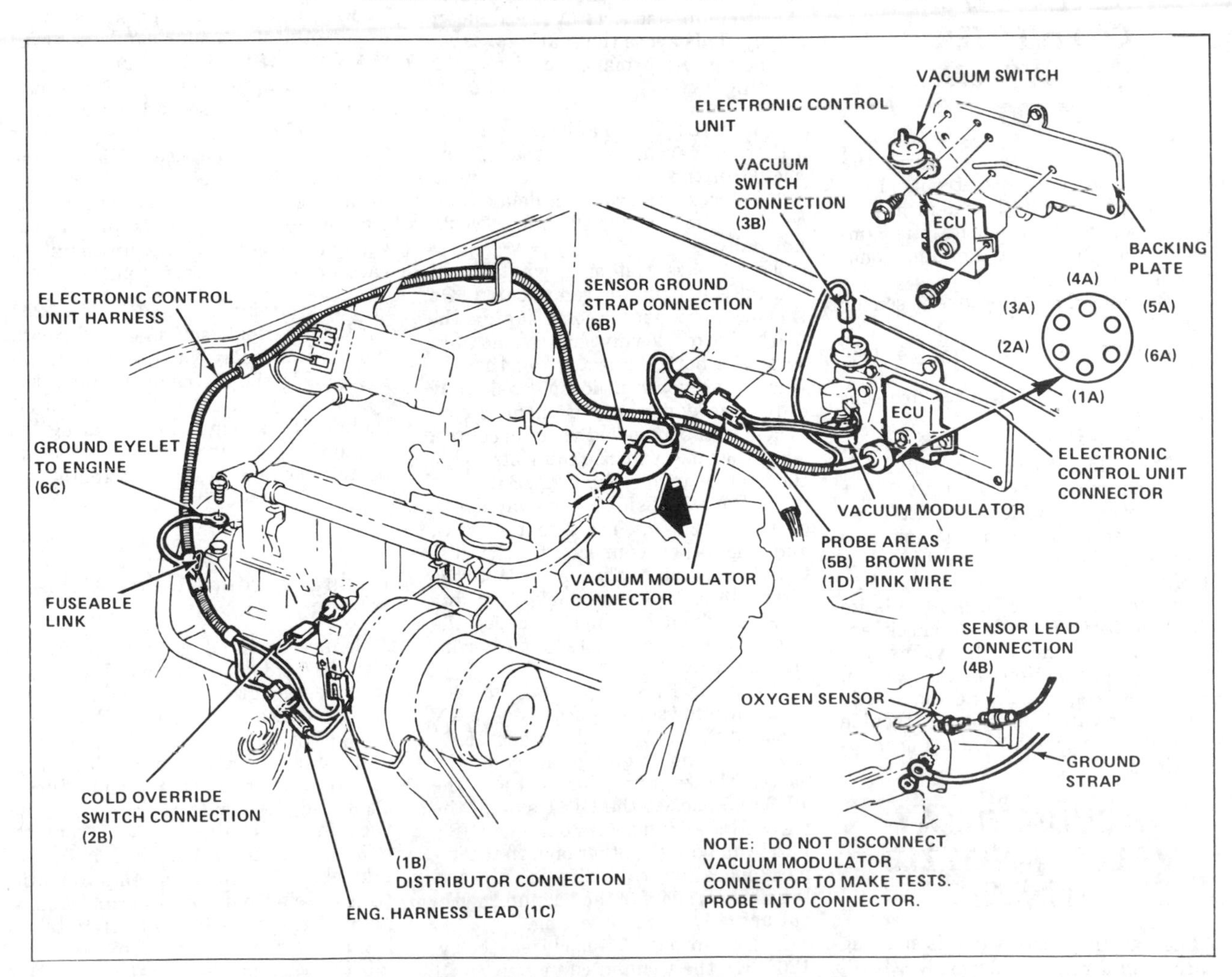

EFC electrical system (© Chevrolet Div. GM Corp.)

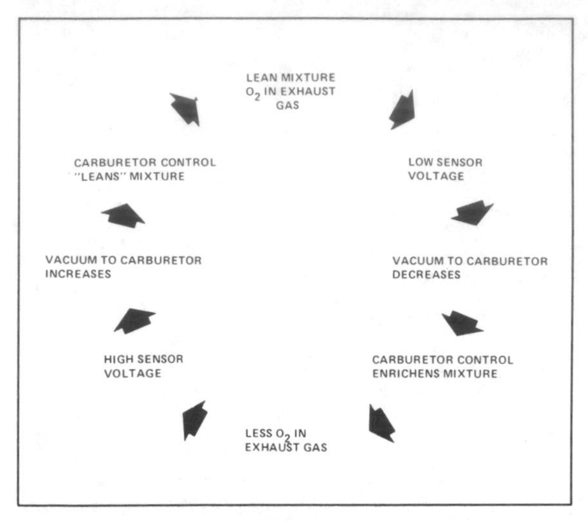

Cycle of Operations (© Chevrolet Div. GM Corp.)

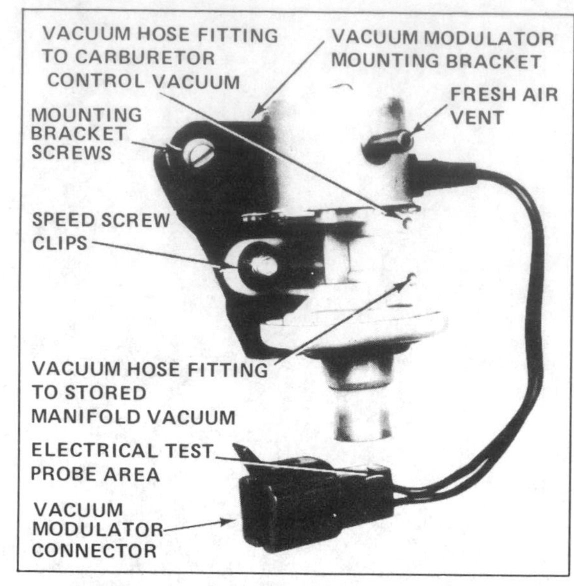

Vacuum modulator (© Chevrolet Div. GM Corp.)

vacuum checks, which need to be done only at time of carburetor overhaul, no maintenance is required.

COMPUTER CONTROLLED CATALYTIC CONVERTER C-4 SYSTEM 1980

The C-4 system is basically the same as the 1979 Electronic Fuel Control. The new C-4 system has an oxygen sensor in the catalytic converter and the following additional controls.

1. Barometric pressure sensor BAR
2. Engine temperature sensor ETS
3. Manifold absolute pressure sensor MAP
4. Lean authority limit switch
5. Throttle position sensor TPS
6. Vacuum control switch VCS
7. Electronic spark timing EST

The electronic control unit is called the electronic control module ECM.

Another difference in the C-4 is its self diagnostic feature. A check engine light in the dash warns the driver of malfunctions in the system. The same lamp, with the diagnostic system activated, flashes a trouble code which assists in locating malfunctions.

VACUUM DELAY VALVE 4-NOZZLE 1979-80

The vacuum delay valve is nothing more than a retard delay valve. It allows the manifold vacuum to act on the distributor diaphragm without any restriction. But when the manifold vacuum drops to zero, as during heavy acceleration or if the engine should die at idle, the valve "traps" the vacuum and lets it bleed down slowly. This keeps the spark advanced for better performance, or for easier starting.

Above 120°F. coolant temperature, a thermal vacuum switch that is connected in the system by hoses, opens and bypasses the vacuum delay valve so that the vacuum advance works normally.

The bypass system is what makes the hose hookup so complicated. On the 4-cyl. without air conditioning, the MAN side of the valve has two nozzles. They are both connected together inside the valve, so it doesn't make any difference which one of them is connected to the manifold vacuum source. The other MAN nozzle connects to the thermal vacuum switch (TVS).

On the DIST side of the valve, one of the nozzles connects to the TVS, and the other connects to the distributor modular valve, or direct to the distributor. It doesn't make any difference which is which, because the two nozzles are connected together inside the valve.

When the TVS is open (above 120° F.) manifold vacuum goes in one MAN nozzle and right out the other MAN nozzle, without going through the valve. The vacuum then goes to the TVS and back to the DIST side of the valve. But all it does is go in one DIST nozzle and out the other one. In the bypass operation, the valve is only acting as a carrier for the vacuum, without actually doing anything.

But when the TVS is closed (below 120° F.) the vacuum can't bypass. It then goes in one MAN nozzle, through the valve, and out the DIST nozzle to the distributor. Then, the retard delay is in operation.

On the 4-cylinder with air conditioning, a different bypass system is used. Manifold vacuum connects to the MAN side of the valve, but the DIST side connects to the upper nozzle on a 3-nozzle TVS. The second nozzle on the DIST side of the vacuum delay valve is capped, and not used. The second nozzle on the MAN side is connected to another system, which doesn't affect the distributor. It is just a convenient way of hooking up manifold vacuum to the other system.

When the engine is cold (below 120° F.) the manifold vacuum goes through the vacuum delay valve and through the TVS to the distributor. When the engine warms up, the TVS shuts off the manifold vacuum, eliminating the vacuum delay valve entirely and sends ported vacuum to the distributor.

Testing Vacuum Delay Valve 1979-80

Connect a hand vacuum pump to the DIST side of the valve. Use a cap or a short length of hose and a bolt to plug the second nozzle on the DIST side. Plug both of the MAN nozzles with your finger. Pump up 15 in. Hg. vacuum and then remove your finger from the MAN side of the valve. The vacuum should drop slowly from 15 to 5 in. Hg. On the 151 4-cyl. it should take about 7 seconds on an automatic transmission car, or 3 seconds on a manual transmission car. If it takes longer, or shorter, or doesn't drop at all, the vacuum delay valve is defective, and must be replaced.

When the manifold vacuum goes through the DS-VDV it is not affected. The two nozzles on the MAN side are connected together inside the valve, without going through the valve mechanism. So the choke delay valve receives manifold vacuum just as if it were hooked up directly to the manifold.

Manifold vacuum connects to the 2-nozzle side of the choke delay valve. The second nozzle on the same side connects to a thermal vacuum switch again just for convenience. The single nozzle side of the delay valve connects to the choke vacuum break.

Testing Choke Vacuum Break Vacuum Delay Valve 1979-80

Use a short length of hose and a "T" to insert a vacuum gauge between the delay valve and the vacuum break unit. Disconnect the second hose on the 2-nozzle side of the delay valve, and cap the nozzle. Leave the manifold vacuum hose hooked up. Start the engine and let it idle, then slip the manifold vacuum hose off the 2-nozzle side of the delay valve. The vacuum gauge should drop to zero immediately, without any delay. If it doesn't, the delay valve is defective, or the hose is restricted. When slipping the hose off the delay valve, hold your finger over the end of the hose to keep the engine from dying.

Reconnect the manifold vacuum hose to the delay valve. The vacuum gauge should rise slowly, taking about 40 seconds to reach full manifold vacuum. Note the highest reading you get, then disconnect the hose from the intake manifold and use a separate hose to attach the vacuum gauge to check manifold vacuum. Both vacuum readings should be within one inch of each other. If not, there is a leak somewhere.

NOTE: *Because so many systems are connected together, a leak in one system can affect another. A single manifold vacuum connection supplies the distributor, the exhaust heat valve, and the vacuum break. When searching for leaks, always isolate each system before changing any parts.*

Repairs are limited to replacement.

CHOKE VACUUM BREAK VACUUM DELAY, VALVE 1979-80

This valve has three nozzles. It connects between manifold vacuum and the choke vacuum break unit. When the engine starts, the delay valve restricts the vacuum so the vacuum breaks opens the choke slowly. If the engine is stopped, the vacuum is released through an internal check valve.

The operation of the valve is very simple, but the vacuum hookup is something else. At first glance, it doesn't make any sense at all, because the manifold vacuum does not run directly from the intake manifold to the delay valve. For convenience in manufacturing, the hose from manifold vacuum runs first to the distributor vacuum delay valve (DS-VDV), which has four nozzles. Then a second hose runs from the DS-VDV to the choke delay valve.

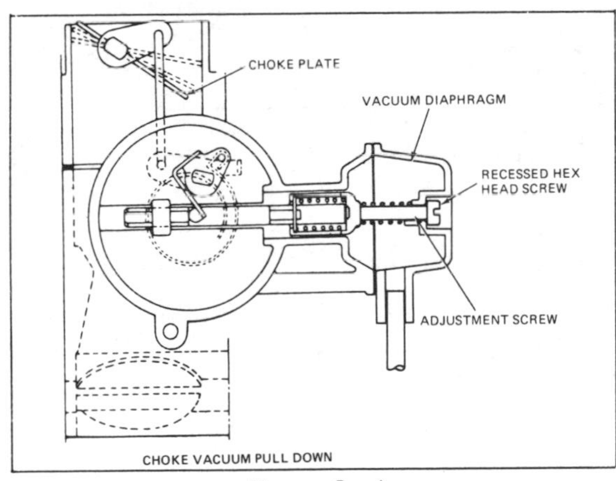

Vacuum Break

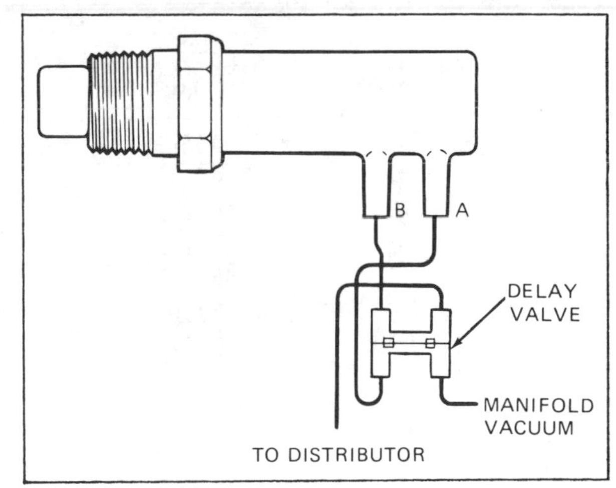

2-nozzle TVS switch—Vega
(© GM Corporation)

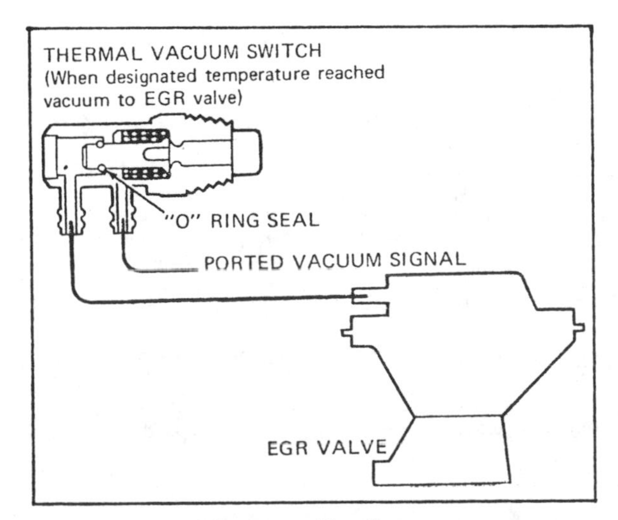

EGR cold override—Vega
(© GM Corporation)

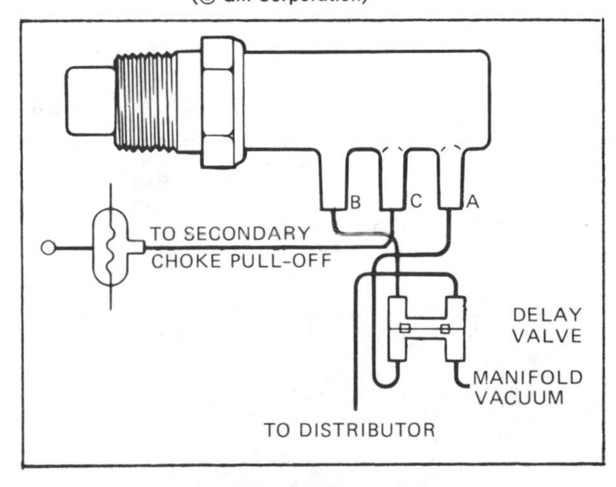

3-nozzle TVS switch—Vega
(© GM Corporation)

TUNE-UP SPECIFICATIONS

ENGINE CODE, 5th CHARACTER OF THE VIN NUMBER
MODEL YEAR CODE, 6th CHARACTER OF THE VIN NUMBER

Year	Eng. V.I.N. Code	Eng. No. Cyl. Disp. (cu in)	Eng.* Mfg.	Carb Bbl	Spark Plugs Orig. Type	Spark Plugs Gap (in)	Dist.	Ignition Timing (deg BTDC) Man Trans	Ignition Timing (deg BTDC) Auto Trans	Valves Intake Opens (deg BTDC)	Fuel Pump Pres. (psi)	Idle Speed (rpm) Man Trans	Idle Speed (rpm) Auto Trans In Drive
1979	E	4/98	C	2	AC-R42TS	.035	EI	①	①	28	5-6.5	②	②
	J	4/98	C	2	AC-R42TS	.035	EI	①	①	31	5-6.5	②	②
1980	9	4/98	C	2	AC-R42TS	.035	EI	①	①	22.5	2.5-6.5	②	②
	0	4/98	C	2	AC-R42TS	.035	EI	①	①	32	2.5-6.5	②	②

NOTE: Should the information in this manual deviate from the specifications on the underhood tune-up decal, the decal specifications should be used as they may reflect production changes.

① See "Initial Timing" section.
② See "Idle Speed & Mixture Adjustments" section.

DISTRIBUTOR SPECIFICATIONS

Year	Distributor Identification①	Centrifugal Advance Start Crank. Deg. @ Eng. RPM	Centrifugal Advance Finish Crank. Deg. @ Eng. RPM	Vacuum Advance Start @ In. Hg.	Vacuum Advance Finish Crank. Deg. @ In. Hg.
1979	1110741	0 @ 1510	16 @ 5250	4	14 @ 8
	1110742	0 @ 1200	20 @ 4800	5	16 @ 11.5
	1110743	0 @ 1200	24 @ 5700	4	30 @ 10.0
	1110744	0 @ 1200	20 @ 4800	4	30 @ 10.0
	1110759	0 @ 1200	24 @ 5700	—	—
	1110760	0 @ 1510	16 @ 5250	5	16 @ 11.5
	1110778②	0 @ 1200	24 @ 5700	5	24 @ 12.0
	1110778③	0 @ 1200	20 @ 4800	5	24 @ 12.0

① Stamped on dist. housing
② W/vac. no. 604
③ W/vac. no. 608

Model 5210-C

Chevrolet Chevette

Year	Carb. Part No. ① ②	Float Level (Dry) (in.)	Float Drop (in.)	Pump Position	Fast Idle Cam (in.)	Choke Plate Pulldown* (in.)	Secondary Vacuum Break (in.)	Fast Idle Setting (rpm)	Choke Unloader (in.)	Choke Setting
1979	466361, 466363, 466369, 466371	0.50	—	—	0.110	0.245	—	2500	0.350	2 Rich
	466364, 466362, 466370, 466372	0.50	—	—	0.110	0.250	—	2500	0.350	2 Rich
	466365, 466366, 466367, 466368, 466373, 466374, 466375, 466376	0.50	—	—	0.130	0.300	—	2500	0.350	1 Rich
1980	All	0.50	—	—	0.110	0.120	—	⑤	0.350	Fixed

1ME CARBURETOR SPECIFICATIONS

CHEVROLET PRODUCTS, CHEVETTE

Rochester Carburetor

Year	Carburetor Identification① Number	Float Level (in.)	Metering Rod (in.)	Fast Idle Speed (rpm)	Fast Idle Cam (in.)	Vacuum Break (in.)	Choke Unloader (in.)	Choke Setting (notches)
1979	17059014	3/8	0.095	2000	0.180	0.200	0.400	Index
	17059020	3/8	0.095	2000	0.180	0.200	0.400	Index
	17059013	3/8	0.095	1800	0.180	0.200	0.400	Index
	17059314	3/8	0.100	2000	0.190	0.245	0.400	Index

① Stamped on float bowl, next to fuel inlet nut
② 2200 rpm for the first two numbers
③ 2200 rpm for the last two numbers

CAR SERIAL NUMBER AND ENGINE IDENTIFICATION

1979

The vehicle identification number is mounted on a plate behind the windshield on the driver's side. The sixth character is the model year, with 9 for 1979. The fifth character is the engine code, as follows:

E . . . . 1.6 2-bbl. 4-cyl. L17 Chev.
J . . . . 1.6 2-bbl. 4-cyl. L18 Chev.

NOTE: *1.6 litres is the same as 98 cubic inches.*

1980

The vehicle identification number is mounted on a plate behind the windshield on the driver's side of the vehicle. The sixth character is the model year, with A for 1980. The fifth character is the engine code, as follows:

9 . . . . 1.6 2-bbl. 4-cyl. L17 Chev.
0 . . . . 1.6 2-bbl. 4-cyl. L18 Chev.

NOTE: *1.6 liters is the same as 98 cubic inches.*

EMISSION EQUIPMENT

1979-80

All Models

Closes positive crankcase ventilation
Emission calibrated carburetor
Emission calibrated distributor
Heated air cleaner
Vapor control, canister storage
Exhaust gas recirculation
Catalytic converter, single
Electric choke
Pulse air system
Early fuel evaporation—(1980)
Computer controlled catalytic converter (C-4 system)—(1980)

IDLE SPEED AND MIXTURE ADJUSTMENTS

1979-80

CAUTION: *Emission control adjustment changes are noted on the Vehicle Emission Information Label by the manufacture. Refer to the label before any adjustments are made.*

Air Cleaner . . . . In place
Air cond. . . . . Off
Hoses . . . . See engine label
Mixture adj. . . . . . . See propane rpm below

49 States (1979-80)
1.6 liter (98 cid) Manual trans.
With or w/o AC . . . . 800 (N)
Propane enriched 925
Fast idle . . . . 2500 (N)
1.6 liter (98 cid) Auto. trans.
With or w/o AC . . . . 750 (D)
Propane enriched 800
Fast idle . . . . 2500 (N)

California (1979)
1.6 liter (98 cid) Manual trans.
With or w/o AC . . . . 800 (N)
Propane enriched . . 900 (925-L18 engine)
Fast idle . . . . 2500 (N)
1.6 liter (98 cid) Auto. trans.
With or w/o AC . . . . 750 (D)
Propane enriched 800
Fast idle . . . . 2500 (N)

California (1980)
1.6 liter (98 cid) Manual trans.
With or w/o AC . . . . 800 (N)
Propane enriched . . . . N A
Fast idle . . . . 2600 (N)
1.6 liter (98 cid) Auto. trans.
With or w/o AC . . . . 800 (D)
Propane enriched . . . . N A
Fast idle . . . . 2500 (N)

High Altitude (1979)
1.6 liter (98 cid) Manual trans.
With or w/o AC . . . . 800 (N)
Propane enriched 925
Fast idle . . . . 2500 (N)
1.6 liter (98 cid) Auto. trans.
With or w/o AC . . . . 750 (D)
Propane enriched 825
Fast idle . . . . 2500 (N)

SPARK PLUGS

1979-80

1.6 4-cyl. . . . . . . AC-R42 TS . . .035

VACUUM ADVANCE

1979-80

Diaphragm type . . . . Single
Vacuum source . . . . Ported

INITIAL TIMING

1979-80

1.6 liter (L17) Fed./MT . . 12° BTDC
1.6 liter (L17) Calif./MT
. . . . 12° BTDC
1.6 liter (L17) Fed./AT . . 18° BTDC
1.6 liter (L17) Calif./AT
(1979) . . . . 16° BTDC
1.6 liter (L17) Calif./AT
(1980) . . . . 18° BTDC
1.6 liter (L18) Fed./MT . . 12° BTDC
1.6 liter (L18) Calif./MT
. . . . 12° BTDC
1.6 liter (L18) Fed./AT . . 18° BTDC
1.6 liter (L18) Calif./AT
. . . . 12° BTDC

EMISSION CONTROL SYSTEMS

CLOSED POSITIVE CRANKCASE VENTILATION SYSTEM (PCV) 1979-80

The closed PCV system is used on all Chevette engines. Conditions indicative of malfunctioning PCV systems are rough idle, oil present in the air cleaner, engine oil leakage and excessive oil sludging or diluting.

The PCV system must be sealed so that the system operates properly. If the system is suspected of improper operation, check for the possible causes and correct so that the system functions as designed.

THERMOSTATIC AIR CLEANER 1979-80

The Thermostatic Air Cleaner is used on all Chevette engines. A vacuum motor controls the damper assembly, located in the air cleaner inlet. The damper assembly controls the amount of preheated and non-preheated air entering the air cleaner, to maintain a controled air temperature into the carburetor. The damper assembly motor vacuum is modulated by a temperature sensor in the air cleaner.

PULSE AIR INJECTION SYSTEM (PAIR) 1979-80

The pulse air injection system consists of a pulse air valve which has four check valves, one for each exhaust port. The firing operation of the engine creates a pulsating flow of gases, which have positive or negative pressures, due to the opening and closing of the exhaust valves. When positive pressure is present in the exhaust manifold, the check valve is forced closed and no exhaust gas will flow past the valve and into the fresh air supply line. When negative pressure is present, the fresh air supply line is opened to allow fresh air to enter the exhaust manifold to mix with the exhaust gases, and to further the burning of the gases.

EVAPORATION CONTROL SYSTEM (ECS) 1979-80

The ECS system is designed to limit the release of fuel vapors into the atmosphere from the carburetor and fuel tank. The system consists of a charcoal canister to absorb the fuel vapors and to store them until the engine is in operation. Hoses and lines are used to direct the vapors from the fuel tank and carburetor, to the charcoal canister. A purge office is located in the carburetor and is connected by a hose to the canister, and during periods of part throttle, the vapors are routed to the PCV system, to be burned along with the crankcase gases.

EXHAUST GAS RECIRCULATION SYSTEM (EGR) 1979-80

The E.G.R. system introduces exhaust gases into the engine induction system through passages cast into the intake manifold.

Introduction of these gases into the air/fuel mixture lowers the combustion temperatures to reduce the formation of oxides of nitrogen. The E.G.R. control valves are at a normally closed position during curb idle (no recirculation) and will open to "mix" gases at throttle plate angles just slightly "off-idle."

There are three different kinds of EGR valves used in Chevettes. The appearance of the valves is similar. The major difference is the method used to control how far each valve opens.

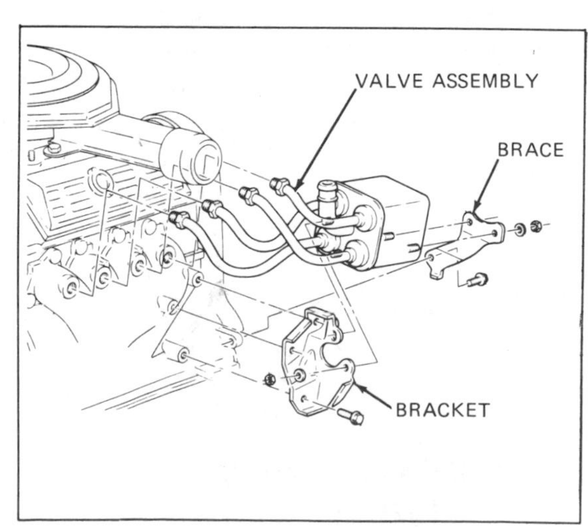

Pulse air valve mounting—1977-79 Chevette (California and High Altitude models)

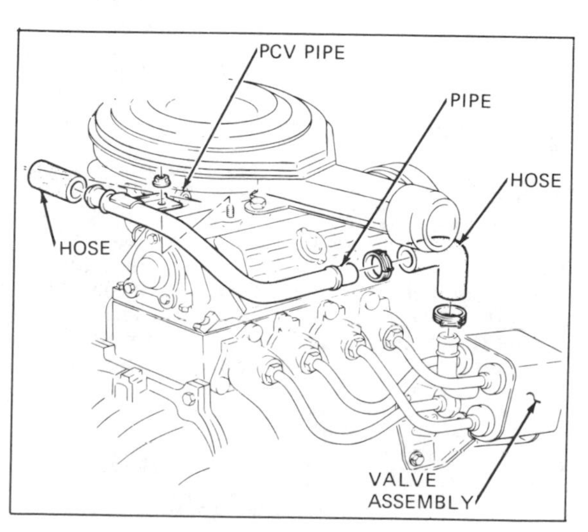

Pulse air pipe and hoses—1977-79 Chevette (California and High Altitude models)

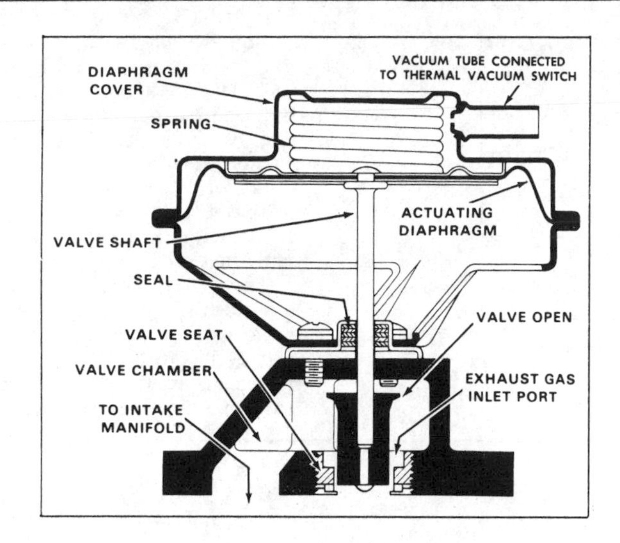

Cross section of EGR valve—1977-79 Chevette

A ported single vacuum valve is used on some engines. The valve opening is determined by the amount of vacuum received from a ported source on the carburetor.

Two different types of backpressure EGR valves are used on most Federal, California and high altitude engines. The valve opening is determined by the amount of vacuum received from a ported source on the carburetor and the amount of backpressure in the exhaust system. One is called a positive transducer valve and the other is called a negative transducer valve. The negative transduced backpressure valve is used on engines with a relatively low backpressure to provide the desired opening point and flow rate.

CATALYTIC CONVERTER 1979-80

A catalytic converter is used as an emission control device and installed as part of the exhaust system. The purpose of the converter is to reduce the hydrocarbons and carbon monoxide pollutants from the exhaust gases, with the use of pellets coated with catalytic materials, containing platinum and palladium. Only unleaded gasoline is to be used for fuel in vehicles equipped with the catalytic converter. Some vehicles produced in 1980 may be equipped with another type of catalytic converter called a Monolith converter. The catalyst in this type of converter is not serviceable.

On 1980 vehicles equipped with the C-4 system the catalytic material used in the three way catalytic converter contains platinum, palladium and rhodium. This three way converter controls the levels of oxides of nitrogen as well as hydrocarbon and carbon monoxide.

ELECTRIC CHOKE 1979-80

An electric choke is used to assist the heating of the choke coil to aid in quicker opening of the choke valve, while the engine is at higher temperatures. The electrical circuits to the choke is routed through the oil pressure switch, when the engine is operating. The current flows to a ceramic resistor that is divided into two sections. A small section is used for the gradual heating of the choke coil, while a large section is used for quick heating of the choke coil. The resistor warms the choke coil to provide proper choke valve opening for good engine warm up performance.

NOTE: *When repairing the choke mechanism, do not install a gasket between the electric choke assembly and the choke housing. To do this, would interrupt the grounding circuits.*

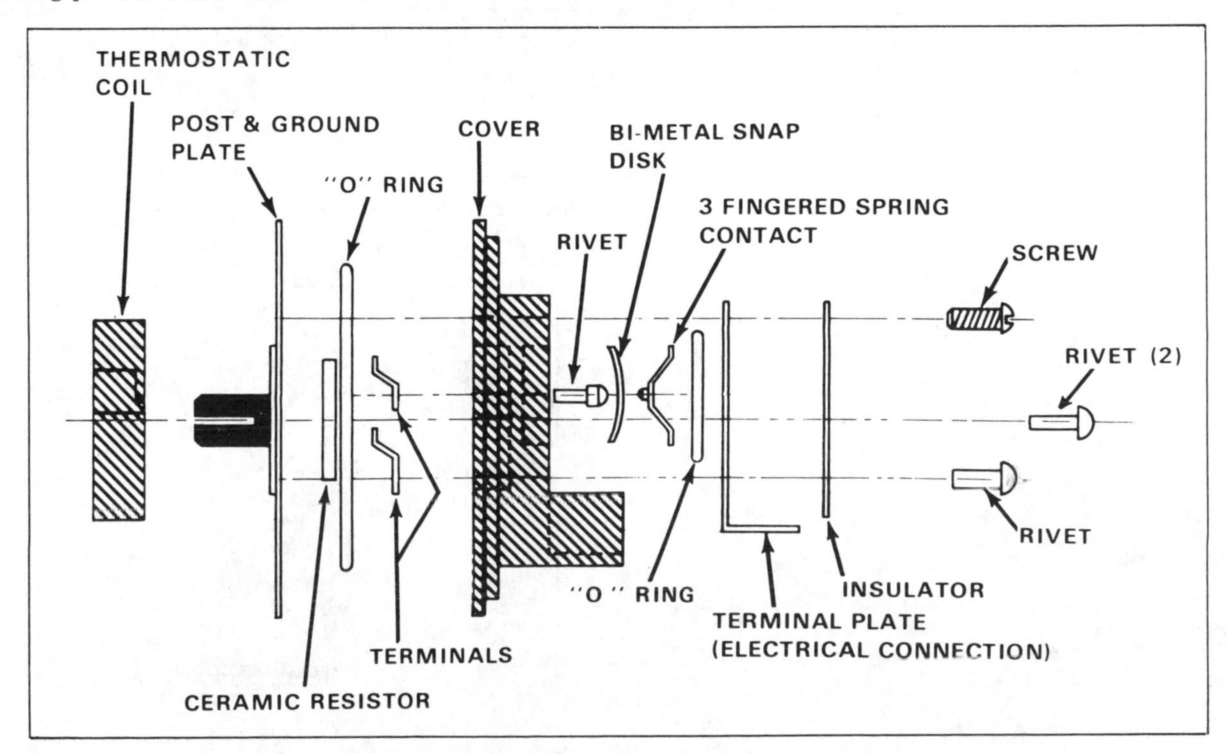

Exploded view of electric choke components—1977-79 Chevette

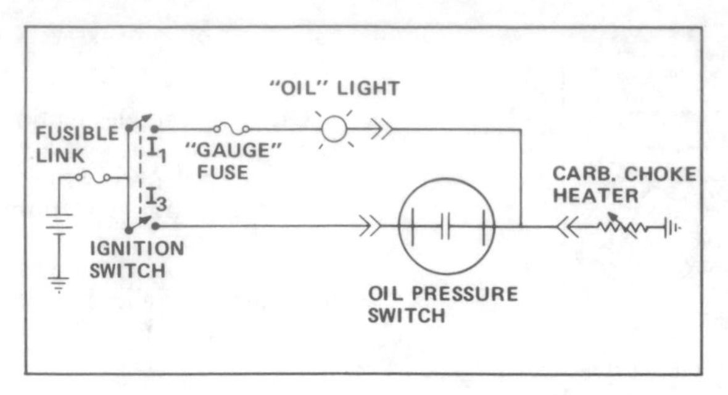

Electric choke circuit schematic—1977-79 Chevette

The system features a "Check Engine" warning lamp which will light in case of a system malfunction and will remain on as long as the engine runs with the malfunction uncorrected. This same lamp, with the ECM diagnostic system activated, will flash a trouble code which will assist in locating the cause of the system malfunction.

The C-4 system controls emissions by close regulation of the air-fuel ratio and by the use of a three way Catalytic Converter which lowers the levels of oxides of Nitrogen, Hydrocarbons and Carbon Monoxide.

EARLY FUEL EVAPORATION SYSTEM (EFE) 1980

The electric EFE system utilizes a ceramic heater grid located underneath the primary bore of the carburetor as part of the carburetor insulator gasket. It heads the incoming fuel-air charge for improved vaporization and driveability on cold driveaway.

COMPUTER CONTROLLED CATALYTIC CONVERTER (C-4 SYSTEM)

The C-4 System used on some 1980 engines is an electronically controlled exhaust emissions system. The major components are an Exhaust Gas Oxygen Sensor (OS), and Electronic Control Module (ECM), a controlled air-fuel ratio carburetor and a three-way Catalytic Converter (ORC).

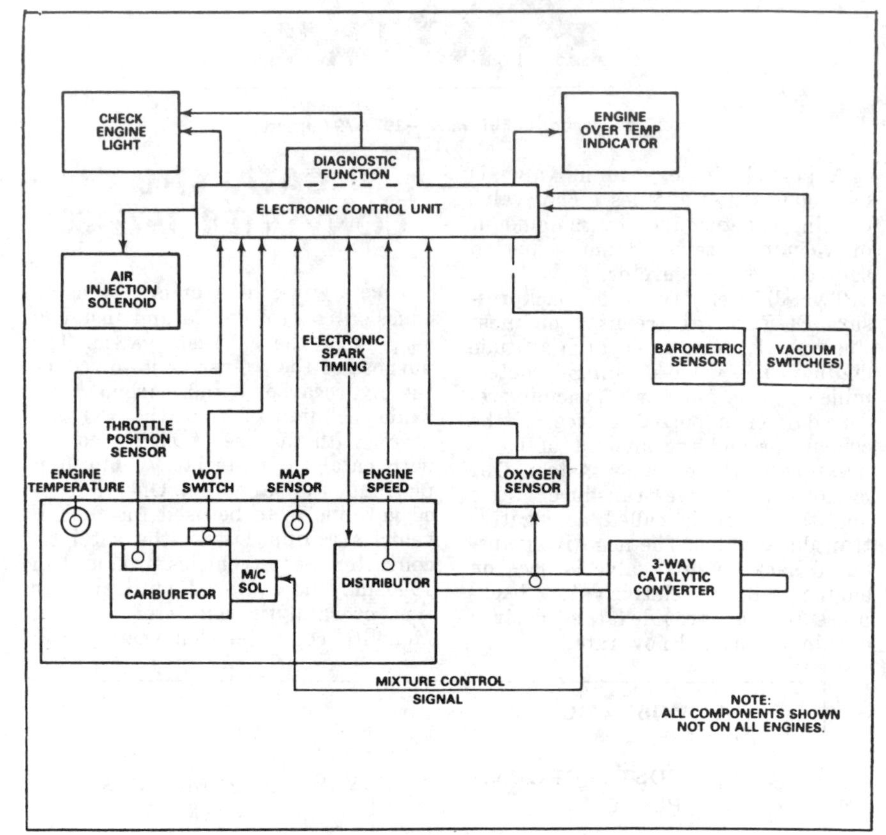

Computer Controlled Catalytic Converter System Schematic

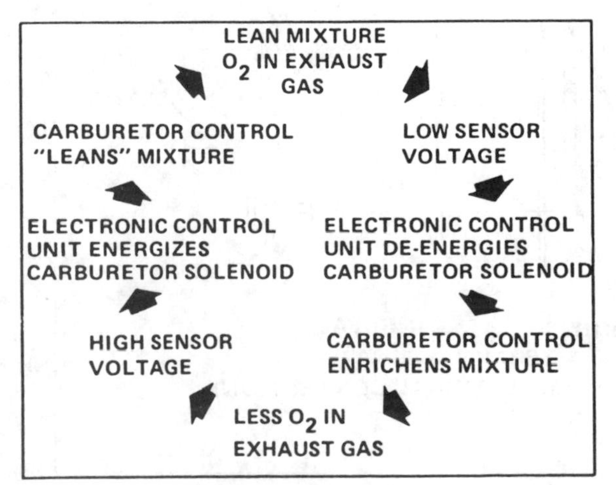

Basic Cycle of Operation

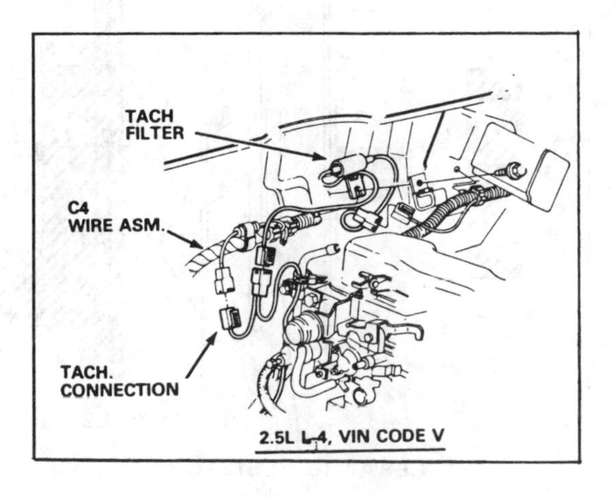

Tachometer Test Leads

VACUUM CIRCUITS

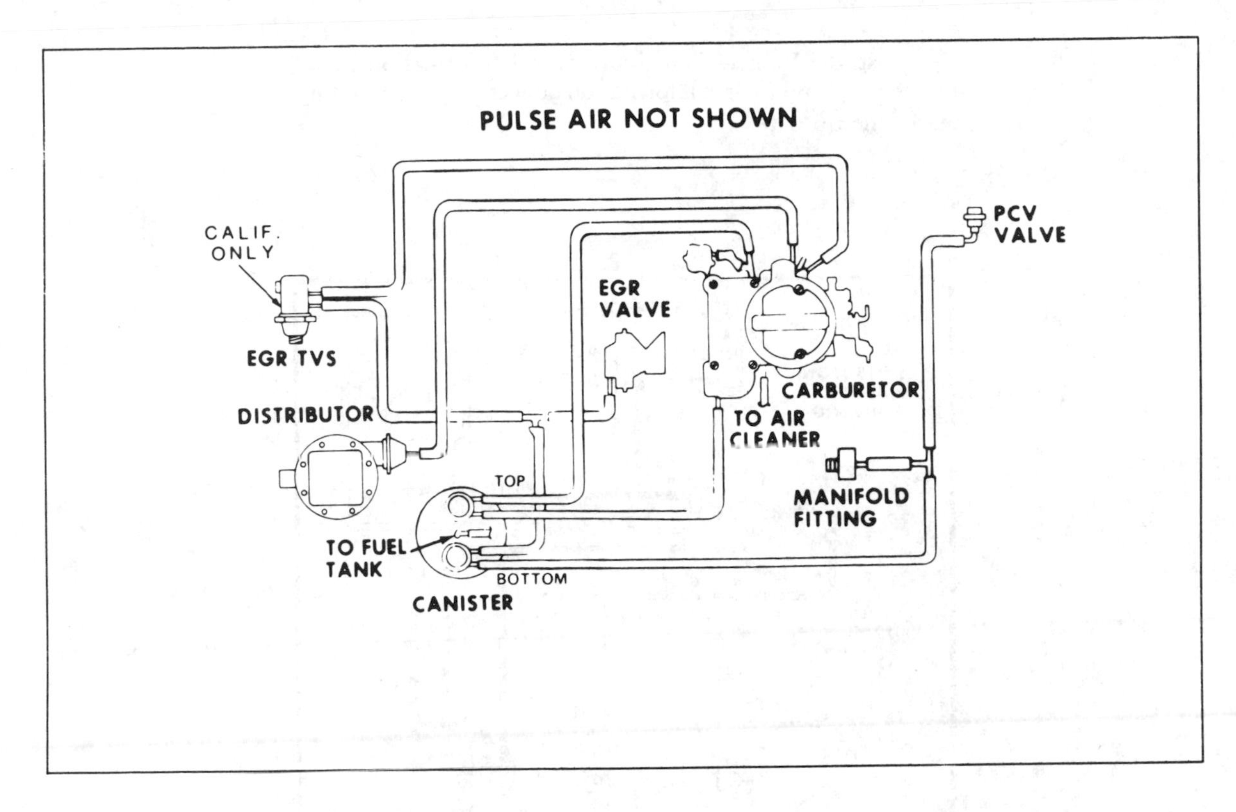

Vacuum hose routing—1978-79 Chevette (California and High Altitude models)

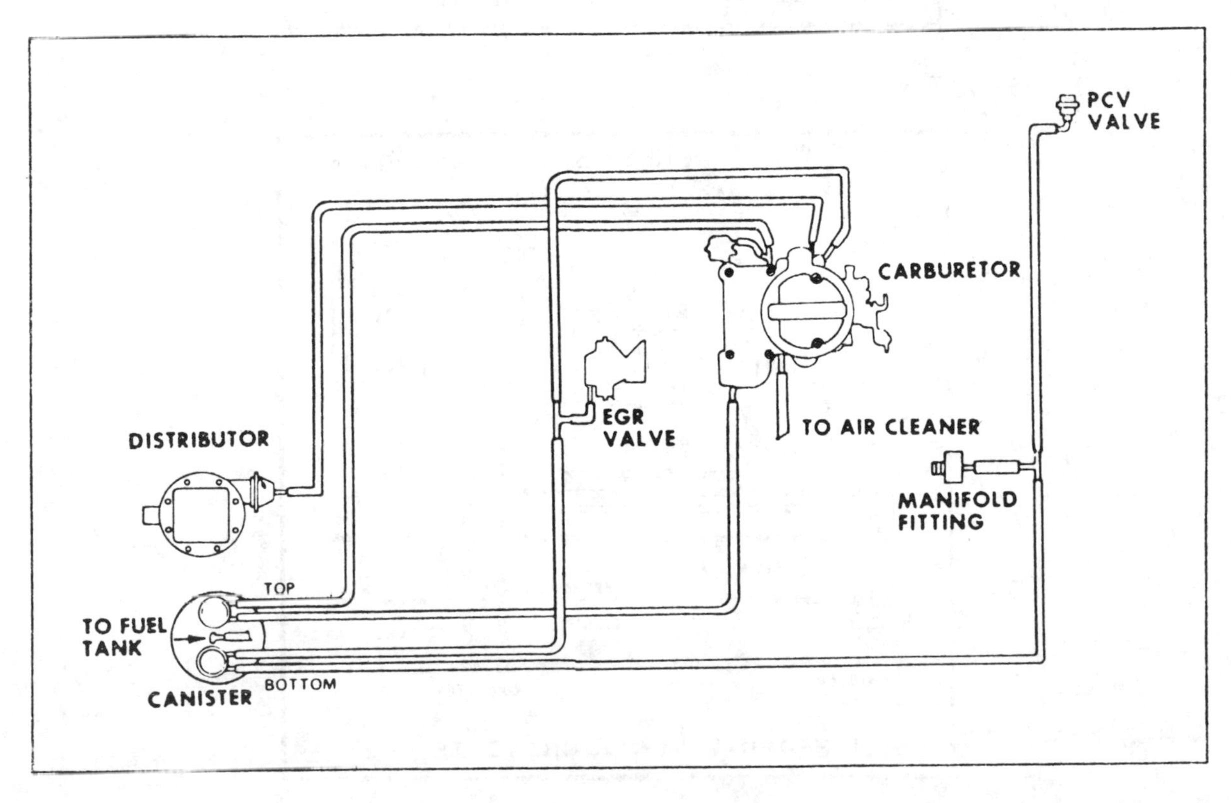

Vacuum hose routing—1978-79 Chevette (Low Altitude models)

VACUUM CIRCUITS

Use the Vehicle Emission Control Information Label for the most up to date information concerning hose routing and tune up data.

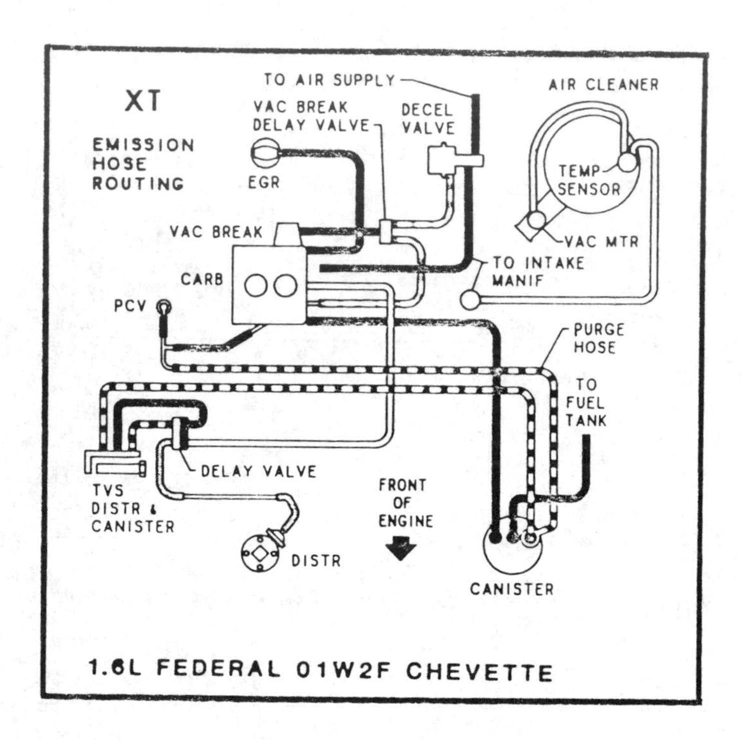

1.6L FEDERAL 01W2F CHEVETTE

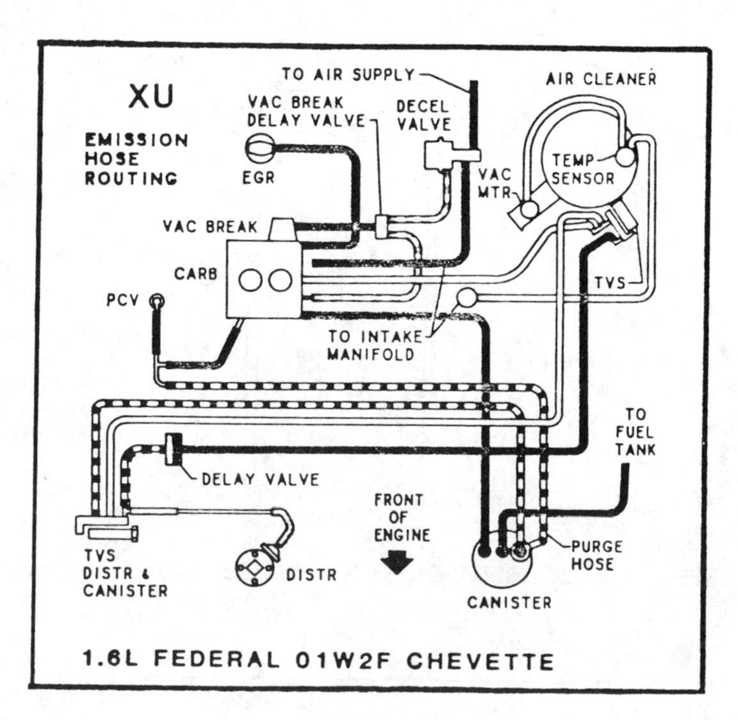

1.6L FEDERAL 01W2F CHEVETTE

1980 Chevette

TUNE-UP SPECIFICATIONS

(Except Starfire & '80 Omega)

ENGINE CODE, 5th CHARACTER OF THE VIN NUMBER
MODEL YEAR CODE, 6th CHARACTER OF THE VIN NUMBER

Year	ENG. V.I.N. Code	ENGINE No. Cyl. Disp. (cu in)	SPARK PLUGS* Orig. Type	Gap (in)	IGNITION TIMING (deg B.T.D.C. @ rpm) Lo. Alt. Manual	Lo. Alt. Auto.	Calif. Hi. Alt. Manual	Calif. Hi. Alt. Auto.	CARBURETION Hot Idle Speed Federal Manual	Federal Auto.(D)	Calif. Hi. Alt Manual	Calif. Hi. Alt Auto.(D)
1979	A	6-231	R46TSX	.060	15 @ 600	15 @ 800	15 @ 600	15 @ 600	800	800	800	800
	2	6-231	①	①	①	①	①	①	—	—	—	—
	F	8-260	R46SZ	.060	18 @ 1100	20 @ 1100	—	18 @ 1100	800	625	—	625•
	P	8-260	DIESEL	—	—	①	—	①	—	①	—	①
	Y	8-301	R46TSX	.060	—	12 @ 650	—	20 @ 1100	650	650	—	650
	G	8-305	R45TS	.045	4 @ 600	4 @ 500	—	4 @ 500	—	600	—	650•
	H	8-305	R45TS	.045	4 @ 500	4 @ 500	—	4 @ 500⑥	—	—	—	—
	L	8-350	R43TS	—	—	—	—	8 @ 500	—	650	—	600⑥
	R	8-350	R46SZ	.060	—	20 @ 1100	—	20 @ 1100	—	650	—	600⑥
	N	8-350	DIESEL	—	—	①	—	①	—	①	—	①
	X	8-350	R46TSX	.060	—	15 @ 550	—	—	—	650	—	—
	K	8-403	R46SZ	.060	—	20 @ 1100	—	20 @ 1100	—	650	—	600⑥
1980	A	6-231	R45TS	.040	15 @ 550	15 @ 550	15 @ 550	15 @ 550	550	670	①	①
	F	8-260	R46SX	.080	20 @ 1100	20 @ 1100	20 @ 1100	20 @ 1100	500	500	500	500
	P	8-260	DIESEL	—	—	⑧	—	—	—	650	—	650
	S	8-265	R45TSX	.060	10 @ 600	10 @ 600	—	10 @ 600	700	625	①	①
	H	8-305	R45TS	.045	4 @ 500	4 @ 500	4 @ 500	4 @ 500	600	600	①	①
	Y	8-307	R46SX	.080	20 @ 1100	20 @ 1100	—	20 @ 1100	—	600	①	①
	R	8-350	R46SX	.080	18 @ 1100	18 @ 1100	16 @ 1100	16 @ 1100	—	600	—	650
	N	8-350	DIESEL	—	—	⑧	—	—	—	650	—	650

— Not applicable
N Transmission in neutral
① See engine decal
⑥ Hi. Alt.—idle 600 rpm
⑦ Calif., Tornado—12 B.T.C. @ 1100 rpm
⑧ Int. timing 16 BTC-38 A.T.D.C.
Ext. timing 64 BTC-17 A.T.D.C.

TUNE-UP SPECIFICATIONS

ENGINE CODE, 5th CHARACTER OF THE VIN NUMBER
MODEL YEAR CODE, 6th CHARACTER OF THE VIN NUMBER

STARFIRE

Year	ENGINE V.I.N. Code	No. Cyl. Disp. (cu in)	SPARK PLUGS* Orig. Type	Gap (in)	IGNITION DIST.	Timing① (deg B.T.D.C. @ rpm) Man. Trans.	Auto Trans.	CARBURETION Idle Speed② (r.p.m.) Man. Trans.	Fed.	Auto. Trans. Cal.	Hi. Alt.
1979	1	4-151	R43TSX	.060	E.I.	14	14	1000/1200	—	650/850	650/850
	V	4-151	R43TSX	.060	E.I.	12	12	900/1250	650/850	—	650/850
	9	4-151	R43TSX	.060	E.I.	12	12	900/1250	650/850	650/850	650/850
	C	6-196	R46TSX	.060	E.I.	15	15	800	550/670	—	—
	A	6-231	R46TSX	.060	E.I.	15	15	800	—	600	600
	2	6-231	R46TSX	.060	E.I.	—	15	—	—	580/670	—
	G	8-305	R45TS	.045	E.I.	4	4②	600/700	500/600	600/650	—
1980	V	4-151	R43TSX	.060	E.I.	14	15	1000	③	650/800	650/800
	A	6-231	R46TSX	.060	E.I.	15	15	800	③	600	—

① Time at curb idle speed unless otherwise indicated. Figure in parentheses is California application.
② Set idle speed with automatic transmission in Drive; manual transmission in Neutral. Where two figures appear, the lower figure indicates idle speed with solenoid deenergized.
③ See underhood certification/specification decal.

DISTRIBUTOR SPECIFICATIONS

(Except Starfire & '80 Omega)

Year	Distributor Identification	Vacuum Model Identification	Centrifugal Advance Start Crank. Deg. @ Eng. RPM	Centrifugal Advance Finish Crank. Deg. @ Eng. RPM	Vacuum Advance Start @ In Hg	Vacuum Advance Finish Crank. Deg. @ In Hg
1979	1103281	1973621	0 @ 1000	20 @ 3800	4	18 @ 12
	1103282	1973624	0 @ 1000	20 @ 3800	4	20 @ 10
	1103285	1973626	0 @ 1200	2 @ 4200	4	24 @ 8
	1103314	1973635	0 @ 825	21.5 @ 3400	4	25 @ 12
	1103320	1973610	0 @ 910	26 @ 4465	4	30 @ 11
	1103322	1973597	0 @ 600	29 @ 4000	6	24 @ 13
	1103323	1973603	0 @ 1000	19 @ 4000	5	16 @ 11
	1103324	1973597	0 @ 600	23 @ 3600	6	24 @ 13
	1103325	1973603	0 @ 1000	13 @ 3600	5	16 @ 11
	1103346	1973597	0 @ 1000	19 @ 4000	6	24 @ 13
	1103347	1973597	0 @ 1000	13 @ 3600	6	24 @ 13
	1103353	1973624	0 @ 1250	11-12 @ 4500	3-6	20 @ 9-12
	1103355	1973662	0 @ 910	26 @ 4465	4	30 @ 9
	1103368	1973626	0 @ 1000	20 @ 3800	4	10 @ 8
	1103379	1973691	0 @ 1000	20 @ 3800	3	20 @ 7.5
	1103396	1973686	0 @ 910	26 @ 4465	5	30 @ 12
	110766	1973688	0 @ 1680	15 @ 3600	4	24 @ 11
	110767	1973687	0 @ 1680	15 @ 3600	3	20 @ 12
	110768	1973687	0 @ 1000	15 @ 3600	3	20 @ 12
	110769	1973688	0 @ 1000	15 @ 3600	4	24 @ 11
	110770	1973689	0 @ 1680	15 @ 3600	3	20 @ 9
	110779	1973698	0 @ 1680	15 @ 3600	3	24 @ 9.5
1980	1110555	1973688	0 @ 1000	15 @ 3600	4	24 @ 11
	1110554	1973718	0 @ 1680	15 @ 3600	3	24 @ 11
	1110552	1973709	0 @ 925	14 @ 3400	3	21 @ 8
	1103419	1973610	0 @ 910	23 @ 3600	4	30 @ 11
	1103426	1973716	0 @ 1000	19 @ 3200	4	20 @ 10
	1103384	1973620	0 @ 800	20 @ 4000	4	15 @ 12
	1103386	1973693	0 @ 1000	20 @ 3800	4	16 @ 7.5
	1103412	197314	0 @ 600	29 @ 4000	4	30 @ 12.5
	1103417	197306	0 @ 1900	17 @ 4400	4	30 @ 12.4
	1103398	1973715	0 @ 600	23 @ 3600	5	30 @ 13.7
	1103413 (1)	1973708	0 @ 6000	13 @ 3500	6	30 @ 13
	1103414 (2)	1973597 (2)	0 @ 600	23 @ 3600	6	24 @ 13

DISTRIBUTOR SPECIFICATIONS

STARFIRE

Year	Distributor Identification	Centrifugal Advance Start Crank. Deg. @ Eng. RPM	Centrifugal Advance Finish Crank. Deg. @ Eng. RPM	Vacuum Advance Start @ In Hg	Vacuum Advance Finish Crank. Deg. @ In Hg
1979	1103281	0 @ 1000	20 @ 3800	4	12 @ 12
	1103285	0 @ 1200	22 @ 4200	4	10 @ 8
	1103365	0 @ 1200	20 @ 4400	5	22 @ 11
	1103379	0 @ 1000	20 @ 3800	3	20 @ 7.5
	1110726	0 @ 1200	18 @ 4000	3.5	22 @ 11
	1110757	0 @ 1200	18 @ 4000	3.5	22 @ 11
	1110766	0 @ 1700	15 @ 3600	4	20 @ 11
	1110767	0-4 @ 2000	12-16 @ 3600	3	20 @ 12
	1110768	2 @ 1250	15 @ 3600	3.5	22 @ 12.5
	1110770	0 @ 1600	15 @ 3600	3	20 @ 9

DISTRIBUTOR SPECIFICATIONS

		CENTRIFUGAL ADVANCE		VACUUM ADVANCE	
Year	DISTRIBUTOR IDENTIFICATION	Start Crank. Deg. @ Eng. RPM	Finish Crank. Deg. @ Eng. RPM	Start @ In Hg	Finish Crank. Deg. @ In Hg
1980	1110558	0 @ 1000	14 @ 4000	3	15 @ 5
	1110559	0 @ 1200	14 @ 4000	3	15 @ 5
	1110560	0 @ 1000	14 @ 4400	4	20 @ 10

2MC, M2MC, M2ME, E2ME CARBURETOR SPECIFICATIONS

OLDSMOBILE — **Rochester Carburetor**

Year	Carburetor Identification[1]	Float Level (in.)	Choke Rod (in.)	Choke Unloader (in.)	Vacuum Break Lean or Front (in.)	Vacuum Break Rich or Rear (in.)	Pump Rod (in.)	Choke Coil Lever (in.)	Automatic Choke (notches)
1979	17059134	15/32	0.243	0.243	0.157	—	1/4[2]	0.120	1 Lean
	17059135	15/32	0.243	0.243	0.157	—	1/4[2]	0.120	1 Lean
	17059136	15/32	0.243	0.243	0.157	—	1/4[2]	0.120	1 Lean
	17059137	15/32	0.243	0.243	0.157	—	1/4[2]	0.120	1 Lean
	17059150	3/8	0.071	0.220	0.195	0.129	1/4[2]	0.120	2 Rich
	17059151	3/8	0.071	0.220	0.243	0.142	11/32[3]	0.120	2 Rich
	17059152	3/8	0.071	0.220	0.195	0.129	1/4[2]	0.120	2 Rich
	17059154	3/8	0.071	0.220	0.157	0.260	11/32[3]	0.120	2 Rich
	17059160	11/32	0.110	0.195	0.129	0.187	1/4[2]	0.120	2 Rich
	17059430	9/32	0.243	0.243	0.157	—	9/32	0.120	1 Lean
	17059432	9/32	0.243	0.243	0.157	—	9/32	0.120	1 Lean
	17059450	3/8	0.071	0.220	0.157	—	11/32[3]	0.120	2 Rich
	17059180	11/32	0.039	0.243	0.103	0.090	1/4[2]	0.120	2 Rich
	17059190	11/32	0.039	0.243	0.103	0.090	1/4[2]	0.120	2 Rich
	17059191	11/32	0.039	0.243	0.103	0.090	9/32[2]	0.120	2 Rich
	17059196	11/32	0.039	0.277	0.129	0.117	1/4[2]	0.120	1 Rich
	17059491	11/32	0.039	0.277	0.129	0.117	9/32[2]	0.120	1 Rich
	17059492	11/32	0.039	0.277	0.129	0.117	9/32[2]	0.120	1 Rich
	17059498	11/32	0.039	0.277	0.129	0.117	9/32[2]	0.120	2 Rich
1980	17080150	3/8	0.071	0.220	0.243	0.157	11/32[3]	0.120	Fixed
	17080152	3/8	0.071	0.220	0.243	0.157	11/32[3]	0.120	Fixed
	17080153	3/8	0.071	0.220	0.243	0.157	11/32[3]	0.120	Fixed
	17080190	9/32	0.139	0.243	0.123	0.110	1/4[2]	0.120	Fixed
	17080191	11/32	0.139	0.243	0.096	0.096	1/4[2]	0.120	Fixed
	17080192	9/32	0.139	0.243	0.123	0.110	1/4[2]	0.120	Fixed
	17080195	9/32	0.139	0.243	0.103	0.071	1/4[2]	0.120	Fixed
	17080197	9/32	0.139	0.243	0.103	0.071	1/4[2]	0.120	Fixed
	17080491	5/16	0.139	0.243	0.117	0.220	Fixed	0.120	Fixed
	17080493	5/16	0.139	0.243	0.117	0.179	Fixed	0.120	Fixed
	17080495	5/16	0.139	0.243	0.117	0.179	Fixed	0.120	Fixed
	17080496	5/16	0.139	0.243	0.117	0.203	Fixed	0.120	Fixed
	17080498	5/16	0.139	0.243	0.117	0.203	Fixed	0.120	Fixed

2SE, E2SE CARBURETOR ADJUSTMENTS

Oldsmobile (except Omega)

Rochester Carburetor

Year	Carburetor Identification	Float Level (in.)	Pump Rod (in.)	Fast Idle (rpm)	Choke Coil Lever (in.)	Fast Idle Cam (deg./in.)	Air Valve Rod (in.)	Primary Vacuum Break (deg./in.)	Choke Setting (notches)	Secondary Vacuum Break (deg./in.)	Choke Unloader (deg./in.)	Secondary Lockout (in.)
1979	17059674	13/64	1/2	2400	.085	18/0.096	.025	22/.123	2 Rich	—	32/.195	.030
	17059675	13/64	17/32	2200	.085	18/0.096	.025	22/.123	1 Rich	—	32/.195	.030
	17059676	13/64	1/2	2400	.085	18/0.096	.025	22/.123	2 Rich	—	32/.195	.030
	17059677	13/64	17/32	2200	.085	18/0.096	.025	22/.123	1 Rich	—	32/.195	.030
1980	17080674	3/16	1/2	2600	.085	18/0.096	.018	19/.103	Fixed	—	32/.195	.025
	17080675	3/16	1/2	2600	.085	18/0.096	.018	21/.117	Fixed	—	32/.195	.025
	17080676	3/16	1/2	2600	.085	18/0.096	.018	19/.103	Fixed	—	32/.195	.025
	17080677	3/16	1/2	2600	.085	18/0.096	.018	21/.117	Fixed	—	32/.195	.025
	17059774	5/32	1/2	①	.085	18/0.096	.018	19/.103	Fixed	—	32/.195	.025
	17059775	5/32	17/32	①	.085	18/0.096	.018	21/.117	Fixed	—	32/.195	.025
	17059776	5/32	1/2	①	.085	18/0.096	.018	19/.103	Fixed	—	32/.195	.025
	17059777	5/32	17/32	①	.085	18/0.096	.018	21/.117	Fixed	—	32/.195	.025

① See Underhood Decal

QUADRAJET CARBURETOR SPECIFICATIONS

OLDSMOBILE

Rochester Carburetor

Year	Carburetor Identification①	Float Level (in.)	Air Valve Spring (turn)	Pump Rod (in.)	Primary Vacuum Break (in.)	Secondary Vacuum Break (in.)	Secondary Opening (in.)	Choke Rod (in.)	Choke Unloader (in.)	Fast Idle Speed ④ (rpm)
1979	17059202	1/2	7/8	1/4	0.164	—	④	0.314	0.243	⑥
	17059207	15/32	7/8	1/4	0.157	—	④	0.243	0.243	⑥
	17059216	15/32	7/8	1/4	0.157	—	④	0.243	0.243	⑥
	17059217	15/32	7/8	1/4	0.157	—	④	0.243	0.243	⑥
	17059218	15/32	7/8	9/32	0.164	—	④	0.243	0.243	⑥
	17059222	15/32	7/8	9/32	0.164	—	①	0.243	0.243	⑥
	17059250	13/32	1/2	9/32	0.129	0.183	④	0.096	0.220	⑥
	17059251	13/32	1/2	9/32	0.129	0.183	④	0.096	0.220	⑥
	17059253	13/32	1/2	9/32	0.129	0.183	④	0.096	0.220	⑥
	17059256	13/32	1/2	9/32	0.136	0.195	④	0.103	0.220	⑥
	17059258	13/32	1/2	9/32	0.136	0.195	④	0.103	0.220	⑥
	17059502	15/32	7/8	1/4	0.164	—	④	0.243	0.243	⑥
	17059504	15/32	7/8	1/4	0.164	—	④	0.243	0.243	⑥
	17059553	13/32	1/2	9/32	0.136	0.230	④	0.103	0.220	⑥
	17059554	13/32	1/2	9/32	0.136	0.230	④	0.103	0.220	⑥
	17059582	15/32	7/8	11/32	0.203	—	④	0.243	0.314	⑥
	17059584	15/32	7/8	11/32	0.203	—	④	0.243	0.314	⑥
1980	17080202	7/16	7/8	1/4⑦	0.157	—	④	0.110	0.243	⑤
	17080204	7/16	7/8	1/4⑦	0.157	—	④	0.110	0.243	⑤
	17080250	13/32	1/2	9/32⑦	0.149	0.211	④	0.090	0.220	⑤
	17080251	13/32	1/2	9/32⑦	0.149	0.211	④	0.090	0.220	⑤
	17080252	13/32	1/2	9/32⑦	0.149	0.211	④	0.090	0.220	⑤
	17080253	13/32	1/2	9/32⑦	0.149	0.211	④	0.090	0.220	⑤
	17080259	13/32	1/2	9/32⑦	0.149	0.211	④	0.090	0.220	⑤
	17080260	13/32	1/2	9/32⑦	0.149	0.211	④	0.090	0.220	⑤

QUADRAJET CARBURETOR SPECIFICATIONS (Cont'd)

OLDSMOBILE

Year	Carburetor Identification①	Float Level (in.)	Air Valve Spring (turn)	Pump Rod (in.)	Primary Vacuum Break (in.)	Secondary Vacuum Break (in.)	Secondary Opening (in.)	Choke Rod (in.)	Choke Unloader (in.)	Fast Idle Speed ④ (rpm)
	17080504	1/2	7/8	Fixed	0.136	0.179	④	0.110	0.243	⑤
	17080553	15/32	1/2	Fixed	0.142	0.220	④	0.090	0.220	⑤
	17080554	15/32	1/2	Fixed	0.142	0.211	④	0.090	0.220	⑤

④ No measurement necessary on two point linkage.
⑤ 3 turns after contacting lever for preliminary setting.
⑥ 2 turns after contacting lever for preliminary setting.
⑦ Inner hole

CAR SERIAL NUMBER AND ENGINE IDENTIFICATION

1979

The vehicle identification number is mounted on a plate behind the windshield on the driver's side. The sixth character is the model year, with 9 for 1979. The fifth character is the engine code, as follows:

A ....231 2-bbl. V-6 LD-5 Buick
F ....260 2-bbl. V-8 LV-8 Olds
H ....305 4-bbl. V-8 LG-4 Chev.
K ....403 4-bbl. V-8 L-80 Olds
L ....350 4-bbl. V-8 LM-1 Chev.
N ....350 Diesel V-8 LF-9 Olds
R ....350 4-bbl. V-8 L-34 Olds
G ....305 2-bbl. V-8 LG-3 Chev.
V ....151 2-bbl. 4-cyl. LX-8 Pont.
1 ....151 2-bbl. 4-cyl. LS-6 Pont.
Y ....301 2-bbl. V-8 L-27 Pont.
P ....260 Diesel V-8 LF-7 Olds
X ....350 2-bbl. V-8 L-77 Chev.

1980

A ....231 2-bbl. V-6 Buick
F ....260 2-bbl. V-8 Olds
H ....305 4-bbl. V-8 Chev.
N ....350 Diesel V-8 Olds
P ....260 Diesel V-8 Olds
R ....350 4 bbl. V 8 Olds
S ....265 2-bbl. V-8 Pont.
V ....151 2-bbl. L4 Pont.
Y ....307 4-bbl. V-8 Olds

EMISSION EQUIPMENT

1979

All Carburetor Models

Closed positive crankcase ventilation
Emission calibrated distributor
Emission calibrated carburetor
Heated air cleaner
Vapor control, canister storage
Exhaust gas recirculation
Catalytic converter
Air Pump
 49-States
  Not used
 Calif.
  V-8 and V-6
 Altitude
  231 V-6
  305 4-bbl. V-8
  350 V-8 "L"
Early fuel evaporation
 4-cyl. (Vin I)
 V-6
 V-8 Chev. only ("H" "L" "G" "X")
 V-8 Pont.
Choke thermal vacuum switch
 49-States
  260 V-8
  301 V-8
  350 V-8 "R"
  403 V-8
 Calif.
  260 V-8
  350 V-8 "R"
  403 V-8
 Altitude
  260 V-8
  350 V-8 "R"
  403 V-8
Spark delay valve
 305 V-8 all except Calif. and altitude
Retard delay valve
 231 V-6 Calif.
 260 V-8 Calif.

Diesel Models

Positive crankcase ventilation

1980

The following devices are on all GM gasoline engines:

Back pressure exhaust gas recirculation BP-EGR
Catalytic converter CC
Evaporative emission control EEC
Positive crankcase ventilation valve PCV
Thermostatic air cleaner TAC
Air injection reactor AIR[1]
 231 V-6 (A)
 260 V-8 (F)
 305 V-8 (H)
 307 V-8 (Y)
 350 V-8 (R)

[1]NOTE: *Only some of the engines will have the AIR on them.*

Barometric pressure sensor BARO C-4
 231 V-6 (A) California
Computer controlled catalytic converter system C-4
 151 L-4 (V) California
 231 V-6 (A) California
 305 V-8 (H) California
 350 V-8 (R) California
Canister purge thermal vacuum switch CP-TVS
 231 V-6 (A) California
 260 V-8 (F)
 305 V-8 (H)
 307 V-8 (Y)
 350 V-8 (R)
Choke vacuum break-distributor thermal vacuum switch CVB-DTVS
 260 V-8 (F)
 305 V-8 (H) California
 307 V-8 (Y)
 350 V-8 (R)
Cold delay spark thermal vacuum switch CDS-TVS
 151 L-4 (V)
Distributor thermal vacuum switch DTVS
 151 L-4 (V)
 260 V-8 (F)
 305 V-8 (H)
 307 V-8 (Y) California
 350 V-8 (R) California
Deceleration valve DV
 151 L-4 (V)
Distributor vacuum delay valve DVDV
 305 V-8 (H)
Distributor vacuum solenoid valve DVSV
 151 L-4 (V)
Delay valve-trapped spark DV-TS
 260 V-8 (F)
 307 V-8 (Y)
 350 V-8 (R)
Electronic control module ECM C-4 California
 151 L-4 (V)
 231 V-6 (A)

EMISSION EQUIPMENT

305 V-8 (H)
350 V-8 (R)
Manifold absolute pressure sensor MAP
231 V-6 (A) California
Oxidation reduction catalyst ORC
151 L-4 (V) California
231 V-6 (A) California
350 V-8 (R) California
Oxygen sensor OS
151 L-4 (V) California
231 V-6 (A) California
305 V-8 (H) California
350 V-8 (R) California
Pulse air injection reactor PAIR
151 L-4 (V) 49 States
Spark advance vacuum modulator SAVM
151 L-4 (V) California
260 V-8 (F)
307 V-8 (Y)
350 V-8 (R) California
Spark retard delay valve SRDV
151 L-4 (V)
Thermostatic air cleaner-distributor thermal vacuum switch TAC-DTVS
305 V-8 (H) 49 States
Vacuum regulator valve VRV
350 V-8 (N) diesel
Engine coolant sensor ECS
231 V-6 (A)
Early fuel evaporator/check valve/delay valve/thermal vacuum switch EFE/-CV/-DV/-TVS
231 V-6 California
260 V-8
305 V-8
EGR/canister purge-TVS EGR/CP-TVS
305 V-8
EGR-distributor thermal vacuum switch EGR-DTVS
350 V-8 (R) California
EGR/EFE/Electronic module retard-DTVS EGR/EFE/EMR-DTVS
260 V-8 (F)
EGR/EFE-thermal vacuum switch EGR/EFE-TVS
231 V-6 California
EGR/EMR-distributor thermal vacuum switch EGR/EMR-DTVS
307 V-8 (Y)
350 V-8 (R) 49 States
EGR-thermal vacuum switch EGR-TVS
151 L-4 (V)
Electronic module retard EMR
307 V-8 (Y)
350 V-8 (R) 49 States
EMR-vacuum switch EMR-VS
307 V-8 (Y)
350 V-8 (R) 49 States
Engine temperature sensor ETS
350 V-8 (N) diesel

Emission Equipment Abbreviations:

AIR—Air Injection Reaction System
BP-EGR—Back Pressure Exhaust Gas Recirculation
CEAB-TVS—Cold Engine Airbleed—TVS
CP-TVS—Canister Purge Thermal Vacuum Switch
CTVS—Choke Thermal Vacuum Switch
CVB-VDV—Choke Vacuum Break-Vacuum Delay Valve
DCV—Distributor Check Valve
DTCV—Distributor Thermal Control Valve
DTVS—Distributor Thermal Vacuum Switch
DVDV—Distributor Vacuum Delay Valve
EEF—Evaporative Emission Control
EFE—Early Fuel Evaporation Valve and Actuator
EFE-CV—EFE Check Valve
EFE/TVS—EFE Thermal Vacuum Switch
EFE-DTVS—EFE Distributor Thermal Vacuum Switch
EGR—Exhaust Gas Recirculation
EGR/EFE-TVS—EGR/Early Fuel Evaporation-TVS
EGR/CP-TVS—EGR Canister Purge Thermal Vacuum Switch
EGR-DTVS—EGR Distributor Thermal Vacuum Switch
EGR-TCV—EGR Thermal Control Valve
EGR-TVS—EGR Thermal Vacuum Switch
OS—Oxygen Sensor
PCV—Postive Crankcase Ventilation
SAVM—Spark Advance Vacuum Modulator
SDV—Spark Delay Valve
SRDV—Spark Retard Delay Valve
TAC—Thermostatic Air Cleaner
TACS—Thermostatic Air Cleaner Sensor
VIS—Vacuum Input Switch
VM—Vacuum Modulator
VM-CV—Vacuum Modulator Check Valve

IDLE SPEED AND MIXTURE ADJUSTMENTS

1979

NOTE: *Refer to Vehicle Emission Information Label before making any adjustments.*

Air cleaner .... In place
Air cond. .... Off
Auto. trans. .... Drive
Idle CO .... Not used
Mixture adj. See propane speed below
Vacuum hoses ..... See engine label

151 4-cyl. 49-States Vin code (V)
Auto. trans. and AC
Curb or "on" idle 850
Slow or "off" idle 650
Propane enriched idle 695
Manual trans. and AC
Curb or "on" idle 1250
Slow or "off" idle 900
Propane enriched idle 1040
Auto. trans. no AC
Curb or "on" idle 650
Slow or "off" idle 500
Propane enriched idle 695
Manual trans. no AC
Curb or "on" idle 900
Slow or "off" idle 500
Propane enriched idle 1040

151 4-cyl. Vin code (1)
Auto. trans. AC
Curb or "on" idle 850
Slow or "off" idle 650
Propane enriched idle .....
Manual trans. AC
Curb or "on" idle 1200
Slow or "off" idle 1000
Propane enriched idle .....
Auto. trans. no AC
Curb or "on" idle 650
Slow or "off" idle 500
Propane enriched idle .....
Manual trans. no AC
Curb or "on" idle 1000
Slow or "off" idle 500
Propane enriched idle .....

231 V-6 49-States Vin code (A)
Auto. trans.
Curb or "on" idle 670
Slow or "off" idle 550
Propane enriched idle 575
Manual trans.
Curb or "on" idle 800
Slow or "off" idle 600
Propane enriched idle 1000

231 V-6 California Vin code (A)
Auto trans.
Curb or "on" idle .... none
Slow or "off" idle 600
Propane enriched idle 615
Manual trans.
Curb or "on" idle 800
Slow or "off" idle 600
Propane enriched idle 840

231 V-6 Altitude Vin code (A)
Auto trans.
Curb or "on" idle .... none
Slow or "off" idle 600
Propane enriched idle 615

260 V-8 49-States Vin code (F)
Auto. trans.
Curb or "on" idle 625
Slow or "off" idle 500
Propane enriched idle .. 560-580
Manual trans.
Curb or "on" idle 800
Slow or "off" idle 650
Propane enriched idle .. 780-800

260 V-8 California Vin code (F)
Auto. trans.
Curb or "on" idle 625
Slow or "off" idle 500
Propane enriched idle .. 530-550

IDLE SPEED AND MIXTURE ADJUSTMENTS

260 V-8 Altitude Vin code (F)
Auto. trans.
Curb or "on" idle 650
Slow or "off" idle 550
Propane enriched idle 575
260 Diesel All Vin code (P)
Auto. trans.
Curb or "on" idle 650
Slow or "off" idle 590
Manual trans.
Curb or "on" idle 659
Slow or "off" idle 575
301 V-8 All Vin code (Y)
Auto. trans.
Curb or "on" idle 650
Slow or "off" idle 500
Propane enriched idle 530
305 V-8 49-States Vin code (G)
Auto trans.
Curb or "on" idle 600
Slow or "off" idle 500
Propane enriched idle . . 520-540
Manual trans.
Curb or "on" idle ....none/ 700 Omega
Slow or "off" idle ..... 600 (N)
Propane enriched idle . . 710-750
305 V-8 California Vin code (G)
Auto. trans.
Curb or "on" idle 650
Slow or "off" idle ..... 600 (N)
Propane enriched idle . . 640-660
305 V-8 4-bbl. 49-States Vin (H)
Auto. trans.
Curb or "on" idle 600
Slow or "off" idle 500
Propane enriched idle . . 530-570
Manual trans.
Curb or "on" idle ....none
Slow or "off" idle 700
Propane enriched idle . . 800-850
305 V-8 4-bbl. California Vin (H)
Auto. trans.
Curb or "on" idle 600
Slow or "off" idle 600
Propane enriched idle . . 525-560
305 V-8 4-bbl. Altitude Vin code (H)
Auto. trans.
Curb or "on" idle 650
Slow or "off" idle 600
Propane enriched idle . . 630-670
350 V-8 California Vin code (L)
Auto. trans.
Curb or "on" idle 600
Slow or "off" idle 500
Propane enriched idle . . 525-560
350 V-8 Altitude Vin code (L)
Auto. trans.
Curb or "on" idle 600
Slow or "off" idle 500
Propane enriched idle . . 630-670
350 Diesel Vin code (N)
Auto. trans.
Curb or "on" idle 650
Slow or "off" idle 575
350 V-8 49-States Vin code (R)
Auto. trans.
Curb or "on" idle 650
Slow or "off" idle 550
Propane enriched idle .625-645/ Toronado 585-590

350 V-8 California Vin code (R)
Auto. trans.
Curb or "on" idle 600
Slow or "off" idle 500
Propane enriched idle . . 565-585
350 V-8 Altitude Vin code (R)
Auto. trans.
Curb or "on" idle 650
Slow or "off" idle 550
Propane enriched idle 590
403 V-8 49-States Vin code (K)
Auto. trans.
Curb or "on" idle 650
Slow or "off" idle 550
Propane enriched idle . . 625-645
403 V-8 California Vin code (K)
Auto. trans.
Curb or "on" idle 600
Slow or "off" idle 500
Propane enriched idle . . 565-585
403 V-8 Altitude Vin code (K)
Auto. trans.
Curb or "on" idle 650
Slow or "off" idle 550
Propane enriched idle 590

NOTE: *On cars equipped with automatic transmission the idle adjustments are made with the transmission selector lever in the (DR) range, unless otherwise noted.*

On cars equipped with manual transmission adjustments are made with transmission selector lever in the (N) range, unless otherwise noted.

When making Curb or "on" idle adjustments turn the idle solenoid screw in or out to adjust RPM (AC on) clutch wires disconnected and wheels blocked. w/o AC, idle stop solenoid energized—if used.

1980

NOTE: *Refer to Vehicle Emission Information Label before making any adjustments.*

Air cleaner ....In place
Air cond. ....Off
Auto trans. ....In drive
151 4-cyl. 49-states VIN code (V)
Auto trans and AC
Carb screw 550
Fast idle ....2,600 (P)
Solenoid on 650
Propane enriched 700
Auto trans and no AC
Carb screw 650
Fast idle ....2,600 (P)
Solenoid on 850
Propane enriched ...... —
Manual and AC
Carb screw ....1,000 (N)
Fast idle ....2,600 (N)
Solenoid on ....1,250 (N)
Propane enriched 1,150
Manual and no AC
Carb screw ....550 (N)
Fast idle ....2,600 (N)
Solenoid on ....1,000 (N)
Propane enriched ...... —

151 4-cyl. California VIN code (V)
Auto trans and AC
Carb screw 650
Fast idle ....2,600 (P)
Solenoid on 850
Idle mixture set by C-4 computer
Auto trans no AC
Carb screw 500
Fast idle ....2,600 (P)
Solenoid on 650
Idle mixture set by C-4 computer
Manual and AC
Carb screw ....1,000 (N)
Fast idle ....2,400 (N)
Solenoid ....1,200 (N)
Idle mixture set by C-4 computer
Manual no AC
Carb screw ....500 (N)
Fast idle ....2,400 (N)
Solenoid on ....1,000 (N)
Idle mixture set by C-4 computer

231 V-6 49 states VIN code (A)
Auto trans and AC
Carb screw 550
Fast idle ....2,000 (P)
Solenoid on 670
Propane enriched 610①
① Cutlass only 600 rpm
Manual and AC
Carb screw ....600 (N)
Fast idle ....2,200 (N)
Solenoid on ....800 (N)
Propane enriched .....830 (N)

231 V-6 California VIN code (A)
Auto trans and AC
Carb screw 550
Fast idle ....2,200 (P)
Solenoid on 620
Idle mixture set by C-4 computer

260 V-8 49 states VIN code (F)
Auto trans and AC
Carb screw 500
Fast idle 700
Solenoid on 625
Propane enriched 530-550

265 V-8 49 states VIN code (S)
Auto trans and AC
Carb screw 500
Fast idle 700
Solenoid on 625
Propane enriched 530-550
305 V-8 49 states VIN code (H)
Auto trans and AC
Carb screw 500
Fast idle ....1,850 (P)
Solenoid on 600
Propane enriched 530-570

305 V-8 California VIN code (H)
Auto trans and AC
Carb screw 550
Fast idle ....2,200 (P)
Solenoid on 650
Idle mixture set by C-4 computer
307 V-8 49 states VIN code (Y)
Auto trans and AC
Carb screw 500
Fast idle 700
Solenoid on 600
Propane enriched 530

IDLE SPEED AND MIXTURE ADJUSTMENTS

350 V-8 9 states VIN code (R)
Auto trans and AC
Carb screw 500
Fast idle 700
Solenoid on 600
Propane enriched 565-585

350 V-8 California VIN code (R)
Auto trans and AC
Carb screw 550
Fast idle 700
Solenoid on 650
Idle mixture set by C-4 computer

350 V-8 Diesel 50 states VIN code (N)
Pump idle 600
Solenoid on 750①

① Fast idle relay disconnected (single green wire at front of dash).

INITIAL TIMING

1979

NOTE: *Engine timing information is noted on the Vehicle Emission Information Label by the manufacture. Refer to this information before any adjustments are made.*

151 4-cyl. .... 14° BTDC
231 V-6 .... 15° BTDC
260 V-8 (at 1100 rpm)
49-States auto. trans. .. 20° BTDC
49-States man. trans. .. 18° BTDC
Calif. auto. trans. ..... 18° BTDC
Altitude auto. trans. ... 18° BTDC
301 V-8 .... 12° BTDC
305 2-bbl. V-8 "G"
49-States .... 4° BTDC
Calif.
Starfire .... 2° BTDC
305 4-bbl. V-8 "H"
49-States .... 4° BTDC
Altitude .... 8° BTDC
350 4-bbl. V-8 "L" ...... 8° BTDC
350 4-bbl. V-8 "R"
(at 1100 rpm) .... 20° BTDC
260 & 350 Diesel V-8 ... Align pump marks with engine stopped
403 V-8 .... 20° BTDC

1980

151 4-cyl. .... 12° BTDC
151 4-cyl. Calif. manual
(at 1100 rpm) .... 12° BTDC
231 V-6 .... 15° BTDC
231 V-6 Manual
(at 800 rpm) .... 15° BTDC
260 V-8 (at 1,100 rpm) .... 20° BTDC①
260 V-8 (at 1,100 rpm) .... 18° BTDC①
305 V-8 49 st.
(at 500 rpm) .... 4° BTDC
305 V-8 Calif.
(at 550 rpm) .... 4° BTDC
307 V-8 (at 1,100 rpm) .. 20° BTDC
350 V-8 (at 1,100 rpm) .. 18° BTDC
350 V-8 Diesel .... —

① See emission label for exact information.

VACUUM ADVANCE

1979

Diaphragm type .... Single
Vacuum source
151 4-cyl. .... Ported
231 V-6
49-State .... Manifold
Calif. .... Ported
Altitude .... Manifold
260 V-8 49-States
Auto. trans. .... Modulated
Man. trans. .... Manifold
260 V-8 Calif. .... Ported
260 V-8 Altitude ..... Modulated
305 2-bbl. V-8
49-States .... Manifold
Calif. .... Ported
Altitude .... Manifold
305 4-bbl. V-8 .... Manifold
350 4-bbl. V-8 "L" ..... Manifold
350 4-bbl. V-8 "R"
49-States .... Manifold
Calif. .... Ported
Altitude .... Ported
403 V-8
49-States .... Manifold
Calif. .... Ported
Altitude .... Ported

1980

NOTE: *Models with Electronic Spark Timing EST have timing advance controlled by computer. All other models see the under hood label.*

SPARK PLUGS

1979

151 4-cyl. ...... AC-R43TSX .. .060
231 V-6 ...... AC-R46TSX .. .060
260 V-8 ...... AC-R46SZ .. .060
301 2-bbl. ..... AC-R46TSX .. .060
305 2-bbl. ...... AC-R45TS .. .045
305 4-bbl. ...... AC-45TS .. .045
350 4-bbl.
"L" ...... AC-R45TS .. .045
"R" ...... AC-R46SZ .. .060
"X" ...... AC-R46TSX .. .060
403 4-bbl. ...... AC-R46SZ .. .060

1980

151 4-cyl. .... R44TSX .060
231 V-6 .... R45TS .040
260 V-8 .... R46SX .080
305 V-8 .... R45TS .045
307 V-8 .... R46SX .080
350 V-8 .... R46SX .080
350 V-8 Diesel .... None

EMISSION CONTROL SYSTEMS

CLOSED POSITIVE CRANKCASE VENTILATION SYSTEM 1979-80

The closed PCV system is continued for all engines installed by Oldsmobile Division. The operation of the system remains the same.

NOTE: *PCV Systems are used with the Diesel Engines.*

EMISSION CALIBRATED DISTRIBUTORS 1979-80

The emission calibrated distributors are used to provide the engine with the correct timing to fire the air/fuel mixture for good engine performance and fuel economy, and to remain within the allowable emission levels. Various type valves and switches are used with the vacuum advance units to assist in controlling the emissions and driveability.

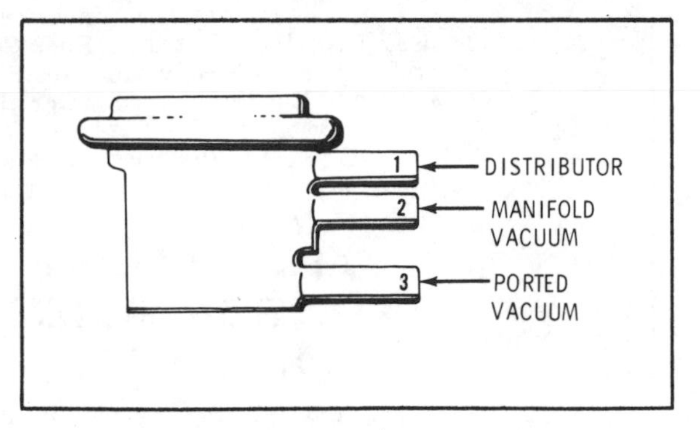

1978-79 Spark Advance Vacuum Modulator (SAVM)

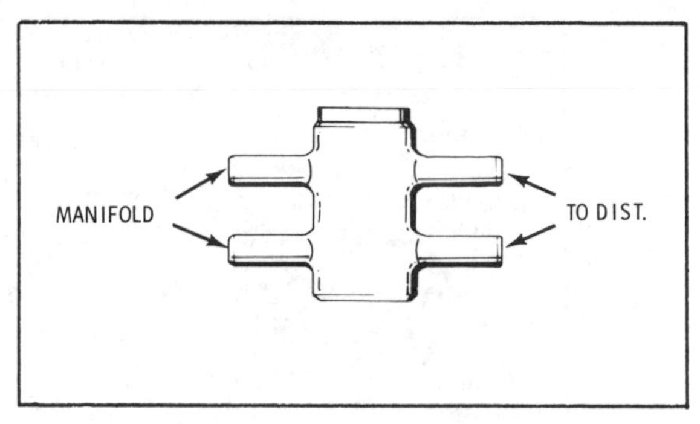

1978-79 Spark Retard Delay Valve (SRDV) Vin code V

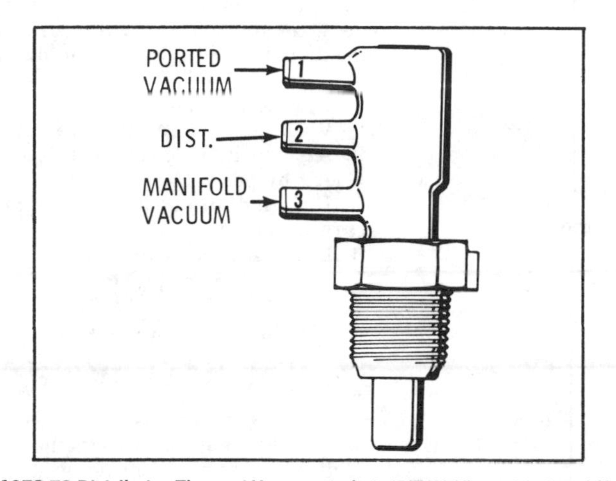

1978-79 Distributor Thermal Vacuum Switch (DTVS) Vin codes R and K

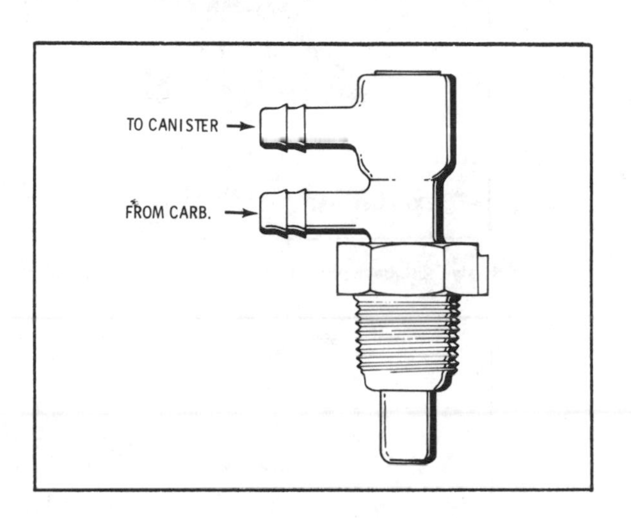

1978-79 Canister purge—Thermal Vacuum Switch (CP-TVS) Vin code A

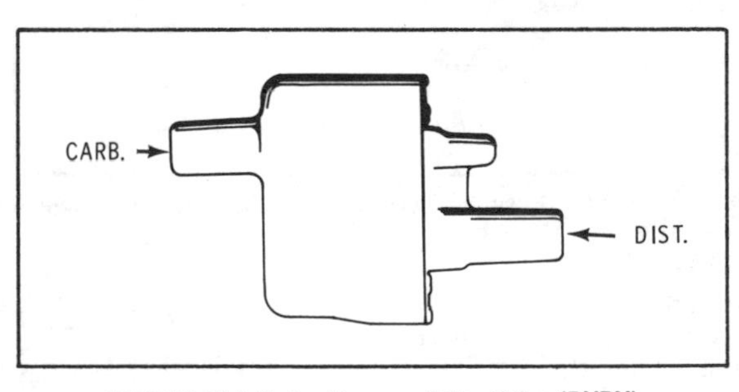

1978-79 Distributor Vacuum Delay Valve (DVDV)

The switches and valves are as follows:

Distributor Check Valve (DCV) Calif. Vin code F
Distributor Thermal Control Valve (DTCV) Vin code F
Distributor Thermal Vacuum Switch (DTVS) Vin code 5
Distributor Thermal Vacuum Switch (DTVS) Vin code R&K
Distributor Thermal Vacuum Switch (DTVS) Vin code V – Non/AC
Distributor Thermal Vacuum Switch (DTVS) Vin code V – W/AC
Distributor Vacuum Delay Valve (DVDW) Calif. Vin code – R & K
Spark Advance Vacuum Modulator (SAVM)
Spark Delay Valve (SDV)
Spark Retard Delay Valve (SRDV)
Electronic Spark Timing/Exhaust Gas Recirculation – Thermal Vacuum Switch (EST/EGR-TVS)

EMISSION CALIBRATED CARBURETORS 1979-80

Emission calibrated carburetors are used to provide the engine with the proper air/fuel mixture for good engine performance and fuel economy. Various switches and valves are used in conjunction with the fuel delivery system.

The switches and valves are as follows:

Vacuum Modulator Check Valve (VMCV)
Choke Thermal Vacuum Switch (CTVS) three port – Vin code F, R, & K
Choke Thermal Vacuum Switch (CTVS) two port – Vin code A
Choke Vacuum Break – Vacuum Delay Valve (CVB-VDV)
Cold Engine Air Bleed – Thermal Vacuum Switch – (CEAB-TVS) Vin code 1

EVAPORATIVE EMISSION CONTROL SYSTEM (EEC) 1979-80

The EEC system used is the same as that used on the 1978 models. Safeguards are built into the system to avoid the possibility of liquid fuel being drawn into the system.

The safeguards are:

1. A fuel tank overfill protector is installed on all models to provide adequate room for expansion of the fuel during temperature changes.
2. A fuel tank venting system is provided on all models to assure that the tank will be vented during any vehicle attitude. A domed fuel tank is used on sedans and coupes.
3. A pressure-vacuum relief valve, located in the gas cap, controls the fuel tank internal pressure.

EXHAUST GAS RECIRCULATION SYSTEM (EGR) 1979-80

The EGR system is used on all en-

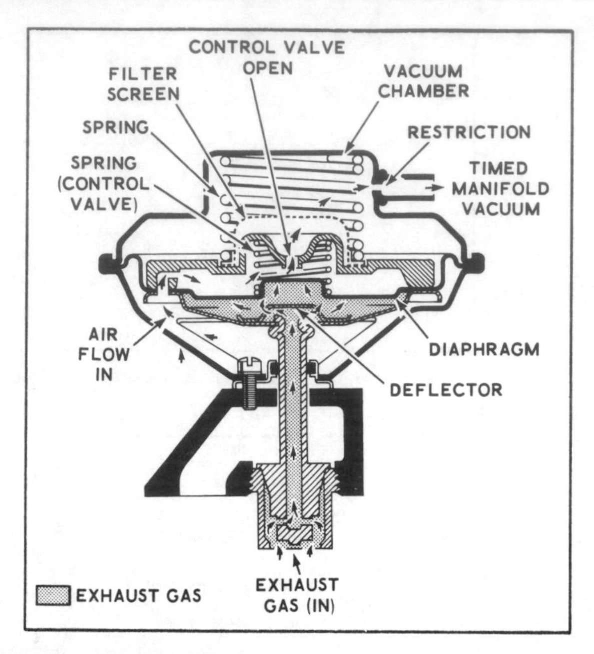

EGR valve with back pressure transducer—typical (Control valve open)

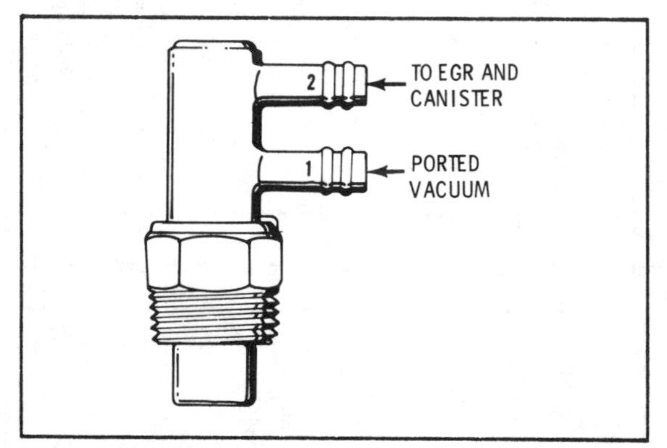

1978-79 Exhaust Gas Recirculation/Canister Purge—Thermal Vacuum Switch (EGR/CP-TVS) Vin code U, H and L

gines to meter exhaust gases into the induction air system, so that during the combustion cycle, the Oxides of Nitrogen (NO_x), emission can be reduced.

Ported and Back pressure types of EGR systems are used. The ported system uses a timed vacuum port in the carburetor to regulate the EGR valve, while the back pressure modulated system regulates the timed vacuum according to the exhaust back pressure level.

Two types of EGR valves are used, one without a transducer and one with a back pressure transducer, which is built into the internal components of the EGR valve. The transducer acts as a pressure regulator to control the flow of exhaust gases more evenly in certain engines.

NOTE: *Various Switches and Check Valves are used in the vacuum circuits.*

CATALYTIC CONVERTER (CC) 1979-80

The Catalytic Converter is used on all vehicle models to lower the levels of Hydrocarbons (HC) and Carbon Monoxide (CO) from the exhaust emission with the use of materials coated with Platinum and Palladium, to act as a catalyst.

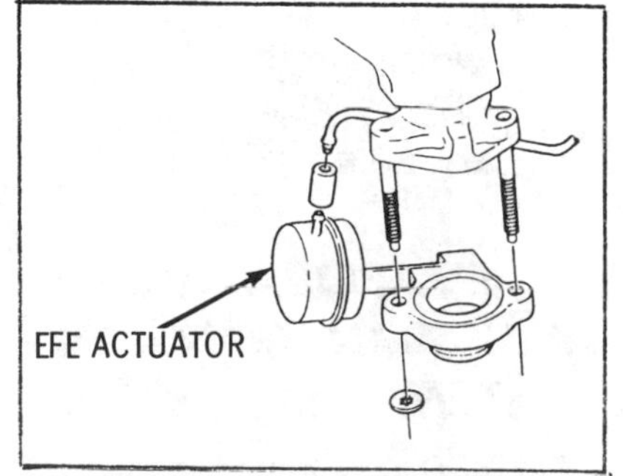

EFE actuator attached to valve—typical

A Phase II catalytic converter is used with the Electronic Fuel Control system. This converter uses Platinum and Rhodium plated material as the catalyst.

NOTE: *The 1980 system is called computer controlled catalytic converter system.*

AIR INJECTION REACTION SYSTEM (AIR) 1979-80

The Air Injection system is used on some engines to provide additional oxygen to the combustion process, after the exhaust gases leave the cylinders, to continue the combustion in the exhaust manifolds. The following components are used in the system:

1. Engine driven air pump
2. Vacuum differential valve
3. Air flow and control hoses
4. Air by-pass valve
5. Differential vacuum delay and separator valve
6. Exhaust check valve

EARLY FUEL EVAPORATION SYSTEM (EFE) 1979-80

The EFE system is designed to direct rapid heat to the engine air induction system to provide quick fuel evaporation and more uniform fuel distribution during cold driveaway. A vacuum operated valve is located between the exhaust manifold and the exhaust pipe is controlled by a thermal vacuum switch (TVS) to pass vacuum to the valve when temperatures are below a specific calibration point. As the vacuum is applied to the valve, the exhaust gas flow is increased under the intake manifold to preheat the manifold during cold engine operation. The system is used on most models.

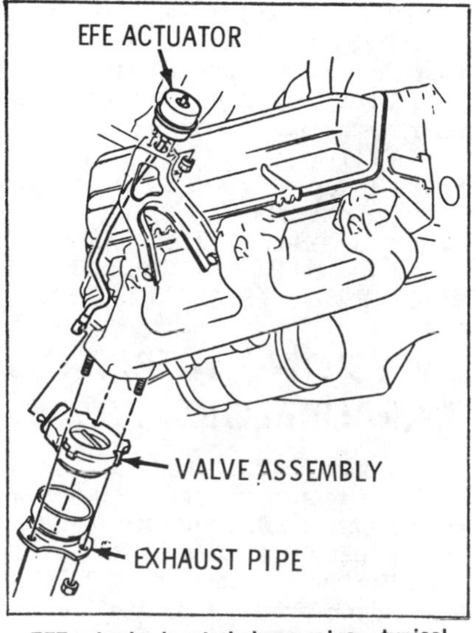

EFE actuator located above valve—typical

ELECTRIC CHOKE 1979-80

The Electric Choke is used on selected models of 1979. At temperatures below 50° F., the electrical current is directed to the small segment resistor of the choke to allow gradual opening of the choke valve. At temperatures of 70° F. and above, the electrical current is directed to both the small and the large segment resistors, to allow quicker opening of the choke valve.

ELECTRONIC FUEL CONTROL SYSTEM (EFC) 1979-80

The EFC system is designed to control the exhaust emissions by regulating the air/fuel mixture with the use of the following components:

1. Exhaust gas oxygen sensor
2. Electric control unit
3. Vacuum modulator
4. Controlled air-fuel ratio carburetor
5. Phase II catalytic converter

The exhaust gas sensor is placed in the exhaust gas stream and generates a voltage which varies with the oxygen content of the exhaust gases. As the exhaust gas oxygen content rises, indicating a lean mixture, the voltage falls. As the exhaust gas oxygen content decreases, indicating a rich mixture, the voltage rises.

The voltage signal is sent to the electronic control unit, where the voltage signal is monitored. The electronic control unit transmits a signal to the vacuum modulator. This signal is of a constant current and is continually cycling on and off.

To assist in maintaining proper electrical signal regulation during periods of unstable engine operation, a bi-metal switch is located in the cylinder head and provides an open circuit or a ground circuit for the electrical signal, when the temperature is above or below the calibration level value.

If below the calibration point, the electronic control unit uses this signal to restrict the amount of leanness that the carburetor can go lean. A temperature sensitive diode within the ECU, assists in the determination of the strength of the electrical signal. When the engine warms up, this feature drops out, indicating the engine is warm and the drivability will not be impaired while the carburetor is in the maximum lean range. the same as for the 1978 models. The EFC system should be included in any diagnosis of the following problens:

1. Detonation
2. Stalls or rough idle – cold
3. Stalls or rough idle – hot
4. Missing
5. Hesitation
6. Surges
7. Sluggish or spongy operation
8. Poor gas mileage
9. Hard starting – cold
10. Hard starting – hot
11. Foul exhaust odor
12. Engine cuts out

The 1980 Electronic Fuel Control System is called Computer Controlled Catalytic Converter System and is designated by C-4. The systems computer has a self diagnostic feature, that gives a trouble code for a failure by flashing a light under the dash.

EMISSION EQUIPMENT

Diesel Models
Exhaust gas recirculation valve
Engine temperature switch
Vacuum regulator valve
EGR—vacuum switch

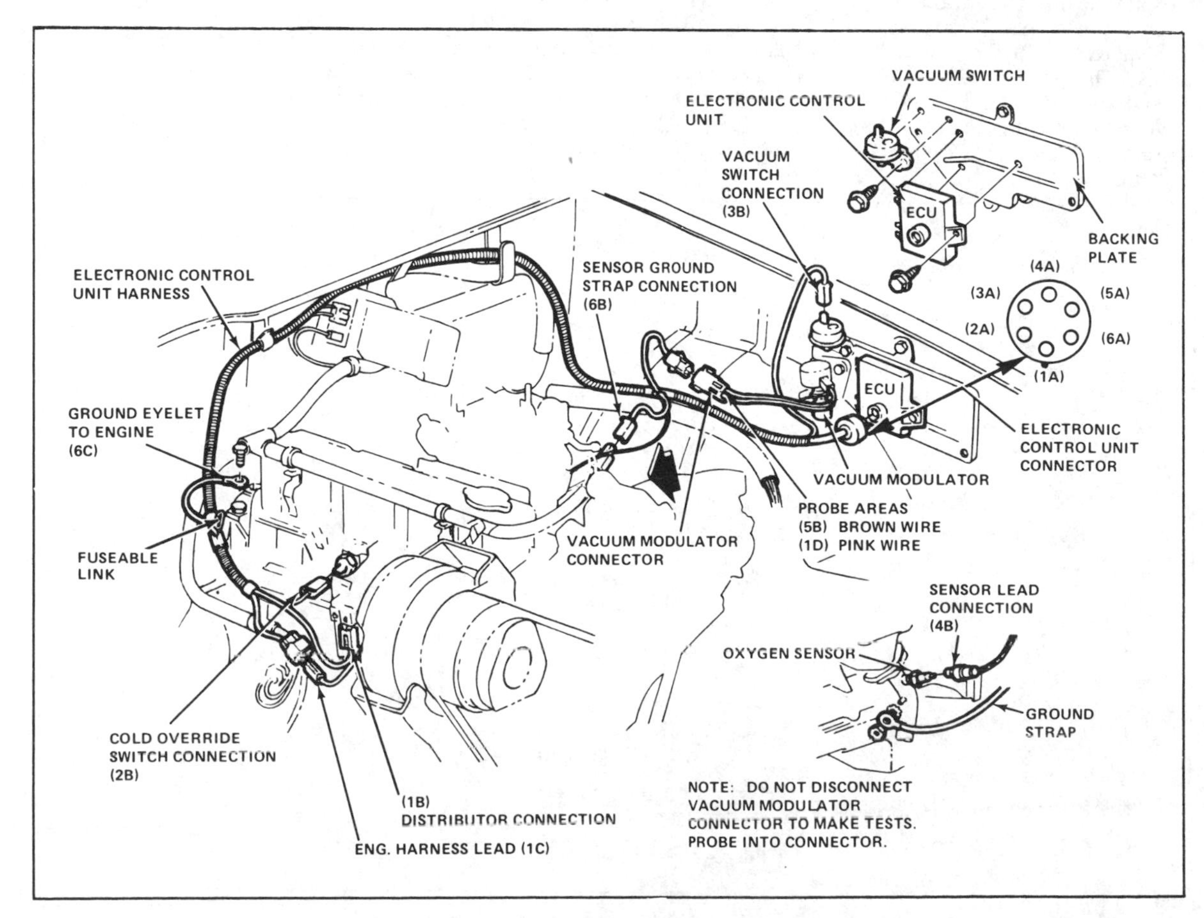

Electrical wiring connections—1978-79 Electronic Fuel Control system

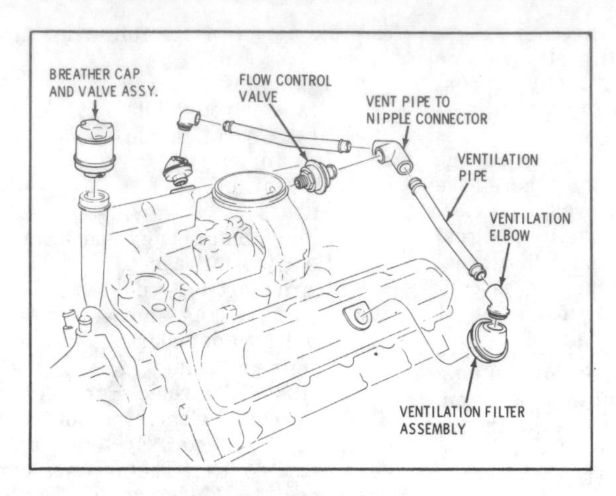

Diesel ventilation system

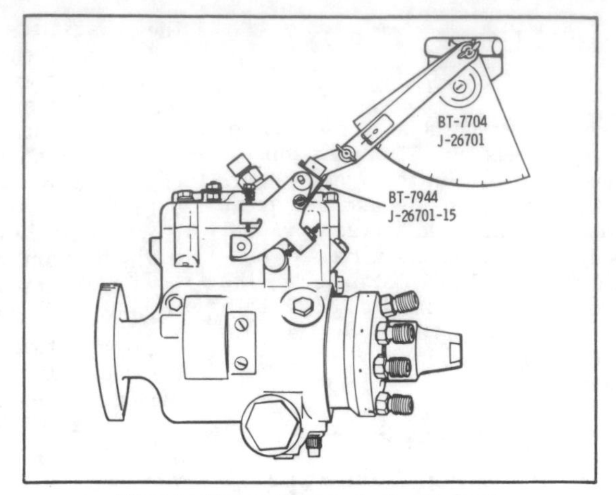

Attaching the angle gauge to throttle lever

DIESEL ENGINE 1979-80

The diesel engine is basically a clean burning engine. The vehicles are passing California specifications with only PCV, EGR, ETS, VRV, and EGR-VS.

The PCV system uses a flow control valve. This valve will close down the flow from the crankcase if it gets large enough to keep the engine running.

Diesel EGR Systems are similar to the gasoline engines. Exhaust gases are recirculated into the intake manifold to cool the combustion and reduce nitrous oxides in the exhaust.

The diesel engine creates very little vacuum and a vacuum pump is needed to operate vacuum controls. The Vacuum Regulator Valve (VRV) is used to control vacuum to vary with position of the throttle lever.

An EGR-Vacuum switch cuts out vacuum to the EGR at idle speeds and cuts it in at higher speeds.

Vacuum Regulator Valve Adjustment

1. Remove the air crossovers and install safety covers J-26996-2 or equivalent on the air intakes.
2. Remove the throttle rod from the throttle lever.
3. Loosen the vacuum regulator valve injection pump bolts.
4. Install an adapter BT-7944 or J-26701-15 or equivalent and angle gauge BT-7704 or J-26701 on the throttle lever.
5. Rotate the throttle lever to the wide open position and set the angle gauge to zero.
6. Center the bubble in the gauges level.
7. Set the angle gauge to 50 degrees.
8. Rotate the throttle lever so that the level bubble is centered.
9. Attach an outside vacuum source, to the inboard port of the vacuum valve and install a vacuum gauge on the outside port. Apply 18-22 inches of vacuum.
10. Rotate the vacuum valve clockwise until the vacuum gauge reads 7-8 inches.
11. Tighten the vacuum valve bolts. Remove the vacuum source and gauge.
12. Install the throttle rod to the bellcrank.
13. Remove the covers and install the crossover.

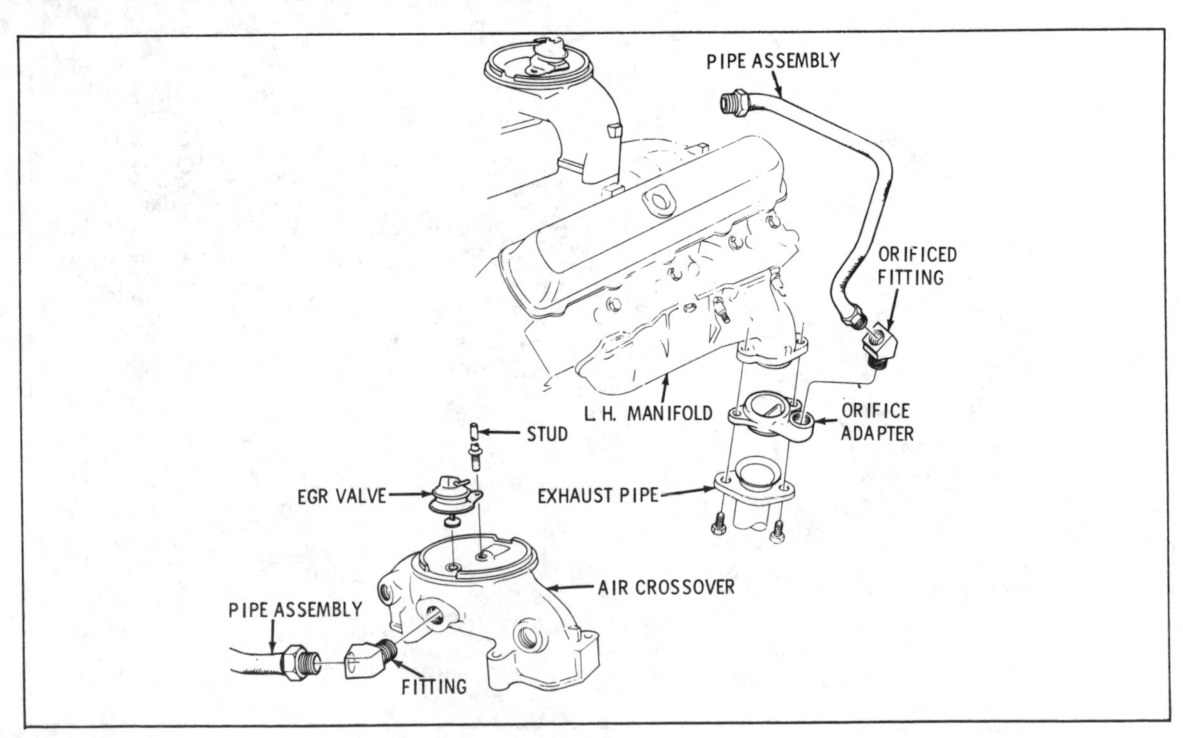

Location of EGR Valve and EGR Orifice (© General Motors Corp.)

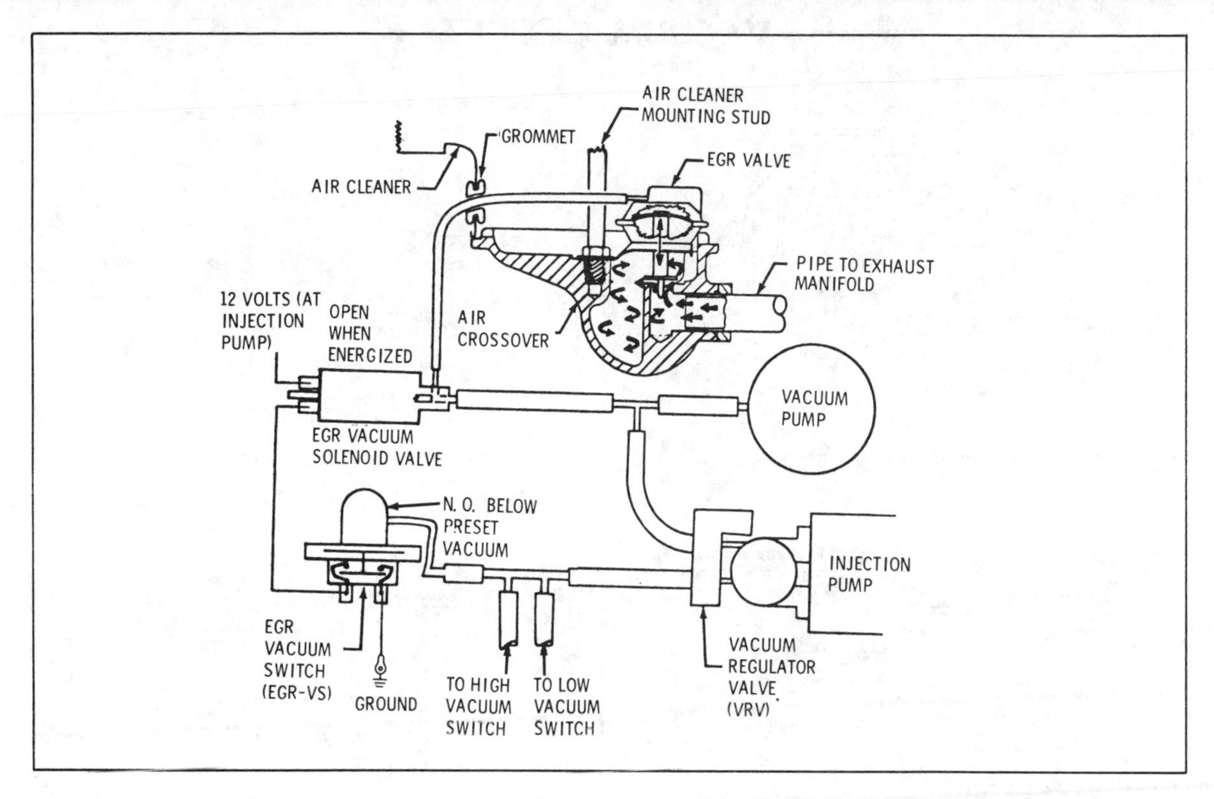

Diesel EGR System (© General Motors Corp.)

Slow Idle Adjustment

1. Bring engine to normal operating temperature.

2. Insert magnetic pick-up tachometer J-26925 or equivalent into the timing indicator hole.
3. With the driving wheels blocked and the parking brake set, adjust the slow idle screw on the injection pump to the specification shown on the emission label.

Fast Idle Adjustment

1. With the engine off disconnect the single green wire from the idle relay located in front of the dash.
2. With the driving wheels blocked and the parking brake set, start the engine and adjust the energized solenoid to the specification listed on the emission label.
3. Turn the ignition off and reconnect the green wire to the relay.

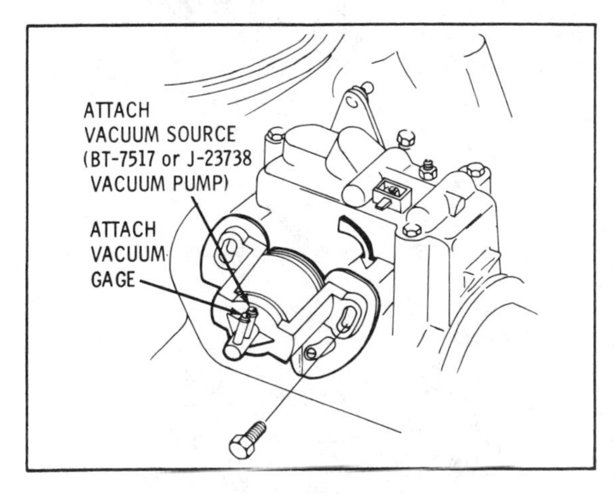

Vacuum ports of the Vacuum Regulator Valve

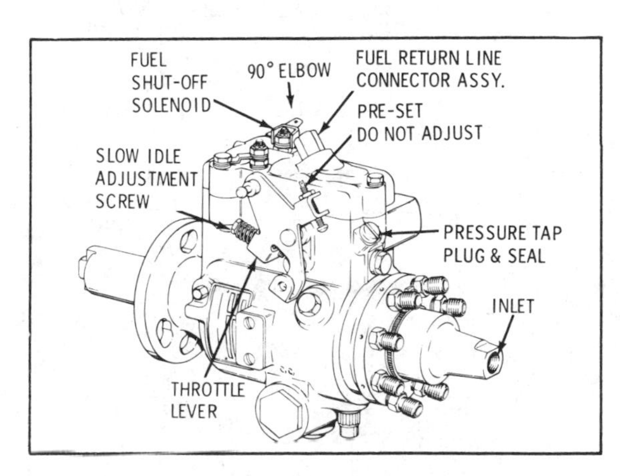

Idle Adjustment Screw

VACUUM CIRCUITS

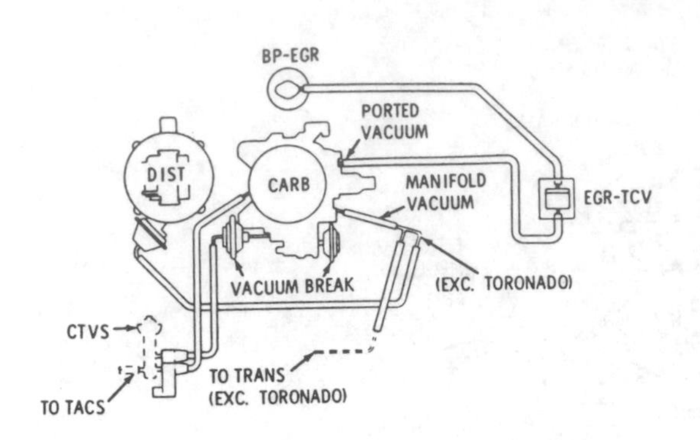

1979 350 V-8 Vin code R and 403 V-8 Vin code K—Except Calif. and Alt.

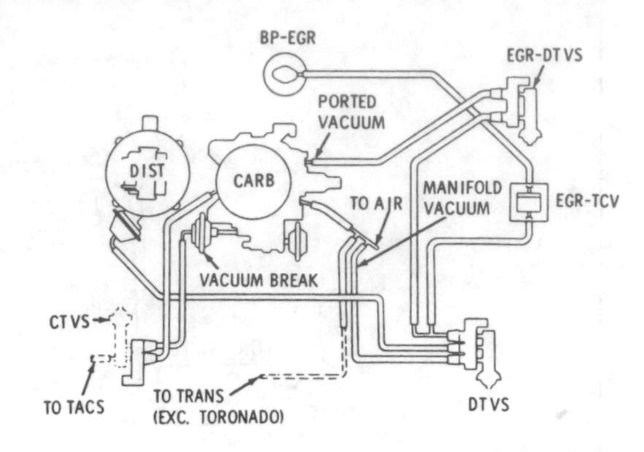

1979 350 V-8 Vin code R and 403 V-8 Vin code K—Calif.

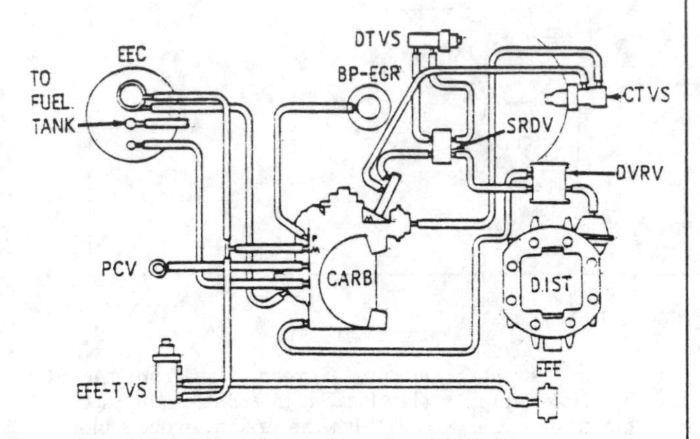

1979 301 V-8—Except Calif. and Alt. (with AC)

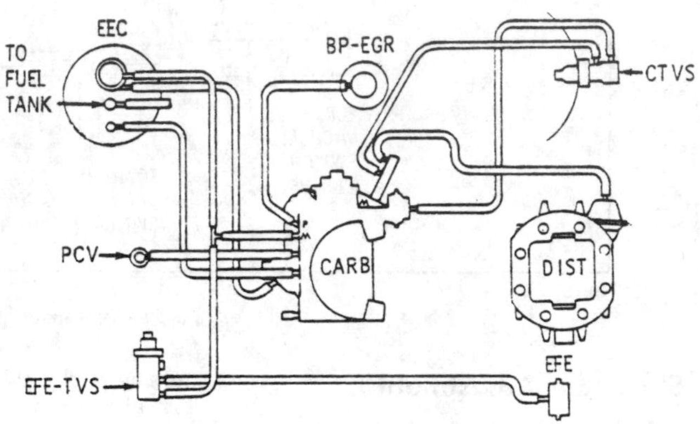

1979 301 V-8—Except Calif. and Alt. (without AC)

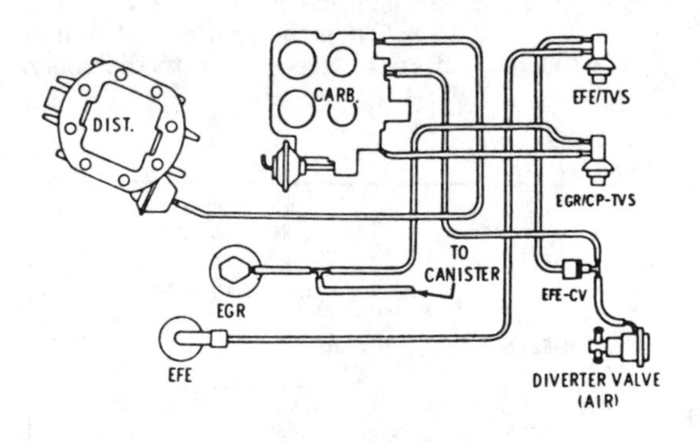

1978 49-States—305 V-8

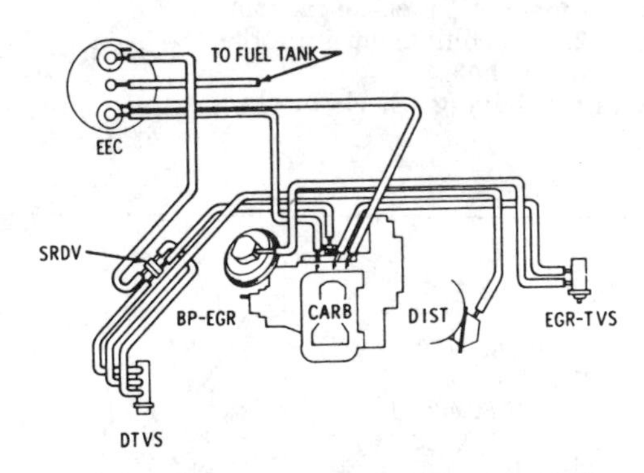

1979 151 4-cyl.—Except Calif. (w/o AC, MT)

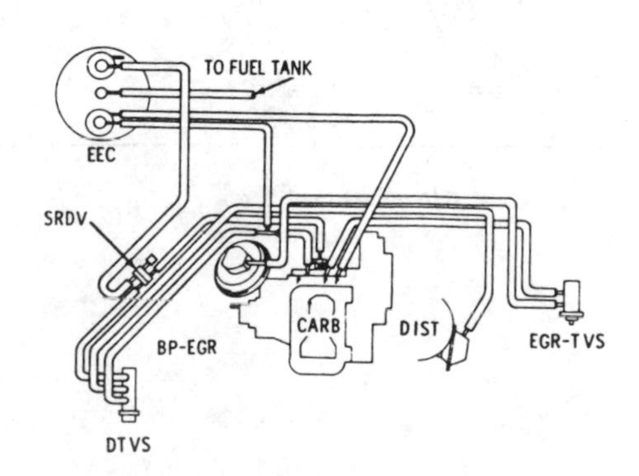

1979 151 4-cyl.—Except Calif. (with AC, AT)

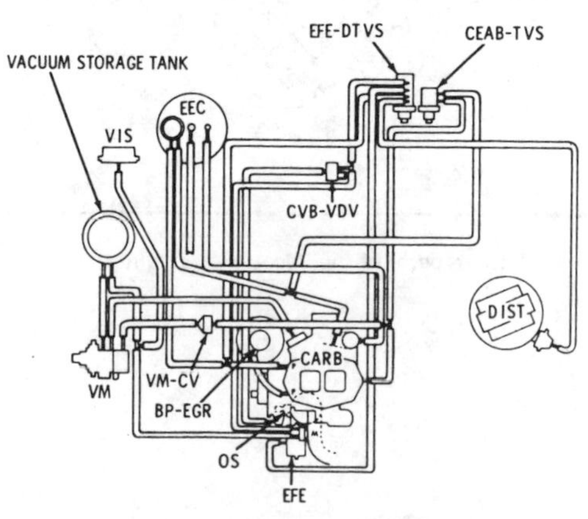

1979 151 4-cyl.—Calif. (with EFC)

VACUUM CIRCUITS

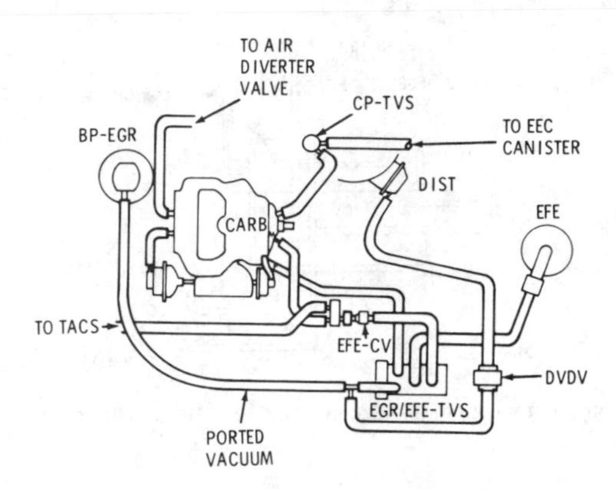

1979 231 V-6—Calif. Starfire (with MT)

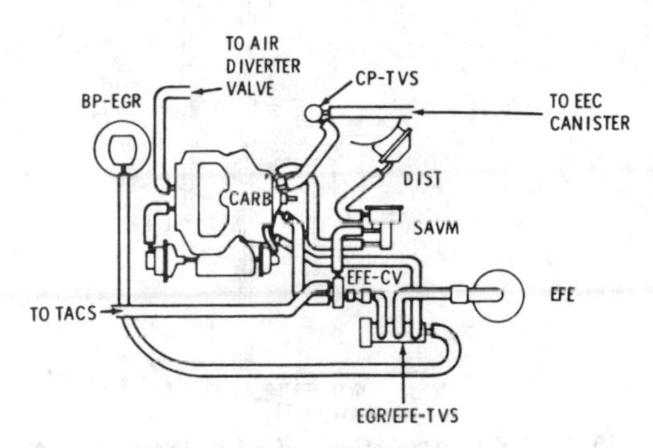

1979 231 V-6—Calif. Omega and 88 models, Alt. Starfire and Cutlass (with AT)

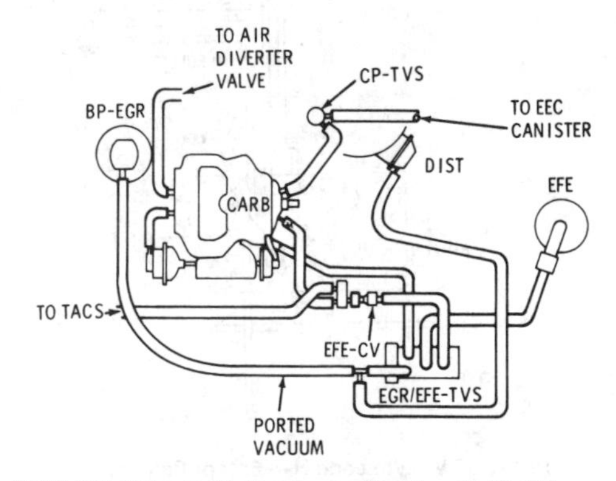

1979 231 V-6—Calif. Starfire and Cutlass (with AT)

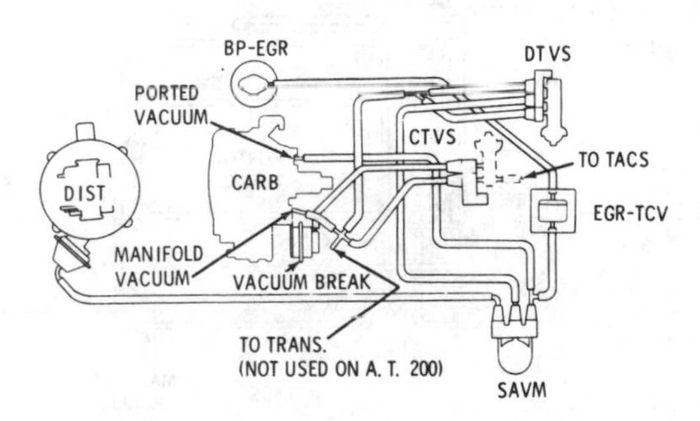

1979 260 V-8—Except Calif. and Alt. (with AT)

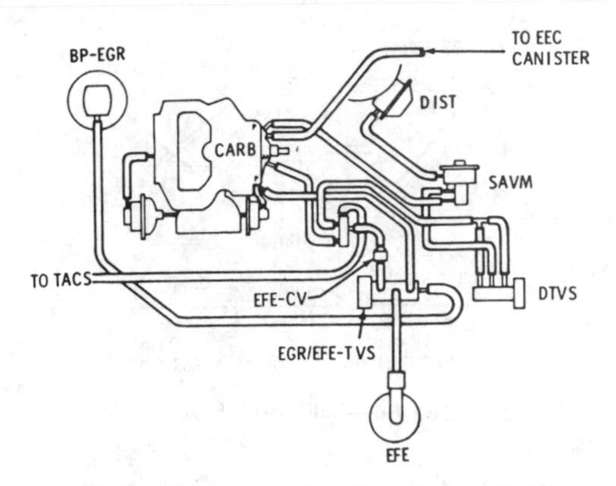

1979 231 V-6—Except Calif. and Alt. (with AT)

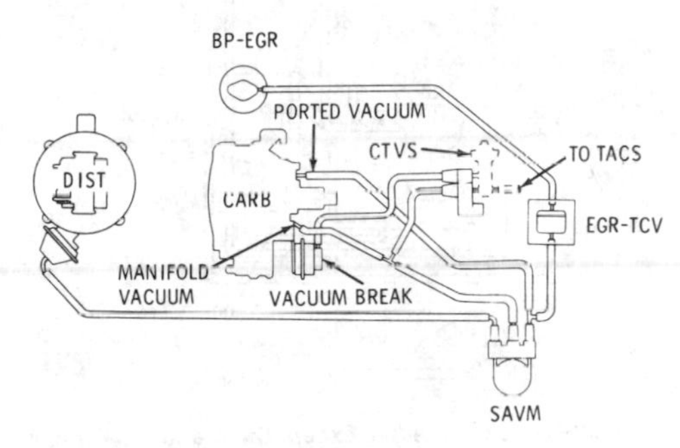

1979 260 V-8--Except Calif. and Alt. (with MT)

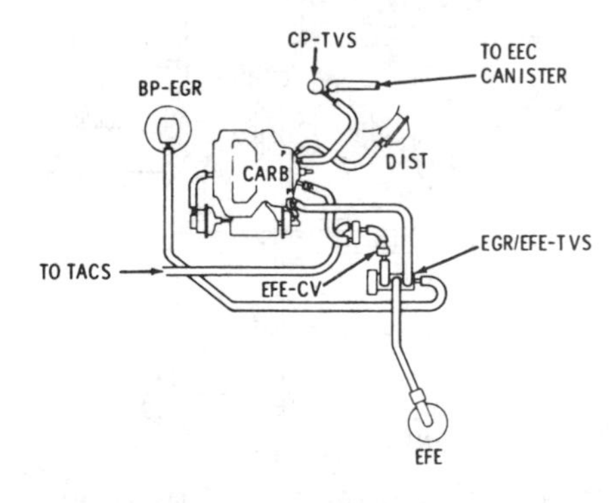

1979 231 V-6—Except Calif. and Alt. (with MT)

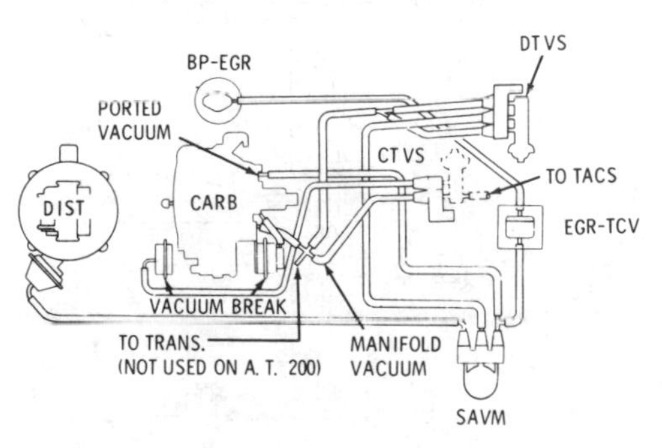

1979 260 V-8—Alt. (With AT)

VACUUM CIRCUITS

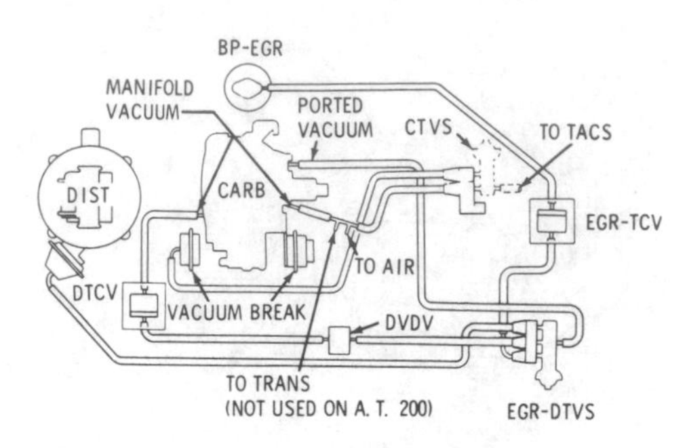

1979 260 V-8—Calif. (with AT)

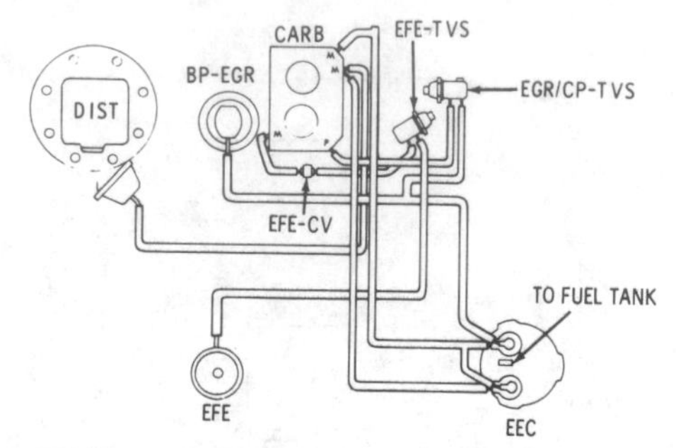

1979 305 V-8 Vin code G—Except Calif. and Alt. Starfire models

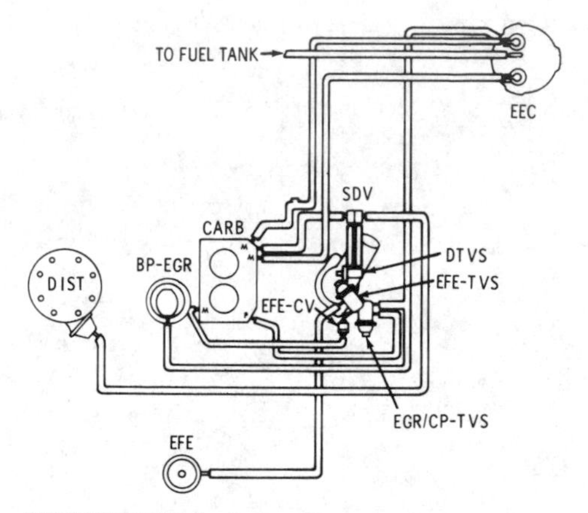

1979 305 V-8 Vin code G—Except Calif. and Alt. Omega and 88 models, except SW

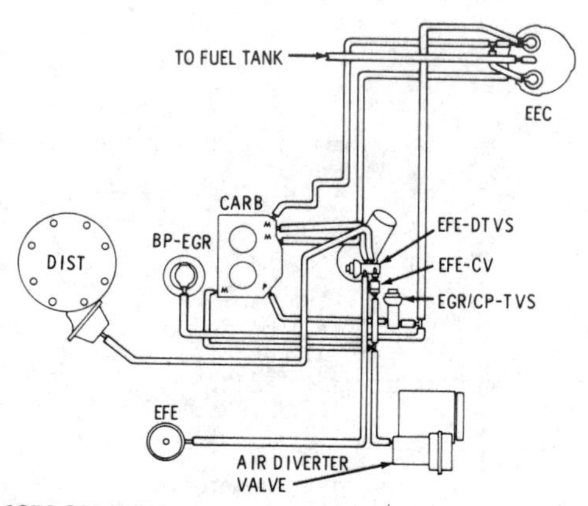

1979 305 V-8 Vin code G—Calif. 88 models, except SW

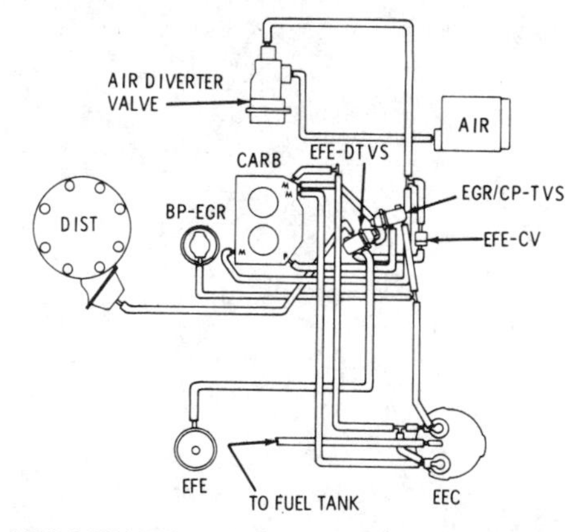

1979 305 V-8 Vin code G—Calif. Starfire models

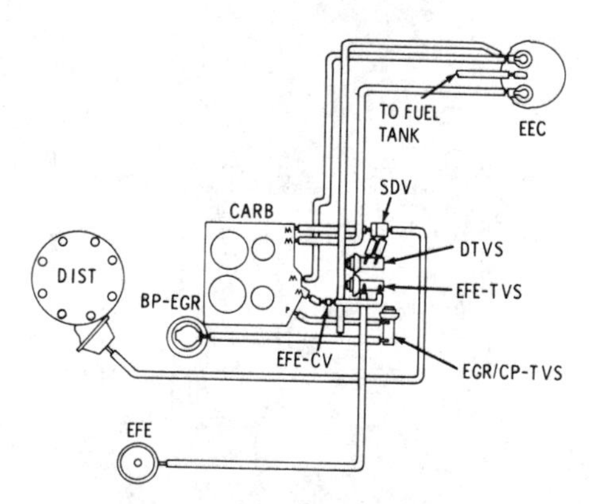

1979 305 V-8 Vin code H—Except Calif. and Alt.

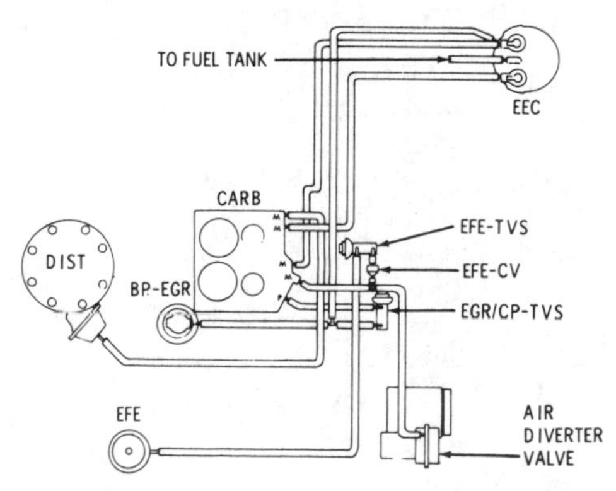

1979 305 V-8 Vin code H and 350 V-8 Vin code L—Calif. and Alt.

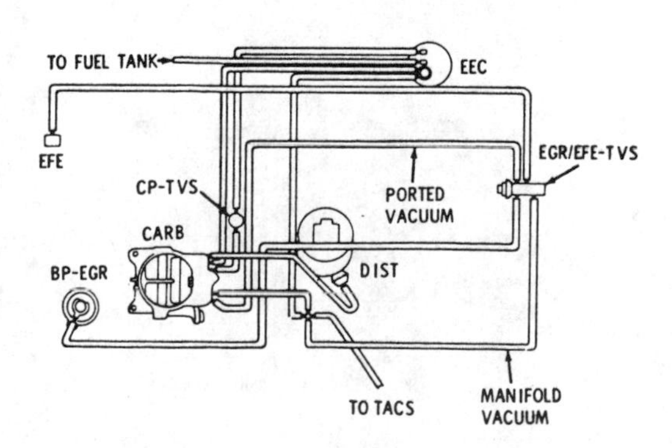

1979 350 V-8 Vin code X—except Calif. and Alt.

VACUUM CIRCUITS

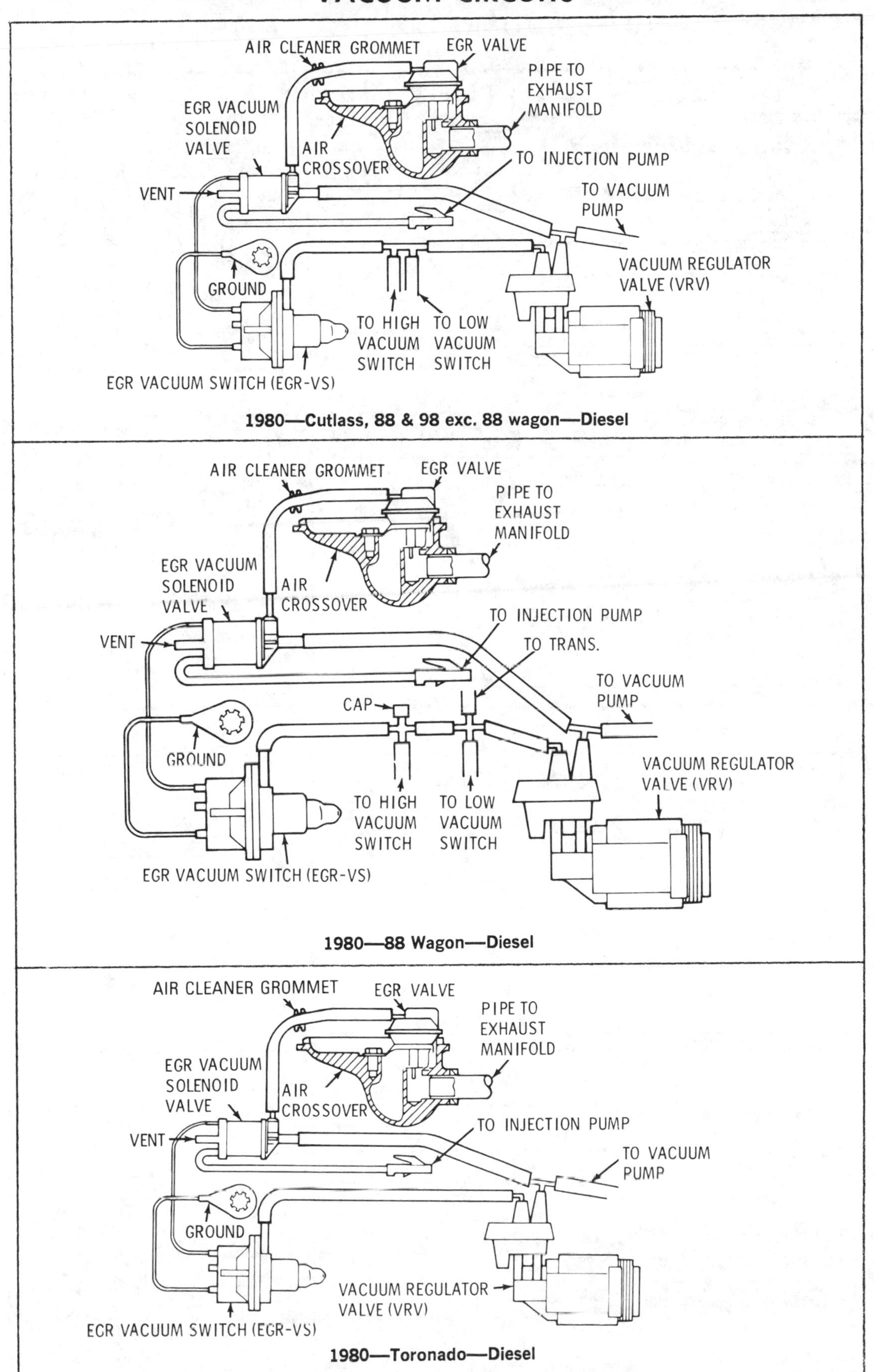

1980—Cutlass, 88 & 98 exc. 88 wagon—Diesel

1980—88 Wagon—Diesel

1980—Toronado—Diesel

VEHICLE EMISSION CONTROL LABELS AND VACUUM HOSE ROUTINGS

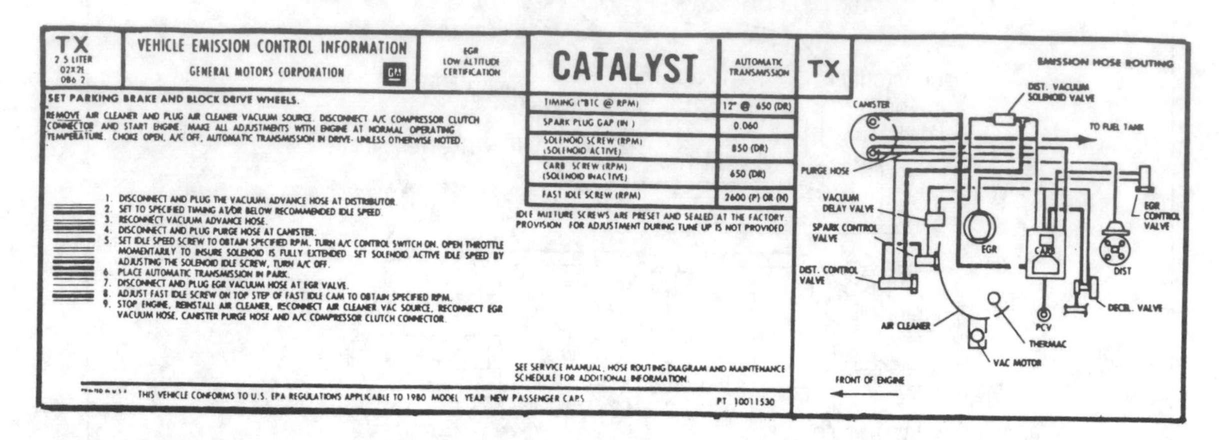

TX
2.5 LITER
02X2E
086-2

VEHICLE EMISSION CONTROL INFORMATION
GENERAL MOTORS CORPORATION

EGR LOW ALTITUDE CERTIFICATION

CATALYST

AUTOMATIC TRANSMISSION

SET PARKING BRAKE AND BLOCK DRIVE WHEELS.

REMOVE AIR CLEANER AND PLUG AIR CLEANER VACUUM SOURCE. DISCONNECT A/C COMPRESSOR CLUTCH CONNECTOR AND START ENGINE. MAKE ALL ADJUSTMENTS WITH ENGINE AT NORMAL OPERATING TEMPERATURE. CHOKE OPEN, A/C OFF, AUTOMATIC TRANSMISSION IN DRIVE- UNLESS OTHERWISE NOTED.

1. DISCONNECT AND PLUG THE VACUUM ADVANCE HOSE AT DISTRIBUTOR.
2. SET TO SPECIFIED TIMING AT/OR BELOW RECOMMENDED IDLE SPEED.
3. RECONNECT VACUUM ADVANCE HOSE.
4. DISCONNECT AND PLUG PURGE HOSE AT CANISTER.
5. SET IDLE SPEED SCREW TO OBTAIN SPECIFIED RPM. TURN A/C CONTROL SWITCH ON. OPEN THROTTLE MOMENTARILY TO INSURE SOLENOID IS FULLY EXTENDED. SET SOLENOID ACTIVE IDLE SPEED BY ADJUSTING THE SOLENOID IDLE SCREW, TURN A/C OFF.
6. PLACE AUTOMATIC TRANSMISSION IN PARK.
7. DISCONNECT AND PLUG EGR VACUUM HOSE AT EGR VALVE.
8. ADJUST FAST IDLE SCREW ON TOP STEP OF FAST IDLE CAM TO OBTAIN SPECIFIED RPM.
9. STOP ENGINE, REINSTALL AIR CLEANER, RECONNECT AIR CLEANER VAC SOURCE, RECONNECT EGR VACUUM HOSE, CANISTER PURGE HOSE AND A/C COMPRESSOR CLUTCH CONNECTOR.

TIMING (°BTC @ RPM)	12° @ 650 (DR)
SPARK PLUG GAP (IN.)	0.060
SOLENOID SCREW (RPM) (SOLENOID ACTIVE)	850 (DR)
CARB SCREW (RPM) (SOLENOID INACTIVE)	650 (DR)
FAST IDLE SCREW (RPM)	2600 (P) OR (N)

IDLE MIXTURE SCREWS ARE PRESET AND SEALED AT THE FACTORY. PROVISION FOR ADJUSTMENT DURING TUNE UP IS NOT PROVIDED.

SEE SERVICE MANUAL, HOSE ROUTING DIAGRAM AND MAINTENANCE SCHEDULE FOR ADDITIONAL INFORMATION.

THIS VEHICLE CONFORMS TO U.S. EPA REGULATIONS APPLICABLE TO 1980 MODEL YEAR NEW PASSENGER CARS.

PT 10011530

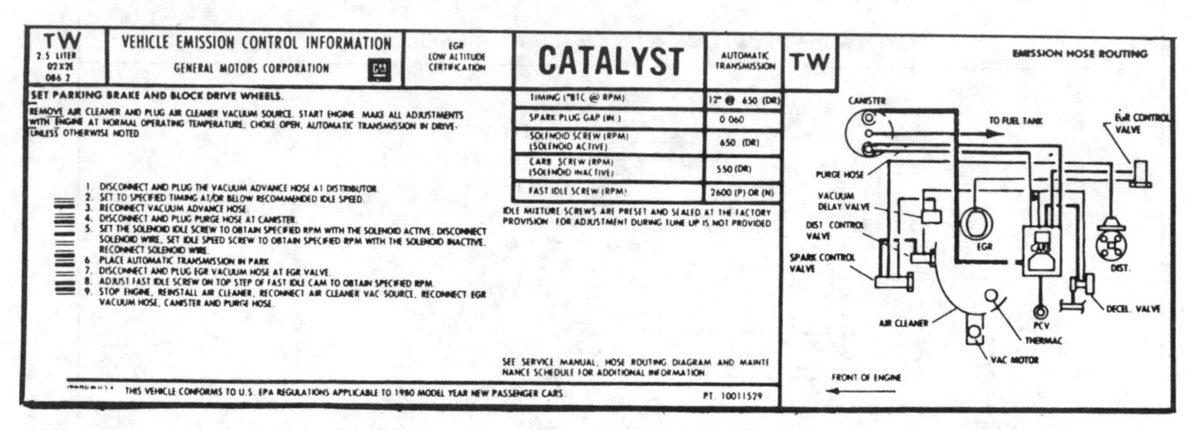

TW
2.5 LITER
02X2E
086-2

VEHICLE EMISSION CONTROL INFORMATION
GENERAL MOTORS CORPORATION

EGR LOW ALTITUDE CERTIFICATION

CATALYST

AUTOMATIC TRANSMISSION

SET PARKING BRAKE AND BLOCK DRIVE WHEELS.

REMOVE AIR CLEANER AND PLUG AIR CLEANER VACUUM SOURCE. START ENGINE. MAKE ALL ADJUSTMENTS WITH ENGINE AT NORMAL OPERATING TEMPERATURE, CHOKE OPEN, AUTOMATIC TRANSMISSION IN DRIVE-UNLESS OTHERWISE NOTED.

1. DISCONNECT AND PLUG THE VACUUM ADVANCE HOSE AT DISTRIBUTOR.
2. SET TO SPECIFIED TIMING AT/OR BELOW RECOMMENDED IDLE SPEED.
3. RECONNECT VACUUM ADVANCE HOSE.
4. DISCONNECT AND PLUG PURGE HOSE AT CANISTER.
5. SET THE SOLENOID IDLE SCREW TO OBTAIN SPECIFIED RPM WITH THE SOLENOID ACTIVE. DISCONNECT SOLENOID WIRE, SET IDLE SPEED SCREW TO OBTAIN SPECIFIED RPM WITH THE SOLENOID INACTIVE. RECONNECT SOLENOID WIRE.
6. PLACE AUTOMATIC TRANSMISSION IN PARK.
7. DISCONNECT AND PLUG EGR VACUUM HOSE AT EGR VALVE.
8. ADJUST FAST IDLE SCREW ON TOP STEP OF FAST IDLE CAM TO OBTAIN SPECIFIED RPM.
9. STOP ENGINE, REINSTALL AIR CLEANER, RECONNECT AIR CLEANER VAC SOURCE, RECONNECT EGR VACUUM HOSE, CANISTER AND PURGE HOSE.

TIMING (°BTC @ RPM)	12° @ 650 (DR)
SPARK PLUG GAP (IN.)	0.060
SOLENOID SCREW (RPM) (SOLENOID ACTIVE)	650 (DR)
CARB SCREW (RPM) (SOLENOID INACTIVE)	550 (DR)
FAST IDLE SCREW (RPM)	2600 (P) OR (N)

IDLE MIXTURE SCREWS ARE PRESET AND SEALED AT THE FACTORY. PROVISION FOR ADJUSTMENT DURING TUNE UP IS NOT PROVIDED.

SEE SERVICE MANUAL, HOSE ROUTING DIAGRAM AND MAINTENANCE SCHEDULE FOR ADDITIONAL INFORMATION.

THIS VEHICLE CONFORMS TO U.S. EPA REGULATIONS APPLICABLE TO 1980 MODEL YEAR NEW PASSENGER CARS.

PT. 10011529

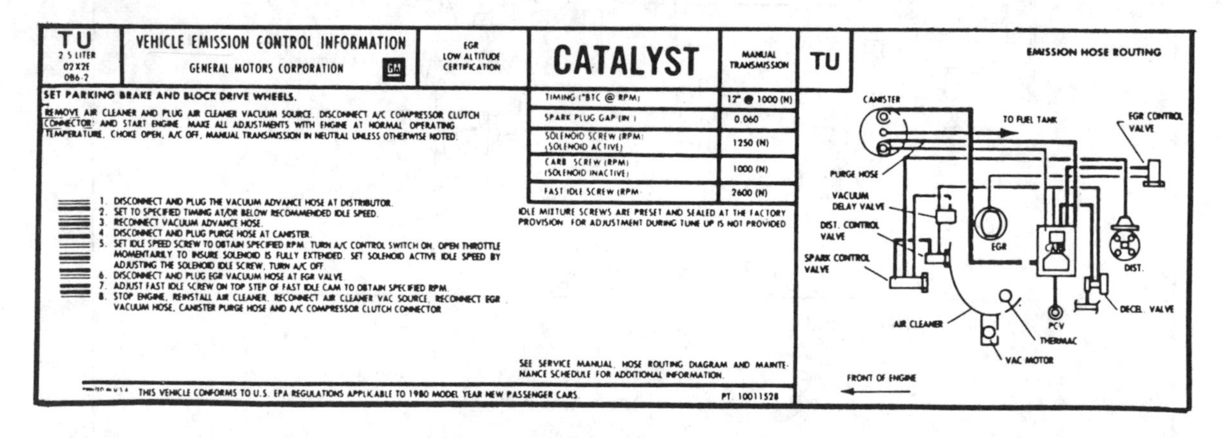

TU
2.5 LITER
02X2E
086-2

VEHICLE EMISSION CONTROL INFORMATION
GENERAL MOTORS CORPORATION

EGR LOW ALTITUDE CERTIFICATION

CATALYST

MANUAL TRANSMISSION

SET PARKING BRAKE AND BLOCK DRIVE WHEELS.

REMOVE AIR CLEANER AND PLUG AIR CLEANER VACUUM SOURCE. DISCONNECT A/C COMPRESSOR CLUTCH CONNECTOR AND START ENGINE. MAKE ALL ADJUSTMENTS WITH ENGINE AT NORMAL OPERATING TEMPERATURE. CHOKE OPEN, A/C OFF, MANUAL TRANSMISSION IN NEUTRAL UNLESS OTHERWISE NOTED.

1. DISCONNECT AND PLUG THE VACUUM ADVANCE HOSE AT DISTRIBUTOR.
2. SET TO SPECIFIED TIMING AT/OR BELOW RECOMMENDED IDLE SPEED.
3. RECONNECT VACUUM ADVANCE HOSE.
4. DISCONNECT AND PLUG PURGE HOSE AT CANISTER.
5. SET IDLE SPEED SCREW TO OBTAIN SPECIFIED RPM. TURN A/C CONTROL SWITCH ON. OPEN THROTTLE MOMENTARILY TO INSURE SOLENOID IS FULLY EXTENDED. SET SOLENOID ACTIVE IDLE SPEED BY ADJUSTING THE SOLENOID IDLE SCREW, TURN A/C OFF.
6. DISCONNECT AND PLUG EGR VACUUM HOSE AT EGR VALVE.
7. ADJUST FAST IDLE SCREW ON TOP STEP OF FAST IDLE CAM TO OBTAIN SPECIFIED RPM.
8. STOP ENGINE, REINSTALL AIR CLEANER, RECONNECT AIR CLEANER VAC SOURCE, RECONNECT EGR VACUUM HOSE, CANISTER PURGE HOSE AND A/C COMPRESSOR CLUTCH CONNECTOR.

TIMING (°BTC @ RPM)	12° @ 1000 (N)
SPARK PLUG GAP (IN.)	0.060
SOLENOID SCREW (RPM) (SOLENOID ACTIVE)	1250 (N)
CARB SCREW (RPM) (SOLENOID INACTIVE)	1000 (N)
FAST IDLE SCREW (RPM)	2600 (N)

IDLE MIXTURE SCREWS ARE PRESET AND SEALED AT THE FACTORY. PROVISION FOR ADJUSTMENT DURING TUNE UP IS NOT PROVIDED.

SEE SERVICE MANUAL, HOSE ROUTING DIAGRAM AND MAINTENANCE SCHEDULE FOR ADDITIONAL INFORMATION.

THIS VEHICLE CONFORMS TO U.S. EPA REGULATIONS APPLICABLE TO 1980 MODEL YEAR NEW PASSENGER CARS.

PT. 10011528

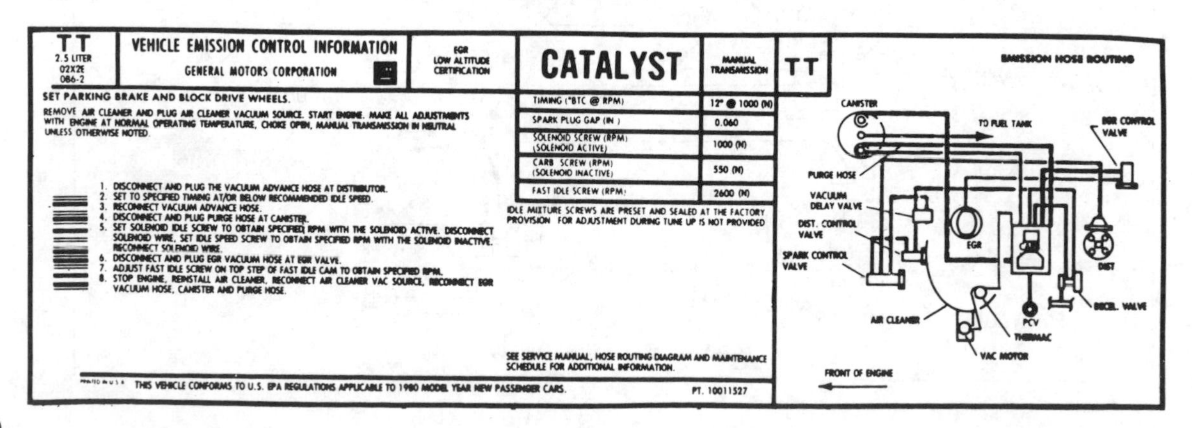

TT
2.5 LITER
02X2E
086-2

VEHICLE EMISSION CONTROL INFORMATION
GENERAL MOTORS CORPORATION

EGR LOW ALTITUDE CERTIFICATION

CATALYST

MANUAL TRANSMISSION

SET PARKING BRAKE AND BLOCK DRIVE WHEELS.

REMOVE AIR CLEANER AND PLUG AIR CLEANER VACUUM SOURCE. START ENGINE. MAKE ALL ADJUSTMENTS WITH ENGINE AT NORMAL OPERATING TEMPERATURE, CHOKE OPEN, MANUAL TRANSMISSION IN NEUTRAL UNLESS OTHERWISE NOTED.

1. DISCONNECT AND PLUG THE VACUUM ADVANCE HOSE AT DISTRIBUTOR.
2. SET TO SPECIFIED TIMING AT/OR BELOW RECOMMENDED IDLE SPEED.
3. RECONNECT VACUUM ADVANCE HOSE.
4. DISCONNECT AND PLUG PURGE HOSE AT CANISTER.
5. SET SOLENOID IDLE SCREW TO OBTAIN SPECIFIED RPM WITH THE SOLENOID ACTIVE. DISCONNECT SOLENOID WIRE, SET IDLE SPEED SCREW TO OBTAIN SPECIFIED RPM WITH THE SOLENOID INACTIVE. RECONNECT SOLENOID WIRE.
6. DISCONNECT AND PLUG EGR VACUUM HOSE AT EGR VALVE.
7. ADJUST FAST IDLE SCREW ON TOP STEP OF FAST IDLE CAM TO OBTAIN SPECIFIED RPM.
8. STOP ENGINE, REINSTALL AIR CLEANER, RECONNECT AIR CLEANER VAC SOURCE, RECONNECT EGR VACUUM HOSE, CANISTER AND PURGE HOSE.

TIMING (°BTC @ RPM)	12° @ 1000 (N)
SPARK PLUG GAP (IN.)	0.060
SOLENOID SCREW (RPM) (SOLENOID ACTIVE)	1000 (N)
CARB SCREW (RPM) (SOLENOID INACTIVE)	550 (N)
FAST IDLE SCREW (RPM)	2600 (N)

IDLE MIXTURE SCREWS ARE PRESET AND SEALED AT THE FACTORY. PROVISION FOR ADJUSTMENT DURING TUNE UP IS NOT PROVIDED.

SEE SERVICE MANUAL, HOSE ROUTING DIAGRAM AND MAINTENANCE SCHEDULE FOR ADDITIONAL INFORMATION.

THIS VEHICLE CONFORMS TO U.S. EPA REGULATIONS APPLICABLE TO 1980 MODEL YEAR NEW PASSENGER CARS.

PT. 10011527

VEHICLE EMISSION CONTROL LABELS AND VACUUM HOSE ROUTINGS

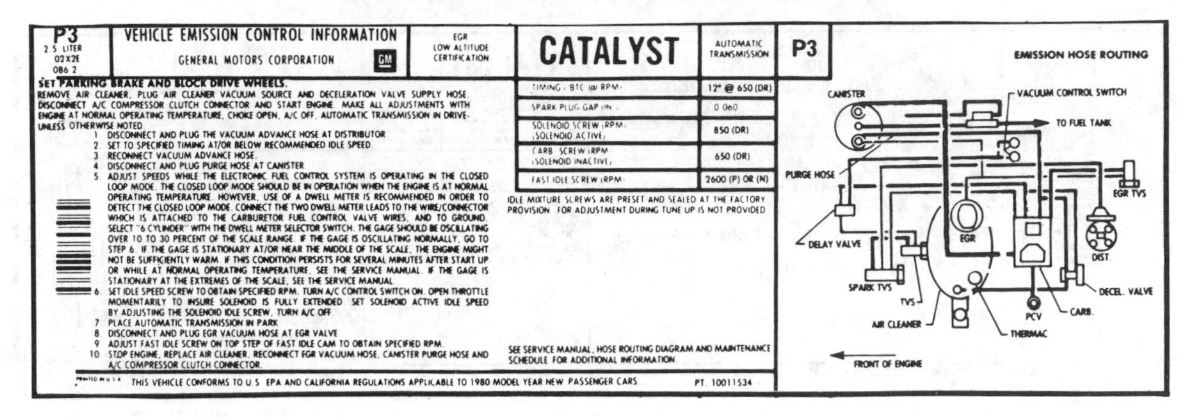

P3 2.5 LITER 02X2E 0B6 2 — VEHICLE EMISSION CONTROL INFORMATION — GENERAL MOTORS CORPORATION — EGR LOW ALTITUDE CERTIFICATION — **CATALYST** — AUTOMATIC TRANSMISSION — **P3**

SET PARKING BRAKE AND BLOCK DRIVE WHEELS.
REMOVE AIR CLEANER, PLUG AIR CLEANER VACUUM SOURCE AND DECELERATION VALVE SUPPLY HOSE. DISCONNECT A/C COMPRESSOR CLUTCH CONNECTOR AND START ENGINE. MAKE ALL ADJUSTMENTS WITH ENGINE AT NORMAL OPERATING TEMPERATURE, CHOKE OPEN, A/C OFF, AUTOMATIC TRANSMISSION IN DRIVE-UNLESS OTHERWISE NOTED.

1. DISCONNECT AND PLUG THE VACUUM ADVANCE HOSE AT DISTRIBUTOR.
2. SET TO SPECIFIED TIMING AT/OR BELOW RECOMMENDED IDLE SPEED.
3. RECONNECT VACUUM ADVANCE HOSE.
4. DISCONNECT AND PLUG PURGE HOSE AT CANISTER.
5. ADJUST SPEEDS WHILE THE ELECTRONIC FUEL CONTROL SYSTEM IS OPERATING IN THE CLOSED LOOP MODE. THE CLOSED LOOP MODE SHOULD BE IN OPERATION WHEN THE ENGINE IS AT NORMAL OPERATING TEMPERATURE. HOWEVER, USE OF A DWELL METER IS RECOMMENDED IN ORDER TO DETECT THE CLOSED LOOP MODE. CONNECT THE TWO DWELL METER LEADS TO THE WIRE/CONNECTOR WHICH IS ATTACHED TO THE CARBURETOR FUEL CONTROL VALVE WIRES, AND TO GROUND. SELECT "6 CYLINDER" WITH THE DWELL METER SELECTOR SWITCH. THE GAGE SHOULD BE OSCILLATING OVER 10 TO 30 PERCENT OF THE SCALE RANGE. IF THE GAGE IS OSCILLATING NORMALLY, GO TO STEP 6. IF THE GAGE IS STATIONARY AT/OR NEAR THE MIDDLE OF THE SCALE, THE ENGINE MIGHT NOT BE SUFFICIENTLY WARM. IF THIS CONDITION PERSISTS FOR SEVERAL MINUTES AFTER START UP OR WHILE AT NORMAL OPERATING TEMPERATURE, SEE THE SERVICE MANUAL. IF THE GAGE IS STATIONARY AT THE EXTREMES OF THE SCALE, SEE THE SERVICE MANUAL.
6. SET IDLE SPEED SCREW TO OBTAIN SPECIFIED RPM. TURN A/C CONTROL SWITCH ON. OPEN THROTTLE MOMENTARILY TO INSURE SOLENOID IS FULLY EXTENDED. SET SOLENOID ACTIVE IDLE SPEED BY ADJUSTING THE SOLENOID IDLE SCREW. TURN A/C OFF.
7. PLACE AUTOMATIC TRANSMISSION IN PARK.
8. DISCONNECT AND PLUG EGR VACUUM HOSE AT EGR VALVE.
9. ADJUST FAST IDLE SCREW ON TOP STEP OF FAST IDLE CAM TO OBTAIN SPECIFIED RPM.
10. STOP ENGINE, REPLACE AIR CLEANER, RECONNECT EGR VACUUM HOSE, CANISTER PURGE HOSE AND A/C COMPRESSOR CLUTCH CONNECTOR.

TIMING (°BTC @ RPM)	12° @ 650 (DR)
SPARK PLUG GAP (IN.)	0.060
SOLENOID SCREW (RPM) (SOLENOID ACTIVE)	850 (DR)
CARB. SCREW (RPM) (SOLENOID INACTIVE)	650 (DR)
FAST IDLE SCREW (RPM)	2600 (P) OR (N)

IDLE MIXTURE SCREWS ARE PRESET AND SEALED AT THE FACTORY. PROVISION FOR ADJUSTMENT DURING TUNE UP IS NOT PROVIDED.

SEE SERVICE MANUAL, HOSE ROUTING DIAGRAM AND MAINTENANCE SCHEDULE FOR ADDITIONAL INFORMATION.

PRINTED IN U.S.A. THIS VEHICLE CONFORMS TO U.S. EPA AND CALIFORNIA REGULATIONS APPLICABLE TO 1980 MODEL YEAR NEW PASSENGER CARS. PT. 10011534

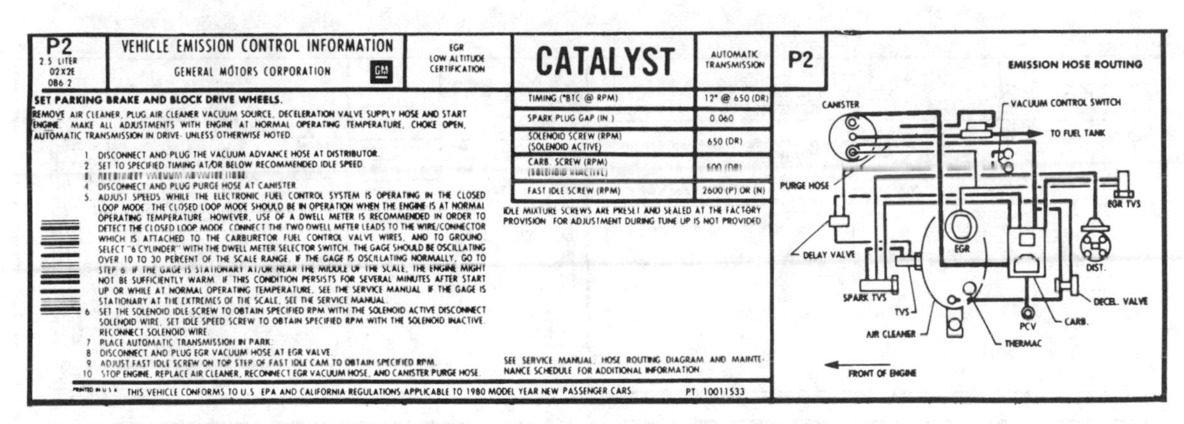

P2 2.5 LITER 02X2E 0B6 2 — VEHICLE EMISSION CONTROL INFORMATION — GENERAL MOTORS CORPORATION — EGR LOW ALTITUDE CERTIFICATION — **CATALYST** — AUTOMATIC TRANSMISSION — **P2**

SET PARKING BRAKE AND BLOCK DRIVE WHEELS.
REMOVE AIR CLEANER, PLUG AIR CLEANER VACUUM SOURCE, DECELERATION VALVE SUPPLY HOSE AND START ENGINE. MAKE ALL ADJUSTMENTS WITH ENGINE AT NORMAL OPERATING TEMPERATURE, CHOKE OPEN, AUTOMATIC TRANSMISSION IN DRIVE- UNLESS OTHERWISE NOTED.

1. DISCONNECT AND PLUG THE VACUUM ADVANCE HOSE AT DISTRIBUTOR.
2. SET TO SPECIFIED TIMING AT/OR BELOW RECOMMENDED IDLE SPEED.
3. RECONNECT VACUUM ADVANCE HOSE.
4. DISCONNECT AND PLUG PURGE HOSE AT CANISTER.
5. ADJUST SPEEDS WHILE THE ELECTRONIC FUEL CONTROL SYSTEM IS OPERATING IN THE CLOSED LOOP MODE. THE CLOSED LOOP MODE SHOULD BE IN OPERATION WHEN THE ENGINE IS AT NORMAL OPERATING TEMPERATURE. HOWEVER, USE OF A DWELL METER IS RECOMMENDED IN ORDER TO DETECT THE CLOSED LOOP MODE. CONNECT THE TWO DWELL METER LEADS TO THE WIRE/CONNECTOR WHICH IS ATTACHED TO THE CARBURETOR FUEL CONTROL VALVE WIRES, AND TO GROUND. SELECT "6 CYLINDER" WITH THE DWELL METER SELECTOR SWITCH. THE GAGE SHOULD BE OSCILLATING OVER 10 TO 30 PERCENT OF THE SCALE RANGE. IF THE GAGE IS OSCILLATING NORMALLY, GO TO STEP 6. IF THE GAGE IS STATIONARY AT/OR NEAR THE MIDDLE OF THE SCALE, THE ENGINE MIGHT NOT BE SUFFICIENTLY WARM. IF THIS CONDITION PERSISTS FOR SEVERAL MINUTES AFTER START UP OR WHILE AT NORMAL OPERATING TEMPERATURE, SEE THE SERVICE MANUAL. IF THE GAGE IS STATIONARY AT THE EXTREMES OF THE SCALE, SEE THE SERVICE MANUAL.
6. SET THE SOLENOID IDLE SCREW TO OBTAIN SPECIFIED RPM WITH THE SOLENOID ACTIVE DISCONNECT SOLENOID WIRE. SET IDLE SPEED SCREW TO OBTAIN SPECIFIED RPM WITH THE SOLENOID INACTIVE. RECONNECT SOLENOID WIRE.
7. PLACE AUTOMATIC TRANSMISSION IN PARK.
8. DISCONNECT AND PLUG EGR VACUUM HOSE AT EGR VALVE.
9. ADJUST FAST IDLE SCREW ON TOP STEP OF FAST IDLE CAM TO OBTAIN SPECIFIED RPM.
10. STOP ENGINE, REPLACE AIR CLEANER, RECONNECT EGR VACUUM HOSE, AND CANISTER PURGE HOSE.

TIMING (°BTC @ RPM)	12° @ 600 (DR)
SPARK PLUG GAP (IN.)	0.060
SOLENOID SCREW (RPM) (SOLENOID ACTIVE)	650 (DR)
CARB. SCREW (RPM) (SOLENOID INACTIVE)	600 (DR)
FAST IDLE SCREW (RPM)	2600 (P) OR (N)

IDLE MIXTURE SCREWS ARE PRESET AND SEALED AT THE FACTORY. PROVISION FOR ADJUSTMENT DURING TUNE UP IS NOT PROVIDED.

SEE SERVICE MANUAL, HOSE ROUTING DIAGRAM AND MAINTENANCE SCHEDULE FOR ADDITIONAL INFORMATION.

PRINTED IN U.S.A. THIS VEHICLE CONFORMS TO U.S. EPA AND CALIFORNIA REGULATIONS APPLICABLE TO 1980 MODEL YEAR NEW PASSENGER CARS. PT. 10011533

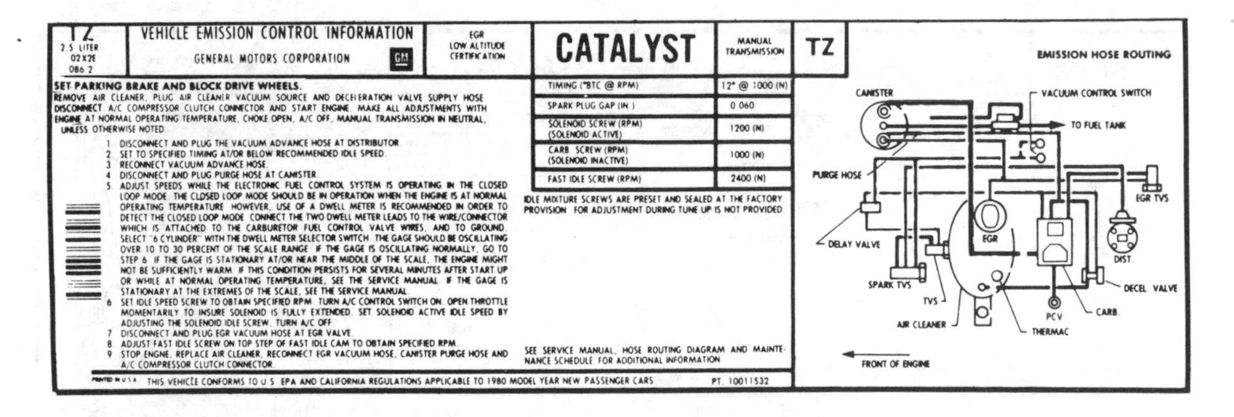

TZ 2.5 LITER 02X2E 086 2 — VEHICLE EMISSION CONTROL INFORMATION — GENERAL MOTORS CORPORATION — EGR LOW ALTITUDE CERTIFICATION — **CATALYST** — MANUAL TRANSMISSION — **TZ**

SET PARKING BRAKE AND BLOCK DRIVE WHEELS.
REMOVE AIR CLEANER, PLUG AIR CLEANER VACUUM SOURCE AND DECELERATION VALVE SUPPLY HOSE. DISCONNECT A/C COMPRESSOR CLUTCH CONNECTOR AND START ENGINE. MAKE ALL ADJUSTMENTS WITH ENGINE AT NORMAL OPERATING TEMPERATURE, CHOKE OPEN, A/C OFF, MANUAL TRANSMISSION IN NEUTRAL, UNLESS OTHERWISE NOTED.

1. DISCONNECT AND PLUG THE VACUUM ADVANCE HOSE AT DISTRIBUTOR.
2. SET TO SPECIFIED TIMING AT/OR BELOW RECOMMENDED IDLE SPEED.
3. RECONNECT VACUUM ADVANCE HOSE.
4. DISCONNECT AND PLUG PURGE HOSE AT CANISTER.
5. ADJUST SPEEDS WHILE THE ELECTRONIC FUEL CONTROL SYSTEM IS OPERATING IN THE CLOSED LOOP MODE. THE CLOSED LOOP MODE SHOULD BE IN OPERATION WHEN THE ENGINE IS AT NORMAL OPERATING TEMPERATURE. HOWEVER, USE OF A DWELL METER IS RECOMMENDED IN ORDER TO DETECT THE CLOSED LOOP MODE. CONNECT THE TWO DWELL METER LEADS TO THE WIRE/CONNECTOR WHICH IS ATTACHED TO THE CARBURETOR FUEL CONTROL VALVE WIRES, AND TO GROUND. SELECT "6 CYLINDER" WITH THE DWELL METER SELECTOR SWITCH. THE GAGE SHOULD BE OSCILLATING OVER 10 TO 30 PERCENT OF THE SCALE RANGE. IF THE GAGE IS OSCILLATING NORMALLY, GO TO STEP 6. IF THE GAGE IS STATIONARY AT/OR NEAR THE MIDDLE OF THE SCALE, THE ENGINE MIGHT NOT BE SUFFICIENTLY WARM. IF THIS CONDITION PERSISTS FOR SEVERAL MINUTES AFTER START UP OR WHILE AT NORMAL OPERATING TEMPERATURE, SEE THE SERVICE MANUAL. IF THE GAGE IS STATIONARY AT THE EXTREMES OF THE SCALE, SEE THE SERVICE MANUAL.
6. SET IDLE SPEED SCREW TO OBTAIN SPECIFIED RPM. TURN A/C CONTROL SWITCH ON. OPEN THROTTLE MOMENTARILY TO INSURE SOLENOID IS FULLY EXTENDED. SET SOLENOID ACTIVE IDLE SPEED BY ADJUSTING THE SOLENOID IDLE SCREW. TURN A/C OFF.
7. DISCONNECT AND PLUG EGR VACUUM HOSE AT EGR VALVE.
8. ADJUST FAST IDLE SCREW ON TOP STEP OF FAST IDLE CAM TO OBTAIN SPECIFIED RPM.
9. STOP ENGINE, REPLACE AIR CLEANER, RECONNECT EGR VACUUM HOSE, CANISTER PURGE HOSE AND A/C COMPRESSOR CLUTCH CONNECTOR.

TIMING (°BTC @ RPM)	12° @ 1000 (N)
SPARK PLUG GAP (IN.)	0.060
SOLENOID SCREW (RPM) (SOLENOID ACTIVE)	1200 (N)
CARB. SCREW (RPM) (SOLENOID INACTIVE)	1000 (N)
FAST IDLE SCREW (RPM)	2400 (N)

IDLE MIXTURE SCREWS ARE PRESET AND SEALED AT THE FACTORY. PROVISION FOR ADJUSTMENT DURING TUNE UP IS NOT PROVIDED.

SEE SERVICE MANUAL, HOSE ROUTING DIAGRAM AND MAINTENANCE SCHEDULE FOR ADDITIONAL INFORMATION.

PRINTED IN U.S.A. THIS VEHICLE CONFORMS TO U.S. EPA AND CALIFORNIA REGULATIONS APPLICABLE TO 1980 MODEL YEAR NEW PASSENGER CARS. PT. 10011532

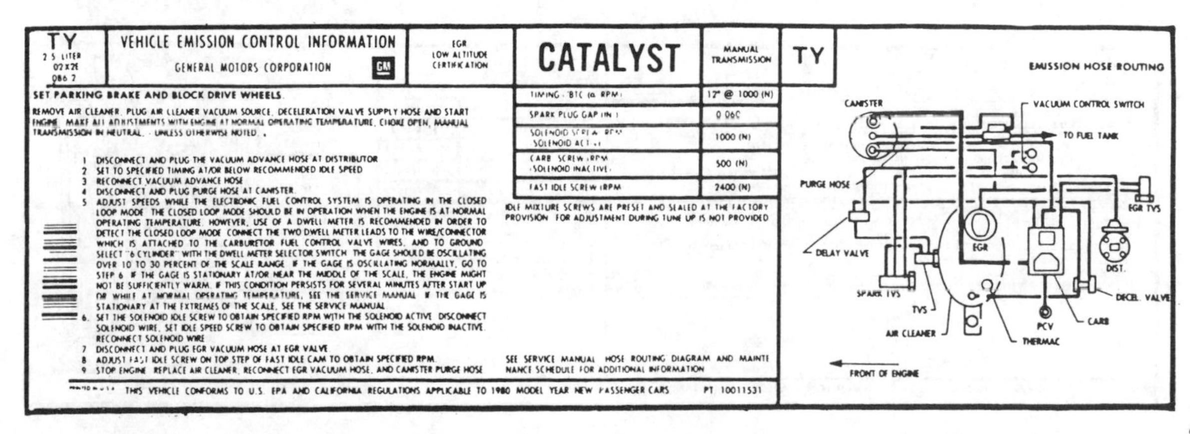

TY 2.5 LITER 02X2E 086 2 — VEHICLE EMISSION CONTROL INFORMATION — GENERAL MOTORS CORPORATION — EGR LOW ALTITUDE CERTIFICATION — **CATALYST** — MANUAL TRANSMISSION — **TY**

SET PARKING BRAKE AND BLOCK DRIVE WHEELS.
REMOVE AIR CLEANER, PLUG AIR CLEANER VACUUM SOURCE, DECELERATION VALVE SUPPLY HOSE AND START ENGINE. MAKE ALL ADJUSTMENTS WITH ENGINE AT NORMAL OPERATING TEMPERATURE, CHOKE OPEN, MANUAL TRANSMISSION IN NEUTRAL - UNLESS OTHERWISE NOTED.

1. DISCONNECT AND PLUG THE VACUUM ADVANCE HOSE AT DISTRIBUTOR.
2. SET TO SPECIFIED TIMING AT/OR BELOW RECOMMENDED IDLE SPEED.
3. RECONNECT VACUUM ADVANCE HOSE.
4. DISCONNECT AND PLUG PURGE HOSE AT CANISTER.
5. ADJUST SPEEDS WHILE THE ELECTRONIC FUEL CONTROL SYSTEM IS OPERATING IN THE CLOSED LOOP MODE. THE CLOSED LOOP MODE SHOULD BE IN OPERATION WHEN THE ENGINE IS AT NORMAL OPERATING TEMPERATURE. HOWEVER, USE OF A DWELL METER IS RECOMMENDED IN ORDER TO DETECT THE CLOSED LOOP MODE. CONNECT THE TWO DWELL METER LEADS TO THE WIRE/CONNECTOR WHICH IS ATTACHED TO THE CARBURETOR FUEL CONTROL VALVE WIRES, AND TO GROUND. SELECT "6 CYLINDER" WITH THE DWELL METER SELECTOR SWITCH. THE GAGE SHOULD BE OSCILLATING OVER 10 TO 30 PERCENT OF THE SCALE RANGE. IF THE GAGE IS OSCILLATING NORMALLY, GO TO STEP 6. IF THE GAGE IS STATIONARY AT/OR NEAR THE MIDDLE OF THE SCALE, THE ENGINE MIGHT NOT BE SUFFICIENTLY WARM. IF THIS CONDITION PERSISTS FOR SEVERAL MINUTES AFTER START UP OR WHILE AT NORMAL OPERATING TEMPERATURE, SEE THE SERVICE MANUAL. IF THE GAGE IS STATIONARY AT THE EXTREMES OF THE SCALE, SEE THE SERVICE MANUAL.
6. SET THE SOLENOID IDLE SCREW TO OBTAIN SPECIFIED RPM WITH THE SOLENOID ACTIVE. DISCONNECT SOLENOID WIRE. SET IDLE SCREW TO OBTAIN SPECIFIED RPM WITH THE SOLENOID INACTIVE. RECONNECT SOLENOID WIRE.
7. DISCONNECT AND PLUG EGR VACUUM HOSE AT EGR VALVE.
8. ADJUST FAST IDLE SCREW ON TOP STEP OF FAST IDLE CAM TO OBTAIN SPECIFIED RPM.
9. STOP ENGINE. REPLACE AIR CLEANER, RECONNECT EGR VACUUM HOSE, AND CANISTER PURGE HOSE.

TIMING (°BTC @ RPM)	12° @ 1000 (N)
SPARK PLUG GAP (IN.)	0.060
SOLENOID SCREW (RPM) (SOLENOID ACT[illegible])	1000 (N)
CARB. SCREW (RPM) (SOLENOID INACTIVE)	500 (N)
FAST IDLE SCREW (RPM)	2400 (N)

IDLE MIXTURE SCREWS ARE PRESET AND SEALED AT THE FACTORY. PROVISION FOR ADJUSTMENT DURING TUNE UP IS NOT PROVIDED.

SEE SERVICE MANUAL, HOSE ROUTING DIAGRAM AND MAINTENANCE SCHEDULE FOR ADDITIONAL INFORMATION.

PRINTED IN U.S.A. THIS VEHICLE CONFORMS TO U.S. EPA AND CALIFORNIA REGULATIONS APPLICABLE TO 1980 MODEL YEAR NEW PASSENGER CARS. PT. 10011531

VEHICLE EMISSION CONTROL LABELS AND VACUUM HOSE ROUTINGS

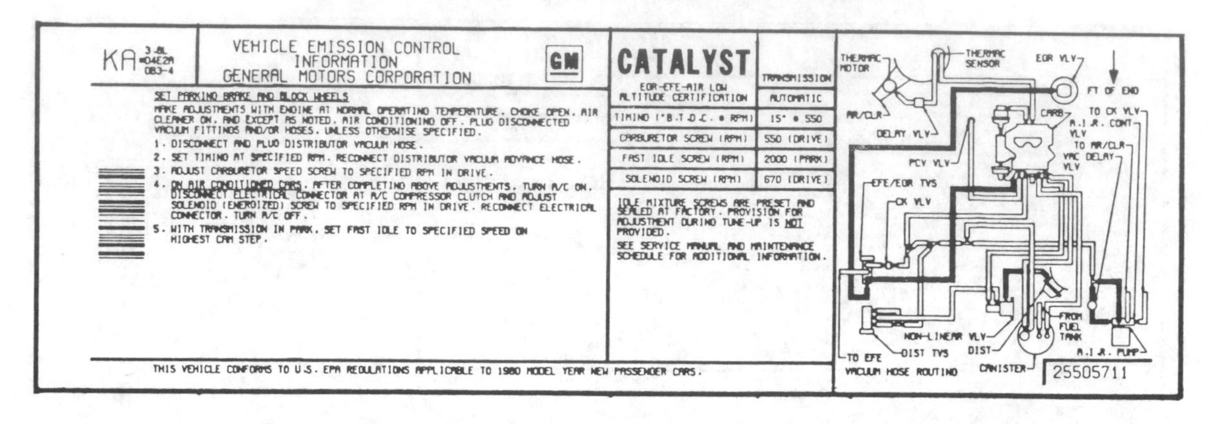

KA 3.8L =04E2A 083-4

VEHICLE EMISSION CONTROL INFORMATION
GENERAL MOTORS CORPORATION — GM

SET PARKING BRAKE AND BLOCK WHEELS

MAKE ADJUSTMENTS WITH ENGINE AT NORMAL OPERATING TEMPERATURE, CHOKE OPEN, AIR CLEANER ON, AND EXCEPT AS NOTED, AIR CONDITIONING OFF. PLUG DISCONNECTED VACUUM FITTINGS AND/OR HOSES, UNLESS OTHERWISE SPECIFIED.

1. DISCONNECT AND PLUG DISTRIBUTOR VACUUM HOSE.
2. SET TIMING AT SPECIFIED RPM. RECONNECT DISTRIBUTOR VACUUM ADVANCE HOSE.
3. ADJUST CARBURETOR SPEED SCREW TO SPECIFIED RPM IN DRIVE.
4. ON AIR CONDITIONED CARS, AFTER COMPLETING ABOVE ADJUSTMENTS, TURN A/C ON. DISCONNECT ELECTRICAL CONNECTOR AT A/C COMPRESSOR CLUTCH AND ADJUST SOLENOID (ENERGIZED) SCREW TO SPECIFIED RPM IN DRIVE. RECONNECT ELECTRICAL CONNECTOR. TURN A/C OFF.
5. WITH TRANSMISSION IN PARK, SET FAST IDLE TO SPECIFIED SPEED ON HIGHEST CAM STEP.

CATALYST

EGR-EFE-AIR LOW ALTITUDE CERTIFICATION	TRANSMISSION AUTOMATIC
TIMING (°B.T.D.C. @ RPM)	15° @ 550
CARBURETOR SCREW (RPM)	550 (DRIVE)
FAST IDLE SCREW (RPM)	2000 (PARK)
SOLENOID SCREW (RPM)	670 (DRIVE)

IDLE MIXTURE SCREWS ARE PRESET AND SEALED AT FACTORY. PROVISION FOR ADJUSTMENT DURING TUNE-UP IS NOT PROVIDED.

SEE SERVICE MANUAL AND MAINTENANCE SCHEDULE FOR ADDITIONAL INFORMATION.

THIS VEHICLE CONFORMS TO U.S. EPA REGULATIONS APPLICABLE TO 1980 MODEL YEAR NEW PASSENGER CARS.

25505711

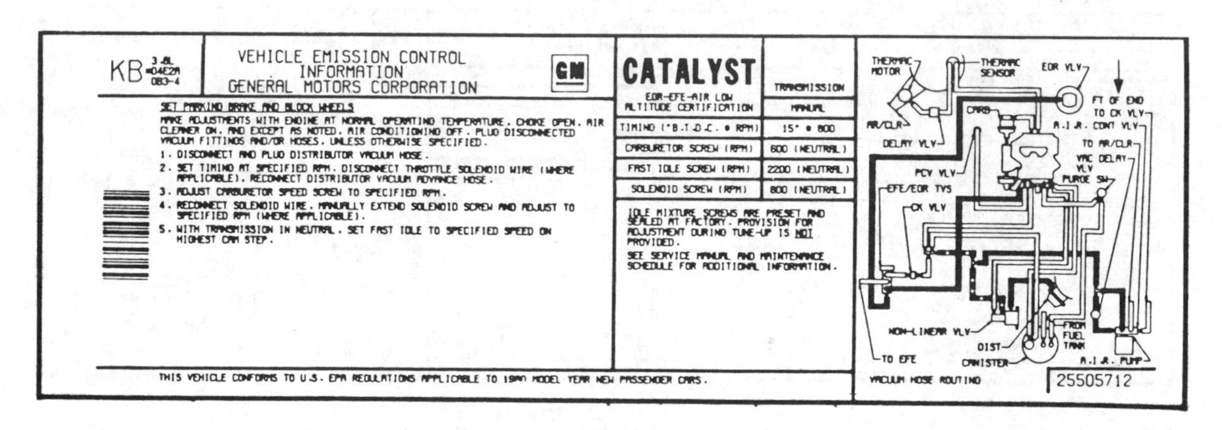

KB 3.8L =04E2A 083-4

VEHICLE EMISSION CONTROL INFORMATION
GENERAL MOTORS CORPORATION — GM

SET PARKING BRAKE AND BLOCK WHEELS

MAKE ADJUSTMENTS WITH ENGINE AT NORMAL OPERATING TEMPERATURE, CHOKE OPEN, AIR CLEANER ON, AND EXCEPT AS NOTED, AIR CONDITIONING OFF. PLUG DISCONNECTED VACUUM FITTINGS AND/OR HOSES, UNLESS OTHERWISE SPECIFIED.

1. DISCONNECT AND PLUG DISTRIBUTOR VACUUM HOSE.
2. SET TIMING AT SPECIFIED RPM. DISCONNECT THROTTLE SOLENOID WIRE (WHERE APPLICABLE). RECONNECT DISTRIBUTOR VACUUM ADVANCE HOSE.
3. ADJUST CARBURETOR SPEED SCREW TO SPECIFIED RPM.
4. RECONNECT SOLENOID WIRE. MANUALLY EXTEND SOLENOID SCREW AND ADJUST TO SPECIFIED RPM (WHERE APPLICABLE).
5. WITH TRANSMISSION IN NEUTRAL, SET FAST IDLE TO SPECIFIED SPEED ON HIGHEST CAM STEP.

CATALYST

EGR-EFE-AIR LOW ALTITUDE CERTIFICATION	TRANSMISSION MANUAL
TIMING (°B.T.D.C. @ RPM)	15° @ 800
CARBURETOR SCREW (RPM)	600 (NEUTRAL)
FAST IDLE SCREW (RPM)	2200 (NEUTRAL)
SOLENOID SCREW (RPM)	800 (NEUTRAL)

IDLE MIXTURE SCREWS ARE PRESET AND SEALED AT FACTORY. PROVISION FOR ADJUSTMENT DURING TUNE-UP IS NOT PROVIDED.

SEE SERVICE MANUAL AND MAINTENANCE SCHEDULE FOR ADDITIONAL INFORMATION.

THIS VEHICLE CONFORMS TO U.S. EPA REGULATIONS APPLICABLE TO 1980 MODEL YEAR NEW PASSENGER CARS.

25505712

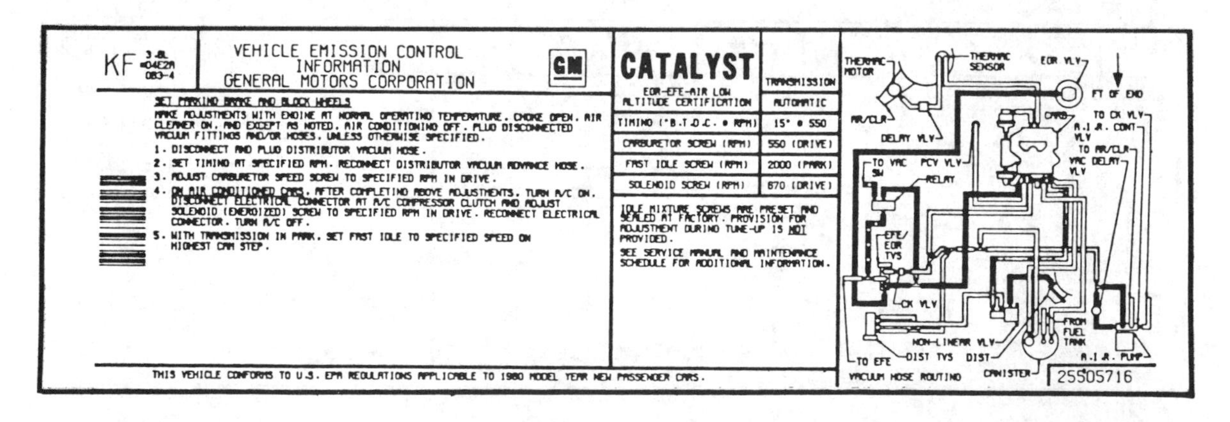

KF 3.8L =04E2A 083-4

VEHICLE EMISSION CONTROL INFORMATION
GENERAL MOTORS CORPORATION — GM

SET PARKING BRAKE AND BLOCK WHEELS

MAKE ADJUSTMENTS WITH ENGINE AT NORMAL OPERATING TEMPERATURE, CHOKE OPEN, AIR CLEANER ON, AND EXCEPT AS NOTED, AIR CONDITIONING OFF. PLUG DISCONNECTED VACUUM FITTINGS AND/OR HOSES, UNLESS OTHERWISE SPECIFIED.

1. DISCONNECT AND PLUG DISTRIBUTOR VACUUM HOSE.
2. SET TIMING AT SPECIFIED RPM. RECONNECT DISTRIBUTOR VACUUM ADVANCE HOSE.
3. ADJUST CARBURETOR SPEED SCREW TO SPECIFIED RPM IN DRIVE.
4. ON AIR CONDITIONED CARS, AFTER COMPLETING ABOVE ADJUSTMENTS, TURN A/C ON. DISCONNECT ELECTRICAL CONNECTOR AT A/C COMPRESSOR CLUTCH AND ADJUST SOLENOID (ENERGIZED) SCREW TO SPECIFIED RPM IN DRIVE. RECONNECT ELECTRICAL CONNECTOR. TURN A/C OFF.
5. WITH TRANSMISSION IN PARK, SET FAST IDLE TO SPECIFIED SPEED ON HIGHEST CAM STEP.

CATALYST

EGR-EFE-AIR LOW ALTITUDE CERTIFICATION	TRANSMISSION AUTOMATIC
TIMING (°B.T.D.C. @ RPM)	15° @ 550
CARBURETOR SCREW (RPM)	550 (DRIVE)
FAST IDLE SCREW (RPM)	2000 (PARK)
SOLENOID SCREW (RPM)	670 (DRIVE)

IDLE MIXTURE SCREWS ARE PRESET AND SEALED AT FACTORY. PROVISION FOR ADJUSTMENT DURING TUNE-UP IS NOT PROVIDED.

SEE SERVICE MANUAL AND MAINTENANCE SCHEDULE FOR ADDITIONAL INFORMATION.

THIS VEHICLE CONFORMS TO U.S. EPA REGULATIONS APPLICABLE TO 1980 MODEL YEAR NEW PASSENGER CARS.

25505716

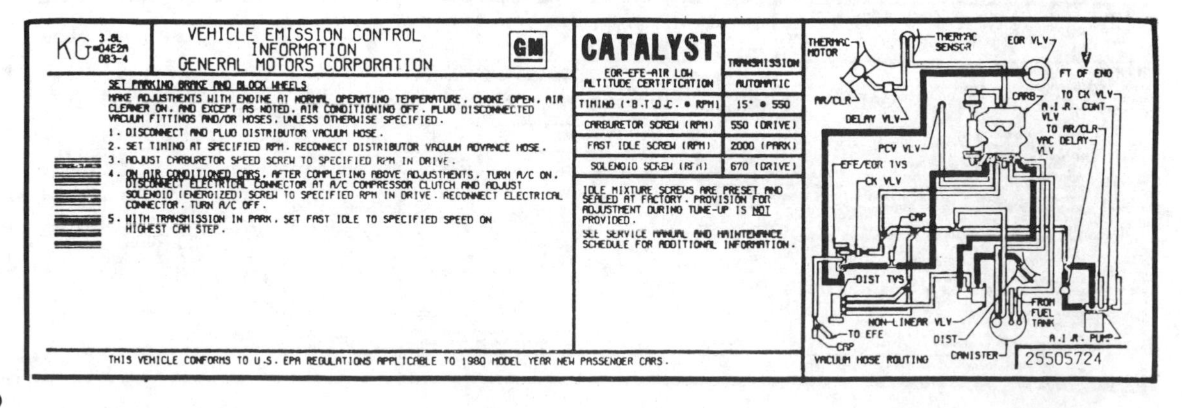

KG 3.8L =04E2A 083-4

VEHICLE EMISSION CONTROL INFORMATION
GENERAL MOTORS CORPORATION — GM

SET PARKING BRAKE AND BLOCK WHEELS

MAKE ADJUSTMENTS WITH ENGINE AT NORMAL OPERATING TEMPERATURE, CHOKE OPEN, AIR CLEANER ON, AND EXCEPT AS NOTED, AIR CONDITIONING OFF. PLUG DISCONNECTED VACUUM FITTINGS AND/OR HOSES, UNLESS OTHERWISE SPECIFIED.

1. DISCONNECT AND PLUG DISTRIBUTOR VACUUM HOSE.
2. SET TIMING AT SPECIFIED RPM. RECONNECT DISTRIBUTOR VACUUM ADVANCE HOSE.
3. ADJUST CARBURETOR SPEED SCREW TO SPECIFIED RPM IN DRIVE.
4. ON AIR CONDITIONED CARS, AFTER COMPLETING ABOVE ADJUSTMENTS, TURN A/C ON. DISCONNECT ELECTRICAL CONNECTOR AT A/C COMPRESSOR CLUTCH AND ADJUST SOLENOID (ENERGIZED) SCREW TO SPECIFIED RPM IN DRIVE. RECONNECT ELECTRICAL CONNECTOR. TURN A/C OFF.
5. WITH TRANSMISSION IN PARK, SET FAST IDLE TO SPECIFIED SPEED ON HIGHEST CAM STEP.

CATALYST

EGR-EFE-AIR LOW ALTITUDE CERTIFICATION	TRANSMISSION AUTOMATIC
TIMING (°B.T.D.C. @ RPM)	15° @ 550
CARBURETOR SCREW (RPM)	550 (DRIVE)
FAST IDLE SCREW (RPM)	2000 (PARK)
SOLENOID SCREW (RPM)	670 (DRIVE)

IDLE MIXTURE SCREWS ARE PRESET AND SEALED AT FACTORY. PROVISION FOR ADJUSTMENT DURING TUNE-UP IS NOT PROVIDED.

SEE SERVICE MANUAL AND MAINTENANCE SCHEDULE FOR ADDITIONAL INFORMATION.

THIS VEHICLE CONFORMS TO U.S. EPA REGULATIONS APPLICABLE TO 1980 MODEL YEAR NEW PASSENGER CARS.

25505724

VEHICLE EMISSION CONTROL LABELS AND VACUUM HOSE ROUTINGS

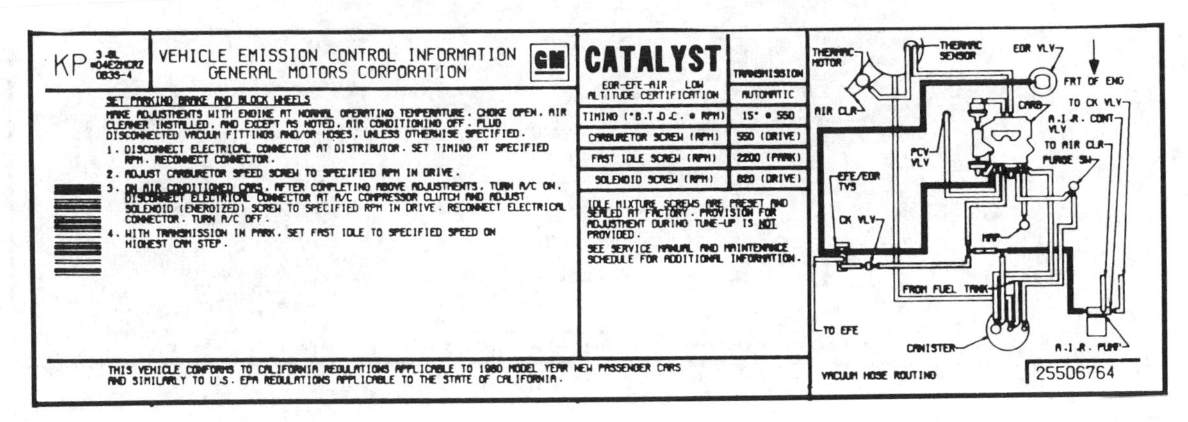

KP 3.8L #04E2HCRZ 0835-4

VEHICLE EMISSION CONTROL INFORMATION
GENERAL MOTORS CORPORATION
GM

SET PARKING BRAKE AND BLOCK WHEELS

MAKE ADJUSTMENTS WITH ENGINE AT NORMAL OPERATING TEMPERATURE. CHOKE OPEN. AIR CLEANER INSTALLED. AND EXCEPT AS NOTED. AIR CONDITIONING OFF. PLUG DISCONNECTED VACUUM FITTINGS AND/OR HOSES. UNLESS OTHERWISE SPECIFIED.

1. DISCONNECT ELECTRICAL CONNECTOR AT DISTRIBUTOR. SET TIMING AT SPECIFIED RPM. RECONNECT CONNECTOR.
2. ADJUST CARBURETOR SPEED SCREW TO SPECIFIED RPM IN DRIVE.
3. ON AIR CONDITIONED CARS. AFTER COMPLETING ABOVE ADJUSTMENTS. TURN A/C ON. DISCONNECT ELECTRICAL CONNECTOR AT A/C COMPRESSOR CLUTCH AND ADJUST SOLENOID (ENERGIZED) SCREW TO SPECIFIED RPM IN DRIVE. RECONNECT ELECTRICAL CONNECTOR. TURN A/C OFF.
4. WITH TRANSMISSION IN PARK. SET FAST IDLE TO SPECIFIED SPEED ON HIGHEST CAM STEP.

CATALYST
EGR-EFE-AIR LOW ALTITUDE CERTIFICATION

	TRANSMISSION
	AUTOMATIC
TIMING (°B.T.D.C. @ RPM)	15° @ 550
CARBURETOR SCREW (RPM)	550 (DRIVE)
FAST IDLE SCREW (RPM)	2200 (PARK)
SOLENOID SCREW (RPM)	820 (DRIVE)

IDLE MIXTURE SCREWS ARE PRESET AND SEALED AT FACTORY. PROVISION FOR ADJUSTMENT DURING TUNE-UP IS NOT PROVIDED.

SEE SERVICE MANUAL AND MAINTENANCE SCHEDULE FOR ADDITIONAL INFORMATION.

THIS VEHICLE CONFORMS TO CALIFORNIA REGULATIONS APPLICABLE TO 1980 MODEL YEAR NEW PASSENGER CARS AND SIMILARLY TO U.S. EPA REGULATIONS APPLICABLE TO THE STATE OF CALIFORNIA.

VACUUM HOSE ROUTING 25506764

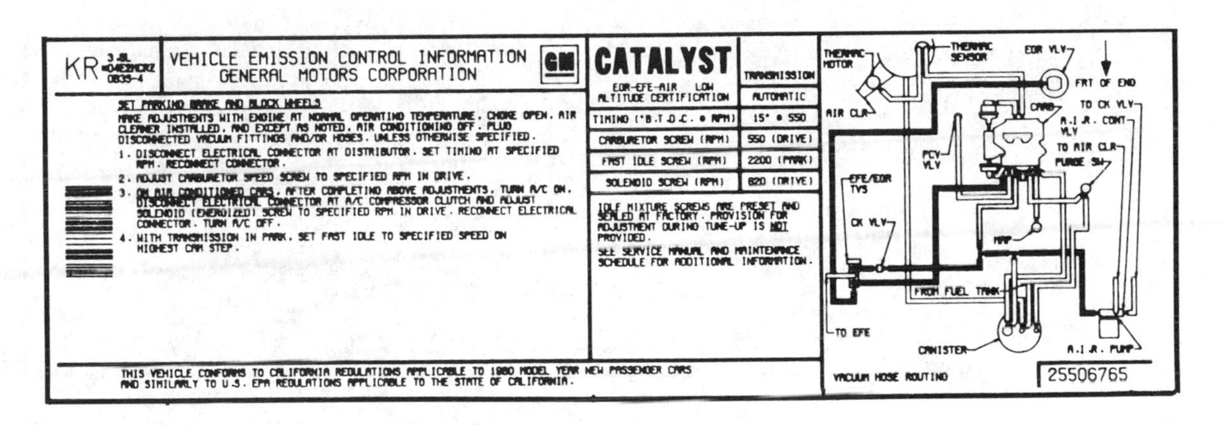

KR 3.8L #04E2HCRZ 0835-4

VEHICLE EMISSION CONTROL INFORMATION
GENERAL MOTORS CORPORATION
GM

SET PARKING BRAKE AND BLOCK WHEELS

MAKE ADJUSTMENTS WITH ENGINE AT NORMAL OPERATING TEMPERATURE. CHOKE OPEN. AIR CLEANER INSTALLED. AND EXCEPT AS NOTED. AIR CONDITIONING OFF. PLUG DISCONNECTED VACUUM FITTINGS AND/OR HOSES. UNLESS OTHERWISE SPECIFIED.

1. DISCONNECT ELECTRICAL CONNECTOR AT DISTRIBUTOR. SET TIMING AT SPECIFIED RPM. RECONNECT CONNECTOR.
2. ADJUST CARBURETOR SPEED SCREW TO SPECIFIED RPM IN DRIVE.
3. ON AIR CONDITIONED CARS. AFTER COMPLETING ABOVE ADJUSTMENTS. TURN A/C ON. DISCONNECT ELECTRICAL CONNECTOR AT A/C COMPRESSOR CLUTCH AND ADJUST SOLENOID (ENERGIZED) SCREW TO SPECIFIED RPM IN DRIVE. RECONNECT ELECTRICAL CONNECTOR. TURN A/C OFF.
4. WITH TRANSMISSION IN PARK. SET FAST IDLE TO SPECIFIED SPEED ON HIGHEST CAM STEP.

CATALYST
EGR-EFE-AIR LOW ALTITUDE CERTIFICATION

	TRANSMISSION
	AUTOMATIC
TIMING (°B.T.D.C. @ RPM)	15° @ 550
CARBURETOR SCREW (RPM)	550 (DRIVE)
FAST IDLE SCREW (RPM)	2200 (PARK)
SOLENOID SCREW (RPM)	820 (DRIVE)

IDLE MIXTURE SCREWS ARE PRESET AND SEALED AT FACTORY. PROVISION FOR ADJUSTMENT DURING TUNE-UP IS NOT PROVIDED.

SEE SERVICE MANUAL AND MAINTENANCE SCHEDULE FOR ADDITIONAL INFORMATION.

THIS VEHICLE CONFORMS TO CALIFORNIA REGULATIONS APPLICABLE TO 1980 MODEL YEAR NEW PASSENGER CARS AND SIMILARLY TO U.S. EPA REGULATIONS APPLICABLE TO THE STATE OF CALIFORNIA.

VACUUM HOSE ROUTING 25506765

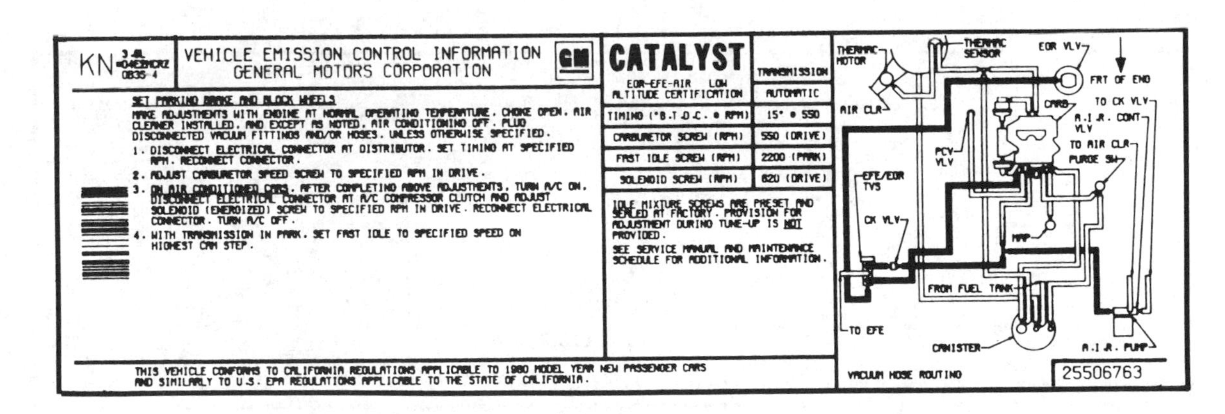

KN 3.8L #04E3HCRZ 0835-4

VEHICLE EMISSION CONTROL INFORMATION
GENERAL MOTORS CORPORATION
GM

SET PARKING BRAKE AND BLOCK WHEELS

MAKE ADJUSTMENTS WITH ENGINE AT NORMAL OPERATING TEMPERATURE. CHOKE OPEN. AIR CLEANER INSTALLED. AND EXCEPT AS NOTED. AIR CONDITIONING OFF. PLUG DISCONNECTED VACUUM FITTINGS AND/OR HOSES. UNLESS OTHERWISE SPECIFIED.

1. DISCONNECT ELECTRICAL CONNECTOR AT DISTRIBUTOR. SET TIMING AT SPECIFIED RPM. RECONNECT CONNECTOR.
2. ADJUST CARBURETOR SPEED SCREW TO SPECIFIED RPM IN DRIVE.
3. ON AIR CONDITIONED CARS. AFTER COMPLETING ABOVE ADJUSTMENTS. TURN A/C ON. DISCONNECT ELECTRICAL CONNECTOR AT A/C COMPRESSOR CLUTCH AND ADJUST SOLENOID (ENERGIZED) SCREW TO SPECIFIED RPM IN DRIVE. RECONNECT ELECTRICAL CONNECTOR. TURN A/C OFF.
4. WITH TRANSMISSION IN PARK. SET FAST IDLE TO SPECIFIED SPEED ON HIGHEST CAM STEP.

CATALYST
EGR-EFE-AIR LOW ALTITUDE CERTIFICATION

	TRANSMISSION
	AUTOMATIC
TIMING (°B.T.D.C. @ RPM)	15° @ 550
CARBURETOR SCREW (RPM)	550 (DRIVE)
FAST IDLE SCREW (RPM)	2200 (PARK)
SOLENOID SCREW (RPM)	820 (DRIVE)

IDLE MIXTURE SCREWS ARE PRESET AND SEALED AT FACTORY. PROVISION FOR ADJUSTMENT DURING TUNE-UP IS NOT PROVIDED.

SEE SERVICE MANUAL AND MAINTENANCE SCHEDULE FOR ADDITIONAL INFORMATION.

THIS VEHICLE CONFORMS TO CALIFORNIA REGULATIONS APPLICABLE TO 1980 MODEL YEAR NEW PASSENGER CARS AND SIMILARLY TO U.S. EPA REGULATIONS APPLICABLE TO THE STATE OF CALIFORNIA.

VACUUM HOSE ROUTING 25506763

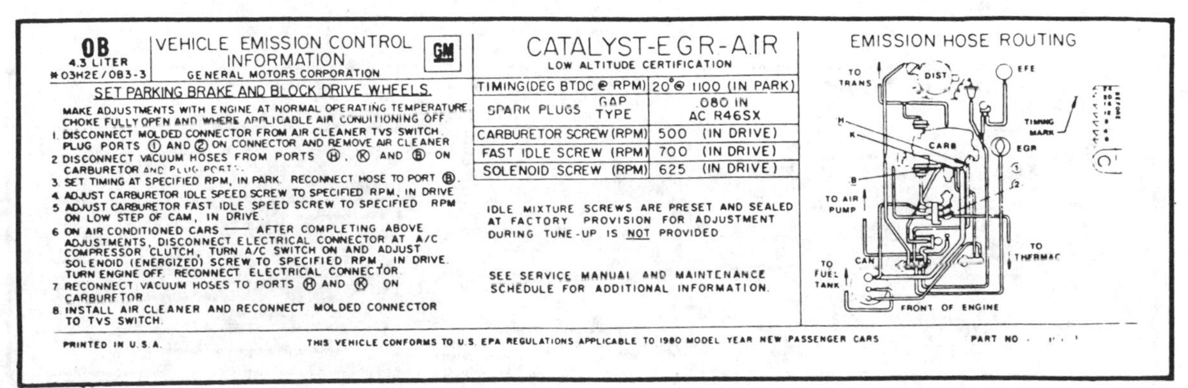

OB 4.3 LITER #03H2E/0B3-3

VEHICLE EMISSION CONTROL INFORMATION
GENERAL MOTORS CORPORATION
GM

SET PARKING BRAKE AND BLOCK DRIVE WHEELS.

MAKE ADJUSTMENTS WITH ENGINE AT NORMAL OPERATING TEMPERATURE. CHOKE FULLY OPEN AND WHERE APPLICABLE AIR CONDITIONING OFF.

1. DISCONNECT MOLDED CONNECTOR FROM AIR CLEANER TVS SWITCH. PLUG PORTS ① AND ② ON CONNECTOR AND REMOVE AIR CLEANER
2. DISCONNECT VACUUM HOSES FROM PORTS Ⓗ, Ⓚ AND Ⓑ ON CARBURETOR AND PLUG PORTS.
3. SET TIMING AT SPECIFIED RPM, IN PARK. RECONNECT HOSE TO PORT Ⓑ.
4. ADJUST CARBURETOR IDLE SPEED SCREW TO SPECIFIED RPM, IN DRIVE
5. ADJUST CARBURETOR FAST IDLE SPEED SCREW TO SPECIFIED RPM ON LOW STEP OF CAM, IN DRIVE.
6. ON AIR CONDITIONED CARS — AFTER COMPLETING ABOVE ADJUSTMENTS, DISCONNECT ELECTRICAL CONNECTOR AT A/C COMPRESSOR CLUTCH, TURN A/C SWITCH ON AND ADJUST SOLENOID (ENERGIZED) SCREW TO SPECIFIED RPM, IN DRIVE. TURN ENGINE OFF. RECONNECT ELECTRICAL CONNECTOR.
7. RECONNECT VACUUM HOSES TO PORTS Ⓗ AND Ⓚ ON CARBURETOR
8. INSTALL AIR CLEANER AND RECONNECT MOLDED CONNECTOR TO TVS SWITCH.

CATALYST-EGR-AIR
LOW ALTITUDE CERTIFICATION

TIMING (DEG BTDC @ RPM)		20° @ 1100 (IN PARK)
SPARK PLUGS	GAP	.080 IN
	TYPE	AC R46SX
CARBURETOR SCREW (RPM)		500 (IN DRIVE)
FAST IDLE SCREW (RPM)		700 (IN DRIVE)
SOLENOID SCREW (RPM)		625 (IN DRIVE)

IDLE MIXTURE SCREWS ARE PRESET AND SEALED AT FACTORY PROVISION FOR ADJUSTMENT DURING TUNE-UP IS NOT PROVIDED

SEE SERVICE MANUAL AND MAINTENANCE SCHEDULE FOR ADDITIONAL INFORMATION.

EMISSION HOSE ROUTING

PRINTED IN U.S.A.

THIS VEHICLE CONFORMS TO U.S. EPA REGULATIONS APPLICABLE TO 1980 MODEL YEAR NEW PASSENGER CARS

PART NO

VEHICLE EMISSION CONTROL LABELS AND VACUUM HOSE ROUTINGS

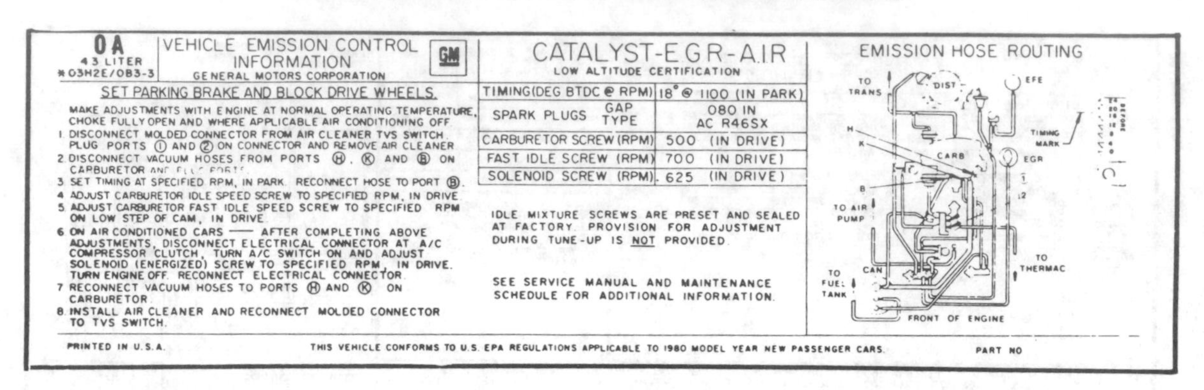

OA
4.3 LITER
#03H2E/0B3-3

VEHICLE EMISSION CONTROL INFORMATION
GENERAL MOTORS CORPORATION

SET PARKING BRAKE AND BLOCK DRIVE WHEELS.

MAKE ADJUSTMENTS WITH ENGINE AT NORMAL OPERATING TEMPERATURE, CHOKE FULLY OPEN AND WHERE APPLICABLE AIR CONDITIONING OFF.

1. DISCONNECT MOLDED CONNECTOR FROM AIR CLEANER TVS SWITCH. PLUG PORTS ① AND ② ON CONNECTOR AND REMOVE AIR CLEANER.
2. DISCONNECT VACUUM HOSES FROM PORTS Ⓗ, Ⓚ AND Ⓑ ON CARBURETOR AND PLUG PORTS.
3. SET TIMING AT SPECIFIED RPM, IN PARK. RECONNECT HOSE TO PORT Ⓑ.
4. ADJUST CARBURETOR IDLE SPEED SCREW TO SPECIFIED RPM, IN DRIVE.
5. ADJUST CARBURETOR FAST IDLE SPEED SCREW TO SPECIFIED RPM ON LOW STEP OF CAM, IN DRIVE.
6. ON AIR CONDITIONED CARS —— AFTER COMPLETING ABOVE ADJUSTMENTS, DISCONNECT ELECTRICAL CONNECTOR AT A/C COMPRESSOR CLUTCH, TURN A/C SWITCH ON AND ADJUST SOLENOID (ENERGIZED) SCREW TO SPECIFIED RPM, IN DRIVE. TURN ENGINE OFF. RECONNECT ELECTRICAL CONNECTOR.
7. RECONNECT VACUUM HOSES TO PORTS Ⓗ AND Ⓚ ON CARBURETOR.
8. INSTALL AIR CLEANER AND RECONNECT MOLDED CONNECTOR TO TVS SWITCH.

CATALYST-EGR-AIR
LOW ALTITUDE CERTIFICATION

TIMING (DEG BTDC @ RPM)	18° @ 1100 (IN PARK)
SPARK PLUGS GAP / TYPE	.080 IN / AC R46SX
CARBURETOR SCREW (RPM)	500 (IN DRIVE)
FAST IDLE SCREW (RPM)	700 (IN DRIVE)
SOLENOID SCREW (RPM)	625 (IN DRIVE)

IDLE MIXTURE SCREWS ARE PRESET AND SEALED AT FACTORY. PROVISION FOR ADJUSTMENT DURING TUNE-UP IS NOT PROVIDED.

SEE SERVICE MANUAL AND MAINTENANCE SCHEDULE FOR ADDITIONAL INFORMATION.

EMISSION HOSE ROUTING

PRINTED IN U.S.A. THIS VEHICLE CONFORMS TO U.S. EPA REGULATIONS APPLICABLE TO 1980 MODEL YEAR NEW PASSENGER CARS. PART NO

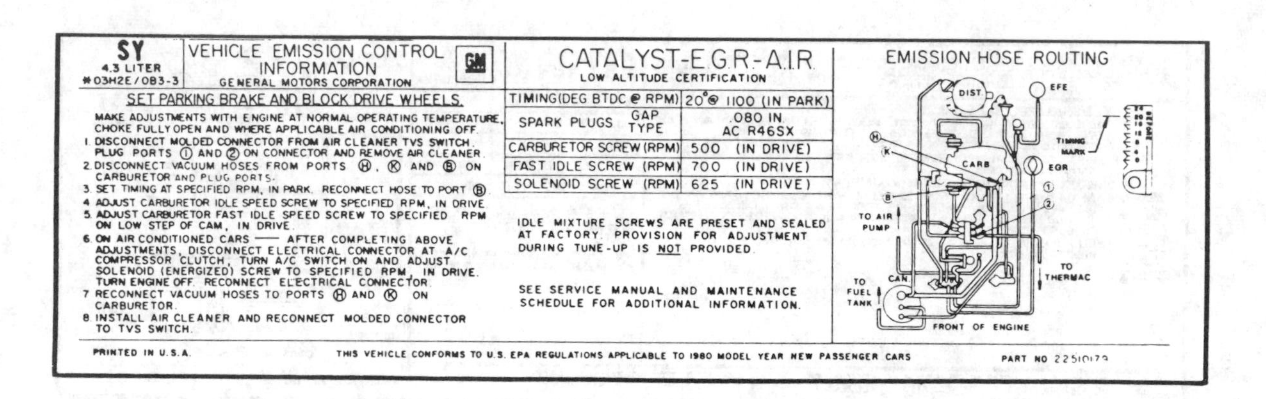

SY
4.3 LITER
#03H2E/0B3-3

VEHICLE EMISSION CONTROL INFORMATION
GENERAL MOTORS CORPORATION

SET PARKING BRAKE AND BLOCK DRIVE WHEELS.

MAKE ADJUSTMENTS WITH ENGINE AT NORMAL OPERATING TEMPERATURE, CHOKE FULLY OPEN AND WHERE APPLICABLE AIR CONDITIONING OFF.

1. DISCONNECT MOLDED CONNECTOR FROM AIR CLEANER TVS SWITCH. PLUG PORTS ① AND ② ON CONNECTOR AND REMOVE AIR CLEANER.
2. DISCONNECT VACUUM HOSES FROM PORTS Ⓗ, Ⓚ AND Ⓑ ON CARBURETOR AND PLUG PORTS.
3. SET TIMING AT SPECIFIED RPM, IN PARK. RECONNECT HOSE TO PORT Ⓑ.
4. ADJUST CARBURETOR IDLE SPEED SCREW TO SPECIFIED RPM, IN DRIVE.
5. ADJUST CARBURETOR FAST IDLE SPEED SCREW TO SPECIFIED RPM ON LOW STEP OF CAM, IN DRIVE.
6. ON AIR CONDITIONED CARS —— AFTER COMPLETING ABOVE ADJUSTMENTS, DISCONNECT ELECTRICAL CONNECTOR AT A/C COMPRESSOR CLUTCH, TURN A/C SWITCH ON AND ADJUST SOLENOID (ENERGIZED) SCREW TO SPECIFIED RPM, IN DRIVE. TURN ENGINE OFF. RECONNECT ELECTRICAL CONNECTOR.
7. RECONNECT VACUUM HOSES TO PORTS Ⓗ AND Ⓚ ON CARBURETOR.
8. INSTALL AIR CLEANER AND RECONNECT MOLDED CONNECTOR TO TVS SWITCH.

CATALYST-EGR-AIR
LOW ALTITUDE CERTIFICATION

TIMING (DEG BTDC @ RPM)	20° @ 1100 (IN PARK)
SPARK PLUGS GAP / TYPE	.080 IN / AC R46SX
CARBURETOR SCREW (RPM)	500 (IN DRIVE)
FAST IDLE SCREW (RPM)	700 (IN DRIVE)
SOLENOID SCREW (RPM)	625 (IN DRIVE)

IDLE MIXTURE SCREWS ARE PRESET AND SEALED AT FACTORY. PROVISION FOR ADJUSTMENT DURING TUNE-UP IS NOT PROVIDED.

SEE SERVICE MANUAL AND MAINTENANCE SCHEDULE FOR ADDITIONAL INFORMATION.

EMISSION HOSE ROUTING

PRINTED IN U.S.A. THIS VEHICLE CONFORMS TO U.S. EPA REGULATIONS APPLICABLE TO 1980 MODEL YEAR NEW PASSENGER CARS. PART NO 22510179

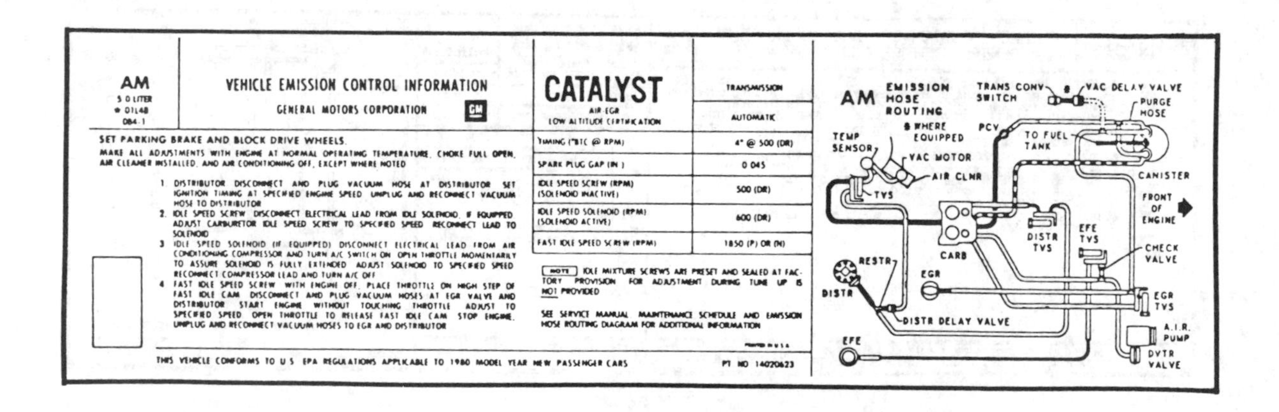

AM
5.0 LITER
* 01L48
084-1

VEHICLE EMISSION CONTROL INFORMATION
GENERAL MOTORS CORPORATION

SET PARKING BRAKE AND BLOCK DRIVE WHEELS.

MAKE ALL ADJUSTMENTS WITH ENGINE AT NORMAL OPERATING TEMPERATURE, CHOKE FULL OPEN, AIR CLEANER INSTALLED, AND AIR CONDITIONING OFF, EXCEPT WHERE NOTED.

1. DISTRIBUTOR: DISCONNECT AND PLUG VACUUM HOSE AT DISTRIBUTOR. SET IGNITION TIMING AT SPECIFIED ENGINE SPEED. UNPLUG AND RECONNECT VACUUM HOSE TO DISTRIBUTOR.
2. IDLE SPEED SCREW: DISCONNECT ELECTRICAL LEAD FROM IDLE SOLENOID, IF EQUIPPED. ADJUST CARBURETOR IDLE SPEED SCREW TO SPECIFIED SPEED. RECONNECT LEAD TO SOLENOID.
3. IDLE SPEED SOLENOID (IF EQUIPPED): DISCONNECT ELECTRICAL LEAD FROM AIR CONDITIONING COMPRESSOR AND TURN A/C SWITCH ON. OPEN THROTTLE MOMENTARILY TO ASSURE SOLENOID IS FULLY EXTENDED. ADJUST SOLENOID TO SPECIFIED SPEED. RECONNECT COMPRESSOR LEAD AND TURN A/C OFF.
4. FAST IDLE SPEED SCREW: WITH ENGINE OFF, PLACE THROTTLE ON HIGH STEP OF FAST IDLE CAM. DISCONNECT AND PLUG VACUUM HOSES AT EGR VALVE AND DISTRIBUTOR. START ENGINE WITHOUT TOUCHING THROTTLE. ADJUST TO SPECIFIED SPEED. OPEN THROTTLE TO RELEASE FAST IDLE CAM. STOP ENGINE. UNPLUG AND RECONNECT VACUUM HOSES TO EGR AND DISTRIBUTOR.

CATALYST
AIR-EGR
LOW ALTITUDE CERTIFICATION

	TRANSMISSION AUTOMATIC
TIMING (°BTC @ RPM)	4° @ 500 (DR)
SPARK PLUG GAP (IN.)	0.045
IDLE SPEED SCREW (RPM) (SOLENOID INACTIVE)	500 (DR)
IDLE SPEED SOLENOID (RPM) (SOLENOID ACTIVE)	600 (DR)
FAST IDLE SPEED SCREW (RPM)	1850 (P) OR (N)

NOTE: IDLE MIXTURE SCREWS ARE PRESET AND SEALED AT FACTORY. PROVISION FOR ADJUSTMENT DURING TUNE UP IS NOT PROVIDED.

SEE SERVICE MANUAL, MAINTENANCE SCHEDULE AND EMISSION HOSE ROUTING DIAGRAM FOR ADDITIONAL INFORMATION.

PRINTED IN U.S.A.

THIS VEHICLE CONFORMS TO U.S. EPA REGULATIONS APPLICABLE TO 1980 MODEL YEAR NEW PASSENGER CARS. PT NO 14020623

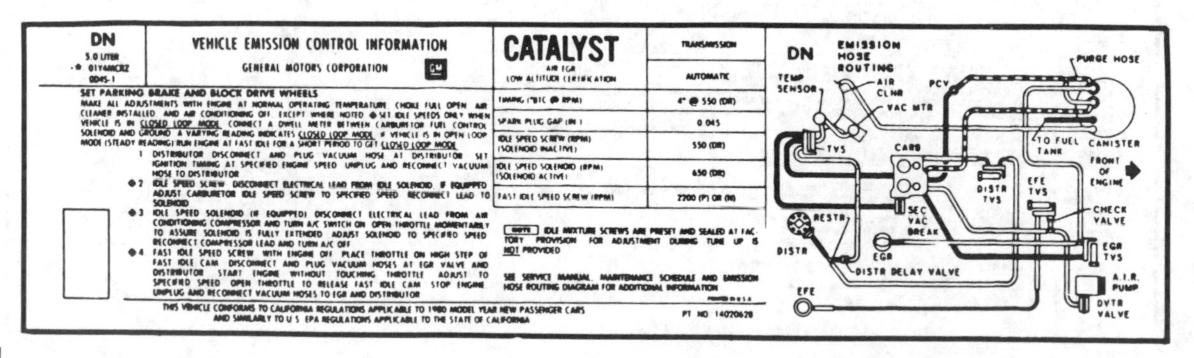

DN
5.0 LITER
* 01Y4MCRZ
0D45-1

VEHICLE EMISSION CONTROL INFORMATION
GENERAL MOTORS CORPORATION

SET PARKING BRAKE AND BLOCK DRIVE WHEELS

MAKE ALL ADJUSTMENTS WITH ENGINE AT NORMAL OPERATING TEMPERATURE, CHOKE FULL OPEN, AIR CLEANER INSTALLED, AND AIR CONDITIONING OFF, EXCEPT WHERE NOTED. ◆ SET IDLE SPEEDS ONLY WHEN VEHICLE IS IN CLOSED LOOP MODE. CONNECT A DWELL METER BETWEEN CARBURETOR FUEL CONTROL SOLENOID AND GROUND. A VARYING READING INDICATES CLOSED LOOP MODE. IF VEHICLE IS IN OPEN LOOP MODE (STEADY READING) RUN ENGINE AT FAST IDLE FOR A SHORT PERIOD TO GET CLOSED LOOP MODE.

1. DISTRIBUTOR: DISCONNECT AND PLUG VACUUM HOSE AT DISTRIBUTOR. SET IGNITION TIMING AT SPECIFIED ENGINE SPEED. UNPLUG AND RECONNECT VACUUM HOSE TO DISTRIBUTOR.

◆2. IDLE SPEED SCREW: DISCONNECT ELECTRICAL LEAD FROM IDLE SOLENOID, IF EQUIPPED. ADJUST CARBURETOR IDLE SPEED SCREW TO SPECIFIED SPEED. RECONNECT LEAD TO SOLENOID.

◆3. IDLE SPEED SOLENOID (IF EQUIPPED): DISCONNECT ELECTRICAL LEAD FROM AIR CONDITIONING COMPRESSOR AND TURN A/C SWITCH ON. OPEN THROTTLE MOMENTARILY TO ASSURE SOLENOID IS FULLY EXTENDED. ADJUST SOLENOID TO SPECIFIED SPEED. RECONNECT COMPRESSOR LEAD AND TURN A/C OFF.

◆4. FAST IDLE SPEED SCREW: WITH ENGINE OFF, PLACE THROTTLE ON HIGH STEP OF FAST IDLE CAM. DISCONNECT AND PLUG VACUUM HOSES AT EGR VALVE AND DISTRIBUTOR. START ENGINE WITHOUT TOUCHING THROTTLE. ADJUST TO SPECIFIED SPEED. OPEN THROTTLE TO RELEASE FAST IDLE CAM. STOP ENGINE. UNPLUG AND RECONNECT VACUUM HOSES TO EGR AND DISTRIBUTOR.

CATALYST
AIR-EGR
LOW ALTITUDE CERTIFICATION

	TRANSMISSION AUTOMATIC
TIMING (°BTC @ RPM)	4° @ 550 (DR)
SPARK PLUG GAP (IN.)	0.045
IDLE SPEED SCREW (RPM) (SOLENOID INACTIVE)	550 (DR)
IDLE SPEED SOLENOID (RPM) (SOLENOID ACTIVE)	650 (DR)
FAST IDLE SPEED SCREW (RPM)	2200 (P) OR (N)

NOTE: IDLE MIXTURE SCREWS ARE PRESET AND SEALED AT FACTORY. PROVISION FOR ADJUSTMENT DURING TUNE UP IS NOT PROVIDED.

SEE SERVICE MANUAL, MAINTENANCE SCHEDULE AND EMISSION HOSE ROUTING DIAGRAM FOR ADDITIONAL INFORMATION.

PRINTED IN U.S.A.

THIS VEHICLE CONFORMS TO CALIFORNIA REGULATIONS APPLICABLE TO 1980 MODEL YEAR NEW PASSENGER CARS AND SIMILARLY TO U.S. EPA REGULATIONS APPLICABLE TO THE STATE OF CALIFORNIA. PT NO 14020628

VEHICLE EMISSION CONTROL LABELS AND VACUUM HOSE ROUTINGS

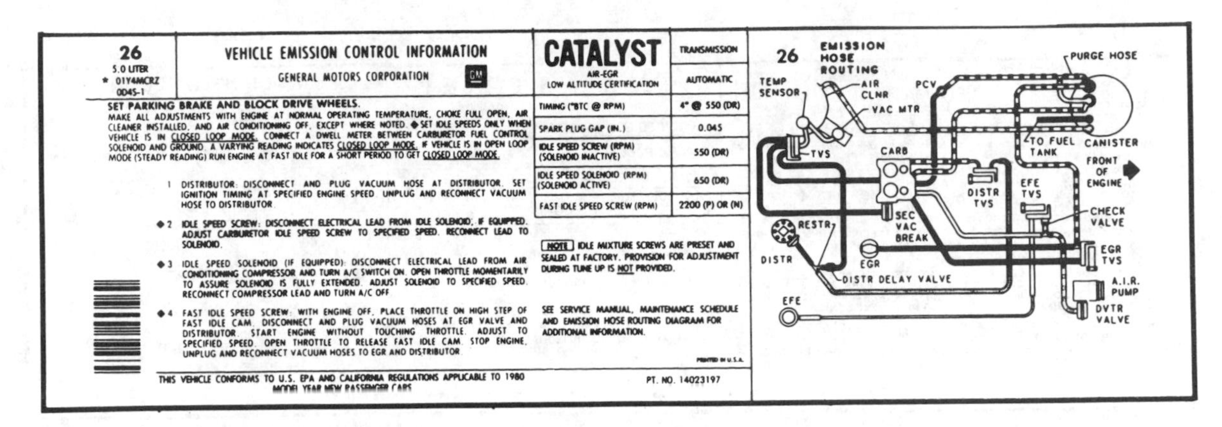

26
5.0 LITER
★ 01Y4MCRZ
0D4S-1

VEHICLE EMISSION CONTROL INFORMATION
GENERAL MOTORS CORPORATION

SET PARKING BRAKE AND BLOCK DRIVE WHEELS.
MAKE ALL ADJUSTMENTS WITH ENGINE AT NORMAL OPERATING TEMPERATURE, CHOKE FULL OPEN, AIR CLEANER INSTALLED, AND AIR CONDITIONING OFF, EXCEPT WHERE NOTED. ◆ SET IDLE SPEEDS ONLY WHEN VEHICLE IS IN CLOSED LOOP MODE. CONNECT A DWELL METER BETWEEN CARBURETOR FUEL CONTROL SOLENOID AND GROUND. A VARYING READING INDICATES CLOSED LOOP MODE. IF VEHICLE IS IN OPEN LOOP MODE (STEADY READING) RUN ENGINE AT FAST IDLE FOR A SHORT PERIOD TO GET CLOSED LOOP MODE.

1 DISTRIBUTOR: DISCONNECT AND PLUG VACUUM HOSE AT DISTRIBUTOR. SET IGNITION TIMING AT SPECIFIED ENGINE SPEED. UNPLUG AND RECONNECT VACUUM HOSE TO DISTRIBUTOR.

◆ 2 IDLE SPEED SCREW: DISCONNECT ELECTRICAL LEAD FROM IDLE SOLENOID, IF EQUIPPED. ADJUST CARBURETOR IDLE SPEED SCREW TO SPECIFIED SPEED. RECONNECT LEAD TO SOLENOID.

◆ 3 IDLE SPEED SOLENOID (IF EQUIPPED): DISCONNECT ELECTRICAL LEAD FROM AIR CONDITIONING COMPRESSOR AND TURN A/C SWITCH ON. OPEN THROTTLE MOMENTARILY TO ASSURE SOLENOID IS FULLY EXTENDED. ADJUST SOLENOID TO SPECIFIED SPEED. RECONNECT COMPRESSOR LEAD AND TURN A/C OFF.

◆ 4 FAST IDLE SPEED SCREW: WITH ENGINE OFF, PLACE THROTTLE ON HIGH STEP OF FAST IDLE CAM. DISCONNECT AND PLUG VACUUM HOSES AT EGR VALVE AND DISTRIBUTOR. START ENGINE WITHOUT TOUCHING THROTTLE. ADJUST TO SPECIFIED SPEED. OPEN THROTTLE TO RELEASE FAST IDLE CAM. STOP ENGINE, UNPLUG AND RECONNECT VACUUM HOSES TO EGR AND DISTRIBUTOR.

THIS VEHICLE CONFORMS TO U.S. EPA AND CALIFORNIA REGULATIONS APPLICABLE TO 1980 MODEL YEAR NEW PASSENGER CARS

CATALYST AIR-EGR LOW ALTITUDE CERTIFICATION	TRANSMISSION AUTOMATIC
TIMING (°BTC @ RPM)	4° @ 550 (DR)
SPARK PLUG GAP (IN.)	0.045
IDLE SPEED SCREW (RPM) (SOLENOID INACTIVE)	550 (DR)
IDLE SPEED SOLENOID (RPM) (SOLENOID ACTIVE)	650 (DR)
FAST IDLE SPEED SCREW (RPM)	2200 (P) OR (N)

NOTE IDLE MIXTURE SCREWS ARE PRESET AND SEALED AT FACTORY. PROVISION FOR ADJUSTMENT DURING TUNE UP IS NOT PROVIDED.

SEE SERVICE MANUAL, MAINTENANCE SCHEDULE AND EMISSION HOSE ROUTING DIAGRAM FOR ADDITIONAL INFORMATION.

PRINTED IN U.S.A.

PT. NO. 14023197

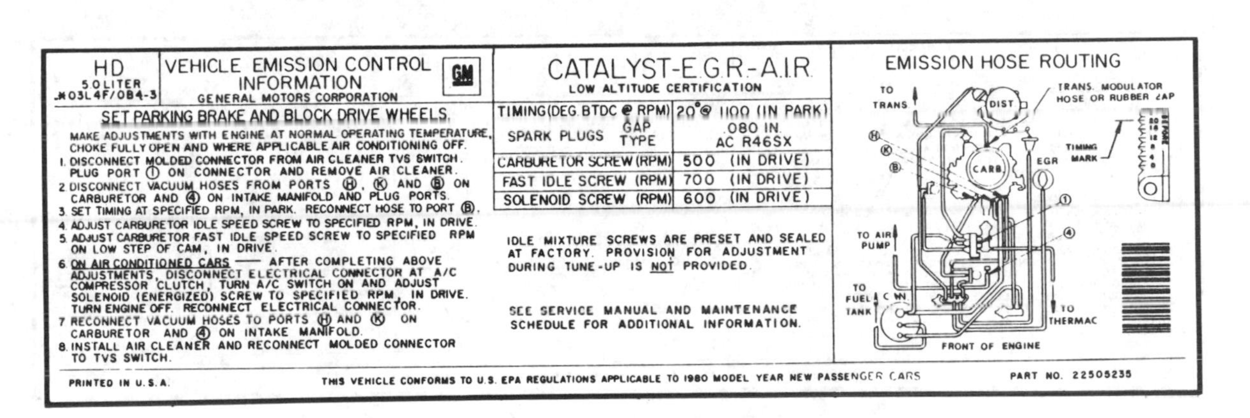

HD
5.0 LITER
★03L4F/0B4-3

VEHICLE EMISSION CONTROL INFORMATION
GENERAL MOTORS CORPORATION

SET PARKING BRAKE AND BLOCK DRIVE WHEELS.
MAKE ADJUSTMENTS WITH ENGINE AT NORMAL OPERATING TEMPERATURE, CHOKE FULLY OPEN AND WHERE APPLICABLE AIR CONDITIONING OFF.
1. DISCONNECT MOLDED CONNECTOR FROM AIR CLEANER TVS SWITCH. PLUG PORT Ⓘ ON CONNECTOR AND REMOVE AIR CLEANER.
2. DISCONNECT VACUUM HOSES FROM PORTS Ⓗ, Ⓚ AND Ⓑ ON CARBURETOR AND ④ ON INTAKE MANIFOLD AND PLUG PORTS.
3. SET TIMING AT SPECIFIED RPM, IN PARK. RECONNECT HOSE TO PORT Ⓑ.
4. ADJUST CARBURETOR IDLE SPEED SCREW TO SPECIFIED RPM, IN DRIVE.
5. ADJUST CARBURETOR FAST IDLE SPEED SCREW TO SPECIFIED RPM ON LOW STEP OF CAM, IN DRIVE.
6. ON AIR CONDITIONED CARS —— AFTER COMPLETING ABOVE ADJUSTMENTS, DISCONNECT ELECTRICAL CONNECTOR AT A/C COMPRESSOR CLUTCH, TURN A/C SWITCH ON AND ADJUST SOLENOID (ENERGIZED) SCREW TO SPECIFIED RPM, IN DRIVE. TURN ENGINE OFF. RECONNECT ELECTRICAL CONNECTOR.
7. RECONNECT VACUUM HOSES TO PORTS Ⓗ AND Ⓚ ON CARBURETOR AND ④ ON INTAKE MANIFOLD.
8. INSTALL AIR CLEANER AND RECONNECT MOLDED CONNECTOR TO TVS SWITCH.

CATALYST-E.G.R.-A.I.R.
LOW ALTITUDE CERTIFICATION

TIMING(DEG BTDC @ RPM)	20° @ 1100 (IN PARK)
SPARK PLUGS GAP TYPE	.080 IN. AC R46SX
CARBURETOR SCREW (RPM)	500 (IN DRIVE)
FAST IDLE SCREW (RPM)	700 (IN DRIVE)
SOLENOID SCREW (RPM)	600 (IN DRIVE)

IDLE MIXTURE SCREWS ARE PRESET AND SEALED AT FACTORY. PROVISION FOR ADJUSTMENT DURING TUNE-UP IS NOT PROVIDED.

SEE SERVICE MANUAL AND MAINTENANCE SCHEDULE FOR ADDITIONAL INFORMATION.

PRINTED IN U.S.A. THIS VEHICLE CONFORMS TO U.S. EPA REGULATIONS APPLICABLE TO 1980 MODEL YEAR NEW PASSENGER CARS PART NO. 22505235

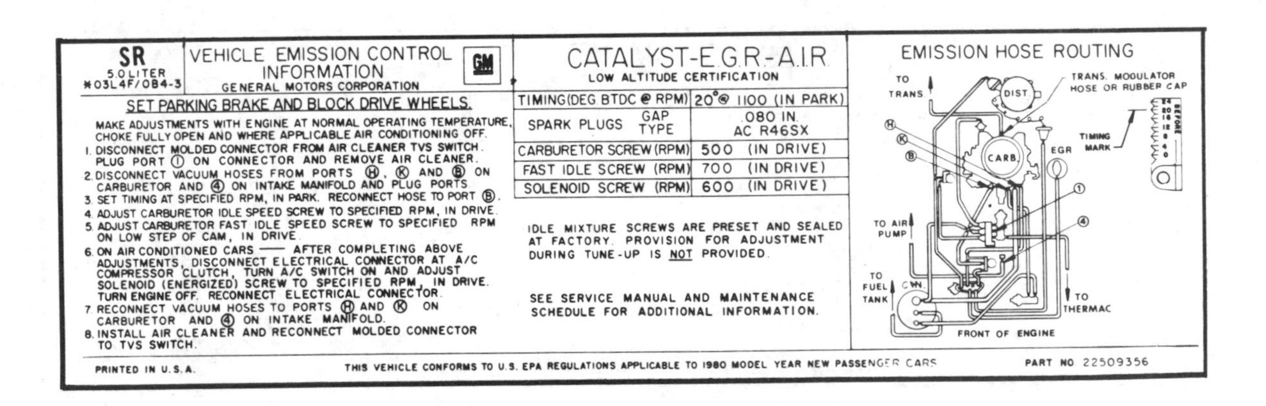

SR
5.0 LITER
★03L4F/0B4-3

VEHICLE EMISSION CONTROL INFORMATION
GENERAL MOTORS CORPORATION

SET PARKING BRAKE AND BLOCK DRIVE WHEELS.
MAKE ADJUSTMENTS WITH ENGINE AT NORMAL OPERATING TEMPERATURE, CHOKE FULLY OPEN AND WHERE APPLICABLE AIR CONDITIONING OFF.
1. DISCONNECT MOLDED CONNECTOR FROM AIR CLEANER TVS SWITCH. PLUG PORT Ⓘ ON CONNECTOR AND REMOVE AIR CLEANER.
2. DISCONNECT VACUUM HOSES FROM PORTS Ⓗ, Ⓚ AND Ⓑ ON CARBURETOR AND ④ ON INTAKE MANIFOLD AND PLUG PORTS.
3. SET TIMING AT SPECIFIED RPM, IN PARK. RECONNECT HOSE TO PORT Ⓑ.
4. ADJUST CARBURETOR IDLE SPEED SCREW TO SPECIFIED RPM, IN DRIVE.
5. ADJUST CARBURETOR FAST IDLE SPEED SCREW TO SPECIFIED RPM ON LOW STEP OF CAM, IN DRIVE.
6. ON AIR CONDITIONED CARS —— AFTER COMPLETING ABOVE ADJUSTMENTS, DISCONNECT ELECTRICAL CONNECTOR AT A/C COMPRESSOR CLUTCH, TURN A/C SWITCH ON AND ADJUST SOLENOID (ENERGIZED) SCREW TO SPECIFIED RPM, IN DRIVE. TURN ENGINE OFF. RECONNECT ELECTRICAL CONNECTOR.
7. RECONNECT VACUUM HOSES TO PORTS Ⓗ AND Ⓚ ON CARBURETOR AND ④ ON INTAKE MANIFOLD.
8. INSTALL AIR CLEANER AND RECONNECT MOLDED CONNECTOR TO TVS SWITCH.

CATALYST-E.G.R.-A.I.R.
LOW ALTITUDE CERTIFICATION

TIMING(DEG BTDC @ RPM)	20° @ 1100 (IN PARK)
SPARK PLUGS GAP TYPE	.080 IN. AC R46SX
CARBURETOR SCREW (RPM)	500 (IN DRIVE)
FAST IDLE SCREW (RPM)	700 (IN DRIVE)
SOLENOID SCREW (RPM)	600 (IN DRIVE)

IDLE MIXTURE SCREWS ARE PRESET AND SEALED AT FACTORY. PROVISION FOR ADJUSTMENT DURING TUNE-UP IS NOT PROVIDED.

SEE SERVICE MANUAL AND MAINTENANCE SCHEDULE FOR ADDITIONAL INFORMATION.

PRINTED IN U.S.A. THIS VEHICLE CONFORMS TO U.S. EPA REGULATIONS APPLICABLE TO 1980 MODEL YEAR NEW PASSENGER CARS PART NO. 22509356

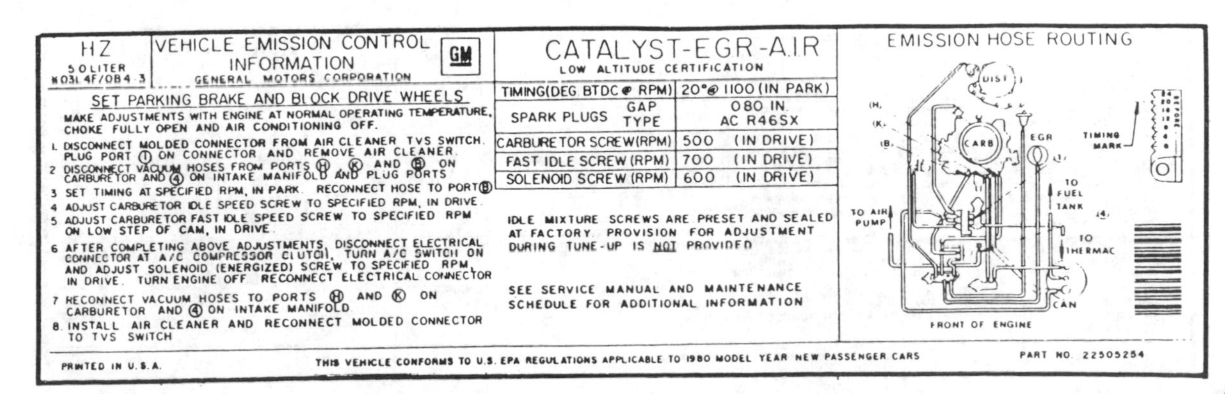

HZ
5.0 LITER
★03L4F/0B4-3

VEHICLE EMISSION CONTROL INFORMATION
GENERAL MOTORS CORPORATION

SET PARKING BRAKE AND BLOCK DRIVE WHEELS
MAKE ADJUSTMENTS WITH ENGINE AT NORMAL OPERATING TEMPERATURE, CHOKE FULLY OPEN AND AIR CONDITIONING OFF.
1. DISCONNECT MOLDED CONNECTOR FROM AIR CLEANER TVS SWITCH. PLUG PORT Ⓘ ON CONNECTOR AND REMOVE AIR CLEANER.
2. DISCONNECT VACUUM HOSES FROM PORTS Ⓗ, Ⓚ AND Ⓑ ON CARBURETOR AND ④ ON INTAKE MANIFOLD AND PLUG PORTS.
3. SET TIMING AT SPECIFIED RPM, IN PARK. RECONNECT HOSE TO PORT Ⓑ.
4. ADJUST CARBURETOR IDLE SPEED SCREW TO SPECIFIED RPM, IN DRIVE.
5. ADJUST CARBURETOR FAST IDLE SPEED SCREW TO SPECIFIED RPM ON LOW STEP OF CAM, IN DRIVE.
6. AFTER COMPLETING ABOVE ADJUSTMENTS, DISCONNECT ELECTRICAL CONNECTOR AT A/C COMPRESSOR CLUTCH, TURN A/C SWITCH ON AND ADJUST SOLENOID (ENERGIZED) SCREW TO SPECIFIED RPM, IN DRIVE. TURN ENGINE OFF. RECONNECT ELECTRICAL CONNECTOR.
7. RECONNECT VACUUM HOSES TO PORTS Ⓗ AND Ⓚ ON CARBURETOR AND ④ ON INTAKE MANIFOLD.
8. INSTALL AIR CLEANER AND RECONNECT MOLDED CONNECTOR TO TVS SWITCH.

CATALYST-EGR-A.I.R.
LOW ALTITUDE CERTIFICATION

TIMING(DEG BTDC @ RPM)	20° @ 1100 (IN PARK)
SPARK PLUGS GAP TYPE	.080 IN. AC R46SX
CARBURETOR SCREW (RPM)	500 (IN DRIVE)
FAST IDLE SCREW (RPM)	700 (IN DRIVE)
SOLENOID SCREW (RPM)	600 (IN DRIVE)

IDLE MIXTURE SCREWS ARE PRESET AND SEALED AT FACTORY. PROVISION FOR ADJUSTMENT DURING TUNE-UP IS NOT PROVIDED.

SEE SERVICE MANUAL AND MAINTENANCE SCHEDULE FOR ADDITIONAL INFORMATION.

PRINTED IN U.S.A. THIS VEHICLE CONFORMS TO U.S. EPA REGULATIONS APPLICABLE TO 1980 MODEL YEAR NEW PASSENGER CARS PART NO. 22505254

VEHICLE EMISSION CONTROL LABELS AND VACUUM HOSE ROUTINGS

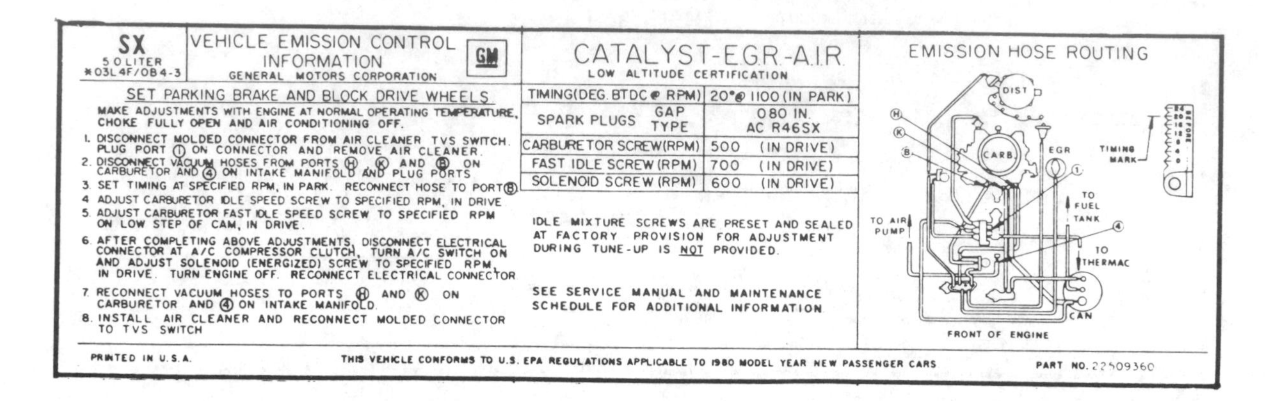

SX
5.0 LITER
*03L4F/0B4-3

VEHICLE EMISSION CONTROL INFORMATION
GENERAL MOTORS CORPORATION
GM

SET PARKING BRAKE AND BLOCK DRIVE WHEELS.

MAKE ADJUSTMENTS WITH ENGINE AT NORMAL OPERATING TEMPERATURE, CHOKE FULLY OPEN AND AIR CONDITIONING OFF.

1. DISCONNECT MOLDED CONNECTOR FROM AIR CLEANER TVS SWITCH. PLUG PORT ① ON CONNECTOR AND REMOVE AIR CLEANER.
2. DISCONNECT VACUUM HOSES FROM PORTS Ⓗ, Ⓚ AND Ⓑ ON CARBURETOR AND ④ ON INTAKE MANIFOLD AND PLUG PORTS.
3. SET TIMING AT SPECIFIED RPM, IN PARK. RECONNECT HOSE TO PORT Ⓑ.
4. ADJUST CARBURETOR IDLE SPEED SCREW TO SPECIFIED RPM, IN DRIVE.
5. ADJUST CARBURETOR FAST IDLE SPEED SCREW TO SPECIFIED RPM ON LOW STEP OF CAM, IN DRIVE.
6. AFTER COMPLETING ABOVE ADJUSTMENTS, DISCONNECT ELECTRICAL CONNECTOR AT A/C COMPRESSOR CLUTCH, TURN A/C SWITCH ON AND ADJUST SOLENOID (ENERGIZED) SCREW TO SPECIFIED RPM, IN DRIVE. TURN ENGINE OFF. RECONNECT ELECTRICAL CONNECTOR.
7. RECONNECT VACUUM HOSES TO PORTS Ⓗ AND Ⓚ ON CARBURETOR AND ④ ON INTAKE MANIFOLD.
8. INSTALL AIR CLEANER AND RECONNECT MOLDED CONNECTOR TO TVS SWITCH.

CATALYST-EGR-AIR
LOW ALTITUDE CERTIFICATION

TIMING(DEG. BTDC @ RPM)	20° @ 1100 (IN PARK)
SPARK PLUGS GAP / TYPE	.080 IN. / AC R46SX
CARBURETOR SCREW(RPM)	500 (IN DRIVE)
FAST IDLE SCREW(RPM)	700 (IN DRIVE)
SOLENOID SCREW (RPM)	600 (IN DRIVE)

IDLE MIXTURE SCREWS ARE PRESET AND SEALED AT FACTORY. PROVISION FOR ADJUSTMENT DURING TUNE-UP IS NOT PROVIDED.

SEE SERVICE MANUAL AND MAINTENANCE SCHEDULE FOR ADDITIONAL INFORMATION.

EMISSION HOSE ROUTING

PRINTED IN U.S.A. THIS VEHICLE CONFORMS TO U.S. EPA REGULATIONS APPLICABLE TO 1980 MODEL YEAR NEW PASSENGER CARS. PART NO. 22509360

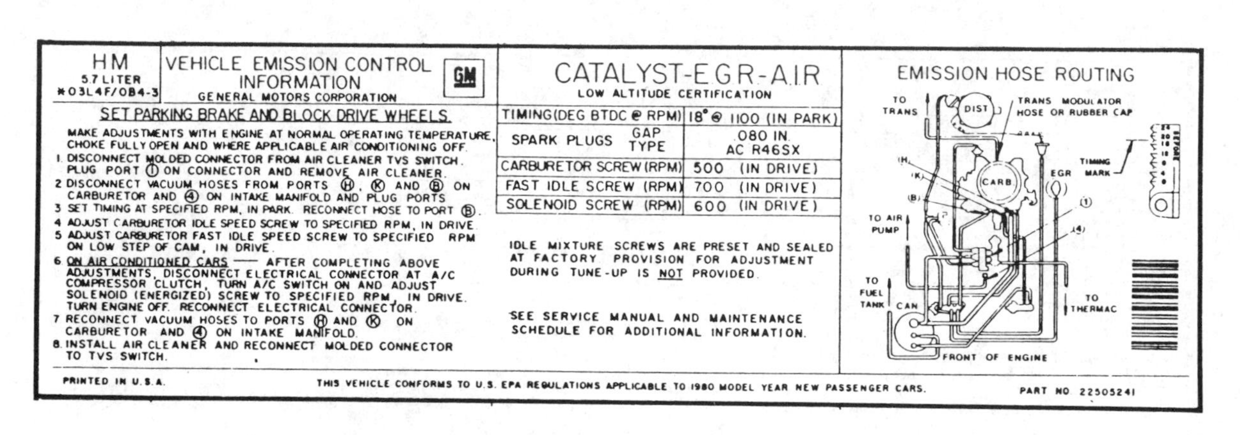

HM
5.7 LITER
*03L4F/0B4-3

VEHICLE EMISSION CONTROL INFORMATION
GENERAL MOTORS CORPORATION
GM

SET PARKING BRAKE AND BLOCK DRIVE WHEELS.

MAKE ADJUSTMENTS WITH ENGINE AT NORMAL OPERATING TEMPERATURE, CHOKE FULLY OPEN AND WHERE APPLICABLE AIR CONDITIONING OFF.

1. DISCONNECT MOLDED CONNECTOR FROM AIR CLEANER TVS SWITCH. PLUG PORT ① ON CONNECTOR AND REMOVE AIR CLEANER.
2. DISCONNECT VACUUM HOSES FROM PORTS Ⓗ, Ⓚ AND Ⓑ ON CARBURETOR AND ④ ON INTAKE MANIFOLD AND PLUG PORTS.
3. SET TIMING AT SPECIFIED RPM, IN PARK. RECONNECT HOSE TO PORT Ⓑ.
4. ADJUST CARBURETOR IDLE SPEED SCREW TO SPECIFIED RPM, IN DRIVE.
5. ADJUST CARBURETOR FAST IDLE SPEED SCREW TO SPECIFIED RPM ON LOW STEP OF CAM, IN DRIVE.
6. ON AIR CONDITIONED CARS —— AFTER COMPLETING ABOVE ADJUSTMENTS, DISCONNECT ELECTRICAL CONNECTOR AT A/C COMPRESSOR CLUTCH, TURN A/C SWITCH ON AND ADJUST SOLENOID (ENERGIZED) SCREW TO SPECIFIED RPM, IN DRIVE. TURN ENGINE OFF. RECONNECT ELECTRICAL CONNECTOR.
7. RECONNECT VACUUM HOSES TO PORTS Ⓗ AND Ⓚ ON CARBURETOR AND ④ ON INTAKE MANIFOLD.
8. INSTALL AIR CLEANER AND RECONNECT MOLDED CONNECTOR TO TVS SWITCH.

CATALYST-EGR-AIR
LOW ALTITUDE CERTIFICATION

TIMING(DEG BTDC @ RPM)	18° @ 1100 (IN PARK)
SPARK PLUGS GAP / TYPE	.080 IN. / AC R46SX
CARBURETOR SCREW(RPM)	500 (IN DRIVE)
FAST IDLE SCREW (RPM)	700 (IN DRIVE)
SOLENOID SCREW (RPM)	600 (IN DRIVE)

IDLE MIXTURE SCREWS ARE PRESET AND SEALED AT FACTORY. PROVISION FOR ADJUSTMENT DURING TUNE-UP IS NOT PROVIDED.

SEE SERVICE MANUAL AND MAINTENANCE SCHEDULE FOR ADDITIONAL INFORMATION.

EMISSION HOSE ROUTING

PRINTED IN U.S.A. THIS VEHICLE CONFORMS TO U.S. EPA REGULATIONS APPLICABLE TO 1980 MODEL YEAR NEW PASSENGER CARS. PART NO 22505241

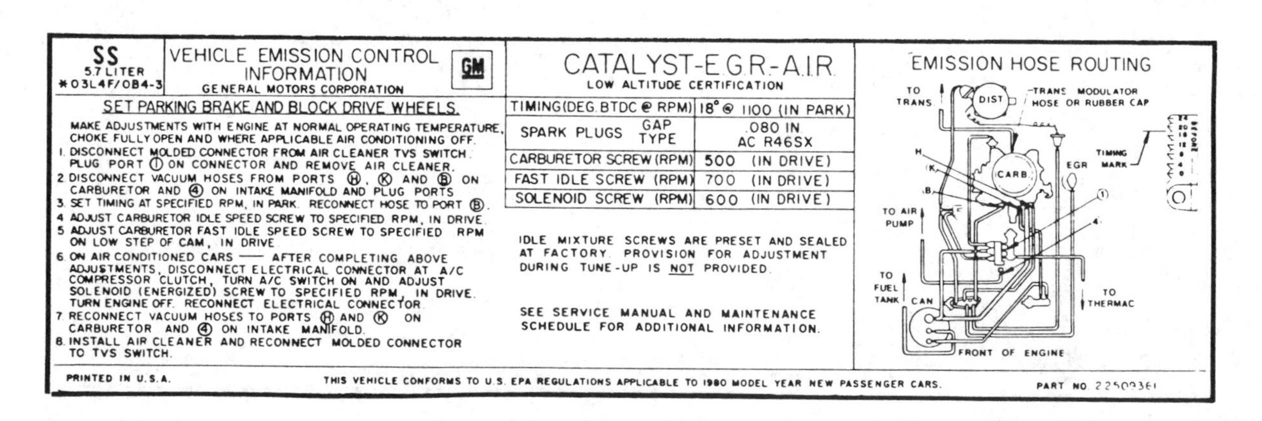

SS
5.7 LITER
*03L4F/0B4-3

VEHICLE EMISSION CONTROL INFORMATION
GENERAL MOTORS CORPORATION
GM

SET PARKING BRAKE AND BLOCK DRIVE WHEELS.

MAKE ADJUSTMENTS WITH ENGINE AT NORMAL OPERATING TEMPERATURE, CHOKE FULLY OPEN AND WHERE APPLICABLE AIR CONDITIONING OFF.

1. DISCONNECT MOLDED CONNECTOR FROM AIR CLEANER TVS SWITCH. PLUG PORT ① ON CONNECTOR AND REMOVE AIR CLEANER.
2. DISCONNECT VACUUM HOSES FROM PORTS Ⓗ, Ⓚ AND Ⓑ ON CARBURETOR AND ④ ON INTAKE MANIFOLD AND PLUG PORTS.
3. SET TIMING AT SPECIFIED RPM, IN PARK. RECONNECT HOSE TO PORT Ⓑ.
4. ADJUST CARBURETOR IDLE SPEED SCREW TO SPECIFIED RPM, IN DRIVE.
5. ADJUST CARBURETOR FAST IDLE SPEED SCREW TO SPECIFIED RPM ON LOW STEP OF CAM, IN DRIVE.
6. ON AIR CONDITIONED CARS —— AFTER COMPLETING ABOVE ADJUSTMENTS, DISCONNECT ELECTRICAL CONNECTOR AT A/C COMPRESSOR CLUTCH, TURN A/C SWITCH ON AND ADJUST SOLENOID (ENERGIZED) SCREW TO SPECIFIED RPM, IN DRIVE. TURN ENGINE OFF. RECONNECT ELECTRICAL CONNECTOR.
7. RECONNECT VACUUM HOSES TO PORTS Ⓗ AND Ⓚ ON CARBURETOR AND ④ ON INTAKE MANIFOLD.
8. INSTALL AIR CLEANER AND RECONNECT MOLDED CONNECTOR TO TVS SWITCH.

CATALYST-EGR-AIR
LOW ALTITUDE CERTIFICATION

TIMING(DEG. BTDC @ RPM)	18° @ 1100 (IN PARK)
SPARK PLUGS GAP / TYPE	.080 IN / AC R46SX
CARBURETOR SCREW (RPM)	500 (IN DRIVE)
FAST IDLE SCREW (RPM)	700 (IN DRIVE)
SOLENOID SCREW (RPM)	600 (IN DRIVE)

IDLE MIXTURE SCREWS ARE PRESET AND SEALED AT FACTORY. PROVISION FOR ADJUSTMENT DURING TUNE-UP IS NOT PROVIDED.

SEE SERVICE MANUAL AND MAINTENANCE SCHEDULE FOR ADDITIONAL INFORMATION.

EMISSION HOSE ROUTING

PRINTED IN U.S.A. THIS VEHICLE CONFORMS TO U.S. EPA REGULATIONS APPLICABLE TO 1980 MODEL YEAR NEW PASSENGER CARS. PART NO 22509361

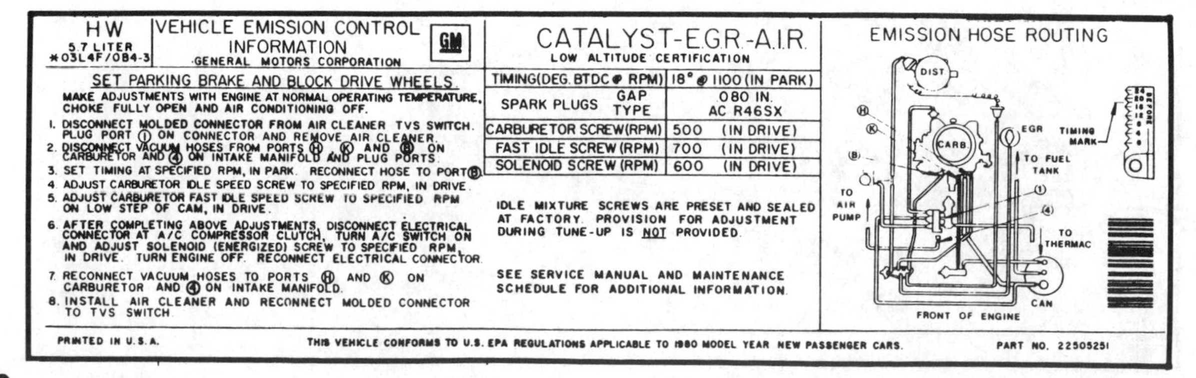

HW
5.7 LITER
*03L4F/0B4-3

VEHICLE EMISSION CONTROL INFORMATION
GENERAL MOTORS CORPORATION
GM

SET PARKING BRAKE AND BLOCK DRIVE WHEELS.

MAKE ADJUSTMENTS WITH ENGINE AT NORMAL OPERATING TEMPERATURE, CHOKE FULLY OPEN AND AIR CONDITIONING OFF.

1. DISCONNECT MOLDED CONNECTOR FROM AIR CLEANER TVS SWITCH. PLUG PORT ① ON CONNECTOR AND REMOVE AIR CLEANER.
2. DISCONNECT VACUUM HOSES FROM PORTS Ⓗ, Ⓚ AND Ⓑ ON CARBURETOR AND ④ ON INTAKE MANIFOLD AND PLUG PORTS.
3. SET TIMING AT SPECIFIED RPM, IN PARK. RECONNECT HOSE TO PORT Ⓑ.
4. ADJUST CARBURETOR IDLE SPEED SCREW TO SPECIFIED RPM, IN DRIVE.
5. ADJUST CARBURETOR FAST IDLE SPEED SCREW TO SPECIFIED RPM ON LOW STEP OF CAM, IN DRIVE.
6. AFTER COMPLETING ABOVE ADJUSTMENTS, DISCONNECT ELECTRICAL CONNECTOR AT A/C COMPRESSOR CLUTCH, TURN A/C SWITCH ON AND ADJUST SOLENOID (ENERGIZED) SCREW TO SPECIFIED RPM, IN DRIVE. TURN ENGINE OFF. RECONNECT ELECTRICAL CONNECTOR.
7. RECONNECT VACUUM HOSES TO PORTS Ⓗ AND Ⓚ ON CARBURETOR AND ④ ON INTAKE MANIFOLD.
8. INSTALL AIR CLEANER AND RECONNECT MOLDED CONNECTOR TO TVS SWITCH.

CATALYST-EGR-AIR
LOW ALTITUDE CERTIFICATION

TIMING(DEG. BTDC @ RPM)	18° @ 1100 (IN PARK)
SPARK PLUGS GAP / TYPE	.080 IN. / AC R46SX
CARBURETOR SCREW(RPM)	500 (IN DRIVE)
FAST IDLE SCREW (RPM)	700 (IN DRIVE)
SOLENOID SCREW (RPM)	600 (IN DRIVE)

IDLE MIXTURE SCREWS ARE PRESET AND SEALED AT FACTORY. PROVISION FOR ADJUSTMENT DURING TUNE-UP IS NOT PROVIDED.

SEE SERVICE MANUAL AND MAINTENANCE SCHEDULE FOR ADDITIONAL INFORMATION.

EMISSION HOSE ROUTING

PRINTED IN U.S.A. THIS VEHICLE CONFORMS TO U.S. EPA REGULATIONS APPLICABLE TO 1980 MODEL YEAR NEW PASSENGER CARS. PART NO. 22505251

VEHICLE EMISSION CONTROL LABELS AND VACUUM HOSE ROUTINGS

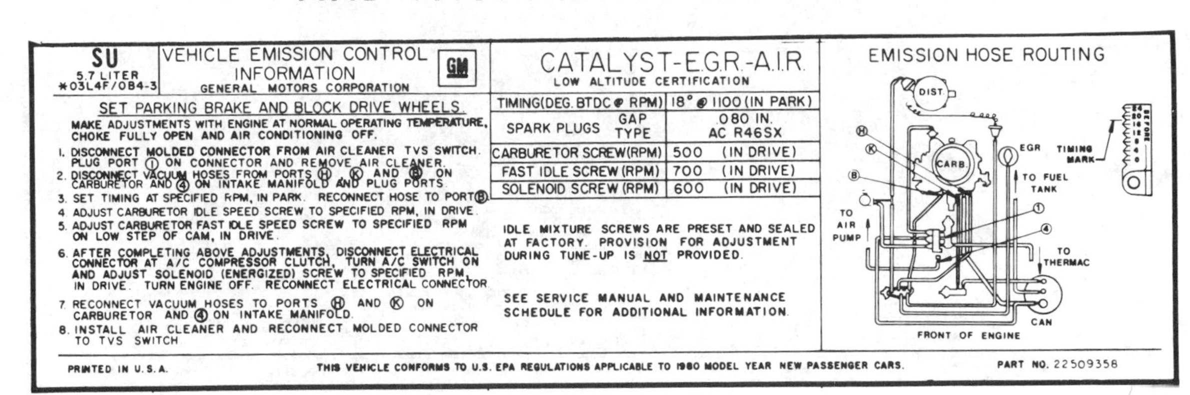

SU
5.7 LITER
*03L4F/084-3

VEHICLE EMISSION CONTROL INFORMATION
GENERAL MOTORS CORPORATION
GM

SET PARKING BRAKE AND BLOCK DRIVE WHEELS.
MAKE ADJUSTMENTS WITH ENGINE AT NORMAL OPERATING TEMPERATURE, CHOKE FULLY OPEN AND AIR CONDITIONING OFF.
1. DISCONNECT MOLDED CONNECTOR FROM AIR CLEANER TVS SWITCH. PLUG PORT ① ON CONNECTOR AND REMOVE AIR CLEANER.
2. DISCONNECT VACUUM HOSES FROM PORTS Ⓗ, Ⓚ AND Ⓑ ON CARBURETOR AND ④ ON INTAKE MANIFOLD AND PLUG PORTS.
3. SET TIMING AT SPECIFIED RPM, IN PARK. RECONNECT HOSE TO PORT Ⓑ.
4. ADJUST CARBURETOR IDLE SPEED SCREW TO SPECIFIED RPM, IN DRIVE.
5. ADJUST CARBURETOR FAST IDLE SPEED SCREW TO SPECIFIED RPM ON LOW STEP OF CAM, IN DRIVE.
6. AFTER COMPLETING ABOVE ADJUSTMENTS, DISCONNECT ELECTRICAL CONNECTOR AT A/C COMPRESSOR CLUTCH, TURN A/C SWITCH ON AND ADJUST SOLENOID (ENERGIZED) SCREW TO SPECIFIED RPM, IN DRIVE. TURN ENGINE OFF. RECONNECT ELECTRICAL CONNECTOR.
7. RECONNECT VACUUM HOSES TO PORTS Ⓗ AND Ⓚ ON CARBURETOR AND ④ ON INTAKE MANIFOLD.
8. INSTALL AIR CLEANER AND RECONNECT MOLDED CONNECTOR TO TVS SWITCH.

CATALYST-E.G.R.-A.I.R.
LOW ALTITUDE CERTIFICATION

TIMING (DEG. BTDC @ RPM)	18° @ 1100 (IN PARK)
SPARK PLUGS GAP / TYPE	.080 IN. / AC R46SX
CARBURETOR SCREW (RPM)	500 (IN DRIVE)
FAST IDLE SCREW (RPM)	700 (IN DRIVE)
SOLENOID SCREW (RPM)	600 (IN DRIVE)

IDLE MIXTURE SCREWS ARE PRESET AND SEALED AT FACTORY. PROVISION FOR ADJUSTMENT DURING TUNE-UP IS NOT PROVIDED.

SEE SERVICE MANUAL AND MAINTENANCE SCHEDULE FOR ADDITIONAL INFORMATION.

EMISSION HOSE ROUTING

PRINTED IN U.S.A. THIS VEHICLE CONFORMS TO U.S. EPA REGULATIONS APPLICABLE TO 1980 MODEL YEAR NEW PASSENGER CARS. PART NO. 22509358

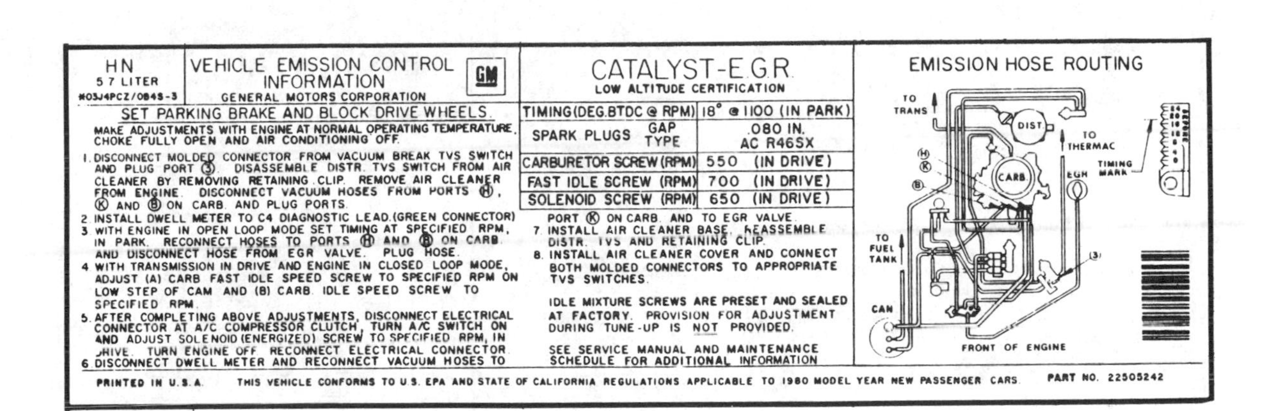

HN
5.7 LITER
*03J4PCZ/084S-3

VEHICLE EMISSION CONTROL INFORMATION
GENERAL MOTORS CORPORATION
GM

SET PARKING BRAKE AND BLOCK DRIVE WHEELS.
MAKE ADJUSTMENTS WITH ENGINE AT NORMAL OPERATING TEMPERATURE, CHOKE FULLY OPEN AND AIR CONDITIONING OFF.
1. DISCONNECT MOLDED CONNECTOR FROM VACUUM BREAK TVS SWITCH AND PLUG PORT ③. DISASSEMBLE DISTR. TVS SWITCH FROM AIR CLEANER BY REMOVING RETAINING CLIP. REMOVE AIR CLEANER FROM ENGINE. DISCONNECT VACUUM HOSES FROM PORTS Ⓗ, Ⓚ AND Ⓑ ON CARB. AND PLUG PORTS.
2. INSTALL DWELL METER TO C4 DIAGNOSTIC LEAD (GREEN CONNECTOR).
3. WITH ENGINE IN OPEN LOOP MODE SET TIMING AT SPECIFIED RPM, IN PARK. RECONNECT HOSES TO PORTS Ⓗ AND Ⓑ ON CARB. AND DISCONNECT HOSE FROM EGR VALVE. PLUG HOSE.
4. WITH TRANSMISSION IN DRIVE AND ENGINE IN CLOSED LOOP MODE, ADJUST (A) CARB FAST IDLE SPEED SCREW TO SPECIFIED RPM ON LOW STEP OF CAM AND (B) CARB. IDLE SPEED SCREW TO SPECIFIED RPM.
5. AFTER COMPLETING ABOVE ADJUSTMENTS, DISCONNECT ELECTRICAL CONNECTOR AT A/C COMPRESSOR CLUTCH, TURN A/C SWITCH ON AND ADJUST SOLENOID (ENERGIZED) SCREW TO SPECIFIED RPM, IN DRIVE. TURN ENGINE OFF. RECONNECT ELECTRICAL CONNECTOR.
6. DISCONNECT DWELL METER AND RECONNECT VACUUM HOSES TO PORT Ⓚ ON CARB. AND TO EGR VALVE.
7. INSTALL AIR CLEANER BASE. REASSEMBLE DISTR. TVS AND RETAINING CLIP.
8. INSTALL AIR CLEANER COVER AND CONNECT BOTH MOLDED CONNECTORS TO APPROPRIATE TVS SWITCHES.

CATALYST-E.G.R.
LOW ALTITUDE CERTIFICATION

TIMING (DEG. BTDC @ RPM)	18° @ 1100 (IN PARK)
SPARK PLUGS GAP / TYPE	.080 IN. / AC R46SX
CARBURETOR SCREW (RPM)	550 (IN DRIVE)
FAST IDLE SCREW (RPM)	700 (IN DRIVE)
SOLENOID SCREW (RPM)	650 (IN DRIVE)

IDLE MIXTURE SCREWS ARE PRESET AND SEALED AT FACTORY. PROVISION FOR ADJUSTMENT DURING TUNE-UP IS NOT PROVIDED.

SEE SERVICE MANUAL AND MAINTENANCE SCHEDULE FOR ADDITIONAL INFORMATION.

EMISSION HOSE ROUTING

PRINTED IN U.S.A. THIS VEHICLE CONFORMS TO U.S. EPA AND STATE OF CALIFORNIA REGULATIONS APPLICABLE TO 1980 MODEL YEAR NEW PASSENGER CARS. PART NO. 22505242

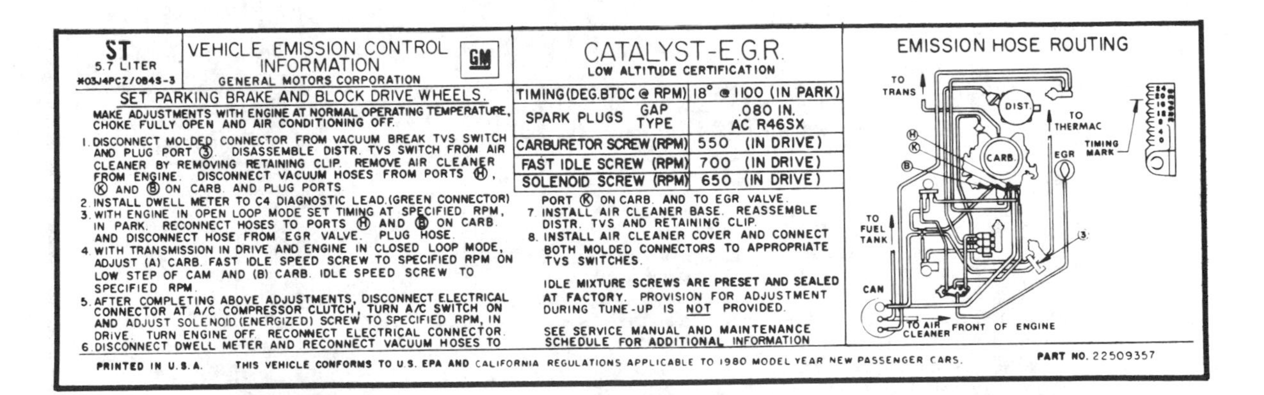

ST
5.7 LITER
*03J4PCZ/084S-3

VEHICLE EMISSION CONTROL INFORMATION
GENERAL MOTORS CORPORATION
GM

SET PARKING BRAKE AND BLOCK DRIVE WHEELS.
MAKE ADJUSTMENTS WITH ENGINE AT NORMAL OPERATING TEMPERATURE, CHOKE FULLY OPEN AND AIR CONDITIONING OFF.
1. DISCONNECT MOLDED CONNECTOR FROM VACUUM BREAK TVS SWITCH AND PLUG PORT ③. DISASSEMBLE DISTR. TVS SWITCH FROM AIR CLEANER BY REMOVING RETAINING CLIP. REMOVE AIR CLEANER FROM ENGINE. DISCONNECT VACUUM HOSES FROM PORTS Ⓗ, Ⓚ AND Ⓑ ON CARB. AND PLUG PORTS.
2. INSTALL DWELL METER TO C4 DIAGNOSTIC LEAD (GREEN CONNECTOR).
3. WITH ENGINE IN OPEN LOOP MODE SET TIMING AT SPECIFIED RPM, IN PARK. RECONNECT HOSES TO PORTS Ⓗ AND Ⓑ ON CARB. AND DISCONNECT HOSE FROM EGR VALVE. PLUG HOSE.
4. WITH TRANSMISSION IN DRIVE AND ENGINE IN CLOSED LOOP MODE, ADJUST (A) CARB. FAST IDLE SPEED SCREW TO SPECIFIED RPM ON LOW STEP OF CAM AND (B) CARB. IDLE SPEED SCREW TO SPECIFIED RPM.
5. AFTER COMPLETING ABOVE ADJUSTMENTS, DISCONNECT ELECTRICAL CONNECTOR AT A/C COMPRESSOR CLUTCH, TURN A/C SWITCH ON AND ADJUST SOLENOID (ENERGIZED) SCREW TO SPECIFIED RPM, IN DRIVE. TURN ENGINE OFF. RECONNECT ELECTRICAL CONNECTOR.
6. DISCONNECT DWELL METER AND RECONNECT VACUUM HOSES TO PORT Ⓚ ON CARB. AND TO EGR VALVE.
7. INSTALL AIR CLEANER BASE. REASSEMBLE DISTR. TVS AND RETAINING CLIP.
8. INSTALL AIR CLEANER COVER AND CONNECT BOTH MOLDED CONNECTORS TO APPROPRIATE TVS SWITCHES.

CATALYST-E.G.R.
LOW ALTITUDE CERTIFICATION

TIMING (DEG. BTDC @ RPM)	18° @ 1100 (IN PARK)
SPARK PLUGS GAP / TYPE	.080 IN. / AC R46SX
CARBURETOR SCREW (RPM)	550 (IN DRIVE)
FAST IDLE SCREW (RPM)	700 (IN DRIVE)
SOLENOID SCREW (RPM)	650 (IN DRIVE)

IDLE MIXTURE SCREWS ARE PRESET AND SEALED AT FACTORY. PROVISION FOR ADJUSTMENT DURING TUNE-UP IS NOT PROVIDED.

SEE SERVICE MANUAL AND MAINTENANCE SCHEDULE FOR ADDITIONAL INFORMATION.

EMISSION HOSE ROUTING

PRINTED IN U.S.A. THIS VEHICLE CONFORMS TO U.S. EPA AND CALIFORNIA REGULATIONS APPLICABLE TO 1980 MODEL YEAR NEW PASSENGER CARS. PART NO. 22509357

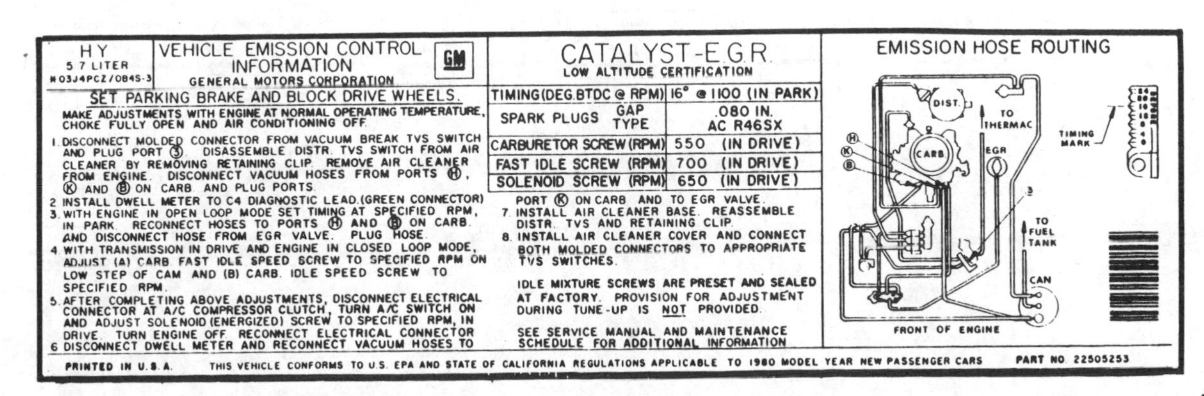

HY
5.7 LITER
*03J4PCZ/084S-3

VEHICLE EMISSION CONTROL INFORMATION
GENERAL MOTORS CORPORATION
GM

SET PARKING BRAKE AND BLOCK DRIVE WHEELS.
MAKE ADJUSTMENTS WITH ENGINE AT NORMAL OPERATING TEMPERATURE, CHOKE FULLY OPEN AND AIR CONDITIONING OFF.
1. DISCONNECT MOLDED CONNECTOR FROM VACUUM BREAK TVS SWITCH AND PLUG PORT ③. DISASSEMBLE DISTR. TVS SWITCH FROM AIR CLEANER BY REMOVING RETAINING CLIP. REMOVE AIR CLEANER FROM ENGINE. DISCONNECT VACUUM HOSES FROM PORTS Ⓗ, Ⓚ AND Ⓑ ON CARB. AND PLUG PORTS.
2. INSTALL DWELL METER TO C4 DIAGNOSTIC LEAD (GREEN CONNECTOR).
3. WITH ENGINE IN OPEN LOOP MODE SET TIMING AT SPECIFIED RPM, IN PARK. RECONNECT HOSES TO PORTS Ⓗ AND Ⓑ ON CARB. AND DISCONNECT HOSE FROM EGR VALVE. PLUG HOSE.
4. WITH TRANSMISSION IN DRIVE AND ENGINE IN CLOSED LOOP MODE, ADJUST (A) CARB FAST IDLE SPEED SCREW TO SPECIFIED RPM ON LOW STEP OF CAM AND (B) CARB. IDLE SPEED SCREW TO SPECIFIED RPM.
5. AFTER COMPLETING ABOVE ADJUSTMENTS, DISCONNECT ELECTRICAL CONNECTOR AT A/C COMPRESSOR CLUTCH, TURN A/C SWITCH ON AND ADJUST SOLENOID (ENERGIZED) SCREW TO SPECIFIED RPM, IN DRIVE. TURN ENGINE OFF. RECONNECT ELECTRICAL CONNECTOR.
6. DISCONNECT DWELL METER AND RECONNECT VACUUM HOSES TO PORT Ⓚ ON CARB. AND TO EGR VALVE.
7. INSTALL AIR CLEANER BASE. REASSEMBLE DISTR. TVS AND RETAINING CLIP.
8. INSTALL AIR CLEANER COVER AND CONNECT BOTH MOLDED CONNECTORS TO APPROPRIATE TVS SWITCHES.

CATALYST-E.G.R.
LOW ALTITUDE CERTIFICATION

TIMING (DEG. BTDC @ RPM)	16° @ 1100 (IN PARK)
SPARK PLUGS GAP / TYPE	.080 IN. / AC R46SX
CARBURETOR SCREW (RPM)	550 (IN DRIVE)
FAST IDLE SCREW (RPM)	700 (IN DRIVE)
SOLENOID SCREW (RPM)	650 (IN DRIVE)

IDLE MIXTURE SCREWS ARE PRESET AND SEALED AT FACTORY. PROVISION FOR ADJUSTMENT DURING TUNE-UP IS NOT PROVIDED.

SEE SERVICE MANUAL AND MAINTENANCE SCHEDULE FOR ADDITIONAL INFORMATION.

EMISSION HOSE ROUTING

PRINTED IN U.S.A. THIS VEHICLE CONFORMS TO U.S. EPA AND STATE OF CALIFORNIA REGULATIONS APPLICABLE TO 1980 MODEL YEAR NEW PASSENGER CARS. PART NO. 22505253

VEHICLE EMISSION CONTROL LABELS AND VACUUM HOSE ROUTINGS

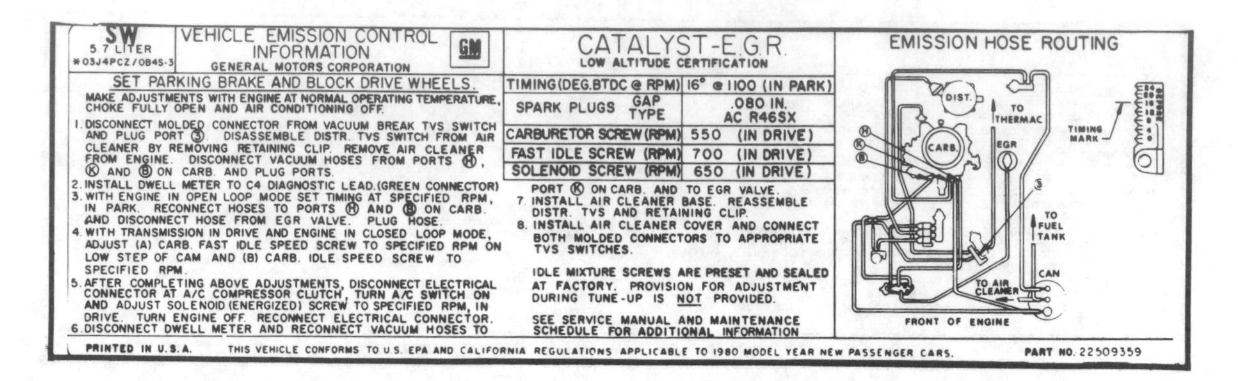

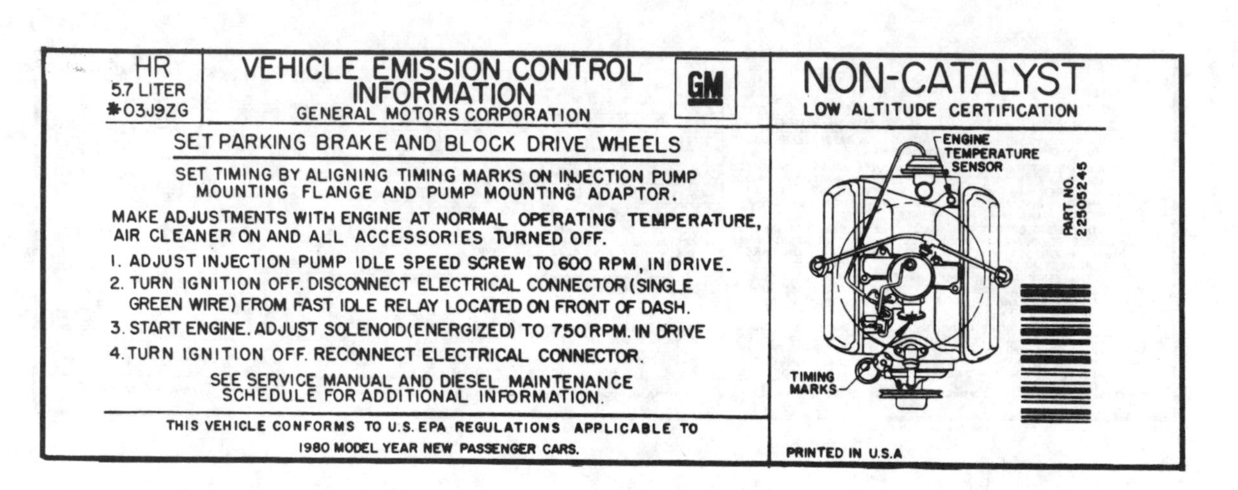

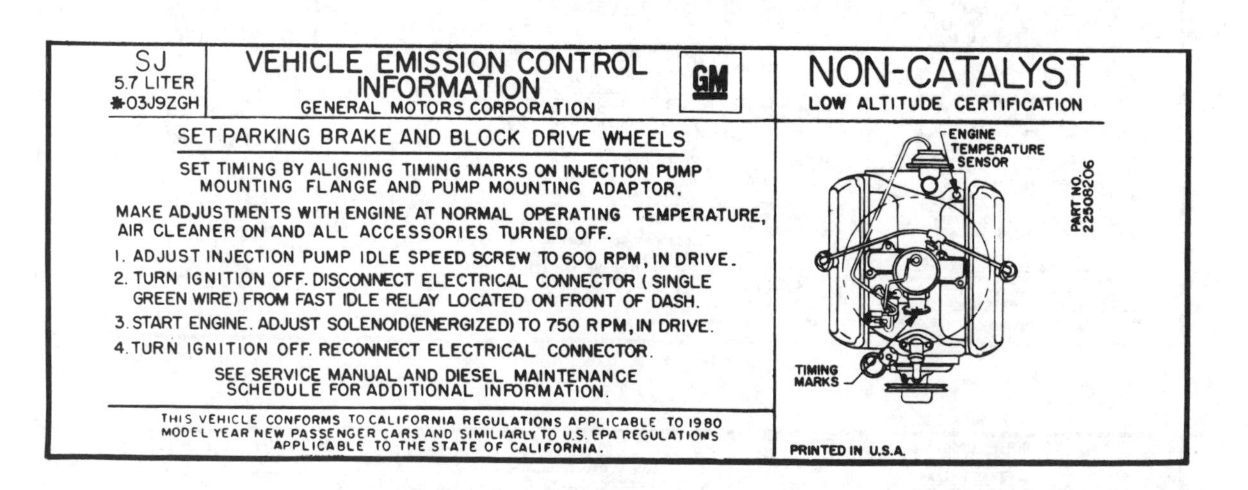

TUNE-UP SPECIFICATIONS

EXCEPT ASTRE, SUNBIRD & '80 PHOENIX

YEAR	ENG. V.I.N. Code	ENGINE No. Cyl Disp. (cu in)	SPARK PLUGS* Orig. Type	SPARK PLUGS* Gap (in)	DIST.	IGNITION TIMING (deg BTDC) Man Trans.	IGNITION TIMING (deg BTDC) Auto Trans.	CARBURETION HOT IDLE SPEED Lo-Aft Man. Trans In (N)	Lo-Aft Auto Trans In (Drive)	Calif Man Trans N	Calif Auto Trans In Drive	Hi-Alt Man Trans N	Hi-Alt Auto Trans In Drive
1979	V	4-151	R43TSX	.060	E.I.	13 C 14	12 C 14	900	600/675	1000	850	—	—
	2	6-231	R46TSX	.060	E.I.	15	15	800	600	800	600	800	600
	A	6-231	R46TSX	.060	E.I.	15	15	800	600	800	600	800	600
	Y	8-301	R46TSX	.060	E.I.	12	14	800	650	800	650	—	—
	W	8-301	R45TSX	.060	E.I.	—	12	—	650	—	—	—	—
	G	8-305	R45TS	.045	E.I.	4	4	700	600/675	—	600/700	—	—
	H	8-305	R46SZ	.080	E.I.	4	4	—	—	—	600	—	500
	L	8-350	R46SZ	.080	E.I.	—	15②	—	600	—	600	—	600
	R	8-350	R45TSX	.060	E.I.	—	20@ 1100	—	600/675	—	—	—	—
	X	8-350	R46TSX	.080	E.I.	—	15	—	600/675	—	—	—	—
	Z	8-400	R45TSX	.060	E.I.	18	18	775	650	—	—	—	—
	K	8-403	R46SZ	.080	E.I.	—	18 C 20	—	650	—	600	—	—
1980	K	6-229	R45TSX	.045	E.I.	④	①	750	600	800	600	800	600
	A	6-231	R46TSX	.060	E.I.	15	15	800	670	800	600	800	600
	S	8-265	R45TSX	.060	E.I.	④	④	700	650	700	600	—	—
	W	8-301	R45TSX	.060	E.I.	—	12	—	600/675	—	600/700	—	—
	T	8-301	R45TSX	.060	E.I.	Turbo-	④		600/675	—	600/700	—	—
	H	8-305	R45TSX	.045	E.I.	4	4	700	600/675	—	600/700	—	500
	X	8-350	R46TSX	.080	E.I.	—	15	—	670	—	—	—	—
	R	8-350	R45TSX	.060	E.I.	—	20@ 1100	—	600/675	—	600/700	—	—
	N	8-350	DIESEL	—	E.I.	—	—	—	575	—	—	—	—

NOTE: Should the information provided in this manual deviate from the specifications on the underhood tune-up label, the label specifications should be used, as they may reflect production changes.

NOTE: Where two idle speeds are listed, the higher speed is with the air conditioner on.

① Cal. and high alt. R45TSX
② Cal. 20°—high alt. 8°
③ Calif.—12
A 6 cylinder, dwell 32½°, gap .019
8 cylinder, dwell 30°, gap .019
C California
④ See engine decal

DISTRIBUTOR SPECIFICATIONS

EXCEPT ASTRE, SUNBIRD & '80 PHOENIX

YEAR	DIST. IDENT.	CENTRIFUGAL ADVANCE Start Crank. Deg. @ Eng. RPM	CENTRIFUGAL ADVANCE Finish Crank. Deg. @ Eng. RPM	VACUUM ADVANCE Start @ In hg	VACUUM ADVANCE Finish Crank. Deg. @ In hg
1979	1103281	0 @ 1000	20 @ 3800	4	18 @ 12
	1103282	0 @ 1000	20 @ 3800	4	20 @ 10
	1103285	0 @ 1200	22 @ 4200	4	20 @ 8
	1103310	0 @ 1000	14 @ 4400	4	25 @ 12
	1103314	0 @ 825	21.5 @ 3400	4	25 @ 12
	1103315	0 @ 1000	20 @ 4400	5	25 @ 11
	1103323	0 @ 1000	19 @ 4000	5	16 @ 11
	1103325	0 @ 1000	13 @ 3600	5	16 @ 11
	1103337	0 @ 1100	16 @ 2400	4	24 @ 10
	1103346	0 @ 1000	19 @ 4000	6	24 @ 13

DISTRIBUTOR SPECIFICATIONS

EXCEPT ASTRE, SUNBIRD & '80 PHOENIX

YEAR	DIST. IDENT.	CENTRIFUGAL ADVANCE Start Crank. Deg. @ Eng. RPM	CENTRIFUGAL ADVANCE Finish Crank. Deg. @ Eng. RPM	VACUUM ADVANCE Start @ In hg	VACUUM ADVANCE Finish Crank. Deg. @ In hg
1980	1110752	2 @ 1550	14 @ 4100	4.0	16 @ 7.5
	1110554	2 @ 1950	15 @ 5000	3.5	24 @ 13.5
	1110576	2 @ 1200	14 @ 3400	3.5	24 @ 8.5
	1110552	2 @ 1200	14 @ 3400	3.5	24 @ 8.5
	1110769	2 @ 1200	15 @ 3600	4.5	24 @ 9.5
	1110555	2 @ 1200	15 @ 3600	4.5	24 @ 9.5
	1110784	Electronic Spark Timing (EST)			
	1103450	2 @ 1050	18 @ 5000	4.5	20 @ 10.5
	1103425	2 @ 1250	18 @ 4600	4.5	20 @ 10.5
	1103407	2 @ 1200	23 @ 4400	4.5	20 @ 10.5
	1103444	1 @ 1200	14 @ 4400	7.5	19 @ 13.5
	1103386	2 @ 1150	20 @ 3800	5.0	20.5 @ 8.5
	1103447	2 @ 2200	17 @ 4400	7.0	24 @ 14.6
	1103417	2 @ 2200	17 @ 4400	6.0	24 @ 13.5
	1103413	2 @ 1400	12 @ 5000	6.5	30 @ 15.8

QUADRAJET CARBURETOR SPECIFICATIONS

PONTIAC

Rochester Carburetor

Year	Carburetor Identification①	Float Level (in.)	Air Valve Spring (turn)	Pump Rod (in.)	Primary Vacuum Break (in.)	Secondary Vacuum Break (in.)	Secondary Opening (in.)	Choke Rod (in.)	Choke Unloader (in.)	Fast Idle Speed② (rpm)
1979	17058263	17/32	5/8	3/8	0.164	0.243	④	0.129	0.220	⑤
	17059250, 253	13/32	1/2	9/32	0.129	0.183	④	0.096	0.220	⑤
	17059241	5/16	3/4	3/8	0.120	0.113	④	0.096	0.243	⑤
	17059271	9/16	5/8	3/8	0.142	0.227	④	0.110	0.203	⑤
	17059272	15/32	5/8	3/8	0.136	0.195	④	0.074	0.220	⑤
	17059502, 504	15/32	7/8	1/4	0.164	—	④	0.243	0.243	⑤
	17059553	13/32	1/2	9/32	0.136	0.230	④	0.103	0.220	⑤
	17059582, 584	15/32	7/8	11/32	0.203	—	④	0.243	0.314	⑤
1980	17080249	7/16	3/4	9/32⑥	0.129	0.114	④	0.096	0.243	③
	17080270	15/32	5/8	3/8⑦	0.149	0.211	④	0.074	0.220	③
	17080272	15/32	5/8	3/8⑦	0.129	0.175	④	0.074	0.203	③
	17080274	15/32	5/8	5/16⑥	0.110	0.164	④	0.083	0.203	③
	17080502	1/2	7/8	Fixed	0.136	0.179	④	0.110	0.243	③
	17080504	1/2	7/8	Fixed	0.136	0.179	④	0.110	0.243	③
	17080553	15/32	1/2	Fixed	0.142	0.220	④	0.090	0.220	③

① The carburetor identification number is stamped on the float bowl, near the secondary throttle lever.
② On low step.
③ 1½ turns after contacting lever for preliminary setting
④ No measurement necessary on two point linkage; see text.
⑤ 2 turns after contacting lever for preliminary setting.
⑥ Inner hole
⑦ Outer hole

2SE, E2SE CARBURETOR ADJUSTMENTS

Pontiac (except Phoenix)

Year	Carburetor Identification	Float Level (in.)	Pump Rod (in.)	Fast Idle (rpm)	Choke Coil Lever (in.)	Fast Idle Cam (deg./in.)	Air Valve Rod (in.)	Primary Vacuum Break (deg./in.)	Choke Setting (notches)	Secondary Vacuum Break (deg./in.)	Choke Unloader (deg./in.)	Secondary Lockout (in.)
1979	17059674	3/16	1/2	2400	.120	18/0.096	.025	19/.103	2 Rich	—	32/.195	.01-.04
	17059675	3/16	17/32	2200	.120	18/0.096	.025	21/.117	1 Rich	—	32/.195	.01-.04
	17059676	3/16	1/2	2400	.120	18/0.096	.025	19/.103	2 Rich	—	32/.195	.01-.04
	17059677	3/16	17/32	2200	.120	18/0.096	.025	21/.117	1 Rich	—	32/.195	.01-.04
1980	17080674	3/16	1/2	①	.085	18/0.096	.018	19/.103	Fixed	—	32/.195	.012
	17080675	3/16	1/2	①	.085	18/0.096	.018	21/.117	Fixed	—	32/.195	.012
	17080676	3/16	1/2	①	.085	18/0.096	.018	19/.103	Fixed	—	32/.195	.012
	17080677	3/16	1/2	①	.085	18/0.096	.018	21/.117	Fixed	—	32/.195	.012
	17059774	5/32	1/2	①	.085	18/0.096	.018	19/.103	Fixed	—	32/.195	.012
	17059775	5/32	17/32	①	.085	18/0.096	.018	21/.117	Fixed	—	32/.195	.012
	17059776	5/32	1/2	①	.085	18/0.096	.018	19/.103	Fixed	—	32/.195	.012
	17059777	5/32	17/32	①	.085	18/0.096	.018	21/.117	Fixed	—	32/.195	.012

① See Underhood Decal

2MC, M2MC, M2ME, E2ME CARBURETOR SPECIFICATIONS

PONTIAC

Rochester Carburetor

Year	Carburetor Identification①	Float Level (in.)	Choke Rod (in.)	Choke Unloader (in.)	Vacuum Break Lean or Front (in.)	Vacuum Break Rich or Rear (in.)	Pump Rod (in.)	Choke Coil Lever (in.)	Automatic Choke (notches)
1979	17059134, 135, 136, 137	13/32	0.243	0.243	0.157	—	9/32②	0.120	1 Lean
	17059180, 190, 191	11/32	0.139	0.243	0.103	0.090	1/4②	0.120	2 Rich
	17059160	11/32	0.110	0.195	0.129	0.203	9/32②	0.120	2 Rich
	17059196	11/32	0.139	0.277	0.129	0.117	1/4②	0.120	1 Rich
	17059434, 436	13/32	0.243	0.243	0.164	—	9/32②	0.120	2 Lean
	17059492, 498	11/32	0.139	0.277	0.129	0.117	9/32②	0.120	2 Rich
	17059430, 432	9/32	0.243	0.243	0.171	—	9/32②	0.120	1 Lean
	17059491	11/32	0.139	0.277	0.129	0.117	9/32②	0.120	1 Rich
1980	17080130, 131, 132, 133, 146, 147, 148, 149	11/32	0.110	0.243	0.142	—	1/4②	0.120	Fixed
	17080160	5/16	0.110	0.243	0.168	0.207	1/4②	0.120	Fixed
	17080190	9/32	0.074	0.243	0.123	0.110	1/4②	0.120	Fixed
	17080191	11/32	0.139	0.243	0.096	0.096	1/4②	0.120	Fixed
	17080192	9/32	0.139	0.243	0.096	0.110	1/4②	0.120	Fixed
	17080195	9/32	0.139	0.243	0.103	0.071	1/4②	0.120	Fixed
	17080197	9/32	0.139	0.243	0.103	0.071	1/4②	0.120	Fixed
	17080490	5/16	0.139	0.243	0.117	0.203	1/4②	0.120	Fixed
	17080491	5/16	0.139	0.243	0.117	0.220	1/4②	0.120	Fixed
	17080492	5/16	0.139	0.243	0.117	0.203	1/4②	0.120	Fixed
	17080493	5/16	0.139	0.243	0.117	0.179	3/8	0.120	Fixed
	17080494	5/16	0.139	0.243	0.117	0.179	1/4②	0.120	Fixed

2MC, M2MC, M2ME, E2ME CARBURETOR SPECIFICATIONS (Cont'd)

PONTIAC

Rochester Carburetor

Year	Carburetor Identification①	Float Level (in.)	Choke Rod (in.)	Choke Unloader (in.)	Vacuum Break Lean or Front (in.)	Vacuum Break Rich or Rear (in.)	Pump Rod (in.)	Choke Coil Lever (in.)	Automatic Choke (notches)
1980	17080495	5/16	0.139	0.243	0.117	0.179	3/8	0.120	Fixed
	17080496	5/16	0.139	0.243	0.117	0.203	3/8	0.120	Fixed
	17080498	5/16	0.139	0.243	0.117	0.203	3/8	0.120	Fixed

① The carburetor identification number is stamped on the float bowl, next to the fuel inlet nut.
② Inner hole

CAR SERIAL NUMBER AND ENGINE IDENTIFICATION

1979

The vehicle identification number is mounted behind the windshield on the driver's side. The sixth character is the model year, with 9 for 1979. The fifth character is the engine code, as follows:

A ....231 2-bbl. V-6 LD-5 Buick
H ....305 4-bbl. V-8 LG-4 Chev.
K ....403 4-bbl. V-8 L-80 Olds.
L ....350 4-bbl. V-8 LM-1 Chev.
R ....350 4-bbl. V-8 L-34 Olds.
V ......151 2-bbl. 4-cyl. LX-8 Pont.
W ....301 4-bbl. V-8 L-37 Pont.
X ....350 4-bbl. V-8 L-77 Buick
Y ....301 2-bbl. V-8 L-27 Pont.
Z ....400 4-bbl. V-8 L-78 Pont. (w/Perf. Pkg. W-72)
Z ....400 4-bbl. V-8 W-72 Pont.
1 ....151 2-bbl. 4-cyl. LS-6 Pont.

Vin code "V" represents the 151 CID, 4-cyl. engine used in the 49-States, while the Vin code "1" represents the 151 CID engine used in the state of California.

1980

The vehicle identification number is mounted behind the windshield on the driver's side. The sixth character is the model year, with A for 1980. The fifth character is the engine code, as follows:

A ......231 (3.8L) 2bbl. V-6 LD-5 Buick
H ......305 (5.0L) 4bbl. V-8 LG-4 Chev.
K ......229 (3.8L) 2bbl. V-6 LC-3 Chev.
N ......350 (5.7L) Diesel V-8 LF-9 Olds.
R ......350 (5.7L) 4bbl. V-8 L-34 Olds.
S ......265 (4.3L) 2bbl. V-8 LS-5 Pont.
T ......301 (4.9L) Turbo V-8 LU-8 Pont.
W ......301 (4.9L) 4 bbl. V-8 L-37 Pont.
X ......350 (5.7L) 4bbl. V-8 L-77 Pont.

NOTE: *The 400 cu. in. V-8 has been cancelled. A new turbo-charged version of the 301 V-8 is available as well as the new 265 V-8 which is derived from the 301 engine.*

EMISSION EQUIPMENT

All Models

Closed positive crankcase ventilation
Emission calibrated distributor
Emission calibrated carburetor
Heated air cleaner
Vapor control, canister storage
Exhaust gas recirculation system w/back pressure transducer
Catalytic converter
Air pump
49-States
Not used
Calif.
All V-6 and V-8 models
Not used on 4-cyl.
Altitude
Series H—231 CID—2-bbl. V-6 w/auto. trans.
Series A & G—305 CID—4 bbl. V-8 w/auto. trans.
Series A Wagon, F & X—350 CID—4-bbl. V-8 w/auto. trans.
Early fuel evaporation
All models, except Olds-built engines and 151 CID
Electric choke
Idle solenoid
Choke vacuum break
Pulsair system
Computer Controlled Catalyic Converter System (C-4)
4-cyl. engines for 49-States use
Temperature Vacuum Switch (TVS)
Distributor Spark-Vacuum Modulated Valve (DS-VMV)
Distributor Spark-Thermal Vacuum Switch (DS-TVS)

IDLE SPEED AND MIXTURE ADJUSTMENTS

1979

The idle mixture screws have been preset and sealed at the factory. Idle mixture should only be adjusted in cases of major carburetor overhaul, throttle body replacement or high idle CO levels, as determined by state or local inspections, to obtain the correct idle speed with the use of the propane enrichment method. Adjustment of the idle speed by any other method, may violate emission laws of the Federal, States or Provincial agencies.

Refer to the Emission Control Information label to obtain the correct idle speed, timing adjustment, disconnection of hoses or tubes and the location of the propane enrichment hose to the carburetor. The ECI Label reflects any manufacturing changes to the engine, concerning the emission levels.

Air cleaner ....In place
Air cond. ....Off
Auto. trans. ....Drive
Vapor hose to carb. ....Plugged
EGR hose to carb. ....Plugged
Idle CO ....Not used
Mixture adj. See propane speed below

IDLE SPEED AND MIXTURE ADJUSTMENTS

151 CID 4-cyl. Low Alt. Vin code V
w/Auto. trans.
Curb or "on" idle 850D
Slow or "off" idle 650D
(500 D—w/ AC)
Propane enriched idle 695
w/Manual trans.
Curb or "on" idle 1250N
Slow or "off" idle 900N
(500 N—w/AC)
Propane enriched idle 1040
151 CID 4-cyl. Calif. Vin code 1
w/auto. trans.
Curb or "on" idle 850D
Slow or "off" idle 650D
(500 D—w/AC)
Propane enriched idle ... ②
w/Manual trans.
Curb or "on" idle 1000N
Slow or "off" idle 500N
Propane enriched idle ... ②
231 CID V-6 Low Alt. Vin code A
w/Auto. trans.
Curb or "on" idle 670D
Slow or "off" idle 550D
Propane enriched idle 575
w/Manual trans.
Curb or "on" idle 800N
Slow or "off" idle 600N
Propane enriched idle 1000
231 CID V-6 Calif. Vin Code A
w/Auto. trans.
Curb or "on" idle .... N/A
Slow or "off" idle 600D
Propane enriched idle 615
w/Manual trans.
Curb or "on" idle 800N
Slow or "off" idle 600N
Propane enriched idle 840
231 CID V-6 High Alt. Vin code A
w/Auto. trans.
Curb or "on" idle .... N/A
Slow or "off" idle 600D
Propane enriched idle 615
301 CID 2-bbl. V-8 Low Alt. Vin code Y
w/Auto. trans.
Curb or "on" idle 650D
Slow or "off" idle 500D
Propane enriched idle 530
301 CID 4-bbl. V-8 Low Alt. Vin code W
w/Auto. trans.
Curb or "on" idle 650D
Slow or "off" idle 500D
Propane enriched idle 540
w/Manual trans.
Curb or "on" idle 800N
Slow or "off" idle 700N
Propane enriched idle ... 810N
305 CID 2-bbl. V-8 Low Alt. Vin code G
w/Auto. trans.
Curb or "on" idle 600D
Slow or "off" idle 500D
Propane enriched idle ..520-540
w/Manual trans.
Curb or "on" idle 700N
Slow or "off" idle 600N
Propane enriched idle ..710-750
305 CID 2-bbl. V-8 Calif. Vin code G
w/Auto trans.
Curb or "on" idle 650D
Slow or "off" idle 600D
Propane enriched idle ..640-660
305 CID 4-bbl. V-8 Calif. Vin code H
w/Auto trans.
Curb or "on" idle 600D
Slow or "off" idle 500D
Propane enriched idle ..540-560
305 CID 4-bbl. V-8 High Alt. Vin code H
w/Auto. trans.
Curb or "on" idle 600D
Slow or "off" idle 500D
Propane enriched idle ..640-660
350 CID 4-bbl. V-8 Low Alt. Vin code X
w/Auto. trans.
Curb or "on" idle .... N/A
Slow or "off" idle 550D
Propane enriched idle 800
350 CID 4-bbl. V-8 Calif. Vin code R & L
w/Auto. trans.
Curb or "on" idle 600D
Slow or "off" idle 500D
Propane enriched idle ... R-①, L-640-660
350 CID 4-bbl. V-8 High Alt. Vin code R & L
w/Auto. trans.
Curb or "on" idle 650D
Slow or "off" idle 600D
Propane enriched idle ... R-①, L-640-660
400 CID 4-bbl. V-8 Low Alt. Vin code Z
w/Manual trans.
Curb or "on" idle .... N/A
Slow or "off" idle 775N
Propane enriched idle ... 800
403 CID 4-bbl. V-8 Low Alt. Vin code K
w/Auto. trans.
Curb or "on" idle 650D
Slow or "off" idle 550D
Propane enriched idle ... ①
403 CID 4-bbl. V-8 Calif. Vin code K
w/Auto. trans.
Curb or "on" idle 600D
Slow or "off" idle 500D
Propane enriched idle ... ①
403 CID 4-bbl. V-8 High Alt. Vin code K
w/Auto. trans.
Curb or "on" idle .... ①
Slow or "off" idle .... ①
Propane enriched idle ... ①

N/A—Not Applicable
① Refer to the Emission Control Information Label
② Refer to the 1979 Emission specifications

1980

The idle mixture screws have been preset and sealed at the factory. Idle mixture should only be adjusted in cases of major carburetor overhaul, throttle body replacement or high idle CO levels, as determined by state or local inspections, to obtain the correct idle speed with the use of the propane enrichment method. Adjustment of the idle speed by any other method, may violate emission laws of the Federal, States or Provincial agencies.

Refer to the Emission Control Information label to obtain the correct idle speed, timing adjustment, disconnection of hoses or tubes and the location of the propane enrichment hose to the carburetor. The ECI Label reflects any manufacturing changes to the engine, concerning the emission levels.

Air cleaner ....In place
Air cond. ....Off
Auto. trans. ....Drive
Vapor hose to carb. ....Plugged
EGR hose to carb. ....Plugged
Idle CO ....Not used
Mixture adj. See propane speed below
229 CID Vin code K
w/Auto. trans.
Curb or "on' idle 675D
Slow or "off" idle 600D
Propane enriched idle 630-650
w/Manual trans.
Curb or "on" idle 750N
Slow or "off" idle 700N
Propane enriched idle 850-900
231 CID Vin code A
w/Auto. trans.
Curb or "on" idle 670D
Slow or "off" idle 550D
Propane enriched idle 600①-610
w/Manual trans.
Curb or "on" idle 800N
Slow or "off" idle 600N
Propane enriched idle 830
w/C4 system
Curb or "on" idle 620D
Slow or "off" idle 550D
Propane enriched idle .... ②
26 CID Vin code S
w/Auto. trans.
Curb or "on" idle 650D
Slow or "off" idle 550D
Propane enriched idle ..580-590
301 CID Vin code W
w/Auto. trans.
Curb or "on" idle 650D
Slow or "off" idle 500D
Propane enriched idle 640-650
w/Turbocharger Vin code T
Curb or "on" idle 650D
Slow or "off" idle 600D
Propane enriched idle 640-650
305 CID Vin code H
w/Auto. trans., C 4 system
Curb or "on" idle 650D
Slow or "off" idle 550D
Propane enriched idle .... ②
350 CID Vin code X
w/Auto. trans.
Curb or "on" idle 670D
Slow or "off" idle 550D
Propane enriched idle 590
w/C4 system Vin code R
Curb or "on" idle 650D
Slow or "off" idle 550D
Propane enriched idle .... ②

① A-body (Lemans)
② Propane enrichment method not to be used with C-4 system. ECM unit will respond to mixture adjustment with an opposite "correction" to the air/fuel mixture.

INITIAL TIMING

1979

NOTE: *Distributor vacuum hose must be disconnected and plugged. Set timing at idle speed unless shown otherwise.*

151 CID
- Low Alt.
  - w/Auto. trans. ....12° @ 650D
  - w/Manual trans. ...12° @ 900N
- Calif.
  - All ....14° @ 1000

231 CID
- Low Alt.
  - w/Auto trans. ......15° @ 600
  - w/Manual trans. ....15° @ 800
- Calif.
  - w/Auto trans. ......15° @ 600
  - w/Manual trans. ....15° @ 800
- High Alt.
  - w/Auto. trans. ......15° @ 600

301 CID
- Low Alt.
  - w/Auto. trans. (2 & 4-bbl.) .....12° @ 650D
  - w/Manual trans. (4-bbl.) ....14° @ 750N

305 CID
- Low Alt.
  - w/Auto. trans. (2-bbl.) .... 4° @ 500D
  - w/Manual trans. (2-bbl.) .... 4° @ 600N
- Calif.
  - w/Auto. trans. (2-bbl., series F, X) 4° @ 500D
  - w/Auto. trans. (2-bbl., series H) . 2° @ 600D
  - w/Auto. trans. (4-bbl) .... 4° @ 500D
- High Alt.
  - w/Auto. trans. .... 4° @ 600D

350 CID
- Low Alt.
  - w/Auto. trans. (4-bbl.) ....15° @ 550
- Calif.
  - w/Auto. trans. (4-bbl., series X) . 8° @ 500D
  - (Series B) ...20° @ 1100P
- High Alt.
  - w/Auto. trans. (4-bbl.) .... 8° @ 600D

400 CID
- Low Alt.
  - w/Manual trans. ...18° @ 775N

403 CID
- Low Alt.
  - w/Auto. trans. ...18° @ 1100P
- Calif.
  - w/Auto. trans. ...20° @ 1100P

1980

229 CID VIN Code K
- w/Auto. trans. ....10° @ 600D
- w/Manual trans. ..10° @ 700N

231 CID VIN Code A
- w/Auto. trans. ....15° @ 550D
- w/Manual trans. ..15° @ 800N
- w/C4 system
  - w/Auto. trans. ....15° @ 550D
  - w/Manual trans. ..See label

265 CID VIN Code S
- w/Auto. trans. ....10° @ 700D

301 CID VIN Code W (Code T-Turbo)
- w/Auto. trans. ....12° @ 500D
- w/Auto. trans. (YR, YN) ......12° @ 550D
- w/Turbocharger ... 8° @ 600D

305 CID VIN Code H
- w/Auto. trans. & C4 4° @ 550D

350 CID VIN Code X (Code R-w/C4)
- w/Auto. trans. ....15° @ 550D
- w/Auto. trans. & C4 18° @ 1100D

SPARK PLUGS

1979

151 4-cyl. ......AC-R46TSX.. .060
231 V-6 ....AC-R46TSX.. .060
301 2-bbl. ......AC-R46TSX.. .060
301 4-bbl. ......AC-R45TSX.. .060
305 2-bbl. ....AC-R45TS.. .045
350 V-8 "L" .....AC-R45TS.. .045
350 V-8 "R" .....AC-R46SZ.. .060
350 V-8 "X" ...AC-R46TSX.. .060
400 V-8 ....AC-R45TSX.. .060
403 V-8 ....AC-R46SZ.. .060

1980

229 V-6 ....R45TS....... .045
231 V-6 ....R45TS....... .040
231 V-6 w/C4, M.T. See label
265 V-8 ....R45TSX...... .060
301 V-8 ....R45TSX...... .060
305 V-8 ....R45TS....... .045
350 V-8 ....R45TSX...... .060
350 V-8 w/C4, A.T.R46SX.... .080

VACUUM ADVANCE
1979

Diaphragm Type ....Single
Vacuum Source
- 151 4-cyl.
  - Low Alt. (w/AC) ......Ported
  - Calif. ....Ported
  - Low Alt. (without AC) .Manifold
- 231 V-6 ....Ported
- 301 2-bbl. V-8 ....Manifold
- 301 4-bbl. V-8 ....Manifold
- 305 2-bbl. V-8 ....Manifold
- 350 V-8 ....Manifold
- 400 V-8 ....Manifold
- 403 V-8
  - Low and High Alt. ....Manifold
  - Calif. ....Ported

VACUUM SOURCE
1980

There are two types of vacuum source.
1. Ported vacuum
2. Manifold vacuum

Ported vacuum advance systems have a timed port in the carburetor throttle body above the throttle valve. The timed port provides vacuum during open throttle operations only. Manifold vacuum advance systems use manifold vacuum from either a manifold vacuum port on the carburetor or a fitting in the intake manifold.

Vacuum is provided whenever the engine is running. Various types of vacuum controls are used in the emission system to modify or control the operation of the different emission control components. Therefore, in many cases, manifold vacuum is used until the engine warms to the point where ported vacuum applied to a valve is greater than the calibrated vacuum from the manifold. When the ported vacuum is greater than the calibration vacuum, a full ported signal is provided to the distributor.

All of this will help provide good cold engine response and performance.

EMISSION CONTROL SYSTEMS

1980

Distributors used with the C-4 emission system have all spark timing done electronically. The vacuum and centrifugal advance mechanisms have been removed from the Electronic Spark Timing High Energy Ignition (EST-HEI) distributor.

Electronic Spark Control (ESC) is used on Vin code T turbocharged engines and Vin code W performance engines and it is a closed loop system that controls engine detonation by adjusting spark timing. The HEI's distributor electronic module is modified so it can respond to the controller signal. This command is delayed if detonation is occurring, thus providing the amount of retard needed as the amount of retard is a function of the degree of detonation.

In addition, a modified module is used with the Electronic Module Retard (EMR) system. This is a spark control system with an HEI module that has built in a timing retard feature. When the retard circuit is grounded, the firing of the spark plugs is delayed for a certain, calibrated number of crankshaft degrees. The grounding circuit is controlled by the Electronic Control Module (ECM). When the retard circuit is open, there is no delay and the plugs fire as controlled by engine speed and vacuum. If the EMR-HEI module is removed or replaced for any reason, the ignition timing must be checked and set to specifications.

Due to the addition of the three new HEI modules, it should be noted that replacement will require the exact equal of the removed part. A seven terminal module is used with EST units, a five terminal unit with the turbocharged engine, and another five terminal unit for the 5.7 engine (Vin code R) equipped with EMR. The two five terminal modules *cannot* be interchanged.

CLOSED POSITIVE CRANKCASE VENTILATION SYSTEM 1979-80

The closed PCV system is continued on all engines used by Pontiac Motor Division. The operation of the system remains the same.

ENGINE MODIFICATION 1979-80

The engine modification system is basically the same for 1980 as was used on the 1979 models.

Minor carburetor calibration changes are made to comply with the Federal and States mandated changes in the emission standards.

Before any adjustments are made, check the emission control information label, attached to the vehicle, for up-to-date information.

THERMOSTATIC AIR CLEANER (TAC) 1979-80

The TAC system is continued in use with little or no change.

EVAPORATION EMISSION CONTROL SYSTEM (EEC) 1979-80

The EEC system used is the same as that used on the 1979 models. Safeguards are built into the system to avoid the possibility of liquid fuel being drawn into the system.

The safeguards are;

1. A fuel tank overfill protector is installed on all models to provide adequate room for expansion of the fuel during temperature changes.
2. A fuel tank venting system is provided on all models to assure that the tank will be vented during any vehicle attitude. A domed fuel tank is used on sedans and coupes.
3. A pressure-vacuum relief valve, located in the gas cap, controls the fuel tank internal pressure.

EXHAUST GAS RECIRCULATION SYSTEM (EGR) 1979-80

The EGR system remains in use.

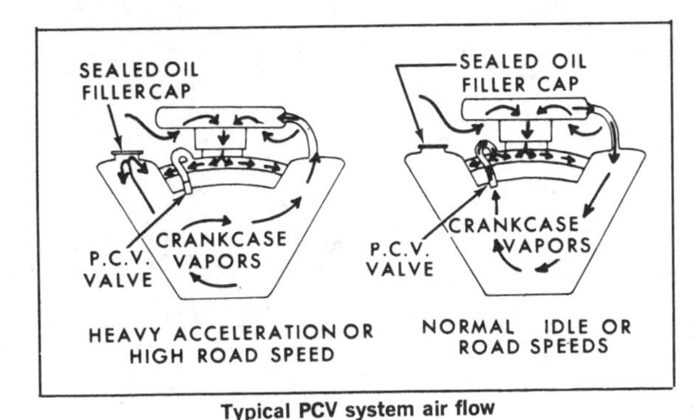

Typical PCV system air flow

TEMPERATURE SENSOR INSIDE AIR CLEANER (SHOWN OPEN)
FULL VACUUM SOURCE
AIR CLEANER VACUUM MOTOR (VALVE SHOWN OPEN)

Thermostatically controlled air cleaner

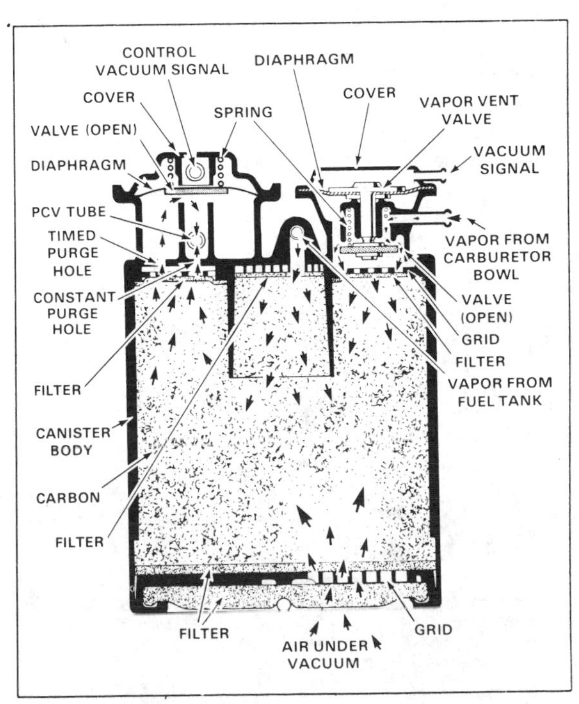

Typical vapor storage canister (© General Motors Corp.)

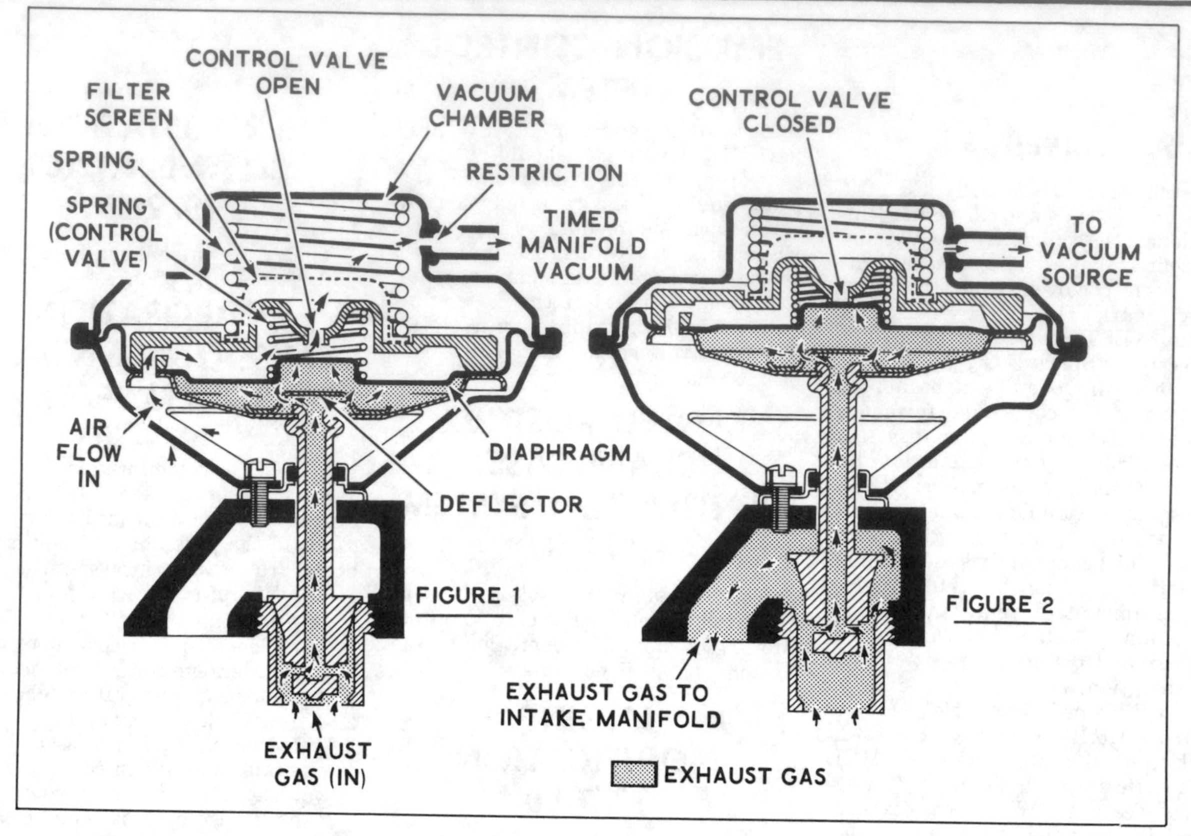

Cross section of ported EGR valve—Both on and off positions

System operation remains the same as used on the 1979 engine models. The EGR valve is designed to be closed during periods of deceleration and idle, to avoid rough idle and stalling from the dilution of the air/fuel mixture by the exhaust gases.

AIR INJECTION REACTION SYSTEM (AIR) 1979-80

The Air Injection Reaction system is carried over to most of the 1980 models and operates in the same manner as the 1979 models.

1980

A new family of lightweight, smaller nylon AIR management valves are being used on the 1980 engines. In addition to the standard diverter valve function, and elimination of backfires during deceleration, these new valves have the potential to provide fuel economy improvement, catalytic converter protection and if desired, electronic control.

EARLY FUEL EVAPORATION SYSTEM (EFE) 1979-80

The Early Fuel Evaporation Sys-

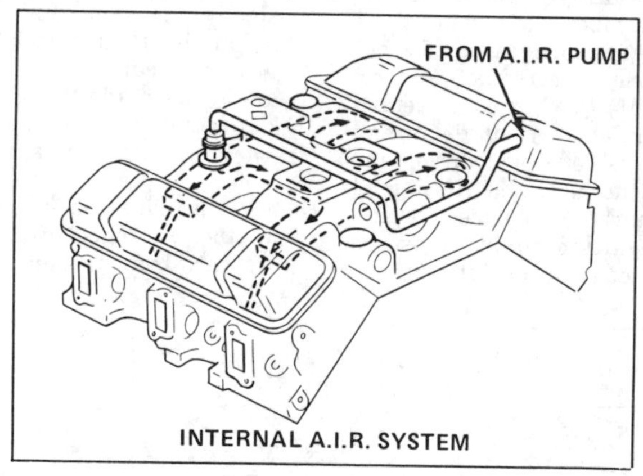

Internal AIR system—V-6 engines

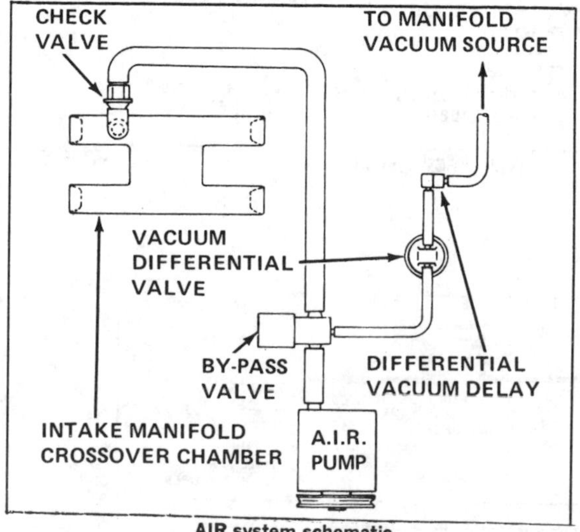

AIR system schematic

tem is used on most 1980 models. The operation of the system remains the same as the 1979 models. Refer to the Emission Control Information label for usage clarification.

ELECTRIC CHOKE 1979

The Electric Choke is used on select models in 1980 and operates in the same manner as the 1979 models. At temperatures below 50° F., the electrical current is directed to the small segment resistor of the choke to allow gradual opening of the choke valve. At temperatures of 70° F. and above, the electrical current is directed to both the small and the large segment resistors, to allow quicker opening of the choke valve.

ELECTRONIC FUEL CONTROL SYSTEM (EFC) 1979-80

The EFC system is continued for 1980 models. The operation remains the same as for the 1979 models. The EFC system should be included in any diagnosis of the following problems;

1. Detonation
2. Stalls or rough idle—cold
3. Stalls or rough idle—hot
4. Missing
5. Hesitation
6. Surges
7. Sluggish or spongy operation
8. Poor gas milage
9. Hard starting—cold
10. Hard starting—hot
11. Foul exhaust odor
12. Engine cuts out

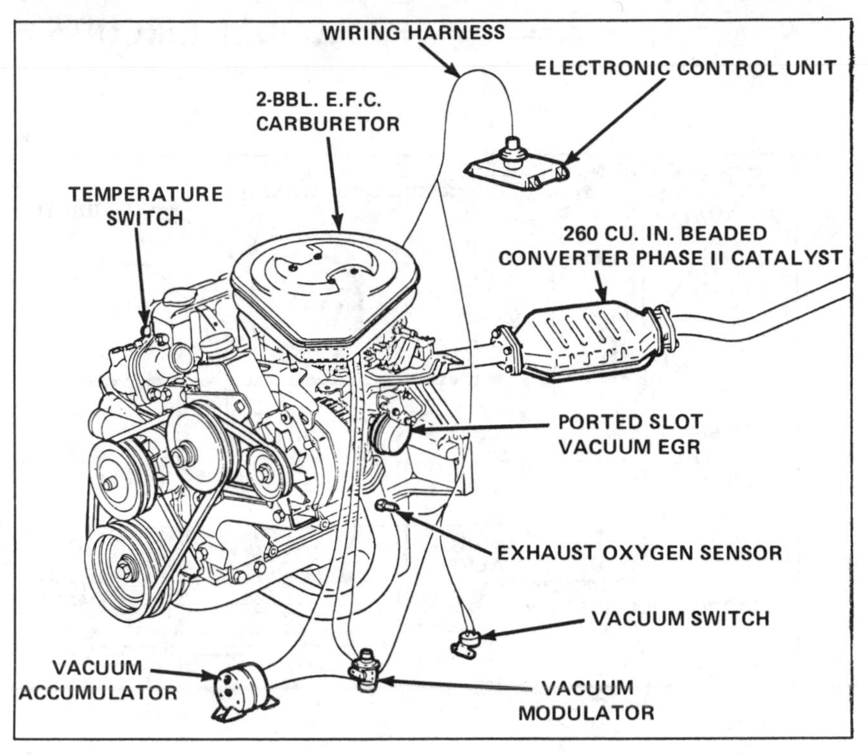

Electronic fuel control components

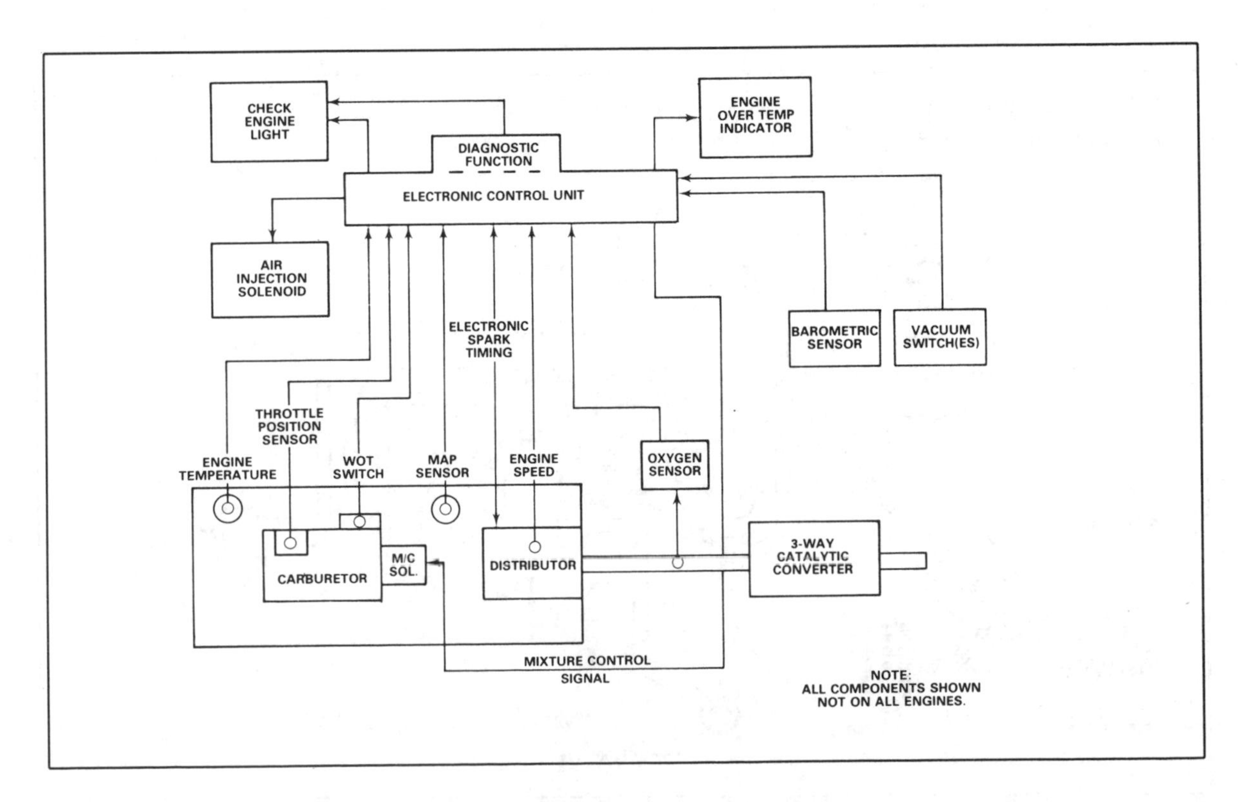

C-4 Schematic (© General Motors Corp.)

VACUUM CIRCUITS

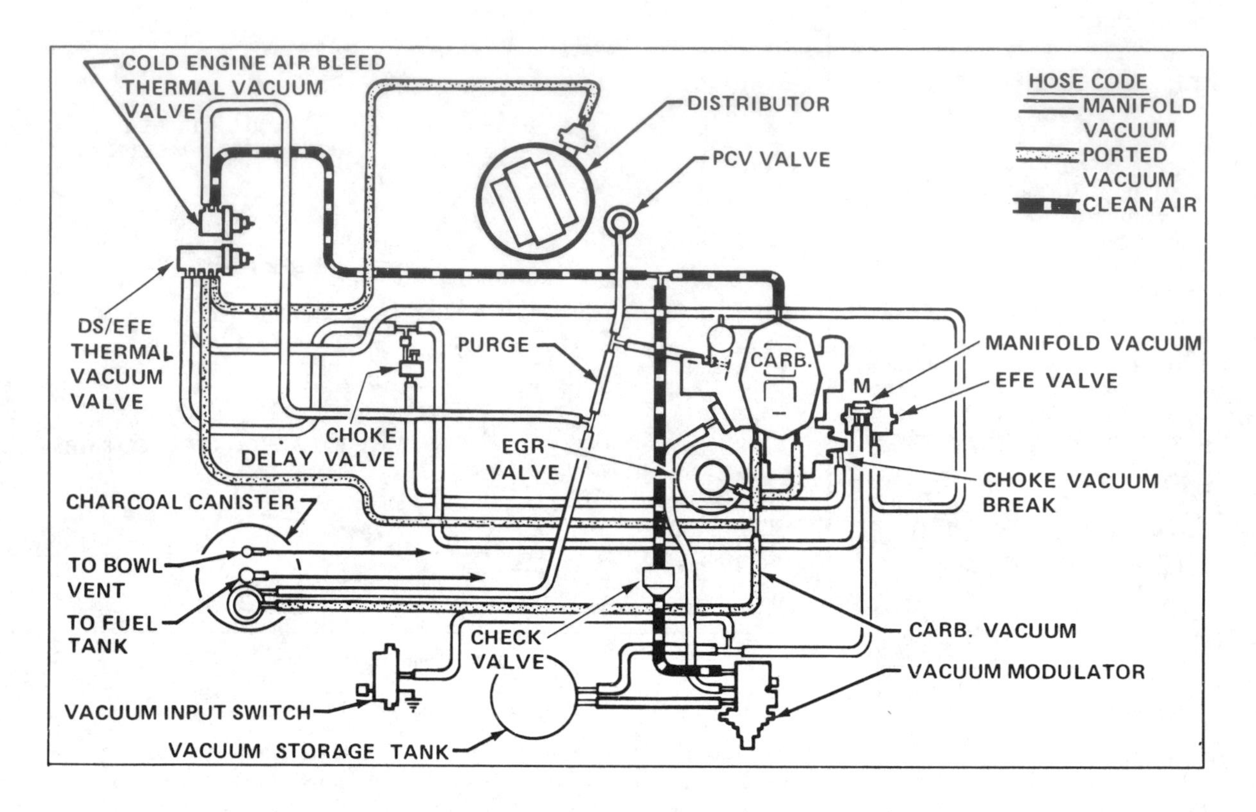

1979 151 CID 4 cyl.—Vin code 1—Calif.

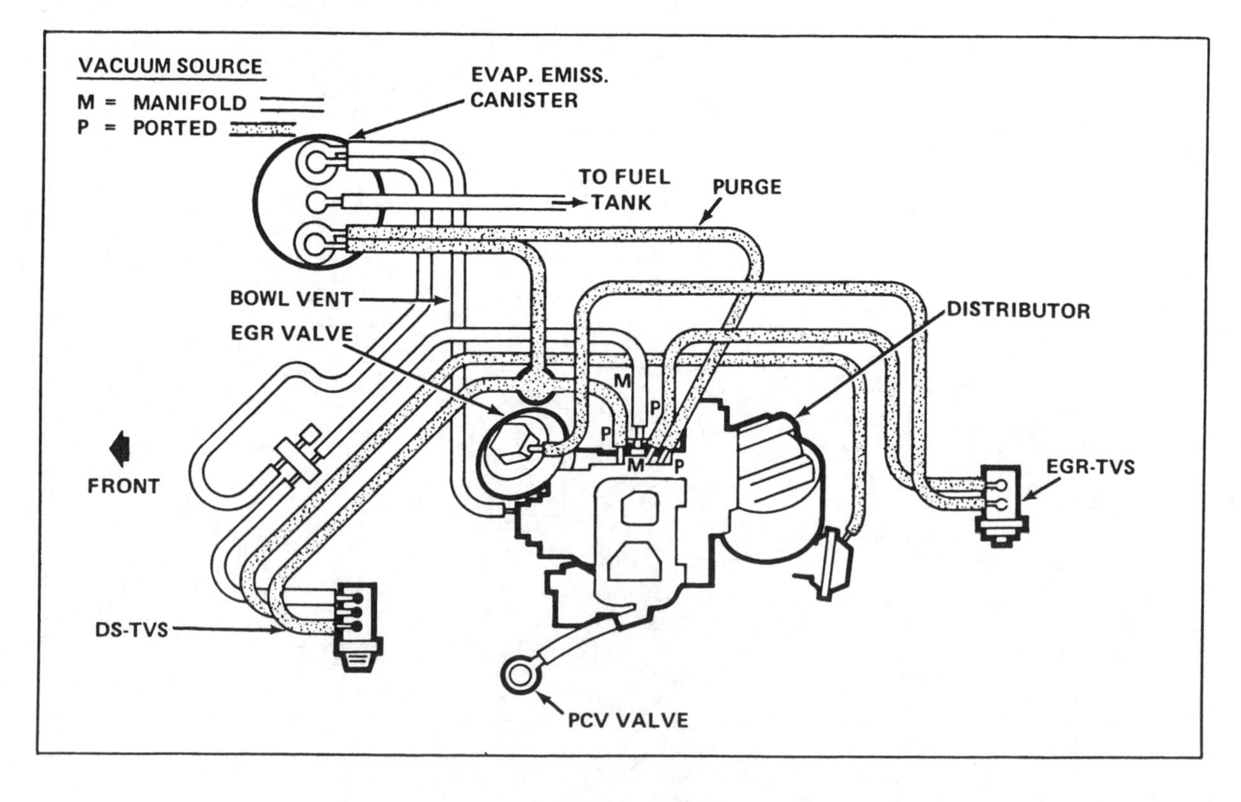

1979 151 CID 4 cyl.—Vin code V—Low Alt. (Auto. trans. w/AC)

VACUUM CIRCUITS

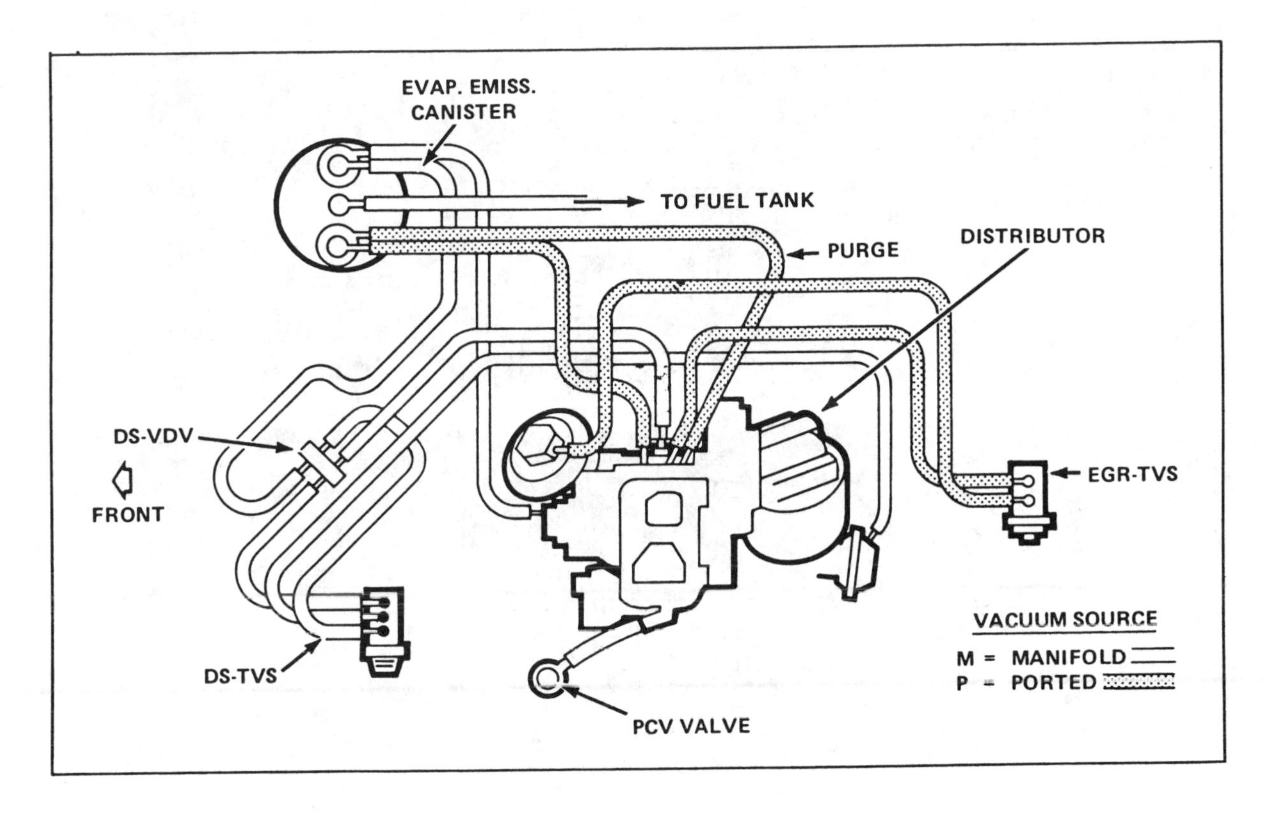

1979 151 CID 4 cyl.—Vin code V—Low Alt. (Auto. and Manual trans. without AC)

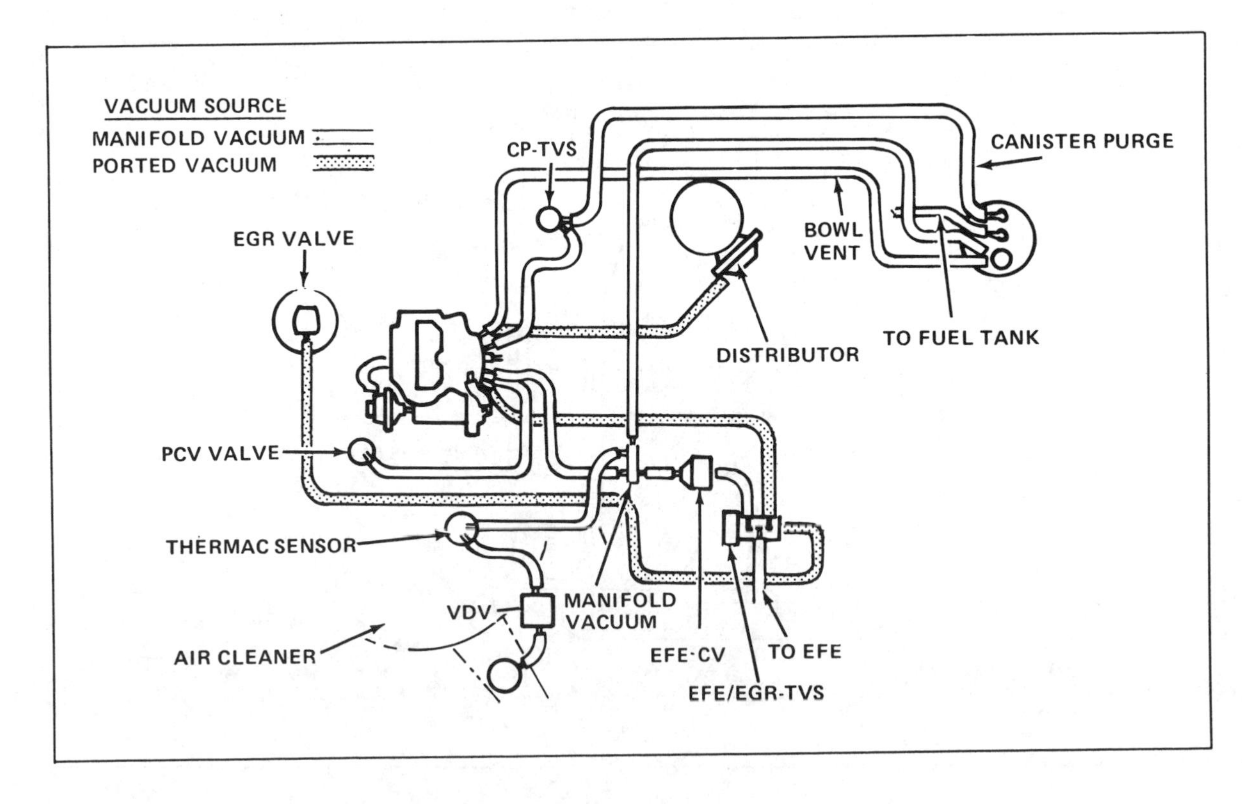

1979 231 CID V-6—Vin code A—Low Alt. (w/Manual trans.)

VACUUM CIRCUITS

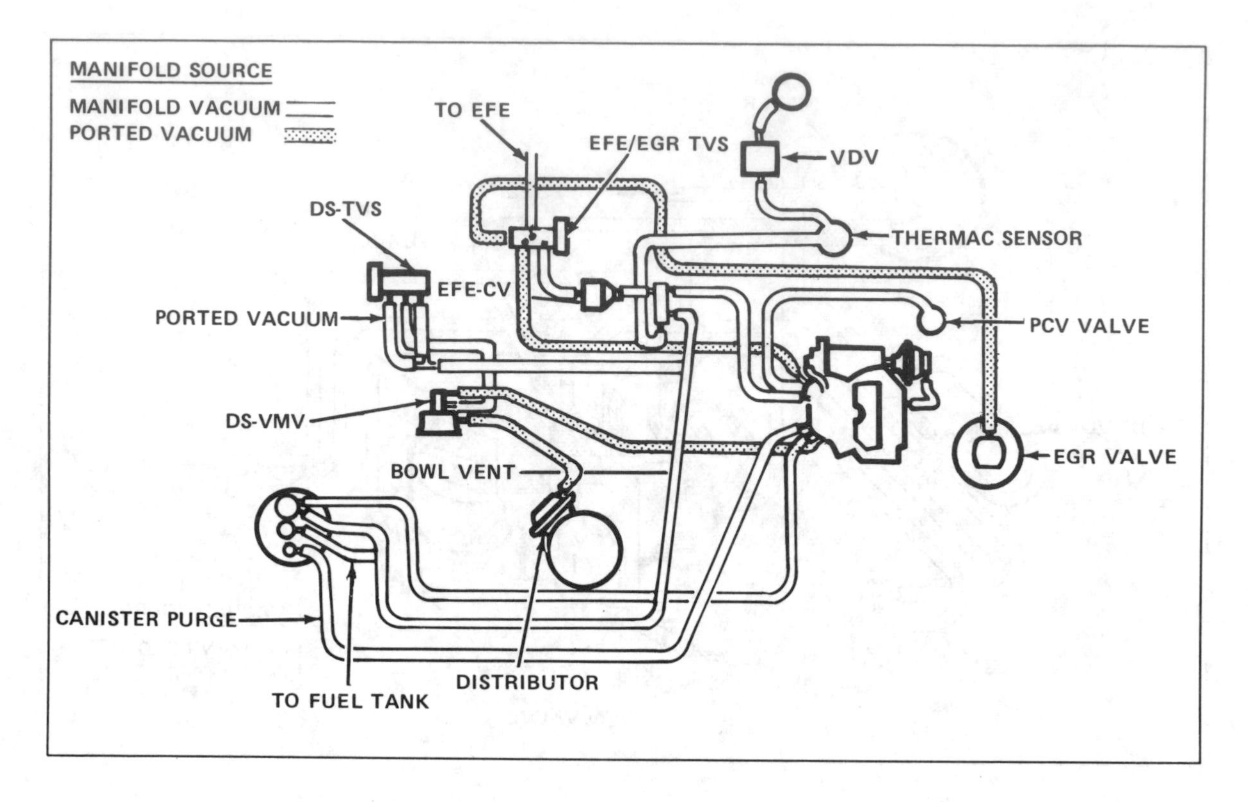

1979 231 CID V-6—Vin code A—Low Alt. (w/Auto. trans.)

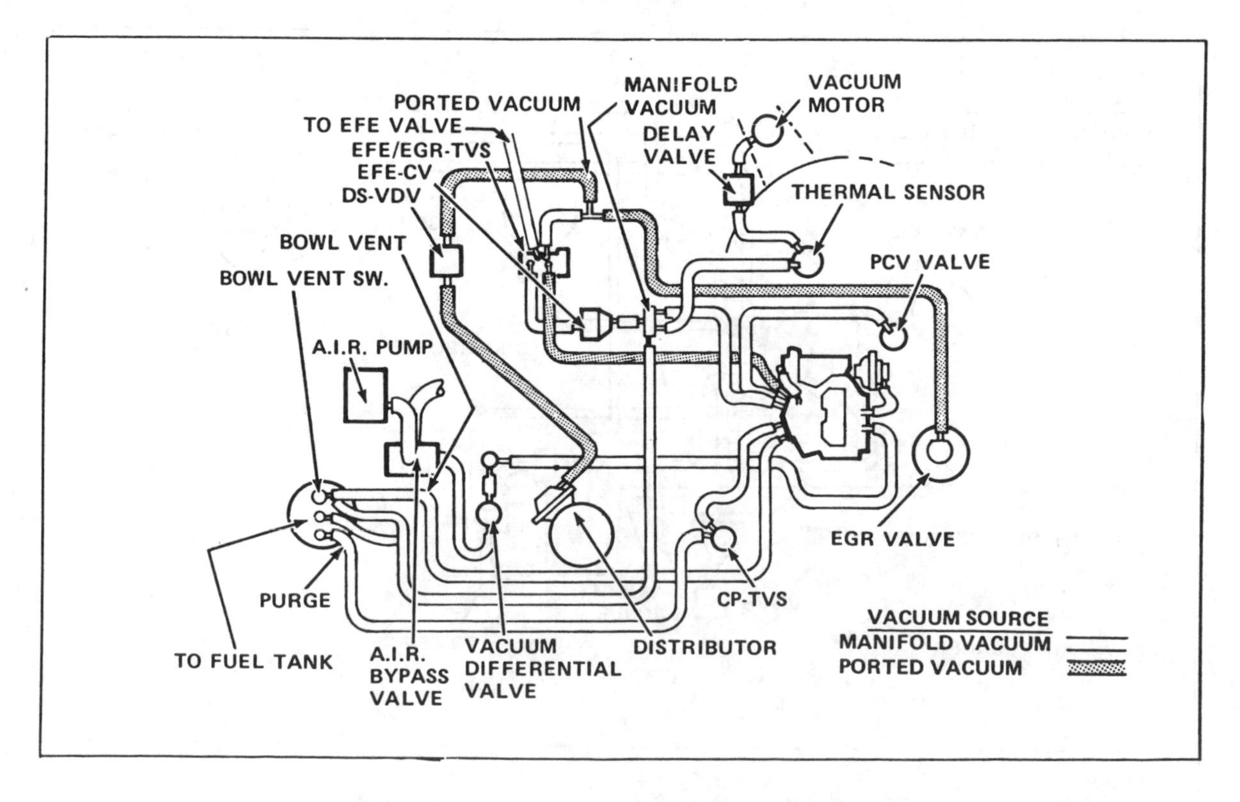

1979 231 CID V-6—Vin code A—Calif. (w/Manual trans.)

VACUUM CIRCUITS

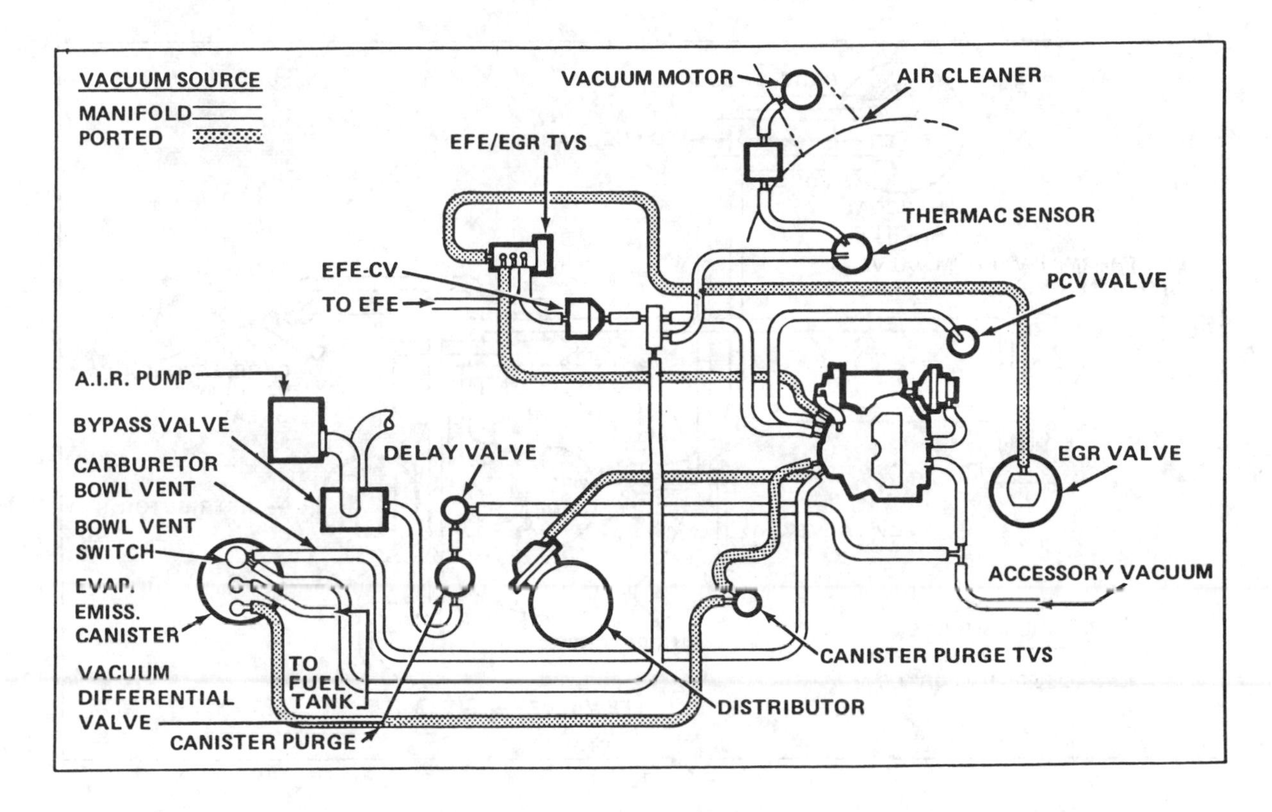

1979 231 CID V-6—Vin code A—Calif. and High Alt. (w/Auto. trans.)

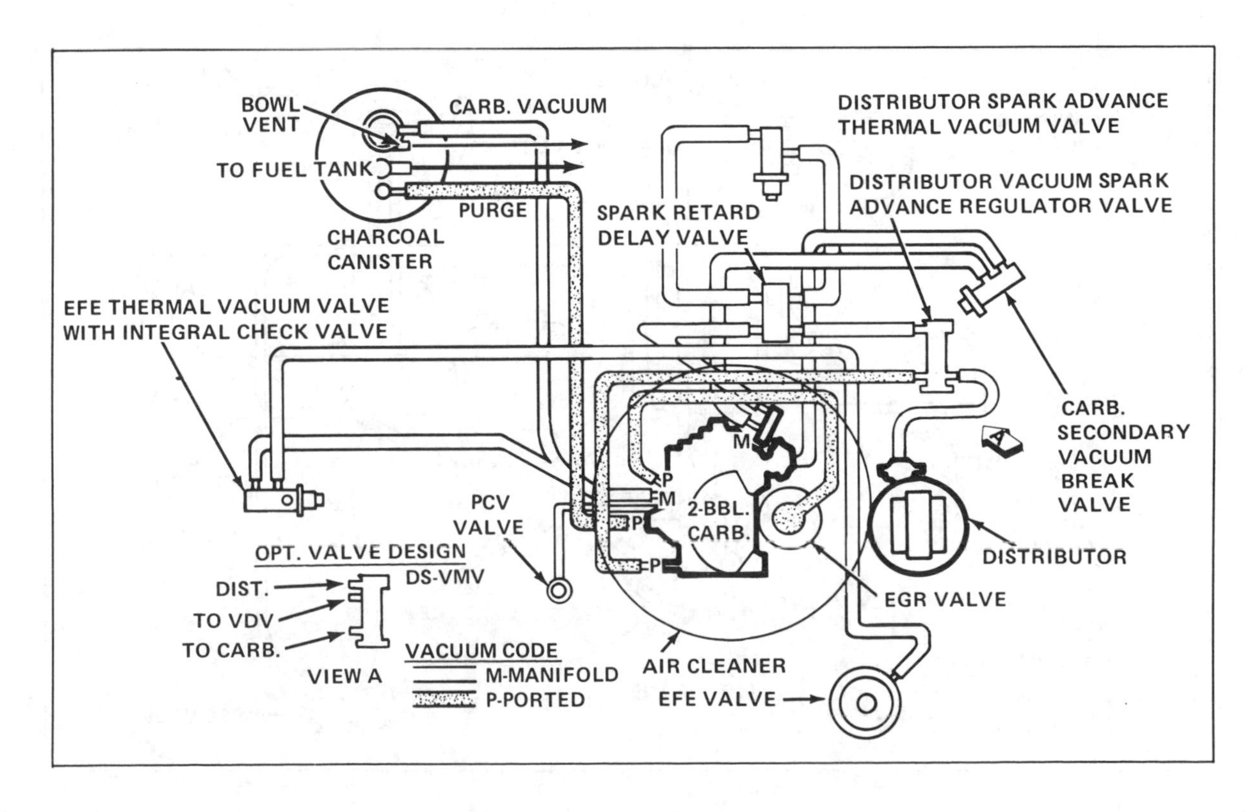

1979 301 CID V-8—Vin code Y (2-bbl. w/Auto. trans. and AC)

VACUUM CIRCUITS

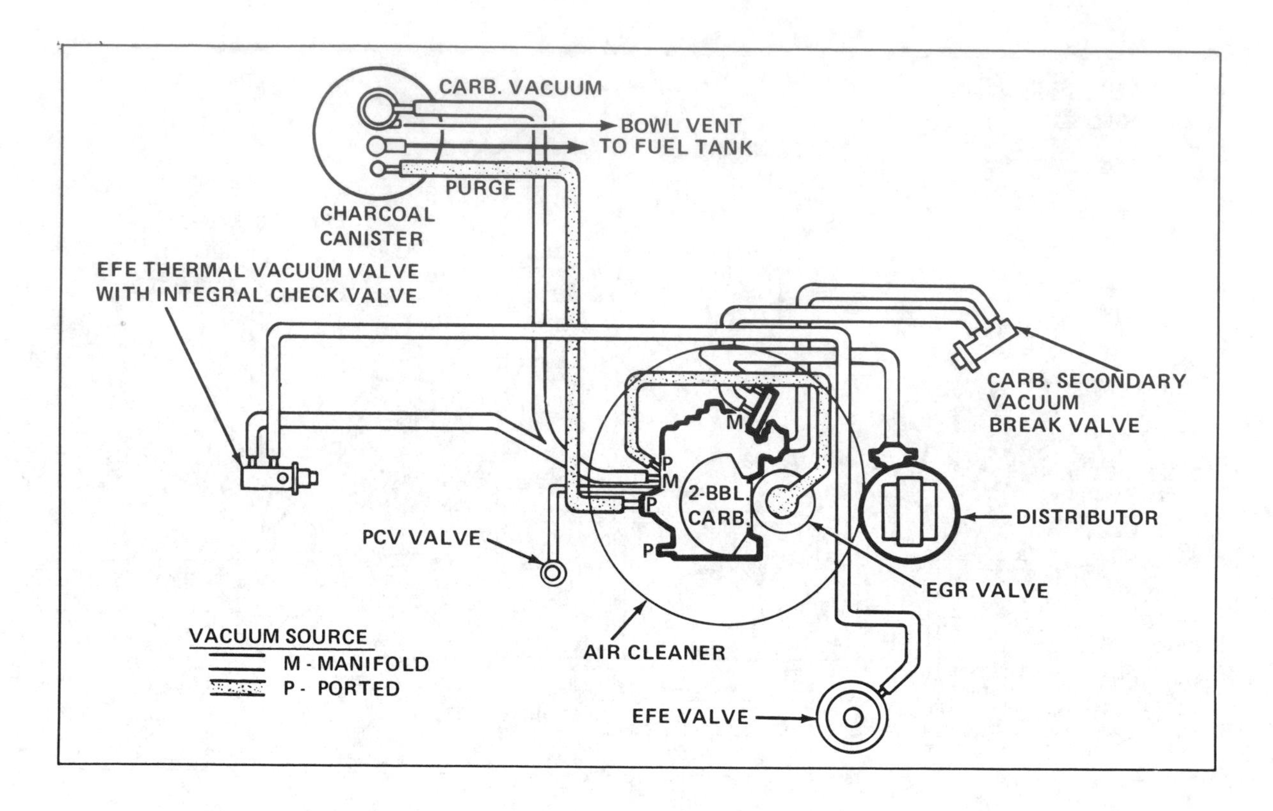

1979 301 CID V-8—Vin code Y (2-bbl. w/Auto. trans. and without AC)

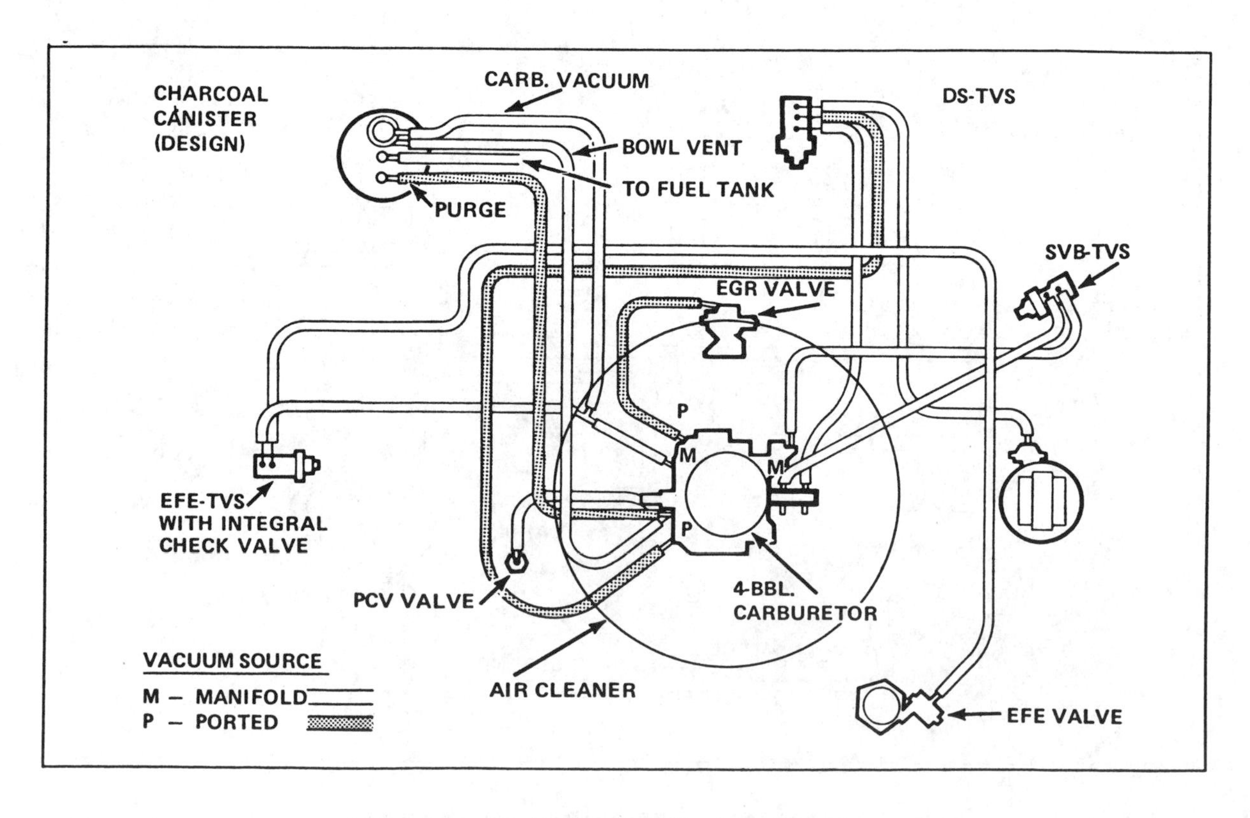

1979 301 CID V-8—Vin code W (4-bbl, w/Manual trans.)

VACUUM CIRCUITS

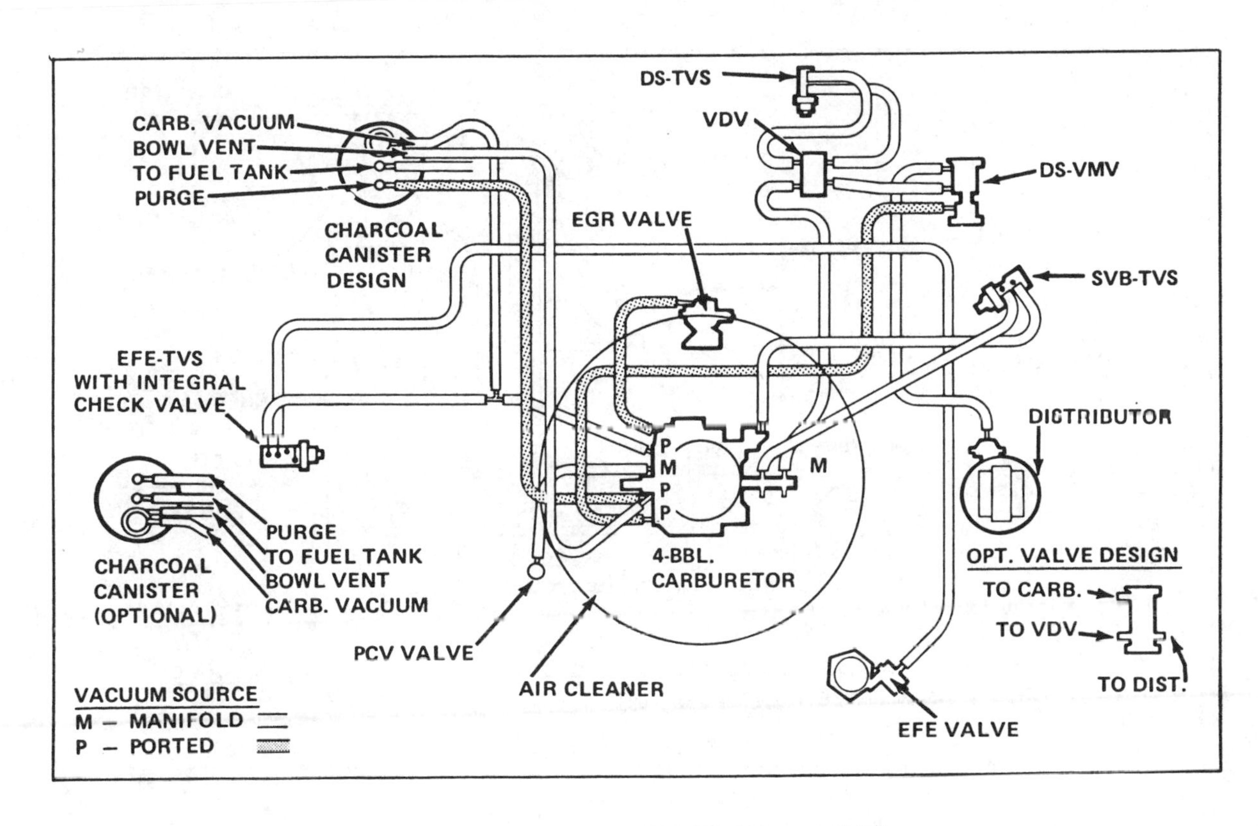

1979 301 CID V-8—Vin code W (4-bbl. w/Auto. trans. and AC)

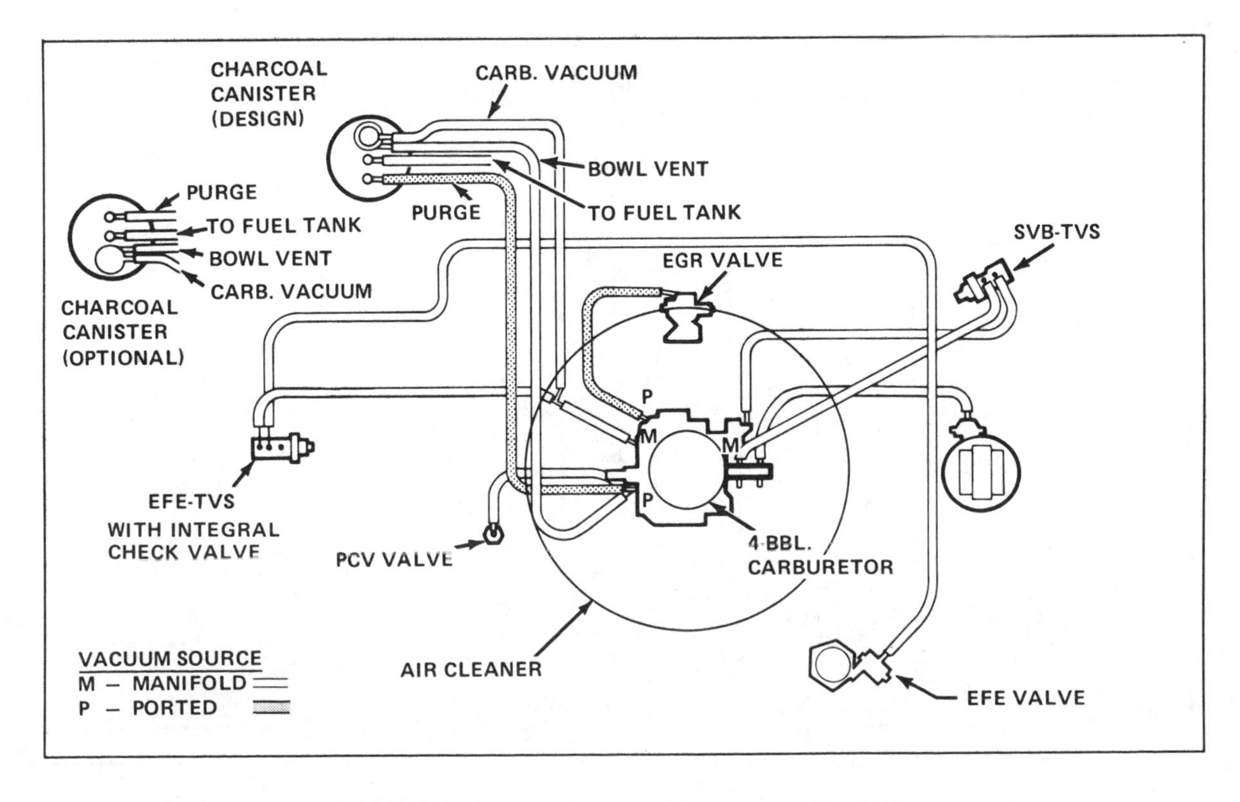

1979 301 CID V-8—Vin code W (4-bbl, w/Auto. trans and without AC)

VACUUM CIRCUITS

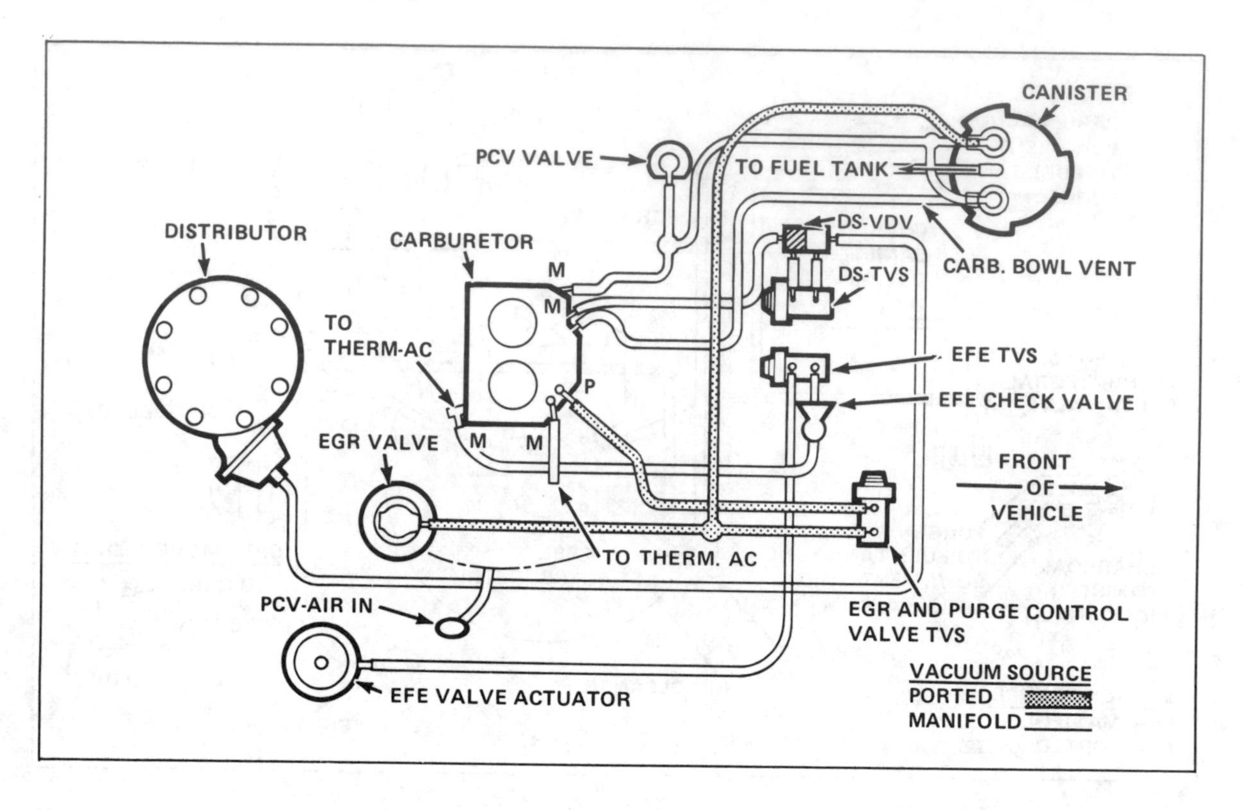

1979 305 CID V-8—Vin code G—Low Alt. (2-bbl. except Series H)

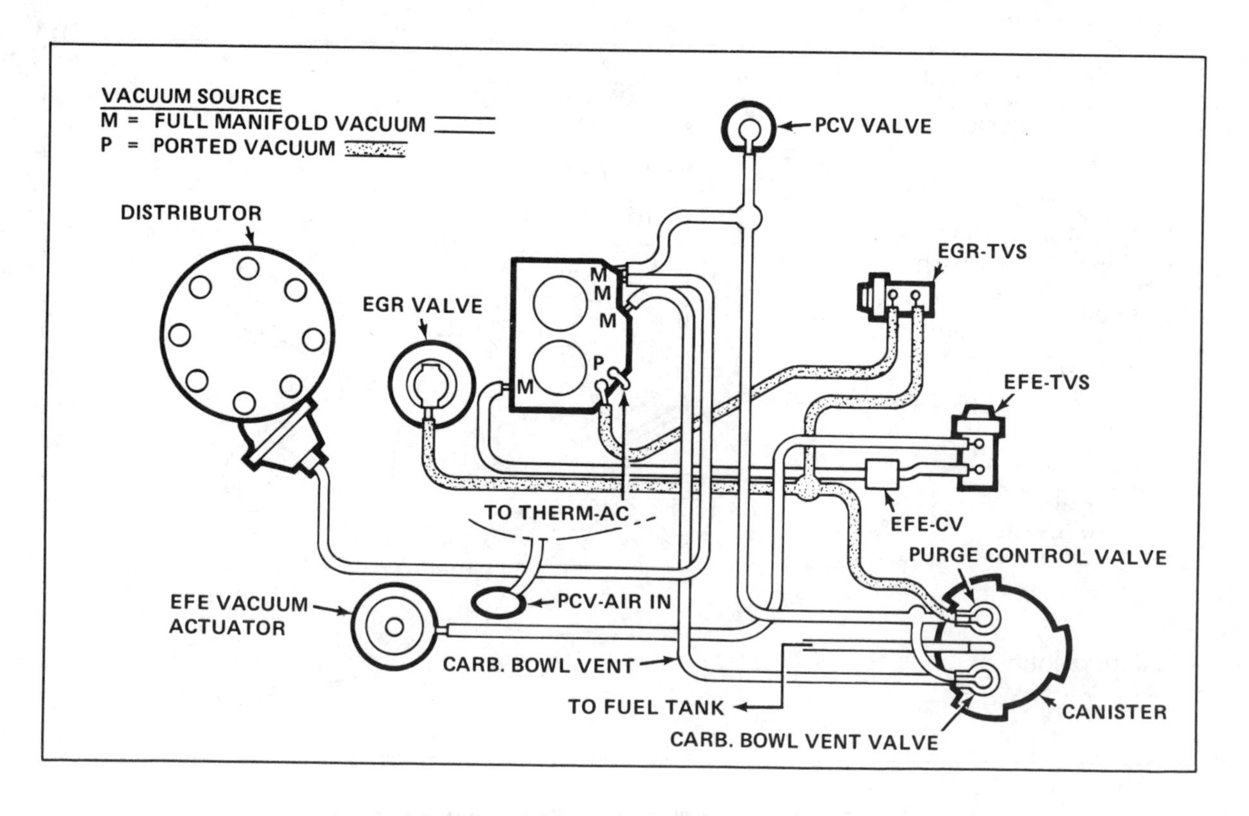

1979 305 CID V-8—Vin code G—Low Alt. (2-bbl, Series H only)

VACUUM CIRCUITS

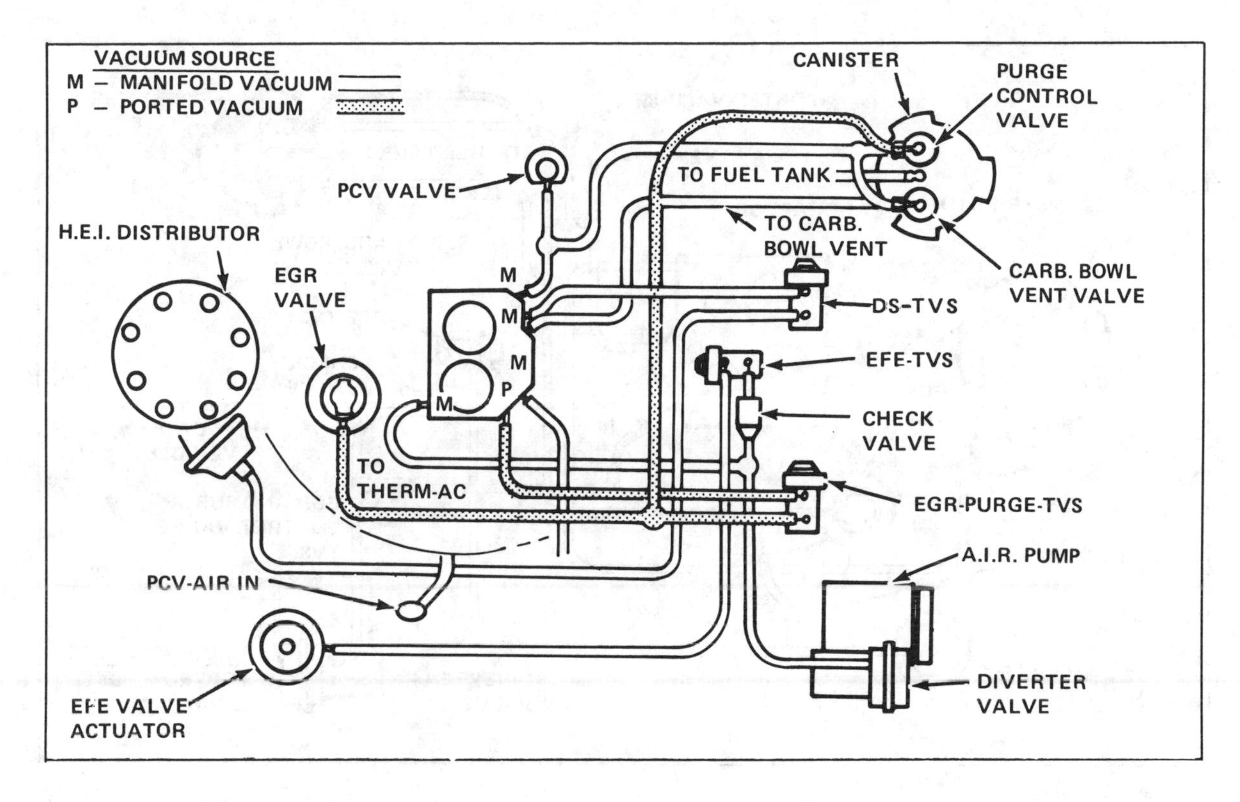

1979 305 CID V-8—Vin code G—Calif. (2-bbl. except Series H)

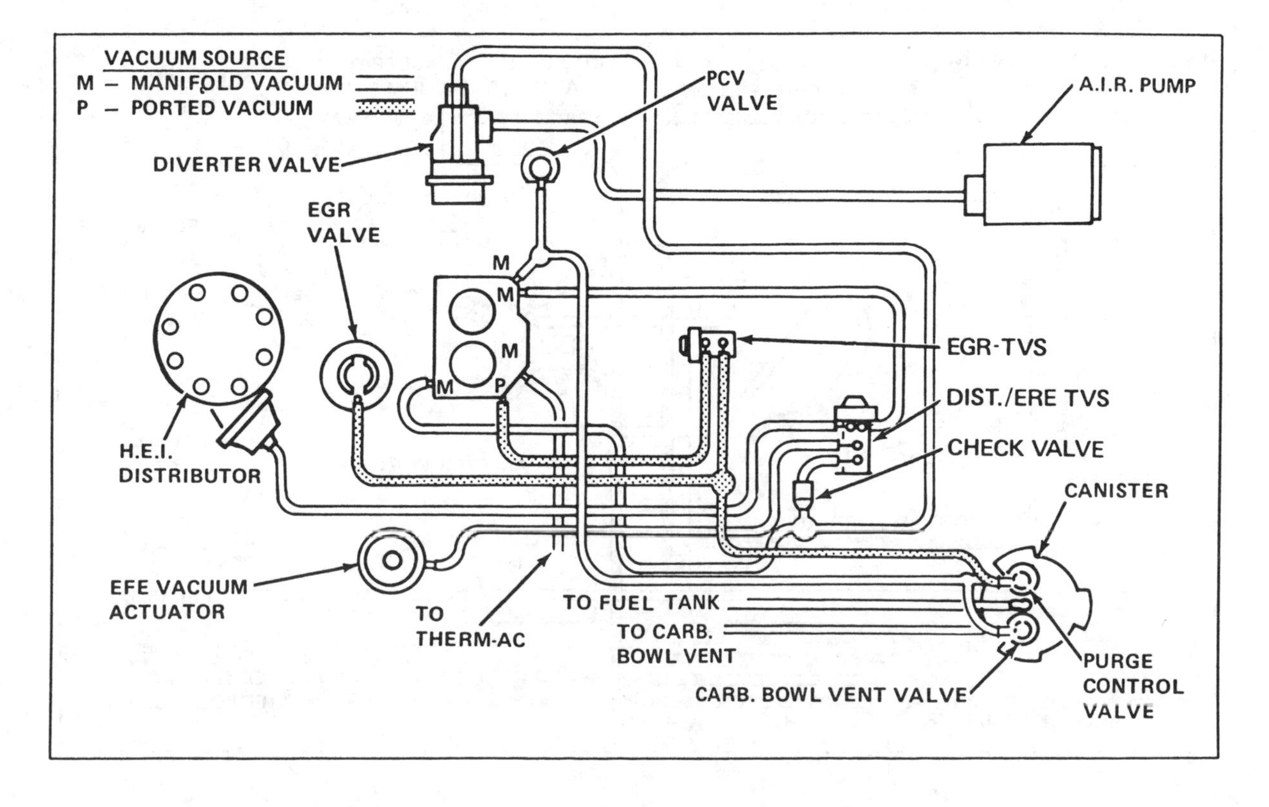

1979 305 CID V-8—Vin code G—Calif. (2-bbl, Series H only)

VACUUM CIRCUITS

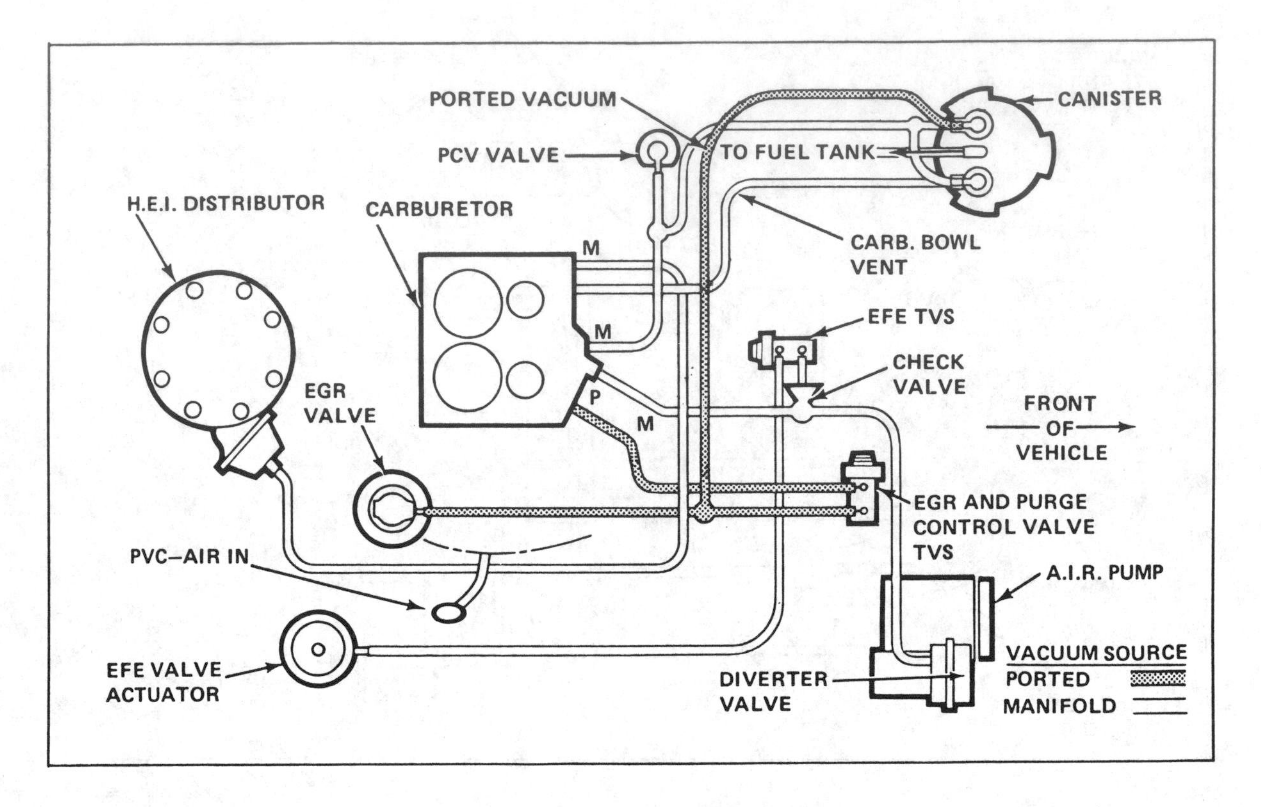

1979 305 CID V-8—Vin code H—Calif. and High Alt. (4-bbl.)
350 CID V-8—Vin code L—Calif. and High Alt. (4-bbl.)

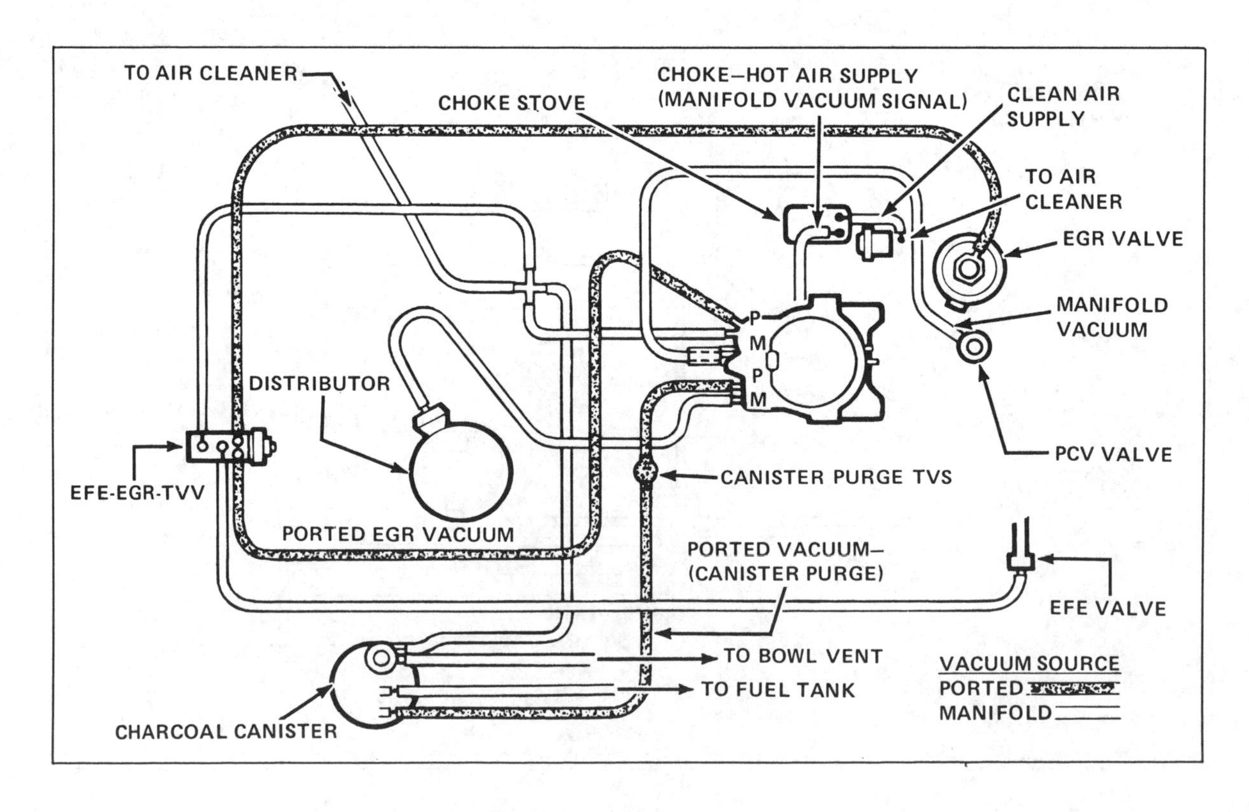

1979 350 CID V-8—Vin code X (4-bbl.)

VACUUM CIRCUITS

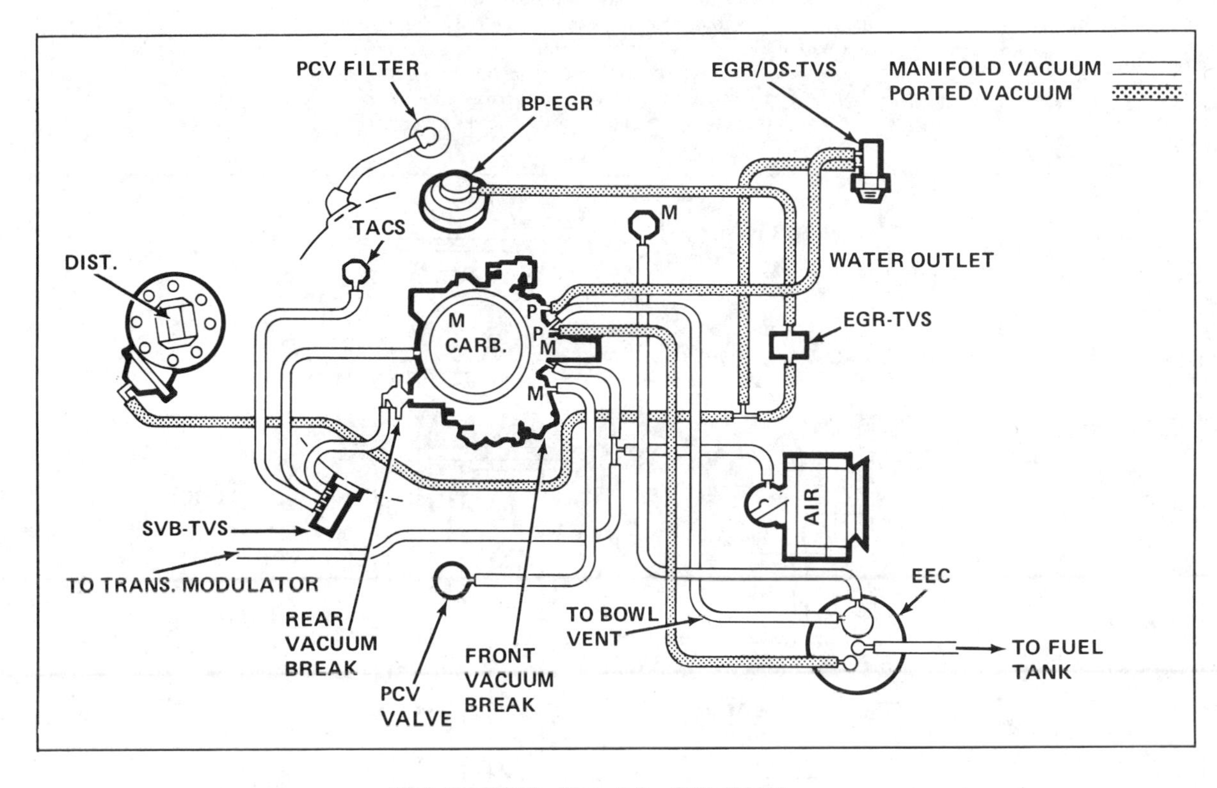

1979 350 CID V-8—Vin code R—Calif. (4-bbl.)
403 CID V-8—Vin code K—Calif. (4-bbl.)

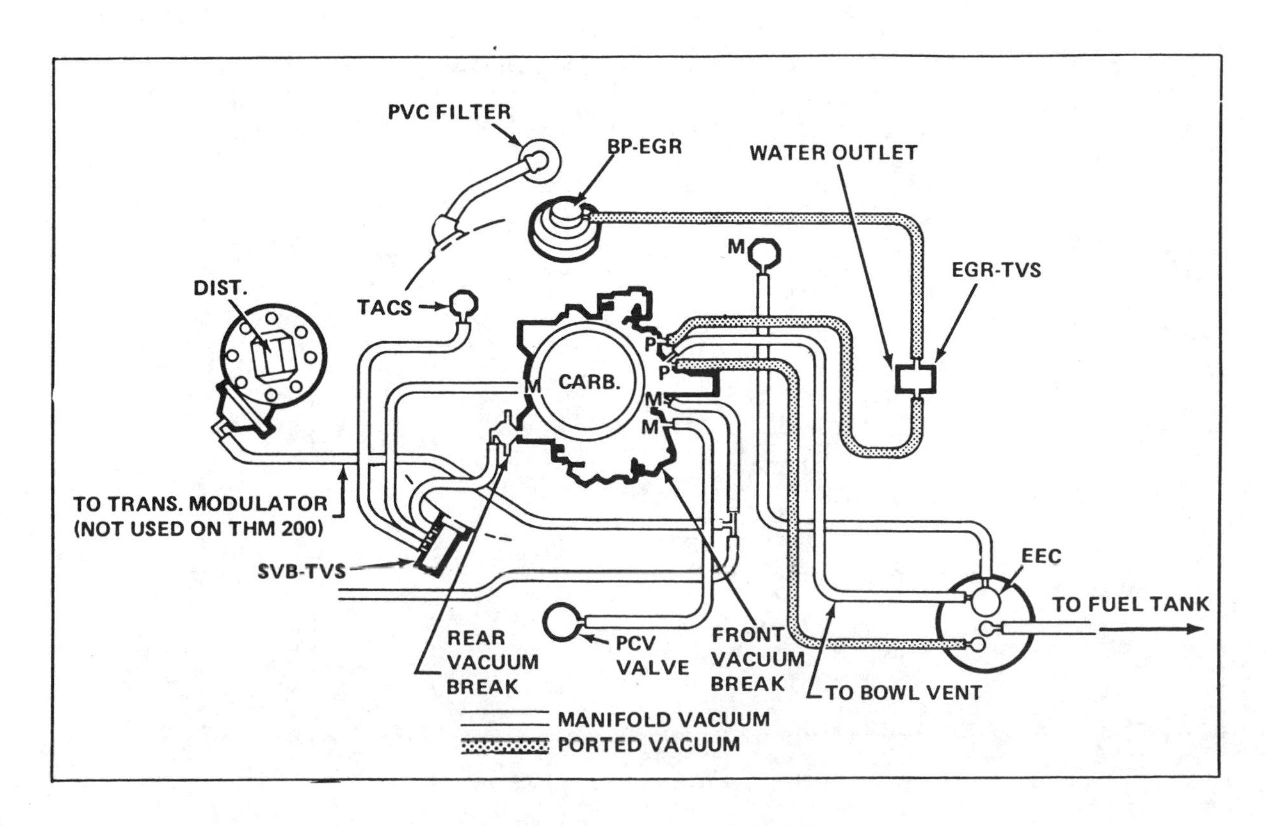

1979 350 CID V-8—Vin code R—High Alt. (4-bbl.)
403 CID V-8—Vin code K—High Alt. (4-bbl.)

VACUUM CIRCUITS

NOTICE: Vacuum hose schematics shown in this section show start of production configurations. It is possible, due to a design change, that the schematic shown does not agree in all details to the schematic shown on the Vehicles Emission Control Information Label. The Vehicle Emission Control Label attached to the radiator support or fan shroud should always be checked for up to date information.

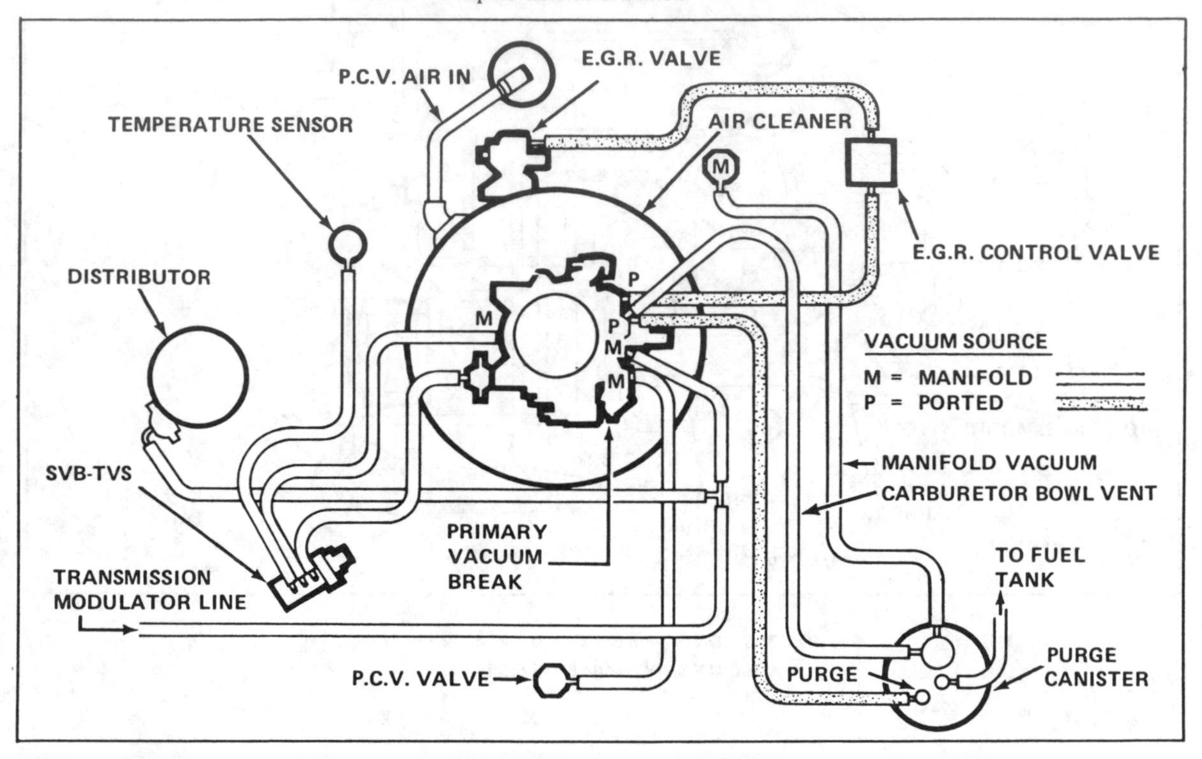

1979 403 CID V-8—Vin code K—Low Alt. (4-bbl.)

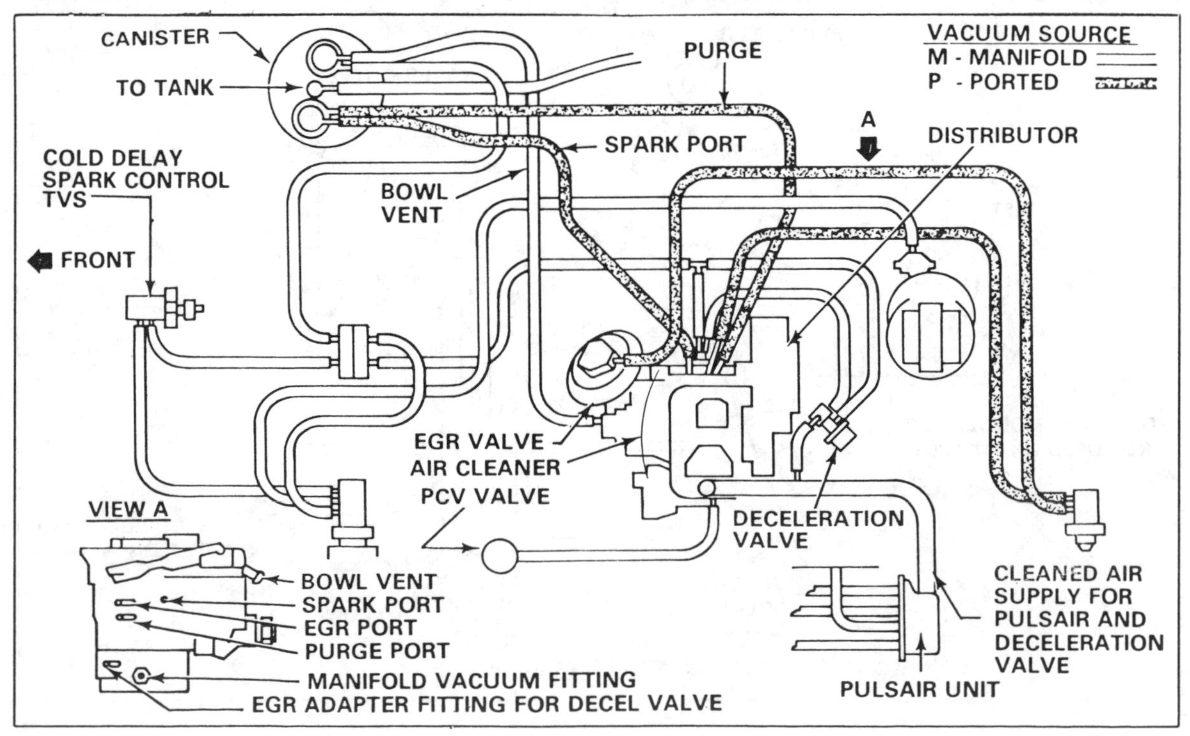

1980 2.5 liter L-4, Code V, w/o C-4, all except automatic transmission w/A.C.
(© General Motors Corp.)

VACUUM CIRCUITS

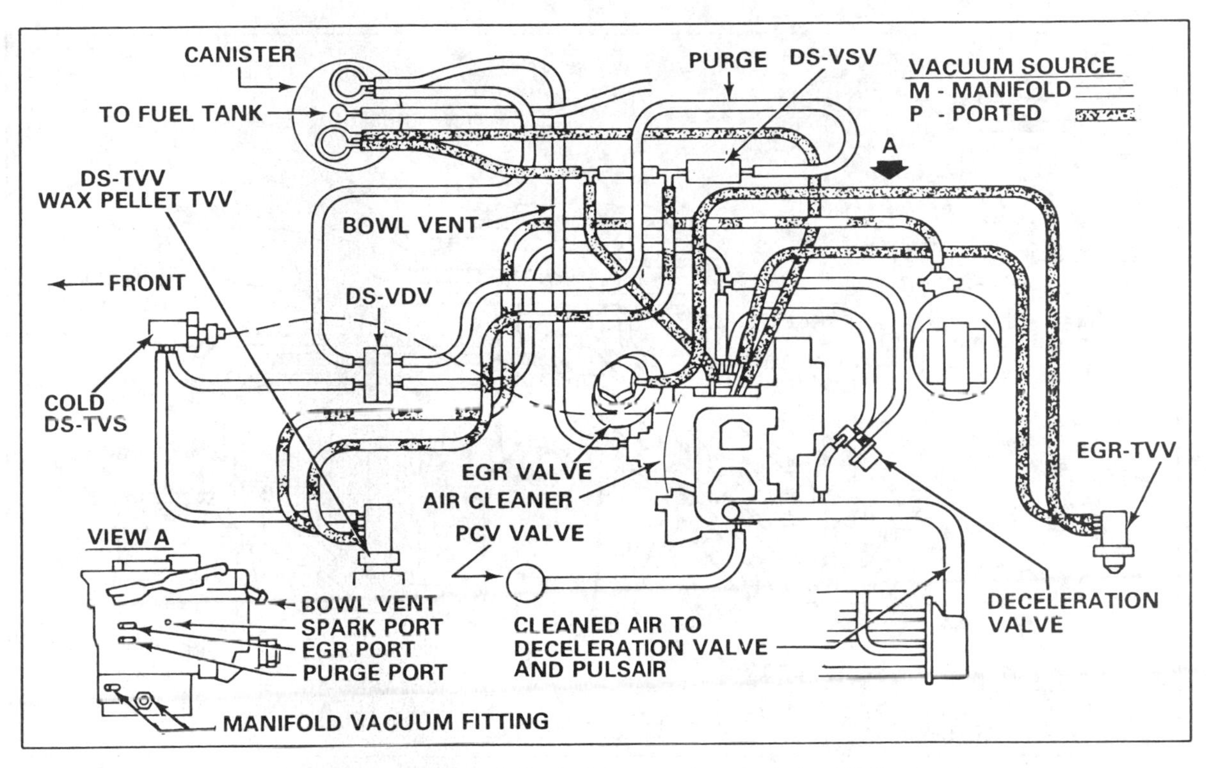

1980 2.5 liter L-4, Vin Code V, w/o C-4, automatic transmission w/A.C. only
(© General Motors Corp.)

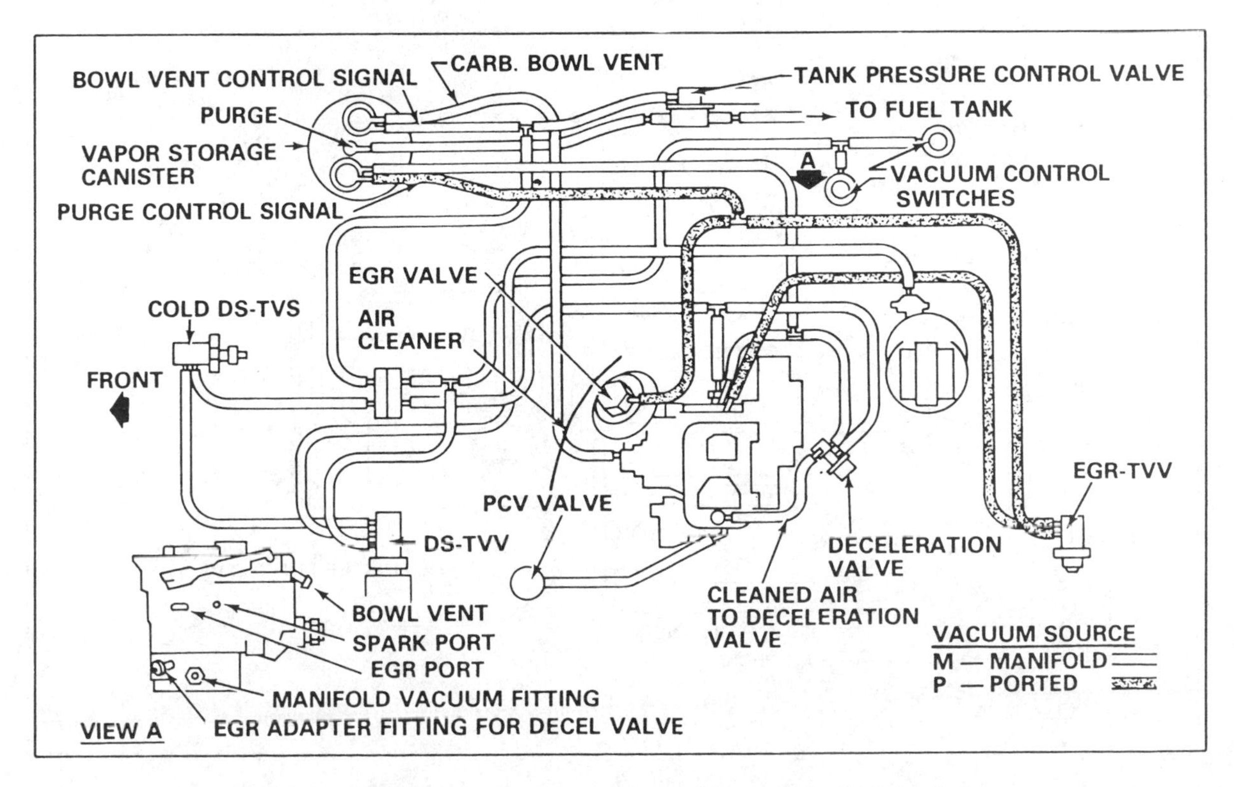

1980 2.5 liter L-4, Vin Code V, with C-4, manual transmission (© General Motors Corp.)

VACUUM CIRCUITS

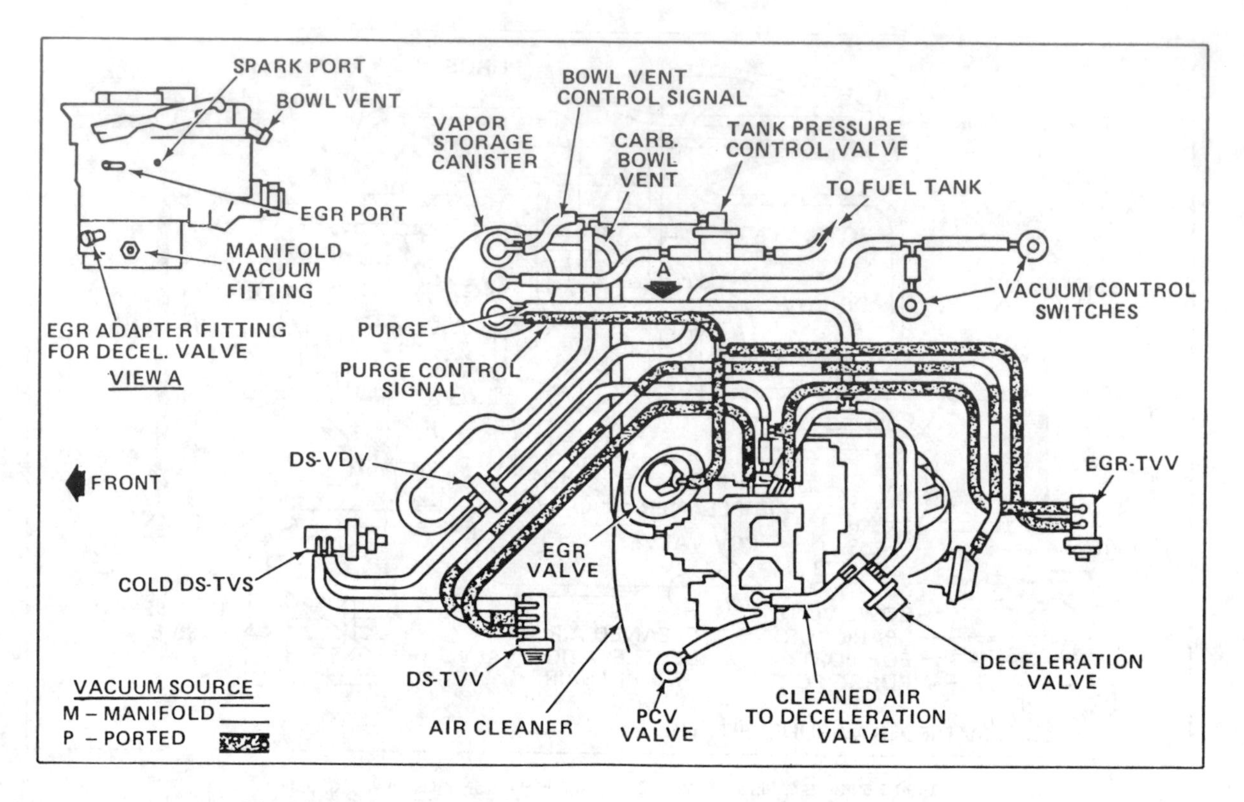

1980 2.5 liter L-4, Vin Code V, with C-4, automatic transmission (© General Motors Corp.)

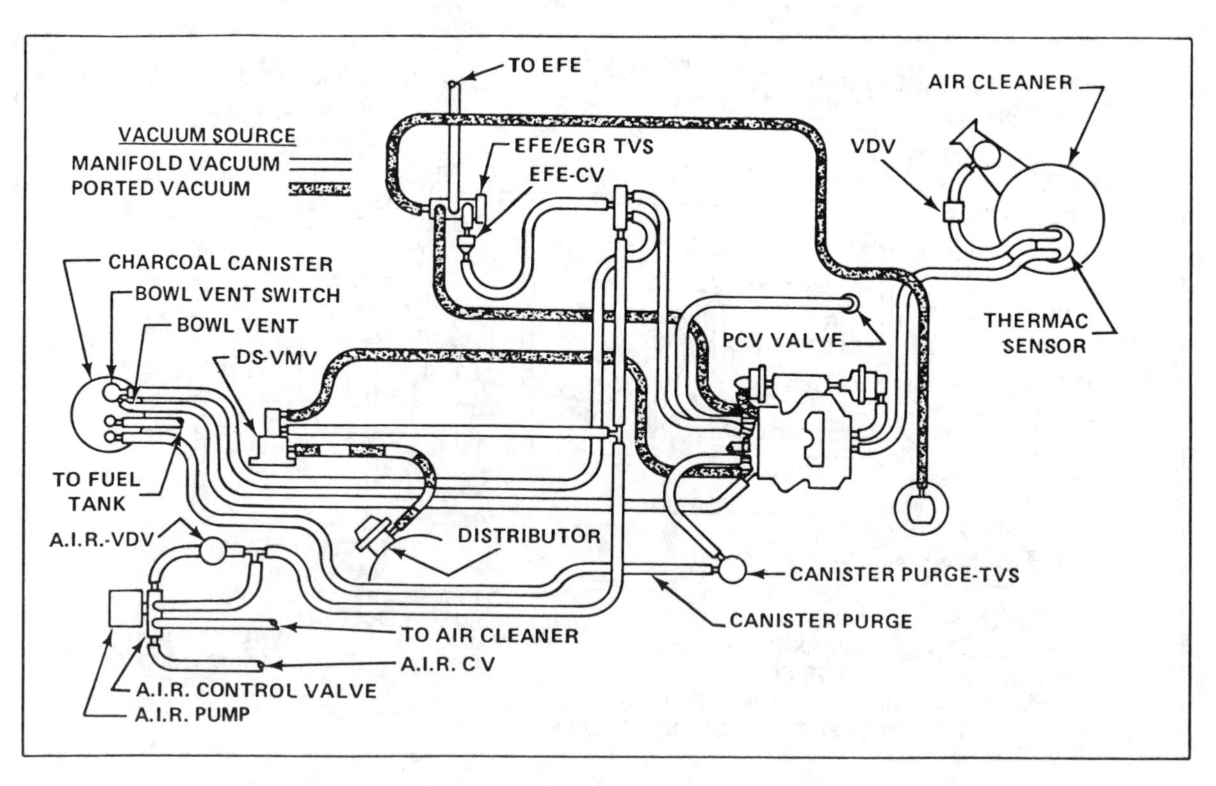

1980 3.8 liter V-6, Vin Code A, w/o C-4, manual transmission (© General Motors Corp.)

VACUUM CIRCUITS

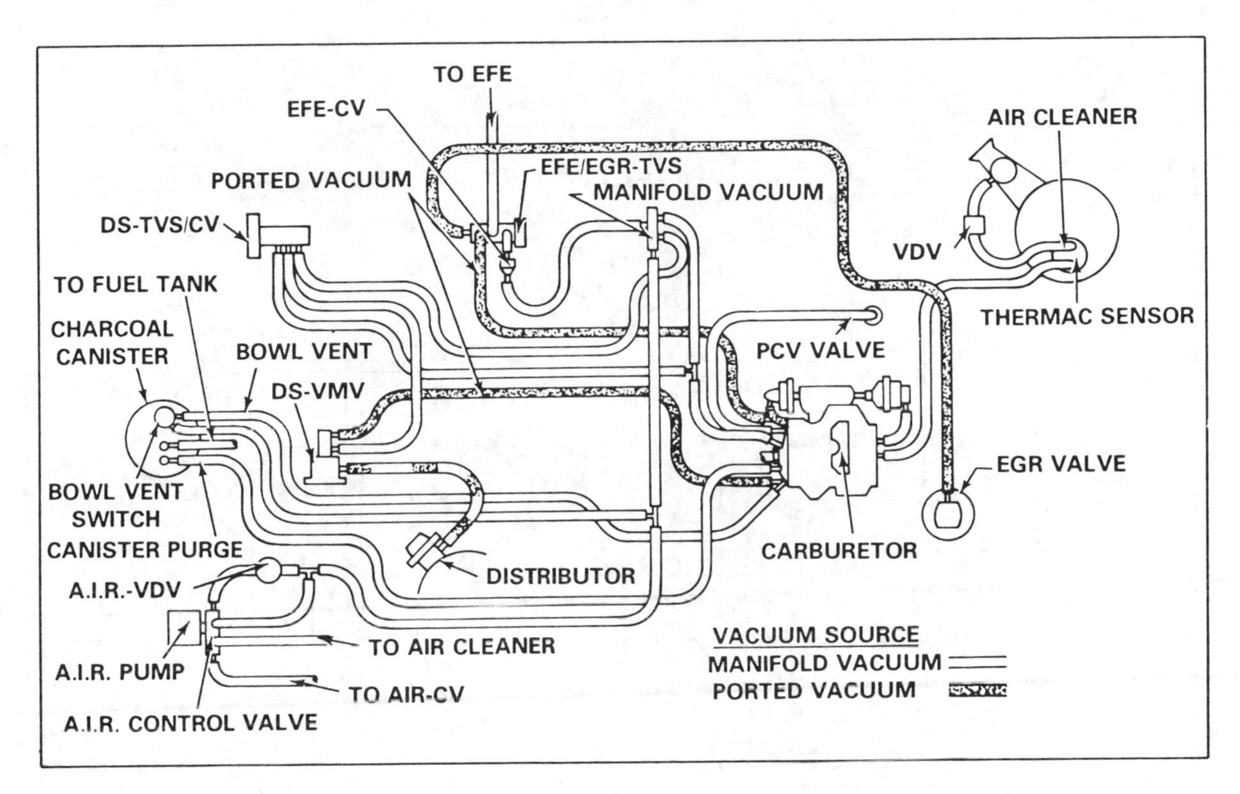

1980 3.8 liter V-6, Vin Code A, w/o C-4, automatic transmission (© General Motors Corp.)

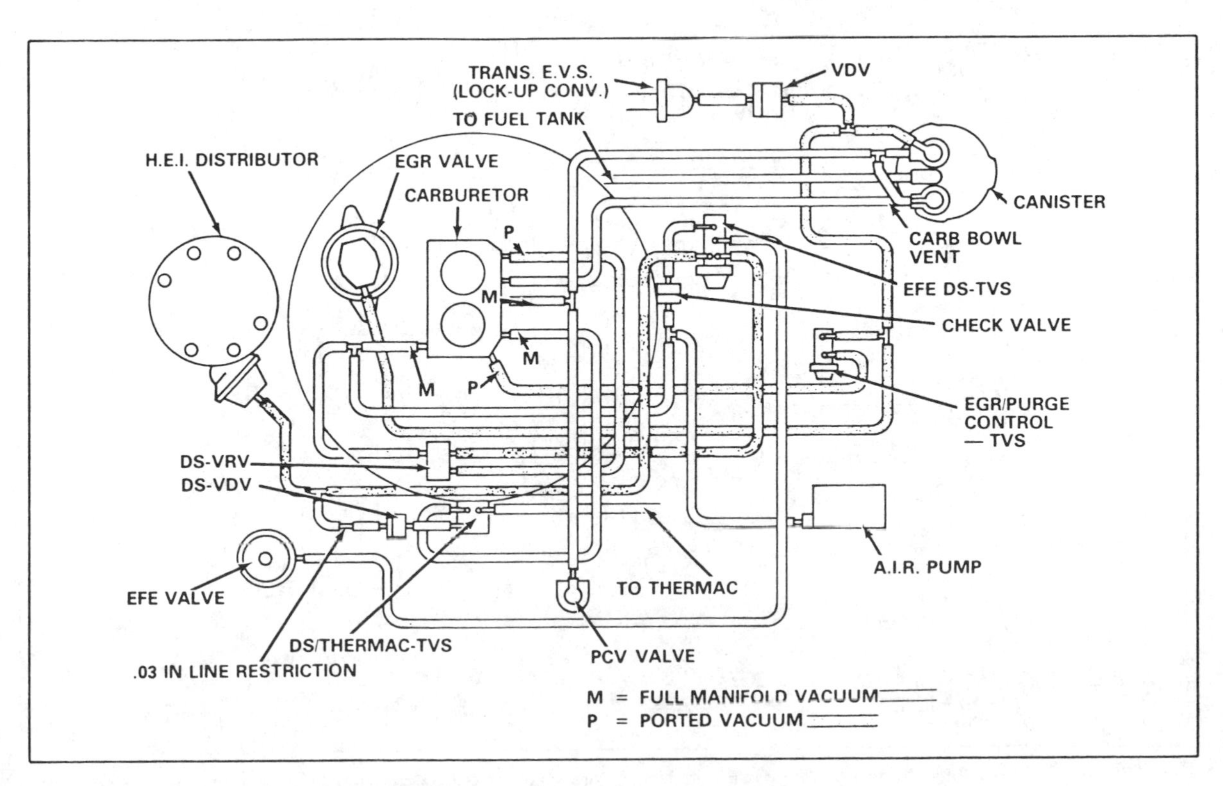

1980 3.8 liter V-6, Vin Code K (© General Motors Corp.)

VACUUM CIRCUITS

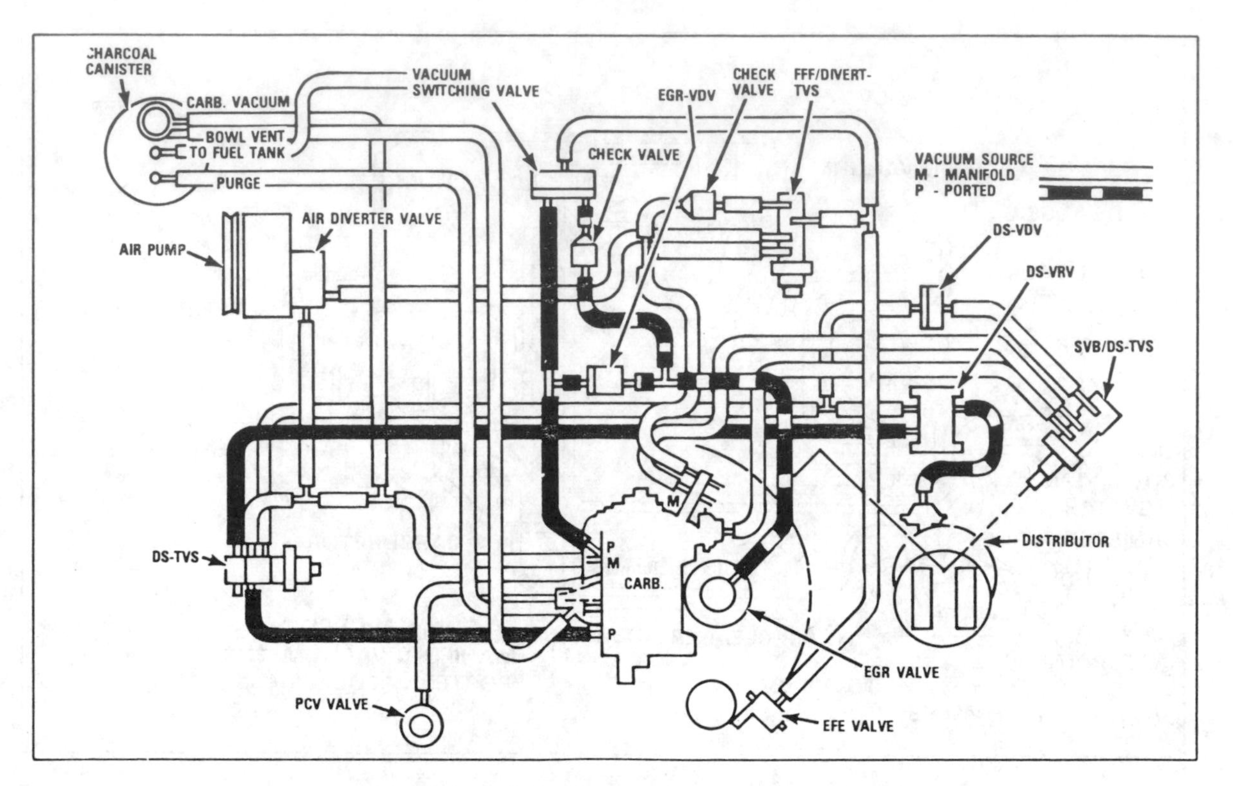

1980 4.3 liter V-8, Vin Code S (© General Motors Corp.)

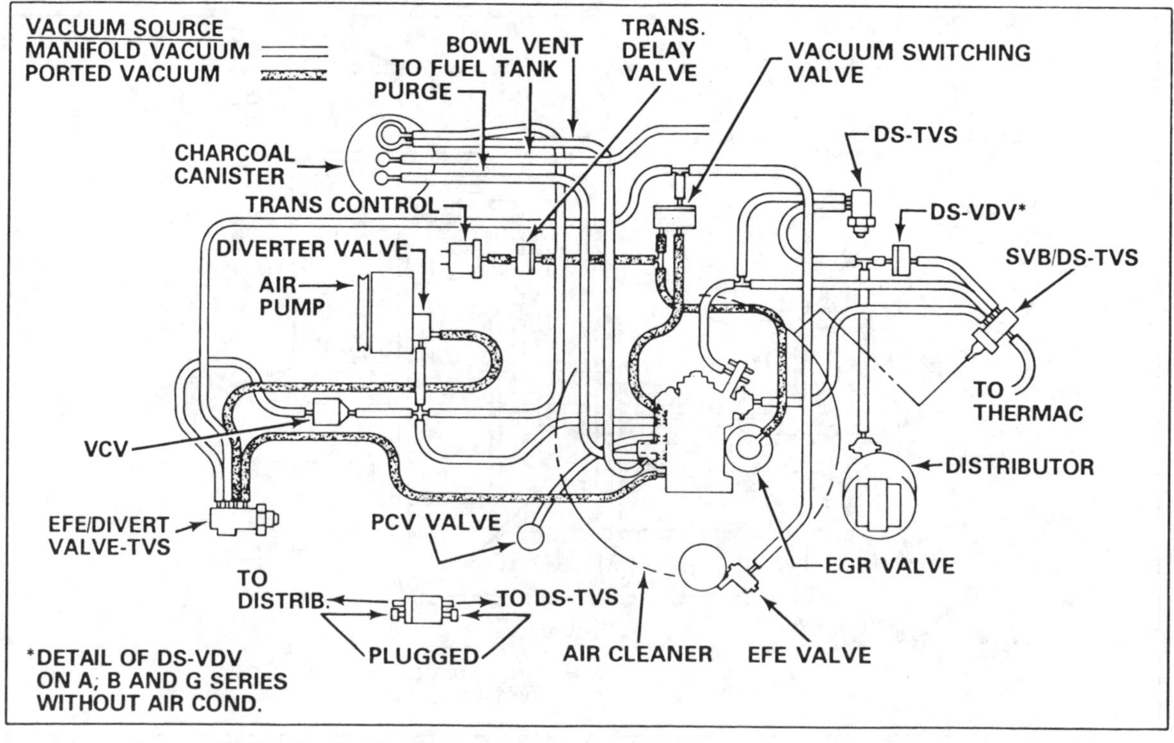

1980 4.3 liter V-8, Vin Code S, with lock-up torque converter (© General Motors Corp.)

VACUUM CIRCUITS

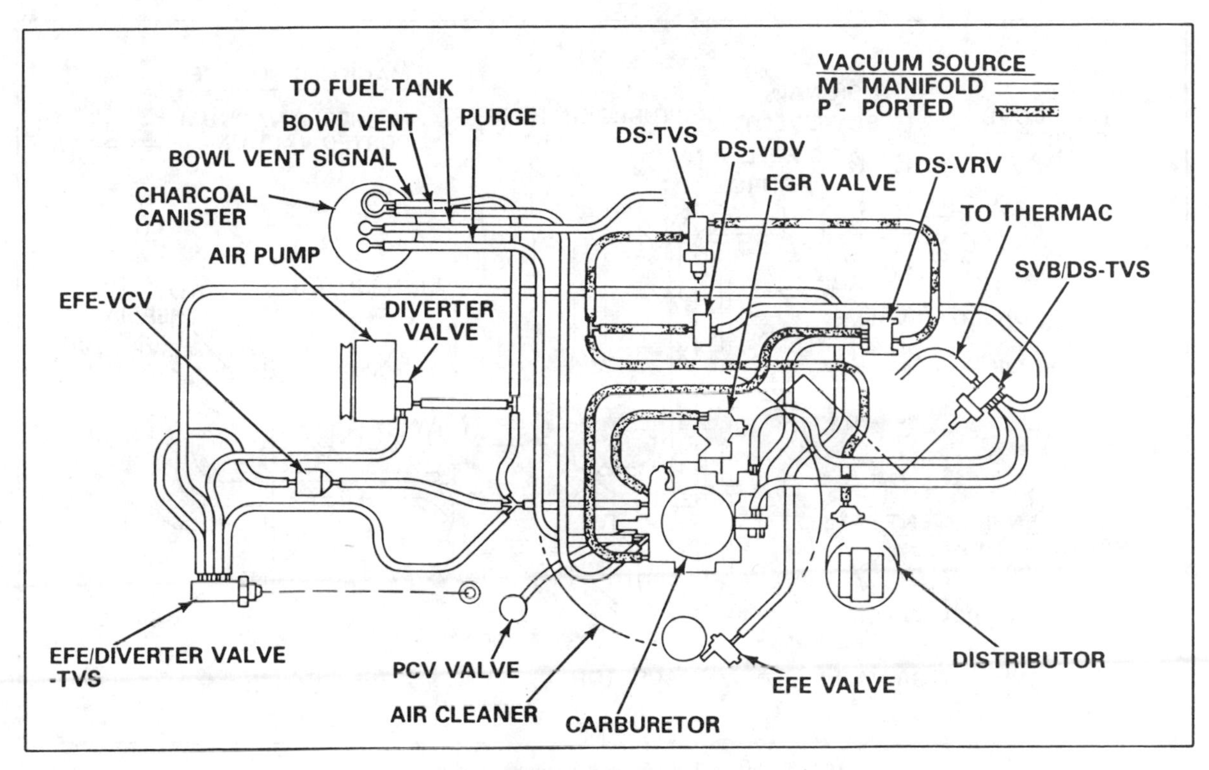

1980 4.9 liter V-8, Vin Code W (© General Motors Corp.)

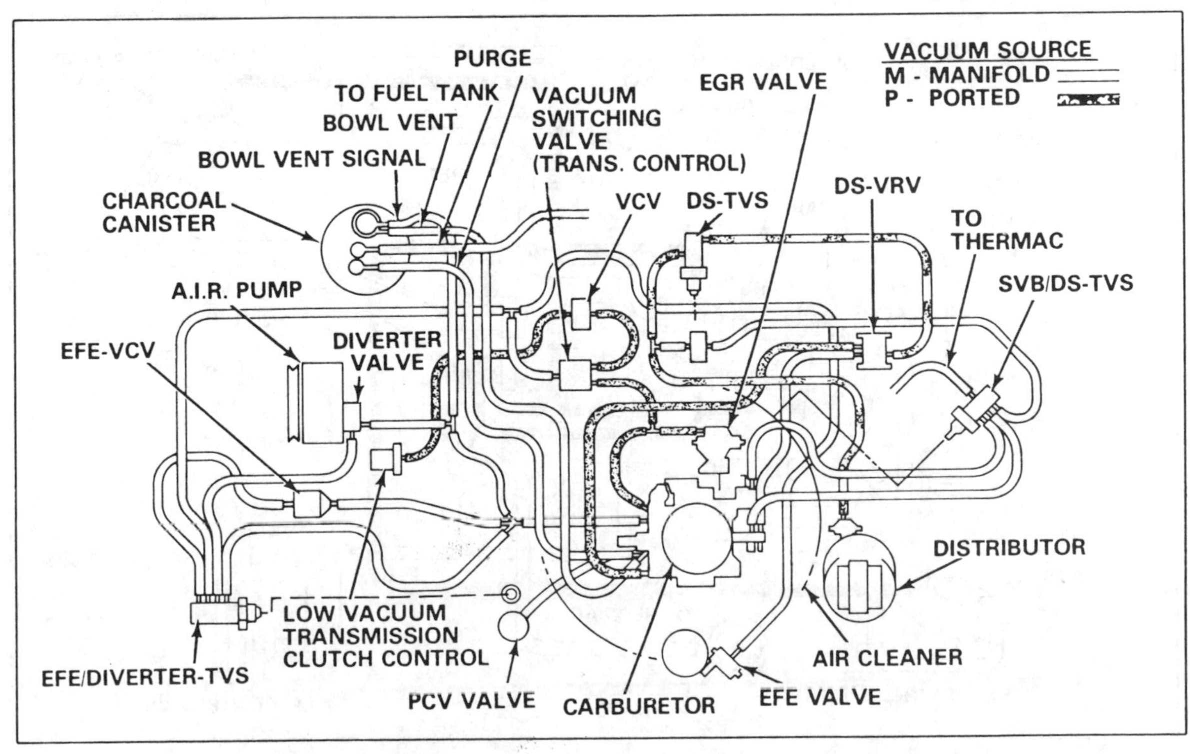

1980 4.9 liter V-8, Vin Code W, with lock-up torque converter (© General Motors Corp.)

VACUUM CIRCUITS

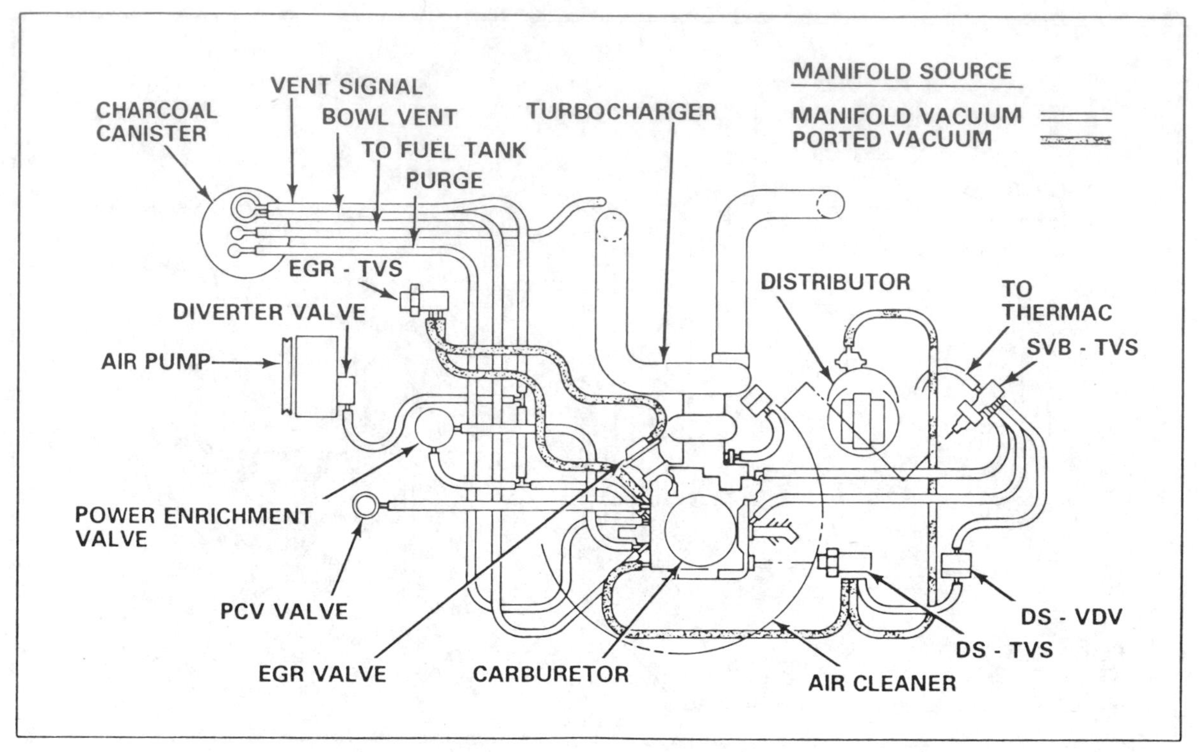

1980 4.9 liter V-8, Vin Code T (© General Motors Corp.)

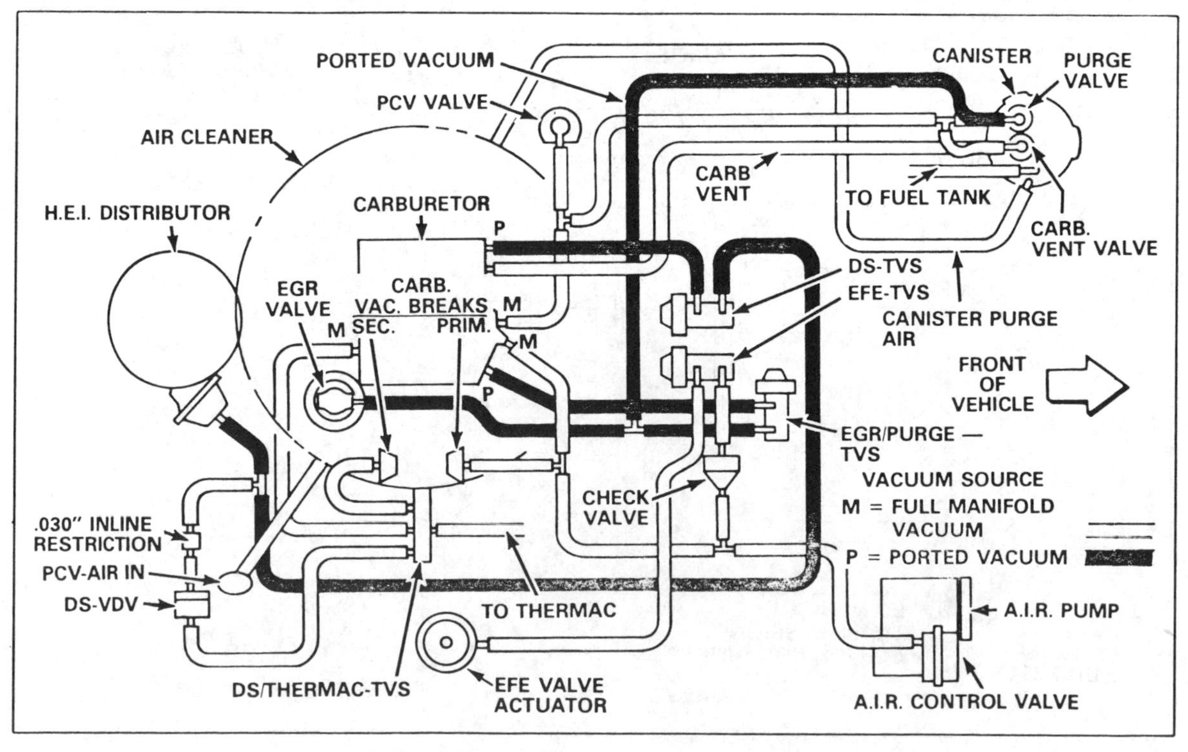

1980 5.0 liter V-8, Vin Code H (© General Motors Corp.)

VACUUM CIRCUITS

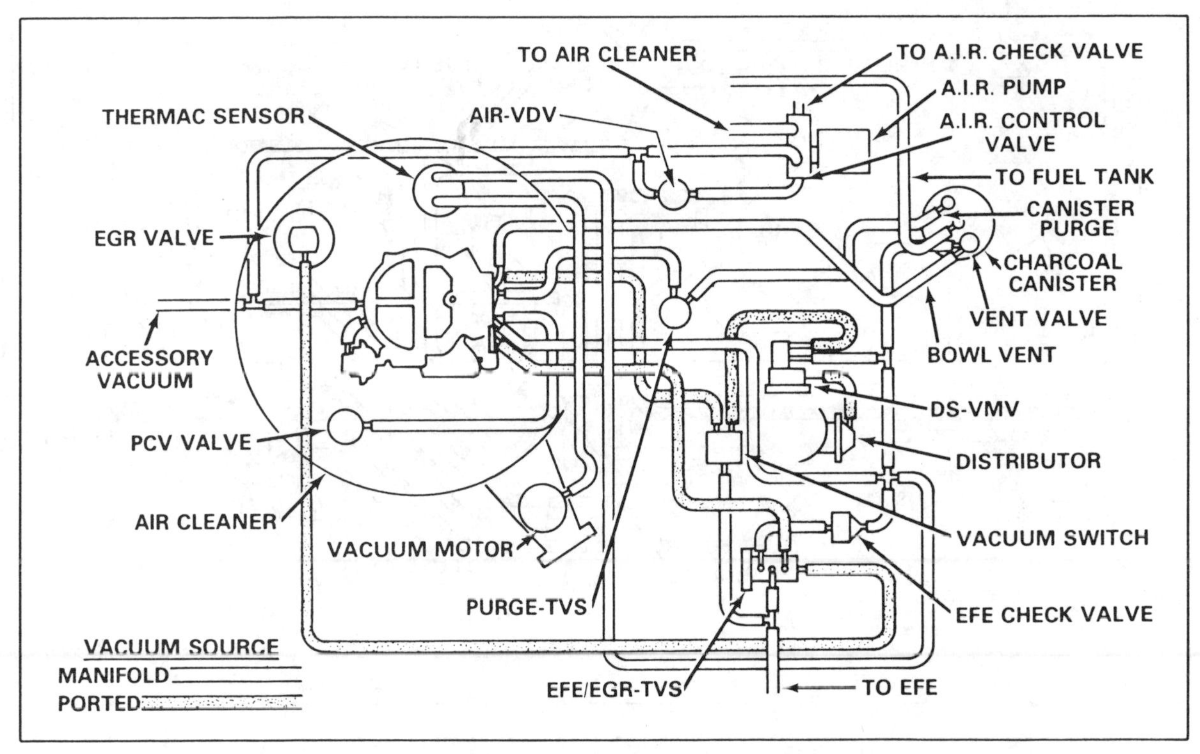

1980 5.7 liter V-8, Vin Code X (© General Motors Corp.)

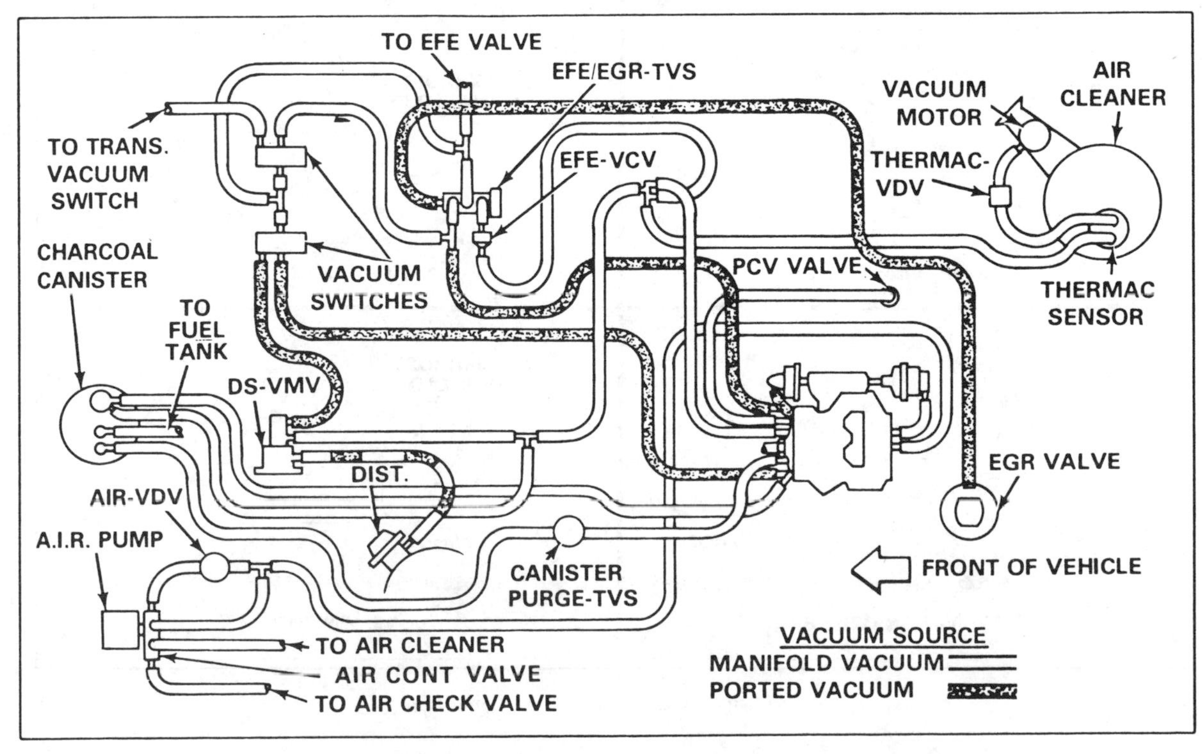

1980 5.7 liter V-8, Vin Code X, with lock-up torque converter (© General Motors Corp.)

VACUUM CIRCUITS

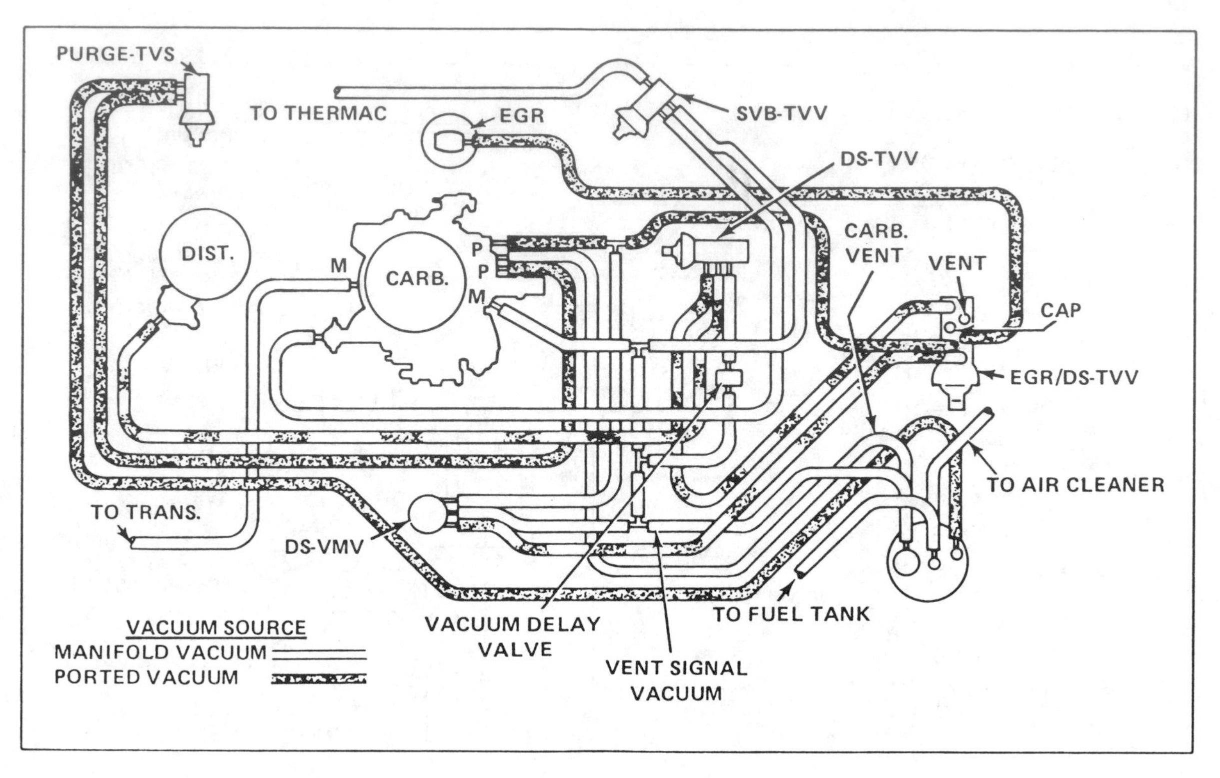

1980 5.7 liter V-8, Vin Code R (© General Motors Corp.)

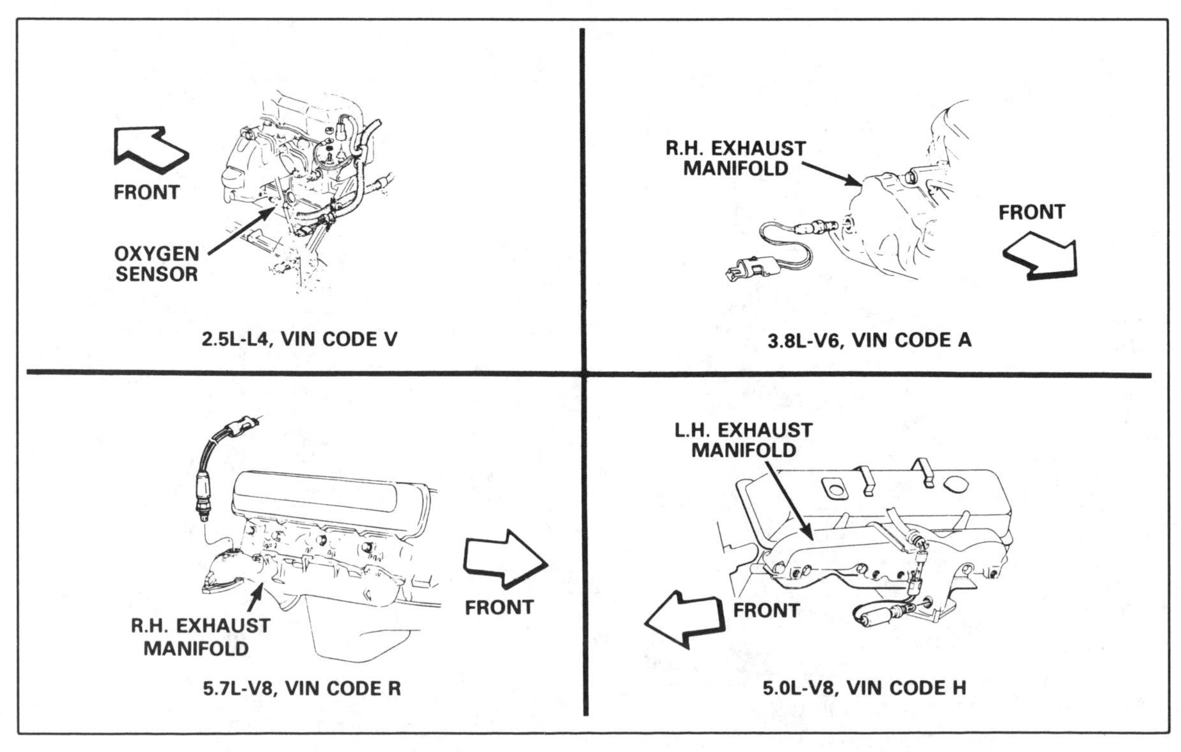

Oxygen Sensor locations (© General Motors Corp.)

TUNE-UP SPECIFICATIONS

CITATION, OMEGA, PHOENIX & SKYLARK

YR.	ENG. V.I.N. Code	ENGINE No. Cyl. Disp. (cu in)	APPLIC. & TRANS.	SPARK PLUGS* Orig. Type	Gap (in)	TIMING deg BTDC @ rpm	IDLE SPEEDS Solenoid Screw A/C	Non A/C	Base	Curb	Fast
1980	5	4-151	Lo.Alt./Man.	R43TSX	.060	10 @ 1000	1300N	1000N	500N	1000N	2400N①
			Lo.Alt./Auto.	R43TSX	.060	10 @ 650	900D	650D	500D	650D	2600N
			Calif./Man.	R43TSX	.060	10 @ 1000	1200N	1000N	500N	1000N	2200N
			Calif./Auto.	R43TSX	.060	10 @ 650	900D	650D	500D	650D	2600N
	7	6-173	Lo.Alt./Man.	R44TS	.045	2 @ 750	1200N	1200N	750N	—	1900N
			Lo.Alt./Auto.	R44TS	.045	6 @ 700	850D	—	—	700D	2000P
			Calif./Man.	R44TS	.045	6 @ 750	—	—	750N	—	2000N
			Calif./Auto.	R44TS	.045	10 @ 700	800D	—	—	700D	2000P

NOTE: The underhood certification/specification decal is the authority for performance specifications affecting vehicle emissions. Use this manual's information only when that decal is not available.

— Not applicable
N Neutral
D Drive
P Park
① Carburetor numbers 17059621 and 17059619—2600

2SE, E2SE CARBURETOR ADJUSTMENTS

Citation, Omega, Phoenix, Skylark

Rochester Carburetor

Year	Carburetor Identification	Float Level (in.)	Pump Rod (in.)	Fast Idle (rpm)	Choke Coil Lever (in.)	Fast Idle Cam (deg./in.)	Air Valve Rod (in.)	Primary Vacuum Break (deg./in.)	Choke Setting (notches)	Secondary Vacuum Break (deg./in.)	Choke Unloader (deg./in.)	Secondary Lockout (in.)
1980	17059614	3/16	1/2	2600	.085	18/.096	.025	17/.090	Fixed	—	36/.227	.120
	17059615	3/16	5/32	2600	.085	18/.096	.025	19/.103	Fixed	—	36/.227	.120
	17059616	3/16	1/2	2600	.085	18/.096	.025	17/.090	Fixed	—	36/.227	.120
	17059617	3/16	5/32	2600	.085	18/.096	.025	19/.103	Fixed	—	36/.227	.120
	17059650	3/16	3/32	2000	.085	27/.157	.025	30/.179	Fixed	38/.243	30/.179	.120
	17059651	3/16	3/32	1900	.085	27/.157	.025	22/.123	Fixed	23/.120	30/.179	.120
	17059652	3/16	3/32	2000	.085	27/.157	.025	30/.179	Fixed	38/.243	30/.179	.120
	17059653	3/16	3/32	1900	.085	27/.157	.025	22/.123	Fixed	23/.120	30/.179	.120
	17059714	11/16	5/32	2600	.085	18/.096	.025	23/.129	Fixed	—	32/.195	.120
	17059715	11/16	3/32	2200	.085	18/.096	.025	25/.142	Fixed	—	32/.195	.120
	17059716	11/16	5/32	2600	.085	18/.096	.025	23/.129	Fixed	—	32/.195	.120
	17059717	11/16	3/32	2200	.085	18/.096	.025	25/.142	Fixed	—	32/.195	.120
	17059760	1/8	5/64	2000	.085	17.5/.093	.025	20/.110	Fixed	33/.203	35/.220	.120
	17059762	1/8	5/64	2000	.085	17.5/.093	.025	20/.110	Fixed	33/.203	35/.220	.120
	17059763	1/8	5/64	2000	.085	17.5/.093	.025	20/.110	Fixed	33/.203	35/.220	.120
	17059618	3/16	1/2	2600	.085	18/.096	.025	17/.090	Fixed	—	36/.227	.120
	17059619	3/16	5/32	2600	.085	18/.096	.025	19/.103	Fixed	—	36/.227	.120
	17059620	3/16	1/2	2600	.085	18/.096	.025	17/.090	Fixed	—	36/.227	.120
	17059621	3/16	5/32	2600	.085	18/.096	.025	19/.103	Fixed	—	36/.227	.120

DISTRIBUTOR SPECIFICATIONS

YEAR	DISTRIBUTOR IDENTIFICATION	CENTRIFUGAL ADVANCE Start Crankshaft Deg. @ Eng. RPM	CENTRIFUGAL ADVANCE Finish Crankshaft Deg. @ Eng. RPM	VACUUM ADVANCE Start @ In. Hg.	VACUUM ADVANCE Finish Crankshaft Deg @ In. Hg.
1980	1103361	0-4.5 @ 1100	20-24 @ 4800	3.5	8.5-11.5 @ 20
	1103362	0-4.5 @ 1100	24-28 @ 4800	3.5	8.5-11.5 @ 20
	1110782	0-4 @ 1050	19-23 @ 4000	3.5	16.5-22 @ 8
	1110783	0-3 @ 1400	19-23 @ 4000	4	19-20 @ 10
	1110786	0-3 @ 1050	19-23 @ 4000	3.5	22 @ 9
	1110787	0 @ 1050	22 @ 4000	3.5	20 @ 9

CAR SERIAL NUMBER AND ENGINE IDENTIFICATION

1980

The vehicle identification number is mounted behind the windshield on the driver's side. The sixth character is the model year, with A for 1980. The fifth character is the engine code, as follows:

5 ....2.5L (151) 2-bbl. L-4 Pont.
7 ......2.8L (171) 2-bbl. V-6 Chev.

EMISSION EQUIPMENT

1980

Emission calibrated carburetor
Emission calibrated distributor
Catalyic converter
Early Fuel Evaporation System (EFE)
Exhaust Gas Recirculation System (EGR)
Closed Positive Crankcase Ventilation System (PCV)
Thermostic Air Cleaner
Pulsair Injection System
Deceleration Valve
Computer Controlled Catalytic Converter System (C-4)
Distributor Spark Thermal Vacuum Switch (DS-TVS)
Distributor Spark Vacuum Delay Valve (DS-VDV)
Distributor Spark Vacuum Regulator Valve (DS-VRV)
Distributor Spark Vacuum Modulated Valve (DS-VMV)
Vapor Storage Canister
Electric Choke

NOTE: *Not all vehicles will have all equipment listed, depending on application.*

IDLE SPEED AND MIXTURE ADJUSTMENTS

All engines are equipped with a two-barrel carburetor. Two models are used; a 2SE unit and an E2SE unit. The 2SE models are used in most applications, while the E2SE unit is the California engine version. The E2SE is used in conjunction with the Computer Controlled Catalytic Converter System (C-4). The C-4 system is sensitive to any change in mixture control adjustment. If it is improperly set, the system may not be able to correct or control the air/fuel mixture. For this reason, plugs are installed in the airhorn, float bowl and over the idle mixture adjusting screws in the throttle body to discourage tampering with factory settings. Mixture adjustment settings should never be changed from the original factory settings, unless there is a physical change in the carburetor or major components requiring replacement parts, or there is a definite emission irregularity.

Propane Enrichment or Lean Drop procedures for idle mixture adjustment used on conventional carburetors must not be used when adjusting the C-4 system carburetor because the Electronic Control Module (ECM) will respond to any mixture change with an opposite "correction" to the air/fuel mixture. For idle speed specifications, and adjustments, it is always best to refer to the emission control underhood label.

2.5 Liter 4-cyl (151)
w/Auto. trans.
Curb or "on" idle 650D
Slow or "off" idle 500D
Propane enriched idle 700
w/Manual trans.
Curb or "on" idle 1000N
Slow or "off" idle 500N
Propane enriched idle ..1150
California w/Auto. trans.
Curb or "on" idle 650D
Slow or "off" idle 500D
Propane enriched idle .... ①
California w/Manual trans.
Curb or "on" idle 1000N
Slow or "off" idle 500N
Propane enriched idle ... ①
2.8 Liter V-6 (173)
w/Auto. trans.
Curb or "on" idle .... ②
Slow or "off" idle .... ②
w/Manual trans.
Curb or "on" idle .... ②
Slow or "off" idle .... ②

① Propane enrichment method is not to be used to adjust the carburetors in the C-4 system.
② See underhood label.

INITIAL TIMING

Timing specifications for each engine are found on the tune-up label under the hood of the vehicle and these specifications should be strictly followed. When using a timing light, connect an adapter between the No. 1 spark plug and the plug wire, or use an inductive type pick-up. It is important to not pierce the plug lead.

INITIAL TIMING

Once the insulation of the spark plug cable has been broken, voltage will jump to the nearest ground and the plug will not fire properly. A magnetic timing probe hole is built in for use with special timing equipment. Follow the manufacturer's instructions for use of the equipment. If a change is necessary, loosen the distributor hold-down bolt at the base of the distributor. On the L-4 engines, loosen the nut and slide the clamp off. Do not remove the mounting bolt.

2.5 Liter (151) Vin code 5, All
w/Auto. trans. ....10° @ 650D
w/Manual trans. ..10° @ 1000N
2.8 Liter (171) Vin code 7
w/Auto. trans. .... 6° @ 700P
w/Manual trans. .. 2° @ 750N
California
w/Auto. trans. ....10° @ 700P
California
w/Manual trans. .. 6° @ 750N

SPARK PLUGS 1980

2.5 Liter (151) 4-cyl.,
All....R43TSX.... .060

2.8 Liter (171) V-6
All.....R-44TS.... .045

VACUUM SOURCE 1980

There are two types of vacuum source.

1. Ported vacuum
2. Manifold vacuum

Ported vacuum advance systems have a timed port in the carburetor throttle body above the throttle valve. The timed port provides vacuum during open throttle operations only. Manifold vacuum advance systems use manifold vacuum from either a manifold vacuum port on the carburetor or a fitting in the intake manifold.

Vacuum is provided whenever the engine is running. Various types of vacuum controls are used in the emission system to modify or control the operation of the different emission control components. Therefore, in many cases, manifold vacuum is used until the engine warms to the point where ported vacuum applied to a valve is greater than the calibrated vacuum from the manifold. When the ported vacuum is greater than the calibration vacuum, a full ported signal is provided to the distributor.

All of this will help provide good cold engine response and performance.

EMISSION CONTROL SYSTEMS

POSITIVE CRANKCASE VENTILATION SYSTEM 1980

All engines are equipped with a Positive Crankcase Ventilation System (PCV) to help produce a more complete scavenging of crankcase vapors.

To help with a smooth idle, a PCV valve is used which restricts the ventilation system whenever the manifold vacuum is high.

The usual checks can be made, such as placing the thumb over the end of the valve to check for vacuum and shaking the valve to check for rattle, indicating the check needle is free and in operating order.

Typical Thermac system (© General Motors Corp.)

THERMOSTATIC AIR CLEANER 1980

The Thermostatic Air Cleaner (Thermac) is used on all engines. It consists of a damper assembly in the air cleaner inlet, controlled by a vacuum motor to mix pre-heated air with non pre-heated air, as it enters the air cleaner. This maintains a controlled air temperature going into the carburetor. This, in turn, allows a leaner carburetor and choke setting for lower emissions. Some engines use a delay check valve to keep the door from opening too soon.

EVAPORATION EMISSION CONTROL SYSTEM (EEC) 1980

The Evaporative Emission Control System is a closed system that prevents the vapor from the gasoline in the fuel tank and the carburetor from entering the atmosphere. The usual vapor storage canister is used along with purge valves and vacuum lines as required by engine and application.

EXHAUST GAS RECIRCULATION SYSTEM (EGR) 1980

The Exhaust Gas Recirculation System (EGR) is used on all engines. It sends a metered amount of exhaust gas into the engine intake system

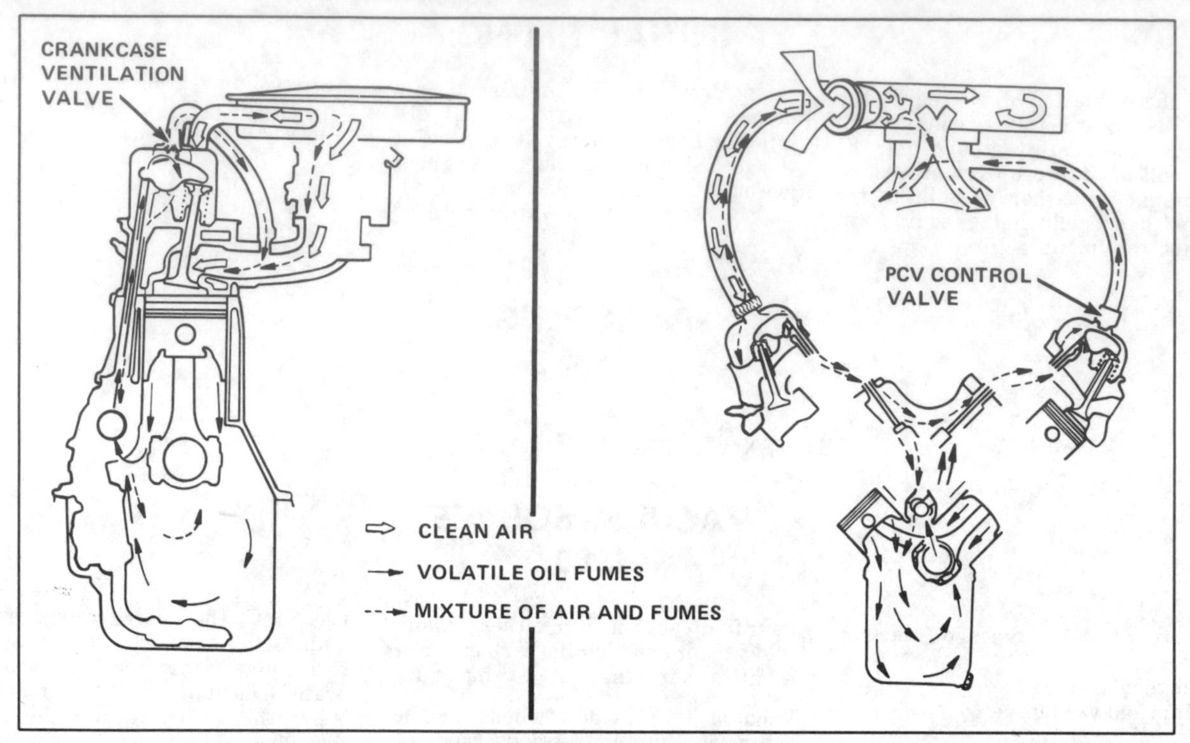

Typical PCV system (© General Motors Corp.)

through either passages cast into the intake manifold or carburetor-to-manifold spacer plate. The purpose of this system is to lower the combustion temperature and thereby reduce the amount of NOX that is formed. The EGR valve regulates the amount of exhaust admitted depending on engine operation.

Two types of EGR systems are used: Vacuum Modulated and Exhaust Back Pressure Modulated. The major difference between the valves is the method used to control how far each valve opens.

With a Vacuum Modulated EGR System, a ported vacuum signal, controlled by throttle position, determines the amount of exhaust gas admitted to the intake manifold. When the throttle is closed, as it would be idle or deceleration, there is no vacuum signal to the EGR valve because the vacuum port is above the closed throttle valve.

The Exhaust Back Pressure Modulated EGR Valve uses a transducer mounted inside the EGR Valve to control the vacuum signal. The vacuum signal is generated in the same manner as for the Vacuum Modulated EGR system. However, the transducer uses exhaust gas pres-

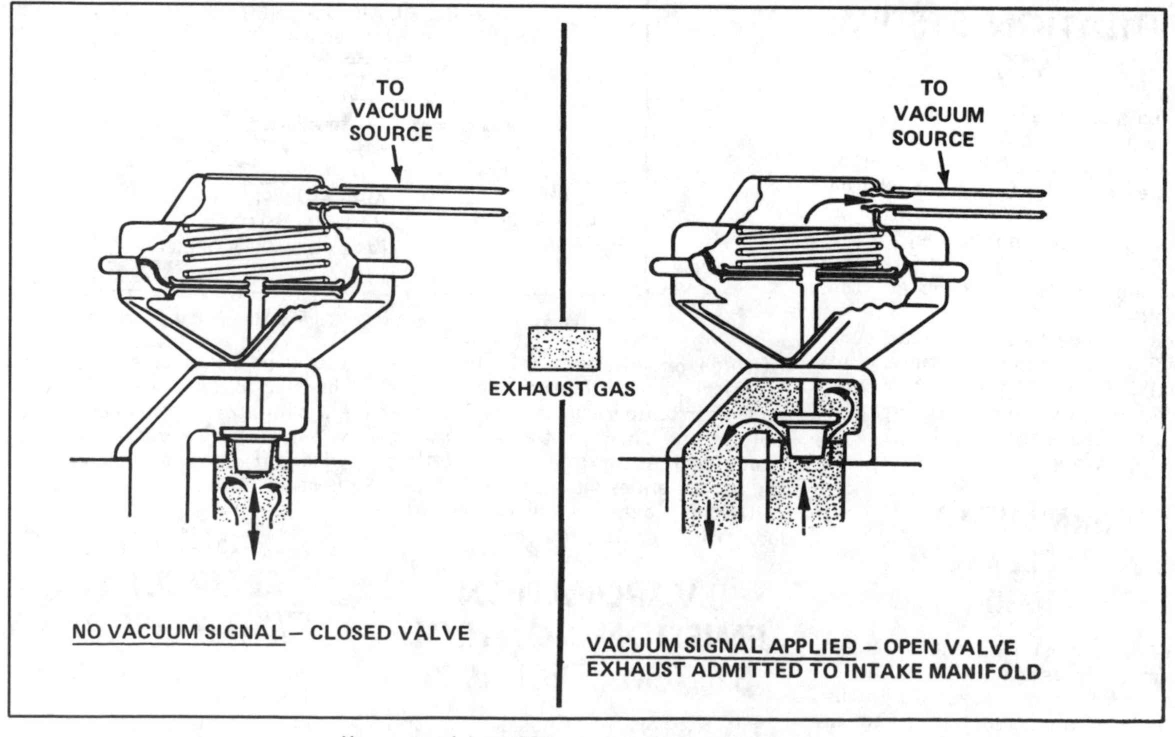

Vacuum modulated EGR valve (© General Motors Corp.)

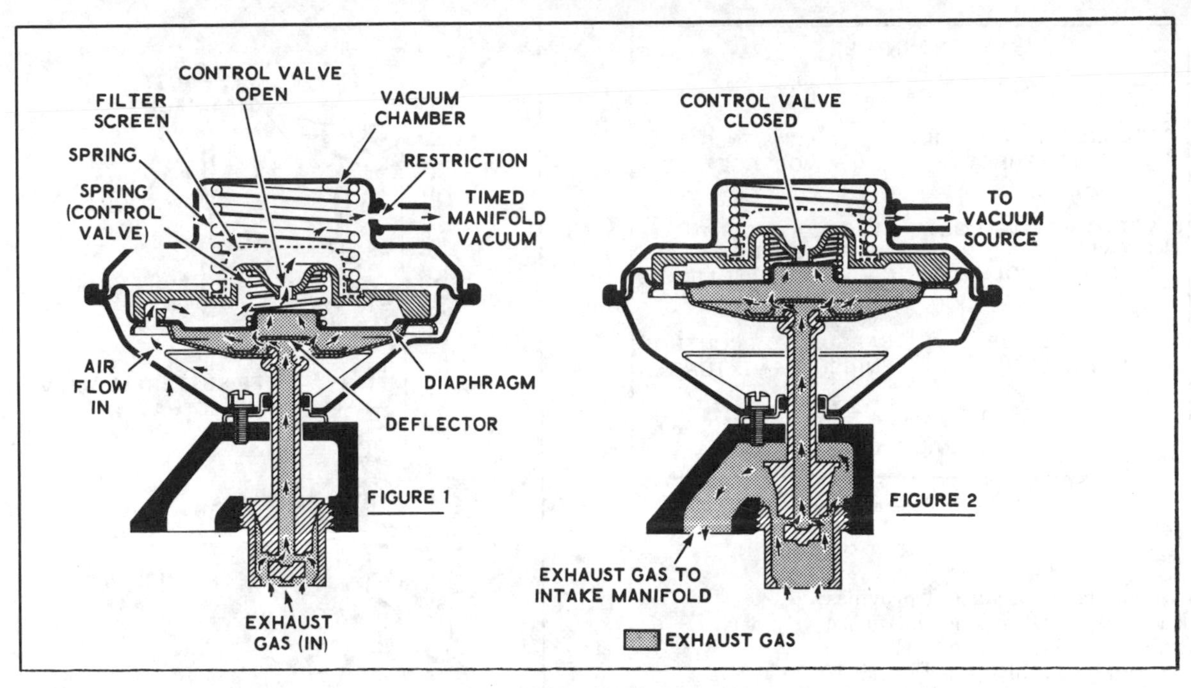

Exhaust gas modulated EGR valve (© General Motors Corp.)

sure to control an air bleed within the valve to modify the vacuum signal from the carburetor.

COMPUTER CONTROLLED CATALYTIC CONVERTER SYSTEM (C-4) 1980

The C-4 system used on 1980 California engines, as a forerunner of equipment to be found on all gas-burning 1981 GM cars, is an electronically controlled exhaust emission system. It includes electronic spark control and automatic altitude compensation. The major components are an Exhaust Gas Oxygen Sensor (OS), an Electronic Control Module (ECM), a controlled air-fuel ratio carburetor and a three-way catalytic converter.

The system features a "Check Engine" warning light which will come on in case of a problem in the system. It will remain on as long as the engine runs with the problem uncorrected. This same lamp, when the ECM diagnostic mode is activated, will flash a trouble code number on the instrument panel to assist in locating the problem. The technican compares the code number with a code list to determine the area of the malfunction. This feature saves much time when diagnosing trouble in the C-4 system.

PULSAIR SYSTEM 1980

The Pulsair system is used on some engines and it is a system of distribution pipes and check valves. The pulses of the exhaust system siphons air into the exhaust system, into the exhaust port near the exhaust valve, to help lower the emissions.

EARLY FUEL EVAPORATION SYSTEM (EFE) 1980

The EFE system can be found on the V-6 engines and its purpose is to provide a rapid source of heat to the intake system during cold driveaway.

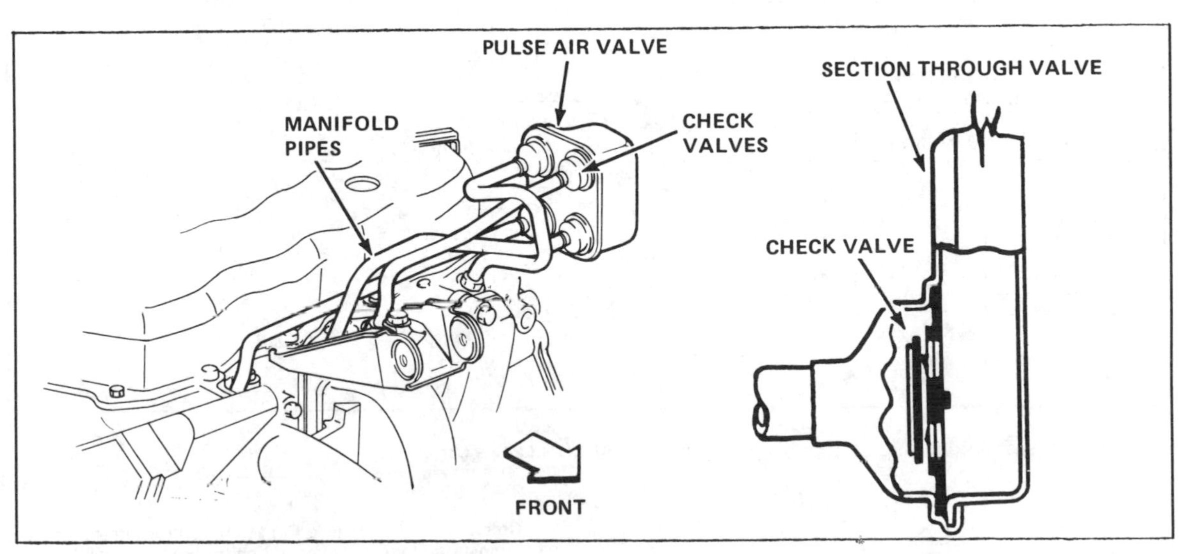

Typical Pulsair system (© General Motors Corp.)

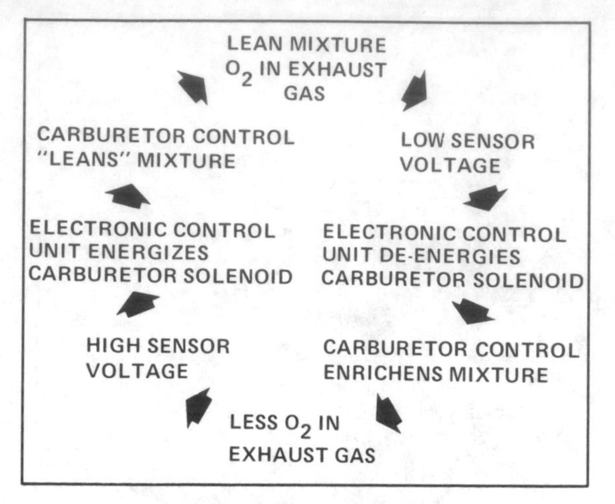

C-4 theory (© General Motors Corp.)

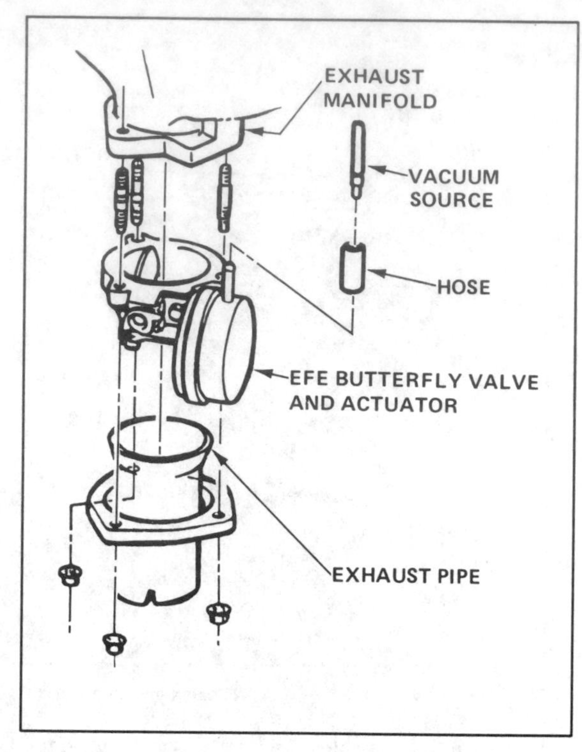

E.F.E. system (© General Motors Corp.)

This is important because it provides quick fuel evaporation and thus more uniform distribution to help the engine run smoothly when cold. It also reduces the amount of time that the choke is required thus reducing the amount of emissions.

The EFE system uses a butterfly valve in the exit of the exhaust manifold to increase the flow of exhaust gas to the intake manifold during cold operation. The valve is operated and controlled by a vacuum actuator which is in turn, operated by a thermal vacuum switch (TVS) which allows vacuum to the valve when the coolant temperature is below the calibration level.

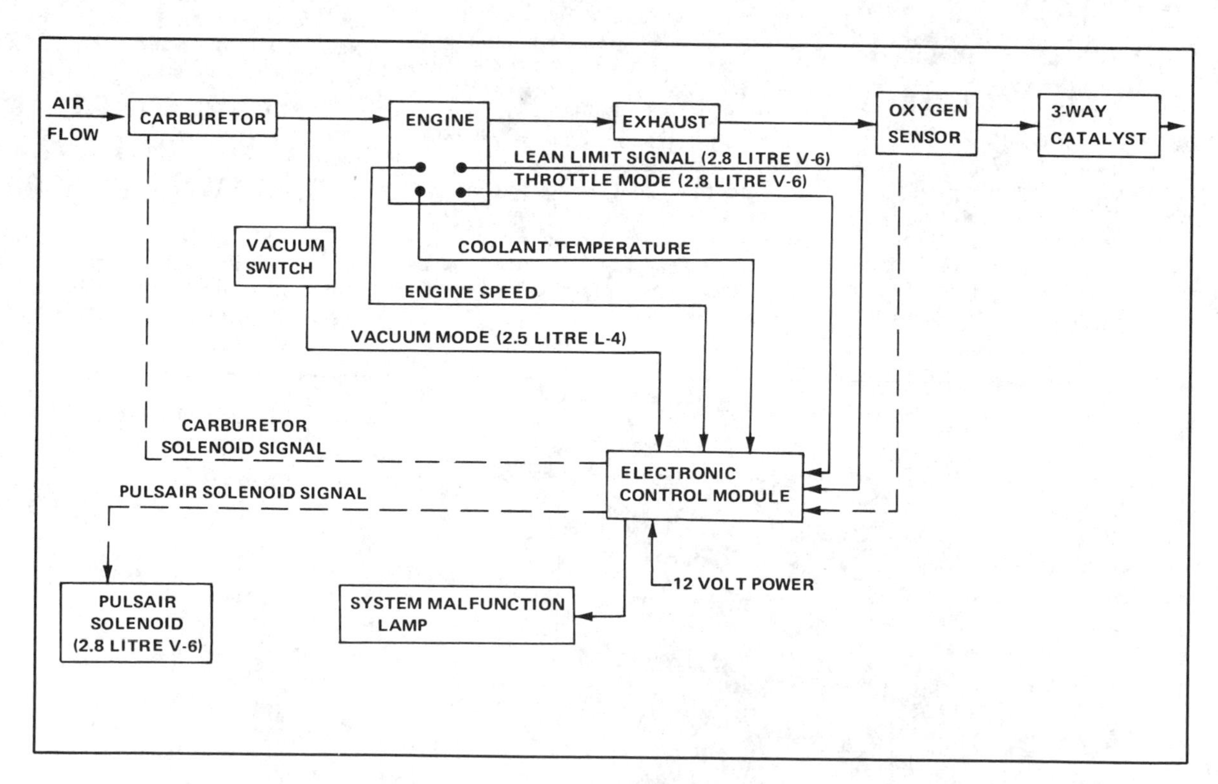

C-4 schematic (© General Motors Corp.)

Note: —Use the Vehicle Emission Control Information Label for the most up to date information concerning hose routing and tune up data.

TUNE-UP SPECIFICATIONS

Engine Code, 5th character of the VIN number

Year	ENG. V.I.N. Code	ENGINE No. Cyl. Disp. (cu in)	SPARK PLUGS Orig. Type	Gap (in)	IGNITION TIMING① (deg B.T.D.C.) Federal Man	Federal Auto	California Man	California Auto	CARBURETION① HOT IDLE SPEED Federal Man	Federal Auto	California Man	California Auto
1979	C	6-225	RBL-16Y	.035	12	12	—	8	675	675	—	750
	D	6-225	RN-12Y	.035	12	12	—	8	675	675	—	750
	G	8-318	RN-12Y	.035	16	16	16	16	725	725	—	—
	H	8-318	RN-12Y	.035	—	16	16	16	750	750	—	750
	K	8-360	RN-12Y	.035	—	12	—	16	750	750	—	750
	J	8-360	RN-12Y	.035	—	16	—	16	750	750	—	750
	L	8-360	RN-12Y	.035	16	16	—	—	750	750	—	—
1980	C	6-225	RBL-16Y	.035	12	12	—	8	675	675	—	750
	D	6-225	RN-12Y	.035	12	12	—	8	675	675	—	750
	G	8-318	RN-12Y	.035	16	16	—	16	725	725	—	750
	H	8-318	RN-12Y	.035	—	16	—	16	750	750	—	750
	K	8-360	RN-12Y	.035	—	12	—	12	750	750	—	750
	L	8-360	RN-12Y	.035	16	16	—	16	750	750	—	750

① Refer to the Emissions Control Label and Specific Sections in this Manual for additional information

DISTRIBUTOR SPECIFICATIONS

Year	DISTRIBUTOR IDENTIFICATION	CENTRIFUGAL ADVANCE Start Dist Deg @ Dist RPM	CENTRIFUGAL ADVANCE Finish Dist Deg @ Dist RPM	VACUUM ADVANCE Start @ In Hg	VACUUM ADVANCE Finish Dist Deg @ In Hg
1979-80	3874876	.2-2.2 @ 600	3.7-5.7 @ 2500	7	7.3-9.8 @ 11.5
1979	4091101	.1-1.2 @ 600	5.8-7.8 @ 2060	9	7.5-9.5 @ 12.5
1980	4111501	.0-1.5 @ 700	7.5-10.0 @ 2200	7	10.0-12.0 @ 1.5

1979 ELECTRONIC SPARK CONTROL SYSTEM

Custom I.C. Spark Control Computer Part Number		**4111656**	**4111657**	**4111674**	**4111750**
Engine Application		360-2 E57	360-4 E58	318-4 E47	318-2 E45
Vacuum Advance (Range)		4°-14°	0°-12°	4°-13°	4°-12°
Crank + Electrical = (Basic Timing)		12° + 0°	10° + 0°	10° + 6°	10° + 6°
Speed Advance (Ground Carb Switch and Disconnect Throttle Transducer Before Checking)	@1100 RPM	3°-7°	1°-5°	0°-1°	0°-1°
	@2000	8°-12°	11°-15°	1°-5°	1°-5°
Zero Time Offset		18°-22°	0°-3°	0°-3°	18°-22°
Accumulator Clock Up (in minutes)		0	8	8	0
Vacuum Advance—Full	@1100 RPM	13°-17°	17°-21°	16°-22°	12°-16°
Accumulator	@1500	18°-22°	26°-30°	23°-27°	18°-22°
Throttle Maximum Advance (Throttle Open 20°)		0°	3°-7°	3°-7°	3°-7°

NOTE: The following specifications are published from the latest information available at the time of publication. If anything differs from those on the Emission Control Information Label, use the specification on the label.

1979 ELECTRONIC SPARK CONTROL SYSTEM

Custom I.C. Spark Control Computer Part Number		4111373	4111392	4111439	4111440
Engine Application		225 E-24-25	318-4 E46	360-4 E58	318-2 E44
Vacuum Advance (Range)		5″-11″	4″-13″	4″-12″	4″-12″
Crank + Electrical = (Basic Timing)		10° + 5°	10° + 6°	10° + 6°	10° + 6°
Speed Advance (Ground Carb Switch and Disconnect Throttle Transducer Before Checking)	@1100 RPM	0°-2°	0°-1°	2°-6°	0°-1°
	@2000	1°-4°	1°-5°	6°-10°	1°-5°
Zero Time Offset		0°-3°	8°-12°	0°-3°	8°-12°
Accumulator Clock Up (in minutes)		8	8	8	8
Vacuum Advance—Full	@1100 RPM	11°-15°	16°-20°	8°-12°	12°-16°
Accumulator	@1500	18°-22°	23°-27°	18°-22°	18°-22°
Throttle Maximum Advance (Throttle Open 20°)		6°-10°	3°-7°	6°-10°	3°-7°

1979 ELECTRONIC SPARK CONTROL SYSTEM

Custom I.C. Spark Control Computer Part Number		4111441	4111442	4111492	4111540
Engine Application		360-2 E57	360-4 E58	318-4 E47	318-2 E45
Vacuum Advance (Range)		4″-12″	4″-12″	4″-13″	4″-12″
Crank + Electrical = (Basic Timing)		12° + 0°	10° + 6°	10° + 6°	10° + 6°
Speed Advance (Ground Carb Switch and Disconnect Throttle Transducer Before Checking)	@1100 RPM	3°-7°	2°-6°	0°-1°	0°-1°
	@2000	8°-12°	6°-10°	1°-5°	1°-5°
Zero Time Offset		6°-10°	6°-10°	8°-12°	8°-12°
Accumulator Clock Up (in minutes)		8	8	8	8
Vacuum Advance—Full	@1100 RPM	11°-17°	8°-12°	16°-20°	12°-16°
Accumulator	@1500	18°-22°	18°-22°	23°-27°	18°-22°
Throttle Maximum Advance (Throttle Open 20°)		0°	6°-10°	3°-7°	3°-7°

1979 ELECTRONIC SPARK CONTROL SYSTEM

Custom I.C. Spark Control Computer Part Number		4111574	4111575	4111650	4111652
Engine Application		318-4 E46	360-4 E56	318-2 E44	360-2 E57
Vacuum Advance (Range)		4°-13°	4°-10°	4°-12°	4°-14°
Crank + Electrical = (Basic Timing)		10° + 6°	10° + 6°	10° + 6°	12° + 0°
Speed Advance (Ground Carb Switch and Disconnect Throttle Transducer Before Checking)	@1100 RPM	0°-1°	2°-6°	0°-1°	0°-1°
	@2000	1°-5°	6°-10°	1°-5°	11°-15°
Zero Time Offset		0°-3°	2°-6°	18°-22°	0°-3°
Accumulator Clock Up (in minutes)		8	8	8	8
Vacuum Advance—Full	@1100 RPM	16°-20°	11°-15°	12°-16°	17°-21°
Accumulator	@1500	23°-27°	18°-22°	18°-22°	26°-30°
Throttle Maximum Advance (Throttle Open 20°)		3°-7°	6°-10°	3°-7°	3°-7°

1980 ELECTRONIC SPARK CONTROL SYSTEM

Custom I.C. Spark Control Computer Part Number		4145000	4145003	4145004	4145005
Engine Application		360-4 Fed	318-4 Cal	360-2 Fed	360-2 Can
Crank + RUN = (Basic Timing)		8° + 8°	8° + 8°	12° + 0°	12° + 0°
Vacuum Advance (Range)		4″-14″	4″-14″	4″-14″	4″-14″
Zero Time Advance		23°-27°	28°-32°	23°-27°	23°-27°
Accumulator Clock Up (in minutes)		0	0	0	0
Vacuum Advance—Full Accumulator	@1100 RPM	10°-14°	14°-18°	0°-2°	16°-20°
	@2500	23°-27°	28°-32°	23°-27°	23°-27°
Speed Advance (Ground Carb Switch and Disconnect Throttle Transducer Before Checking)	@1100 RPM	1°-4°	0°-2°	0°-4°	4°-6°
	@2000 RPM	4°-8°	1°-5°	4°-8°	7°-10°
	@4800	12°-16°	12°-16°	16°-20°	16°-20°
O_2 Feed Back Air Switching Solenoid Warm Up Schedule		No	Yes 25 Sec	No	No

1980 ELECTRONIC SPARK CONTROL SYSTEM

Custom I.C. Spark Control Computer Part Number		4105006	4105007	4145087
Engine Application		318-4 Can	225-1 Cal	360-4 Fed
Crank + RUN = (Basic Timing)		10° + 0°	12° + 0°	8° + 8°
Vacuum Advance (Range)		4″-14″	0″-10″	4″-14″
Zero Time Advance		0°-2°	0°-2°	21°-25°
Accumulator Clock Up (in minutes)		8 Min.	1 Min.	0
Vacuum Advance—Full Accumulator	@1100 RPM	19°-23°	0°-2°	0°-2°
	@2500	28°-32°	20°-24°	21°-25°
Speed Advance (Ground Carb Switch and Disconnect Throttle Transducer Before Checking)	@1100 RPM	0°-2°	0°-2°	0°-2°
	@2000 RPM	10°-14°	1°-5°	4°-8°
	@4800	18°-22°	1°-5°	12°-16°
O_2 Feed Back Air Switching Solenoid Warm Up Schedule		No	Yes* 100 Sec.	No

* Has Purge Solenoid in Parallel

Model 2280

Chrysler Corporation

Year	Carb. Part No.	Float Level (in.)	Accelerator Pump Adjustment (in.)	Bowl Vent Clearance (in.)	Fast Idle (rpm)	Choke Unloader Clearance (in.)	Vacuum Kick (in.)	Fast Idle Cam Position (In.)	Choke
1978	R-7990-A	.313	Flush	.030	1600	.310	.150	.070	Fixed
1979	R-8448-A	.313	Flush	.030	1600	.310	.150	.070	Fixed

CARTER TQ SPECIFICATIONS

CHRYSLER PRODUCTS

CARTER—Carburetors

Year	Model ①	Float Setting (in.)	Secondary Throttle Linkage (in.)	Secondary Air Valve Opening (in.)	Secondary Air Valve Spring (turns)	Accelerator Pump (in.)	Choke Control Lever (in.)	Choke Unloader (in.)	Vacuum Kick (in.)	Fast Idle Speed (rpm)
1979	9195S	29/32	②	3/8	2	33/64	3 3/8	0.310	0.100	1600
	9197S	29/32	②	1/2	1 1/2	33/64	3 3/8	0.310	0.100	1600
	9196S, 9198S, 9202S	29/32	②	1/2	2	33/64	3 3/8	0.310	0.100	1600
1980	9236S	29/32	②	1/2	3	11/32③	3 3/8	0.310	0.100	1600
	9243S	29/32	②	1/2	2 5/8	11/32④	3 3/8	0.310	0.100	1600
	9244S	29/32	②	1/2	2 1/2	11/32④	3 3/8	0.310	0.100	1200

① Model numbers located on the tag or on the casting
② Adjust link so primary and secondary stops both contact at same time
③ Slot #1
④ Slot #2
NOTE: All choke settings are fixed.

Model 1945

Chrysler Corporation

Holley Carburetor

Year	Carb. Part No. ②	Float Level (in.)	Accelerator Pump Adjustment (in.)	Bowl Vent Clearance (in.)	Fast Idle (rpm)	Choke Unloader Clearance (in.)	Vacuum Kick (in.)	Fast Idle Cam Position (in.)	Choke
1979	R-8523-A	①	1.70③	1/16	1400	.250	.110	.080	Fixed
	R-8452-A	①	1.615④	1/16	1600	.250	.110	.080	Fixed
	R-8555-A	①	1.70③	1/16	1400	.250	.110	.080	Fixed
	R-8727-A	①	1.615④	1/16	1600	.250	.110	.080	Fixed
	R-8680-A	①	1.615④	1/16	1500	.250	.130	.080	Fixed
1980	R-8718-A	①	1.70③	1/16	1400	.250	.150	.090	Fixed
	R-8831-A	①	1.615④	1/16	1600	.250	.140	.090	Fixed
	R-8832-A	①	1.70③	1/16	1400	.250	.110	.090	Fixed
	R-8833-A	①	1.615④	1/16	1600	.250	.110	.090	Fixed

① Flush with the top of the bowl cover gasket, plus or minus 1/32
② Located on a tag attached to the carburetor.
③ Position #1
④ Position #2

Model 2245

Chrysler Corporation

Holley Carburetor

Year	Carb.★ Part No.	Float Level (in.)	Accelerator Pump Adjustment (in.)	Bowl Vent Clearance (in.)	Fast Idle (rpm)	Choke Unloader Clearance (in.)	Vacuum Kick (in.)	Fast Idle Cam Position (in.)	Choke
1978	R-7991-A	.188	.265	.025	1600	.170	.110	.110	Fixed
	R-8326-A	.188	.265	.025	1600	.170	.110	.110	Fixed
1979	R-8450-A	.188	.266	.025	1600	.170	.110	.110	Fixed
	R-8774-A	.188	.266	.025	1600	.170	.110	.110	Fixed

★ Located on a tag attached to the carburetor.

Model 5220

Chrysler Corporation — Holley Carburetor

Year	Carb. Part No.	Accelerator Pump	Dry Float Level (in.)	Vacuum Kick (in.)	Fast Idle RPM (w/fan)	Throttle Position Transducer (in.)	Throttle Stop Speed RPM	Choke
1979	R-8524A, 8526A,8532A, 8534A,8528A, 8530A	#2 hole	.480	.040	1700	—	700	2 Rich
	R-8525A, 8541A,8531A, 8533A,8527A, 8529A	#2 hole	.480	.070	1400	—	700	2 Rich
1980	R8838A, 8839A, 9110A, 9111A, 9325A, 9327A	#2 hole	.480	.040	1700	—	700	Fixed
	R8726A, 8727A, 8837A, 9108A, 9321A, 9323A	#2 hole	.480	.070	1400	—	700	Fixed
	R9109A	#2 hole	.480	.100	1400	—	700	Fixed

CARTER BBD SPECIFICATIONS

CHRYSLER PRODUCTS — CARTER—Carburetors

Year	Model ④	Float Level (in.)	Accelerator Pump Travel (in.)	Bowl Vent (in.)	Choke Unloader (in.)	Choke Vacuum Kick	Fast Idle Cam Position	Fast Idle Speed (rpm)	Automatic Choke Adjustment
1979	8198S	¼	0.500③	0.080	0.280	0.100	0.070	1600	Fixed
	8199S	¼	0.500③	0.080	0.280	0.100	0.070	1600	Fixed
1980	8233S	¼	0.500③	0.080	0.280	0.130	0.070	1500	Fixed
	8235S	¼	0.500③	0.080	0.280	0.130	0.070	1700	Fixed
	8237S	¼	0.500③	0.080	0.280	0.110	0.070	1500	Fixed
	8239S	¼	0.500③	0.080	0.280	0.110	0.070	1500	Fixed
	8286S	¼	0.500③	0.080	0.280	0.100	0.070	1400	Fixed

CAR SERIAL NUMBER AND ENGINE IDENTIFICATON

1979-80

The vehicle identification number is mounted on a plate behind the windshield on the driver's side. The sixth character is the model year, with 9 for 1979 and A for 1980. The fifth chara cter is the engine code, as follows:

C .... 225 1-bbl. 6-cyl.
D .... 225 2-bbl. 6-cyl.
G 318 2bbl. V-8
H 318 4-bbl. V-8
J (1979) 360 4-bbl. V-8
K 360 2bbl. V-8
L 360 4-bbl. V-8

EMISSION EQUIPMENT

1979-80

All Models

Closed positive crankcase ventilation
Heated air cleaner
Vapor control, canister storage
Exhaust gas recirculation
Emission calibrated carburetor
Emission calibrated distributor
Catalytic converter
Fresh air intake
Electric assist choke
Orifice spark advance control
Air injection system
Air aspirator system

IDLE SPEED AND MIXTURE ADJUSTMENTS

1979

EXHAUST ANALYZER METHOD

NOTE: *This is an alternate method for California cars only. The preferred method is to use propane enrichment.*

Air cleaner ....In place
Air cond. ....Off
Auto. trans. ....Park

Air pump ....Disconnect for mixture setting
Vac. adv. hose ....Plugged
EGR vac. hose ....Plugged
Lean burn engine .....Ground carb. switch with jumper

Mixture adj. ....Set idle CO
Idle CO
225 1-bbl. 6-cyl. .....Probe 0.3%
318 4-bbl. V-8 ....Tail 0.5%
360 4-bbl. V-8 ....Tail 0.5%

Idle Speed
225 6-cyl. 750
318 4-bbl. V-8 750
360 4-bbl. V-8 750

IDLE SPEED AND PROPANE ENRICHED MIXTURE ADJUSTMENTS 1979

NOTE: *The following specifications are obtained from the latest information at the time of publication. Should this data differ from the data on the emissions label, use the information on the vehicle emissions label. All measurements are in RPM's.*

225 1-bbl. Manual trans.
Idle set 675
Fast idle 1400
Propane enriched 845

225 1-bbl. Auto. trans.
Idle set .... 675 (750 Calif.)
Fast idle ......1600 (1500 Calif.)
Propane enriched .830 (925 Calif.)

225 2-bbl. Auto. trans.
Idle set ....750 (725 Canada)
Fast idle 1600
Propane enriched ....855 (890 Canada)

318 2-bbl. Auto. trans.
Idle set 730
Fast idle 1600
Propane enriched 850

318 4-bbl. Auto. trans.
Idle set 750
Fast idle 1600
Propane enriched 830

360 2-bbl. Auto. trans.
Idle set 750
Fast idle 1600
Propane enriched 890

360 4-bbl. Auto. trans.
Idle set 750
Fast idle 1600
Propane enriched .... 830 (870 Federal)

1980

NOTE: *The following specifications are obtained from the latest information at the time of publication. Should this data differ from the data on the emissions label, use the data on the label. All measurements are in RPM's*

225 1-bbl. Manual trans.
Idle set 725
Fast idle 1400
Propane enriched .... 880 (860 Canada)

225 1-bbl. Auto. trans.
Idle set 725
Fast idle 1600
Propane enriched .... 835 (860 Canada)

225 2-bbl. Auto. trans.
Idle set 750
Fast idle 1600
Propane enriched 885

318 2-bbl. Auto. trans.
Idle set ....700 (730 Canada)
Fast idle ....1500 (1700 Canada)
Propane enriched ....800 (760 Canada)

318 4-bbl. Auto. trans.
Idle set 750
Fast idle 1600
Propane enriched 800

360 2-bbl. Auto. trans.
Idle set 700
Fast idle 1500
Propane enriched 800

360 4-bbl. Auto. trans.
Idle set ....750 (730 Canada)
Fast idle ......(12 (1600 Canada)
Propane enriched ....870 (830 Canada)

INITIAL TIMING

1979

NOTE: *The following specifications are published from the latest information available at the time of publication. If this information disagrees with the information on the vehicle emissions label, use the data on the emissions label.*

225 1-bbl. 6-cyl.
49-States ....12° BTDC
Calif. .... 8° BTDC

225 2-bbl. 6-cyl. ....12° BTDC
318 2-bbl. V-8 ....16° BTDC
318 4-bbl. V-8 ....16° BTDC
360 2-bbl. V-8 ....12° BTDC
360 4-bbl. V-8
49-States ....16° BTDC
Calif. ....16° BTDC

1980

NOTE: *The following specifications are published from the latest information available at the time of publication. If this information differs from the information on the vehicle emissions label, use the data on the emissions label.*

225 1-bbl. 6 cyl. ....12 BTDC
225 2-bbl. 6 cyl. ....12 BTDC
318 2-bbl. V-8 ....12 BTDC
318 4-bbl. V-8 (Canada) ..10 BTDC
360 2-bbl. V-8 ....12 BTDC
360 4-bbl. V-8 (49 States) .16 BTDC
360 4-bbl. V-8 (Canada) ..10 BTDC

SPARK PLUGS 1979-80

225 6-cyl. .....CH-RBL16Y.. .035
318 V-8 ....CH-RN12Y.. .035
360 V-8 ....CH-RN12Y.. .035

VACUUM ADVANCE 1979

Diaphragm type ....Single
Vacuum source ....Manifold

VAPOR CONTROL SYSTEM 1979-80

Some models still have the normal 3-hose carbon canister, with one hose bringing vapors from the tank, the second hose bringing vapors from the carburetor bowl vent, and the third hose for purging. The purge hose on 3-hose canisters is connected to a carburetor port that is above the throttle blade. At idle the port is not subject to engine vacuum, so the canister is not purged. Above idle the port is uncovered and purging takes place.

Some engines use a 4-hose canister. Two hoses bring vapors to the canister, one from the tank and the other from the carburetor bowl. The other two hoses are for purging. One of them is connected to ported vacuum at the carburetor. It turns on a purge valve on top of the canister. When the purge valve opens, a hose from the canister to manifold vacuum purges the canister.

Some of the larger V-8's use two carbon canisters. One canister is the 3-hose type. It connects to the carburetor bowl vent and to a carburetor port. The third outlet on the canister is capped. Purging takes place only above idle when the port is uncovered.

EMISSION CONTROL SYSTEMS

The second canister is connected to the fuel tank. It is the purge valve 4-outlet type, but the outlet that normally connects to the bowl vent is capped off. With this system, one canister takes only the carburetor bowl vent vapors, and the other canister takes only the fuel tank vapors. This reduces the chance that any vapors will get into the atmosphere.

Both canisters have filters in the bottom, which must be replaced every 30,000 miles.

HEATED INLET AIR SYSTEM 1979-80

The purpose of the heated inlet air system is to control the temperature of the air entering the carburetor when ambient temperatures are low. By increasing the temperature of the air being introduced into the carburetor, a much leaner calibration is made possible, thereby reducing hydrocarbon emissions. In addition to a refined fuel air mixture, other benefits derived as a result of the heated air include smoother engine warm-up operation (a more volatile mixture as a result of heat) and a minimal amount of carburetor icing.

ELECTRIC ASSIST CHOKE SYSTEM 1979-80

This system basically consists of an electric heating element, located next to a bimetal spring inside the choke housing, which assists engine heat to control choke duration. The electric current necessary to generate heat in the heater is routed through the oil pressure switch. A minimum pressure of 4 psi is required to close the switch before current can begin flowing through the switch and to the heater.

In addition to the electric heating element, two different electrical control units are used to regulate choke duration. A single stage control shortens only summer choke operation above approximately 80 degrees F. A duel stage control functions during summer temperatures in a manner similar to that of the single stage control, but also stabilizes choke duration in the winter by reducing current flow to the heater with the aid of a resistor permanently connected to both terminals of the control. The resistor is effective at temperatures below approximately 55 degrees F.

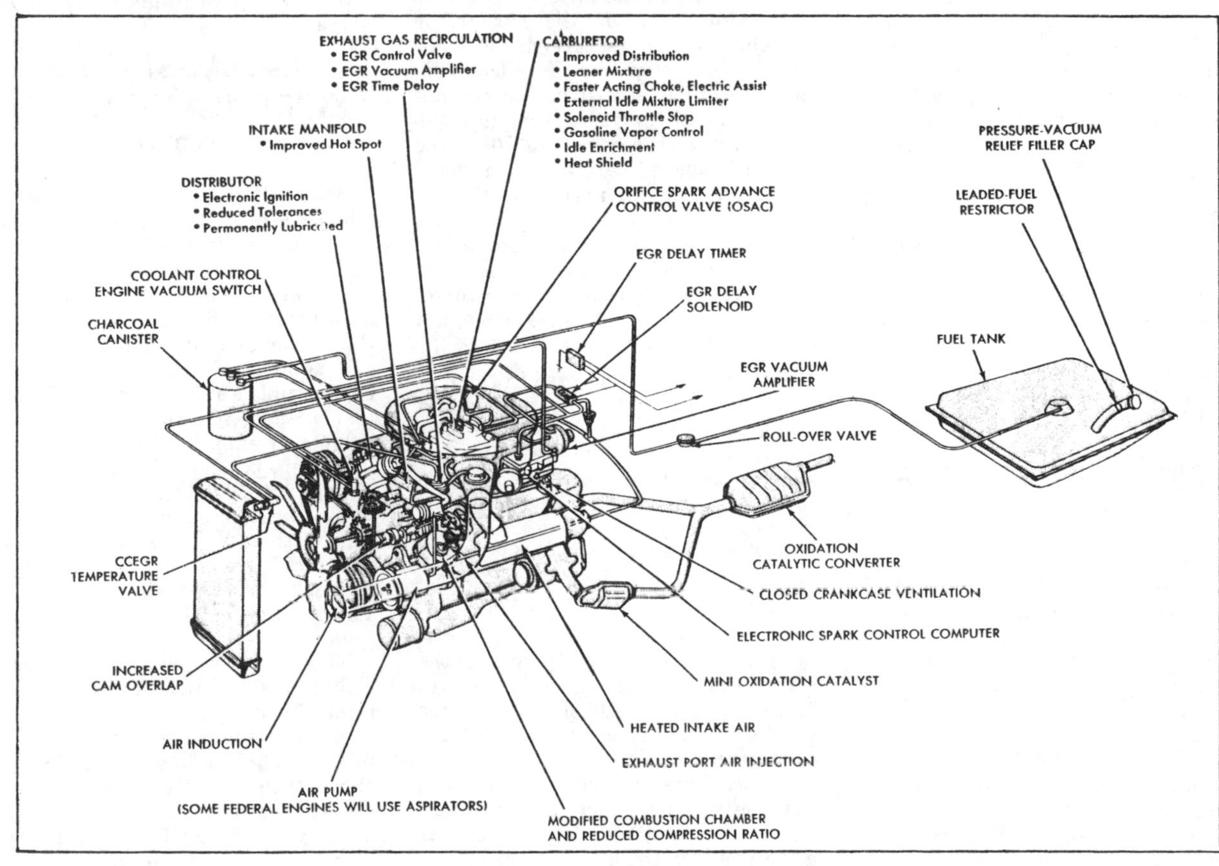

Typical emission control system and components

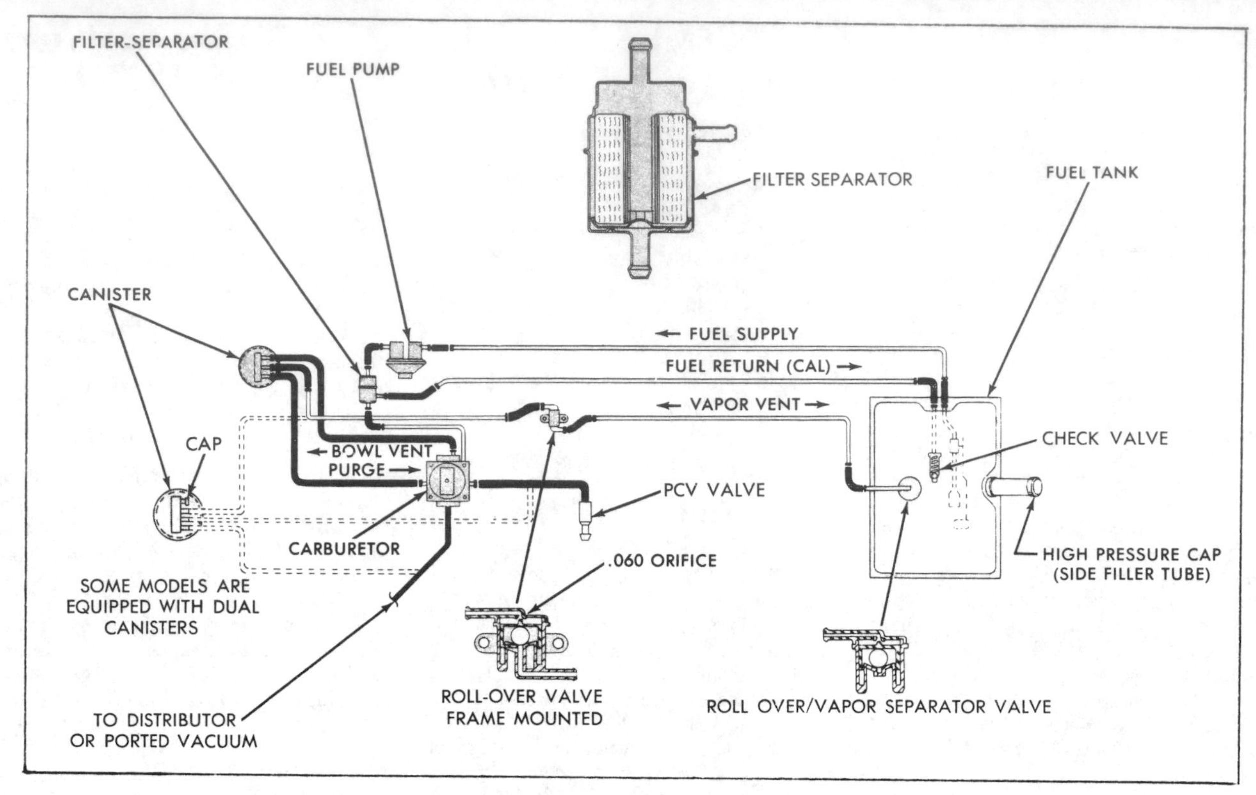

Evaporation control system—typical

IGNITION SYSTEM 1979-80

The purpose or design function of the ignition system is to allow a lean air fuel mixture to be burned under various operating conditions. The Electronic Spark Control system is comprised of a spark control computer, five engine sensors, and a specially calibrated carburetor.

The spark control computer is capable of receiving several and various signals from the five engine sensors simultaneously, computing these signals within milliseconds, thereby monitoring engine performance, and advancing or retarding the ignition timing by signaling the ignition coil to produce the electrical impulses which fire the spark plugs.

ENGINE SENSORS 1979-80

There are five engine sensors: vacuum transducer, pick-up coil, throttle position transducer, carburetor switch, and coolant switch.

The vacuum transducer, located on the spark control computer, senses intake manifold vacuum, and signals the computer which, in turn, regulates ignition timing.

A high vacuum reading will command an additional advance in timing. A low vacuum reading will command the timing to retard.

In order to achieve maximum advance for any inch of vacuum, the carburetor switch must remain open for a specified amount of time. During this state, ignition advance is not immediate; it is gradual, and develops at a slow rate. If the carburetor switch closes prematurely, the additional advance will be cancelled. This information is stored in the computer's memory, and the computer will slowly return the advance to 0. If the switch is opened before the computer can return the advance to 0, the advance will again gradually build from the point recorded in the computer's memory. If the switch is opened after the advance has been allowed to return to 0, ignition advance will again develop gradually.

The pick-up coil, located in the distributor, senses engine speed and crankshaft location, and transmits this information to the computer. The signal transmitted from the pick-up coil is reference signal. When the computer receives this signal, maximum timing advance is made available.

The function of the throttle position transducer is to sense the throttle position and rate of change. As the throttle plates start to open, regardless of position, additional advance will be directed on command from the computer.

The carburetor switch, located on the idle stop solenoid or air conditioning solenoid, signals the computer as to whether the engine is at idle or off idle.

The coolant switch, located on the thermostat housing, senses engine coolant temperature. Its function is to signal the computer when the engine coolant temperature is below 150 degrees.

EXHAUST GAS RECIRCULATION (EGR) 1979

Recirculated exhaust gas, when introduced into the combustion chamber, has the effect of minimizing the oxides of nitrogen emitted into the air by lowering combustion flame temperatures during engine operation. Only a modulated amount of exhaust gas is allowed to dilute the highly combustible air fuel mixture. The system responsible for regulating the volume of exhaust gas available during firing is called Exhaust Gas Recirculation (EGR).

Two alternate systems are used to control the rate of exhaust gas recirculation, depending on the engine model. These two systems are: Ported Vacuum Control and Venturi Vacuum Control.

The ported vacuum control system utilizes a slot type port in the carburetor throttle body which is exposed to an increasing percentage of manifold vacuum as the throttle blade opens. This throttle bore port is connected through an external nipple directly to the EGR valve. The flow rate is dependent on manifold vacuum throttle position, and exhaust

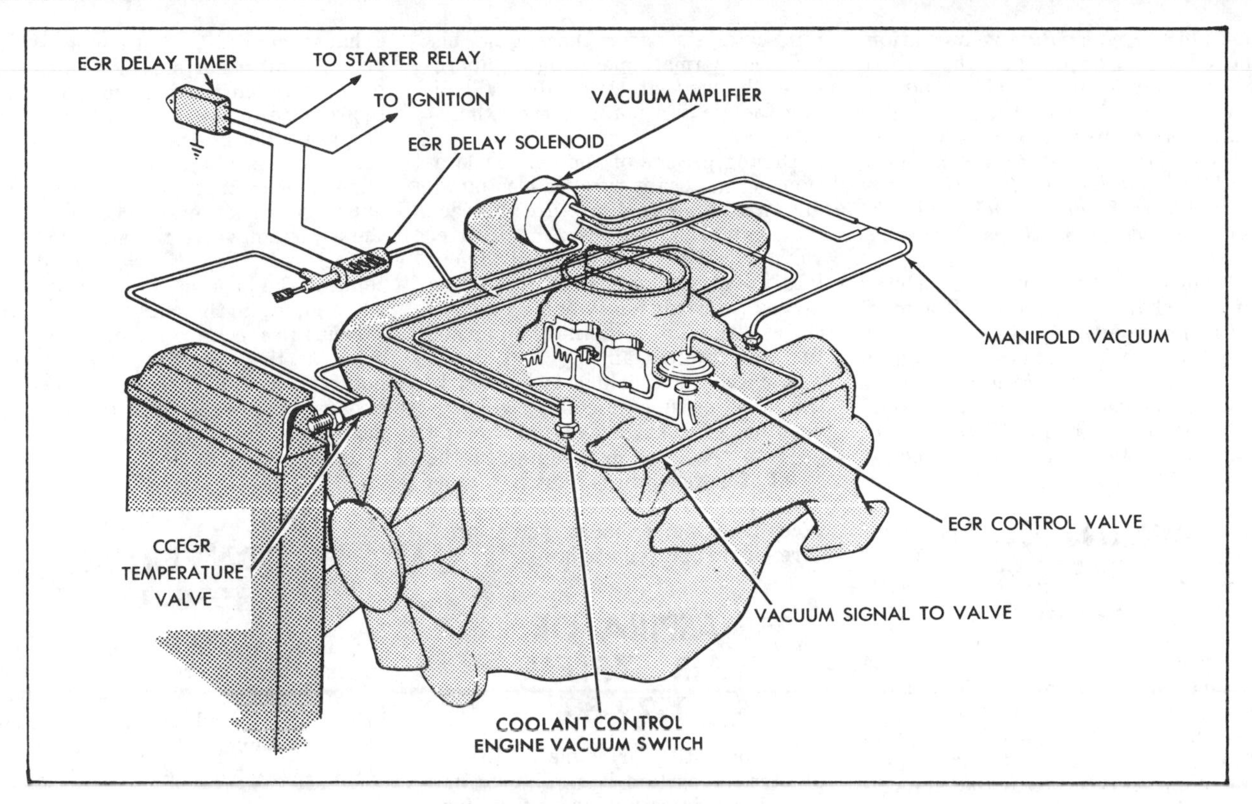

E. G. R. system—typical

gas back pressure. Recycle at wide open throttle is eliminated by calibrating the valve opening point above manifold vacuums available at wide open throttle as port vacuum cannot exceed manifold vacuum. Elimination of wide open throttle recycle provides maximum performance.

The venturi vacuum control system utilizes a vacuum tap at the throat of the carburetor venturi to provide a control signal. Because of the low amplitude of this signal, it is necessary to use a vacuum amplifier to increase the vacuum to the level required to operate the EGR control valve. Elimination of recycle at wide open throttle is accomplished by a dump diaphragm which compares venturi and manifold vacuum to determine when wide open throttle is achieved. At wide open throttle, the internal reservoir is "dumped," limiting output to the EGR valve to manifold vacuum. The valve opening point is set above the manifold vacuums available at wide open throttle, permitting the valve to be closed at wide open throttle. This system is dependent primarily on engine intake airflow as indicated by the venturi signal, and is also affected by intake vacuum and exhaust gas back pressure.

During engine warm-up, when combustion flame temperature is relatively low, control vacuum is prevented from reaching the EGR valve by use of a coolant control exhaust gas recirculation valve (CCEGR). The CCEGR, mounted in the engine and/or radiator top tank, remains closed until the coolant reaches a specified temperature, plus or minus 5 degrees F.

Valve Color Code	Opening Temperature (degrees F.)
Blue	75
Black	98
Yellow	125
Red	150

When the coolant reaches the calibration temperature, the valve opens thereby permitting vacuum to reach the EGR valve.

1980

The E.G.R. system is basically the same as the system used in 1979 with the addition of a Charge Temperature Switch (CTS). The charge temperature switch is installed in the intake manifold No. 6 branch runner on 6 cylinder engines and No. 8 branch runner on 8 cylinder engines. When the intake charge temperature is below 60 degrees F, the CTS will be closed allowing no EGR timer

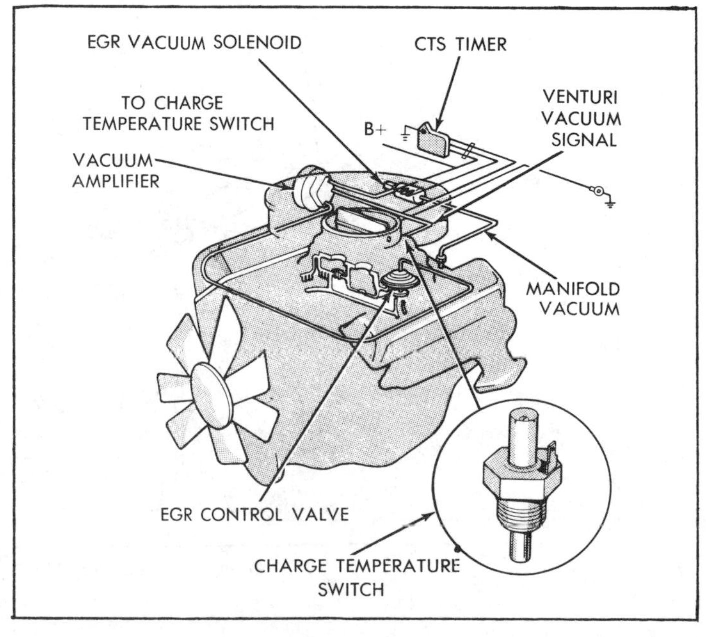

EGR-CTS system

function and no EGR valve operation. The CTS will open when the intake charge temperature is above 60 degrees F, thus allowing the EGR valve and timer to operate.

NOTE: *Do not overtorque the CTS as you will break off the nylon threads in the intake manifold. The torque requirement is 60 inch-pounds.*

Because of function and wiring harness changes, three EGR timers will be used on the CTS-EGR system. They are black, orange, and red in color. They are not interchangeable with timers used in previous years.

The Ported Vacuum System is not used on vehicles produced in 1980.

AIR INJECTION SYSTEM 1979-80

A very effective way of reducing the level of carbon monoxide and hydrocarbons present in the exhaust gases is to inject a controlled amount of air directly into the exhaust system at a point very close to the exhaust valve. The injected air assists in further oxidation of the exhaust gases thereby holding the level of undesirable gases at a minimum. The system consists basically of a belt driven injection pump, a diverter valve, a switching valve, and a check valve.

Filtered air from the clean side of the air cleaner is drawn into the pump and delivered to an air switching valve. During engine warm-up, the air is directed to the exhaust ports to assist the oxidation process in the mini-catalyst. After the engine has reached normal operating temperature, the air switching valve will direct the air to a point in the exhaust pipe.

During periods of sudden deceleration, when abrupt throttle closing allows a rich air fuel mixture to reach the combustion chamber, the diverter valve will direct the injected air out into the atmosphere, thus preventing a backfire condition which would occur as a result of combining unburned fuel and injected air.

A check valve, which is located in the injection tube assembly, will prevent hot exhaust gases from backing up into the hose and pump in the event of a system break down, such as would be caused by pump belt failure, excessively high exhaust system pressure, or a ruptured air hose.

ASPIRATOR AIR SYSTEM 1979-80

The operating principle of the aspirator air system is similar to that of the air injection system in that both systems rely on the introduction of air into the exhaust manifold to reduce carbon monoxide and hydrocarbon emissions to an acceptable level. Though similar, both systems are distinctly different in design and operation.

The aspirator air system is a relatively simple system to diagnose and repair when compared to the air injection system. An aspirator valve, located between the air cleaner and exhaust manifold, and connected by hoses, contains a spring loaded diaphragm which is sensitive to changes in pressure. Whenever a vacuum pulse occurs in the exhaust ports and manifold passages, the diaphragm opens to allow fresh air to mix with the exhaust gases. At high engine speeds, the aspirator valve remains closed.

To test if the valve has failed, disconnect the hose at the aspirator inlet, and, with the engine at idle and in neutral, check for escaping exhaust gases at the inlet. If the pulses can not be felt, and there is evidence of exhaust gases present at the inlet, the valve has failed and should be replaced.

MINI CATALYTIC CONVERTER 1979-80

In addition to the conventional type (underfloor converter) catalytic converter, a mini-converter is also used to initiate the exhaust gas oxidation before the exhaust gases reach the main underfloor catalytic converter.

CRANKCASE VENTILATION SYSTEM 1979-80

This system, which utilizes a conventional PCV valve, is similar to the systems used on most vehicles marketed in the United States in that it directs crankcase vapors through the air cleaner and into the combustion area for complete burning.

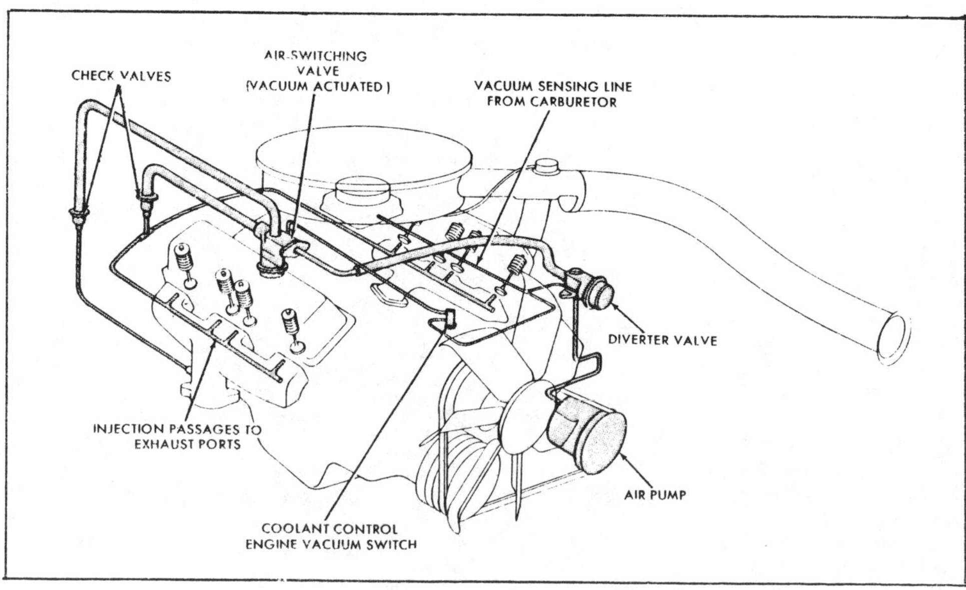

Air Injection System—Typical

VACUUM CIRCUITS

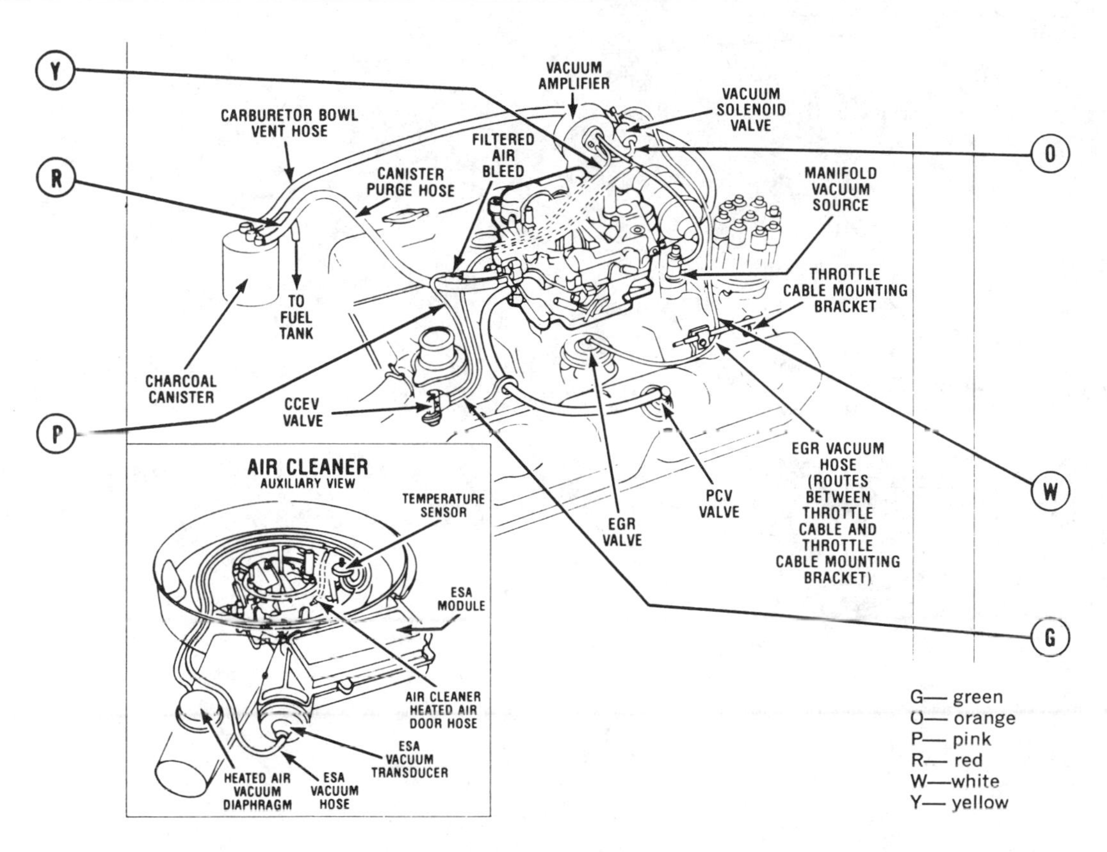

1979 Vacuum house diagram, 360 4 bbl., 49 states, (automatic transmission)

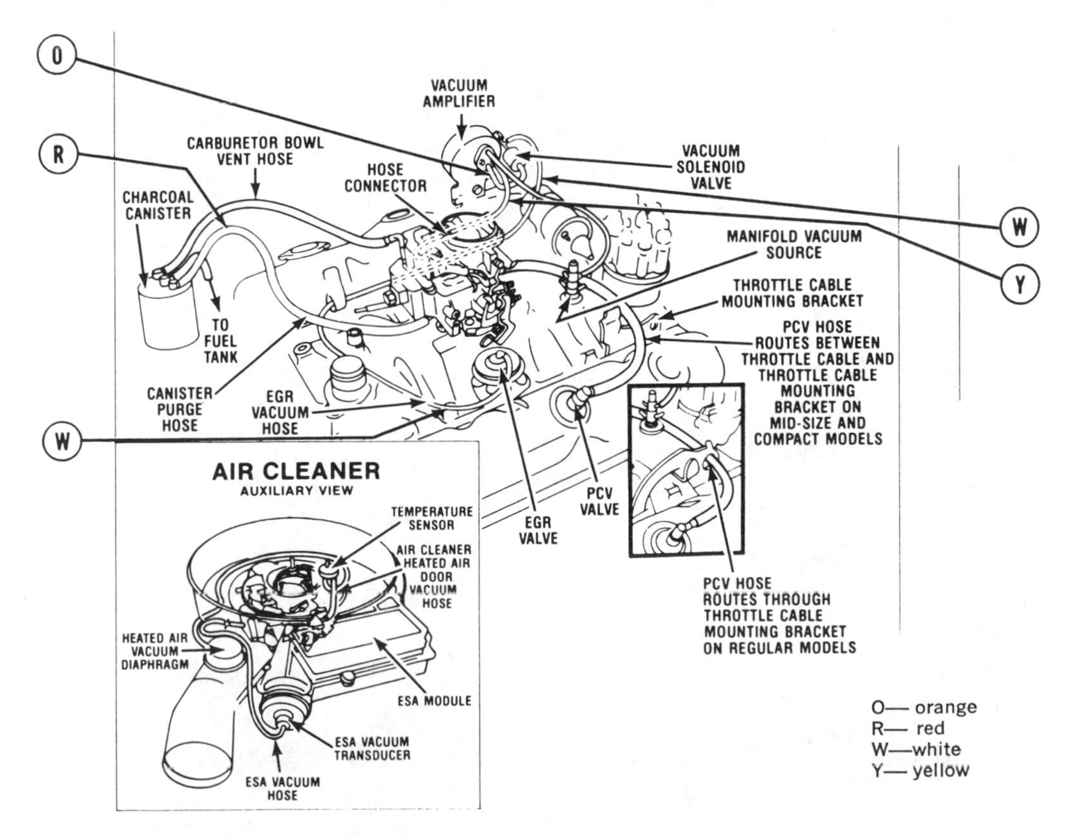

1979 Vacuum hose diagram 318 2 bbl., 49 states and Canada (automatic transmission)

VACUUM CIRCUITS

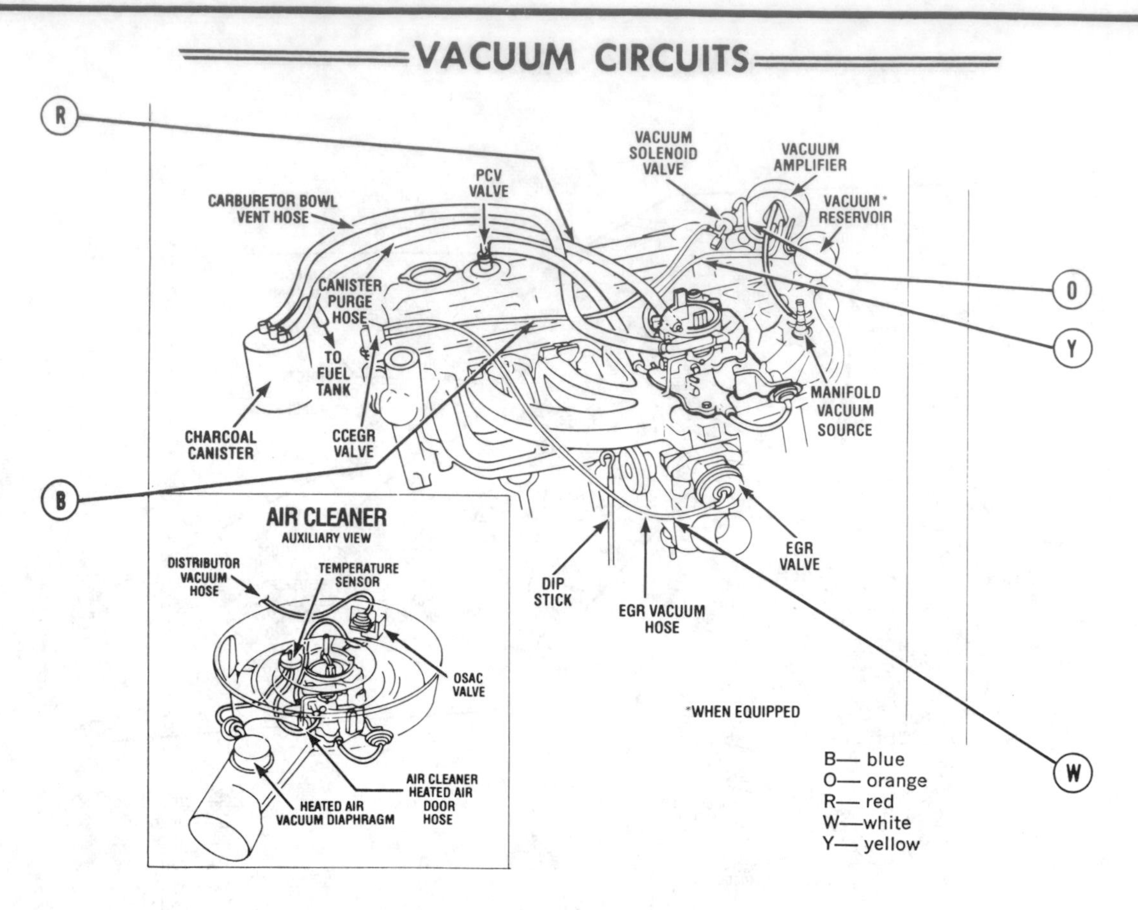

1979 Vacuum hose diagram, 225 bbl., 49 states and Canada

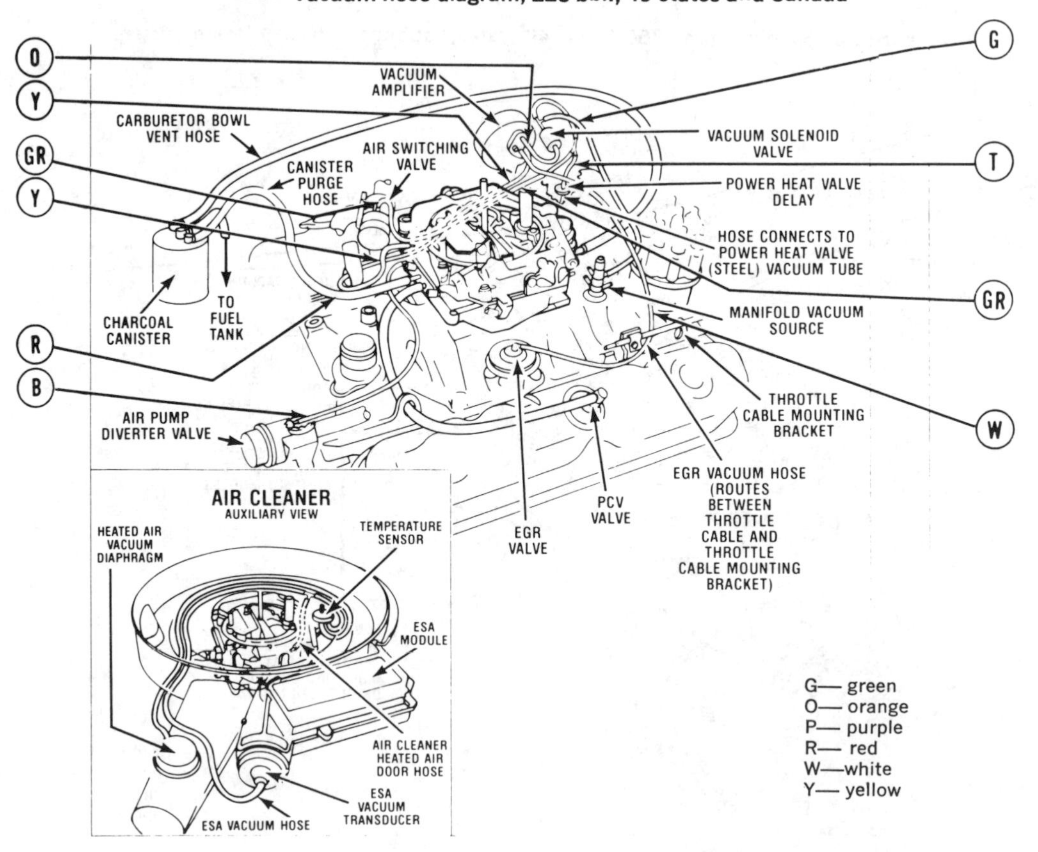

1979 Vacuum hose diagram, 318 4 bbl., California (automatic transmission)

VACUUM CIRCUITS

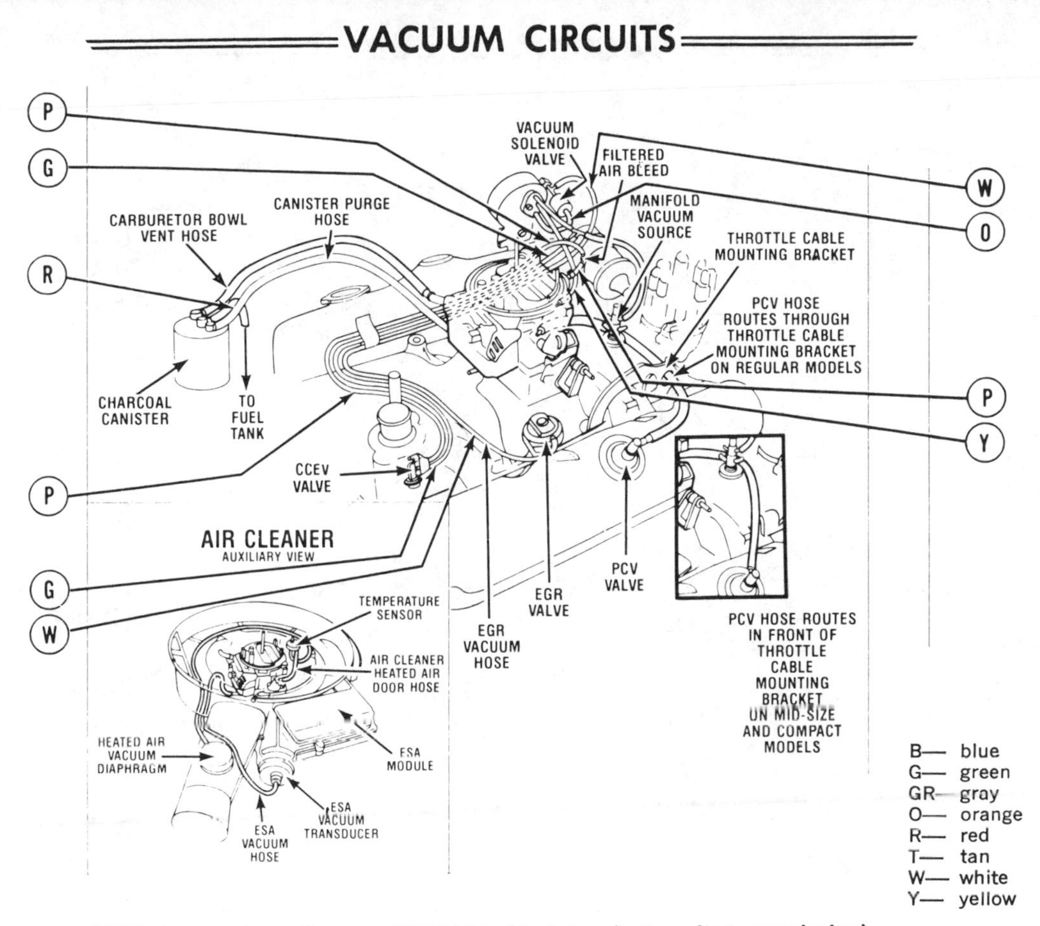

1979 Vacuum hose diagram, 360 2 bbl., 49 states, (automatic transmission)

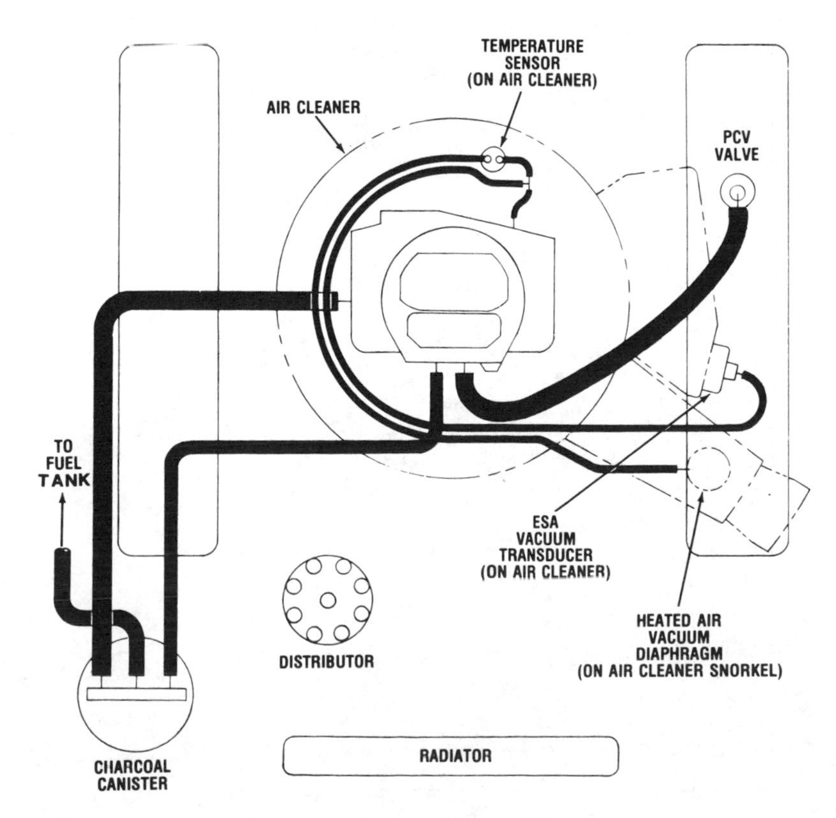

1979 49-State 400 4-Bbl. Electronic Lean Burn

VACUUM CIRCUITS

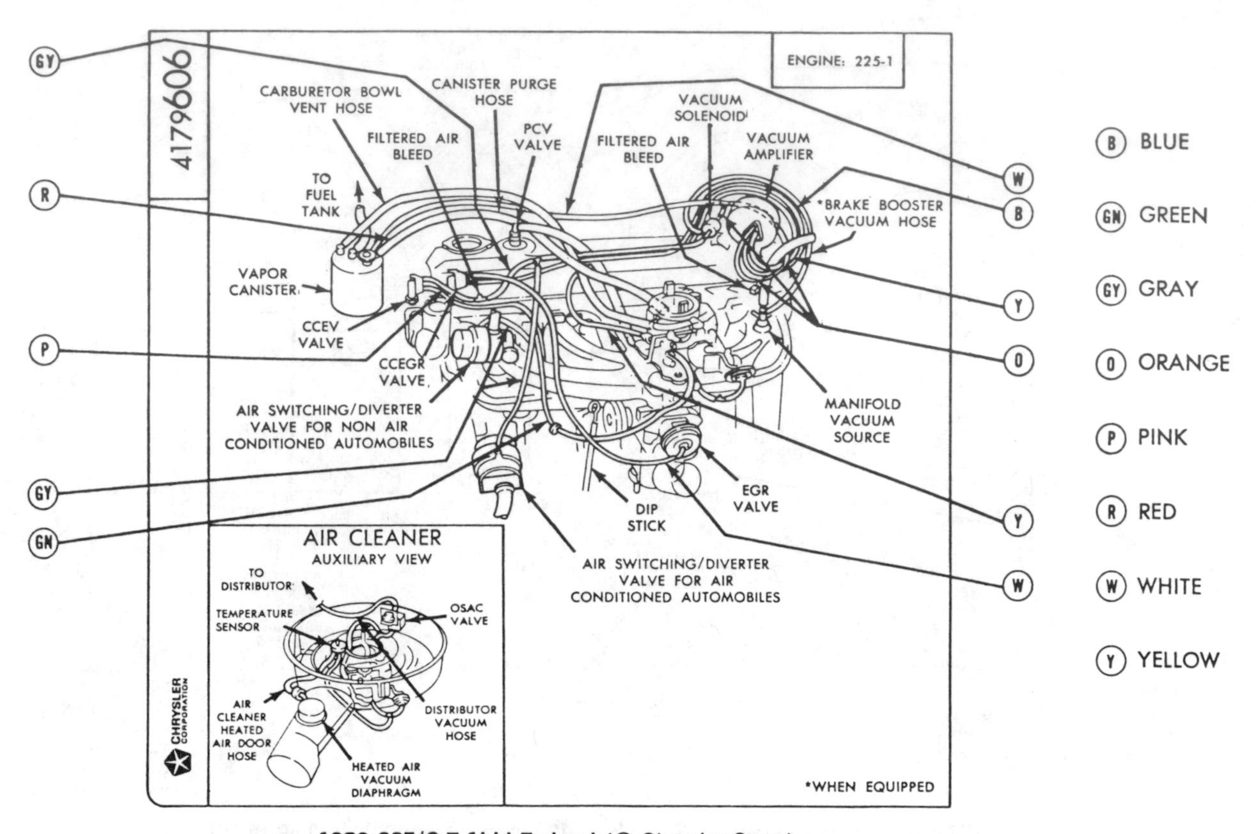

1980 225/3.7-1bbl Federal (© Chrysler Corp.)

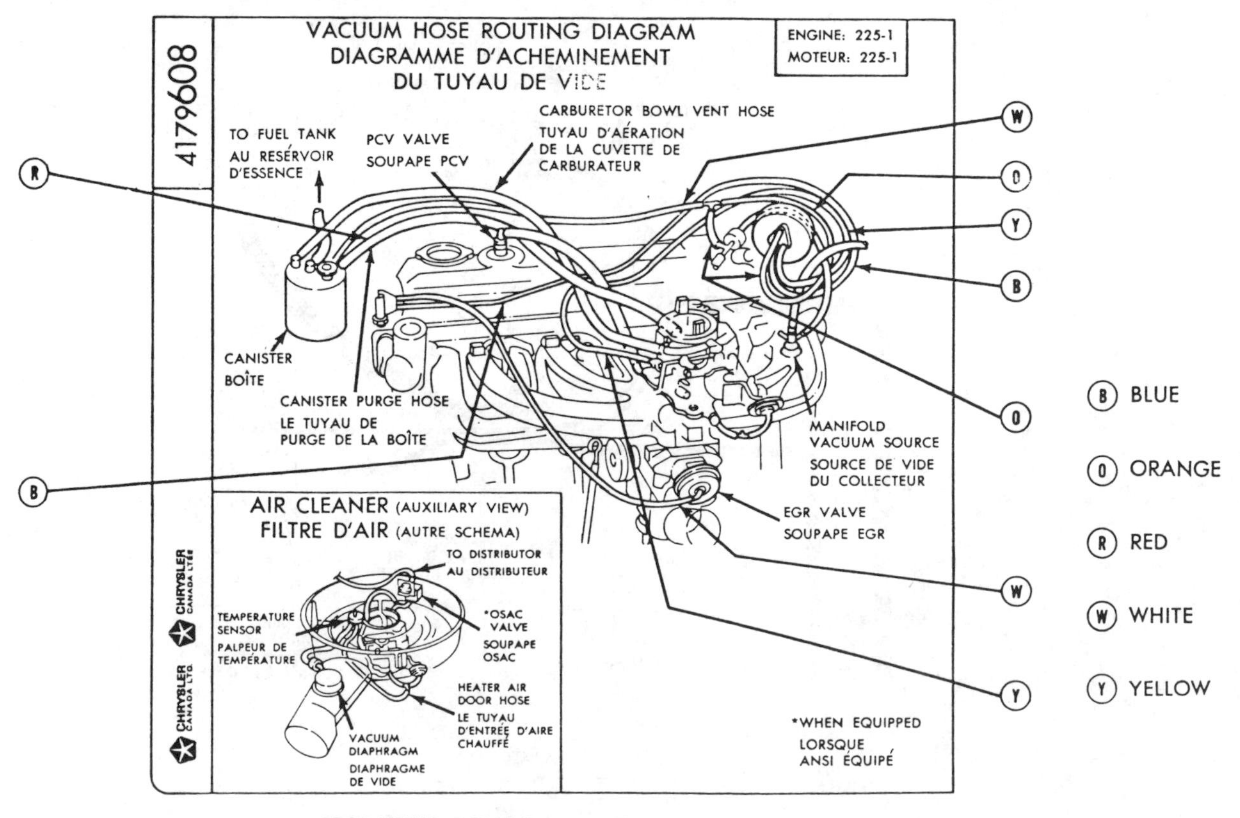

1980 225/3.7-1bbl Canada (© Chrysler Corp.)

VACUUM CIRCUITS

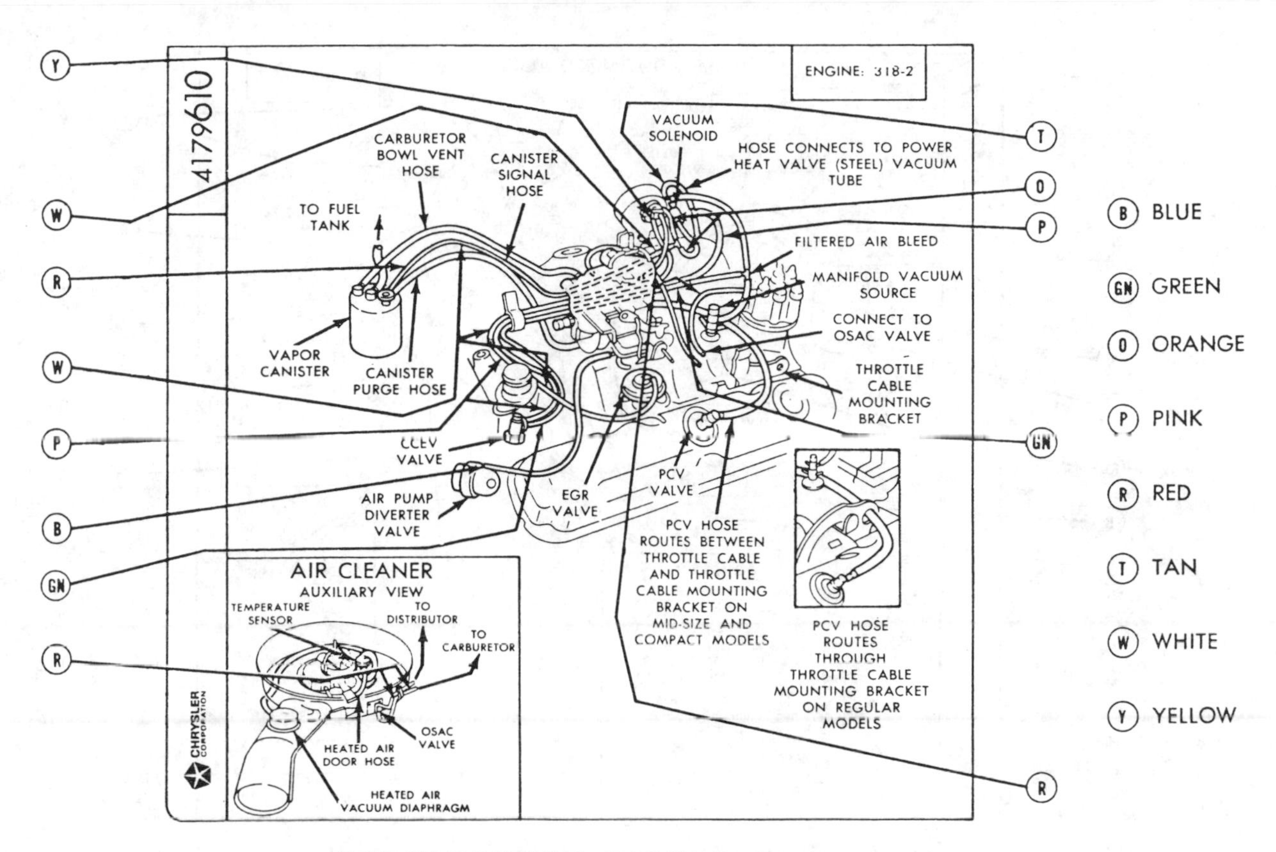

1980 318/5.2-2bbl Federal (© Chrysler Corp.)

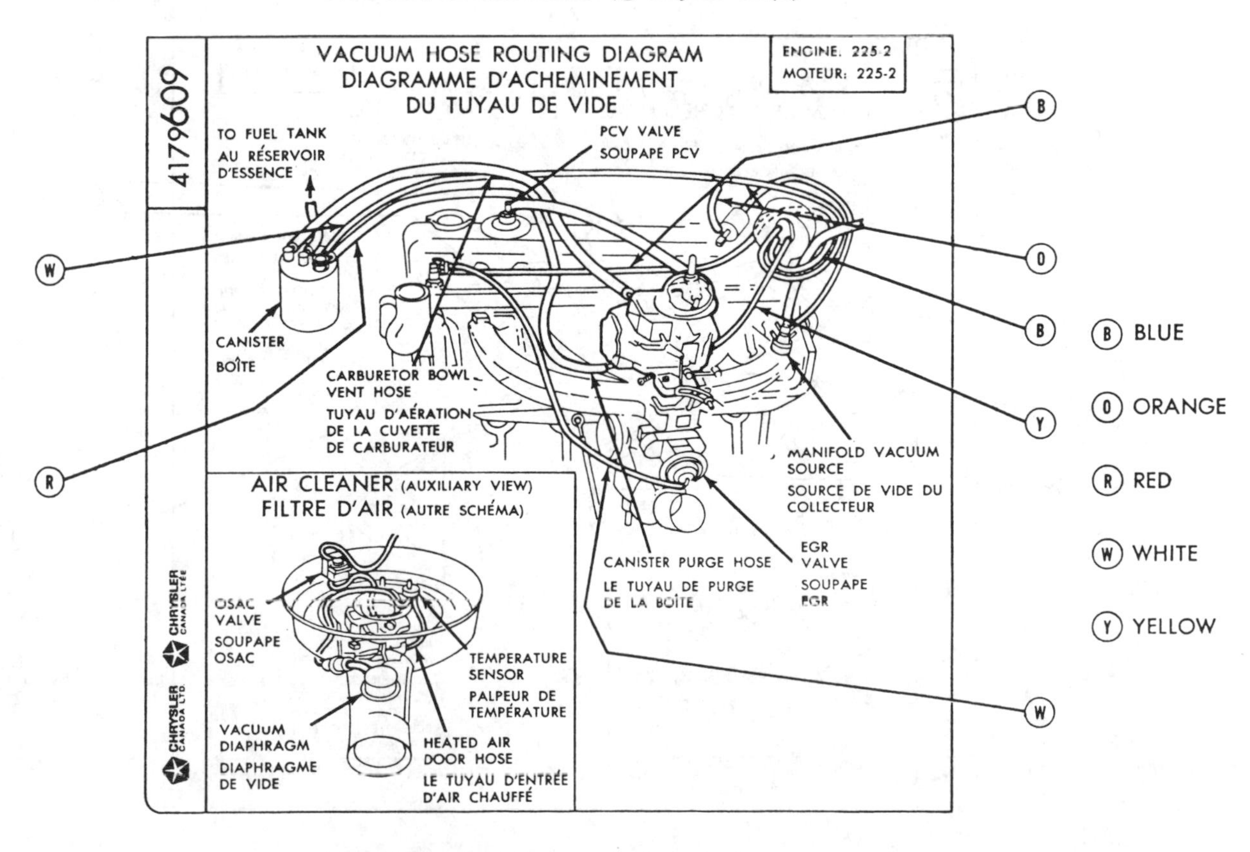

1980 318/5.2-2bbl Canada (© Chrysler Corp.)

VACUUM CIRCUITS

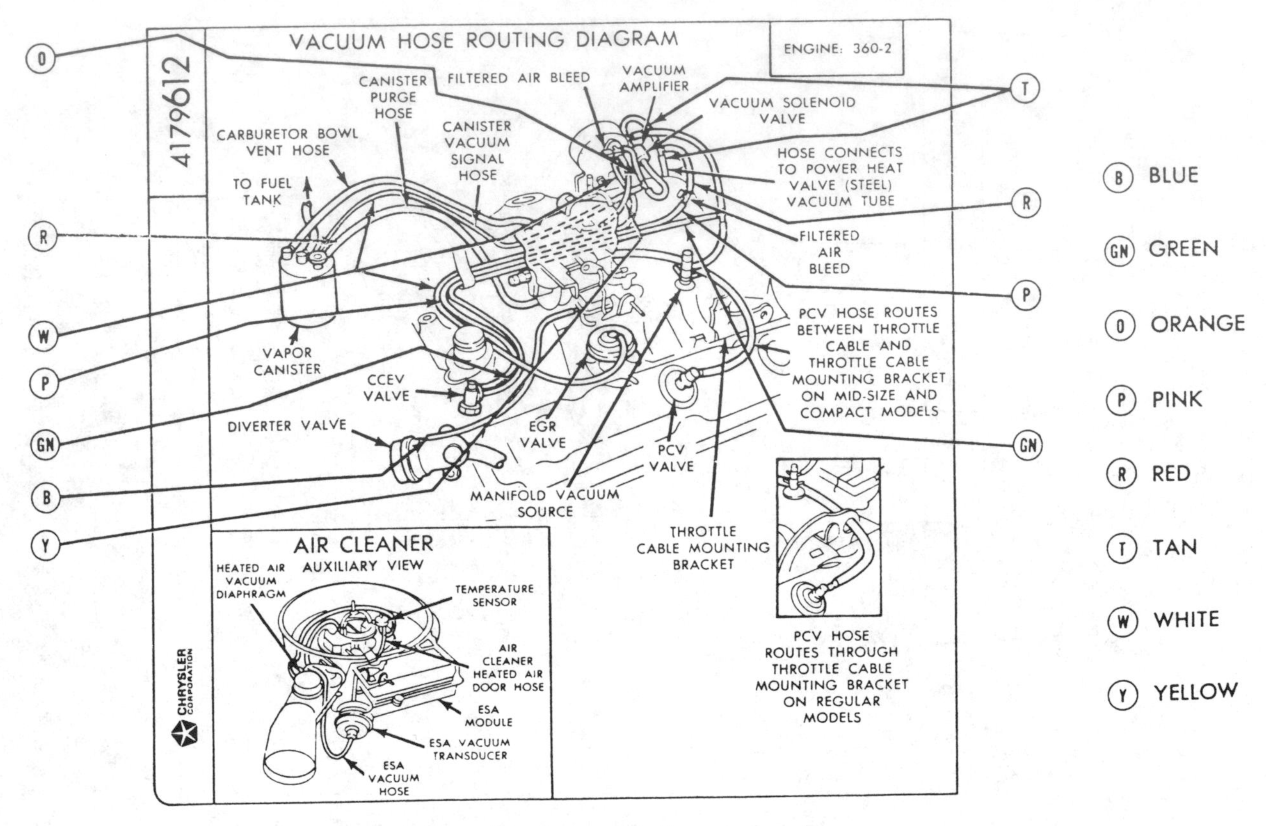

1980 360/5.9-2bbl Federal (© Chrysler Corp.)

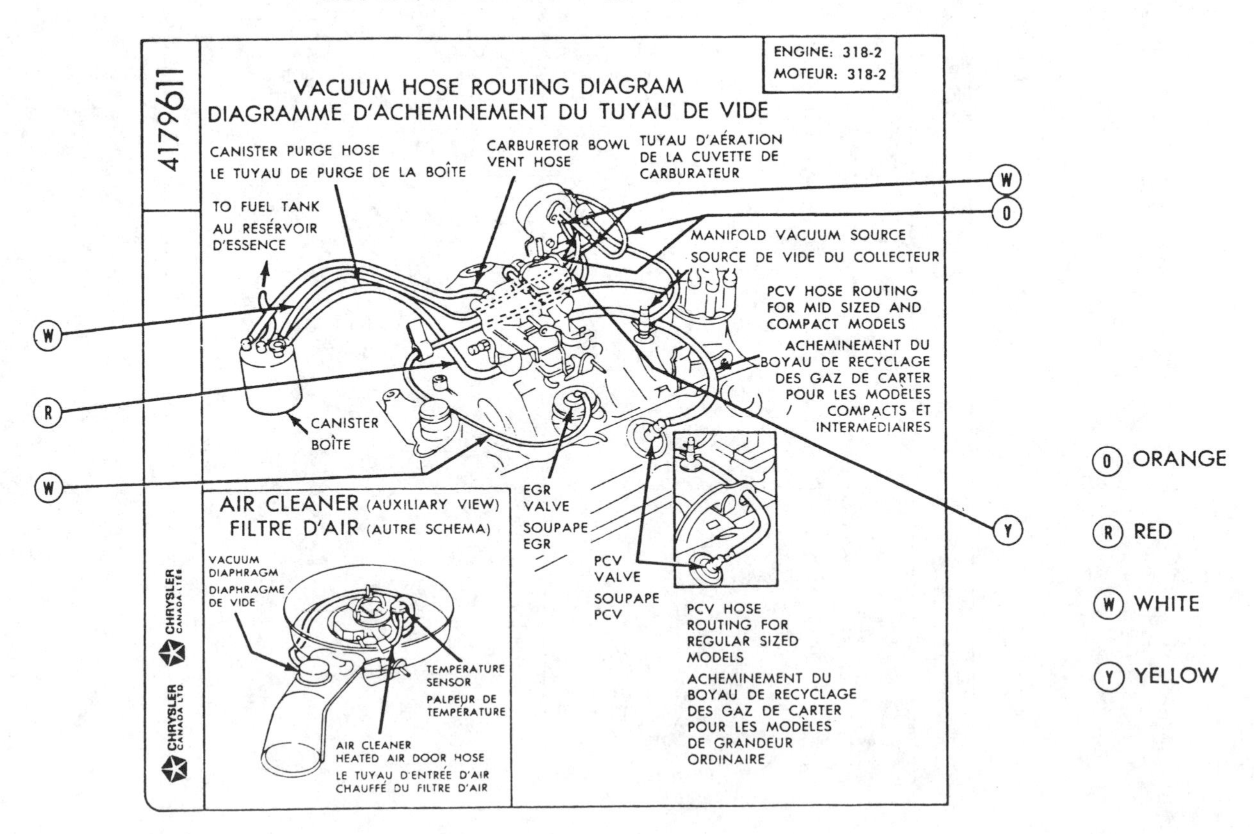

1980 360/5.9-2bbl Canada (© Chrysler Corp.)

VACUUM CIRCUITS

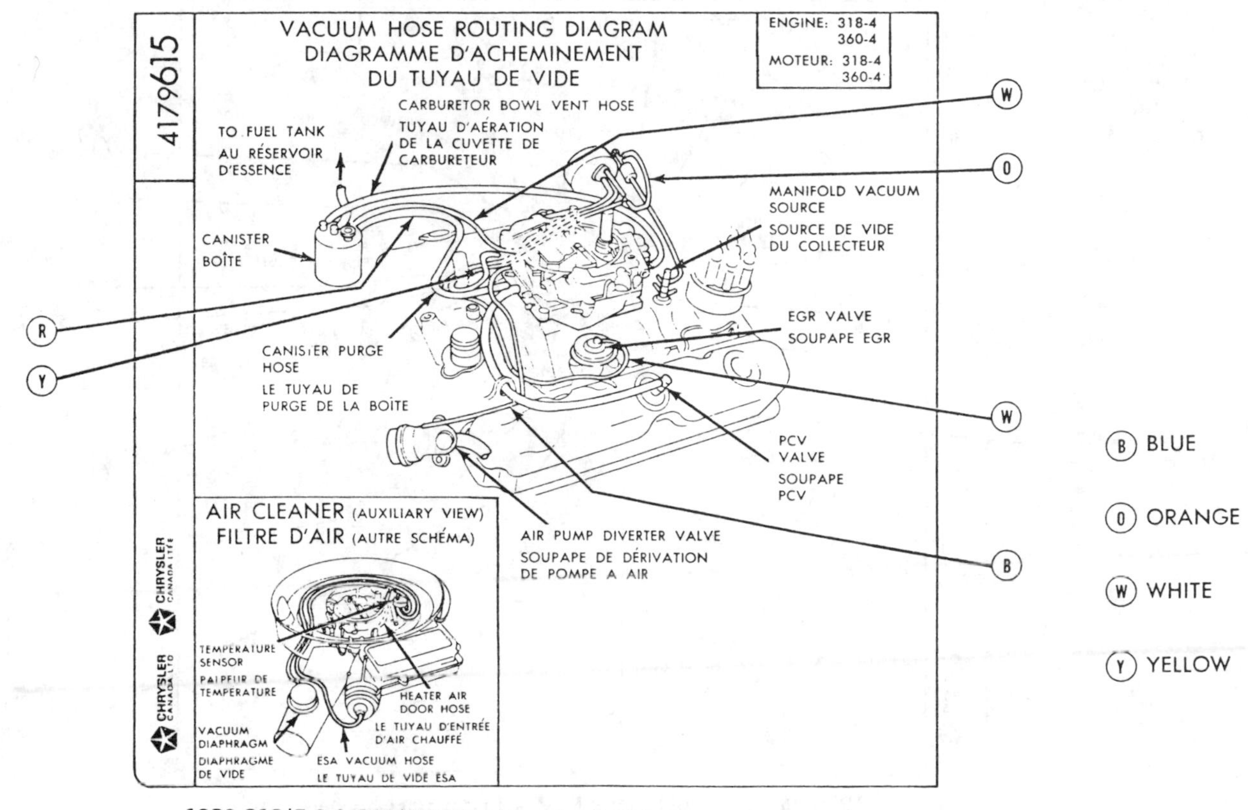

1980 318/5.2 & 360/5.9-4bbl Canada, F, M & J body (© Chrysler Corp.)

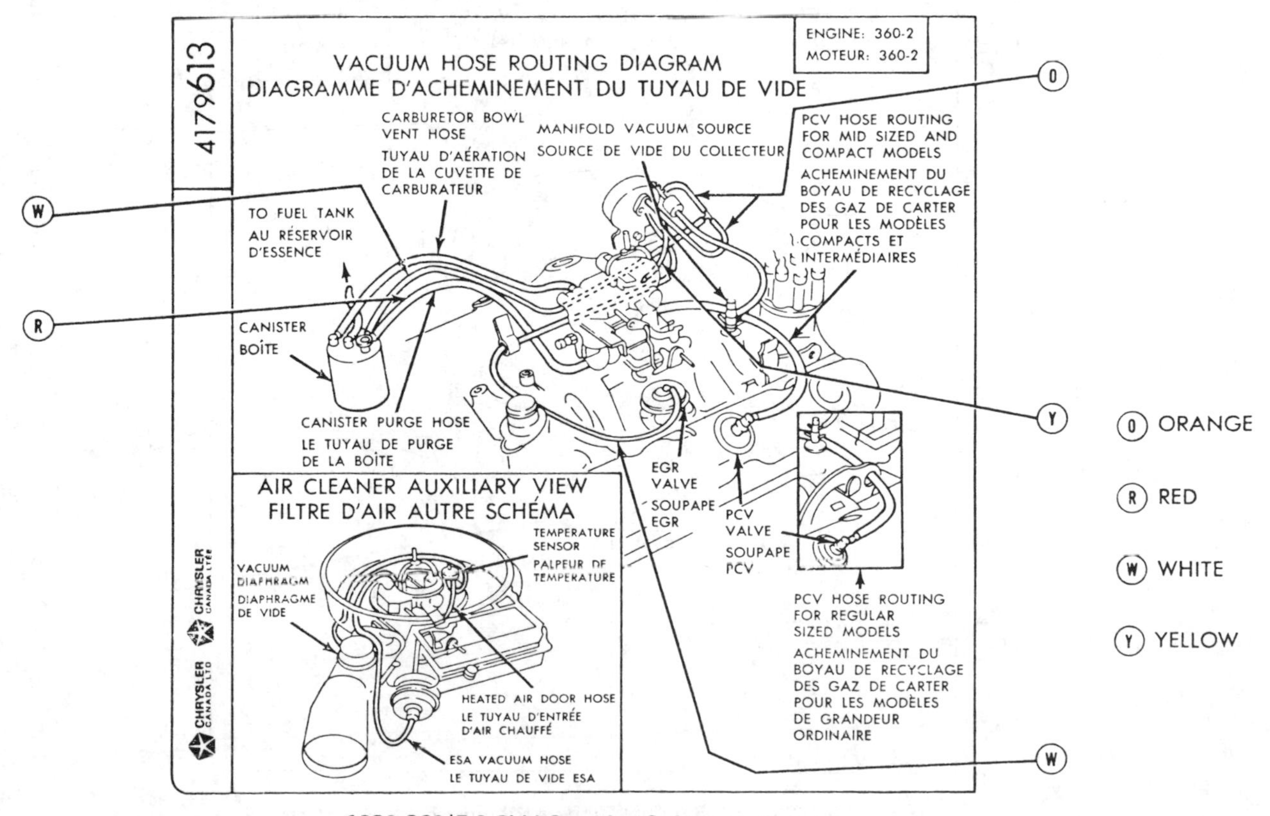

1980 360/5.9-2bbl Canada (© Chrysler Corp.)

VACUUM CIRCUITS

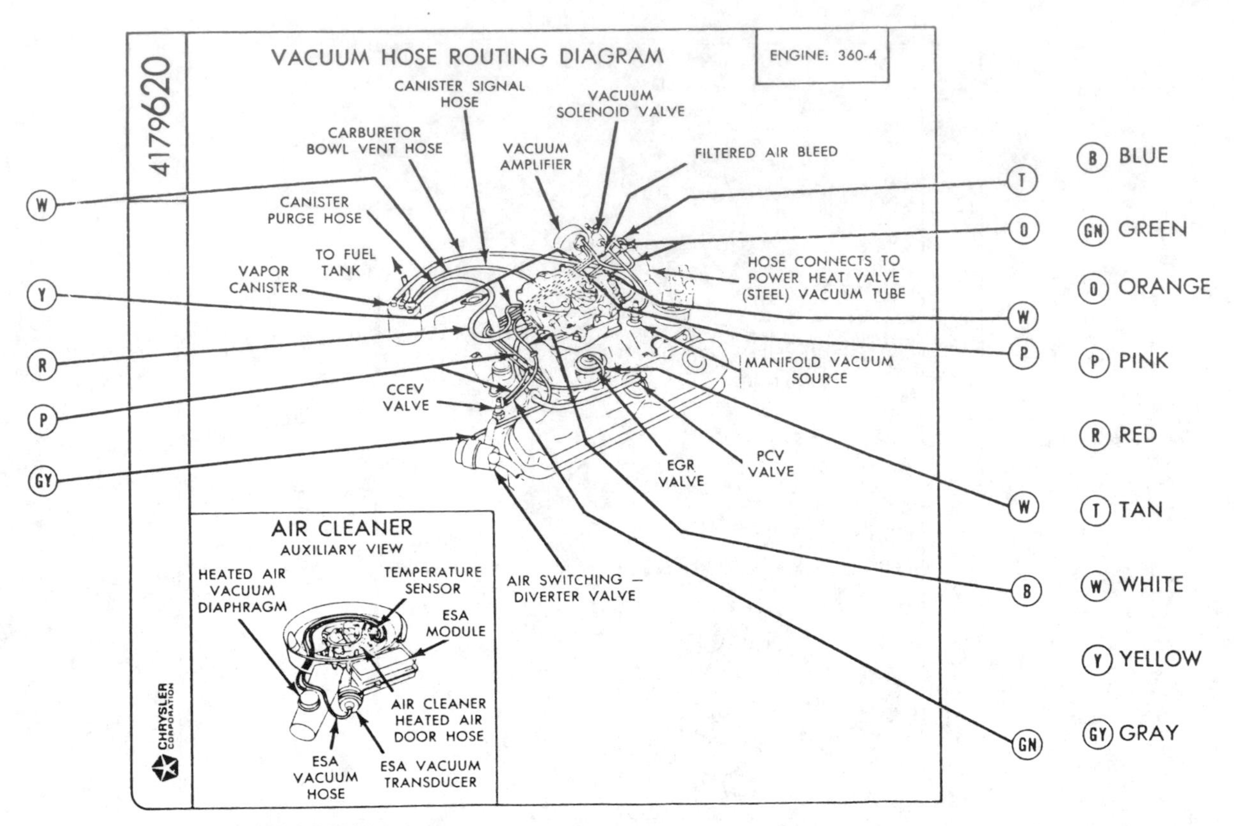

1980 360/5.2-4bbl Federal F, M & J body (© Chrysler Corp.)

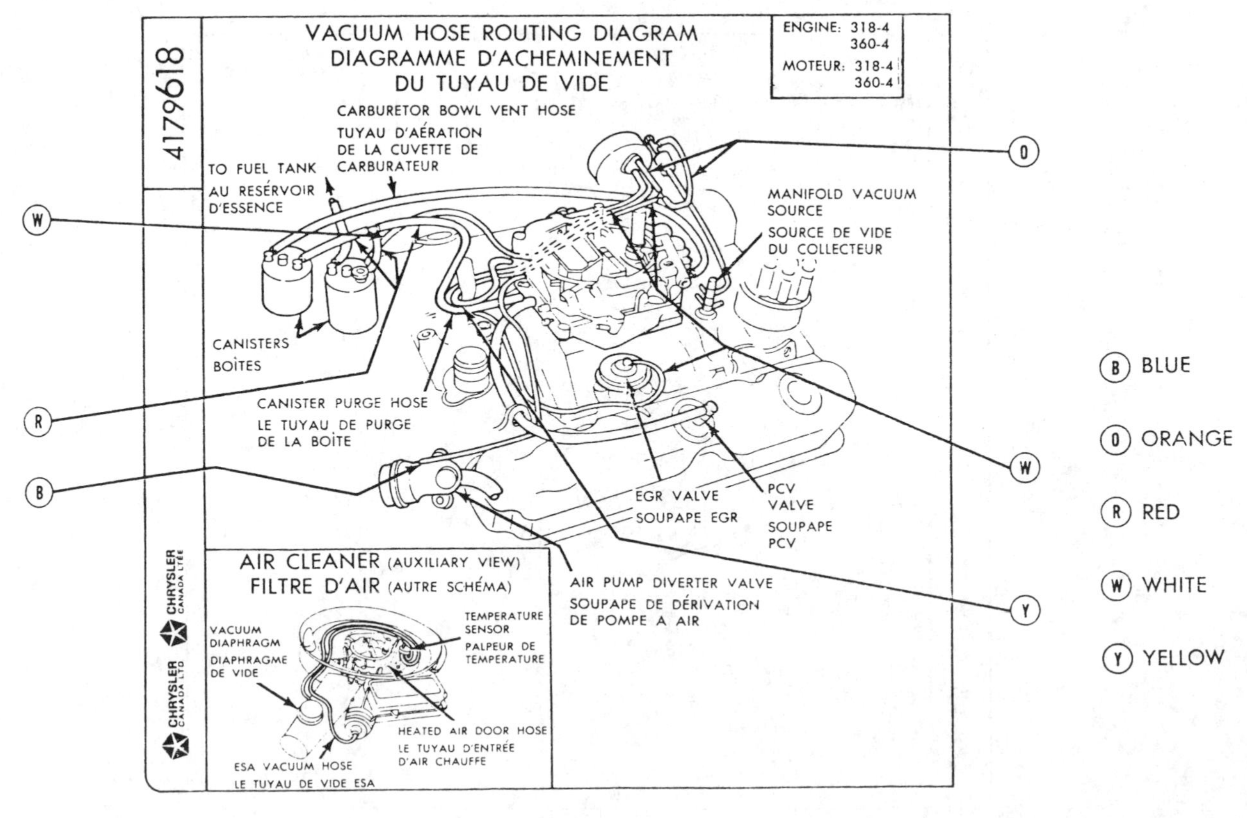

1980 318/5.2 & 360/5.9-4bbl Canada R body (© Chrysler Corp.)

VACUUM CIRCUITS

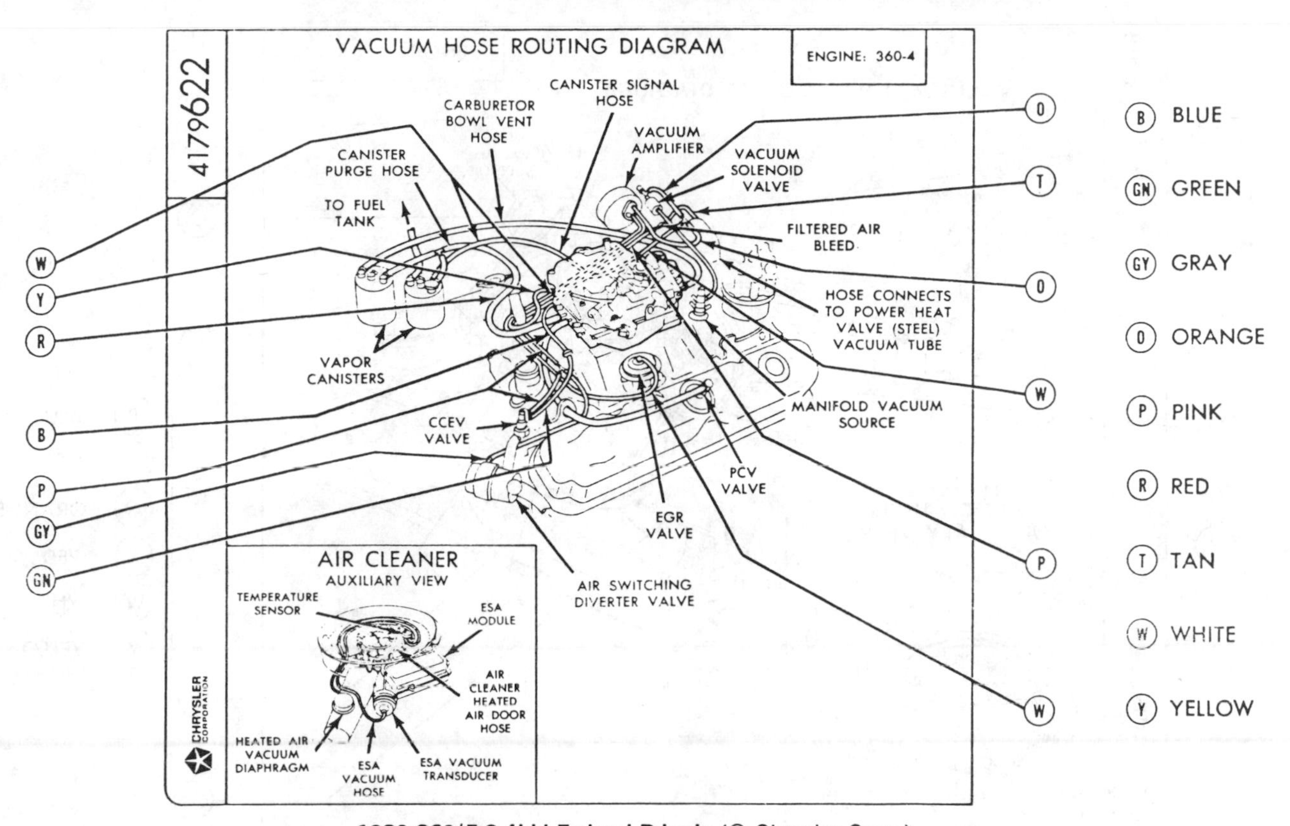

1980 360/5.9-4bbl Federal R body (© Chrysler Corp.)

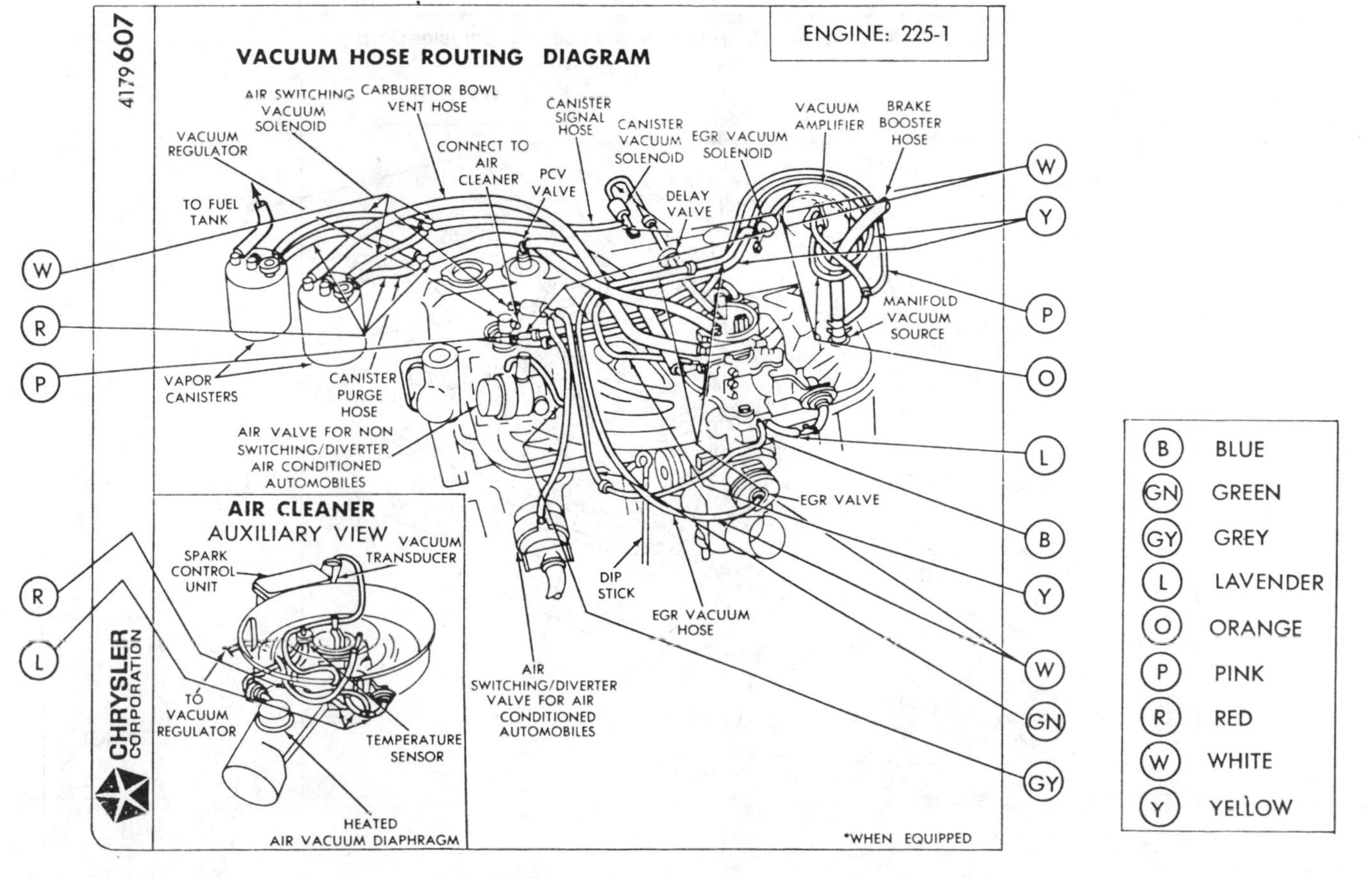

1980 225/3.7-1bbl California (© Chrysler Corp.)

VACUUM CIRCUITS

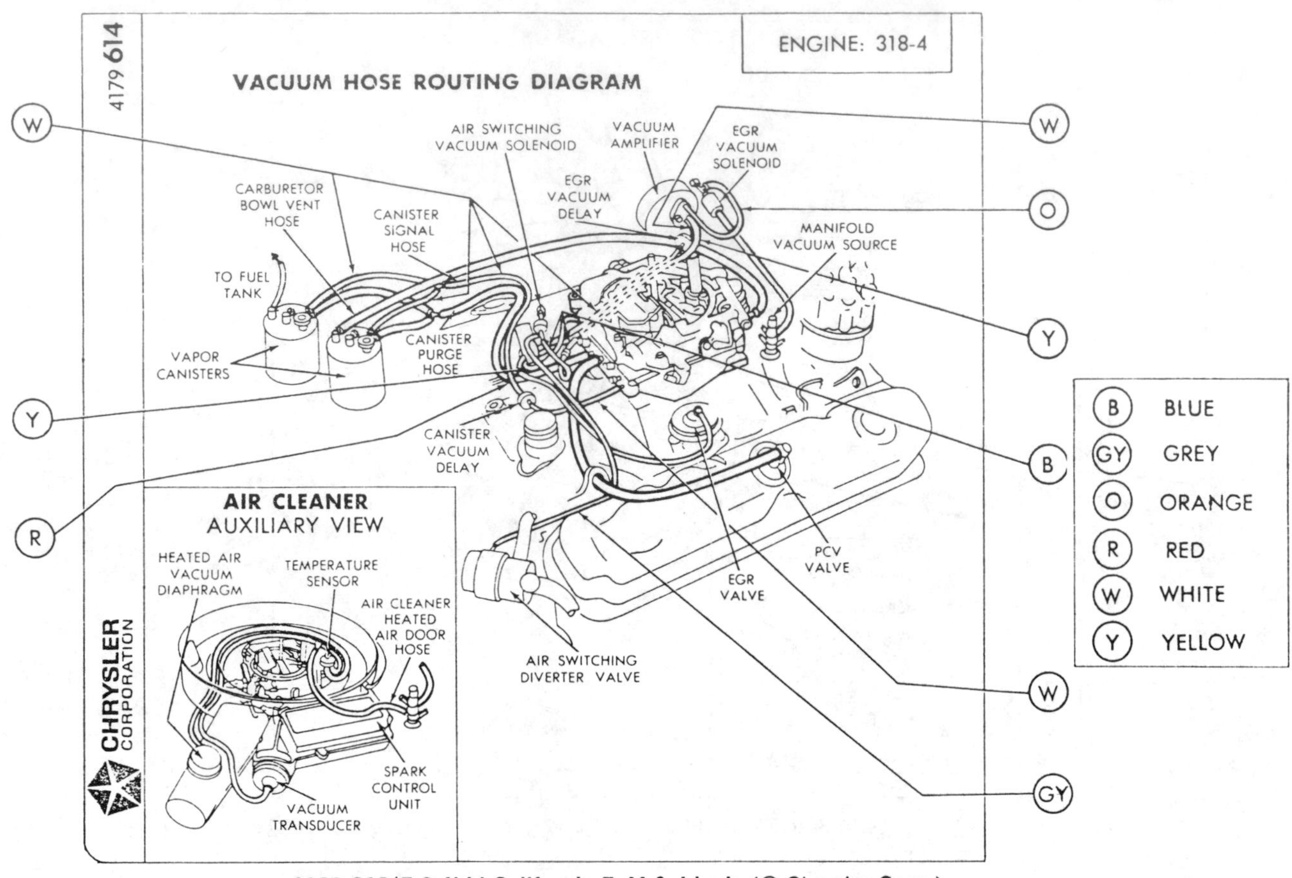

1980 318/5.2-4bbl California F, M & J body (© Chrysler Corp.)

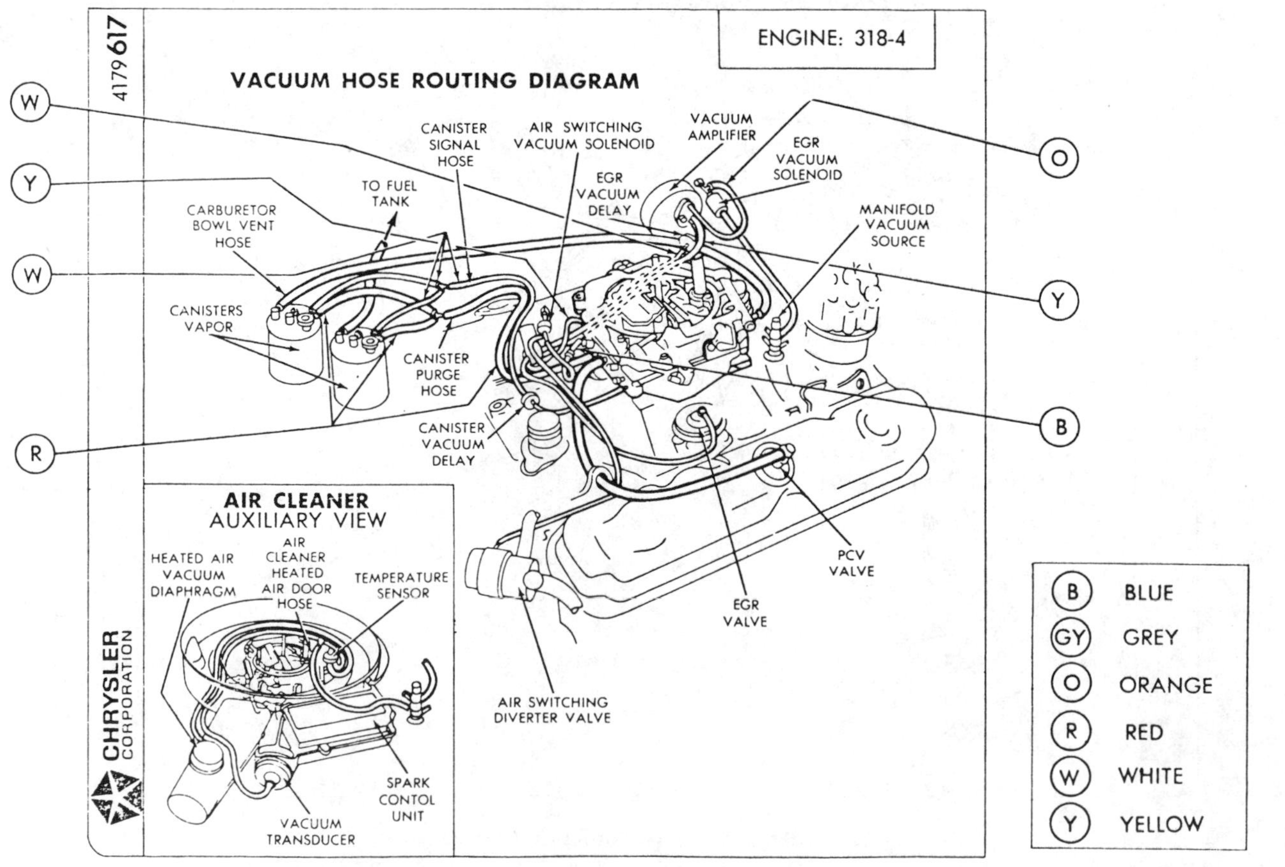

1980 318/5.2-4bbl California R body (© Chrysler Corp.)

VACUUM CIRCUITS

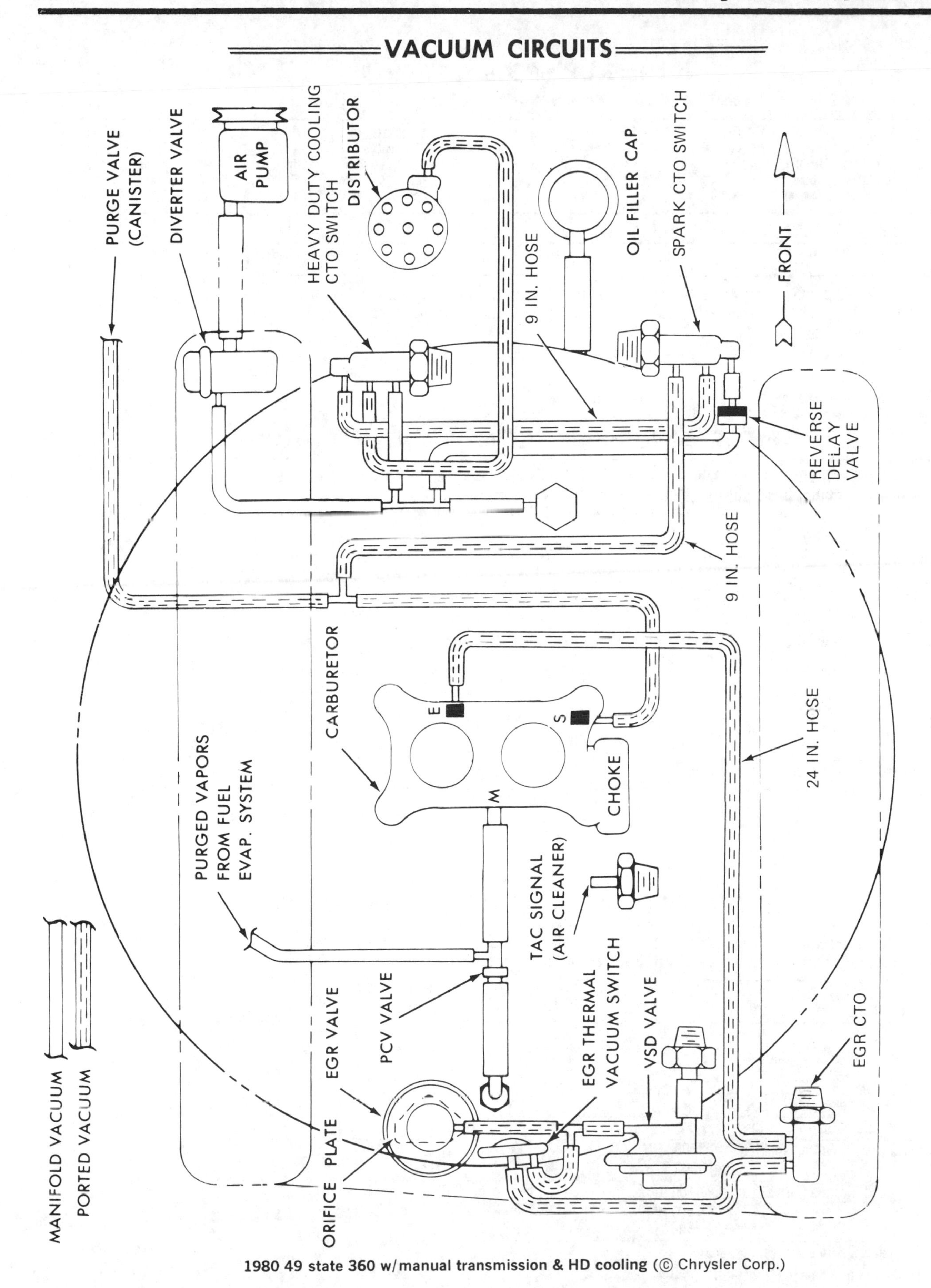

1980 49 state 360 w/manual transmission & HD cooling (© Chrysler Corp.)

TUNE-UP SPECIFICATIONS

ENGINE CODE, 5th CHARACTER OF THE VIN NUMBER

Year	Eng. V.I.N. Code	ENGINE No. Cyl. Disp. (cu in)	H.P.	SPARK PLUGS Orig. Type	SPARK PLUGS Gap (in)	DIST.	IGNITION TIMING (deg) Man Trans	IGNITION TIMING (deg) Auto Trans	VALVES Intake Opens (deg) B.	FUEL PUMP Pres. (psi)	IDLE SPEED (rpm) Man Trans	IDLE SPEED (rpm) Auto Trans in Drive
1979	A	104.7	75	RN-12Y	0.35	ESC	15B	15B	14B	4-6	②	②
1980	A	104.7	75	RN-12Y	0.35	ESC	12B	12B	14B	4-6	②	②

NOTE: Should the specifications in this manual deviate from the specifications on the engine compartment decal, the specifications on the decal should be used.

ESC Electronic Spark Control
② See Idle Speed section

ELECTRONIC SPARK CONTROL SYSTEM—1979

The following specifications are published from the latest information available at the time of publication. If anything differs from those on the Emission Control Information Label, use the specification on the label.

Custom I.C. Spark Control Computer Part Number	5206721	5206784	5206785	5206790	5206793
Engine Application	E-12 1.7L	E-12 1.7L	E-12 1.7L	E-12 1.7L	E-12 1.7L
Vacuum Advance (Range)	0″-10″	0″-10″	0″-10″	0″-10″	0″-10″
Crank + Electrical = (Basic Timing)	10° + 5°	10° + 5°	10° + 5°	10° + 5°	10° + 0°
Speed Advance (Ground Carb Switch) @ 1100 RPM	0°-3°	0°-3°	0°-3°	0°-3°	0°-3°
@ 2000	8°-12°	8°-12°	8°-12°	8°-12°	8°-12°
Zero Time Offset	6°-10°	2°-6°	3°-7°	0°-3°	0°-3°
Accumulator Clock Up (in minutes)	8	8	8	8	8
Vacuum Advance—Full Accumulator @ 2000 RPM	18°-22°	18°-22°	18°-22°	18°-22°	18°-22°
@ 3000	23°-27°	23°-27°	28°-32°	28°-32°	23°-27°
Throttle Maximum Advance (Throttle Open 20°)	0°	0°	0°	0°	0°

ELECTRONIC SPARK CONTROL SYSTEM—1980

The following specifications are published from the latest information available at the time of publication. If anything differs from those on the Emission Control Information Label, use the specification on the label.

Custom I.C. Spark Control Computer Part Number	5213008	5213012
Engine Application	E-12 1.7L	E-12 1.7L
Crank + Run = (Basic Timing)	10° + 0°	10° + 0°
Vacuum Advance (Range)	0″-10″	0″-10″
Zero Time Advance	18°-22°	23°-27°
Accumulator Clock Up (in minutes)	0	0
Vacuum Advance—Full Accumulator @ 2000 RPM	5°-9°	5°-9°
@ 3000	18°-22°	23°-27°
Speed Advance (Ground Carb Switch) @ 1100 RPM	0°-3°	0°-3°
@ 2000 RPM	8°-12°	13°-17°
@ 4800	18°-22°	20°-24°
Warm Up Schedule (in seconds)	25	25
Throttle Maximum Advance (Throttle Open 20°)	0°	0°

CAR SERIAL NUMBER AND ENGINE IDENTIFICATION

1979-80

Behind the windshield on the driver's side is the vehicle identification number. The sixth character is the model year, with 9 for 1979 and "A" for 1980. The fifth character is the engine code, as follows:

A .... 1.7 liter 4-cyl.

EMISSION EQUIPMENT

All Models

Closed positive crankcase ventilation
Heated air cleaner
Vapor control, canister storage
Exhaust gas recirculation
Emission calibrated carburetor
Emission calibrated distributor
Catalytic converter
Fresh air intake
Full electric choke
Air injection system
Air aspirator system (1979)

IDLE SPEED AND MIXTURE ADJUSTMENTS EXHAUST ANALYZER METHOD

1979

NOTE: *This is an alternate method for California cars only. The preferred method is to use propane enrichment.*

Air cleaner .... In place
Air cond. .... Off
Auto. trans. .... Park
Engine fan .... Operating
Air pump .... Disconnect for mixture setting
Vac. adv. hose .... Plugged
EGR vac. hose .... Plugged
Lean burn engine ..... Ground carb. switch with jumper
Mixture adj. .... Set idle CO

NOTE: *After disconnecting the air pump for mixture settings, the opening to the exhaust must be plugged to prevent air from entering the exhaust. Do not plug the pump output.*

Idle CO
1.7 4-cyl. .... Probe 0.5%

Idle Speed
Auto. trans. with air cond. .....
AC idle speedup ....
Auto. trans. no air cond.
Solenoid connected ....
Solenoid disconnected ....
Man. trans. with air cond. 900
AC idle speedup 850
Man. trans. no air cond.
Solenoid connected 900
Solenoid disconnected 700

NOTE: *All engines have a solenoid on the carburetor. On engines without air conditioning, the solenoid is used for anti-dieseling. It is energized whenever the ignition switch is on. Curb idle speed is adjusted by turning the screw on top of the solenoid. The "solenoid disconnected" speed is set with the throttle stop screw.*

On engines with air conditioning, the solenoid is used for air conditioning idle speedup. It is energized only when the air conditioning is on. The curb idle speed is adjusted by turning the screw on top of the solenoid. The "AC idle speedup" speed is adjusted with an Allen screw underneath the idle speed screw. To reach the Allen screws, the idle speed screw must be unscrewed and removed from the solenoid. There is no throttle stop screw on air conditioned engines.

IDLE SPEED AND PROPANE ENRICHED MIXTURE ADJUSTMENTS

1979

NOTE: *This is the preferred method for adjusting all cars.*

NOTE: *The vehicle emissions sticker sometimes reflects specification changes made in production. Sticker figures must be used if they disagree with this information.*

Make all adjustments with the engine fully warmed up. All measurements are in RPM's.

1.7 liter Manual trans. No AC
Idle set (w/fan) 900
Fast idle 1400
Propane enriched .... 1050 (975 Calif.)
Throttle stop speed ... 700

1.7 liter Manual trans. with AC
Idle set (w/fan) 900
Fast idle 1400
Propane enriched .... 1050 (975 Calif.)
AC idle speed (AC on) 850

1.7 liter Auto. trans. No AC
Idle set (w/fan) 900
Fast idle 1700
Propane enriched .... 1050 (975 Calif.)
Throttle stop speed ... 700

1.7 liter Auto. trans. with AC
Idle set (w/fan) 900
Fast idle 1700
Propane enriched .... 1050 (975 Calif.)
AC idle speed (AC on) .... 750(D)

1980

NOTE: *The vehicle emissions sticker sometimes reflects specification changes made in production. Sticker figures must be used if they disagree with this information.*

Make all adjustments with the engine fully warmed up. All measurements are in RPM's.

1.7 liter Manual trans. No AC
Idle set 900
Fast idle 1400
Propane enriched 1000
Throttle stop speed 700

1.7 liter Manual trans. with AC
Idle set 900
Fast idle 1400

IDLE SPEED AND PROPANE ENRICHED MIXTURE ADJUSTMENTS

Propane enriched 1000
AC idle speed 850
1.7 liter Auto. trans. No AC
Idle set 900
Fast idle 1700
Propane enriched 1000
Throttle stop speed 700
1.7 liter Auto. trans. with AC
Idle set 900
Fast idle 1700
Propane enriched 1000
AC idle speed 750

INITIAL TIMING

1979

1.7 4-cyl. ....15° BTDC

1980

1.7 4-cyl. ....12° BTDC

SPARK PLUGS 1979-80

1.7 4-cyl. ....RN12Y .035

VACUUM ADVANCE 1979-80

Diaphragm type ....Single
Vacuum source ....Ported

EMISSION CONTROL SYSTEMS

CORRESPONDING FUNCTIONS 1979-80

SYSTEM	FUNCTION
Air pumps (AIR)	Reduces exhaust HC and CO
Carburetor Calibration	Controls exhaust HC and CO
Catalytic Converter	Reduces exhaust HC and CO
Distributor Calibration	Controls exhaust HC and CO
Electric Choke	Reduces exhaust HC and CO
Evaporation Control System	Vapor control
Exhaust Gas Recirculation	Controls NOx
Heated Inlet Air	Reduces HC
Initial Engine Timing	Controls exhaust HC, CO, and Nox
Positive Crankcase Ventilation	Controls Crankcase HC

CARBURETOR 1979-80

Carburetion on the OMNI and HORIZON is accomplished through the use of a Holly model 5220 carburetor, a staged, dual venturi carburetor on which the primary venturi is smaller than the secondary venturi.

The primary stage is comprised of an idle circuit, transfer system, diaphragm type accelerator pump system, main metering system, and a power enrichment system.

HEATED INLET AIR SYSTEM 1979-80

The purpose of the heated inlet air system is to control the temperature of the air entering the carburetor when ambient temperatures are low. By increasing the temperature of the air being introduced into the carburetor, a much leaner calibration is made possible, thereby reducing hydrocarbon emissions. In addition to a refined fuel air mixture, other benefits derived as a result of the heated air include smoother engine warm-up operation (a more volatile mixture as a result of heat) and a minimal amount of carburetor icing.

FULL ELECTRIC CHOKE SYSTEM 1979-80

A full-electric-choke is an electrically heated choke comprised of an electric heater and switch which are sealed within the choke housing. During cold weather the switch remains open to reduce heater output until the choke area reaches a sufficiently warm temperature. If the choke area is at a sufficiently warm temperature, then the switch closes and full heater output moves the choke to the open position.

IGNITION SYSTEM 1979-80

The electronic ignition system is used on all vehicles. This system eliminates the need for breaker points by using a Hall Effect switch to inform the electronic control unit when to trigger the delivery of electrical energy to the spark plugs. With worn breaker points eliminated as a cause of engine misfiring and increased emissions, better control of exhaust emissions can be achieved.

EXHAUST GAS RECIRCULATION (EGR) 1979-80

Recirculated exhaust gas, when introduced into the combustion chamber, has the effect of minimizing the oxides of nitrogen emitted into the air by lowering combustion flame temperatures during engine operation. Only a modulated amount of exhaust gas is allowed to dilute the highly combustible air fuel mixture. The system responsible for regulating the volume of exhaust gas available during firing is called Exhaust Gas Recirculation (EGR).

During engine warm-up, when combustion flame temperature is relatively low, control vacuum is prevented from reaching the EGR valve by

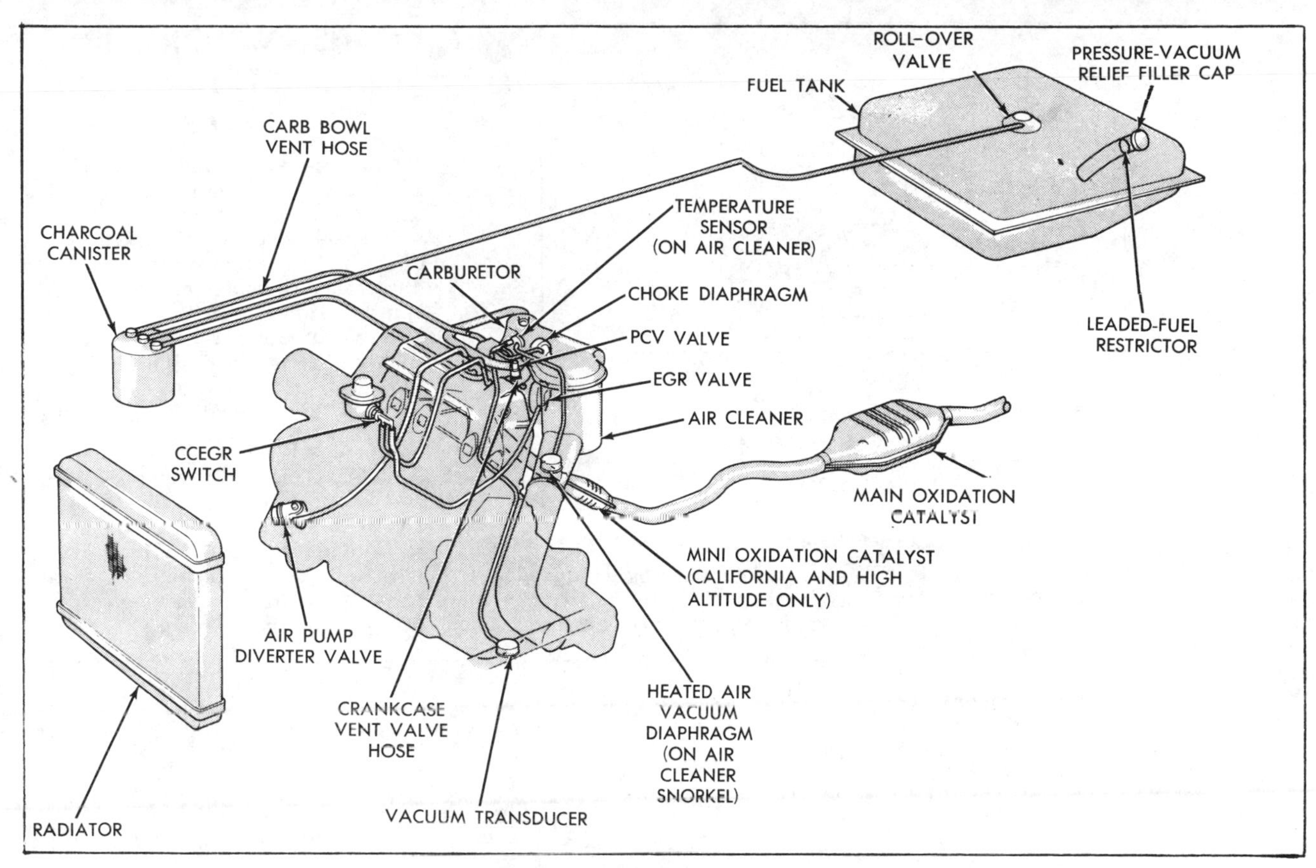

Emission control system—Omni and Horizon

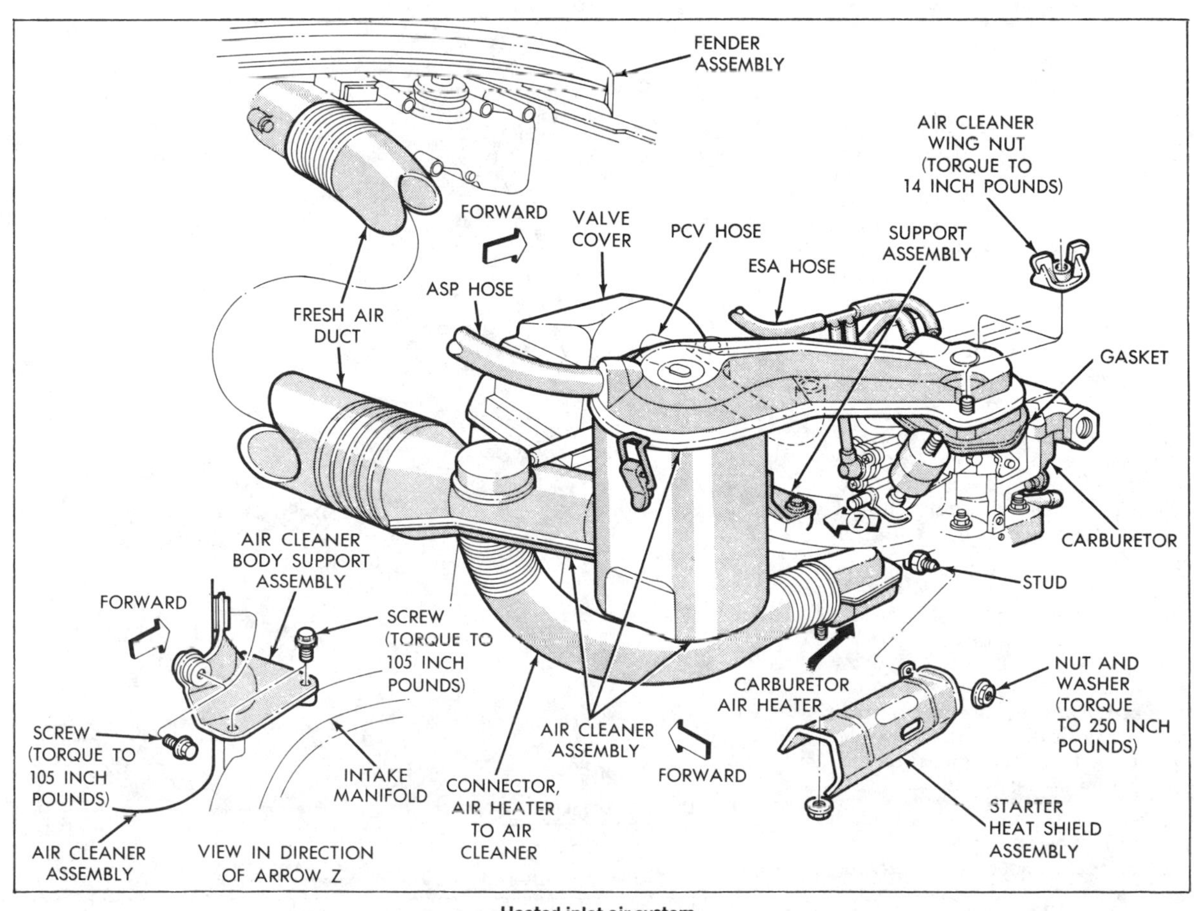

Heated inlet air system

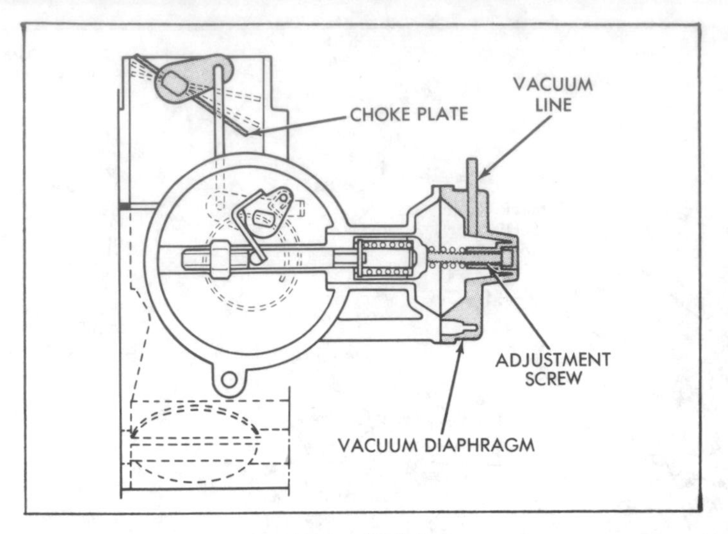

Full electric choke

use of a coolant control exhaust gas recirculation valve (CCEGR). The CCEGR, mounted in the thermostat housing, remains closed until the coolant temperature exceeds 125 degrees F. When the coolant temperature reaches the calibration temperature of 125 degrees F., plus or minus 5 degrees, the valve opens thereby permitting vacuum to reach the EGR valve.

AIR INJECTION SYSTEM 1979-80

One very effective way of reducing the level of carbon monoxide and hydrocarbons present in the exhaust gases is to inject a controlled amount of air directly into the exhaust system at a point very close to the exhaust valve. The injected air assists in further oxidation of the exhaust gases thereby holding the level of undesirable gases at a minimum. The system consists basically of a belt driven injection pump, a diverter valve, and a check valve.

Filtered air from the clean side of the air cleaner is drawn into the pump and delivered to an air injection manifold. Exactly where the air is injected depends on the emission control code.

During periods of sudden deceleration, when abrupt throttle closing allows a rich air fuel mixture to reach the combustion chamber, the diverter valve will direct the injected air out into the atmosphere, thus preventing a backfire condition which would occur as a result of combining unburned fuel and injected air.

A check valve, which is located in the injection tube assembly, will prevent hot exhaust gases from backing up into the hose and pump in the event of a system break down, such as would be caused by pump belt failure, excessively high exhaust system pressure, or a ruptured air hose.

ASPIRATOR AIR SYSTEM 1979

The operating principle of the aspirator air system is similar to that of the air injection system in that both systems rely on the introduction of air into the exhaust manifold to reduce carbon monoxide and hydrocarbon emissions to an acceptable level. Though similar, both systems are distinctly different in design and operation.

NOTE: *Refer to "Air Injection System."*

The aspirator air system is a relatively simple system to diagnose and repair when compared to the air injection system. An aspirator valve, located between the air cleaner and exhaust manifold, and connected by hoses, contains a spring loaded di-

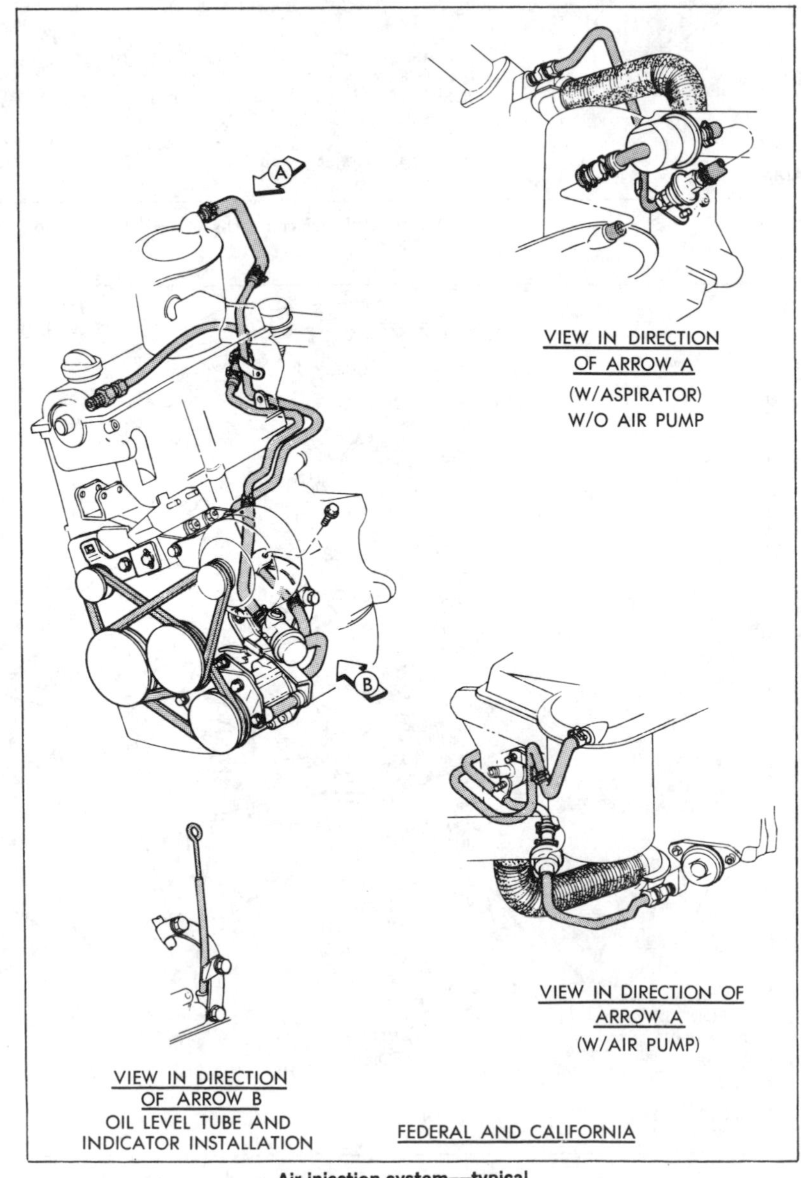

Air injection system—typical

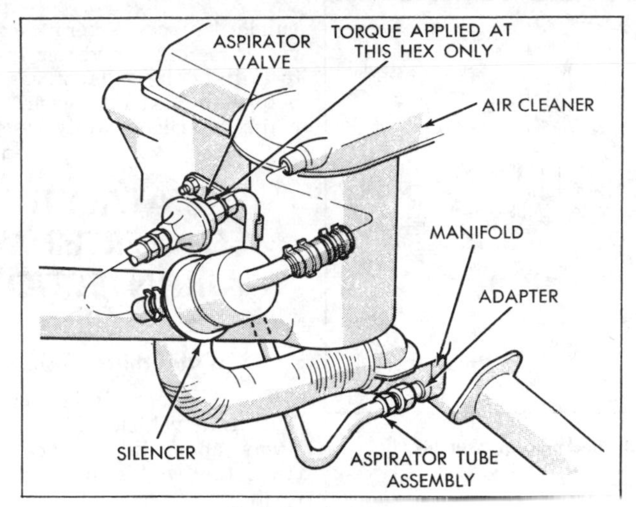

Aspirator air system

aphragm which is sensitive to changes in pressure. Whenever a vacuum pulse occurs in the exhaust ports and manifold passages, the diaphragm opens to allow fresh air to mix with the exhaust gases. At high engine speeds, the aspirator valve remains closed.

To test if the valve has failed, disconnect the hose at the aspirator inlet, and, with the engine at idle and in neutral, check for escaping exhaust gases at the inlet. If the pulses can not be felt, and there is evidence of exhaust gases present at the inlet, the valve has failed and should be replaced.

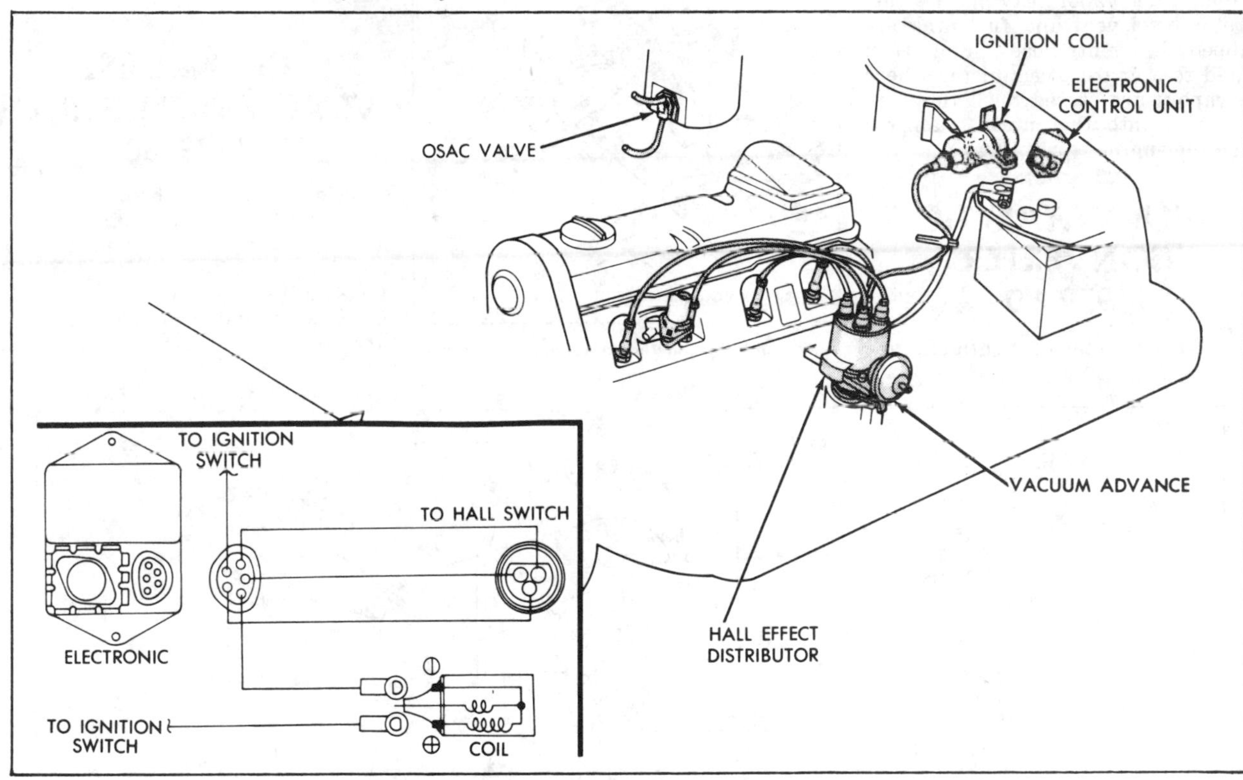

Electronic ignition system and major components—typical

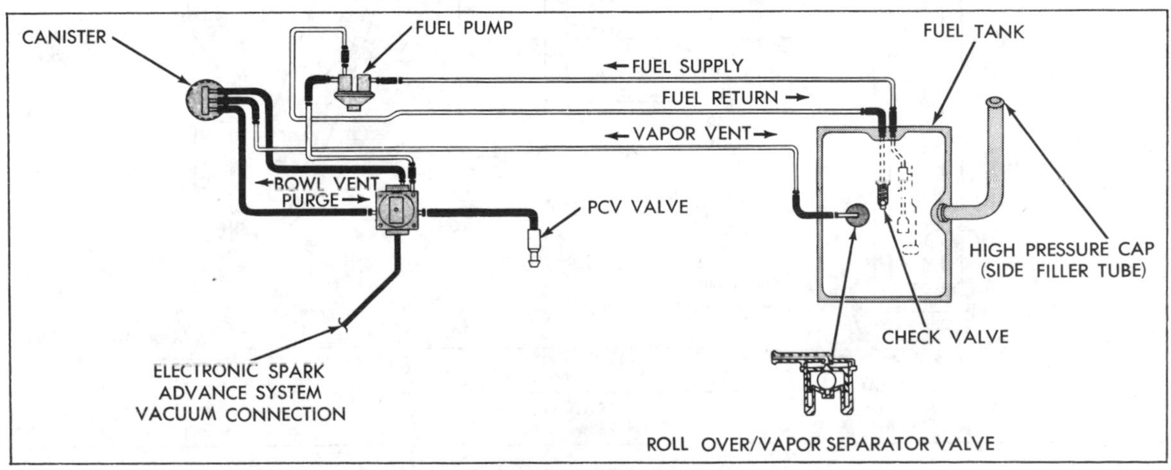

Evaporation control system

EVAPORATION CONTROL SYSTEM 1979-80

Pollution can occur even when an engine is not running. This is because gasoline will continue to evaporate as long as the temperature of the fuel is high enough to permit evaporation. When the fuel is in a state of expansion, it will separate and rise away from the liquid fuel into the surrounding atmosphere. This constitutes pollution because gasoline fumes are hydrocarbons, an undesirable pollutant.

A method of controlling the emission of fuel vapors is to contain the fuel system in an evaporation control system. Fuel vapor leaving the carburetor bowl vent and fuel tank are trapped in vent hoses, and then routed to a charcoal cannister where the vapors are stored until they can be drawn into the intake manifold when the engine is running.

MINI CATALYTIC CONVERTER 1979-80

In addition to the conventional type (underfloor converter) catalytic converter, a mini-converter is also used to initiate the exhaust gas oxidation before the exhaust gases reach the main underfloor catalytic converter.

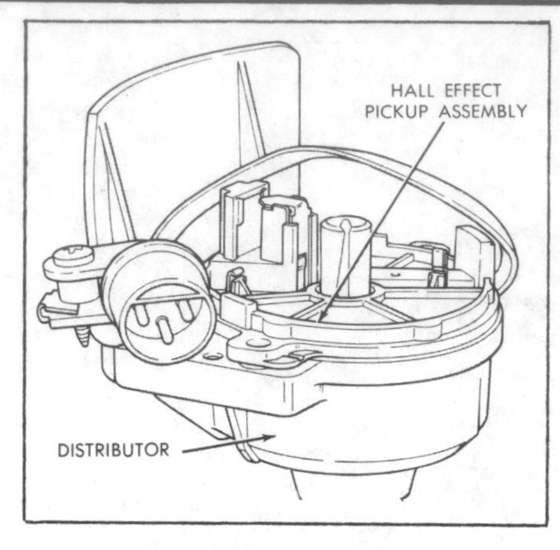

Hall effect pick-up assembly

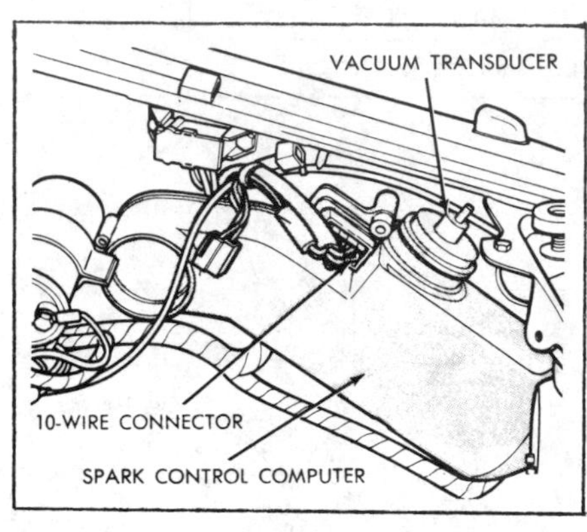

Spark control computer and vacuum transducer

CATALYTIC CONVERTER WITH AIR INJECTION 1980

(California vehicles only)

The front catalytic converter for California vehicles is a combination 3-way and oxidation catalyst. Between the two biscuits, air is injected so that the second biscuit acts as a full oxidation catalyst.

CRANKCASE VENTILATION SYSTEM 1979-80

This system, which utilizes a conventional PCV valve, is similar to the systems used on most vehicles marketed in the United States in that it directs crankcase vapors through the air cleaner and into the combustion area for complete burning.

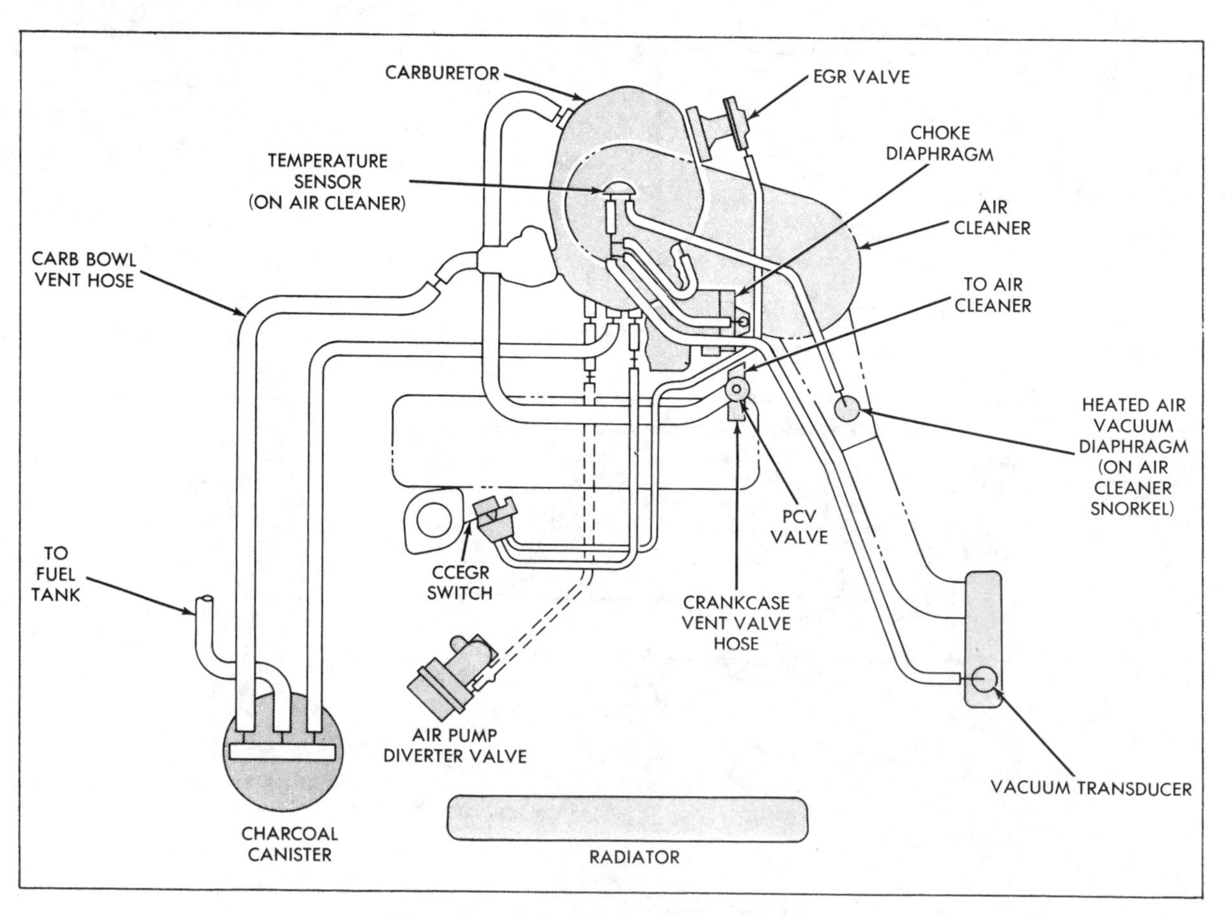

Vacuum hose routing diagram—Omni and Horizon

TUNE-UP SPECIFICATIONS

Engine Code, 5th character of the VIN number

FORD, LINCOLN, MERCURY

Year	ENG. V.I.N. CODE	No. Cyl. Disp. (cu in)	IGNITION Spark Plugs* Orig. Type	Gap (in)	Timing (deg. BTDC @ rpm) Man. Trans.	Auto. Trans.	CARBURETION Idle Speed (r.p.m.) Manual	Auto Cal.	Fed.
1979	Y	4-140	AWSF-42	.035	6 @ 550	20 @ 600	850N	750D/600N	800D/600N
	W	4-140	AWSF-42	.035	2 @ 650	—	900N	—	—
	Z	6-170	AWSF-42	.035	9 @ 650	9 @ 650	850N	750D/600N	650D
	T	6-200	BSF-82	.050	8 @ 750	10 @ 750	700N	650D	
	L	6-250	BSF-82	.050	4 @ 750	10 @ 750	800N	600D	
	F	8-302	ASF-52	.050	12 @ 500	8 @ 500	800N	600D	
	H	8-351	ASF-52	.050	—	15 @ 800	—	600D	
	Q	8-351	ASF-42	.050	—	12 @ 800	—	600D	
	S	8-400	ASF-52	.050	—	14 @ 800	—	650D	600D
1980	A	4-140	AWSF-42	.035	To obtain 1980 Specifications, refer to Vehicle Emission Control Information Label for Calibration Number, and relate to Calibration Specifications Charts.				
	B	6-200	BSF-82	.050					
	C	6-250	BSF-82	.050					
	D	8-255	ASF-52	.050					
	F	8-302	ASF-52	.050					
	G	8-351	ASF-52	.050					

The underhood certification/specification decal is the authority for performance specifications affecting vehicle emissions. Use this manual's information only when that decal is not available.

DISTRIBUTOR SPECIFICATIONS

FORD FULL-SIZE

Year	DISTRIBUTOR IDENTIFICATION	CENTRIFUGAL ADVANCE Start Dist. Deg. @ Dist. RPM	Finish Dist. Deg. @ Dist. RPM	VACUUM ADVANCE Start In. Hg.	Finish Dist. Deg. @ In. Hg.
1979	D9AE-AAA	0-2.7 @ 500	10.6-13.2 @ 2,500	2.3	12.7-15.2 @ 25
	D9AE-ZA	0-3 @ 500	12.2-14.7 @ 2,500	1.8	14.7-17.2 @ 25
	D9AE-ABA	0-2 @ 580	9.2-12.2 @ 2,500	2.8	6.7-9.2 @ 25①
	D9SE-AA	0-2 @ 500	3.6-6.1 @ 2,500	2.8	14.7-17.7 @ 25
1980	D9AE-TA (49s)②	0-6 @ 980	20-26 @ 5,000	2.2	25.5-30.5 @ 25

① Retard maximum—2.4 @ 9.3

② All California 302 V-8 Engines Use Electronic Engine Control on Spark Advance.
All 50 States 351 V-8 Engines Use Electronic Engine Control on Spark Advance.

DISTRIBUTOR SPECIFICATIONS

LINCOLN CONTINENTAL

Year	DISTRIBUTOR IDENTIFICATION	CENTRIFUGAL ADVANCE Start Dist. Deg. @ Dist. RPM	Finish Dist. Deg. @ Dist. RPM	VACUUM ADVANCE Start In. Hg.	Finish Dist. Deg. @ In. Hg.
1979	D9AE-YA	0.2 @ 610	9.7-12.6 @ 2,500	2	14.7-17.7 @ 25
	D9AE-ACA	0-2 @ 725	7.5-10.2 @ 2,500	2.4	6.9-9.2 @ 25
	D8OE-AA	0-2.7 @ 500	13.6-16.3 @ 2,500	2	12.7-15.2 @ 25
1980①	—	—	—	—	—

① All 1980 Lincoln Applications use the Ford Electronic Engine Control System to Control Timing Advance.

DISTRIBUTOR SPECIFICATIONS

FORD COMPACT AND INTERMEDIATE CARS

Year	DISTRIBUTOR IDENTIFICATION	CENTRIFUGAL ADVANCE Start Dist. Deg. @ Dist. RPM	CENTRIFUGAL ADVANCE Finish Dist. Deg. @ Dist. RPM	VACUUM ADVANCE Start In. Hg.	VACUUM ADVANCE Finish Dist. Deg. @ In. Hg.
1979	D9ZE-CA	0-3 @ 500	10-12.7 @ 2,500	2.8	14.7-17.2 @ 25
	D9SE-AA	0-2 @ 500	3.6-6.1 @ 2,500	2.8	14.7-17.2 @ 25
	D9AE-PA	0-3.6 @ 550	4.4-7.1 @ 2,500	2	14.7-17.2 @ 25
	D7AE-UA	0-2 @ 730	7.5-10.2 @ 2,500	2.3	11.2-13.2 @ 25
	D8OE-AA	0-2.7 @ 500	13.7-16.3 @ 2,500	2	12.7-15.2 @ 25
	D8BE-JA	0-1 @ 550	7.5-10.5	3.5	6.7-9.2 @ 13
	D9DE-CA	0-1 @ 1000	3.5-6.5 @ 2,500	3	8.7-11.2 @ 11.5
	D8DE-CA	0-1 @ 475	6.5-9 @ 2,500	2	10.7-13.2 @ 10.8
	D8DE-EA	0-1 @ 450	9.5-12.2 @ 2,500	3	10.7-13.2 @ 14
	D7DE-AA	0-1 @ 525	9.5-12.5 @ 2,500	3	12.7-15.2 @ 11
	D9BE-CA	0-1 @ 700	7.5 @ 2,500	4	12 @ 14
	D97E-CA	0-1 @ 450	10-13 @ 2,500	4	7.5-8.5 @ 13
	D7EE-DA	0-1 @ 510	11.5-14 @ 2,500	1.75	10.7-13.2 @ 12.4
	D7EE-EA	0-1 @ 525	11.5-14 @ 2,500	2	10.7-13.2 @ 7.9
	D7EE-CA	0-1 @ 800	5-7.5 @ 2,500	2.3	10.7-13.2 @ 7.9
	D7EE-HA	0-2.5 @ 1200	5-7.5 @ 2,500	2.3	10.7-13.2 @ 7.9
	D9TE-BA	0-1 @ 550	7-9.5 @ 2,500	4.5	8.7-11.2 @ 12.5
	D8BE-EA	0-1 @ 550	7-9.5 @ 2,500	4.5	8.7-11.2 @ 12.5
1980	D8BE-JA	0-4 @ 1,320	15.2-21 @ 5,000	2	13.5-18.5 @ 25
	D9DE-CA	.5-4.5 @ 1,120	8-13.4 @ 5,000	2	17.5-22.5 @ 25
	E0SE-CA	0-4 @ 1,270	19.8-25.4 @ 5,000	2.1	21.5-26.5 @ 25
	E0ZE-AA	0-5.2 @ 1,000	9.6-15 @ 5,000	2.3	21.5-26.5 @ 25
	D9ZE-CA	0-6 @ 1,000	20-25.5 @ 5,000	2.8	29.5-34.5 @ 25
	E0ZE-EA	0-5.2 @ 1,080	9.6-15 @ 5,000	2.4	21.5-26.5 @ 25
	E0SE-DA	0-5 @ 1,480	19.4-26.2 @ 5,000	1.8	21.5-26.5 @ 25
	E0EE-CA	0-5 @ 1,160	22.6-28 @ 5,000	1.8	21.5-26.5 @ 25
	E0EE-DA	0-6.4 @ 1,200	23.2-28 @ 5,000	2.1	21.5-26.5 @ 25
	E0EE-FA	0-5.2 @ 2,280	10.2-15.8 @ 5,000	2	13.5-18.5 @ 25
	E0EE-BA	0-3.5 @ 1,080	12.2-22.6 @ 5,000	1.7	17.5-22.5 @ 25
	E0EE-EA	0-6.2 @ 1,000	22.2-27.5 @ 5,000	2	13.5-18.5 @ 25
	D9BE-DA	0-4 @ 1,260	11.8-19.5 @ 5,000	1.8	17.5-22.5 @ 25
	D8BE-EA	0-4 @ 1,270	18.2-23.5 @ 5,000	2.4	17.5-22.5 @ 25

1980 FORD PROPANE ENRICHMENT CHART

①	②	②
①	②	②
①	②	②

① See instruction on Location of Calibration Number
② "O" Gain Turns Lean RPM Drop: 1/4/10

HOW TO DETERMINE VEHICLE CALIBRATION NUMBERS 1980

An emission control decal and an engine code label must be used to determine the vehicle calibration code number.

1. The emission control label is gold or silver and might be located on the valve cover, radiator support bracket, under side of the hood, fresh air inlet tube or the windshield washer bottle.
2. The engine code label is yellow or white and will be located on the valve cover.
3. Determine the model year from the emission label. Use only the last digit. For example, 1980 would be "0."
4. Determine the calibration base and the revision level from the engine code label. For example the calibration base could be "93J" and the revision level "RO."
5. The complete calibration number in the example would be "O-93J-RO."

NOTE: *Some of the calibrations are carried over from prior years. If a number can't be located using prefix "0"; check "9", "8" or "7" prefixes.*

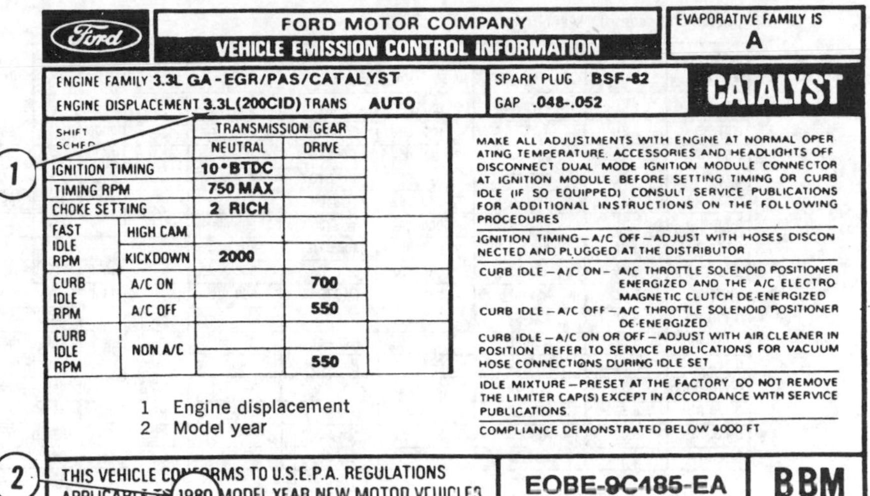

Emission control decal (© Ford Motor Co.)

C 2 XL830AA | C 93J | R 0 | S 26
A B C D E F G H J K L M
1 2 3 80 10°
1 2 3 4 5 6 7 8 9 0

1 Calibration base
2 Revision level

Engine code label (© Ford Motor Co.)

CALIBRATION SERVICE SPECIFICATIONS

Calibration Number	Engine	Spark Plug	Spark Plug Gap	Ignition Timing	Timing RPM	Fast Idle RPM High Cam	Fast Idle RPM Kick Down	Curb Idle RPM A/C On	Curb Idle RPM A/C Off	Curb Idle RPM Non-AC	TSP Off RPM AC	TSP Off RPM Non-AC	Choke Cap Setting	Decel Throttle Control RPM
0-1B-R11	2.3L	AWSF-42	.032-.036	6^0 BTDC	600	–	2000	750	–	750	–	–	2 RICH	–
0-1B-R12	2.3L	AWSF-42	.032-.036	6^0 BTDC	600	–	2000	750	–	750	600	600	2 RICH	–
0-2C-R10	2.3L	AWSF-42	.032-.036	6^0 BTDC	550	–	1800	850	–	850	–	–	2 RICH	–
0-1D-R0	2.3L	AWSF-42	.032-.036	6^0 BTDC	600	–	2000	–	750	750	–	–	1 LEAN	–
0-2D-R0	2.3L	AWSF-42	.032-.036	6^0 BTDC	550	–	1800	850	–	850	–	–	2 RICH	–
0-1G-R11	2.3L	AWSF-42	.032-.036	6^0 BTDC	600	–	2000	750	–	750	–	–	2 RICH	–
0-1G-R12	2.3L	AWSF-42	.032-.036	6^0 BTDC	600	–	2000	750	–	750	600	600	2 RICH	–

Calibration Number	Engine	Spark Plug	Spark Plug Gap	Ignition Timing	Timing RPM	Fast Idle RPM High Cam	Fast Idle RPM Kick Down	Curb Idle RPM A/C On	Curb Idle RPM A/C Off	Curb Idle RPM Non-AC	TSP Off RPM AC	TSP Off RPM Non-AC	Choke Cap Setting	Decel Throttle Control RPM
0-8A-R0	4.1L	BSF-82	.048-.052	4^0 BTDC	750 max.	–	1700	800	700	700	–	–	2 RICH	–
0-29A-R0	4.1L	BSF-82	.048-.052	8^0 BTDC	750 max.	–	1700	700	600	600	–	450	2 RICH	–
0-16A-R0	4.2L	ASF-42	.048-.052	6^0 BTDC	500 max.	2000	–	650	500	500	–	–	3 RICH	–
0-16A-R12	4.2L	ASF-42	.048-.052	6^0 BTDC	550	1800	–	700	550	550	–	–	4 RICH	–
0-16C-R0	4.2L	ASF-42	.048-.052	6^0 BTDC	500 max.	2000	–	650	500	500	–	–	3 RICH	–
0-16C-R13	4.2L	ASF-42	.048-.052	6^0 BTDC	550	1800	–	700	550	550	–	–	4 RICH	–
0-16C-R15	4.2L	ASF-42	.048-.052	6^0 BTDC	500	1800	–	650	500	500	–	–	4 RICH or "V"	–
0-16C-R18	4.2L	ASF-42	.048-.052	6^0 BTDC	500	1800	–	625	550	550	–	–	4 RICH or "V"	–
0-16S-R0	4.2L	ASF-42	.048-.052	6^0 BTDC	550	2100	–	650	500	500	–	–	3 RICH	–
0-16S-R14	4.2L	ASF-42	.048-.052	6^0 BTDC	500	2000	–	650	500	500	–	–	3 RICH or "V"	–
0-16W-R0	4.2L	ASF-42	.048-.052	6^0 BTDC	500	2000	–	650	500	500	–	–	3 RICH or "V"	–
0-16X-R0	4.2L	ASF-42	.048-.052	6^0 BTDC	500	2000	–	650	500	500	–	–	3 RICH or "V"	–
9-11C-R1	5.0L	ASF-52	.048-.052	8^0 BTDC	500	2100	–	675	600	–	–	600	3 RICH	–
0-11A-R10	5.0L	ASF-52	.048-.052	8^0 BTDC	500	2000	–	625	500	500	–	–	3 RICH or "V"	–

1 All timing, choke and RPM specifications shown should be set with the Automatic Transmission in NEUTRAL, except CURB IDLE RPM which is set with Automatic Transmission in DRIVE.

2 All timing, choke and RPM specifications shown should be set with the Manual Transmission in NEUTRAL.

CALIBRATION SERVICE SPECIFICATIONS

REVISED

Calibration Number	Engine	Spark Plug	Spark Plug Gap	Ignition Timing	Timing RPM	High Cam	Kick Down	A/C On	A/C Off	Non-AC	VOTM On	VOTM Off	Choke Cap Setting
0-1P-R0	2.3L	AWSF-42	.032-.036	12° BTD C	600	–	2000	–	750	750	–	–	Index
0-2T-R0	2.3L	AWSF-42	.032-.036	6° BTD C	650	–	2000	–	850	850	–	–	Index
0-6A-R0	3.3L	BSF-82	.048-.052	10° BTD C	750	–	1600	–	–	700	–	–	2 Rich
0-7Q-R1	3.3L	BSF-82	.048-.052	10° BTD C	750 Max.	–	2300	700	600	600	–	–	2 Rich
0-27A-R3	3.3L	BSF-82	.048-.052	7° BTD C	750 Max.	–	2000	700	550	550	–	–	Index
0-11A-R0	5.0L	ASF-52	.048-.052	8° BTD C	500 Max.	2000	–	650	500	500	–	–	3 Rich
0-11B-R0	5.0L	ASF-52	.048-.052	10° BTD C	500 Max.	2000	–	650	500	500	–	–	3 Rich
0-13A-R0	5.0L	ASF-52	.048-.052	6° BTD C	500	–	1850	–	500	500	–	–	1 Rich
0-13F-R0	5.0L	ASF-52	.048-.052	6° BTD C	500	–	1850	–	500	500	–	–	1 Rich
0-13R-R0	5.0L	ASF-52	.048-.052	Not Adj.	–	2000**	–	650	500	–	650	500	Index
0-13T-R0	5.0L	ASF-52	.048-.052	Not Adj.	–	2000**	–	650	500	–	650	500	Index
0-14A-R0	5.0L	ASF-52	.048-.052	Not Adj.	–	2100*	–	–	550	550	–	–	Index
0-12A-R0	5.8L	ASF-52	.048-.052	Not Adj.	–	1650**	–	640	550	–	640	550	Index
0-12B-R0	5.8L	ASF-52	.048-.052	Not Adj.	–	1650**	–	640	550	–	640	550	Index
0-12C-R0	5.8L	ASF-52	.048-.052	Not Adj.	–	1650**	–	640	550	–	640	550	Index
0-12N-R0	5.8L	ASF-52	.048-.052	Not Adj.	–	1650**	–	640	550	–	640	550	Index

*1st High Step
**2nd High Step

(1) All timing, choke and RPM specifications shown should be set with the Automatic Transmission in NEUTRAL, except CURB IDLE RPM which is set with Automatic Transmission in DRIVE.

(2) All timing, choke and RPM specifications shown should be set with the Manual Transmission in NEUTRAL.

CALIBRATION SERVICE SPECIFICATIONS

REVISED

Calibration Number	Engine	Spark Plug	Spark Plug Gap	Ignition Timing	Timing RPM	Fast Idle RPM		Curb Idle RPM			TSP On RPM		Choke Cap Setting	Decel Throttle Control RPM
						High Cam	Kick Down	A/C On	A/C Off	Non-A/C	A/C	Non-A/C		
0-1B-R0	2.3L	AWSF-42	.032-.036	6° BTDC	600	–	2000	–	750	750	–	–	1 Lean	–
0-1G-R0	2.3L	AWSF-42	.032-.036	6° BTDC	600	–	2000	–	750	750	–	–	1 Lean	–
0-2B-R0	2.3L	AWSF-42	.032-.036	6° BTDC	550	–	1800	–	850	850	–	–	2 Rich	–
0-2C-R0	2.3L	AWSF-42	.032-.036	6° BTDC	550	–	1800	–	850	850	–	–	2 Rich	–
0-2E-R0	2.3L	AWSF-32	.032-.036	6° BTDC	550	–	1800	–	850	850	–	–	2 Rich	–
0-6A-R01	3.3L	BSF-82	.048-.052	12° BTDC	750 Max.	–	1600	900	700	–	–	–	2 Rich	–
0-7A-R0	3.3L	BSF-82	.048-.052	10° BTDC	750 Max.	–	2000	700	550	550	–	–	2 Rich	–
0-9A-R0	4.1L	BSF-82	.048-.052	10° BTDC	750 Max.	–	1700	700	600	550	450	450	2 Rich	–
0-11E-R0	5.0L	ASF-52	.048-.052	6° BTDC	500	2100	–	700	550	550	–	–	3 Rich	–
0-11D-R0	5.0L	ASF-52	.048-.052	8° BTDC	500	2000	–	650	500	500	–	–	3 Rich	–
0-11N-R0	5.0L	ASF-52	.048-.052	6° BTDC	500 Max.	2100	–	650	500	500	–	–	3 Rich	–
0-11R-R0	5.0L	ASF-52	.048-.052	6° BTDC	500	2100	–	650	500	500	–	–	3 Rich	–

(1) All timing, choke, and RPM specifications shown should be set with the Automatic Transmission in NEUTRAL, except CURB IDLE RPM which is set with Automatic Transmission in DRIVE.

(2) All timing, choke, and RPM specifications shown should be set with the Manual Transmission in NEUTRAL.

CALIBRATION SERVICE SPECIFICATIONS

Calibration Number	Engine	Spark Plug	Spark Plug Gap	Ignition Timing	Timing RPM	Fast Idle RPM		Curb Idle RPM			TSP Off RPM		Choke Cap Setting	Decel Throttle Control RPM
						High Cam	Kick Down	A/C On	A/C Off	Non-AC	AC	Non-AC		
0-11E-R13	5.0L	ASF-52	048-.052	6° BTDC	500	2200	–	700	550	–	–	–	3 RICH or "V"	–
0-11N-R10	5.0L	ASF-52	048-.052	6° BTDC	500	2100	–	650	500	500	–	–	3 RICH or "V"	–
0-11N-R11	5.0L	ASF-52	.048-.052	6° BTDC	500	2100	–	625	500	500	–	–	3 RICH or "V"	–
0-11N-R12	5.0L	ASF-52	.048-.052	6° BTDC	500	2100	–	625	500	500	–	–	4 RICH or "V"	–
0-11N-R13	5.0L	ASF-52	.048-.052	6° BTDC	500	2100	–	625	500	500	–	–	4 RICH or "V"	–
0-11P-R10	5.0L	ASF-52	.048-.052	8° BTDC	500	2000	–	600	500	–	–	500	3 RICH	–
0-11P-R11	5.0L	ASF-52	.048-.052	8° BTDC	500	2000	–	650	500	500	–	–	3 RICH or "V"	–
0-11R-R10	5.0L	ASF-52	.048-.052	6° BTDC	500	2100	–	650	500	500	–	–	3 RICH or "V"	–
0-11V-R12	5.0L	ASF-52	.048-.052	8° BTDC	500	1800	–	700	550	–	–	–	4 RICH or "V"	–
0-11W-R0	5.0L	ASF-52	.048-.052	8° BTDC	500	2000	–	700	550	–	–	550	3 RICH or "V"	–
0-11W-R10	5.0L	ASF-52	.048-.052	8° BTDC	500	2000	–	625	550	550	–	–	3 RICH or "V"	–
0-11X-R0	5.0L	ASF-52	.048-.052	8° BTDC	500	2000	–	650	500	500	–	–	3 RICH or "V"	–
0-11Y-R0	5.0L	ASF-52	.048-.052	8° BTDC	500	2000	–	700	550	–	–	550	3 RICH or "V"	–
0-13A-R0	5.0L	ASF 52	.048-.052	6° BTDC	500	–	1700	–	500	500	–	–	1 RICH	–
0-13A-R11	5.0L	ASF-52	.048-.052	6° BTDC	500	–	1850	–	500	500	–	–	1 RICH	–
0-13A-R12	5.0L	ASF-52	.048-.052	6° BTDC	500	–	1700	500	–	500	–	–	1 RICH	–

CALIBRATION SERVICE SPECIFICATIONS

Calibration Number	Engine	Spark Plug	Spark Plug Gap	Ignition Timing	Timing RPM	Fast Idle RPM		Curb Idle RPM			TSP Off RPM		Choke Cap Setting	Decel Throttle Control RPM
						High Cam	Kick Down	A/C On	A/C Off	Non-AC	AC	Non-AC		
0-13D-R0	5.0L	ASF-52	.048-.052	8° BTDC	500	2100-1st High Step	1850	–	500	500	–	–	RICH	–
0-13F-R11	5.0L	ASF-52	.048-.052	6° BTDC	500	–	1850	–	500	500	–	–	1 RICH	–
0-13R-R10	5.0L	ASF-52	.048-.052	–	–	2000	–	–	–	–	650	500	–	–
0-13R-R11	5.0L	ASF-52	.048-.052	–	–	–	2000	–	–	–	650	500	–	–
0-13T-R10	5.0L	ASF-52	.048-.052	–	–	2000	–	–	–	–	650	500	–	–
0-13T-R11	5.0L	ASF-52	.048-.052	–	–	–	2000	–	–	–	650	500	–	–
0-14A-R10	5.0L	ASF-52	.048-.052	–	–	2100-1st High Step	–	–	–	550	–	–	INDEX	–
0-14A-R11	5.0L	ASF-52	.048-.052	–	–	2100-1st High Step	–	–	550	–	–	–	INDEX	–
0-14N-R0	5.0L	ASF-52	.048-.052	Not Adj.	–	2100-1st High Step	–	–	550	550	–	–	INDEX	–
0-14N-R10	5.0L	ASF-52	.048-.052	–	–	2100-1st High Step	–	–	–	550	–	–	INDEX	–
0-12A-R5	5.8L	ASF-52	.048-.052	–	–	1650	–	–	–	–	640	550	–	–
0-12C-R5	5.8L	ASF-52	.048-.052		–	1050	–	–	–	–	640	550	–	–
0-12G-R0	5.8L	ASF-52	.048-.052	6° BTDC	800	2200	–	650	550	–	–	–	RICH	–
0-12H-R0	5.8L	ASF-52	.048-.052	17° BTDC	800	2200	–	650	550	500	–	–	RICH	–
0-12J-R0	5.8L	ASF-52	.048-.052	12° BTDC	800	2200	–	650	550	–	–	550	1 RICH	–
0-12P-R0	5.8L	ASF-52	.048-.052	Not Adj.	–	1650	–	640	550	–	640	550	INDEX	–

1 All timing, choke and RPM specifications shown should be set with the Automatic Transmission in NEUTRAL, except CURB IDLE RPM which is set with Automatic Transmission in DRIVE.

2 All timing, choke and RPM specifications shown should be set with the Manual Transmission in NEUTRAL.

Motorcraft Model 2700 VV Specifications

Ford Products

Year	Model	Float Level (in.)	Float Drop (in.)	Fast Idle Cam Setting (notches)	Cold Enrichment Metering Rod (in.)	Control Vacuum (in. H_2O)	Venturi Valve Limiter (in.)	Choke Cap Setting (notches)	Control Vacuum Regulator Setting (in.)
1979	D9ZE-LB	$1\frac{3}{64}$	$1\frac{15}{32}$	1 Rich/2nd step	.125	①	②	Index	.230
	D84E-KA	$1\frac{3}{64}$	$1\frac{15}{32}$	1 Rich/3rd step	.125	5.5	$\frac{61}{64}$	Index	——
1980	All	$1\frac{3}{64}$	$1\frac{15}{32}$	1 Rich/4th step	.125	③	④	⑤	.075

① Venturi Air Bypass 6.8-7.3
Venturi Valve Diaphragm 4.6-5.1
② Limiter Setting .38-.42
Limiter Stop Setting .73-.77
③ See text
④ Opening gap: 0.99-1.01
Closing gap: 0.94-0.98
⑤ See underhood decal

Model 1946

Ford Motor Co.

Year	Part Number	Float Level	Choke Pulldown (in.)	Dechoke (in.)	Fast Idle Cam (in.)	Accelerator Pump Stroke Slot
1978-79	All	①	.026	.250	.080	#2
1980	EOBE-ALA, AMA	①	.100	.150	.070	#2
	EOEE-ANA, APA	①	.100	.150	.070	#2
	EOZE-BBA, BAA	①	.120	.150	.086	#2
	EOZE-DA, EA	①	.110	.150	.070	#2
	EOZE-FA, GA	①	.110	.150	.070	#2
	EOBE-AA, CA	①	.100	.150	.070	#2
	EOBE-ZA, AAA	①	.115	.150	.090	#1

CARTER YF, YFA SPECIFICATIONS

FORD MOTOR CO.

Year	Model ①	Float Level (in.)	Fast Idle Cam (in.)	Unloader (in.)	Choke
1979	D9BE-RA D9DE-CB, DB, AA, BA, CA, EA	25/32	0.140	0.250	1 Rich
1980	DEDE-GA, HA, EODE-JA, NA, LA, MA	25/32	0.140	0.250	2 Rich

① Model number located on the tag or casting

FORD, AUTOLITE, MOTORCRAFT MODEL 5200 SPECIFICATIONS

Ford Products

Year	(9510)* Carburetor Identification①	Dry Float Level (in.)	Pump Hole Setting	Choke Plate Pulldown (in.)	Fast Idle Cam Linkage (in.)	Fast Idle (rpm)	Dechoke (in.)	Choke Setting	Dashpot (in.)
1979	D9ZE-ND	0.460	3	0.236	0.118	1800	0.236	2 Rich	—
	D9BE-AAA, D9BE-ABA, D9EE-AMA	0.460	2	0.236	0.118	1800	0.236	2 Rich	—
	D9EE-ANA, D9EE-ASA, D9EE-AYA	0.460	2	0.236	0.118	1800	0.236	1 Rich	—
1980	D9EE-APA, ANA	0.460	2	0.236	0.118	②	0.236	1 Rich	—
	EOEE-GA, RA	0.460	2	0.196	0.078	②	0.196	②	—
	EOEE-JA, TA	0.460	2	0.196	0.078	②	0.196	②	—
	EOEE-JC, TC	0.460	—	0.196	0.078	②	0.196	②	—
	EOEE-JD, TD	0.460	2	0.177	0.078	②	0.196	②	—
	EOEE-AEA, AFA	0.460	2	0.196	0.078	②	0.196	②	—
	EOZE-ACB	0.460	—	0.275	0.157	②	0.236	②	—
	EOZE-AZA	0.460	2	0.275	0.157	②	0.393	②	—
	EOZE-AAA	0.460	3	0.275	0.157	②	0.236	②	—
	EOZE-ACA	0.460	2	0.275	0.157	②	0.236	②	—
	EOZE-ATA	0.460	2	0.275	0.118	②	0.236	②	—

* Basic carburetor number

① Figure given is for all manual transmissions; for automatic trans. the figures are: (49 states) 2000 RPM; (Calif.) 1800 RPM.

② See underhood decal

MOTORCRAFT MODEL 7200 VV SPECIFICATIONS

Year	Model	Float Level (in.)	Float Drop (in.)	Fast Idle Cam Setting (notches)	Cold Enrichment Metering Rod (in.)	Control Vacuum (in. H_2O)	Venturi Valve Limiter (in.)	Choke Cap Setting (notches)
1979	D9AE-ACA	1 3/64	1 15/32	1 Rich/3rd step	.125	7.5	.73-.77①	Index
	D9ME-AA	1 3/64	1 15/32	1 Rich/3rd step	.125	7.5	.73-.77①	Index
1980	All	1 3/64	1 15/32	1 Rich/3rd step	.125	②	③	④

① Limiter Stop Setting: .99-1.01

② See text

③ Opening gap: 0.99-1.01
Closing gap: 0.39-0.41

④ See underhood decal

FORD, AUTOLITE, MOTORCRAFT MODELS 2100, 2150 SPECIFICATIONS

Ford Products

Year	(9510)* Carburetor Identification	Dry Float Level (in.)	Wet Float Level (in.)	Pump Setting Hole # ①	Choke Plate Pulldown (in.)	Fast Idle Cam Linkage Clearance (in.)	Fast Idle (rpm)	Dechoke (in)	Choke Setting	Dashpot (in.)
1978-79	D84E-EA	7/16	13/16	2	0.110	⑧	⑨	—	3 Rich	—
	D8AE-JA	3/8	3/4	3	0.167	⑧	⑨	—	3 Rich	—
	D8BE-ACA	7/16	3/4	4	0.155	⑧	⑨	—	2 Rich	—
	D8BE-ADA	7/16	13/16	2	0.110	⑧	⑨	—	3 Rich	—
	D8BE-AEA	7/16	13/16	2	0.110	⑧	⑨	—	4 Rich	—
	D8BE-AFA	7/16	13/16	2	0.110	⑧	⑨	—	4 Rich	—
	D8BE-MB	3/8	13/16	3	0.122	⑧	⑨	—	Index	—
	D8DE-HA	19/32	13/16	3	0.157	⑧	⑨	—	Index	—
	D8KE-EA	19/32	13/16	2	0.135	⑧	⑨	—	3 Rich	—
	D8OE-BA	3/8	3/4	3	0.167	⑧	⑨	—	3 Rich	—
	D8OE-EA	19/32	13/16	2	0.136	⑧	⑨	—	Index	—
	D8OE-HA	7/16	13/16	3	0.180	⑧	⑨	—	2 Rich	—
	D8SE-CA	19/32	13/16	3	0.150	⑧	⑨	—	2 Rich	—
	D8ZE-TA	3/8	3/4	4	0.135	⑧	⑨	—	Index	—
	D8ZE-UA	3/8	3/4	4	0.135	⑧	⑨	—	Index	—
	D8WE-DA	7/16	13/16	4	0.143	⑧	⑨	—	1 Rich	—
	D8YE-AB	3/8	13/16	3	0.122	⑧	⑨	—	Index	—
	D8SE-DA, FA	7/16	13/16	3	0.147	⑧	⑨	—	3 Rich	—
	D8SE-FA, GA	3/8	13/16	3	0.147	⑧	⑨	—	3 Rich	—
1980	E04E-PA, RA	—	13/16	2	0.104	⑧	⑨	1/4	⑨	—
	E0BE-AUA	—	13/16	3	0.116	⑧	⑨	1/4	⑨	—
	E0DE-SA, TA	—	13/16	2	0.104	⑧	⑨	1/4	⑨	—
	E0KE-CA, DA	—	13/16	3	0.116	⑧	⑨	1/4	⑨	—
	E0KE-GA, HA	—	13/16	3	0.116	⑧	⑨	1/4	⑨	—
	E0KE-JA, KA	—	13/16	3	0.116	⑧	⑨	1/4	⑨	—
	D84E-TA, UA	—	13/16	2	0.125	⑧	⑨	1/4	⑨	—
	E04E-ADA, AEA	—	13/16	2	0.104	⑧	⑨	1/4	⑨	—
	E04E-CA	—	13/16	2	0.104	⑧	⑨	1/4	⑨	—
	E04E-EA, FA	—	13/16	2	0.104	⑧	⑨	1/4	⑨	—
	E04E-JA, KA	—	13/16	2	0.137	⑧	⑨	1/4	⑨	—
	E04E-SA, TA	—	13/16	2	0.104	⑧	⑨	1/4	⑨	—
	E04E-VA, YA	—	13/16	2	0.104	⑧	⑨	1/4	⑨	—
	E0DE-TA, VA	—	13/16	2	0.104	⑧	⑨	1/4	⑨	—
	E0SE-GA, HA	—	13/16	2	0.104	⑧	⑨	1/4	⑨	—
	E0SE-LA, MA	—	13/16	2	0.104	⑧	⑨	1/4	⑨	—
	E0SE-NA	—	13/16	2	0.104	⑧	⑨	1/4	⑨	—
	E0SE-PA	—	13/16	2	0.137	⑧	⑨	1/4	⑨	—
	E0VE-FA	—	13/16	2	0.104	⑧	⑨	1/4	⑨	—
	E0WE-BA, CA	—	13/16	2	0.137	⑧	⑨	1/4	⑨	—
	D9AE-ANA, APA	—	13/16	3	0.129	⑧	⑨	1/4	⑨	—
	D9AE-AVA, AYA	—	13/16	3	0.129	⑧	⑨	1/4	⑨	—
	E0AE-AGA	—	13/16	3	0.159	⑧	⑨	1/4	⑨	—

* Basic carburetor number for Ford products
① With link in inboard hole of pump lever
⑧ Opposite "V" notch
⑨ See underhood decal

CAR SERIAL NUMBER AND ENGINE IDENTIFICATION

1979

The vehicle identification number is mounted on the driver's side. The first character is the model year, with 8 for 1978. The fifth character is the engine code, as follows.

F 302 2-bbl. V-8
H 351 2-bbl. V-8
L .... 250 1-bbl. 6-cyl.
S 400 2-bbl. V-8
T .... 200 1-bbl. 6-cyl.

351 engines come in two different designs, the "W" and the "M". Both engines use the "H" code letter. Although the bore and stroke of both engines is the same, the internal parts are not interchangeable. The two engines can be identified by looking at the fuel pump mounting bolts. The "W" bolts are on a horizontal line. The "M" bolts are on a vertical line. The "W" engine is the same general design as the 302 V-8. The "M" engine is the same general design as the 400 V-8.

1980

The vehicle identification number is mounted on the driver's side of the dash and is visible through the windshield. The first character indicates the year; with 0 for 1980. The fifth character is the engine code as follows:

B .... 200 1-bbl. 6-cyl.
C .... 250 1-bbl. 6-cyl.
D 255 2-bbl. V-8
F 302 2-bbl. V-8
T .... 140 2-bbl. 4-cyl.

EMISSION EQUIPMENT

1979

All Models

Closed positive crankcase ventilation
Emission calibrated carburetor
Emission calibrated distributor
Heated air cleaner
Vapor control, canister storage
Electric choke
Exhaust gas recirculation
Catalytic converter
Fresh air tube to air cleaner

NOTE: *In addition to the above listed equipment some vehicles will also be equipped with an air pump system.*

Altitude compensation carburetors may be found on high altitude vehicles only.

1980

Air Cleaner Temperature Control System
1. Cold weather modulator
2. TVS Temperature Vacuum System
3. Duct and door in air cleaner

Catalytic Converter
Closed PCV System
CFI-EFI Electronic Fuel Injection
EEC Electronic Engine Control; Includes 8 sensors and a computer to control timing and fuel.
EGR Exhaust Gas Recirculation
PVS Porter Vacuum System
VDV Vacuum Delay Valve
Thermactor System
1. Air Supply Pump
2. By Pass Valve
3. Air Control Valve
4. Check Valve
5. Decel Valve
6. Air Cleaner Temperature Sensor
7. Solenoid Vacuum Valve
8. Thermactor Idle Vacuum Valve
9. Thermactor Vacuum Vent Valve

Evaporative Emission Control System; Charcoal Cannister System.
VOTM Vacuum Operated Throttle Modulator Venturi Valve Carburetor

IDLE SPEED

1979

NOTE: *Make all adjustments with the engine at normal operating temperature. Refer to the Engine Emission Label, located in the engine compartment, for additional instructions before making any adjustments.*

NOTE: *In many instances the information on the underhood emissions label can not be found elsewhere and is special to the vehicle being serviced; therefore, it should be copied and kept with the owner's manual.*

Ford/Mercury 302 V-8 (F) 49 states

Calibration 9-11E-R1
- Auto. trans./air conditioned
  - Fast idle 1750
  - Curb idle
    - AC/ on 625
    - A/C off 550
  - VOTM*
    - A/C off 625

Calibration 9-11E-R1
- Auto. trans./ non A/C
  - Fast idle 1750
  - Curb idle 550
  - VOTM* 625

Ford/Mercury 302 V-8 (F) Police station wgn. 49 states

Calibration 9-11F-R1
- Auto. trans./air conditioned
  - Fast idle 1750
  - Curb idle
    - A/C on 625
    - A/C off 550
  - VOTM*
    - A/C off 625

Calibration 9-11F-R1
- Auto. trans./non A/C
  - Fast idle 1750
  - Curb idle 550
  - VOTM* idle 625

Ford/Mercury 302 V-8 (F) California only

Calibration 9-11Q-R01
- Auto. trans.
  - Fast idle 1800
  - Curb idle
    - A/C on 625
    - A/C off 550
    - Non A/C 550
  - VOTM*
    - A/C 625
    - Non A/C 625

Calibration 9-11Q-R-10
- Auto trans.
  - Fast idle 1800
  - Curb idle
    - A/C on 625
    - A/C off 550
    - Non A/C 550
  - VOTM*
    - A/C 625
    - Non A/C 625

Ford/Mercury 351 V-8 (H) 49 states

Calibration 9-12G-R0
- Auto. trans.
  - Fast idle 2200
  - Curb idle
    - A/C on 650
    - A/C off 600
    - Non A/C 600

Calibration 9-12F-R0
- Auto. trans.
  - Fast idle 2200
  - Curb idle
    - A/C on 650
    - A/C off 600
    - Non A/C 600

Mercury w/EEC II 351 V-8 (H) 49 states

Calibration 9-12A-R0
- Auto. trans.
  - Fast idle 2000
  - Curb idle
    - A/C on 640
    - A/C off 550

IDLE SPEED

Non A/C 550
Calibration 9-12B-R0
Auto. trans.
Fast idle 2000
Curb idle
A/C on 640
A/C off 550
Non A/C 550

Ford/Mercury w/EEC II 351 V-8 (H) California only

Calibration 9-12N-R0
Auto. trans.
Fast idle 2100
Curb idle
A/C 620
Non A/C 620
Calibration 9-12P R0
Auto. trans.
Fast idle 2100
Curb idle
A/C 620
Non A/C 620

Granada/Monarch 302 V-8 (F) 49 states

Calibration 9-11C-R0
Auto. trans.
Fast idle 2100
Curb idle
A/C on 675
A/C off 600
Non A/C 600
Calibration 9-11C-R1
Auto. trans.
Fast idle 2100
Curb idle
A/C on 675
A/C off 600
Non A/C 600
Calibration 9-10C-R0
Manual trans.
Fast idle 2300
Curb idle
A/C on 875
A/C off 800
Non A/C 800
Calibration 9-10C-R10
Manual trans.
Fast idle 2300
Curb idle
A/C on 875
A/C off 800
Non A/C 800

Granada/Monarch 302 V-8 (F) California only

Calibration 9-11S-R0
Auto. trans.
Fast idle 1800
Curb idle
A/C on 675
A/C off 600
Non A/C 600
VOTM*
A/C 700
Non A/C 700

Granada/Monarch 250 I-6 (L) 49 states

Calibration 9-9A-R0
Auto. trans.
Fast idle 1700
Curb idle
A/C on 700
A/C off 600
Non A/C 600
TSP off**
A/C off 500
Non A/C 500
Calibration 9-9A-R12
Auto. trans.
Fast idle 1700
Curb idle
A/C on 700
A/C off 600
Non A/C 600
TSP off**
A/C off 450
Non A/C 450
Calibration 9-8A-R0
Manual trans.
Fast idle 1700
Curb idle
A/C 800
Non A/C 800

Granada/Monarch 250 I-6 (L) California only

Calibration 9-9P-R0
Auto. trans.
Fast idle 2300
Curb idle
A/C on 700
A/C off 600
Non A/C 600
Tsp off**
A/C 450
Non A/C 450

Cougar/LTD II/Thunderbird 302 V-8 (F) 49 states

Calibration 9-11J-R0
Auto. trans.
Fast idle 2100
Curb idle
A/C on 675
A/C off 600
Non A/C 600

Cougar/LTD II/Thunderbird 351 V-8 (H) 49 states

Calibration 9-12E-R0
Auto. trans.
Fast idle 2100
Curb idle
A/C on 650
A/C off 600
Non A/C 600
Calibration 9-14E-R0
Auto. trans.
Fast idle 2200
Curb idle
A/C on 650
A/C off 600
Non A/C 600

Cougar/LTD II/Thunderbird 351 V-8 (H) California only

Calibration R-140-R0
Auto. trans.
Fast idle 2300
Curb idle
A/C on 650
A/C off 600
Non A/C 600

Lincoln Continental/Mark V 400 V-8 (S) 49 states

Calibration 9-17F-R00
Auto. trans.
Fast idle 2200
Curb idle
A/C on 675
A/C off 600

Lincoln Continental/Mark V 400 V-8 (S) California only

Calibration 9-17P-R0
Auto. trans.
Fast idle 2200
Curb idle
A/C on 650
A/C off 650
Calibration 9-17Q-R0
Auto. trans.
Fast idle 2300
Curb idle
A/C on 650
A/C off 600

Versailles 302 V-8 (F) 49 states

Calibration 8-11D-R0
Auto. trans.
Fast idle 1900
Curb idle
A/C 625
Calibration 8-11L-R11
Auto. trans.
Fast idle 3100
Curb idle
A/C 625
VOTM*
A/C 750

Versailles 302 V-8 (F) California only

Calibration 9-11P-R0/9-11P-R1
Auto. trans.
Fast idle 2900
Curb idle
A/C 625
VOTM*
A/C 800
Calibration 9-11V-R0
Fast idle 1800
Curb idle
A/C on 675
A/C off 600
VOTM*
A/C 700

* VOTM (Vacuum Operated Throttle Modulator)
** TSP (Throttle Solenoid Positioner)

NOTE: *All curb idle settings are made at the solenoid with the solenoid either energized or deenergized.*

IDLE SPEED

Various throttle positioners are used depending on vehicle application. Curb idle adjustments are made at the throttle positioners and not at the "off idle" adjusting screw.

A combination of devices can be used including a solenoid-dashpot, a solenoid-diaphragm, a vacuum operated throttle solenoid, and a vacuum operated throttle kicker actuator.

When the A/C-heater control is positioned in the A/C mode the throttle kicker solenoid is energized thereby allowing intake manifold vacuum to reach the throttle kicker actuator which is positioned against the throttle lever. When a vacuum signal is applied to the throttle kicker actuator, it will increase the engine RPM for increased cooling and smoother idle. The throttle kicker actuator is also energized during engine warm-up or if an engine overheat condition exists.

1980

See the Ford Calibration Service Specifications.

IGNITION TIMING AND SPARK PLUG APPLICATION

1979

Ford/Mercury 302 V-8 (F) 49 states

Calibration 9-11E-R1/Auto. trans./ with or without A/C
6° BTDC @ 500 rpm
ASF-52 .048-.052

Ford/Mercury Police station wgn. 302 V-8 (F) 49 states

Calibration 9-11F-R1/Auto. trans./ with or without A/C
6° BTDC @ 500 rpm
ASF-52 .048-.052

Ford/Mercury 302 V-8 (F) California only

Calibration 9-11Q-R01/9-11Q-R10
6° BTDC @ 500 rpm
ASF-52-6 .058-.062

Ford/Mercury 351 V-8 (H) 49 states

Calibration 9-12G-R0/Auto. trans.
10° BTDC @ 800 rpm
ASF-52 .048-.052
Calibration 9-12F-R0/Auto. trans.
15° BTDC @ 800 rpm
ASF-52 .048-052
Calibration 9-12A-R0/9-12B-R0/ Auto. trans.
Timing is not adjustable (equipped with EEC)
ASF-52 .048-.052

Ford/Mercury 351 V-8 (H) California only

Calibration 9-12N-R0/9-12P-R0/ Auto. trans.
Timing is not adjustable (equipped with EEC)
ASF-52 .048-.052

Granada/Monarch 302 V-8 (F) 49 states

Calibration 9-11C-R0/9-11C-R1/ Auto. trans.
8° BTDC @ 500 rpm
ASF-52 .048-.052
Calibration 9-10C-R0/Manual trans.
12° BTDC @ 500 rpm
ASF-52 .048-.052
Calibration 9-10C-R10/Manual trans.
8° BTDC @ 500 rpm
ASF-52 .048-.052

Granada/Monarch 302 V-8 (F) California only

Calibration 9-11S-R0/Auto. trans.
12° BTDC @ 500 rpm
ASF-52-6 .058-.062

Granada/Monarch 250 I-6 (L) 49 states

Calibration 9-9A-R0/9-9A-R12/Auto. trans.
10° BTDC @ 750 rpm
BSF-82 .048-.052
Calibration 9-8A-R0/Manual trans.
4° BTDC @ 750 rpm
BSF-82 .048-.052

Granada/Monarch 250 I-6 (L) California only

Calibration 9-9P-R0/Auto. trans.
6° BTDC @ 750 rpm
BSF-82 .048-.052

Cougar/LTD II/Thunderbird 302 V-8 (F) 49 states

Calibration 9-11J-R0/Auto. trans.
8° BTDC @ 500 rpm
ASF-52 .048-.052

Cougar/LTD II/Thunderbird 351 V-8 (H) 49 states

Calibration 9-12E-R0/Auto. trans.
15° BTDC @ 600 rpm
ASF-42 .048-.052
Calibration 9-14E-R0/Auto. trans.
12° BTDC @ 800 rpm
ASF-52 .048-.052

Cougar/LTD II/Thunderbird 351 V-8 (H) California only

Calibration 9-14Q-R0/Auto. trans.
14° BTDC @ 800 rpm
ASF-52 .048-.052

Lincoln Continental/Mark V 400 V-8 (S) 49 states

Calibration 9-17F-R00/Auto. trans.
14° BTDC @ 800 rpm
ASF-52 .048-.052

Lincoln Continental/Mark V 400 V-8 (S) California only

Calibration 9-17P-R0/Auto. trans.
14° BTDC @ 800 rpm
ASF-52-6 .058-.062
Calibration 9-17Q-R0°Auto. trans.
16° BTDC @ 800 rpm
ASF-52 .048-.052

Versailles 302 V-8 (F) 49 states

Calibration 8-11D-R0/Auto. trans.
30° BTDC @ 625 rpm (not adjustable)
ASF-52 .048-.052
Calibration 8-11L-R11/Auto. trans.
15° BTDC @ 625 rpm (not adjustable)
ASF-52 .048-.052

Versailles 302 V-8 (F) California only

Calibration 9-11P-R0/9-11P-R1/Auto trans.
Timing is not adjustable
ASF-52 .048-.052
Calibration 9-11V-R0°Auto. trans.
12° BTDC @ rpm
ASF-52-6 .058-.062

1980

See Ford service calibration specifications for 1980 timing and spark plug information.

EMISSION CONTROL SYSTEMS

ELECTRONIC ENGINE CONTROL SYSTEM 1979

This system, known as Electronic Engine Control, or EEC, is installed on all 1979 Versailles 302 V-8 engines. The system controls engine timing, exhaust gas recirculation, and air pump airflow. Seven sensors feed information into an electronic box mounted on the passenger side of the firewall near the steering column. The box, called an electronic control assembly, takes all the information, processes it and then electrically controls the timing, EGR, and air pump airflow. First, let's look at the 7 sensors and how they do their job.

Inlet Air Temperature Sensor

Mounted in the air cleaner can, near the air inlet, the temperature sensor has two wire connected to it, so current can flow from the electronic control assembly to the sensor and back. The sensor is a thermistor design, which changes resistance as the temperature changes. This affects the amount of current flowing, so that in effect the electronic control assembly knows at all times the temperature inside the air cleaner. The action of the sensor is opposite to temperature changes. In other words, as the temperature goes up, the sensor resistance goes down, and vice versa.

The information transmitted by the sensor is used by the electronic control assembly to determine the right setting for spark timing and air pump airflow. At high temperatures, the spark advance is less, to help prevent pinging.

The sensor is held in place by a clip inside the air cleaner. Resistance of the sensor should be between 6,500 and 45,000 ohms, as measured at the disconnected plug with an ohmmeter.

Throttle Position Sensor

This sensor is a variable resistor mounted on the carburetor and connected to the end of the throttle shaft. For every position of the throttle, there is a corresponding position of the sensor which gives a different current flow to the electronic control assembly. Using a throttle position sensor is almost the same as using a vacuum reading. The electronic control assembly combines the information on throttle position with information from other sensors, and then controls the spark advance, EGR flow rate, and air pump airflow.

The throttle position sensor is adjusted by loosening the mounting screws and rotating the sensor until

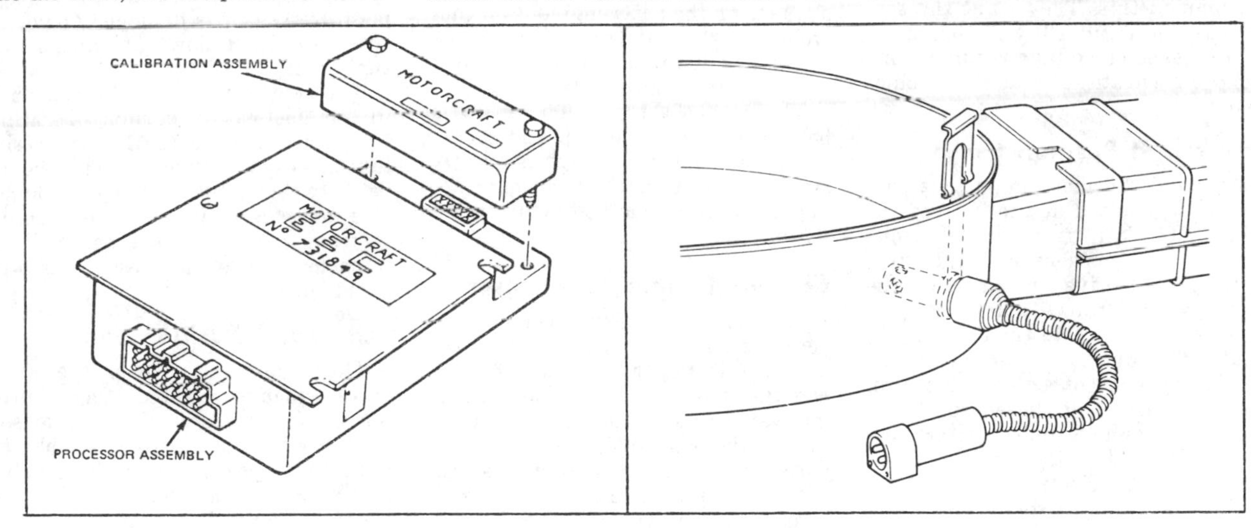

Electronic Control Unit (ECU)

Inlet Air Temperature (IAT) Sensor

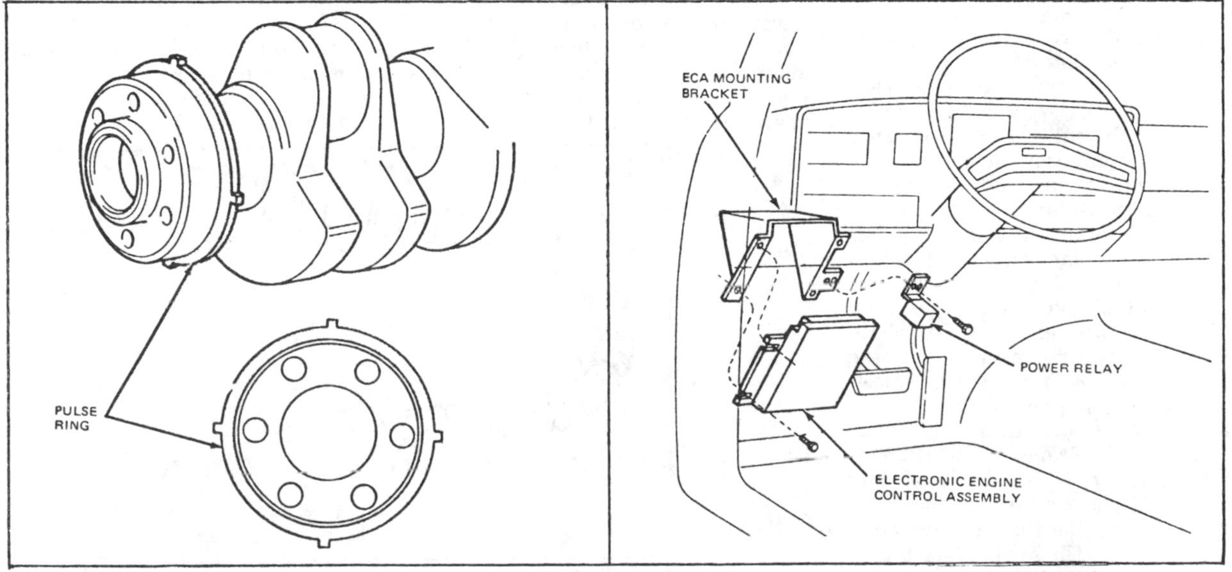

Crankshaft pulse ring

Power relay and ECA attachment

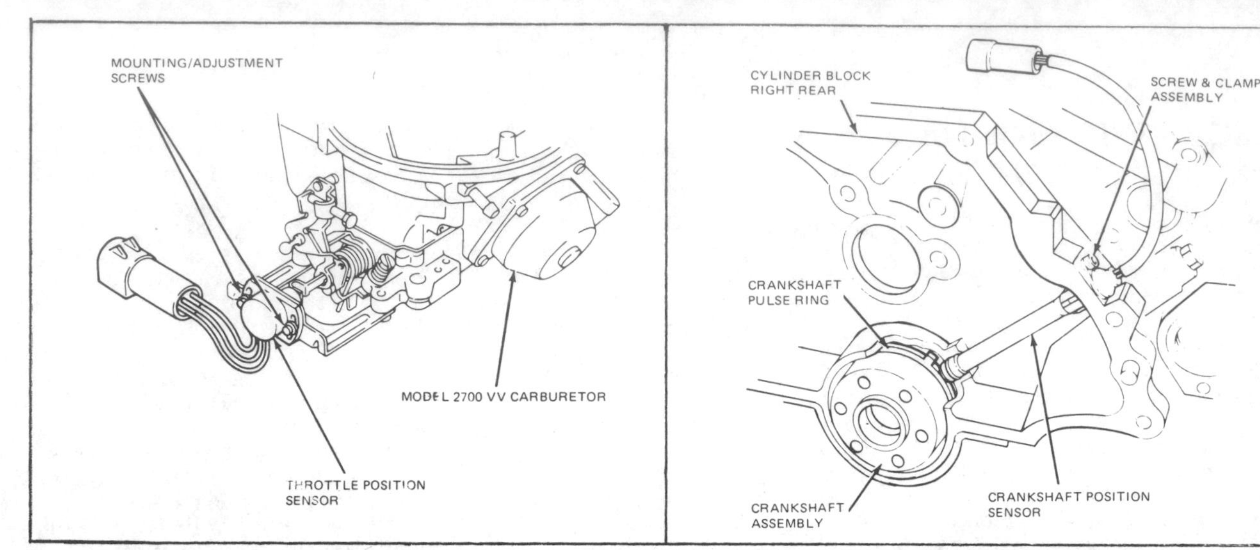

Throttle positioner Sensor

Crankshaft Position (CP) Sensor

the correct readings appear on a system analyzer. Although the sensor is a resistor, and could possibly be set using an ohmmeter, Ford does not recommend this. They want the sensor position adjusted with the analyzer to a specific voltage reading. This is more accurate than using an ohmmeter.

Crankshaft Position Sensor

Probably the most important sensor, it is mounted at the rear of the cylinder block, in front of the torque converter. If this sensor, or the wires to it are defective or damaged, the engine will not run. A pulse ring, permanently attached to the crankshaft, rotates just under the tip of the sensor. Every time one of the lobes on the ring passes by the sensor tip it creates a magnetic pulse in the electromagnet inside the sensor. With this signal, the electronic control assembly knows the engine rpm and the position of the crank, so it can set the spark timing and EGR flow rate to exactly what the engine needs.

The signals from the crankshaft position sensor and the throttle position sensor are combined in the electronic control assembly to determine the load on the engine. This is exactly the same as taking a vacuum reading from the intake manifold, but it eliminates the need for any vacuum diaphragm or connecting hose.

A conventional engine gets its spark timing directly from the distributor, which is driven by the camshaft and timing chain. If the chain wears, the timing retards, so the timing must be periodically reset. This does not happen with the EEC system, because the timing is determined directly from the position of the crankshaft. There is never any need to adjust timing on the EEC system. An engine that is completely worn out will still have perfect timing.

A conventional engine also has problems with wear in the vacuum advance and centrifugal advance systems. The vacuum and centrifugal advance on the EEC engine is all controlled by the electronic control assembly. So it will also remain perfect over the life of the engine.

The crankshaft position sensor is held in place by a clamp and screw. Near the tip of the sensor is an "O" ring that keeps engine oil from escaping. There are no adjustments on the sensor.

Coolant Temperature Sensor

It's just an ordinary coolant temperature sensor with two wires connected to it. But it has a big effect on how the engine runs. It is mounted on the water passage at the left rear of the intake manifold. Its resistance varies with the coolant temperature, with high resistance at low temperatures, and vice versa. The sensor is connected to the electronic control assembly. When the engine is cold, the electronic control assembly cuts back on the amount of exhaust gas that is allowed to recirculate, or maybe eliminates the EGR entirely. If the engine is overheated because of long idling, the electronic control assembly advances the spark timing to increase the idle speed and help cool down the engine. If the coolant temperature sensor is damaged or defective both EGR flow rate and spark timing may be affected.

The sensor is easily checked with an ohmmeter. At normal operating temperature, (160-220°F.) the resistance at the disconnected plug should be between 1,500 and 6,000 ohms. The sensor is removed by unscrewing it from the water passage with a wrench. Relieve the pressure first to avoid burns.

EGR Valve Position Sensor

This sensor is part of the EGR valve, which is mounted on a carburetor spacer. Wires connect the sensor to the electronic control assembly, to indicate the position of the EGR valve metering rod. This indirectly tells the electronic control assembly how much exhaust gas is recirculating. The electronic control assembly compares this information with what it receives from the other sensors, and then increases or decreases the exhaust gas recirculation rate with other controls that wil be explained later. It is im-

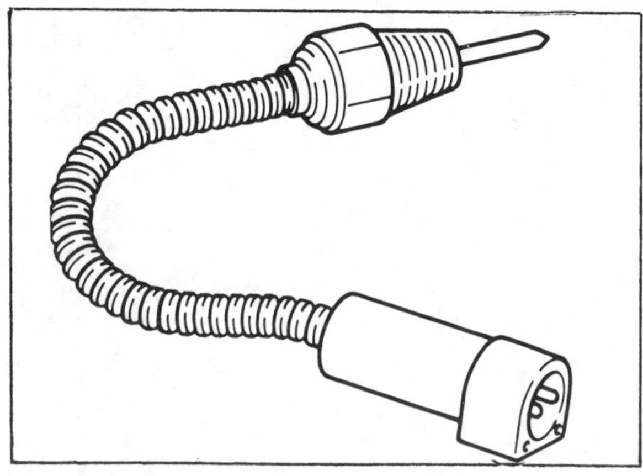

Coolant.Temperature Sensor (ECT)

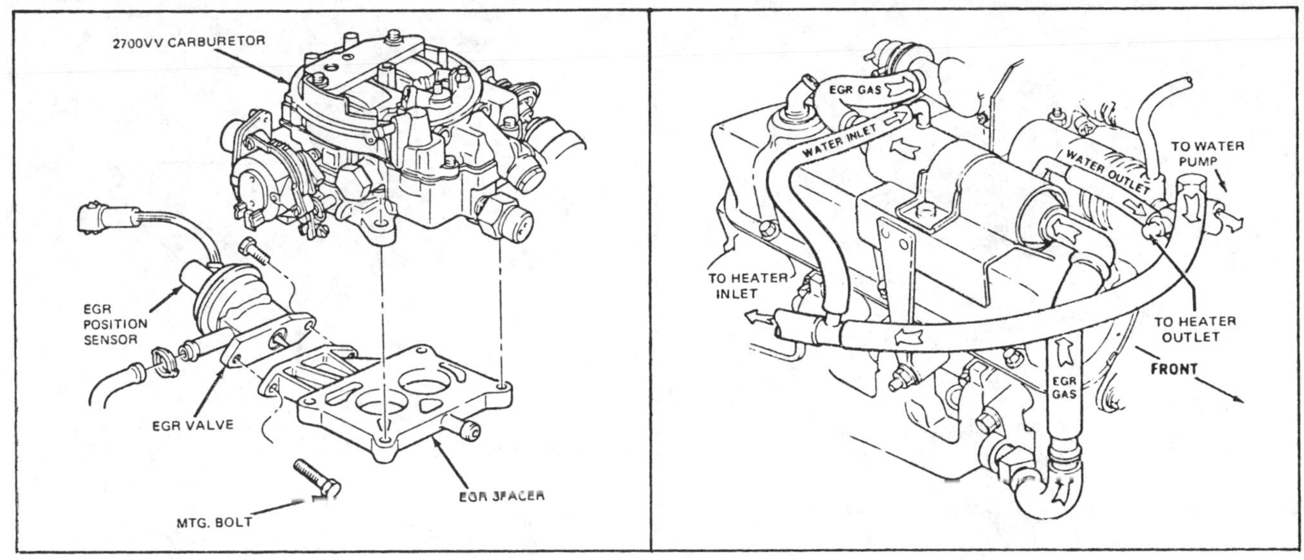

EGR valve and spacer

EGR gas cooler

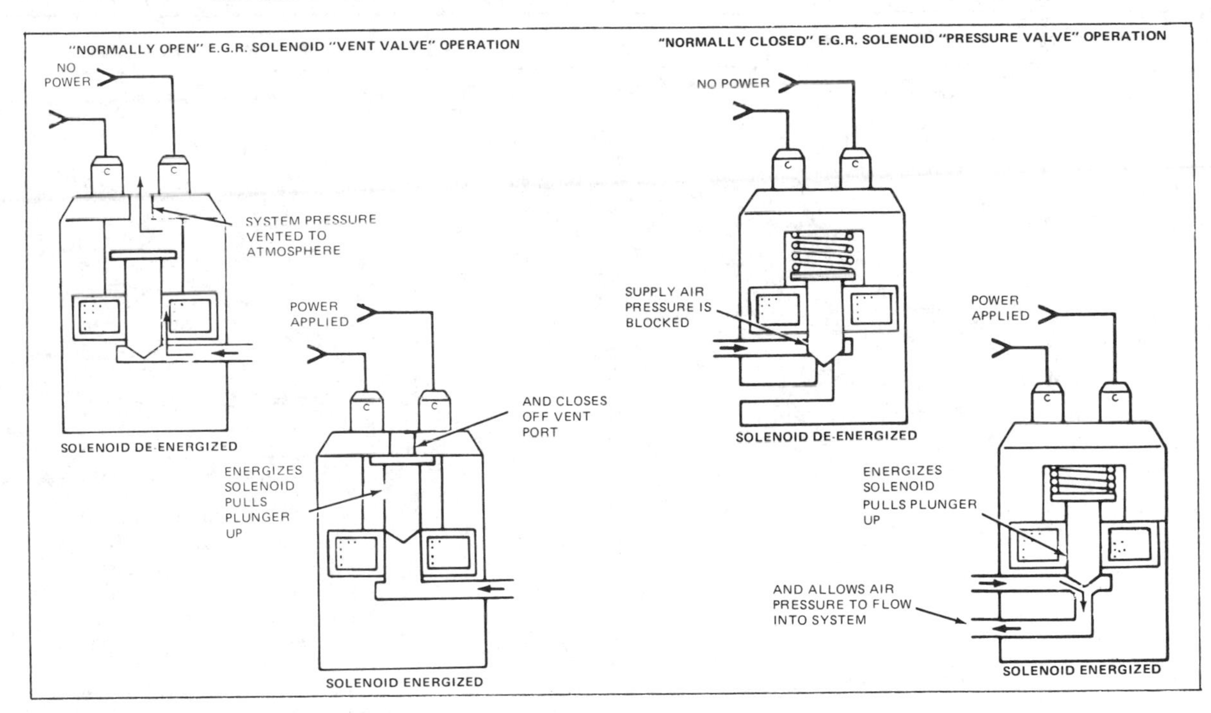

Vent and pressure solenoid

portant to realize that the EGR valve position sensor is only a sensor. It does not actually move the metering rod or directly change the flow rate.

The sensor can be checked with an ohmmeter, after disconnecting the plug. There are three wires connecting to the sensor. The ohmmeter reading between the orange-white and black-and-white wires should be between 2800 and 5300 ohms, with the engine off. Between the orange-white and brown-light green wires, the ohmmeter should read 350 to 940 ohms, with the engine off. If the readings are outside these limits, the entire EGR valve must be replaced. There are no adjustments on the sensor or the valve.

Manifold Absolute Pressure Sensor

Manifold absolute pressure is defined as barometric pressure minus manifold vacuum. The sensor is connected to intake manifold vacuum. In effect, it simply measures intake manifold vacuum and sends a signal through wires to the electronic control assembly. The electronic control assembly uses this information, along with data from other sensors, to set the spark advance and EGR flow rate.

The MAP sensor is at the left rear of the intake manifold, next to the rocker cover. Ford does not give a simple ohmmeter testing procedure, preferring that the system analyzer be used. There are no adjustments to the MAP sensor.

Barometric Pressure Sensor

This sensor has only one purpose, to tell the electronic control assembly the altitude at which the car is operating.

At high altitude, the EGR flow rate is cut back to keep good engine performance. The sensor connects by wires to the electronic control assembly. A simple ohmmeter test is not provided for the barometric sensor. It must be tested with the sytem analyzer. There are no adjustments.

EEC Distributor

Because the electronic control assembly takes care of all timing and

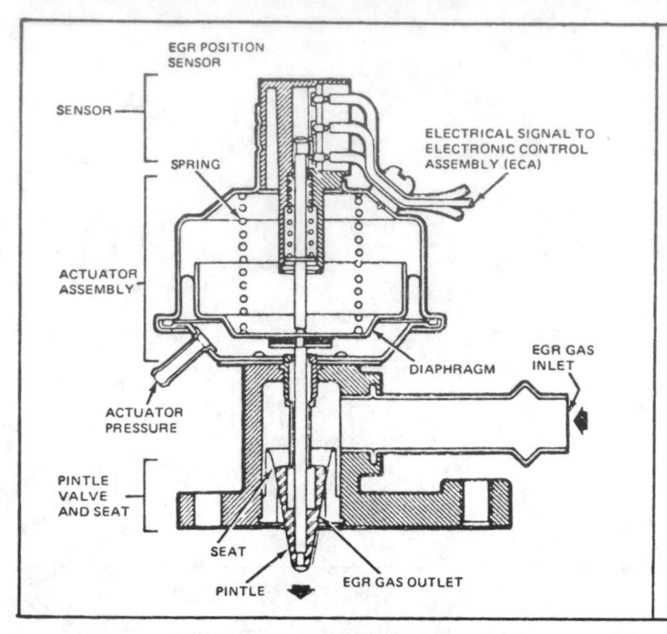

EGR valve used with EEC system

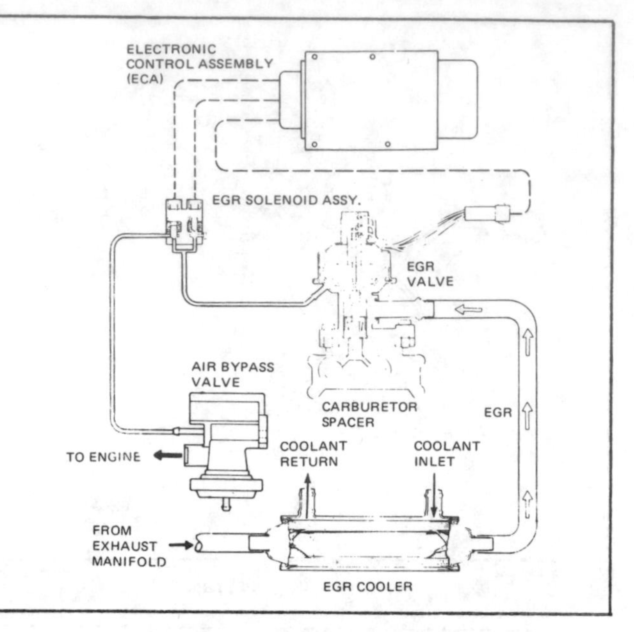

Typical EGR system used with EEC system

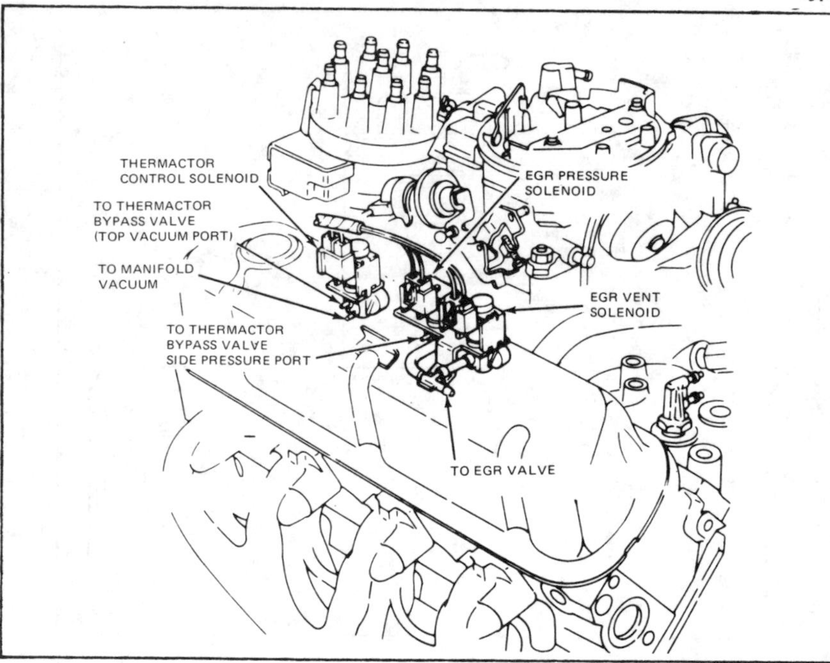

Dual EGR control solenoids

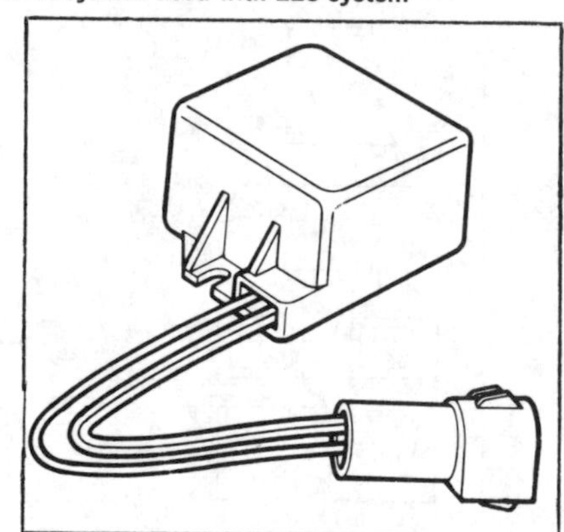

Barametric Pressure (BP) Sensor

advance, the distributor has no vacuum advance, no centrifugal advance, and no timing mechanism. The distributor is nothing more than a rotor and some electrodes that distribute the spark to the correct spark plug wires.

The distributor cap and rotor are a completely new two-level design. Four of the plug wire electrodes are at one level, and the other four are a little bit lower. This puts more distance between the electrodes, and helps prevent crossfiring. In addition, a double-ended rotor is used, so that the spark jumps first from one end of the rotor, and then from the other end. This gives additional separation to prevent crossfiring. The construction of the rotor is such that the spark current actually has to jump twice to get to the plug wire electrode in the cap. It jumps once from the center tower to the pick up drum of the rotor, and a

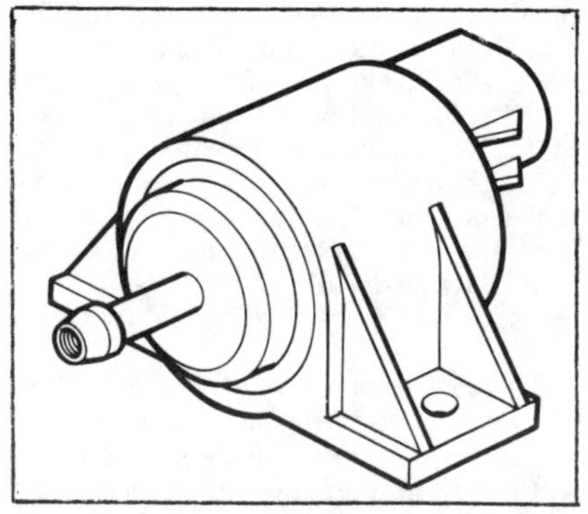

Manifold Absolute Pressure (MAP) Sensor

second time from the top of the rotor to the cap electrode. When you look inside the cap you will see a 4-spoke electrode plate connected to the center tower, and a rotor with two pick-up arms. This plate is positioned so that when the correct rotor pickup arm is over one spoke, the other rotor pickup arm is between the spokes. This was also done to decrease the possibility of crossfire.

Because the distributor has nothing to do with timing, there are no adjustments for distributor position. The clamp and bolt at the base of the distributor fit into a slot, locking the distributor in one position. But there is a rotor position adjustment. A slot in the rotor must be opposite a slot in the housing, when number one cylinder is at top dead center of the compression stroke. This adjustment is made by removing the clamp bolt, lifting up the distributor, and meshing the drive gear teeth in a different position so that the slots line up. After the rotor is as close as possible this way, the rotor mounting screws can be loosened and the rotor moved to bring it into exact alignment with

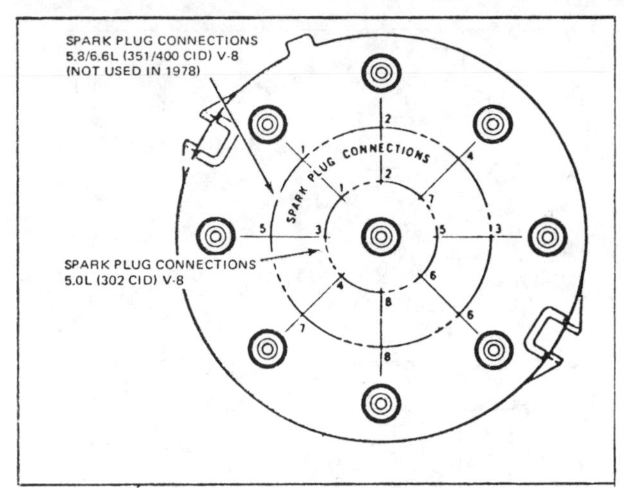

Spark plug wire connections

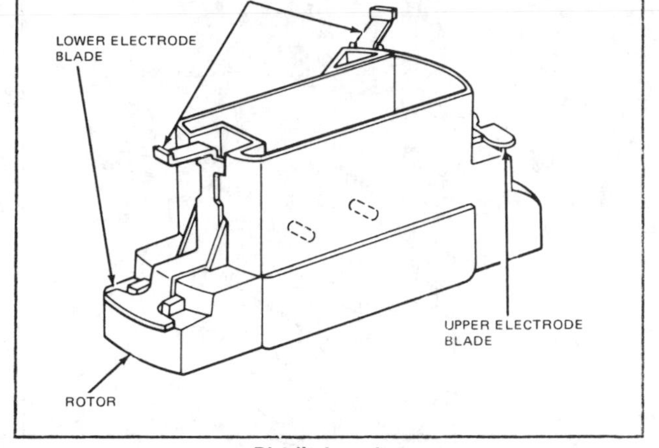

Distributor rotor

the slot. Ford has an official tool T78P-12200-A which fits into the rotor and housing slots at the same time and holds the rotor so it doesn't move while the screws are tightened. The tool is available from Owatonna Tool Company, Owatonna, Minnesota.

Because of the unique 2-level construction of the distributor cap and rotor, the plug wires do not fire in succession around the cap. The rotor fires first from one end, and then the other. This means that although the rotor only turns 45° between firings, the actual firing succession jumps back and forth across the cap. The numbers on the cap are the numbers of the cylinders, and must be connected to those cylinders. If you try to connect the wires in firing order succession around the cap, the engine will not run. Also, there are two sets of numbers on the cap. The inner ring of numbers is for the 302 V-8. The outer ring is for the 351 or 400 V-8, if it ever gets the EEC system. Because four of the numbers are different, the engine will not run if the plug wires are connected according to the wrong number ring.

A surprising thing about this EEC system is that there is no timing setting on the engine. The timing is done by the crankshaft position sensor and the electronic engine control. At one end of the electronic control there is a small trim adjustment, but it only changes the timing a few degrees. The trim adjustment is sealed, and if the seal is broken, free replacement of a defective electronic engine control under warranty may be refused.

Air Pump Airflow Control

A vacuum solenoid, mounted to the left of the intake manifold, is connected by wires to the electronic engine control. The solenoid vacuum path is a normally closed design. The electric coil in the solenoid is energized to open the vacuum passage by current from the electronic engine control. In the open position, the solenoid allows intake manifold vacuum to open the air bypass valve so that air from the pump can flow to each engine exhaust valve. When the solenoid is deenergized, the valve plunger moves to the closed position, shutting off the vacuum to the bypass valve, which then goes into the dump position. Air pump air is then diverted to the atmosphere.

The electronic engine control turns the solenoid on or off according to signals it receives from the inlet air temperature sensor, and the throttle position sensor.

The hose should be connected to the solenoid with the lower outlet to manifold vacuum, and the upper outlet to

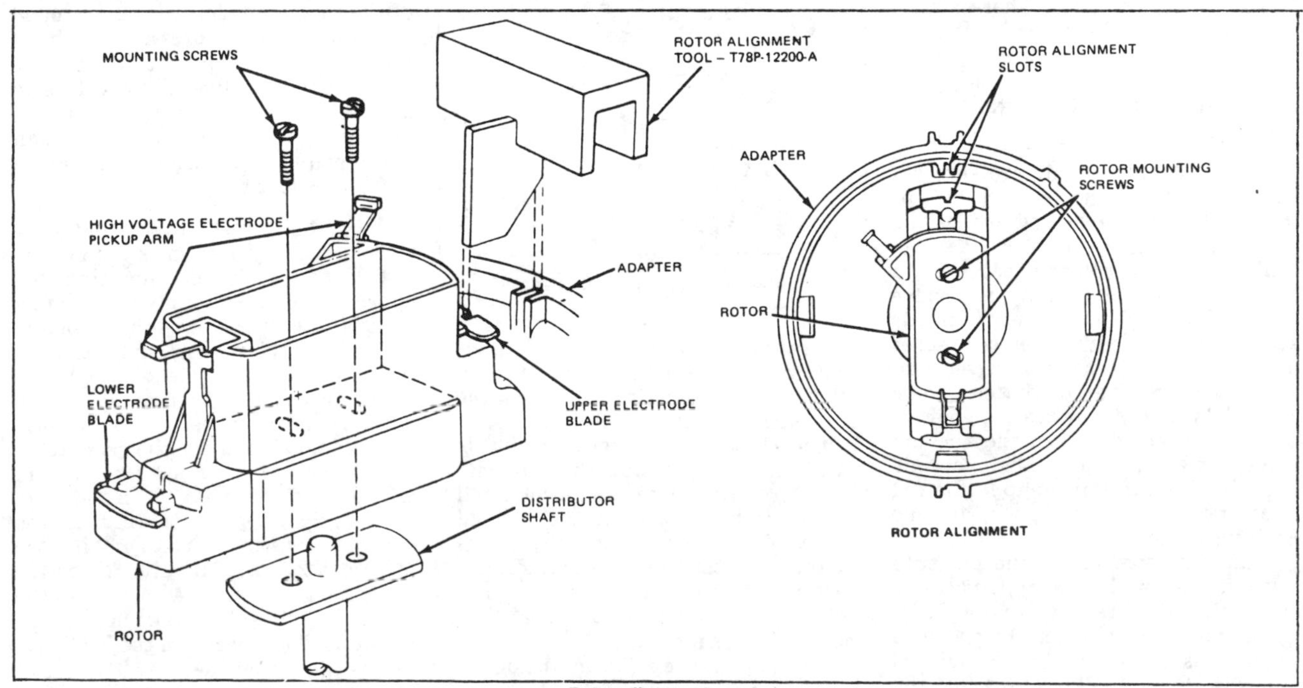

Rotor alignment

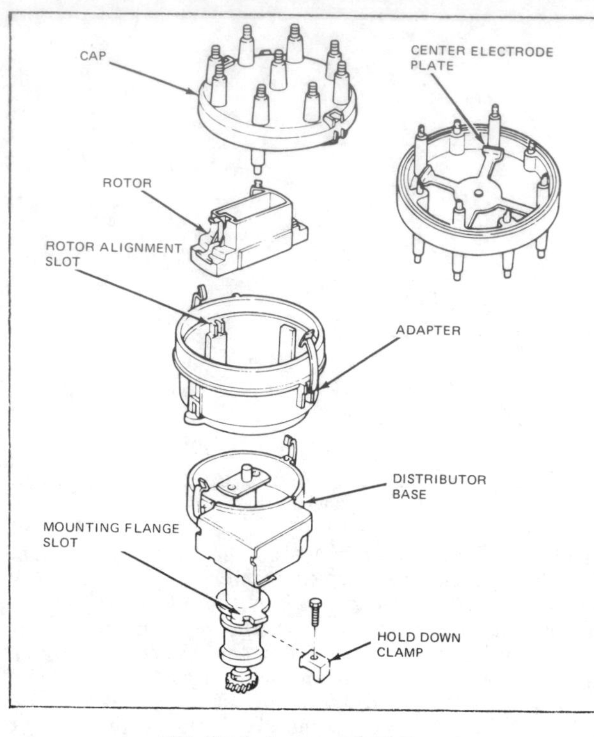

EEC distributor cap and rotor

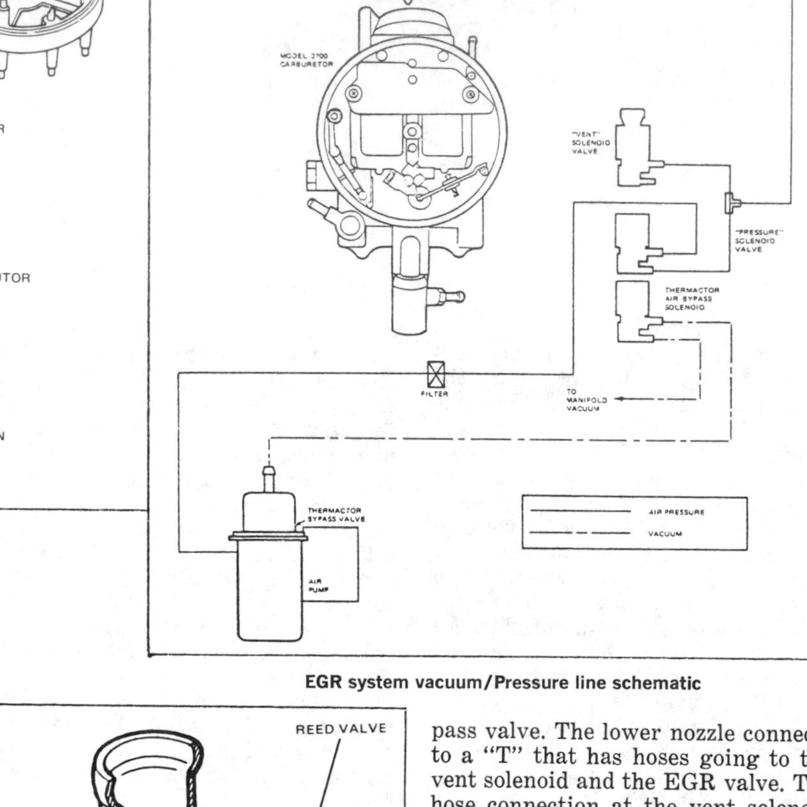

EGR system vacuum/Pressure line schematic

the bypass valve.

Testing the solenoid can be done by idling the engine with a vacuum gauge connected to the upper outlet, after removing the bypass valve hose. When running the engine above idle in PARK, you should get a full manifold vacuum reading on the gauge. Let the engine come down to idle, and in a few seconds the solenoid should close, and the gauge should drop to zero. There is a delay built into the system so the air pump does not pump until the engine has been at idle for a few seconds. There are no adjustments on the solenoid. It must be replaced if it doesn't work right.

Exhaust Gas Recirculation System

The EGR valve is air-pressure operated. A hose connects to the side of the air bypass valve and brings air pump air pressure to the EGR valve. A dual vacuum solenoid setup is in the hose between the bypass valve and the EGR valve. The two solenoids are mounted together on a bracket next to the left rocker arm cover, and connected by wires to the electronic control assembly. The rear solenoid is the vent solenoid. When the vacuum valve inside the solenoid is open, air pressure from the air pump is allowed to escape into the atmosphere and the EGR valve closes.

The front solenoid is the pressure solenoid. When its vacuum valve is open, air pressure goes to the EGR valve and opens it. When the pressure solenoid is closed, the air from the pump is shut off. This also holds the pressure in the hose to the EGR valve, and makes the EGR valve stay in whatever position it was when the pressure solenoid was closed.

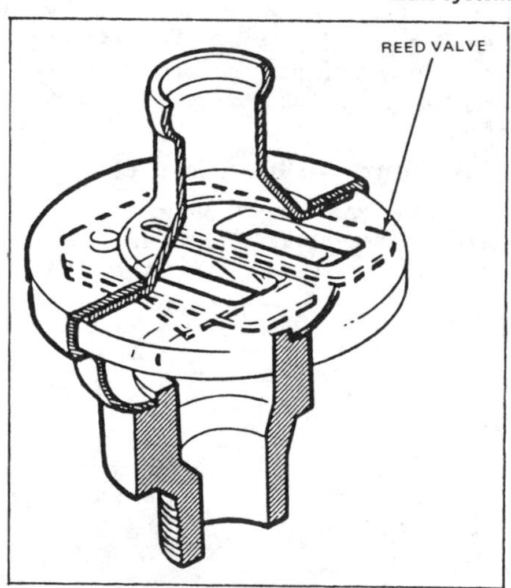

Suction air system inlet valve

The electronic control assembly is continuously analyzing the information is receives from the sensors, and changing the positions of the solenoid valves to keep the flow of exhaust gas at the level the engine needs. Under certain conditions, there can be quite a bit of clicking from the solenoids, but this is normal.

On the pressure solenoid, the upper nozzle connects to the air pump bypass valve. The lower nozzle connects to a "T" that has hoses going to the vent solenoid and the EGR valve. The hose connection at the vent solenoid should be made to the upper nozzle.

The rear, vent solenoid is a normally open design. To test it for leaks, you must first disconnect the electric plug and apply battery voltage to one of the solenoid terminals, and ground the other. Then you can use a hand pressure pump to apply a few pounds pressure to the valve. These valves do not make a perfect seal, but as long as they do not leak any more than half a pound in 5 seconds, they can be considered okay.

The front pressure solenoid is a normally closed design, so it can be tested by simply disconnecting the upper hose and applying a few pounds pressure from a hand pump. If you have the system analyzer, closing or opening the solenoids for testing is simply a matter of pushing a button, which makes it a lot easier.

Spacer plates under the carburetor have been used by Ford for years in EGR systems. A passage in the plate allows exhaust gas to travel from the exhaust crossover passage in the manifold to the EGR valve. From the valve the gas is metered back into the spacer and into the intake part of the manifold. The spacer on the EEC system is a different design. It does not connect to the exhaust crossover pas-

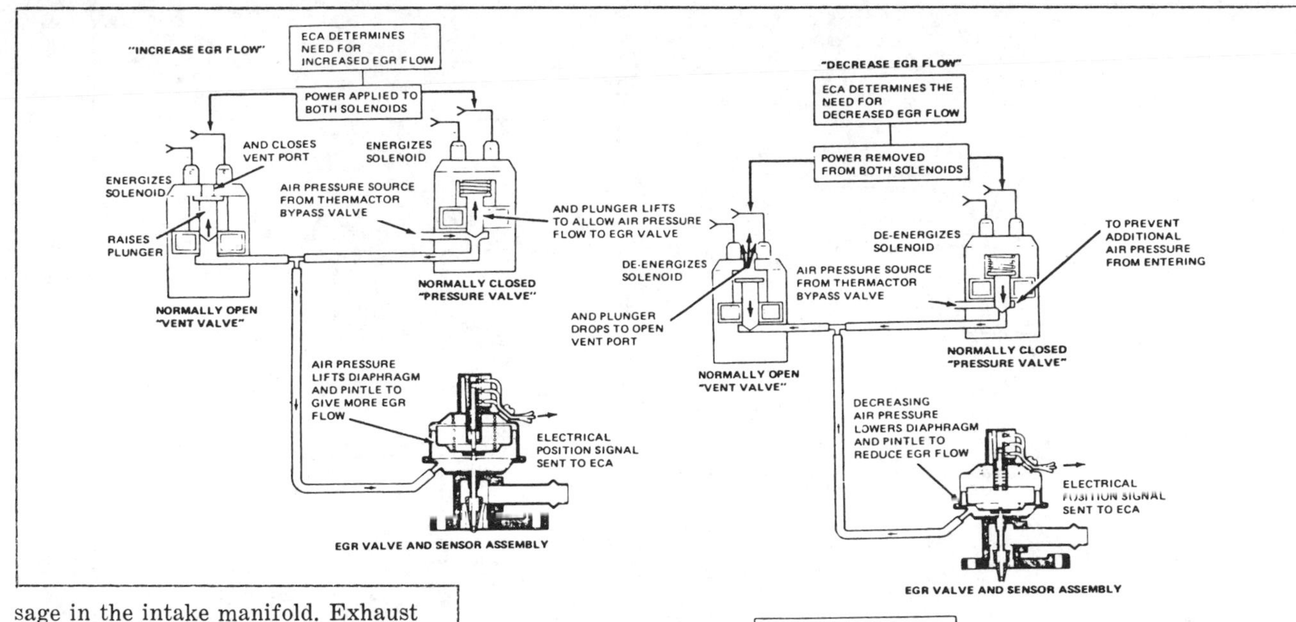

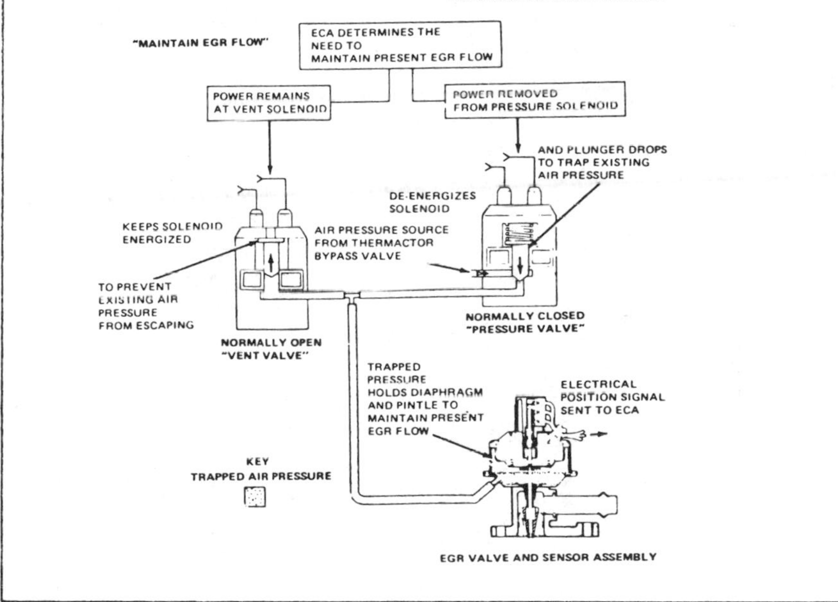

Operation of the EGR system (with EEC)

sage in the intake manifold. Exhaust gases are brought to the EGR valve and allow the gases to flow into the intake part of the manifold.

The tube that carries exhaust gas to the EGR valve is attached to a fitting at the front of the right exhaust manifold. Above the right rocker cover is an EGR cooler, which is connected to engine coolant. Engine coolant flows through the cooler and reduces the temperature of the exhaust gases. From the cooler, a tube takes the cooled exhaust gases to the EGR valve. The EGR valve has two hose connections. The large size hose is for the exhaust gases, and the small size is for pressure to operate the valve.

Diagnostic Testor

This analyzer, available from Owatonna Tool Company, is a plug in unit that will quickly analyze most of the EEC system. Additional equipment needed are an advance-reading timing light, a tachometer, and a vacuum-pressure gauge. The analyzer is made in two parts, and each part must be purchased separately. An advantage is that the voltmeter part can also be used with other Ford analyzers.

EEC Diagnostic Tester
T78L-50-EEC-1
Digital Volt-Ohmmeter
T78L-50-DVOM-1
Owatonna Tool Company
118 Eisenhower Drive
Owatonna, Minnesota 55060

ELECTRONIC ENGINE CONTROL SYSTEM 1980

The 1980 Electronic Engine Control System is basically the same as 1979; however, they added an Ex-

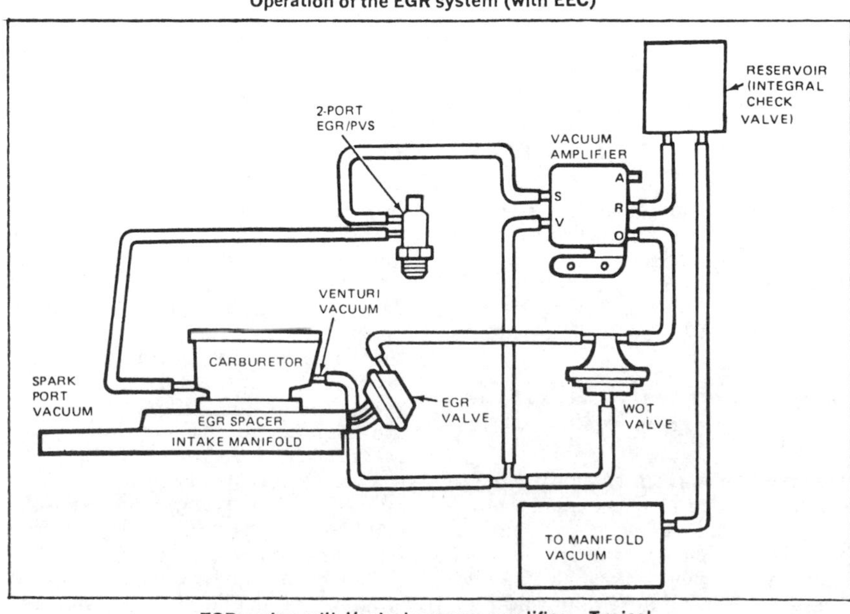

EGR system with Venturi vacuum amplifier—Typical

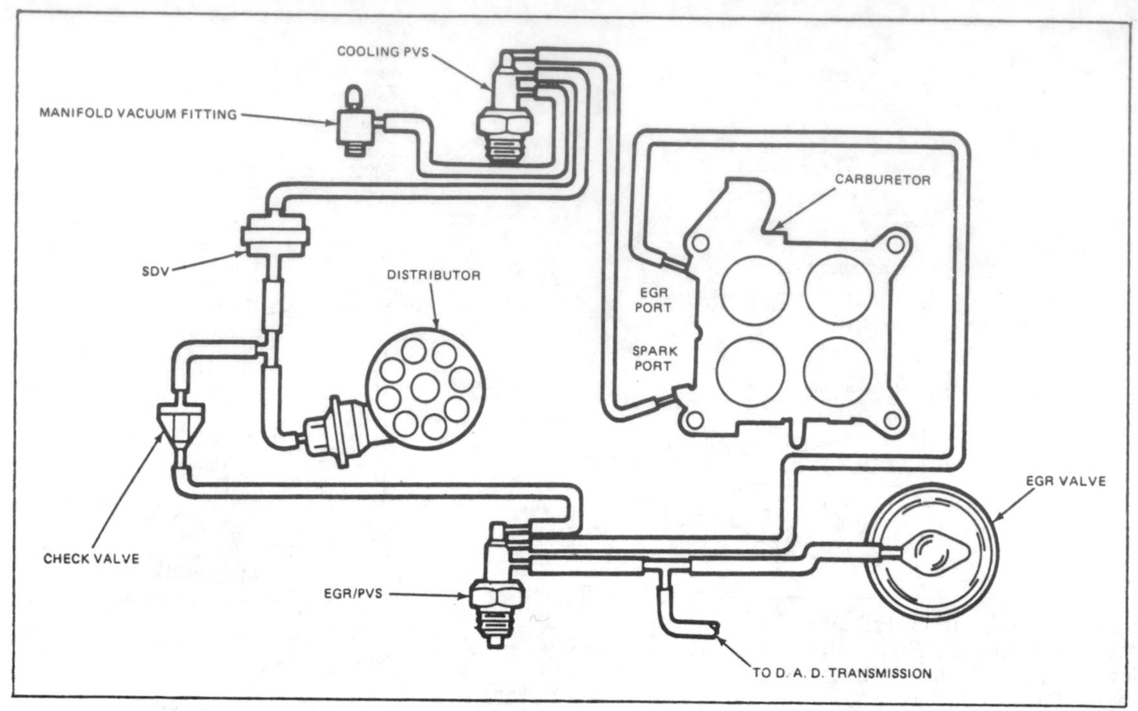

Exhaust gas recirculation and coolant spark control (EGR/CSC) system

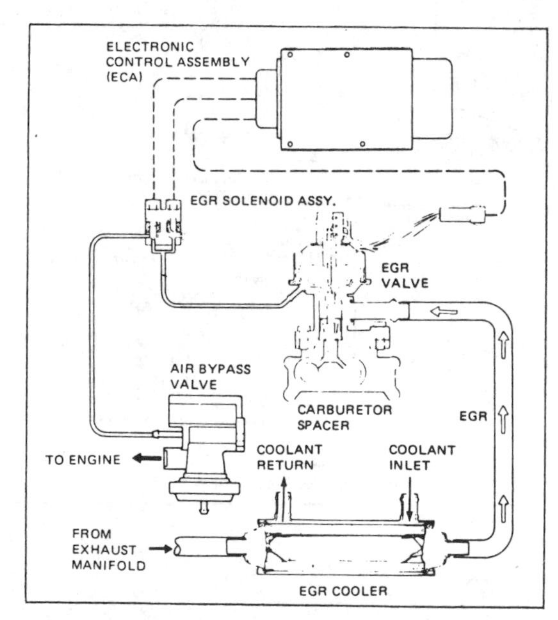

EGR system (with EEC system)—Typical

haust Gas Oxygen Sensor EGO. The system may be combined with Electronic Fuel Injection or Venturi Valve Carburetion. In addition to use on the 302 V-8 in California the EEC System is on all 1980 351 V-8 engines.

SUCTION AIR SYSTEM 1979-80

A hose runs from the clean side of the air cleaner to the exhaust manifold. Mounted about midway in the hose is a silencer and an air inlet valve, which works ilke a check valve. Suction in the exhaust manifold pulls in air from the air cleaner. This air oxidizes the hydrocarbons and carbon monoxides into harmless carbon monoxide and water vapor. The air inlet valve prevents backflow of the exhaust into the air cleaner. Because there is a lot of pulsating in the exhaust, it is normal for the air inlet valve to vibrate, especially at idle. The silencer prevents these vibrations from reaching the air cleaner.

This system is used only on the 200 in. 6-cylinder engine in the Fairmont and Zephyr. To check the system, just remove the silencer and check the air inlet valve to be sure it is vibrating at idle. There should not be any backflow of exhaust out of the valve. The official test for the valve is to use a rubber bulb tester with the engine off. Squeeze the bulb, and if the bulb stays collapsed, it means the valve is okay. If the valve does not vibrate, the system may be clogged, or disconnected.

CATALYTIC CONVERTERS 1979-80

49-State 200 6-cylinder and 302 V-8 engines use a new clamshell type converter that mounts on a bracket attached to the rear engine mount. California 302 V-8 engines also use the clamshell converter except for Granada and Monarch, which use the older design.

Calif. 200 6-cylinder engines use a new mini converter, in the exhaust pipe close to the exhaust manifold. The mini converter gives better control of emissions during warmup, because it gets hot faster.

DUAL MODE IGNITION TIMING 1979-80

A special control assembly takes the place of the usual electronic ignition control. The control assembly retards the spark whenever the vacuum is low on 49-State cars, or the altitude is low on Altitude cars. You can recognize the special control because it has three plug connections, instead of the usual two. Also, there is a vacuum or barometric switch mounted next to

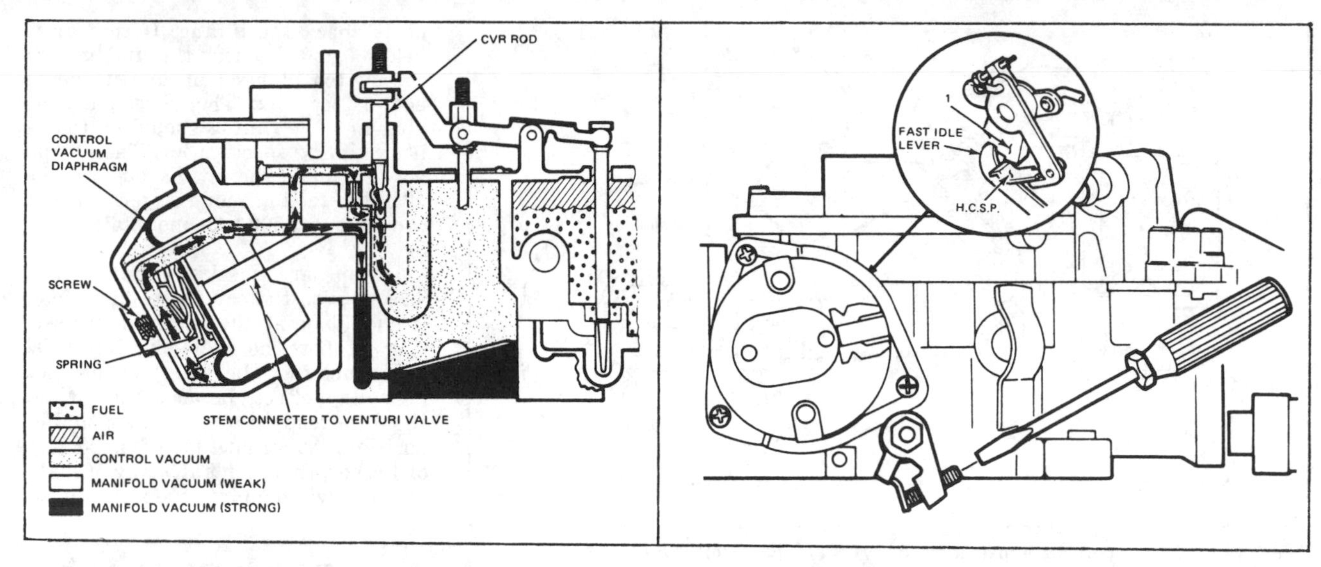

Carburetor control vacuum circuit (© Ford Motor Co.)

Fast idle speed adjustment (© Ford Motor Co.)

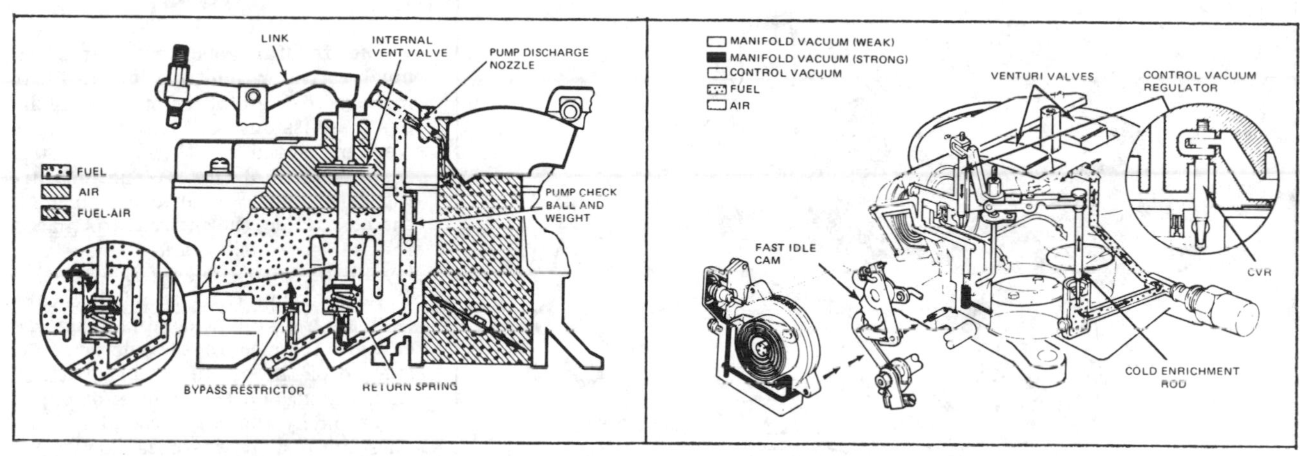

Accelerator pump circuit (© Ford Motor Co.)

Cold enrichment circuit (© Ford Motor Co.)

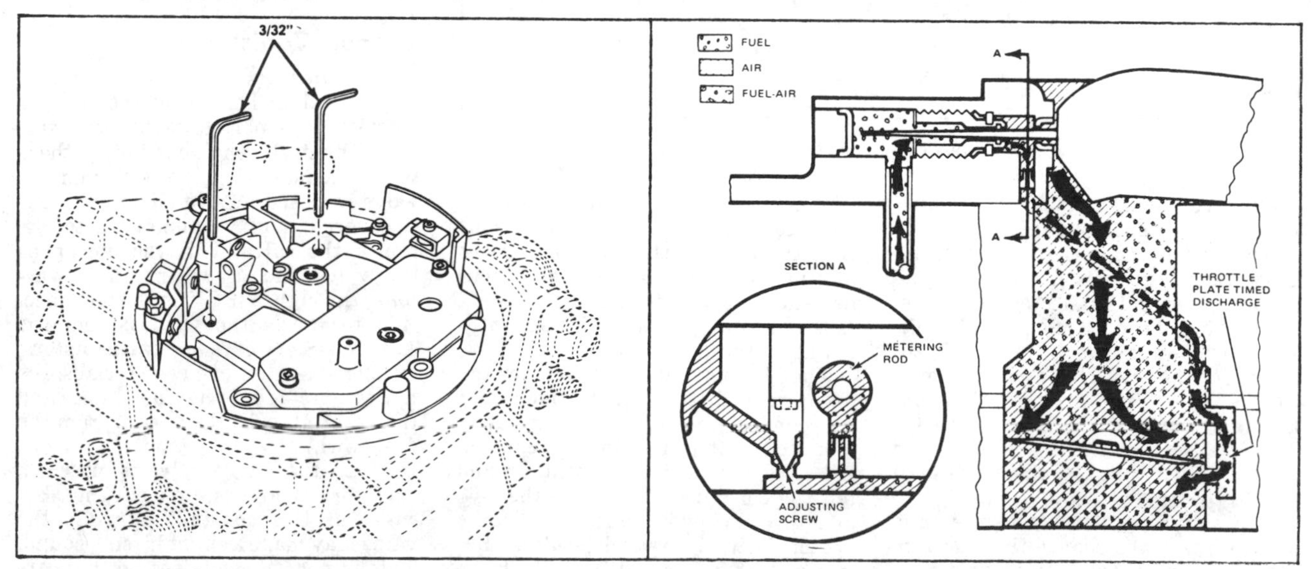

Idle trim adjustment (© Ford Motor Co.)

Idle trim system (© Ford Motor Co.)

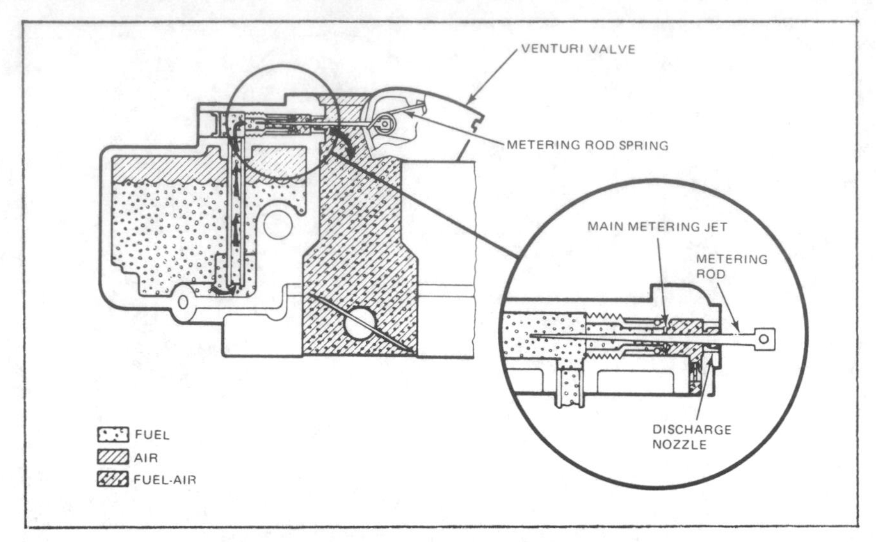

Carburetor main metering system (© Ford Motor Co.)

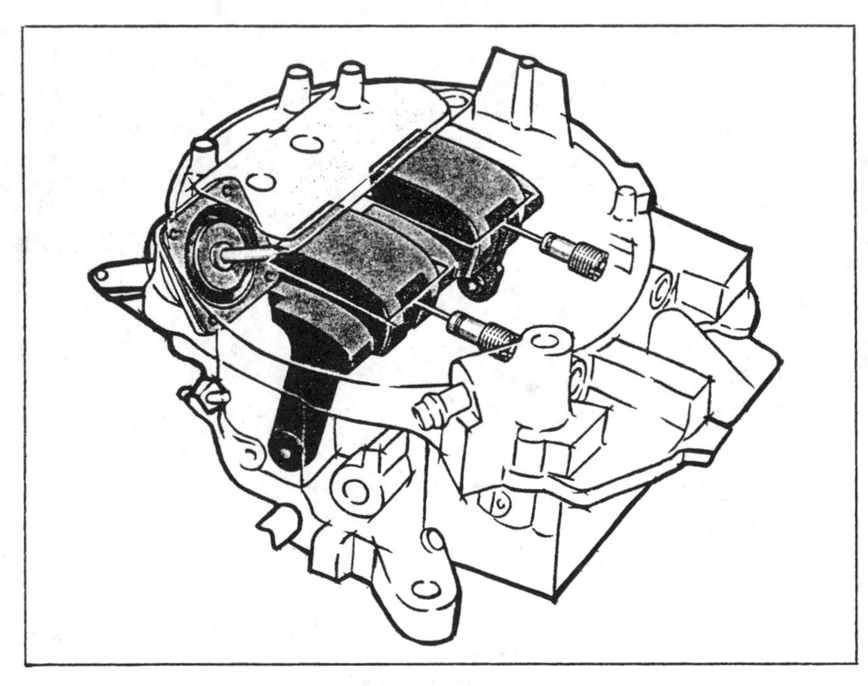

Motorcraft 2700VV carburetor (© Ford Motor Co.)

the control, or in the engine compartment on the fender panel.

The vacuum switch used with 49-State calibrations is mounted on the fender panel and connected to a three-terminal plug on the control unit. When vacuum above 10 in. Hg. is applied to the switch, a signal goes through the wire to the ignition control, and the spark is retarded 3-6°. This means that during acceleration at wide open throttle, the spark will retard to prevent detonation.

A peculiar feature of the system is that some cars have the vacuum switches connected to manifold vacuum, and some to ported vacuum. With a ported vacuum connection, the system will go into 3-6° retard at idle. But with a manifold vacuum connection, there will be normal timing at idle.

When checking the initial timing on these engines, Ford recommends that the 3-terminal plug coming out of the ignition control be disconnected. This will insure that the engine is running at basic timing, without the retard.

The barometric pressure switch used with altitude calibrations does not connect to engine vacuum. It is mounted next to the ignition control. All it does is sense the altitude and tell the control when to give the 3-6° of retard. Below 2,400 ft. altitude, the switch is in the retard position. Between 2,400 and 4,300 ft. it can be in either position. Above 4,300 ft. the switch is in the basic timing position.

Testing the vacuum switch is easy, because you can apply vacuum to it and watch the timing change. But with the altitude switch, all you can do is check the timing. If the car is below 2,400 ft. check the timing with the switch plugged in, and it should be 3-6° retarded. Then disconnect the plug and the timing should go to the basic timing specification. There is no way to make the switch go into the altitude setting, unless you have a convenient 4,300-foot mountain nearby.

If you are checking the altitude switch with the car above 4,300 feet, it should be at the basic timing setting. If not, the switch is defective. Disconnecting the plug should not have any effect, because the switch should already be in the basic timing position. At an altitude of 4,300 feet or higher, there isn't any way of making the switch go into the retard position.

VAPOR CONTROLS 1979-80

The federal government imposed much stricter controls on the car makers 1978 and later models. In the past, emissions of gasoline vapor were measured by attaching hoses to certain parts of the car and measuring how much vapor came through the hose. Now, the entire car is placed in a sealed housing, and the new test is called Sealed Housing Evaporative Determination (SHED). This means that all the vapors coming from the car are trapped in the enclosure. A sample of the air in the enclosure is then taken, and the amount of vapor given off by the car is calculated. To pass the test, it was necessary to redesign the vents on the carburetor and the carbon canister.

Carbon Canister

All canisters now have a purge control valve. The top of the valve is connected to ported carburetor vacuum. When the engine is idling, there is no vacuum from the port, and the valve is closed. Above idle, the port is uncovered, and vacuum goes to the top of the valve. This vacuum opens the valve and allows manifold vacuum through a PCV system connection to purge the vapors from the canister. Some cars use two canisters, arranged so that one canister absorbs the vapors from the fuel tank, and the other absorbs the vapors from the fuel bowl.

When the purge valve is closed, a small orifice allows some purging. Because of this constant purging, the valve may make a slight hissing sound. This is normal, and does not indicate that the valve is leaking. When the valve is open it will make a strong hissing sound if it is lifted off the canister. This hissing sound above idle means that it is working correctly. But if the strong hissing sound is

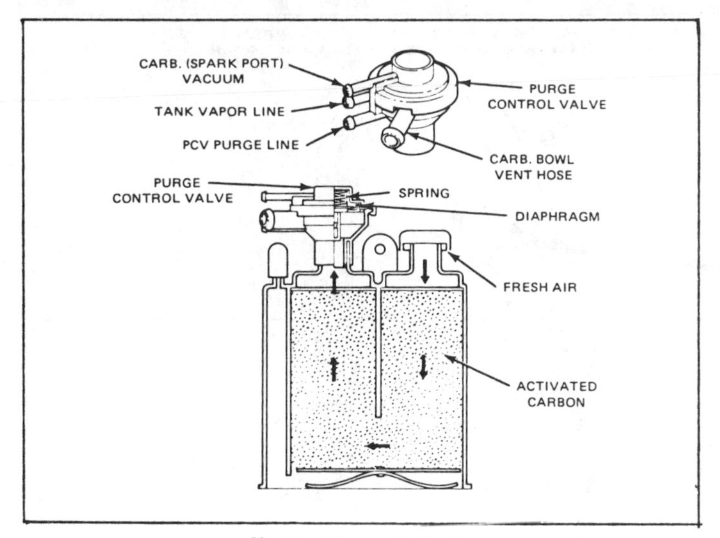

Vapor purge control valve

heard at idle, it usually means the purge valve is stuck open, and should be replaced. An exception to this is those engines that have the top of the purge valve connected to intake manifold vacuum. On those engines the valve is open at idle and above idle, and the canister is purged continuously. To test the valve on those systems, disconnect the hose to the top of the purge valve, and the purging should stop.

Some engines use a ported vacuum switch mounted on a coolant passage. It shuts off the vacuum to the top of the purge valve when the engine is cold. This PVS may be a two-nozzle design, or a 4-nozzle design. If it has 4 nozzles, only two of them are used for the purge valve. The other two are for another emission control on the engine.

Some engines use a purge regulator valve in the purge hose between the carbon canister and intake manifold vacuum. This valve looks like a PCV valve. If intake manifold vacuum increases to where too much air would flow through the purge hose, the valve limits the flow to prevent leaning the mixture.

Some engines use a system of valves and plumbing with a distributor vacuum vent valve, a retard delay valve, a reservoir, and a thermactor idle vacuum valve. The names of these valves include words such as "distributor" and "thermactor" because they were originally designed for another use. Their use in the vapor control system does not mean that they are connected to any other system. The engineers used these valves because they had difficulty preventing rich or lean mixtures while purging the vapors from the canister.

The distributor vacuum vent valve has a 40-second delay built into it. The orifice is so small that it takes that long to build up vacuum and start purging the canister. This gives the engine a change to settle down after starting wihout being upset by the purge vapors. During acceleration the vent valve opens and stops the purging, but the retard continues purging when the engine returns to cruising. The vacuum reservoir helps the retard delay valve maintain the vacuum.

When the engine is started warm, this delay system causes a rich mixture in the engine because it takes several seconds before the air from the canister passes through the delay into the engine. The thermactor idle vacuum valve was added to provide an air bleed. It is a normally open valve, so it bleeds air into the engine as soon as the engine starts. When the vacuum in the system is enough to open the purge valve, it also closes the thermactor idle vacuum valve, and the system purges normally.

Carburetor Bowl Vents

Several types of bowl vent valves are used. Some carburetors have a mechanical vent that opens from the throttle linkage when the engine idles. Others have a vapor operated vent that opens when the vapor pressure builds up. Those valves have been common for several years.

Ford has added a thermal vent valve to some engines. This valve is in the hose connecting the bowl vent to the carbon canister. When the engine compartment is cold, the valve closes. This prevents canister vapors from traveling up the hose to the carburetor, and upsetting the mixture. When the engine compartment warms up, the valve opens, and then the vapors from the carburetor bowl are vented to the carbon canister. If there are any other valves in the system, such as a PVS, those valves also have to be open before the system does any purging.

An electrically operated vent valve, called a solenoid vent valve, is used on some engines. It is a normally open design, so that it vents the fuel bowl to the canister when the engine is not running. When the ignition switch is turned on, the valve closes and no purging takes place during engine operation. This prevents the purging vapors from upsetting the mixture.

With the vent to the canister shut off, some engines gave problems during running because of lack of venting. To solve this problem, the engineers added an auxiliary vent tube on some engines. This tube connects to a "T" in the bowl vent hose, and allows the vapors from the bowl to go into the air cleaner, even though the hose to the canister is shut off.

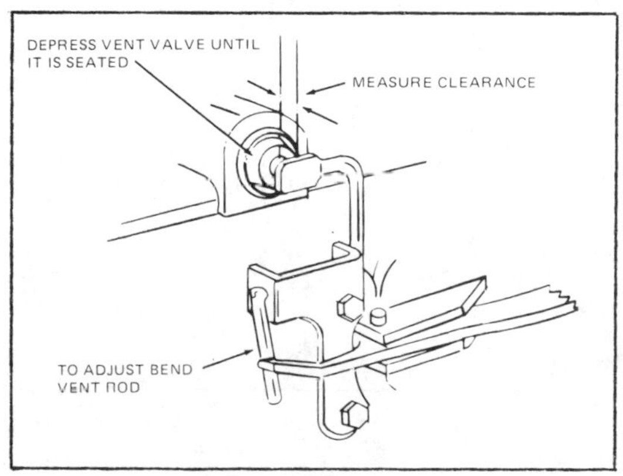

Bowl vent adjustment—Model 2150 2 bbl. carburetor

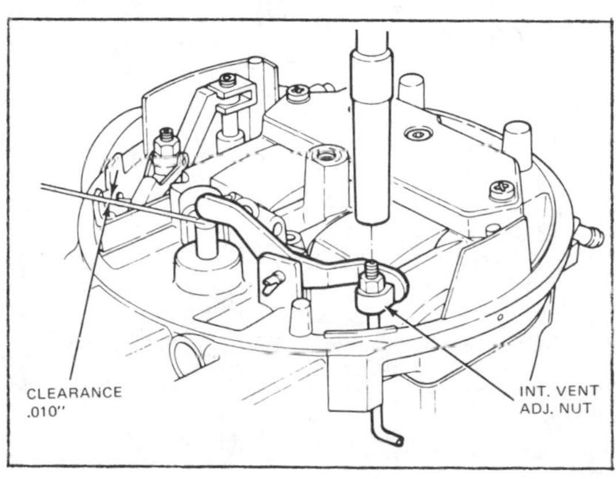

Internal vent system—Model 2700 VV carburetor

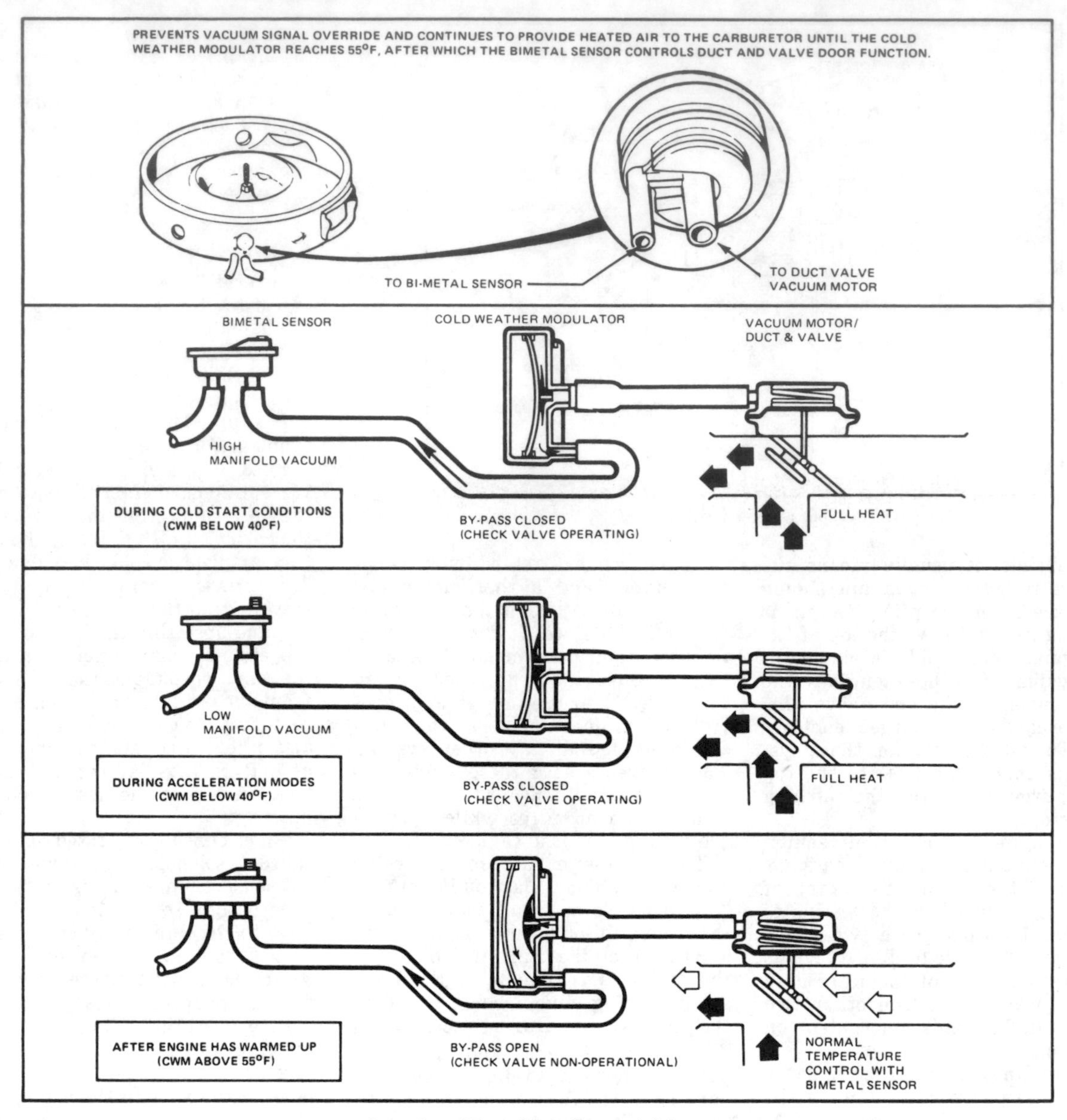

Operation of the cold weather modulator

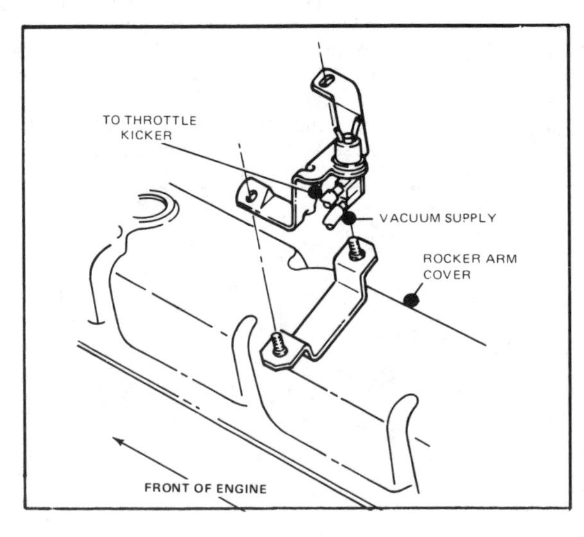

Throttle kicker solenoid

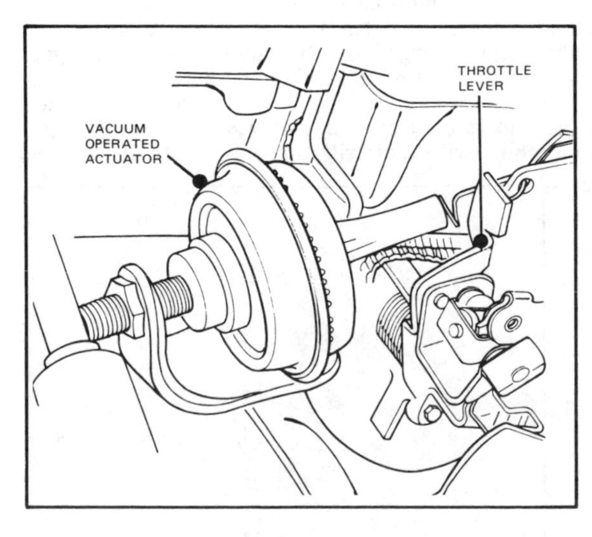

Throttle kicker actuator

Additional escape of vapors was caused by the air intake on the air cleaner. To prevent vapor from spilling out of the air cleaner when the car was parked a vapor dam was installed just inside the air cleaner can opposite the air intake. This keeps the vapors inside the air cleaner. They are purged as soon as the engine starts.

HIGH FLOW PCV VALVE 1979-80

The engine decal on certain engines shows that a high flow PCV valve must be installed according to the maintenance schedule. This means that when the PCV valve is due for replacement, a high flow valve should be installed, instead of the original low-flow valve. After installing the valve, the idle speed and mixture should be checked, because the high flow valve might have changed it. This procedure of installing a high flow valve is not done on all engines. The instructions on the engine decal must be followed.

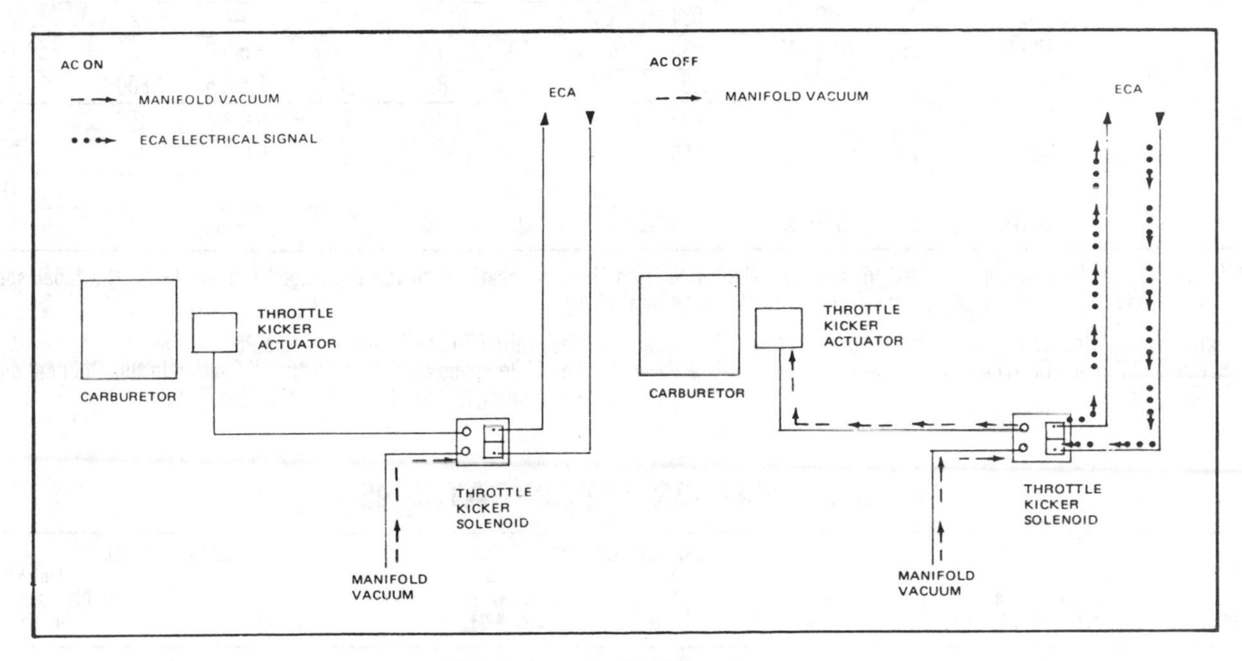

Schematic of the throttle kicker system

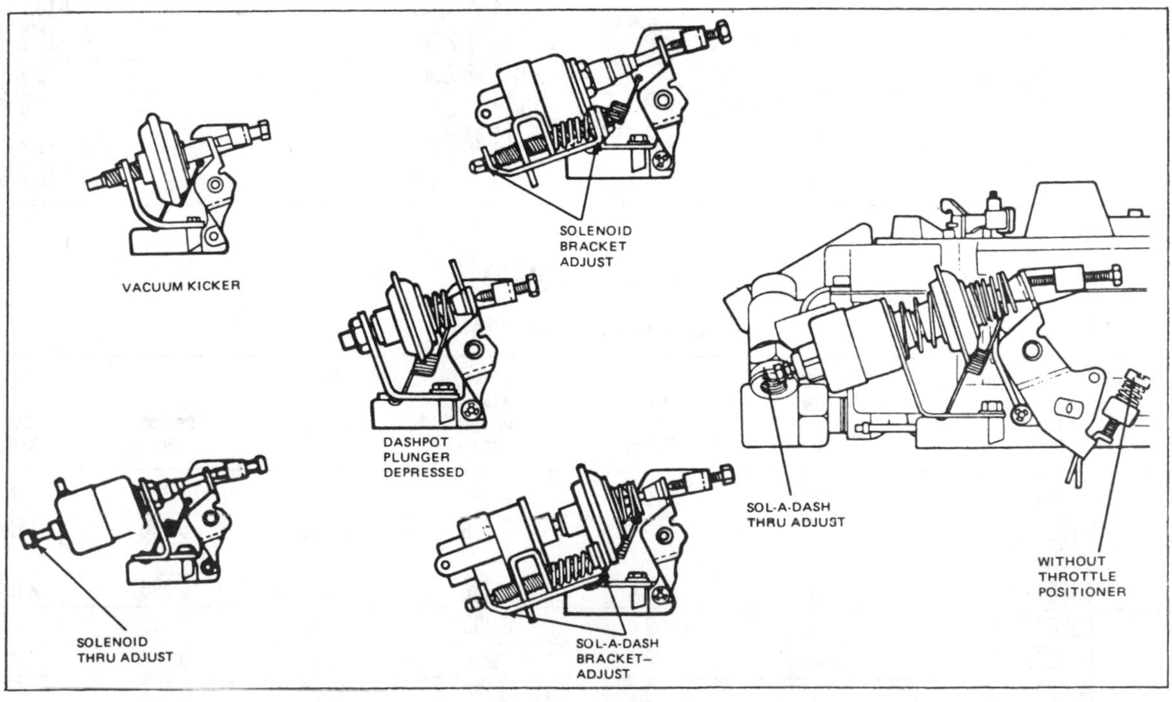

Throttle positioners

TUNE-UP SPECIFICATIONS

Engine Code, 5th character of the VIN number

Yr.	Eng. V.I.N. Code	Engine No. Cyl Disp. (cu in)	Carb Bbl	Spark Plugs Orig. Type	Spark Plugs Gap (in)	Dist. Point Dwell (deg)	Dist. Point Gap (in)	Ignition Timing (deg • BTDC) Man Trans	Ignition Timing (deg • BTDC) Auto Trans	Valves Intake Opens (deg) B	Fuel Pump Pres. (psi)	Idle Speed • (rpm) Man Trans	Idle Speed • (rpm) Auto Trans In Drive
1979	Y	4-140	2v	AWSF-42	.034	E.I.		6	20B(17)	22°	5.5-6.5	850	—
	W	4-140TC	2v	AWSF-32	.034	E.I.		①	①	③	—	②	②
	Z	6-170.8	2v	AWSF-42	.034	E.I.		10	9(6)	20°	3.5-5.8	850	650 (700)
	F	8-302	2v	ASF-52	.050	E.I.		12	8B	16°	5.5-6.5	800	600
1980	A	4-140	2v	AWSF-42	.034	E.I.		①	①	③	5.0-7.0	②	②
	B	6-200	1v	BSF-82	.050	E.I.		①	①	③	5.0-7.0	②	②
	D	8-255	2v	ASF-42	.050	E.I.		①	①	③	6.0-8.0	②	②
	T	4-140TC	2v	AWSF-32	.034	E.I.		①	①	③	—	②	②

NOTE: Should the information provided in this manual deviate from the specifications on the underhood tune-up label, the label specifications should be used, as they may reflect production changes.

• Figure in parentheses indicates California engine
B BEFORE TOP DEAD CENTER
TC Turbo-Charged
E.I. Electronic Ignition

① See "Initial Timing" Section in this Publication
② See "Idle Speed & Mixture Settings" Section in this Publication
③ Information Not Available at Publication Time

DISTRIBUTOR SPECIFICATIONS

YEAR	DISTRIBUTOR IDENTIFICATION	Centrifugal Advance Start Dist. Deg. @ Dist. RPM	Centrifugal Advance Finish Dist. Deg. @ Dist. RPM	Vacuum Advance Start In. Hg.	Vacuum Advance Finish Dist. Deg. @ In. Hg.
1979	D7EE-CA	0-2.1 @ 1235	5-7.5 @ 2,500	2.3	10.7-13.2 @ 16
	D7EE-DA	0-2.1 @ 530	11.5-14 @ 2,500	1.75	10.7-13.2 @ 12.4
	D7EE-EA	0-3 @ 500	11.2-14 @ 2,500	2	10.7-13.2 @ 16
	77TF-CA	0-1 @ 575	9.5-12.5 @ 2,500	3.5	5-7 @ 8.5
	79TF-FA	0-1 @ 600	10-12 @ 2,100	4.5	2-4 @ 10
	D9ZE-FA	0-2.5 @ 485	10.5-13 @ 2,500	1.8	8.7-11.2 @ 7.4
	D9ZE-EA	0-2.5 @ 485	10.5-13 @ 2,500	1.8	10.7-13.2 @ 16.2

Model 6500

Ford Bobcat and Pinto

Year	Carb. Iden.	Dry Float Level (in.)	Pump Hole Setting	Choke Plate Pulldown (in.)	Fast Idle Cam Linkage (in.)	Dechoke (in.)	Choke Setting
1978-79	D9EE-AFC	0.455	2	0.236	0.118	0.236	2 Rich
	D9EE-AJC, D9EE-AKC	0.460	2	0.236	0.118	0.236	1 Rich
	D9EE-AGC	0.460	2	0.236	0.118	0.236	2 Rich
1980	EOEE-NA, VA	0.460	2	0.236	0.118	0.393	①
	EOEE-NC, NV	0.460	2	0.236	0.118	0.157	①
	EOEE-ND, VD	0.460	2	0.236	0.118	0.393	①
	EOZE-AFA, SA	0.460	2	0.236	0.118	0.393	①
	EOZE-AFC· SC	0.460	—	0.236	0.118	0.393	①

① See underhood decal

HOW TO DETERMINE VEHICLE CALIBRATION NUMBERS 1980

An emission control decal and an engine code label must be used to determine the vehicle calibration code number.

1. The emission control label is gold or silver and might be located on the valve cover, radiator support bracket, under side of the hood, fresh air inlet tube or the windshield washer bottle.
2. The engine code label is yellow or white and will be located on the valve cover.
3. Determine the model year from the emission label. Use only the last digit. For example, 1980 would be "O."
4. Determine the calibration base and the revision level from the engine code label. For example the calibration base could be "93J" and the revision level "RO."

Ford — FORD MOTOR COMPANY — VEHICLE EMISSION CONTROL INFORMATION — EVAPORATIVE FAMILY IS A

ENGINE FAMILY 3.3L GA-EGR/PAS/CATALYST — SPARK PLUG BSF-82 — CATALYST
ENGINE DISPLACEMENT 3.3L(200CID) TRANS AUTO — GAP .048-.052

SHIFT SCHED		TRANSMISSION GEAR NEUTRAL	DRIVE
IGNITION TIMING		10°BTDC	
TIMING RPM		750 MAX	
CHOKE SETTING		2 RICH	
FAST IDLE RPM	HIGH CAM		
	KICKDOWN	2000	
CURB IDLE RPM	A/C ON		700
	A/C OFF		550
CURB IDLE RPM	NON A/C		
			550

① ②

MAKE ALL ADJUSTMENTS WITH ENGINE AT NORMAL OPERATING TEMPERATURE, ACCESSORIES AND HEADLIGHTS OFF. DISCONNECT DUAL MODE IGNITION MODULE CONNECTOR AT IGNITION MODULE BEFORE SETTING TIMING OR CURB IDLE (IF SO EQUIPPED). CONSULT SERVICE PUBLICATIONS FOR ADDITIONAL INSTRUCTIONS ON THE FOLLOWING PROCEDURES

IGNITION TIMING—A/C OFF—ADJUST WITH HOSES DISCONNECTED AND PLUGGED AT THE DISTRIBUTOR

CURB IDLE—A/C ON— A/C THROTTLE SOLENOID POSITIONER ENERGIZED AND THE A/C ELECTROMAGNETIC CLUTCH DE-ENERGIZED
CURB IDLE—A/C OFF—A/C THROTTLE SOLENOID POSITIONER DE-ENERGIZED
CURB IDLE—A/C ON OR OFF—ADJUST WITH AIR CLEANER IN POSITION. REFER TO SERVICE PUBLICATIONS FOR VACUUM HOSE CONNECTIONS DURING IDLE SET

IDLE MIXTURE—PRESET AT THE FACTORY. DO NOT REMOVE THE LIMITER CAP(S) EXCEPT IN ACCORDANCE WITH SERVICE PUBLICATIONS

COMPLIANCE DEMONSTRATED BELOW 4000 FT

THIS VEHICLE CONFORMS TO U.S.E.P.A. REGULATIONS APPLICABLE TO 1980 MODEL YEAR NEW MOTOR VEHICLES. — EOBE-9C485-EA — BBM

1980 CALIBRATION SERVICE SPECIFICATIONS

Calibration Number	Engine	Spark Plug	Spark Plug Gap	Ignition Timing	Timing RPM	Fast Idle RPM High Cam	Fast Idle RPM Kick Down	Curb Idle RPM A/C On	Curb Idle RPM A/C Off	Curb Idle RPM Non-AC	TSP Off RPM AC	TSP Off RPM Non-AC	Choke Cap Setting	Decal Throttle Control RPM
0-16A-R0	4.2L	ASF-42	.048-.052	6° BTDC	500 max.	2000	—	650	500	500	—	—	3 Rich	—
0-16C-R0	4.2L	ASF-42	.048-.052	6° BTDC	500 max.	2000	—	650	500	500	—	—	3 Rich	—
0-16S-R0	4.2L	ASF-42	.048-.052	6° BTDC	550	2100	—	650	500	500	—	—	3 Rich	—
0-16W-R0	4.2L	ASF-42	.048-.052	6° BTDC	500	2000	—	650	500	500	—	—	3 Rich or "V"	— —
0-16X-R0	4.2L	ASF-42	.048-.052	6° BTDC	500	2000	—	650	500	500	—	—	3 Rich or "V"	— —
0-1P-R0	2.3L	AWSF-42	.032-.036	12° BTDC	600	—	2000	—	750	750	—	—	Index	—
0-2T-R0	2.3L	AWSF-42	.032-.036	6° BTDC	650	—	2000	—	850	850	—	—	Index	—
0-6A-R0	3.3L	BSF-82	.048-.052	10° BTDC	750	—	1600	—	—	700	—	—	2 Rich	—
0-1B-R0	2.3L	AWSF-42	.032-.036	6° BTDC	600	—	2000	—	750	750	—	—	1 Lean	—
0-1G-R0	2.3L	AWSF-42	.032-.036	6° BTDC	600	—	2000	—	750	750	—	—	1 Lean	—
0-2B-R0	2.3L	AWSF-42	.032-.036	6° BTDC	550	—	1800	—	850	850	—	—	2 Rich	—
0-2C-R0	2.3L	AWSF-42	.032-.036	6° BTDC	550	—	1800	—	850	850	—	—	2 Rich	—
0-2E-R0	2.3L	AWSF-32	.032-.036	6° BTDC	550	—	1800	—	850	850	—	—	2 Rich	—
0-6A-R01	3.3L	BSF-82	.048-.052	12° BTDC	750 max.	—	1600	900	700	—	—	—	2 Rich	—
0-7A-R0	3.3L	BSF-82	.048-.052	10° BTDC	750 max.	—	2000	700	550	550	—	—	2 Rich	—
0-1B-R11	2.3L	AWSF-42	.032-.036	6° BTDC	600	—	2000	750	—	750	—	—	2 Rich	—
0-1B-R12	2.3L	AWSF-42	.032-.036	6° BTDC	600	—	2000	750	—	750	600	600	2 Rich	—
0-2C-R10	2.3L	AWSF-42	.032-.036	6° BTDC	550	—	1800	850	—	850	—	—	2 Rich	—
0-1D-R0	2.3L	AWSF-42	.032-.036	6° BTDC	600	—	2000	—	750	750	—	—	1 Lean	—
0-2D-R0	2.3L	AWSF-42	.032-.036	6° BTDC	550	—	1800	850	—	850	—	—	2 Rich	—
0-1G-R11	2.3L	AWSF-42	.032-.036	6° BTDC	600	—	2000	750	—	750	—	—	2 Rich	—
0-1G-R12	2.3L	AWSF-42	.032-.036	6° BTDC	600	—	2000	750	—	750	600	600	2 Rich	—

① All timing, choke, and RPM specifications shown should be set with the Automatic Transmission in NEUTRAL, except CURB IDLE RPM which is set with Automatic Transmission in DRIVE.

② All timing, choke, and RPM specifications shown should be set with the Manual Transmission in NEUTRAL.

5. The complete calibration number in the example would be "O-93J-RO."

NOTE: *Some of the calibrations are carried over from prior years. If a number can't be located using prefix "0"; check "9", "8" or "7" prefixes.*

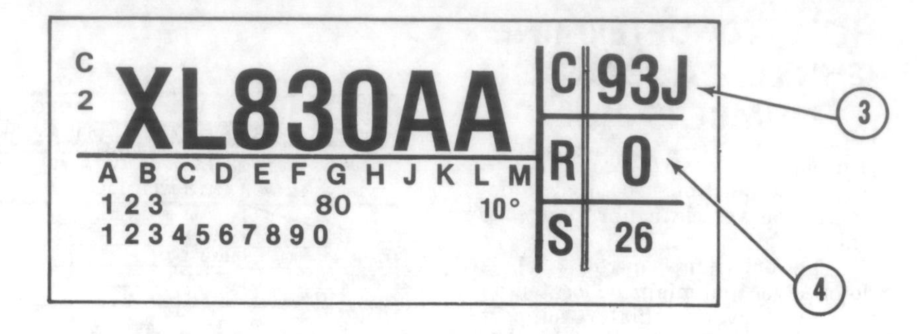

CAR SERIAL NUMBER AND ENGINE IDENTIFICATION

1979

The vehicle identification number is mounted on a plate behind the windshield on the driver's side. The first character is the model year. The fifth character is the engine code, as follows:

Y . . . . .140 2-bbl. 4-cyl.
Z 169 2-bbl. V6
T . . . . .200 1-bbl. 6-cyl.
F 302 2-bbl. V8
W . . . . . . .140 Turbo Charged 4-cyl.

1980

The vehicle identification number is mounted on a plate behind the windshield on the driver's side of the vehicle. The first character is the model year. The fifth character is the engine code, as follows:

A . . . . .140 2-bbl. 4-cyl.
B . . . . .200 1-bbl. 6-cyl.
D . . . . .225 2-bbl. 8-cyl.
T .140 2-bbl. 6-cyl. (Turbo Charged)

EMISSION EQUIPMENT 1979-80

All Models

Closed positive crankcase ventilation
Emission calibrated carburetor
Emission calibrated distributor
Heated air cleaner
Vapor control, canister storage
Exhaust gas recirculation
Electric choke
Catalytic converter
Thermactor System
Feedback carburetor

IDLE SPEED AND IDLE MIXTURES SETTINGS

1979

NOTE: *The underhood specifications sticker sometimes reflects tune-up changes made in production. Sticker figures must be used if they disagree with this information.*

Air cleaner . . . . .In place
Air Cond. .Off
Auto. trans . .Drive unless otherwise noted
Headlights . . . . .Off

NOTE: *On cars equipped with Vacuum Operated Throttle Modulator (VOTM), which provides intermediate fast idle speed during cold engine operation, the vehicle emissions label should be checked before any adjustments are made. Some models also use Throttle Solenoid Positioner (TSP) for curb idle adjustment. Refer to the vehicle emissions label to see if vehicle is so equipped and for the adjustment procedure. Refer to Ford Section for further explanation.*

140 2-bbl. 4-cyl. Vin code (Y) 49 States Pinto and Bobcat
Auto. trans.
Fast idle . . . . .2000 (N)
Curb idle
A/C 800
Non A/C 800
TSP off
A/C 600
Manual trans.
Fast idle 1800
Curb idle
A/C on 1300
A/C off 850
Non A/C 850

140 2-bbl. 4-cyl. Vin code (Y) Calif. Pinto and Bobcat
Manual trans.
Fast idle 1850
Curb idle
Non A/C 850

140 2-bbl. 4-cyl. Vin code (Y) Altitude Pinto and Bobcat
Auto. trans.
Fast idle . . . . .2000 (N)
Curb idle
A/C 800
Non A/C 800
TSP off
A/C 550
Non A/C 550
Manual trans.
Fast idle 1800
Curb idle
A/C 850
Non A/C 850
TSP off
A/C 550
Non A/C 550

169 2-bbl. V6 Vin code (Z) 49 States Pinto and Bobcat
Auto. trans.
Fast idle . .1600, (N), 1800 (N) Bobcat
Curb idle
A/C on 750
A/C off 650
Non A/C 650

169 2-bbl. V6 Vin code (Z) Calif. Pinto and Bobcat
Auto. trans.
Fast idle . . . . .1750 (N)
Curb idle
A/C on 700
Non A/C 700
TSP off
A/C . . . . .600 (N)
Non A/C . . . . .600 (N)

140 2-bbl. 4-cyl. Vin code (Y) 49 States Fairmont and Zephyr
Auto. trans.
Fast idle . . . . .2000 (N)
Curb idle
A/C 800
Non A/C 800
TSP off
A/C 600
Non A/C 600
Manual trans.
Fast idle . . . . . . .1600 Fairmont, 1800 Zephyr

IDLE SPEED AND IDLE MIXTURES SETTINGS

Curb idle
A/C 1300
A/C off 850
Non A/C 850

140 4-cyl. Vin code (Y) Calif. Fairmont and Zephyr
Auto. trans.
Fast idle ...1800 (N), 1300 (N) non A/C
Curb idle
A/C 750
Non A/C 850
TSP off
A/C 600
Non A/C 600
Manual trans.
Fast idle 1800
Curb idle
A/C 850
Non A/C 850
TSP off
A/C 600
Non A/C 600

200 1-bbl. 6-cyl. Vin code (T) 49 States Fairmont and Zephyr
Auto. trans.
Fast idle ....1700 (N)
Curb idle
A/C 650
Non A/C 650
TSP off
A/C ....700 (N)
Non A/C ....700 (N)
Manual trans.
Fast idle 1600
Curb idle
A/C on 850
A/C off 700
Non A/C 700

200 1-bbl. 6-cyl. Vin code (T) Calif. Fairmont and Zephyr
Auto. trans.
Fast idle ....1850 (N)
Curb idle
A/C on 650
A/C off 600
Non A/C 600

200 1-bbl. 6-cyl. Vin code (T) Altitude Fairmont and Zephyr
Auto. trans.
Fast idle ....1700 (N)
Curb idle
A/C 650
Non A/C 650
TSP off
A/C 700
Non A/C 700

302 2-bbl. V8 Vin code (F) 49 States Fairmont and Zephyr
Auto. trans.
Fast idle ....2100 (N)
Curb idle
A/C on 675
A/C off 600
Non A/C 600
Manual trans.
Fast idle 2300
A/C on 875
A/C off 800
Non A/C 800

302 2-bbl. V8 Vin code (F) Calif. Fairmont and Zephyr
Auto. trans.
Fast idle ....1800 N)
A/C on 675
A/C off 600
Non A/C 600
Non A/C 600
VOTM
A/C 700
Non A/C 700

302 2-bbl. V8 Vin code (F) Altitude Fairmont and Zephyr
Auto. trans.
Fast idle ....2100 (N)
Curb idle
A/C on 675
A/C off 600
Non A/C 600

140 2-bbl. 4-cyl. Vin code (Y) 49 States Mustang and Capri
Auto. trans.
Fast idle ....2000 (N)
Curb idle
A/C 800
Non A/C 800
TSP off
A/C 600
Non A/C 600
Manual trans.
Fast idle 1600
Curb idle
A/C on 1300
A/C off 850
Non A/C 850

140 2-bbl. 4-cyl. Vin code (Y) Calif. Mustang and Capri
Auto. trans.
Fast idle ....1800 (N)
Curb idle
A/C 750
Non A/C 850
TSP off
A/C 600
Non A/C 600
Manual trans.
Fast idle 1800
Curb idle
A/C 850
Non A/C 850
TSP off
A/C 600
Non A/C 600

140 2-bbl. 4-cyl. Vin code (Y) Altitude Mustang and Capri
Manual trans.
Fast idle 1800
Curb idle
A/C 850
Non A/C 850
TSP off
A/C 550
Non A/C 550

169 2-bbl. V6 Vin code (Z) 49 States Mustang and Capri
Auto. trans.
Fast idle ...1800 (N) Mustang, 1600 (N) Capri
Curb idle
A/C on 750
A/C off 650
Non A/C 650

169 2-bbl. V6 Vin code (Z) Calif. Mustang and Capri
Auto. trans.
Fast idle ....1750 (N)
Curb idle
A/C 700
Non A/C 700
TSP off
A/C ....600 (N)
Non A/C ....600 (N)

302 2-bbl. V8 Vin code (F) 49 States Mustang and Capri
Auto. trans.
Fast idle ....2100 (N)
Curb
A/C on 675
A/C off 600
Non A/C 600
Manual trans.
Fast idle 2300
Curb idle
A/C on 875
A/C off 800
Non A/C 800

302 2-bbl. V8 Vin code (F) Calif. Mustang and Capri
Auto. trans.
Fast idle ....1800 (N)
Curb idle
A/C on 675
A/C off 600
Non A/C 600
VOTM
A/C 700
Non A/C 700

140 Turbo Charged 4-cyl. Vin code (W) 49 States Mustang and Capri
Manual trans.
Fast idle 1800
Curb idle
A/C on 1300
A/C off 900
Non A/C 900

NOTE: *For explanation of the calibration number see the "Calibration Specification" information at the beginning of this section.*

1980 IDLE MIXTURE ADJUSTMENT—PROPANE ENRICHMENT

Calibration	Gain (RPM)	Reset (RPM)	Selector Lever

NOTE: For the explanation of the calibration number see the "Calibration Specification" information at the beginning of this section.

INITIAL TIMING

1979

NOTE 1: *Before making any adjustments on the ignition timing refer to the Vehicle Emissions Label located in the engine compartment of the vehicle.*

2. *Distributer vacuum hoses must be disconnected and plugged.*

140 2-bbl. 4-cyl.
Auto. trans. . . . . 20° BTDC
Manual trans. . . . . 6° BTDC

169 2-bbl. V6 49 States
Auto. trans. . . . . 9° BTDC
Manual trans. . . . . 6° BTDC

169 2-bbl. V6 California
Auto. trans. . . . . 6° BTDC

200 1-bbl. 6 cyl. 49 States
Auto. trans. . . . . 10° BTDC
Manual trans. . . . . 8° BTDC

200 1-bbl. 6-cyl. California
Auto. trans. . . . . 10° BTDC

302 2-bbl. V8 49 States
Auto. trans. . . . . 8° BTDC
Manual trans. . . . . 12° BTDC

302 2-bbl. V8 California
Auto. trans. . . . . 12° BTDC

140 Turbo Charged 4-cyl. 49 States
Manual trans. . . . . 2° BTDC

1980

NOTE: *For "Initial Timing" information refer to the "Calibration Specification" information at the beginning of this section.*

SPARK PLUGS

1979-80

140 4-cyl. . . . . . AU-AWSF-42 . . . 034
169 V6 . . . . AU-AWSF-42 . . . 034
200 6-cyl. . . . . AU-BSF-82 . . . 050
302 V8 . . . . AU-ASF-52 . . . 050
140 Turbo charged
4-cyl. . . . . . . AU-AWSF-32 . . . 034
255 V-8 . . . . ASF-42 . . . 050

EMISSION CONTROL SYSTEMS 1979-80

FEEDBACK CARBURETOR ELECTRONIC ENGINE CONTROL SYSTEM 1979-80

The 1979 Pinto and Bobcat vehicles manufactured for sale in California and equipped with a 140 Cu. In. engine and equipped with a feedback electronic engine control system. This system consists of three subsystems:

1. Dual catalytic converter
2. Thermactor air control
3. Electronic feedback carburetion

These subsystems work in conjunction with one another to provide improved fuel economy, engine performance and lower exhaust emissions. Basically, the engine control system works as follows; the electronic control unit (ECU) monitors the composition of the exhaust gases via the exhaust gas oxygen (EGO) and commands modification of the carburetor main metering mixture for the proper exhaust emissions.

This system is again being used in the 1980 Pinto, Bobcat, Capri, Mustang, Zephyr, and Fairmont vehicles which are manufactured for sale in California, with the exception that the feedback carburetor electronic engine control system is now also being used on the 140 Cu. In. Turbo Charged engine.

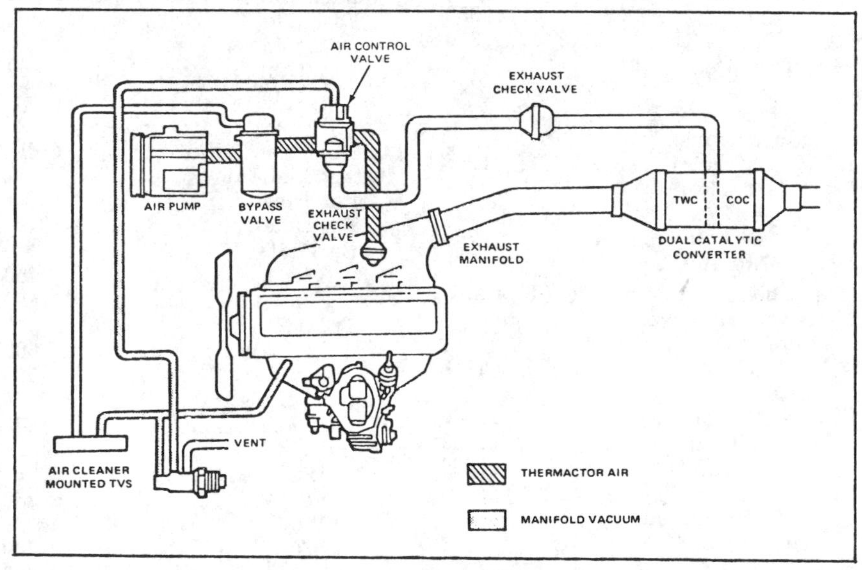

Thermactor system

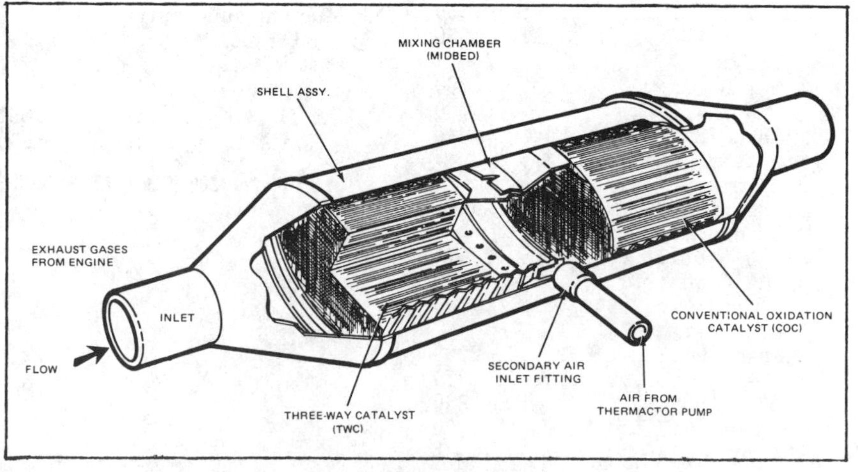

Dual catalytic converter

TURBO CHARGER SYSTEM 1979-80

The operation of a turbo charger is

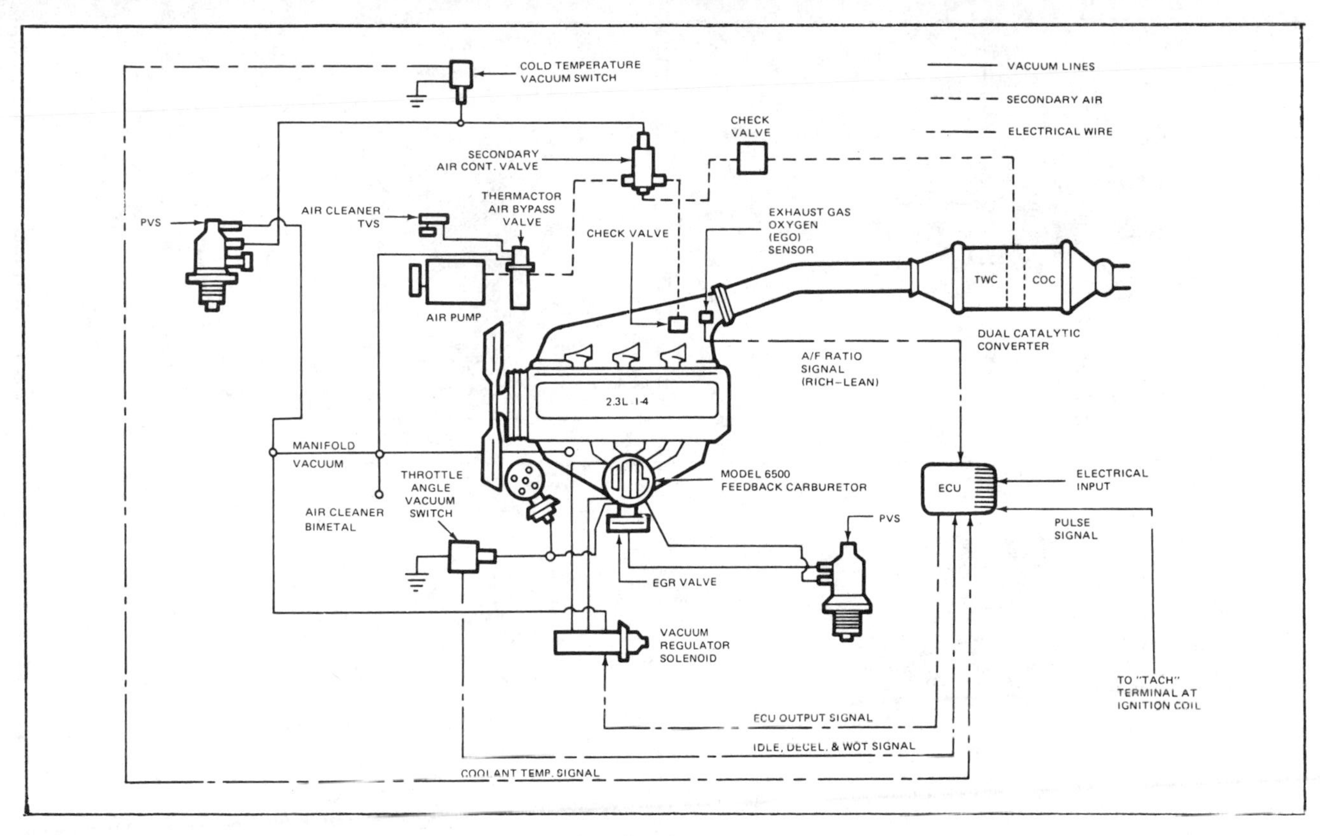

Feedback carburetor electronic engine control system—Schematic

based on using exhaust pressure to turn a turbine on the same shaft with a compressor. The compressor is placed on the manifold side of the carburetor to pull ambient air through the carburetor at a higher volume than normal engine vacuum would. Excess pressure is discharged through the wastegate valve into the exhaust system. The turbo charged system is not effective under light to medium throttle acceleration.

MONOLITH TIMING 1979-80

Some California engines are equipped with a "monolithic" timing system. The system is designed to accept an electronic probe that is connected to digital readout equipment. The probe receptacle is located at the front of all engines, except the 139 4-cyl. engine, which has a boss for monolithic timing in the left rear of the cylinder block.

NOTE: *For additional information with regard to specific emission systems consult the Ford, Mercury section of this manual.*

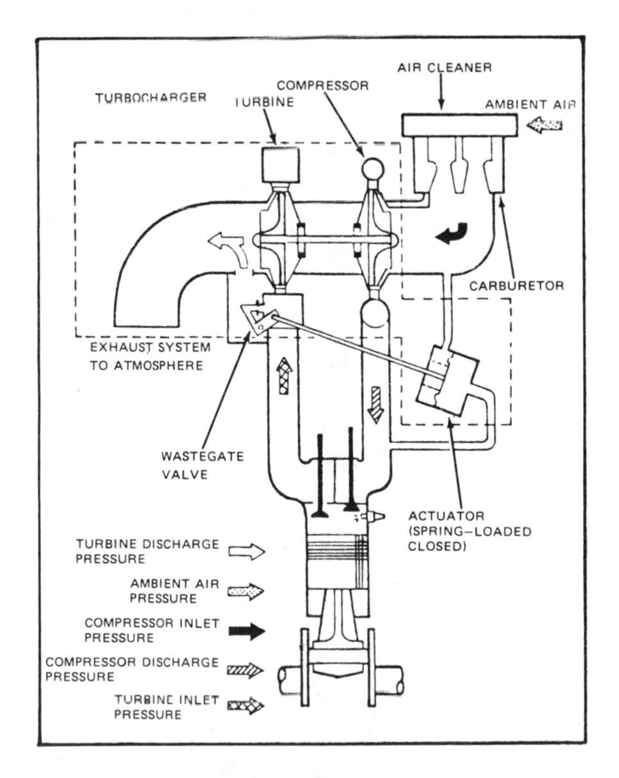

Turbo-charged operation

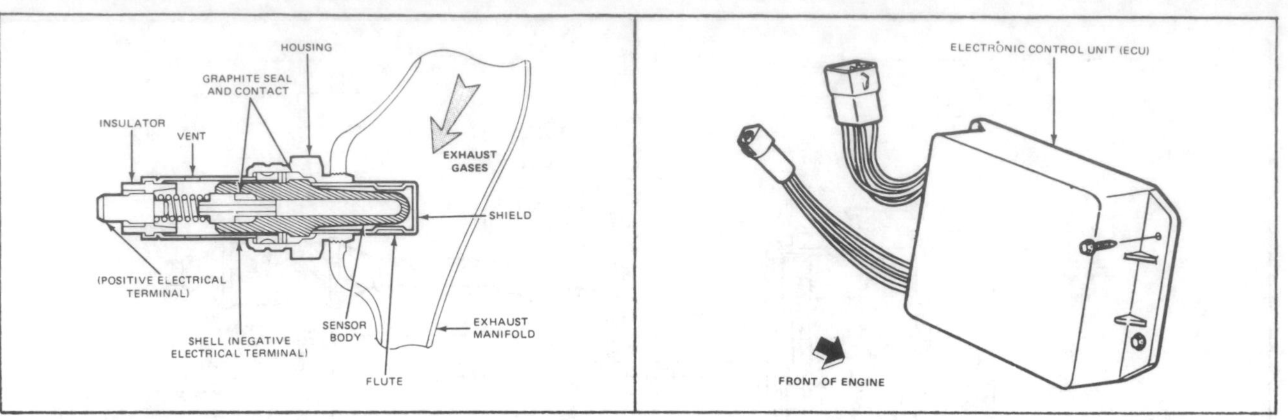

Feedback carburetor, exhaust gas oxygen (EGO) sensor

Electronic Control Unit (ECU)

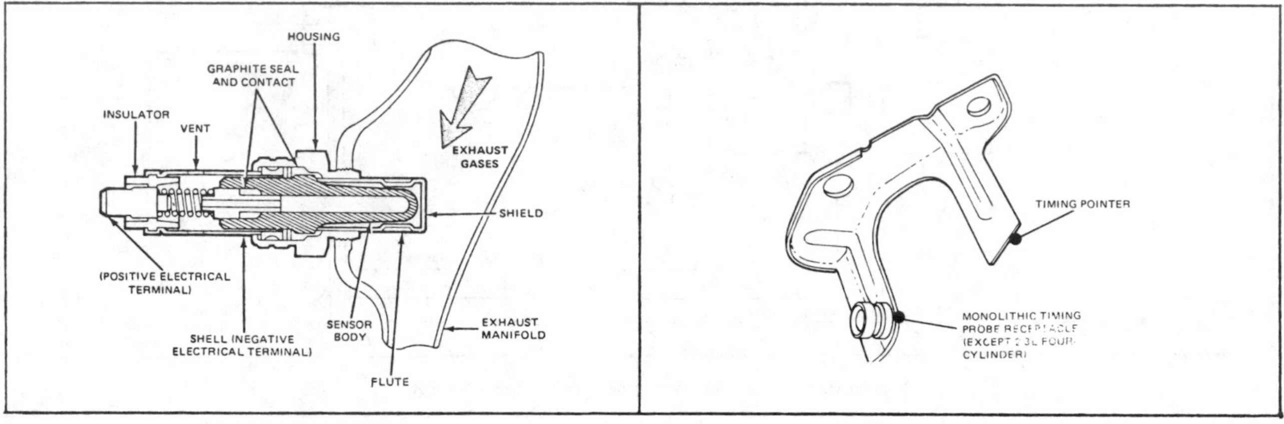

Exhaust Gas Oxygen (EGO) Sensor

Probe receptacle and timing mark

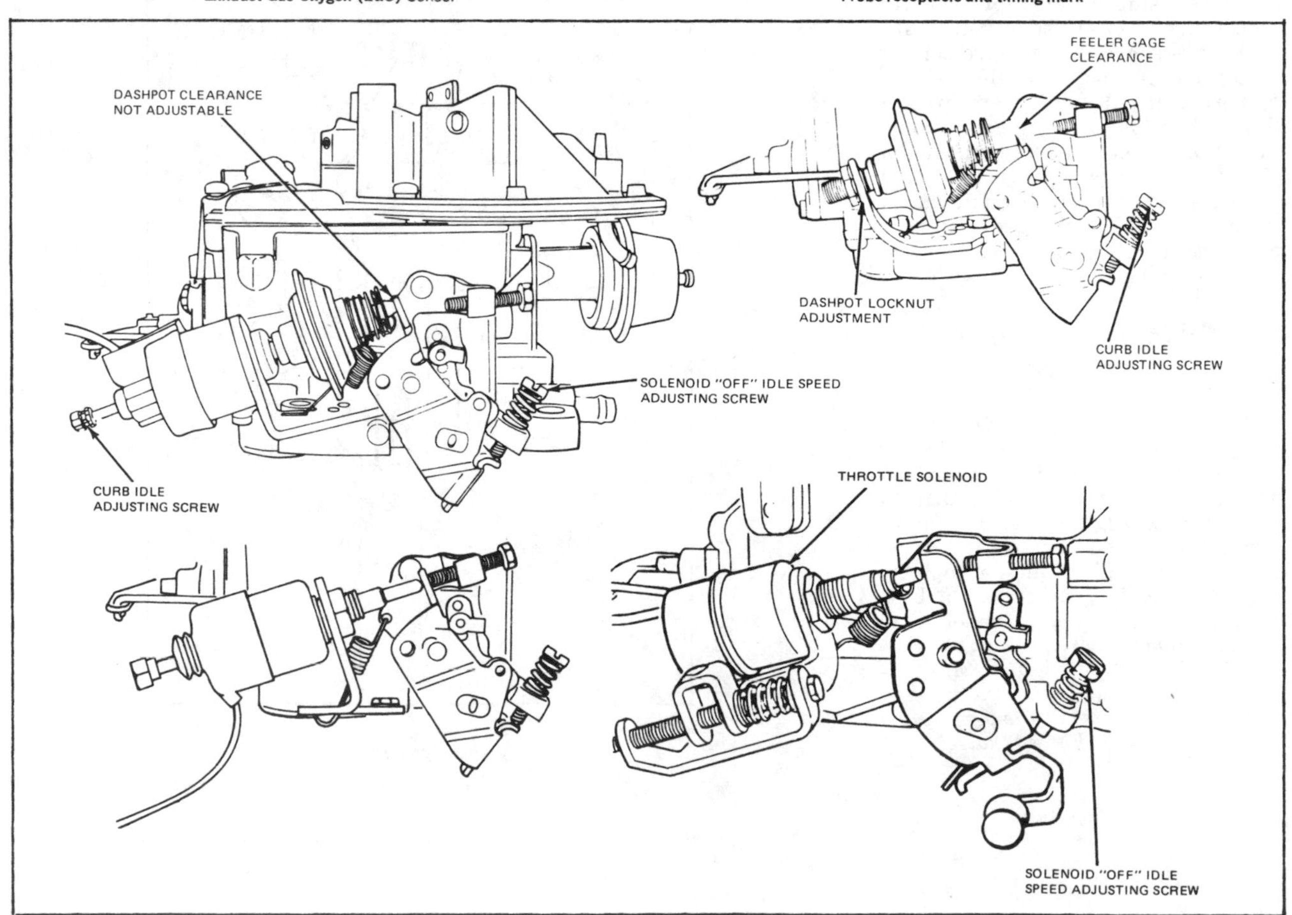

Typical throttle positioners

HOW TO USE THE CALIBRATION SHEETS

Determine the engine displacement and model year from the Exhaust Emission Decal as shown in Figure 1. Obtain the set of vacuum diagrams for the appropriate model year and turn to the correct engine displacement. To find the calibration number installed on a vehicle, refer to the engine code label (not the colored exhaust emission decal). This engine code label is located on the rocker cover as shown below (Figure 2).

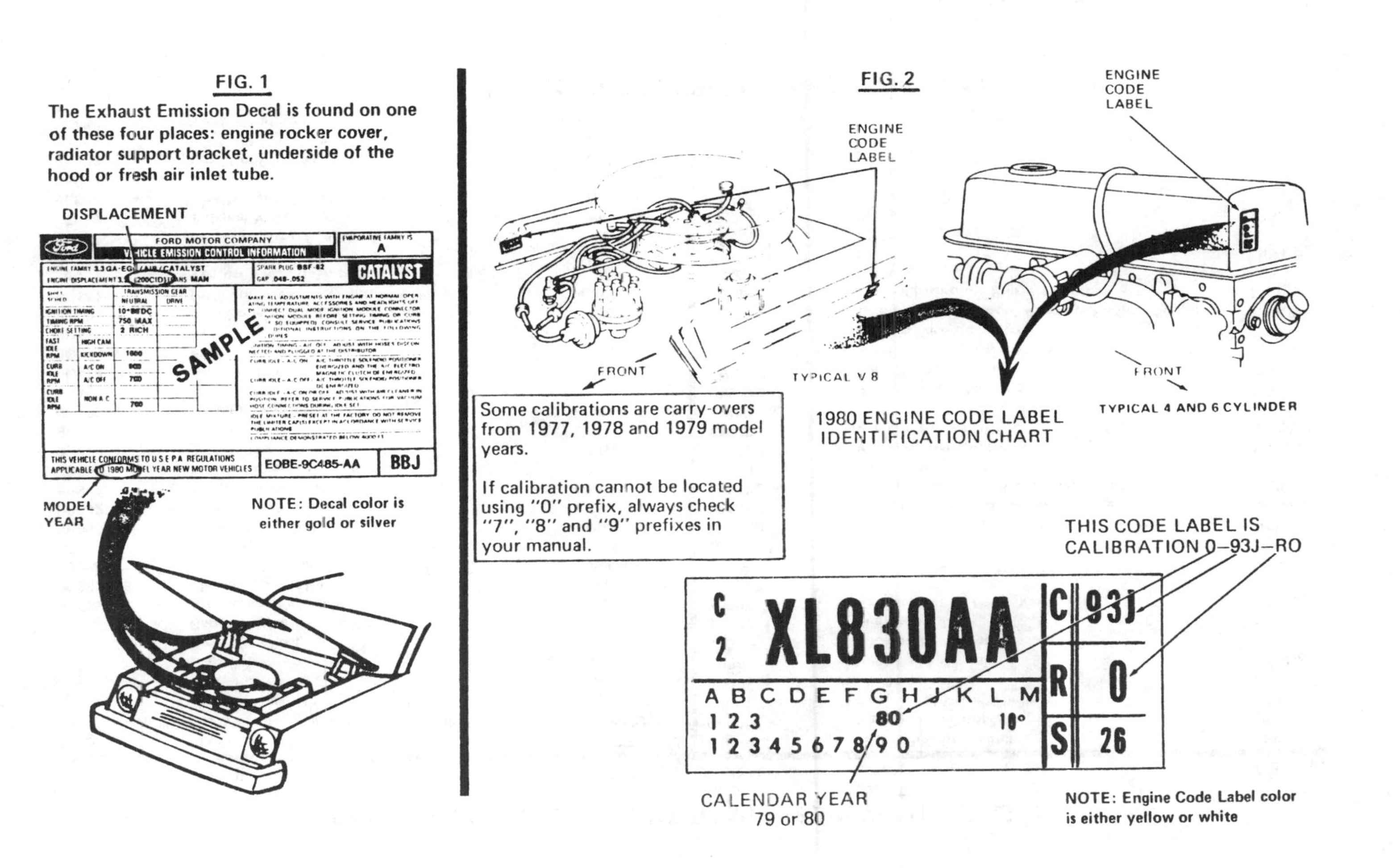

VACUUM SCHEMATIC PART NAME ABBREVIATIONS LIST

BASE PART #	PART NAME	ABBRV.
9510	Carburetor	CARB
	TSP-Vacuum Operated Throttle Modulator	TSP-VOTM
	Vacuum Operated Throttle Modulator	VOTM
	Spark Port	S
	EGR Port	E
	Bowl Vent Port	BV
	Control Valve	CV
9A995	Air Cleaner TVS Valve	TVS
9B289	Thermactor Air By-Pass Valve	Air BPV
9B298	Anti-Backfire Valve	ANTI-BFV
9B981	EGR Valve Actuator	EGR VA
9B963	Purge Control Valve	Purge CV
9B982	Solenoid Valve-Carb Bowl Vent	SV-CBV
9B982	Solenoid Vent Valve	SVV
9B998	Vacuum Regulator/Solenoid	VR/S
9D448 } 9D475 }	EGR Valve	EGR
9D473 } 12A091 }	Vacuum Control Valve	VCV
9D474	Solenoid Valve	Sol V
9D475 } 9D448 }	EGR Valve	EGR
9D604	Air Cleaner Duct & Valve Vacuum Motor	A/CL DV
9E441	Differential Vacuum Control Valve	DVCV
9E451	Venturi Vacuum Amplifier	VVA
9E453	Vacuum Reservoir	VReser
9C369	Separator Assy.-Fuel Vacuum	SA-FV
9E589	Thermal Vent Valve	TVV
9E607	Air Cleaner Bi-Metal Sensor	A/CL Bi Met
9E862	Air Cleaner Cold Weather Modulator	A/CL CWM
9F424	Load Control Valve	LCV
9F452	EGR B/P Transducer	B/P
9F480	Signal Conditioner	SC
9F489	Ported Pressure Switch	PPS
9F490	Vacuum Regulator Valve (3 Port)	VRV
9F491	Air Control Valve (Thermactor)	ACV
9G328	Idle Vacuum Valve	IVV
9H301	Vent Valve Vacuum	VVVac
9K319	Vacuum Restrictor	VRest
9K793	Fuel Decel Valve	FDV
9E897 } 12A208 }	Vacuum Retard Delay Valve	VRDV
12A091 } 9D473 }	Vacuum Control Valve	VCV
12A182	Vacuum Controlled Switch	VCS
12A182	Vacuum Controlled Switch (Cold Temp)	VCS-CT
12A182	Vacuum Controlled Switch (Decel Idle)	VCS-DI
12A189	Vacuum Delay Valve	VDV
12A197	Vacuum Check Valve	V Ck V
12A208 } 9E897 }	Vacuum Retard Delay Valve	VRDV
12A226	Vacuum Vent Valve	VacVV
12A245	Delay Valve - Two Way	DV-TW
12A265	Ignition Timing Vacuum Switch	ITVS
12A268	Ignition Pressure Switch	IPS
12A680	Barometric & Manifold Absolute Pressure Sensor	BMAP
12A680 } 9F479 }	Manifold Absolute Pressure Sensor	MAP
12100 } 12127 }	Distributor	DIST
12127 } 12100 }	Distributor	DIST

DIAGNOSTIC SPECIFICATION ABBREVIATIONS

Calib.	Calibration number	N	Neutral
D	Drive transmission position	N/R	Not required
KD	Kickdown step	PPM	Parts per million
max	Maximum	TBD	Specification to be supplied later
min	Minimum	CALIF. ONLY	California Only
A/C ONLY	Air Conditioning Only	A/T	Automatic transmission
A/C & NON A/C	Air Conditioning & Non Air Conditioning	M/T	Manual transmission
49S	49 States	NON A/C	Non Air Conditioning
50S	50 States	A/C ONLY x–x–x	Shows Air Conditioning Only
MM	Millimeters	NON A/C –II–II–I	Shows Non Air Conditioning Only

VACUUM CIRCUITS

1979—(1.6 L) engines

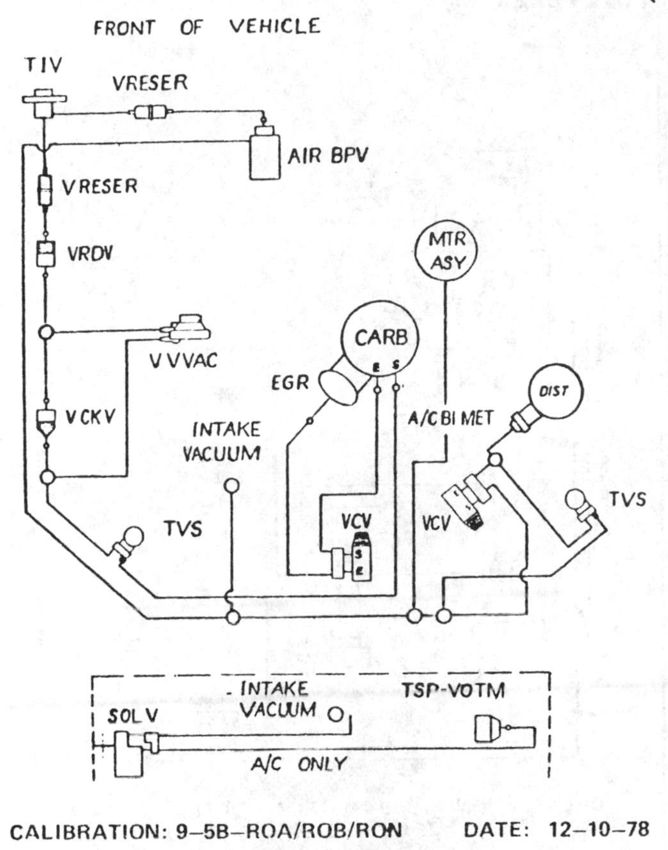

CALIBRATION: 9–5B–ROA/ROB/RON DATE: 12–10–78

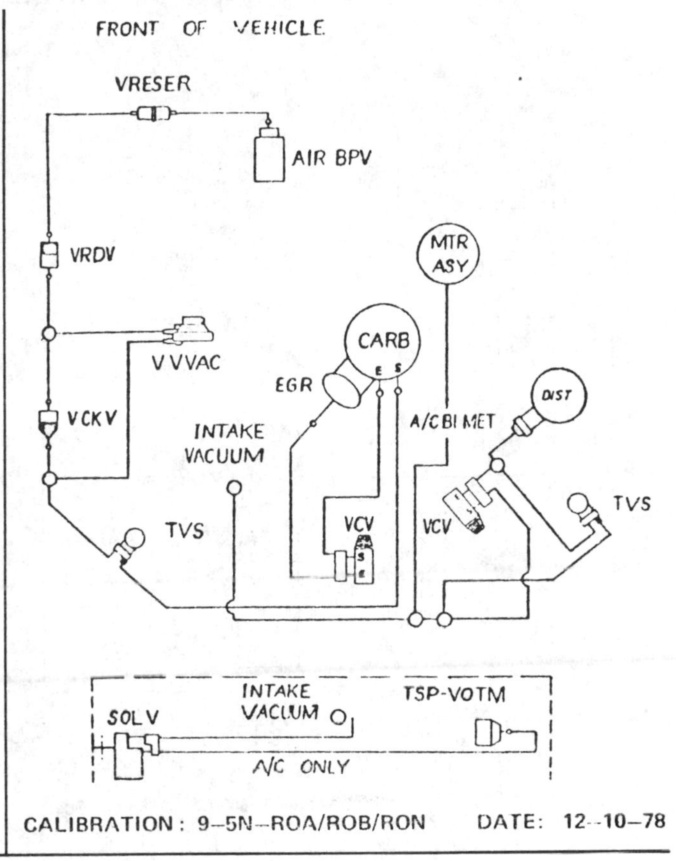

CALIBRATION: 9–5N–ROA/ROB/RON DATE: 12–10–78

1979—140 CID (2.3 L) engines

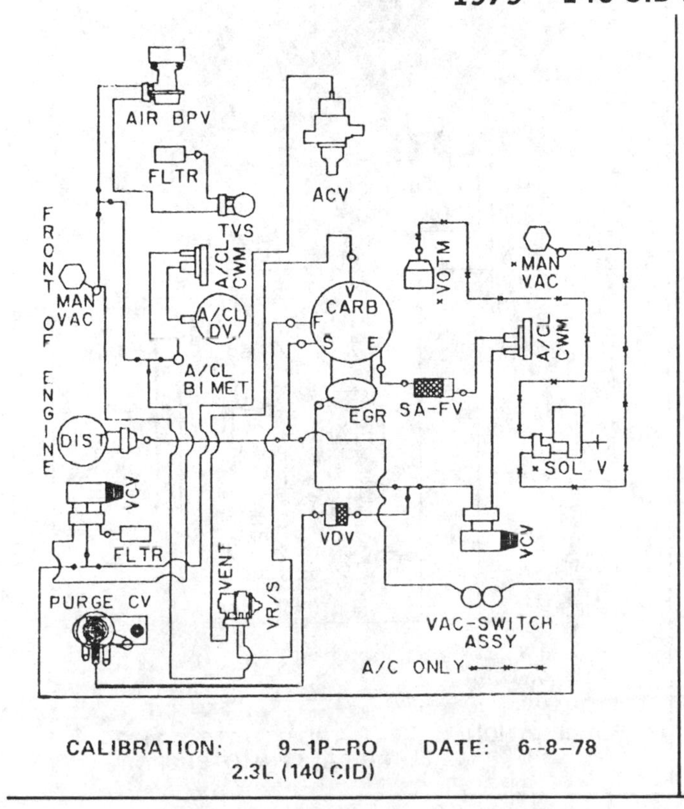

CALIBRATION: 9–1P–RO DATE: 6–8–78
2.3L (140 CID)

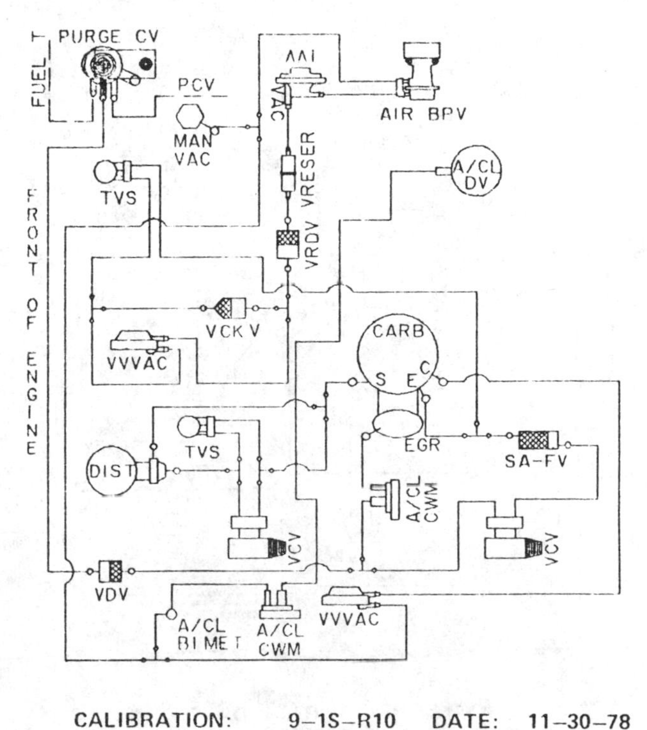

CALIBRATION: 9–1S–R10 DATE: 11–30–78
2.3L (140 CID)

VACUUM CIRCUITS

1979—140 CID (2.3 L) engines

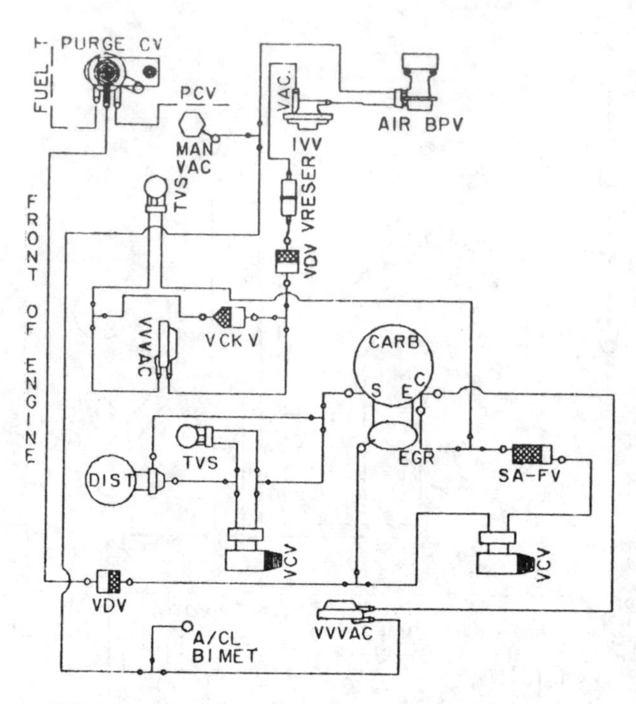

CALIBRATION: 9–1S–RO DATE: 4–14–78
2.3L (140 CID)

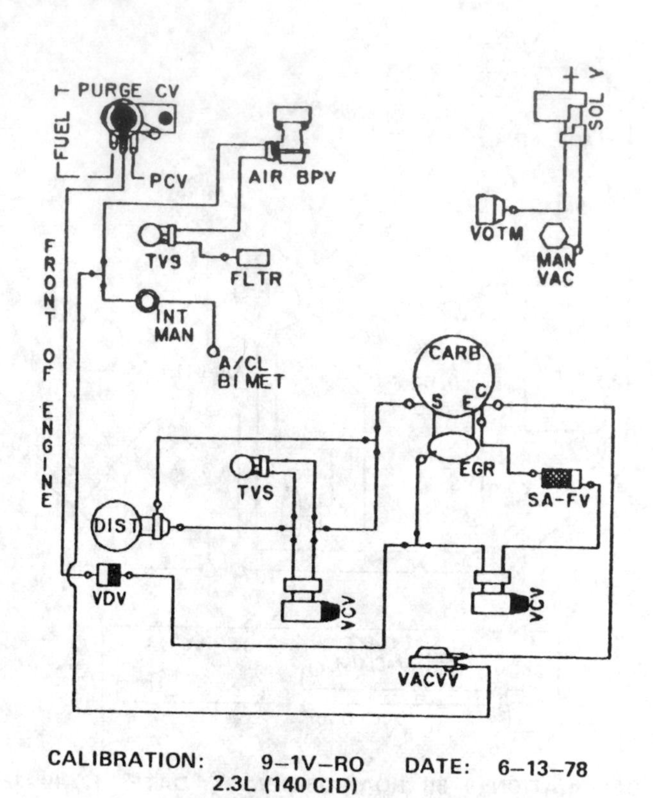

CALIBRATION: 9–1V–RO DATE: 6–13–78
2.3L (140 CID)

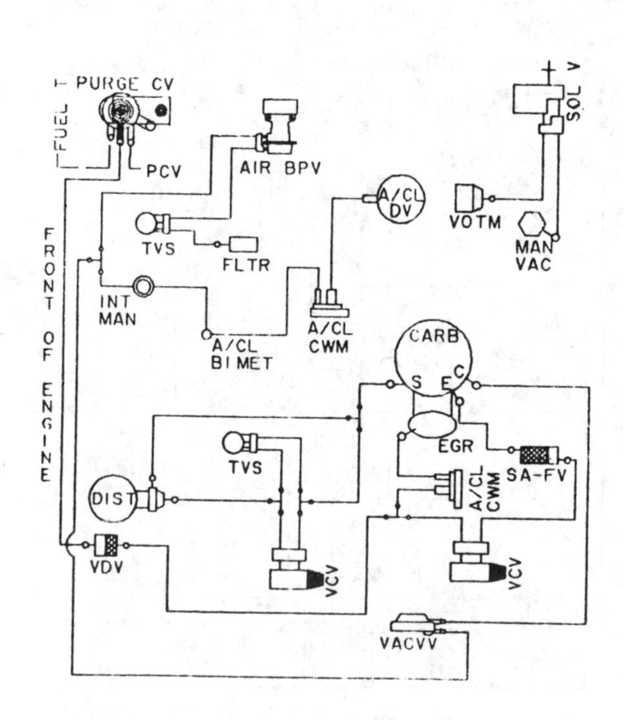

CALIBRATION: 9–1V–R10 DATE: 12–15–78
2.3L (140 CID)

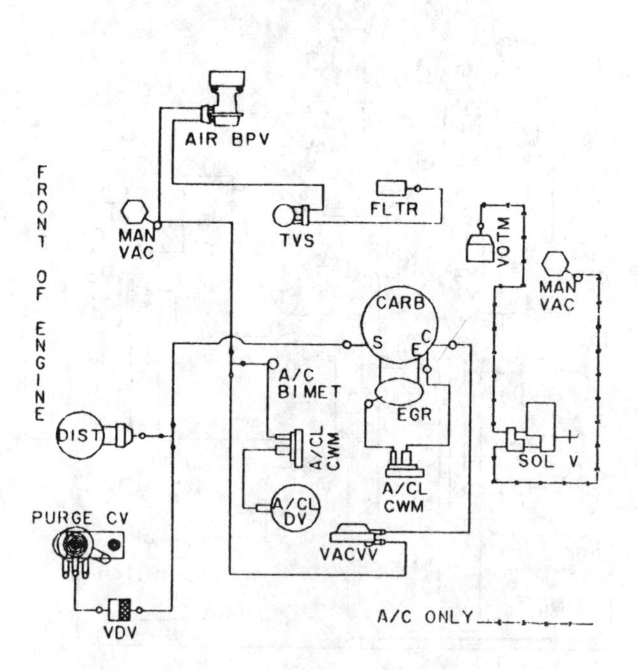

CALIBRATION: 9–1X–RO DATE: 11–7–78
2.3L HIGH ALT PINTO–BOBCAT

VACUUM CIRCUITS

1979—140 CID (2.3 L) engines

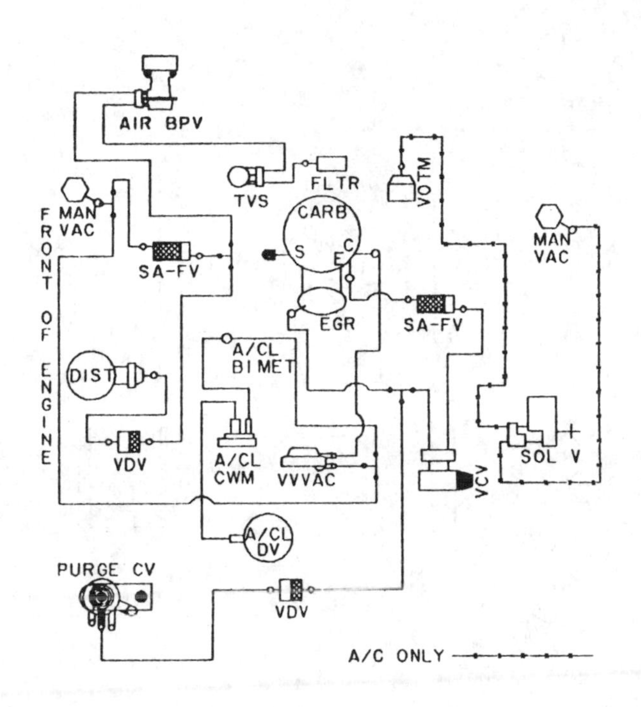

CALIBRATION: 9–2A–RO **DATE:** 10–24–78
2.3L (140 CID)

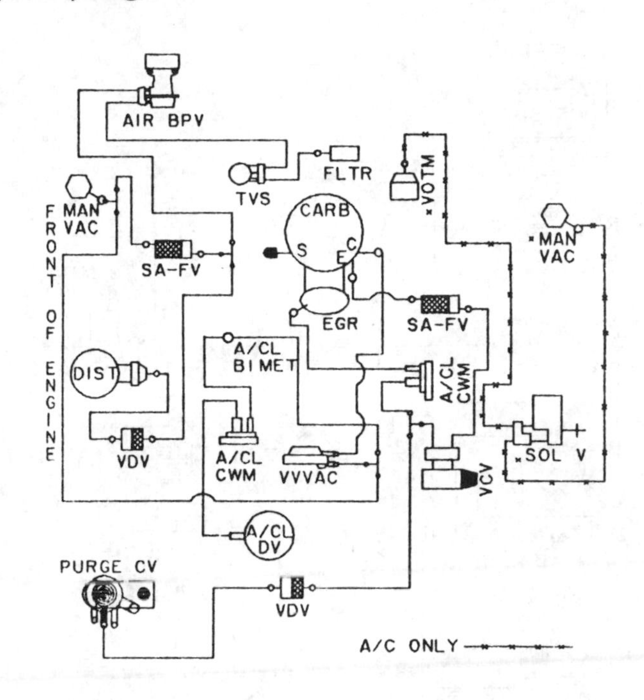

CALIBRATION: 9–2A–R10 **DATE:** 10–20–78
2.3L (140 CID)

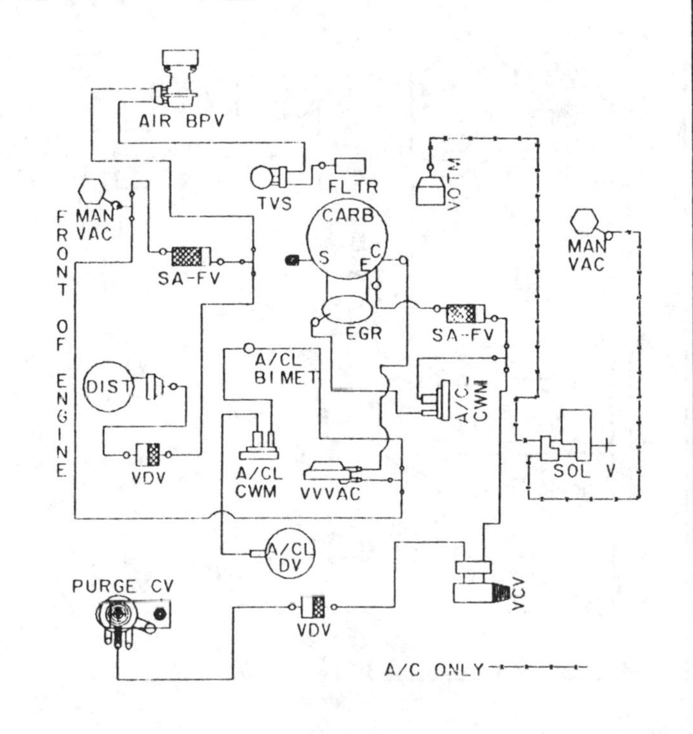

CALIBRATION: 9–2A–R12 **DATE:** 1–3–78
2.3L (140 CID)

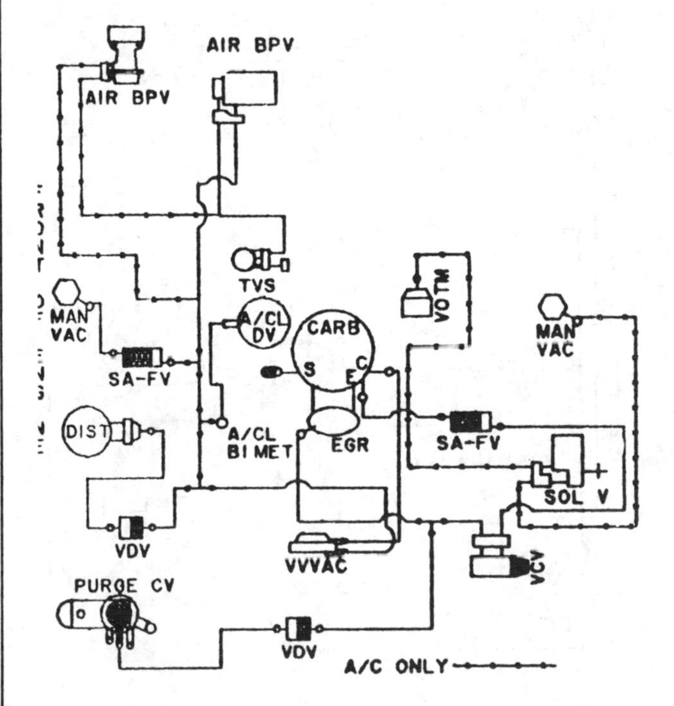

CALIBRATION: 9–2B–RO **DATE:** 7–6–78
2.3L (140 CID)

VACUUM CIRCUITS

1979—140 CID (2.3 L) engines

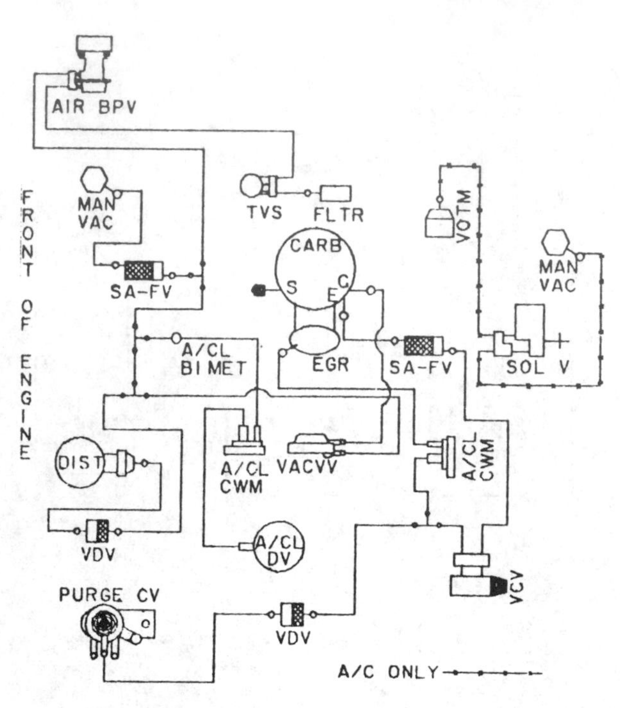

CALIBRATION: 9–2B–R13 DATE: 12–21–78
2.3L (140 CID)

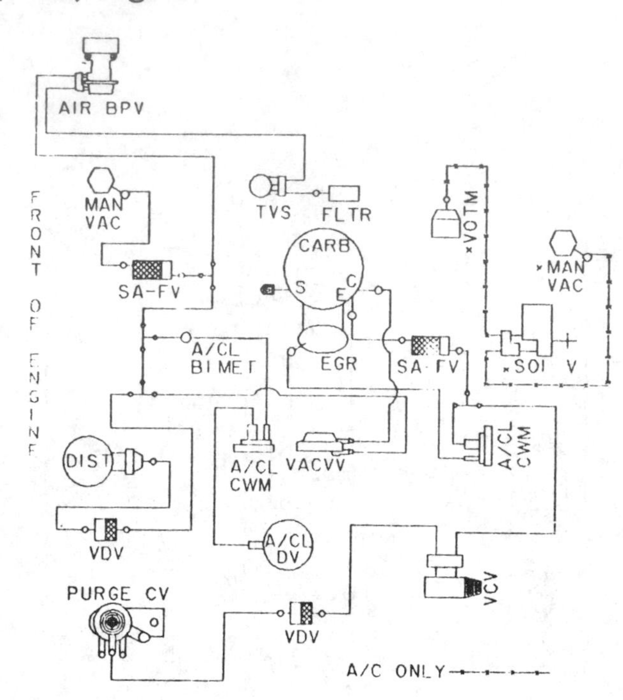

CALIBRATION: 9–2B–R10 DATE: 5–30–78
2.3L (140 CID)

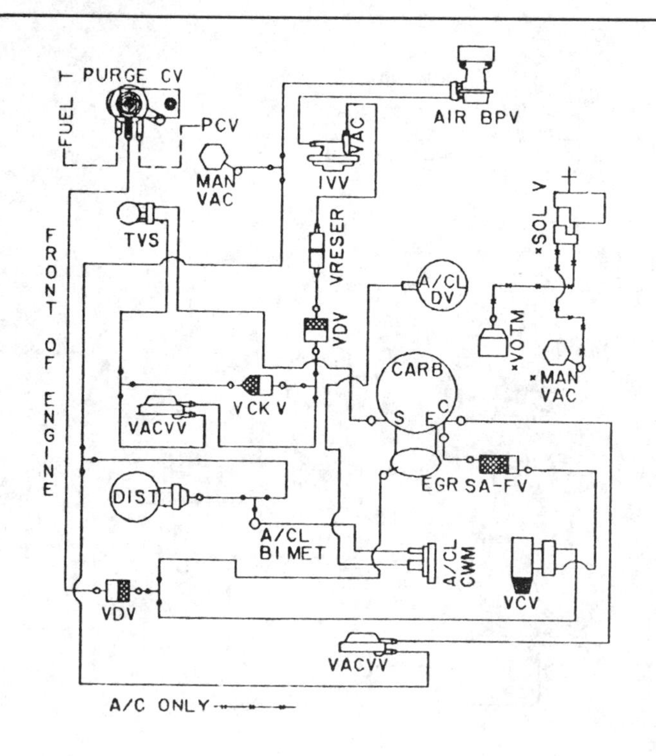

CALIBRATION. 9–2C–RO DATE: 5–19–78
2.3L (140 CID)

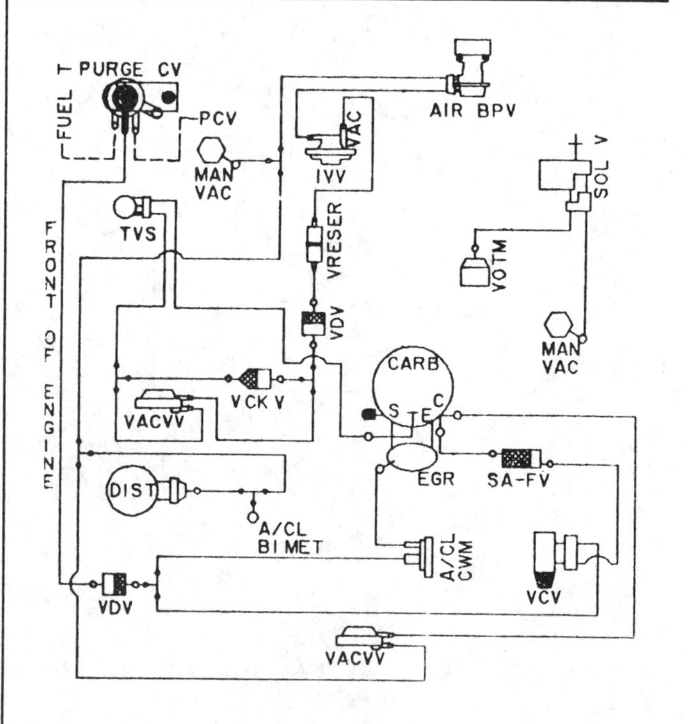

CALIBRATION: 9–2C–R10 DATE: 7–17–78
2.3L (140 CID)

VACUUM CIRCUITS

1979—140 CID (2.3 L) engines

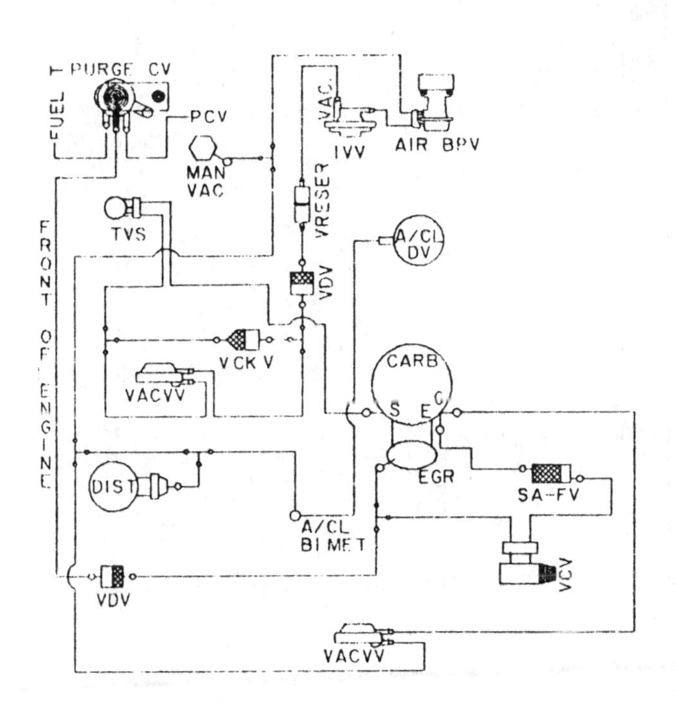

CALIBRATION: 9–2D–RO DATE: 6–14–78
2.3L (140 CID)

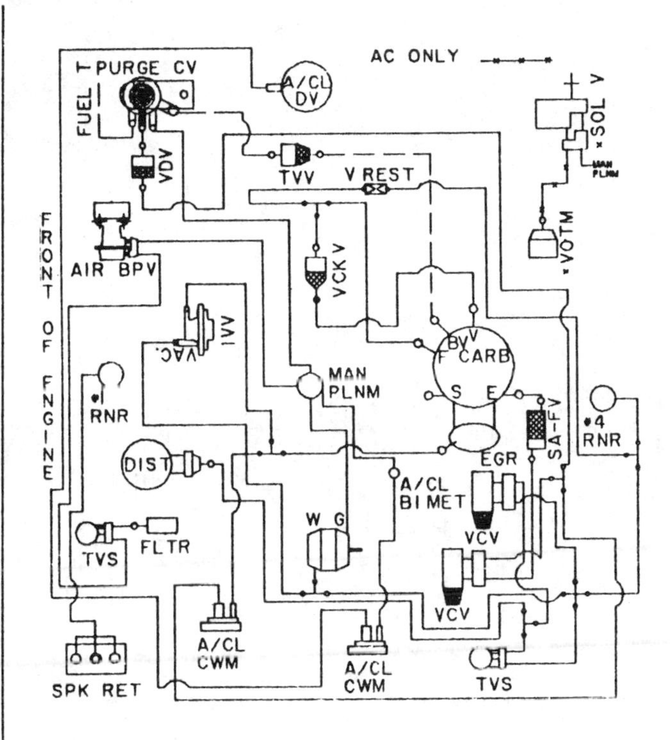

CALIBRATION: 9–2E–RO DATE: 8–7–78
2.3L (140 CID)

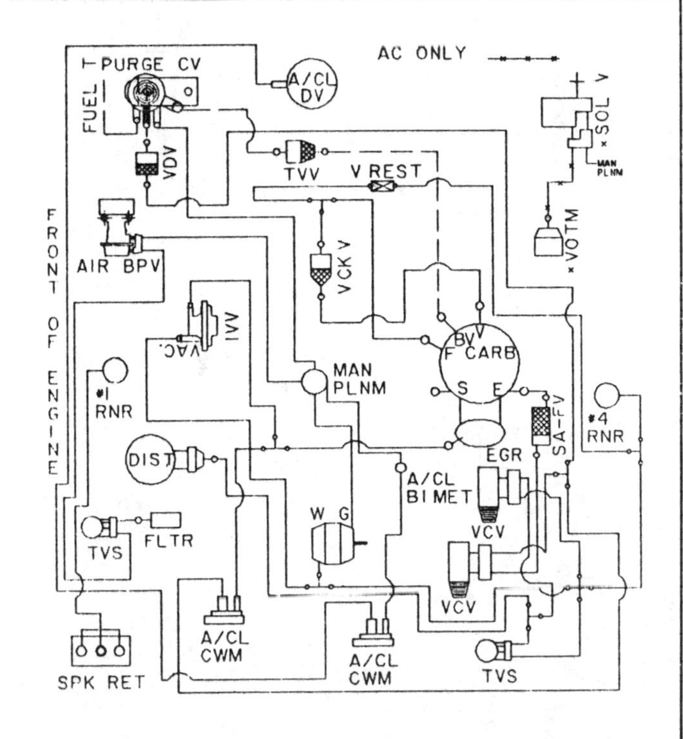

CALIBRATION: 9–2E–R12 DATE: 10–3–78
2.3L (140 CID)

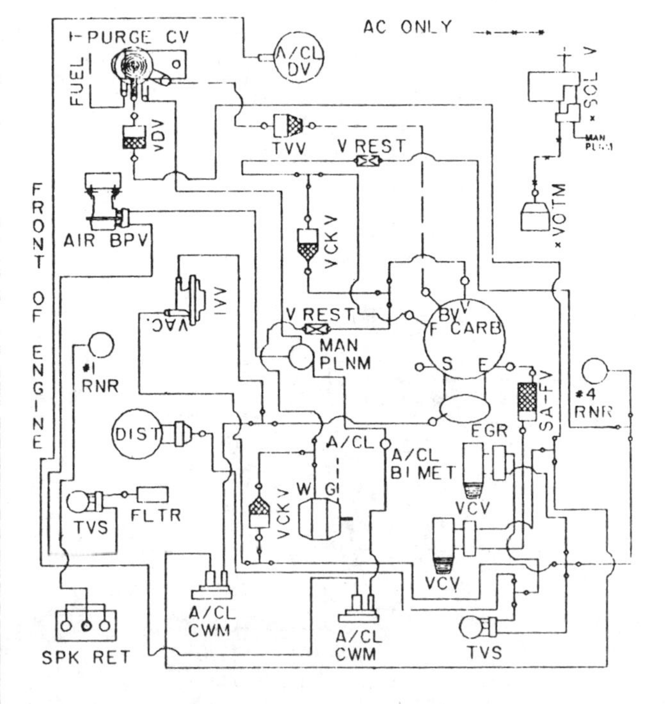

CALIBRATION: 9–2E–R16 DATE: 2–21–79
2.3L (140 CID)

VACUUM CIRCUITS

1979—140 CID (2.3 L) engines

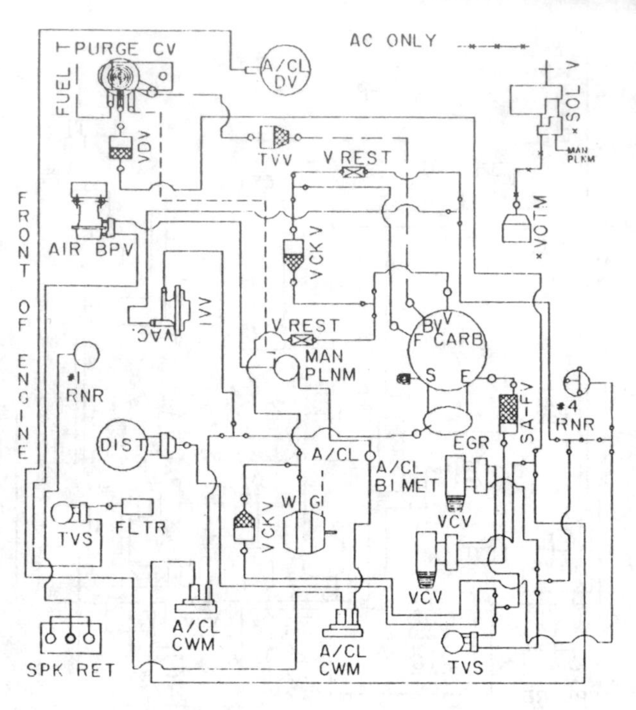

CALIBRATION: 9–2E–R17 DATE: 6–18–79
2.3L (140 CID)

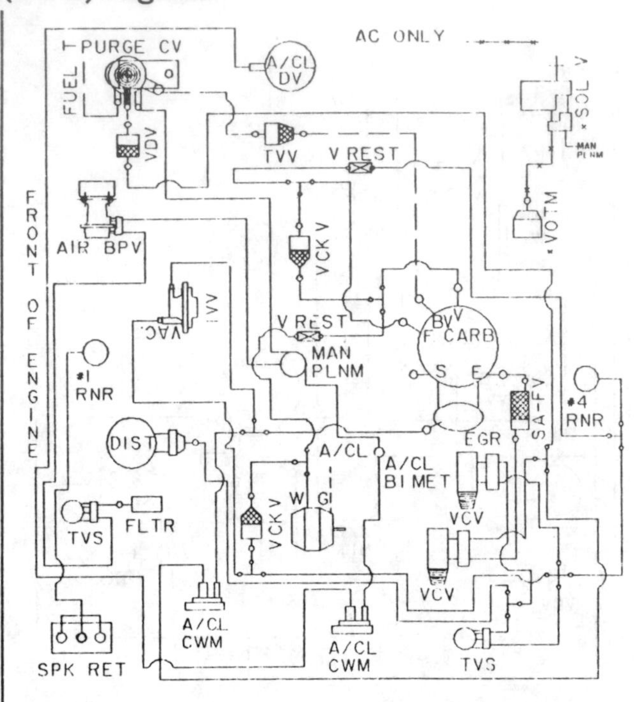

CALIBRATION: 9–2E–R18 DATE: 4–17–79
2.3L (140 CID)

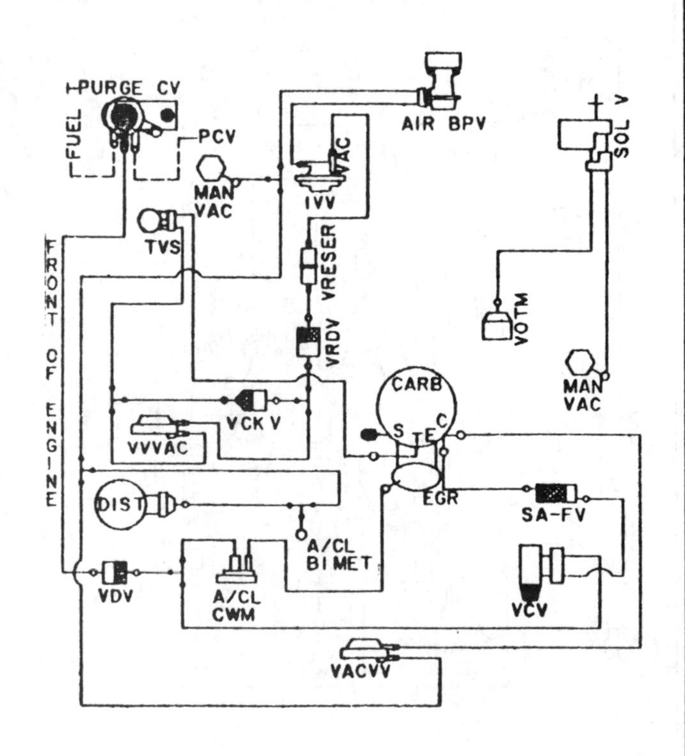

CALIBRATION: 9–2F–RO DATE: 4–10–78
2.3L (140 CID)

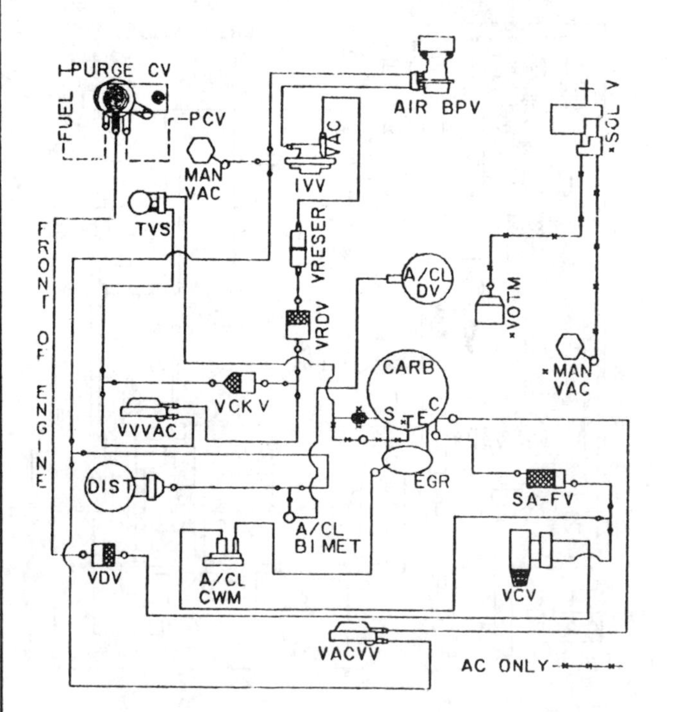

CALIBRATION: 9–2F–R11 DATE: 12–20–78
2.3L (140 CID)

VACUUM CIRCUITS

1979—140 CID (2.3 L) engines

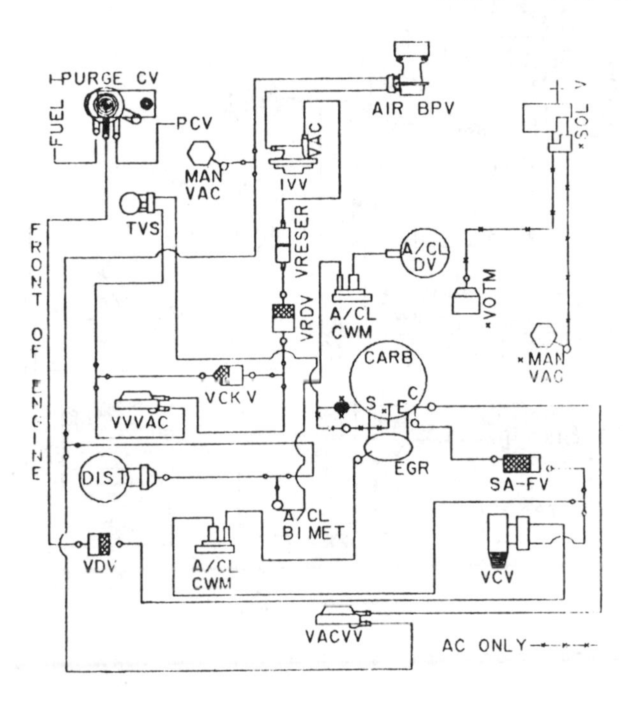

CALIBRATION: 9–2F–R13 DATE: 1–31–79
2.3L (140 CID)

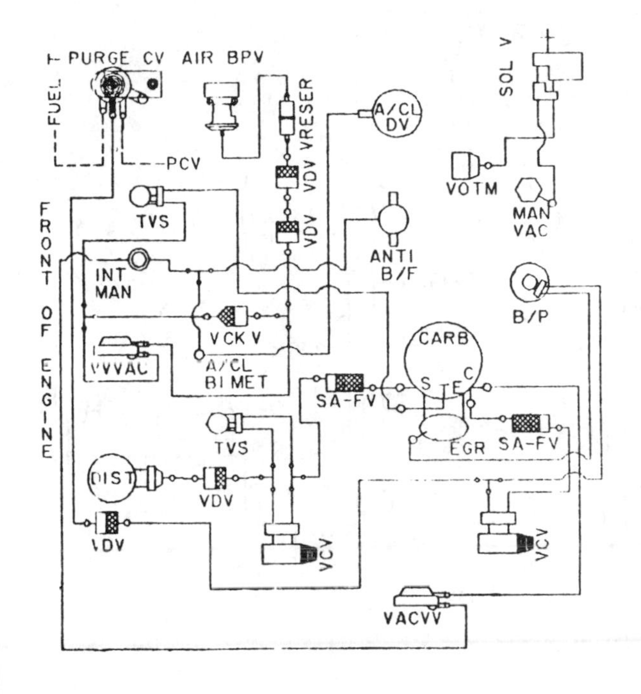

CALIBRATION: 9–2N–RO DATE: 7–11–78
2.3L (140 CID)

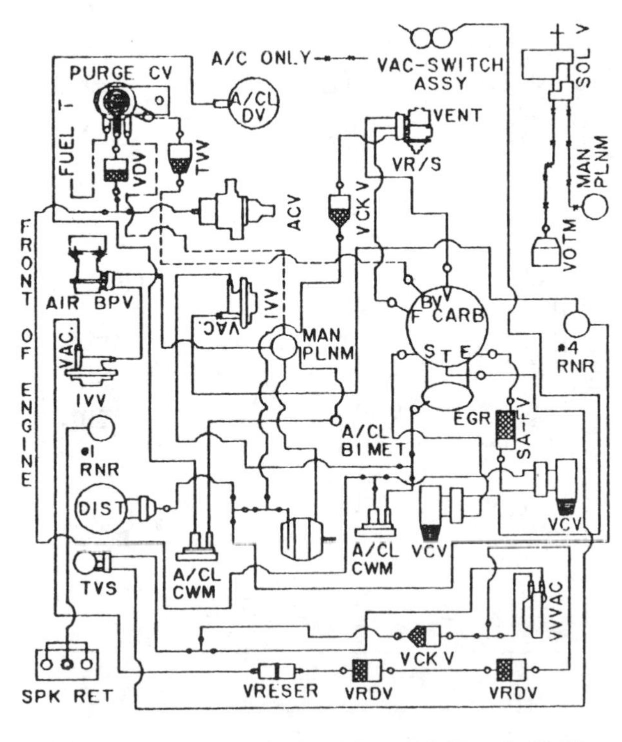

CALIBRATION: 9–2R–RO DATE: 8–7–78
2.3L (140 CID)

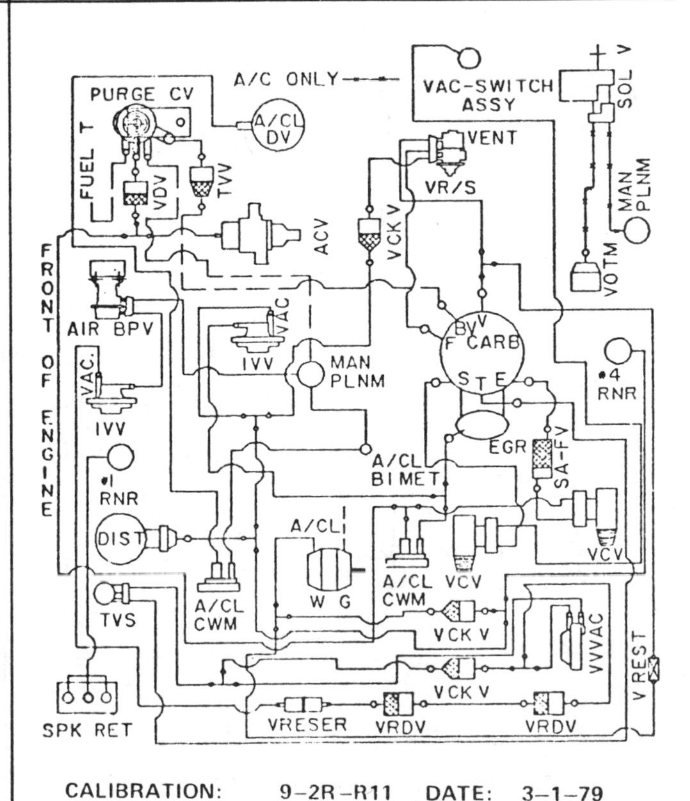

CALIBRATION: 9–2R–R11 DATE: 3–1–79
2.3L (140 CID)

VACUUM CIRCUITS

1979—140 CID (2.3 L) engines

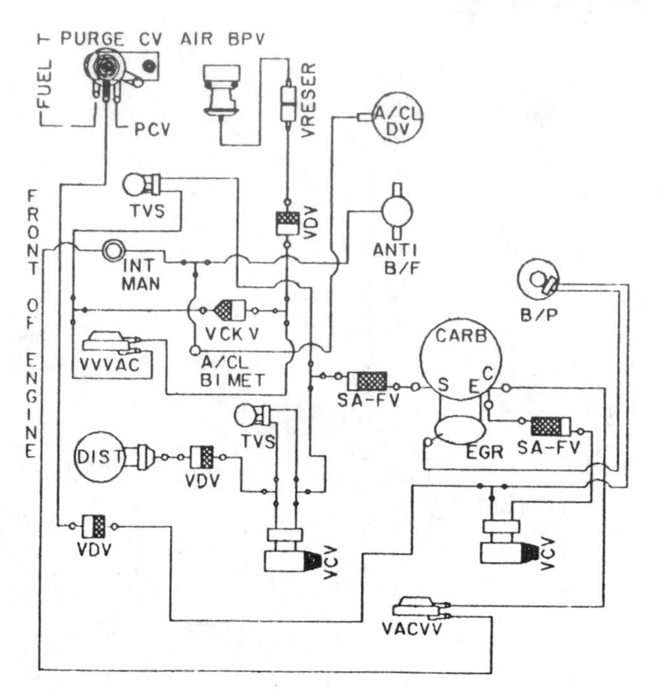

CALIBRATION: 9–2S–RO DATE: 6–9–78
2.3L (140 CID)

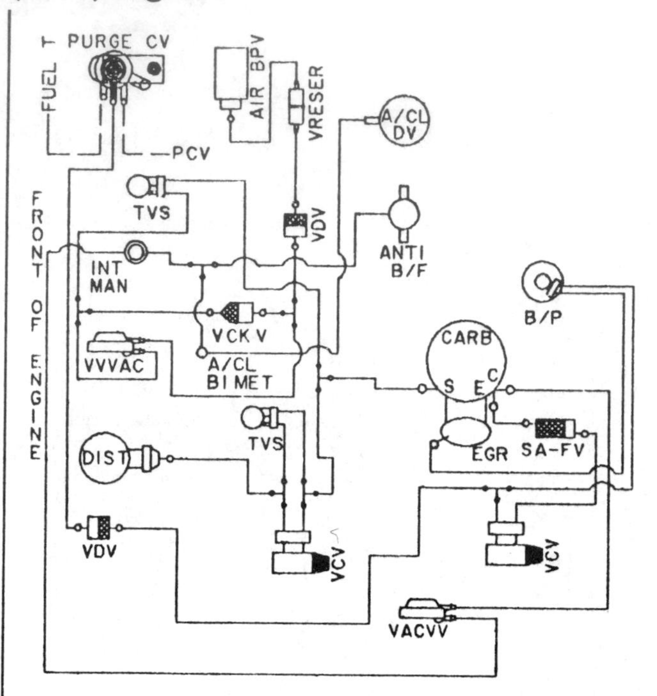

CALIBRATION: 9–2S–R2 DATE: 7–7–78
2.3L (140 CID)

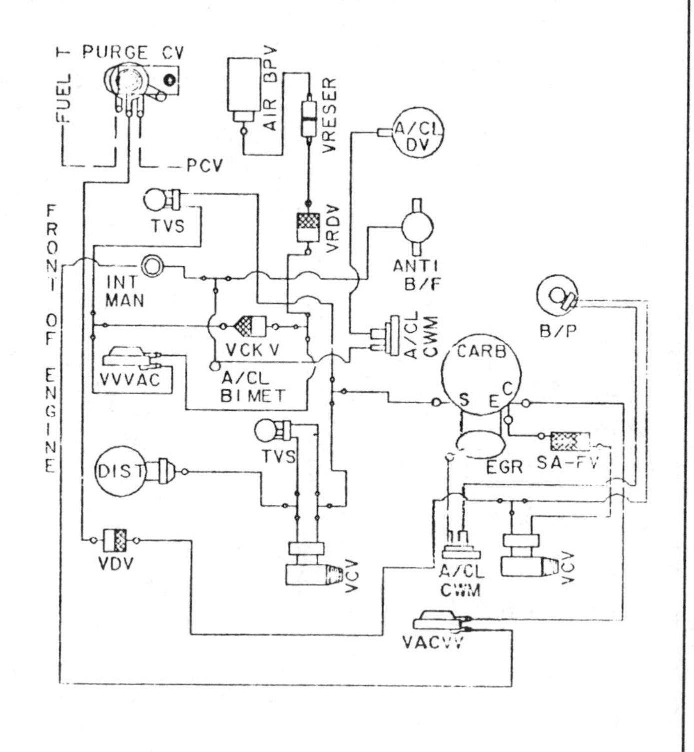

CALIBRATION: 9–2S–R10 DATE: 12–6–78
2.3L (140 CID)

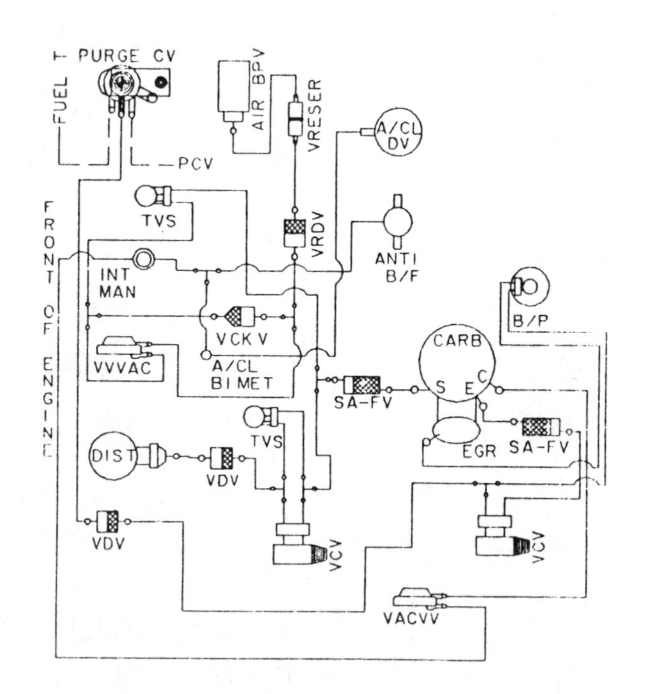

CALIBRATION: 9–2S–R11 DATE: 2–20–79
2.3L (140 CID)

VACUUM CIRCUITS

1979—140 CID (2.3 L) engines

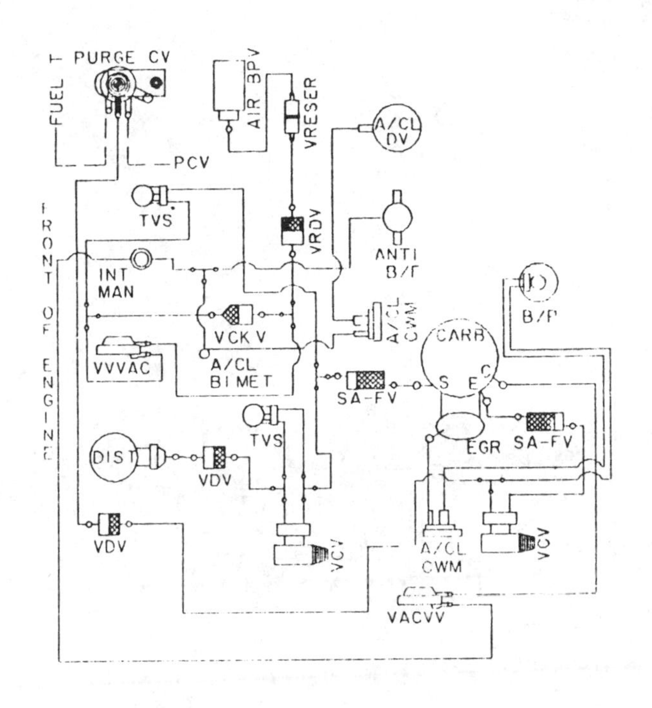

CALIBRATION: 9–2S–R12 DATE: 3–6–79
2.3L (140 CID)

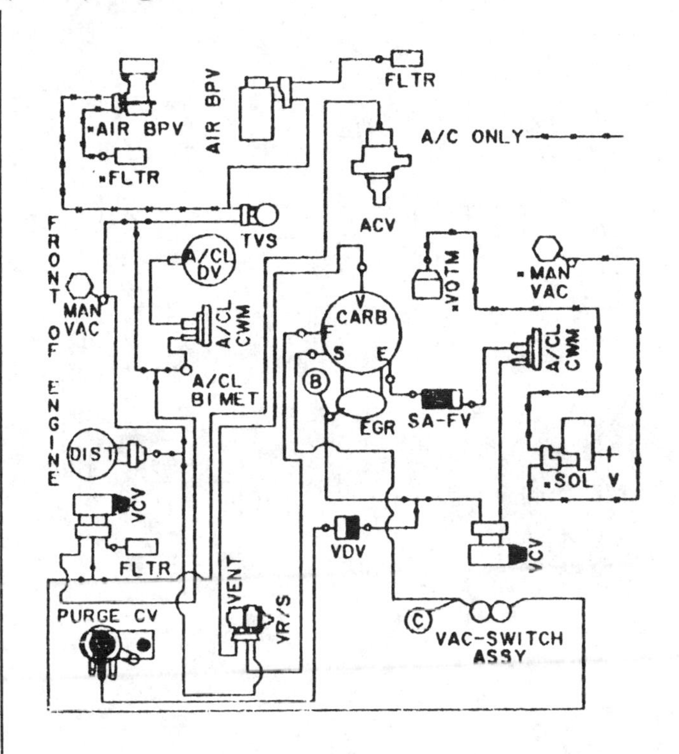

CALIBRATION: 9–2T–RO DATE: 6–2–78
2.3L (140 CID)

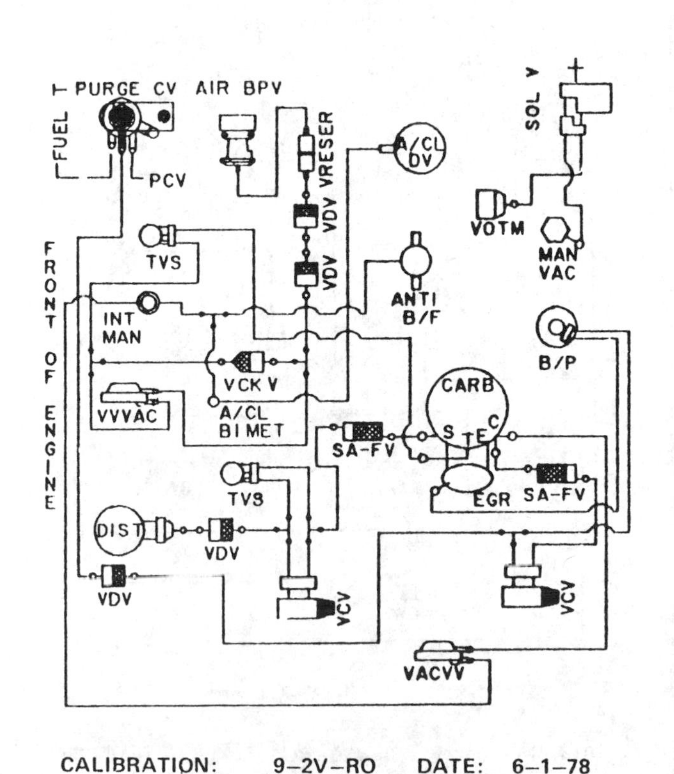

CALIBRATION: 9–2V–RO DATE: 6–1–78
2.3L (140 CID)

PURGE CV AIR BPV
FUEL
PCV
VRESER
VRDV
A/CL DV
SOL V
VOTM
MAN VAC
TVS
VRDV
ANTI B/F
FRONT OF ENGINE
INT MAN
B/P
VCK V
A/CL CWM
CARB
VVVAC
A/CL BI MET
SA-FV
S
E
C
TVS
EGR
SA-FV
DIST
VDV
VDV
VCV
A/CL CWM
VCV
VACVV

CALIBRATION: 9–2V–R10 DATE: 12–19–78
2.3L (140 CID)

VACUUM CIRCUITS

1979—140 CID (2.3 L) engines

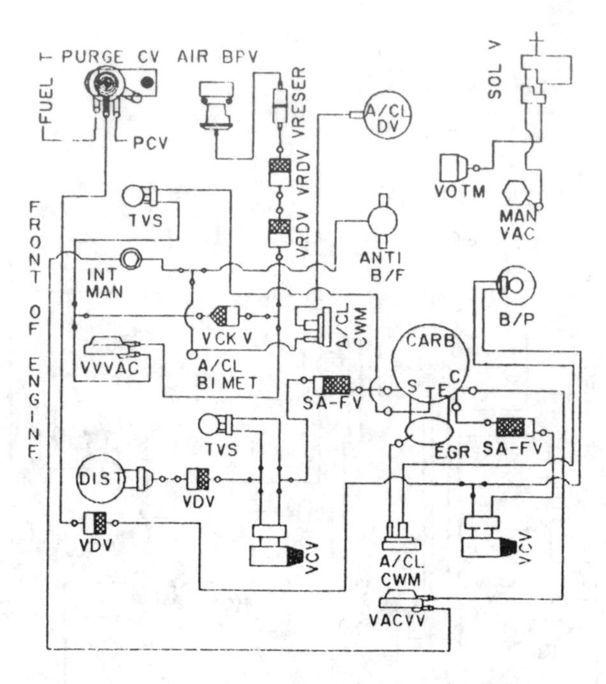

CALIBRATION: 9–2V–R12 DATE: 3–5–79
2.3L (140 CID)

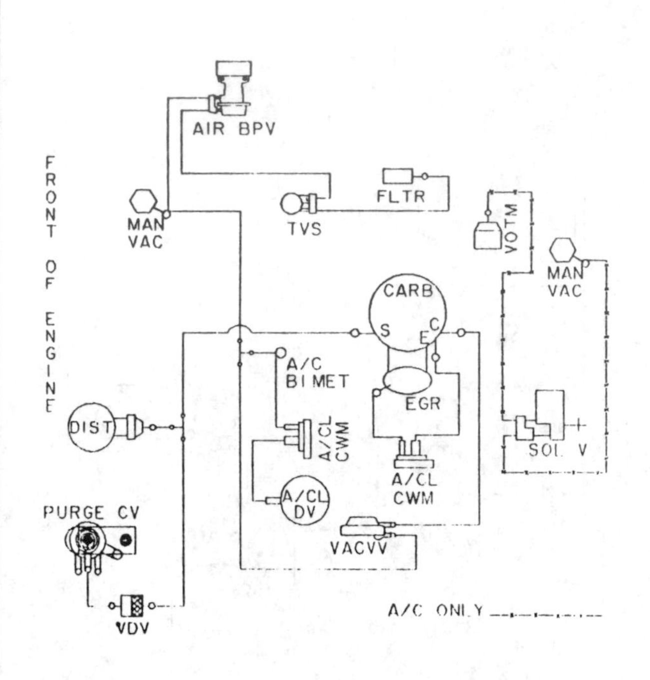

CALIBRATION: 9–2X–RO DATE: 1–22–79
2.3L (140 CID)

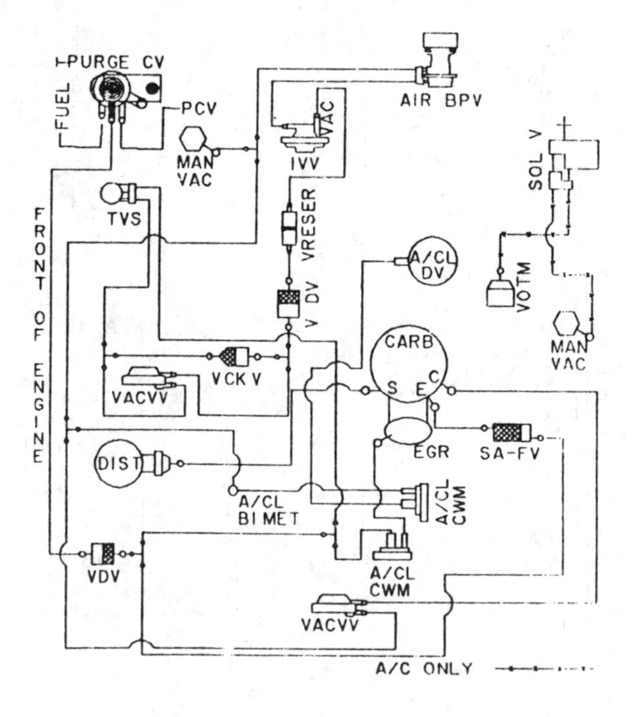

CALIBRATION: 9–2Y–RO DATE: 1–19–79
2.3L (140 CID)

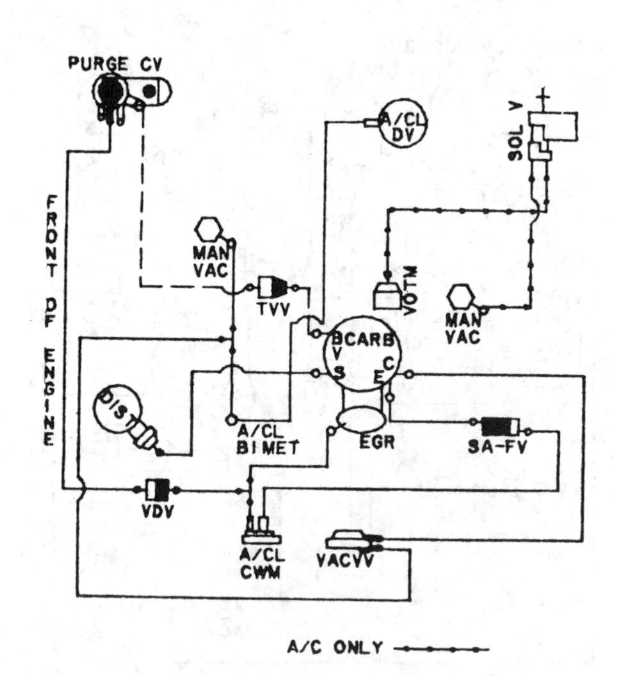

CALIBRATION: 9–21B–RO DATE: 5–18–78
2.3L (140 CID)

VACUUM CIRCUITS

1979—140 CID (2.3 L) engines

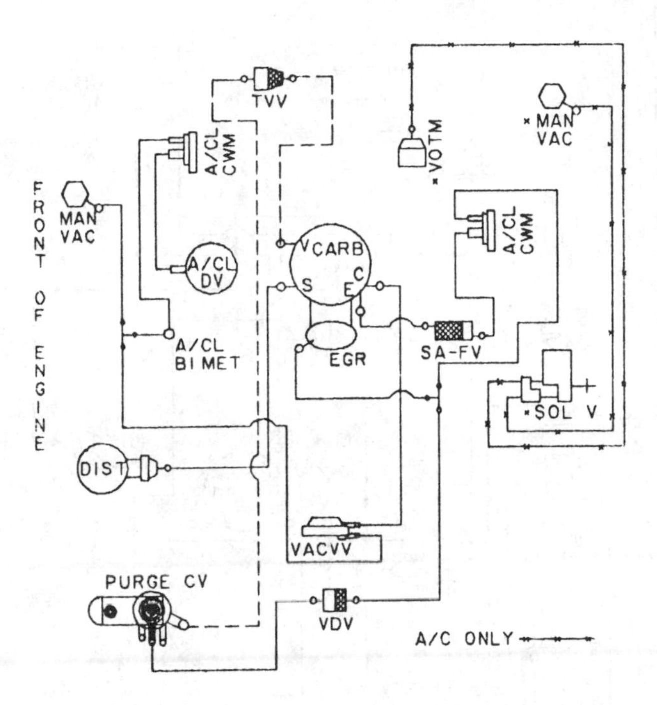

CALIBRATION: 9–21B–R10 DATE: 8–28–78
2.3L (140 CID)

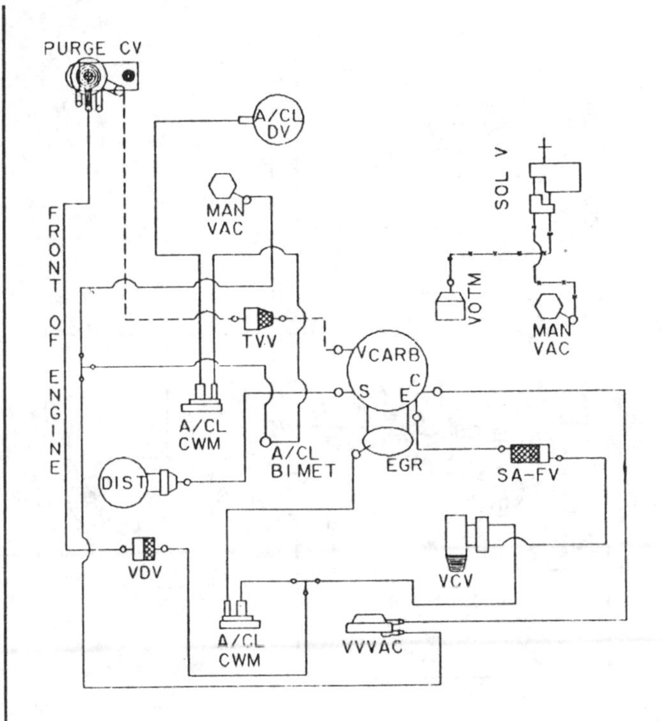

CALIBRATION: 9–21B–R13 DATE: 12–4–78
2.3L (140 CID)

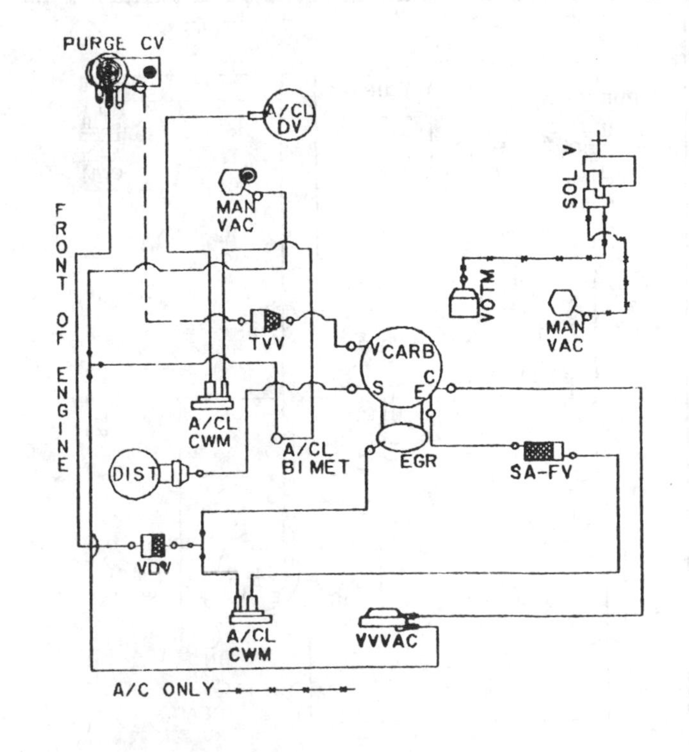

CALIBRATION: 9–21C–RO DATE: 6–7–78
2.3L (140 CID)

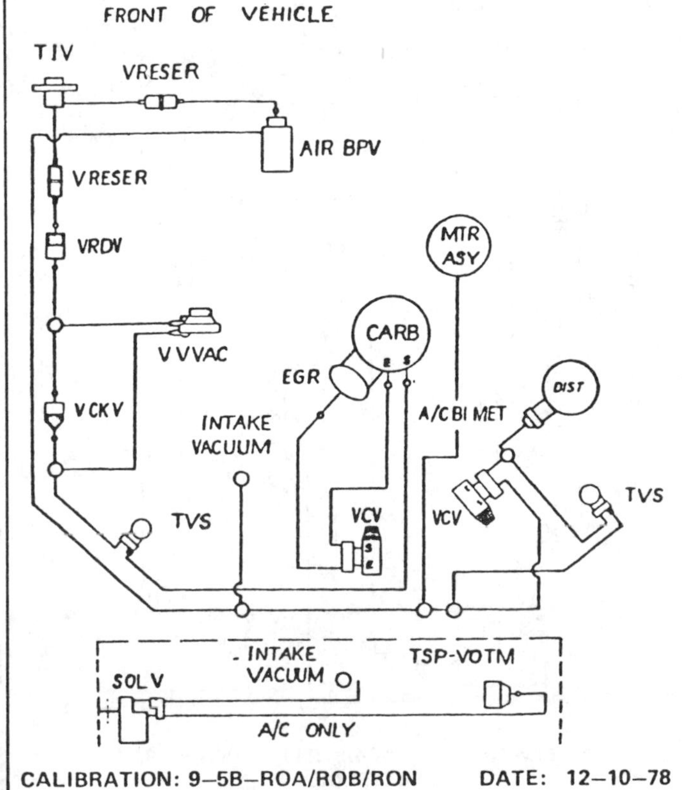

CALIBRATION: 9–5B–ROA/ROB/RON DATE: 12–10–78

VACUUM CIRCUITS

1979—170 CID (2.8 L) engines

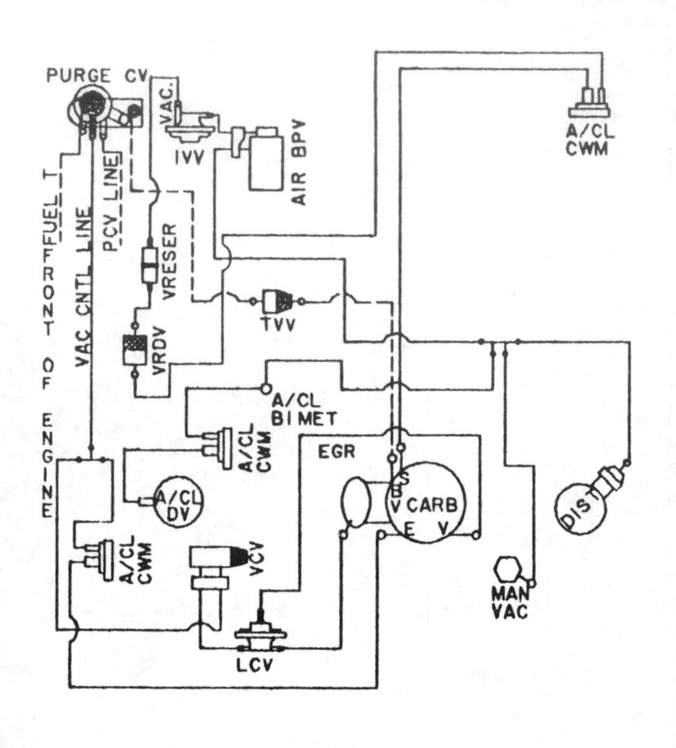

CALIBRATION: 9–3A–RO DATE: 6–12–78
2.8L (171 CID)

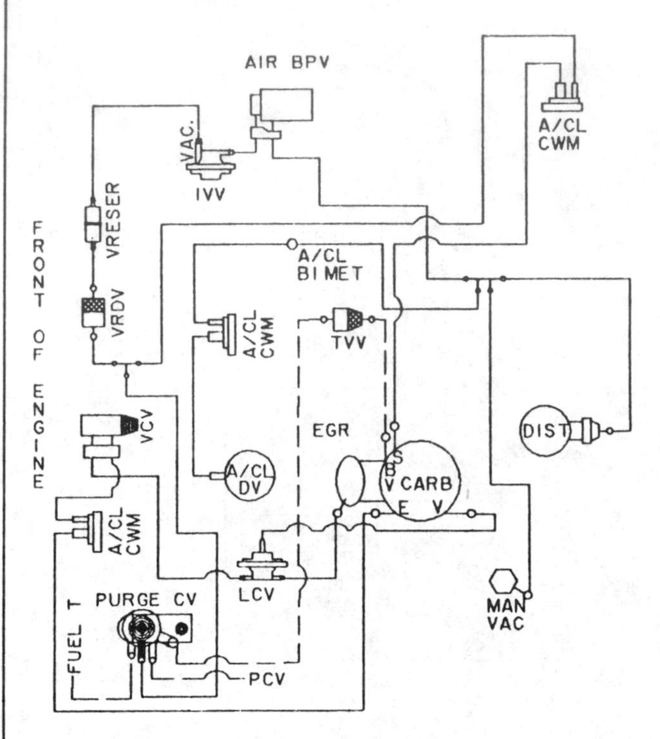

CALIBRATION: 9–4A–RO DATE: 10–18–78
2.8L (171 CID)

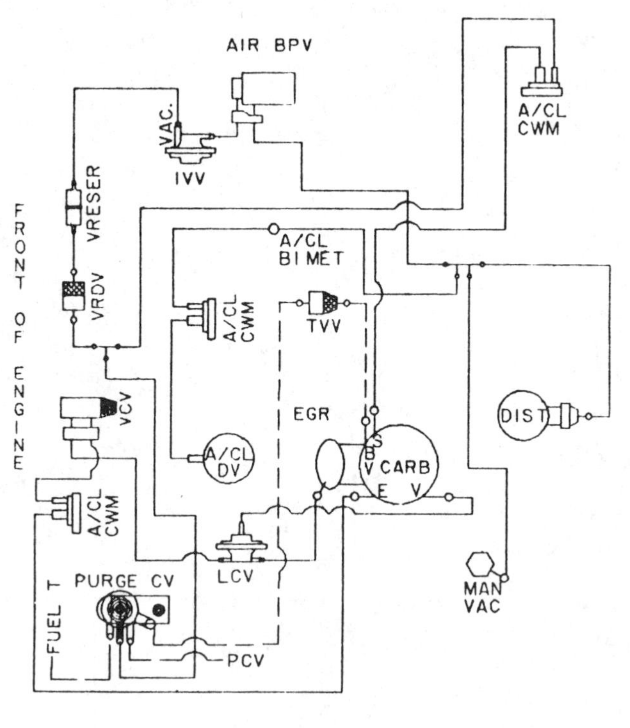

CALIBRATION: 9–4A–R10 DATE: 9–7–78
2.8L (171 CID)

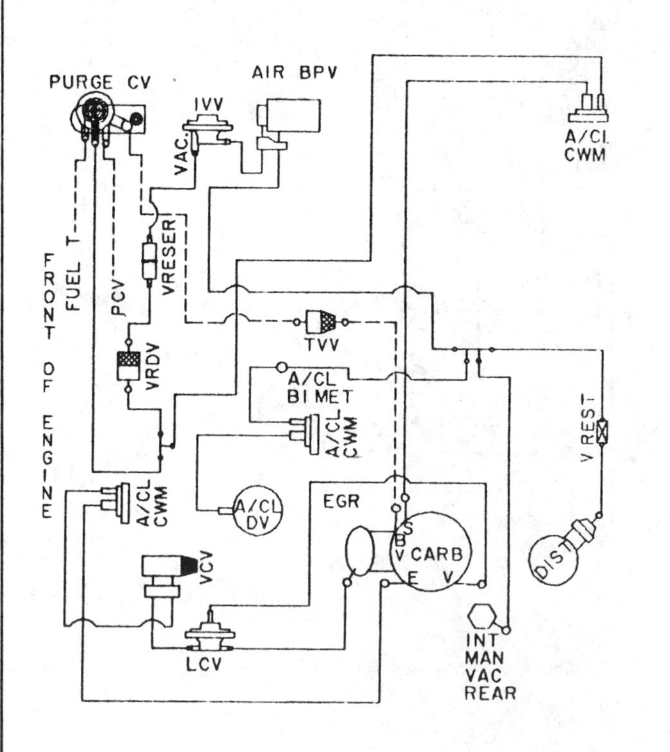

CALIBRATION: 9–4A–R11 DATE: 10–18–78
2.8L (171 CID)

VACUUM CIRCUITS

1979—170 CID (2.8 L) engines

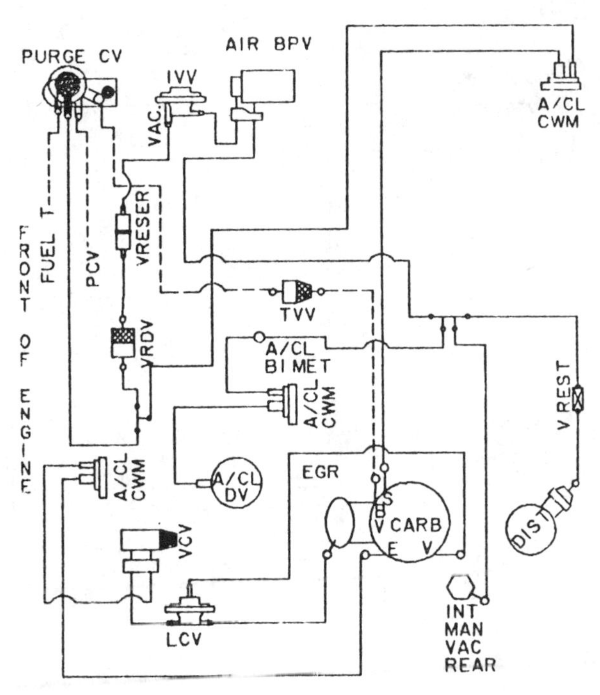

CALIBRATION: 9–4A–R12 DATE: 12–20–78
2.8L (171 CID)

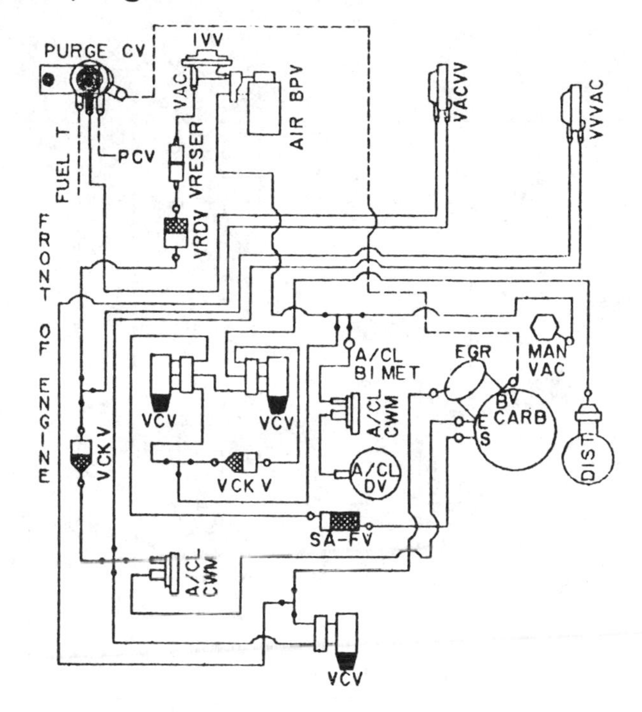

CALIBRATION: 9–4N–RO DATE: 7–11–78
2.8L (171 CID)

1979—200 CID (3.3 L) engines

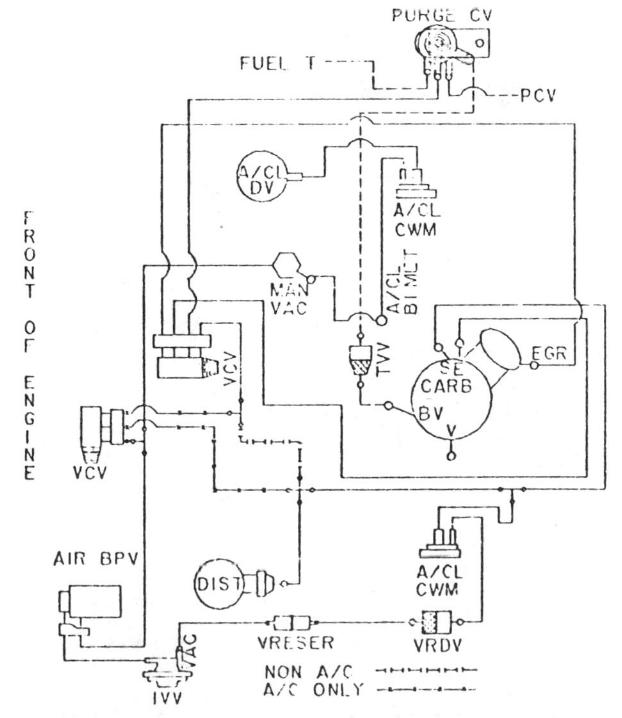

CALIBRATION: 9–6A–RO DATE: 10–25–78
3.3L (200 CID)

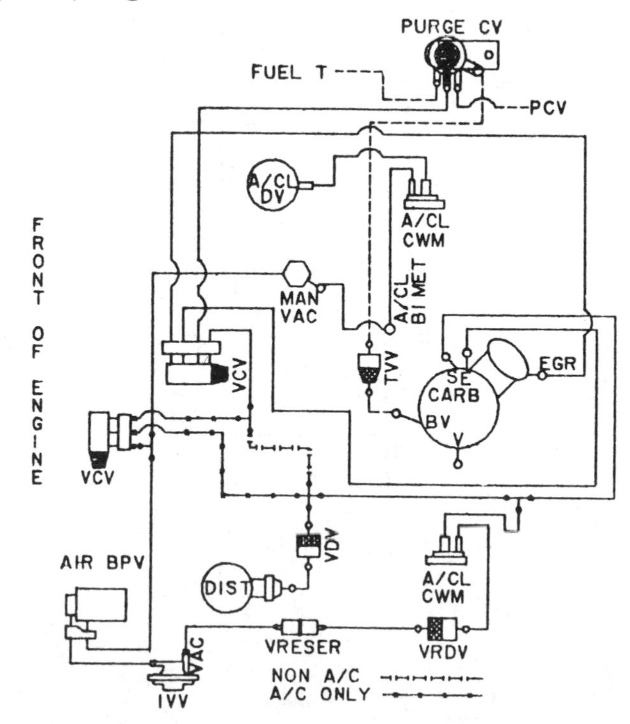

CALIBRATION: 9–6A–R10 DATE: 9–26–78
3.3L (200 CID)

VACUUM CIRCUITS

1979—200 CID (3.3 L) engines

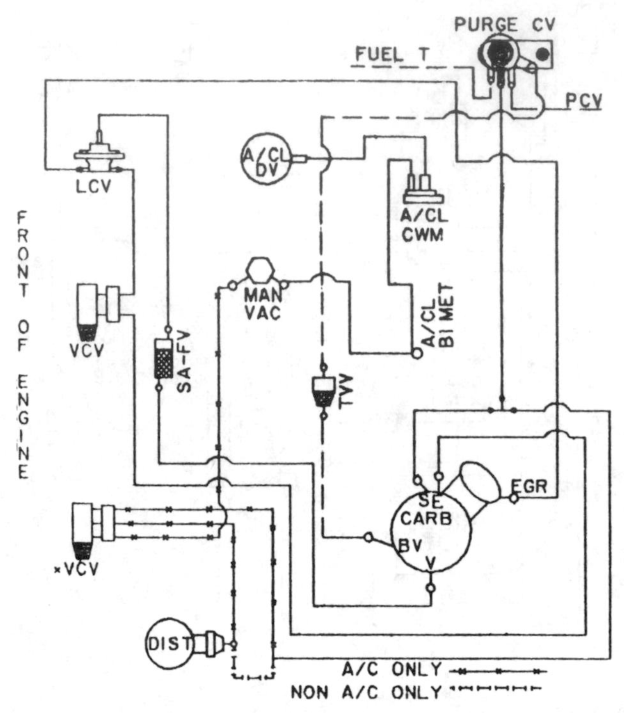

CALIBRATION: 9–7A–RO DATE: 9–26–78
3.3L (200 CID)

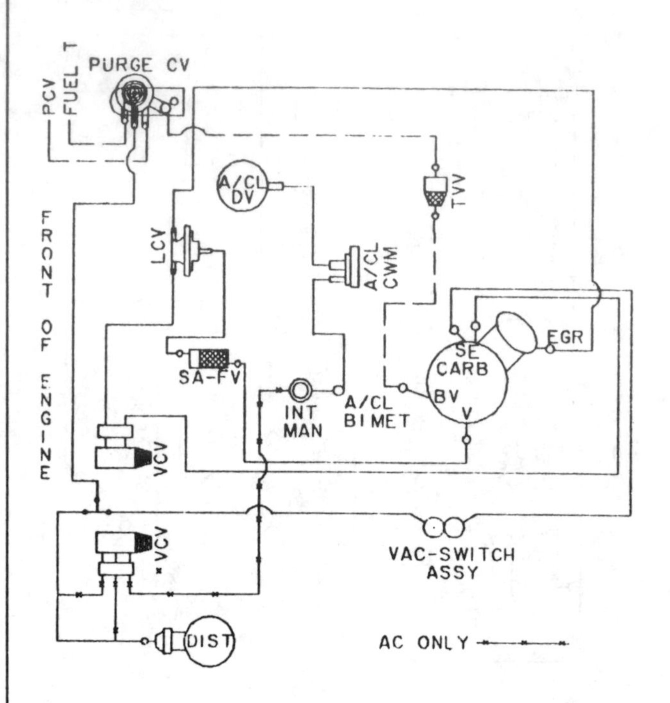

CALIBRATION: 9–7A–R10 DATE: 6–28–78
3.3L (200 CID)

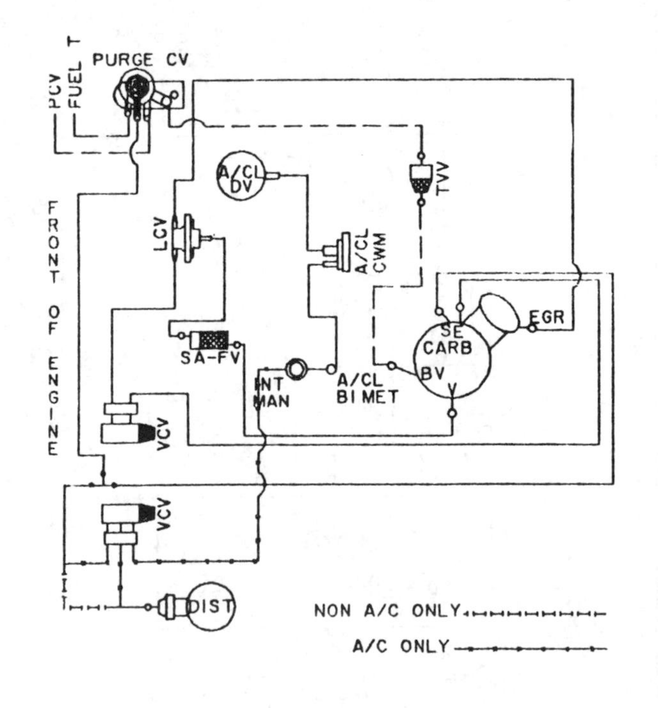

CALIBRATION: 9–7A–R11 DATE: 10–13–78
3.3L (200 CID)

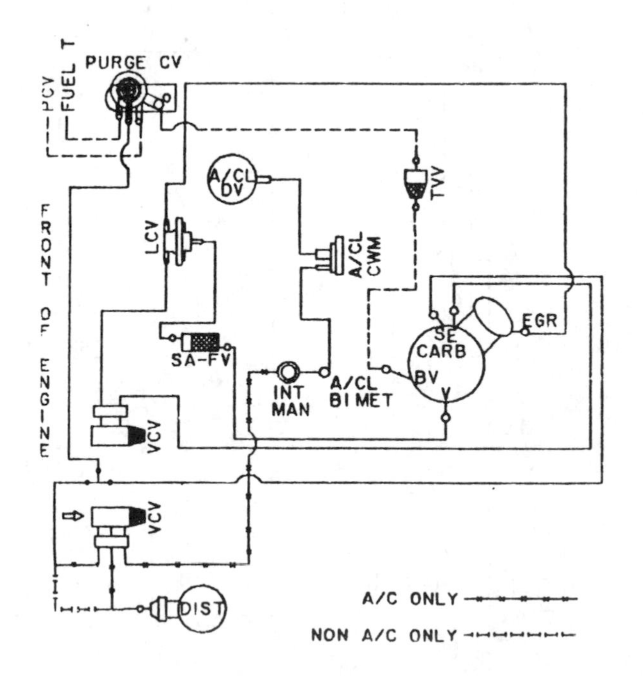

CALIBRATION: 9–7A–R12 DATE: · 10–25–78
3.3L (302 CID)

VACUUM CIRCUITS

1979—200 CID (3.3 L) engines

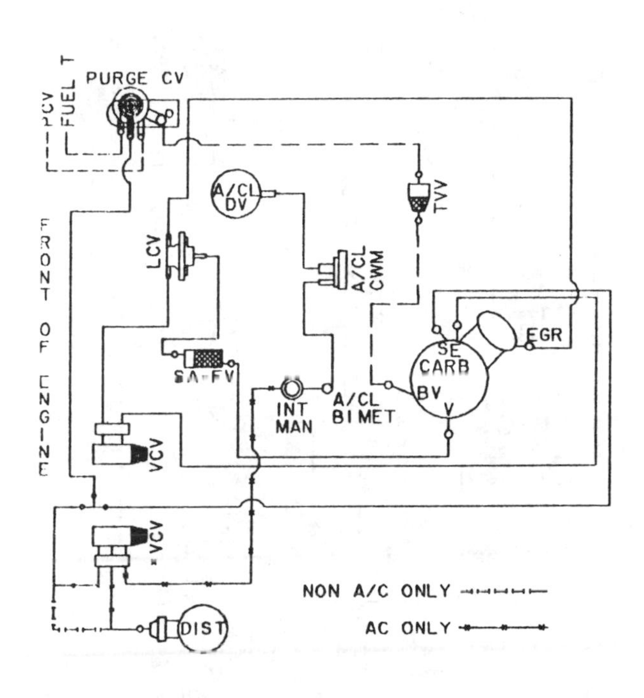

CALIBRATION: 9–7A–R13 DATE: 10–21–78
3.3L (200 CID)

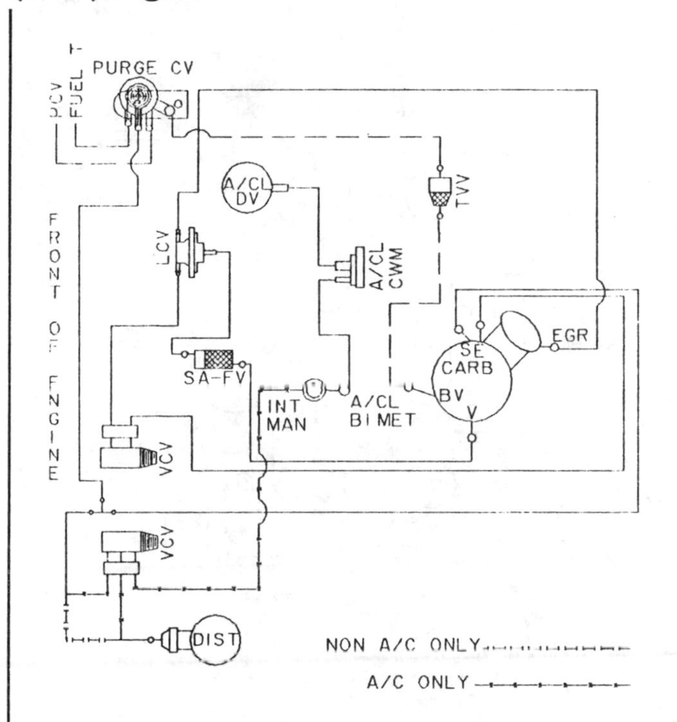

CALIBRATION: 9–7A–R15 DATE: 4–23–79
3.3L (200 CID)

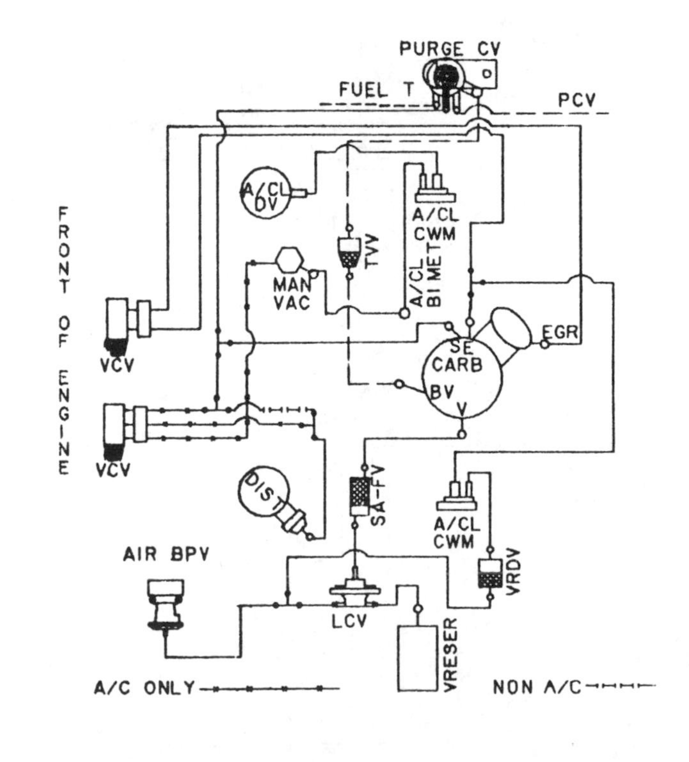

CALIBRATION: 9–7D–RO DATE: 9–29–78
3.3L (200 CID)

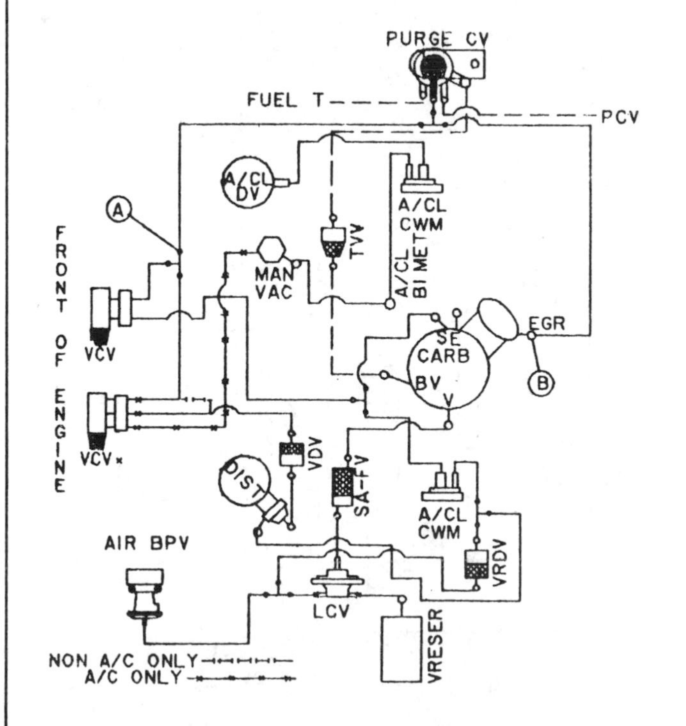

CALIBRATION: 9–7N–R2 DATE: 9–26–78
3.3L (200 CID)

VACUUM CIRCUITS

1979—200 CID (3.3 L) engines

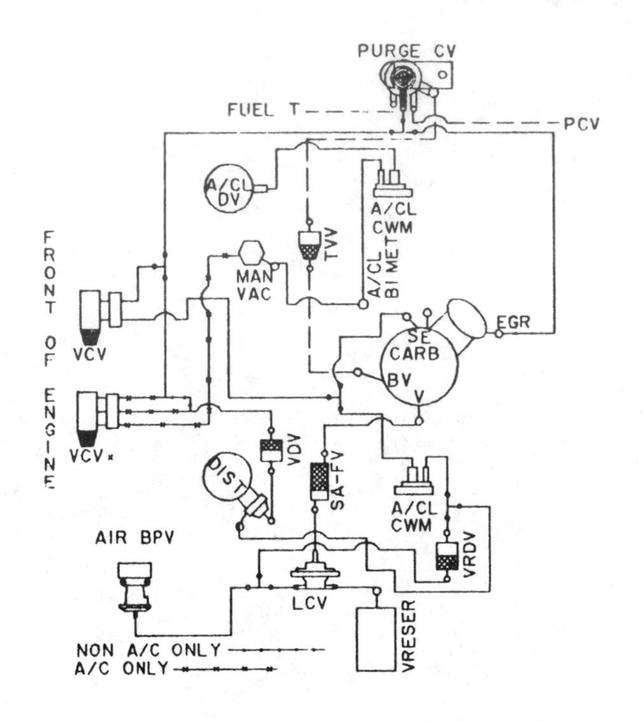

CALIBRATION: 9–7N–R3 DATE: 10–26–78
3.3L (200 CID)

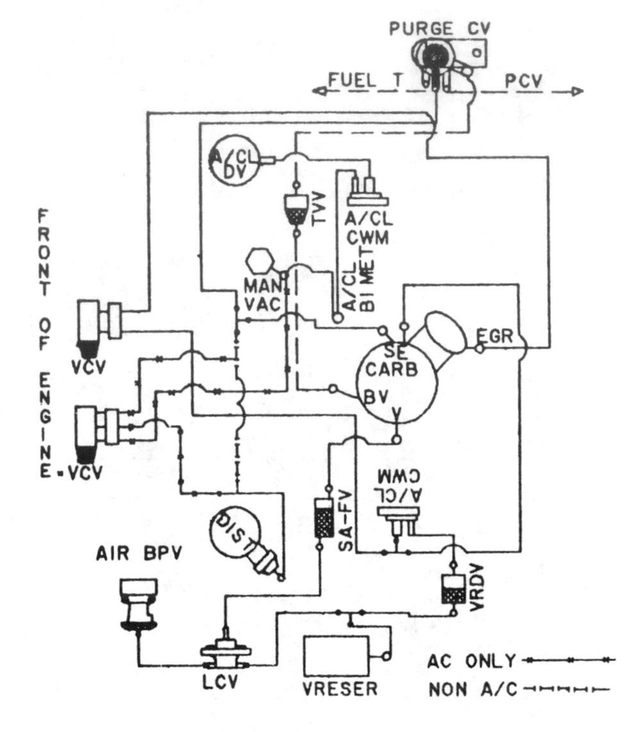

CALIBRATION: 9–7Z–R10 DATE: 9–29–78
3.3L (200 CID)

1979—250 CID (4.1 L) engines

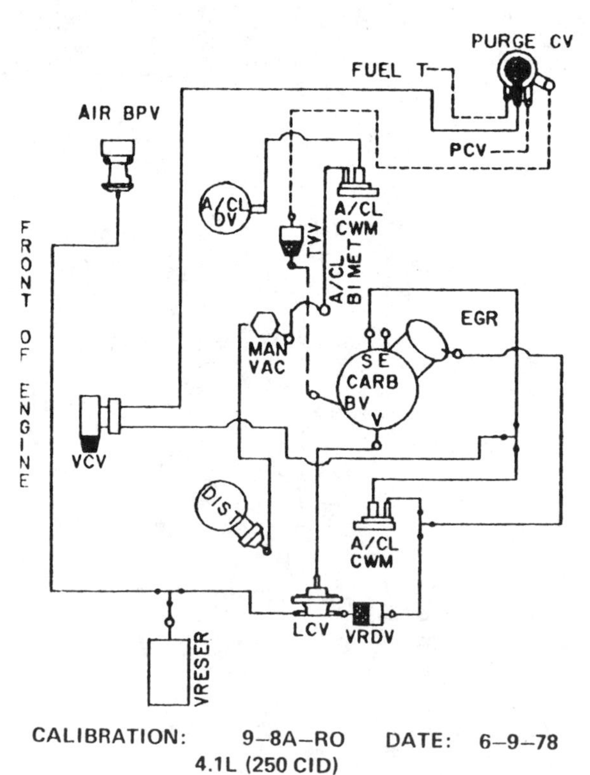

CALIBRATION: 9–8A–RO DATE: 6–9–78
4.1L (250 CID)

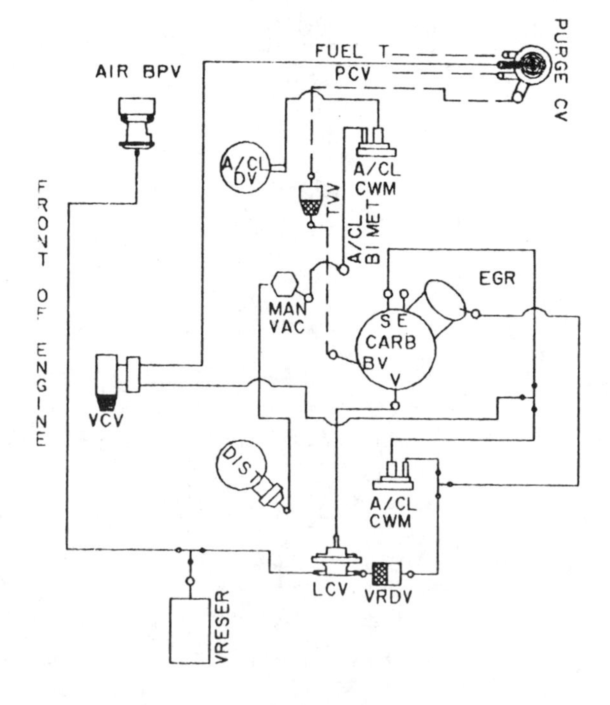

CALIBRATION: 9–8A–R10 DATE: 10–10–78
4.1L (250 CID)

VACUUM CIRCUITS

1979—250 CID (4.1 L) engines

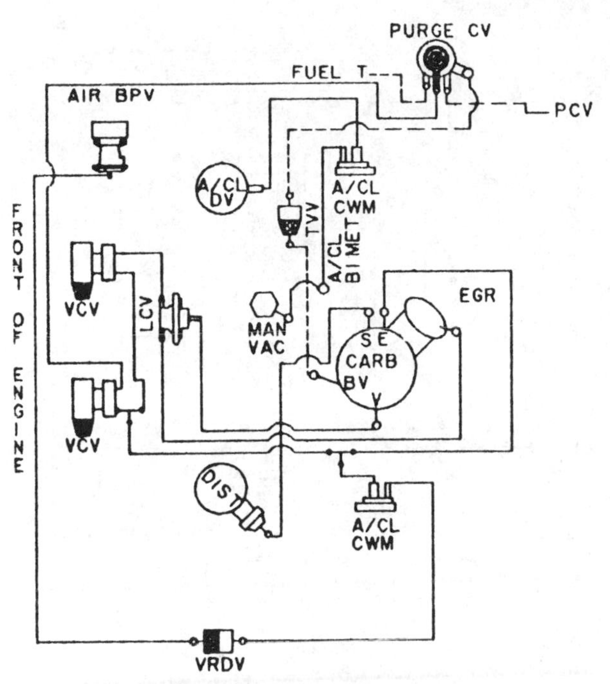

CALIBRATION: 9–9A–RO DATE: 8–30–78
4.1L (250 CID)

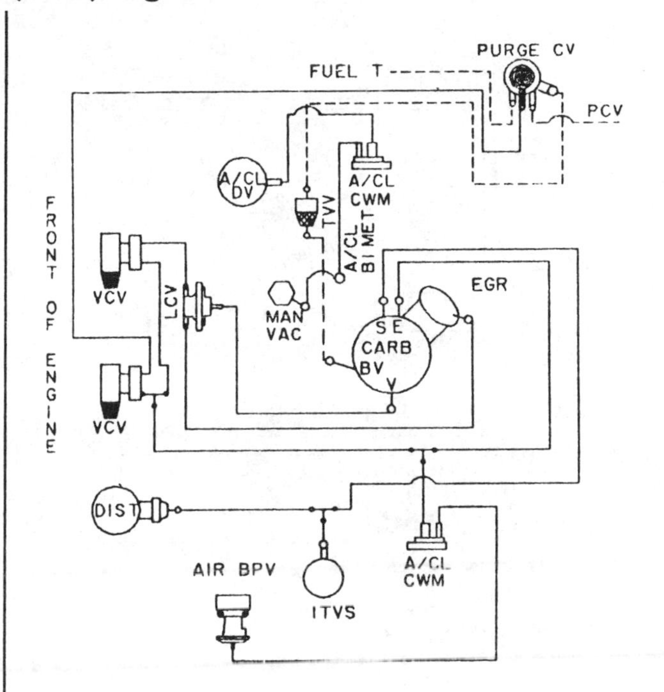

CALIBRATION: 9–9A–R13 DATE: 10–10–78
4.1L (250 CID)

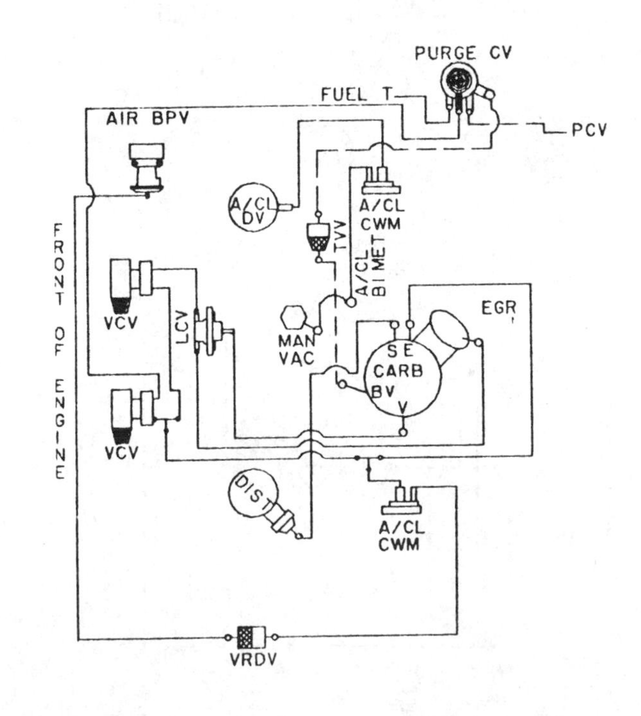

CALIBRATION: 9–9A–R14 DATE: 10–15–78
4.1L (250 CID)

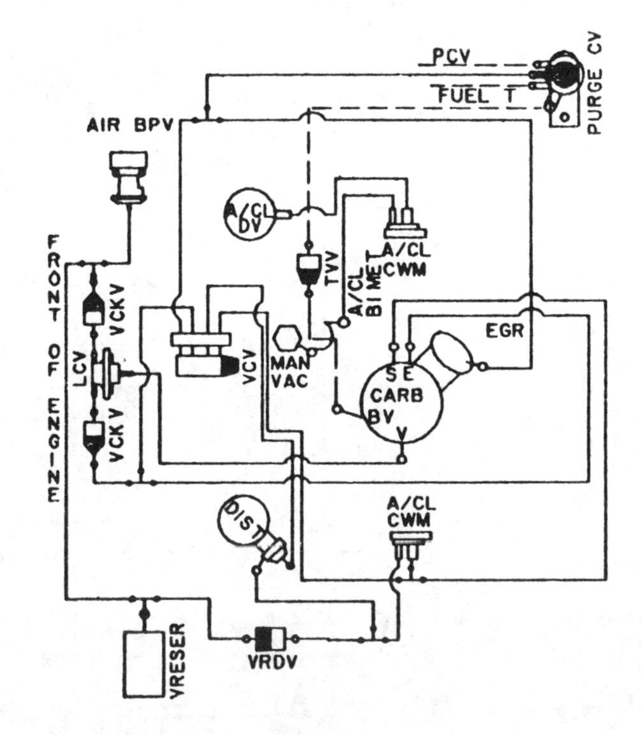

CALIBRATION: 9–9P–RO DATE: 5–26–78
4.1L (250 CID)

VACUUM CIRCUITS

1979—250 CID (4.1 L) engines

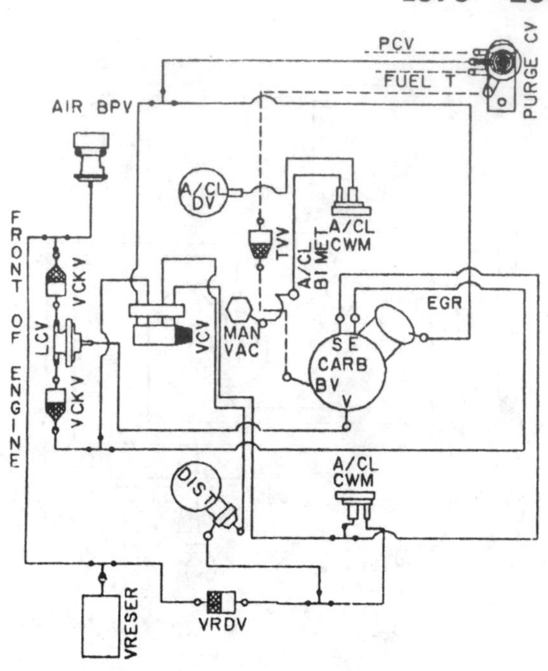

CALIBRATION: 9–9P–R10 DATE: 10–24–78
4.1L (250 CID)

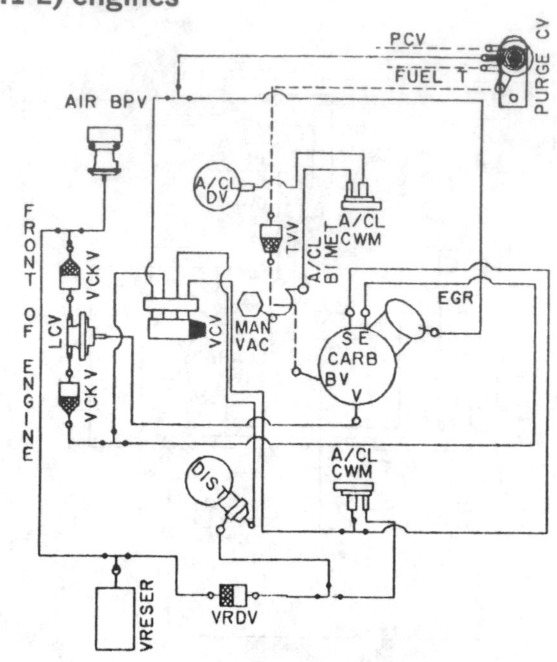

CALIBRATION: 9–9P–R11 DATE: 10–25–78
4.1L (250 CID)

1979—250 CID (4.1 L) engines

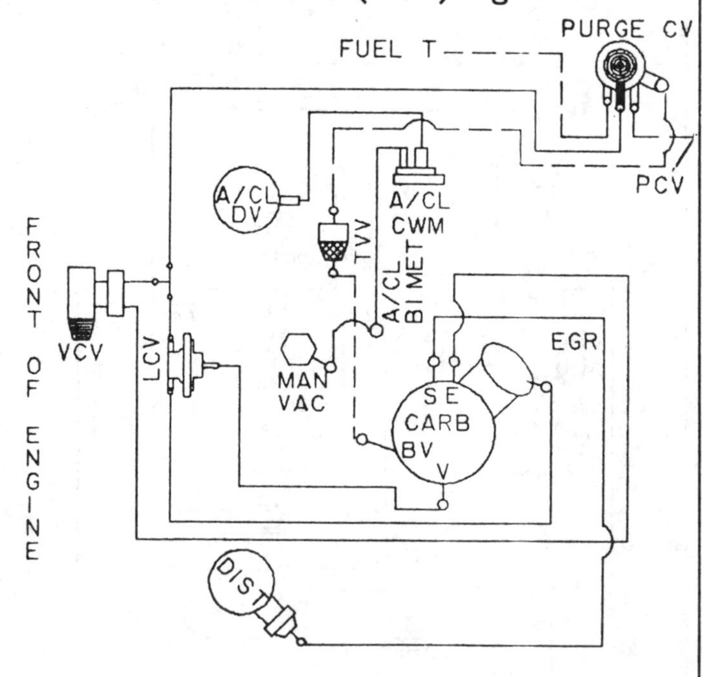

CALIBRATION: 9–29B–R11 DATE: 11–4–78
4.1L (250 CID)

1979—300 CID (4.9 L) engines

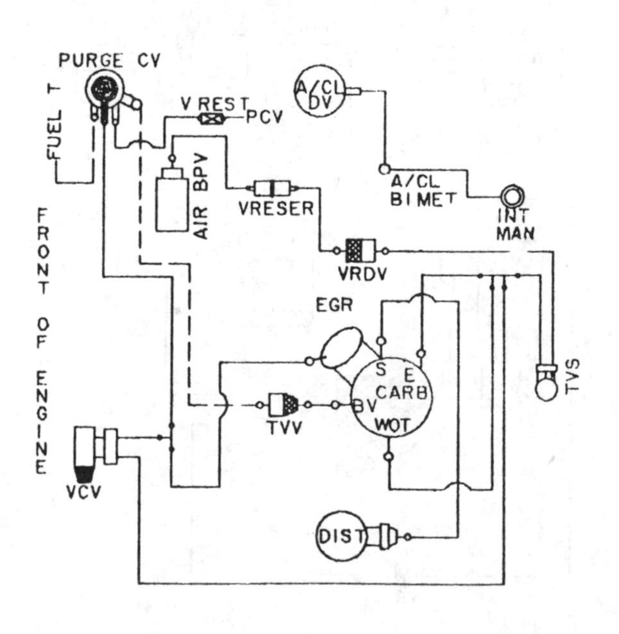

CALIBRATION: 9–51G–RO DATE: 5–8–78
4.9L (300 CID)

VACUUM CIRCUITS

1979—300 CID (4.9 L) engines

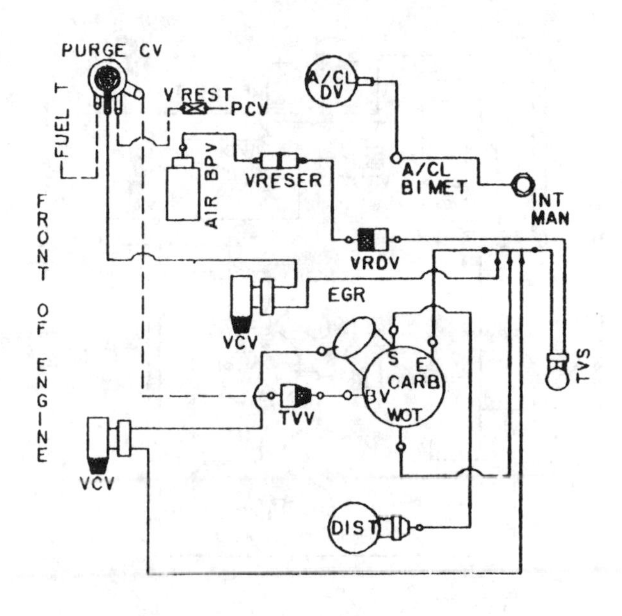

CALIBRATION: 9–51J–RO DATE: 6–13–78
4.9L (300 CID)

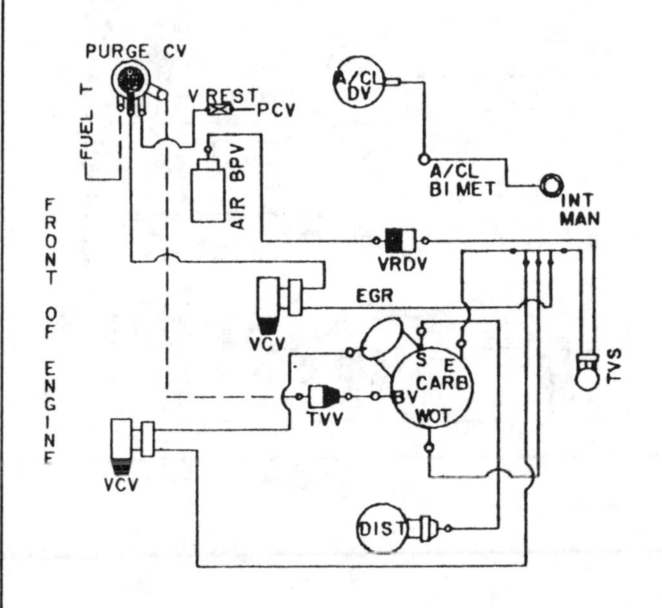

CALIBRATION: 9–51K–R0 DATE: 5–26–78
4.9L (300 CID)

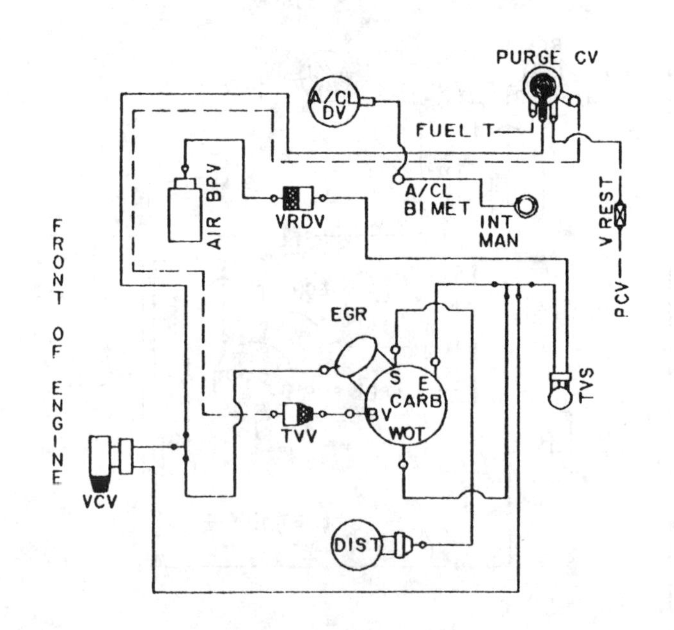

CALIBRATION. 9–51L–RO DATE: 5–8–78
4.9L (300 CID)

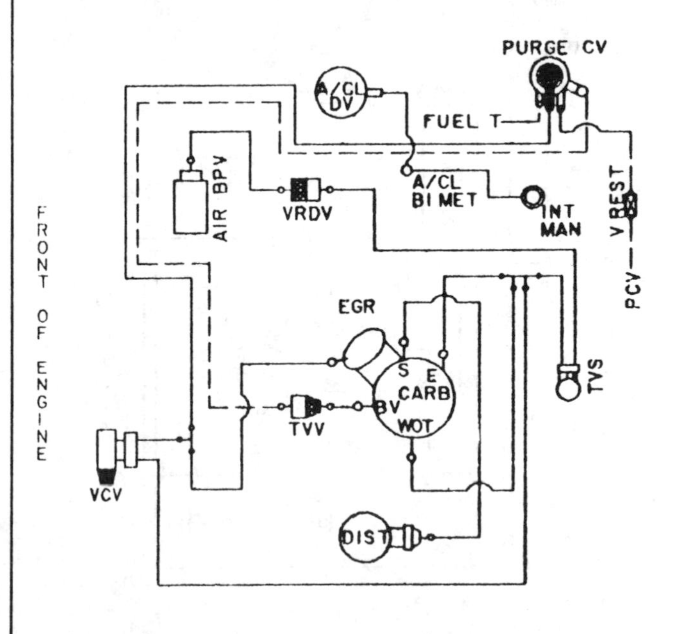

CALIBRATION: 9–51M–RO DATE: 6–16–78
4.9L (300 CID)

VACUUM CIRCUITS

1979—300 CID (4.9 L) engines

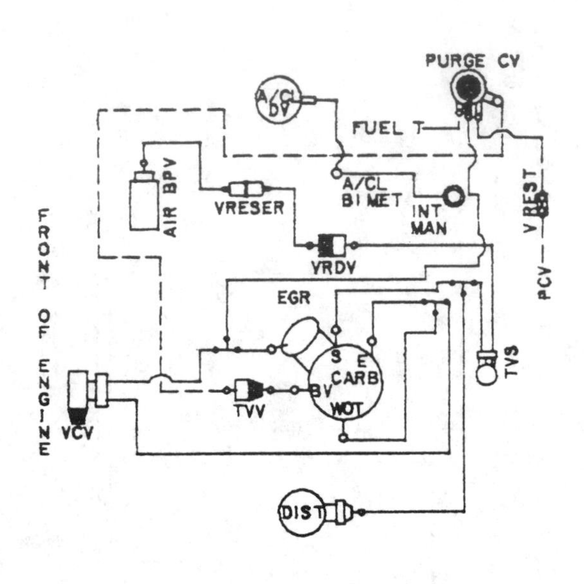

CALIBRATION: 9–51S–RO DATE: 5–11–78
4.9L (300 CID)

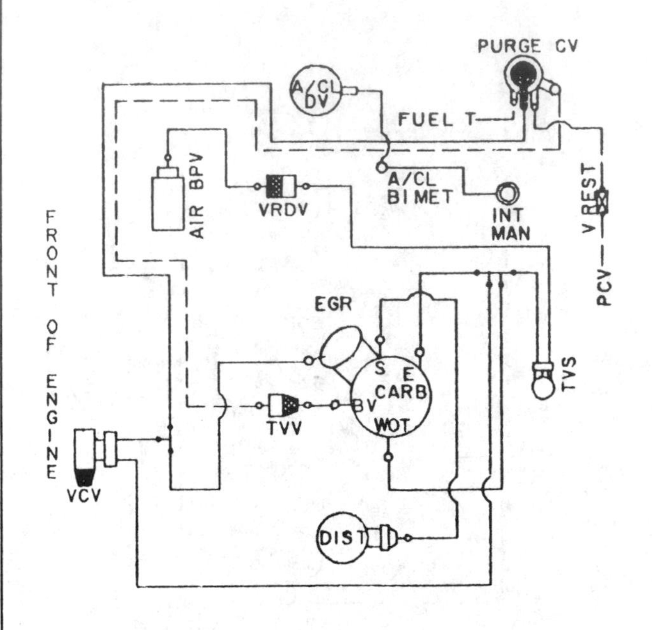

CALIBRATION: 9–51T–RO DATE: 6–6–78
4.9L (300 CID)

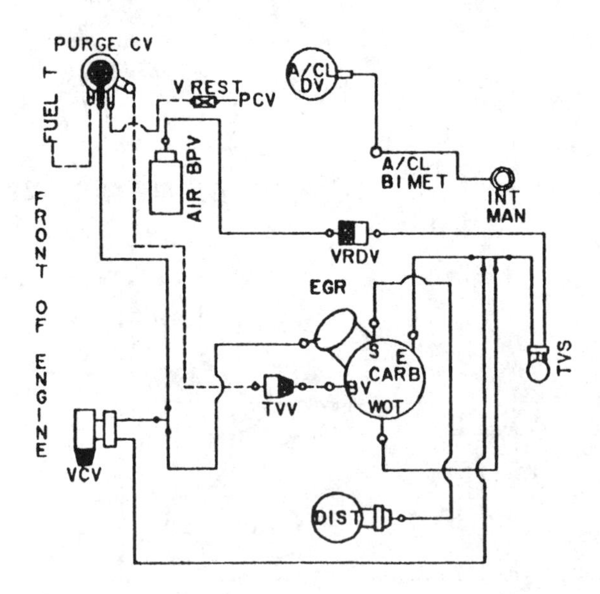

CALIBRATION: 9–52G–RO DATE: 5–8–78
4.9L (300 CID)

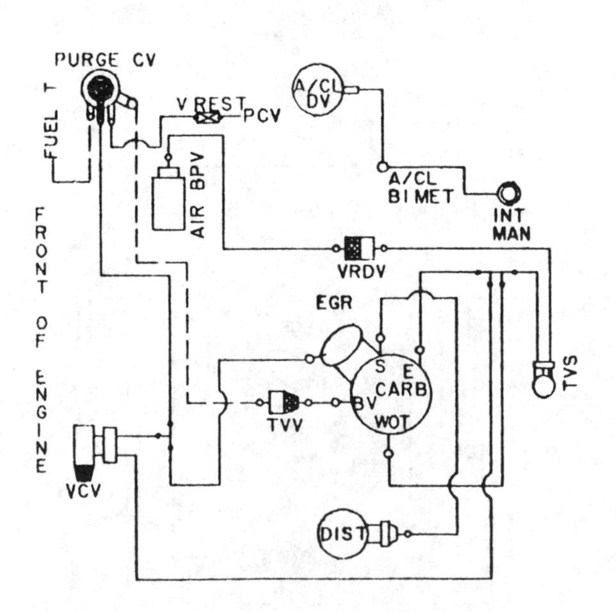

CALIBRATION: 9–52J–RO DATE: 6–5–78
4.9L (300 CID)

VACUUM CIRCUITS

1979—300 CID (4.9 L) engines

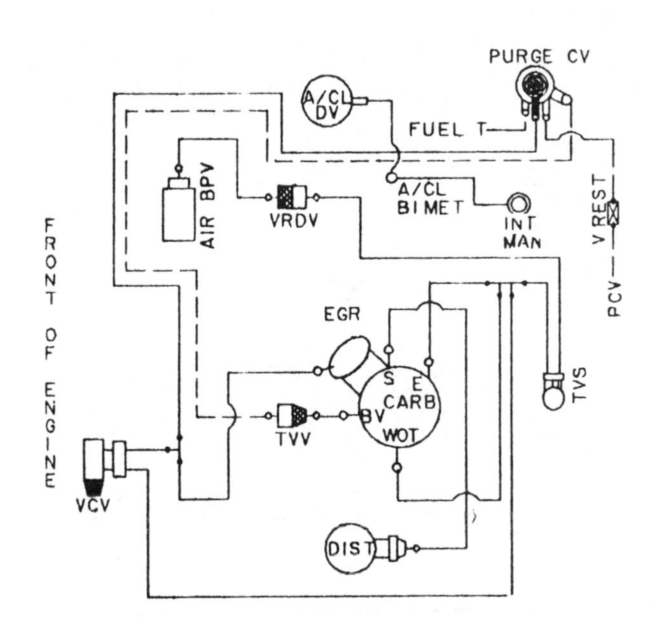

CALIBRATION: 9–52L–RO DATE: 5–8–78
4.9L (300 CID)

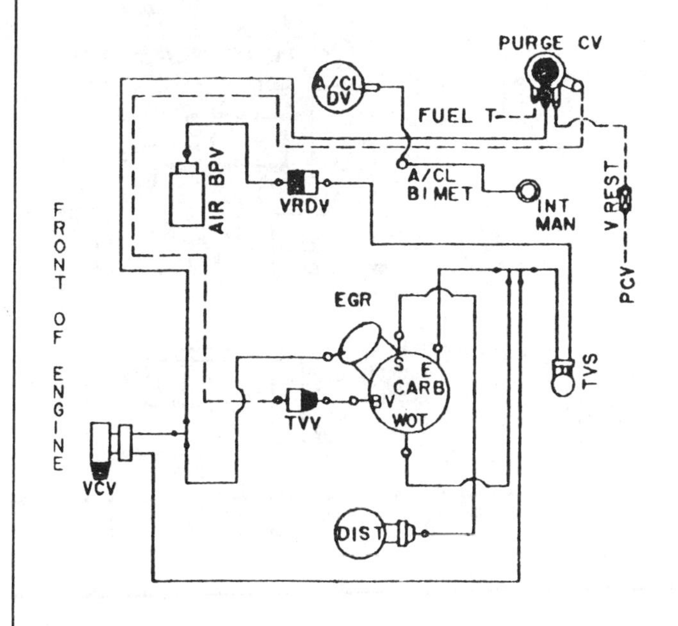

CALIBRATION: 9–52M–RO DATE: 6–12–78
4.9L (300 CID)

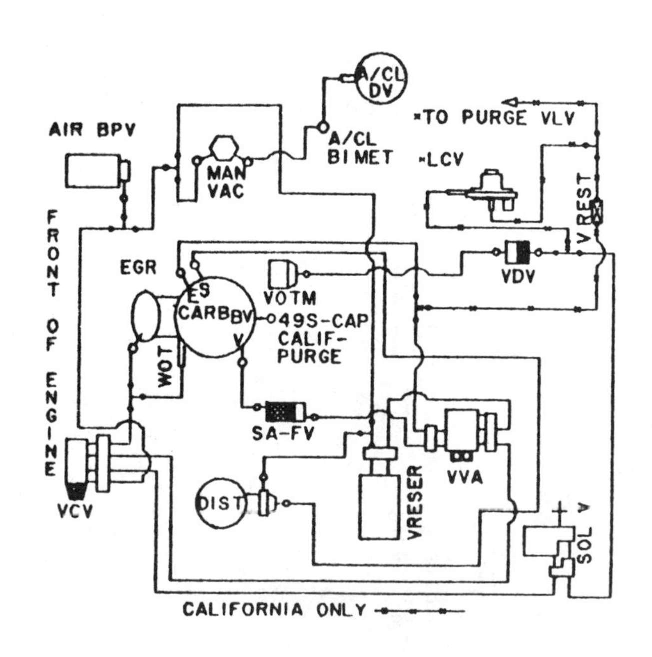

CALIBRATION: 9–77J–RO DATE: 5–3–78
4.9L (300 CID)

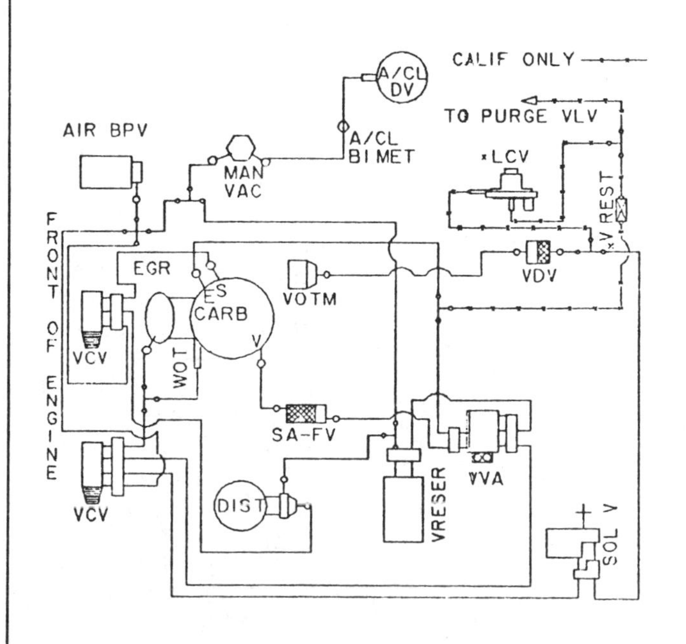

CALIBRATION: 9–77J–R11 DATE: 11–13–78
4.9L (300 CID)

VACUUM CIRCUITS

1979—300 CID (4.9 L) engines

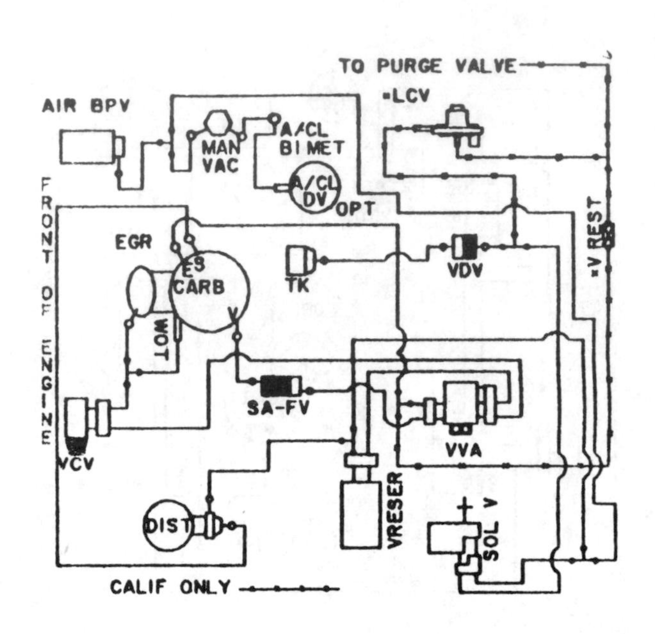

CALIBRATION: 9–77M–RO DATE: 6–12–78
4.9L (300 CID)

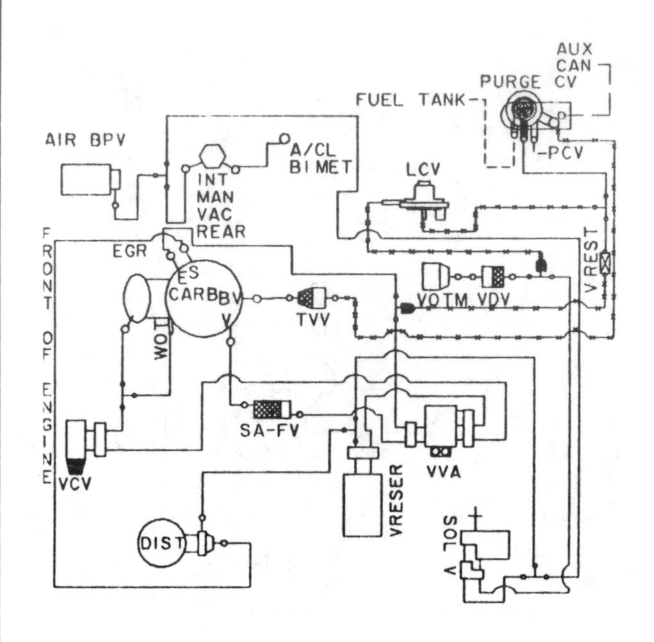

CALIBRATION: 9–77M–R10 DATE: 10–30–78
4.9L (300 CID)

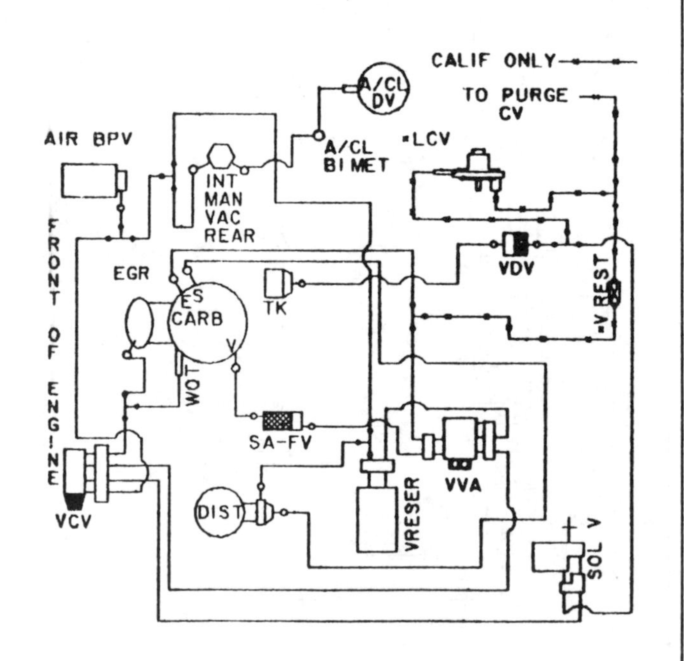

CALIBRATION: 9–78J–RO DATE: 5–3–78
4.9L (300 CID)

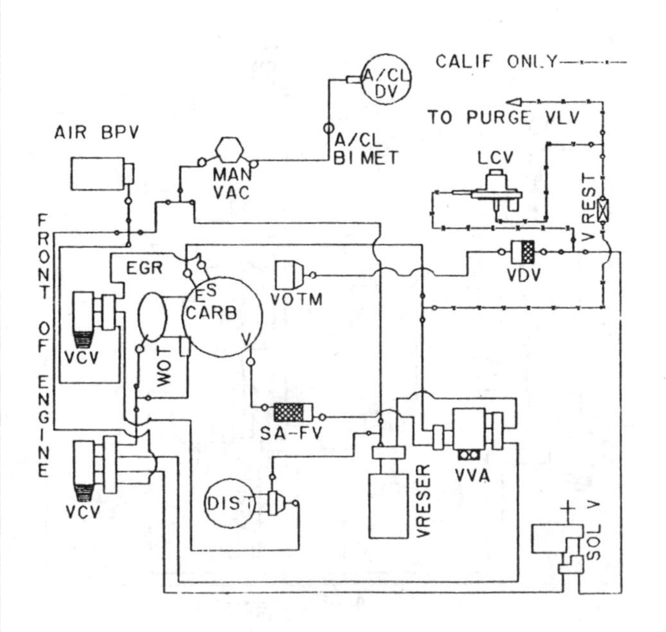

CALIBRATION: 9–78J–R11 DATE: 10–10–78
4.9L (300 CID)

VACUUM CIRCUITS

1979—302 CID (5.0 L) engines

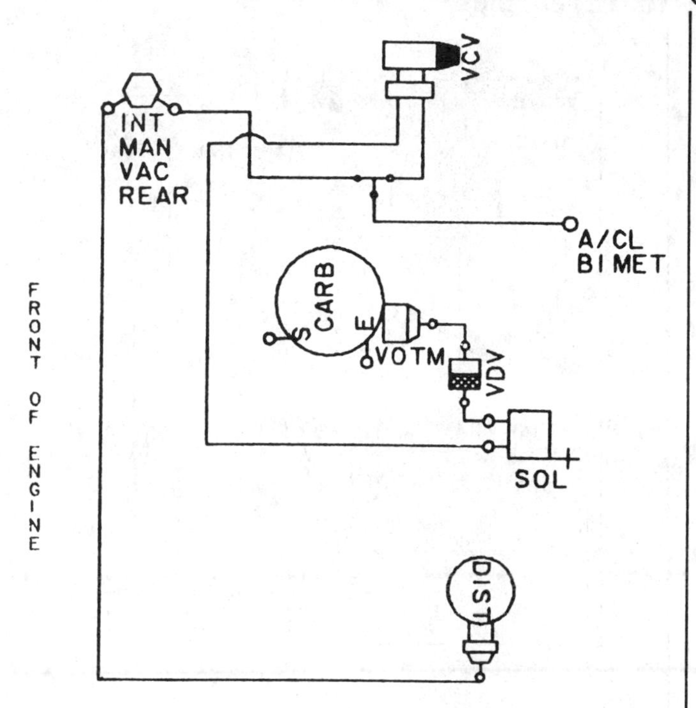

CALIBRATION: * 7–79–R1 DATE: 8–7–78
5.0L (302 CID)

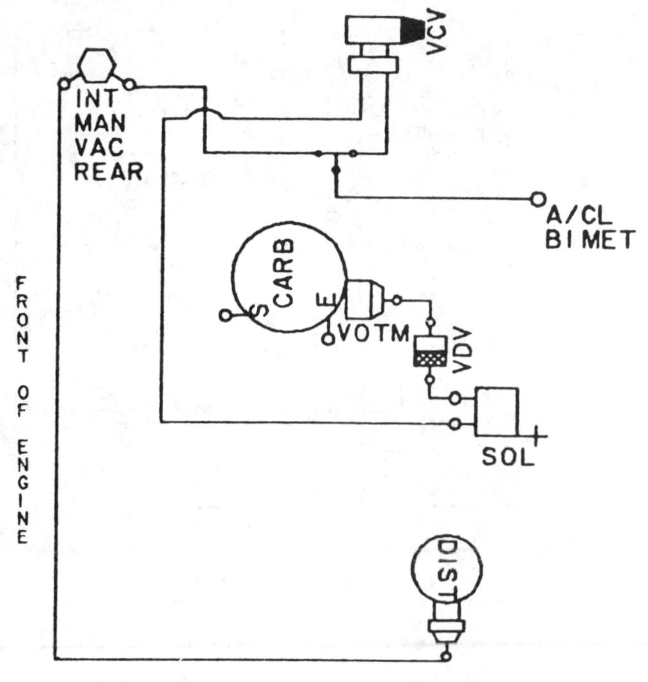

CALIBRATION: * 7–80–RO DATE: 4–25–78
5.0L (302 CID)

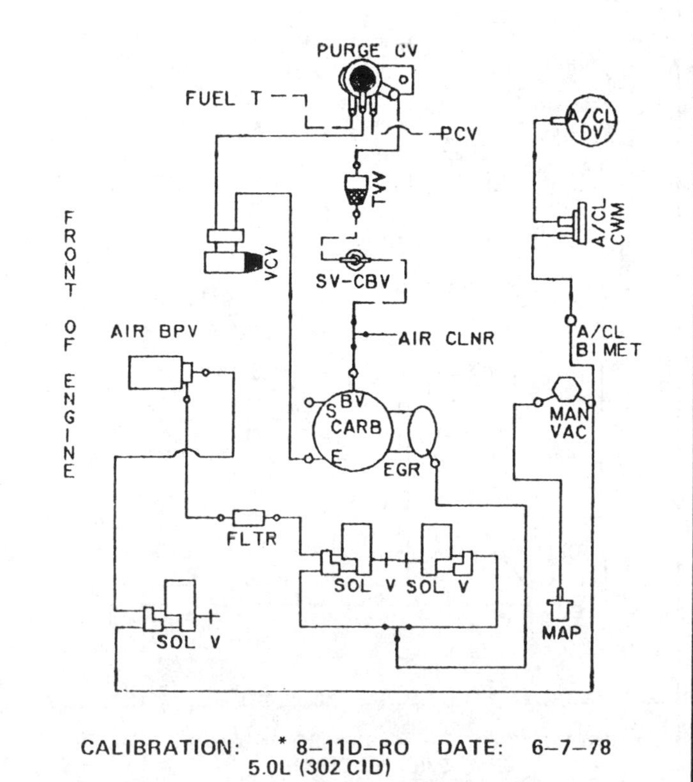

CALIBRATION: * 8–11D–RO DATE: 6–7–78
5.0L (302 CID)

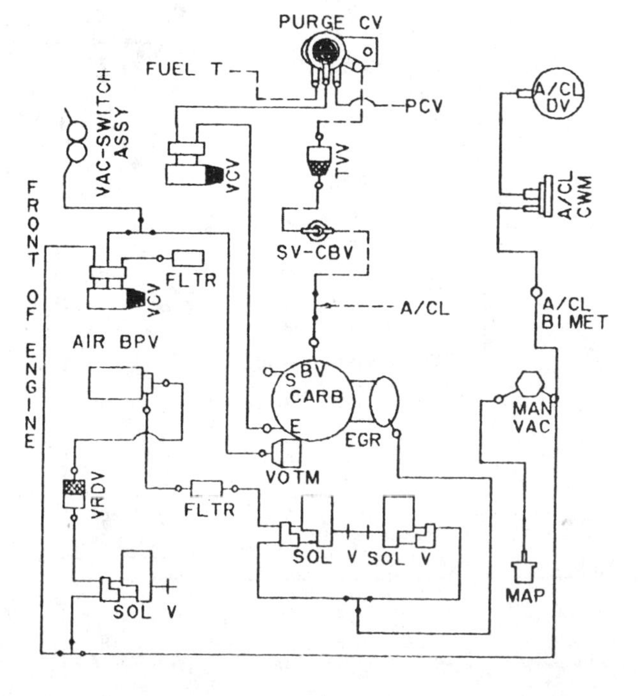

CALIBRATION: * 8–11L–R11 DATE: 5–23–78
5.0L (302 CID)

VACUUM CIRCUITS

1979—302 CID (5.0 L) engines

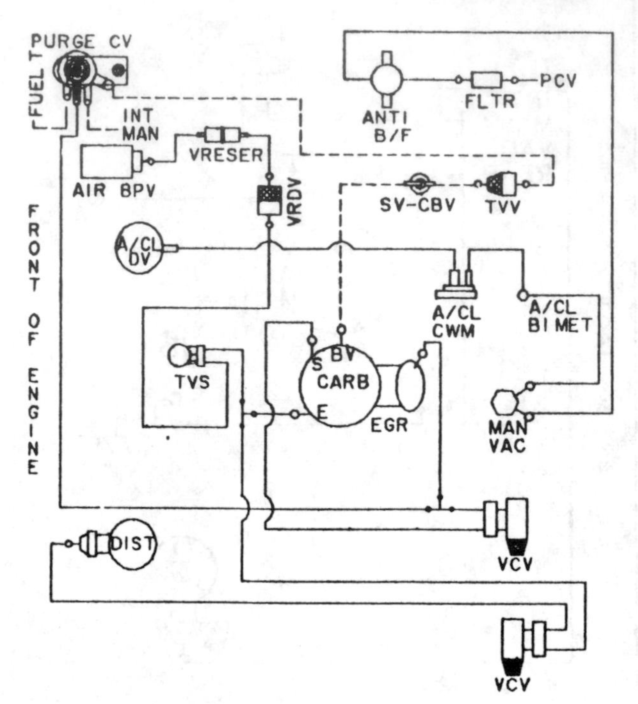

CALIBRATION: 9–10A–RO DATE: 6–7–78
5.0 (302 CID)

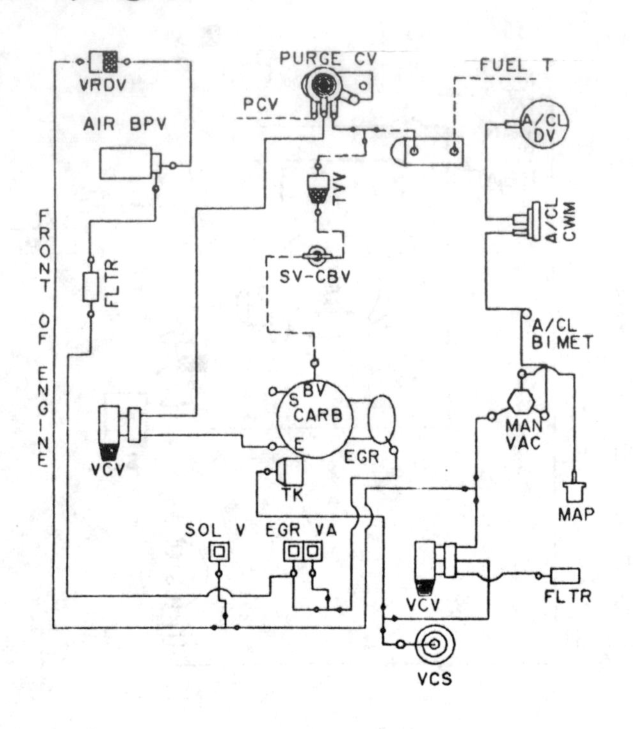

CALIBRATION: *8–11L–R20 DATE: 10–16–78
5.0L (302 CID)

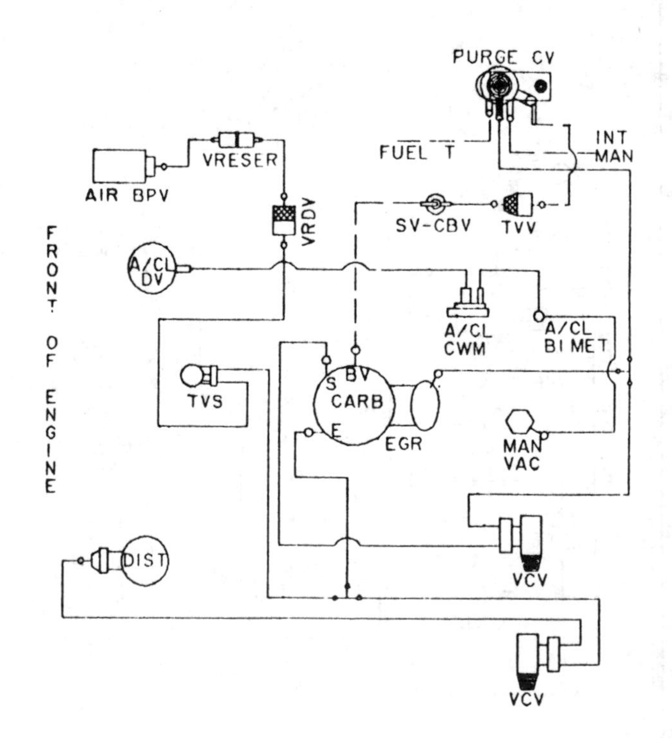

CALIBRATION: 9–10C–RO DATE: 6–15–78
5.0L (302 CID)

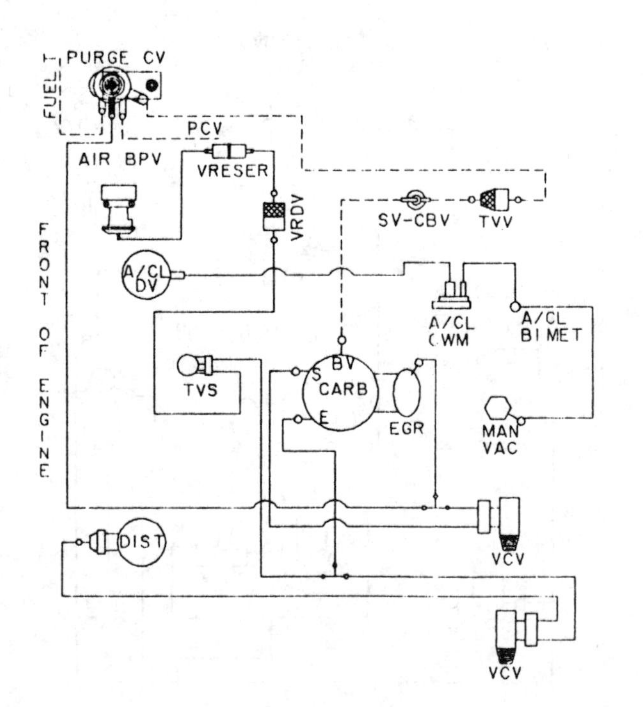

CALIBRATION: 9–10B–RO DATE: 1–18–79
5.0L (302 CID)

VACUUM CIRCUITS

1979—302 CID (5.0 L) engines

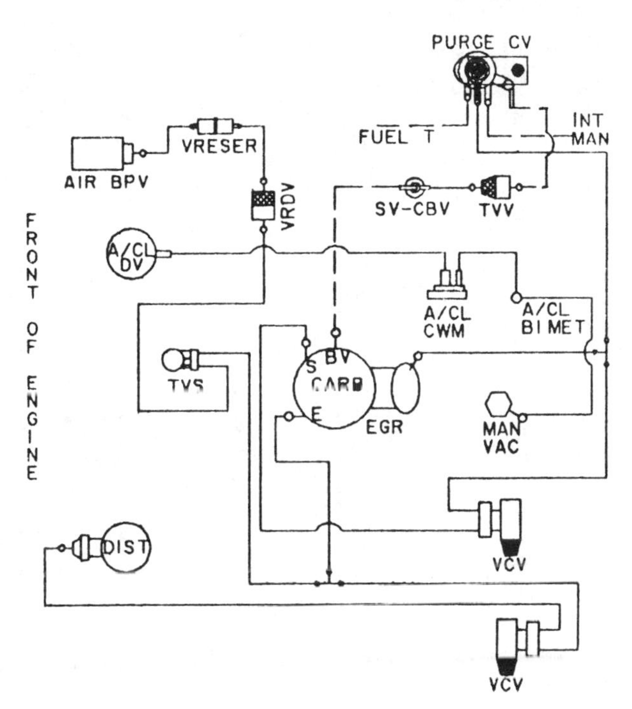

CALIBRATION: 9–10C–R11 DATE: 9–18–78
5.0L (302 CID)

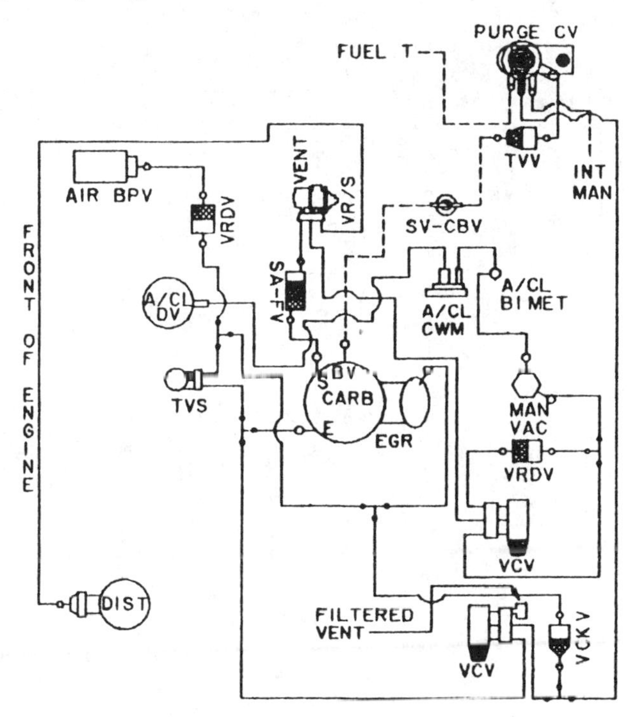

CALIBRATION: 9–11A–RO DATE: 6–7–78
5.0L (302 CID)

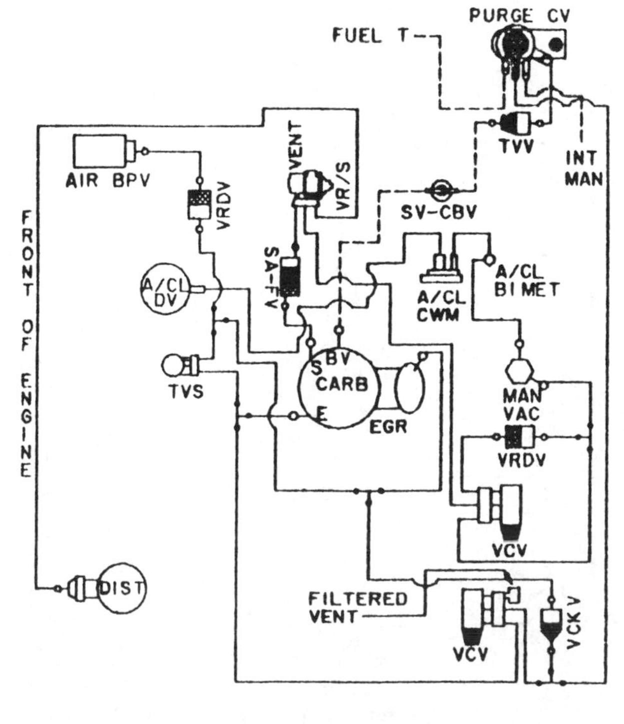

CALIBRATION: 9–11A–R10 DATE: 6–7–78
5.0L (302 CID)

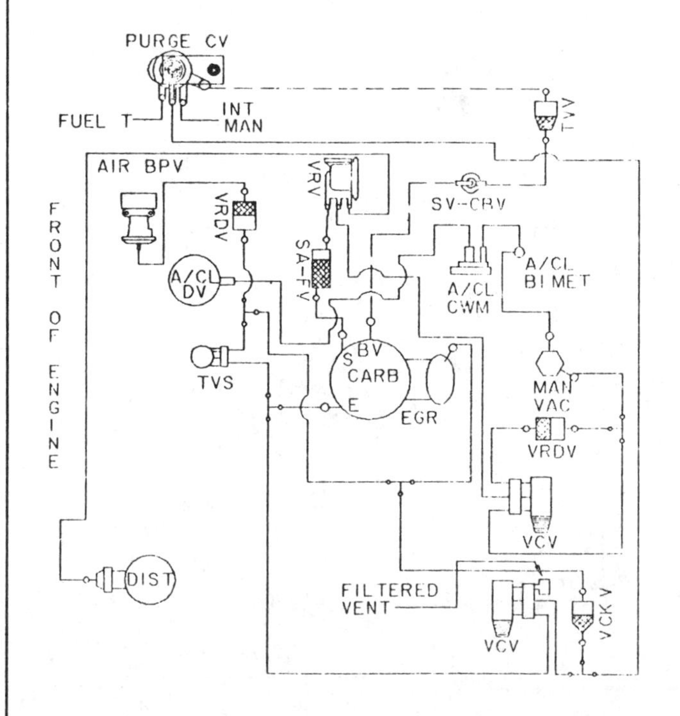

CALIBRATION: 9–11A–R11 DATE: 4–9–79
5.0L (302 CID)

VACUUM CIRCUITS

1979—302 CID (5.0 L) engines

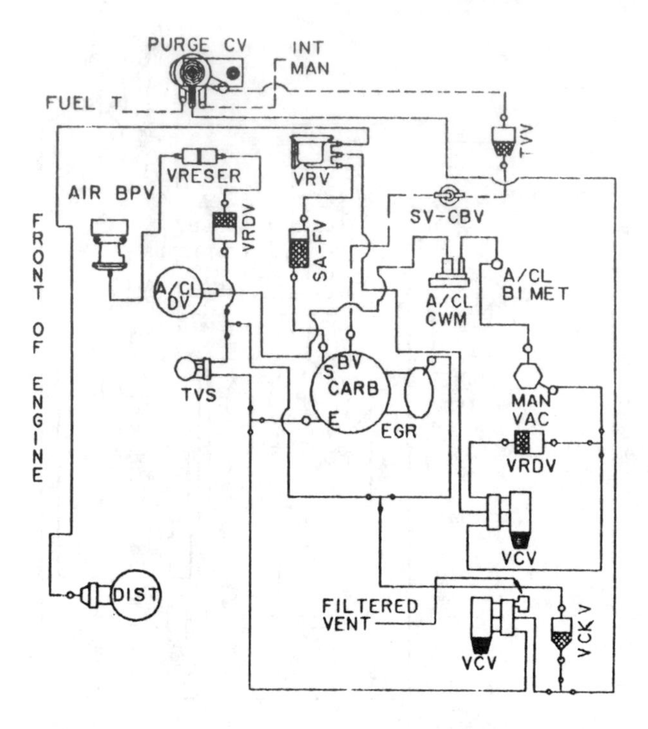

CALIBRATION: 9–11A–R20 DATE: 10–10–78
5.0L (302 CID)

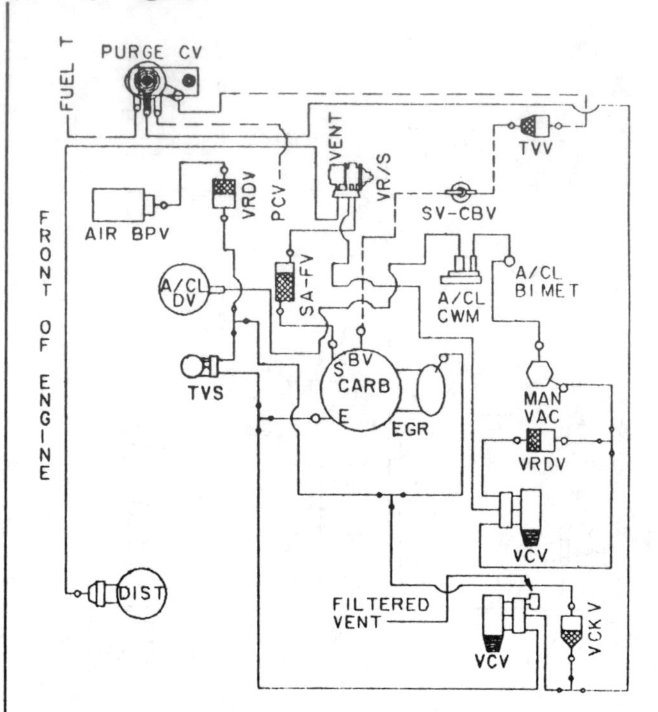

CALIBRATION: 9–11B–RO DATE: 6–12–78
5.0L (302 CID)

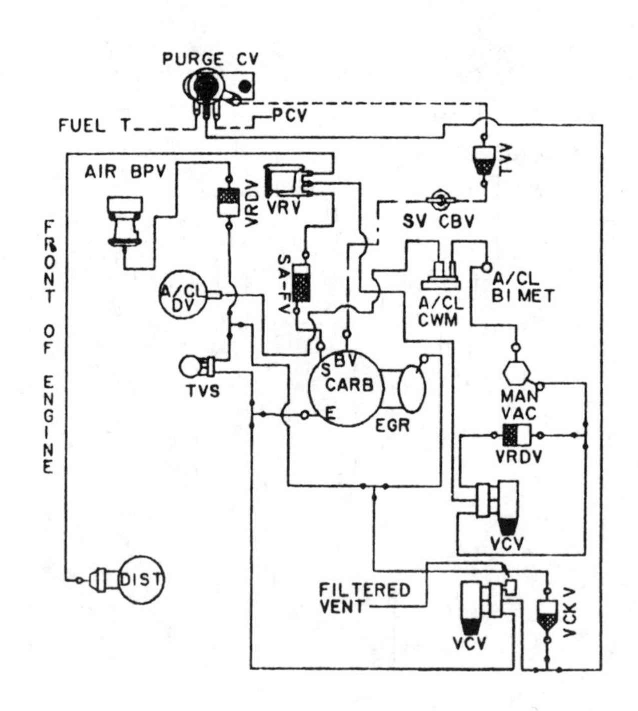

CALIBRATION: 9–11B–R10 DATE: 9–28–78
5.0L (302 CID)

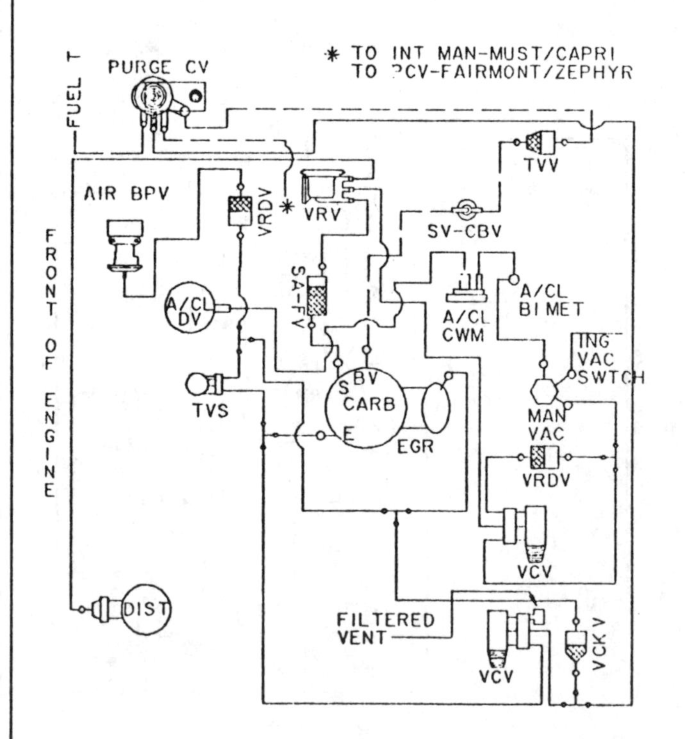

CALIBRATION: 9–11B–R11 DATE: 11–28–78
5.0L (302 CID)

VACUUM CIRCUITS

1979—302 CID (5.0 L) engines

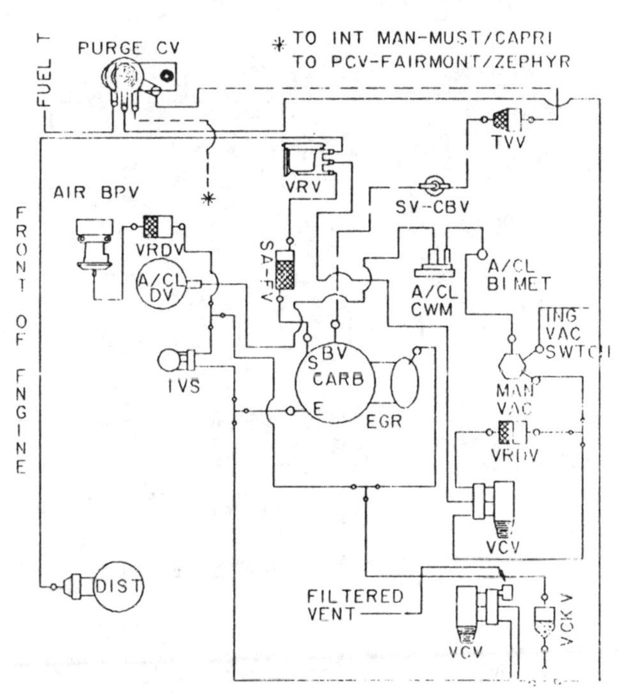

CALIBRATION: 9–11B–R12 DATE: 11–28–78
5.0L (302 CID)

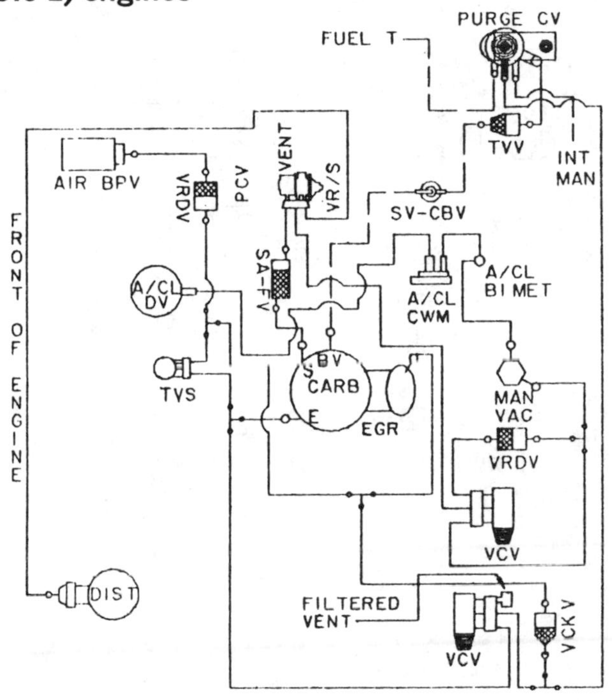

CALIBRATION: 9–11C–R1 DATE: 6–27–78
5.0L (302 CID)

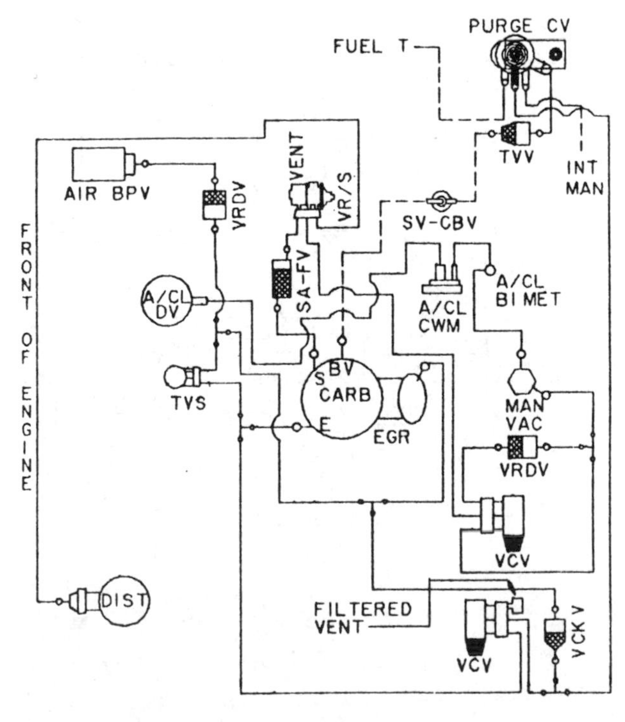

CALIBRATION. 9–11C–R10 DATE: 7–13–78
5.0L (302 CID)

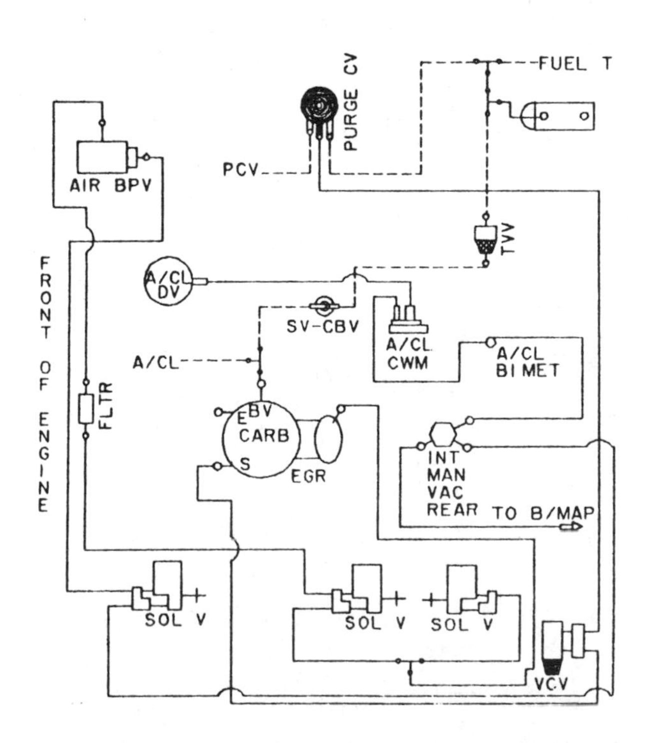

CALIBRATION: 9–11D–R11 DATE: 10–26–78
5.0L (302 CID)

VACUUM CIRCUITS

1979—302 CID (5.0 L) engines

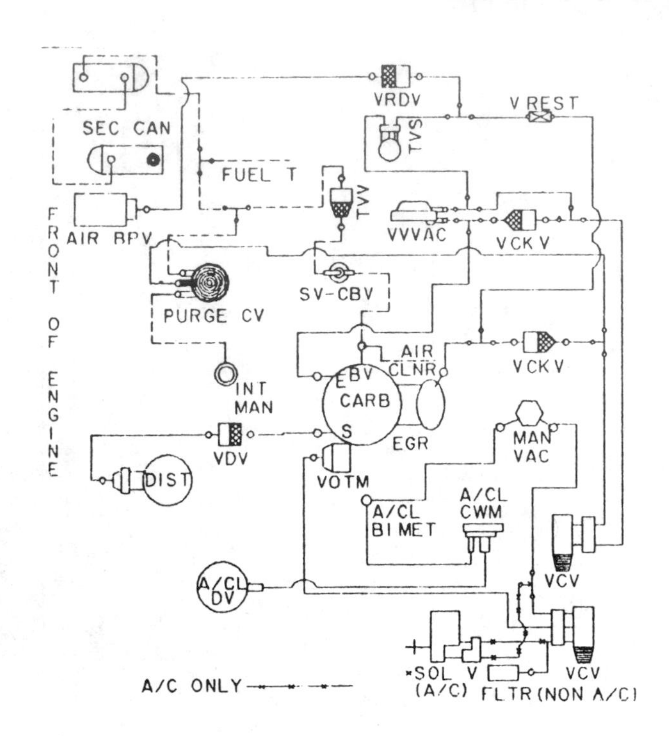

CALIBRATION: 9–11E–R1 DATE: 8–14–78
5.0L (302 CID)

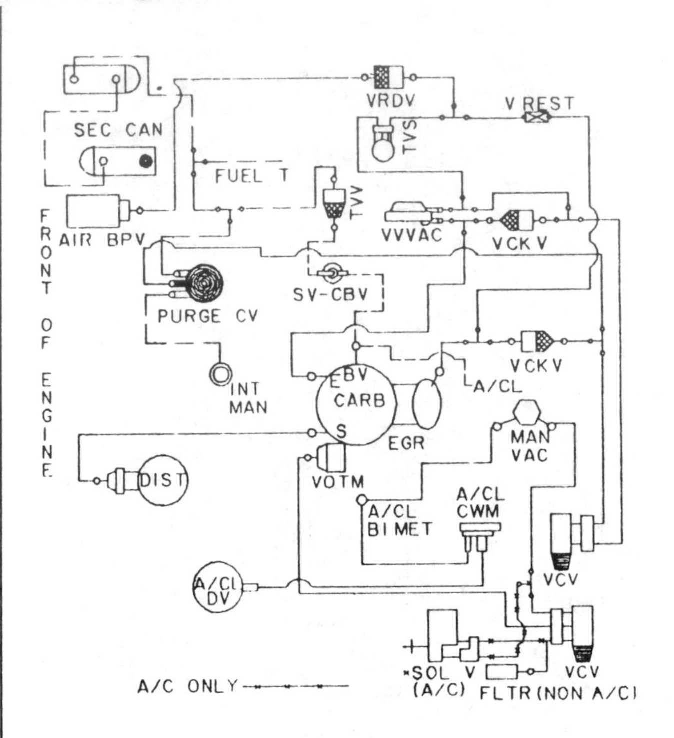

CALIBRATION. 9–11E–R10 DATE: 6–24–78
5.0L (302 CID)

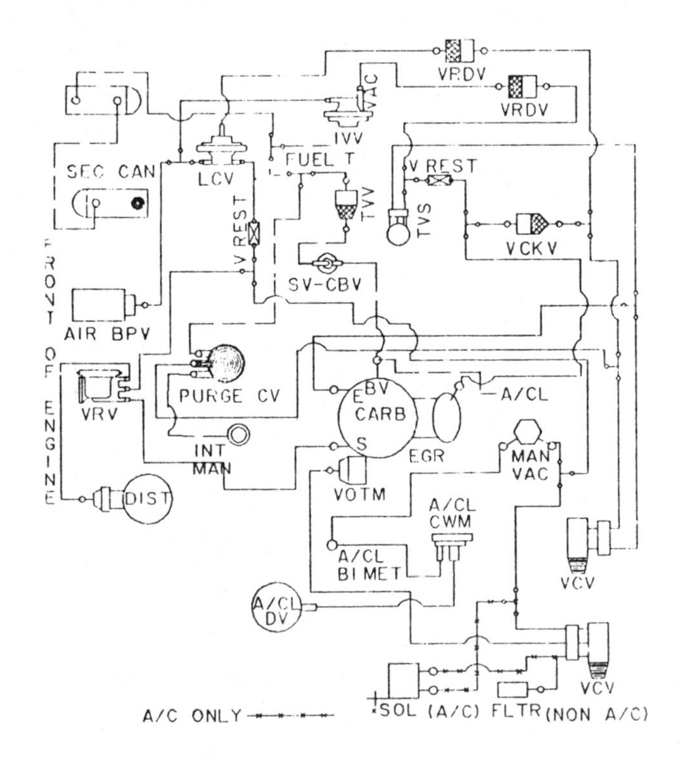

CALIBRATION: 9–11E–R12 DATE: 10–9–78
5.0L (302 CID)

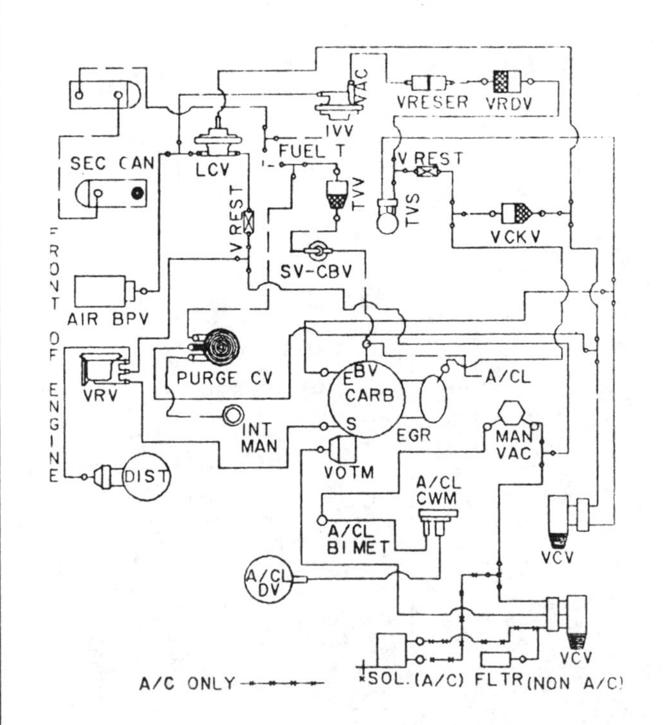

CALIBRATION: 9–11E–R13 DATE: 1–26–79
5.0L (302 CID)

VACUUM CIRCUITS

1979—302 CID (5.0 L) engines

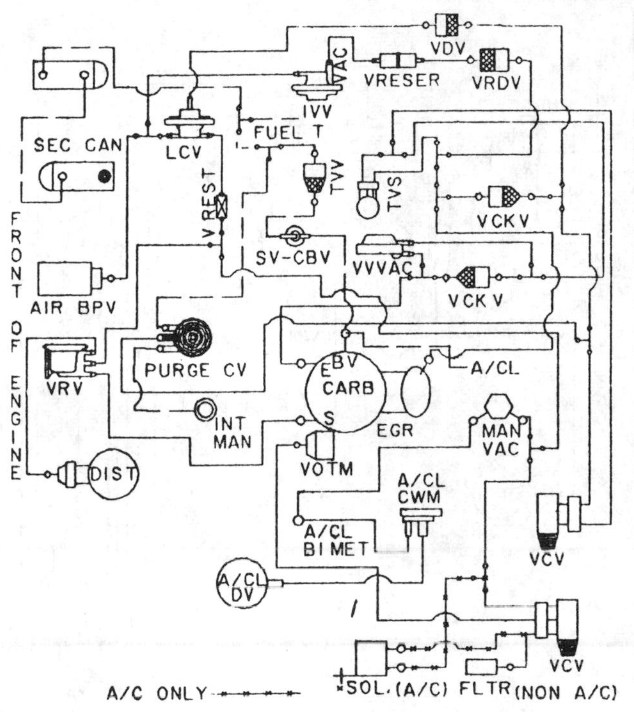

CALIBRATION: 9-11F-R1 DATE: 6-14-78
5.0L (302 CID)

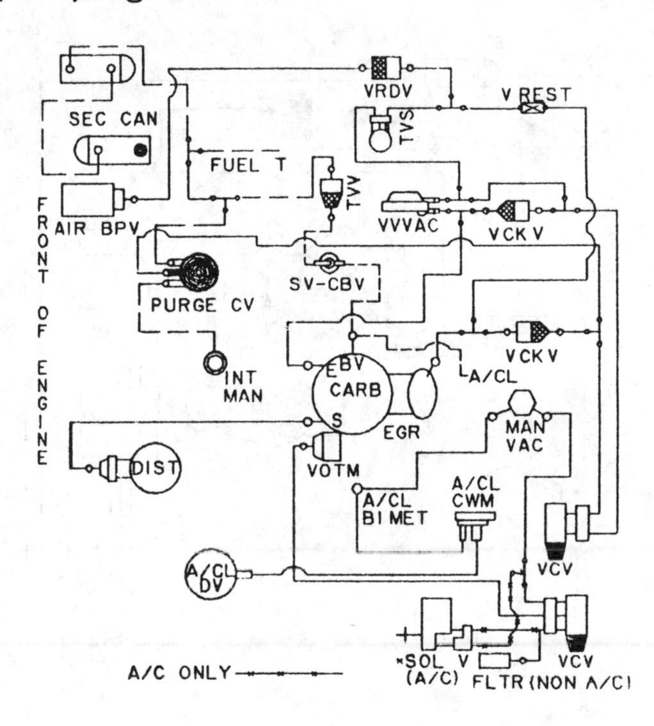

CALIBRATION: 9-11H-R10 DATE: 9-29-78
5.0L (302 CID)

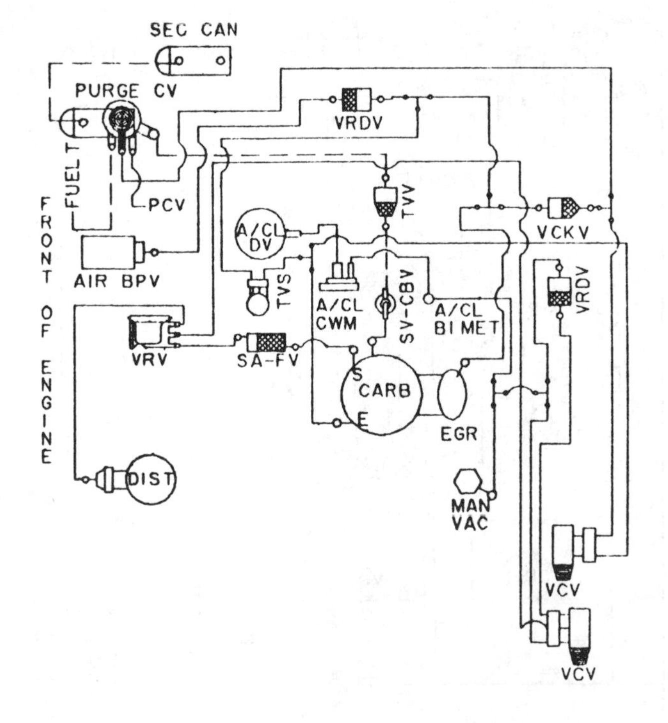

CALIBRATION: 9-11J-RO DATE: 6-8-78
5.0L (302 CID)

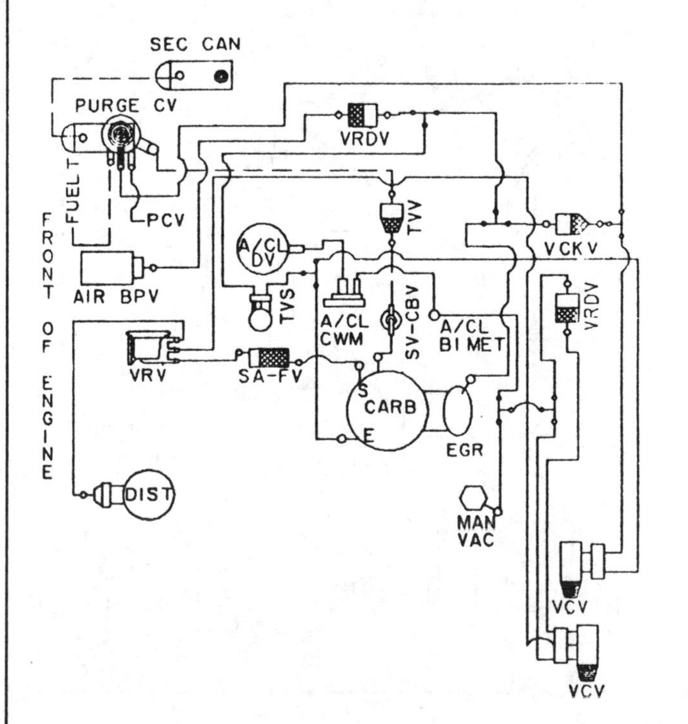

CALIBRATION: 9-11J-R10 DATE: 6-15-78
5.0L (302 CID)

VACUUM CIRCUITS

1979—302 CID (5.0 L) engines

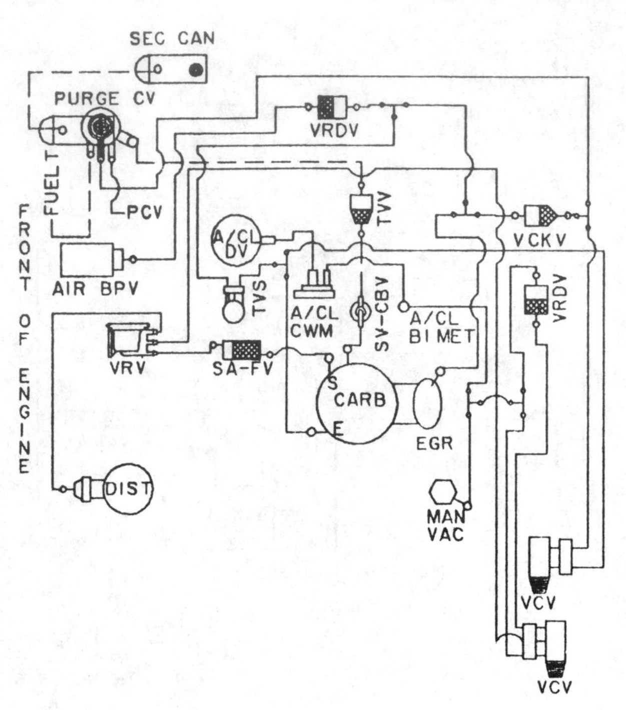

CALIBRATION: 9–IIJ–R13 **DATE:** 7–20–78
5.0L (302 CID)

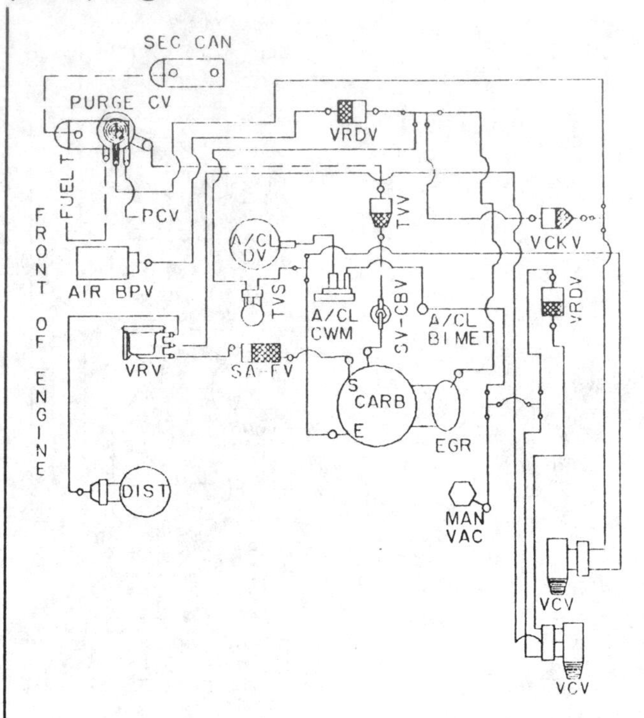

CALIBRATION: 9–11J–R20 **DATE:** 11–21–78
5.0L (302 CID)

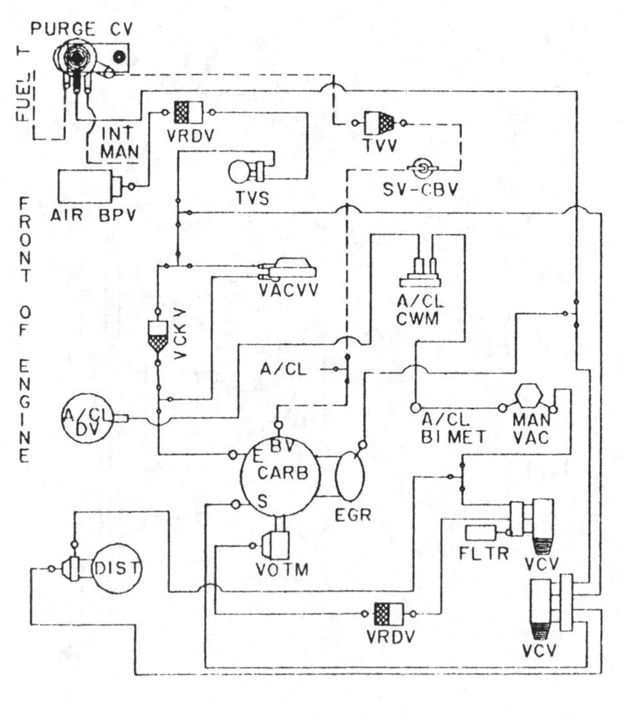

CALIBRATION: 9–11N–RO **DATE:** 6–7–78
5.0L (302 CID)

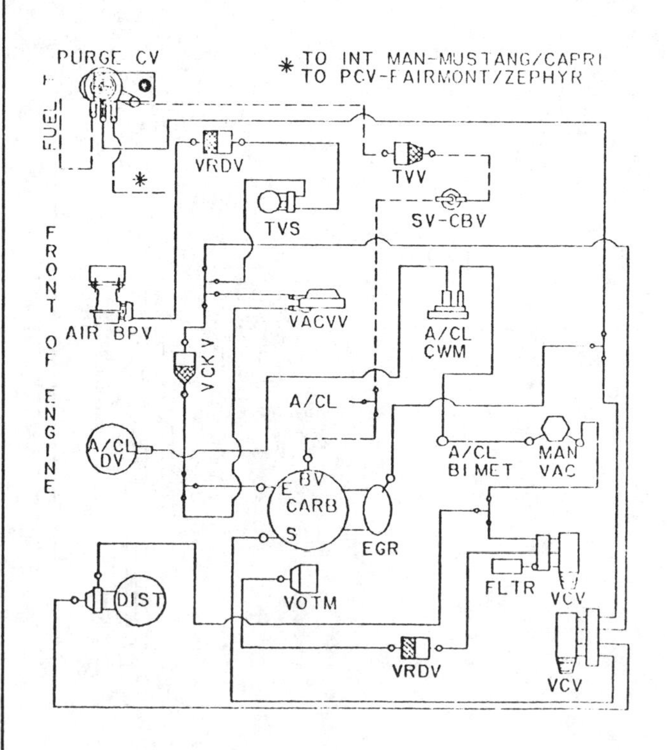

CALIBRATION: 9–11N–R10 **DATE:** 1–12–79
5.0L (302 CID)

VACUUM CIRCUITS

1979—302 CID (5.0 L) engines

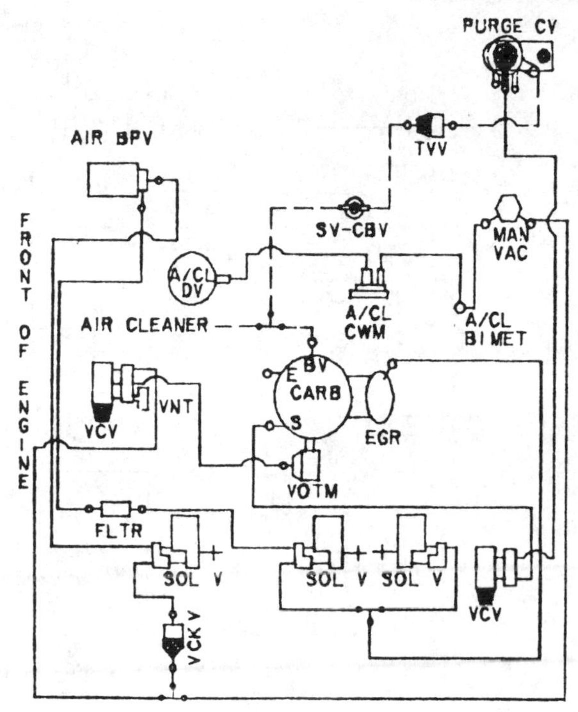

CALIBRATION: 9-11P-RO DATE: 5-24-78
5.0L (302 CID)

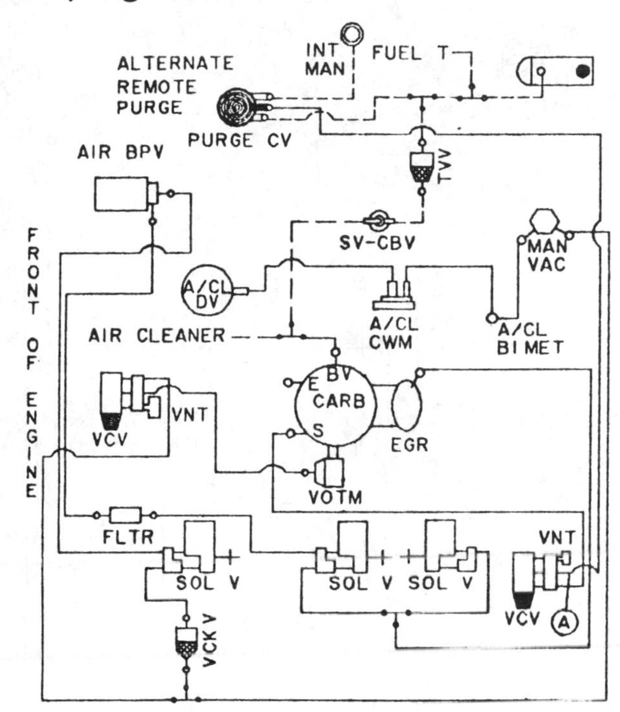

CALIBRATION: 9-11P-R2 DATE: 8-24-78
5.0L (302 CID)

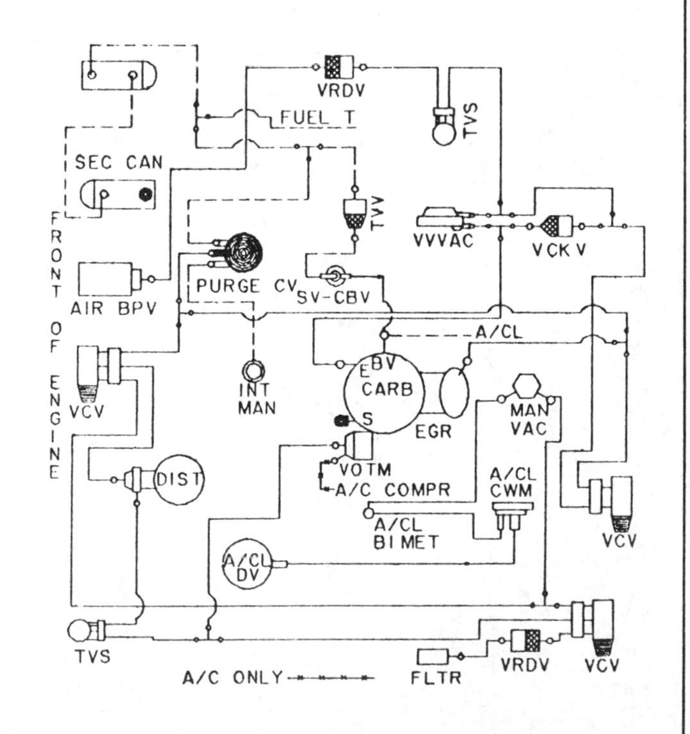

CALIBRATION: 9-11Q-R1 DATE: 6-15-78
5.0L (302 CID)

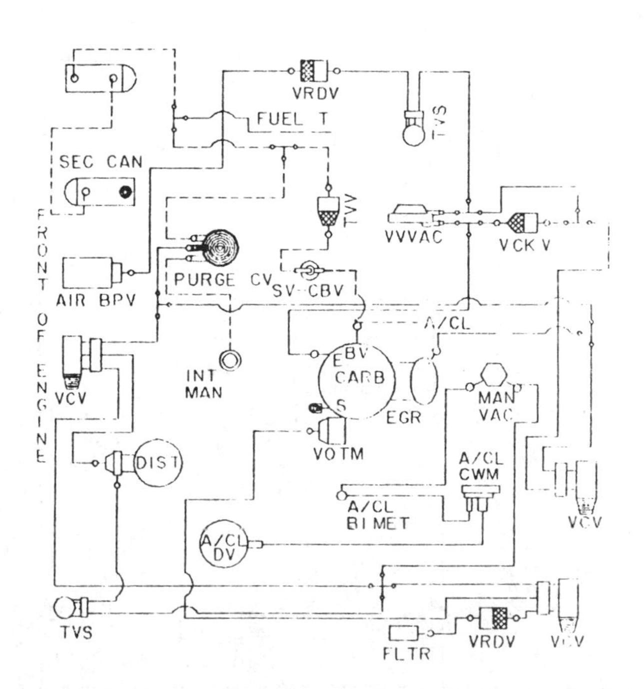

CALIBRATION: 9-11Q-R15 DATE: 1-17-79
5.0L (302 CID)

VACUUM CIRCUITS

1979—302 CID (5.0 L) engines

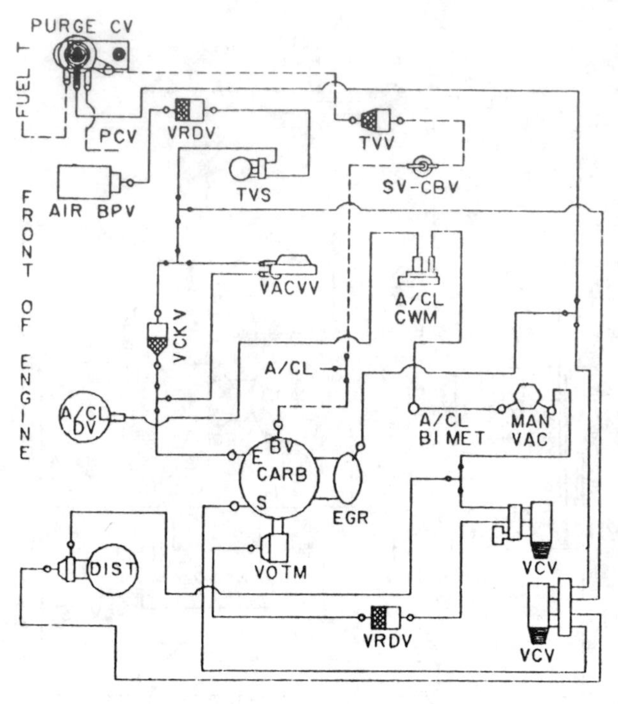

CALIBRATION: 9–11R–RO DATE: 6–7–78
5.0L (302 CID)

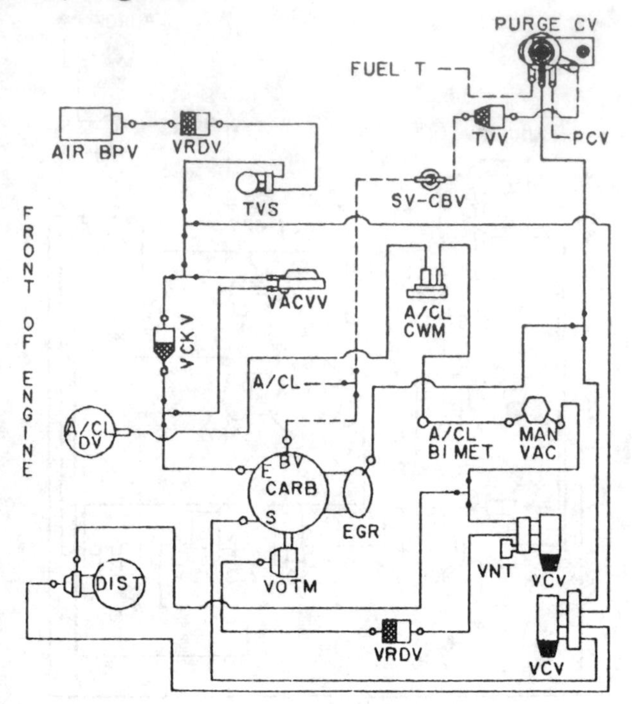

CALIBRATION: 9–11V–RO DATE: 6–8–78
5.0L (302 CID)

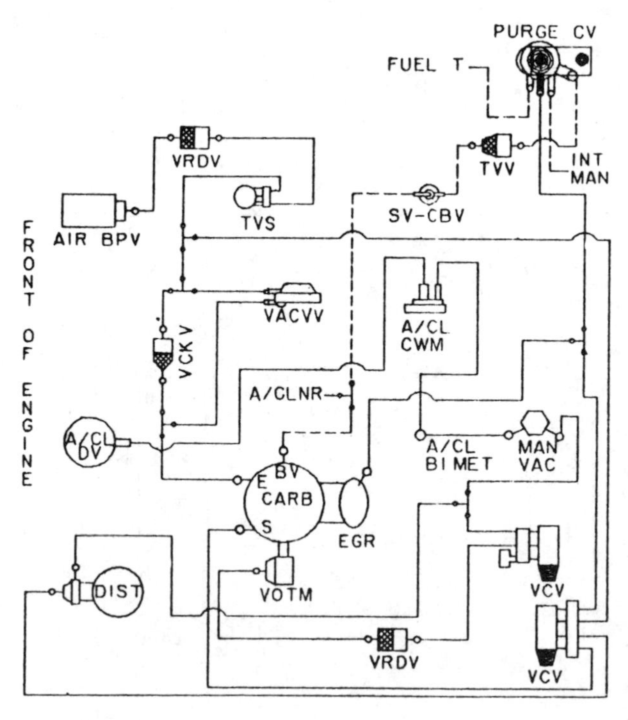

CALIBRATION: 9–11S–RO DATE: 6–7–78
5.0L (302 CID)

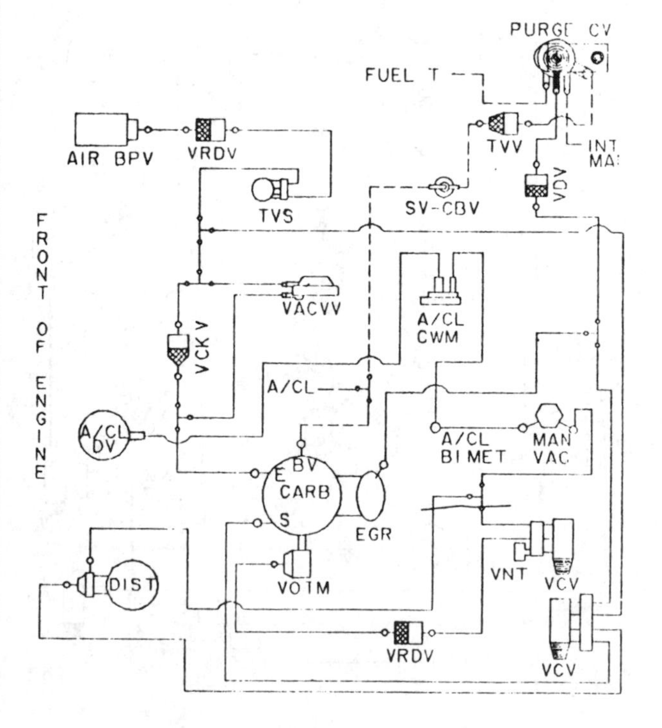

CALIBRATION: 9–IIV–R10 DATE: 8–4–78
5.0L (302 CID)

VACUUM CIRCUITS

1979—302 CID (5.0 L) engines

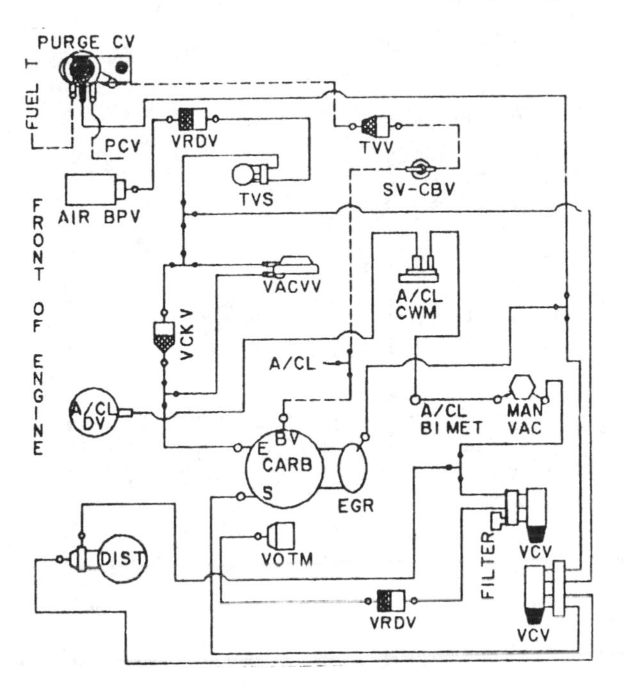

CALIBRATION: 9–11R–R10 DATE: 10–26–78
5.0L (302 CID)

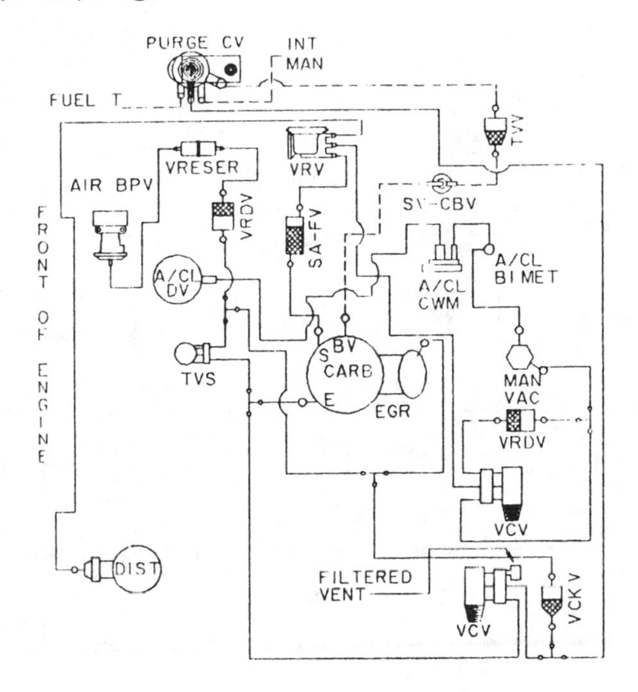

CALIBRATION: 9–11X–R10 DATE: 1–23–79
5.0L (302 CID)

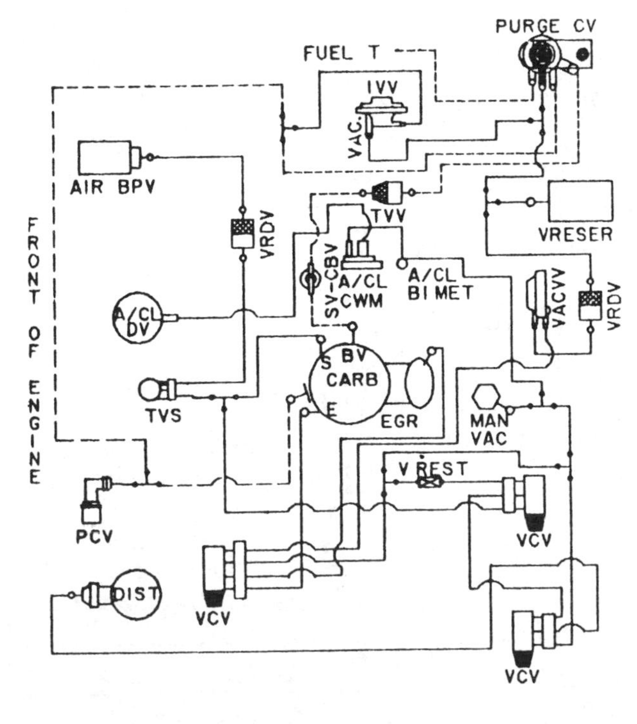

CALIBRATION: 9–11Y–RO DATE: 9–18–78
5.0L (302 CID)

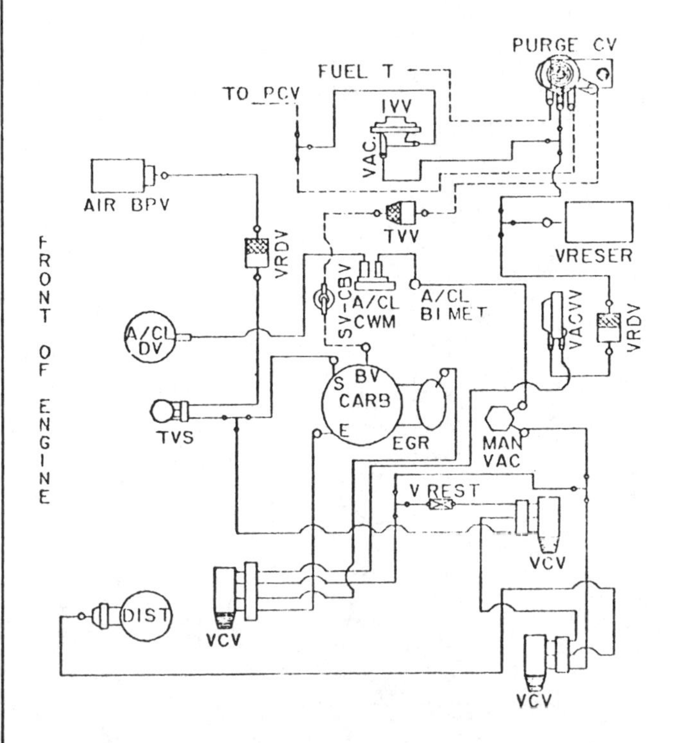

CALIBRATION: 9–11Z–RO DATE: 1–12–79
5.0L (302 CID)

VACUUM CIRCUITS

1979—302 CID (5.0 L) engines

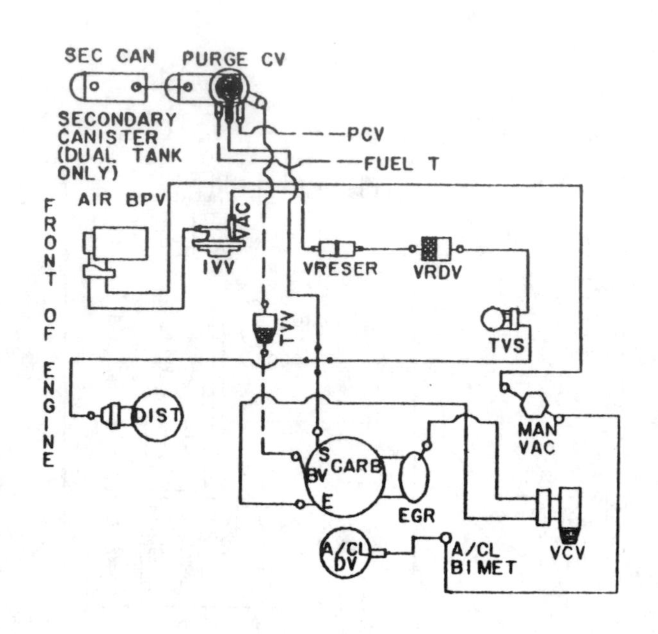

CALIBRATION: 9–53G–RO DATE: 5–12–78
5.0L (302 CID)

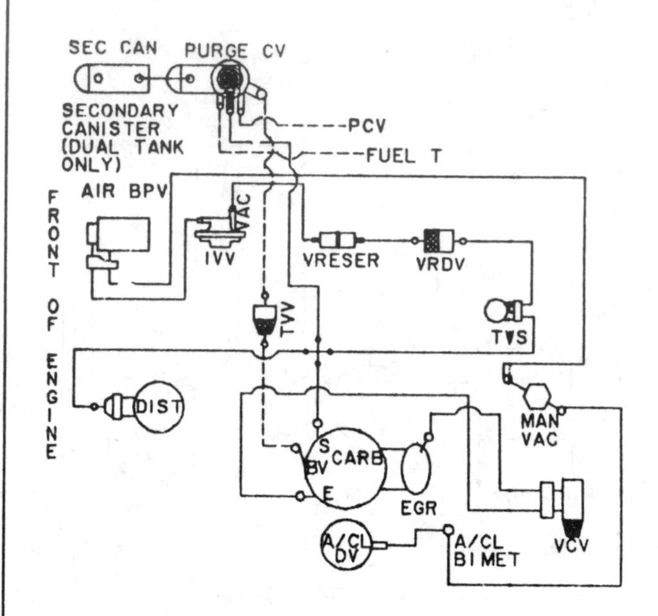

CALIBRATION: 9–53G–R10 DATE: 10–17–78
5.0L (302 CID)

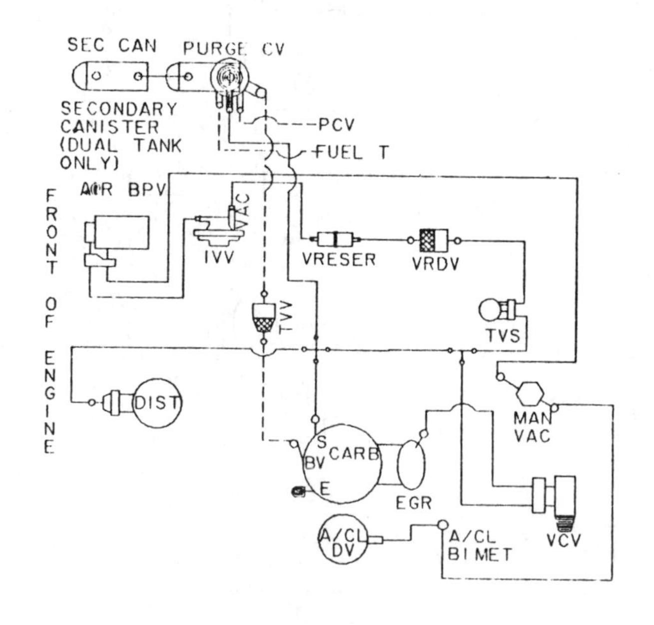

CALIBRATION: 9–53G–R12 DATE: 2–7–79
5.0L (302 CID)

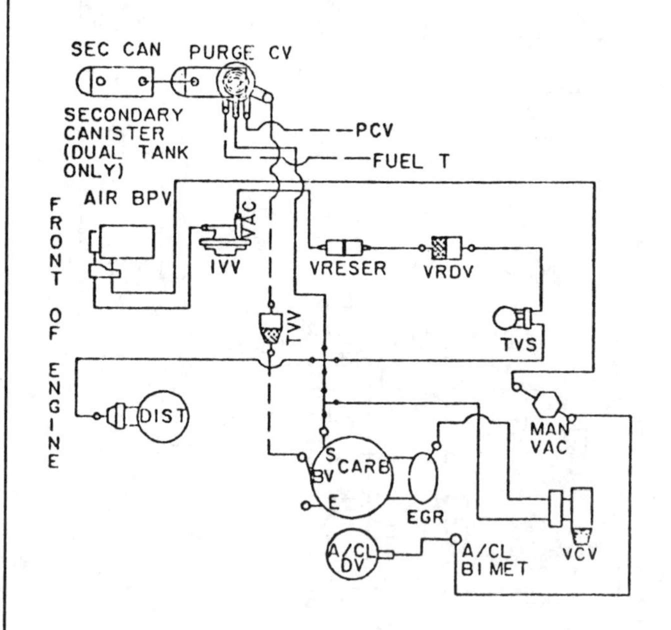

CALIBRATION: 9–53G–R13 DATE: 4–12–79
5.0L (302 CID)

VACUUM CIRCUITS

1979—302 CID (5.0 L) engines

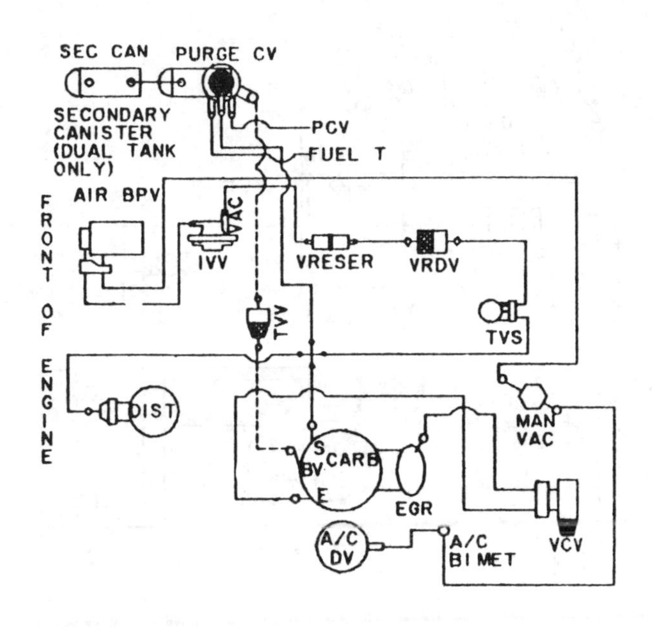

CALIBRATION: 9–53H–RO DATE: 5–12–78
5.0L (302 CID)

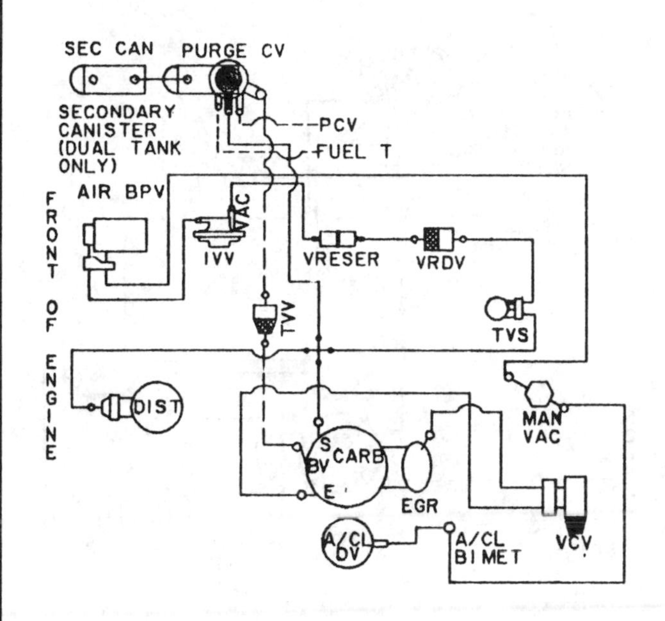

CALIBRATION: 9–53H–R10 DATE: 10–17–78
5.0L (302 CID)

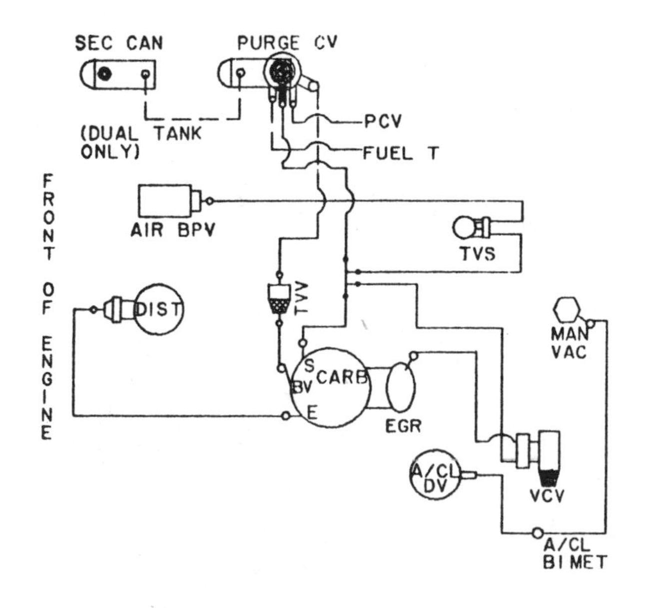

CALIBRATION: 9–54A–RO DATE: 8–10–78
5.0L (302 CID)

FUEL T
PURGE CV
PCV
AIR BPV
TVS
TVV
FRONT OF ENGINE
MAN VAC
S
BV
CARB
E
DIST
EGR
A/CL DV
A/CL BI MET
VCV

CALIBRATION: 9–54G–RO DATE: 5–28–78
5.0L (302 CID)

VACUUM CIRCUITS

1979—302 CID (5.0 L) engines

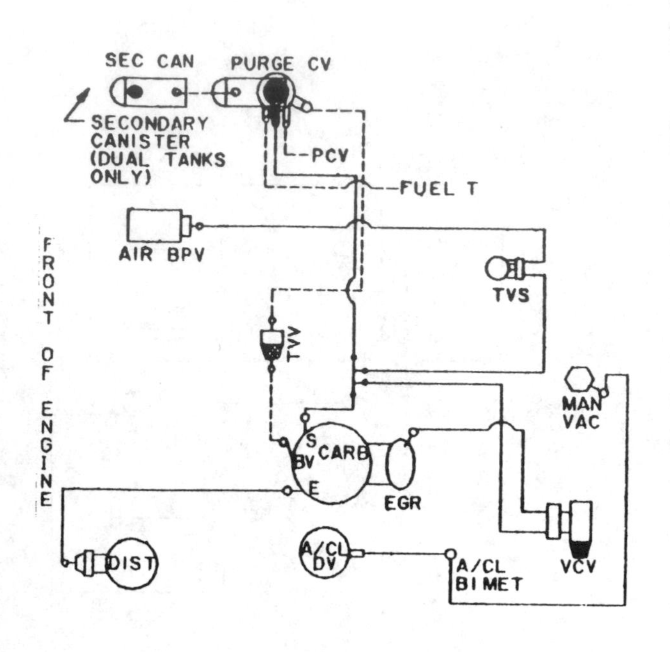

CALIBRATION: 9–54H–RO DATE: 5–31–78
5.0L (302 CID)

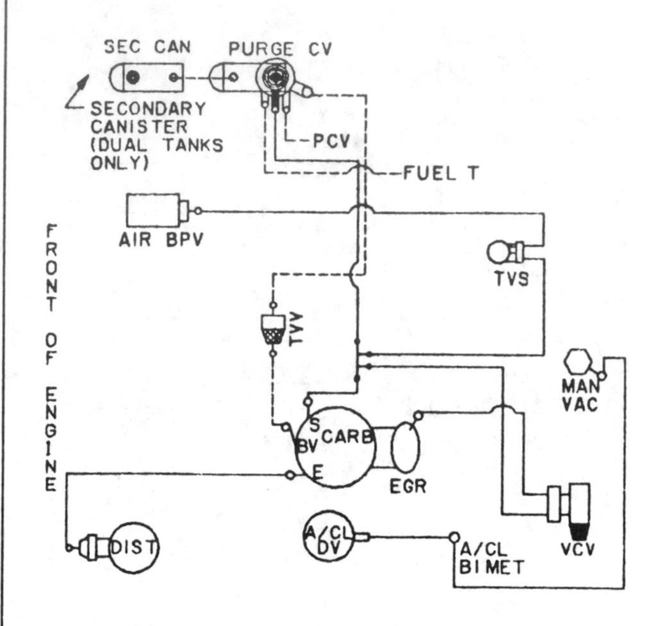

CALIBRATION: 9–54H–R10 DATE: 8–25–78
5.0L (302 CID)

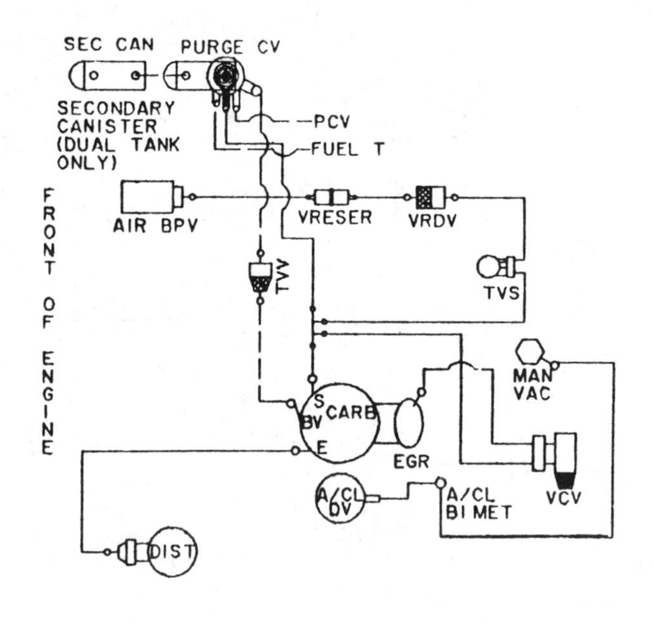

CALIBRATION: 9–54J–R1 DATE: 6–27–78
5.0L (302 CID)

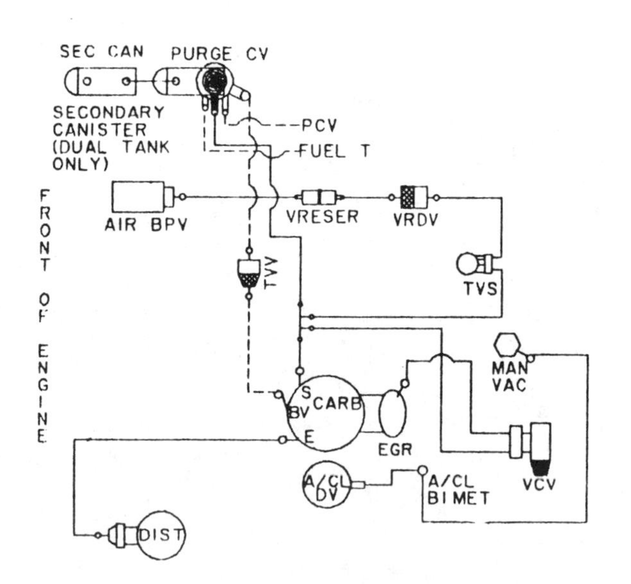

CALIBRATION: 9–54J–R10 DATE: 10–17–78
5.0L (302 CID)

VACUUM CIRCUITS

1979—302 CID (5.0 L) engines

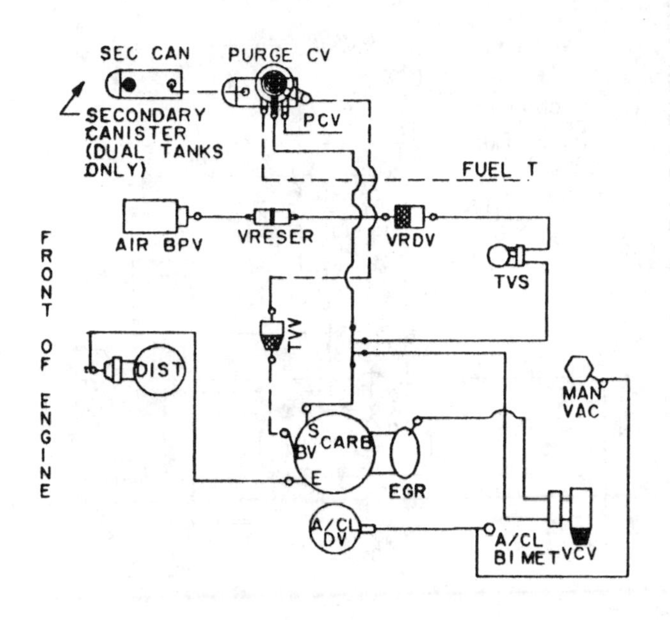

CALIBRATION: 9-54R-RO DATE: 5-28-78
5.0L (302 CID)

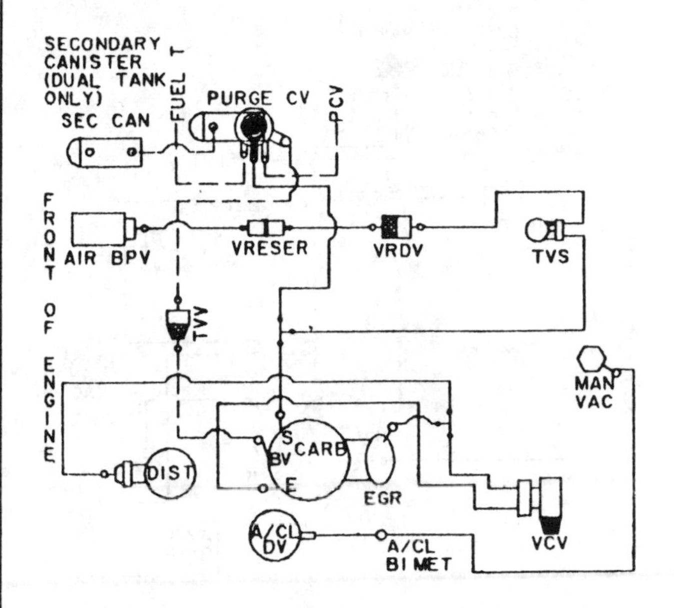

CALIBRATION. 9-54S-RO DATE: 5-25-78
5.0L (302 CID)

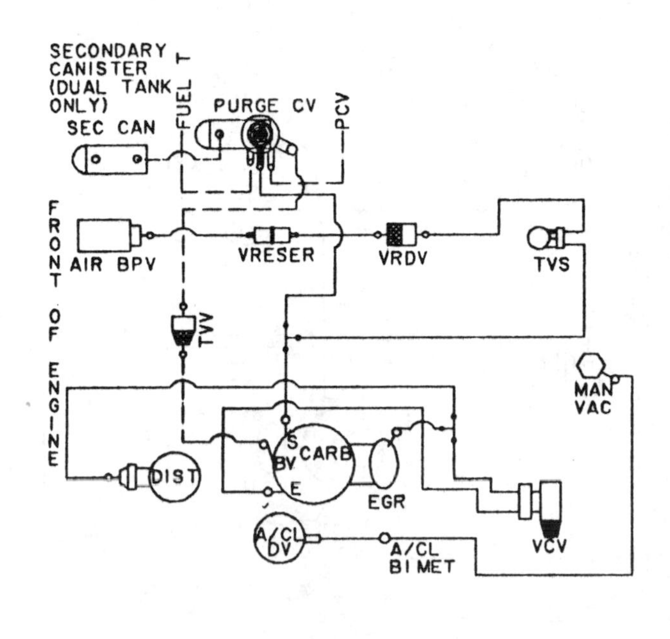

CALIBRATION: 9-54S-R10 DATE: 10-17-78
5.0L (302 CID)

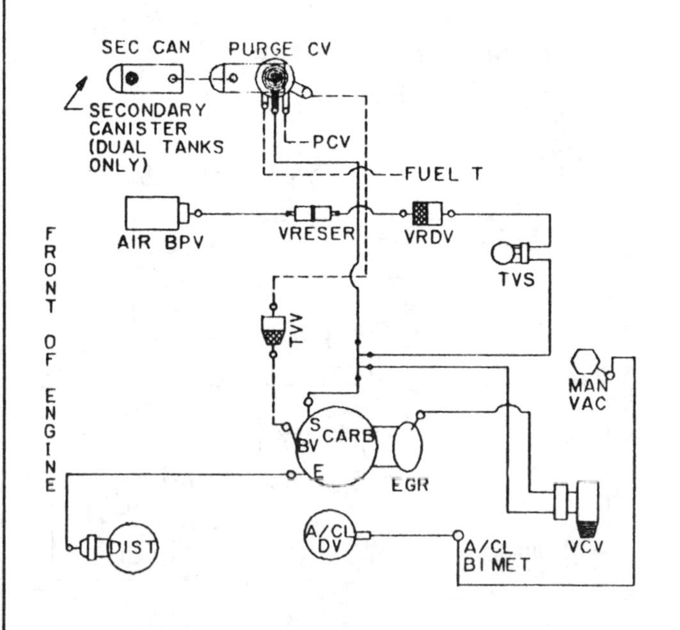

CALIBRATION. 9-54T-RO DATE: 5-31-78
5.0L (302 CID)

VACUUM CIRCUITS

1979—302 CID (5.0 L) engines

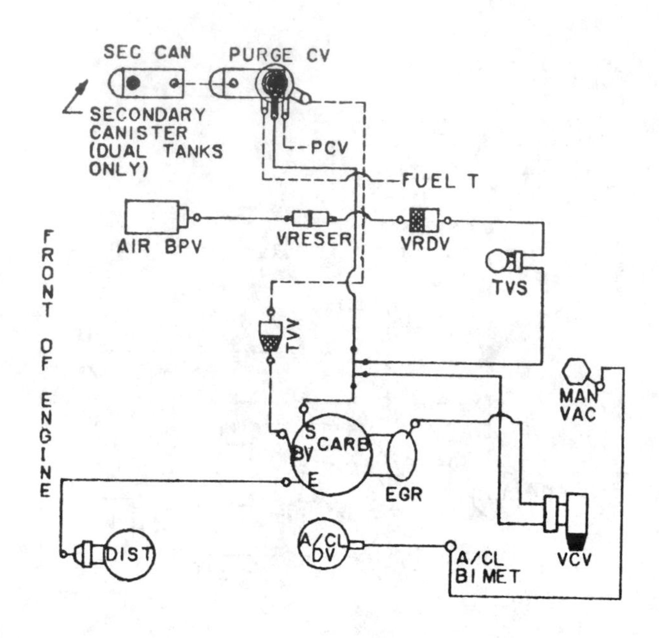

CALIBRATION: 9–54T–R10 DATE: 5–31–78
5.0L (302 CID)

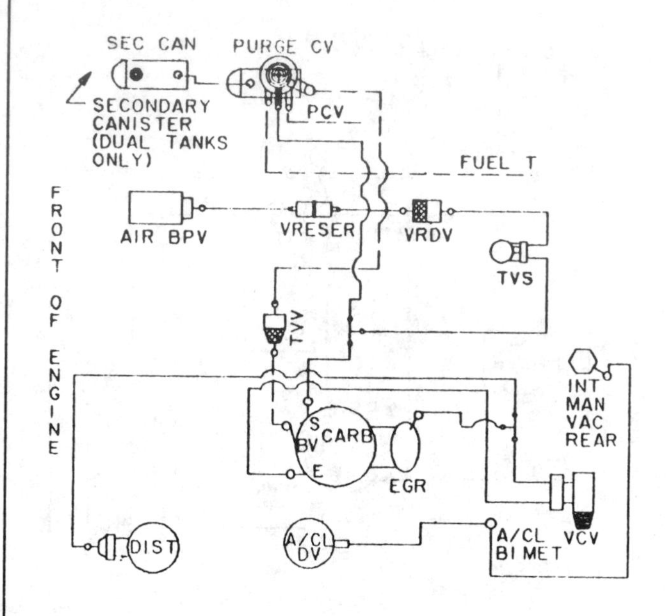

CALIBRATION: 9–54U–RO DATE: 5–30–78
5.0L (302 CID)

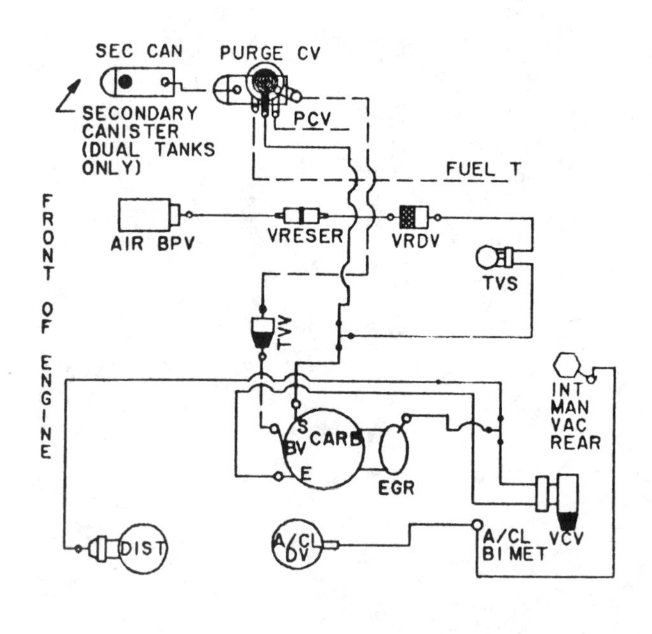

CALIBRATION: 9–54U–R10 DATE: 10–17–78
5.0L (302 CID)

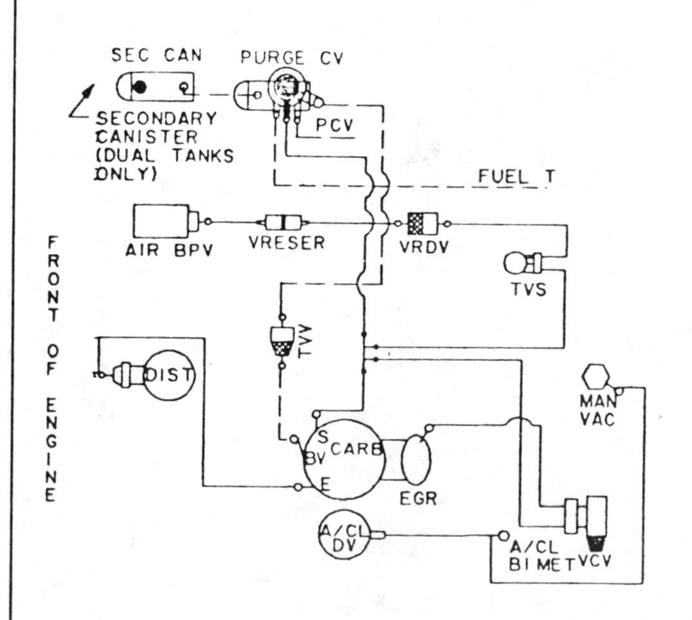

CALIBRATION: 9–54R–RO DATE: 5–28–78
5.0L (302 CID)

VACUUM CIRCUITS

1979—351 CID (5.8 L) engines

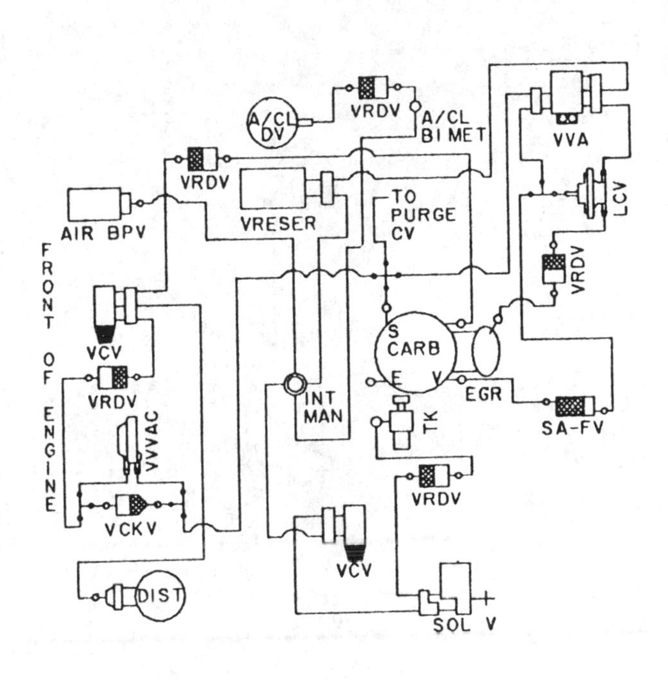

CALIBRATION. 7–72J–R11 DATE: 6–8–78
5.8L (351M CID)

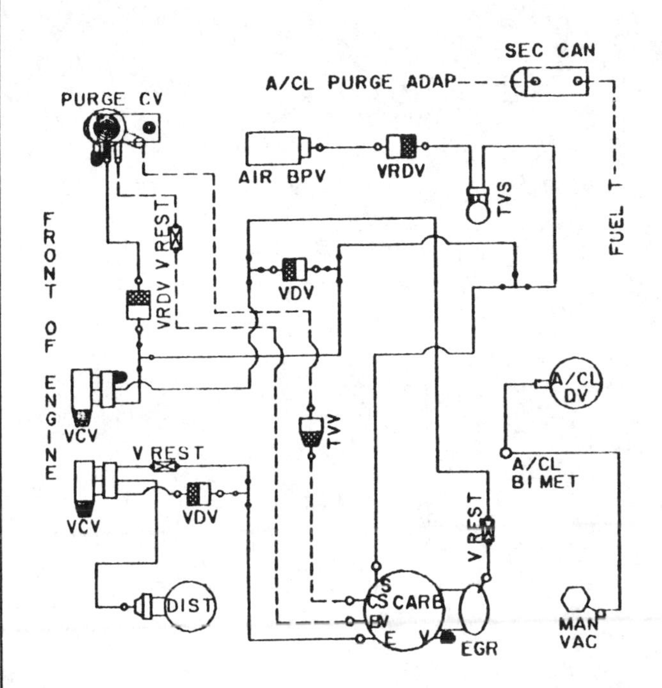

CALIBRATION: 9–14E–R0 DATE: 6–5–78
5.8L (351M CID)

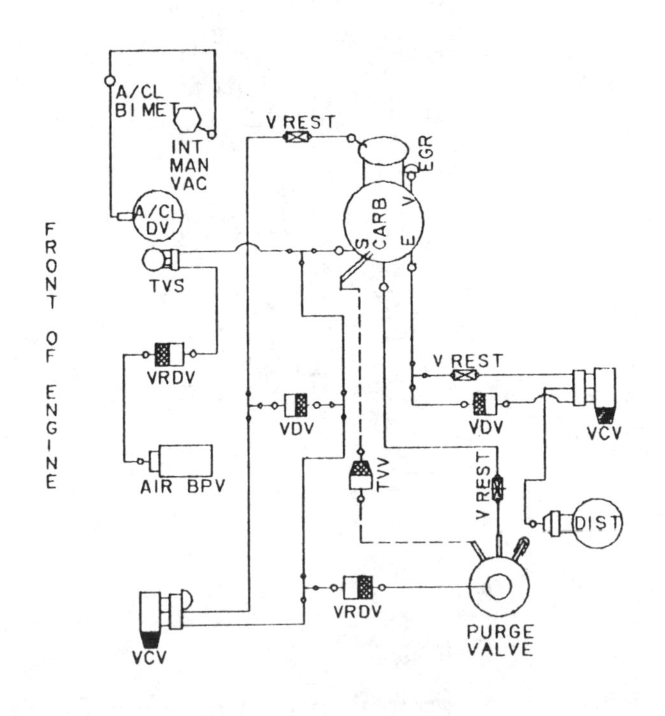

CALIBRATION: 9–14E–R12 DATE: 7-11-78
5.8L (351M CID)

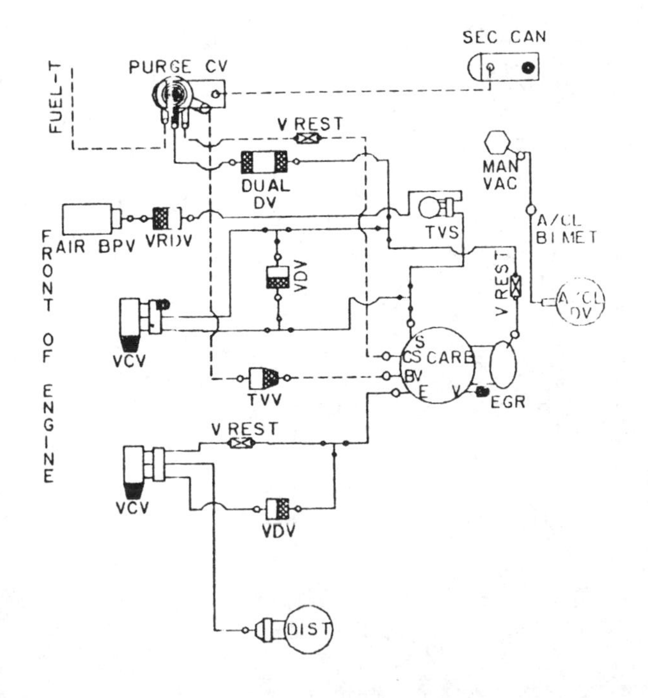

CALIBRATION: 9–14E–R14 DATE: 1–13–79
5.8L (351M CID)

VACUUM CIRCUITS

1979—351 CID (5.8 L) engines

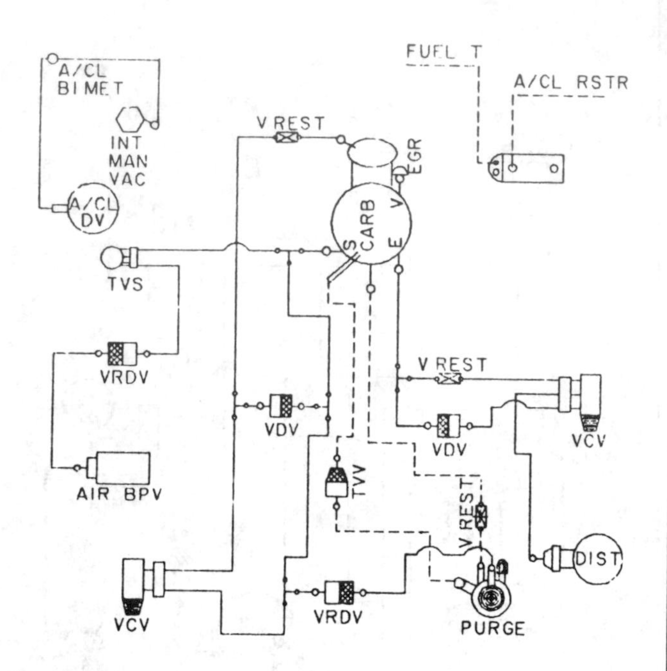

CALIBRATION: 9–14E–R15 DATE: 12–1–78
5.8L (351M CID)

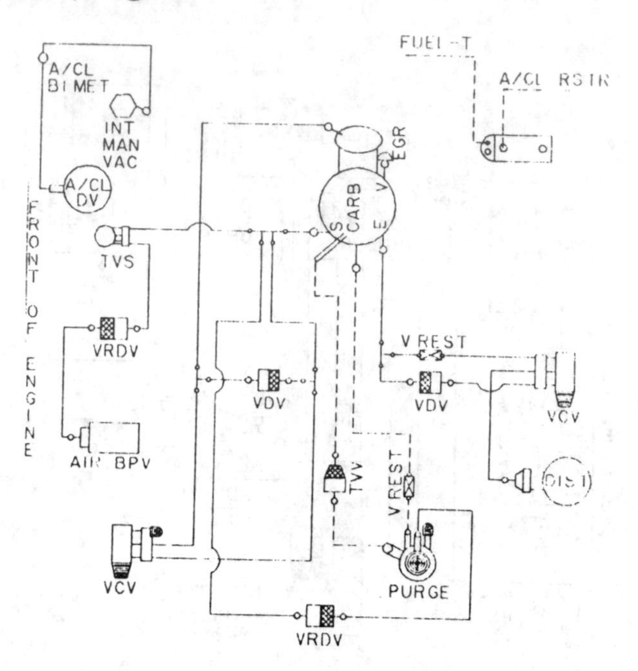

CALIBRATION: 9–14E–R18 DATE: 1–11–79
5.8L (351M CID)

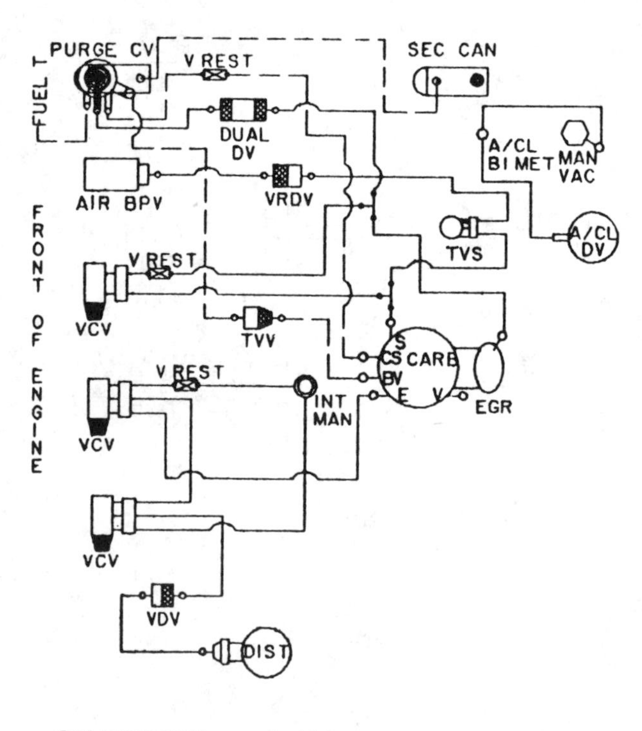

CALIBRATION: 9–14Q–RO DATE: 5–4–78
5.8L (351M CID)

CALIBRATION: 9–14Q–R10 DATE: 2–2–79
5.8L (351M CID)

VACUUM CIRCUITS

1979—351 CID (5.8 L) engines

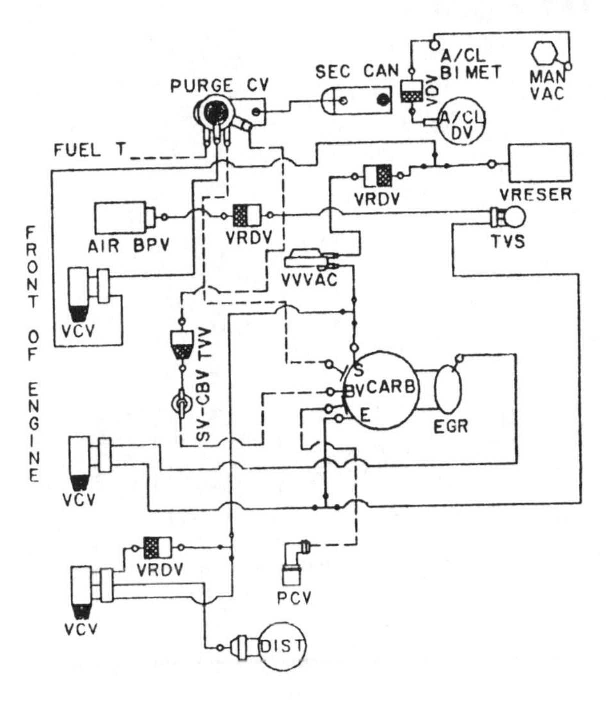

CALIBRATION: 9–14X–RO DATE: 10–10–78
5.8L (351M CID)

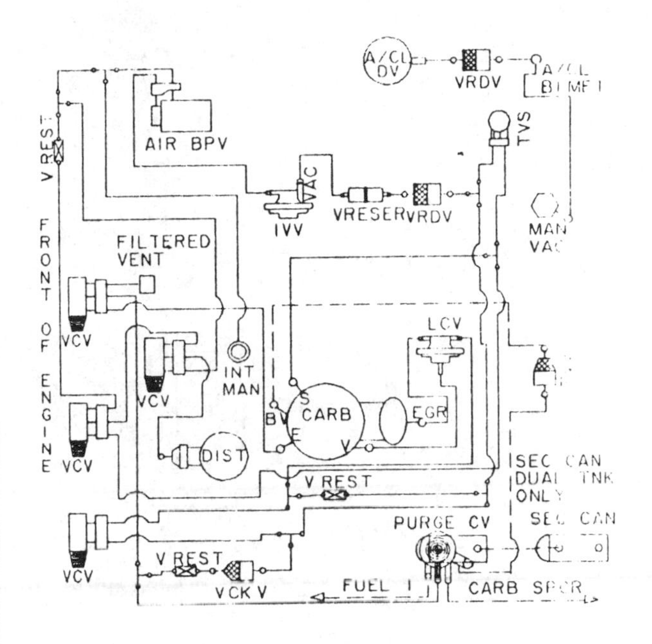

CALIBRATION: 9–59H–ROA DATE: 1–24–79
5.8L (351M CID)

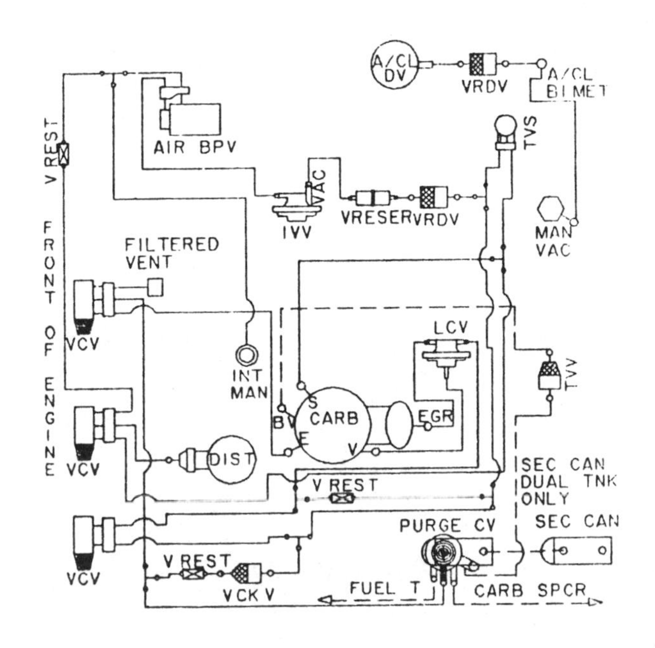

CALIBRATION: 9–59H–RON DATE: 1–24–79
5.8L (351M CID)

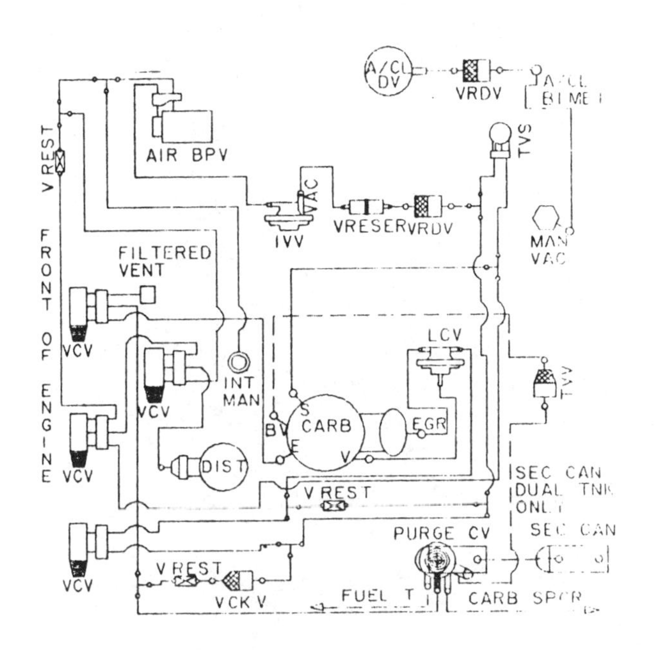

CALIBRATION: 9–59J–ROA DATE: 1–24–79
5.8L (351M CID)

VACUUM CIRCUITS

1979—351 CID (5.8 L) engines

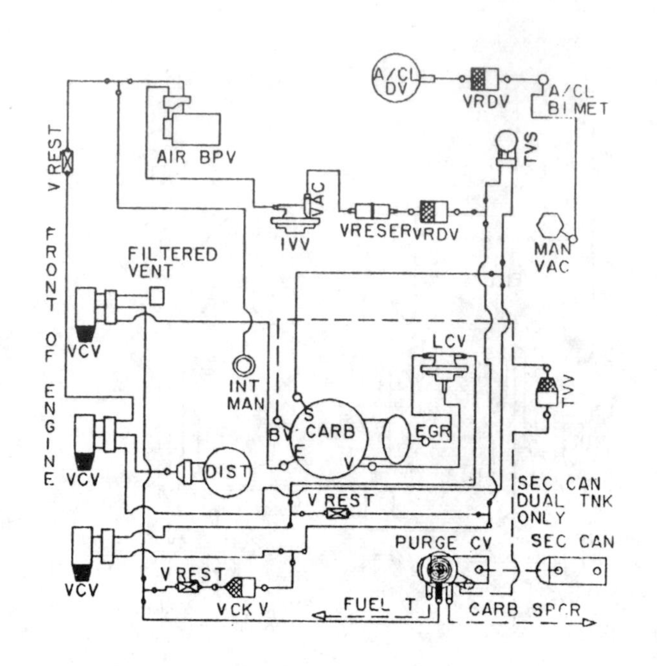

CALIBRATION: 9–59J–RON DATE: 1–24–79
5.8L (351M CID)

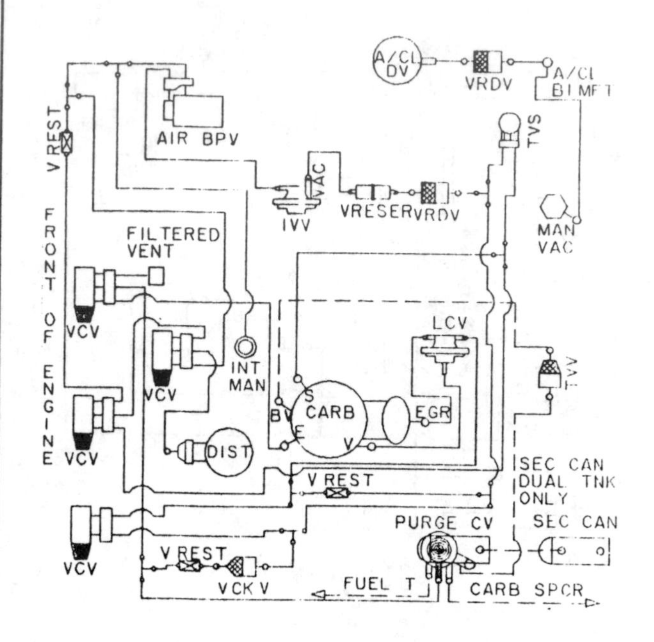

CALIBRATION: 9–59K–ROA DATE: 1–24–79
5.8L (351M CID)

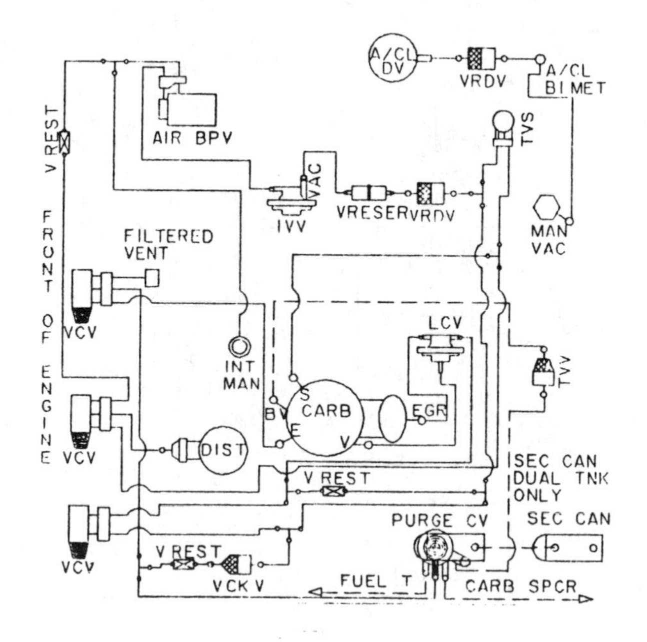

CALIBRATION: 9–59K–RON DATE: 2–26–79
5.8L (351M CID)

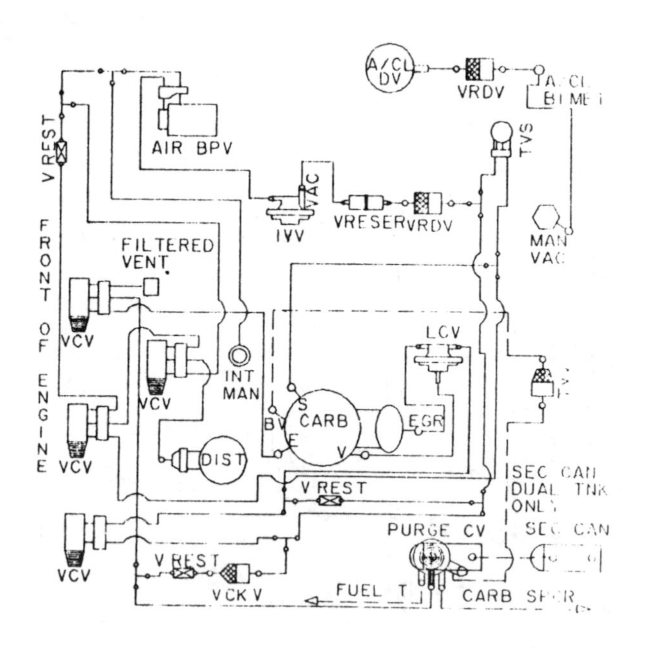

CALIBRATION: 9–59S–ROA DATE: 1–24–79
5.8L (351M CID)

VACUUM CIRCUITS

1979—351 CID (5.8 L) engines

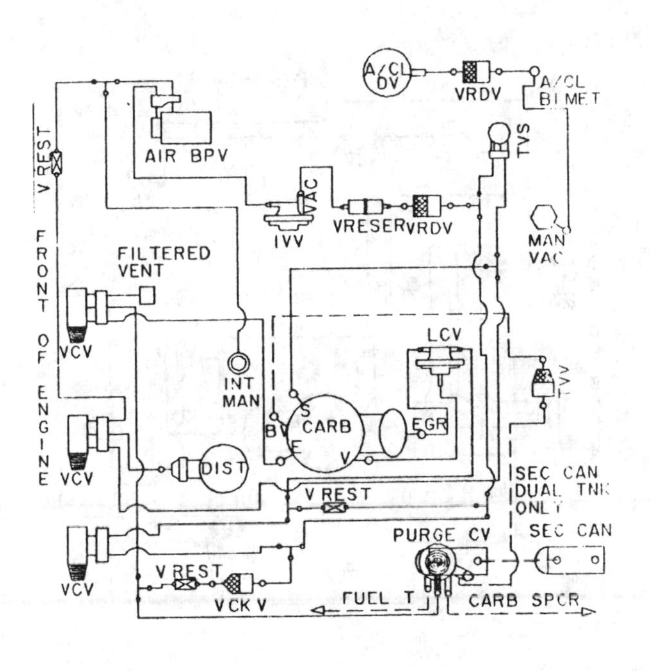

CALIBRATION: 9-59S-RON DATE: 1-24-79
5.8L (351M CID)

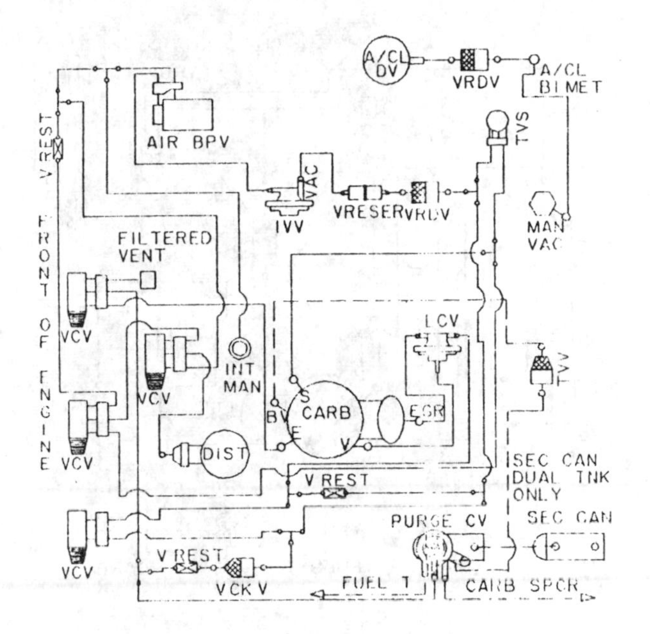

CALIBRATION: 9-59T-ROA DATE: 1-24-79
5.8L (351M CID)

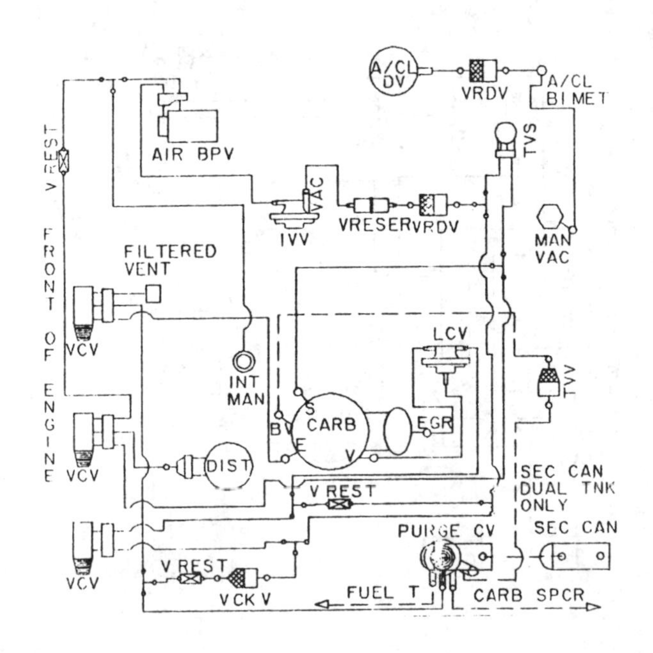

CALIBRATION: 9-59T-RON DATE: 1-24-79
5.8L (351M CID)

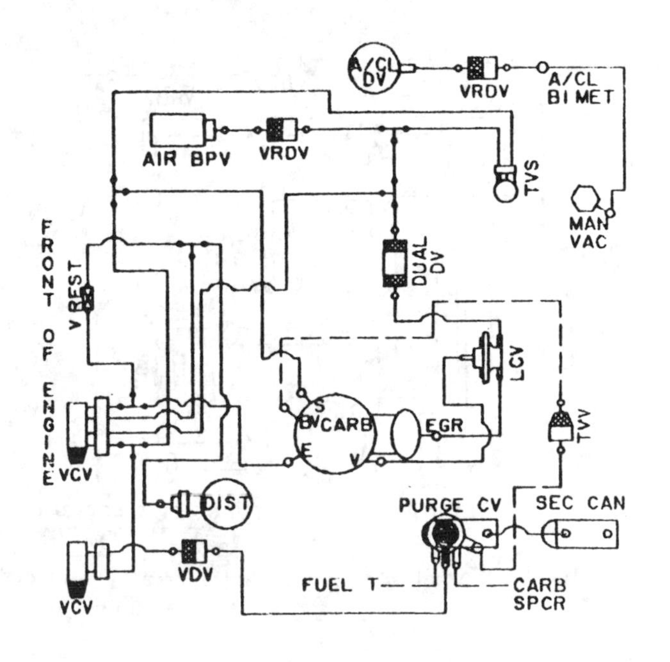

CALIBRATION: 9-60G-RO DATE: 5-16-78
5.8L (351M CID)

VACUUM CIRCUITS

1979—351 CID (5.8 L) engines

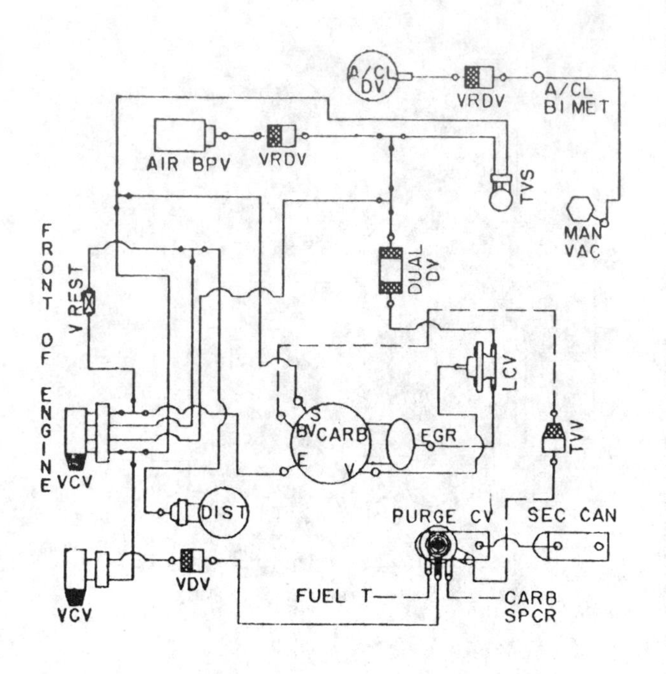

CALIBRATION. 9-60H-R0 DATE: 5-16-78
5.8L (351M CID)

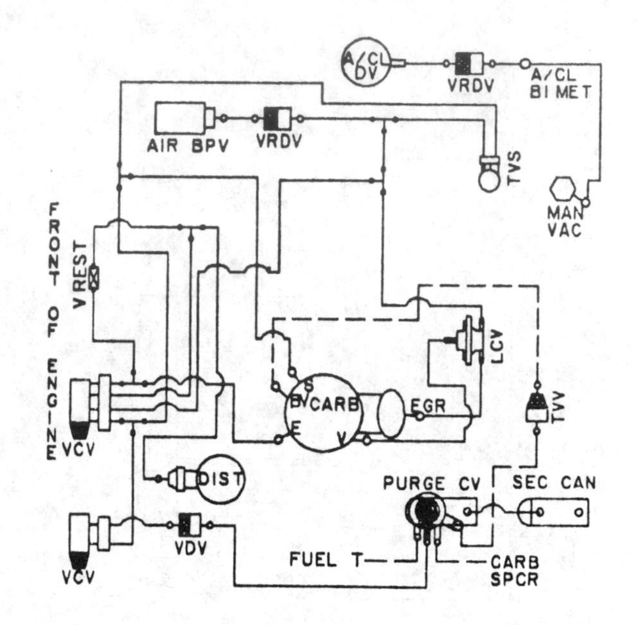

CALIBRATION: 9-60H-R10 DATE: 10-26-78
5.8L (351M CID)

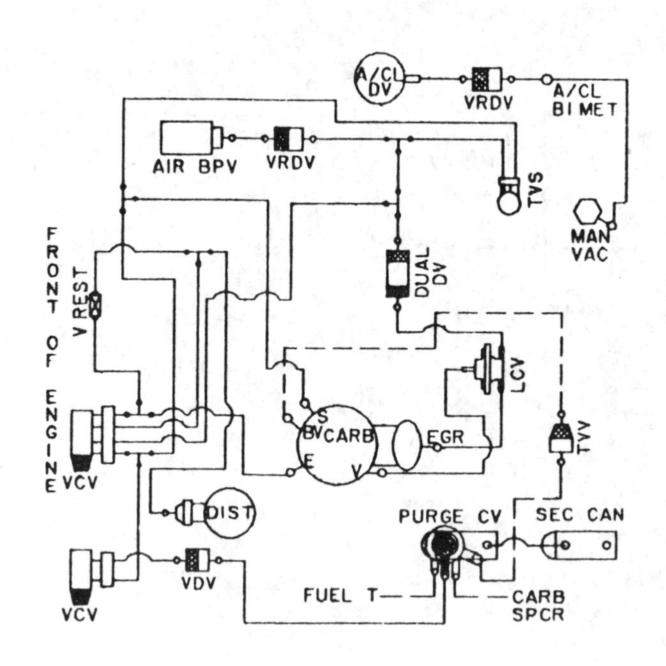

CALIBRATION: 9-60J-RO DATE: 5-18-78
5.8L (351M CID)

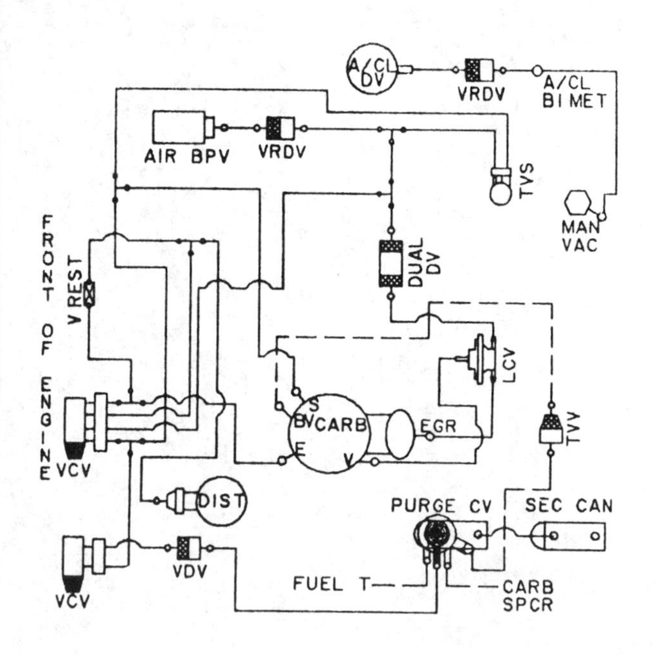

CALIBRATION: 9-60J-R10 DATE: 10-4-78
5.8L (351M CID)

VACUUM CIRCUITS

1979—351 CID (5.8 L) engines

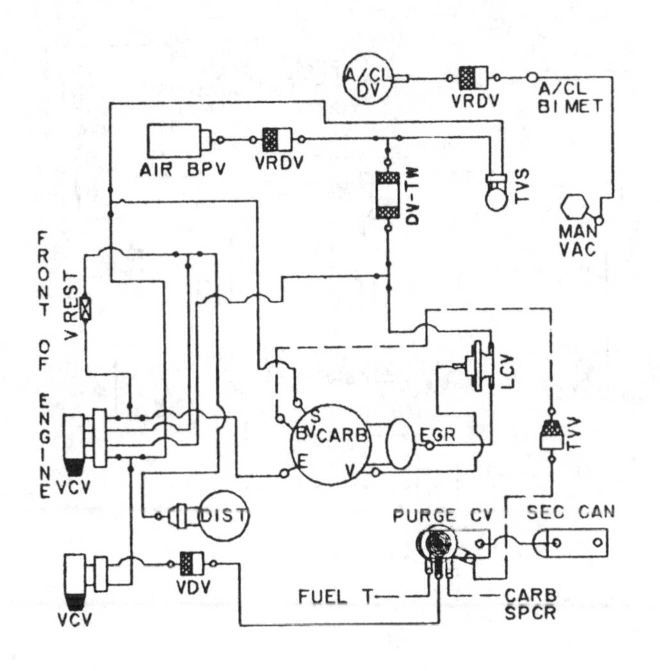

CALIBRATION. 9-60L-RO DATE: 5-18-78
5.8L (351M CID)

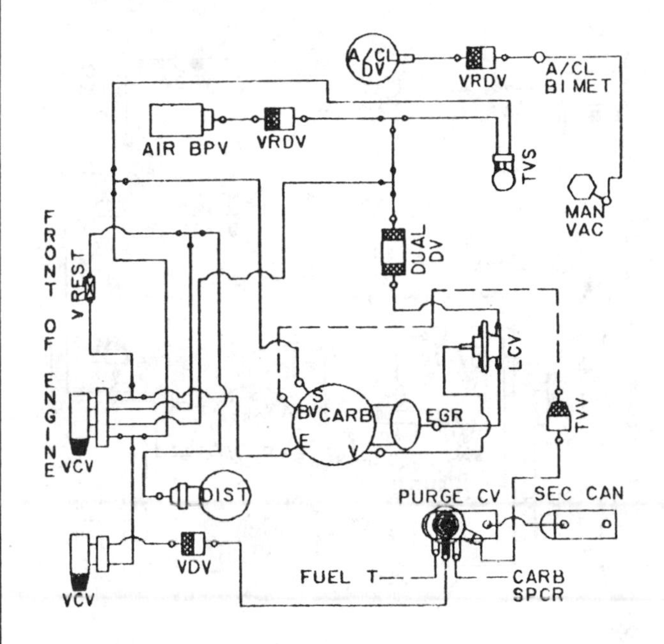

CALIBRATION: 9-60M-RO DATE: 5-16-78
5.8L (351M CID)

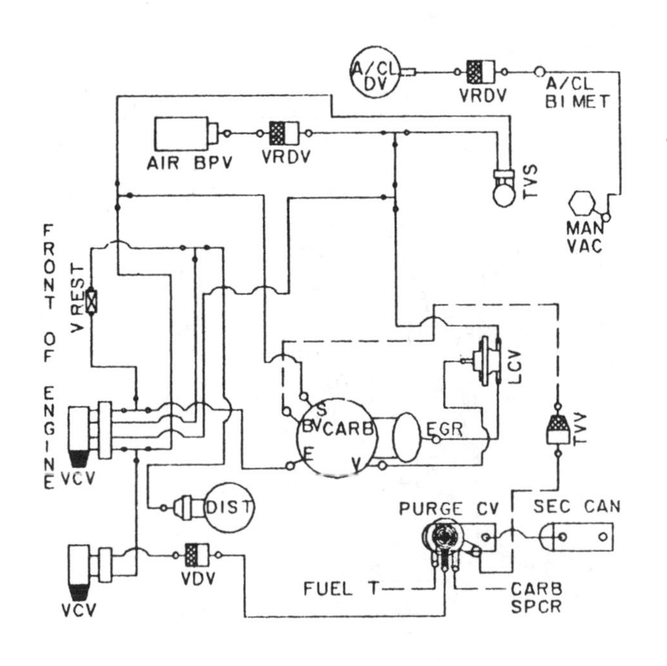

CALIBRATION: 9-60M-R10 DATE: 10-4-78
5.8L (351M CID)

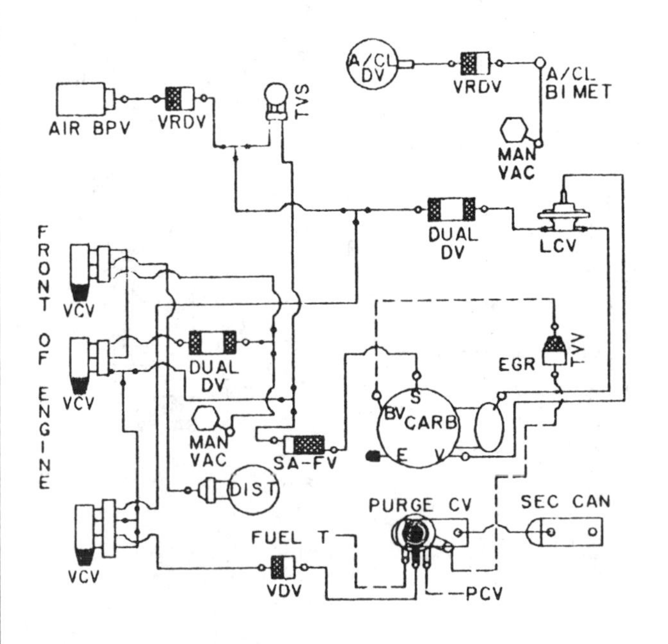

CALIBRATION: 9-60S-RO DATE. 6-15-78
5.8L (351M CID)

VACUUM CIRCUITS

1979—351 CID (5.8 L) engines

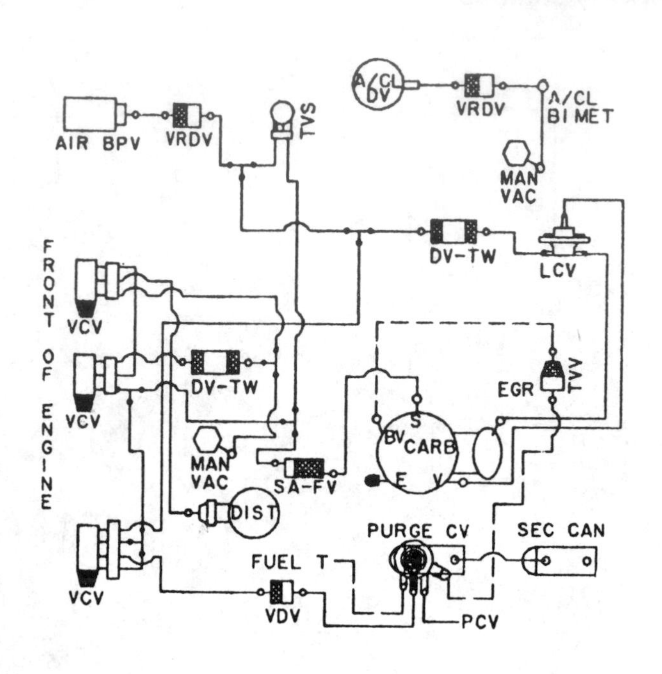

CALIBRATION: 9–60S–R10 DATE: 7–25–78
5.8L (351M CID)

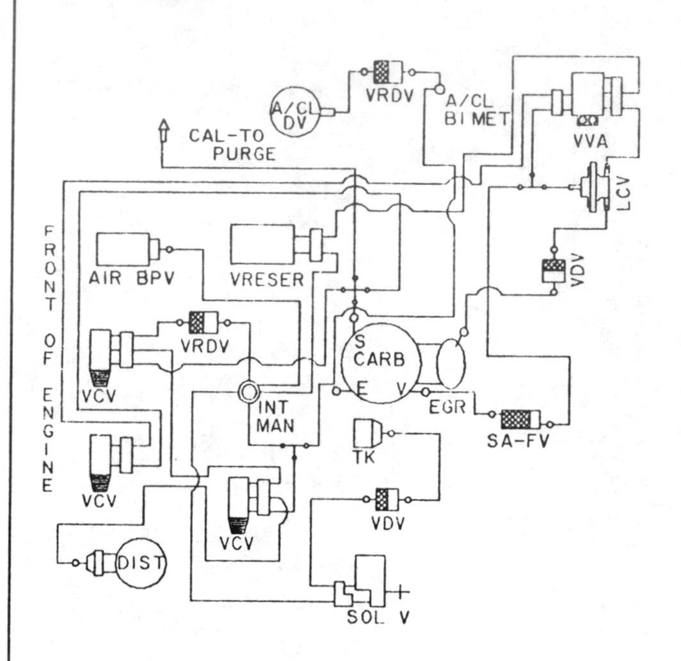

CALIBRATION: 9–71J–R10 DATE: 12–7–78
5.8L (351M CID)

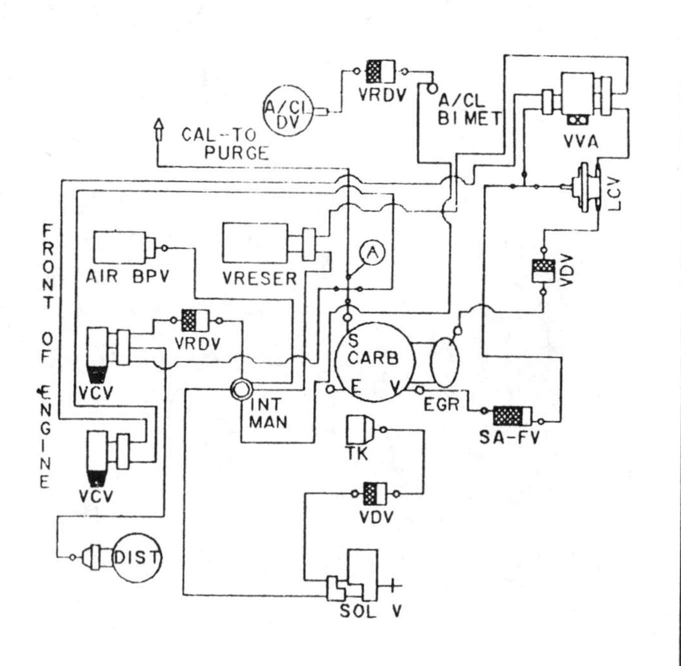

CALIBRATION: 9–72J–RO DATE: 10–13–78
5.8L (351M CID)

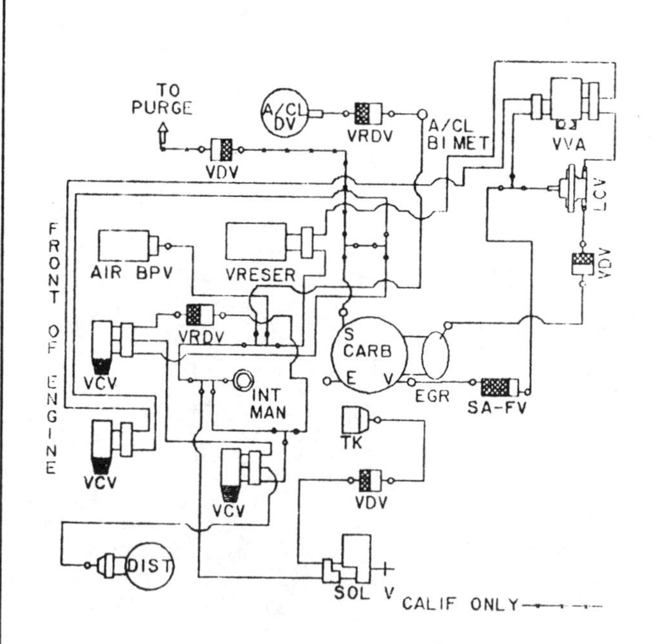

CALIBRATION: 9–72J–R11 DATE: 1–25–79
5.8L (351M CID)

VACUUM CIRCUITS

1979—351 CID (5.8 L) engines

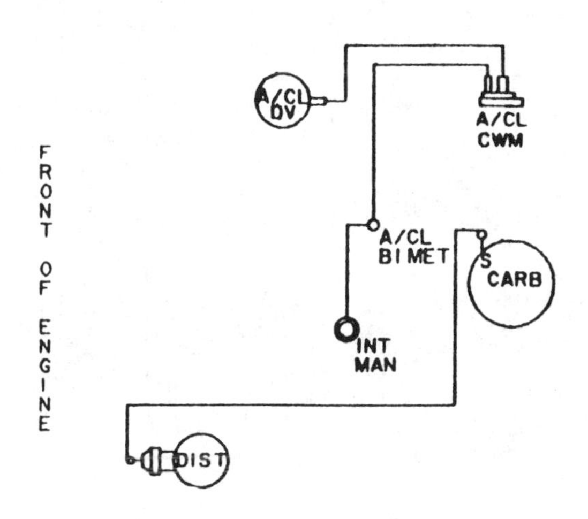

CALIBRATION: 7–75A–R10 DATE: 5–17–78
5.8L (351W CID) CANADIAN ONLY

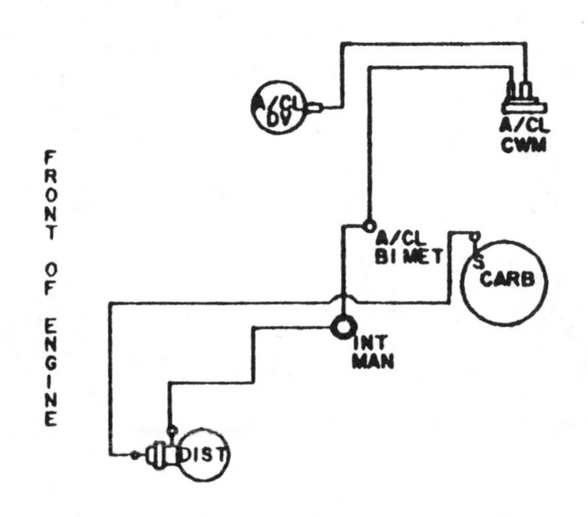

CALIBRATION: 7–76A–R10 DATE: 5–17–78
5.8L (351W CID) CANADIAN ONLY

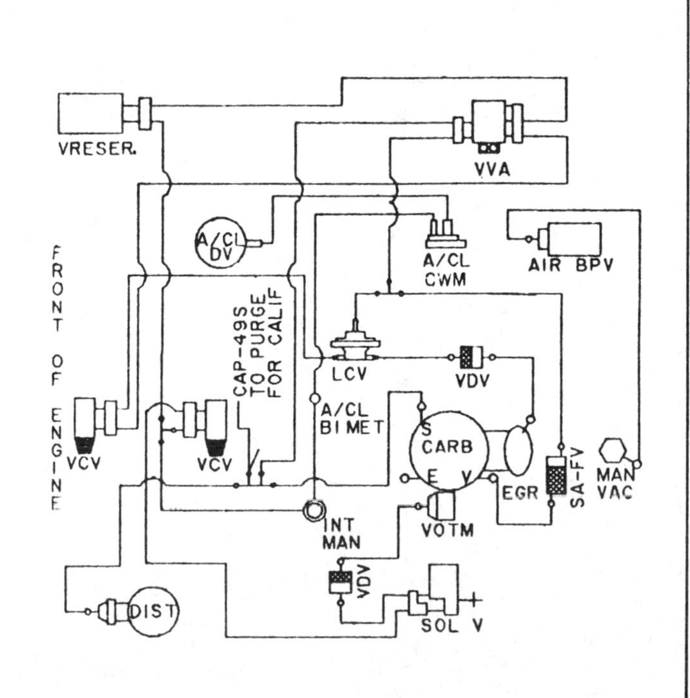

CALIBRATION: 7–76J–R11 DATE: 6–6–78
5.8L (351 CID)

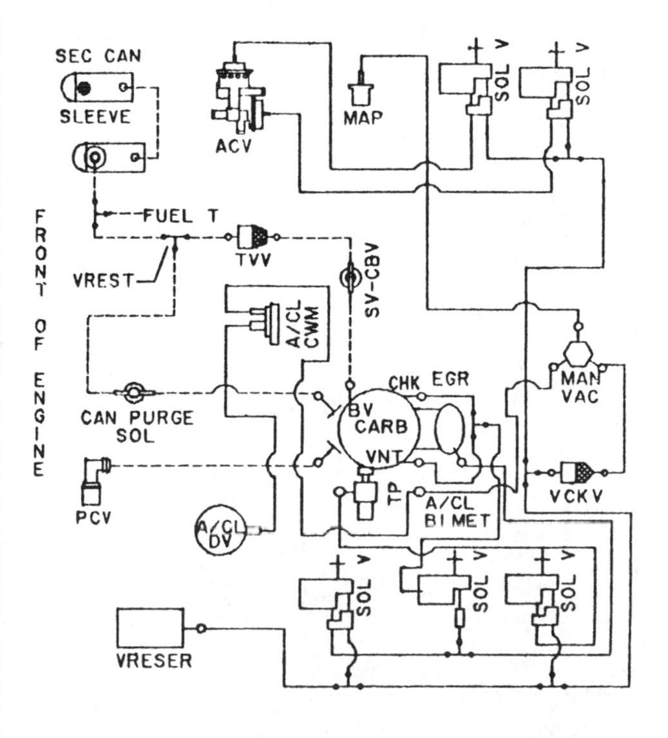

CALIBRATION: 9–12A–RO DATE: 6–21–78
5.8L (351W CID)

VACUUM CIRCUITS

1979—351 CID (5.8 L) engines

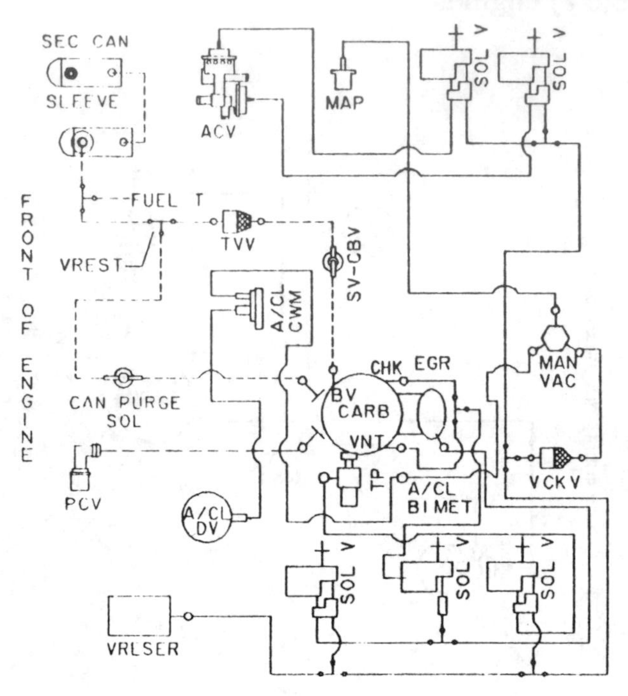

CALIBRATION: 9–12B–RO **DATE:** 6–7–78
5.8L (351W CID)

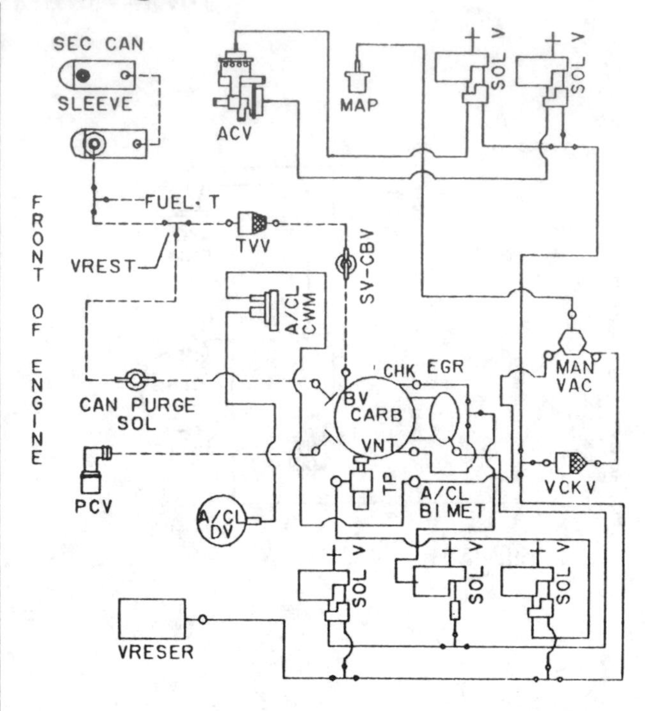

CALIBRATION: 9–12B–R10 **DATE:** 8–12–78
5.8L (351W CID)

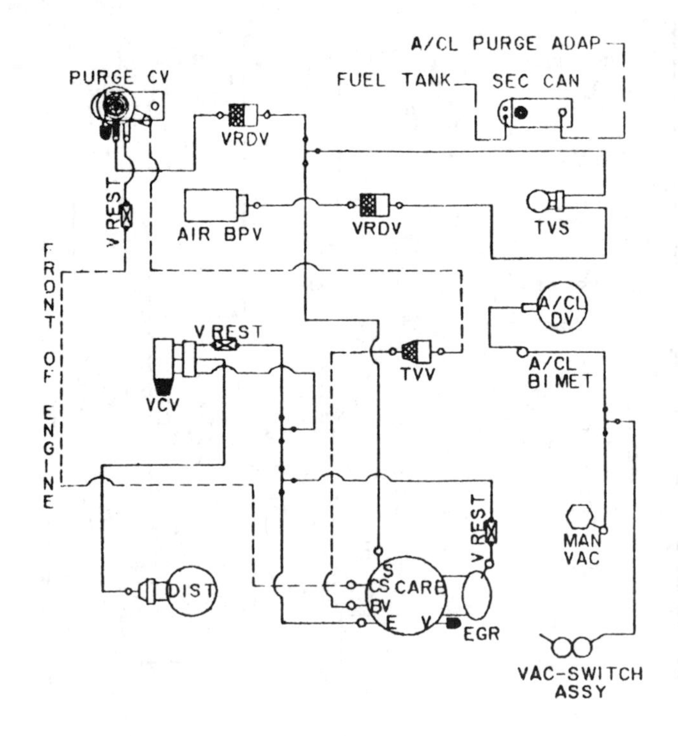

CALIBRATION: 9–12E–RO **DATE:** 6–1–78
5.8L (351W CID)

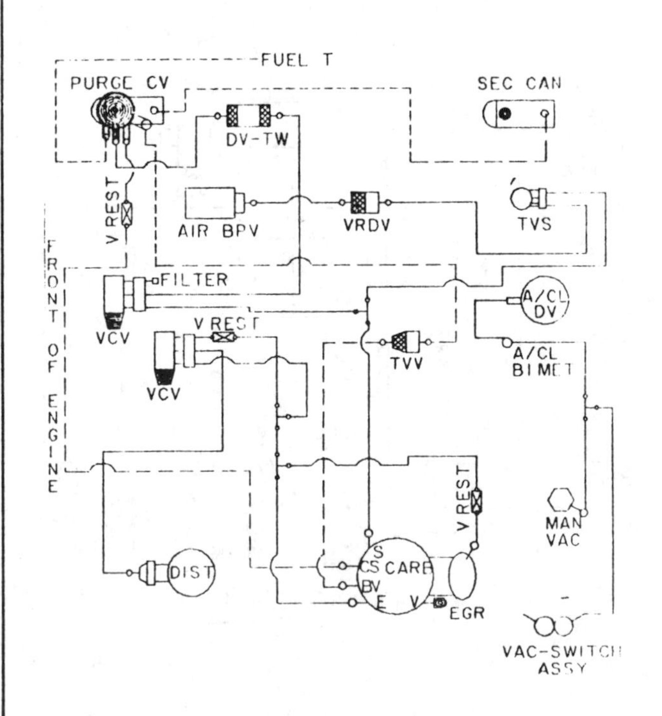

CALIBRATION: 9–12E–R14 **DATE:** 12–20–78
5.8L (351W CID)

VACUUM CIRCUITS

1979—351 CID (5.8 L) engines

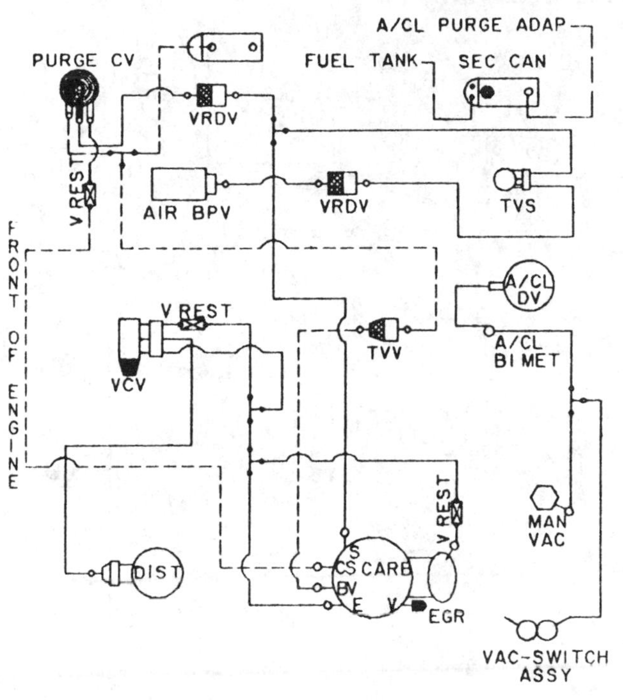

CALIBRATION: 9–12F–RO DATE: 6–1–78
5.8L (351W CID)

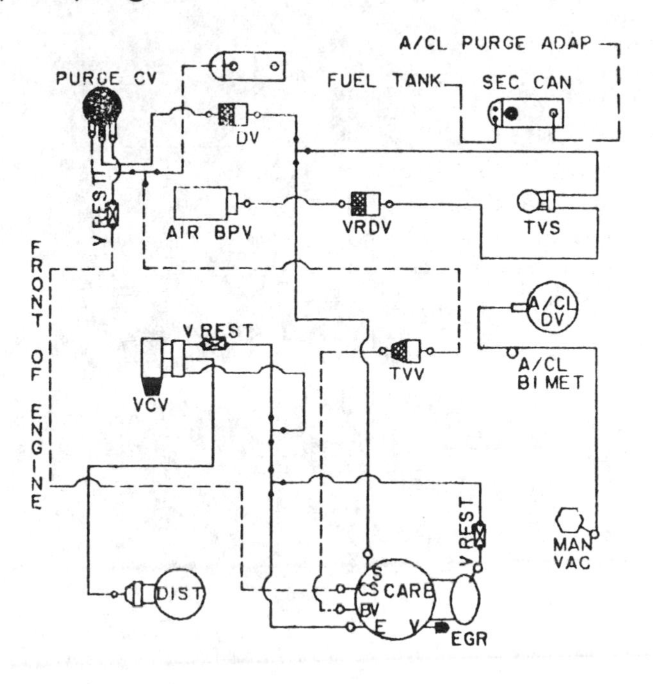

CALIBRATION: 9–12G–RO DATE: 6–1–78
5.8L (351W CID)

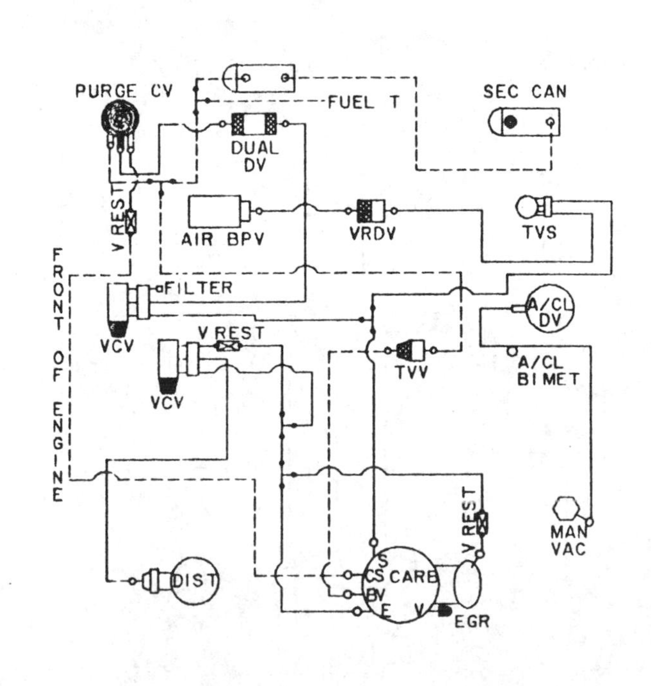

CALIBRATION: 9–12G–R12 DATE: 8–1–78
5.8L (351W CID)

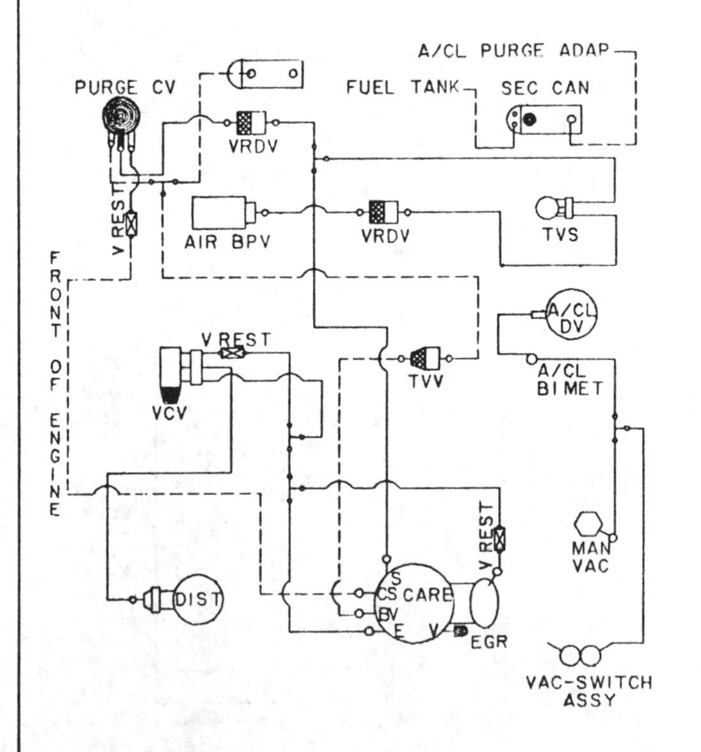

CALIBRATION: 9–12H–R12 DATE: 7–8–78
5.8L (351 W CID)

VACUUM CIRCUITS

1979—351 CID (5.8 L) engines

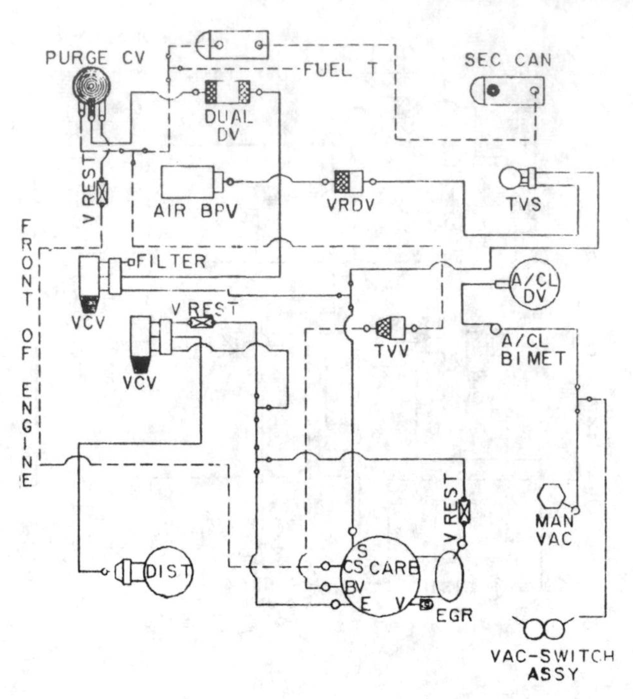

CALIBRATION: 9–12H–R15 DATE: 10–13–78
5.8L (351W CID)

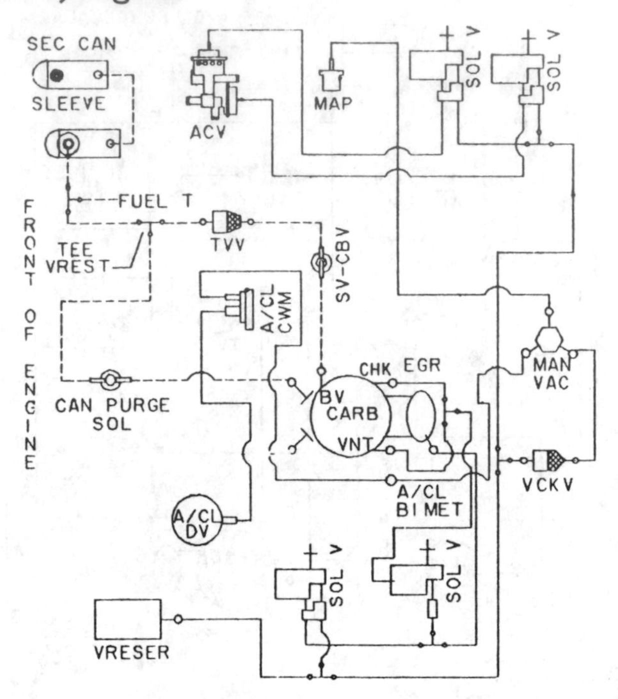

CALIBRATION. 9–12N–RO DATE: 8–14–78
5.8L (351W CID)

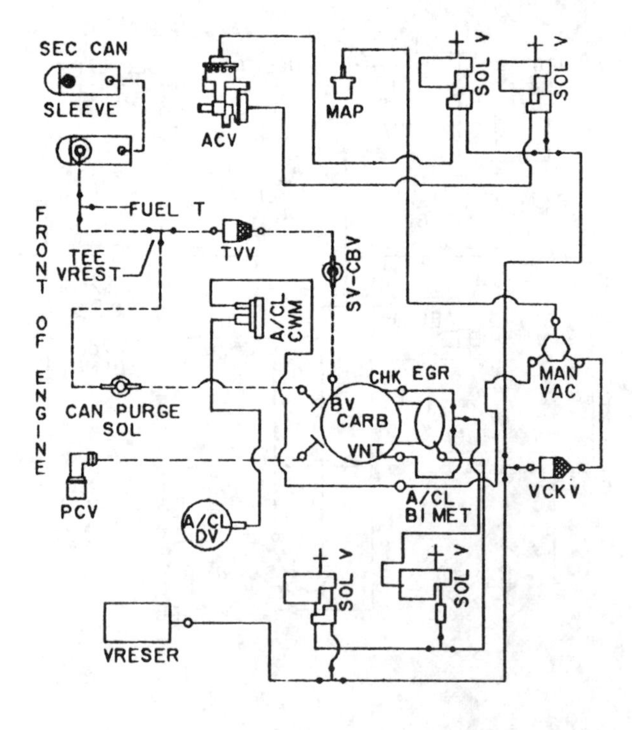

CALIBRATION: 9–12P–R10 DATE: 8–25–78
5.8 (W) (351 CID)

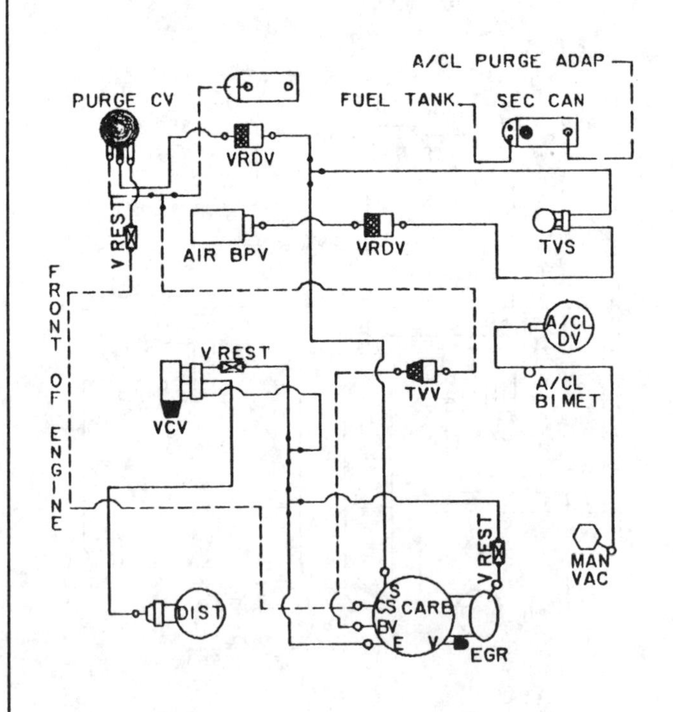

CALIBRATION: 9–12X–RO DATE: 7–13–78
5.8L (351W CID)

VACUUM CIRCUITS

1979—351 CID (5.8 L) engines

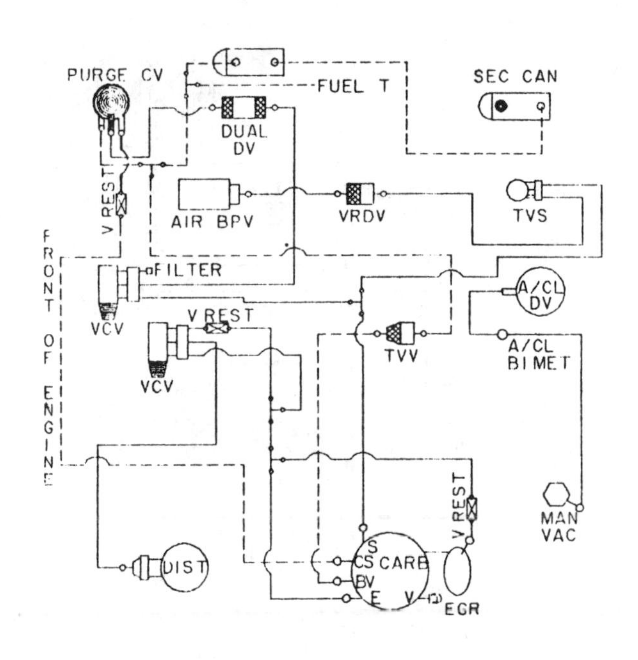

CALIBRATION: 9–12X–R10 DATE: 1–22–79
5.8L (351W CID)

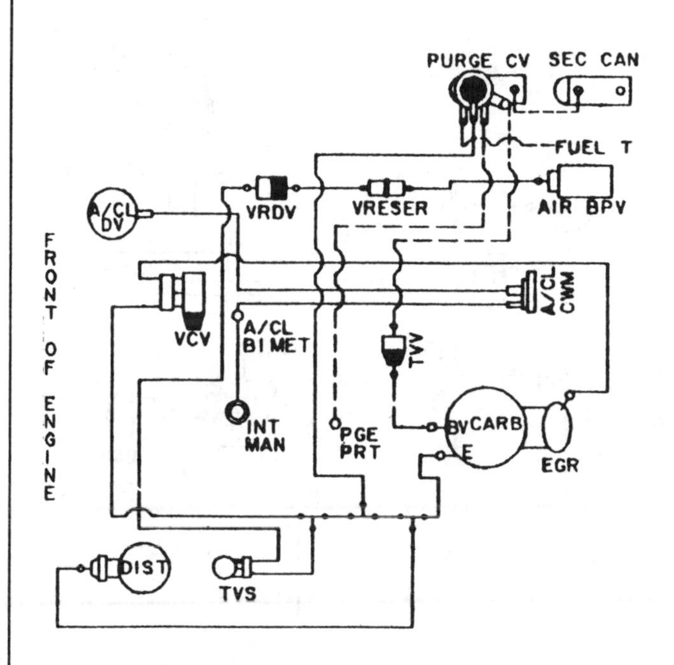

CALIBRATION: 9–63H–RO DATE: 6–13–78
5.8L (351W CID)

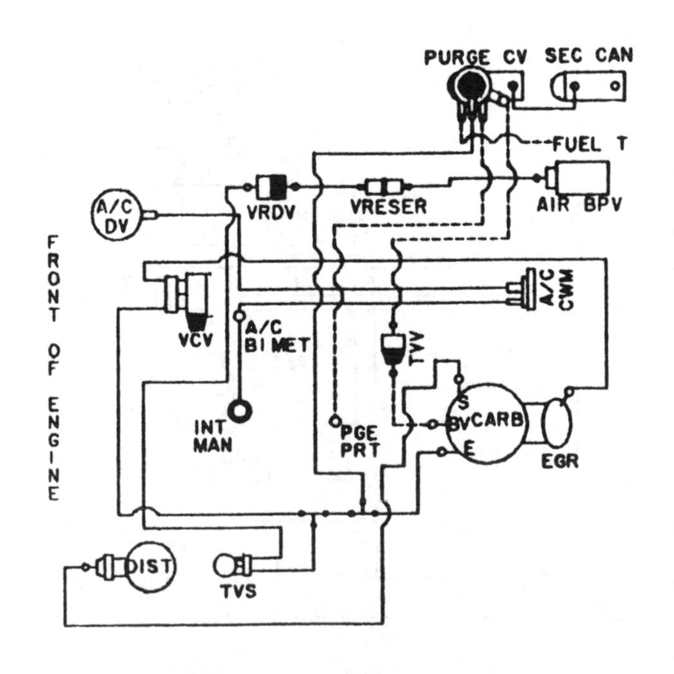

CALIBRATION: 9–64A–RO DATE: 5–30–78
5.8L (351W CID)

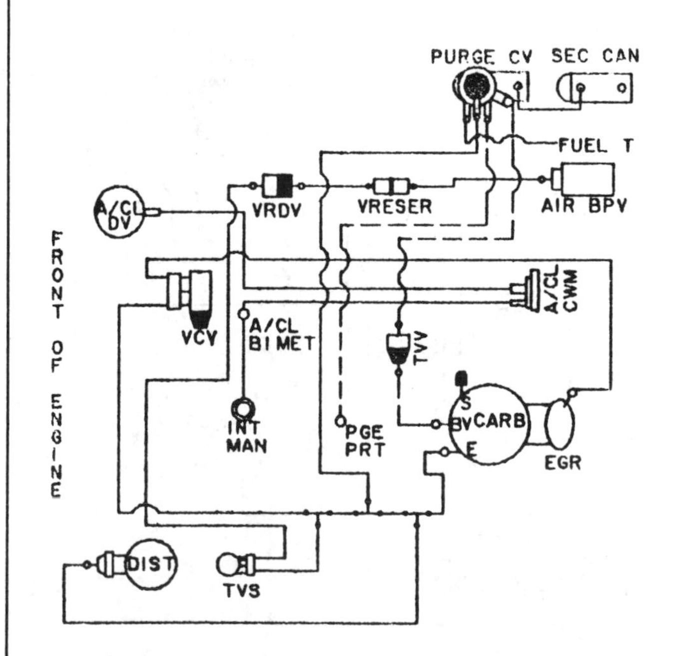

CALIBRATION: 9–64B–RO DATE: 5–30–78
5.8L (351W CID)

VACUUM CIRCUITS

1979—351 CID (5.8 L) engines

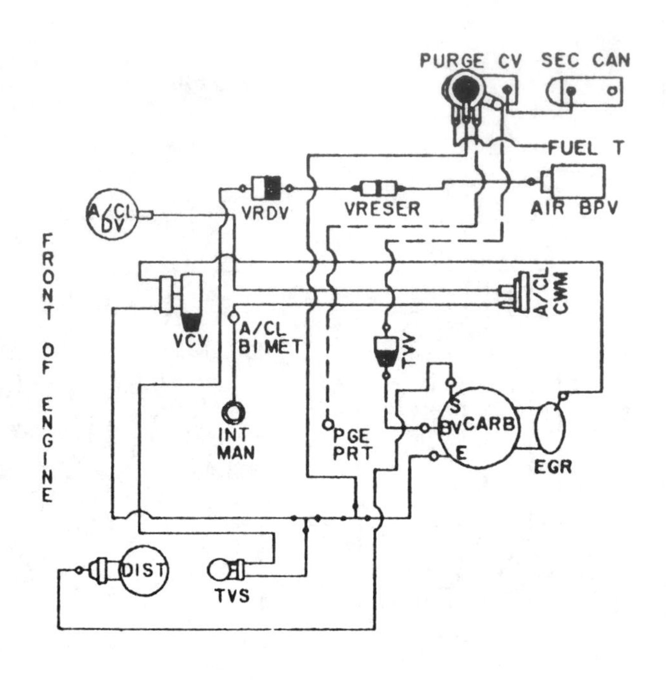

CALIBRATION: 9–64G–RO DATE: 6–7–78
5.8L (351W CID)

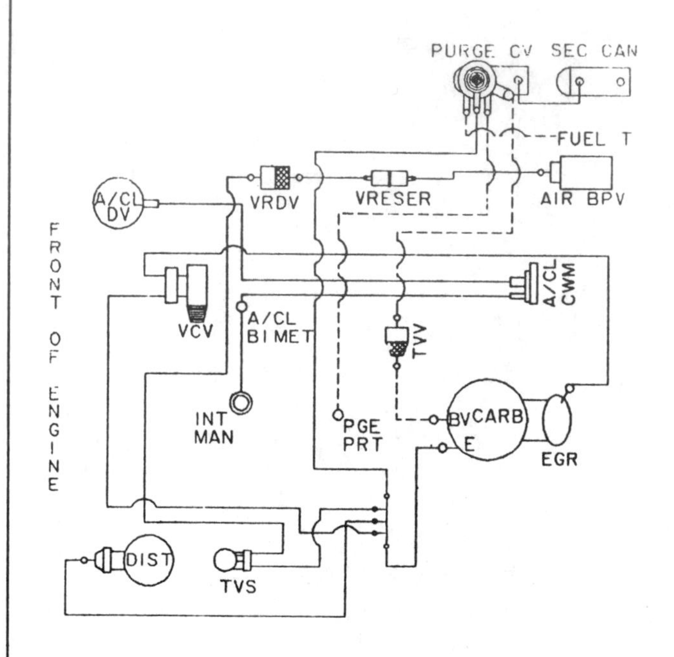

CALIBRATION: 9–64G–R10 DATE: 1–10–79
5.8L (351W CID)

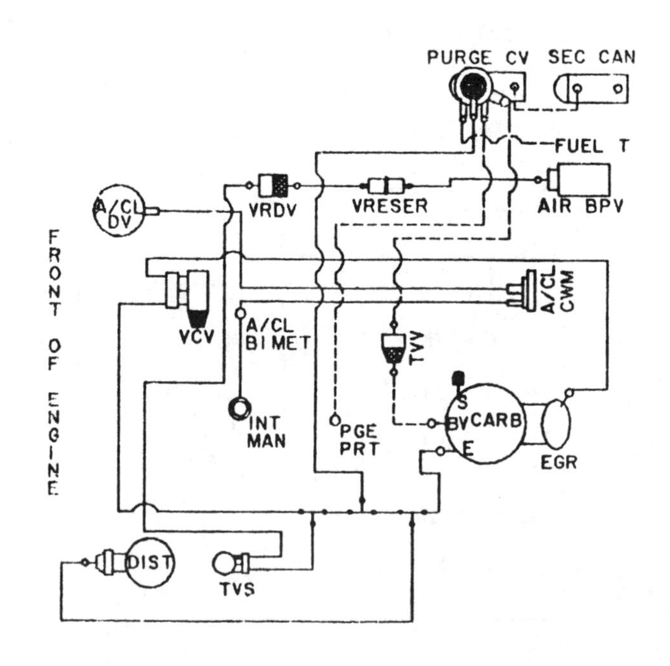

CALIBRATION: 9–64H–RO DATE: 6–16–78
5.8L (351W CID)

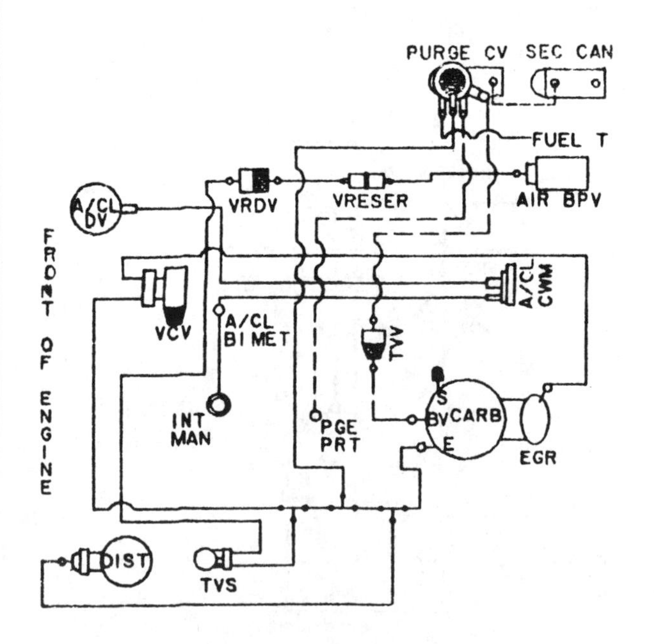

CALIBRATION: 9–64S–RO DATE: 6–4–78
5.8L (351W CID)

VACUUM CIRCUITS

1979—370 CID (6.1 L) engines

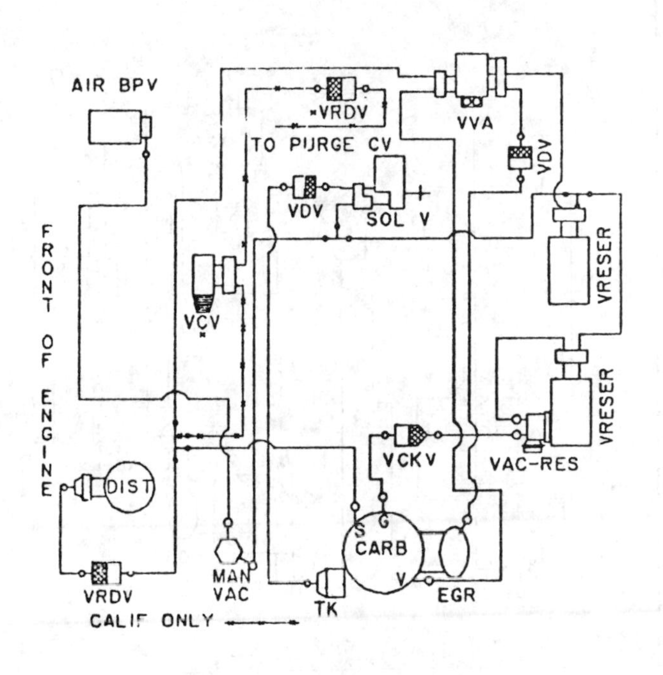

CALIBRATION: 9–83G–RO DATE: 6–14–78
6.1L (370 CID)

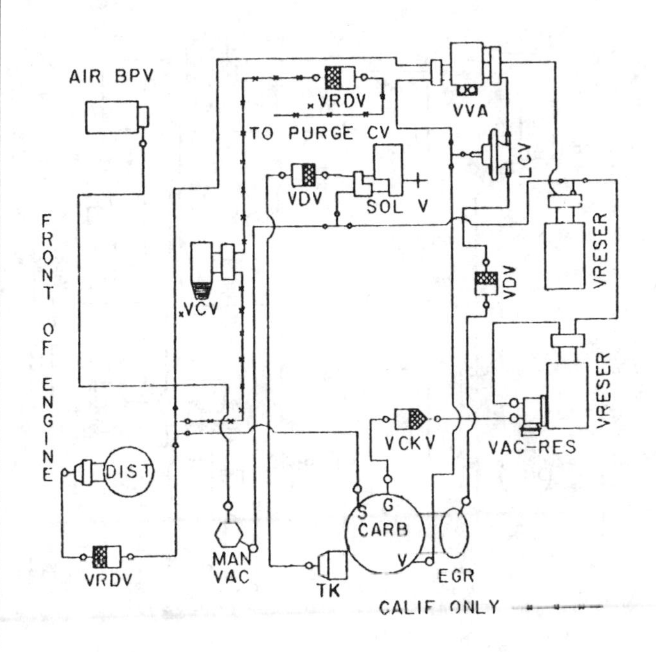

CALIBRATION: 9–83G–R2 DATE: 7–15–78
6.1L (370 CID)

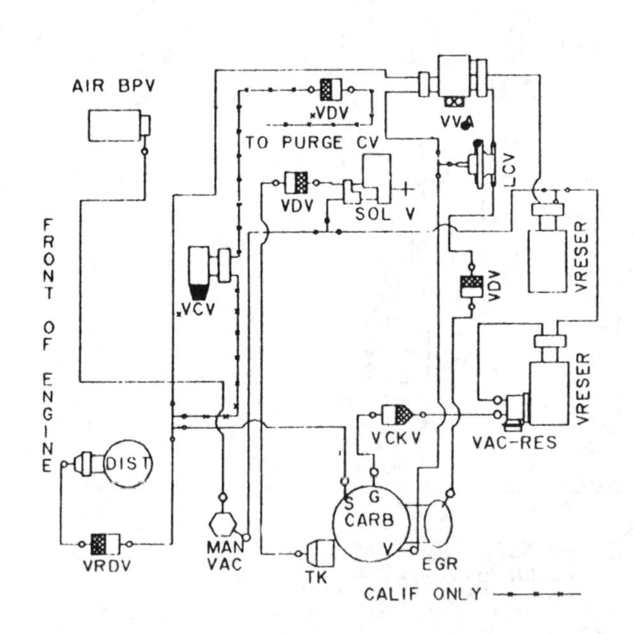

CALIBRATION: 9–83G–R11 DATE: 10–11–78
6.1L (370 CID)

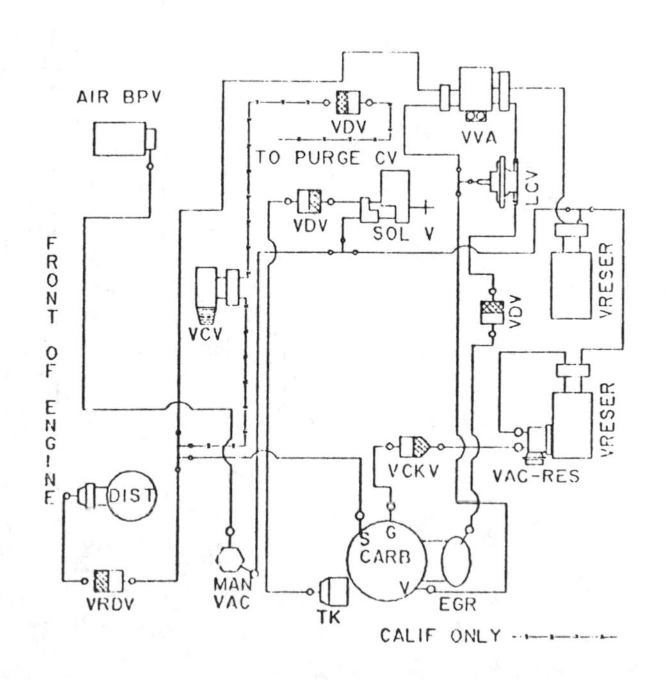

CALIBRATION: 9–83G–R12 DATE: 1–16–79
6.1L (370 CID)

VACUUM CIRCUITS

1979—370 CID (6.1 L) engines

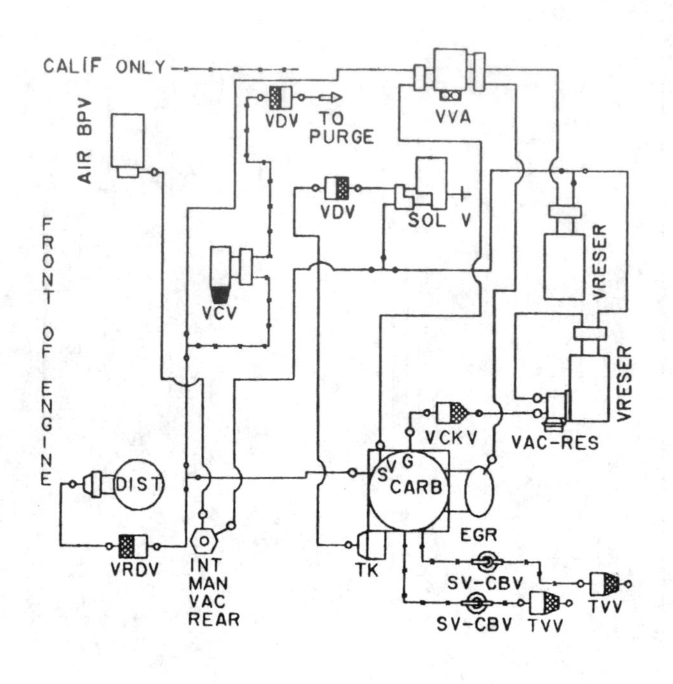

CALIBRATION: 9–83H–R2 DATE: 12–14–78
6.1L (370 CID)

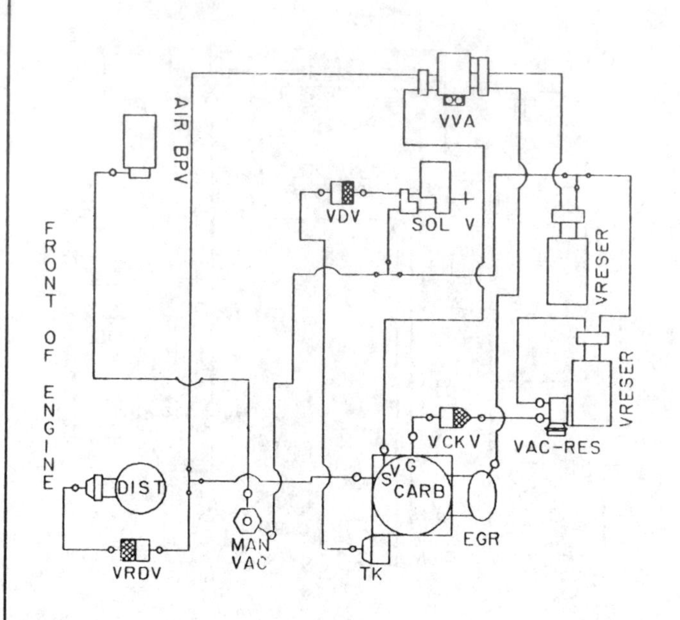

CALIBRATION: 9–83H–R10 DATE: 12–4–78
6.1L (370 CID)

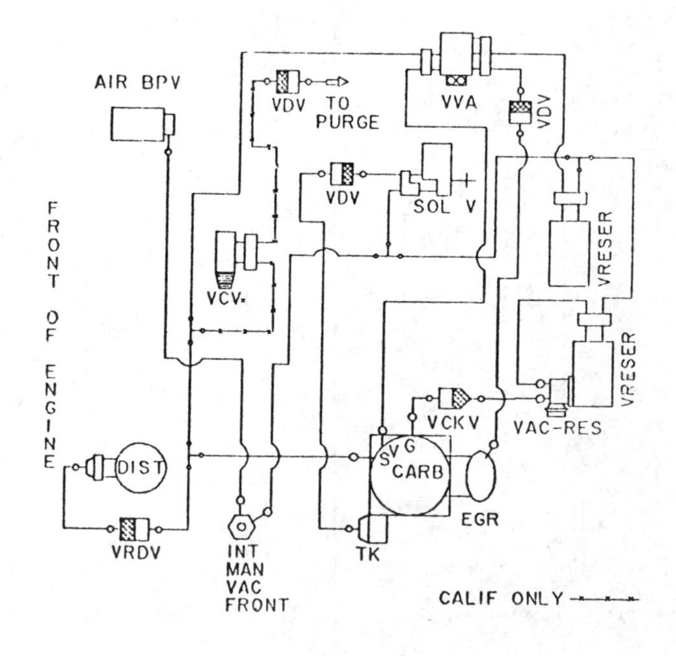

CALIBRATION: 9–83H–R11 DATE: 1–16–79
6.1L (370 CID)

VACUUM CIRCUITS

1979—400 CID (6.6 L) engines

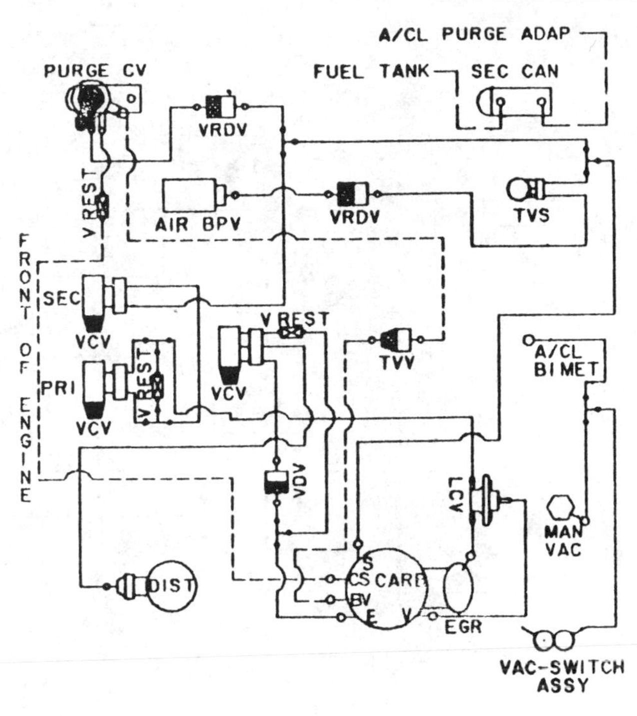

CALIBRATION: 9–17F–RO DATE: 6–6–78
6.6L (400 CID)

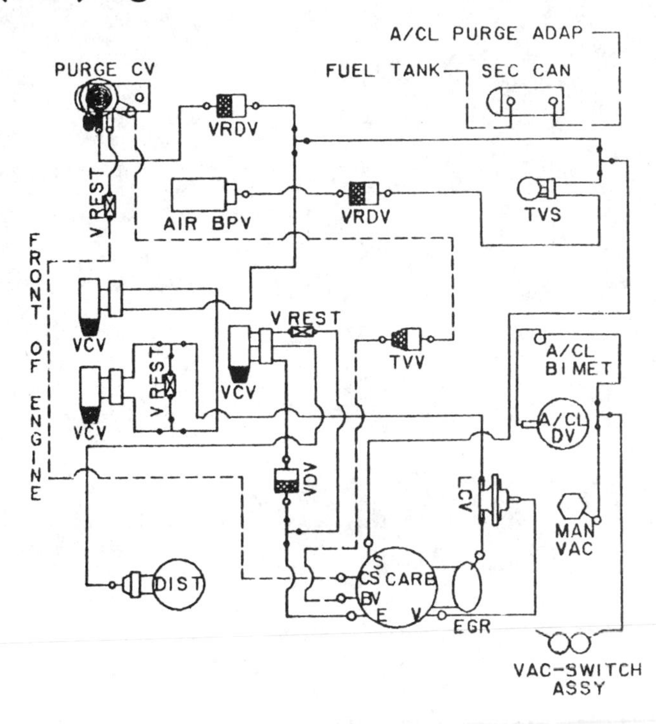

CALIBRATION: 9–17F–R1 DATE: 8–23–78
6.6L (400 CID)

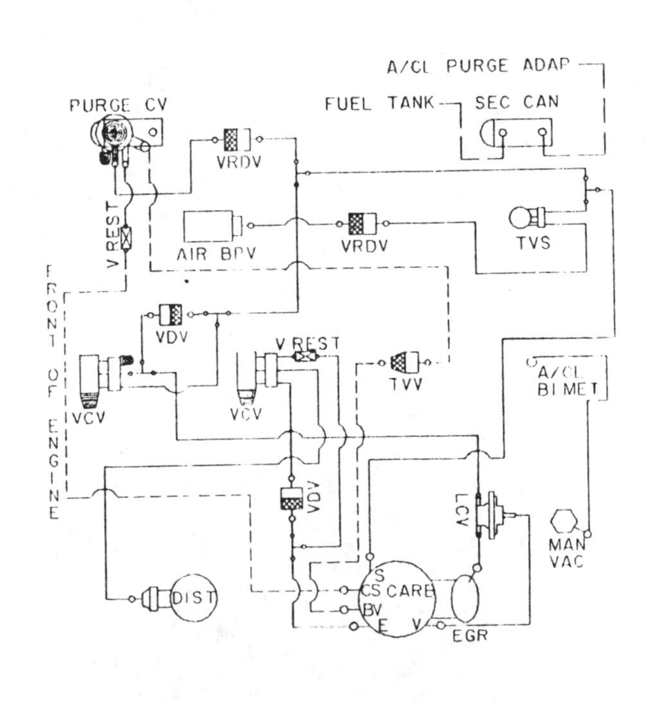

CALIBRATION: 9–17F–R21 DATE: 1–11–79
6.6L (400 CID)

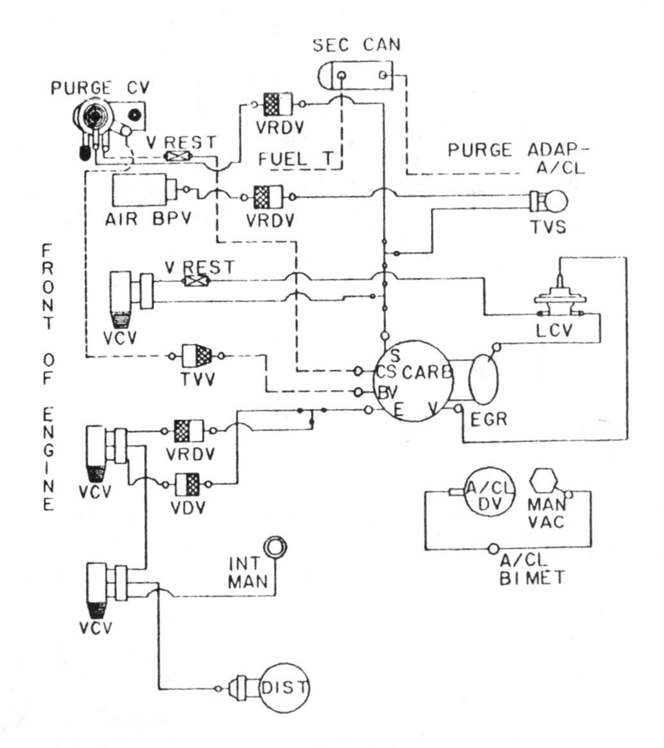

CALIBRATION. 9–17P–RO DATE: 8–16–78
6.6L (400 CID)

VACUUM CIRCUITS

1979—400 CID (6.6 L) engines

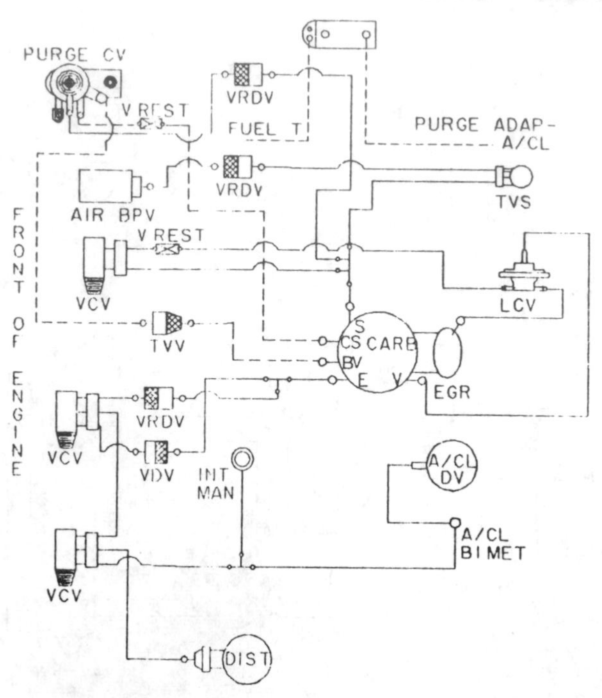

CALIBRATION: 9–17P–R10 DATE: 1–4–79
6.6L (400 CID)

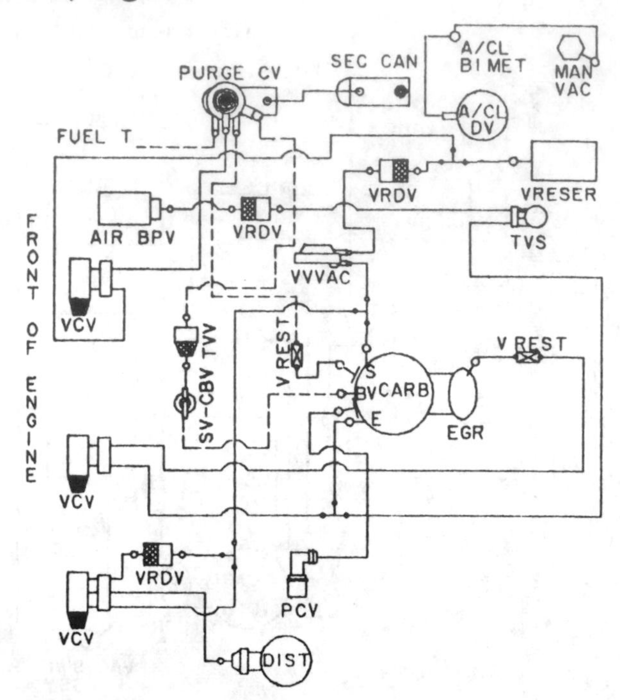

CALIBRATION: 9–17X–RO DATE: 9–28–78
6.6L (400 CID)

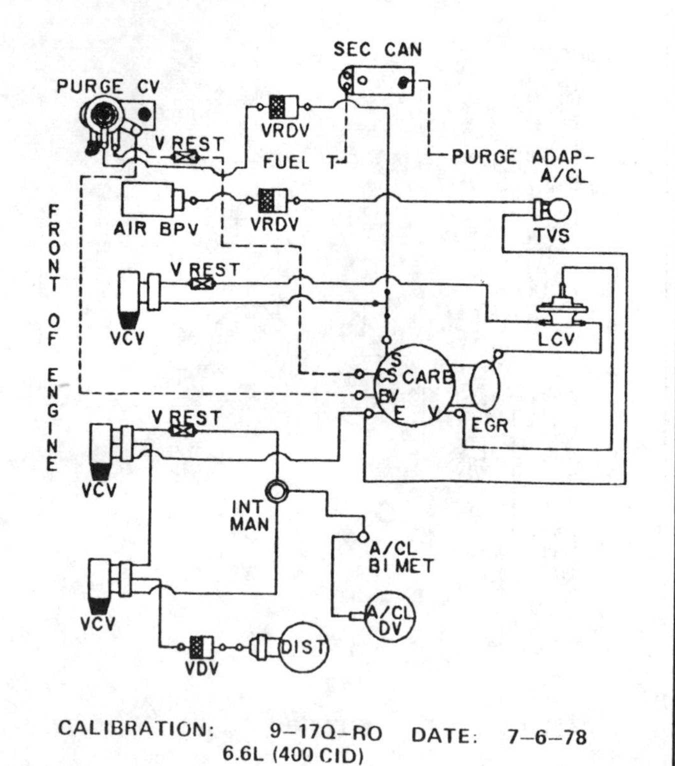

CALIBRATION: 9–17Q–RO DATE: 7–6–78
6.6L (400 CID)

CALIBRATION: 9–61G–ROA DATE: 1–24–79
6.6L (400 CID)

VACUUM CIRCUITS

1979—400 CID (6.6 L) engines

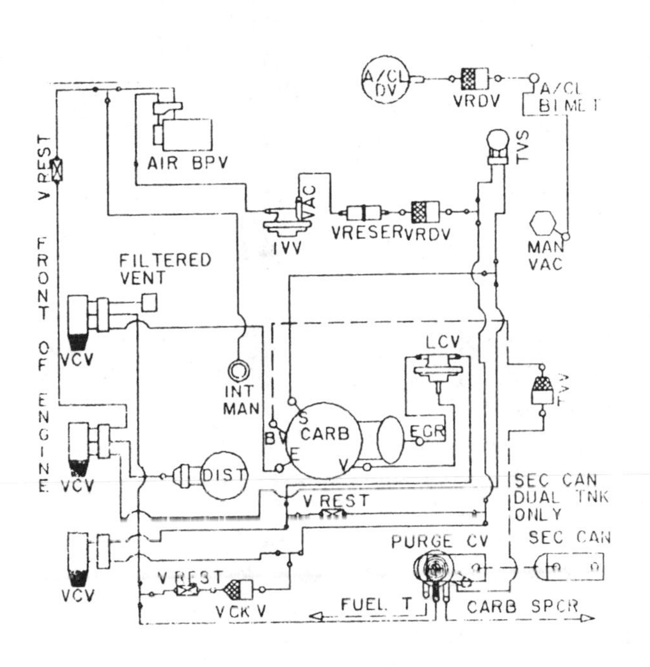

CALIBRATION: 9–61G–RON DATE: 1–24–79
6.6L (400 CID)

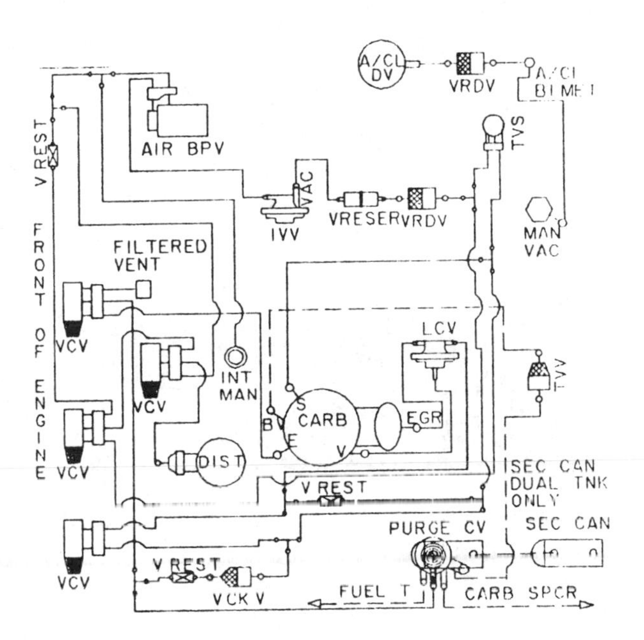

CALIBRATION: 9–61H–ROA DATE: 1–24–79
6.6L (400 CID)

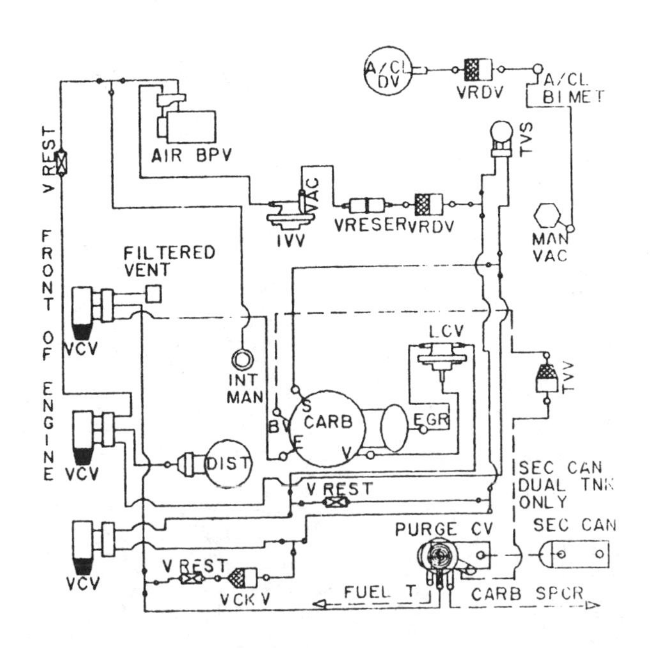

CALIBRATION: 9–61H–RON DATE: 1–24–79
6.6L (400 CID)

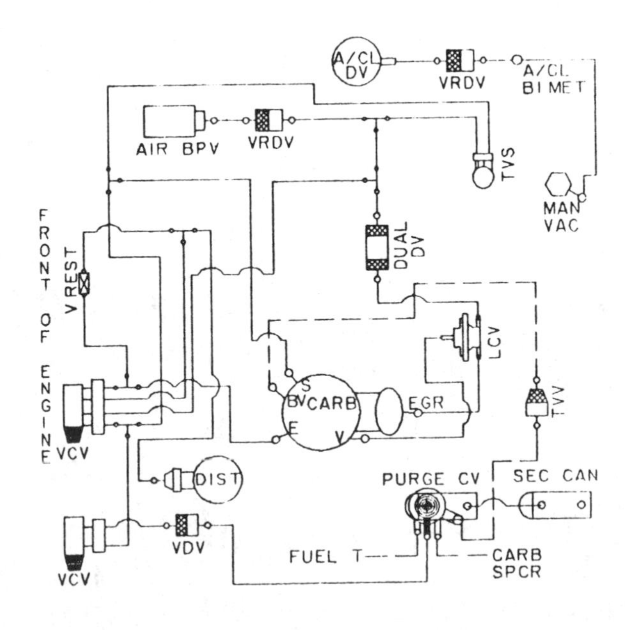

CALIBRATION: 9–62A–RO DATE: 8–10–78
6.6L (400 CID)

VACUUM CIRCUITS

1979—400 CID (6.6 L) engines

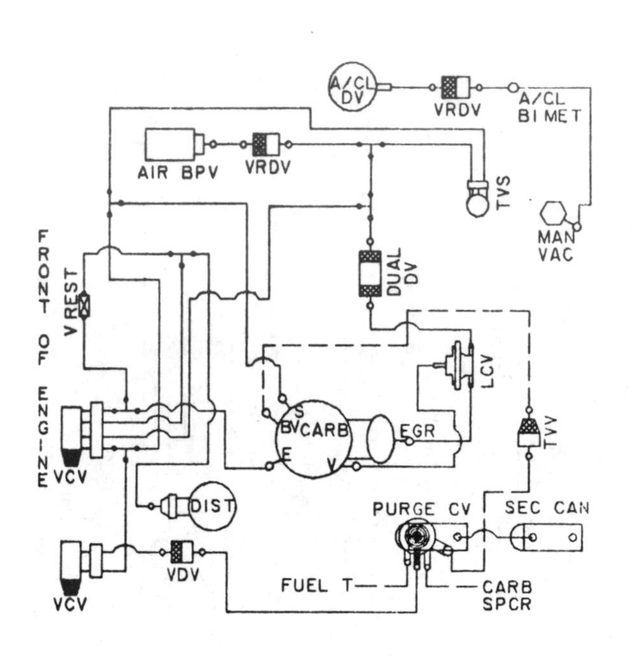

CALIBRATION: 9–62B–RO DATE: 8–10–78
6.6L (400 CID)

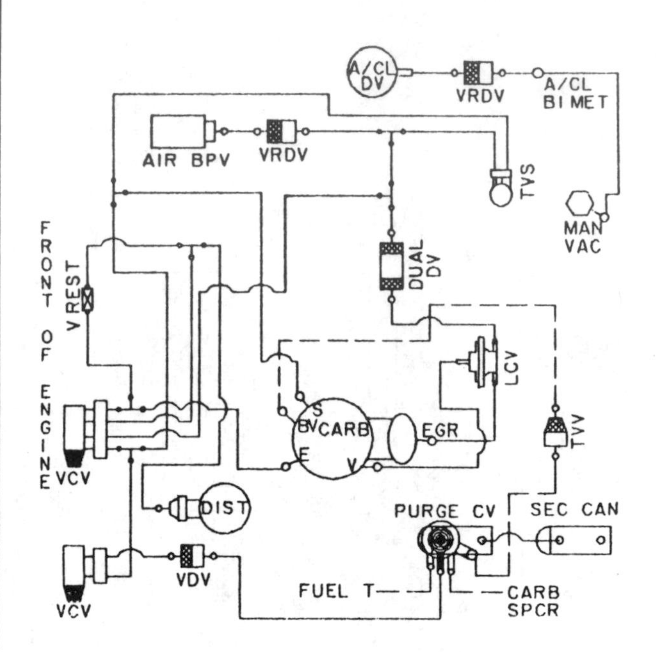

CALIBRATION: 9–62B–R10 DATE: 10–4–78
6.6L (400 CID)

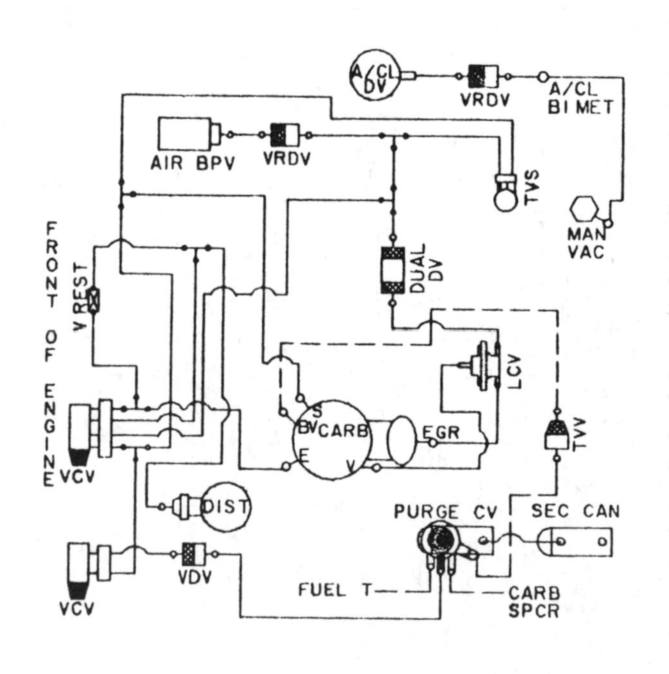

CALIBRATION: 9–62J–RO DATE: 5–16–78
6.6L (400CID)

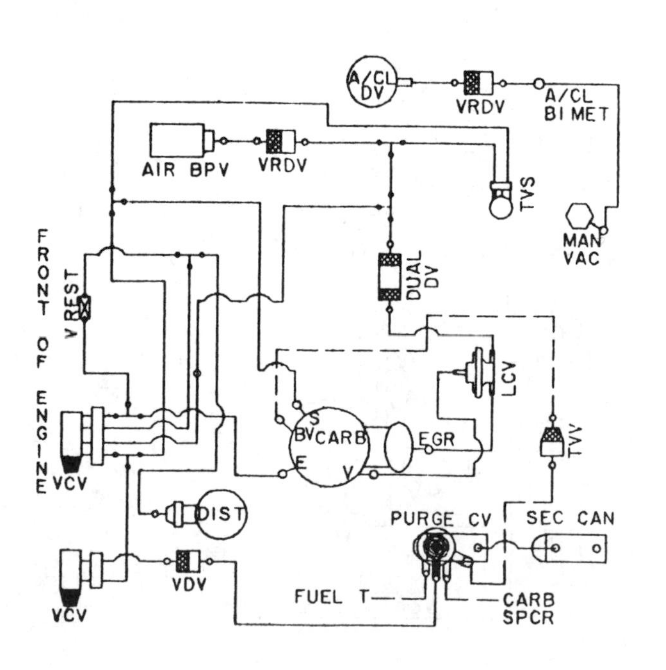

CALIBRATION. 9–62M–RO DATE: 5–24–78
6.6L (400 CID)

VACUUM CIRCUITS

1979—400 CID (6.6 L) engines

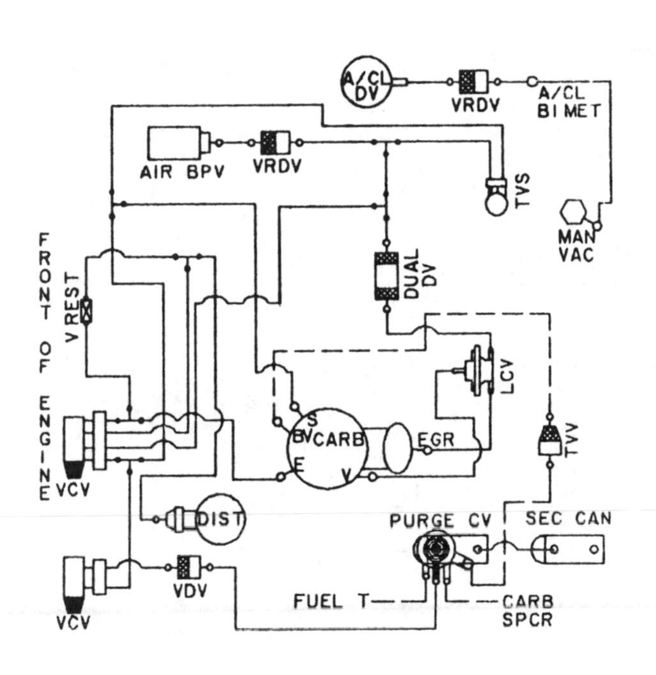

CALIBRATION: 9–62M–R10 DATE: 10–5–78
6.6L (400 CID)

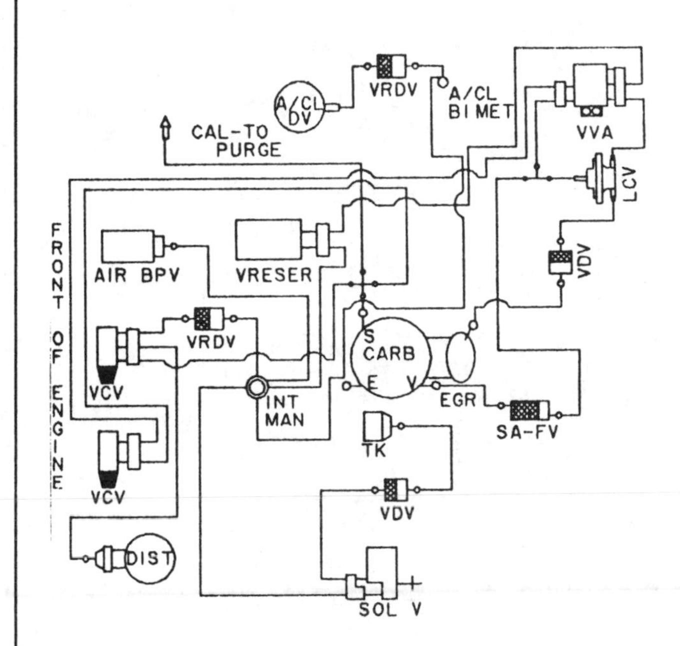

CALIBRATION: 9–73J–RO DATE: 10–2–78
6.6L (400 CID)

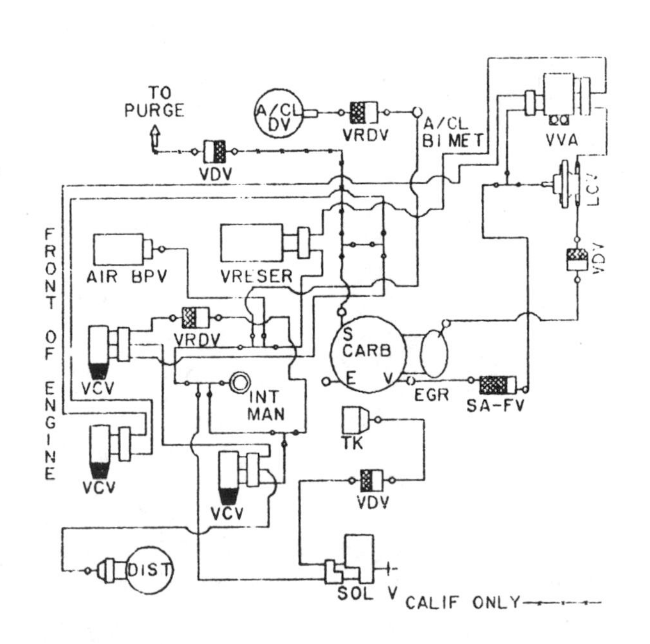

CALIBRATION: 9–73J–R11 DATE: 1–25–79
6.6L (400 CID)

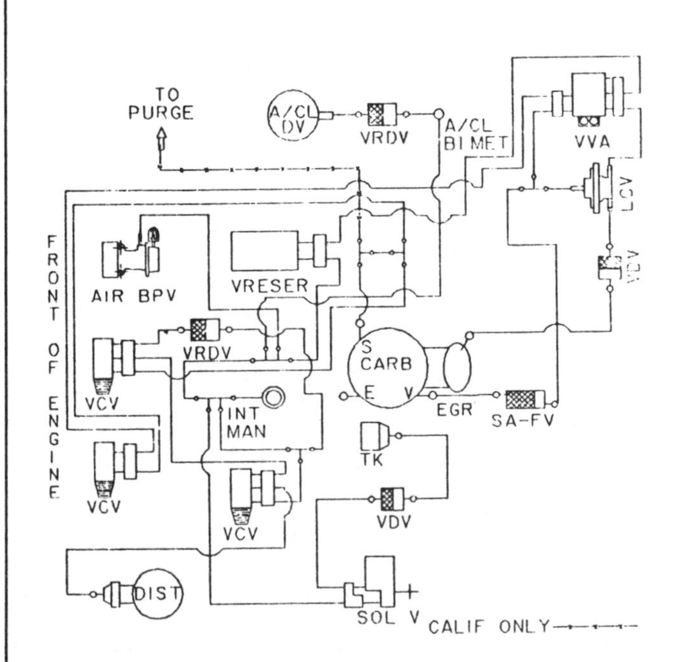

CALIBRATION: 9–73J–R12 DATE: 5–23–79
6.6L (400 CID)

VACUUM CIRCUITS

1979—400 CID (6.6 L) engines

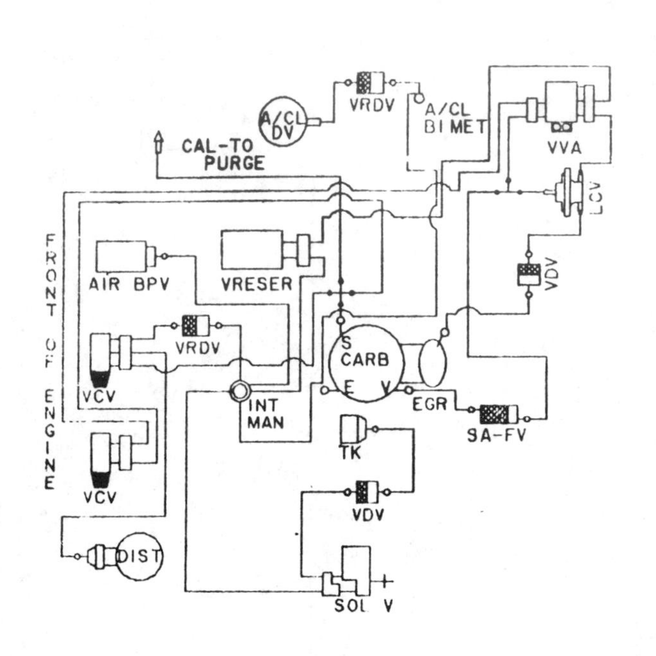

CALIBRATION: 9–74J–RO DATE: 9–29–78
6.6L (400 CID)

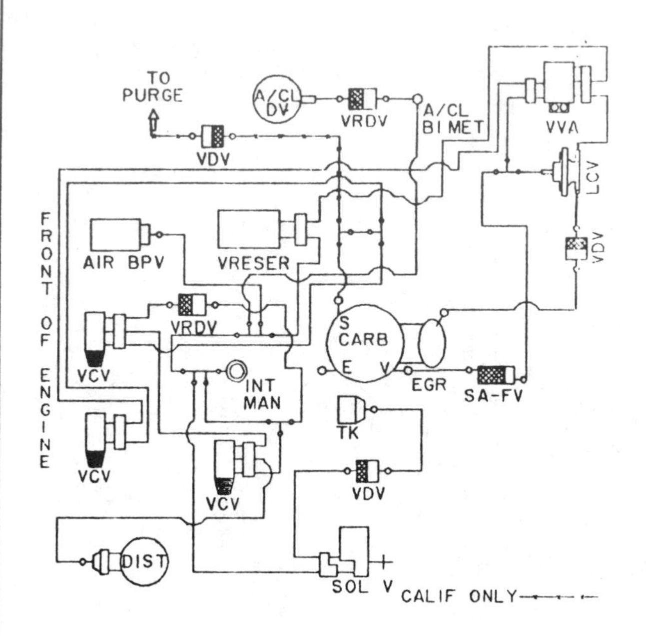

CALIBRATION: 9–74J–R11 DATE: 1–25–79
6.6L (400 CID)

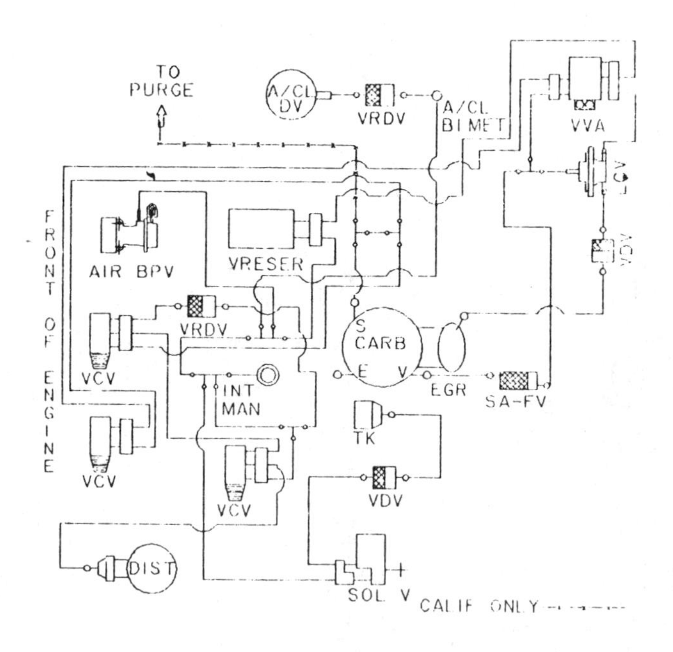

CALIBRATION: 9–74J–R12 DATE: 5–23–79
6.6L (400 CID)

VACUUM CIRCUITS

1979—429 CID (7.0 L) engines

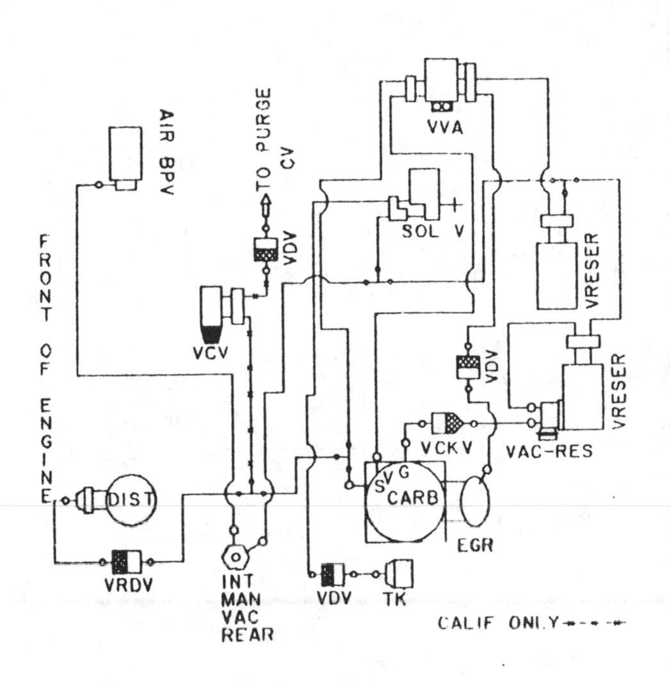

CALIBRATION: 9–87G–RO DATE: 7–11–78
7.0L (429 CID)

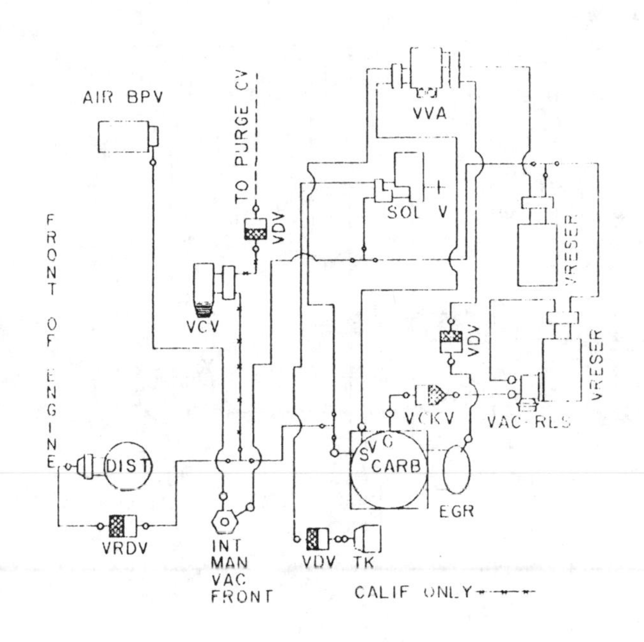

CALIBRATION: 9–87G–R10 DATE: 1–19–79
7.0L (429 CID)

1979—460 CID (7.5 L) engines

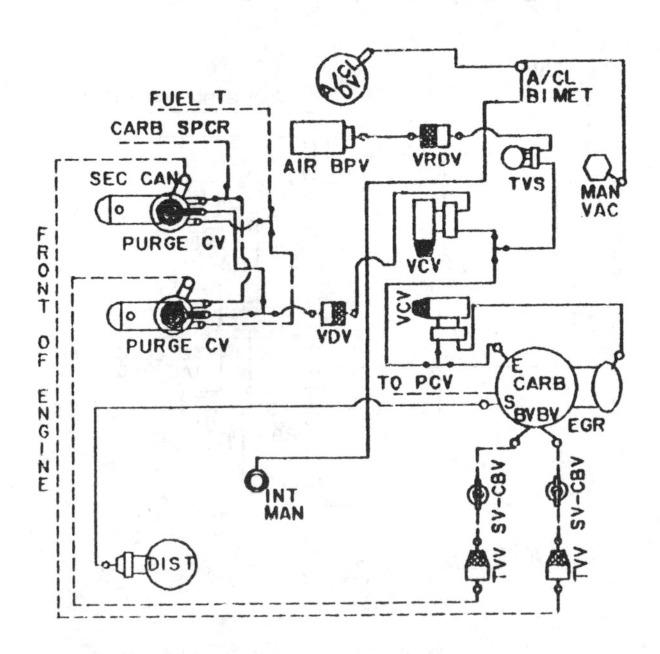

CALIBRATION: 9–66G–RO DATE: 6–13–78
7.5L (460 CID)

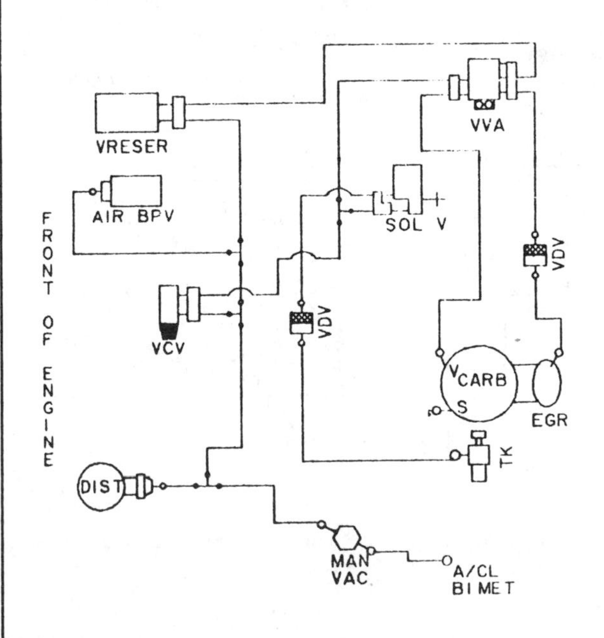

CALIBRATION. 9–97J–RO DATE: 7–10–78
7.5L (460 CID)

VACUUM CIRCUITS

1979—460 CID (7.5 L) engines

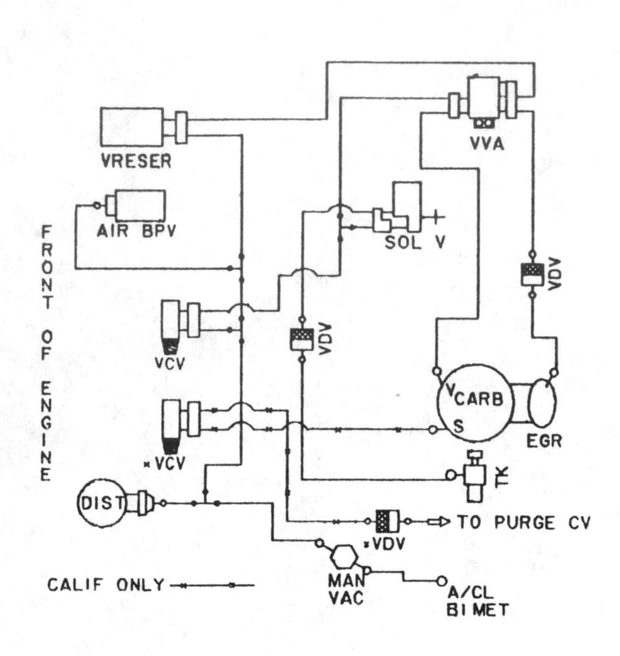

CALIBRATION: 9–97J–R10 DATE: 7–11–78
7.5L (460 CID)

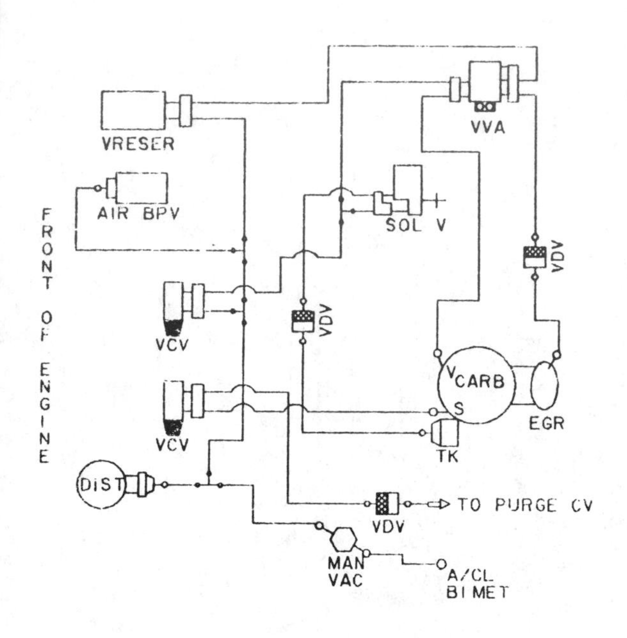

CALIBRATION: 9–97J–R11 DATE: 7–25–78
7.5L (460 CID)

1979—475/477 CID (7.8 L) engines

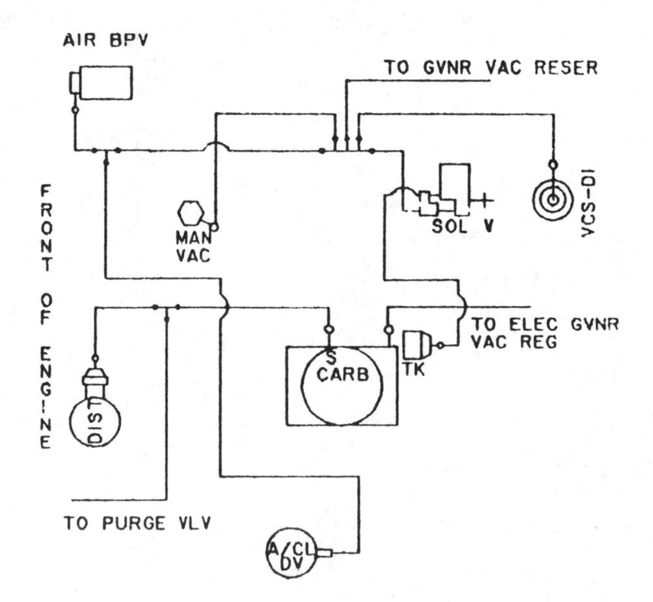

CALIBRATION: 7–93J–RO DATE: 5–5–78
7.8L (475/477)

1979—534 CID (8.8 L) engines

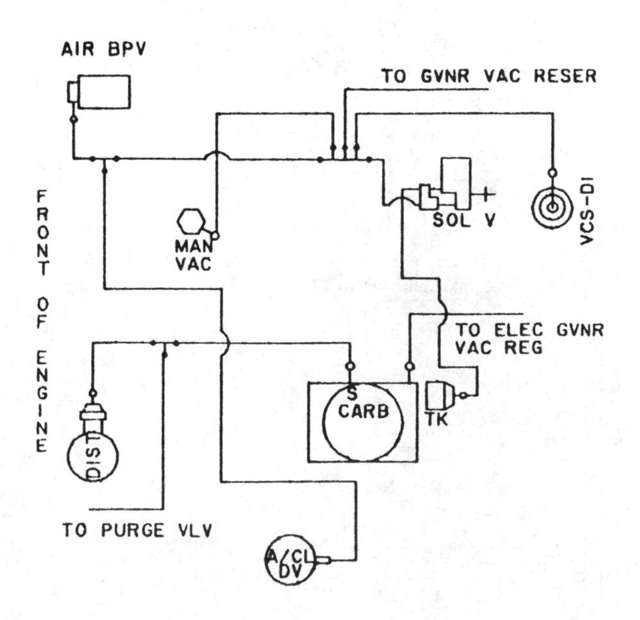

CALIBRATION: 7–95J–RO DATE: 5–25–78
8.8L (534)

VACUUM CIRCUITS

1980—(1.6 L) engines

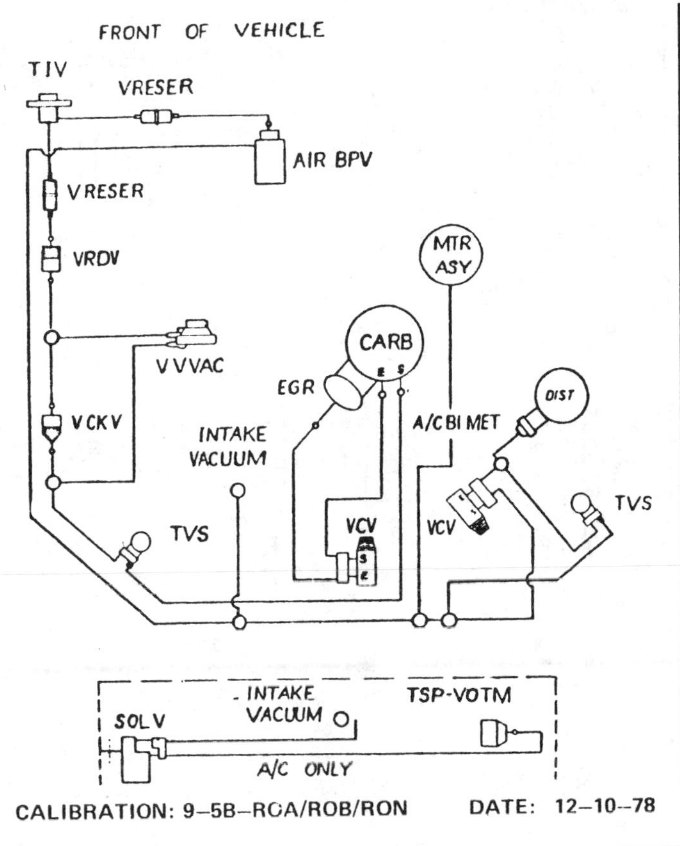

CALIBRATION: 9–5B–ROA/ROB/RON DATE: 12–10–78

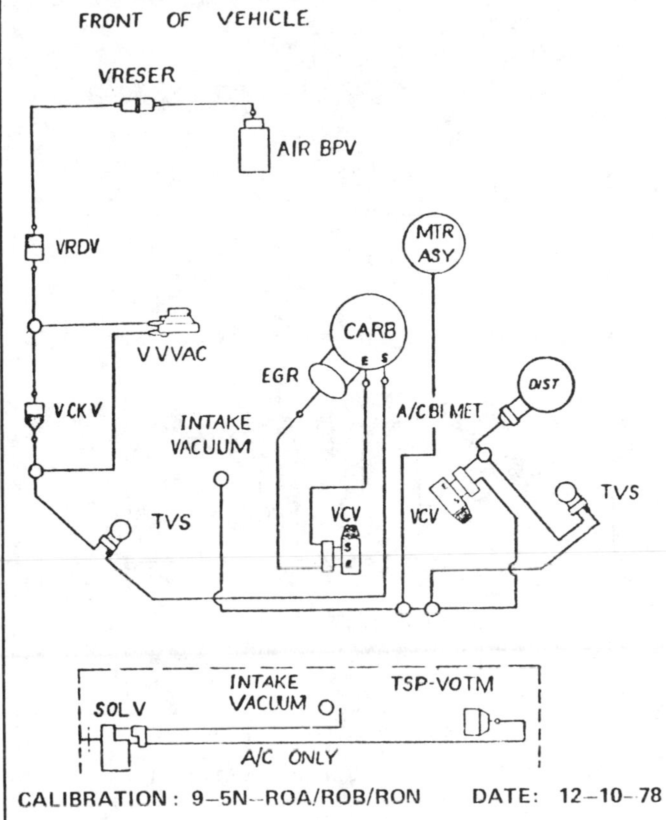

CALIBRATION: 9–5N–ROA/ROB/RON DATE: 12–10–78

1980—140 CID (2.3 L) engines

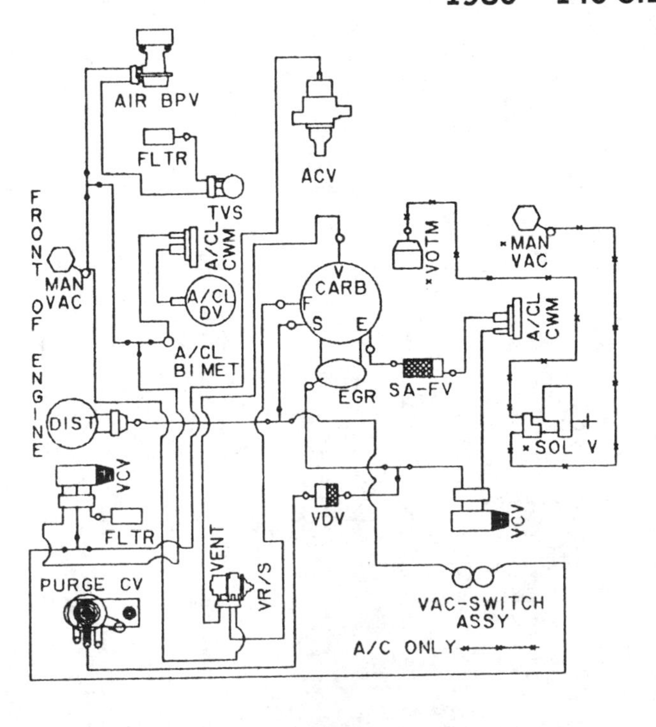

CALIBRATION: 9–1P–RO DATE: 6–8–78
2.3L (140 CID)

CALIBRATION: 9–1S–RO DATE: 4–14–78
2.3L (140 CID)

VACUUM CIRCUITS

1980—140 CID (2.3 L) engines

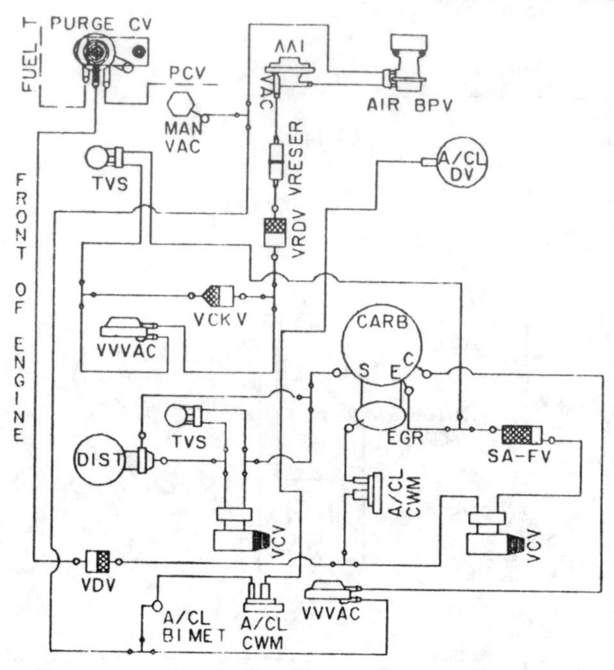

CALIBRATION: 9–1S–R10 DATE: 11–30–78
2.3L (140 CID)

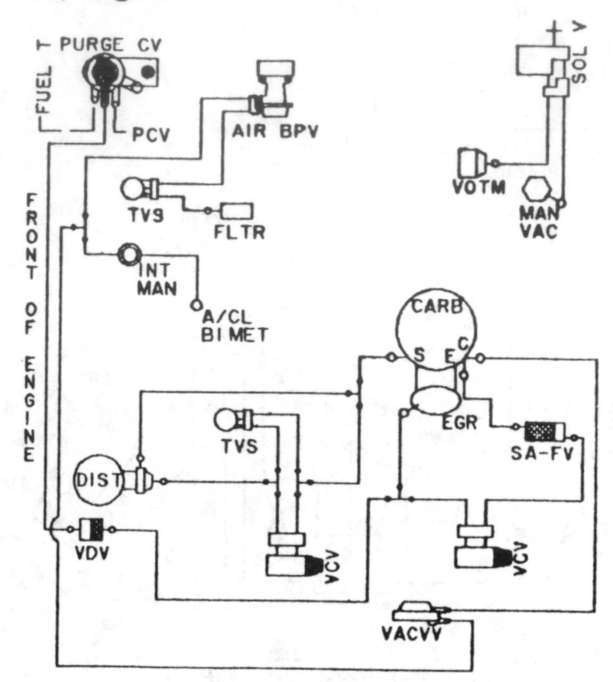

CALIBRATION: 9–1V–RO DATE: 6–13–78
2.3L (140 CID)

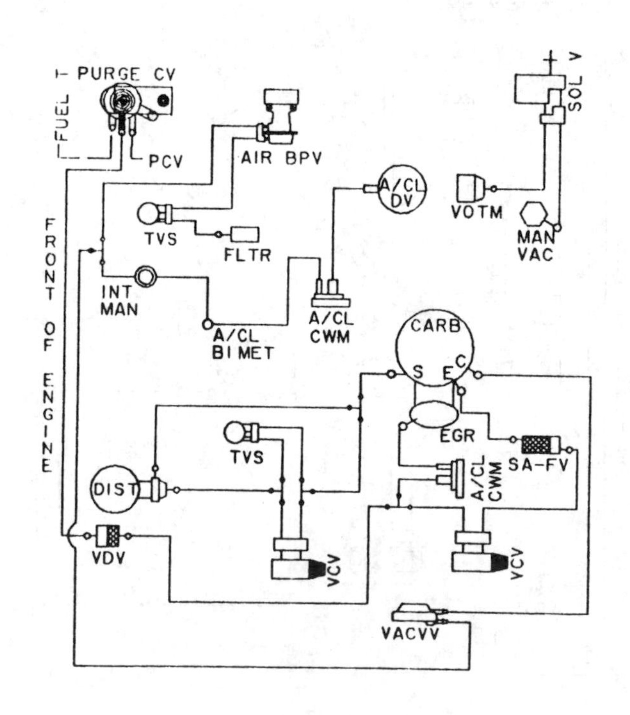

CALIBRATION: 9–1V–R10 DATE: 12–15–78
2.3L (140 CID)

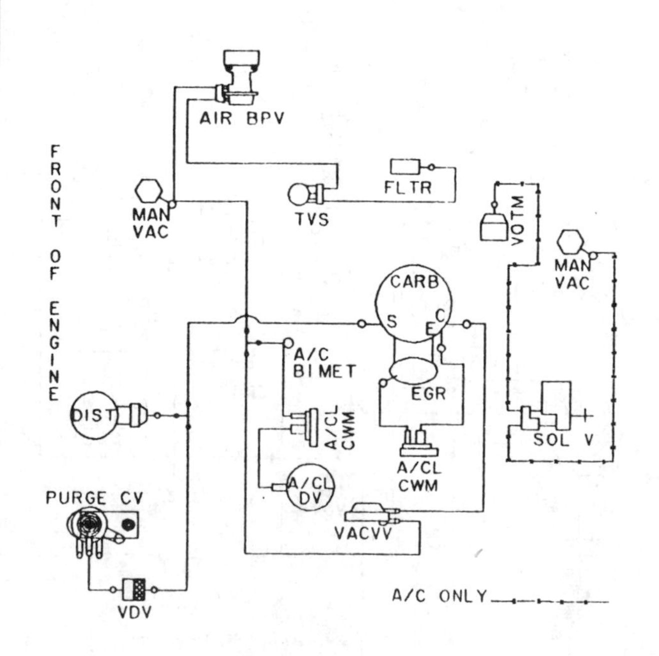

CALIBRATION: 9–1X–RO DATE: 11–7–78
2.3L HIGH ALT PINTO–BOBCAT

VACUUM CIRCUITS

1980—140 CID (2.3 L) engines

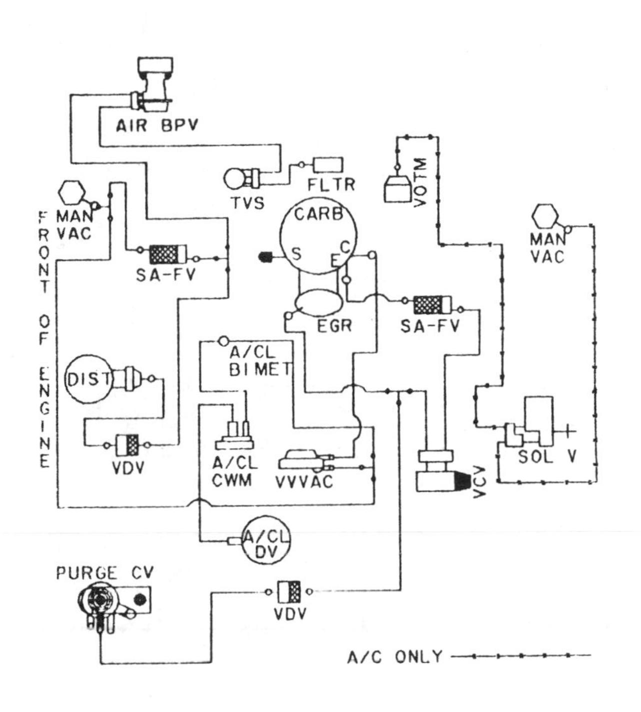

CALIBRATION: 9–2A–RO DATE: 10–24–78
2.3L (140 CID)

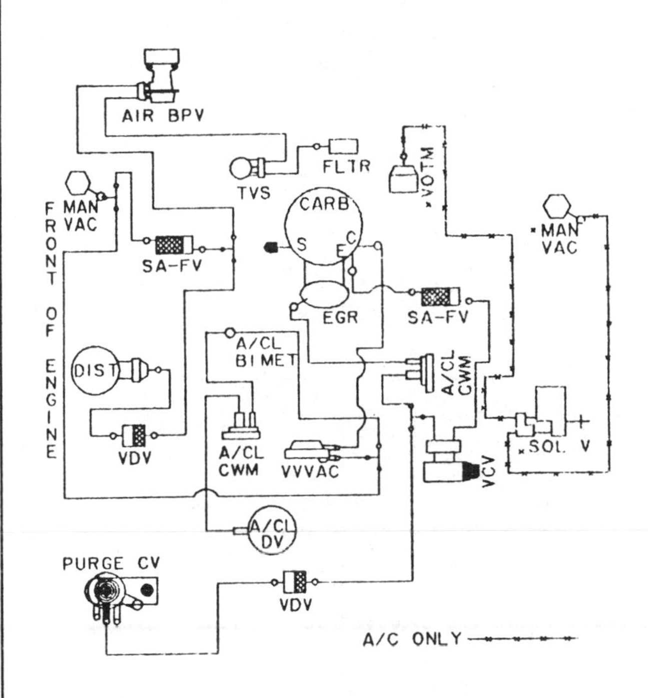

CALIBRATION: 9–2A–R10 DATE: 10–20–78
2.3L (140 CID)

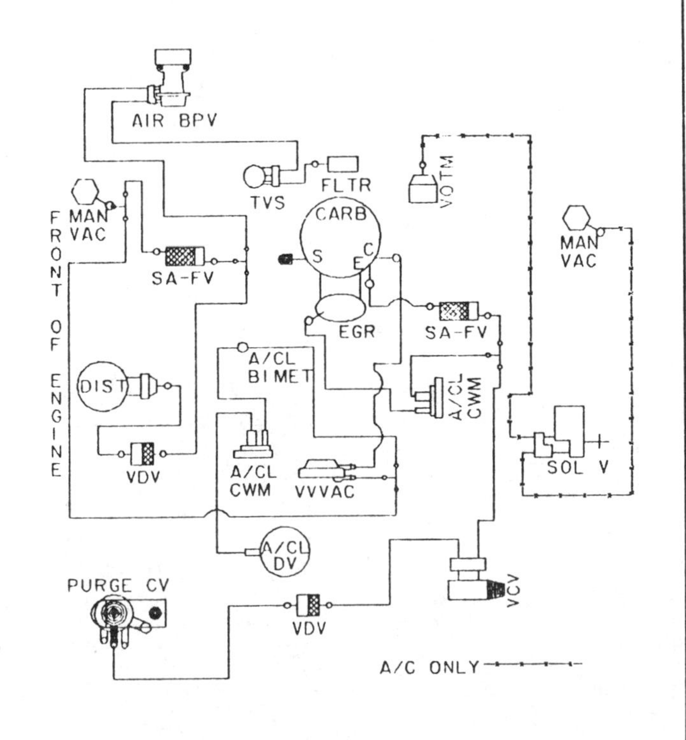

CALIBRATION: 9–2A–R12 DATE: 1–3–78
2.3L (140 CID)

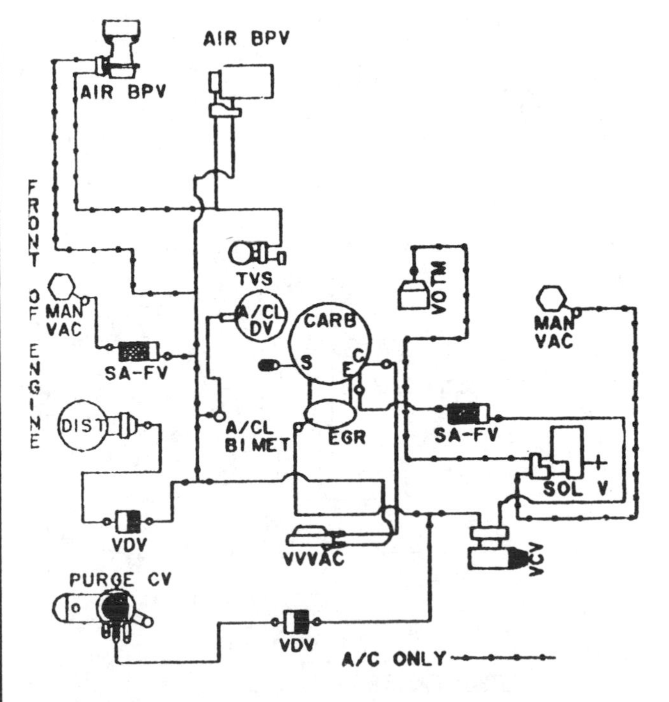

CALIBRATION: 9–2B–RO DATE: 7–6–78
2.3L (140 CID)

VACUUM CIRCUITS

1980—140 CID (2.3 L) engines

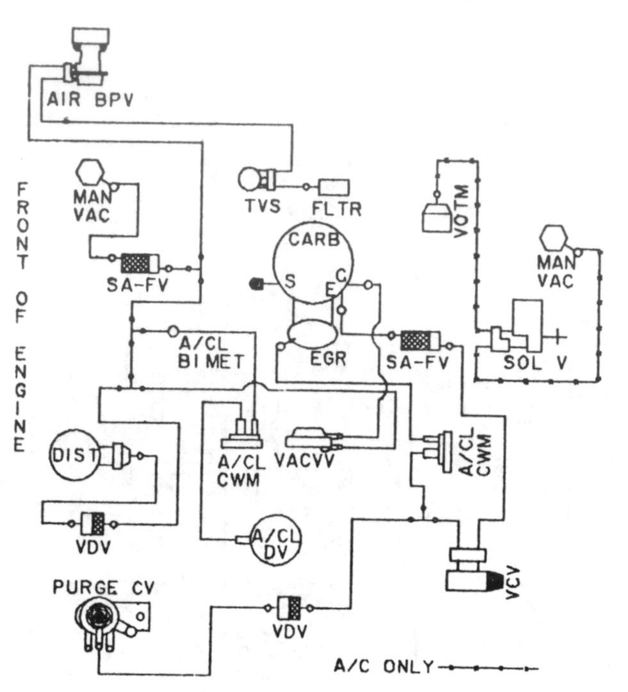

CALIBRATION: 9–2B–R10 **DATE:** 5–30–78
2.3L (140 CID)

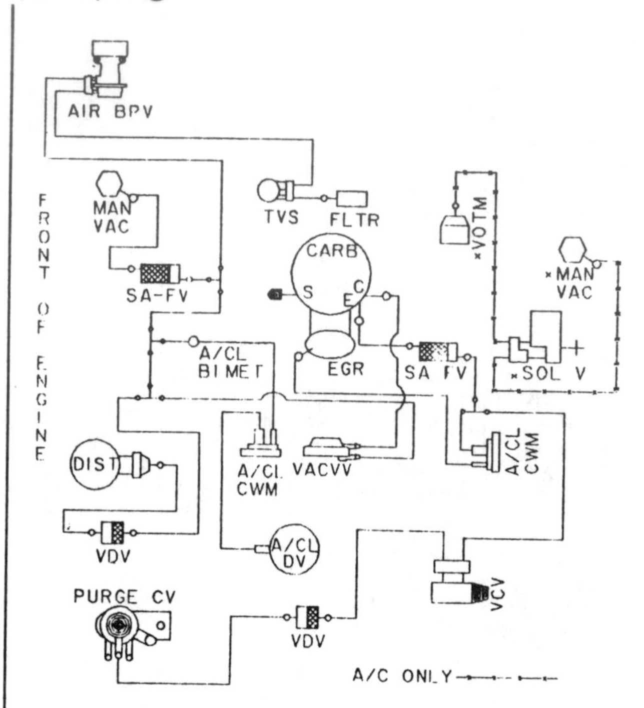

CALIBRATION: 9–2B–R13 **DATE:** 12–21–78
2.3L (140 CID)

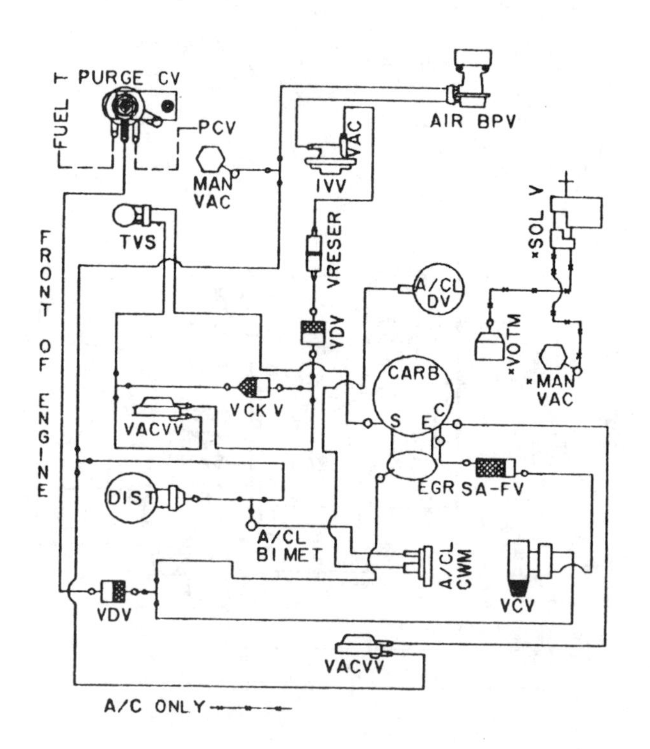

CALIBRATION. 9–2C–RO **DATE:** 5–19–78
2.3L (140 CID)

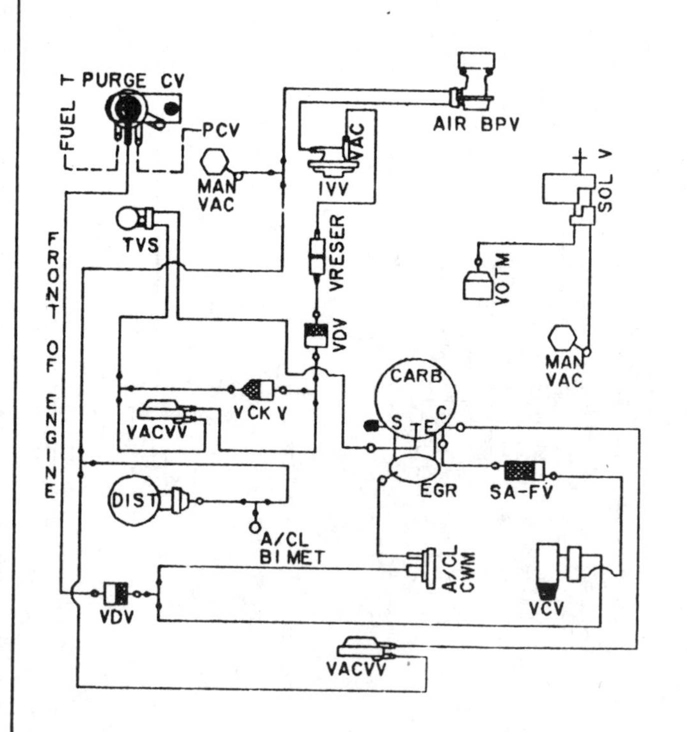

CALIBRATION: 9–2C–R10 **DATE:** 7–17–78
2.3L (140 CID)

VACUUM CIRCUITS

1980—140 CID (2.3 L) engines

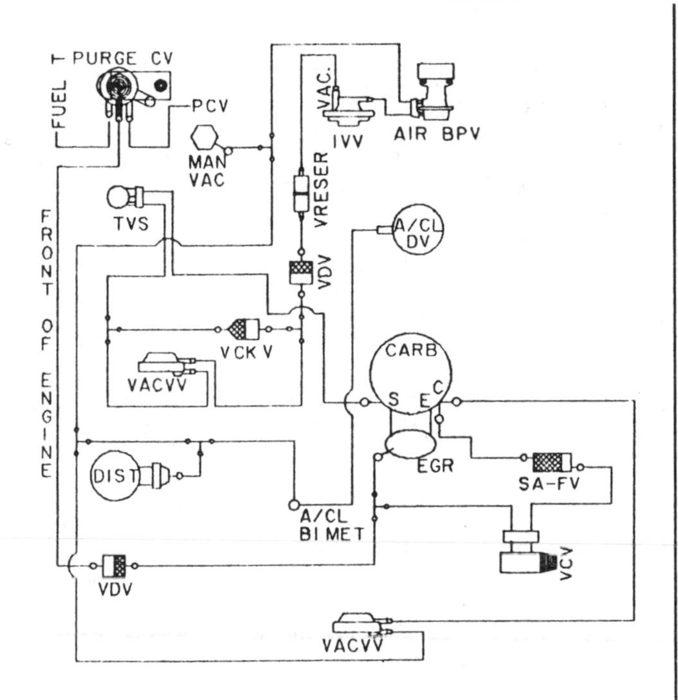

CALIBRATION: 9–2D–RO DATE: 6–14–78
2.3L (140 CID)

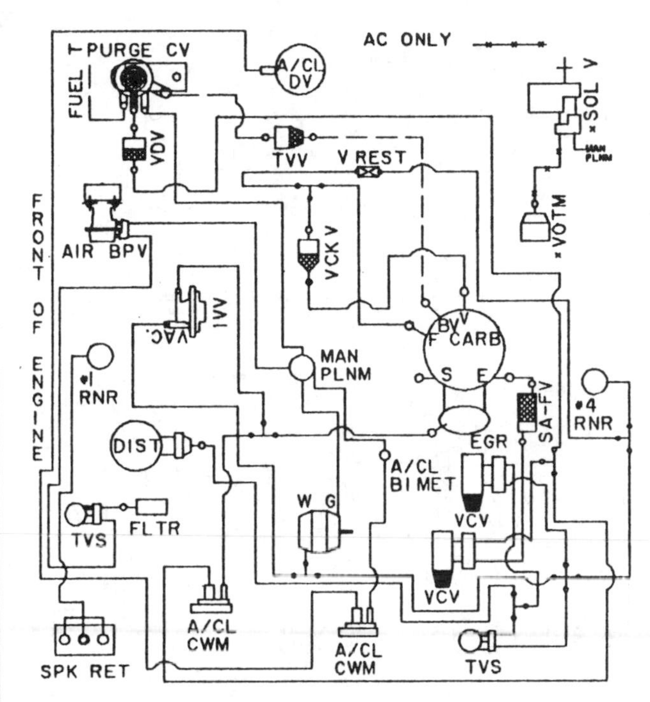

CALIBRATION: 9–2E–RO DATE: 8–7–78
2.3L (140 CID)

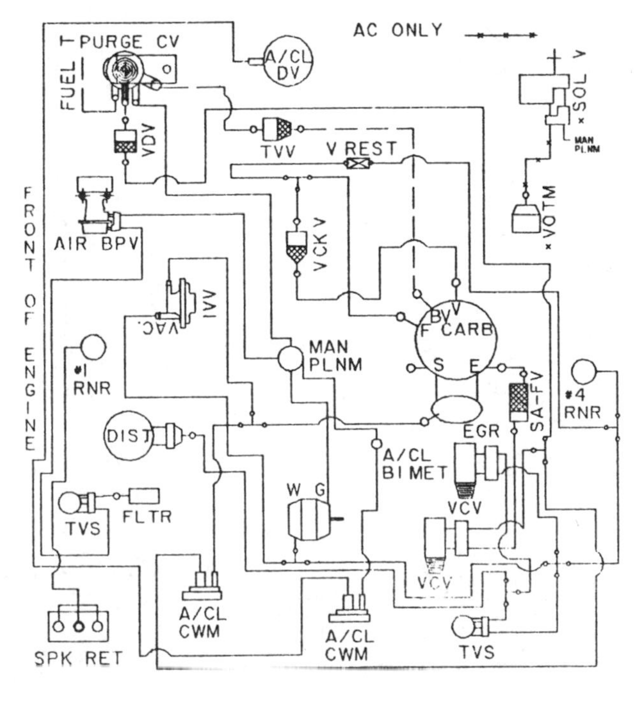

CALIBRATION: 9–2E–R12 DATE: 10–3–78
2.3L (140 CID)

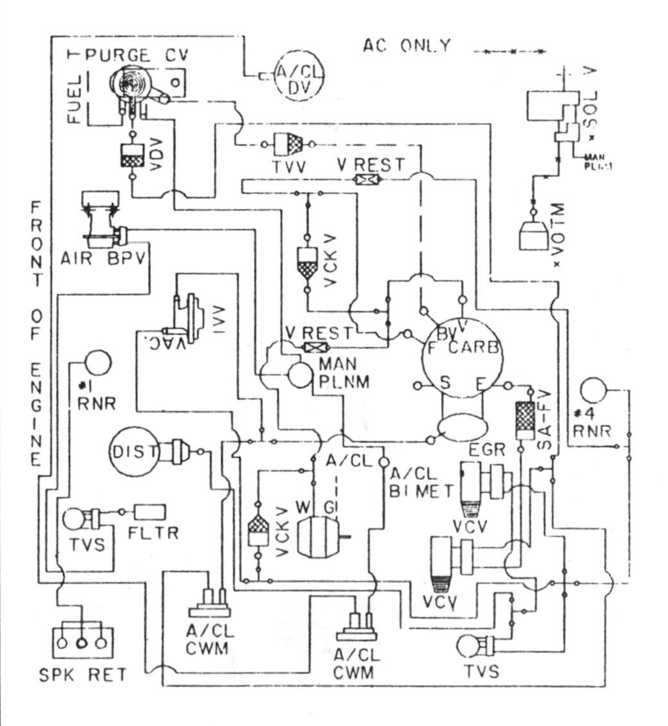

CALIBRATION: 9–2E–R16 DATE: 2–21–79
2.3L (140 CID)

VACUUM CIRCUITS

1980—140 CID (2.3 L) engines

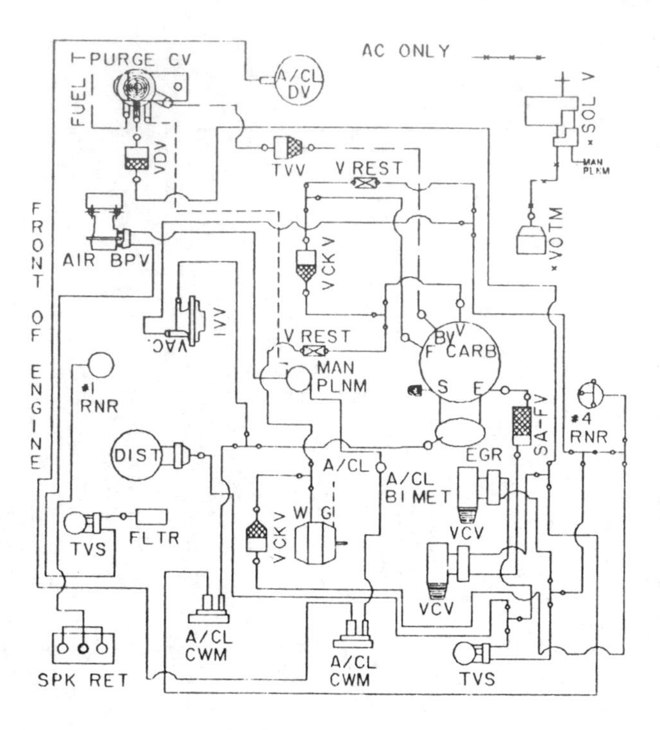

CALIBRATION: 9–2E–R17 DATE: 6–18–79
2.3L (140 CID)

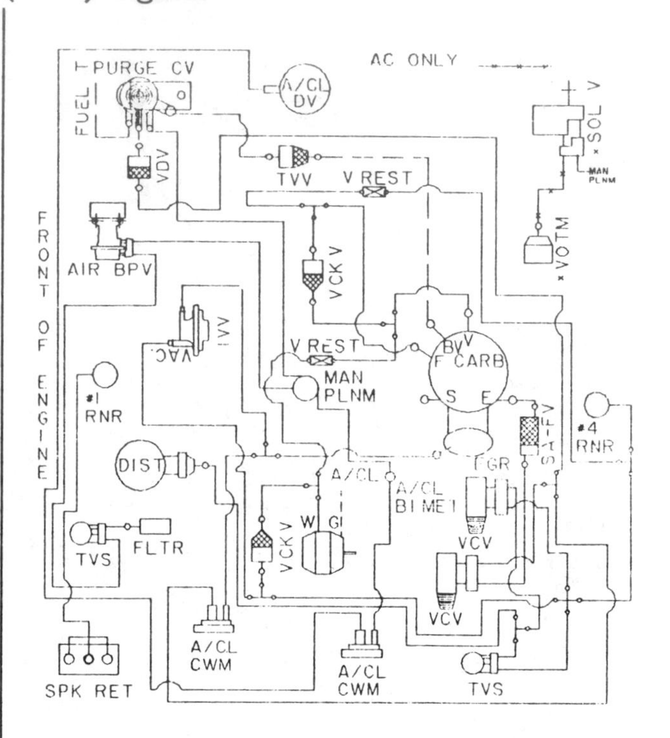

CALIBRATION: 9–2E–R18 DATE: 4–17–79
2.3L (140 CID)

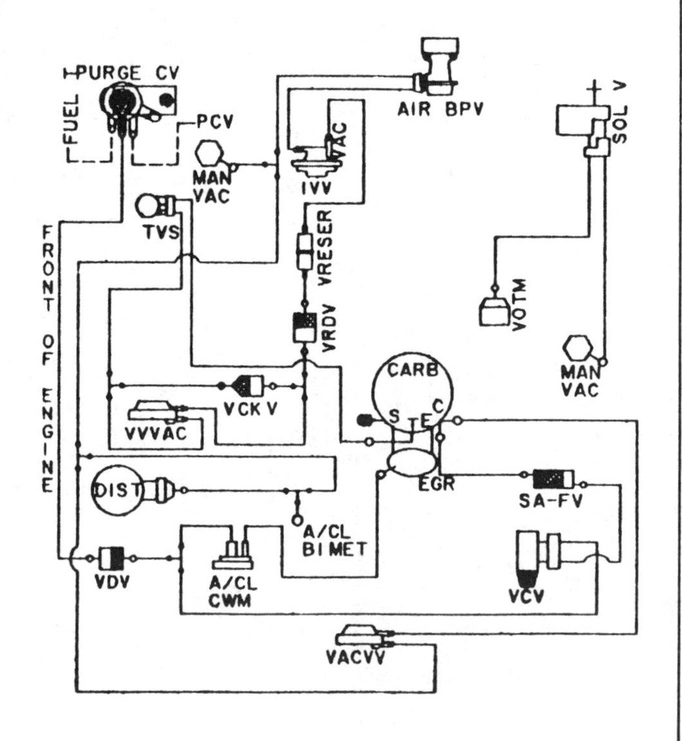

CALIBRATION: 9–2F–RO DATE: 4–10–78
2.3L (140 CID)

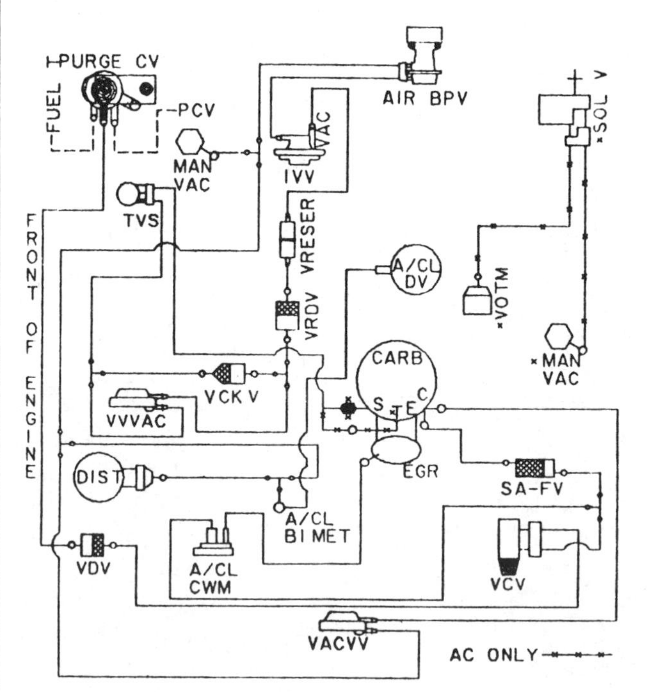

CALIBRATION: 9–2F–R11 DATE: 12–20–78
2.3L (140 CID)

VACUUM CIRCUITS

1980—140 CID (2.3 L) engines

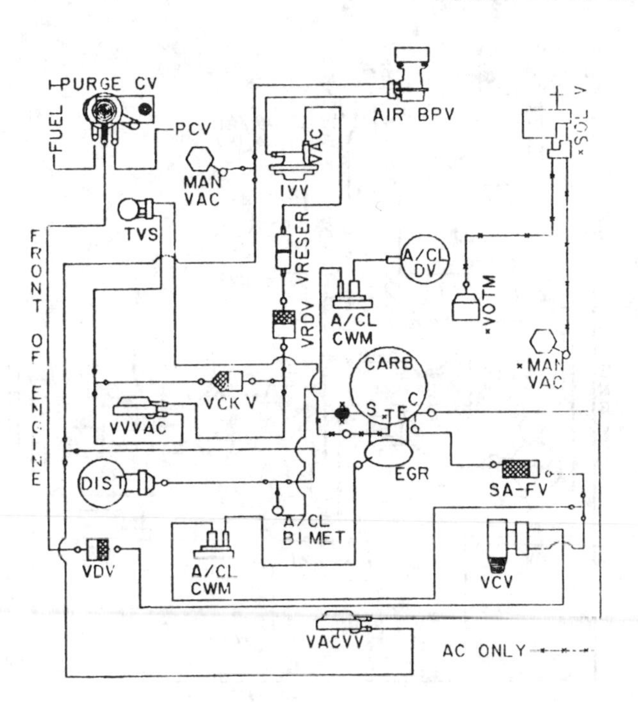

CALIBRATION: 9–2F–R13 DATE: 1–31–79
2.3L (140 CID)

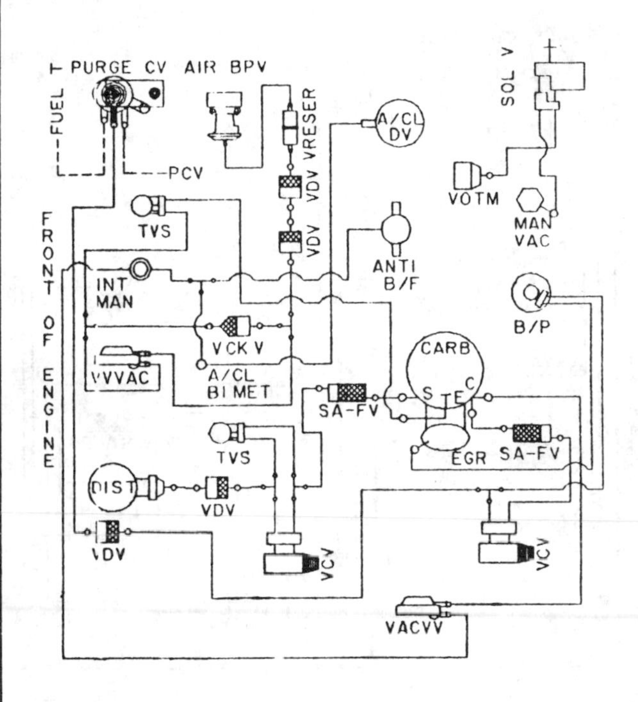

CALIBRATION: 9–2N–RO DATE: 7–11–78
2.3L (140 CID)

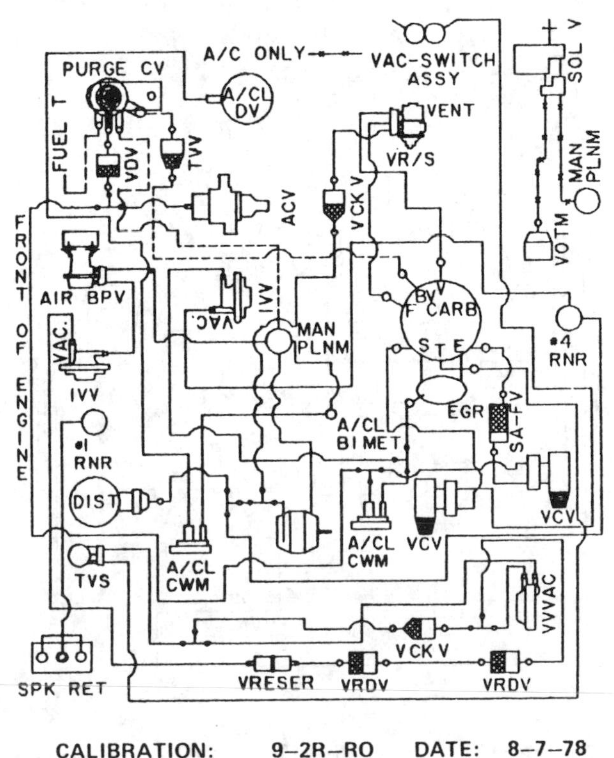

CALIBRATION: 9–2R–RO DATE: 8–7–78
2.3L (140 CID)

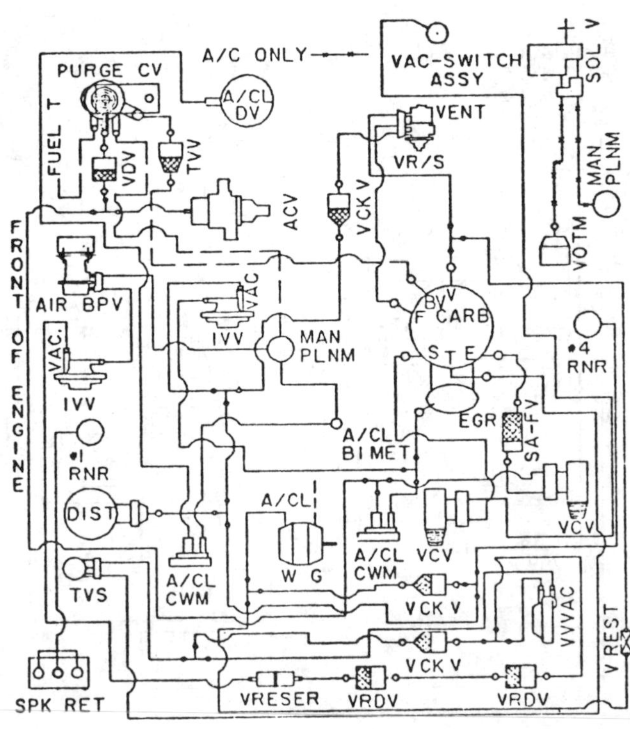

CALIBRATION: 9–2R–R11 DATE: 3–1–79
2.3L (140 CID)

VACUUM CIRCUITS

1980—140 CID (2.3 L) engines

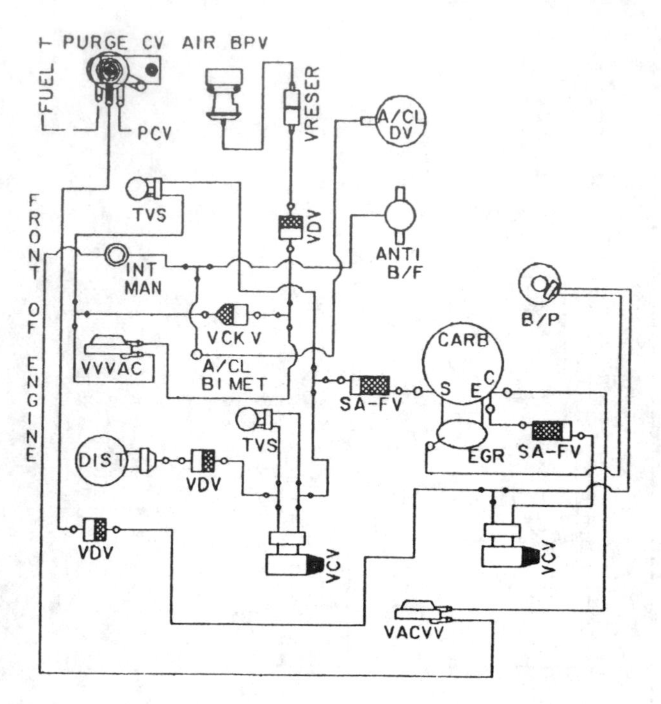

CALIBRATION: 9–2S–RO DATE: 6–9–78
2.3L (140 CID)

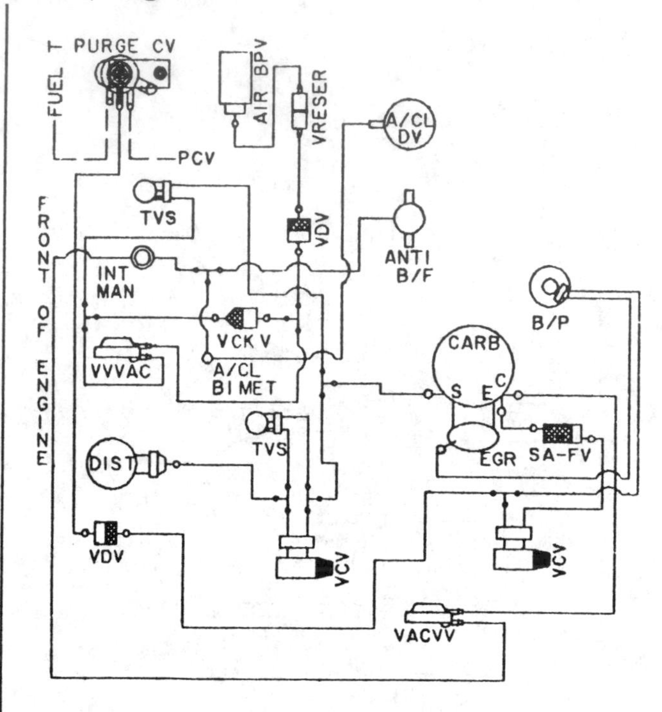

CALIBRATION: 9–2S–R2 DATE: 7–7–78
2.3L (140 CID)

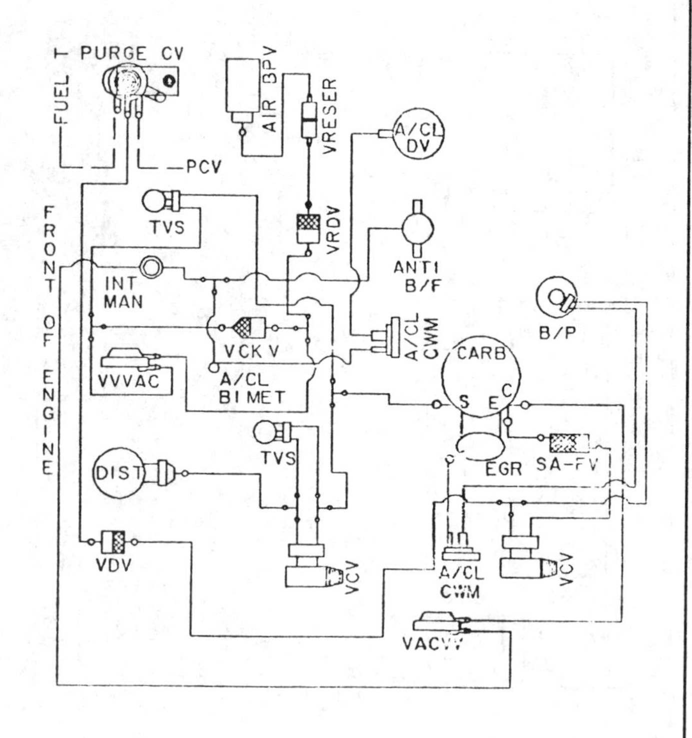

CALIBRATION: 9–2S–R10 DATE: 12–6–78
2.3L (140 CID)

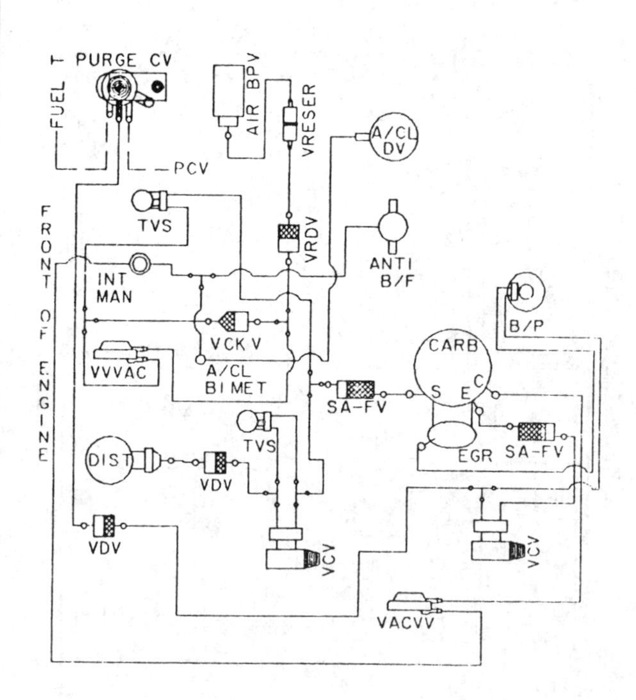

CALIBRATION: 9–2S–R11 DATE: 2–20–79
2.3L (140 CID)

VACUUM CIRCUITS

1980—140 CID (2.3 L) engines

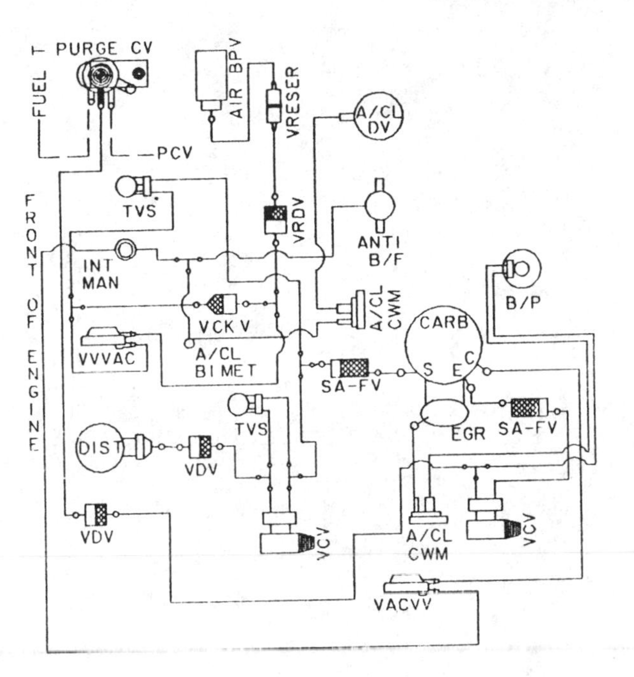

CALIBRATION: 9–2S–R12 DATE: 3–6–79
2.3L (140 CID)

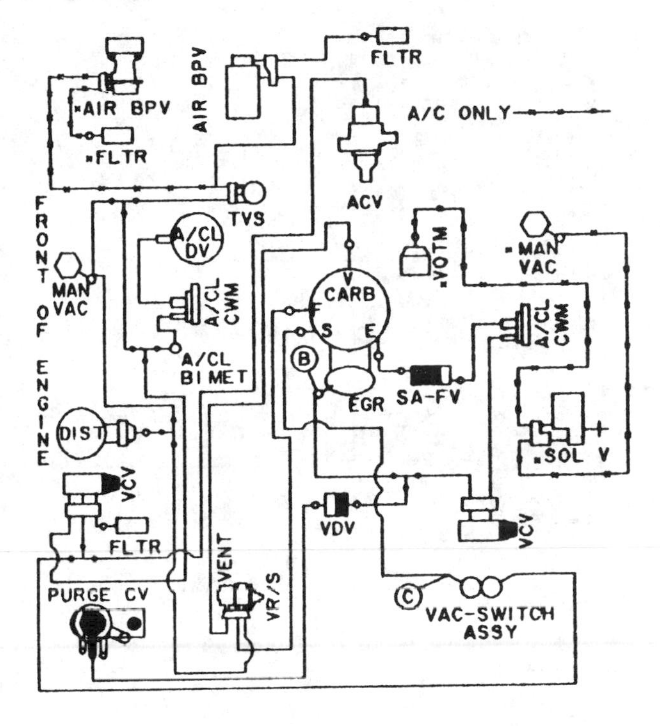

CALIBRATION: 9–2T–RO DATE: 6–2–78
2.3L (140 CID)

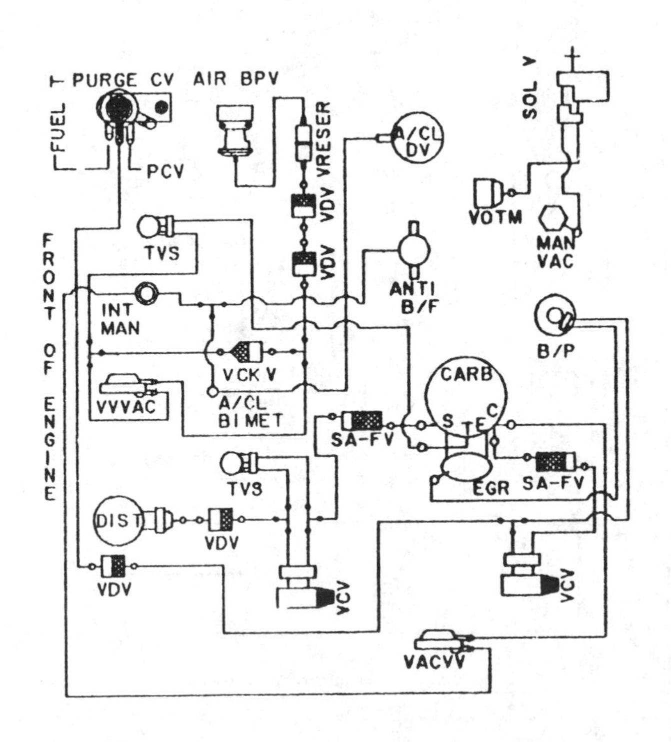

CALIBRATION: 9–2V–RO DATE: 6–1–78
2.3L (140 CID)

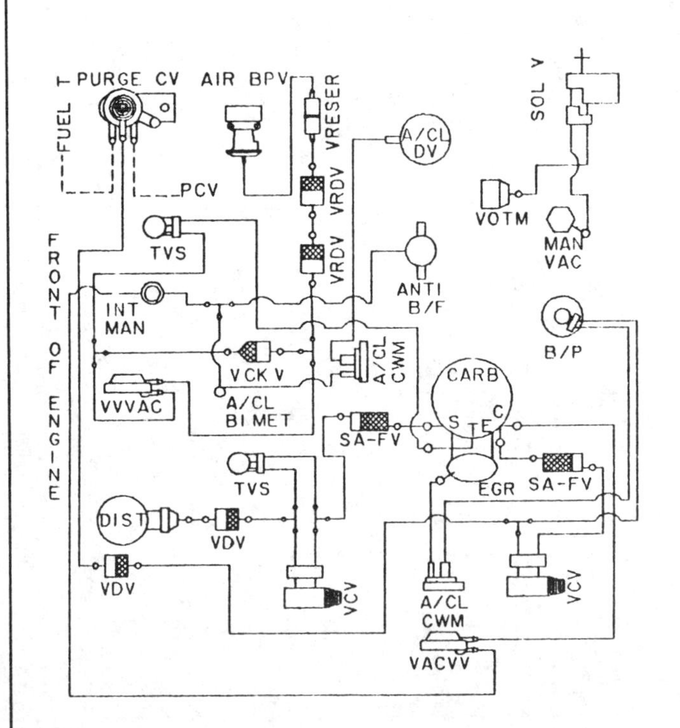

CALIBRATION: 9–2V–R10 DATE: 12–19–78
2.3L (140 CID)

VACUUM CIRCUITS

1980—140 CID (2.3 L) engines

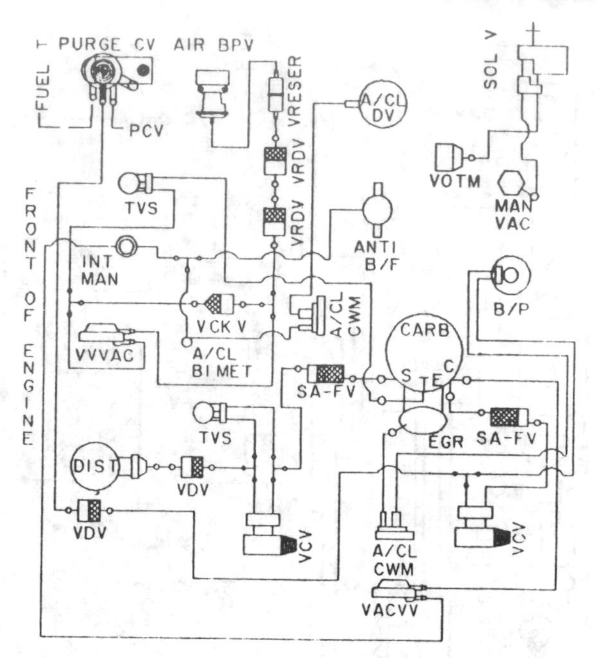

CALIBRATION: 9–2V–R12 DATE: 3–5–79
2.3L (140 CID)

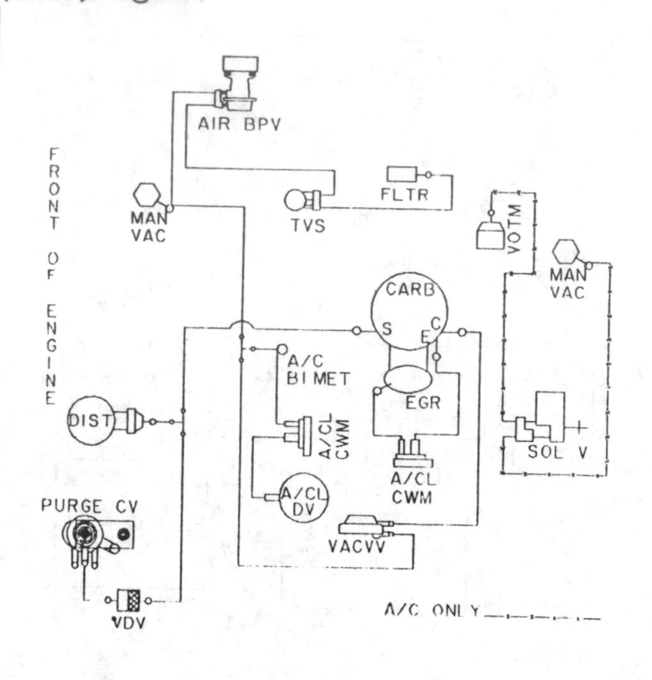

CALIBRATION: 9–2X–RO DATE: 1–22–79
2.3L (140 CID)

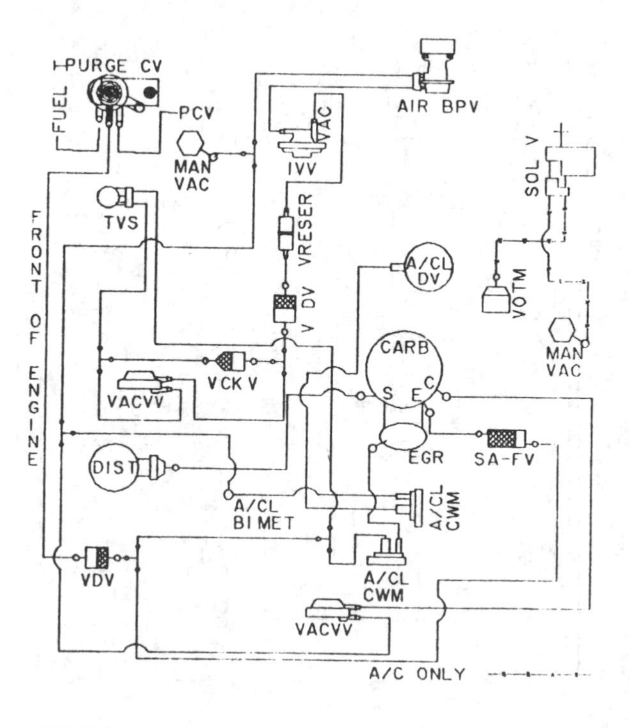

CALIBRATION: 9–2Y–RO DATE: 1–19–79
2.3L (140 CID)

CALIBRATION: 9–21B–RO DATE: 5–18–78
2.3L (140 CID)

VACUUM CIRCUITS

1980—140 CID (2.3 L) engines

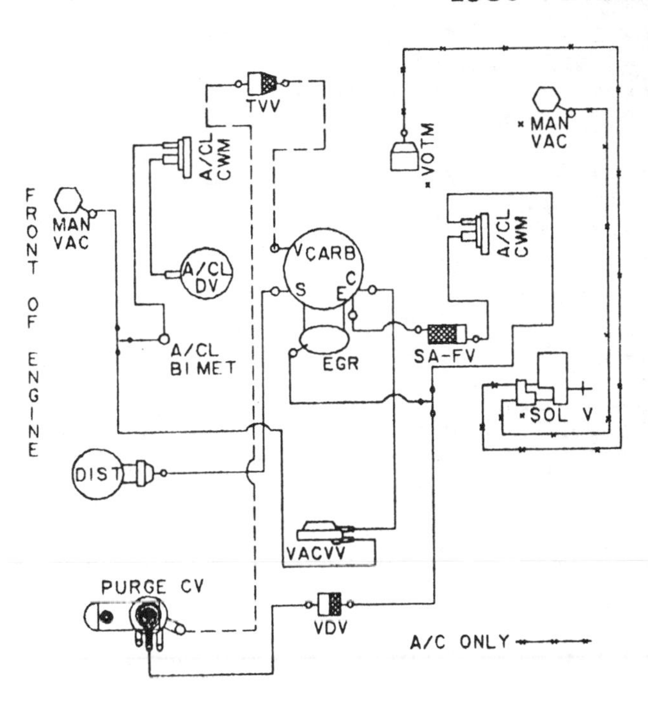

CALIBRATION: 9–21B–R10 DATE: 8–28–78
2.3L (140 CID)

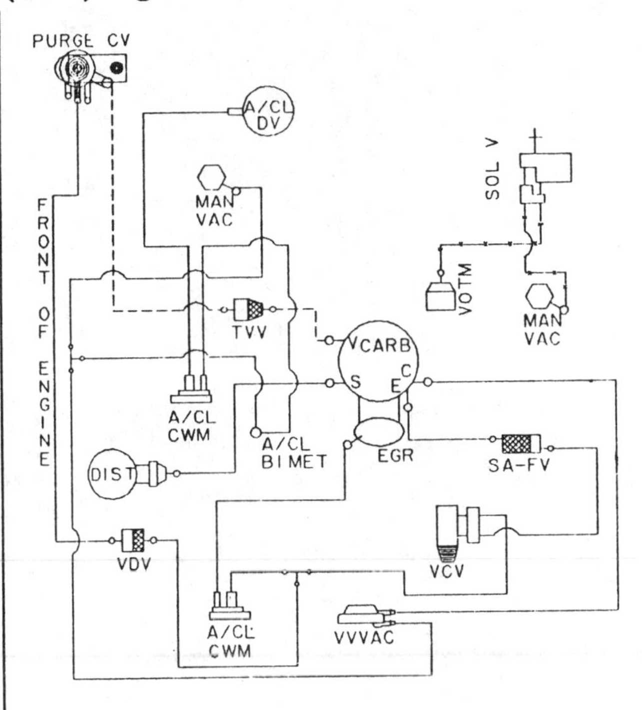

CALIBRATION: 9–21B–R13 DATE: 12–4–78
2.3L (140 CID)

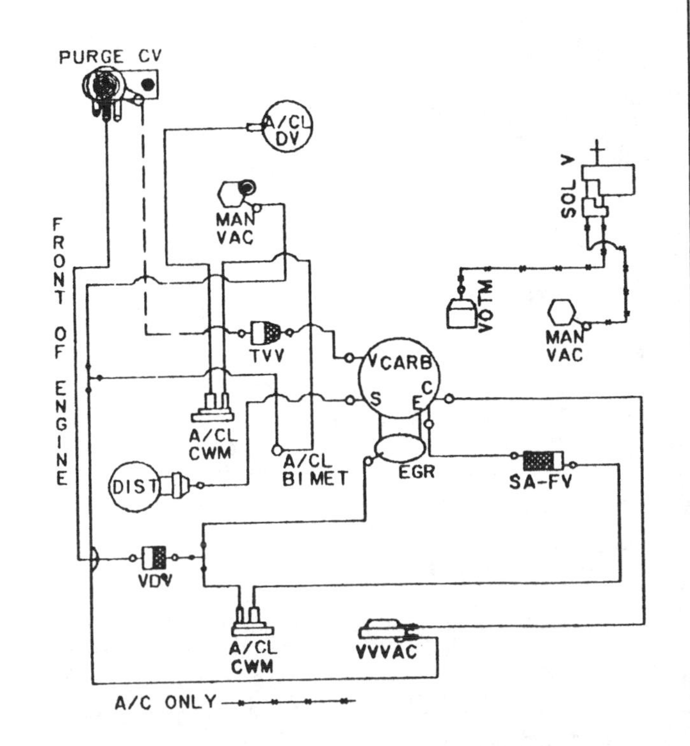

CALIBRATION: 9–21C–RO DATE: 6–7–78
2.3L (140 CID)

49S A/C & NON A/C

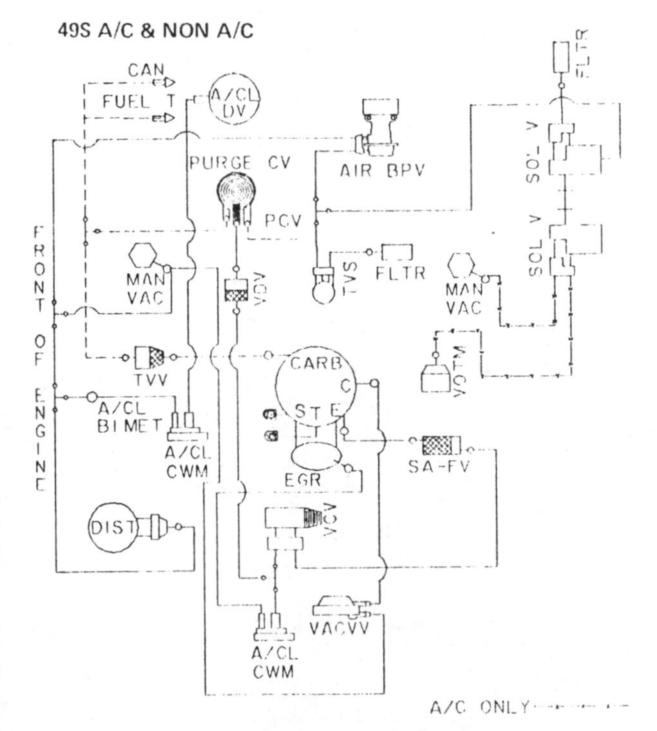

CALIBRATION: 0–1G–RO DATE: 6–22–79
2.3L (140 CID) FAIRMONT/ZEPHYR/
MUSTANG/CAPRI 49S A/T

VACUUM CIRCUITS

1980—140 CID (2.3 L) engines

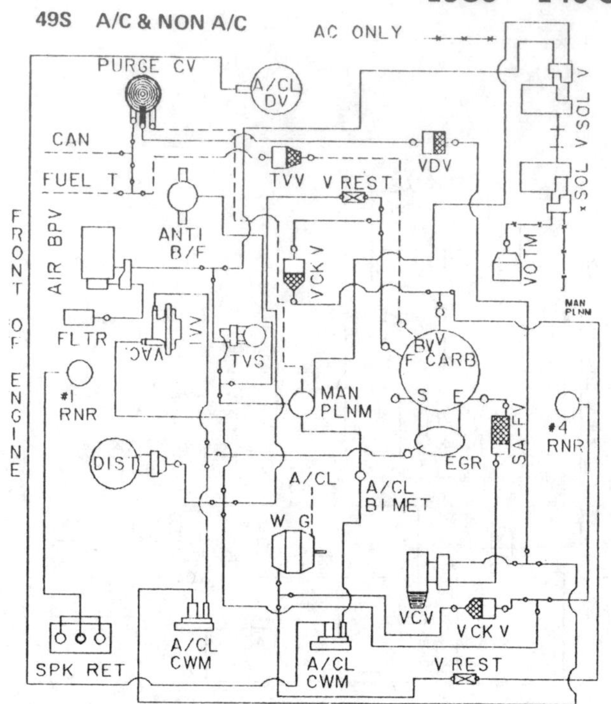

CALIBRATION: 0–2E–RO DATE: 7–26–79
2.3L 49S MUST/CAPRI TURBO M/T

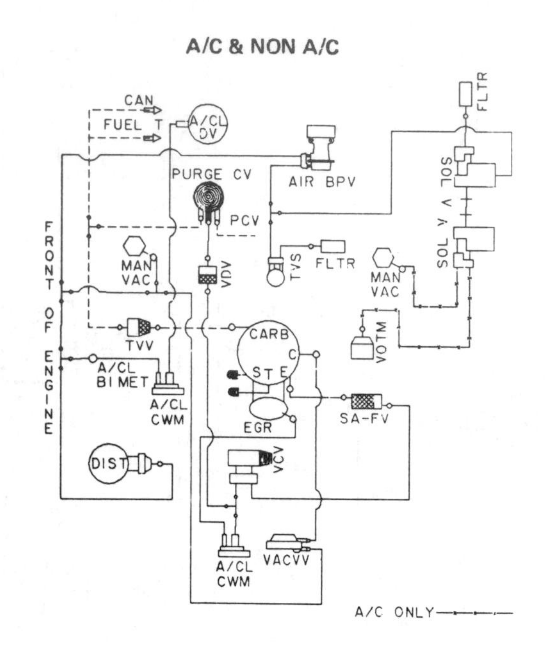

CALIBRATION: 0–1B–RO DATE: 6–20–79
2.3L (140 CID) FAIRMONT/ZEPHYR/
MUST/CAPRI 49S A/T

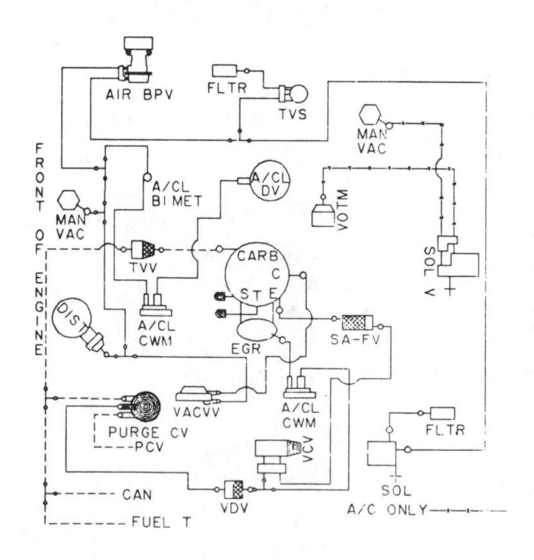

CALIBRATION: 0–1B–RO DATE: 6–20–79
2.3L (140 CID) PINTO/BOBCAT 49S A/T

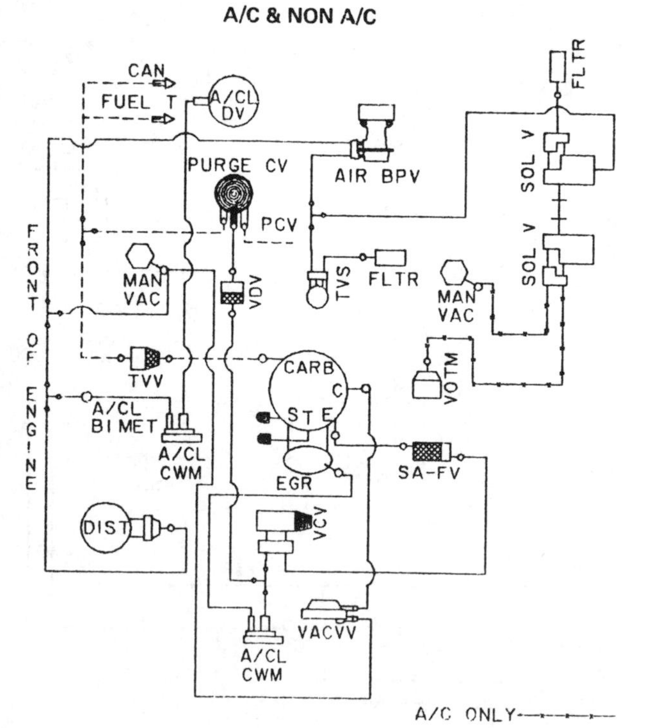

CALIBRATION: 0–1G–RO DATE: 6–22–79
2.3L (140 CID) FAIRMONT/ZEPHYR/
MUSTANG/CAPRI 49S A/T

VACUUM CIRCUITS

1980—140 CID (2.3 L) engines

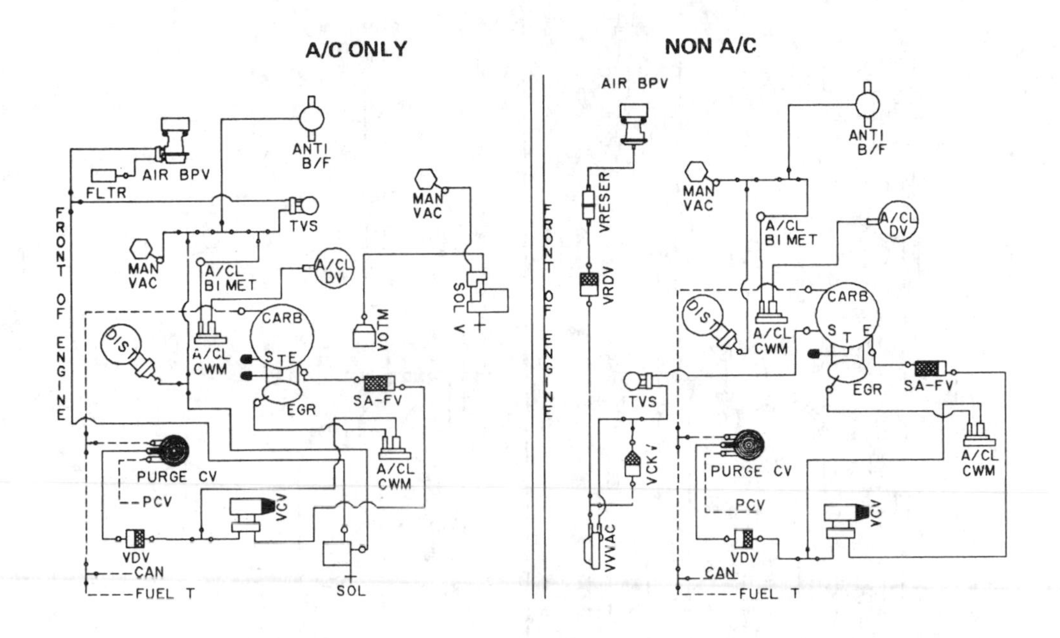

CALIBRATION: 0–2C–RO DATE: 6–25–79
2.3L (140 CID) 49S M/T M-4, M-5 SPD
PINTO/BOBCAT SDN & SWGN

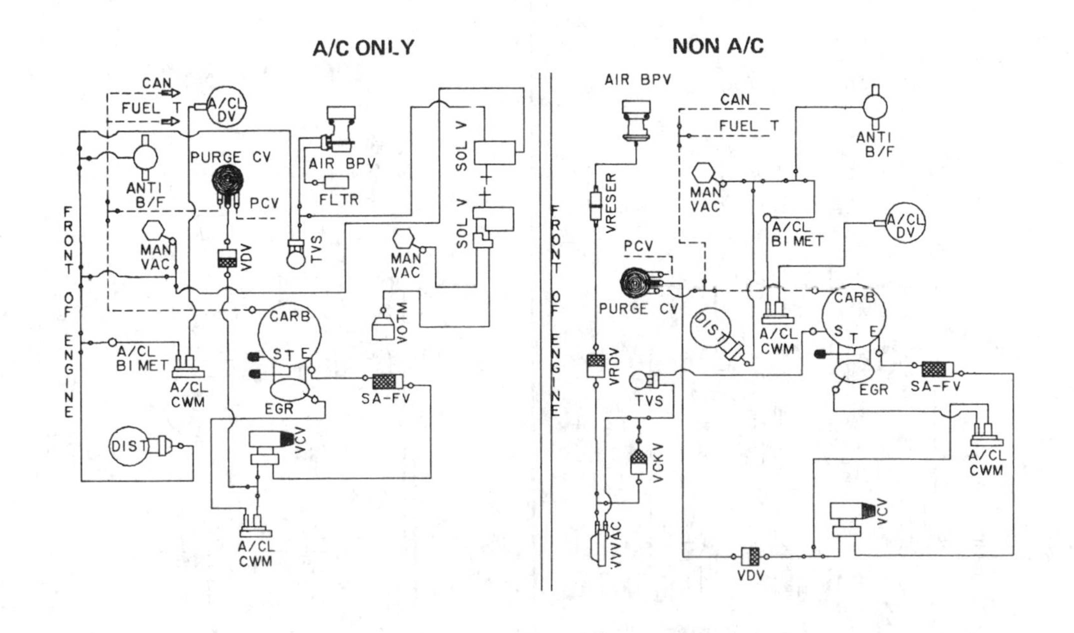

CALIBRATION: 0–2C–RO DATE: 6–25–79
2.3L (140 CID) MUST/CAPRI 49S M/T

VACUUM CIRCUITS

1980—140 CID (2.3 L) engines

49S A/C & NON A/C

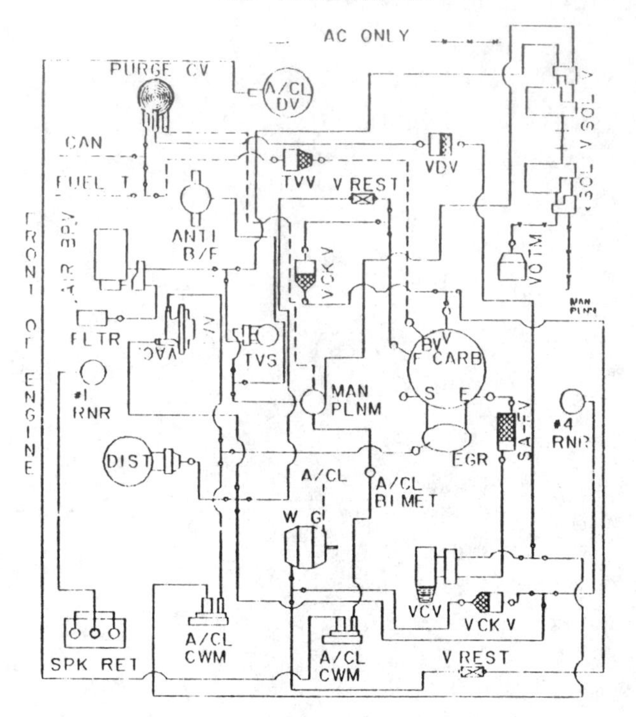

CALIBRATION: 0–2E–RO **DATE:** 7–26–79
2.3L 49S MUST/CAPRI TURBO M/T

49S A/C & NON A/C

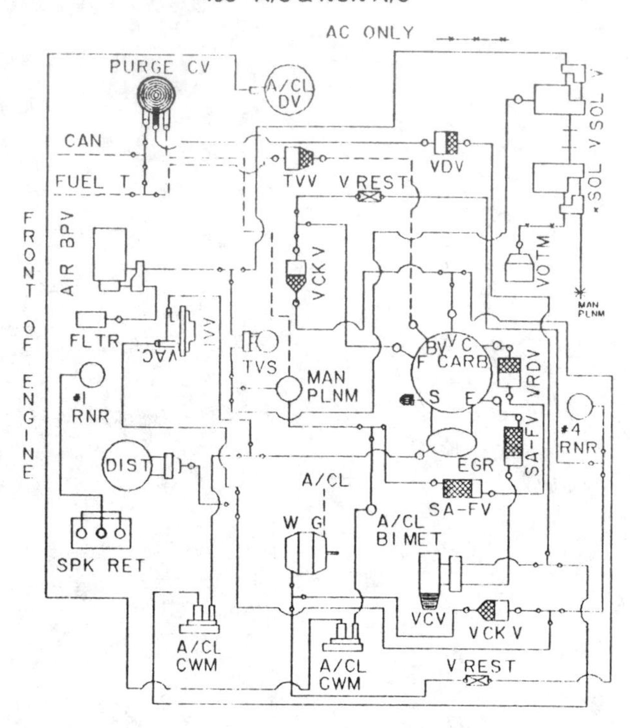

CALIBRATION: 0–01H–R10 **DATE:** 09–17–79
2.3L MUST/CAPRI/FAIRMONT/ZEPHYR 49S TURBO A/T

CALIF. ONLY A/C & NON A/C

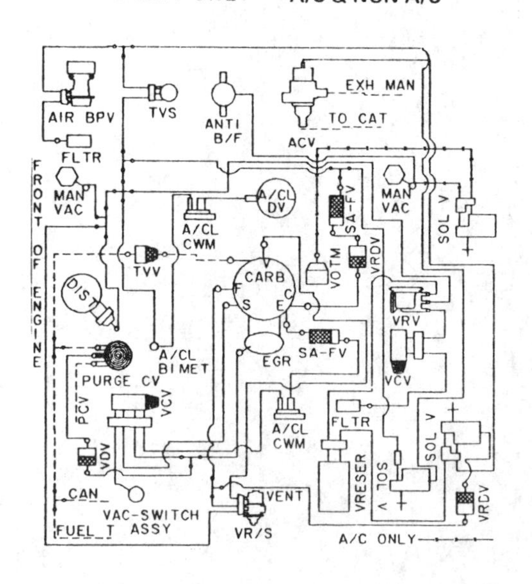

CALIBRATION: 0–2T–RO **DATE:** 7–18–79
2.3L PINTO/BOBCAT CALIF. M/T

CALIF. ONLY A/C & NON A/C

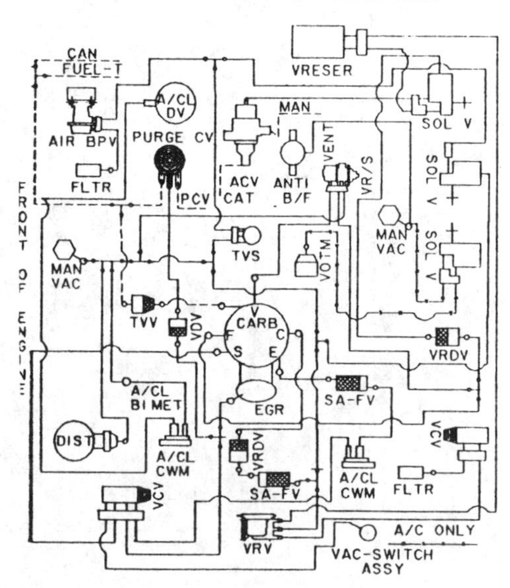

CALIBRATION: 0–2T–RO **DATE:** 7–18–79
2.3L MUST/CAPRI/FAIR/ZEPH CALIF M/T

VACUUM CIRCUITS

1980—140 CID (2.3 L) engines

CALIF. ONLY A/C & NON A/C

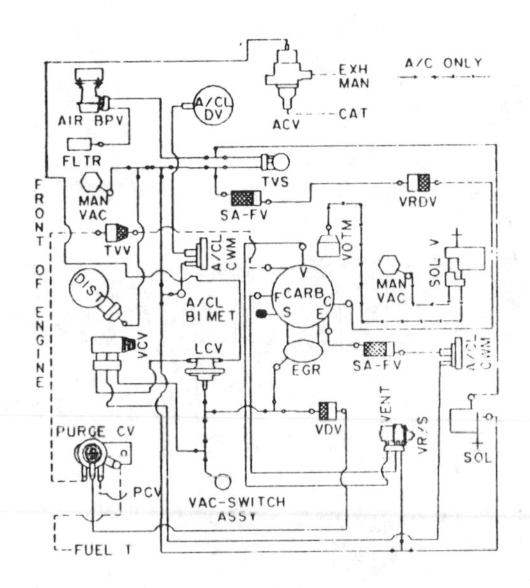

CALIBRATION: 0–1P–RO DATE: 7–19–79
2.3L CALIF A/T PINTO/BOBCAT

CALIF. ONLY A/C & NON A/C

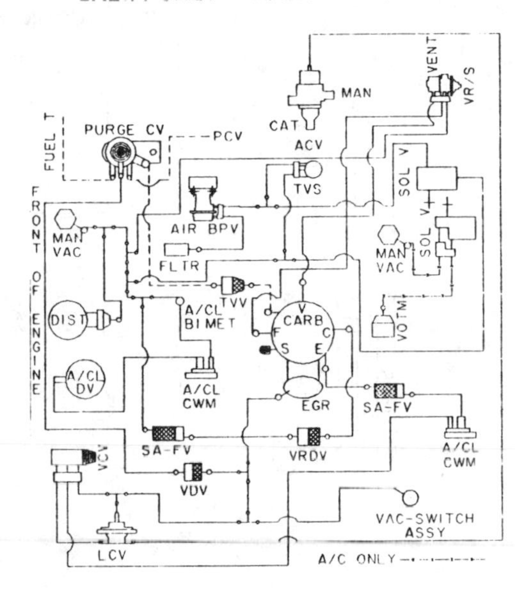

CALIBRATION: 0–1P–RO DATE: 7–20–79
2.3L CALIF A/T FAIRMONT/ZEPHYR
MUST CAPRI

49S A/C ONLY

49S NON A/C

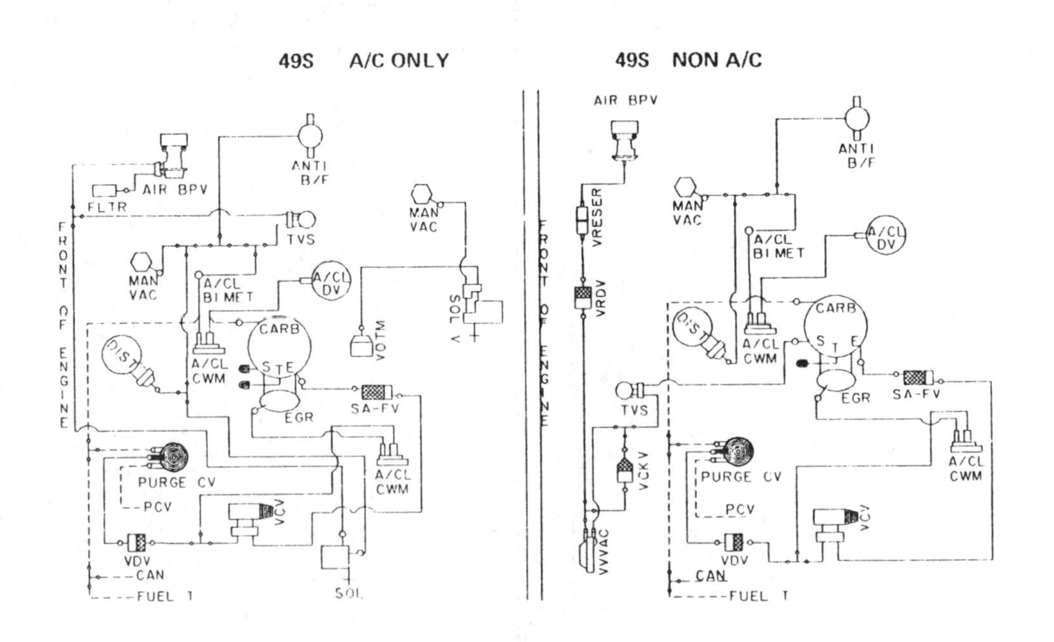

CALIBRATION: 0–2B–RO DATE: 7–31–79
2.3L 49S PINTO/BOBCAT M/T

VACUUM CIRCUITS

1980—140 CID (2.3 L) engines

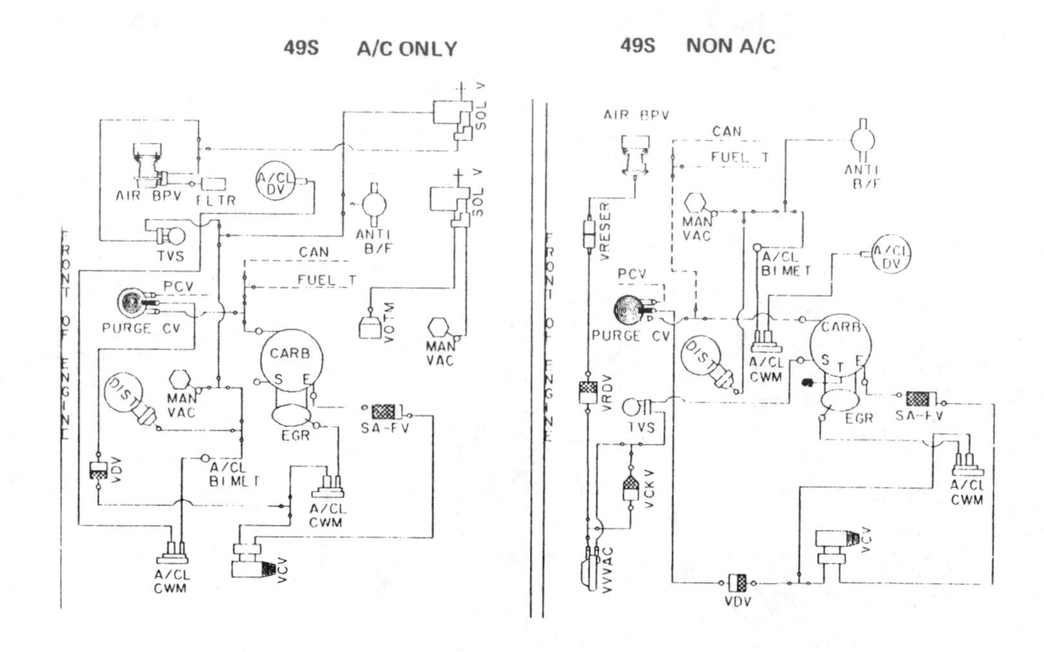

CALIBRATION: 0–2B–RO DATE: 7–31–79
2.3L FAIRMONT/ZEPHYR 49S M/T

49S A/C & NON A/C

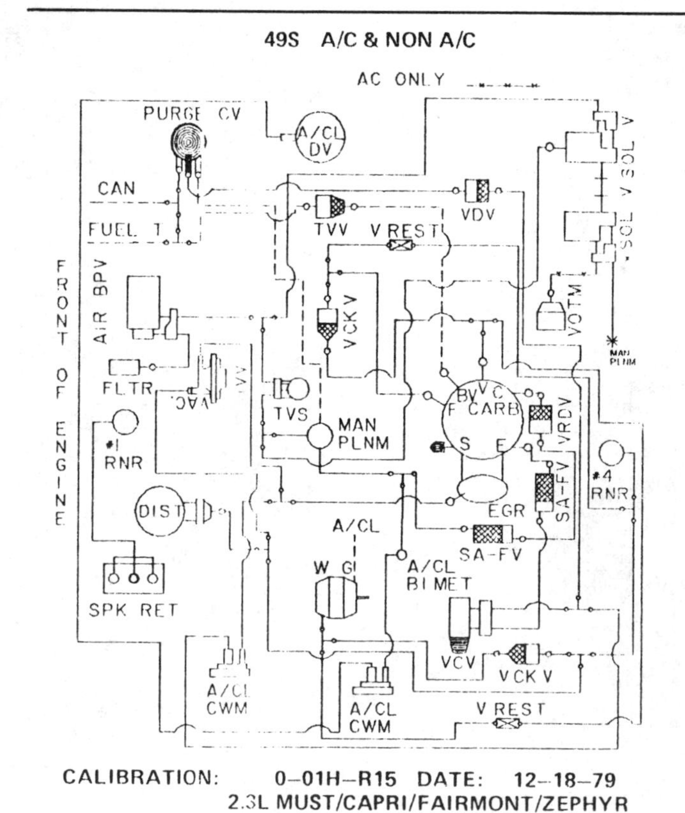

CALIBRATION: 0–01H–R15 DATE: 12–18–79
2.3L MUST/CAPRI/FAIRMONT/ZEPHYR
49S TURBO A/T

CALIF ONLY A/C & NON A/C

CALIBRATION: 0–01S–R11 DATE: 12–12–79
2.3L MUST/CAPRI/FAIRMONT/ZEPHYR
CALIF TURBO A/T

VACUUM CIRCUITS

1980—140 CID (2.3 L) engines

49S A/C & NON A/C

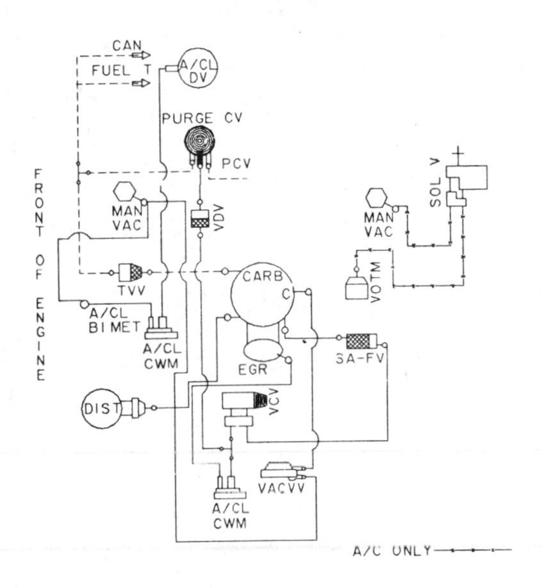

CALIBRATION: 0–21A–R0 DATE: 09–25–79
2.3L FAIRMONT/ZEPHYR/MUST/CAPRI 49S A/T

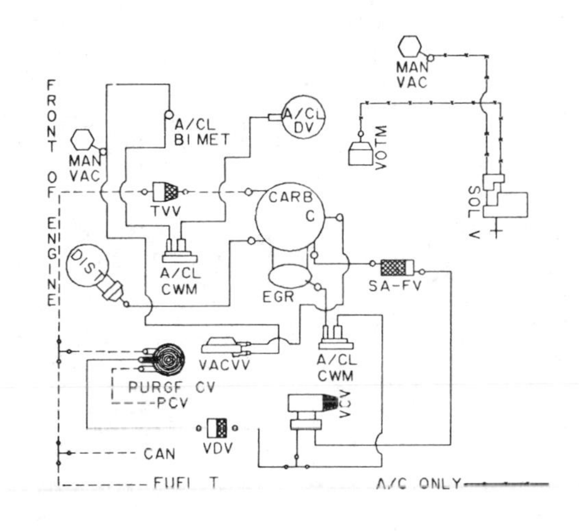

CALIBRATION: 0–21A–R0 DATE: 09–25–79
2.3L PINTO/BOBCAT 49S A/T

49S CANADA A/C & NON A/C

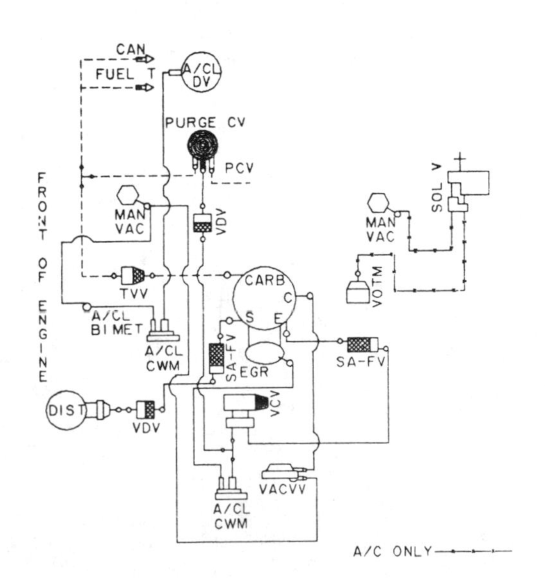

CALIBRATION: 0–21B–R0 DATE: 11–09–79
2.3L FAIRMONT/ZEPHYR 49S/CANADA
A/T

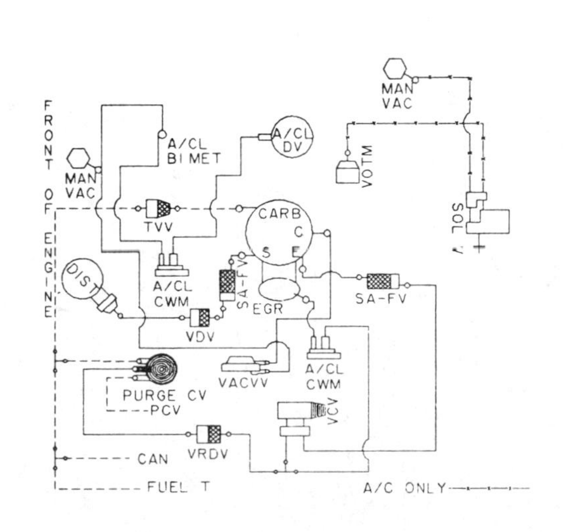

CALIBRATION: 0–21B–R0 DATE: 11–09–79
2.3L PINTO/BOBCAT 49S/CANADA A/T

VACUUM CIRCUITS

1980—140 CID (2.3 L) engines

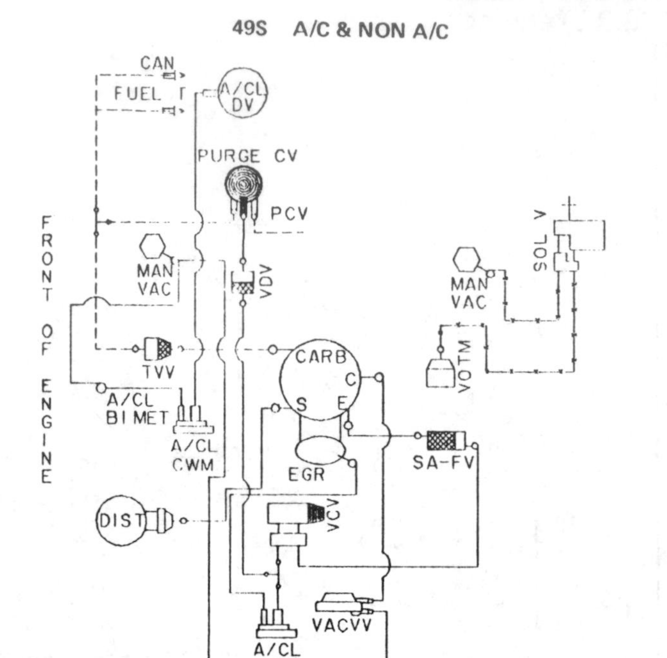

A/C ONLY

CALIBRATION: 0–21B–R10 **DATE:** 12–13–79
2.3L PINTO/BOBCAT 49S A/T

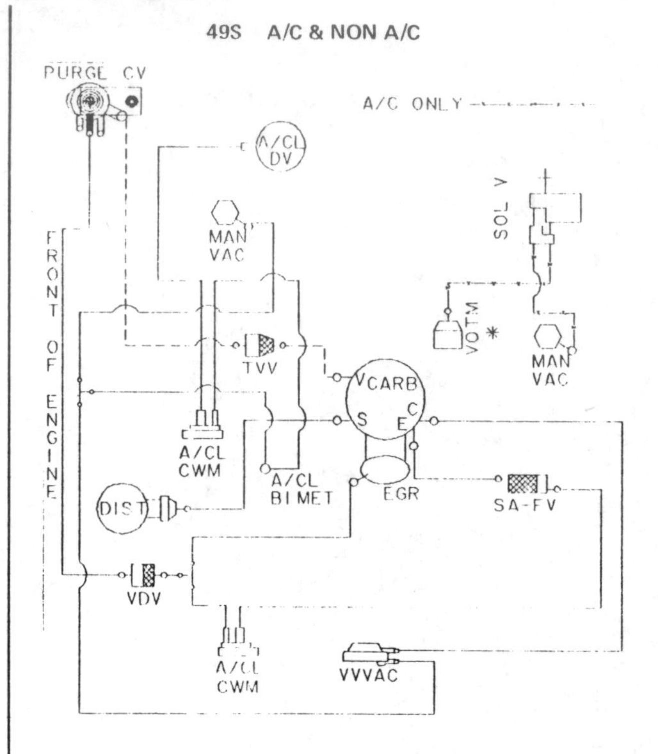

CALIBRATION: 9–21B–R10 **DATE:** 04–30–79
2.3L FAIRMONT/ZEPHYR/MUST/CAPRI
49S A/T

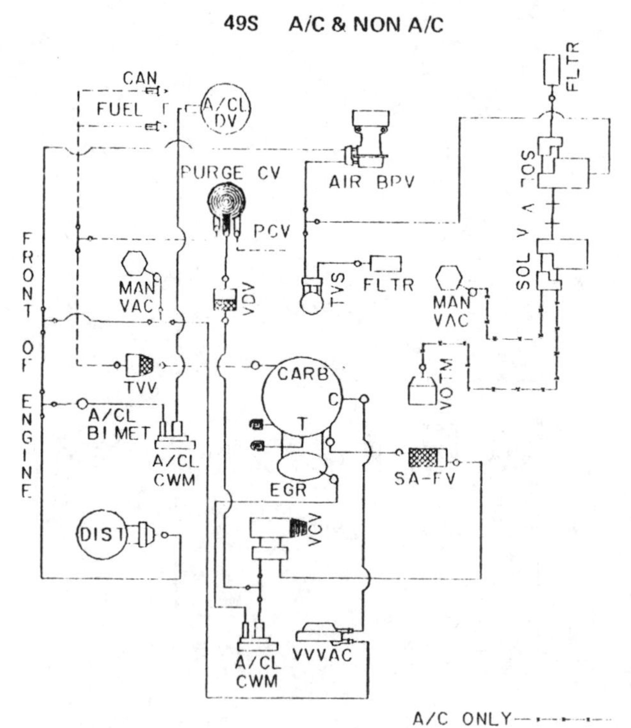

CALIBRATION: 0–1B–R11 **DATE:** 10–03–79
2.3L FAIRMONT/ZEPHYR/MUST/CAPRI
49S A/T

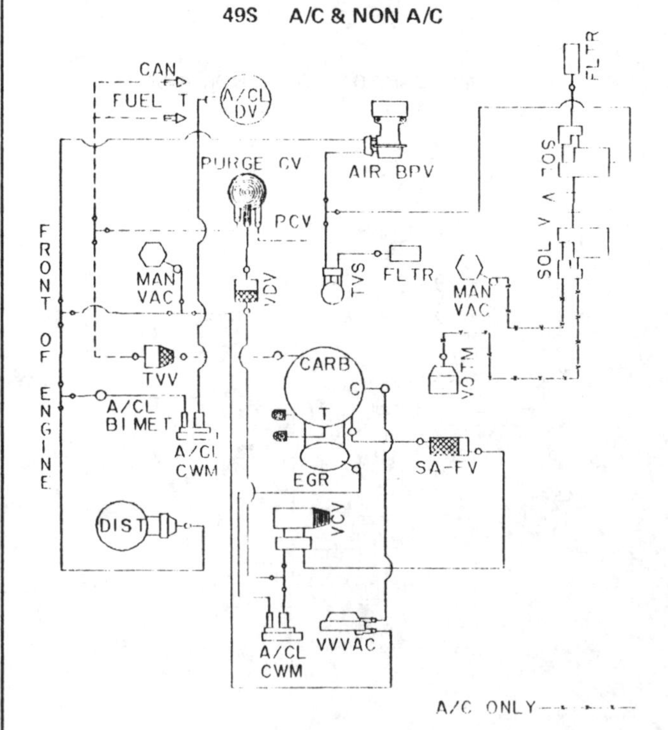

CALIBRATION: 0–1B–R12 **DATE:** 10–11–79
2.3L FAIRMONT/ZEPHYR/MUST/CAPRI
49S A/T

VACUUM CIRCUITS

1980—171 CID (2.8 L) engines

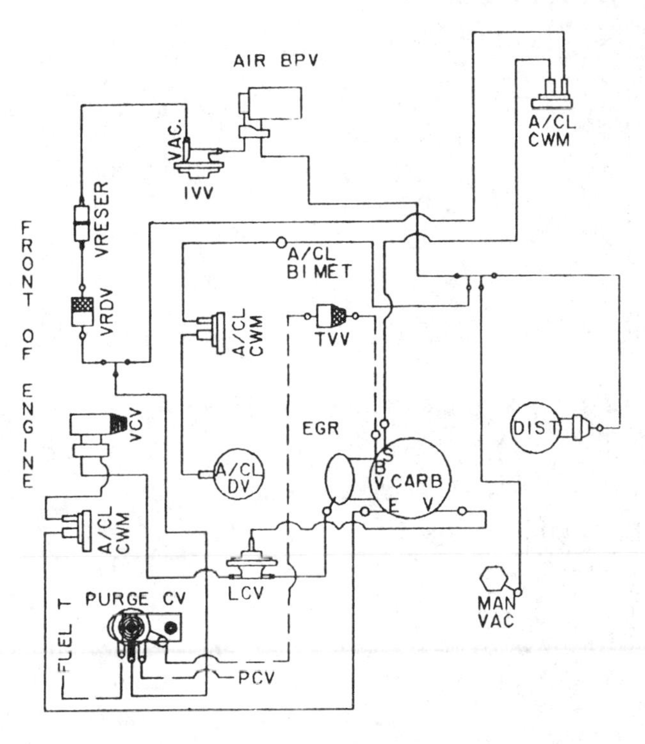

CALIBRATION: 9–4A–R10 DATE: 9–7–78
2.8L (171 CID)

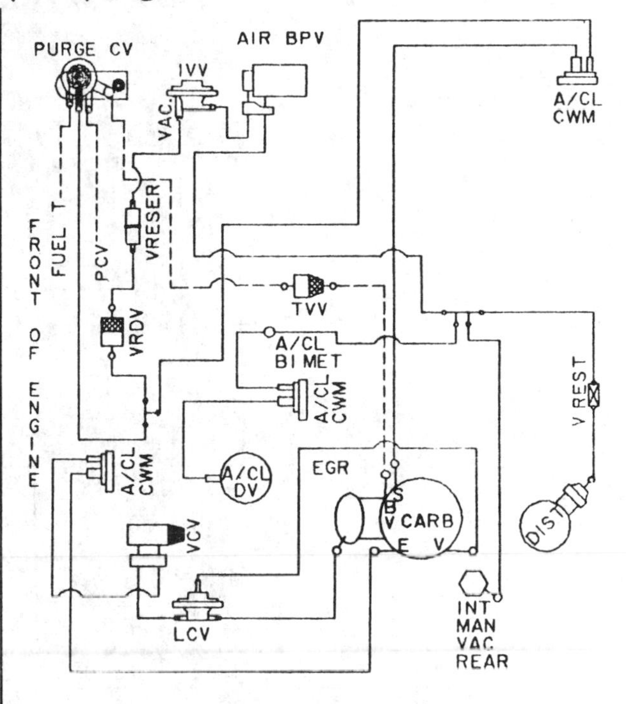

CALIBRATION: 9–4A–R11 DATE: 10–18–78
2.8L (171 CID)

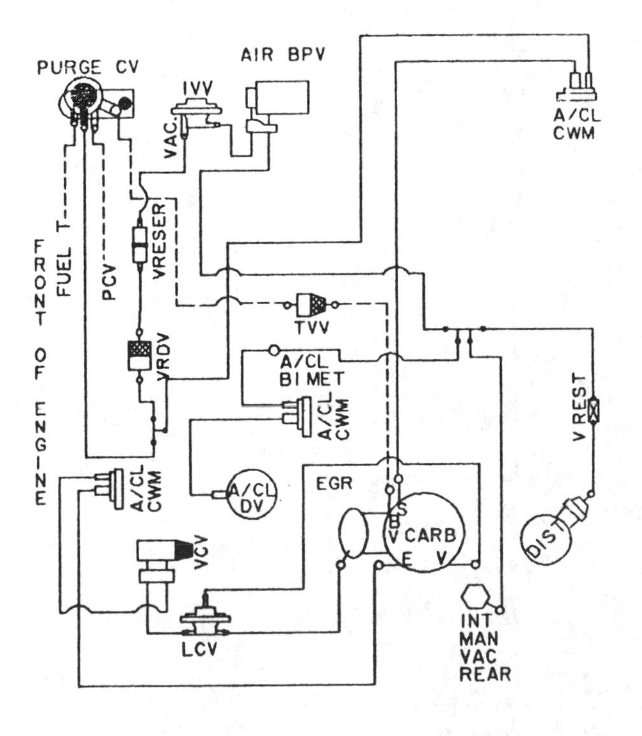

CALIBRATION: 9–4A–R12 DATE: 12–20–78
2.8L (171 CID)

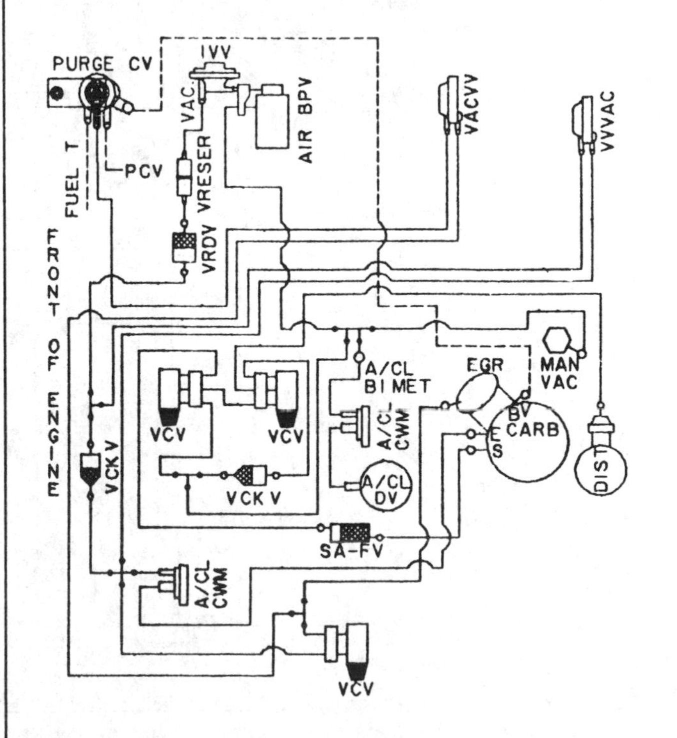

CALIBRATION. 9–4N–RO DATE: 7–11–78
2.8L (171 CID)

VACUUM CIRCUITS

1980—171 CID (2.8 L) engines

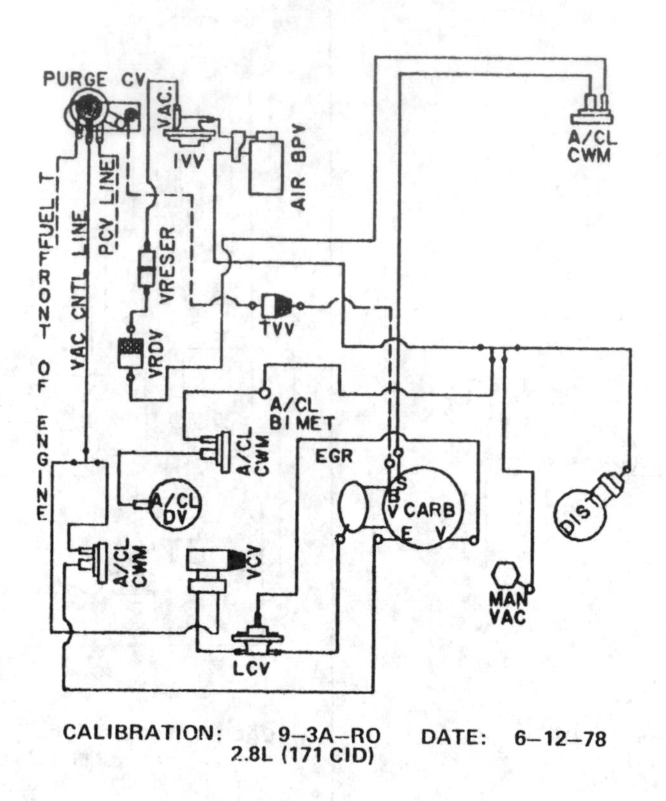

CALIBRATION: 9–3A–RO DATE: 6–12–78
2.8L (171 CID)

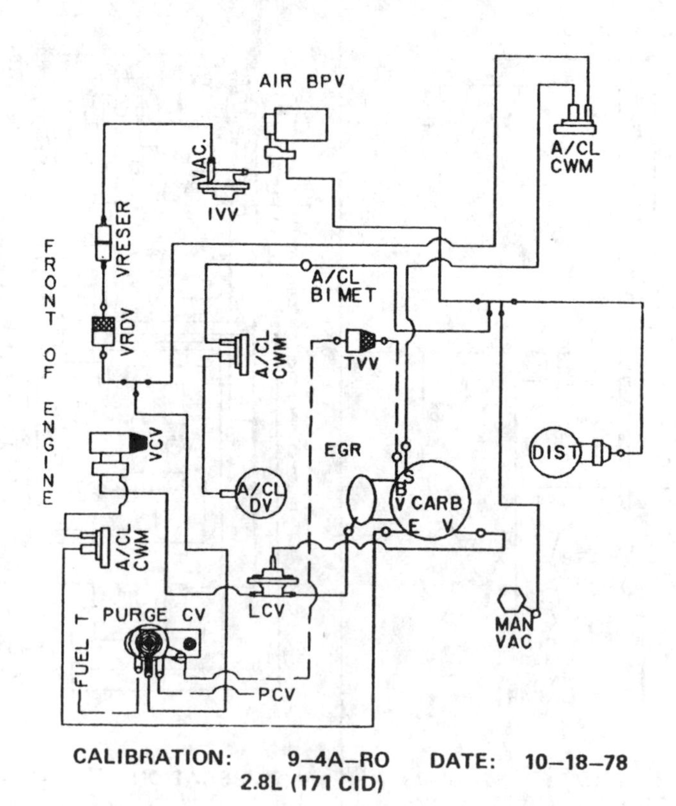

CALIBRATION: 9–4A–RO DATE: 10–18–78
2.8L (171 CID)

VACUUM CIRCUITS

1980—200 CID (3.3 L) engines

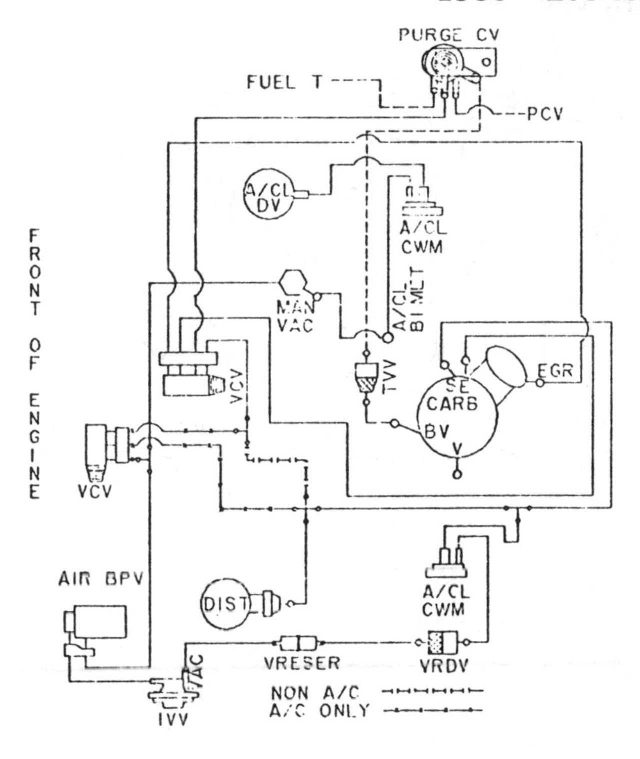

CALIBRATION: 9–6A–RO DATE: 10–25–78
3.3L (200 CID)

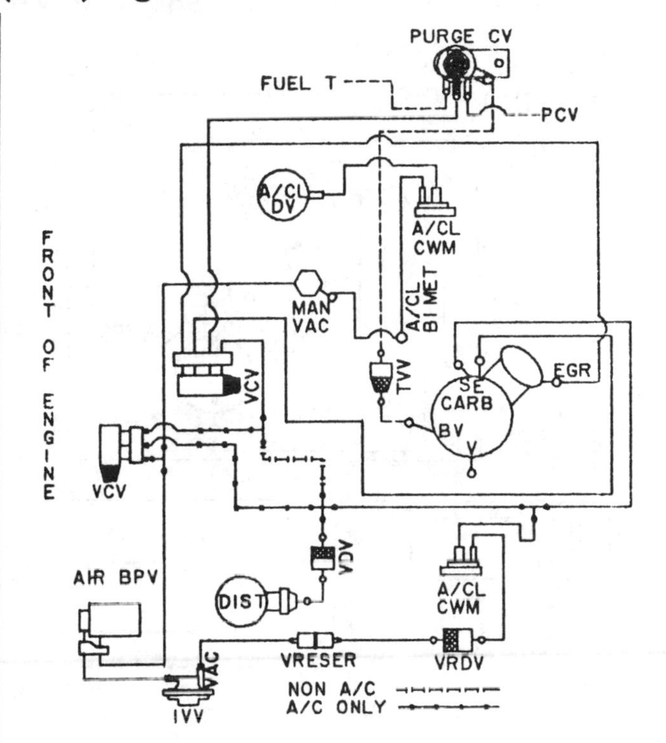

CALIBRATION: 9–6A–R10 DATE: 9–26–78
3.3L (200 CID)

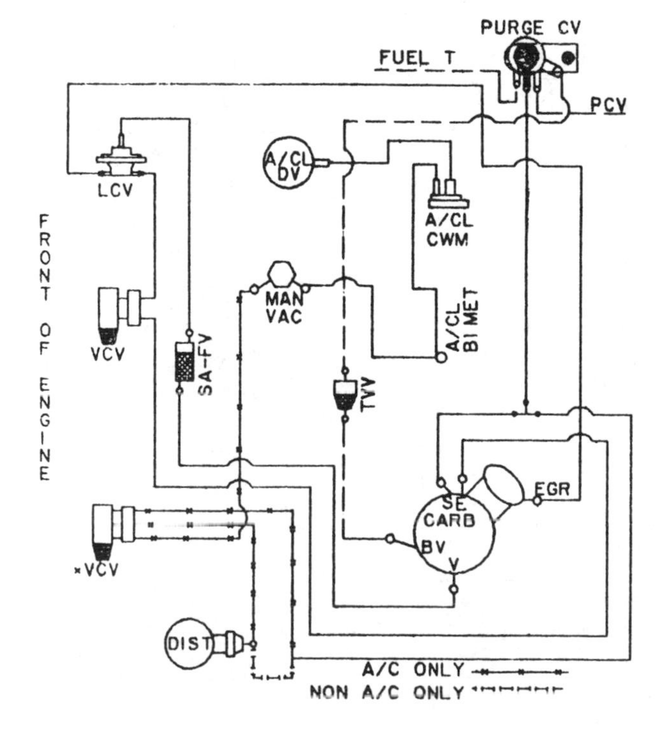

CALIBRATION: 9–7A–RO DATE: 9–26–78
3.3L (200 CID)

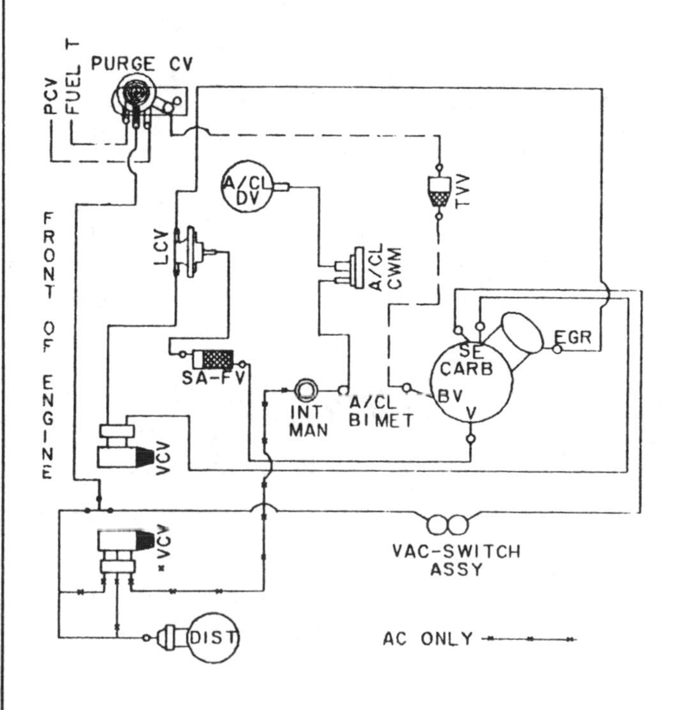

CALIBRATION: 9–7A–R10 DATE: 6–28–78
3.3L (200 CID)

VACUUM CIRCUITS

1980—200 CID (3.3 L) engines

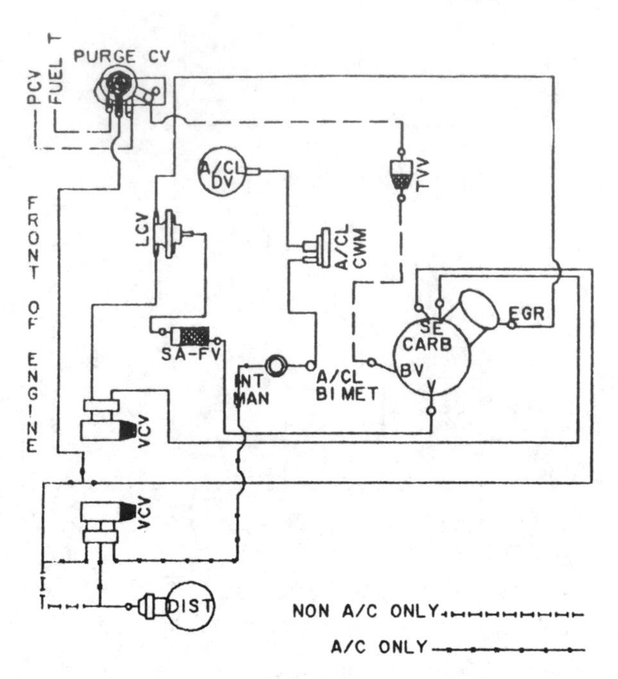

CALIBRATION: 9–7A–R11 DATE: 10–13–78
3.3L (200 CID)

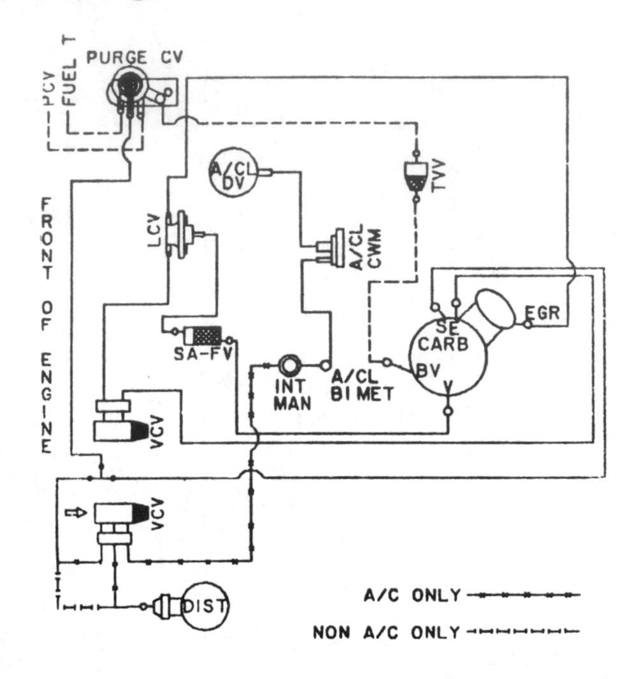

CALIBRATION: 9–7A- 12 DATE: 10–25–78
3.3L (302 D)

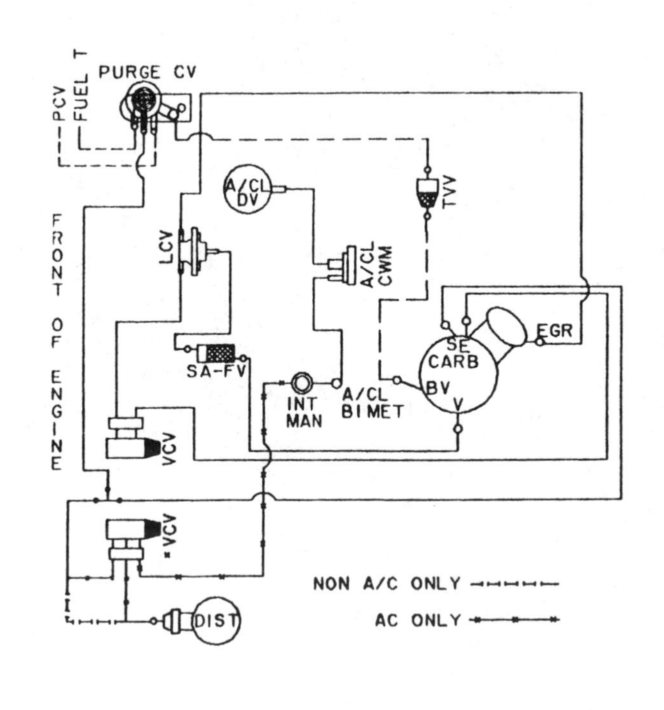

CALIBRATION: 9–7A–R13 DATE: 10–21–78
3.3L (200 CID)

NON A/C ONLY
A/C ONLY

CALIBRATION: 9–7A–R15 DATE: 4–23–79
3.3L (200 CID)

VACUUM CIRCUITS

1980—200 CID (3.3 L) engines

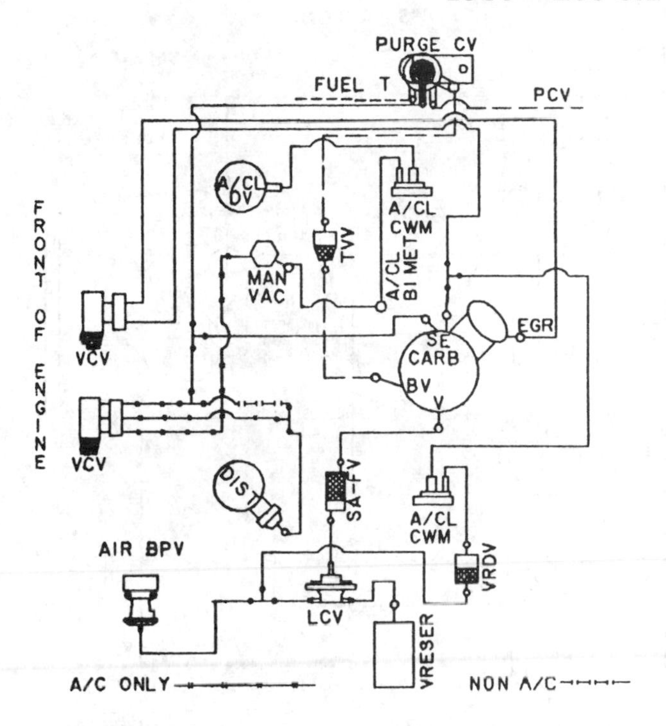

CALIBRATION: 9–7D–RO DATE: 9–29–78
3.3L (200 CID)

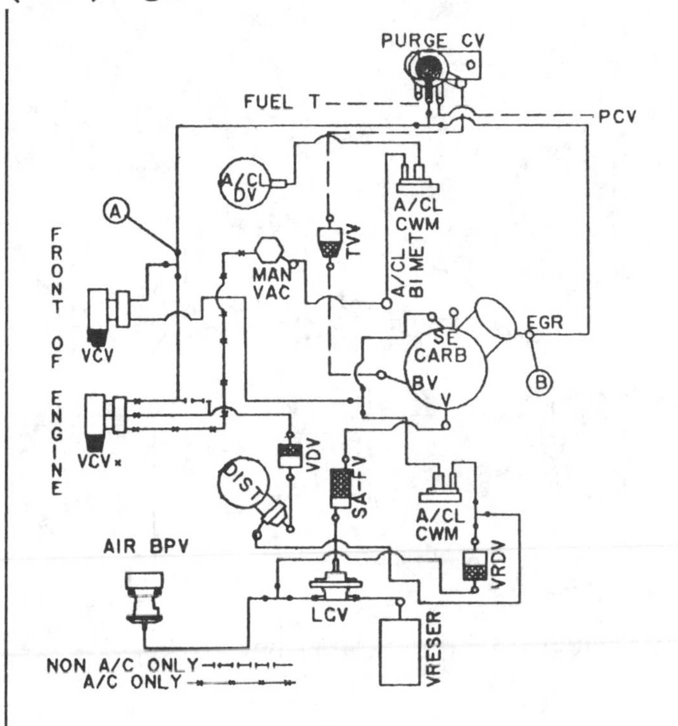

CALIBRATION: 9–7N–R2 DATE: 9–26–78
3.3L (200 CID)

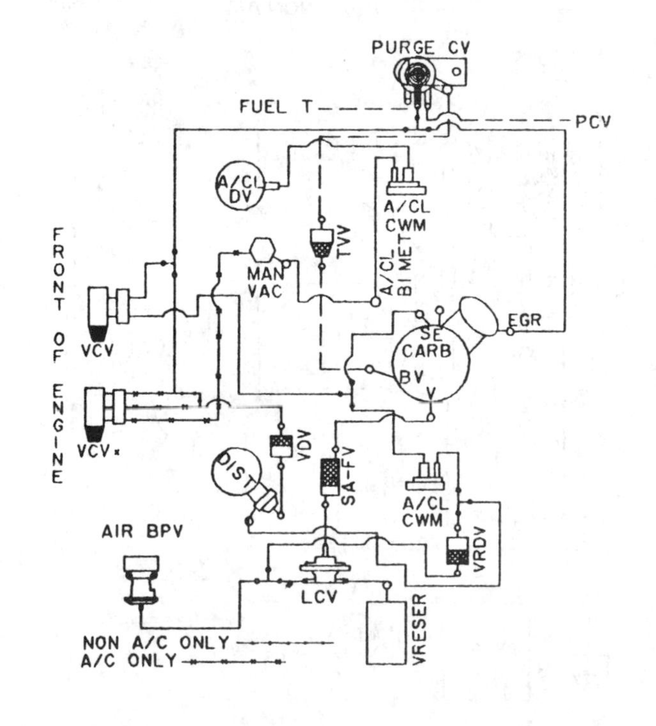

CALIBRATION: 9–7N–R3 DATE: 10–26–78
3.3L (200 CID)

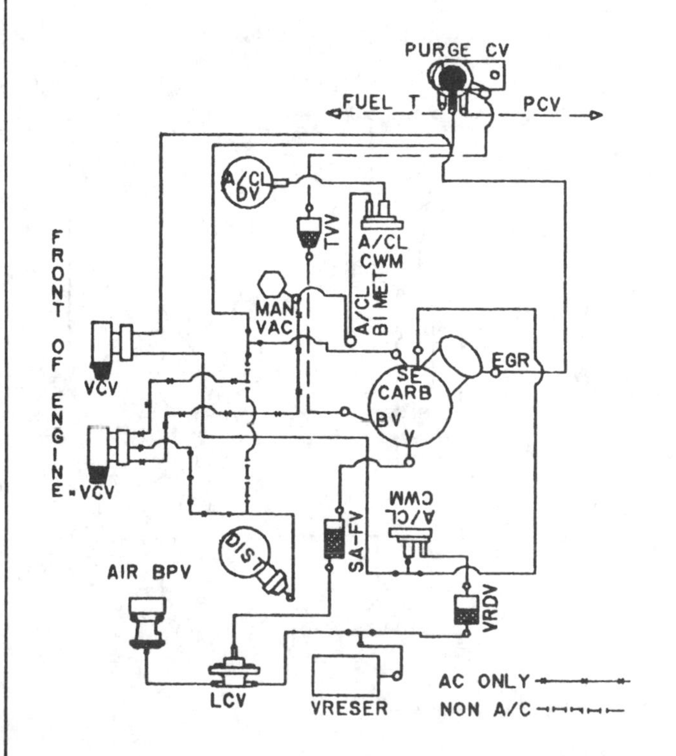

CALIBRATION: 9–7Z–R10 DATE: 9–29–78
3.3L (200 CID)

VACUUM CIRCUITS

1980—200 CID (3.3 L) engines

49S NON A/C

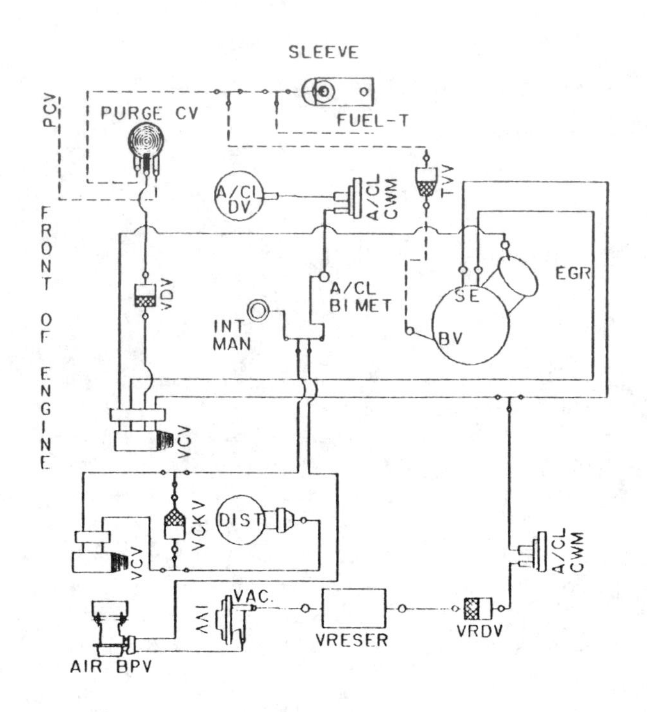

CALIBRATION: 0–6A–RO DATE: 5–9–79
3.3L 49S M/T

49S A/C ONLY

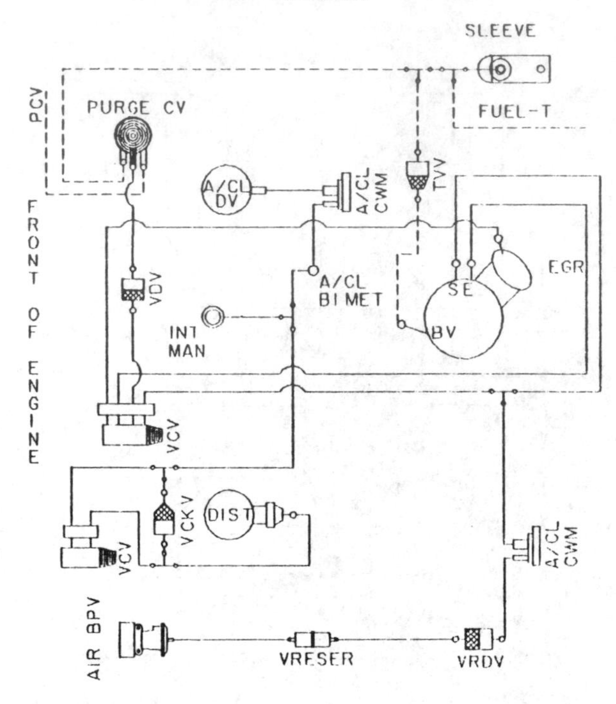

CALIBRATION: 0–6A–R1 DATE: 7–20–79
3.3L 49S M/T

49S A/C & NON A/C

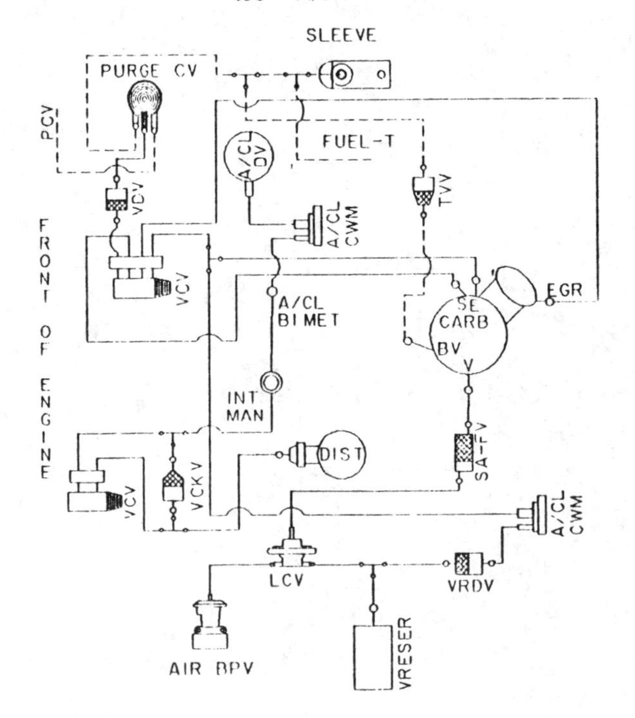

CALIBRATION: 0–7A–RO DATE: 5–10–79
3.3L FAIR/ZEPHYR/MUST/CAPRI 49S A/T

49S A/C & NON A/C

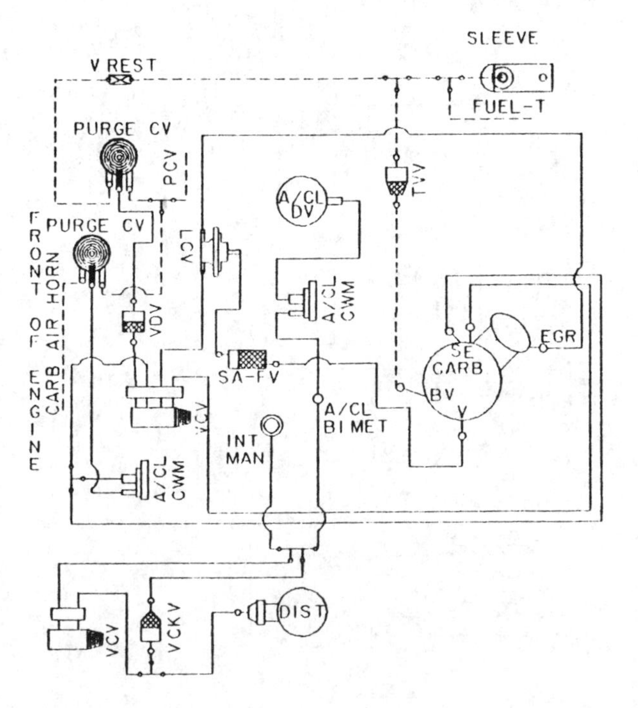

CALIBRATION: 0–27A–R3 DATE: 8–6–79
3.3L FAIR/ZEPH/MUST/CAPRI 49S A/T

VACUUM CIRCUITS

1980—200 CID (3.3 L) engines

CALIF. ONLY A/C | CALIF. ONLY NON A/C

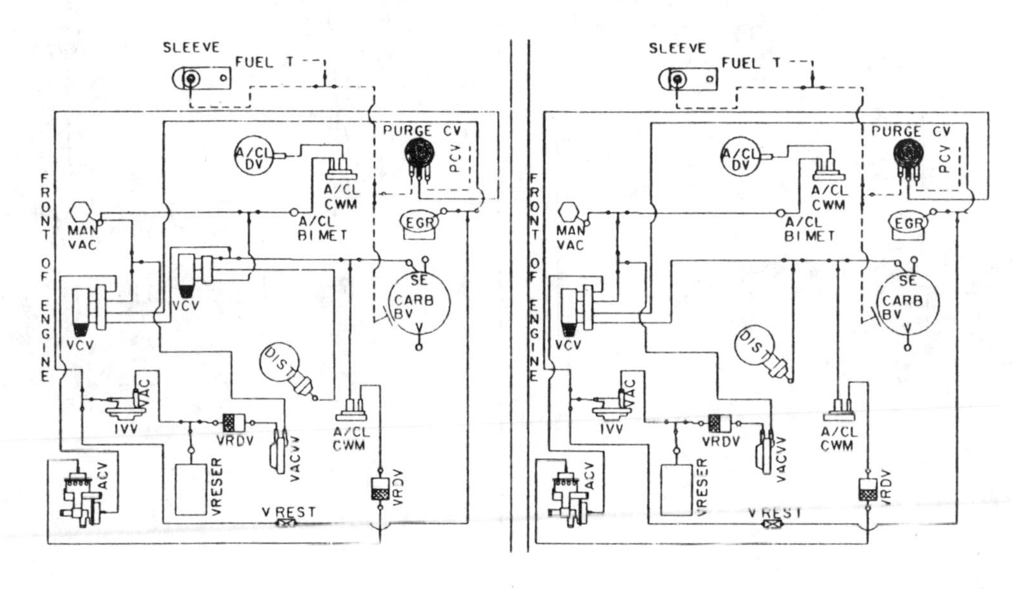

CALIBRATION: 0–7Q–R1 DATE: 8–7–79
3.3L CALIF.

CALIF ONLY A/C & NON A/C

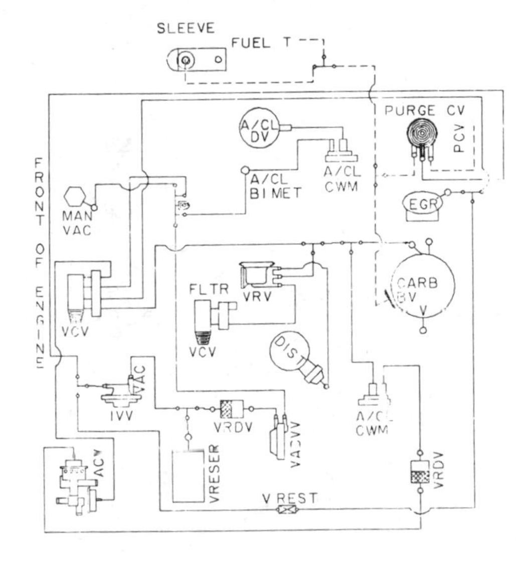

CALIBRATION 0–07P–R11 DATE: 11–05–79
3.3L FAIRMONT/ZEPHYR/MUST/CAPRI
CALIF

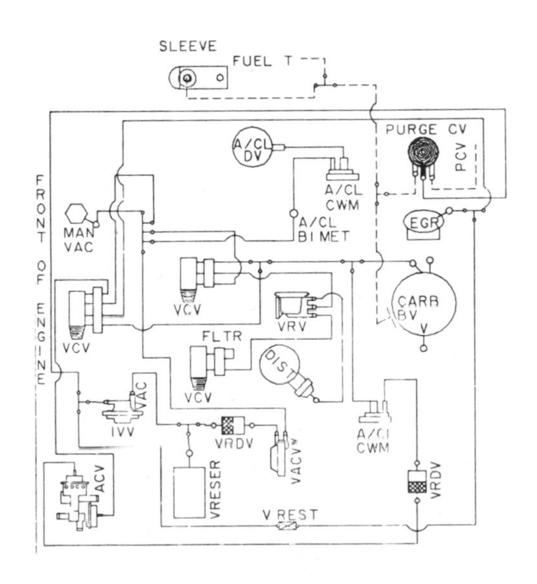

CALIBRATION: 0–07P–R11 DATE: 11–05–79
3.3L FAIRMONT/ZEPHYR/MUST/CAPRI
CALIF

VACUUM CIRCUITS

1980—200 CID (3.3 L) engines

49S A/C & NON A/C

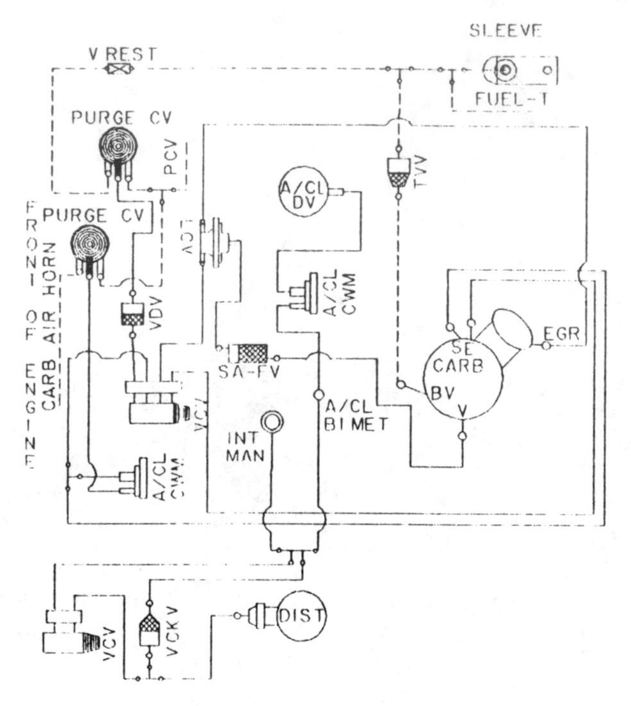

CALIBRATION: 0–27A–R10 DATE: 01–25–80
3.3L FAIRMONT/ZEPHYR/MUST/CAPRI
49 S A/T

CANADA A/C & NON A/C

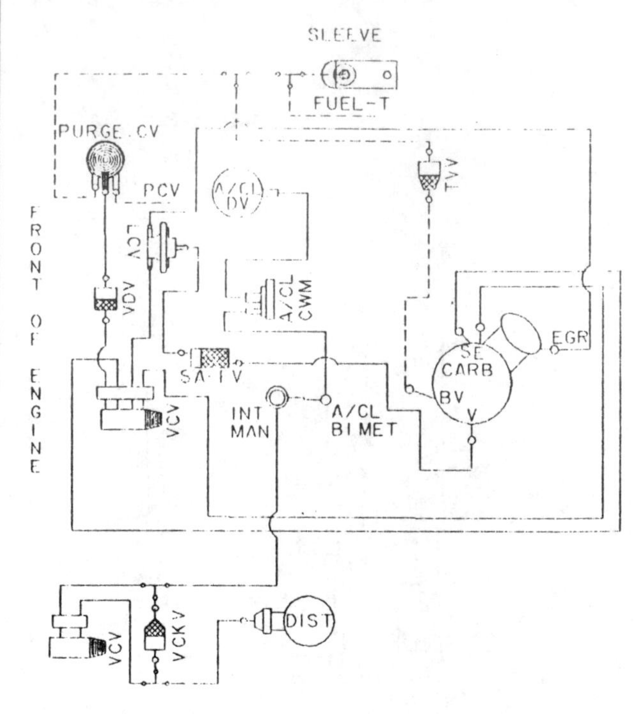

CALIBRATION: 0–27G–R0 DATE: 08–09–79
3.3L FAIRMONT/ZEPHYR CANADA A/T

A/C CALIF ONLY

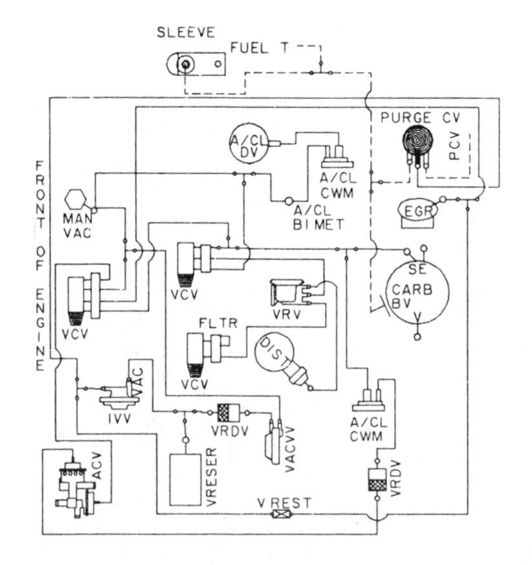

CALIBRATION: 0–7P–R10 DATE: 08–22–79
3.3L FAIRMONT/ZEPHYR/MUST/CAPRI CALIF

NON A/C CALIF ONLY

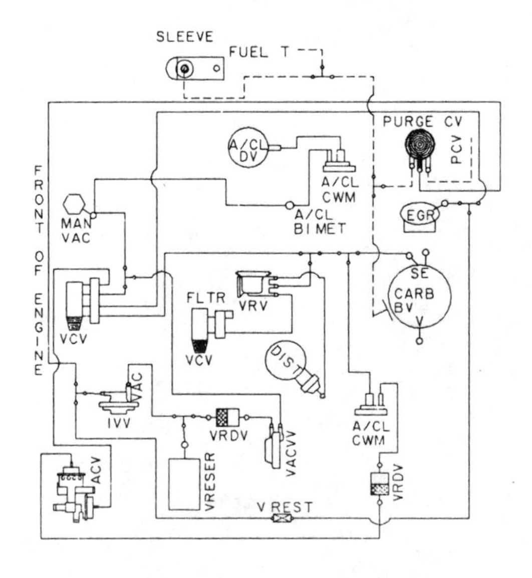

CALIBRATION: 0–7P–R10 DATE: 08–22–79
3.3L FAIRMONT/ZEPHYR/MUST/CAPRI CALIF

VACUUM CIRCUITS

1980—250 CID (4.1 L) engines

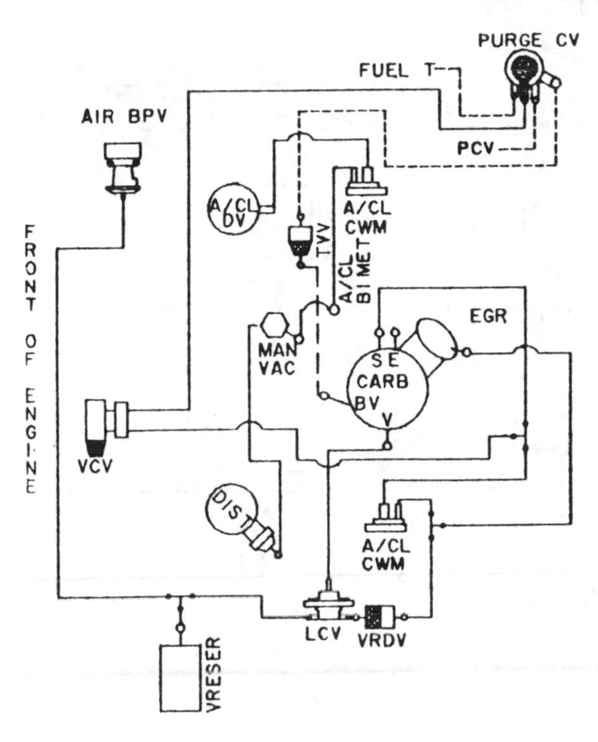

CALIBRATION: 9–8A–RO DATE: 6–9–78
4.1L (250 CID)

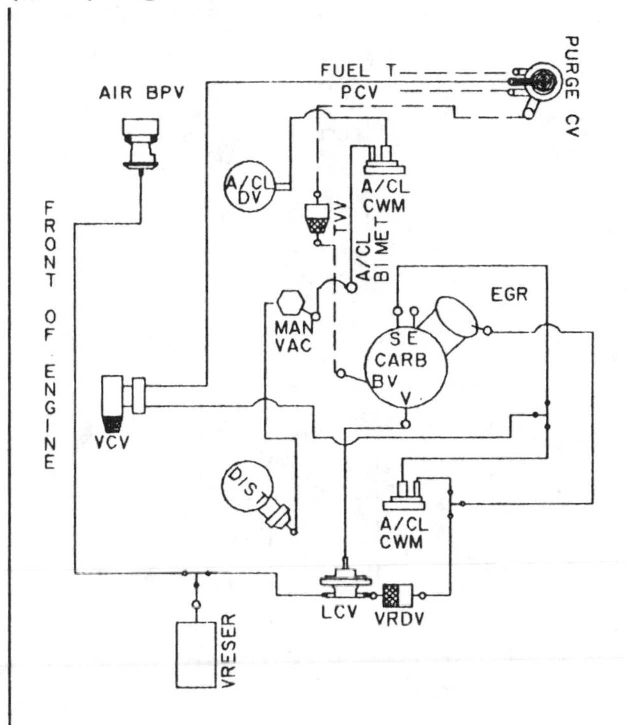

CALIBRATION: 9–8A–R10 DATE: 10–10–78
4.1L (250 CID)

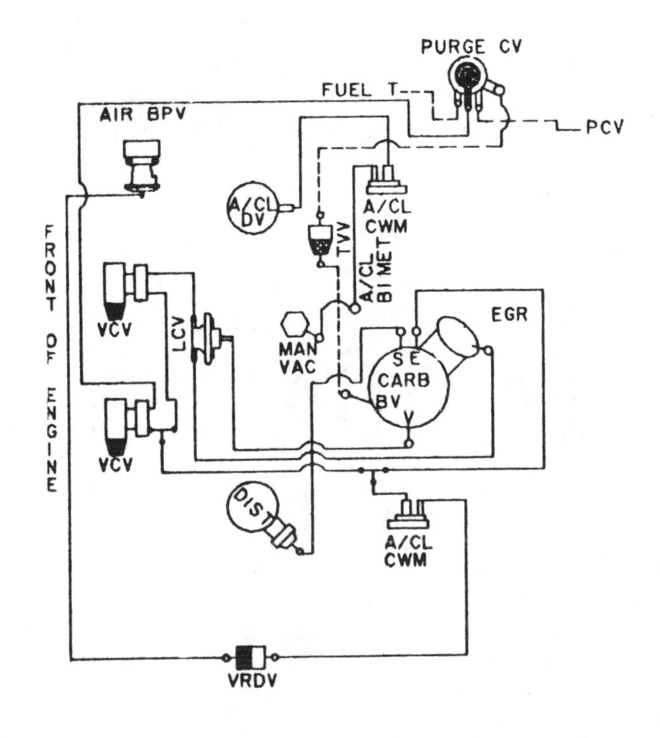

CALIBRATION: 9–9A–RO DATE: 8–30–78
4.1L (250 CID)

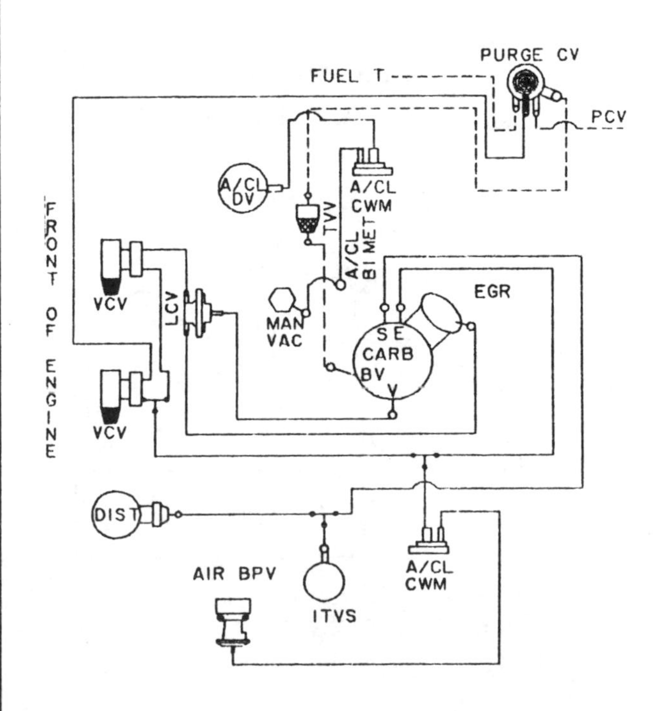

CALIBRATION: 9–9A–R13 DATE: 10–10–78
4.1L (250 CID)

VACUUM CIRCUITS

1980—250 CID (4.1 L) engines

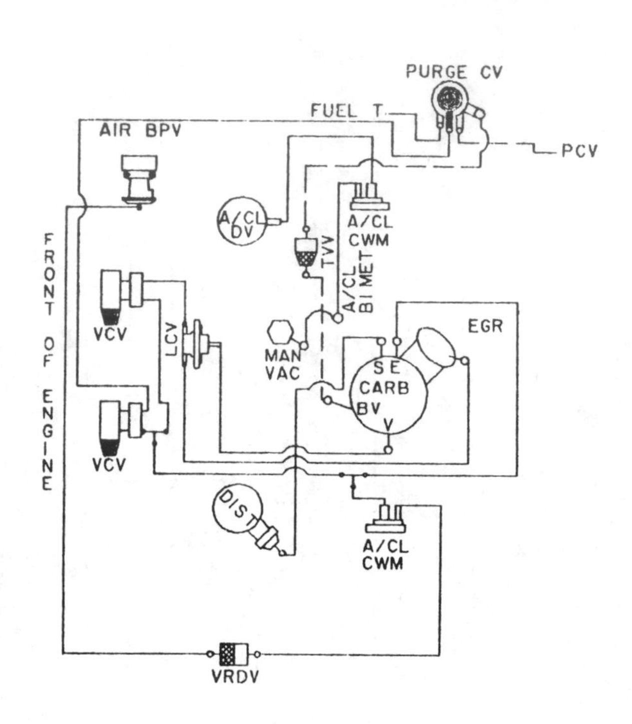

CALIBRATION: 9–9A–R14 DATE: 10–15–78
4.1L (250 CID)

CALIBRATION: 9–9P–RO DATE: 5–26–78
4.1L (250 CID)

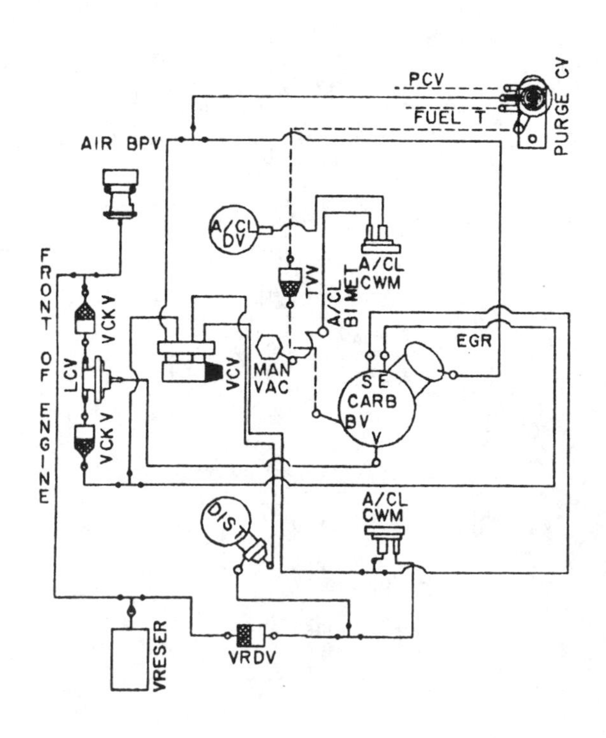

CALIBRATION: 9–9P–R10 DATE: 10–24–78
4.1L (250 CID)

CALIBRATION: 9–9P–R11 DATE: 10–25–78
4.1L (250 CID)

VACUUM CIRCUITS

1980—250 CID (4.1 L) engines

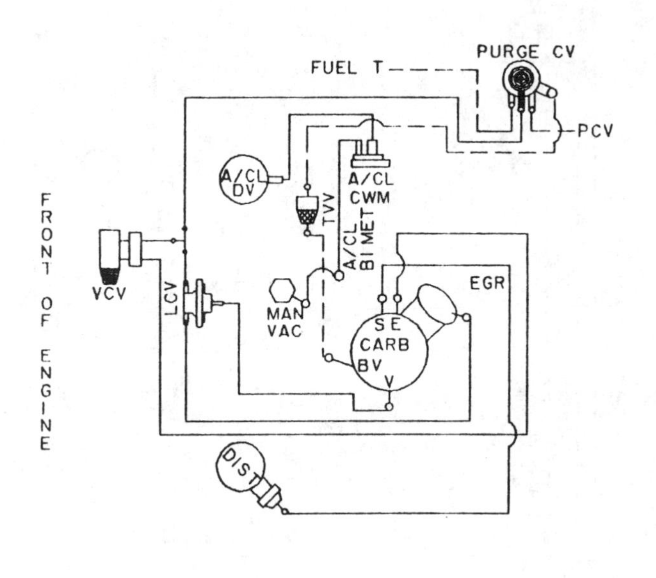

CALIBRATION: 9–29B–R11 DATE: 11–4–78
4.1L (250 CID)

PURGE CV
PCV
FUEL–T
TO CANISTER
A/CL DV
A/CL CWM
TVV
A/CL BI MET
FRONT OF ENGINE
AIR BPV
EGR
MAN VAC
SE
CARB
BV
V
VCV
DIST
A/CL CWM
LCV
VRDV
VRESER

CALIBRATION: 0–8A–RO DATE: 6–1–79
4.1L GRANADA/MONARCH 49S M/T

49S A/C & NON A/C

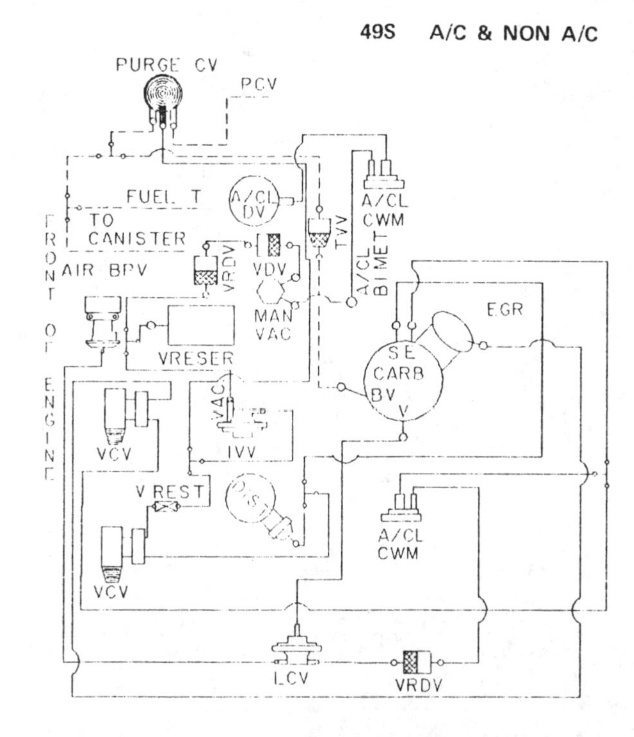

CALIBRATION: 0–9A–RO DATE: 6–1–79
4.1L GRANADA/MONARCH 49S A/T

49S A/C & NON A/C

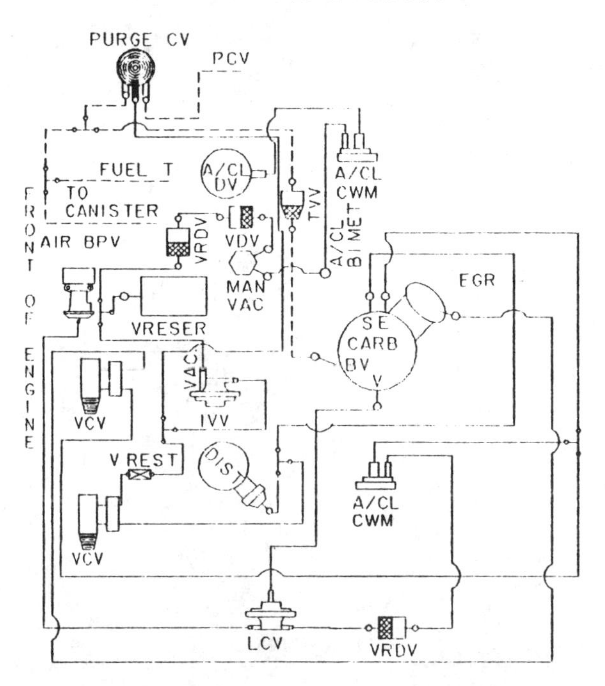

CALIBRATION: 0–9A–RO DATE: 6–1–79
4.1L GRANADA/MONARCH 49S A/T

VACUUM CIRCUITS

1980—250 CID (4.1 L) engines

49S CANADA A/C & NON A/C

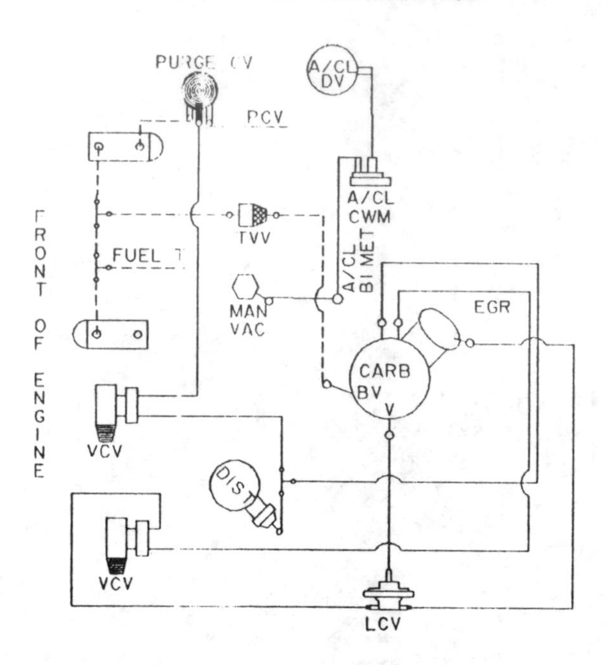

CALIBRATION: 0–29A–R0 **DATE:** 10–1–79
4.1L GRANADA/MONARCH 49S/ CANADA M/T

CALIF ONLY A/C & NON A/C

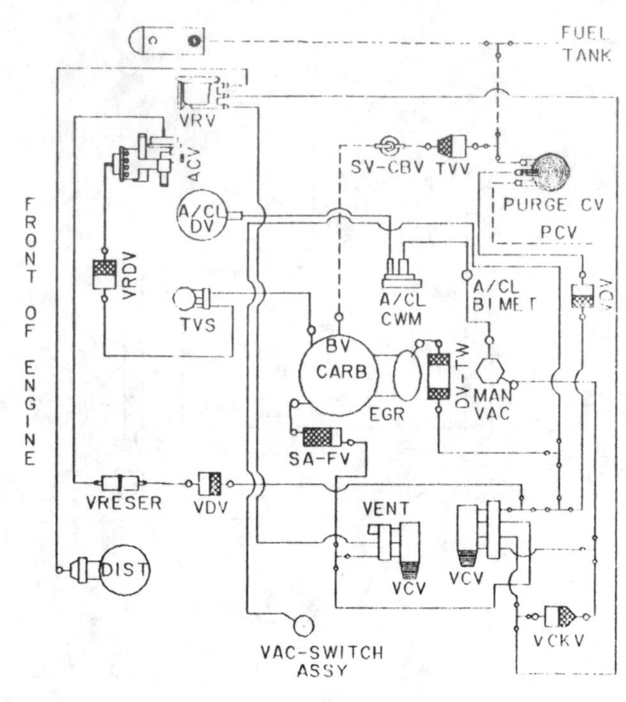

CALIBRATION: 0–16T–R13 **DATE:** 11–19–79
4.2L MUST/CAPRI CALIF

CALIF ONLY A/C & NON A/C

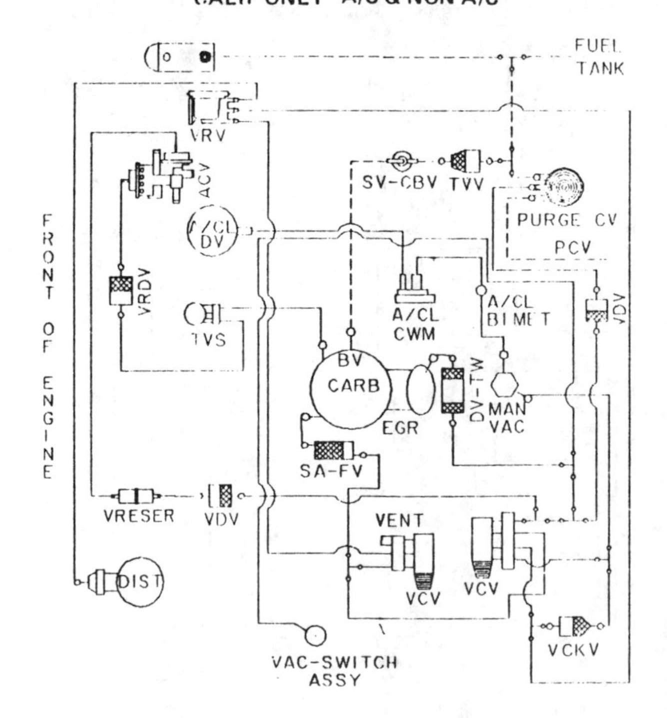

CALIBRATION: 0–16U–R13 **DATE:** 12–05–79
4.2L GRANADA/MONARCH CALIF

A/C & NON A/C

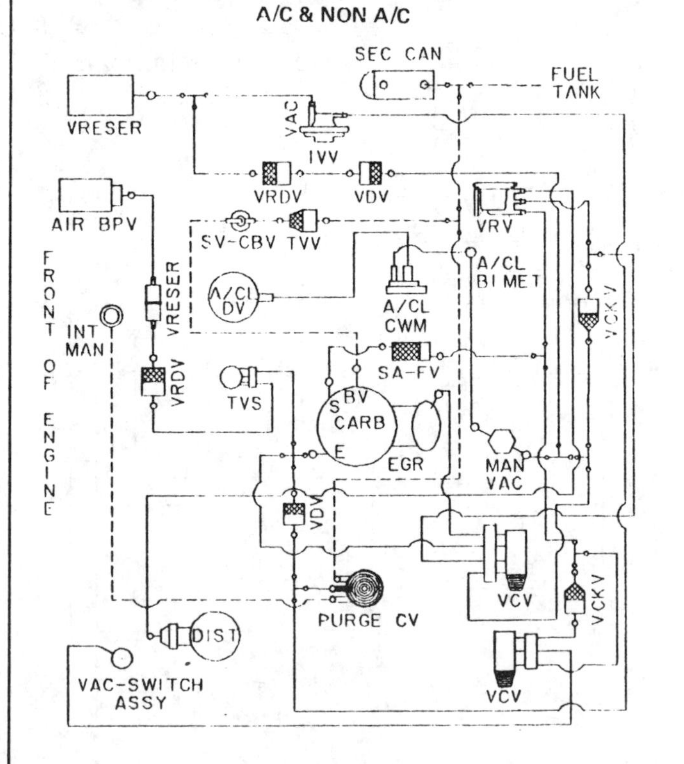

CÁLIBRATION: 0–16C–R19 **DATE:** 10–31–79
4.2L T-BIRD/XR-7

VACUUM CIRCUITS

1980—300 CID (4.9 L) engines

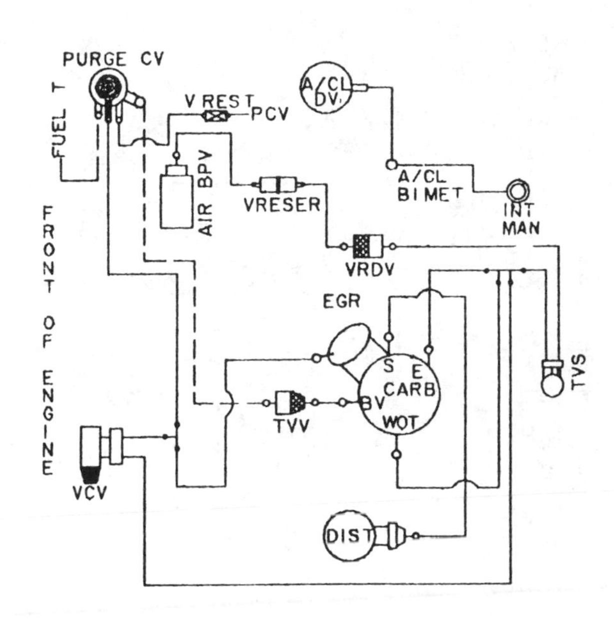

CALIBRATION: 9–51G–RO DATE: 5–8–78
4.9L (300 CID)

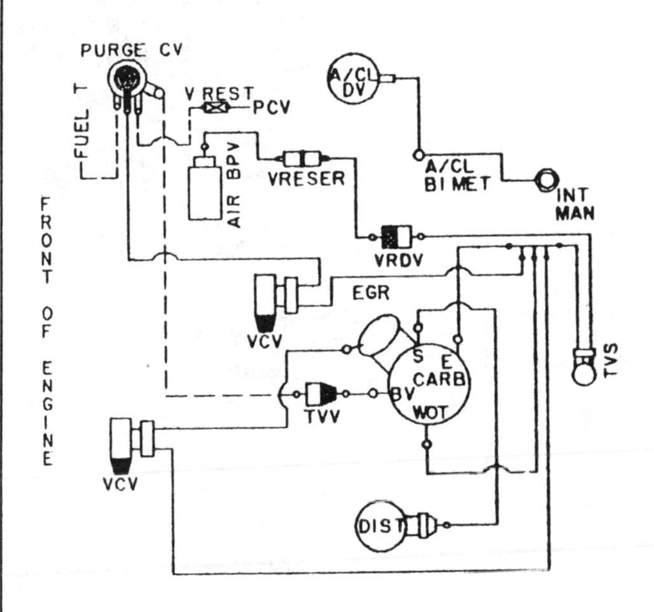

CALIBRATION: 9–51J–RO DATE: 6–13–78
4.9L (300 CID)

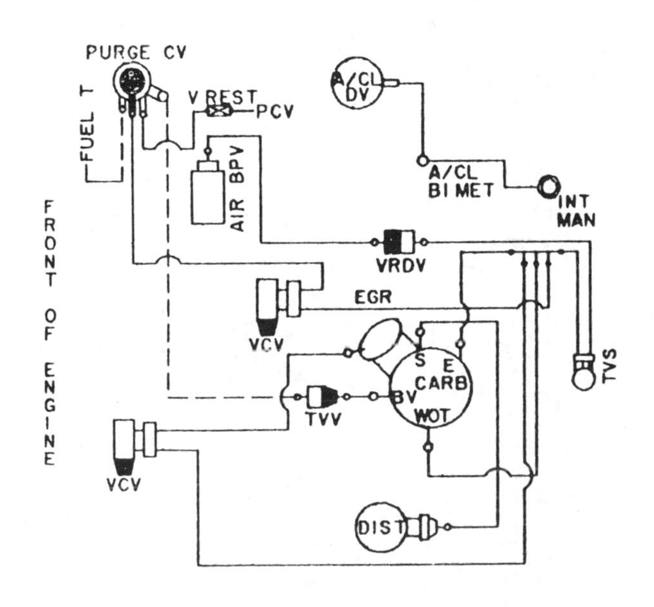

CALIBRATION: 9–51K–RO DATE: 5–26–78
4.9L (300 CID)

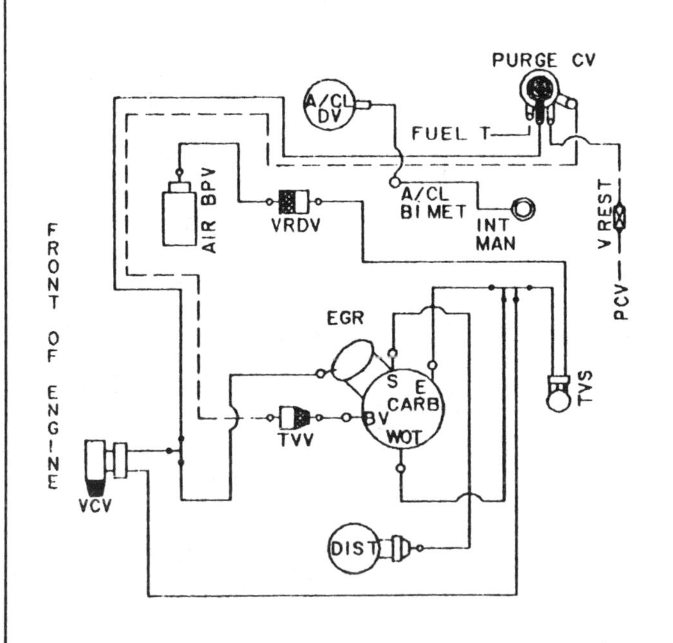

CALIBRATION. 9–51L–RO DATE: 5–8–78
4.9L (300 CID)

VACUUM CIRCUITS

1980—300 CID (4.9 L) engines

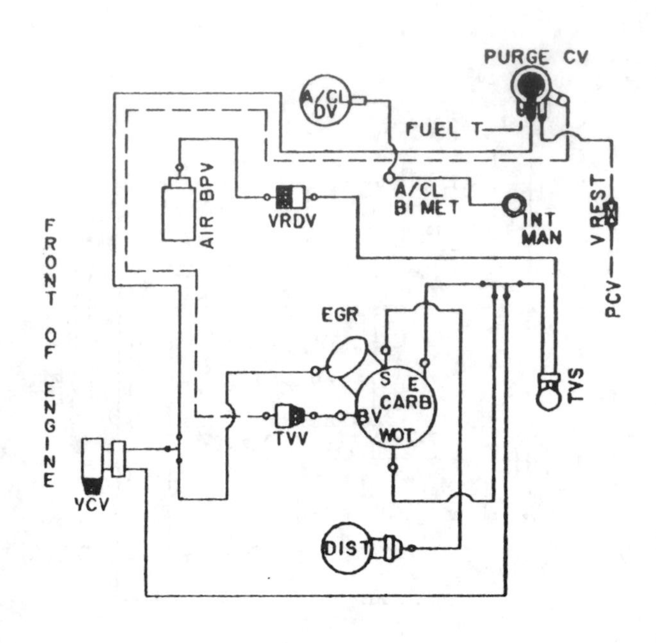

CALIBRATION: 9–51M–RO DATE: 6–16–78
4.9L (300 CID)

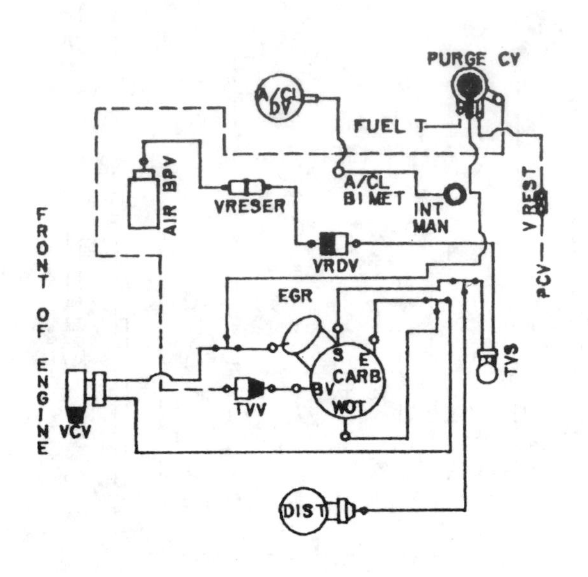

CALIBRATION: 9–51S–RO DATE: 5–11–78
4.9L (300 CID)

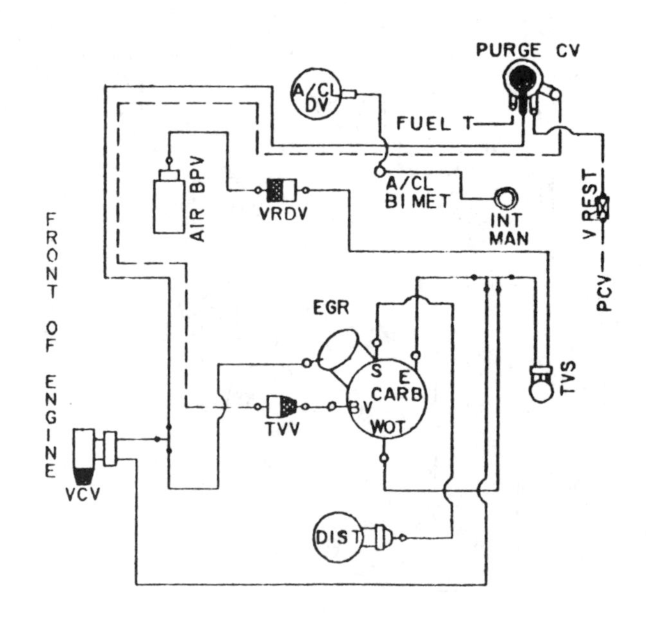

CALIBRATION: 9–51T–RO DATE: 6–6–78
4.9L (300 CID)

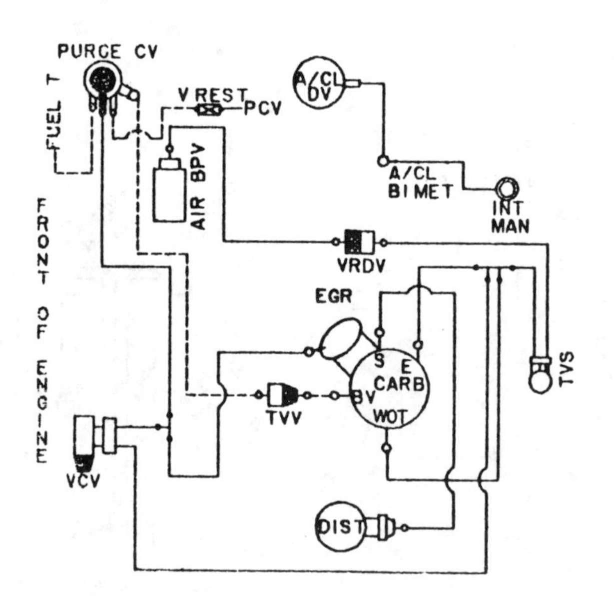

CALIBRATION: 9–52G–RO DATE: 5–8–78
4.9L (300 CID)

VACUUM CIRCUITS

1980—300 CID (4.9 L) engines

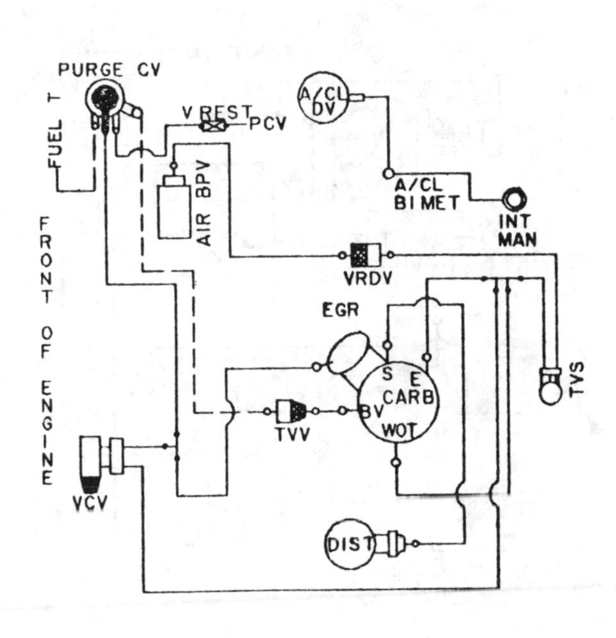

CALIBRATION: 9–52J–RO DATE: 6–5–78
4.9L (300 CID)

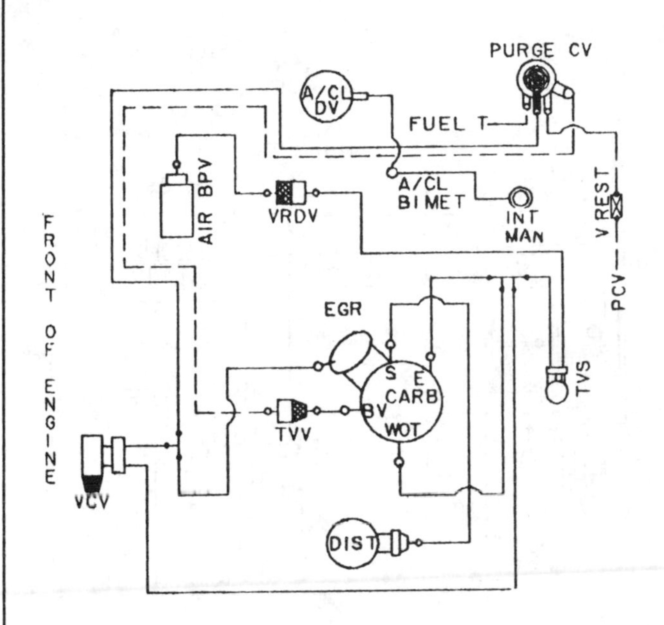

CALIBRATION: 9–52L–RO DATE: 5–8–78
4.9L (300 CID)

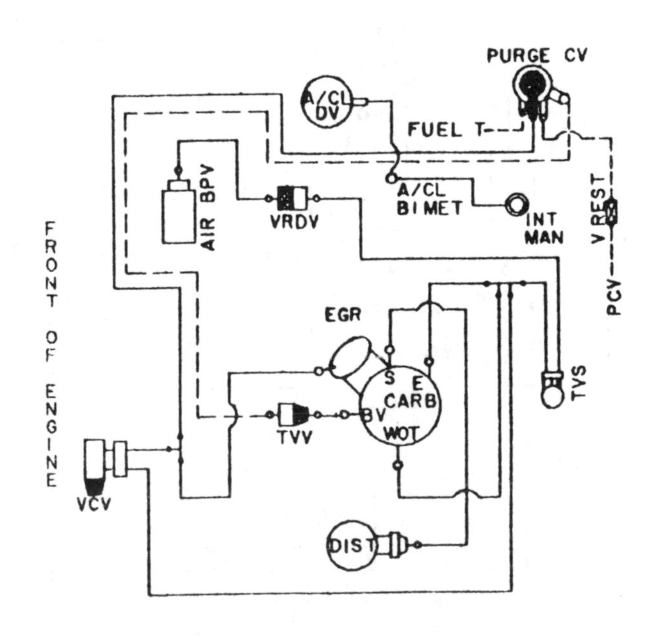

CALIBRATION: 9–52M–RO DATE: 6–12–78
4.9L (300 CID)

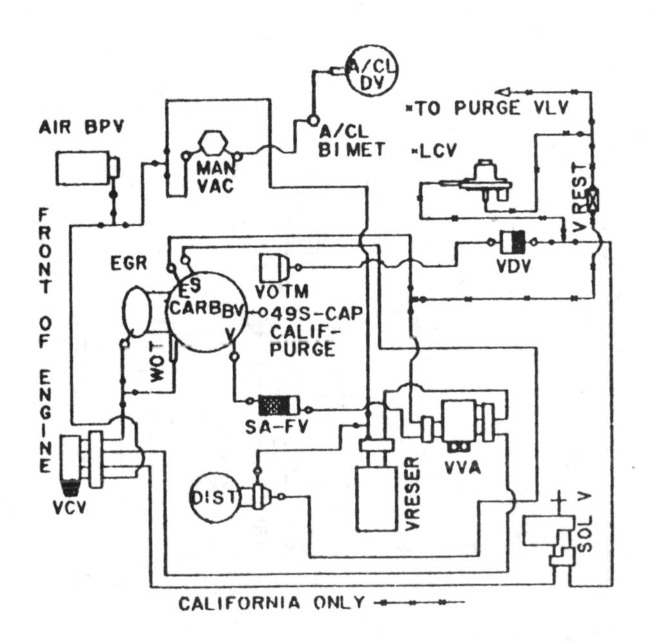

CALIBRATION: 9–77J–RO DATE: 5–3–78
4.9L (300 CID)

VACUUM CIRCUITS

1980—300 CID (4.9 L) engines

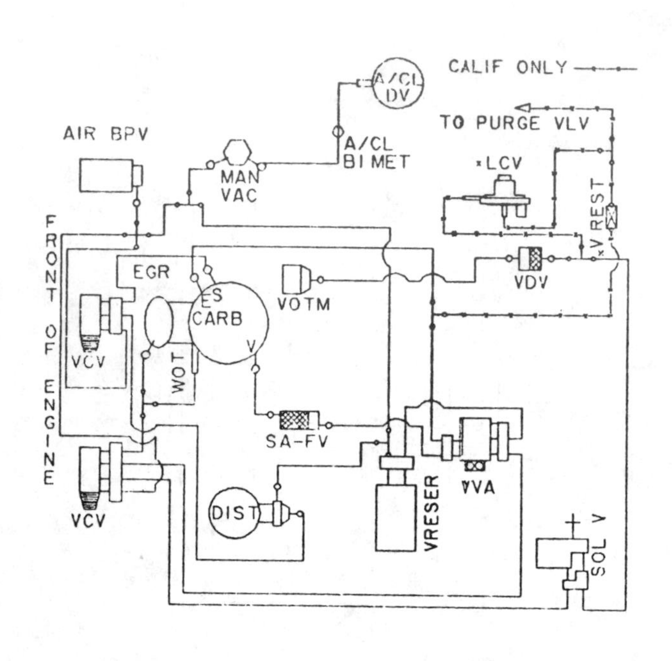

CALIBRATION: 9–77J–R11 DATE: 11–13–78
4.9L (300 CID)

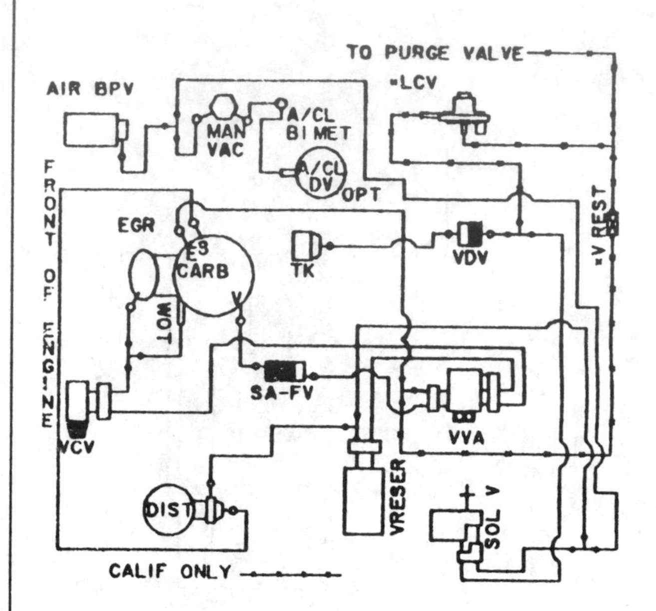

CALIBRATION: 9–77M–RO DATE: 6–12–78
4.9L (300 CID)

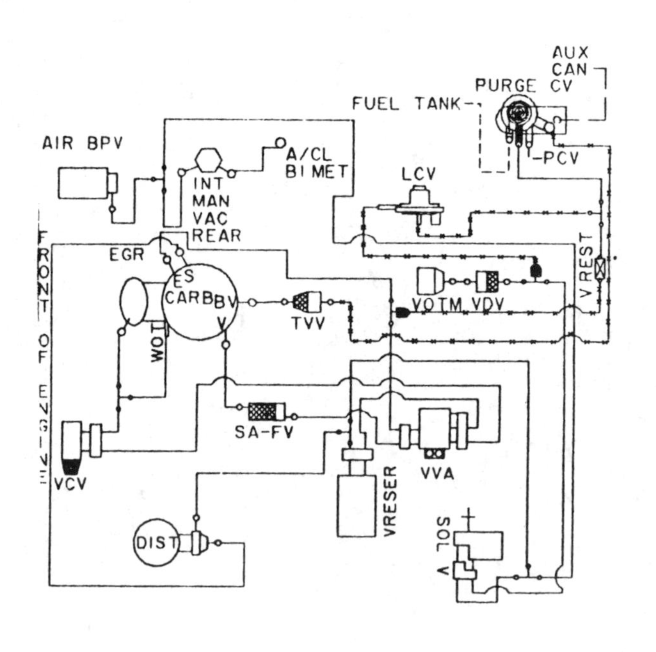

CALIBRATION: 9–77M–R10 DATE: 10–30–78
4.9L (300 CID)

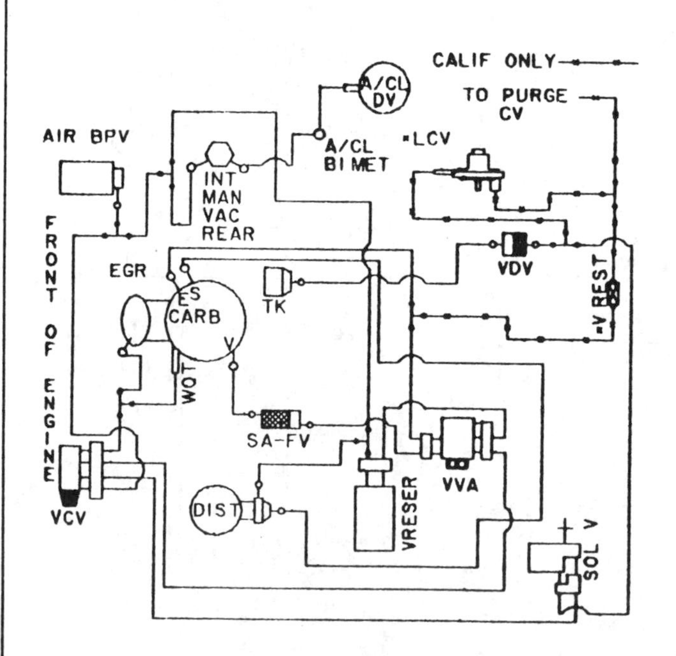

CALIBRATION: 9–78J–RO DATE: 5–3–78
4.9L (300 CID)

VACUUM CIRCUITS

1980—302 CID (5.0 L) engines

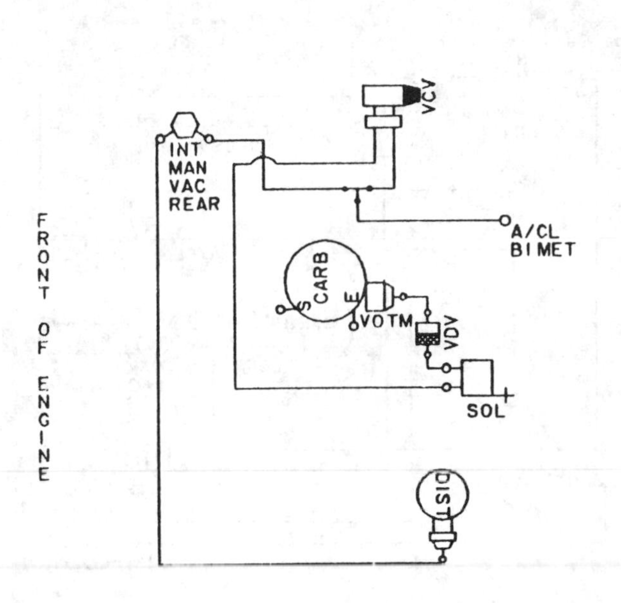

CALIBRATION: * 7–79–R1 DATE: 8–7–78
5.0L (302 CID)

VCV
INT
MAN
VAC
REAR
A/CL
BI MET
CARB
S
E
VOTM
SOL
FRONT OF ENGINE
LSIC

CALIBRATION: * 7–80–RO DATE: 4–25–78
5.0L (302 CID)

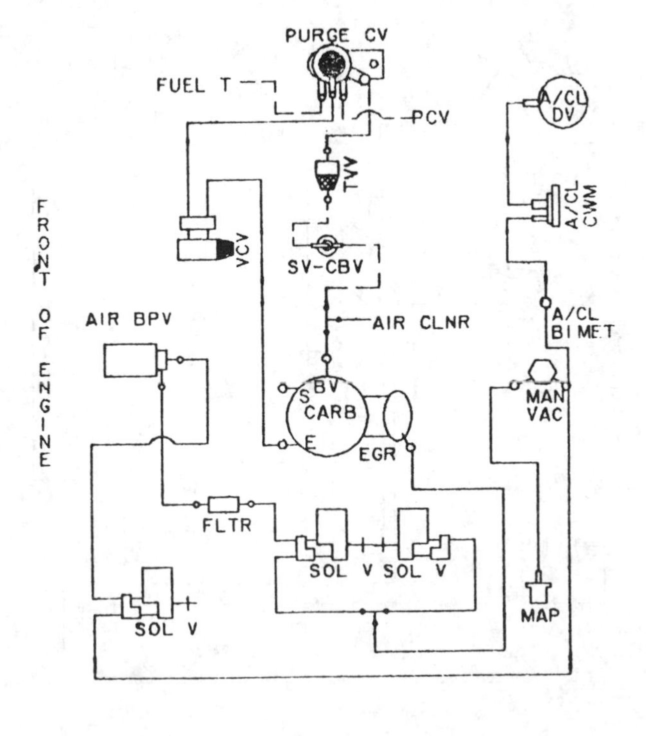

CALIBRATION: * 8–11D–RO DATE: 6–7–78
5.0L (302 CID)

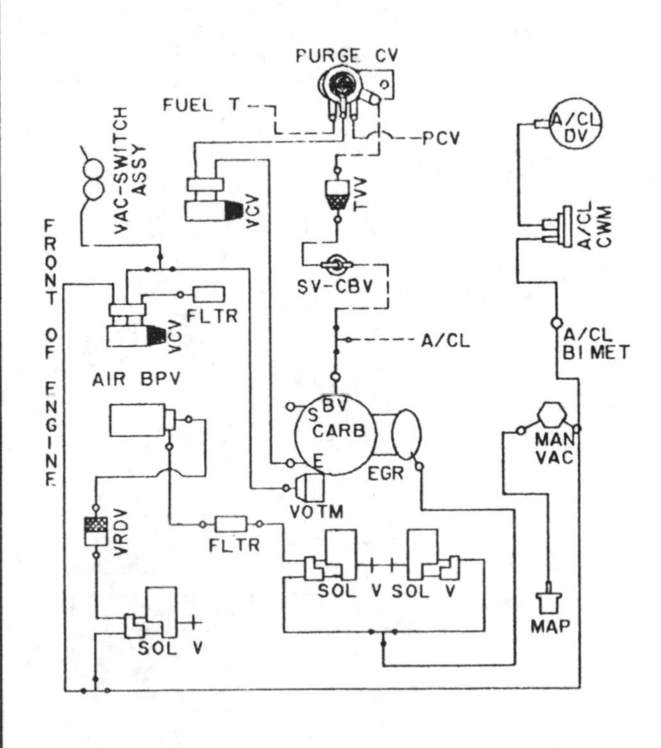

CALIBRATION: * 8–11L–R11 DATE: 5–23–78
5.0L (302 CID)

VACUUM CIRCUITS

1980—302 CID (5.0 L) engines

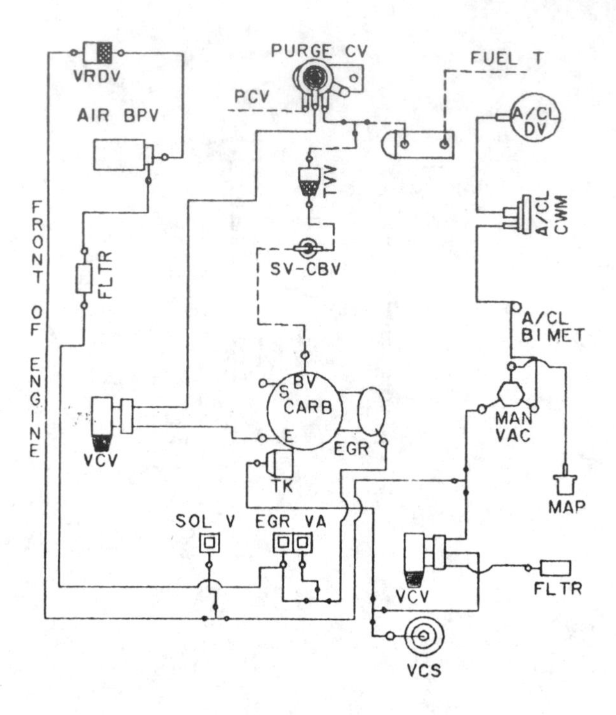

CALIBRATION: *8-11L-R20 DATE: 10-16-78
5.0L (302 CID)

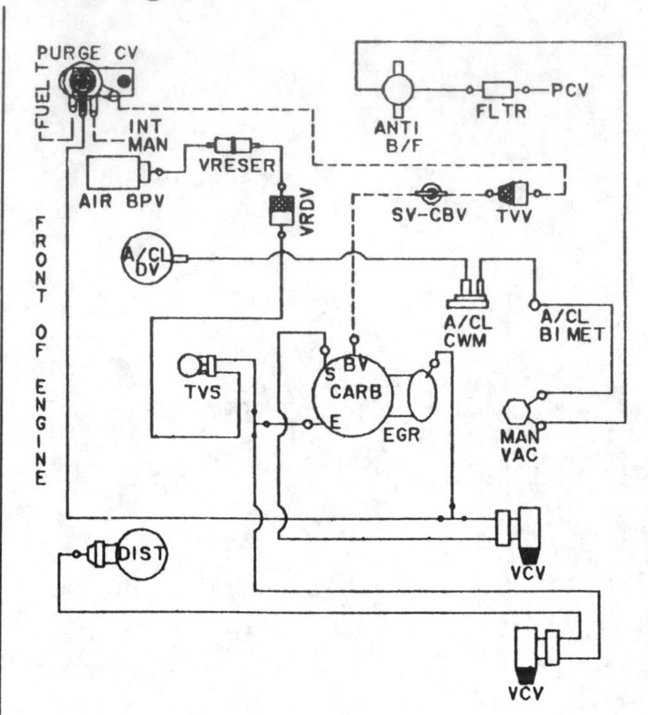

CALIBRATION: 9-10A-RO DATE: 6-7-78
5.0 (302 CID)

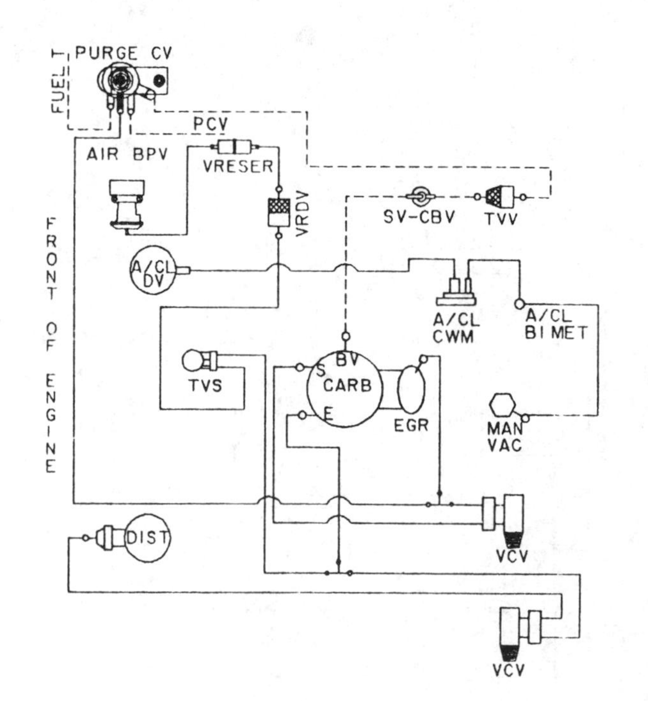

CALIBRATION: 9-10B-RO DATE: 1-18-79
5.0L (302 CID)

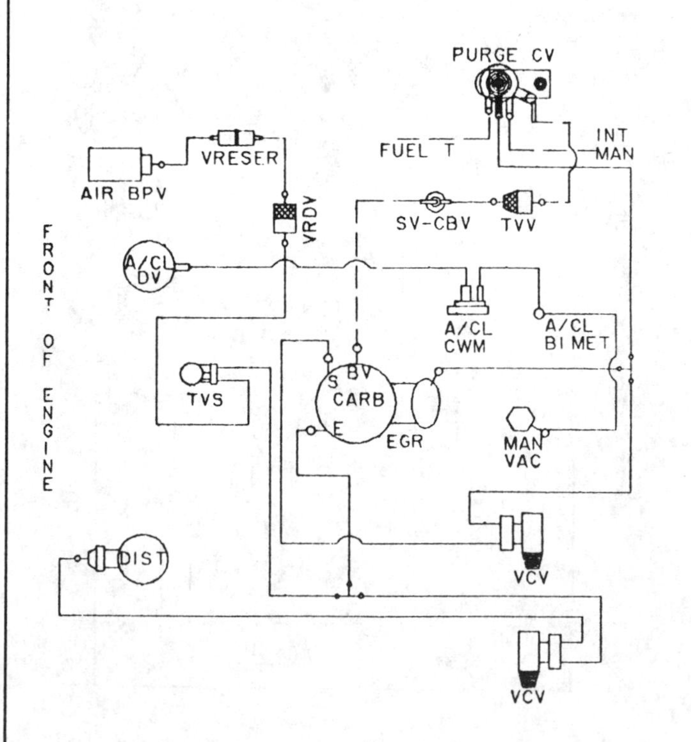

CALIBRATION: 9-10C-RO DATE: 6-15-78
5.0L (302 CID)

VACUUM CIRCUITS

1980—302 CID (5.0 L) engines

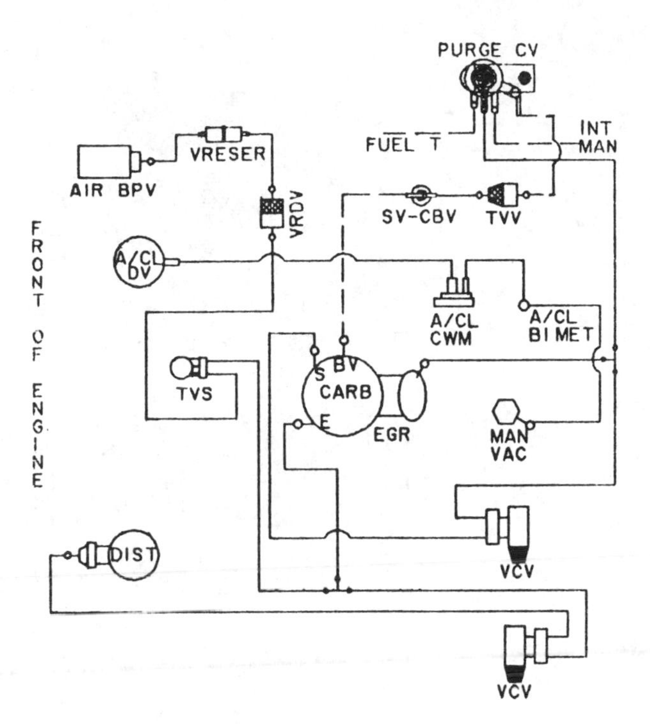

CALIBRATION: 9–10C–R11 DATE: 9–18–78
5.0L (302 CID)

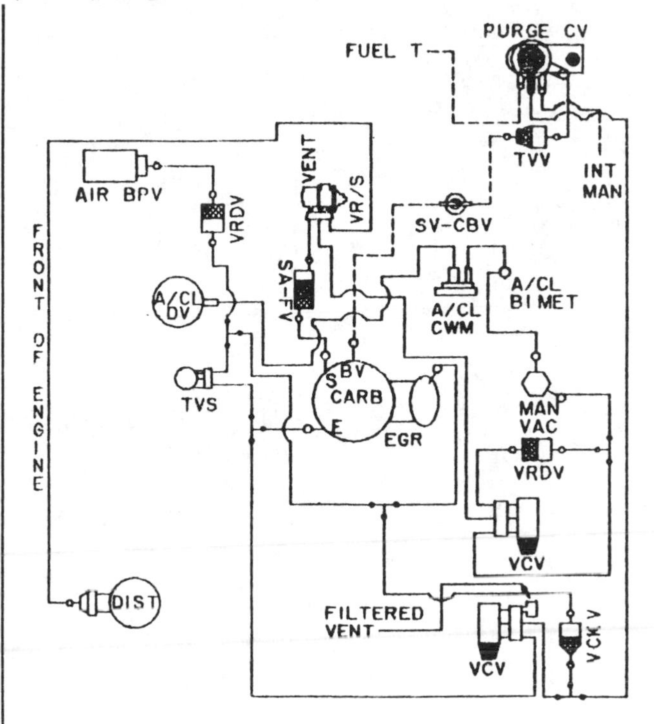

CALIBRATION: 9–11A–RO DATE: 6–7–78
5.0L (302 CID)

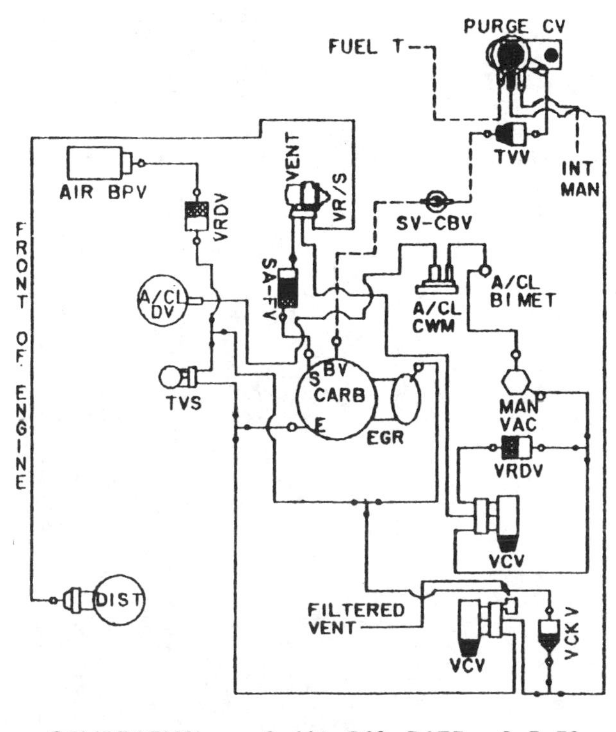

CALIBRATION: 9–11A–R10 DATE: 6–7–78
5.0L (302 CID)

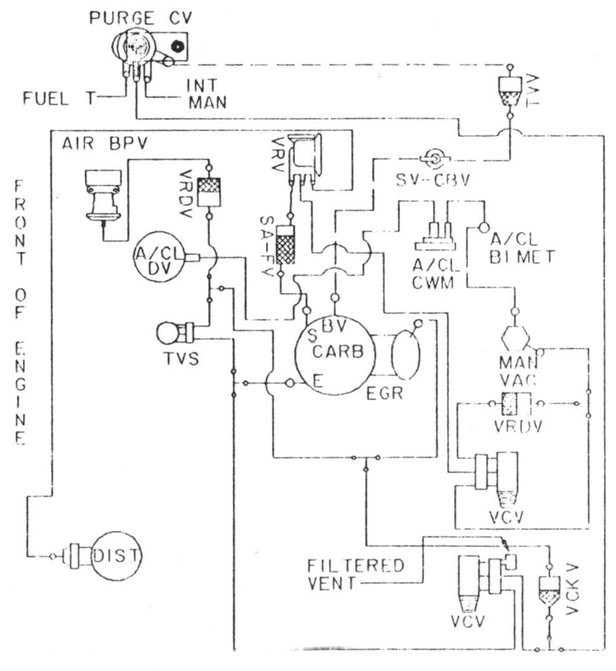

CALIBRATION: 9–11A–R11 DATE: 4–9–79
5.0L (302 CID)

VACUUM CIRCUITS

1980—302 CID (5.0 L) engines

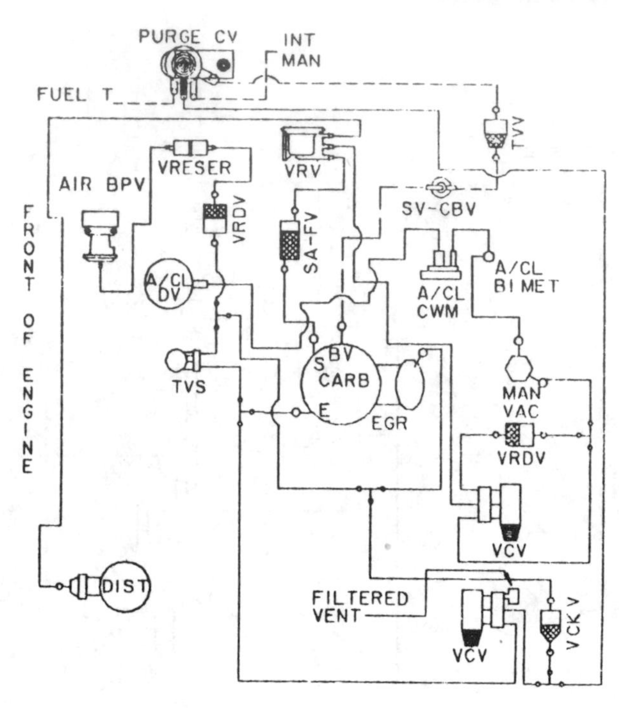

CALIBRATION: 9–11A–R20 **DATE:** 10–10–78
5.0L (302 CID)

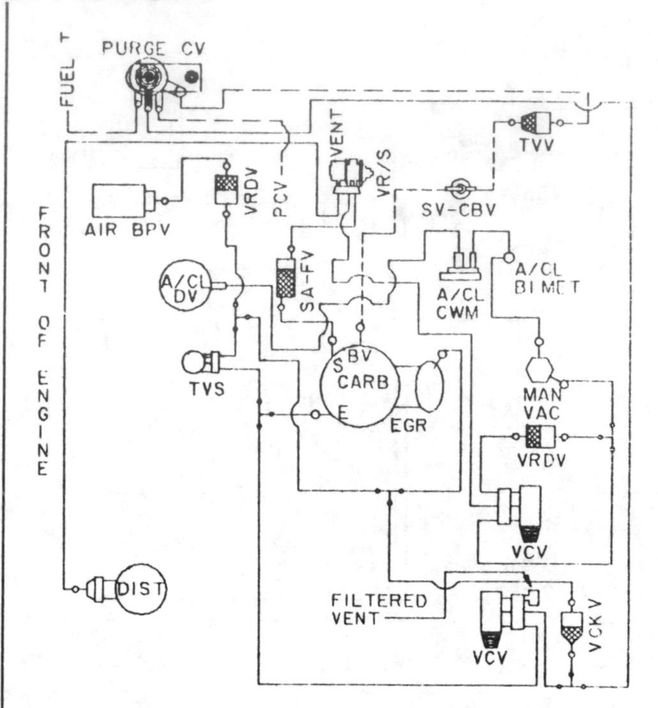

CALIBRATION: 9–11B–RO **DATE:** 6–12–78
5.0L (302 CID)

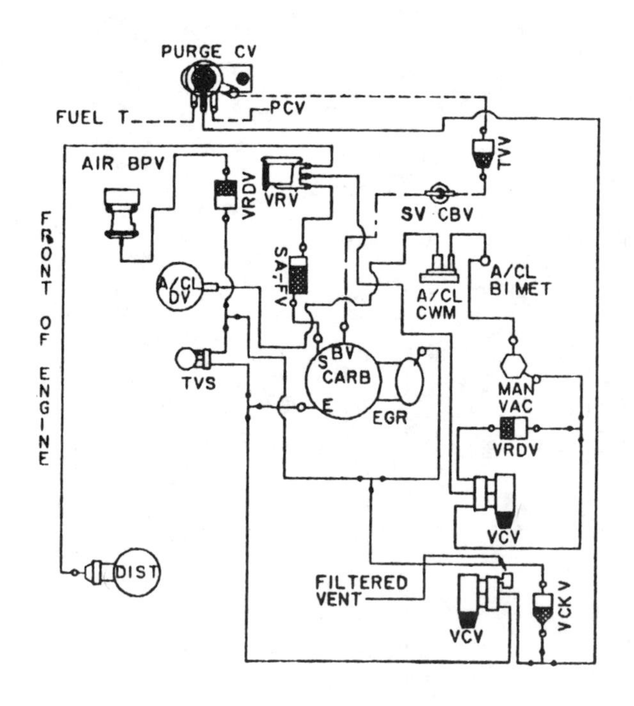

CALIBRATION: 9–11B–R10 **DATE:** 9–28–78
5.0L (302 CID)

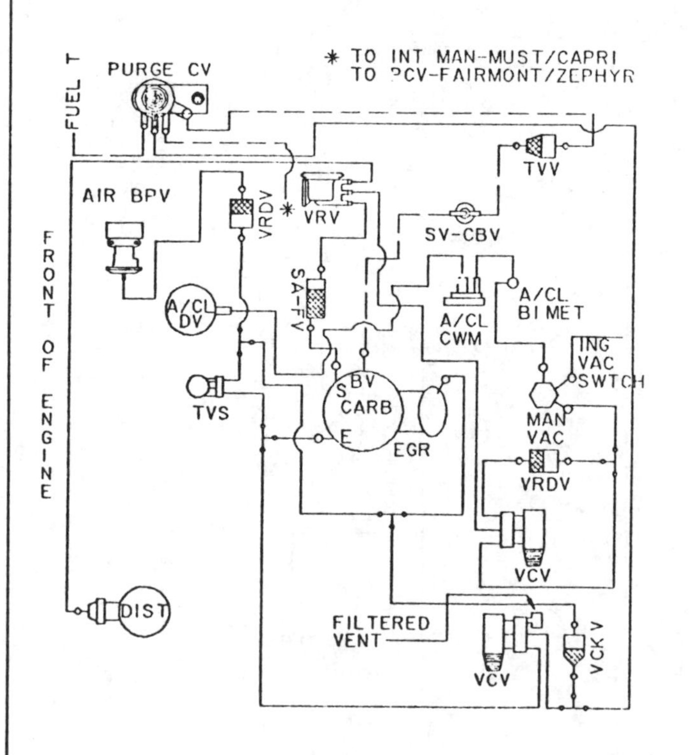

CALIBRATION: 9–11B–R11 **DATE:** 11–28–78
5.0L (302 CID)

VACUUM CIRCUITS

1980—302 CID (5.0 L) engines

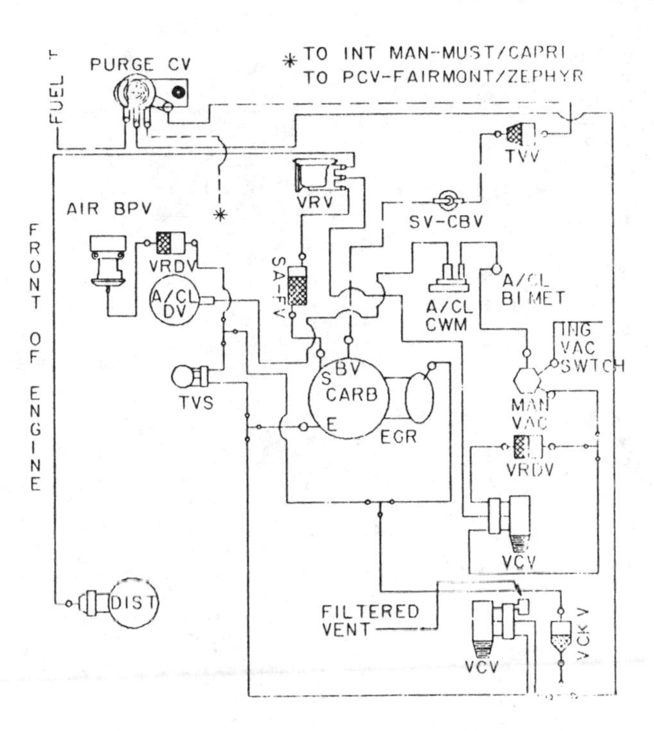

CALIBRATION: 9–11B–R12 DATE: 11–28–78
5.0L (302 CID)

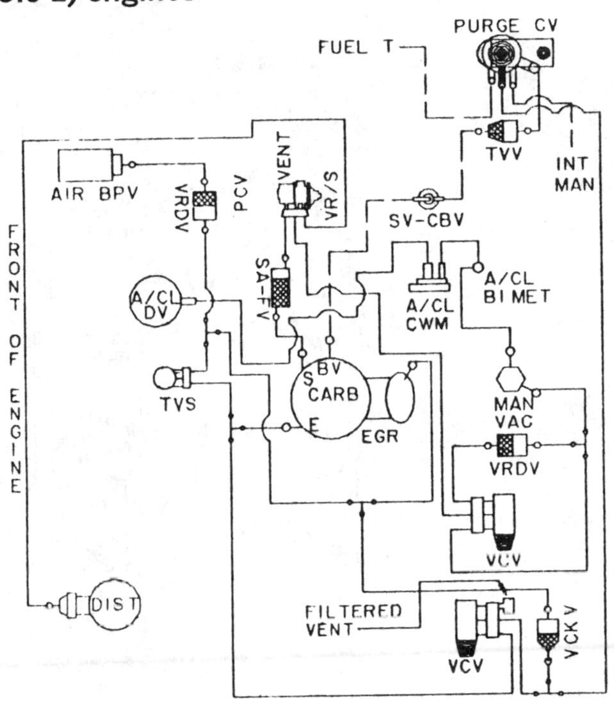

CALIBRATION: 9–11C–R1 DATE: 6–27–78
5.0L (302 CID)

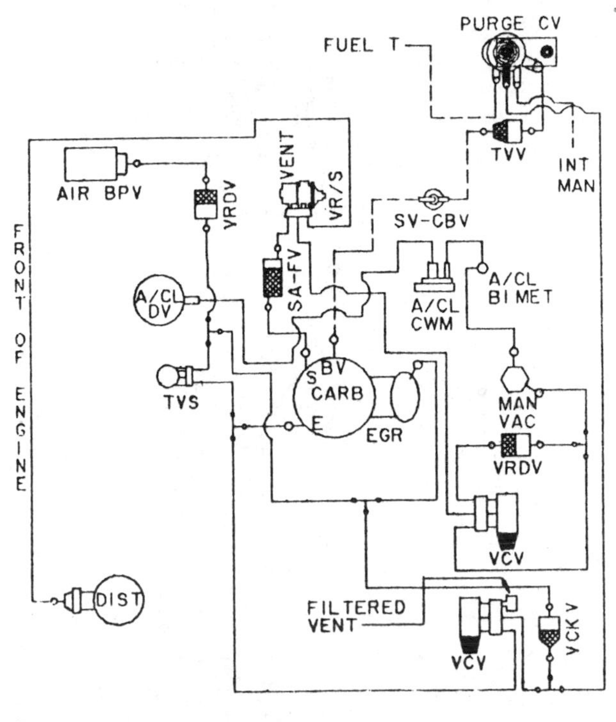

CALIBRATION. 9–11C–R10 DATE: 7–13–78
5.0L (302 CID)

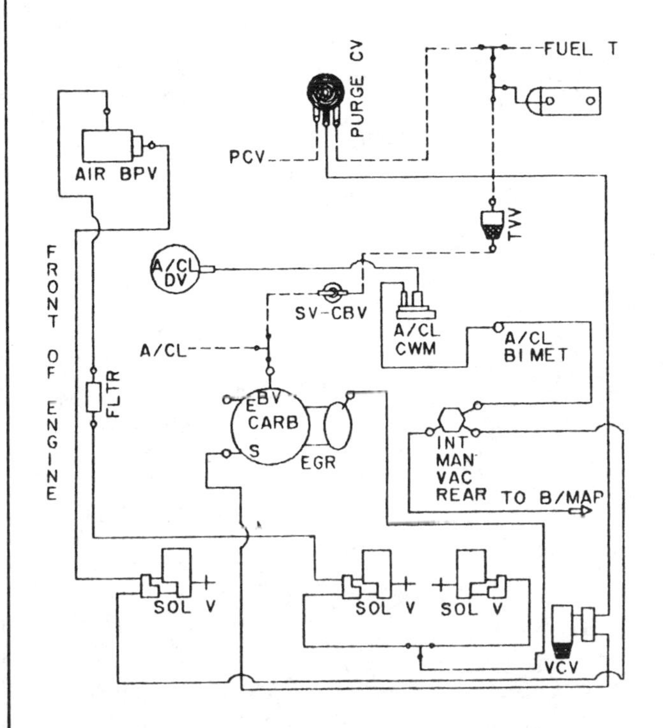

CALIBRATION: 9–11D–R11 DATE: 10–26–78
5.0L (302 CID)

VACUUM CIRCUITS

1980—302 CID (5.0 L) engines

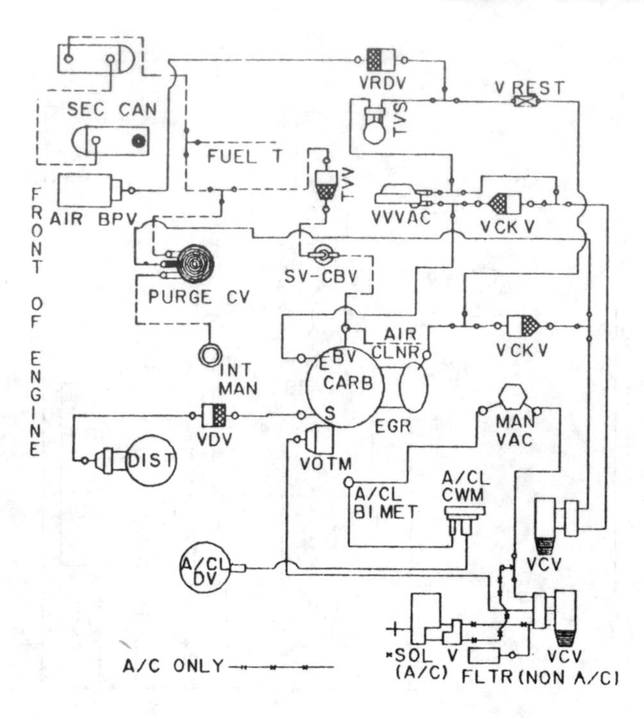

CALIBRATION: 9–11E–R1 DATE: 8–14–78
5.0L (302 CID)

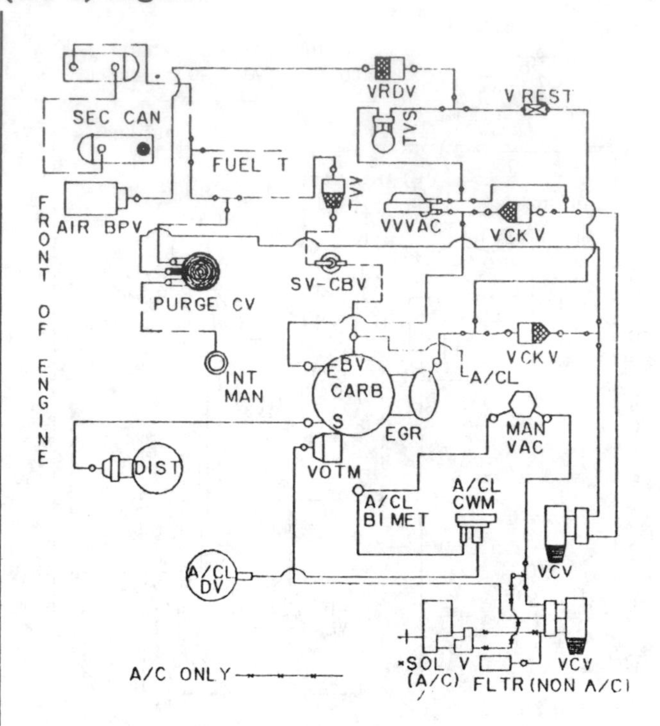

CALIBRATION: 9–11E–R10 DATE: 6–24–78
5.0L (302 CID)

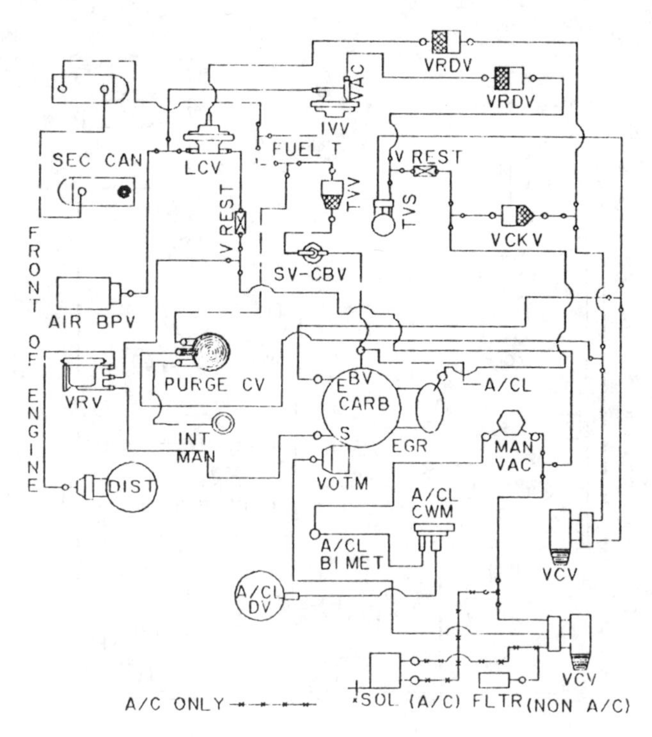

CALIBRATION: 9–11E–R12 DATE: 10–9–78
5.0L (302 CID)

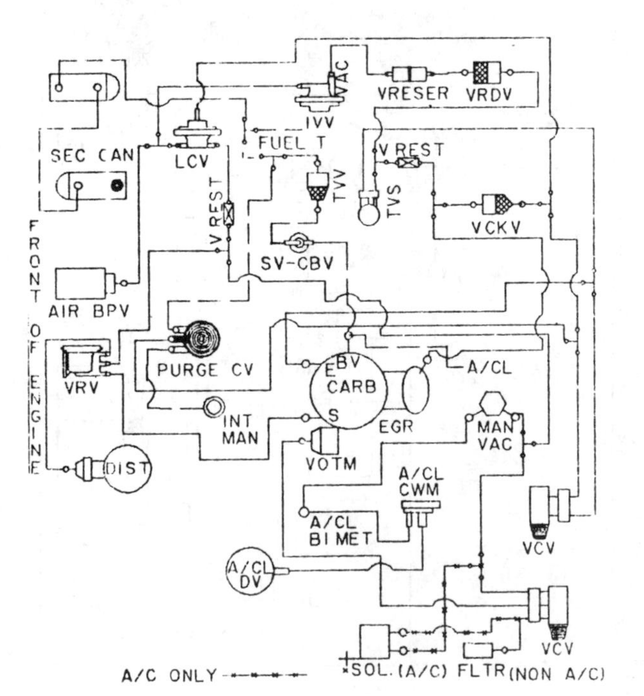

CALIBRATION: 9–11E–R13 DATE: 1–26–79
5.0L (302 CID)

VACUUM CIRCUITS

1980—302 CID (5.0 L) engines

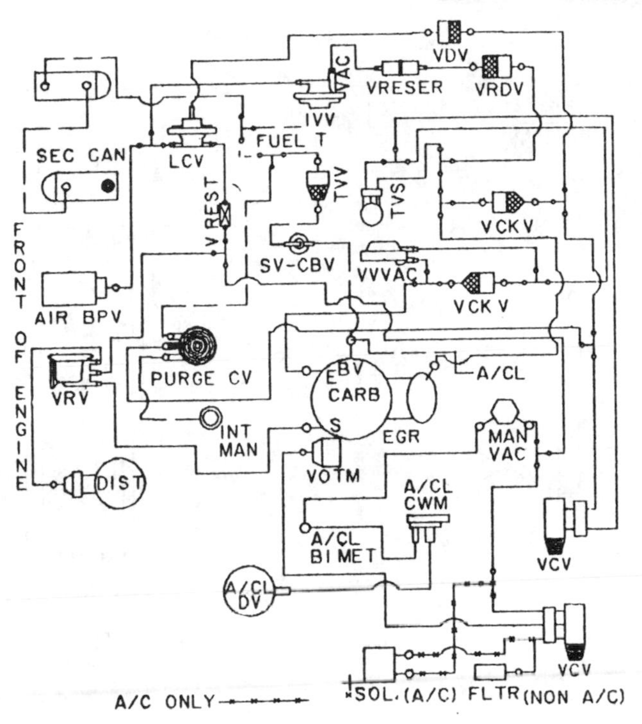

CALIBRATION: 9-11F-R1 DATE: 6-14-78
5.0L (302 CID)

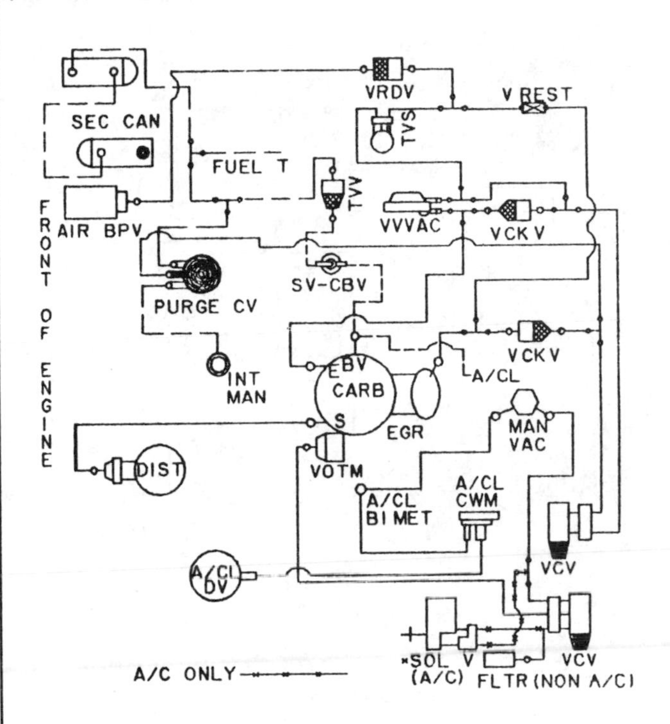

CALIBRATION: 9-11H-R10 DATE: 9-29-78
5.0L (302 CID)

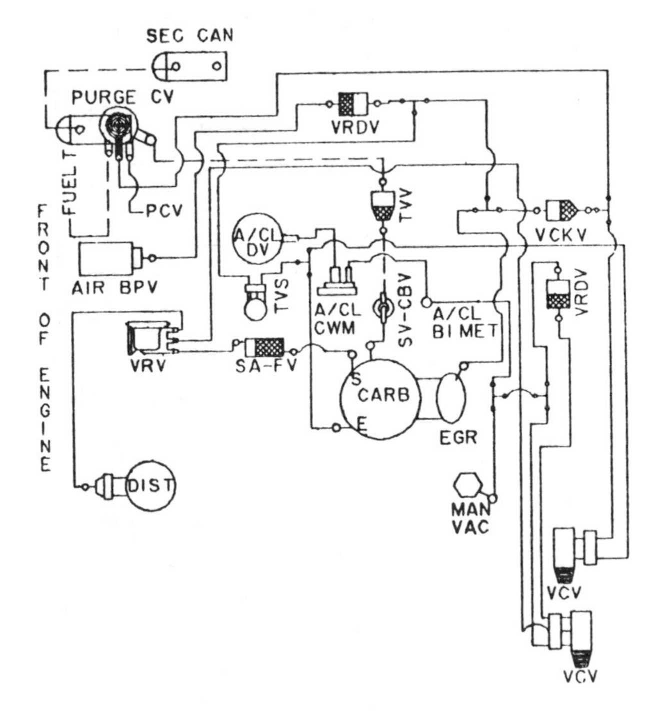

CALIBRATION: 9-11J-RO DATE: 6-8-78
5.0L (302 CID)

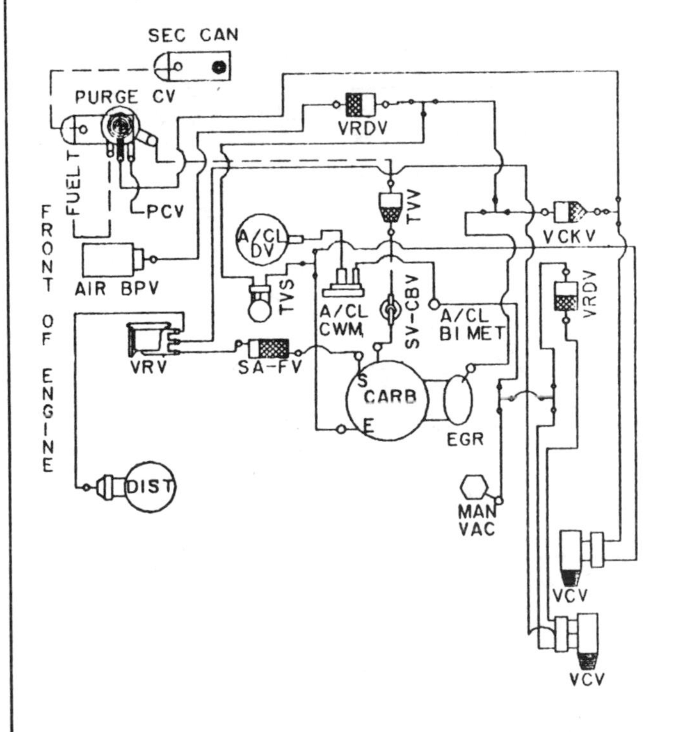

CALIBRATION: 9-11J-R10 DATE: 6-15-78
5.0L (302 CID)

VACUUM CIRCUITS

1980—302 CID (5.0 L) engines

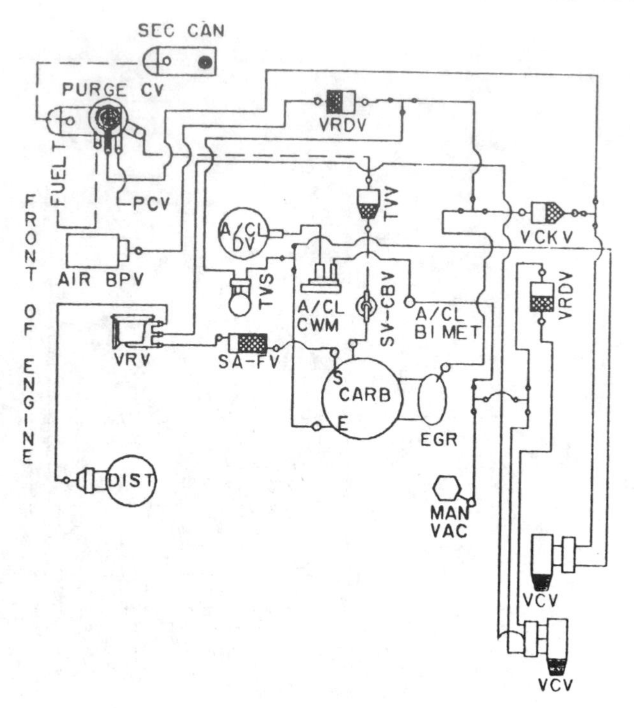

CALIBRATION: 9–IIJ–R13 DATE: 7–20–78
5.0L (302 CID)

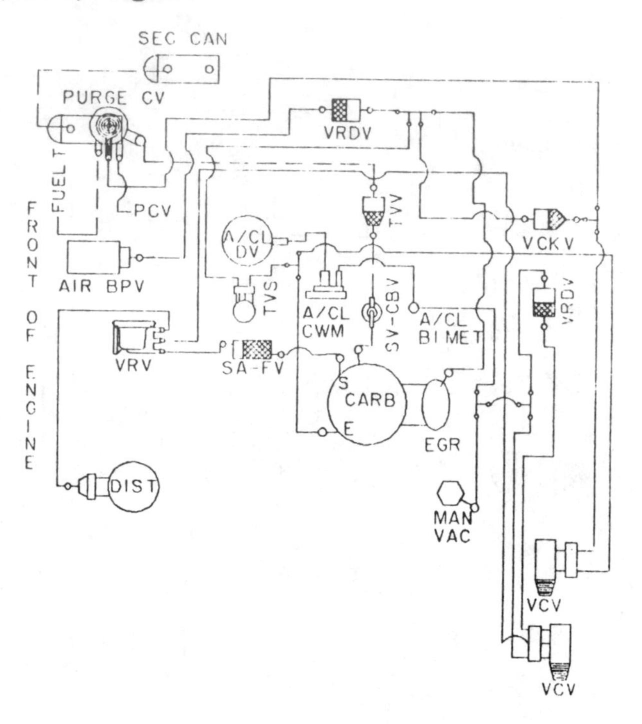

CALIBRATION: 9–11J–R20 DATE: 11–21–78
5.0L (302 CID)

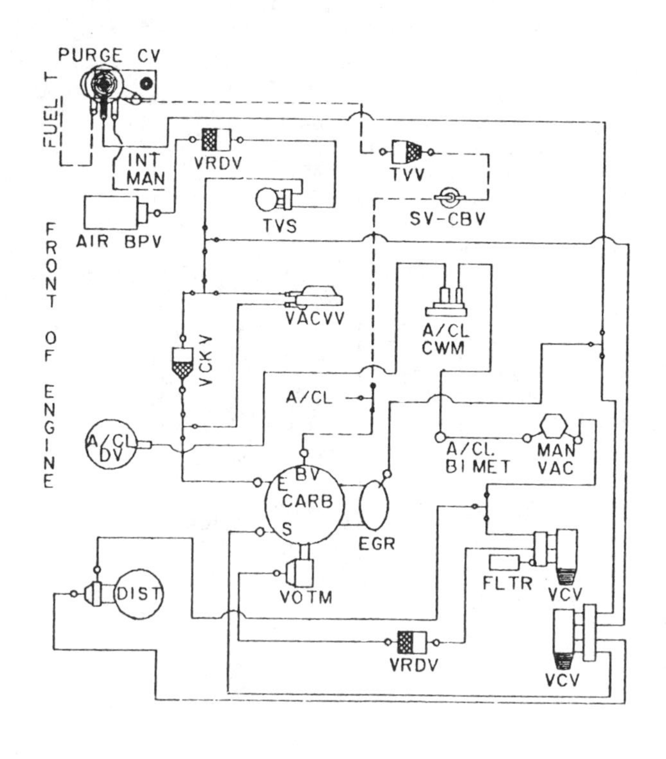

CALIBRATION: 9–11N–RO DATE: 6–7–78
5.0L (302 CID)

PURGE CV

* TO INT MAN-MUSTANG/CAPRI
TO PCV-FAIRMONT/ZEPHYR

FUEL T
VRDV
TVV
*
TVS
SV-CBV
FRONT OF ENGINE
AIR BPV
VACVV
VCKV
A/CL CWM
A/CL
A/CL DV
A/CL BIMET
MAN VAC
E
BV
CARB
S
EGR
DIST
VOTM
FLTR
VCV
VRDV
VCV

CALIBRATION: 9–11N–R10 DATE: 1–12–79
5.0L (302 CID)

VACUUM CIRCUITS

1980—302 CID (5.0 L) engines

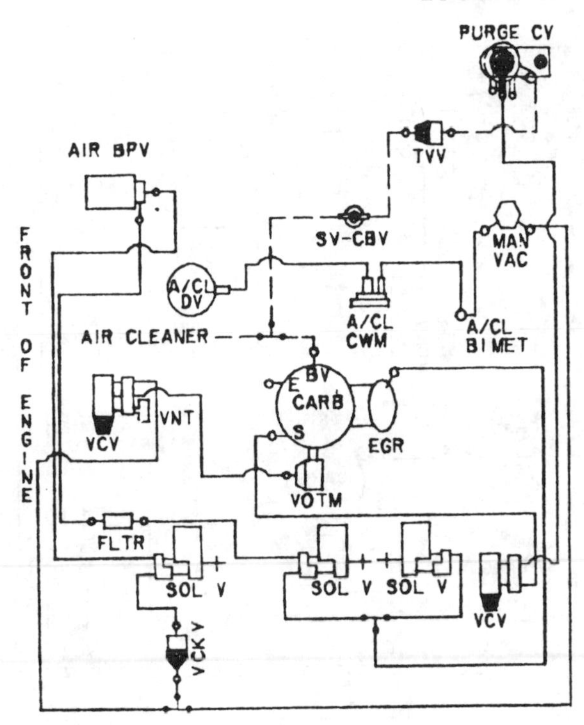

CALIBRATION: 9–11P–RO DATE: 5–24–78
5.0L (302 CID)

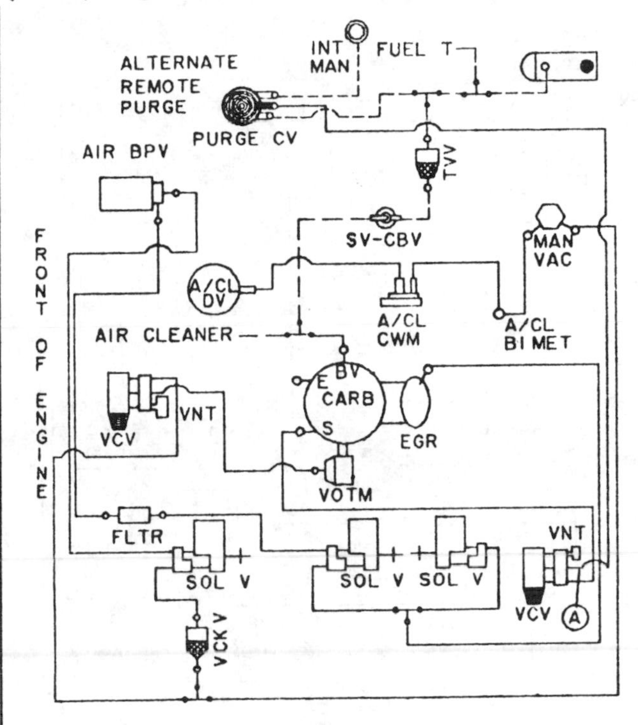

CALIBRATION: 9–11P–R2 DATE: 8–24–78
5.0L (302 CID)

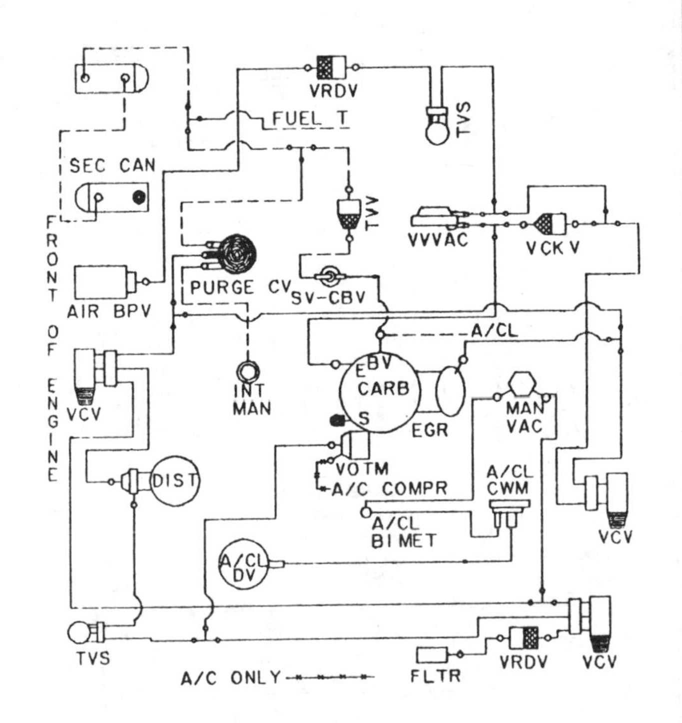

CALIBRATION: 9–11Q–R1 DATE: 6–15–78
5.0L (302 CID)

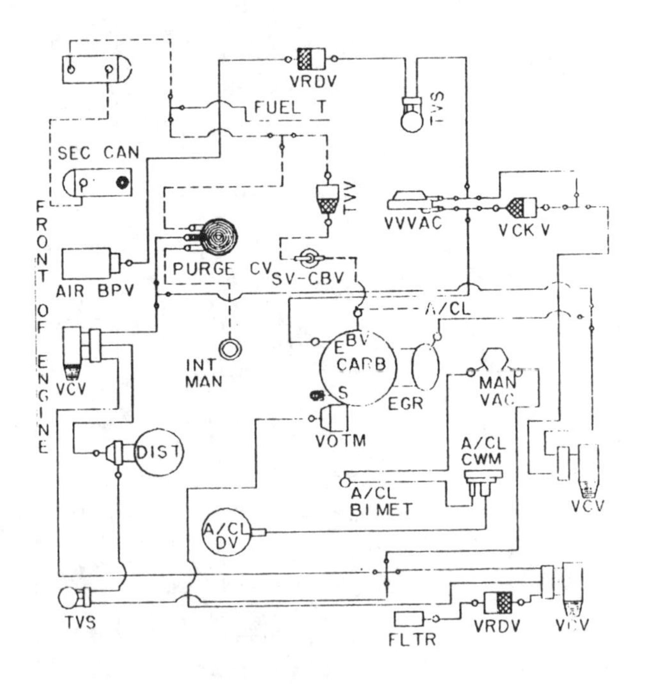

CALIBRATION: 9–11Q–R15 DATE: 1–17–79
5.0L (302 CID)

VACUUM CIRCUITS

1980—302 CID (5.0 L) engines

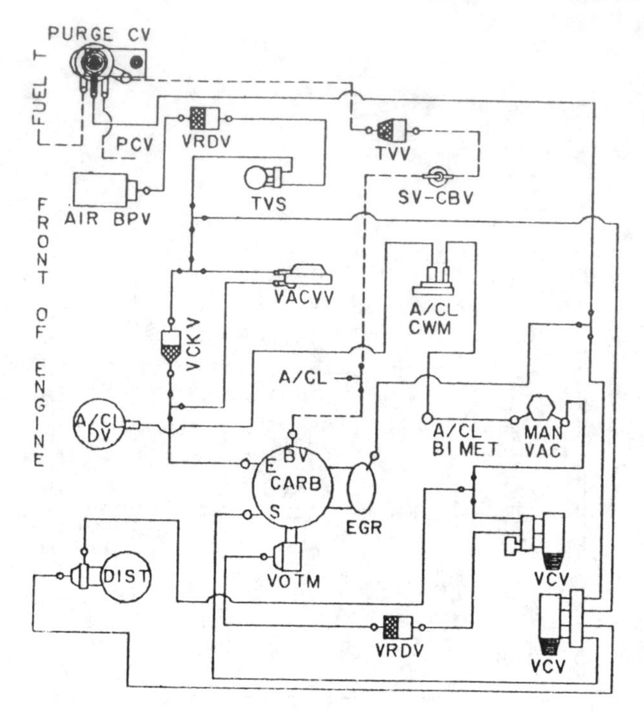

CALIBRATION: 9–11R–RO DATE: 6–7–78
5.0L (302 CID)

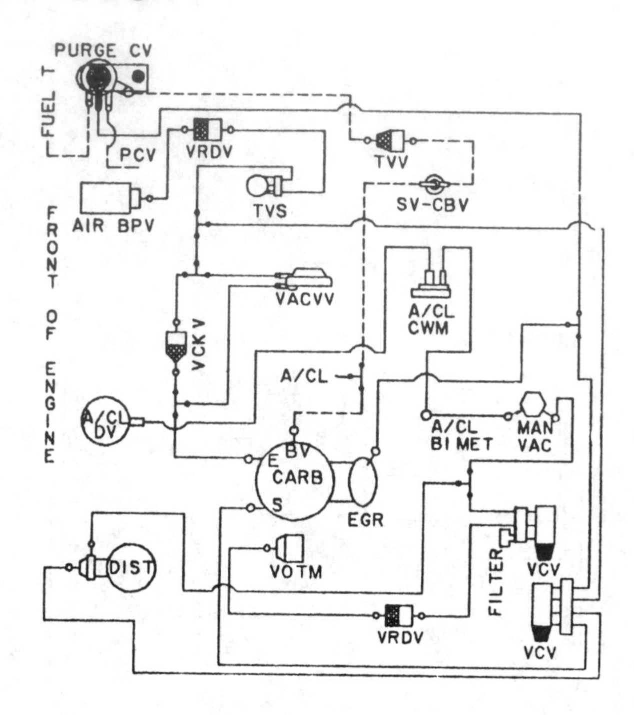

CALIBRATION: 9–11R–R10 DATE: 10–26–78
5.0L (302 CID)

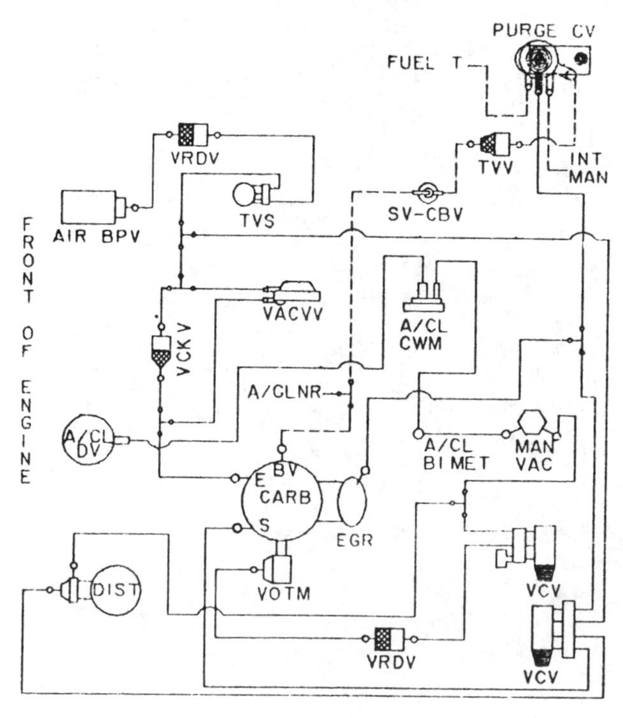

CALIBRATION: 9–11S–RO DATE: 6–7–78
5.0L (302 CID)

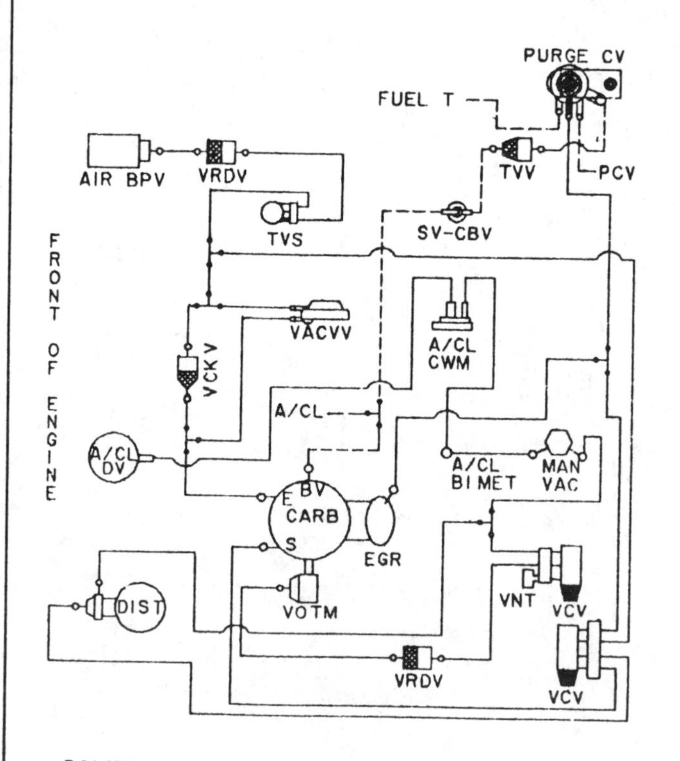

CALIBRATION: 9–11V–RO DATE: 6–8–78
5.0L (302 CID)

VACUUM CIRCUITS

1980—302 CID (5.0 L) engines

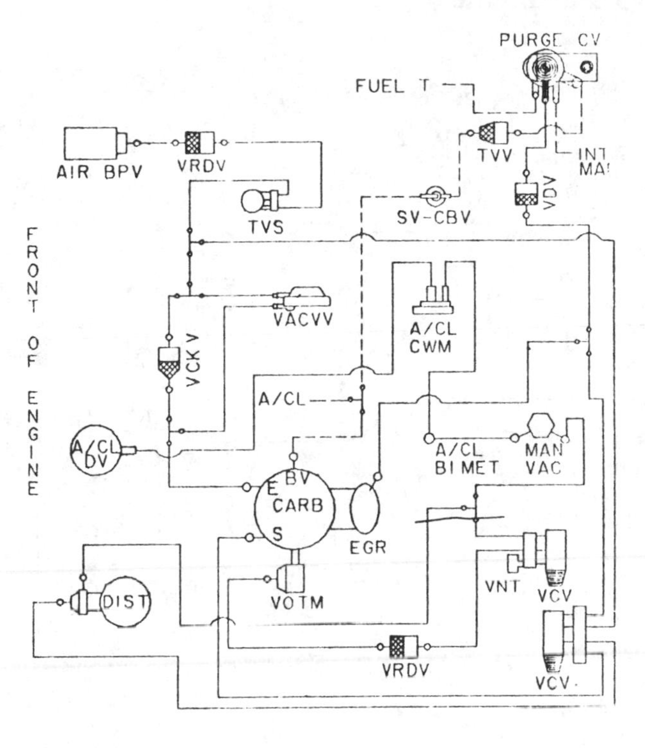

CALIBRATION: 9–IIV–R10 DATE: 8–4–78
5.0L (302 CID)

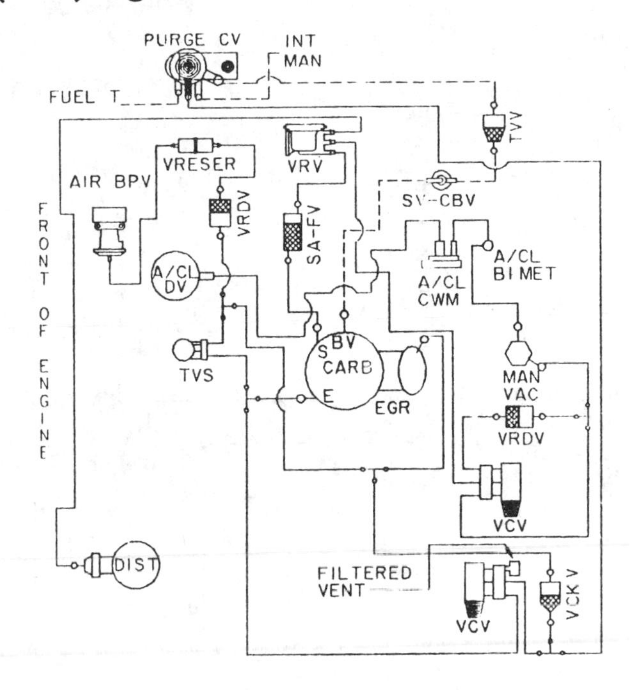

CALIBRATION: 9–11X–R10 DATE: 1–23–79
5.0L (302 CID)

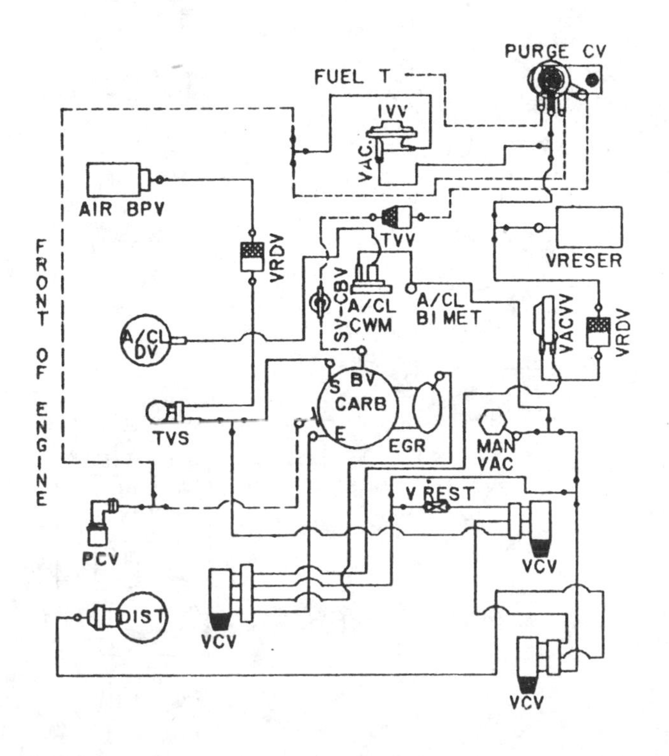

CALIBRATION: 9–11Y–RO DATE: 9–18–78
5.0L (302 CID)

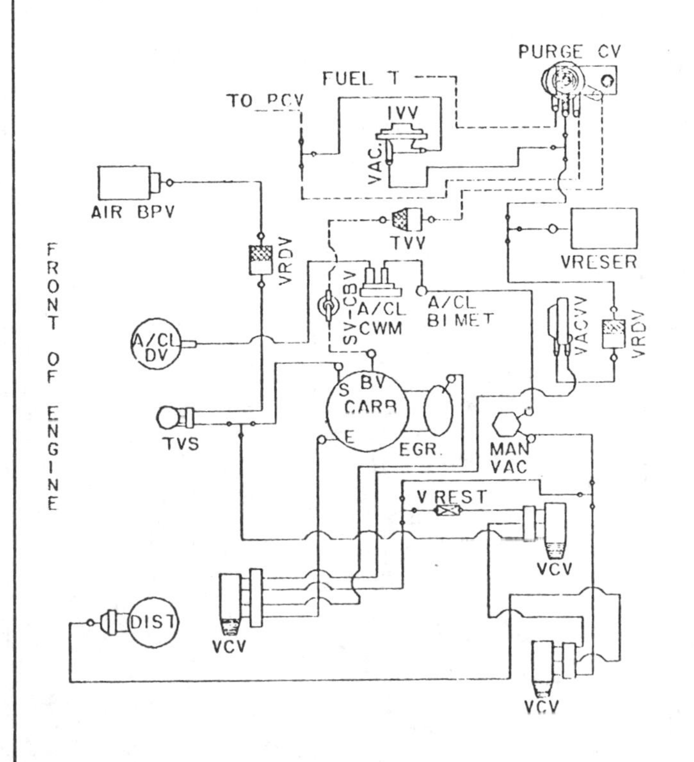

CALIBRATION: 9–11Z–RO DATE: 1–12–79
5.0L (302 CID)

VACUUM CIRCUITS

1980—302 CID (5.0 L) engines

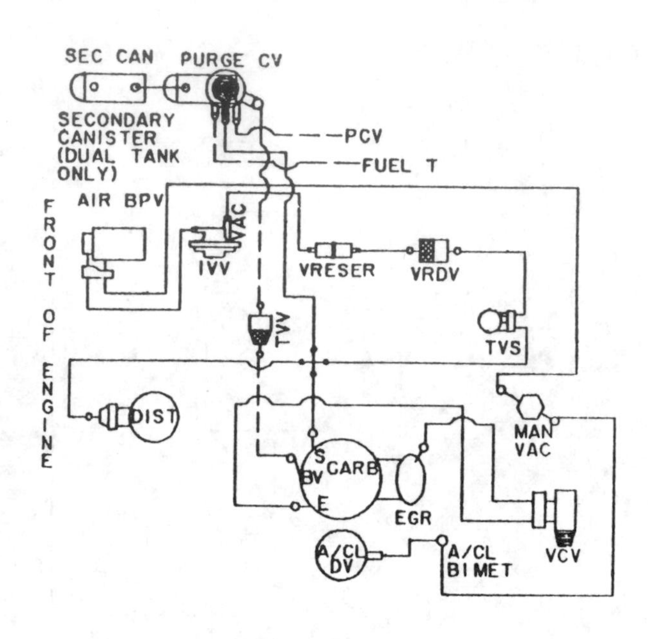

CALIBRATION: 9–53G–RO DATE: 5–12–78
5.0L (302 CID)

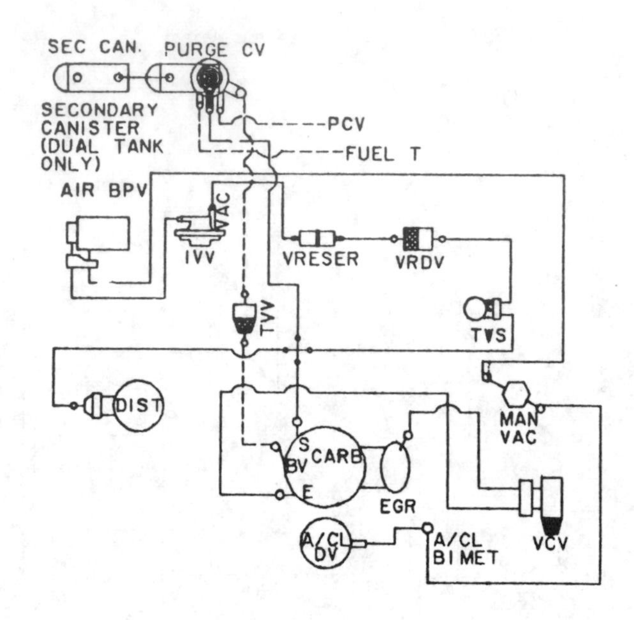

CALIBRATION: 9–53G–R10 DATE: 10–17–78
5.0L (302 CID)

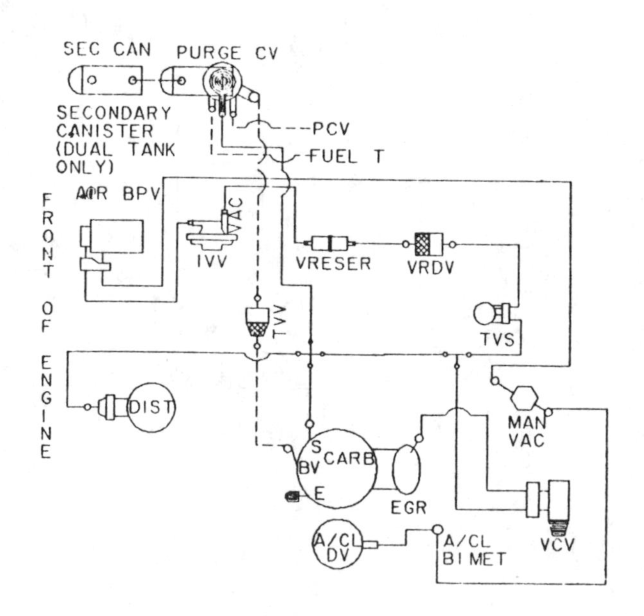

CALIBRATION: 9–53G–R12 DATE: 2–7–79
5.0L (302 CID)

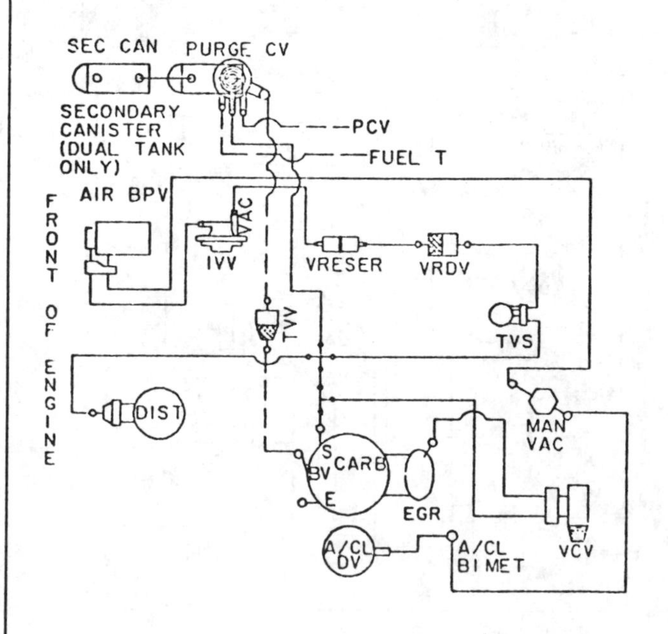

CALIBRATION: 9–53G–R13 DATE: 4–12–79
5.0L (302 CID)

VACUUM CIRCUITS

1980—302 CID (5.0 L) engines

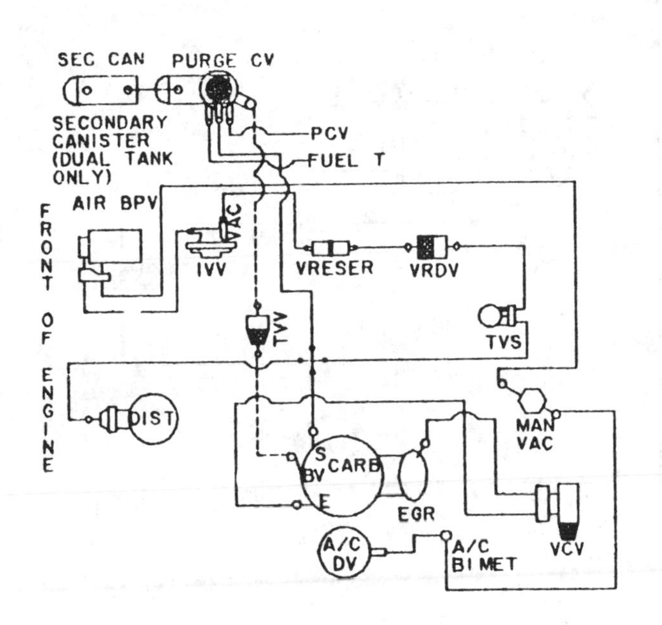

CALIBRATION: 9–53H–RO DATE: 5–12–78
5.0L (302 CID)

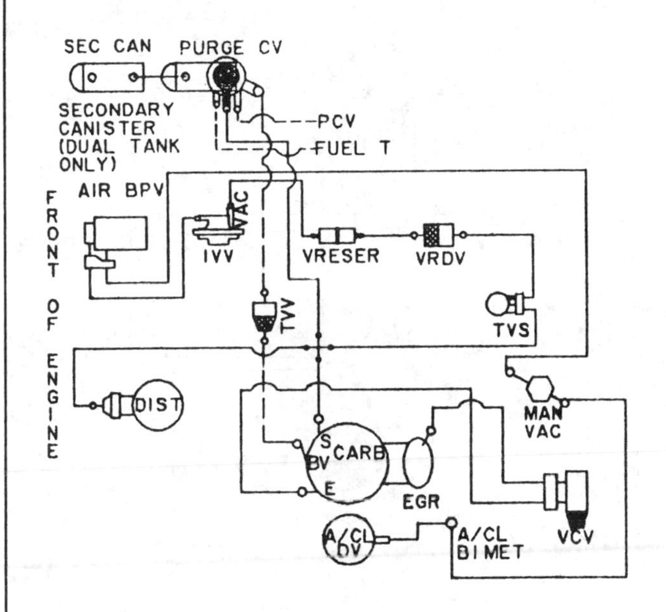

CALIBRATION: 9–53H–R10 DATE: 10–17–78
5.0L (302 CID)

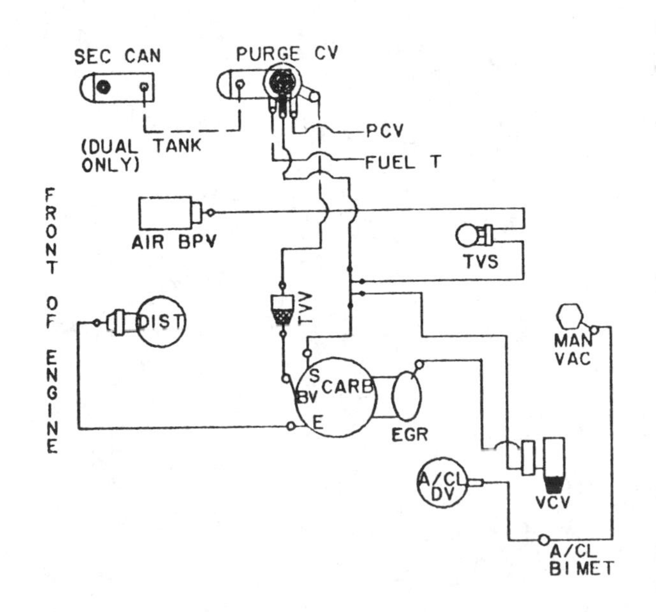

CALIBRATION: 9–54A–RO DATE: 8–10–78
5.0L (302 CID)

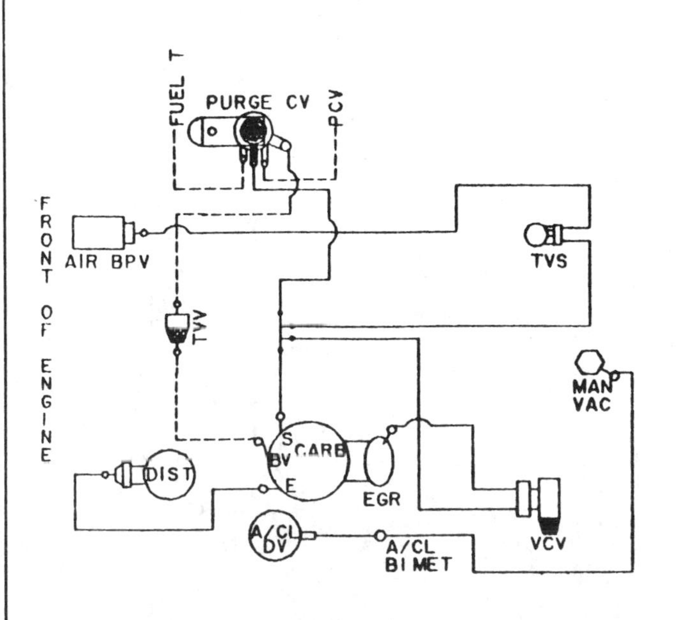

CALIBRATION: 9–54G–RO DATE: 5–28–78
5.0L (302 CID)

VACUUM CIRCUITS

1980—302 CID (5.0 L) engines

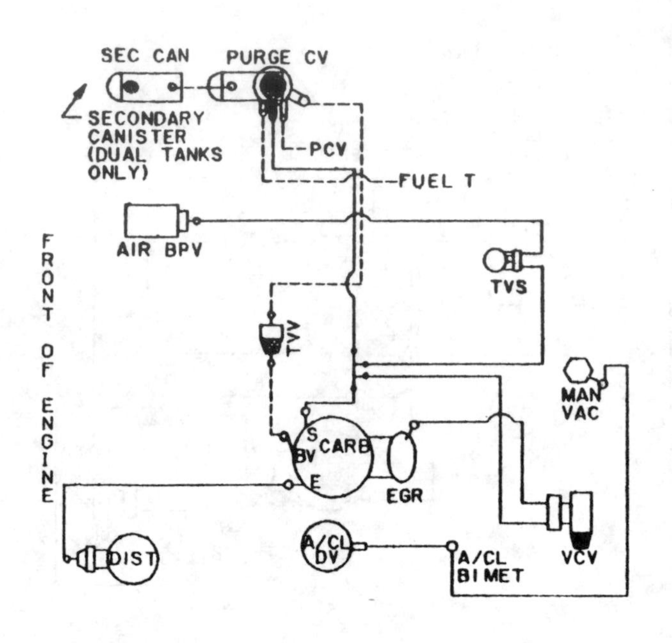

CALIBRATION: 9–54H–RO DATE: 5–31–78
5.0L (302 CID)

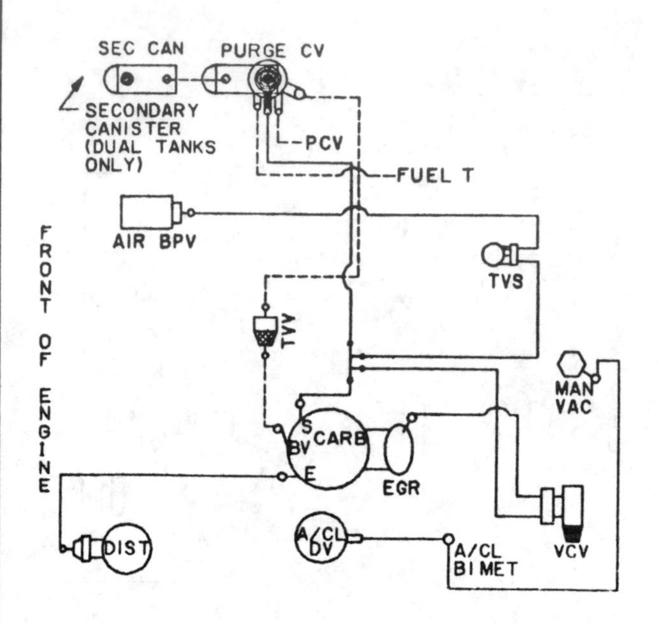

CALIBRATION: 9–54H–R10 DATE: 8–25–78
5.0L (302 CID)

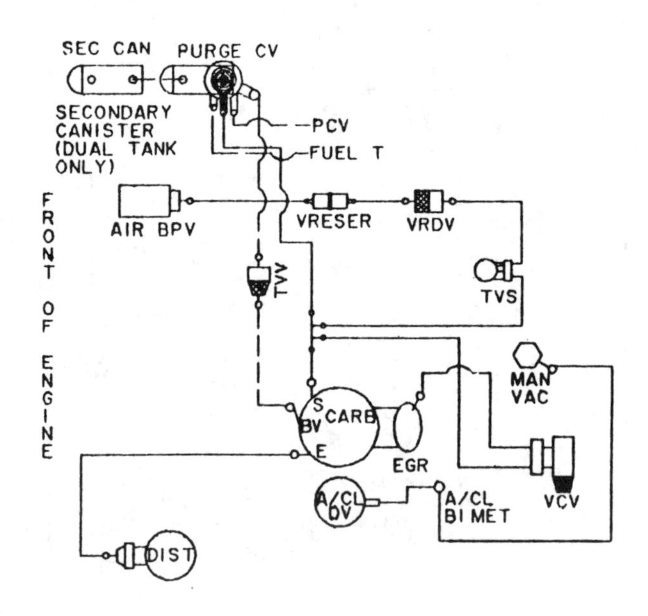

CALIBRATION: 9–54J–R1 DATE: 6–27–78
5.0L (302 CID)

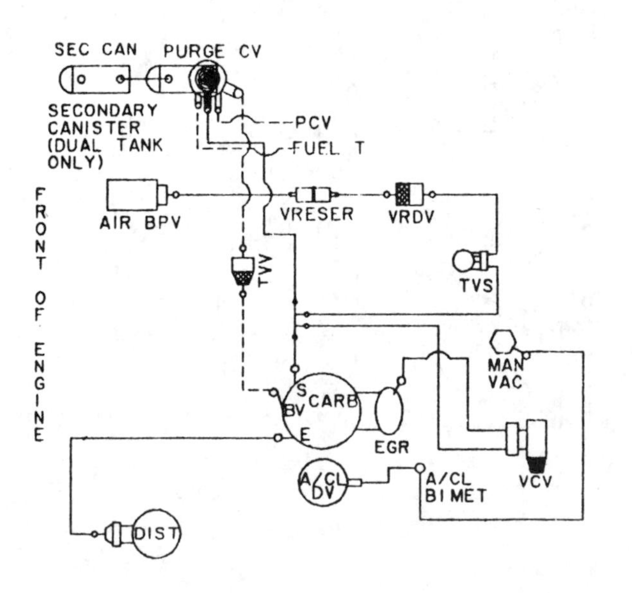

CALIBRATION: 9–54J–R10 DATE: 10–17–78
5.0L (302 CID)

VACUUM CIRCUITS

1980—302 CID (5.0 L) engines

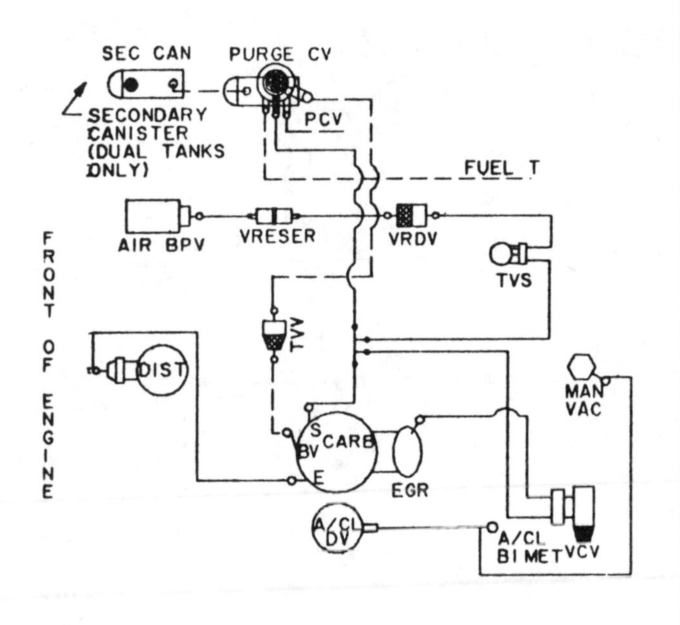

CALIBRATION: 9-54R-RO DATE: 5-28-78
5.0L (302 CID)

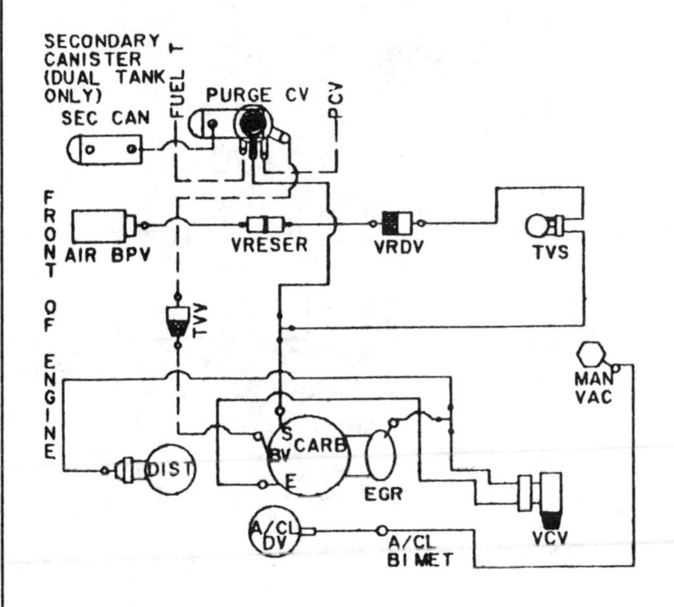

CALIBRATION. 9-54S-RO DATE: 5-25-78
5.0L (302 CID)

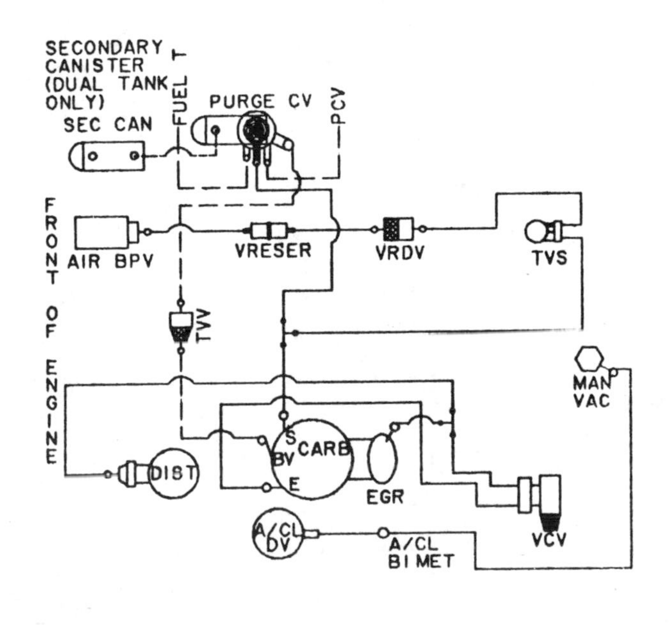

CALIBRATION: 9-54S-R10 DATE: 10-17-78
5.0L (302 CID)

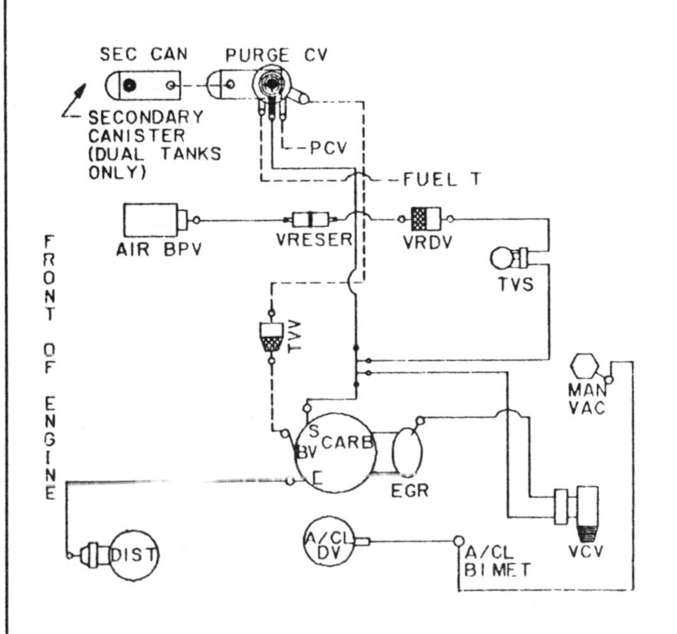

CALIBRATION. 9-54T-RO DATE: 5-31-78
5.0L (302 CID)

VACUUM CIRCUITS

1980—302 CID (5.0 L) engines

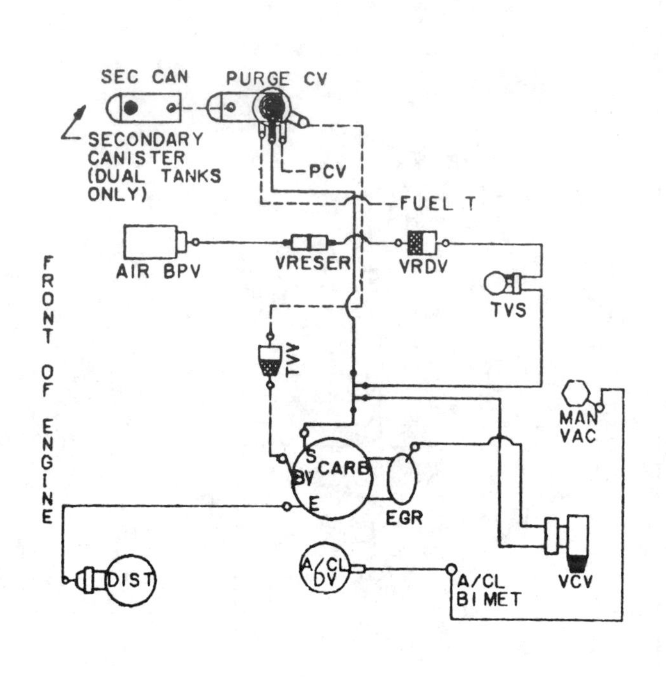

CALIBRATION: 9–54T–R10 DATE: 5–31–78
5.0L (302 CID)

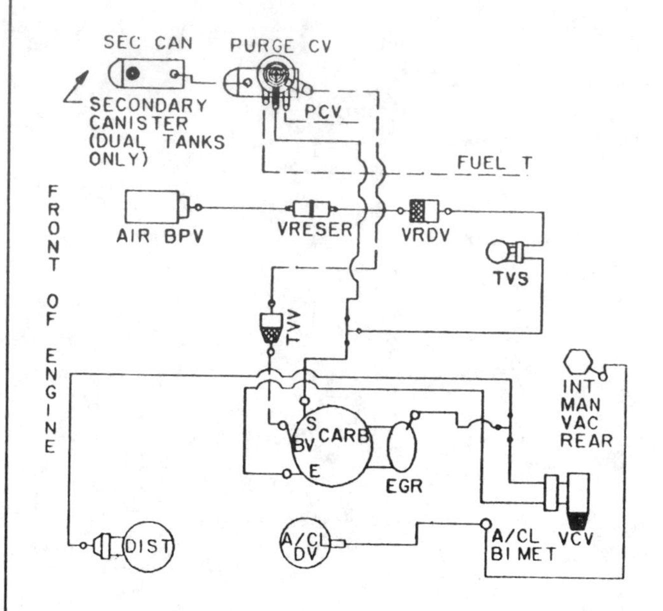

CALIBRATION: 9--54U–RO DATE: 5–30–78
5.0L (302 CID)

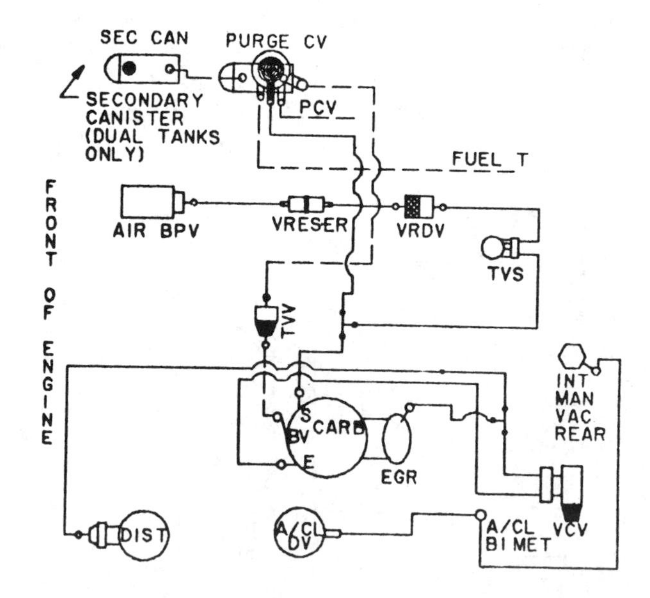

CALIBRATION: 9–54U–R10 DATE: 10–17–78
5.0L (302 CID)

VACUUM CIRCUITS

1980—302 CID (5.0 L) engines

CANADA ONLY A/C & NON A/C

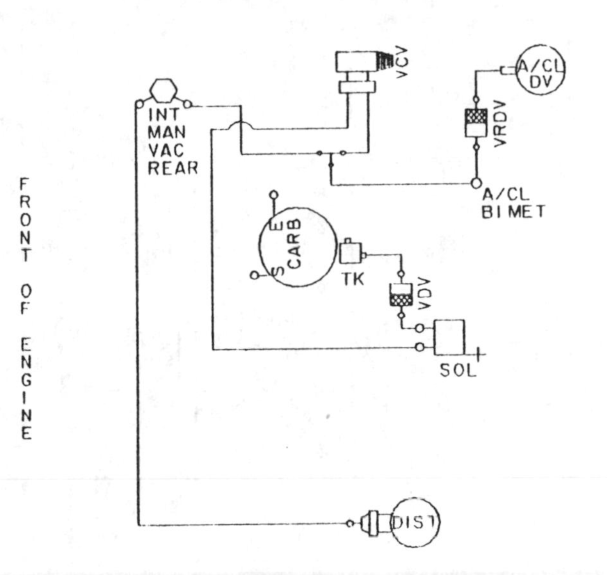

CALIBRATION: 7–79–R1 **DATE:** 7–11–79
5.0L TRK CAN F–150–250 (4X2)
F–150–250–BRO (4X4)

CANADA ONLY A/C & NON A/C

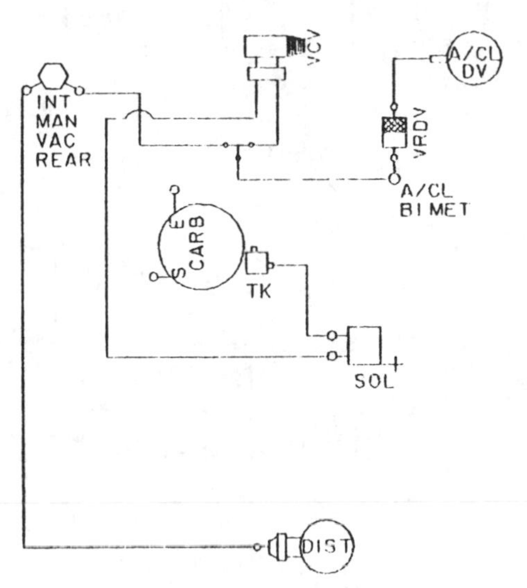

CALIBRATION: 7–80–RO **DATE:** 7–11–79
5.0L TRK CAN F–150–250 (4X2)
F–150–250–BRO (4X4) E–150–250

A/C & NON A/C

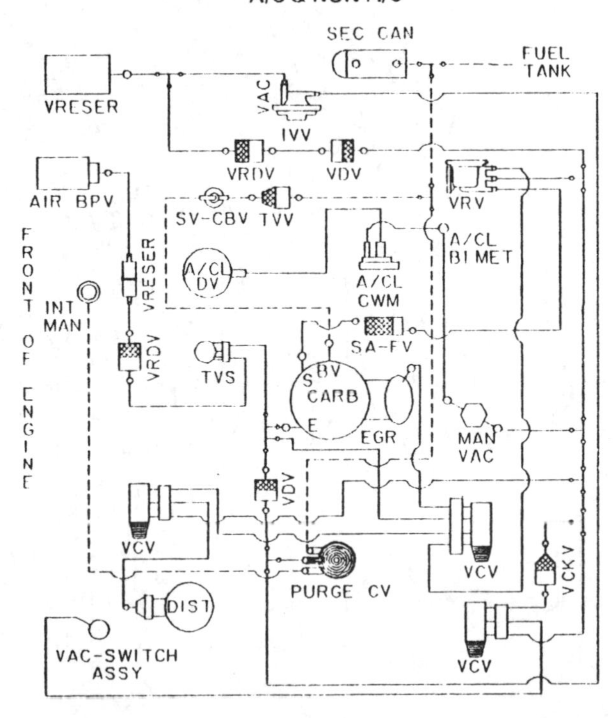

CALIBRATION: 0–11A–RO **DATE:** 8–8–79
5.0L T-BIRD/XR-7

A/C & NON A/C

CALIBRATION: 0–11B–RO **DATE:** 8–8–79
5.0L T-BIRD/XR-7

VACUUM CIRCUITS

1980—302 CID (5.0 L) engines

A/C & NON A/C

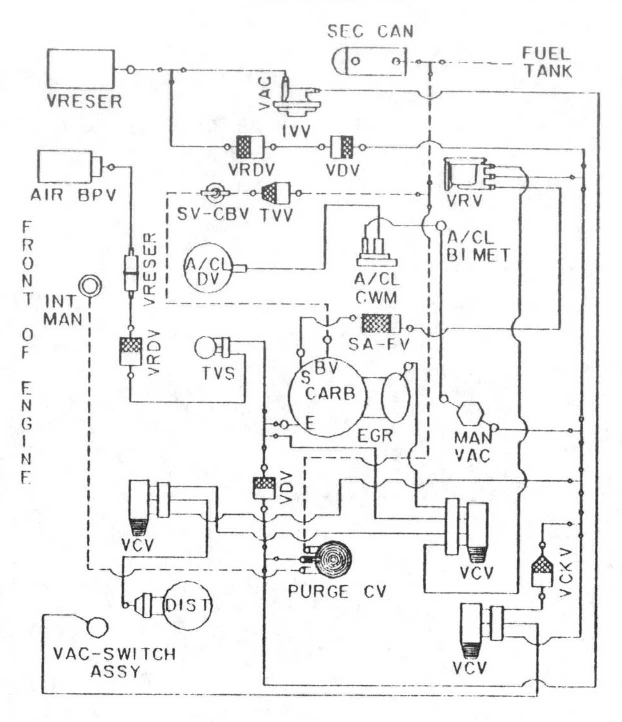

CALIBRATION: 0–11D–RO DATE: 8–8–79
5.0L GRANADA/MONARCH

A/C ONLY

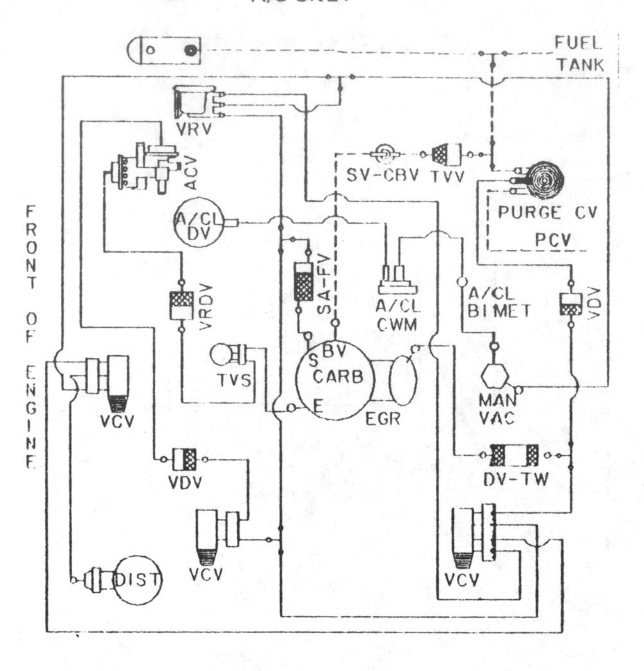

CALIBRATION: 0–11E–RO DATE: 7–19–79
5.0L VERSAILLES 50S A/T

A/C & NON A/C

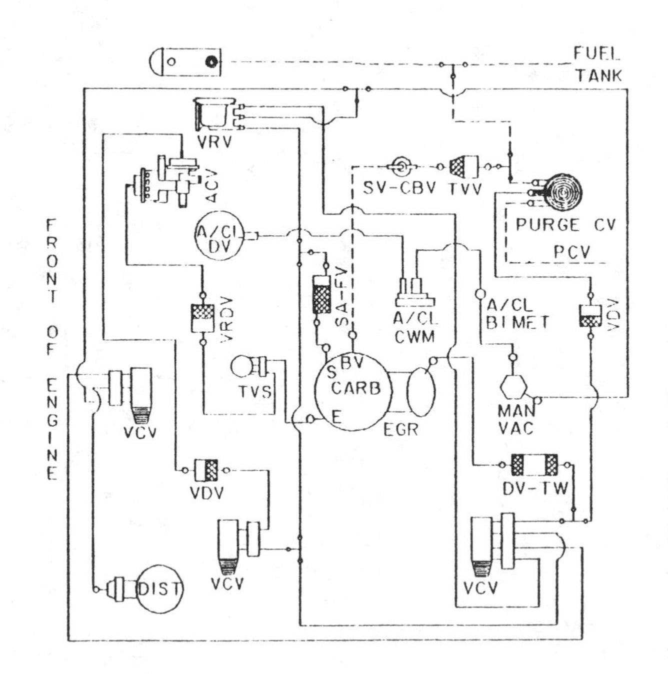

CALIBRATION: 0–11N–RO DATE: 7–16–79
5.0L T'BIRD/XR7 FAIR/ZEPHYR A/T

CALIF. ONLY A/C & NON A/C

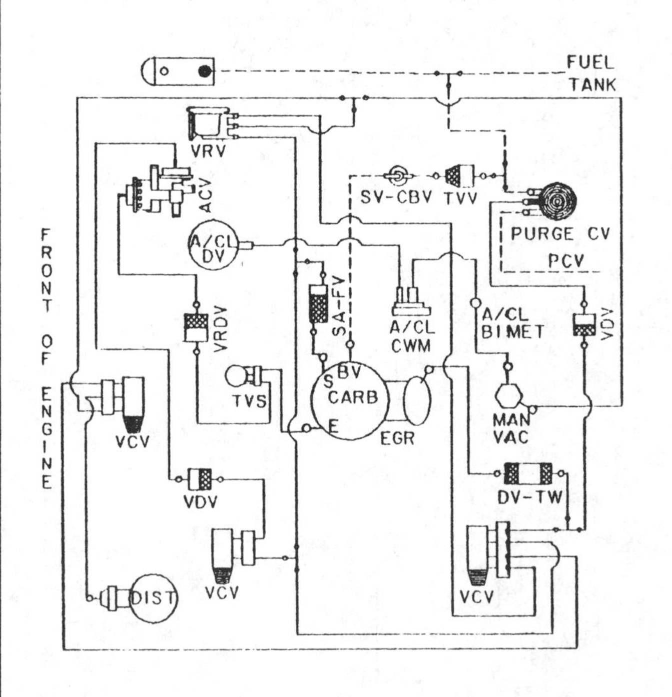

CALIBRATION: 0–11R–RO DATE: 7–12–79
5.0L GRAN/MON CALIF A/T

VACUUM CIRCUITS

1980—302 CID (5.0 L) engines

49S A/C & NON A/C

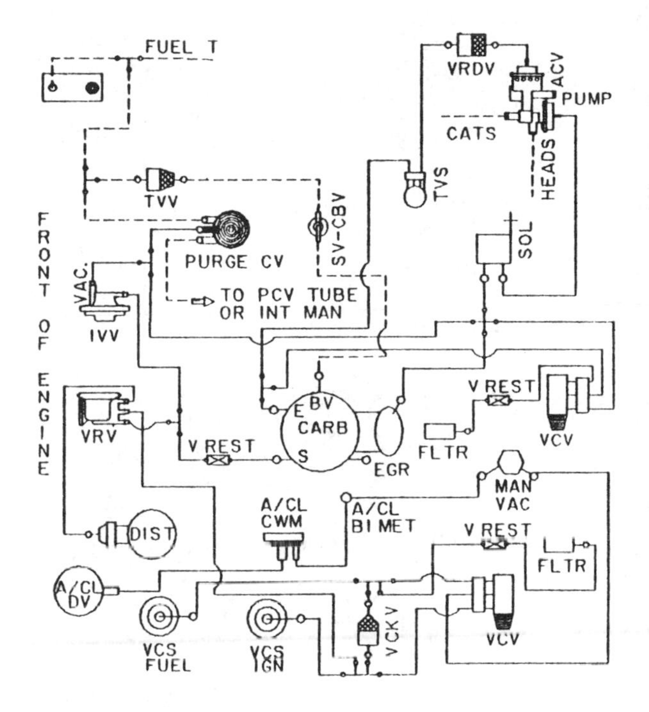

CALIBRATION: 0–13A–RO DATE: 7–18–79
5.0L 49S FORD–MERC SED A/T

49S A/C & NON A/C

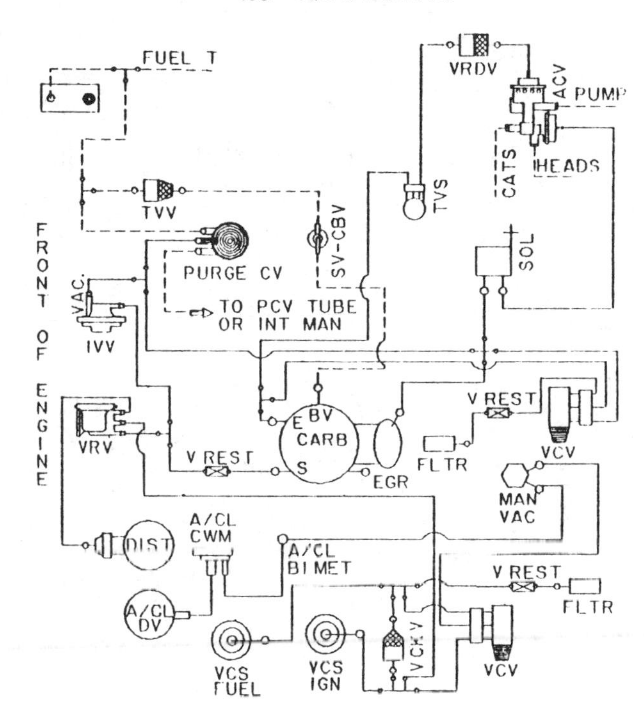

CALIBRATION: 0–13F–RO DATE: 7–18–79
5.0L 49S FORD–MERC S/W A/T

A/C & NON A/C

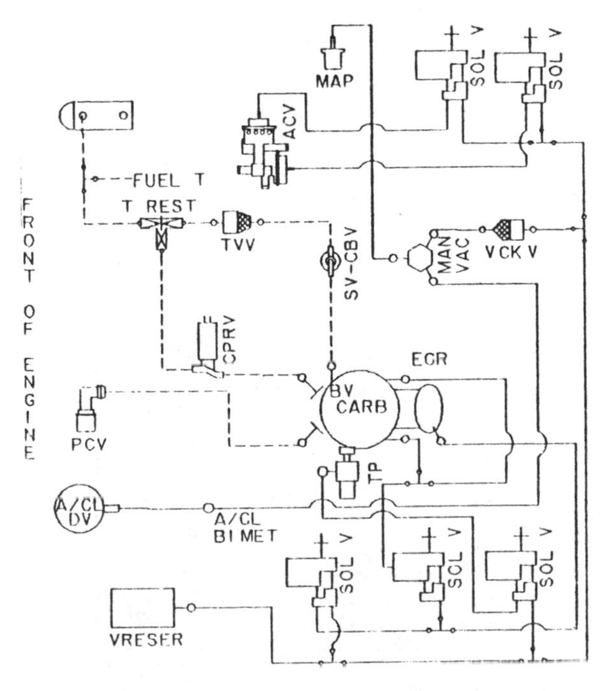

CALIBRATION: 0–13R–RO DATE: 7–16–79
5.0L FORD/MERC CALIF.

CALIF. ONLY A/C & NON A/C

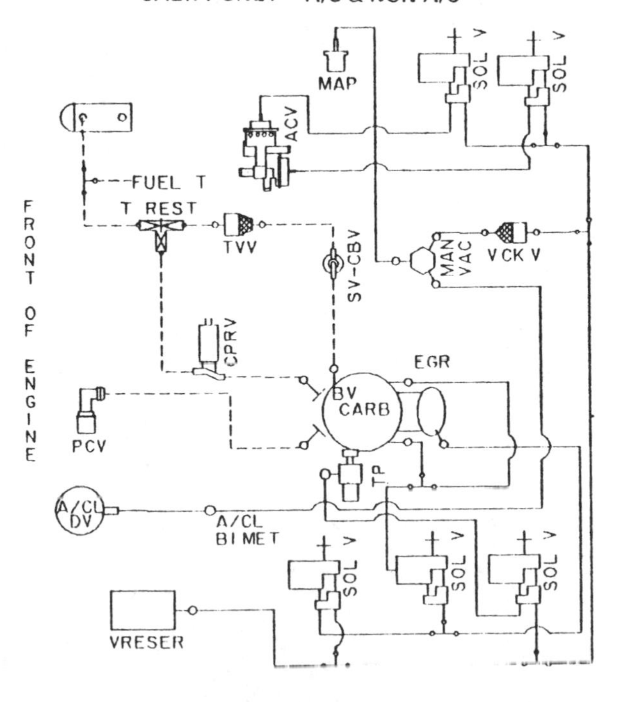

CALIBRATION: 0–13T–RO DATE: 7–16–79
5.0L FORD/MERC CALIF.

VACUUM CIRCUITS

1980—302 CID (5.0 L) engines

49S A/C ONLY

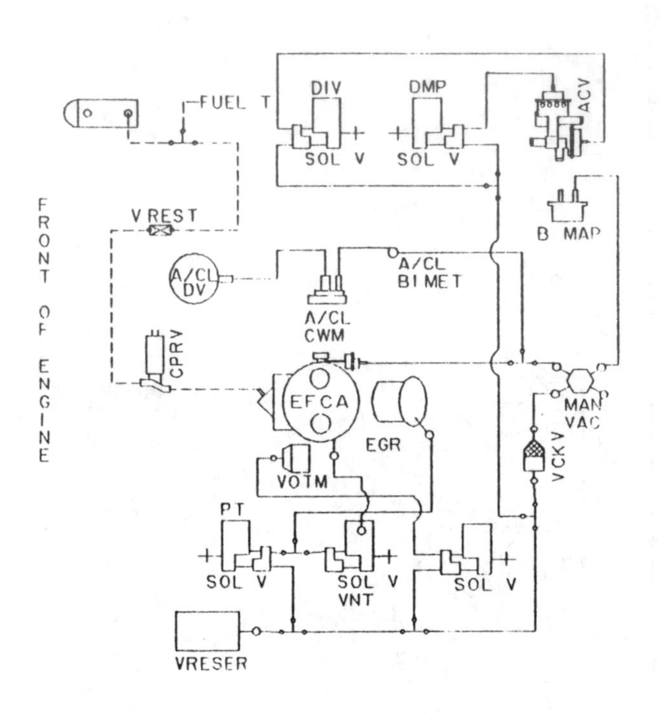

CALIBRATION: 0–14A–RO DATE: 7–17–79
5.0L 49S/CANADA LINC/MARK
FORD/MERC

CALIF. ONLY A/C & NON A/C

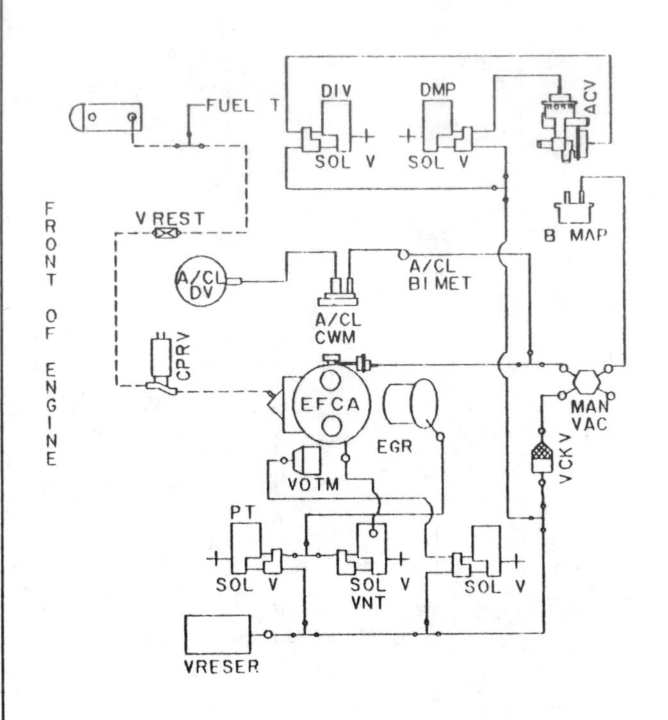

CALIBRATION: 0–14N–RO DATE: 7–23–79
5.0L CALIF LINC/MARK FORD/MERC

CALIF. ONLY A/C & NON A/C

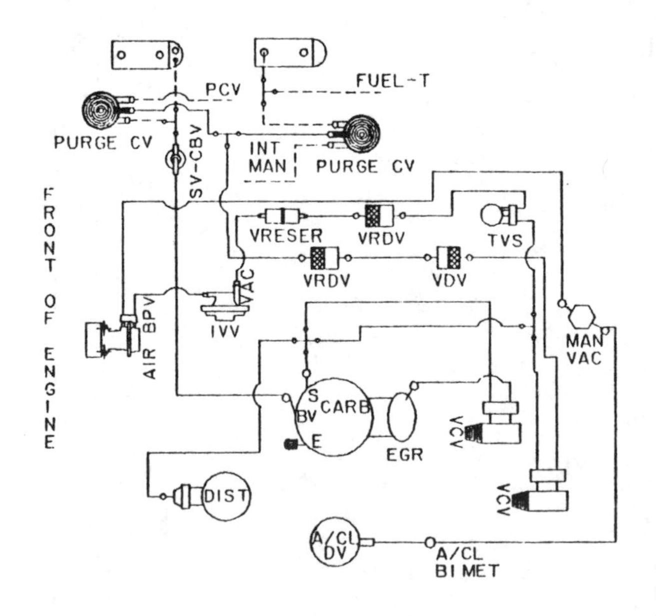

CALIBRATION: 0–53S–RO DATE: 6–28–79
5.0L F–100/F–150 CALIF M3/M4OD

A/C

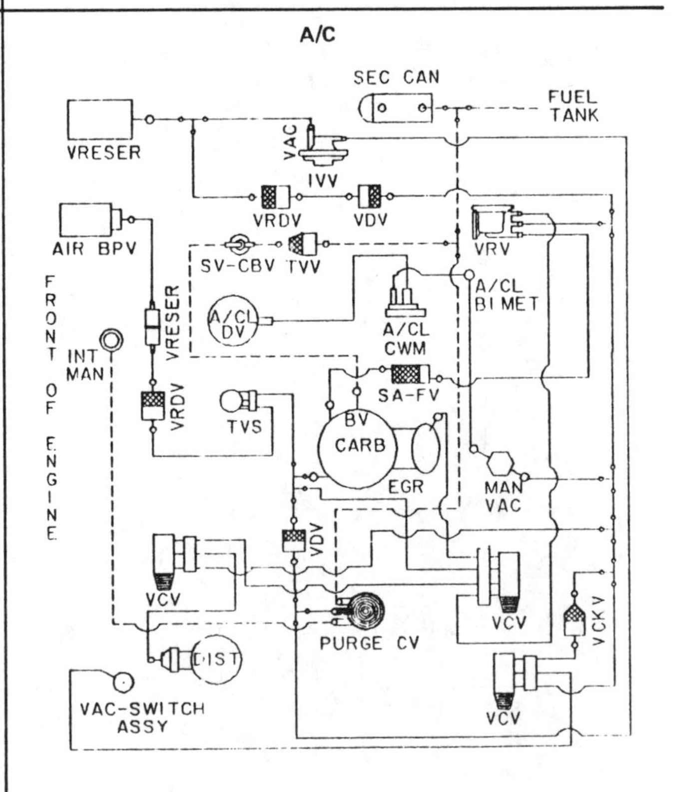

CALIBRATION: 0–11A–R10 DATE: 10–23–79
5.0L T–BIRD/XR7

VACUUM CIRCUITS

1980—302 CID (5.0 L) engines

A/C & NON A/C

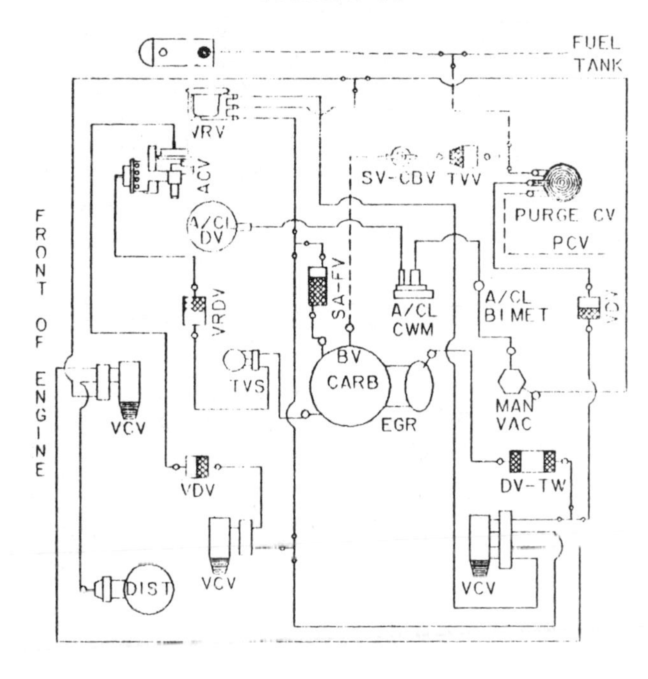

CALIBRATION: 0–11N–R13 DATE: 11–14–79
5.0L T-BIRD/XR7

CALIF ONLY A/C & NON A/C

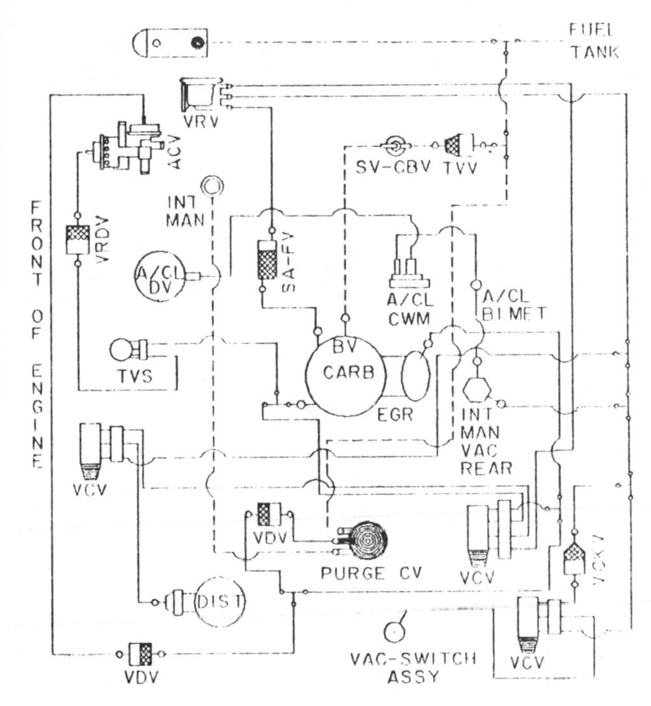

CALIBRATION: 0–11P–R12 DATE: 01–25–80
5.0L T-BIRD/XR7 CALIF

A/C & NON A/C

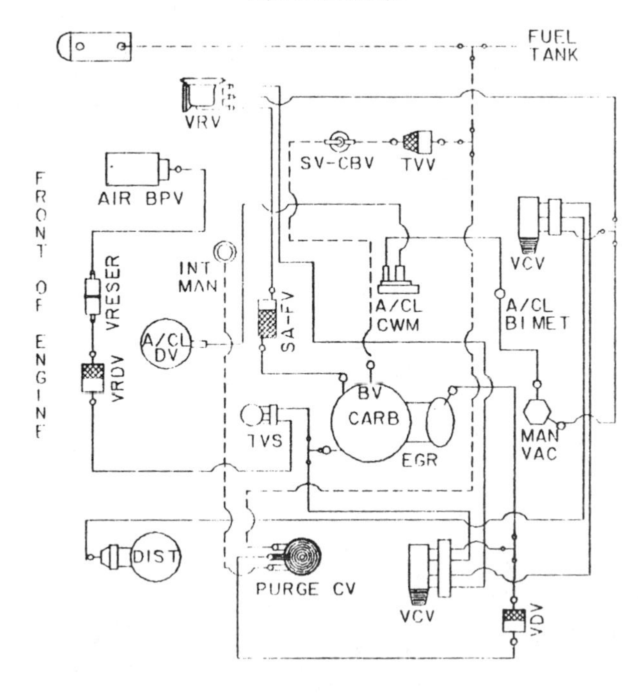

CALIBRATION: 0–11W–R0 DATE: 10–23–79
5.0L T-BIRD/XR7

49S A/C & NON A/C

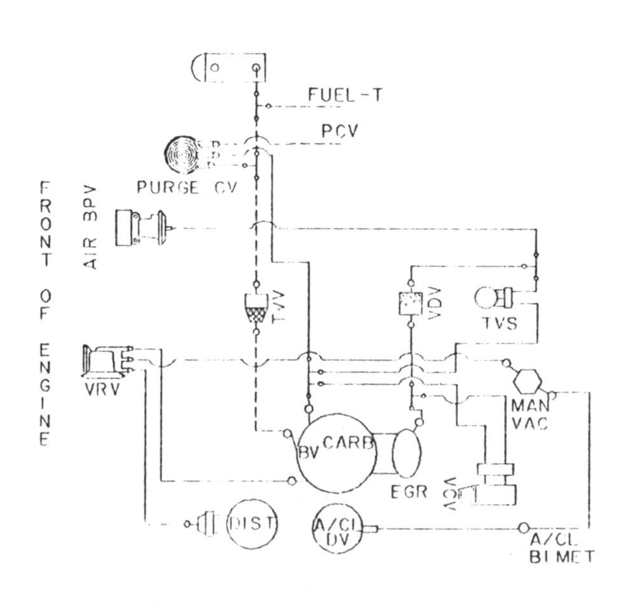

CALIBRATION: 0–54K–R10 DATE: 10–26–79
5.0L F-100/150 49S A/T

VACUUM CIRCUITS

1980—302 CID (5.0 L) engines

49S A/C & NON A/C

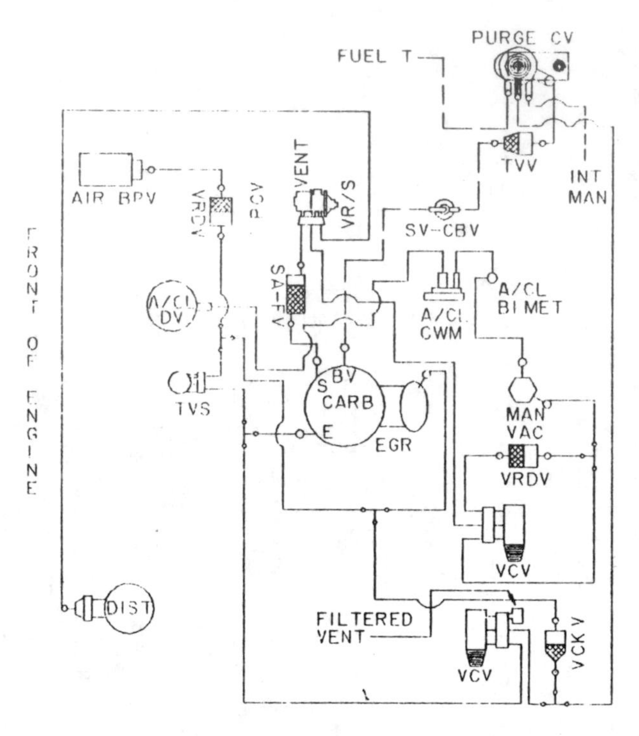

CALIBRATION: 9–11C–R1 DATE: 07–31–79
5.0L GRANADA/MONARCH 49S A/T

CANADA A/C & NON A/C

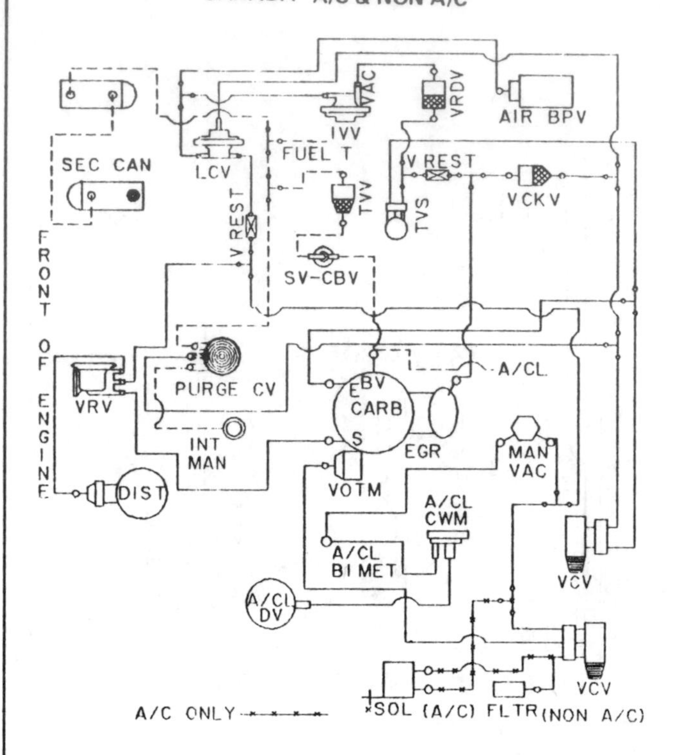

CALIBRATION: 9–11E–R12 DATE: 08–20–79
5.0L FORD/MERC CANADA A/T

CANADA A/C & NON A/C

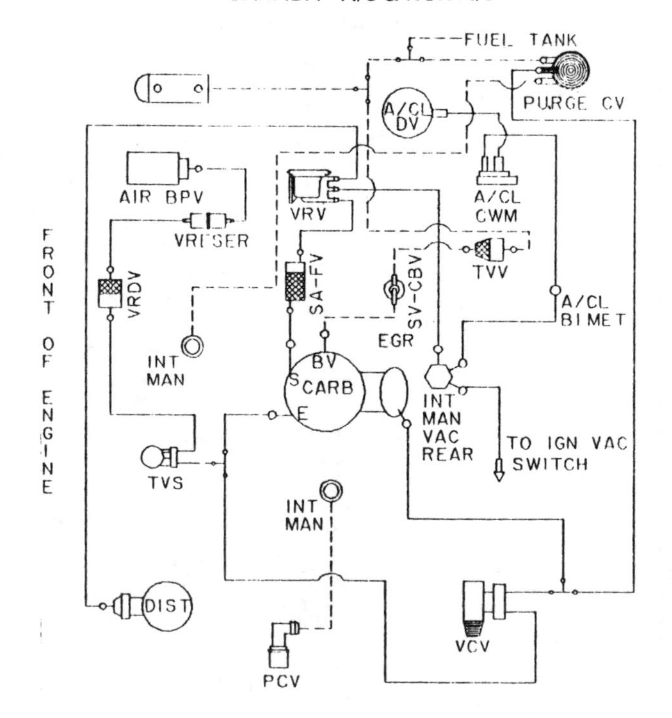

CALIBRATION: 0–11G–R0 DATE: 07–26–79
5.0L T-BIRD/XR-7 CANADA A/T

CANADA A/C & NON A/C

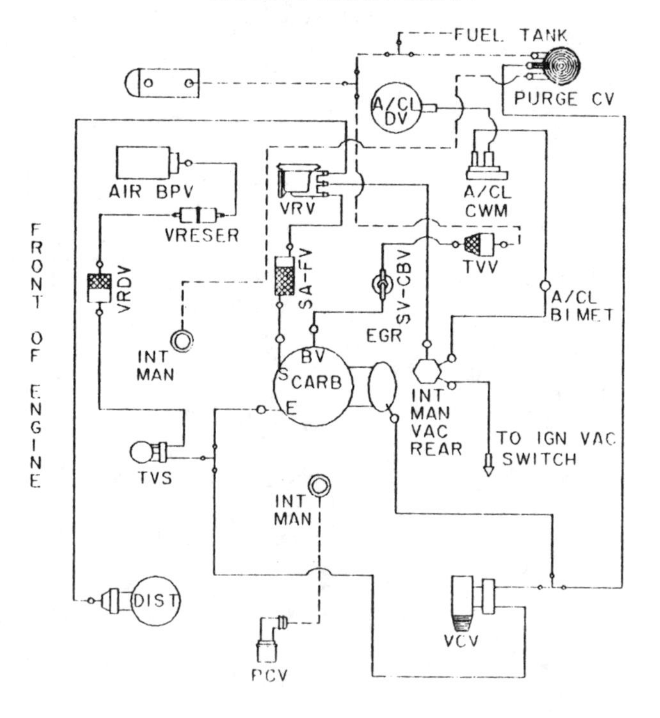

CALIBRATION: 0–11G–R10 DATE: 09–26–79
5.0L T-BIRD/XR-7 CANADA A/T

VACUUM CIRCUITS

1980—302 CID (5.0 L) engines

CANADA A/C & NON A/C

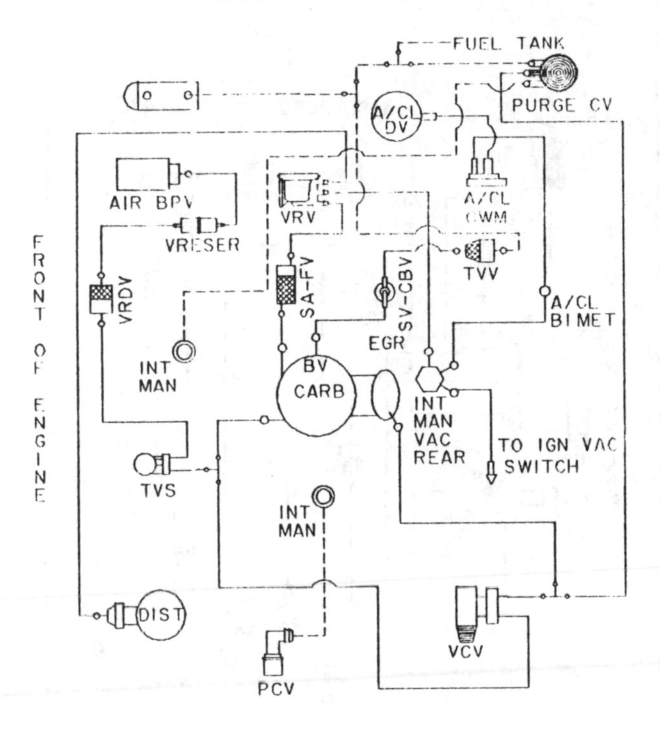

CALIBRATION: 0–11G–R11 DATE: 10–23–79
5.0L T-BIRD/XR-7 CANADA A/T

A/C & NON A/C

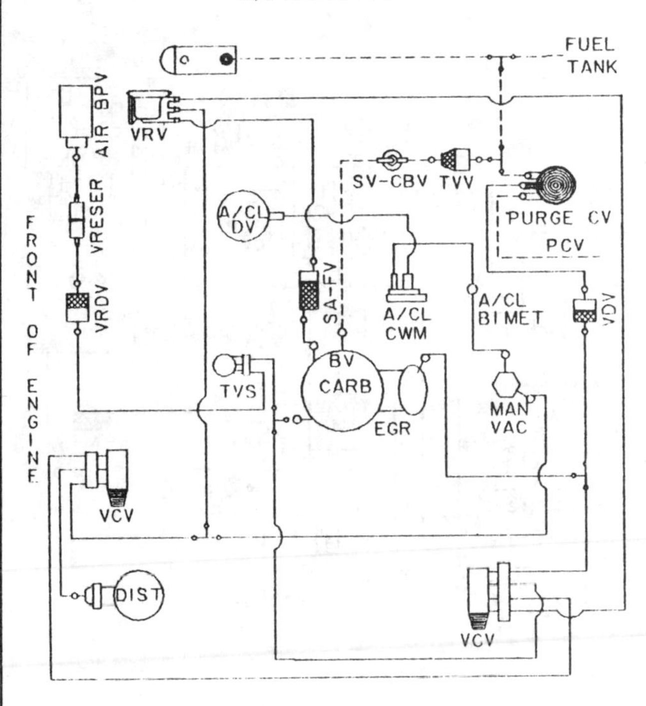

CALIBRATION: 0–11X–R0 DATE: 09–24–79
5.0L GRANADA/MONARCH A/T

49S A/C & NON A/C

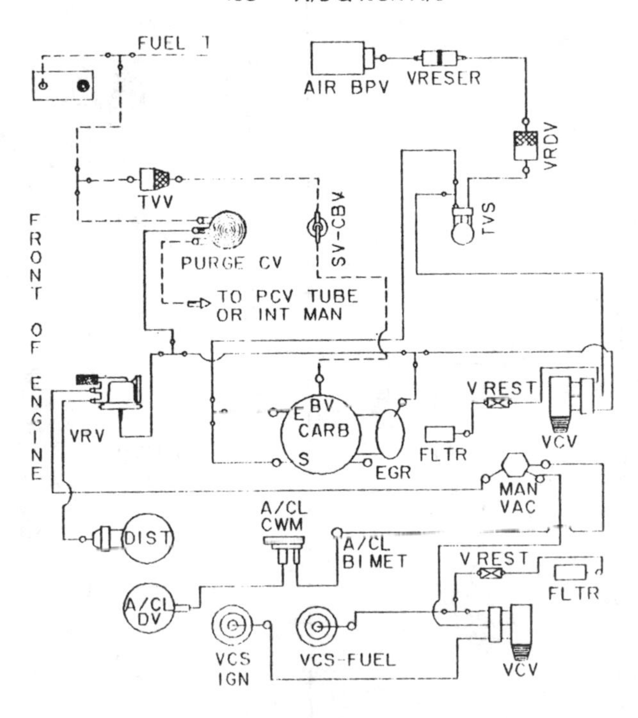

CALIBRATION: 0–13D–R0 DATE: 09–06–79
5.0L FORD SEDAN 49S

VACUUM CIRCUITS

1980—351 CID (5.8 L) engines

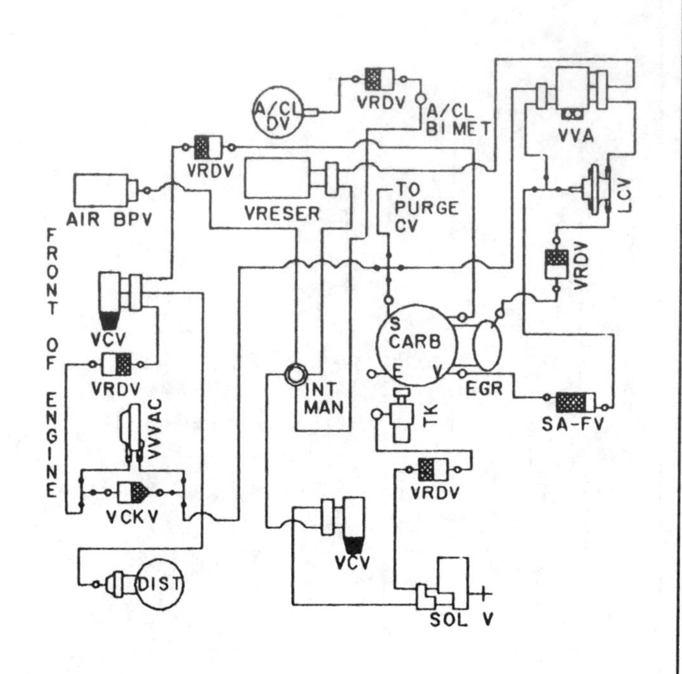

CALIBRATION. 7–72J–R11 DATE: 6–8–78
5.8L (351M CID)

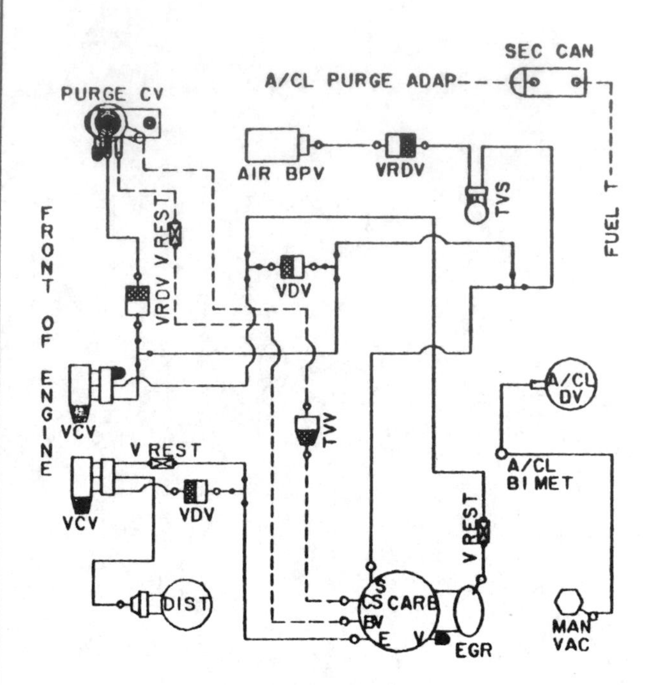

CALIBRATION: 9–14E–RO DATE: 6–5–78
5.8L (351M CID)

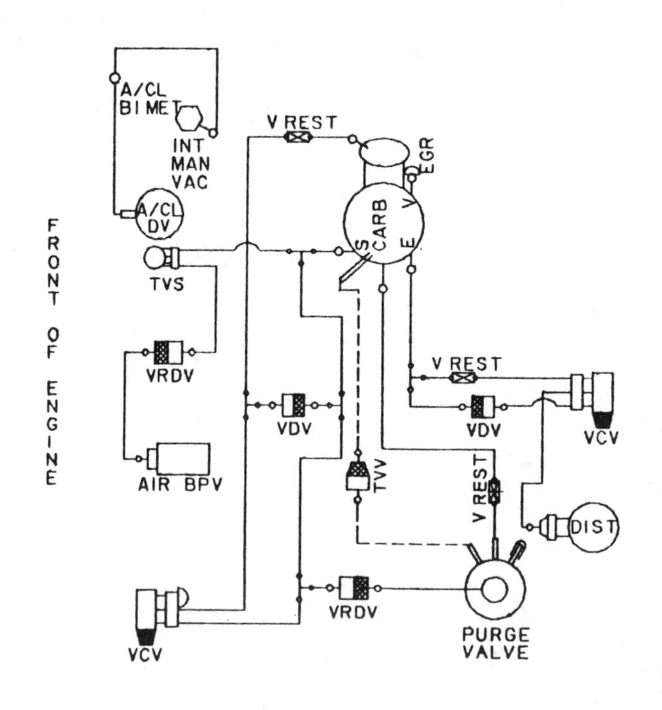

CALIBRATION: 9–14E–R12 DATE: 7–11–78
5.8L (351M CID)

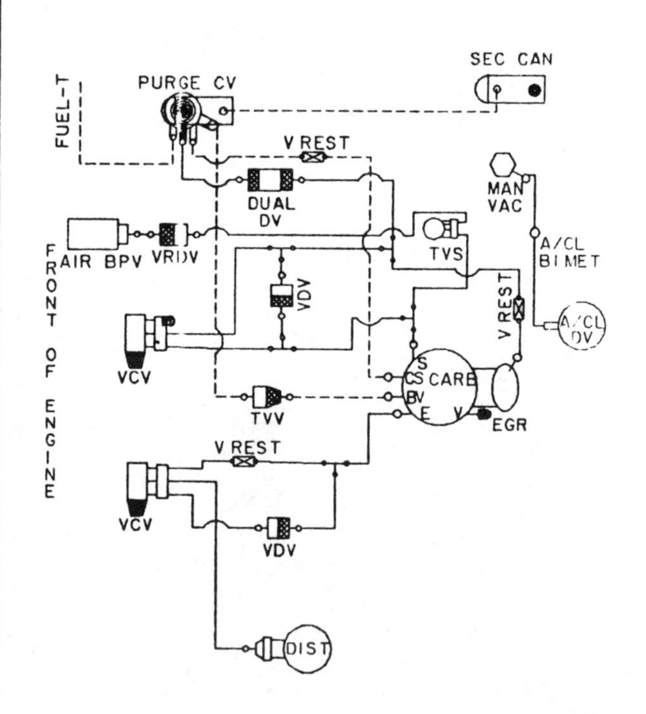

CALIBRATION: 9–14E–R14 DATE: 1–13–79
5.8L (351M CID)

VACUUM CIRCUITS

1980—351 CID (5.8 L) engines

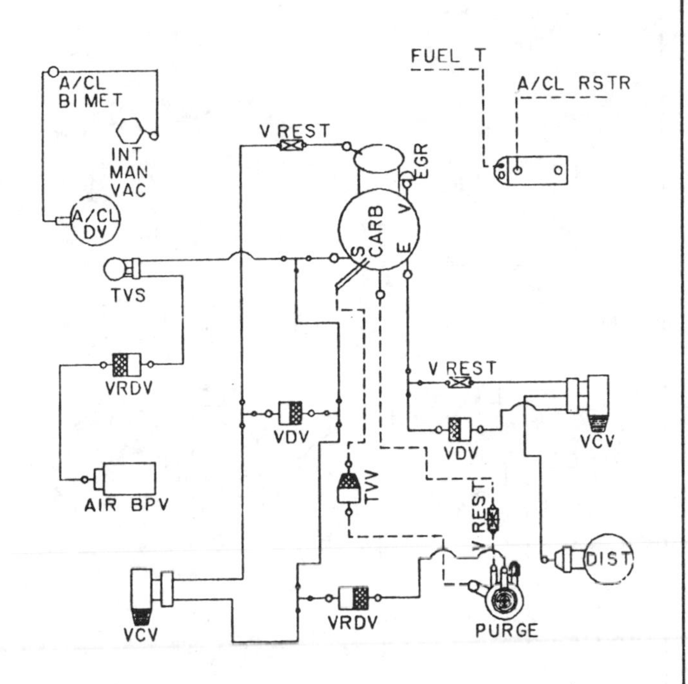

CALIBRATION: 9–14E–R15 DATE: 12–1–78
5.8L (351M CID)

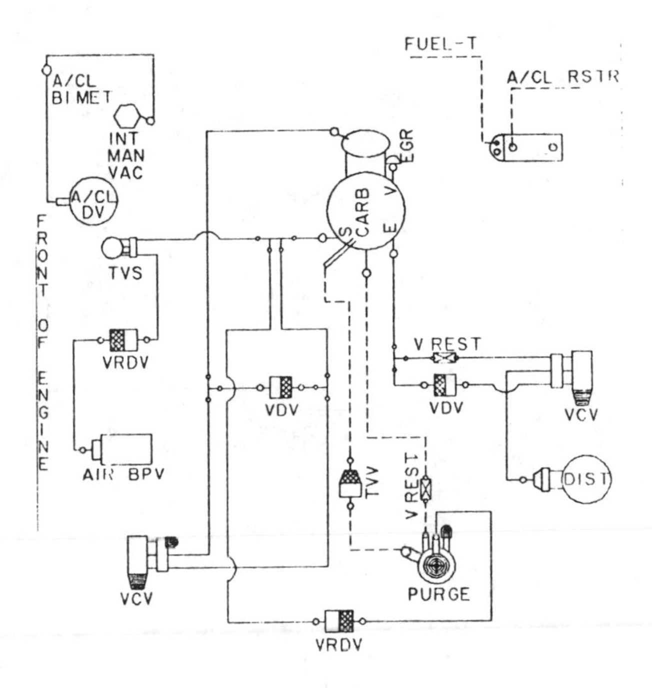

CALIBRATION: 9–14E–R18 DATE: 1–11–79
5.8L (351M CID)

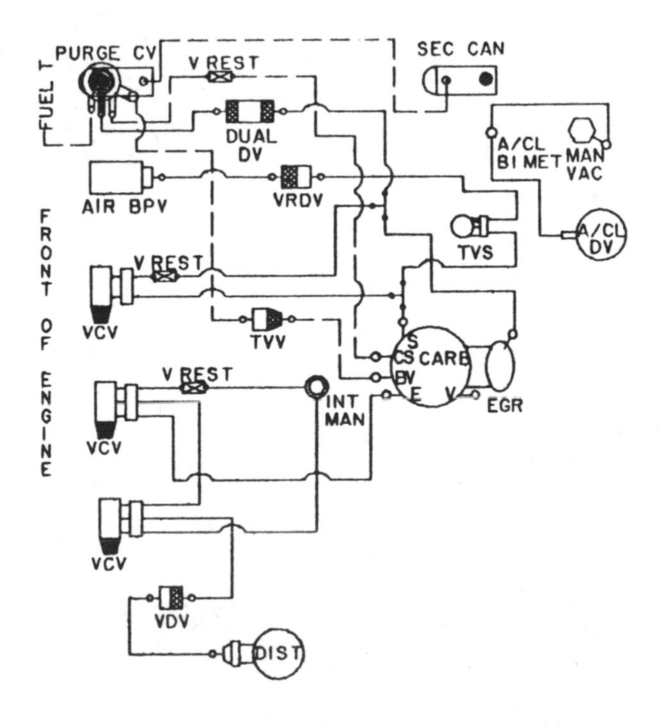

CALIBRATION: 9–14Q–RO DATE: 5-4–78
5.8L (351M CID)

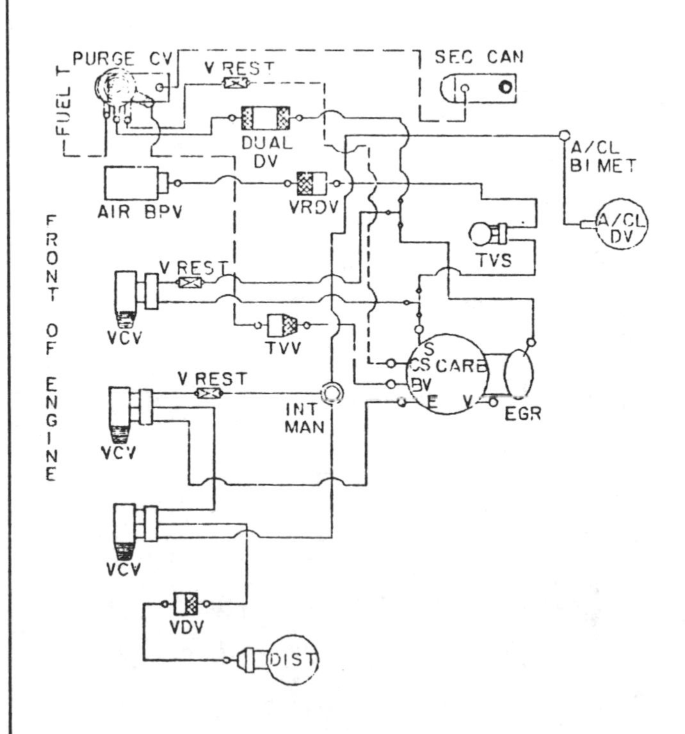

CALIBRATION: 9–14Q–R10 DATE: 2–2–79
5.8L (351M CID)

VACUUM CIRCUITS

1980—351 CID (5.8 L) engines

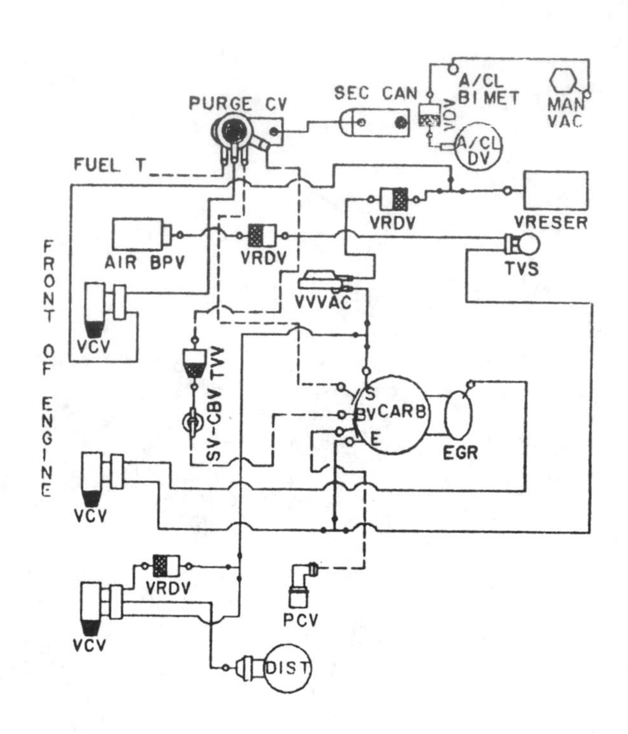

CALIBRATION: 9–14X–RO DATE: 10–10–78
5.8L (351M CID)

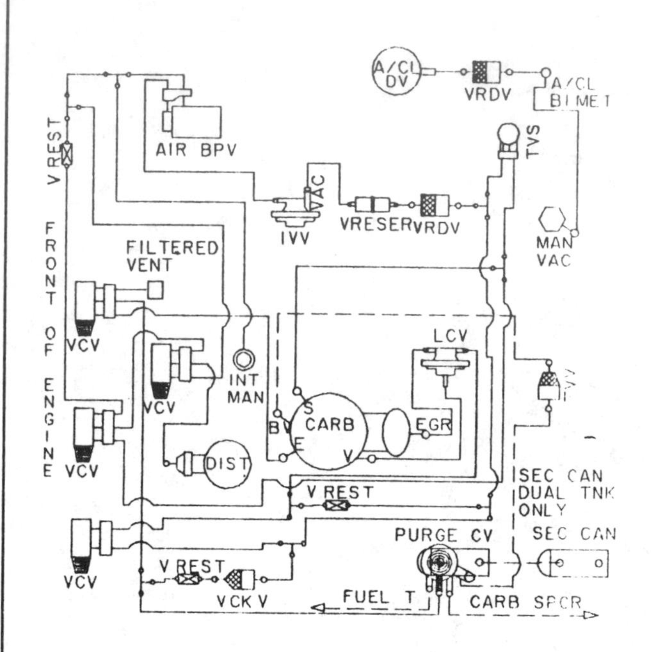

CALIBRATION: 9–59H–ROA DATE: 1–24–79
5.8L (351M CID)

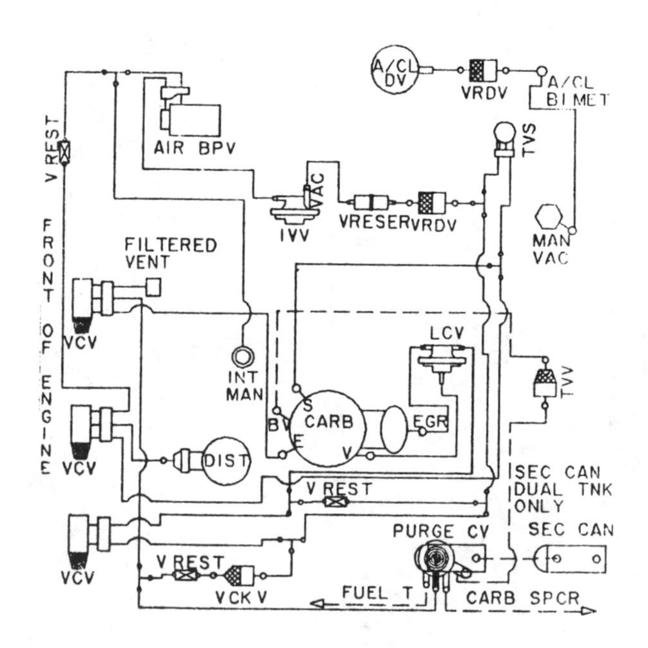

CALIBRATION: 9–59H–RON DATE: 1–24–79
5.8L (351M CID)

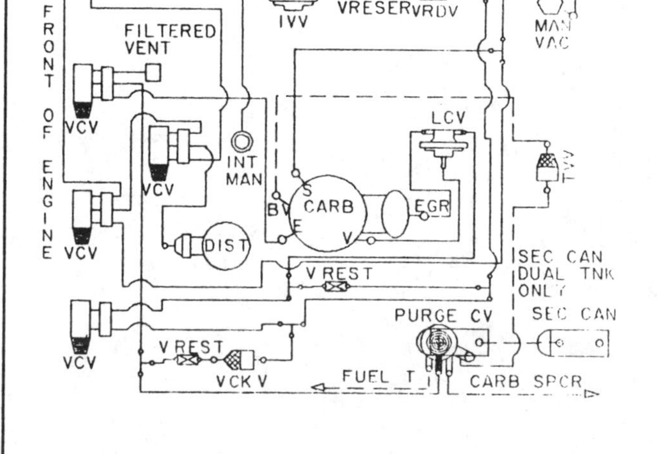

CALIBRATION: 9–59J–ROA DATE: 1–24–79
5.8L (351M CID)

VACUUM CIRCUITS

1980—351 CID (5.8 L) engines

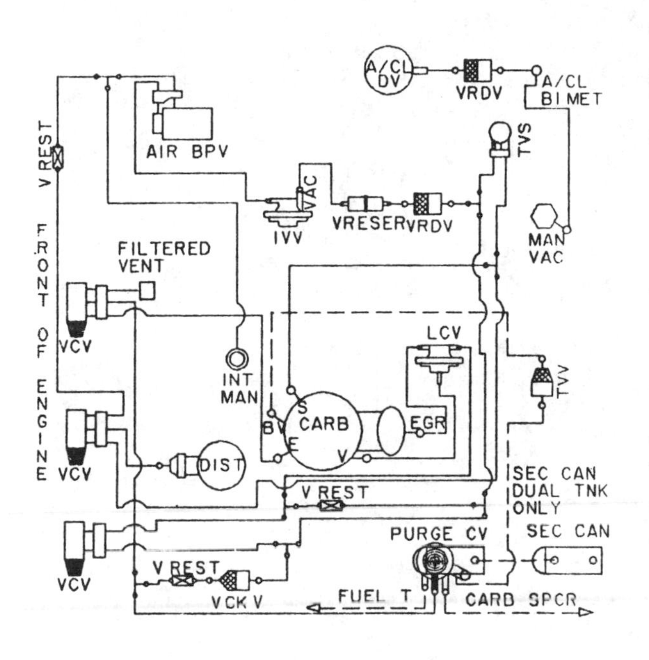

CALIBRATION: 9–59J–RON DATE: 1–24–79
5.8L (351M CID)

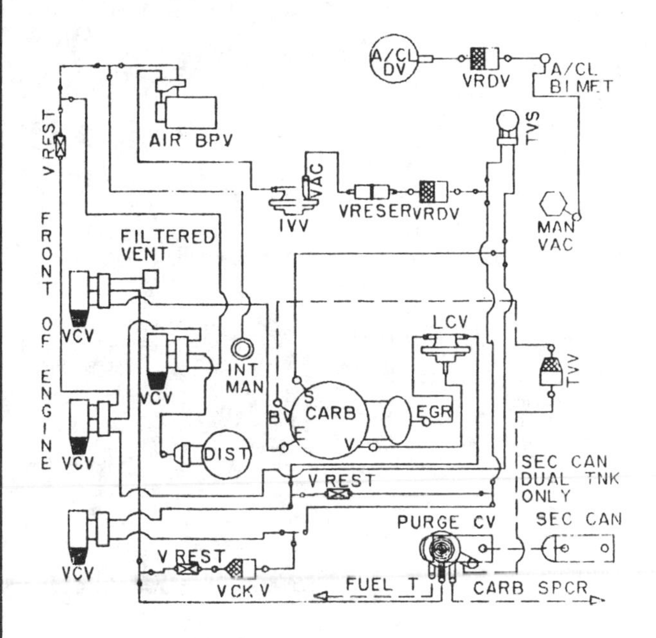

CALIBRATION: 9–59K–ROA DATE: 1–24–79
5.8L (351M CID)

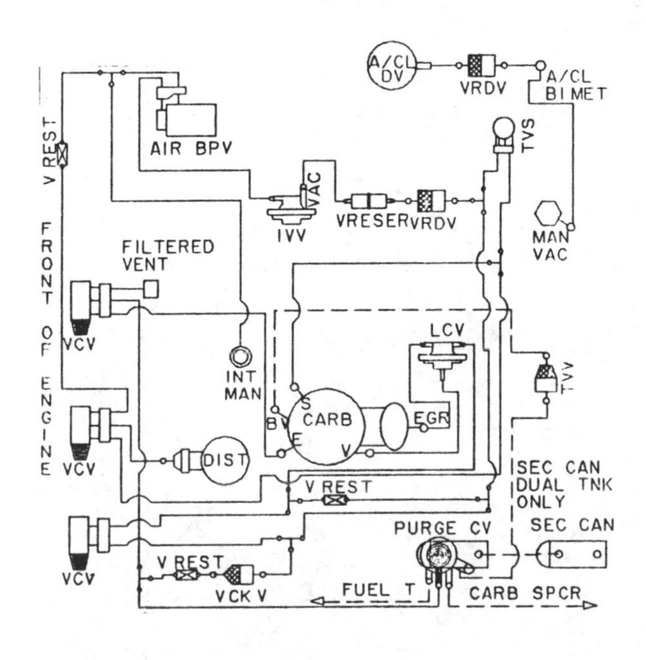

CALIBRATION: 9–59K–RON DATE: 2–26–79
5.8L (351M CID)

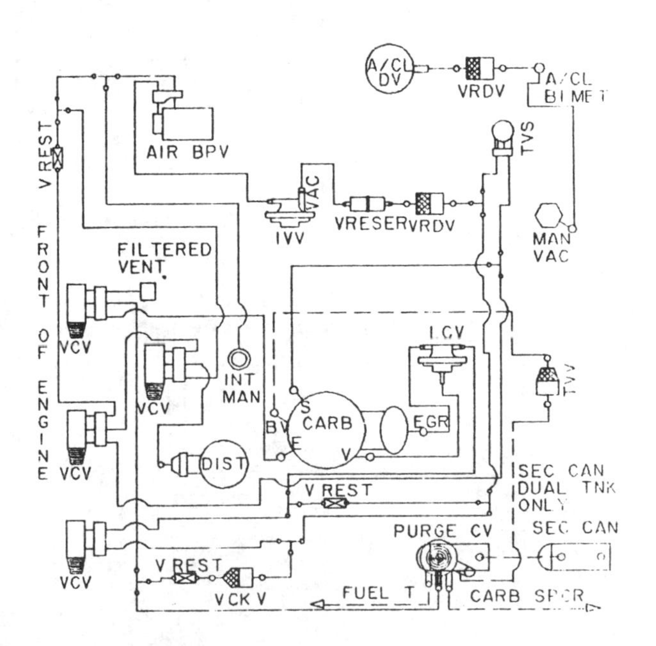

CALIBRATION: 9–59S–ROA DATE: 1–24–79
5.8L (351M CID)

VACUUM CIRCUITS

1980—351 CID (5.8 L) engines

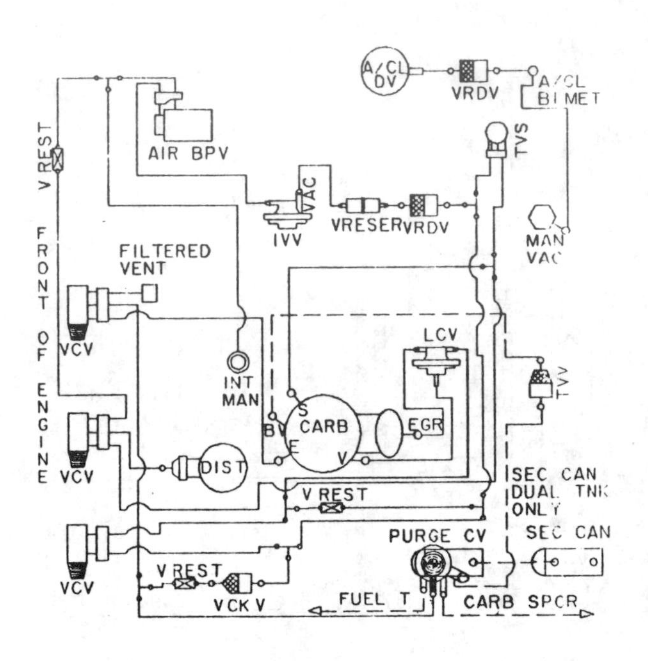

CALIBRATION: 9–59S–RON DATE: 1–24–79
5.8L (351M CID)

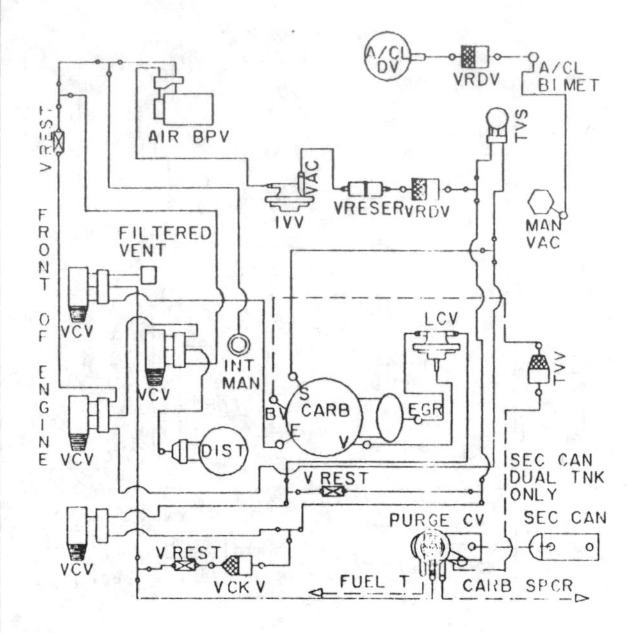

CALIBRATION: 9–59T–ROA DATE: 1–24–79
5.8L (351M CID)

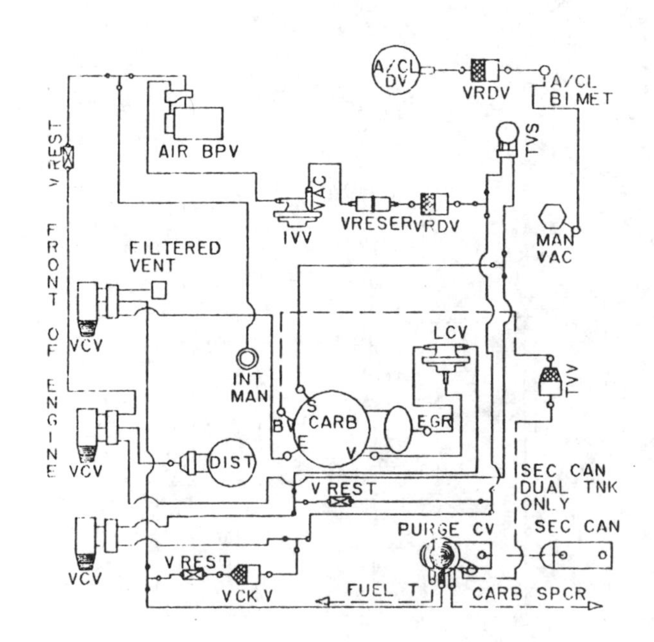

CALIBRATION: 9–59T–RON DATE: 1–24–79
5.8L (351M CID)

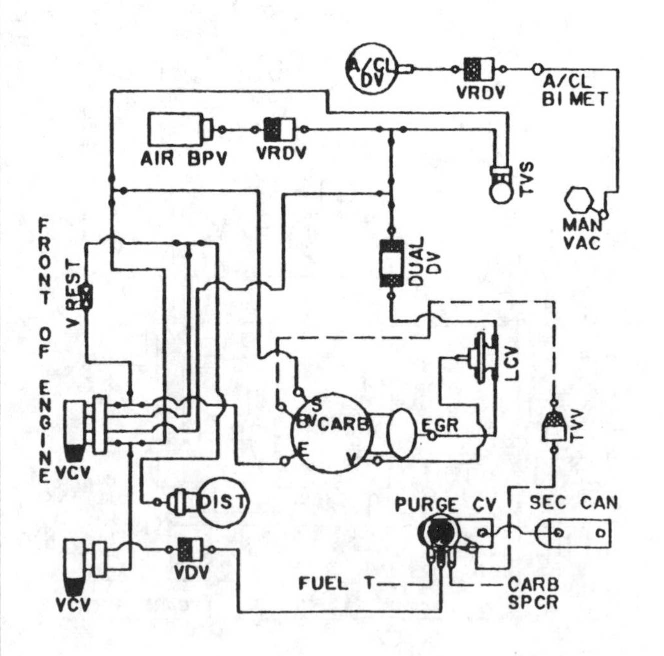

CALIBRATION: 9–60G–RO DATE: 5–16–78
5.8L (351M CID)

VACUUM CIRCUITS

1980—351 CID (5.8 L) engines

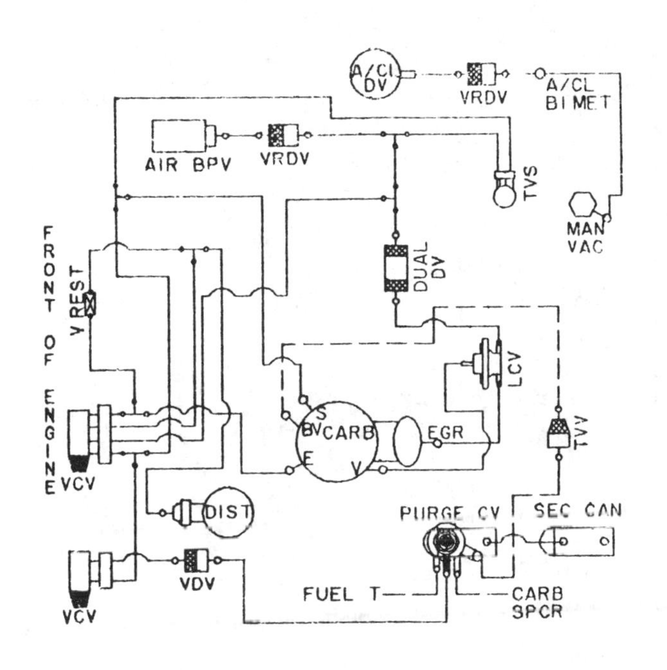

CALIBRATION. 9-60H-R0 DATE: 5-16-78
5.8L (351M CID)

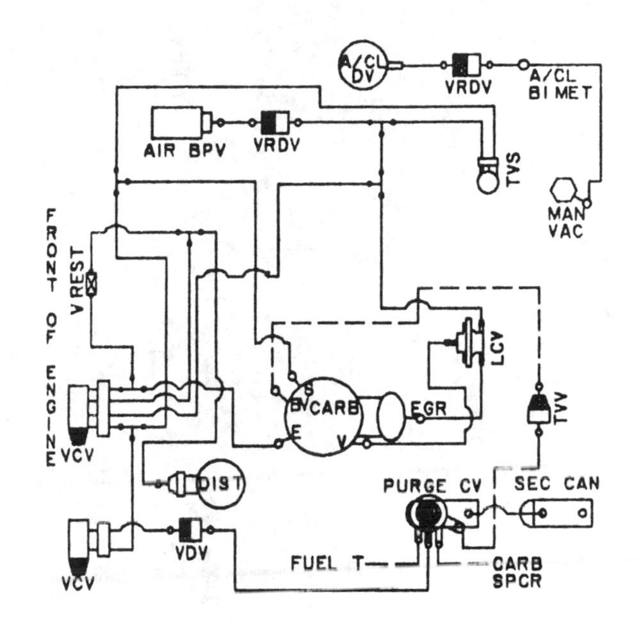

CALIBRATION: 9-60H-R10 DATE: 10-26-78
5.8L (351M CID)

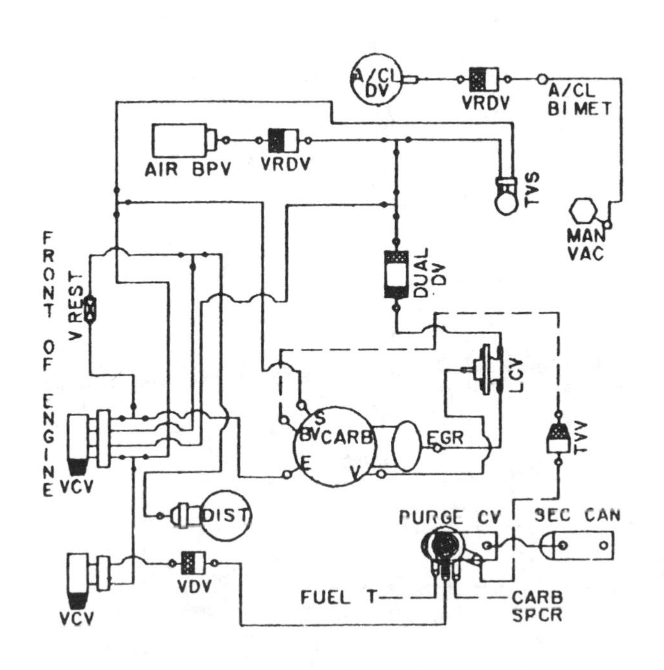

CALIBRATION: 9-60J-RO DATE: 5-18-78
5.8L (351M CID)

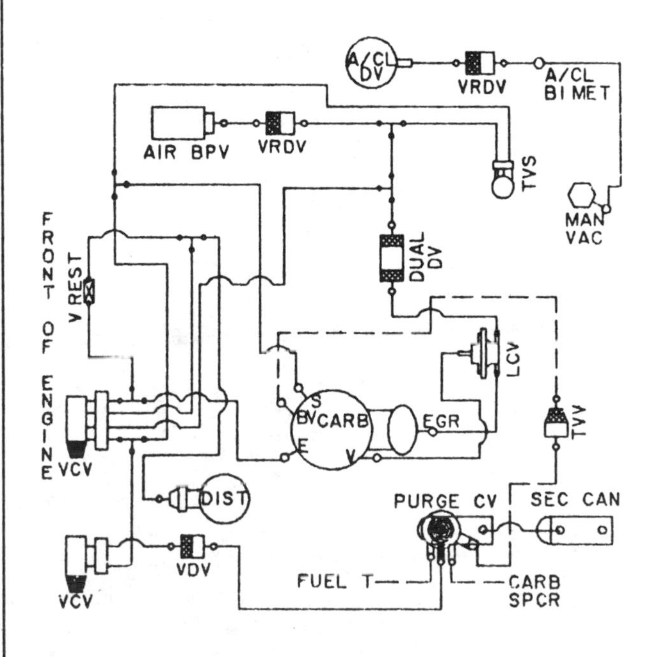

CALIBRATION: 9-60J-R10 DATE: 10-4-78
5.8L (351M CID)

VACUUM CIRCUITS

1980—351 CID (5.8 L) engines

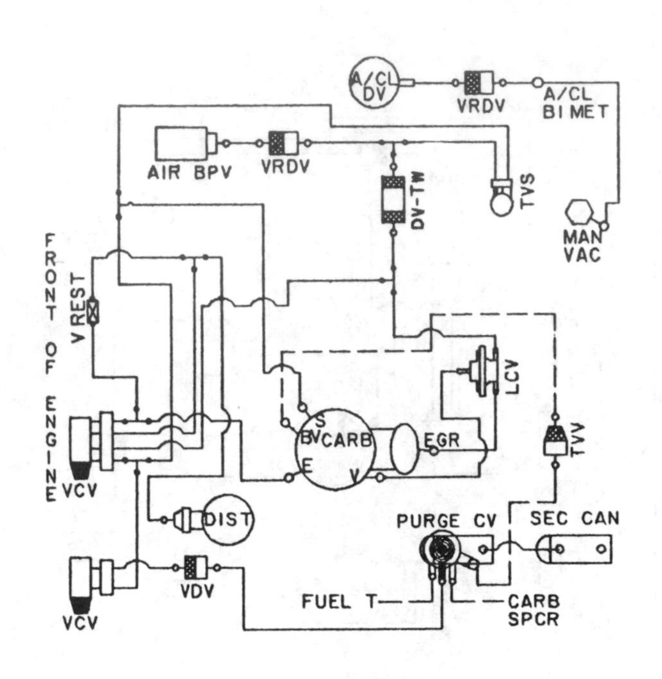

CALIBRATION. 9-60L-RO DATE: 5-18-78
5.8L (351M CID)

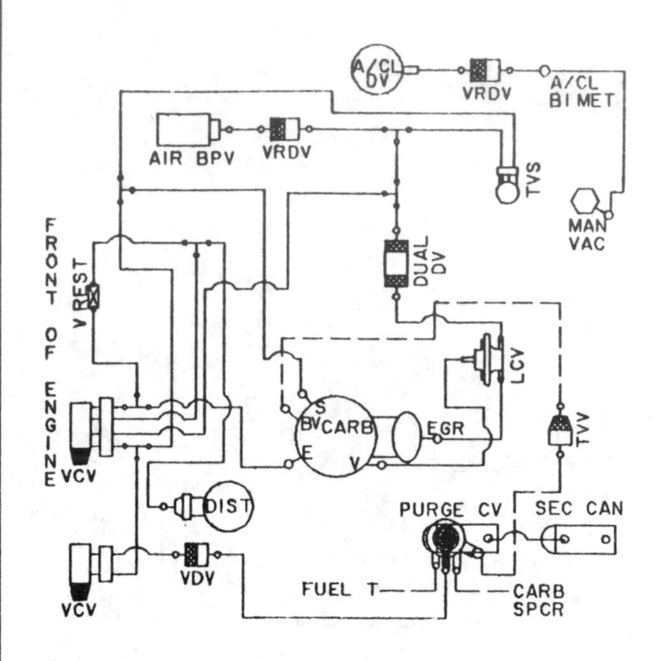

CALIBRATION: 9-60M-RO DATE: 5-16-78
5.8L (351M CID)

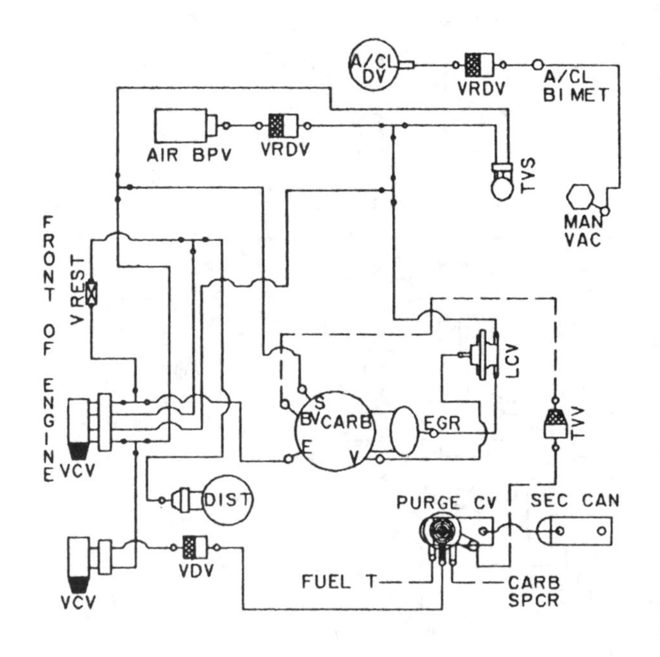

CALIBRATION: 9-60M-R10 DATE: 10-4-78
5.8L (351M CID)

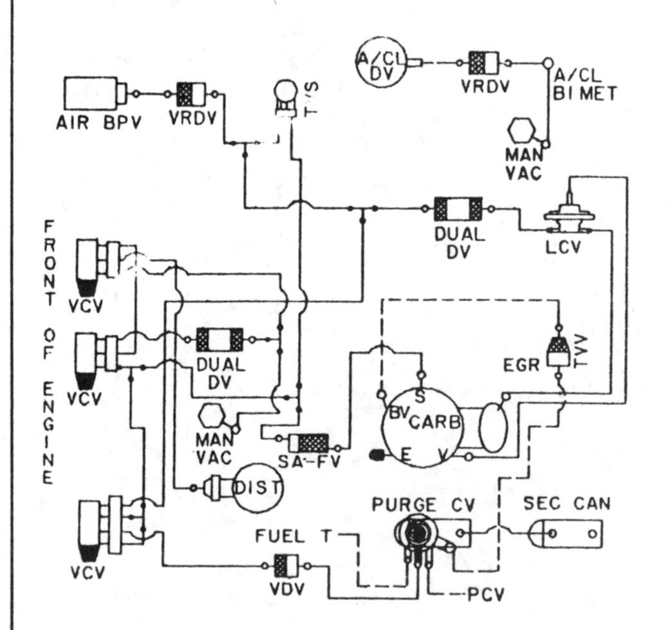

CALIBRATION: 9-60S-RO DATE. 6-15-78
5.8L (351M CID)

VACUUM CIRCUITS

1980—351 CID (5.8 L) engines

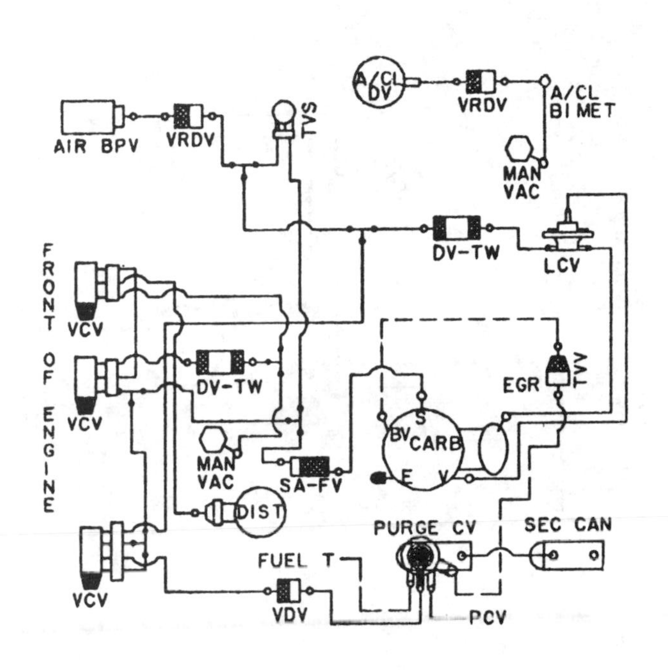

CALIBRATION: 9–60S–R10 DATE: 7–25–78
5.8L (351M CID)

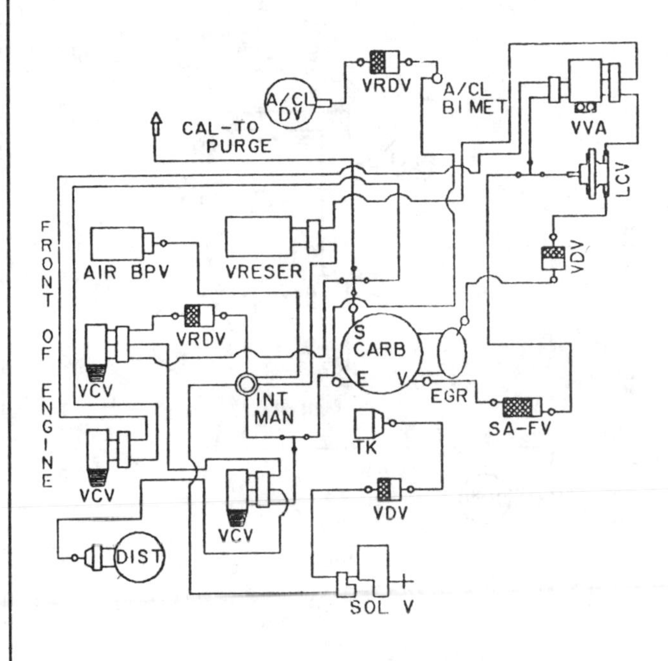

CALIBRATION: 9–71J–R10 DATE: 12–7–78
5.8L (351M CID)

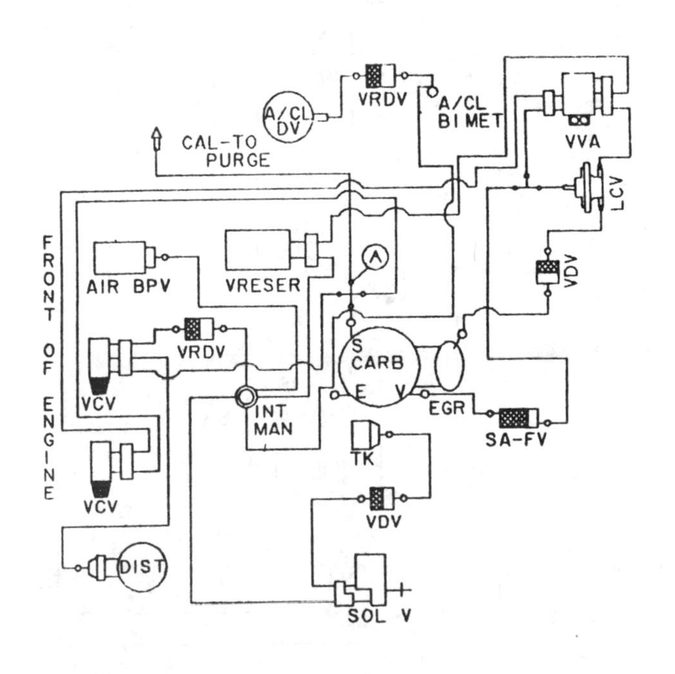

CALIBRATION: 9–72J–RO DATE: 10–13–78
5.8L (351M CID)

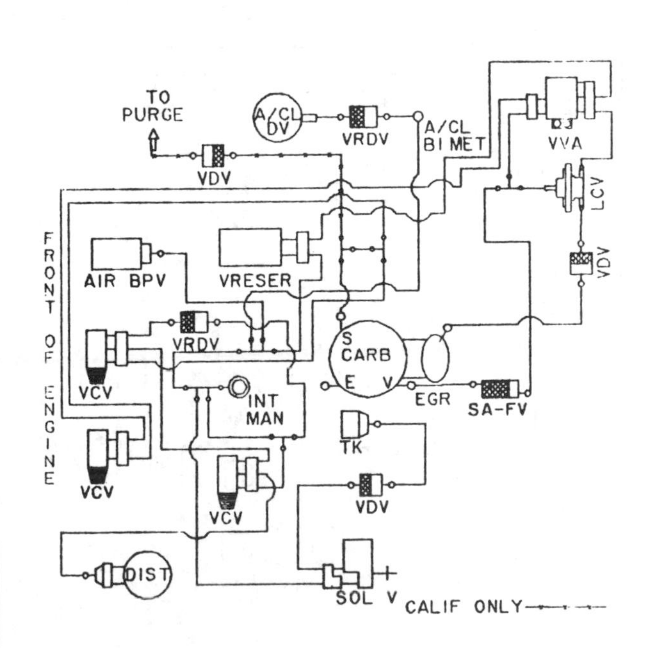

CALIBRATION: 9–72J–R11 DATE: 1–25–79
5.8L (351M CID)

VACUUM CIRCUITS

1980—351 CID (5.8 L) engines

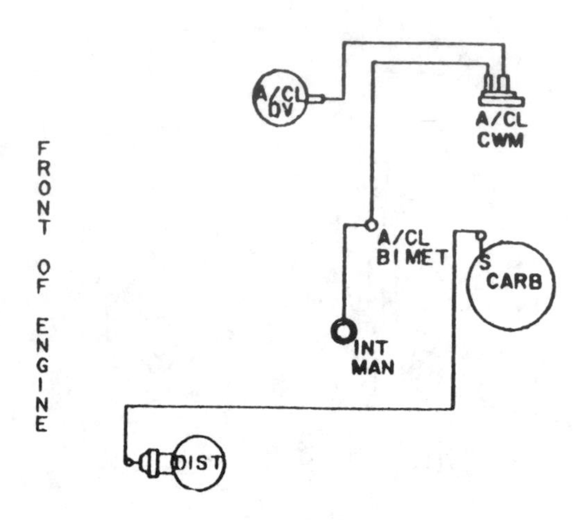

CALIBRATION: 7–75A–R10 DATE: 5–17–78
5.8L (351W CID) CANADIAN ONLY

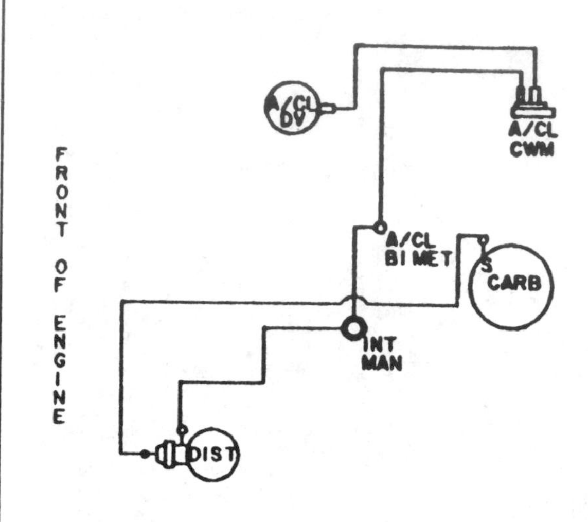

CALIBRATION: 7–76A–R10 DATE: 5–17–78
5.8L (351W CID) CANADIAN ONLY

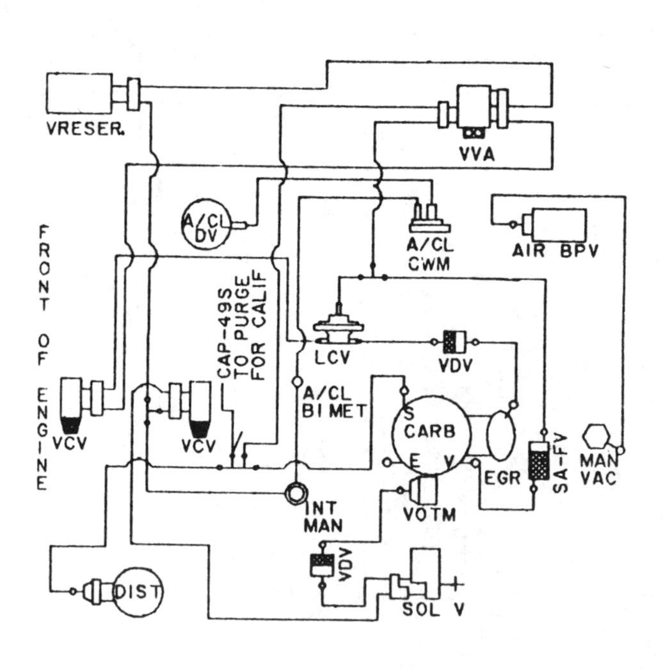

CALIBRATION: 7–76J–R11 DATE: 6–6–78
5.8L (351 CID)

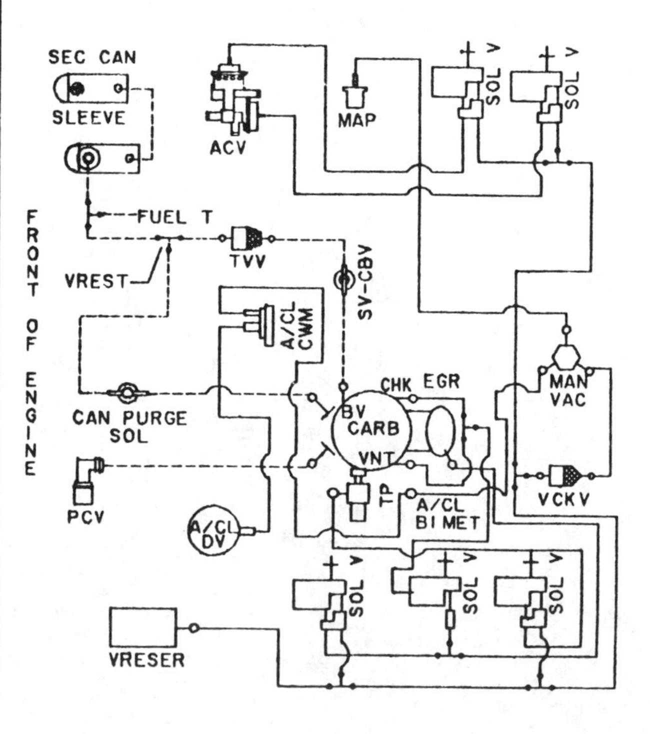

CALIBRATION: 9–12A–RO DATE: 6–21–78
5.8L (351W CID)

VACUUM CIRCUITS

1980—351 CID (5.8 L) engines

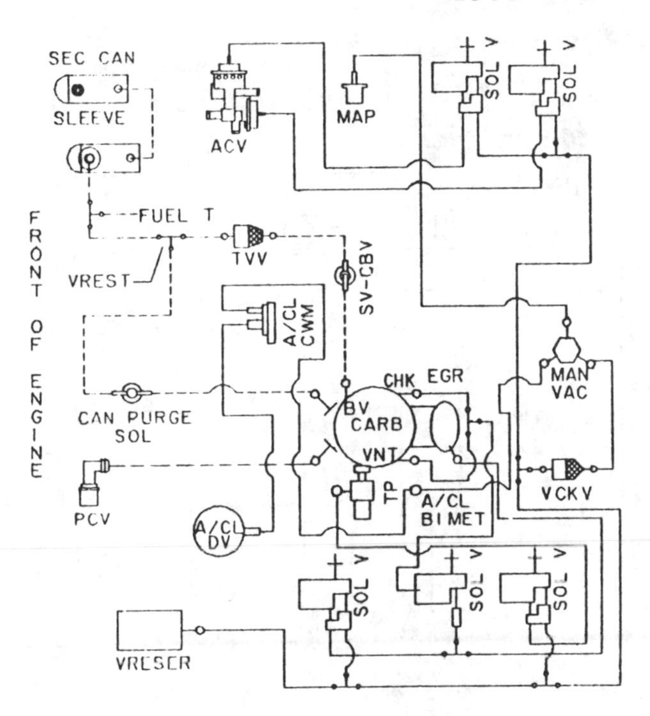

CALIBRATION: 9-12B-RO DATE: 6-7-78
5.8L (351W CID)

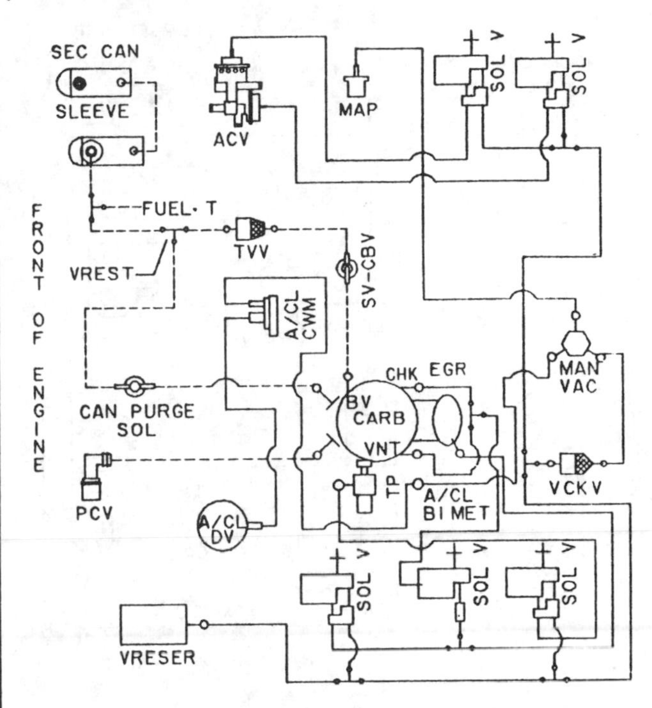

CALIBRATION: 9-12B-R10 DATE: 8-12-78
5.8L (351W CID)

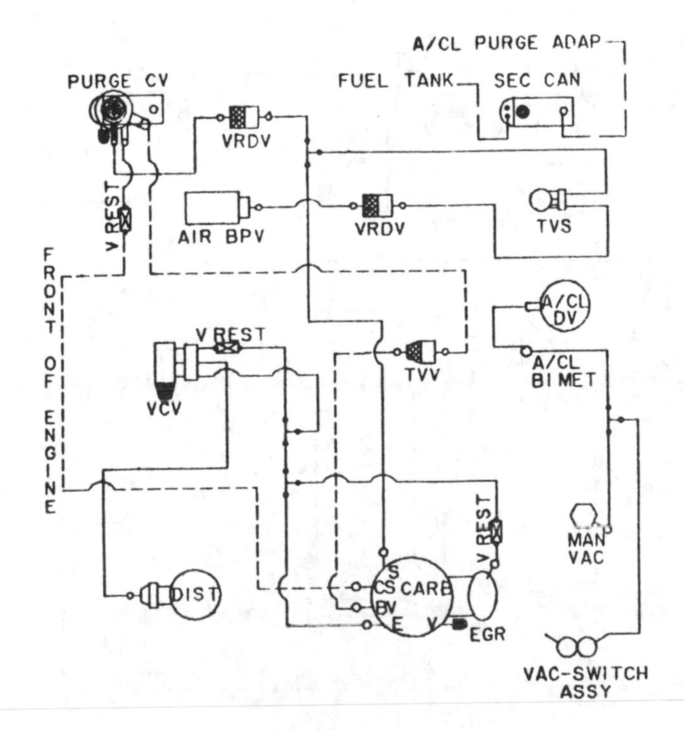

CALIBRATION: 9-12E-RO DATE: 6-1-78
5.8L (351W CID)

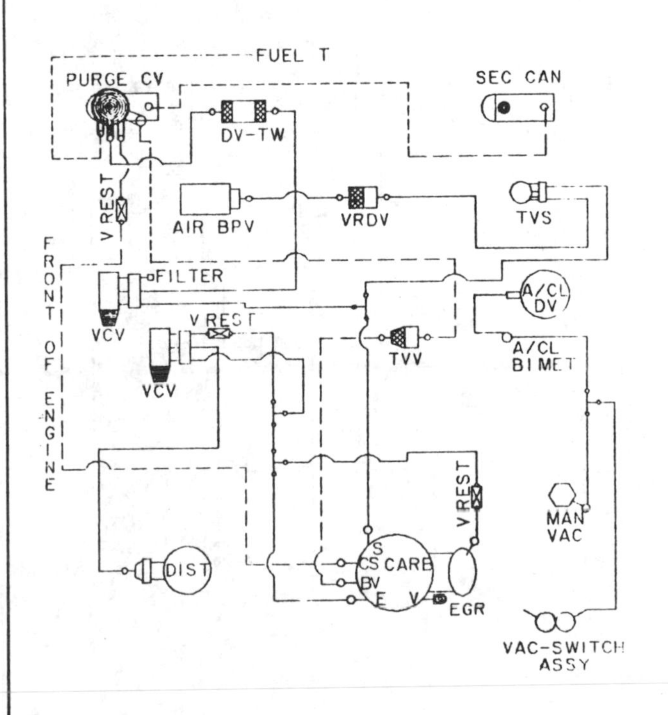

CALIBRATION: 9-12E-R14 DATE: 12-20-78
5.8L (351W CID)

VACUUM CIRCUITS

1980—351 CID (5.8 L) engines

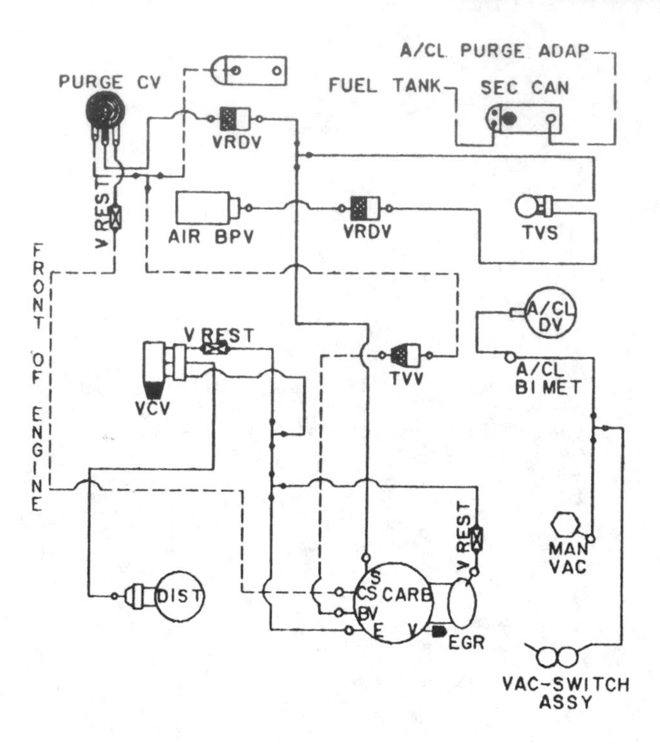

CALIBRATION: 9–12F–RO DATE: 6–1–78
5.8L (351W CID)

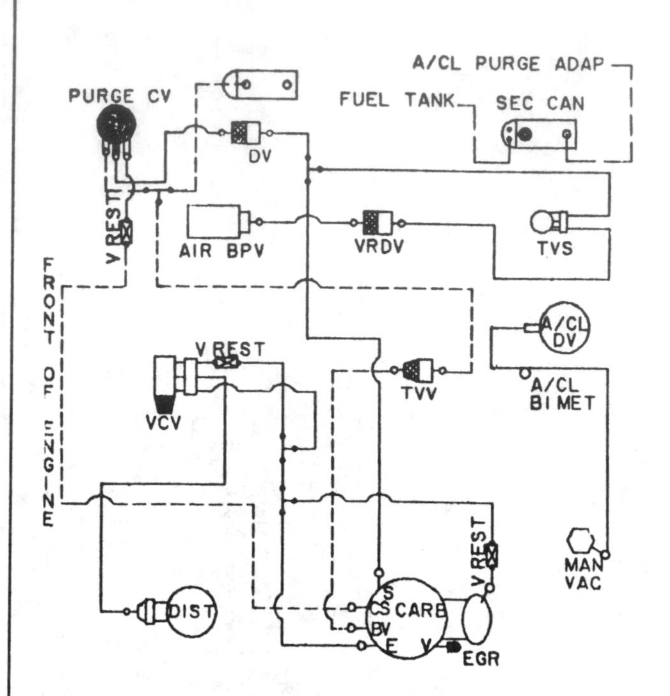

CALIBRATION: 9–12G–RO DATE: 6–1–78
5.8L (351W CID)

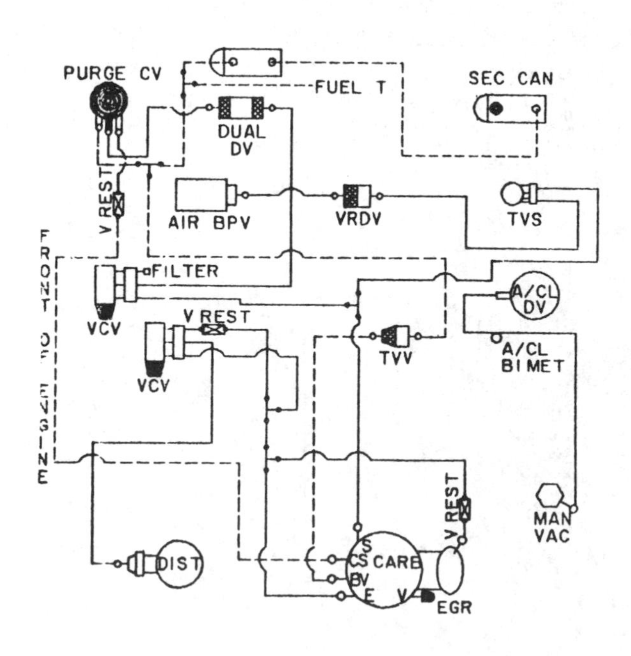

CALIBRATION: 9–12G–R12 DATE: 8–1–78
5.8L (351W CID)

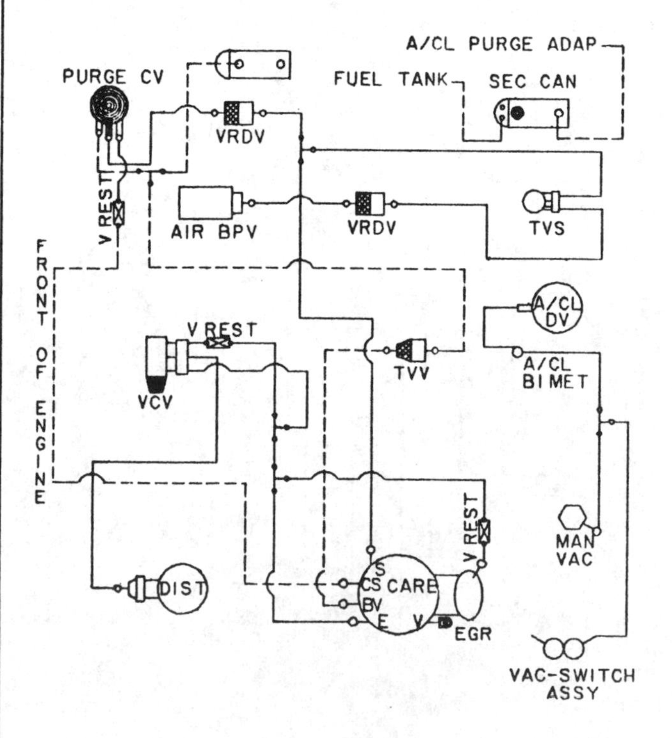

CALIBRATION: 9–12H–R12 DATE: 7–8–78
5.8L (351 W CID)

VACUUM CIRCUITS

1980—351 CID (5.8 L) engines

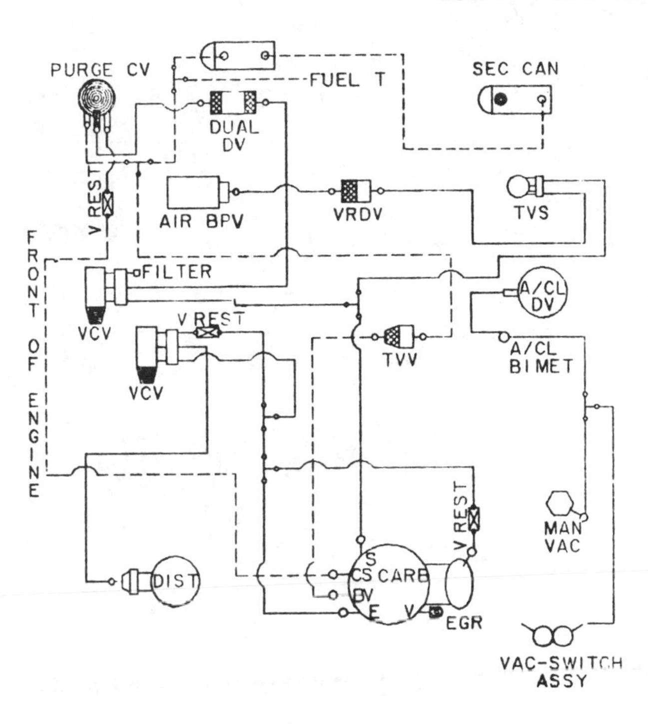

CALIBRATION: 9–12H–R15 DATE: 10–13–78
5.8L (351W CID)

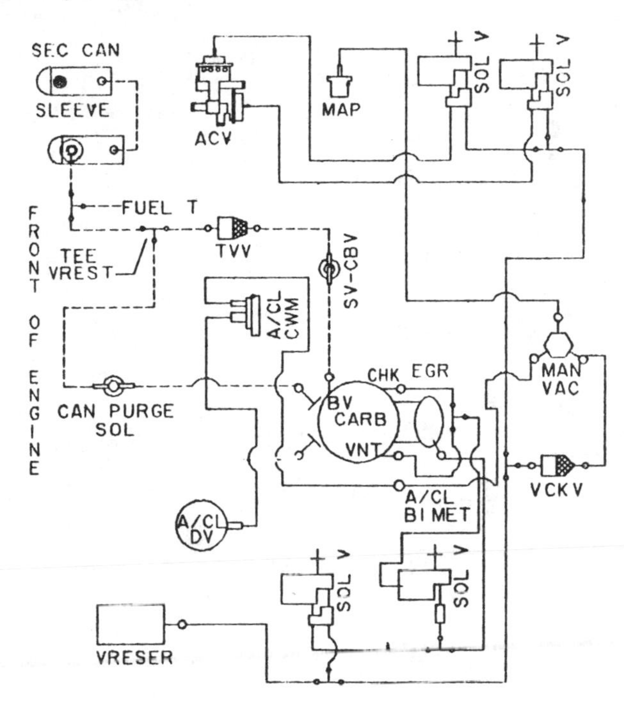

CALIBRATION. 9–12N–RO DATE: 8–14–78
5.8L (351W CID)

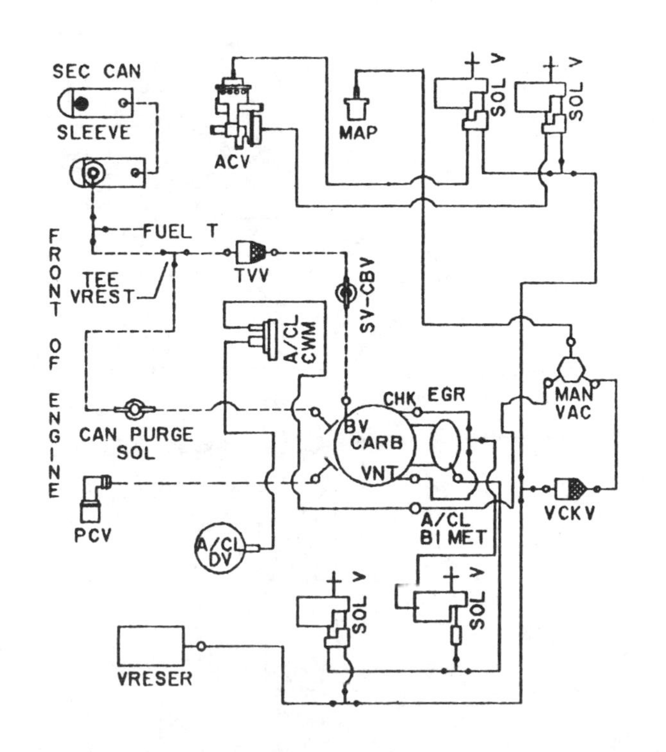

CALIBRATION: 9–12P–R10 DATE: 8–25–78
5.8 (W) (351 CID)

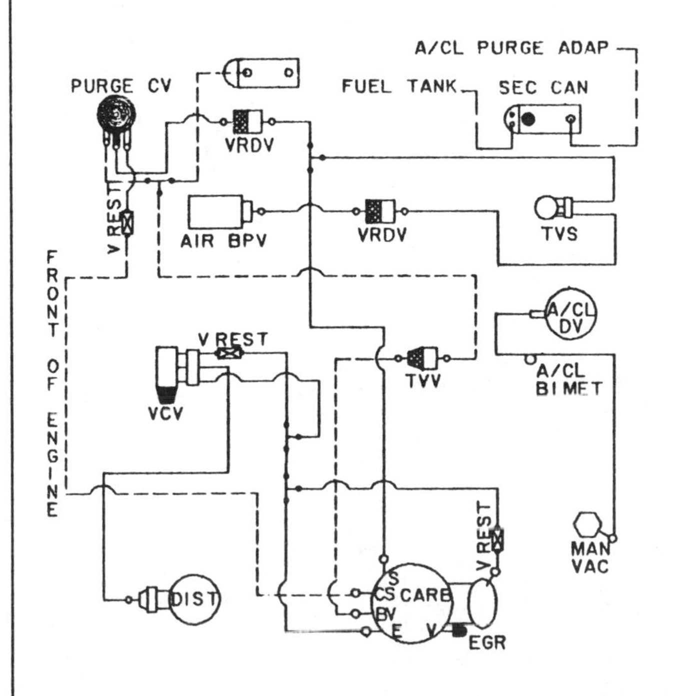

CALIBRATION: 9–12X–RO DATE: 7–13–78
5.8L (351W CID)

VACUUM CIRCUITS

1980—351 CID (5.8 L) engines

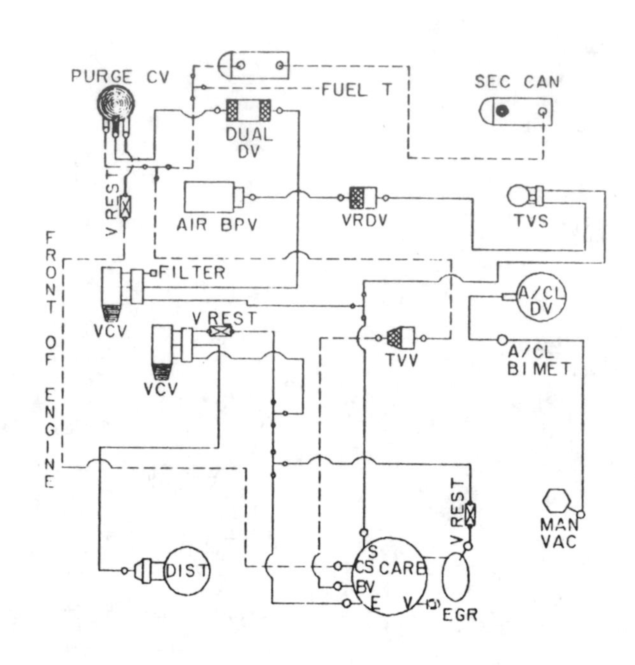

CALIBRATION: 9–12X–R10 DATE: 1–22–79
5.8L (351W CID)

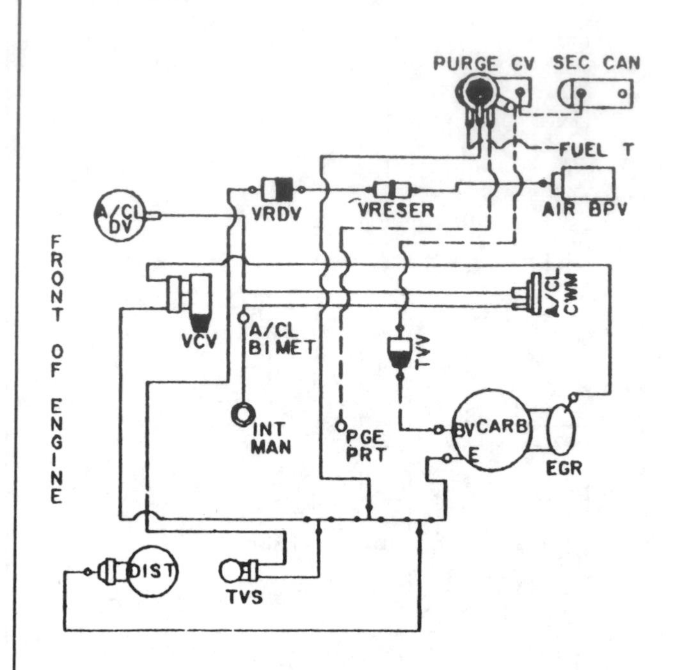

CALIBRATION: 9–63H–RO DATE: 6–13–78
5.8L (351W CID)

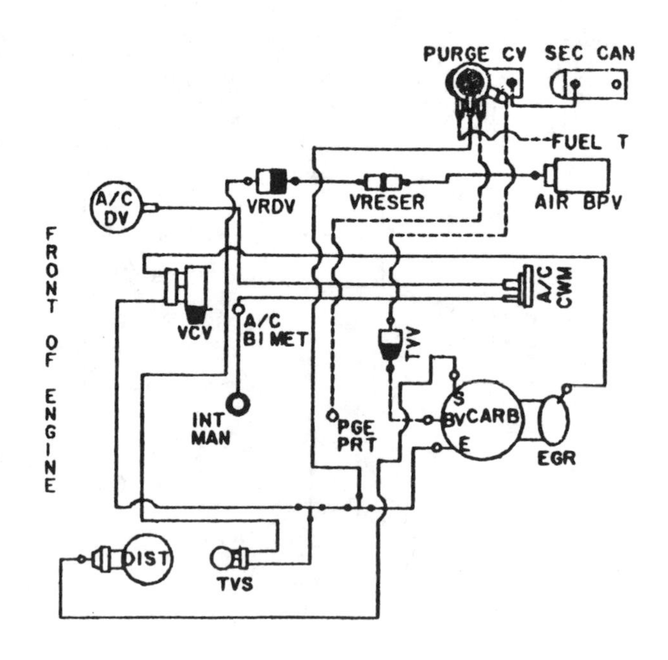

CALIBRATION: 9–64A–RO DATE: 5–30–78
5.8L (351W CID)

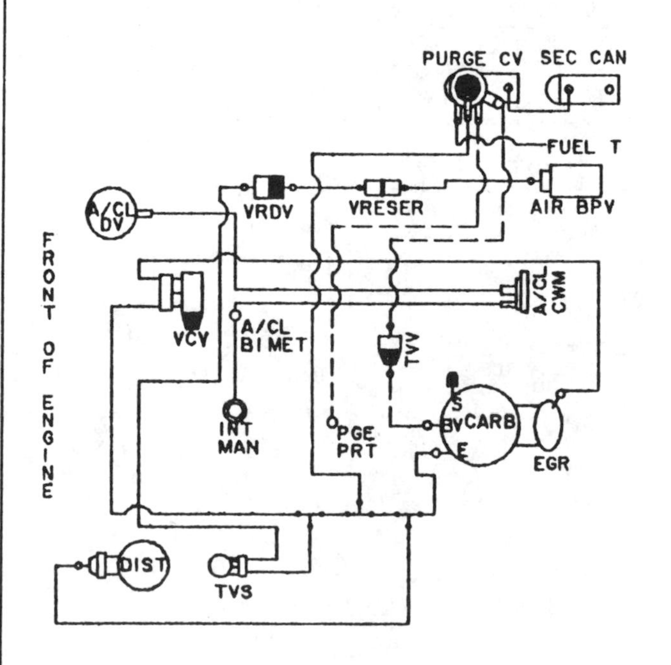

CALIBRATION: 9–64B–RO DATE: 5–30–78
5.8L (351W CID)

VACUUM CIRCUITS

1980—351 CID (5.8 L) engines

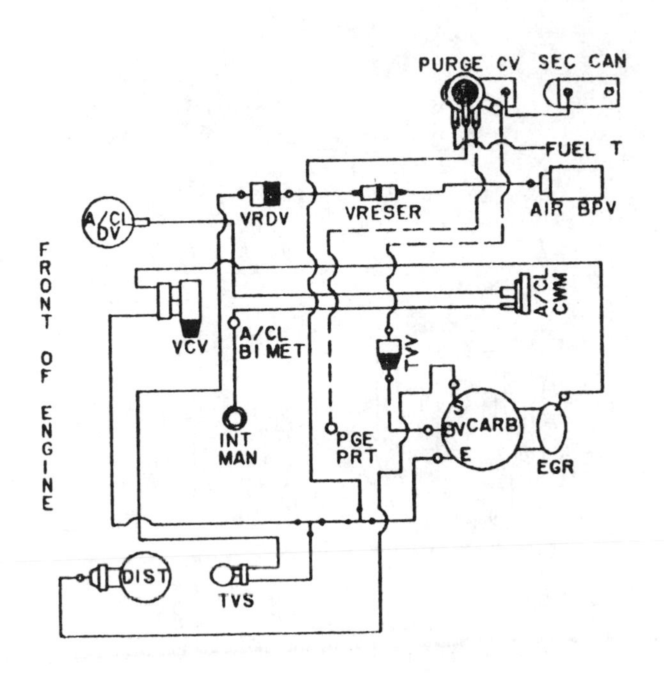

CALIBRATION. 9–64G–RO DATE: 6–7–78
5.8L (351W CID)

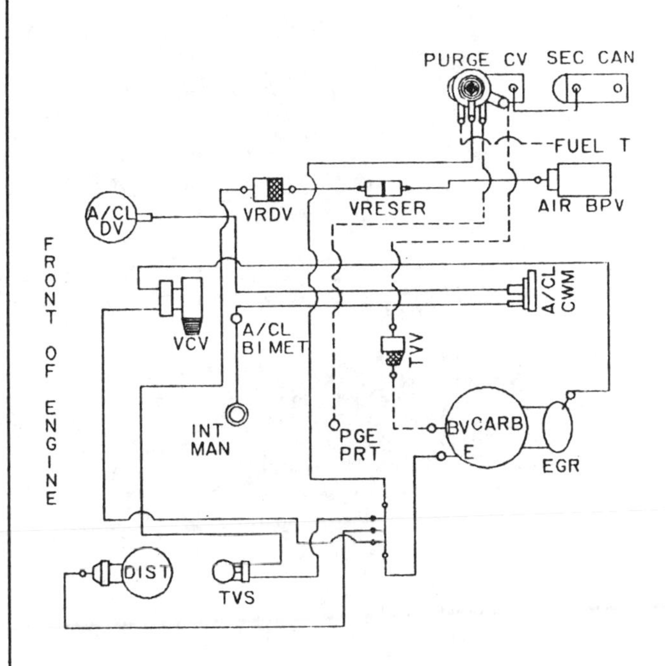

CALIBRATION: 9–64G–R10 DATE: 1–10–79
5.8L (351W CID)

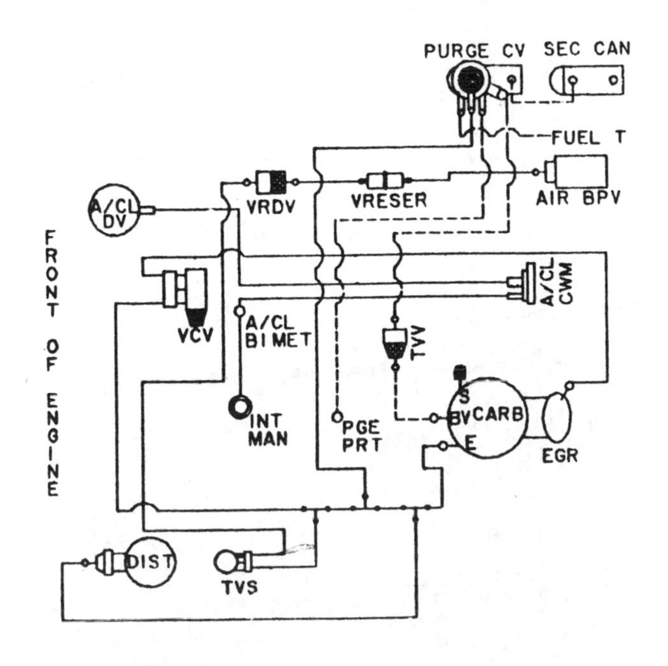

CALIBRATION: 9–64H–RO DATE: 6–16–78
5.8L (351W CID)

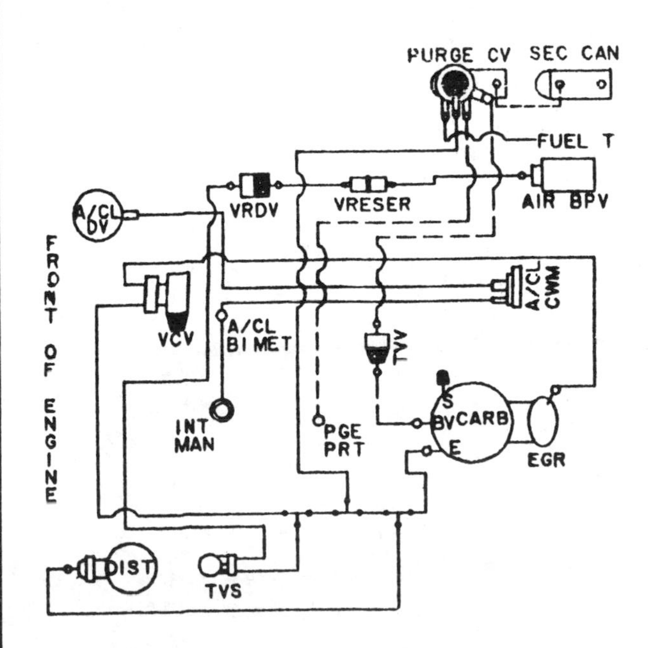

CALIBRATION: 9–64S–RO DATE: 6–4–78
5.8L (351W CID)

VACUUM CIRCUITS

1980—351 CID (5.8 L) engines

49S A/C & NON A/C

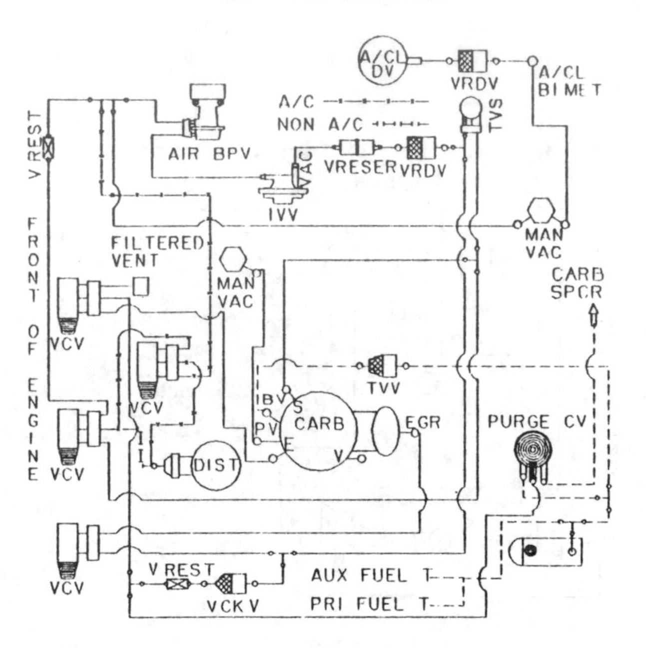

CALIBRATION: 0–59C–RO **DATE:** 8–14–79
5.8L (M) TRUCK 49S M/T

49S A/C & NON A/C

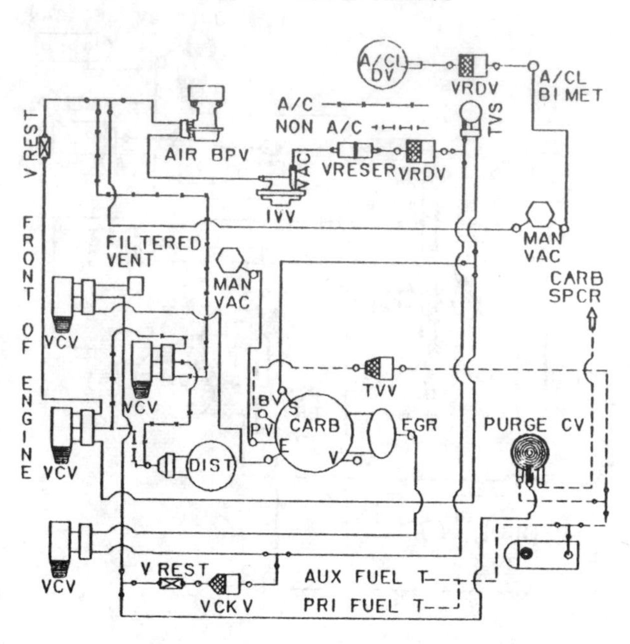

CALIBRATION: 0–59G–RO **DATE:** 8–14–79
5.8L (M) TRUCK 49S M/T

49S A/C & NON A/C

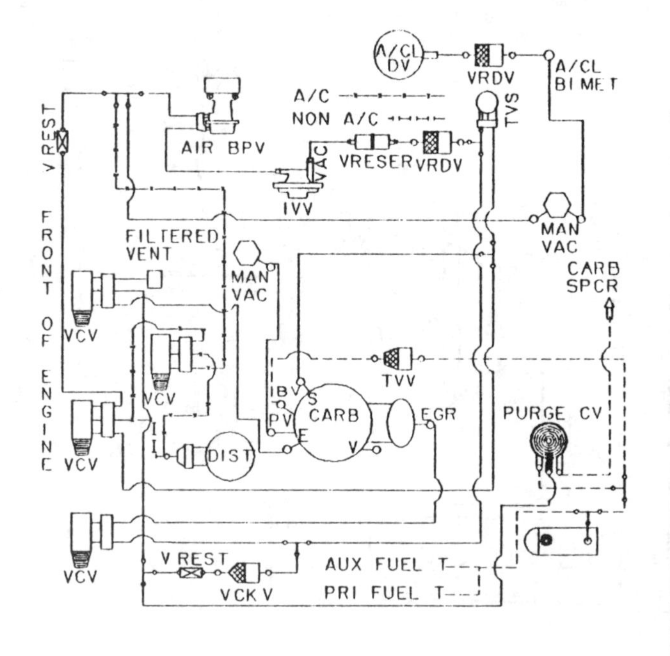

CALIBRATION: 0–59H–RO **DATE:** 8–14–79
5.8L (M) TRUCK 49S M/T

49S A/C & NON A/C

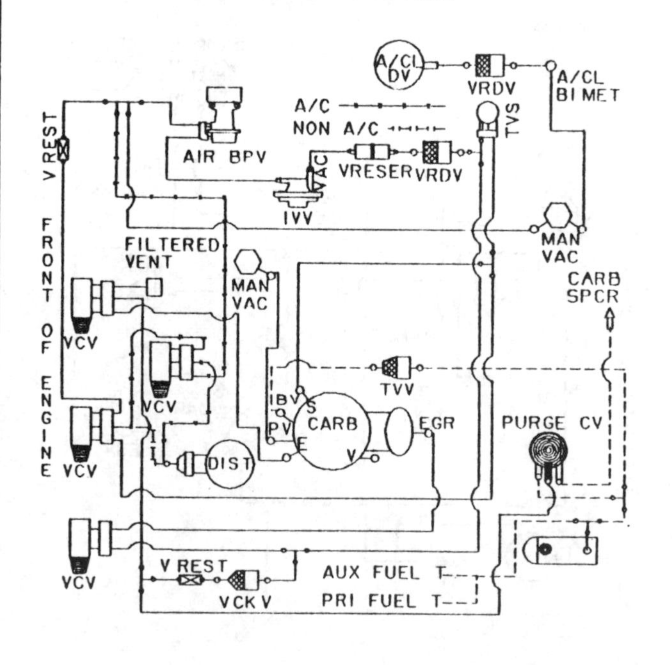

CALIBRATION: 0–59J–RO **DATE:** 8–14–79
5.8L (M) TRUCK 49S M/T

VACUUM CIRCUITS

1980—351 CID (5.8 L) engines

CALIF. ONLY A/C & NON A/C

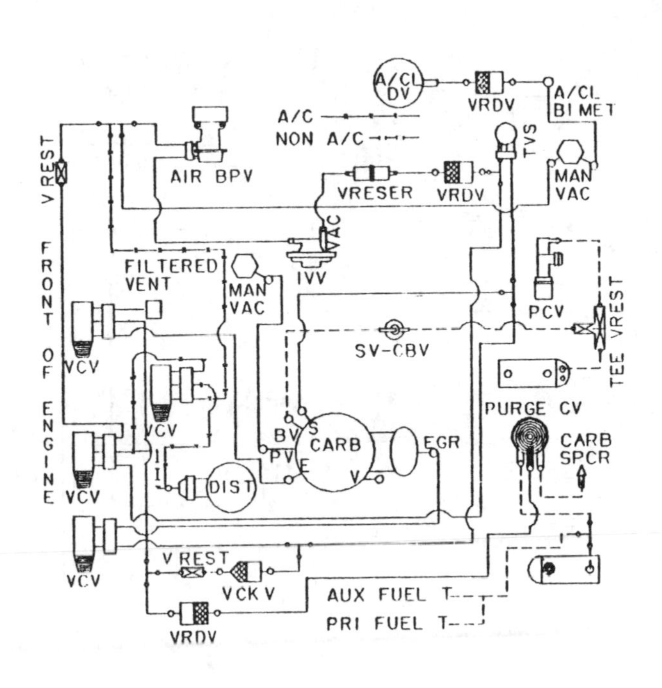

CALIBRATION: 0–59S–RO DATE: 8–14–79
5.8L(M) TRUCK CALIF. M/T

49S A/C & NON A/C

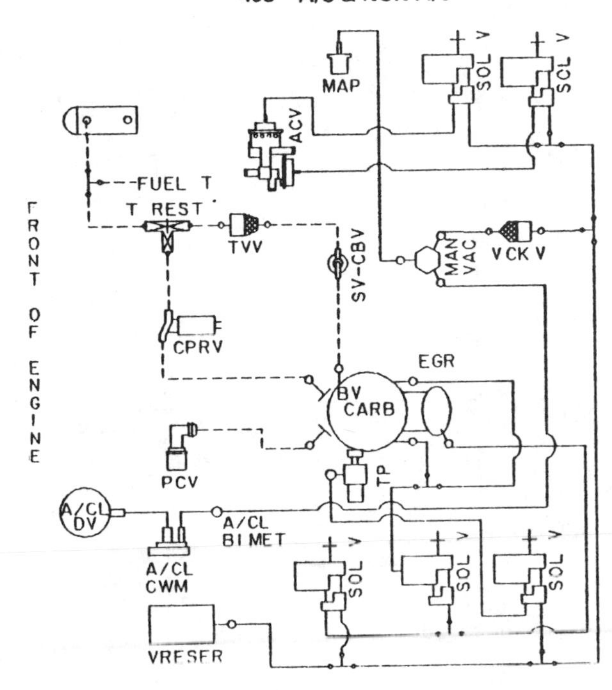

CALIBRATION: 0–12A–RO DATE: 8–7–79
5.8L(W) 49S FORD/MERC/LINC/MARK

49S A/C & NON A/C

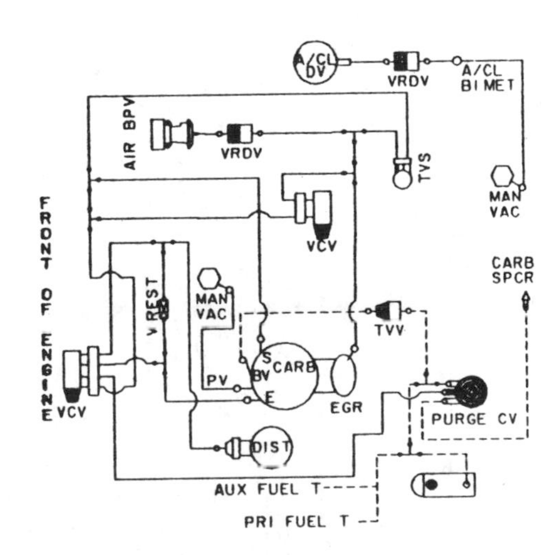

CALIBRATION: 0–60D–RO DATE: 6–15–79
5.8L(M) ECON A/T 49S

CALIF. ONLY A/C & NON A/C

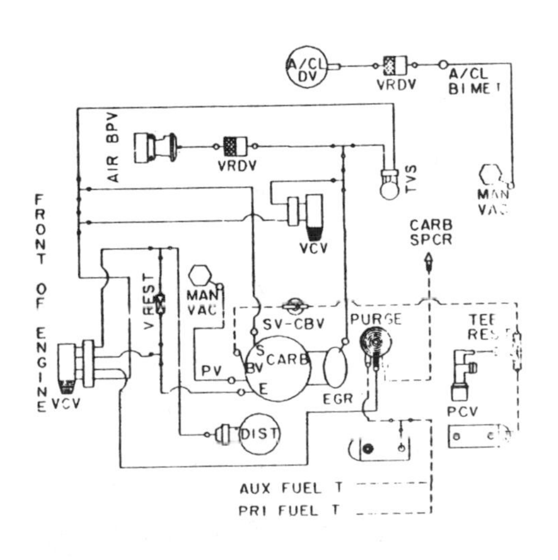

CALIBRATION: 0–60D–RO DATE: 6–15–79
5.8L(M) ECON A/T CALIF.

VACUUM CIRCUITS

1980—351 CID (5.8 L) engines

49S A/C & NON A/C

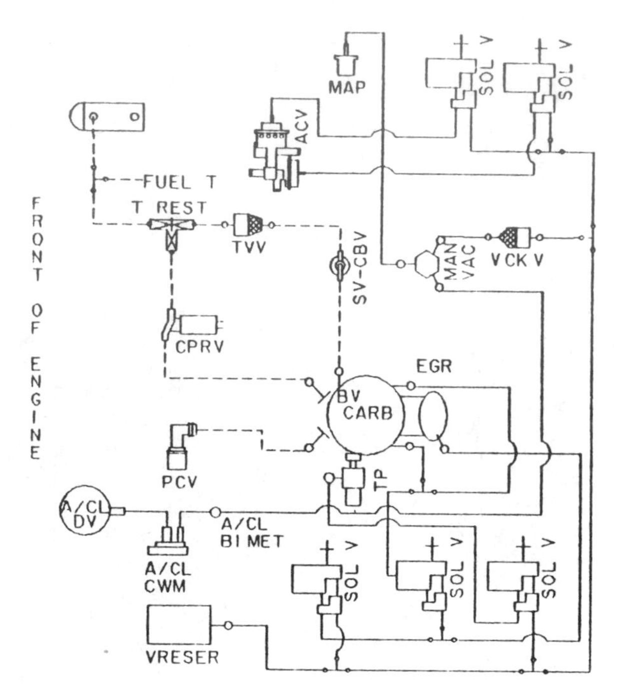

CALIBRATION: 0–12B–RO DATE: 8–7–79
5.8L(W) 49S FORD/MERC/LINC/MARK

49S A/C & NON A/C

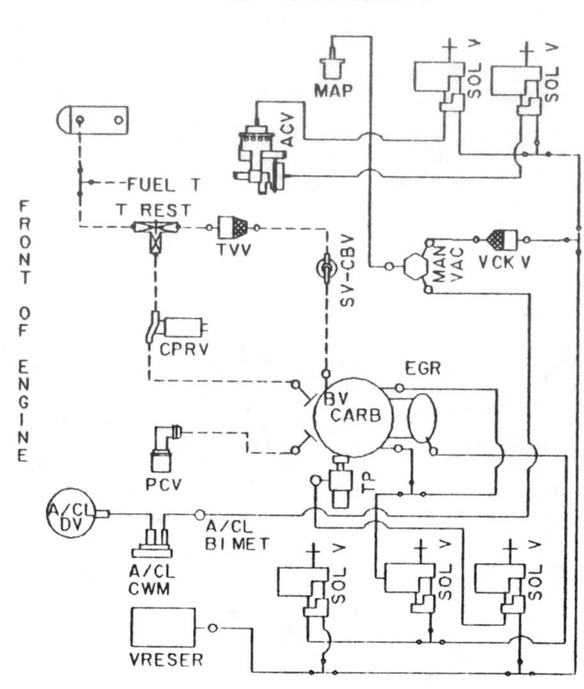

CALIBRATION: 0–12C–RO DATE: 8–7–79
5.8L(W) 49S FORD/MERC SEDAN

A/C & NON A/C

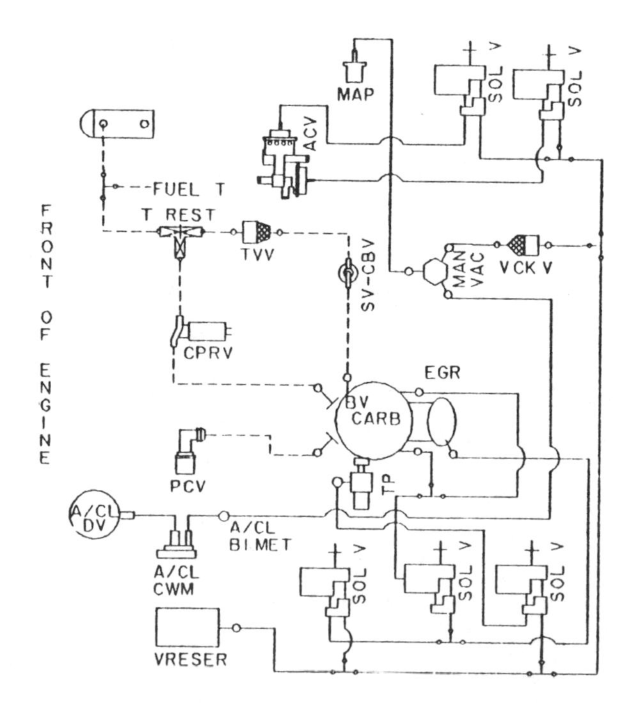

CALIBRATION: 0–12N–RO DATE: 8–7–79
5.8L(W) FORD/MERC/LINC/MARK

49S A/C & NON A/C

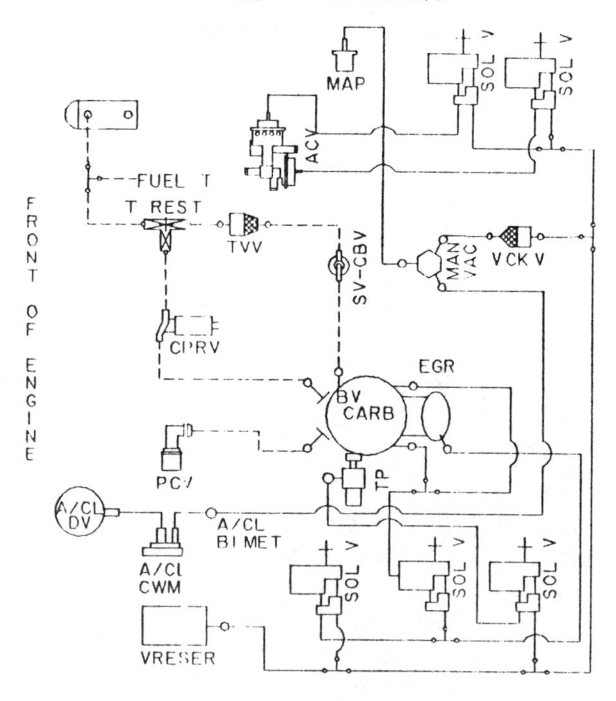

CALIBRATION: 0–12A–R5 DATE: 10–19–79
5.8L (W) FORD/MERC 49S

VACUUM CIRCUITS

1980—351 CID (5.8 L) engines

49S A/C & NON A/C

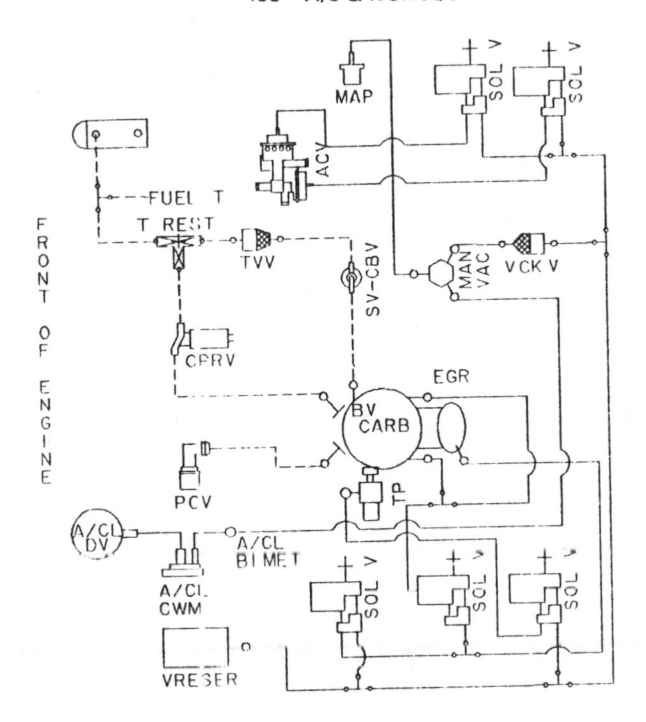

CALIBRATION: 0–12C–R5 DATE: 10–24–79
5.8L (W) FORD/MERC SW/SDN 49S

A/C

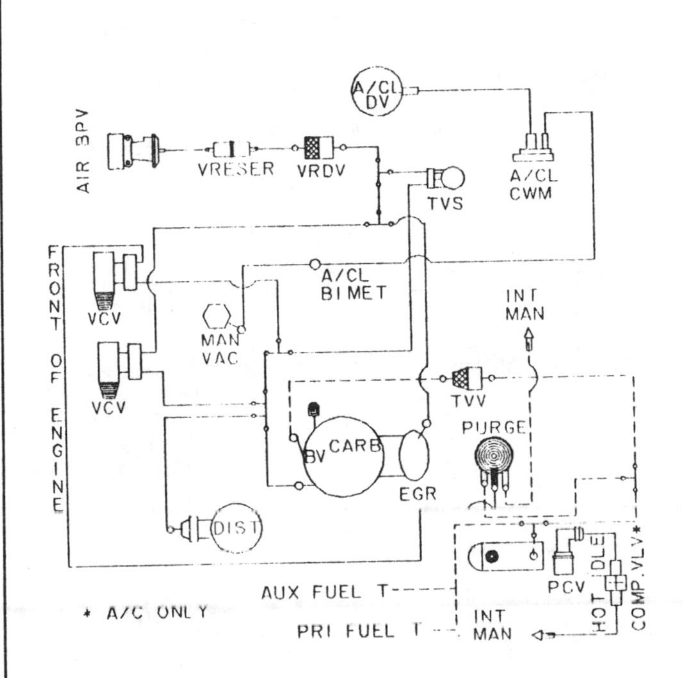

CALIBRATION: 0–64A–R10 DATE: 09–24–79
5.8L (W) ECONOLINE A/T

A/C

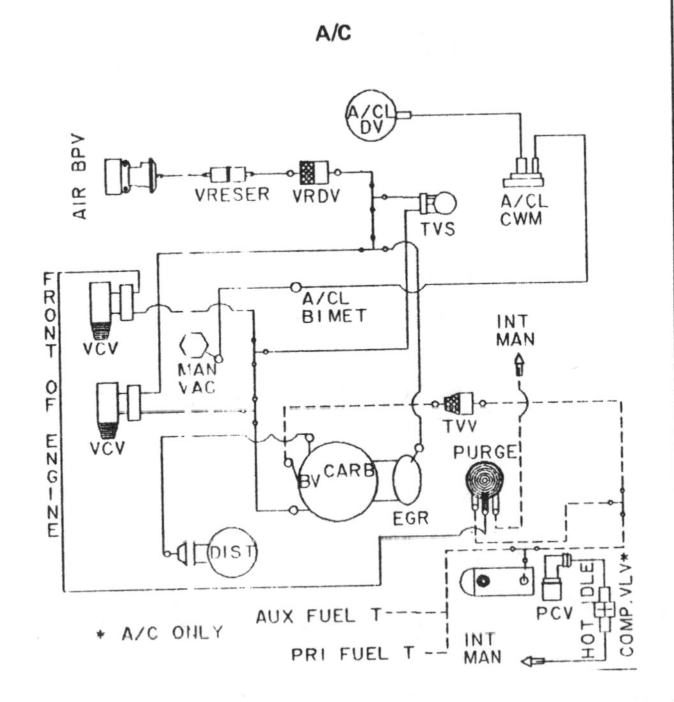

CALIBRATION: 0–64B–R10 DATE: 09–18–79
5.8L (W) ECONOLINE A/T

49S A/C

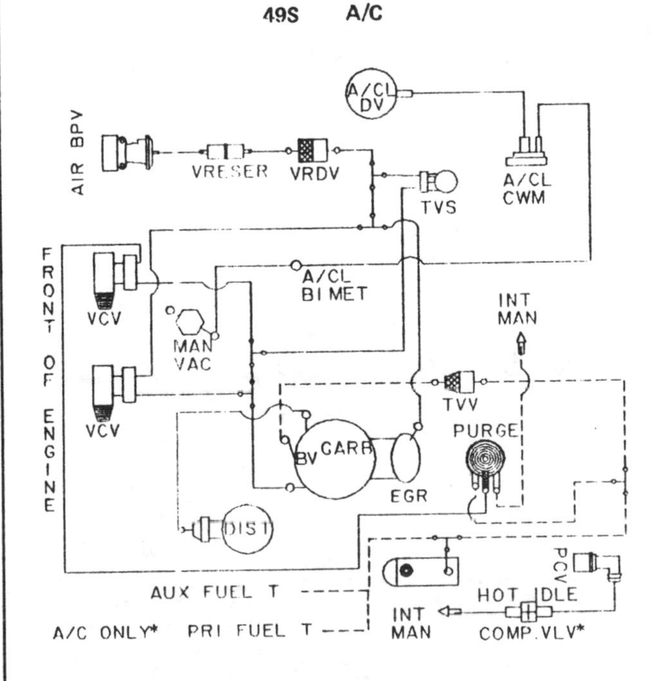

CALIBRATION: 0–64H–R10 DATE: 09–18–79
5.8L (W) ECONOLINE 49S A/T

VACUUM CIRCUITS

1980—351 CID (5.8 L) engines

CALIF ONLY A/C

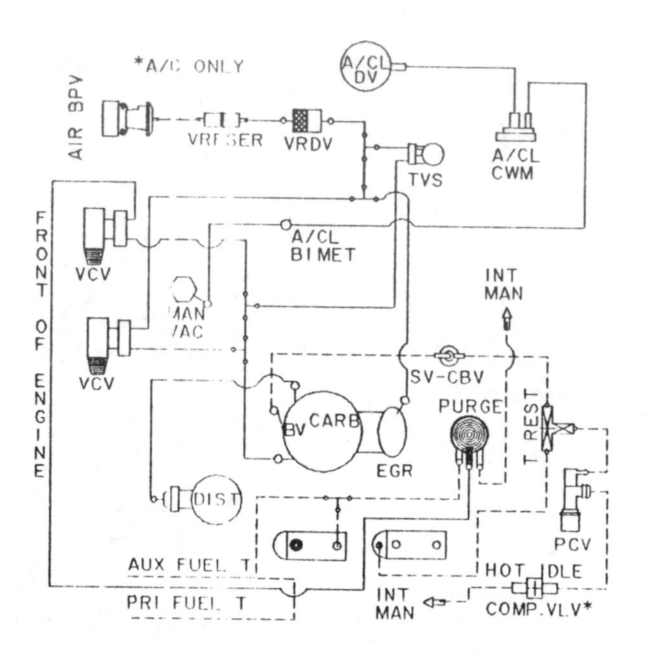

CALIBRATION: 0–64T–R10 DATE: 09–18–79
5.8L (W) ECONOLINE CALIF A/T

CANADA A/C & NON A/C

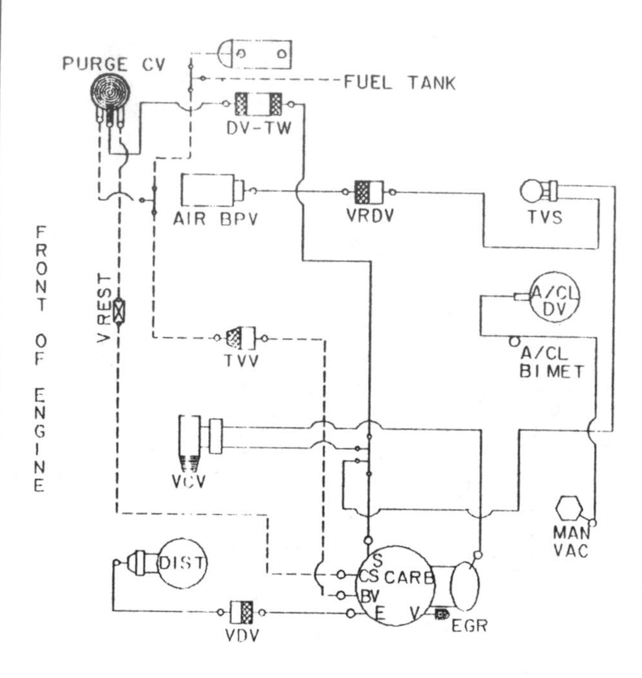

CALIBRATION: 0–12G–R0 DATE: 08–23–79
5.8L (W) LINCOLN/MARK CANADA

49S A/C

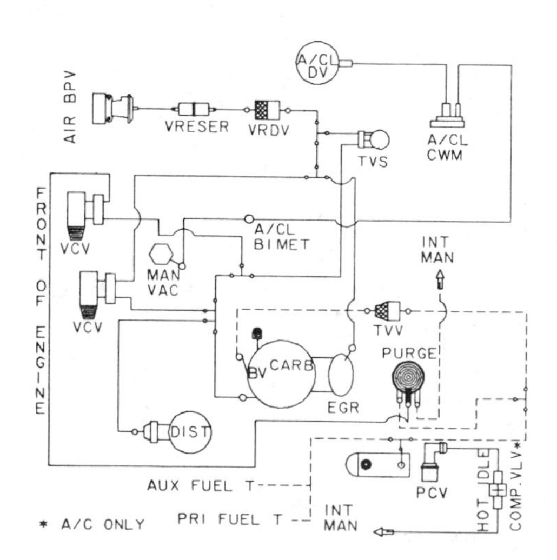

CALIBRATION: 0–64G–R10 DATE: 09–19–79
5.8L (W) ECONOLINE 49S A/T

CALIF ONLY A/C

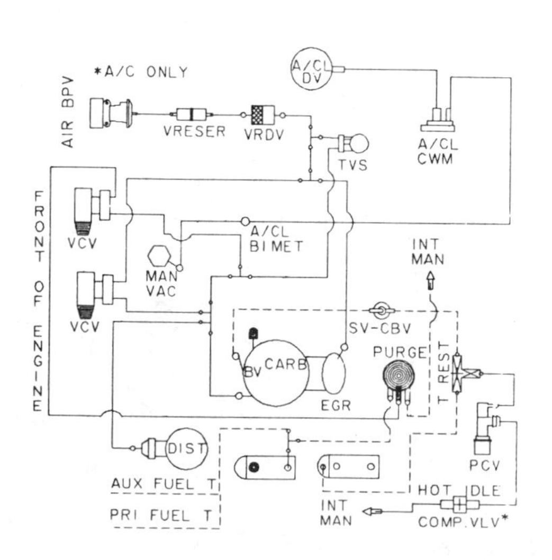

CALIBRATION: 0–64G–R10 DATE: 09–19–79
5.8L (W) ECONOLINE CALIF A/T

VACUUM CIRCUITS

1980—351 CID (5.8 L) engines

A/C & NON A/C

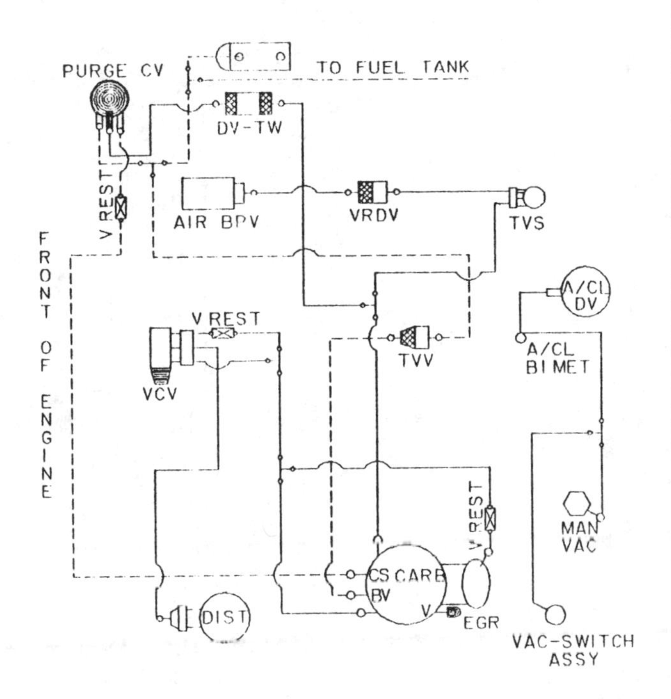

CALIBRATION: 0–12H–R10 DATE: 10–09–79
5.8L (W) FORD/FORD S.W./MERC/MERC S.W.

A/C & NON A/C

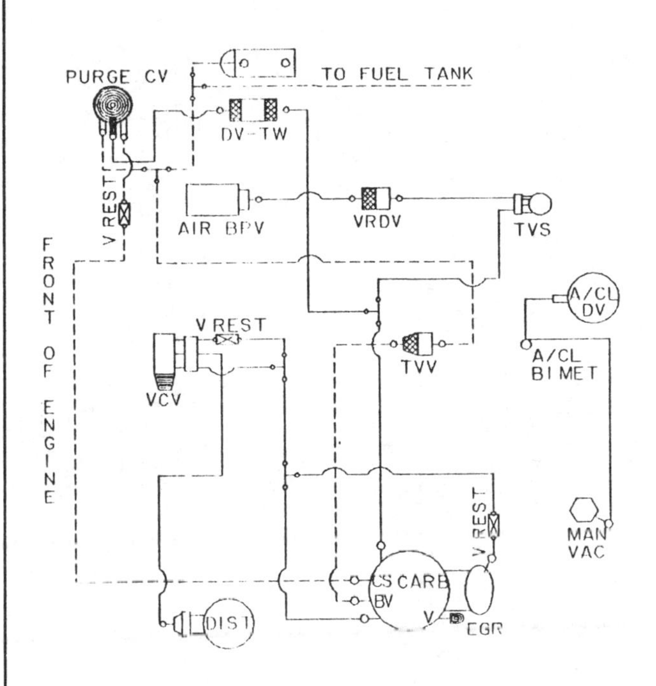

CALIBRATION: 0–12J–R10 DATE: 10–09–79
5.8L (W) FORD/FORD S.W./MERC/MERC S.W.

A/C & NON A/C

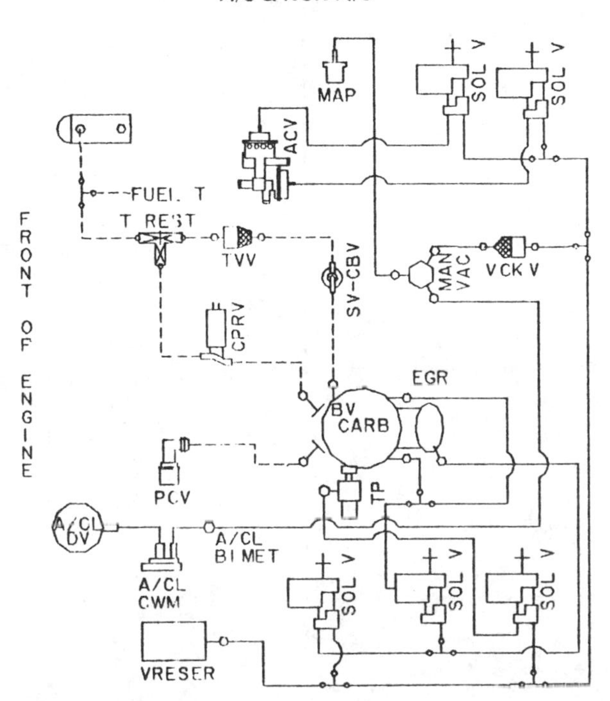

CALIBRATION: 0–12P–R0 DATE: 07–11–79
5.8L (W) FORD/MERC/LINCOLN/MARK

VACUUM CIRCUITS

1980—370 CID (6.1 L) engines

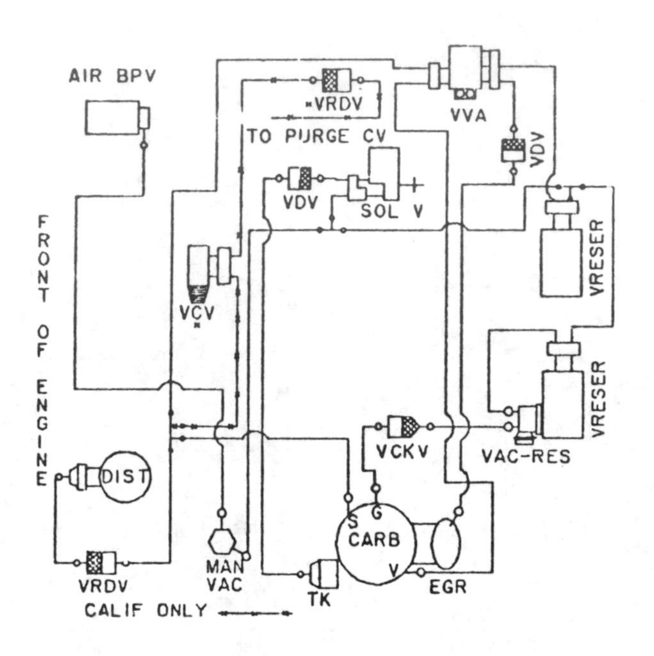

CALIBRATION: 9–83G–RO **DATE:** 6–14–78
6.1L (370 CID)

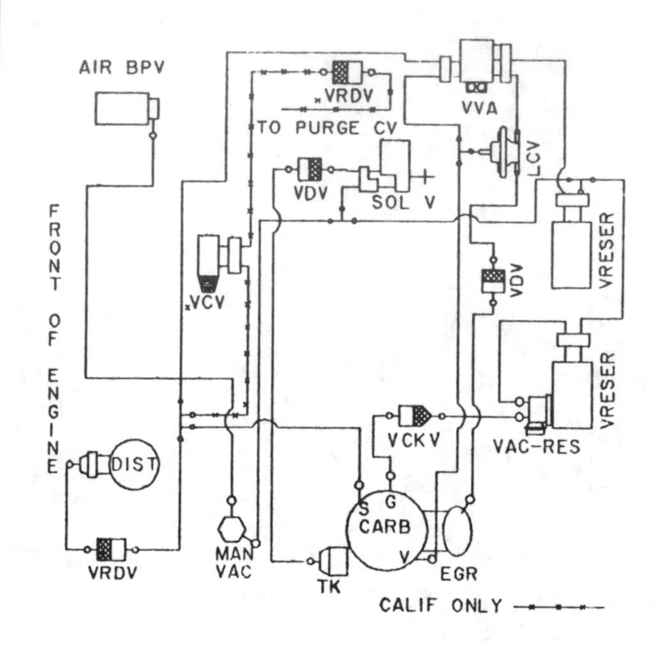

CALIBRATION: 9–83G–R2 **DATE:** 7–15–78
6.1L (370 CID)

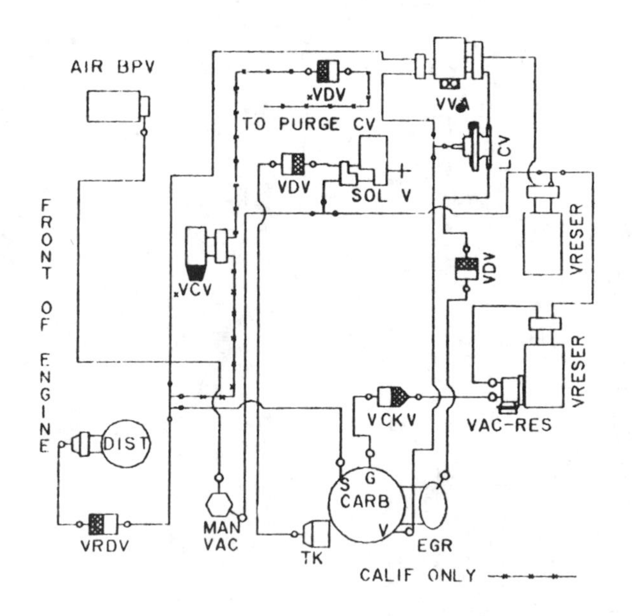

CALIBRATION: 9–83G–R11 **DATE:** 10–11–78
6.1L (370 CID)

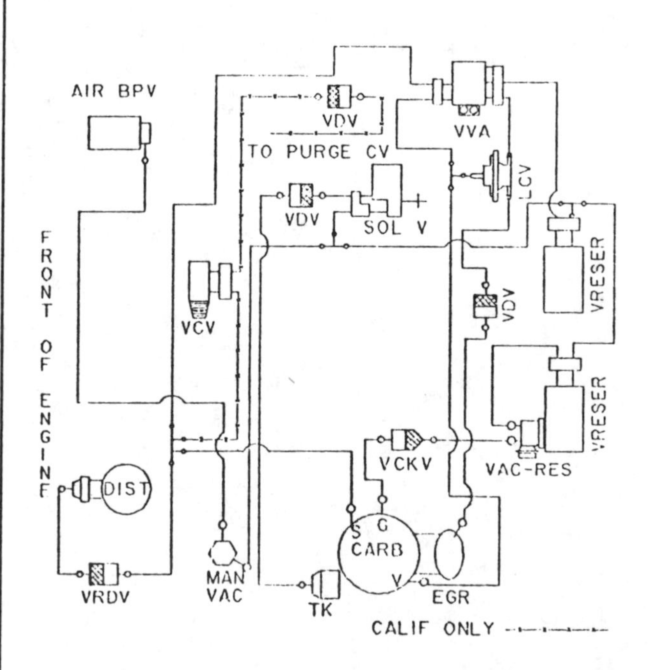

CALIBRATION: 9–83G–R12 **DATE:** 1–16–79
6.1L (370 CID)

VACUUM CIRCUITS

1980—370 CID (6.1 L) engines

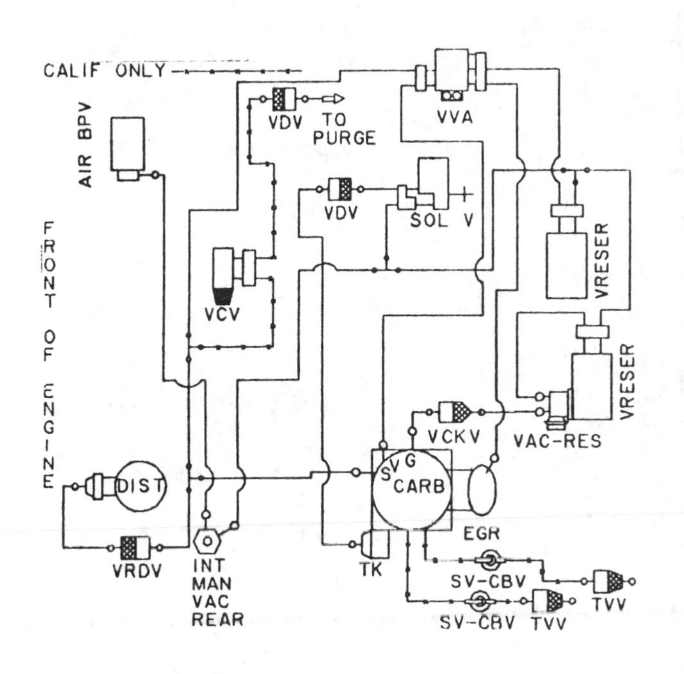

CALIBRATION: 9–83H–R2 **DATE:** 12–14–78
6.1L (370 CID)

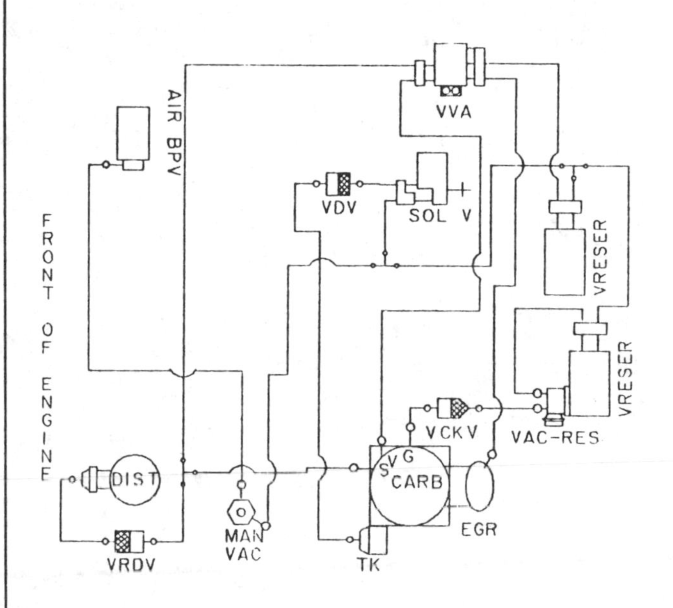

CALIBRATION: 9–83H–R10 **DATE:** 12–4–78
6.1L (370 CID)

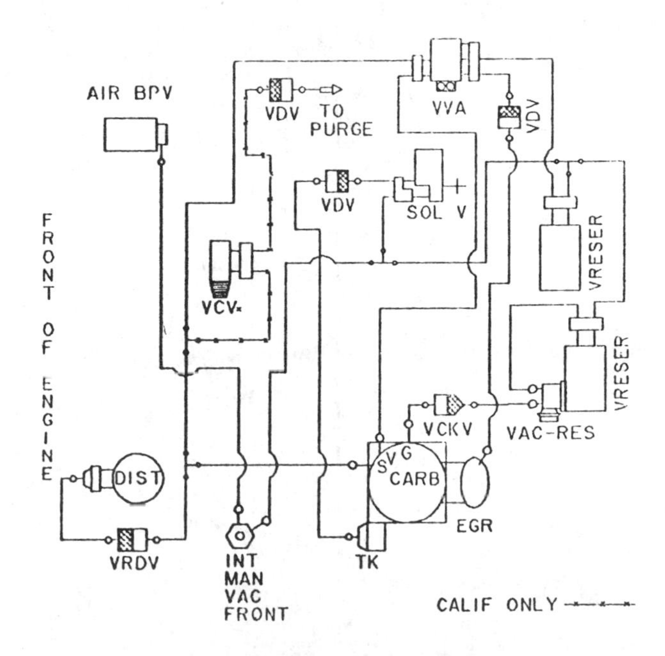

CALIBRATION: 9–83H–R11 **DATE:** 1–16–79
6.1L (370 CID)

VACUUM CIRCUITS

1980—400 CID (6.6 L) engines

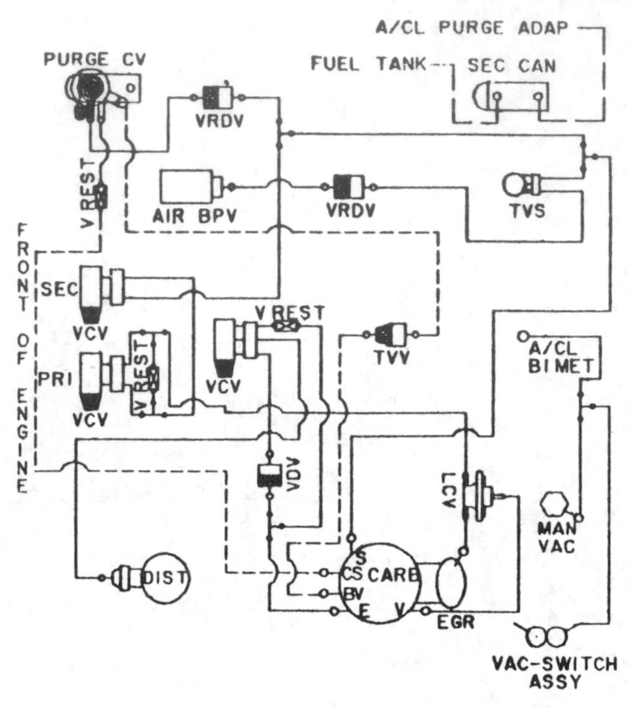

CALIBRATION: 9–17F–RO DATE: 6–6–78
6.6L (400 CID)

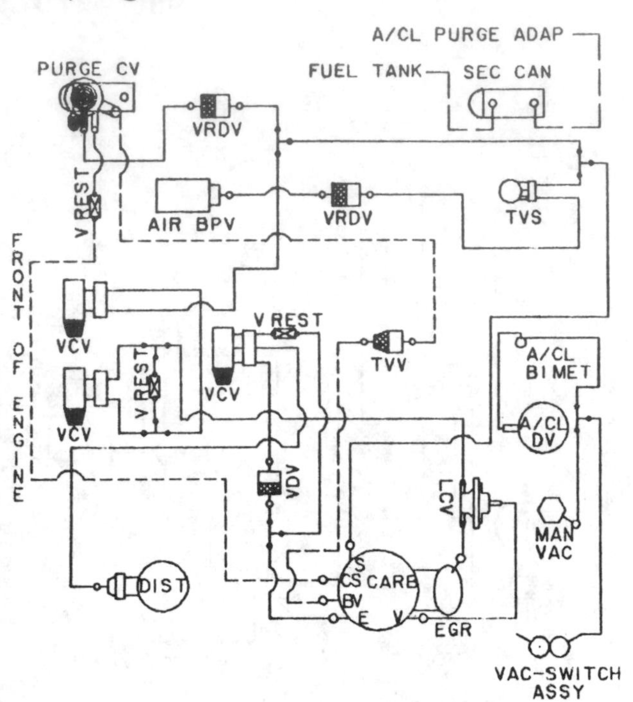

CALIBRATION: 9–17F–R1 DATE: 8–23–78
6.6L (400 CID)

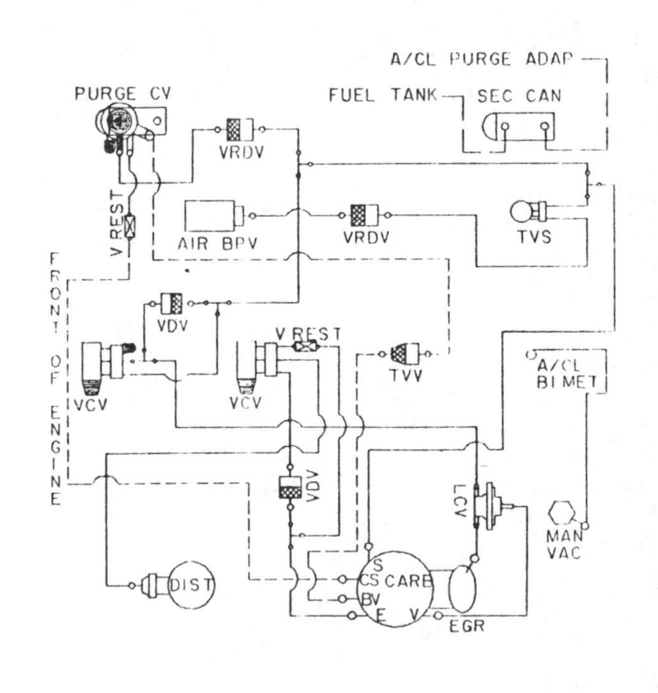

CALIBRATION: 9–17F–R21 DATE: 1–11–79
6.6L (400 CID)

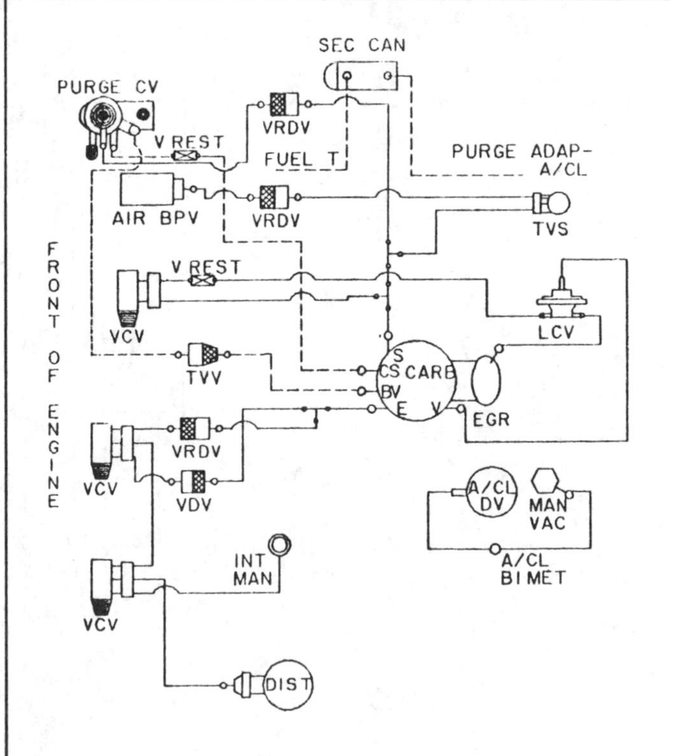

CALIBRATION. 9–17P–RO DATE: 8–16–78
6.6L (400 CID)

VACUUM CIRCUITS

1980—400 CID (6.6 L) engines

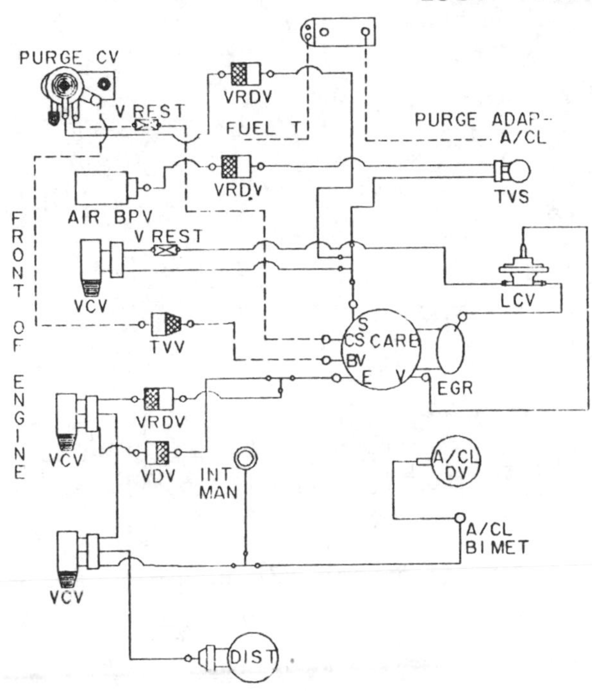

CALIBRATION: 9–17P–R10 DATE: 1–4–79
6.6L (400 CID)

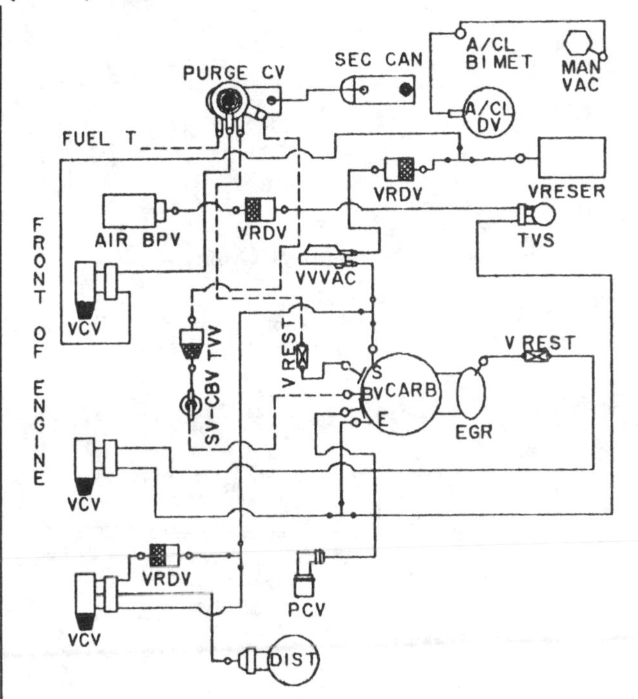

CALIBRATION: 9–17X–RO DATE: 9–28–78
6.6L (400 CID)

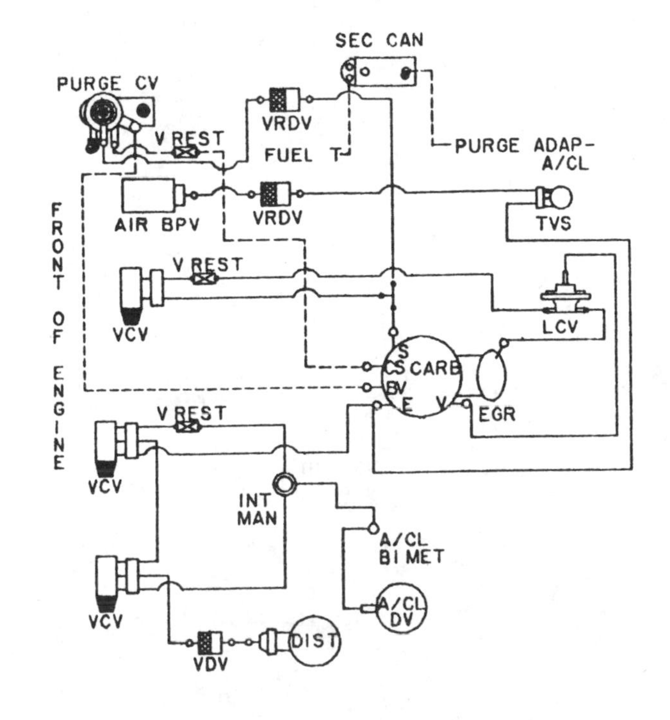

CALIBRATION: 9–17Q–RO DATE: 7–6–78
6.6L (400 CID)

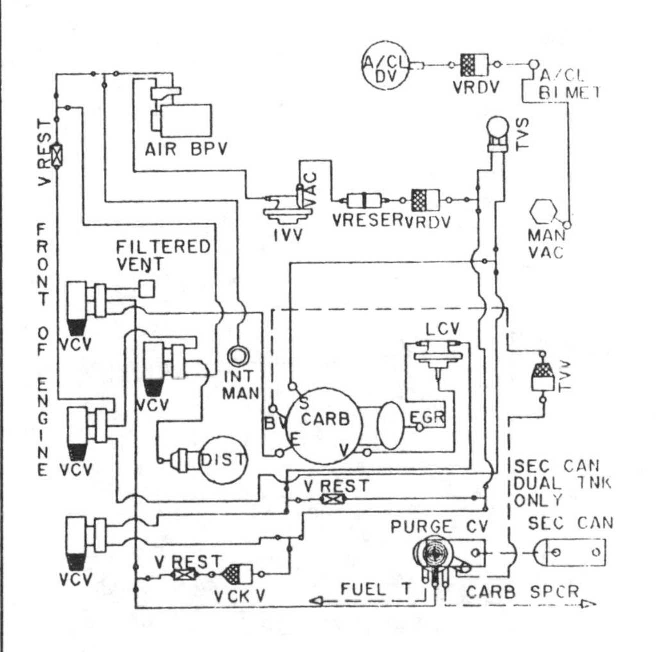

CALIBRATION: 9–61G–ROA DATE: 1–24–79
6.6L (400 CID)

VACUUM CIRCUITS

1980—400 CID (6.6 L) engines

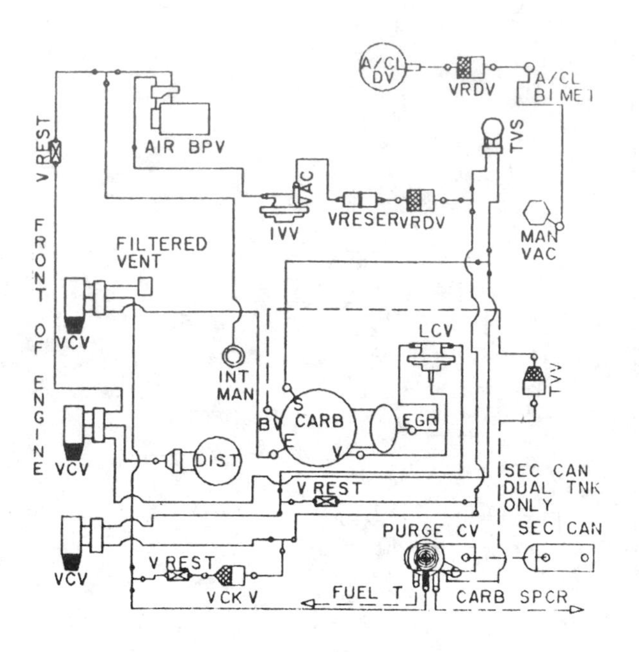

CALIBRATION: 9–61G–RON **DATE:** 1–24–79
6.6L (400 CID)

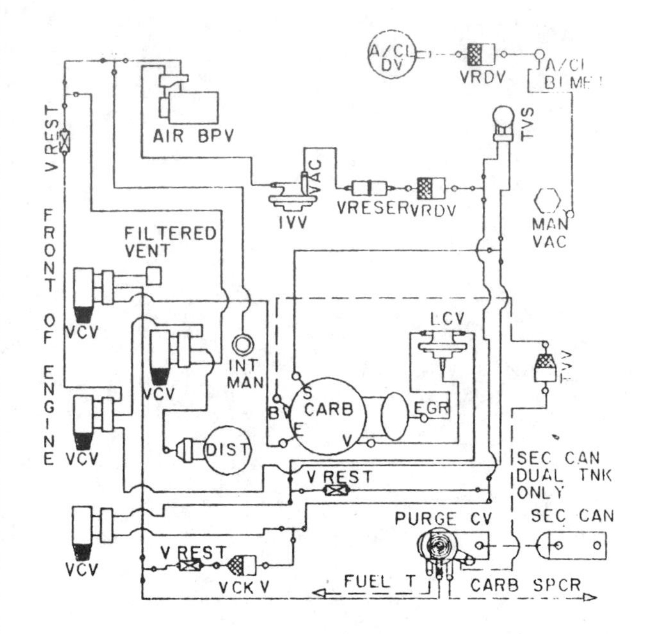

CALIBRATION: 9–61H–ROA **DATE:** 1–24–79
6.6L (400 CID)

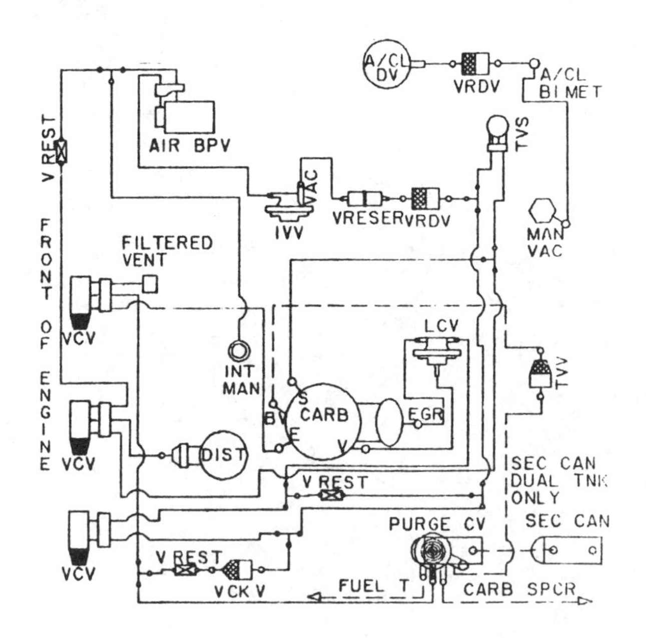

CALIBRATION: 9–61H–RON **DATE:** 1–24–79
6.6L (400 CID)

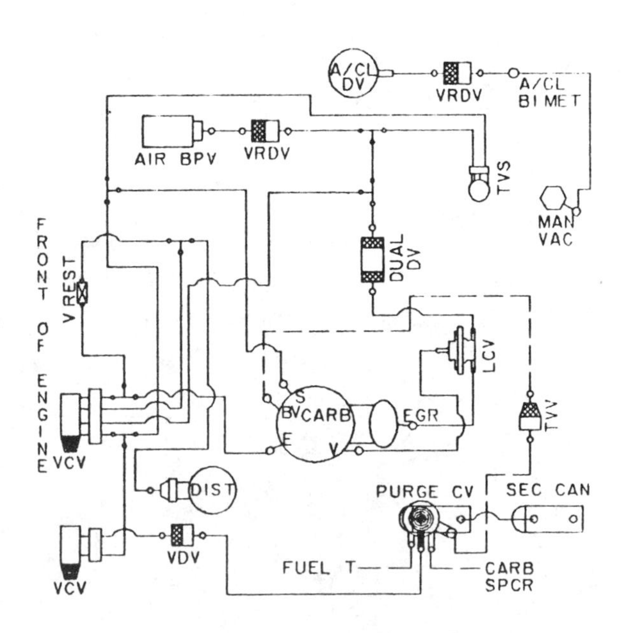

CALIBRATION: 9–62A–RO **DATE:** 8–10–78
6.6L (400 CID)

VACUUM CIRCUITS

1980—400 CID (6.6 L) engines

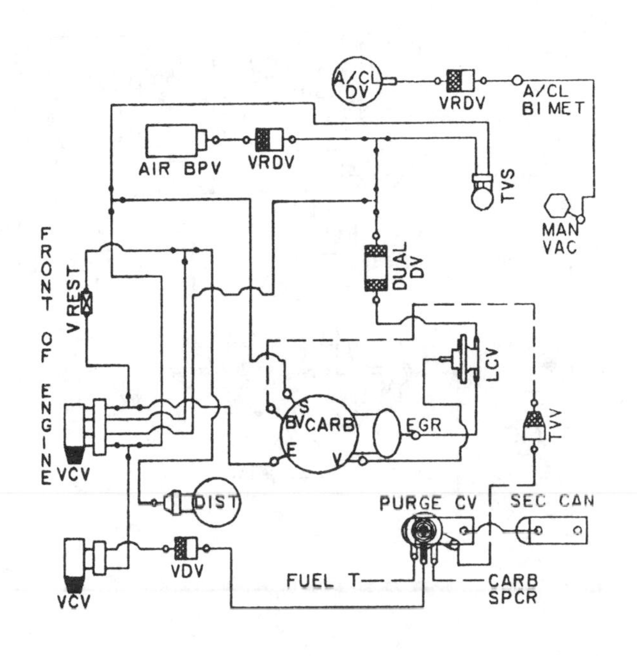

CALIBRATION: 9–62B–RO DATE: 8–10–78
6.6L (400 CID)

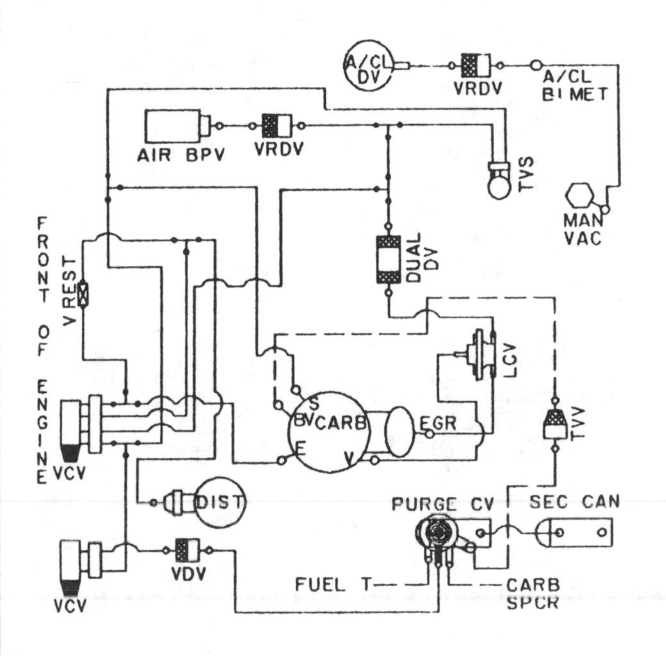

CALIBRATION: 9–62B–R10 DATE: 10–4–78
6.6L (400 CID)

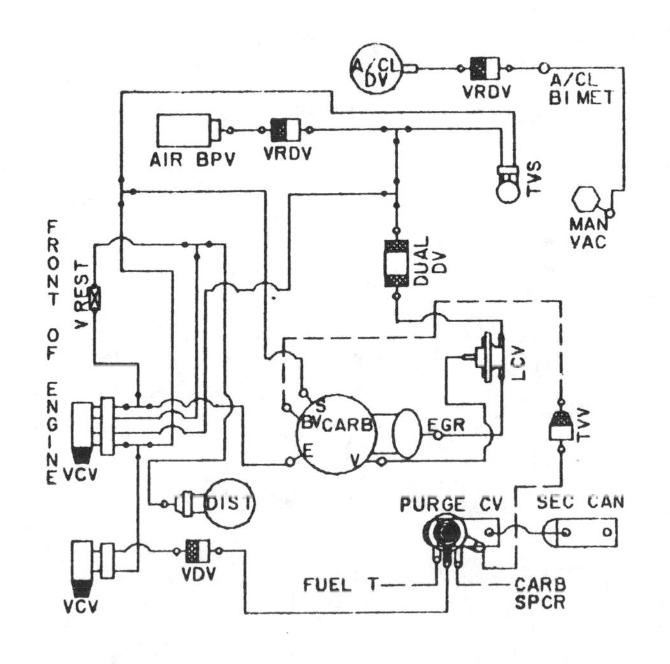

CALIBRATION: 9–62J–RO DATE: 5–16–78
6.6L (400CID)

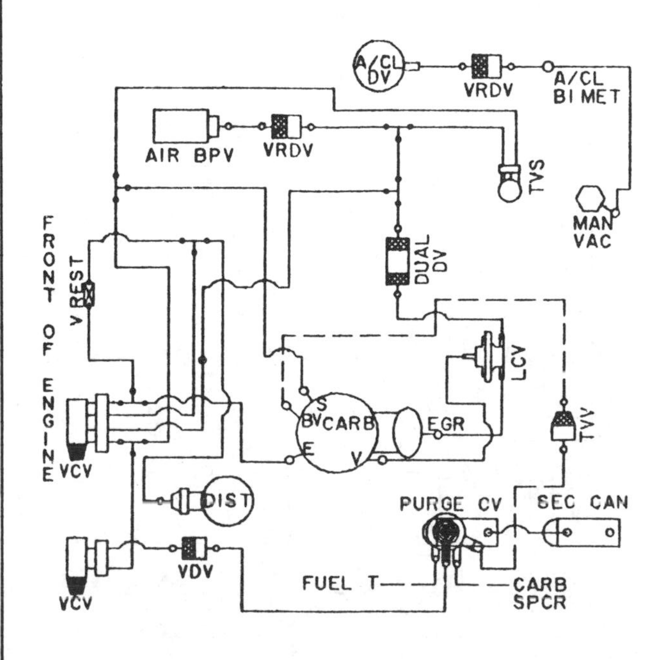

CALIBRATION. 9–62M–RO DATE: 5–24–78
6.6L (400 CID)

VACUUM CIRCUITS

1980—400 CID (6.6 L) engines

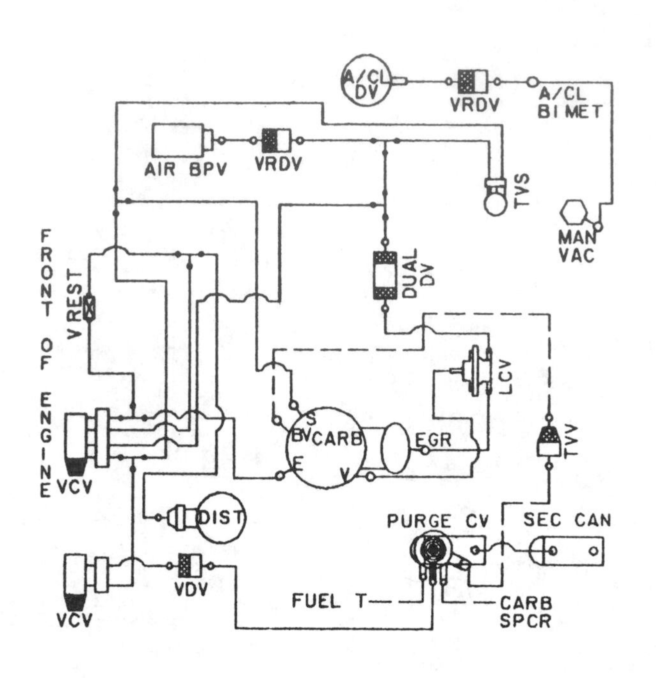

CALIBRATION: 9–62M–R10 DATE: 10–5–78
6.6L (400 CID)

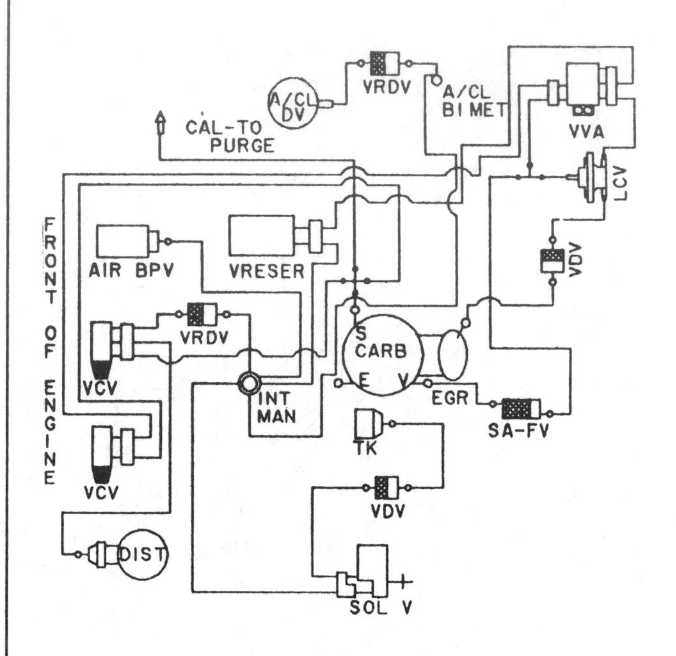

CALIBRATION: 9–73J–RO DATE: 10–2–78
6.6L (400 CID)

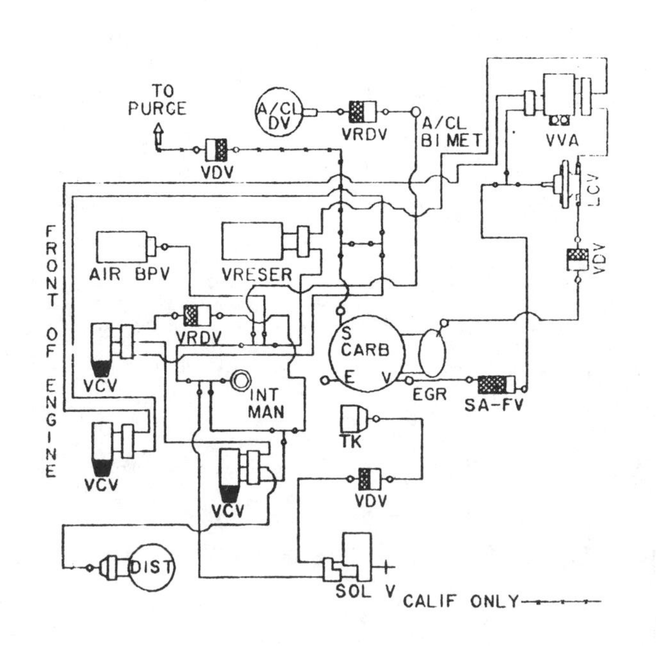

CALIBRATION: 9–73J–R11 DATE: 1–25–79
6.6L (400 CID)

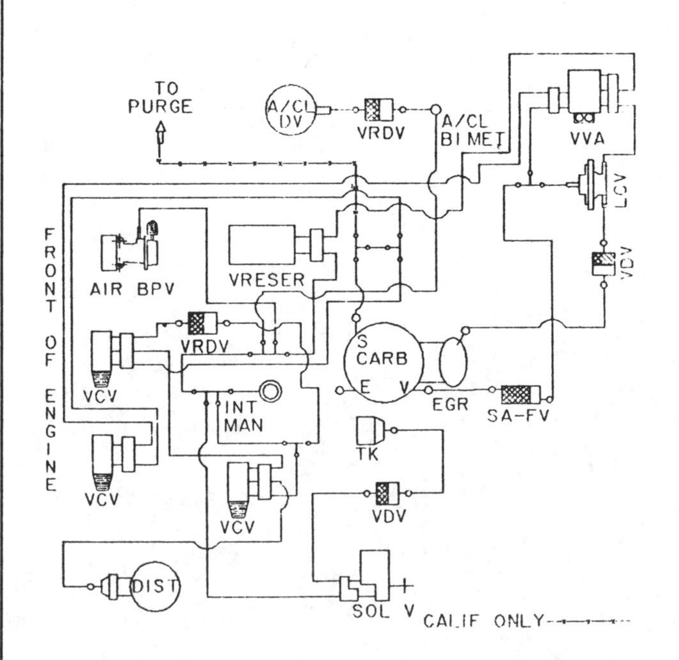

CALIBRATION: 9–73J–R12 DATE: 5–23–79
6.6L (400 CID)

VACUUM CIRCUITS

1980—400 CID (6.6 L) engines

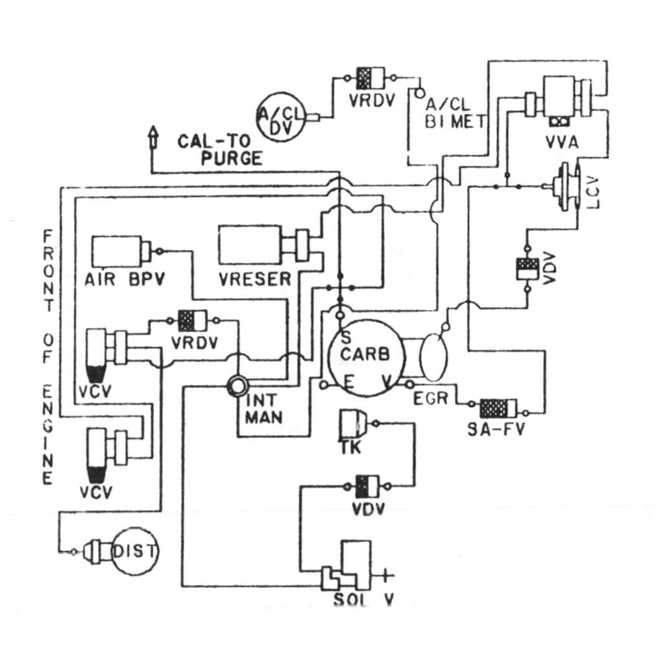

CALIBRATION: 9–74J–RO DATE: 9–29–78
6.6L (400 CID)

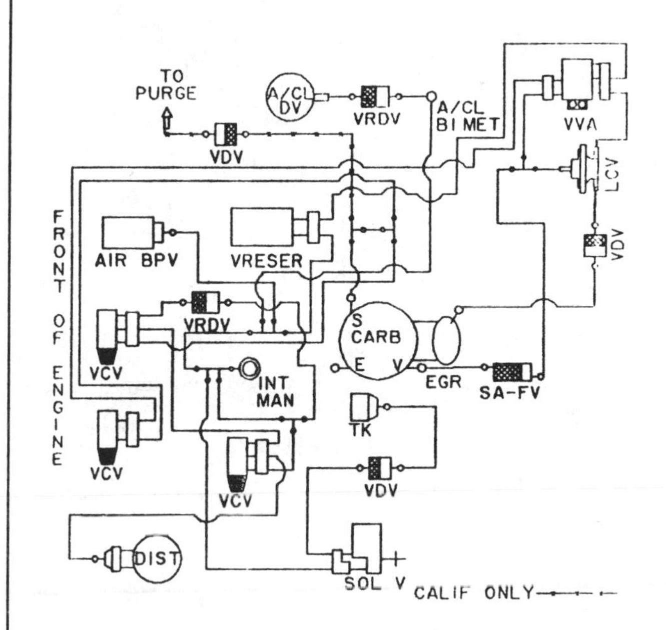

CALIBRATION: 9–74J–R11 DATE: 1–25–79
6.6L (400 CID)

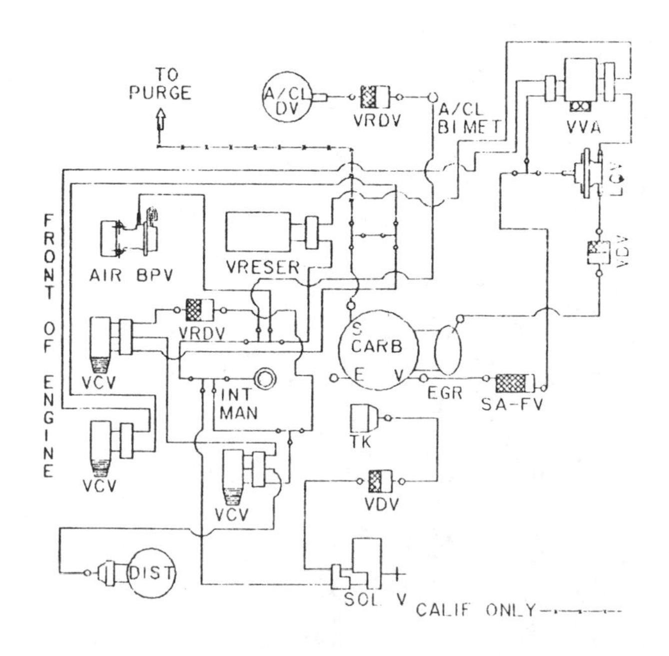

CALIBRATION: 9–74J–R12 DATE: 5–23–79
6.6L (400 CID)

VACUUM CIRCUITS

1980—429 CID (7.0 L) engines

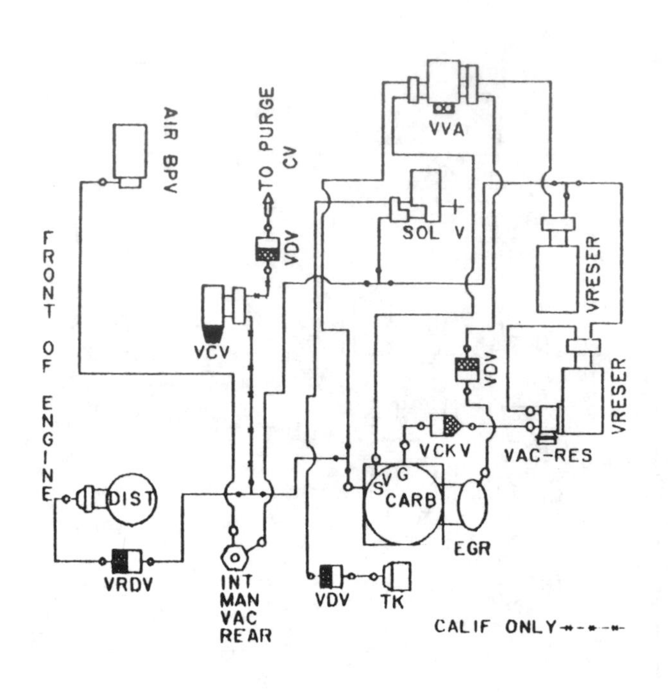

CALIBRATION: 9–87G–RO DATE: 7–11–78
7.0L (429 CID)

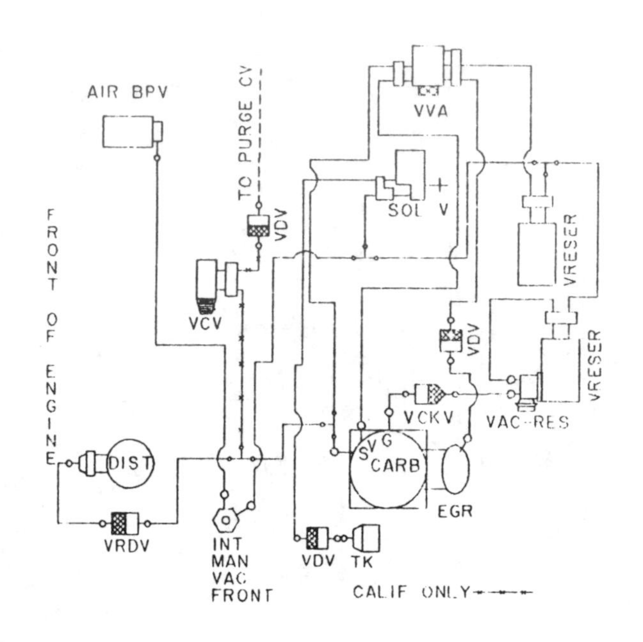

CALIBRATION: 9–87G–R10 DATE: 1–19–79
7.0L (429 CID)

1980—460 CID (7.5 L) engines

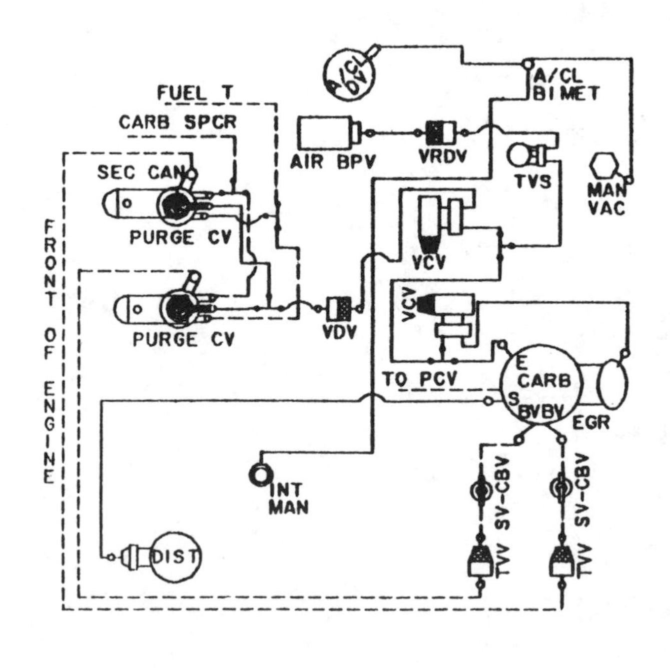

CALIBRATION: 9–66G–RO DATE: 6–13–78
7.5L (460 CID)

VRESER
VVA
AIR BPV
SOL V
FRONT OF ENGINE
VDV
VCV
VDV
CARB
EGR
TK
DIST
MAN VAC
A/CL BI MET

CALIBRATION. 9–97J–RO DATE: 7–10–78
7.5L (460 CID)

VACUUM CIRCUITS

1980—460 CID (7.5 L) engines

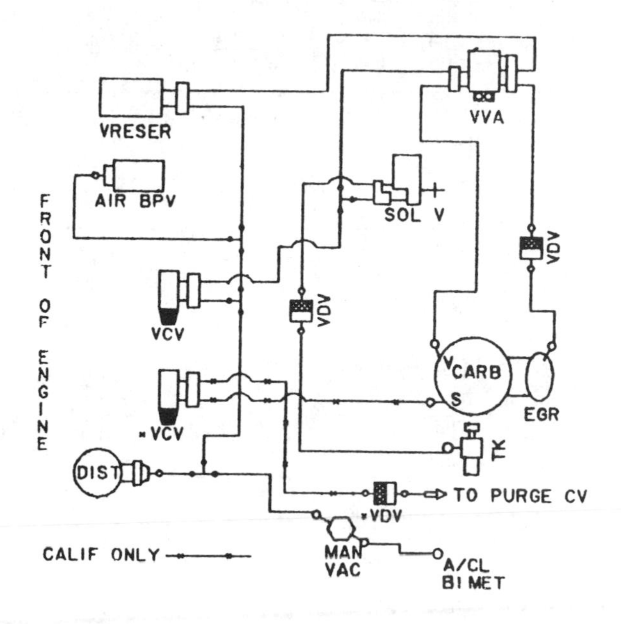

CALIBRATION: 9–97J–R10 DATE: 7–11–78
7.5L (460 CID)

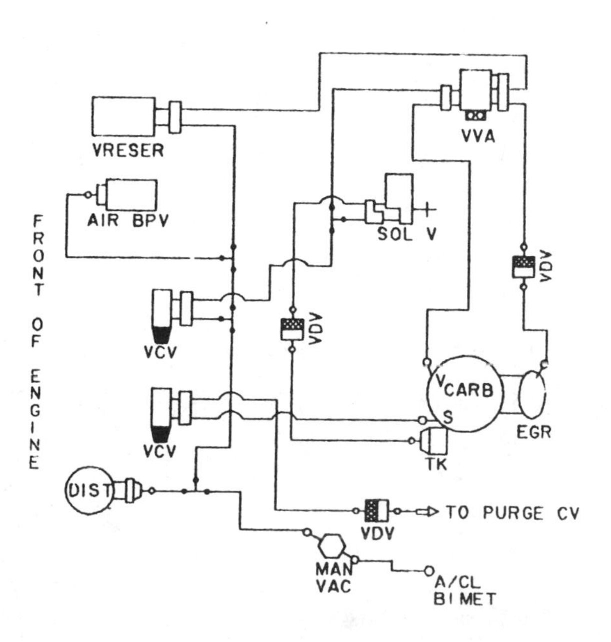

CALIBRATION: 9–97J–R11 DATE: 7–25–78
7.5L (460 CID)

TUNE-UP SPECIFICATIONS

Year	Model	SPARK PLUGS Type	SPARK PLUGS Gap (in.)	DISTRIBUTOR Point Dwell (deg)	DISTRIBUTOR Point Gap (in.)	IGNITION TIMING (deg) MT	IGNITION TIMING (deg) AT	Fuel Pump Pressure (psi)	IDLE SPEED (rpm) MT	IDLE SPEED (rpm) AT①	VALVE CLEARANCE (in.) (Hot) In	VALVE CLEARANCE (in.) (Hot) Ex	Percentage of CO at idle
1975	280 Z Federal	BP-6ES	0.028-0.031	Electronic	⑦	7B⑧	7B⑧	36.3	800	800	0.010	0.012	2.0
1975	280 Z (California)	BP-6ES	0.028-0.031	Electronic	⑦	10B	10B	36.3	800	800	0.010	0.012	2.0
1975	610	BP-6ES	0.031-0.035	49-55	0.017-0.022	12B	12B	3.8	750	650	0.010	0.012	2.0
1975	710	BP-6ES	0.031-0.035	49-55	0.017-0.022	12B	12B	3.8	750	650	0.010	0.012	2.0
1975	B210 (California)	BP-6ES	0.031-0.035	Electronic	⑦	10B	10B	3.8	750	650	0.014	0.014	2.0
1975	710, 610 (California)	BP-6ES	0.031-0.035	Electronic	⑦	12B	12B	3.8	750	650	0.010	0.012	2.0
1976	B-210 (Federal)	BP-5ES	0.031-0.035	49-55	0.017-0.022	10B	10B	3.8	700	650	0.014	0.014	2.0
1976	B-210 (California)	BP-5ES	0.031-0.035	Electronic	⑦	10B	10B	3.8	700	650	0.014	0.014	2.0
1976	610, 710 (Federal)	BP-6ES	0.031-0.035	Electronic	0.018-0.022	12B	12B	3.8	750	650	0.010	0.012	2.0
1976	610, 710 (California)	BP-6ES	0.031-0.035	Electronic	⑦	12B	12B	3.8	750	650	0.010	0.012	2.0
1976	620 (Federal)	BP-6ES	0.031-0.035	49-55	0.018-0.022	12B	12B	3.8	750	650	0.010	0.012	2.0
1976	620 (California)	BP-6ES	0.039-0.043	Electronic	⑦	10B	12B	3.8	750	650	0.010	0.012	2.0
1976	280 Z (Federal)	BP-6ES	0.028-0.031	Electronic	⑦	7B⑧	7B⑧	36.3	800	700	0.010	0.012	2.0
1976	280 Z (California)	BP-6ES	0.028-0.031	Electronic	⑦	10B	10B	36.3	800	700	0.010	0.012	2.0
1977	B-210 (Federal)	BP-5ES	0.039-0.043	49-55	0.018-0.022	10B	8B	3.8	700	650	0.014	0.014	2.0
1977	B-210 (California)	BP-5ES	0.039-0.043	Electronic	⑦	10B	10B	3.8	700	650	0.014	0.014	2.0
1977	610, 710 (Federal)	BP-6ES	0.031-0.035	49-55	0.018-0.022	12B	12B	3.8	750	650	0.010	0.012	2.0
1977	610, 710 (California)	BP-6ES	0.039-0.043	Electronic	⑦	12B	12B	3.8	750	650	0.010	0.012	2.0
1977	620 (Federal)	BP-6ES	0.031-0.035	49-55	0.018-0.022	12B	12B	3.8	750	650	0.010	0.012	2.0
1977	620 (California)	BPR-6ES	0.031-0.035	Electronic	⑦	10B	12B	3.8	750	650	0.010	0.012	2.0
1977-79	280 Z 280 ZX	BP-6ES	0.039-0.043	Electronic	⑦	10B	10B	36.3	800	700	0.010	0.012	2.0 ⑨
1977	F-10 (Federal)	BP-5ES	0.039-0.043	49-55	0.018-0.022	10B	10B	3.8	700	700	0.014	0.014	2.0
1977-78	F-10 (California) (1978 Federal)	BP-5ES	0.039-0.043	Electronic	⑦	10B	10B	3.8	700	700	0.014	0.014	2.0
1978-79	620	BP-6ES-11	0.039-0.043	Electronic	⑦	12B	12B	3.0-3.9⑩	600	600	0.010	0.012	1.0

TUNE-UP SPECIFICATIONS—(Continued)

Year	Model	Spark Plugs Type	Spark Plugs Gap (in.)	Distributor Point Dwell (deg)	Distributor Point Gap (in.)	Ignition Timing (deg) MT	Ignition Timing (deg) AT	Fuel Pump Pressure (psi)	Idle Speed (rpm) MT	Idle Speed (rpm) AT①	Valve Clearance (in.) (Hot) In	Valve Clearance (in.) (Hot) Ex	Percentage of CO at idle
1978-79	510	BP-6ES-11	0.039-0.043	Electronic	⑦	12B	12B	3.0-3.9	600	600	0.010	0.012	1.0
1977-79	810	BP-6ES-11	0.039-0.043	Electronic	⑦	10B	10B	36.3 EFI	700	650	0.010	0.012	1.0 ⑨
1977	200 SX	BP-6ES	0.044	49-55	.020	10B	12B	3.0-3.9	600	600	0.010	0.012	0.3-2.0
1978-79	200 SX	BP-6ES-11	0.039-0.043	Electronic	⑦	12B	12B	3.0-3.9	600	600	0.010	0.012	1.0
1978	B210	BP-5ES-11	0.039-0.043	Electronic	⑦	10B	8B	3.9	700	700	0.010	0.012	2.0 ⑪
1979	210	BP-5ES-11	0.039-0.043	Electronic	⑦	10B	10B	3.9	700	700	0.010	0.012	1.0
1979-80	310	BP-5ES-11⑮	0.039-0.043	Electronic	⑭	8B ⑲	—	3.8	750 ⑱	—	0.014	0.014	2.0 ㉒㉓
1980	280 ZX	BP-6ES-11㉓	0.039-0.043	Electronic	⑭	10B ⑲	10B ⑲	30 ⑯	700 ⑰	700 ⑰	0.010	0.012	1.0 ㉓㉔
1980	210	BP-5ES-11⑮	0.039-0.043	Electronic	⑭	26	26	3.0-3.9	750 ⑱	—	0.014	0.014	2.0 ㉓
1980	620	BP-6ES-11⑮	0.039-0.043	Electronic	⑭	12 ⑳	12	3.0-3.9	600 ⑰	600 ⑰	0.010	0.012	1.0 ㉑
1980	510	BP6-ES ㉗	0.031-0.035	Elec. tronic	⑭	8B ㉕	8B	3.0-3.9	600 ⑰	600 ⑰	0.012	0.012	1.5 ㉒㉓
1980	810	BP-6ES-11	0.039-0.043	Elec. tronic	⑭	10B ⑲	10B	30 ⑯	700 ⑰	650 ⑰	0.010	0.012	1.0 ㉓㉔
1980	200 SX	BP-6ES	0.031-0.035	Electronic	⑭	8B ⑲	8B ⑲	—	700 ⑰	700 ⑰	0.012	0.012	1.3 ㉒㉓

NOTE: **The underhood specifications sticker sometimes reflects tune-up specification changes made in production. Sticker figures must be used if they disagree with this chart.**

① In Drive
② Air pump disconnected
③ Automatic—10B @ 600 below 30°F
④ Automatic—15B @ 600 below 30°F
⑤ Reluctor Gap 0.012-0.016
⑥ 10B—California
⑦ Reluctor Gap 0.008-0.016 in.
⑧ 13 BTDC—Advanced
⑨ 1978-79 1.0%—49 States
0.5%—California
⑩ W/Electric Fuel Pump—4.6 lbs
⑪ Fu Models—1.0%
⑫ Fu Models—5°B @ 700
⑬ California Models—10B @ 650
⑭ 0.012-0.020
⑮ BPR5ES—0.031-0.035
⑯ 37 PSI with Accelerator Fully Depressed
⑰ ± 100 RPM
⑱ ± 50 RPM
⑲ Calif. 6B
⑳ Calif. 12B ± 2 (Standard), 10B ± (Heavy Duty)
㉑ + 1.0 − 0.7
㉒ ± 1.0
㉓ Calif. Cars Idle Mixture Screws are Preset and Sealed
㉔ ± 0.8
㉕ ± 2
㉖ A12A 10 ± 2°
A14 12°
A15 8° ± 2 (D Position)
㉗ BPR-6ES-11

Air Flow Meter Resistance Specifications

Air temperature °C (°F)	Resistance (kΩ)
—30 (—22)	20.3 to 33.0
—10 (—14)	7.6 to 10.8
10 (50)	3.25 to 4.15
20 (68)	2.25 to 2.75
50 (122)	0.74 to 0.94
80 (176)	0.29 to 0.36

Water Temperature Sensor Resistance Specifications

Cooling water temperature °C (°F)	Resistance (kΩ)
—30 (—22)	20.3 to 33.0
—10 (—14)	7.6 to 10.8
10 (50)	3.25 to 4.15
20 (68)	2.25 to 2.75
50 (122)	0.74 to 0.94
80 (176)	0.29 to 0.36

TORQUE SPECIFICATIONS

All readings in ft lbs

Engine Model	Cylinder Head Bolts	Main Bearing Bolts	Rod Bearing Bolts	Crankshaft Pulley Bolt	Flywheel to Crankshaft Bolts
L16	40	33-40	20-24	116-130	69-76
L24	47	33-40	20-24	116-130	101
A12	33-35	36-38	25-26	108-116	47-54
L18	47-62	33-40	33-40	87-116	101-116
A13	54-58	36-43	23-27	108-145	54-61
L20B	47-61	33-40	33-40	87-116	101-116
L26	54-61	33-40	27-31	94-108	94-108
A14, A15	51-54	36-43	23-27	108-145	54-61
L28	54-61	33-40	33-40	94-108	94-108

Engine Identification

Number of Cylinders	Displacement cu. in. (cc)	Type	Engine Model Code
4	97.3 (1,595)	OHC	L16
6	146.0 (2,393)	OHC	L24, L24E
4	71.5 (1,171)	OHV	A12, A12A
4	108.0 (1,770)	OHC	L18
4	78.59 (1,288)	OHV	A13
4	119.1 (1,952)	OHC	L20B
6	156.5 (2,655)	OHC	L26
4	85.24 (1,397)	OHV	A14
6	168.0 (2,753)	OHC	L28, L28E
4	90.80 (1,488)	OHV	A15
4	119.1 (1,952)	OHC	Z20E, Z20S

CAR SERIAL NUMBER AND ENGINE IDENTIFICATION

The car identification number plate is located directly behind the windshield on the right side of the instrument panel. The I.D. number consists of the model and serial numbers.

The engine number is located on the left side of the engine block and consists of the model and serial number.

EMISSION CONTROLS

Emission Control Equipment Applications Table

Year	Model	Engine	Emission Control Systems
1976-77	610	L20B	1,3,4,5,6,7,8,9,10,11
1976-77	710	L20B	1,3,4,5,6,7,8,9,10,11
1976-78	B210	A14,	1,3,4,5,6,7,8,10,11
1979-80	210	A12A A14 A15	1,3,4,5,6,7,8,9,10,11
1976-80	620	L20B	1,2,3,4,5,6,7,8,9,10
1977-78	280Z	L28	1,5,6,7,9,12
1979-80	280ZX	L28E	1,2,3,4,5,6,7,8,9,10,13,14
1977-78	F-10	A14	1,3,4,5,6,7,8,10,11
1978-79	510	L20B	1,3,4,5,6,7,8,9,
1980	510	Z20E Z20S	1,2,4,5,6,7,8,9,10,13
1978-80	810	L24	1,3,5,6,7,9,13①
1978-79	200SX	L20B	1,3,4,5,6,7,8,9
1980	200SX	Z20E	1,2,4,5,6,7,8,9,

1. Closed Crankcase Ventilation System
2. Air Induction System
3. Air Pump System
4. Engine Modification System
5. Fuel Vapor Control System
6. Exhaust Gas Recirculation System
7. Catalytic Converter
8. Early Fuel Evaporation System
9. Boost Controlled Deceleration Device
10. High Altitude Compensator—California
11. TCS—Manual Transmission exc. California
12. Floor Temperature Sensing Device
13. Spark Timing Control
14. Mixture Ratio Feedback System (California)

Various systems are used to control crankcase vapors, exhaust emissions, and fuel vapors. The accompanying chart shows the systems used with various models and engines.

CRANKCASE VENTILATION SYSTEM

The closed crankcase ventilation system is used to route the crankcase vapors (blow-by gases) to the intake manifold (carburetor equipped) or throttle chamber (EPI), to be mixed and burned with the air/fuel mixture.

An air intake hose is connected between the air cleaner assembly or the cover. A return hose is connected between a steel net baffle on the side throttle chamber, to the top engine of the crankcase to the intake manifold or throttle chamber, with a metering Positive Crankcase Valve mounted in the hose.

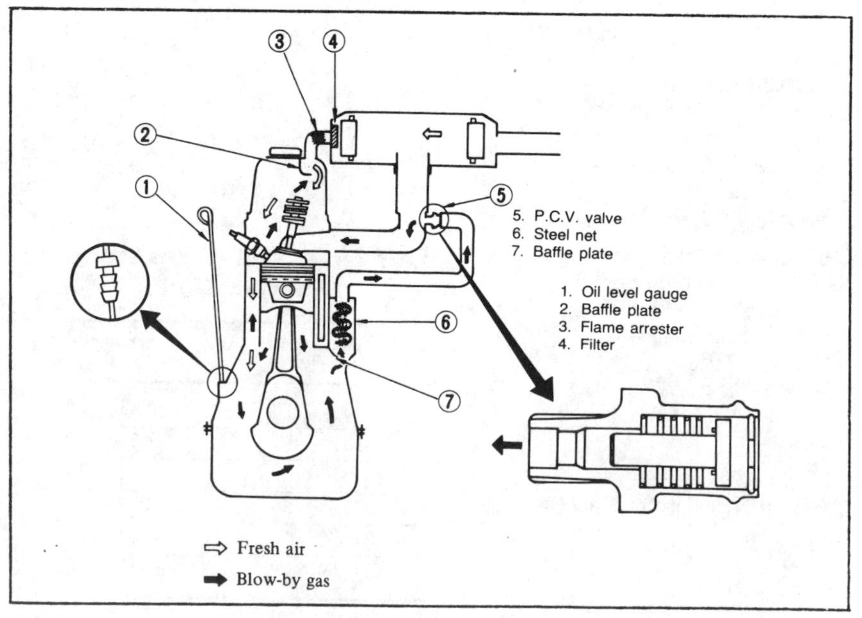

Crankcase ventilation system air flow (PCV)

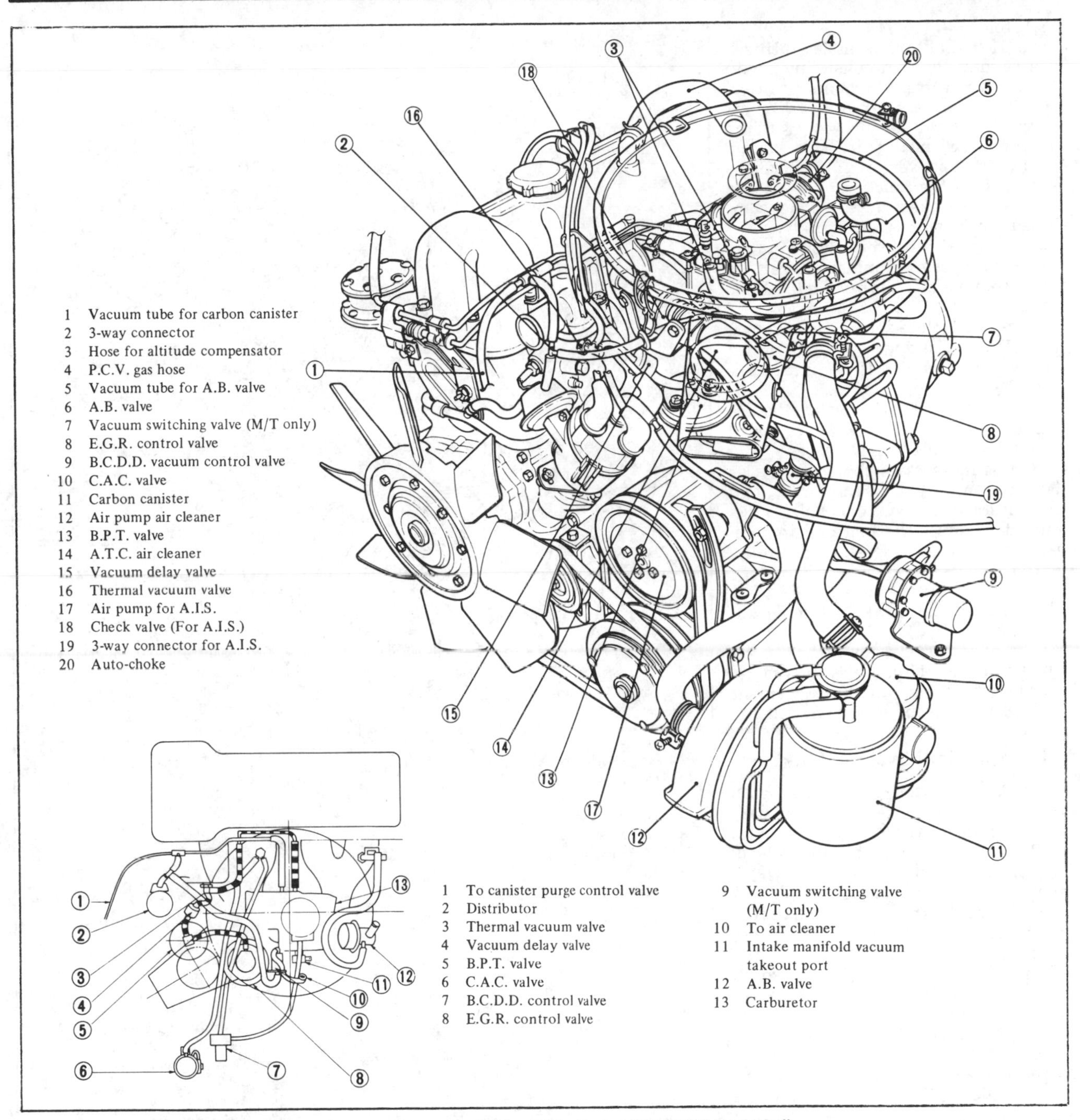

Emission control system—Calif. model—Typical (Model 710 illustrated)

During periods of partial throttle, air is drawn through the air cleaner or throttle chamber, into the top engine cover and through the engine by the engine-developed vacuum. The air mixes with the crankcase vapors and is drawn through the steel net baffle which separates the heavy particles of oil from the vapors. The vapors are then metered through the PCV valve and directed into the intake manifold of the throttle chamber to be burned with the air/fuel mixture.

Under full throttle conditions, when the engine developed vacuum is insufficient to draw the vapors through the PCV valve, the vapors reverse in direction and are drawn into the air cleaner or throttle chamber from the top engine cover by the rush of induction air.

As the engine vacuum is raised, the vapors again flow through the PCV valve and into the combustion chambers. Therefore, no crankcase vapors are allowed to enter the atmosphere.

Air Pump System

In this system, an air injection pump, driven by the engine, compresses, distributes, and injects filtered air into the exhaust port of each cylinder. The air combines with unburned hydrocarbons and carbon monoxide to produce harmless compounds. The system includes an air cleaner, the belt driven air pump, a check valve, and an anti-backfire valve.

The air pump draws air through a hose connected to the carburetor air cleaner or to a separate air cleaner. The pump is a rotary vane unit with an integral pressure regulating valve. The pump outlet pressure passes through a check valve which prevents exhaust gas from entering the pump in case of insufficient pump outlet pressure. An anti-backfire valve admits air from the air pump into the intake manifold on deceleration to prevent backfiring in the exhaust manifold.

In 1976 California models utilized a secondary system consisting of an air control valve which limits injection of secondary air and an emergency relief valve which controls the supply of secondary air. This system protects the converter from overheating. In 1977 the function of these two valves was taken by a single combined air control (C.A.C.) valve.

All engines with the air pump system have a series of minor alterations to accommodate the system. These are:

1. Special close-tolerance carburetor. Most engines, except the L16, require a slightly rich idle mixture adjustment.
2. Distributor with special advance curve. Ignition timing is retarded about 10° at idle in most cases.
3. Cooling system changes such as larger fan, higher fan speed, and thermostatic fan clutch. This is required to offset the increase in temperature caused by retarded timing at idle.
4. Faster idle speed.
5. Heated air intake on some engines.

The only periodic maintenance required on the air pump system is replacement of the air filter element and adjustment of the drive belt.

Air Induction System (A.I.S.)

An air induction system has been introduced for the 1980 model year and is used on the A12A, A14, A15, Z20E and the Z20S engines, replacing the air pump system. This sys tem uses two air induction valves and induces air to the exhaust manifold branches, number 1 and 4, on most models, while inducing air to the cylinder head on the engines used on the B310 model vehicles. The carburetor air cleaner mounted valves incorporate two reed valves to control the secondary air sent to the exhaust area. Two systems are used and can be identified by the system having one air intake tube as being used on non California models (type 1), while the system having two air intake tubes are used on California models (type 2).

Engine Modification System

Engine modifications used on vehicles with the L16 and L18 and L20B, are:

1. A distributor with a secondary set of contact points which are retarded 5°. These secondary points are operational only when cruising or accelerating with a partially open throttle in third gear with manual transmission, or over 13 mph with automatic transmission. A speed sensor is located at the speedometer on automatic transmission models. A temperature sensor in the engine compartment allows retarded timing only when the temperature inside the car is 50°F or above.
2. A solenoid valve in the carburetor opens to supply a lean fuel and air mixture, bypassing the throttle valve, in third gear or over 13 mph as above. The solenoid valve will not open if overridden by a closed throttle switch, a wide open throttle switch, a neutral switch, or a clutch disengaged switch. This arrangement is operational primarily during deceleration, when high intake manifold vacuum is present. For 1976-1979, this system is replaced with a vacuum controlled device in the carburetor to perform the same function.

The engine modification system

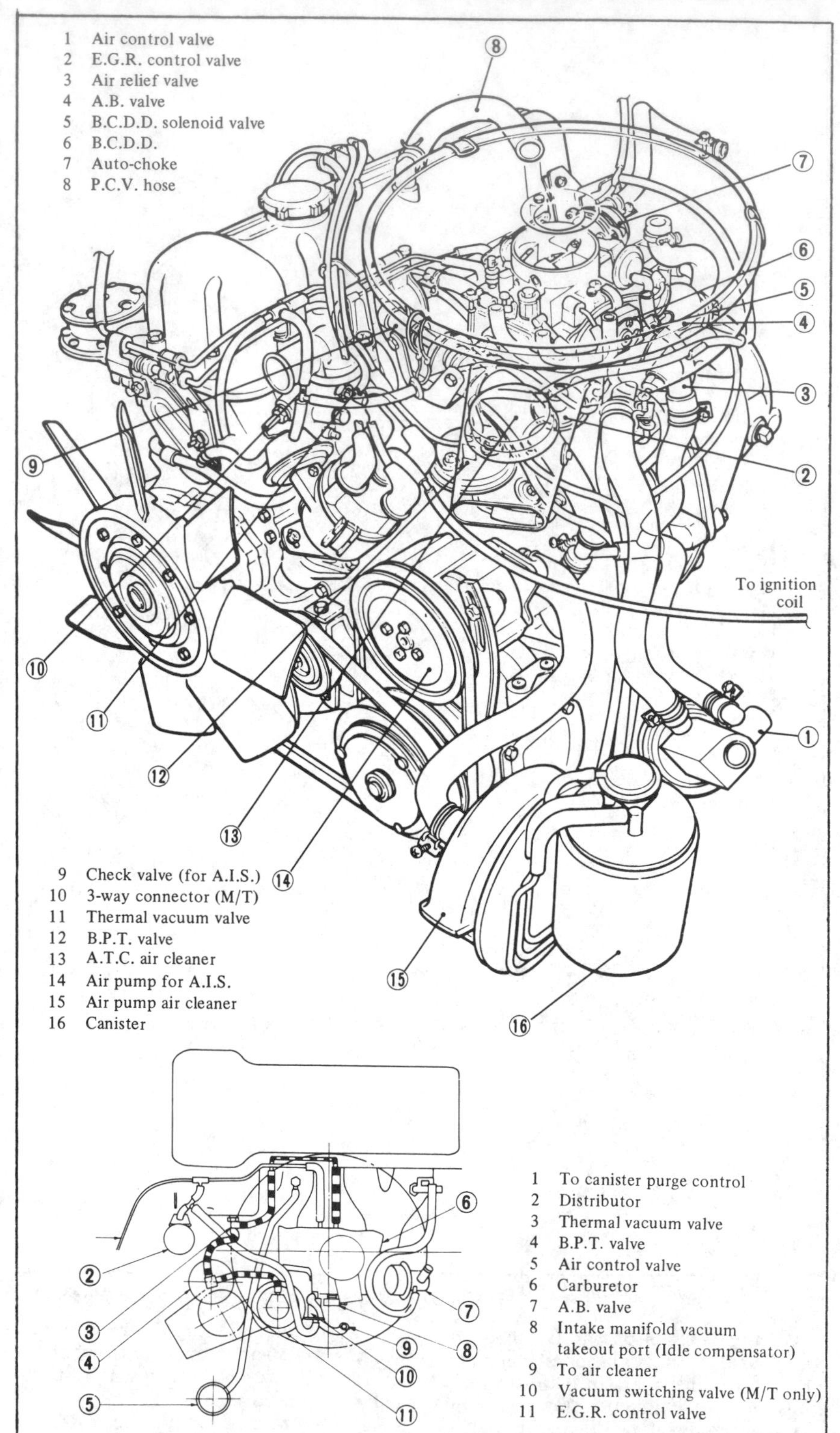

Emission control system—Non-Calif. models—Typical (Model 710 illustrated)

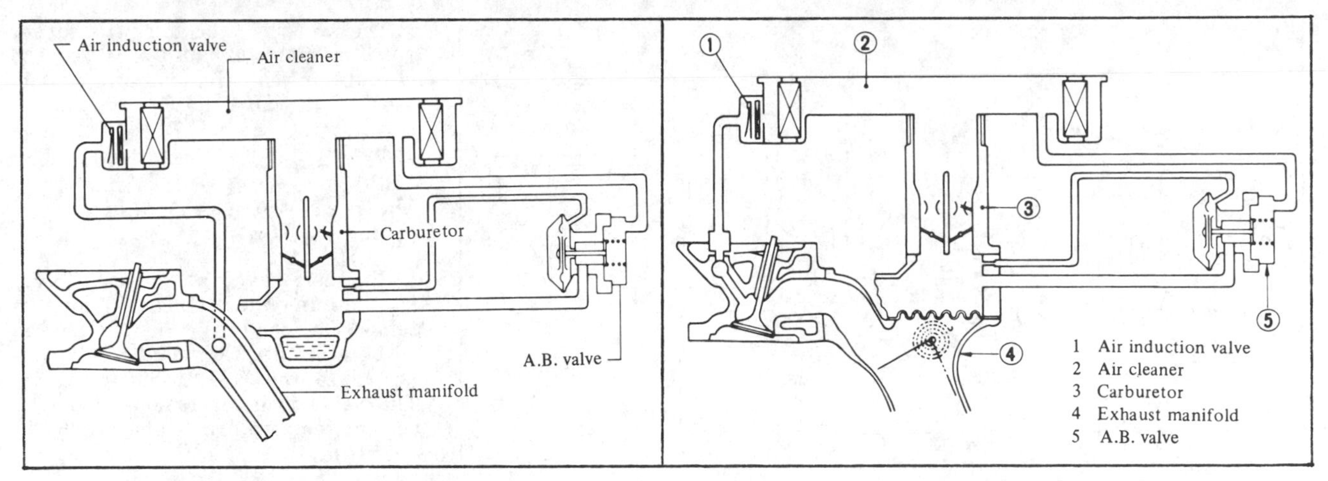

Air inductive system—typical (49 States)

Air inductive system—typical (Canada)

used with the A12, A13, A14, A15 engines are relatively simple. It requires only a throttle positioner which holds the throttle slightly open on deceleration. A vacuum control valve connected to the intake manifold causes a vacuum servo to hold the throttle open slightly during the high vacuum condition of deceleration. The control valve is compensated for the effects of altitude and atmospheric pressure. The carburetor and distributor are specially calibrated for this engine.

The engine modification system for the 240Z sport coupe with the L24 engine and the 260Z with the L26 engine is quite similar to that for the A12, A13 engine, using a vacuum control valve, vacuum servo, and throttle positioner.

The 1976 280Z non-California models use the dual pick-up coil distributor while the California models use the single pick-up coil unit.

Beginning with the 1977 model year, all 280Z models were equipped with the single pick-up coil distributors.

Beginning with the 1979 models, the IC ignition unit has been miniaturized and mounted on the side of the distributor. The pick-up coil has been changed from an arm type of coil to a ring type of coil which eliminates dispersion of the signal waveform.

Z2OE, Z2OS Engines

Based on the block design of the L2OB engine, a new engine has evolved for the 200X and the 510 series. This new engine incorporates an overhead cam, electronic fuel injection or carburetor system, a dual ignition system which utilizes two spark plugs per cylinder for California models, while retaining the conventional one spark plug per cylinder for the non-California models.

The emission controls used on these engine models are as follows:

1. Air induction system
2. Ventri Vacuum Transducer (V.V.T.) type Exhaust Gas Recirculation (E.G.R.) control system
3. Electronic ignition with spark timing control system
4. Redesigned canister for the Evaporator system
5. Catalytic converter
6. Positive Crankcase ventilation (P.C.V.)
7. Deceleration control system

In addition, the following has been added to the carburetor equipped engine (Z2OS).

1. Vacuum controlled fuel shut-off system
2. Coolant temperature controlled spark timing system
3. Carburetor altitude compensator for 510 models

Exhaust Gas Recirculation System (E.G.R.)

For 1976-79 engines, an exhaust gas recirculation (E.G.R.) system is used. This system uses vacuum to actuate a valve which allows a smaller amount of exhaust gases to be drawn into the intake manifold. This results in a decrease in oxides of nitrogen in the exhaust gases. The vacuum required to operate the system is not available at idle or wide throttle openings. A thermostatic switch inside the car shuts off the vacuum to the system when the temperature is below 30°F., thus allowing good cold starting and driveability.

E.G.R. Large-Small and Mixture Ratio Rich-Lean Exchange (California models)

On certain models sold in the state

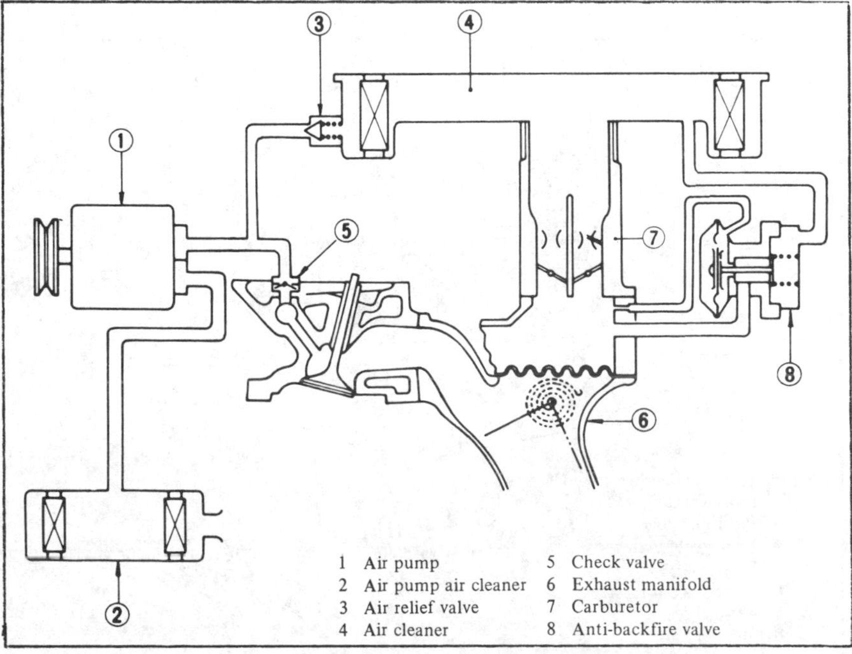

Air injection system—Non-Calif. models—Typical

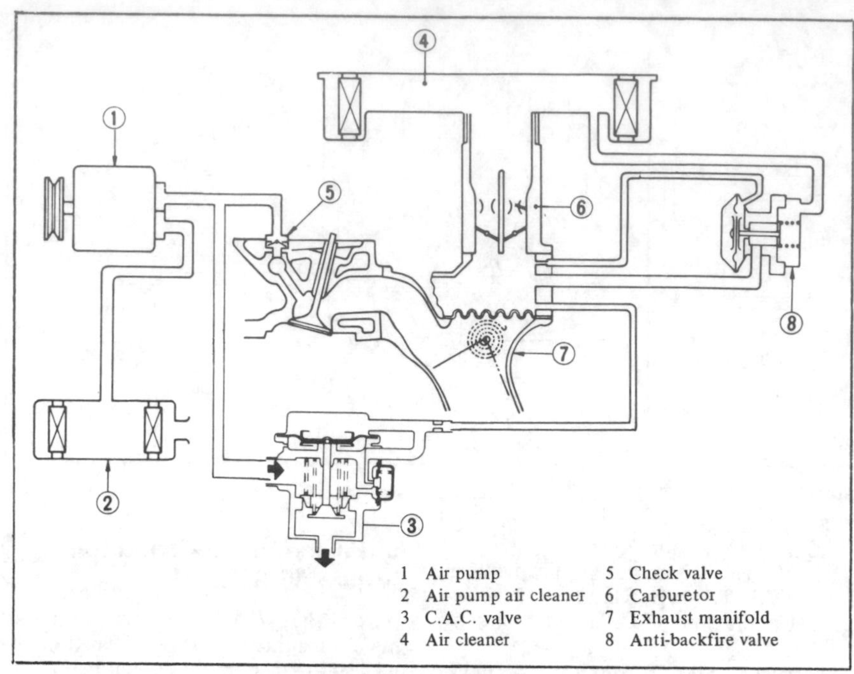

Air injection system—Calif. models—Typical

of California and equipped with manual transmission, an E.G.R. large-small exchange system, which normalizes the E.G.R. rate of flow when the engine temperature is low, such as during warm-up or at high speed operation, has been installed to aid in the driveability. A mixture ratio, rich-lean, exchange system which keeps the air-fuel mixture ratio lean to improve fuel economy, has been installed on the California models and operates in conjunction with the E.G.R. large-small exchange system. A 40 mph speed switch is installed within the speedometer and the vacuum switching switch is controlled by a signal issued and applied by an amplifier attached to the rear of the speedometer-instrument panel.

NOTE: *The E.G.R. system has been discontinued on the 1980 L24E and L28E engines, used in California, but has been retained on the non-California models.*

Venturi Vacuum Transducer, E.G.R. System (V.V.T) California Models 1980

To further reduce the NOx by increasing the E.G.R. flow rate, the system has been changed from a back pressure transducer to a venturi vacuum transducer unit.

The venturi vacuum transducer valve monitors exhaust pressure and carburetor venturi vacuum to control the E.G.R. valve diaphragm and in turn, control the E.G.R. control valve operation.

Thermal Vacuum Valve (2 Port Type)

The thermal vacuum valve (T.V.V.) is used to monitor the cooling system temperature and to open and close the carburetor throttle vacuum line connected to the E.G.R. control valve. The valve is in operation when the coolant temperature is above 203° F. (95° C.).

Vacuum Delay Valve

The vacuum delay valve is located between the throttle vacuum port and the E.G.R. control valve. When the throttle valve is opened rapidly, the vacuum delay valve is used to control the rapid loss of vacuum and to allow the E.G.R. valve to operate for a short time to prevent the rapid deterioration of the NOx emissions. Should the throttle valve be suddenly closed, the valve will instantly open the throttle vacuum line to activate the E.G.R. system.

Early Fuel Evaporation System (E.F.E.)

1976-79 Cars exc. 280Z

In this system, a control valve is welded to the valve shaft and installed on the exhaust manifold through bushing. This heat control valve is actuated by a coil spring, thermostatic spring and counterweight which are assembled on the

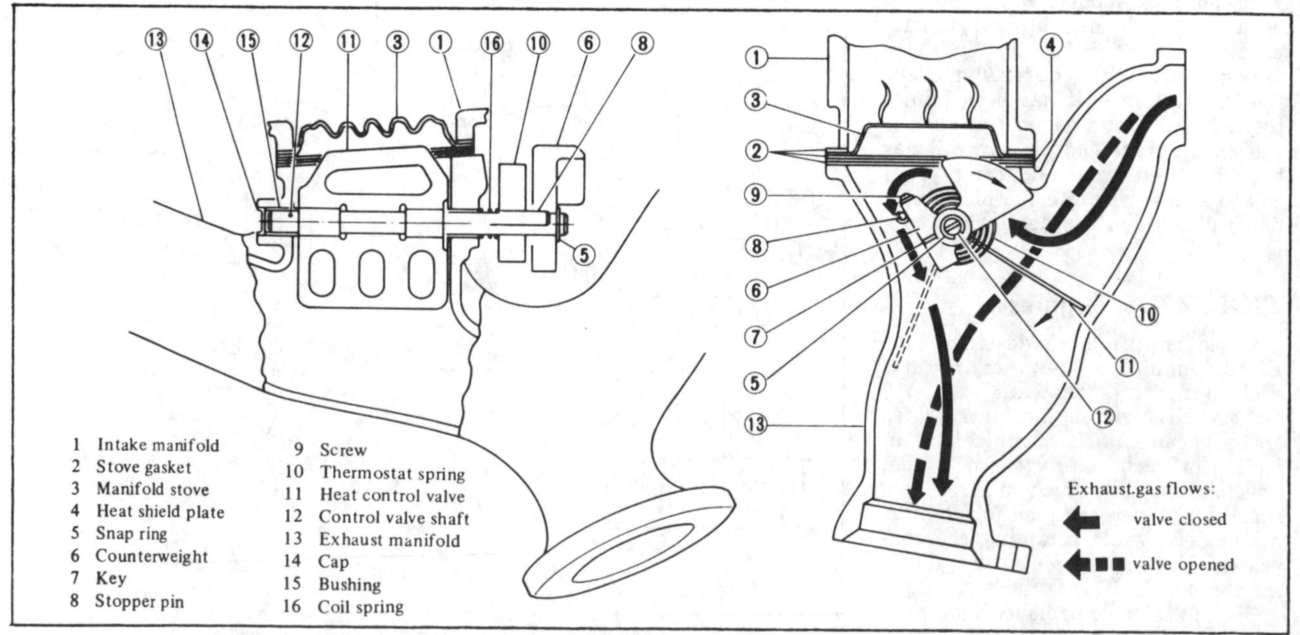

Early fuel evaporative system—Typical

valve shaft projecting at the rear outside of the manifold. The counterweight is secured to the shaft with a key, bolt and snap-ring. A chamber between the intake and exhaust manifolds above the manifold stove heats the air-fuel mixture by means of exhaust gases. This results in better atomization and lower HC content.

Hot Water E.F.E. System

The intake manifolds of the A12A, A14 and A15 engines, used with the 1980 models, utilize a coolant passage under the floor of the manifold, to assist in the vaporization of the air-fuel mixture, to aid in a drivability of the vehicle.

Boost Controlled Deceleration Device (B.C.D.D.)

All 1976-79 Cars exc. B210, F-10

The B.C.D.D. is installed under the throttle chamber as a part of it. It supplies additional air to the intake manifold during coasting to maintain manifold vacuum at the proper operating pressure.

There are two diaphragms in the device. Diaphragm I detects the manifold vacuum and opens the vacuum control valve when vacuum exceeds operating pressure. Diaphragm II operates the air control valve by way of the vacuum transmitted through the vacuum control valve. The air control valve regulates the amount of additional air so that the manifold vacuum can be kept at operating pressure.

On manual transmission models, in addition to the B.C.D.D., the system consists of a vacuum control solenoid valve, speed detecting switch and amplifier.

On automatic transmission models, in addition to the B.C.D.D., the system consists of vacuum control solenoid and inhibitor switch.

A new type B.C.D.D. is used on the 1980 models where applicable. The solenoid valve has been eliminated and is no longer possible to control the unit using the car speed switch and the transmission switch. This new type unit operates when the intake manifold vacuum reaches a high level during decleration, by opening and admitting air.

Altitude Compensator California Models

The altitude compensator is used to provide a proper air-fuel mixture by sending air to the primary and secondary main air bleeds to prevent enrichment in the air-fuel mixture due to the lowering of the atmospheric pressure at high altitudes. The unit is basically the same as the unit used with the 1975 models.

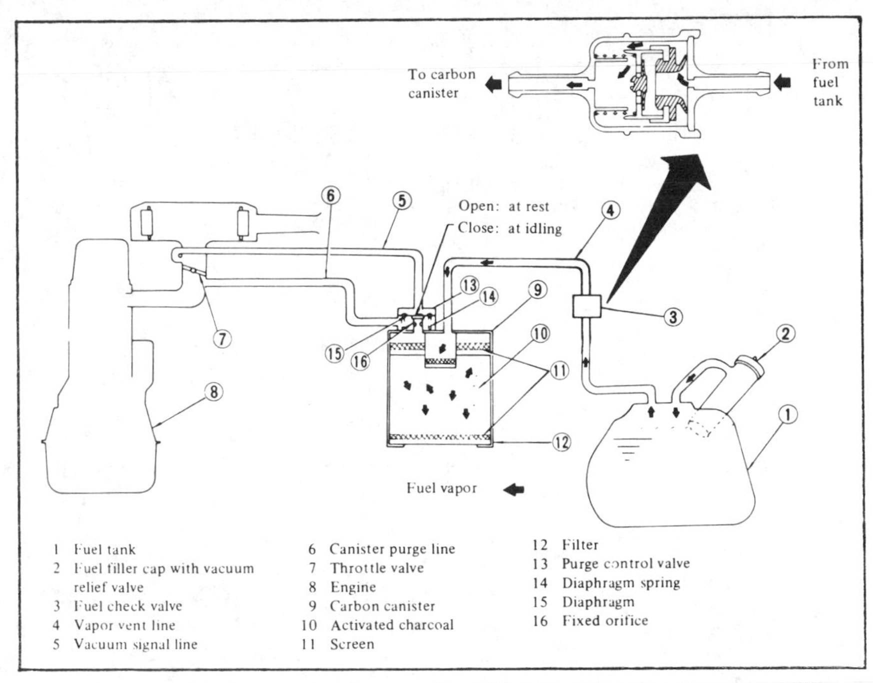

Evaporative emission control system operation with engine stopped or idling—Typical

Carburetor Cooling Fan

A careburetor cooling fan is used on the 310 series, with a timer installed on the units for use on California models.

FUEL VAPOR CONTROL SYSTEM

The fuel vapor control system is used on all vehicles sold in the U.S. It has four major components:

1. A sealed gas tank filler cap to prevent vapors from escaping at this point.
2. A vapor separator which returns liquid fuel to the fuel tank, but allows vapors to pass into the system.
3. A vapor vent line connecting the vapor separator to a flow guide valve.
4. A flow guide valve which allows air into the fuel tank and prevents vapors from the crankcase ventilation system from passing into the vapor vent line and fuel tank.

When the engine is not running, fuel vapors accumulated in the fuel tank, vapor separator, and vapor vent line. When the vapor pressure exceeds 0.4 in. (10 mm) Hg, the flow guide valve opens to allow the vapors to pass into the crankcase ventilation system. Fuel vapors are thus accumulated in the crankcase. When the engine starts, the vapors are disposed of by the crankcase ventilation system. When enough fuel has been used to create a slight vacuum in the fuel tank and fuel vapor control system, the flow guide valve opens to let fresh air from the carburetor air cleaner into the tank.

On engines with sidedraft carburetors, float bowl vapors are routed through the float bowl overflow tubes to the carburetor air cleaner.

Catalytic Converter

All California Cars except Pick-Up

In addition to the air injection system, EGR and the engine modifications, the catalyst further reduces pollutants. Through catalytic action, it changes residual hydrocarbons and carbon monoxide in the exhaust gas into carbon dioxide and water before the exhaust gas is discharged into the atmosphere.

NOTE: *Only unleaded fuel must be used with catalytic converters; lead in fuel will quickly pollute the catalyst and render it useless.*

The emergency air relief valve is used as a catalyst protection device. When the temperature of the catalyst goes above maximum operating temperature, the temperature sensor signals the switching module to activate the emergency air relief valve. This stops air injection into the exhaust manifold and lowers the temperature of the catalyst.

NOTE: *The catalytic converter has been added to all 810 models for 1979 and the floor temperature warning light eliminated.*

Catalytic Converter

All Datsun Models have the catalytic converters installed for 1980. The oval type will be used on most models, with the round converter type used on the remainder models.

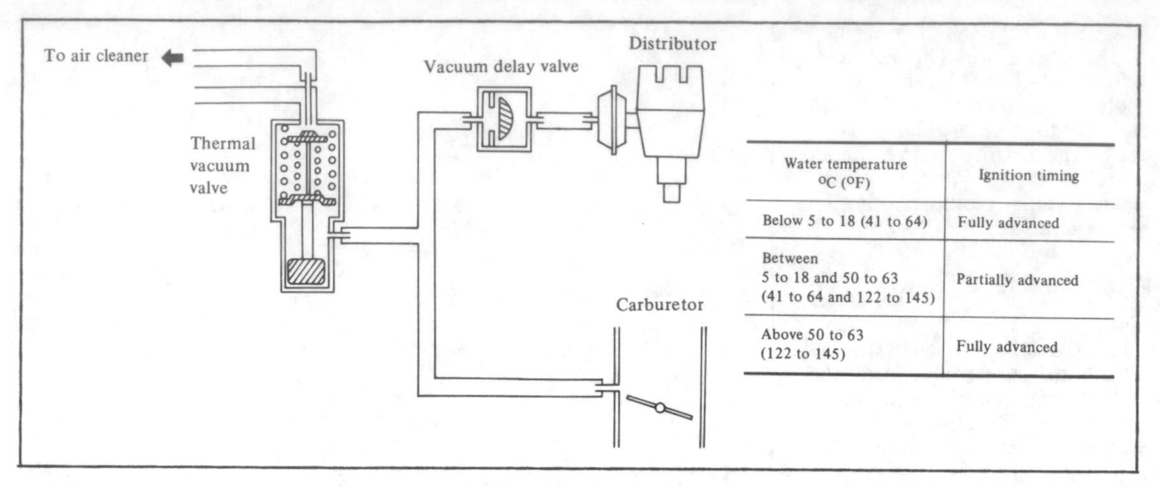

Water temperature °C (°F)	Ignition timing
Below 5 to 18 (41 to 64)	Fully advanced
Between 5 to 18 and 50 to 63 (41 to 64 and 122 to 145)	Partially advanced
Above 50 to 63 (122 to 145)	Fully advanced

Spark timing control system—Non-Calif. models w/automatic transmission—Typical

Three Way Catalytic Converter (T.W.C.)

The three way catalytic converter is used to provide the optimum chemical reaction for converting the exhaust gases, HC, CO, NOx, into harmless matter before it is discharged into the atmosphere. The T.W.C. system is used with the E.F.I. control unit with the feedback circuit controls. The T.W.C. and the regular (oxidation) converters are identical in their outward appearance and identification can be made by checking the following points.

1. The identification marks are the fourth and fifth digits, located on the rear outlet of the converter. 02 represents the regular (oxidation) coverter, while 03 represents the three way converter.
2. A dab of yellow paint is located on the three way converter, while no paint is found on the regular (oxidation) converter.

Floor Temperature Warning System

1976-78 280Z

This system employs temperature sensors to warn of impending catalytic converter overheating. The system consists of a floor sensor located in the luggage compartment, a floor sensor relay located under the front passenger seat and a warning lamp located on the left side of the instrument panel.

Spark Timing Control System

The spark timing control regulates the distributor vacuum advance at low engine coolant temperatures and to achieve good fuel economy levels by using thermal vacuum valves to regulate the passage of vacuum to the vacuum advance.

Thermal Vacuum Valves

Three versions of the T.V.V. are used and are of the wax pellet type, utilizing the wax pellet's thermal expansion force to activate the control shaft, to open or close the vacuum intake or outlet ports. The E.G.R. valve and the distributor vacuum advance operations are controlled by the operation of the thermal vacuum valves, by the monitoring the cooling system temperature.

Mixture Ratio Feedback System (L28E, L24E engines)

The mixture ratio feedback system controls the air-fuel mixture ratio and produces the ideal ratio by means of an exhaust gas sensor and E.F.I. control unit. The exhaust sensor is located in the exhaust manifold and relays the difference in oxygen levels between the exhaust gases and the atmosphere, to the E.F.I. control unit. The control unit controls the amount of fuel being injected into the engine and regulates the air-fuel mixture for the best ratio. The three way converter must be used with this unit.

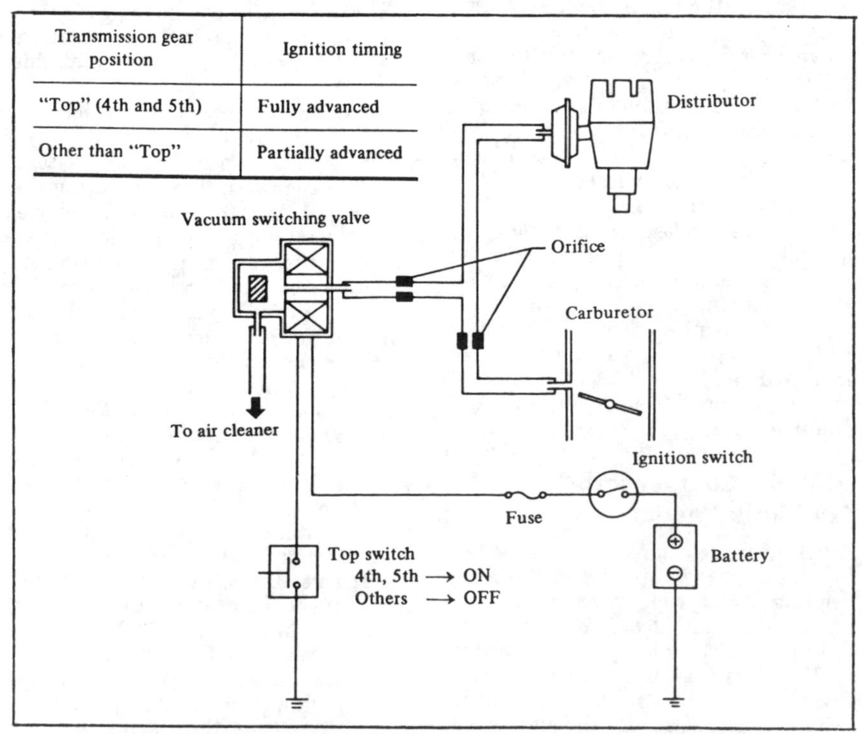

Transmission gear position	Ignition timing
"Top" (4th and 5th)	Fully advanced
Other than "Top"	Partially advanced

Spark timing control system—FU models w/manual transmission—Typical

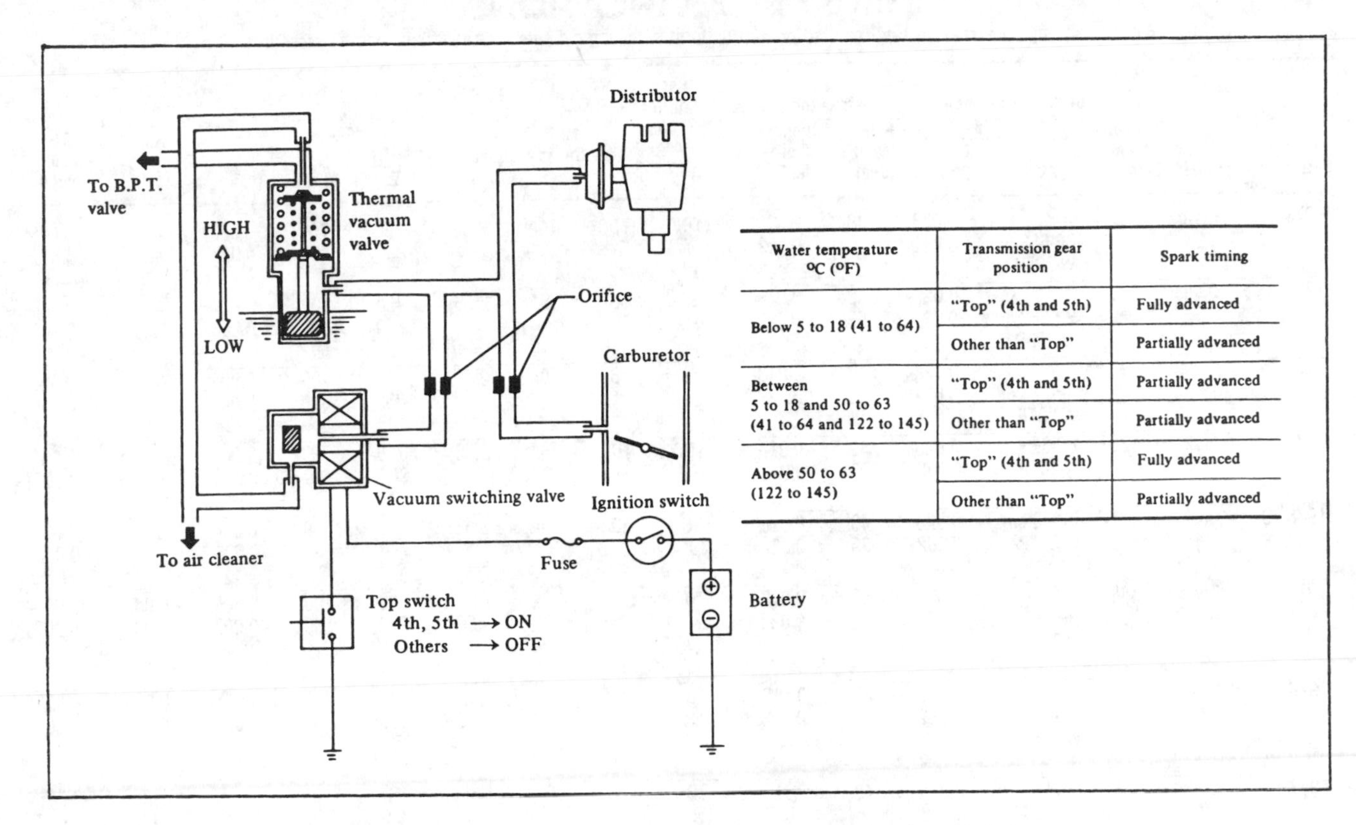

Water temperature °C (°F)	Transmission gear position	Spark timing
Below 5 to 18 (41 to 64)	"Top" (4th and 5th)	Fully advanced
	Other than "Top"	Partially advanced
Between 5 to 18 and 50 to 63 (41 to 64 and 122 to 145)	"Top" (4th and 5th)	Partially advanced
	Other than "Top"	Partially advanced
Above 50 to 63 (122 to 145)	"Top" (4th and 5th)	Fully advanced
	Other than "Top"	Partially advanced

Spark timing control system—USA models w/manual transmission—Typical

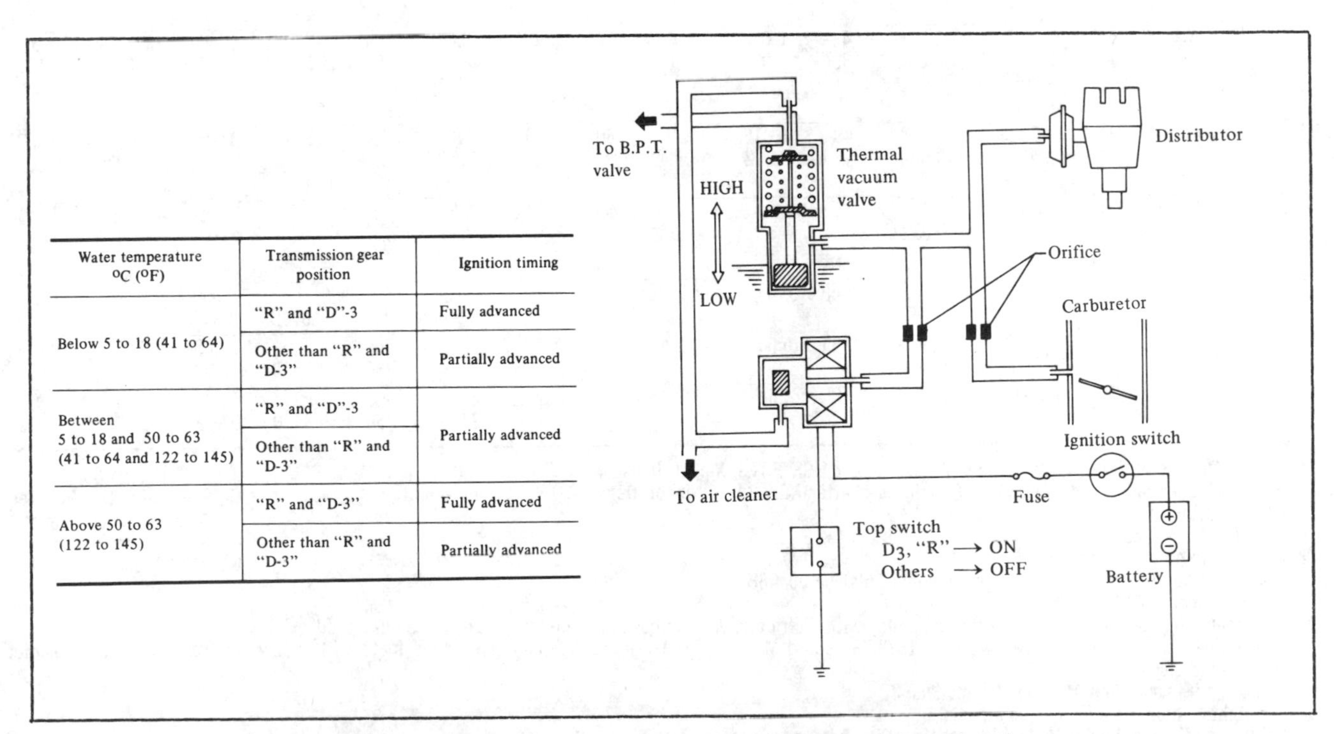

Water temperature °C (°F)	Transmission gear position	Ignition timing
Below 5 to 18 (41 to 64)	"R" and "D"-3	Fully advanced
	Other than "R" and "D-3"	Partially advanced
Between 5 to 18 and 50 to 63 (41 to 64 and 122 to 145)	"R" and "D"-3	Partially advanced
	Other than "R" and "D-3"	
Above 50 to 63 (122 to 145)	"R" and "D-3"	Fully advanced
	Other than "R" and "D-3"	Partially advanced

Spark timing control system—Calif. models w/automatic transmission—Typical

TUNE-UP SPECIFICATIONS

When analyzing compression test results, look for uniformity among cylinders, rather than specific pressures.

Year	Model	Engine Displacement (cc)	Original Equipment Spark Plugs Type	Original Equipment Spark Plugs Gap (in.)	Distributor Point Dwell (deg)	Distributor Point Gap (in.)	Basic Ignition Timing (deg) MT	Basic Ignition Timing (deg) AT	Intake Valve Fully Opens (deg)	Fuel Pump Pressure (psi)	Idle Speed (rpm) MT	Idle Speed (rpm) AT	Valve Clearance (in.) Intake (cold)	Valve Clearance (in.) Auxiliary (cold)	Valve Clearance (in.) Exhaust (cold)
1973	Civic	1170	BP-6ES or W-20EP	0.028-0.031	49-55	0.018-0.022	TDC ③ ⑧	TDC ③ ⑧	10A	2.56	750-850 ④	700-800 ⑤	0.005-0.007	—	0.005-0.007
1974	Civic	1237	BP-6ES or W-20EP	0.028-0.031	49-55	0.018-0.022	5B ⑧	5B ⑧	10A	2.56	750-850 ④	700-800 ⑤	0.004-0.006	—	0.004-0.006
1975-76	Civic	1237	BP-6ES or W-20EP	0.028-0.032	49-55	0.018-0.022	7B ⑧	7B ⑧	10A	2.56	750-850 ④	700-800 ⑤	0.004-0.006	—	0.004-0.006
1977	Civic	1237	BP-6ES or W-20EP	0.028-0.032	49-55	0.018-0.022	TDC ⑧	TDC ⑧	10A	2.56	700-800 ④	700-800 ⑤	0.004-0.006	—	0.004-0.006
1978-79	Civic	1237	BP-6ES or W-20EP	0.028-0.032	49-55	0.018-0.022	2B ⑧	2B ⑧	10A	2.56	650-750 ④	650-750 ⑤	0.004-0.006	—	0.004-0.006
1975	Civic CVCC	1487	BP-6ES or W-20ES ②	0.028-0.032	49-55	0.018-0.022	TDC ⑨	3A ⑨	10A	1.85-2.56	800-900 ④	700-800 ⑤	0.005-0.007	0.005-0.007	0.005-0.007
1976	Civic CVCC	1487	B-6ES or W-20ES ②	0.028-0.032	49-55	0.018-0.022	2B ⑥ ⑨	2B ⑦ ⑨	10A	1.85-2.56	800-900 ④	700-800 ⑤	0.005-0.007	0.005-0.007	0.005-0.007
1977	Civic CVCC	1487	B6EB or W-20ES-L	0.028-0.032	49-55	0.018-0.022	6B ⑩	6B ⑩⑦	10A	1.85-2.56	750-850 ④	650-750 ⑤	0.005-0.007	0.005-0.007	0.005-0.007
1978-79	Civic CVCC	1487	B6EB or W-20ES-L	0.028-0.032	49-55	0.018-0.022	6B ⑩	6B ⑩	10A	1.85-2.56	650-750 ④	600-700 ⑤	0.005-0.007	0.005-0.007	0.007-0.009
1976-77	Accord CVCC	1600	B-6ES or W-20ES ① ②	0.028-0.032	49-55	0.018-0.022	2B ⑨	TDC ⑨	10A	1.85-2.56	750-850 ④	630-730 ⑤	0.005-0.007	0.005-0.007	0.005-0.007
1978	Accord	1600	B6EB or W-20ES-L	0.028-0.032	49-55	0.018-0.022	6B ⑨⑩	6B ⑨⑩	10A	2.13-2.84	750-850 ④	650-750 ⑤	0.005-0.007	0.005-0.007	0.007-0.009
1979	Accord	1751	B7EB	0.028-0.032	Electronic		6B ⑨⑫	4B ⑬⑪	10A	2.13-2.84	650-750 ④	650-750 ⑤	0.005-0.007	0.005-0.007	0.010-0.012
1980	Accord	1751	B7EB⑱	0.028-0.032	Electronic		0°TDC⑲		NA	2.10-2.80	800-850	800-⑳ 850D	0.006-0.007	0.006-0.007	0.011-0.012
	Civic	1300/1500	W-20ES-L B7EB-11⑮	0.040-0.044	Electronic		⑭⑯⑰		NA	2.10-2.80	700-750	700-750D	0.005-0.007	0.005-0.007	0.007-0.008
	Prelude	1751	B7EB⑱	0.028-0.032	Electronic		0°TDC⑲		NA	2.10-2.80	800-850	800-⑳ 850D	0.006-0.007	0.006-0.007	0.011-0.012

① For continuous highway use over 70 mph, use cooler NGK B-7ES, Nippon Denso W-22ES or equivalent
② For continuous low-speed use under 30 mph, use hotter NGK B-5ES, Nippon Denso W-16ES or equivalent
③ Static ignition timing—5B
④ In neutral, with headlights on
⑤ In drive range, with headlights on
⑥ 5-speed sedan (hatchback) from engine number 2500001-up—6B
⑦ Station wagon—TDC
⑧ Aim timing light at red notch on crankshaft pulley with distributor vacuum hose(s) connected at specified idle speed
⑨ Aim timing light at red mark (yellow mark, 1978-79 Accord M/T) on flywheel or torque converter drive plate with distributor vacuum hose connected at specified idle speed
⑩ California (KL) and High Altitude (KH) models: 2B
⑪ Aim light at blue mark (49 States models)
⑫ California (KL) and High Altitude (KH) models: TDC (white mark)
⑬ California (KL) and High Altitude (KH) models: 2 ATDC (black mark)
⑭ w/1300 eng.—2° BTDC
⑮ w/1500 eng.
⑯ w/1500 eng. 49 State Sedan—15° BTDC
⑰ w/1500 eng. Cal./High Altitude—10° BTDC
⑱ B6EB—Cold Climate
⑲ 4° ATDC—1800 cc Cal. 4 dr. Sedan (1751 cc)
⑳ For Calif. 3 spd. Auto, see underhood label

TDC—Top Dead Center
B—Before Top Dead Center
A—After Top Dead Center
— Not Applicable
N.A. Not Available

Note: The underhood specifications sticker often reflects tune-up specification changes made in production. Sticker figures must be used if they disagree with those in this chart.

SERIAL NUMBER IDENTIFICATION CHART

Year	Model	VIN (Chassis Number)	Engine Number
1975	Civic	3300001—4000000	EB2-2000001—2025158
1975	Civic CVCC	1000001—2000000	ED1-1000001—1999999
1975	Civic CVCC Wagon	1000001—2000000	ED2-1000001—1999999
1976	Civic	4000001—5000000	EB2-2025159—2999999
1976	Civic CVCC	2000001—3000000	ED3-2000001—2499999
1976	Civic CVCC①	2000001—3000000	ED3-2500001—2999999
1976	Accord	1000001—2000000	EF1-1000001—2000000
1977	Civic	5000001—6000000	EB2-3000001—4000000
1977	Civic CVCC KL	3000001—4000000	EB3-3000001—3499999
1977	Civic CVCC KA	3000001—4000000	ED3-3500001—3899999
1977	Civic CVCC KH	3000001—4000000	ED3-3900001—4000000
1977	Accord KL	2000001—3000000	EF1-2000001—2499999
1977	Accord KA	2000001—3000000	EF1-2500001—2899999
1977	Accord KH	2000001—3000000	EF1-2900001—3000000
1978	Civic	6000001—7000000	EB3-1000001—1500000
1978	Civic CVCC KL	4000001—5000000	ED3-4000001—4499999
1978	Civic CVCC KA	4000001—5000000	ED3-4500001—4899999
1978	Civic CVCC KH	4000001—5000000	ED3-4900001—5000000
1978	Accord KL	3000001—4000000	EF1-3000001—3499999
1978	Accord KA	3000001—4000000	EF1-3500001—3899999
1978	Accord KH	3000001—4000000	EF1-3900001—4000000
1979	Civic	7000001—8000000	EB3-1500001—2000000
1979	Civic CVCC KL	5000001—6000000	ED3-5000001—5499999
1979	Civic CVCC KA	5000001—6000000	ED3-5500001—5899999
1979	Civic CVCC KH	5000001—6000000	ED3-5900001—6000000
1979	Accord KL	4000001—5000000	EF1-4000001—4499999
1979	Accord KA	4000001—5000000	EF1-4500001—4899999
1979	Accord KH	4000001—5000000	EF1-4900001—5000000
1980	Civic	②1000001—1999999	③EJ1-100001—199999
	Accord/Prelude	2000001—2999999	EK1-200001—299999

Note: Beginning 1976 Civic Wagons have engine serial numbers prefix ED4. KL, KH, and KA designations still apply.

① 5 speed/49 States
KL: California
KH: High Altitude
KA: 49 States
② Note: First digit "1'" indicates 1980 Civic. "2" indicates 1980 Prelude and Accord.
Second digit indicates Emission Group 0-California; 5-49 States Low Altitude 9-High Altitude
③ EJ1-1300cc
EM1-1500cc
EK1-1800cc

TORQUE SPECIFICATIONS

All readings are given in ft lbs

Year	Engine Displacement (cc)	Cylinder Head Bolts	Main Bearing Bolts	Rod Bearing Bolts	Crankshaft Pulley Bolts	Flywheel to Crankshaft Bolts	Manifold In	Manifold Ex	Spark Plugs	Oil Pan Drain Bolt
1973-79	1170, 1237	30-35① 37-42②	27-31	18-21	34-38	34-38	13-17	13-17③	9-12	29-36
1975-79	1487, 1600, 1751 CVCC	40-47	30-35	18-21	58-65	34-48	15-17	15-17	11-18	29-36

① To engine number EB 1-1019949
② From engine number EB 1-1019950
③ 1975-76 models w/AIR—22-33 ft lbs

EMISSION CONTROLS

CRANKCASE EMISSION CONTROL SYSTEM

All engines are equipped with a "Dual Return System" to prevent crankcase vapor emissions. Blow-by gas is returned to the combustion chamber through the intake manifold and carburetor air cleaner. When the throttle is partially opened, blow-by gas is returned to the intake manifold through the breather tubes leading to the tee orifice, located on the outside of the intake manifold. When the throttle is opened wide and the vacuum in the air cleaner rises, blow-by gas is returned to the intake manifold through an additional passage in the air cleaner case.

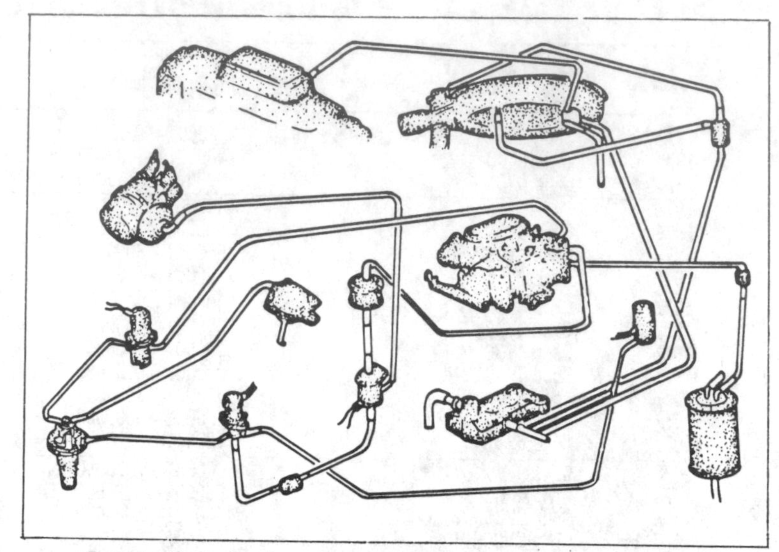

Emission Control system schematic (Accord with manual transmission)

EVAPORATIVE EMISSION CONTROL SYSTEM

Fuel vapor is stored in the expansion chamber, in the fuel tank, and in the vapor line up to the one-way valve. When the vapor pressure becomes higher than the set pressure of the one-way valve, the valve opens and allows vapor into the charcoal canister. While the engine is stopped or idling, the idle cut-off valve in the canister is closed and the vapor is absorbed by the charcoal.

At partially opened throttle, the idle cut-off valve is opened by manifold vacuum. The vapor that was stored in the charcoal canister and in the vapor line is purged into the intake manifold. Any excessive pressure of vacuum which might build up in the fuel tank is relieved by the two-way valve in the filler cap (Civic) or in the engine compartment (Accord).

EXHAUST EMISSION CONTROL SYSTEM

Special control devices are used with engine modifications. Improvements to the combustion chamber, intake manifold, valve timing, carburetor, and distributor comprise the engine modifications. The special control devices consist of the following:

a. Intake air temperature control;
b. Throttle opener;
c. Ignition timing retard unit;
d. Transmission and temperature controlled spark advance (TCS) for the transmission;
e. Temperature controlled spark advance for Hondamatic automatic transmission.

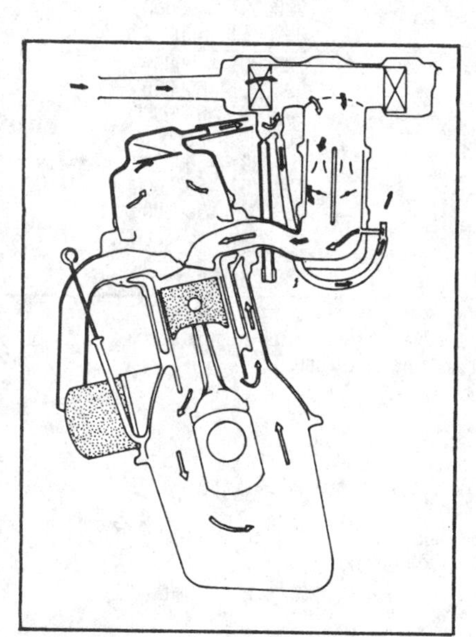

Crankcase ventilation system operation—Typical

1976-79 Models (Except 1976-79 CVCC)

Intake Air Temperature Control

When the temperature in the air cleaner is below approximately 100° F. the air bleed valve, which consists of a bimetallic strip and a rubber seal, remains closed. Intake manifold vacuum is then led into a vacuum motor, on the snorkel of the air cleaner, which moves the air control valve door, allowing only pre-heated air to enter the air cleaner.

When the temperature in the air cleaner becomes higher than approx. 100° F., the air bleed valve opens and the air control valve door returns to the open position allowing only unheated air through the snorkel.

Throttle Opener

The throttle opener is designed to prevent misfiring during deceleration by causing the throttle valve to remain slightly open, allowing better mixture control. The control valve is set to allow the passage of vacuum to the throttle opened diaphragm when the engine vacuum is equal to or greater than the control valve preset vacuum (21.6 ± 1.6 in. Hg) during accleration.

Under running conditions, other than fully closed throttle deceleration, the intake manifold vacuum is less than the control valve set vacuum; therefore the control valve is not actuated. The vacuum remaining in the throttle opener and control valve is returned to atmospheric pressure by the air passage at the valve center.

Ignition Timing Retard Unit

When the engine is idling, the vacuum produced in the carburetor retarder port is communicated to the spark retard unit and the igntion timing, at idle, is retarded.

Transmission Controlled Spark Advance

When the coolant temperature is approximately 120° or higher, and the transmission is in First, Second, or Third gear, the solenoid valve cuts off the vacuum to the spark advance unit.

On later models, the vacuum is cut off to the spark advance unit regardless of temperature when First, Second, or Third gear is selected. Vacuum advance is restored when Fourth gear is selected.

Temperature controlled spark ad-

Emission control system components—Typical of engines with manual transmissions

vance on cars equipped with Hondamatic transmission is designed to reduce Nox emissions. When the coolant temperature is approximately 120° or higher, the solenoid valve is energized, cutting off vacuum to the advance unit.

Ignition Timing Retard Unit

The ignition timing retard unit is used only on Hondamatic models and has no vacuum advance mechanism.

Air Injection System

Beginning with the 1975 model year, an air injection system is used. A belt-driven air pump delivers filtered air under pressure to injection nozzles located at each exhaust port. Here, the additional oxygen supplied by the vane-type pump reacts with any uncombusted fuel mixture, promoting an afterburning effect in the hot exhaust manifold. To prevent a reverse flow in the air injection manifold when exhaust gas pressure exceeds air supply pressure, a non-return check valve is used. To prevent exhaust afterburning or backfiring during deceleration, an anti-afterburn valve delivers air to the intake manifold instead. When manifold vacuum rises above the preset vacuum of the air control valve and/or below that of the air bypass valve, air pump air is returned to the air cleaner.

1976-80 CVCC Models

Throttle Controls

This system controls the closing of the throttle during periods of gear

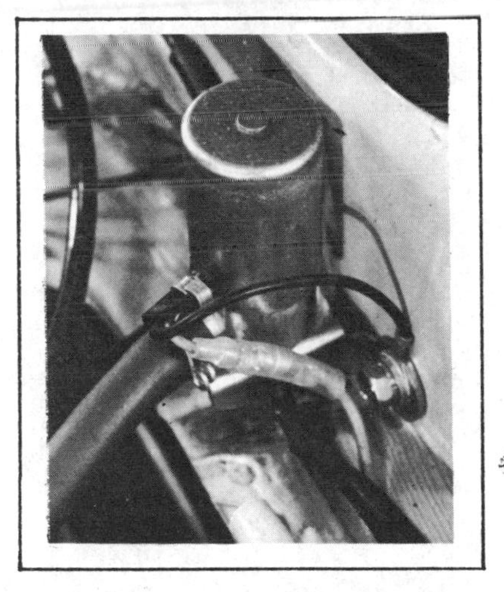

Start control solenoid valve

shifting, deceleration, or anytime the gas pedal is released. This system has two main parts, a dashpot system and a throttle positioner system. The dashpot diaphragm and solenoid valve act to dampen or slow down the throttle return time to 1-4 seconds. The throttle positioner part consists of a speed sensor, a solenoid valve, a control valve and an opener diaphragm which will keep the throttle open a predetermined minimum amount any time the gas pedal is released when the car is traveling 20 mph or faster, and closes it when the car slows to 10 mph.

Ignition Timing Controls

This system uses a coolant temperature sensor to switch distributor vacuum ignition timing controls on or off to reduce hydrocarbon and oxides of nitrogen emissions. The coolant switch is calibrated at 149° F.

1980

Ignition tmiing controls in conjunction with the distributor calibration, affects the time of the ignition of the mixture. The spark is controlled according to the engine's needs and this depends on the engine rpm, the coolant temperature and the load on the engine. Different models and applications have different controls to help get the greatest fuel economy and lowest emissions.

The 1300cc engine, in its 49 state, manual transmission version, uses a coolant temperature switch to activate an ignition solenoid valve. This only comes into play when the coolant temperature is below about 167°F. With the ignition solenoid valve activated, direct manifold vacuum is directed to the distributor vacuum advance unit.

The 1500cc engine, in its 49 state, manual transmission version, sends the vacuum to the distributor directly from the carburetor insulator block by means of a calibrated orifice.

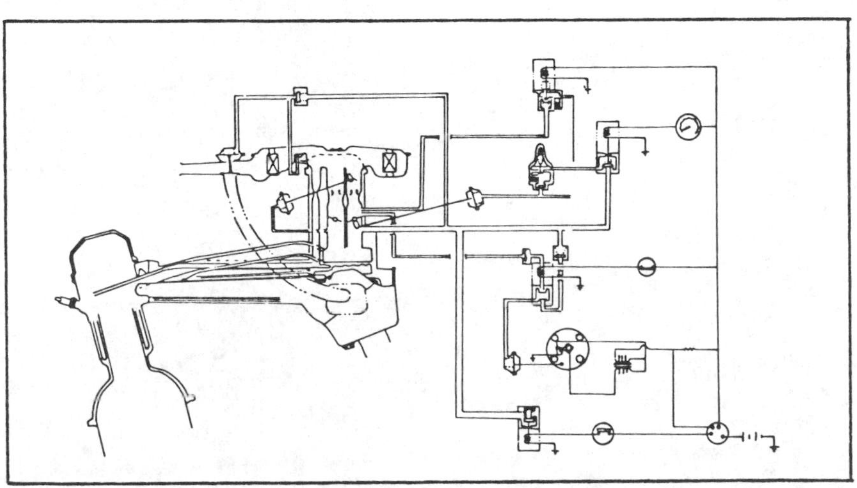

Emission control system schematic (Civic CVCC with manual transmission)

The 1500cc engine with manual transmission, California and High Altitude versions as well as all automatic transmission versions send the vacuum signal only when the engine is above idle since the port is above the throttle plate.

Hot Start Control

This system is designed to prevent an overrich mixture condition in the intake manifold due to vaporization of residual fuel when starting a hot engine. This reduces hydrocarbon and carbon monoxide emissions.

CVCC Engine Modifications

By far, the most important part of the CVCC engine emission control system is the Compound Vortex Controlled Combustion (CVCC) cylinder head itself. Each cylinder has three valves: a conventional intake and conventional exhaust valve, and a smaller auxiliary intake valve. There are actually *two* combustion chambers per cylinder: a precombustion or auxiliary chamber, and the main chamber. During the intake stroke, an extremely lean mixture is drawn in to the main combustion chamber. Simultaneously, a very rich mixture is drawn into the smaller precombustion chamber via the auxiliary intake valve. The spark plug, located in the precombustion chamber, easily ignites the rich premixture, and this combustion spreads out into the main combustion chamber where the lean mixture is ignited. Due to the fact that the volume of the auxiliary chamber is much smaller than the main chamber, the overall mixture is very lean (about 18 parts air to one part fuel).

1980

The 1300cc engine uses a new aluminum block with the CVCC head. A siamesed 2 port exhaust is used which is similar to the 1979 and earlier 1500cc engines. The cylinder head is no longer of the crossflow design.

The 1500cc engine uses a new combustion chamber design, as well as an addditional passage for the EGR to the #4 exhaust port on the California and High Altitude engines.

EMISSION CONTROL EQUIPMENT 1980

CVCC Engine design
Crankcase Control (PCV)
Evaporative Control
Intake Air Temperature Control
Throttle Controls
Exhaust Gas Recirculation (EGR)
Catalytic Converter
Anti-Afterburn Valve
Speed Sensor (in speedo head)
Air Jet Controller (AJC)

NOTE: *Not all vehicles will use all components, depending on application.*

SPEED SENSOR 1980

The speed sensor is located in the speedometer. It can be considered a photointerupter, since it makes use of a notched plate rotating on the speedometer shaft. It is actuated at a predetermined speed. It is used this year on all vehicles.

CATALYTIC CONVERTER 1980

The catalytic converter is of conventional design and it is used on all models except the 1300cc engine, 49 state, manual transmission version. With use of the converter, unleaded fuel is required.

AIR JET CONTROLLER 1980

The air jet controller is an atmospheric pressure sensing device. The application is therefore restricted to the 1500cc engine, California and High Altitude versions. It controls the flow of air to the main and slow air jets of the carburetor auxiliary throat and the secondary slow jet of the main carburetor. When the air pressure outside changes, as it would when the vehicle is subjected to altitude changes in mountainous areas, the bellows in the air jet controller expand (in response to reduced pressure) to open the valve in the AJC thus increasing air flow to the jets to maintain optimum mixture control.

ANTI-AFTERBURN VALVE

The purpose of this valve is to introduce fresh air into the intake manifold whenever the manifold vacuum abruptly increases, as it would on deceleration. This helps reduce backfiring. A spring and diaphragm control the air flow and the valve opening. The valve is sensitive only to changes in vacuum. This valve will be found only on the 1300cc manual transmission, 49 state versions.

TUNE-UP SPECIFICATIONS

Year	Engine Displacement (cc)	Spark Plugs Type	Spark Plugs Gap (in.)	Distributor Point Dwell (deg)	Distributor Point Gap (in.)	Ignition Timing (deg)	Intake Valve Opens (deg)	Cranking Compression Pressure (psi)	Idle Speed (rpm)	Valve Clearance (in) In	Valve Clearance (in) Ex
1973-74	1400	BP-6ES	0.032	49-55	0.020	6B @ 800	24B	178	800	0.011-0.013-	0.011-0.013
1975	1400	BP-6ES	0.030	49-55	0.020	8B @ 800M, 8B @ 900A	24B	178	①	0.012	0.014
1976	1400	BP-6ES	0.032	49-55	0.018	8B @ 900	24B	156	①	0.011	0.015
	1600	BP-6ES	0.032	44-55	0.018	8B @ 900	24B	156	①	0.011	0.015
1977-78	1600	BP-6ES	0.032	49-55③	0.018④	8B @ 850②	24B	156	①	0.010	0.014
1979	1600	BP-6ES	0.032	Electronic	⑤	8B @ 800②	24B	156	①	0.010	0.014
1980	1600	BP-6ES	0.032	Electronic	⑤	8B @ 800②⑥	24B	156	①	0.010	0.014
1980	1800	⑦	⑦	⑦	⑦	⑦	⑦	⑦	⑦	⑦	⑦

NOTE: The underhood specifications sticker often reflects tune-up specification changes made in production. Sticker figures must be used if they disagree with those in this chart.

B Before Top Dead Center
TDC Top Dead Center
M—Manual, A—Automatic
① See Engine Compartment Sticker
② Calif. 900
③ Calif. Electronic Ignition, starting 1977 model year
④ Calif. w/E.I. dir gap Man. trans .008-.016″ Auto trans .012-.016″
⑤ Air Gap Man. trans. exc. 4WD .008-.016″ Auto. trans. and 4WD .012-.016″
⑥ 1980 Hatchback STD, DL, Sedan DL and Hardtop DL 900 rpm
⑦ Information Not Available at Time of Printing

VEHICLE IDENTIFICATION

Year	Model (displacement)	Body Style	Vehicle Identification Number Code
1973-75	1400 GL	Sedan, Coupe, Hardtop	A22L
	1400 GL	Station Wagon	A62L
1975	4 wheel drive	Station Wagon	A64L
1976	1400, 1600 DL	Sedan, Coupe (M.T.)	A22L
	1400, 1600 DL	Sedan (A.T.)	A26L
	1400, 1600 DL	Station Wagon (M.T.)	A62L
	1400, 1600 DL	Station Wagon (A.T.)	A66L
	1400, 1600 GF	Hardtop (M.T.)	A22L
	1400, 1600 GF	Hardtop (A.T.)	A26L
	4 wheel drive	Station Wagon	A64L
1977-79	1600 DL	Sedan, Coupe	A26L
	1600 DL	Station Wagon	A66L
	1600 GF	Hardtop	A26L
	4WD	Station Wagon	A67L
	Brat	4 WD	—
1980	1600 STD, DL	Coupe Hatchback	AF2
	1600 STD, DL- 4WD	Coupe Hatchback	AF3
	1600 DL, GL	Sedan	AB2
	1600 DL, GLF	Hardtop	AW2
	1600 DL, GL	Station Wagon	AM2
	1600 DL, GL- 4WD	Station Wagon	AM3
	1600 4WD	Brat	A69L
	1800 DL	Coupe Hatchback	AF4
	1800 GL	Sedan	AB4
	1800 GLF	Hardtop	AW4
	1800 GL	Station Wagon	AM4

BREAKERLESS DISTRIBUTOR

Beginning with the 1977 model year, vehicles offered for sale in California were equipped with a breakerless distributor. The centrifugal advance, vacuum advance and retard units are the same as used with the conventional distributor.

The air gap between the reluctor and pick-up coil is adjustable and should be 0.008 to 0.016 inch for the standard transmission equipped vehicles and 0.012 to 0.016 inch for automatic transmission equipped vehicles. The ignition timing should be checked if any changes are made to the air gap.

CARBURETOR ADJUSTMENTS

(with or without CO Meter)

1976

With CO meter

(with air injector system connected)
Idle speed—850-950 RPM
CO %—0.15-0.55%
(w/o air injector connected)
Idle speed—850-950 RPM
CO %—0.5-1.5 %

Without CO meter

1. Adjust the engine idle speed to attain 980 RPM.
2. Turn the air/fuel mixture screw clockwise until the engine speed is 900 RPM.
3. The emitted CO should be within the proper range.

1977-78

With CO meter

(with air injected system connected)
49 states and high altitude
Idle speed—850 ± 50 RPM
CO %—0.5 ± 1.5%
California
Idle speed—900 ± 50 RPM
CO %—0.75 ± 0.25%
(w/o air injected system connected)
49 states and high altitude
Idle speed—850 ± 50 RPM
CO %—1.5 ± 0.5%
California
Idle speed—900 ± 50 RPM
CO % 0.75 ± 0.25%

Without CO meter

1. Adjust the engine speed to attain 930 RPM (990 RPM California) by adjusting the air/fuel mixture screw and the throttle screw.
2. Adjust the air/fuel mixture screw clockwise until the engine speed is changed to 850 RPM (900 RPM California).
3. The emitted CO should be within the proper range.

1979-80

Follow the instructions on the Emission Control Information Label, located within the engine compartment.

EMISSION CONTROLS

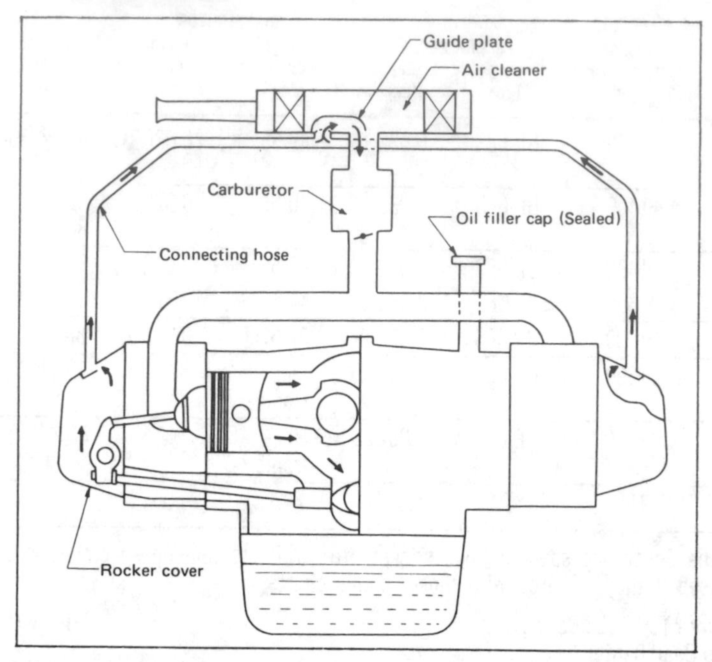

Cross section of crankcase emission control system

EMISSION CONTROLS CRANKCASE EMISSION CONTROL SYSTEM

The sealed crankcase emission control system takes blow-by gas emitted from the crankcase and routes the gas through the air cleaner and into the intake manifold for recombustion.

The system consists of a sealed oil filler cap, a rocker cover with an outlet pipe, an air cleaner with an inlet pipe to receive the connecting hoses and the connecting hoses and clamps.

There are no tests other than making sure that the system is kept clean.

EVAPORATIVE EMISSION CONTROL SYSTEM

1976

Evaporative gas from the fuel tank is not discharged into the atmosphere but conducted to the air cleaner unit and then burned in the combustion chamber. No absorbent is used.

The system consists of a sealed fuel tank and filler cap, two reservoir tanks on the station wagon, an air breather valve or a restriction, breather hoses, breather pipe and the air cleaner.

There is an air breather valve located at the filler cap. When the flap (door) is opened, a spring exerts pressure on the rubber breather hose and pinches it shut.

The vacuum relief valve filler cap relieves any vacuum condition that might arise in the gas tank.

1977-80

This system includes a canister, a check valve, two orifices and on station wagons and 4-wheel drive vehicles, two reserve tanks.

Gasoline vapor evaporated from the fuel in the fuel tank is introduced into the canister located in the engine compartment and absorbed by the activated charcoal particles. As engine speed increases a purge valve is opened and fresh air is sucked in through the bottom filter of the canister purging the absorbed vapor from the activated charcoal.

AIR SUCTION SYSTEM

The Air Suction System is used to reduce the exhaust emissions by supplying secondary air into the exhaust ports without the use of an air pump.

The negative pressure caused by the exhaust gas pulsation and the intake manifold pressure during the valve overlap period, is utilized for the further oxidation of HC and CO and to avoid the use of a catalytic converter or thermal reactor in the exhaust line.

A double reed valve assembly is used to direct fresh air to flow into the exhaust ports under negative pressure and to block the return of the exhaust gas under positive pressure. A reed valve controls air to each side of the engine.

An air suction silencer is used to muffle the sound of the air being admitted into the Air Suction System.

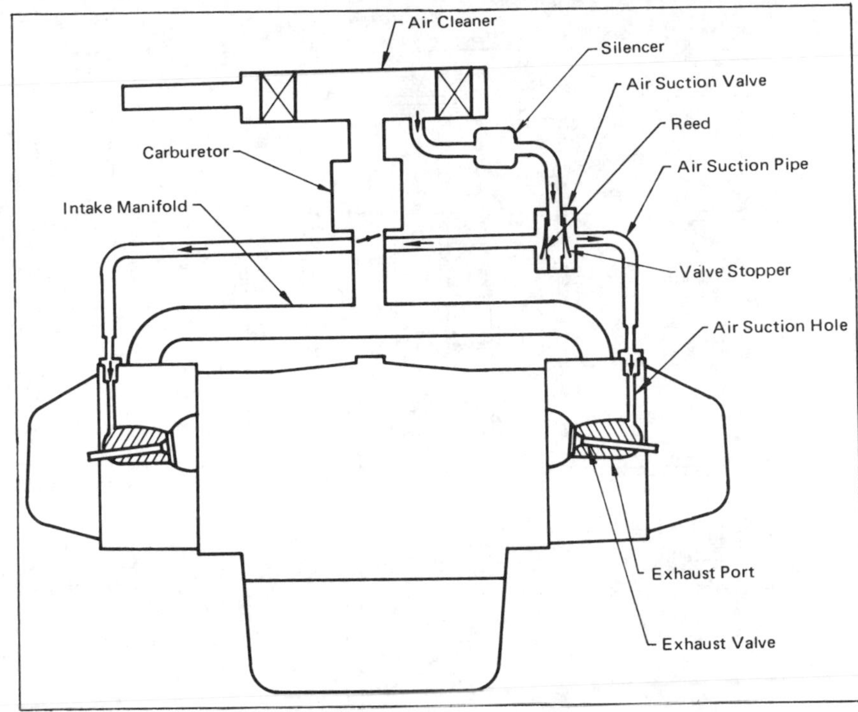

Cross section of air suction system

HOT AIR CONTROL SYSTEM

The hot air control system consists of the air cleaner, the air stove on the exhaust pipe and the air intake hose connecting the air cleaner and air stove. The air cleaner is equipped with an air control valve which maintains the temperature of the air being drawn into the carburetor at 100°-127° F. to reduce HG emission when the underhood temperature is below 100° F. This system should be inspected every 12,000 miles.

COASTING BY-PASS SYSTEM

To control the HC emissions while the vehicle is in the coasting or decelerating mode, a controlled amount of air/fuel mixture is channelled through the coasting by-pass passage in the carburetor to improve the combustion in the cylinders at the periods of high engine vacuum.

The high engine vacuum reacts on a by-pass valve diaphragm, opening a vacuum passage to the servo valve, located on the carburetor, which in turn opens a metered passage for air and fuel, from the carburetor air horn to the section of the throttle bore below the secondary throttle plate.

As the engine vacuum changes to a lower value due to acceleration, the by-pass valve closes the passage to the servo valve and the carburetor returns to its normal function.

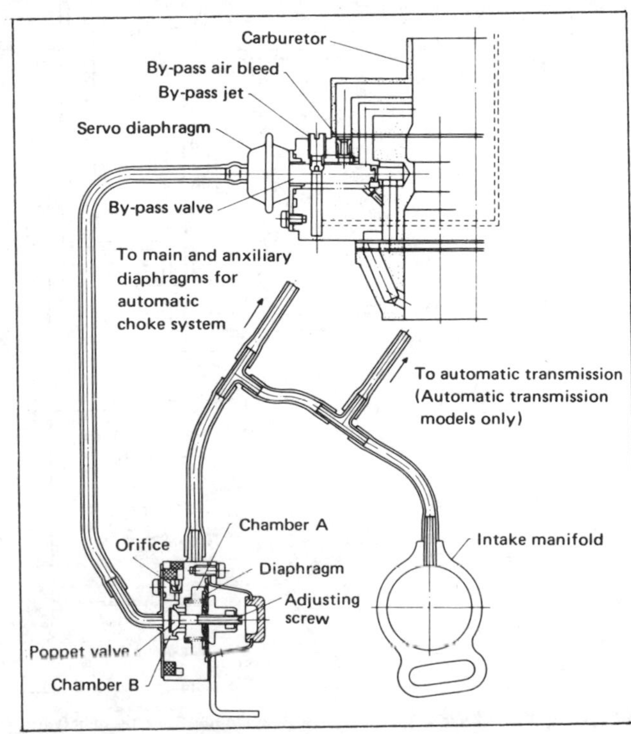

Coasting by-pass system components

ELECTRICALLY ASSISTED AUTOMATIC CHOKE

A vacuum-operated automatic choke replaces the manual chock previously used. The automatic choke uses a chock cap containing a heating element to speed up choke valve opening and reduce CO emissions during warm-up. The heating element gets its power from a special tap on the voltage regulator, when the ignition is on and the engine running.

EXHAUST GAS RECIRCULATION (EGR) SYSTEM

An exhaust gas recirculation (EGR) system is used on 1976 California and all 1977 and later models to reduce NO (oxides of nitrogen) emissions by lowering peak flame temperature during combustion. A small portion of the exhaust gases are routed into the intake manifold via a vacuum-operated EGR control valve.

A solenoid vacuum valve controls the flow of vacuum from a port on the carburetor (above the primary throttle valve) to the EGR valve vacuum diaphragm. The solenoid, in turn, is operated by a coolant temperature switch.

When the coolant temperature is above 122°F., the temperature switch breaks the current flow to the vacuum solenoid valve. The valve closes, permitting the throttle port vacuum to

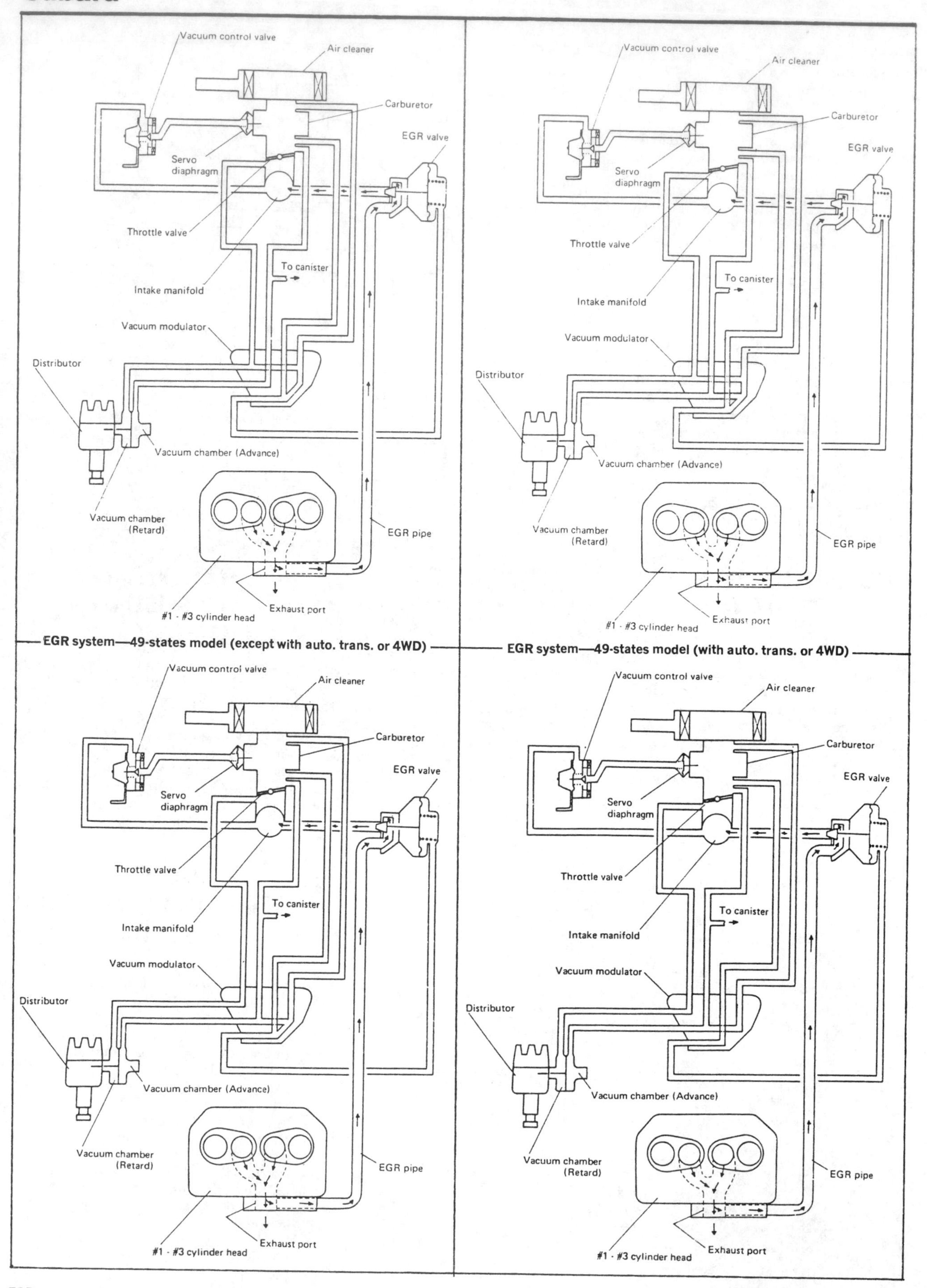

EGR system—49-states model (except with auto. trans. or 4WD)

EGR system—49-states model (with auto. trans. or 4WD)

EGR system—Calif. models (except with auto. trans.), 49-states and high altitude models (except with auto. trans. or 4WD)

EGR system—Calif. models (with auto. trans.)

operate the EGR valve diaphragm. This causes the EGR valve to open under conditions other than idle or wide-open throttle.

Below 122°F., the vacuum solenoid valve is energized to vent the vacuum from the throttle port into the atmosphere through a filter. By preventing exhaust gas recirculation from occurring before the engine has warmed-up, cold driveability is greatly improved.

ENGINE MODIFICATION SYSTEM

The vacuum is routed to the distributor vacuum retard unit. The vacuum unit retards the ignition spark in order to promote complete combustion in the cylinders.

There is an anti-dieseling solenoid mounted opposite the float bowl on the carburetor. This prevents the engine from dieseling when the ignition switch is turned off. When the ignition switch is turned off, an electromagnet in the switch is also cut off. A spring inside the housing forces a plunger into position, blocking the fuel passages leading to the opening below the throttle plates. When the ignition switch is turned on, it energizes the electromagnet in the switch and pulls the plunger out of the fuel passage, allowing fuel to reach the opening below the throttle plates.

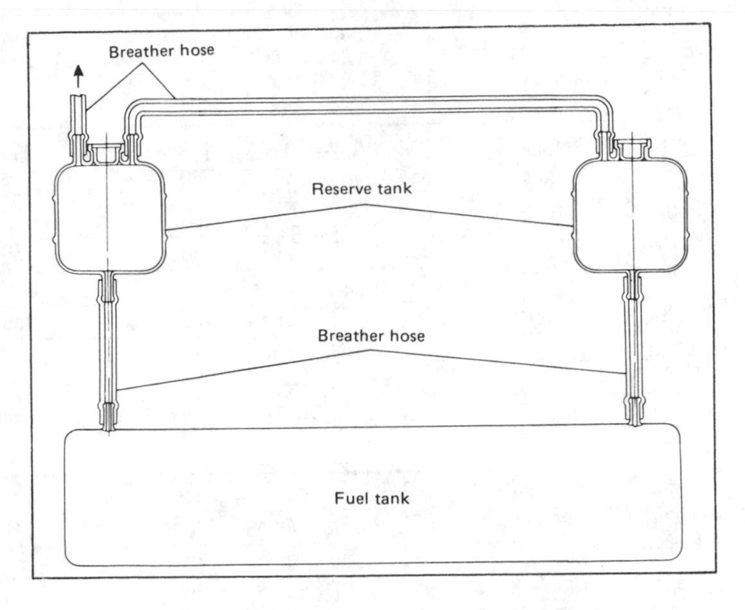

Vapor separators mounted on station wagon—typical

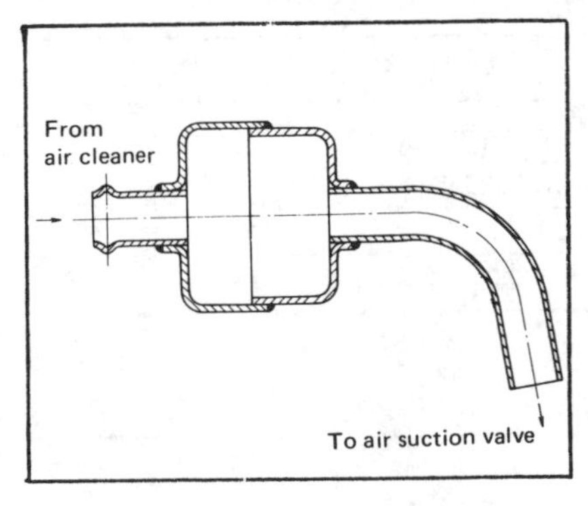

Cross section of air suction system silencer

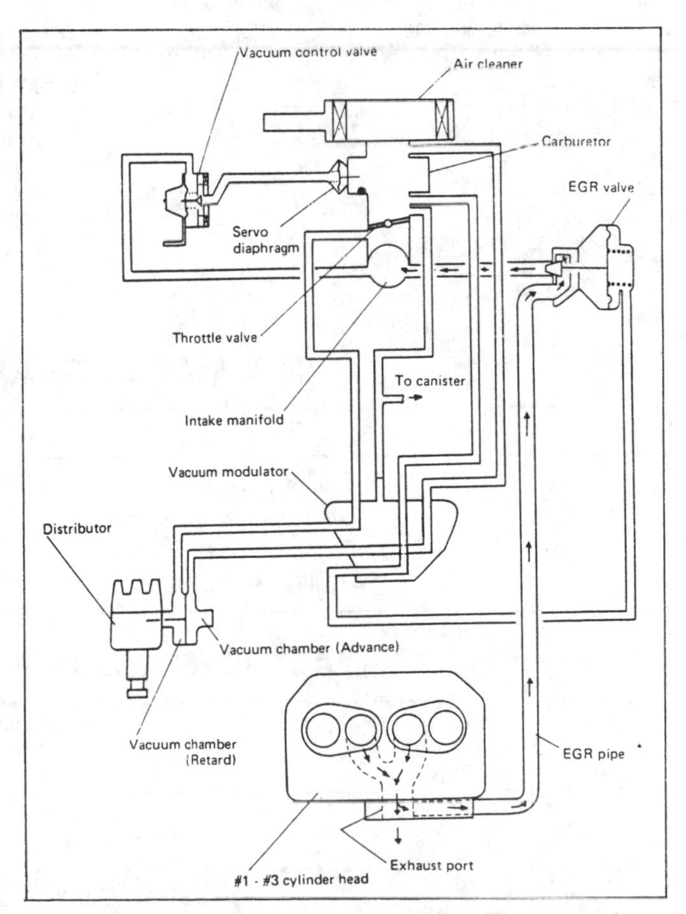

EGR system—49-states and high altitude (with auto. trans.)

TUNE-UP SPECIFICATIONS

Year	Engine Type	Spark Plugs Type (ND)	Spark Plugs Gap (in.)	Distributor Point Dwell (deg)	Distributor Point Gap (in.)	Ignition Timing (deg) ▲ MT	Ignition Timing (deg) ▲ AT	Compression Press.	Fuel Pump Press. **	Idle Speed rpm ▲ MT	Idle Speed rpm ▲ AT	Valve Clearance (in.) Intake	Valve Clearance (in.) Exhaust
1975-77	2T-C	W16EP	0.030	52⑤	0.018	10B⑥	10B⑥	171	2.8-4.3⑦	850	850	0.008	0.013
	20R	W16EP	0.030	52	0.018⑩	8B	8B	156	2.2-4.2	850	850	0.008	0.012
	4M③	W16EP	0.030	41	0.018	10B	10B	156	4.2-5.4	800	750	0.007	0.010
	4M④	W16EP	0.030	41	0.018	5B	5B	156	4.2-5.4	800	750	0.007	0.010
	2F	W14EX	0.037	41	0.018	7B	—	149	3.4-4.7	650	—	0.008	0.014
	3K-C	W20EP	0.031	52	0.018	5B	—	156	2.8-4.3	750	—	0.008	0.012
1979	3K-C	BPR5EA-L	0.031	Electronic		8B	8B	156	3.0-4.5	750	750	0.008	0.012
	2T-C	BP5EA-L	0.031	Electronic		10B⑪	10B⑪	171	3.0-4.5	850	850	0.008	0.013
	4M	BPR5EA-L	0.031	Electronic		10B⑪	10B⑪	156	4.2-5.4	750	750	0.011	0.014
	20R	BP5EA-L⑫	0.031	Electronic		8B	8B	156	2.2-4.2	800	850	0.008	0.012
1980	3K-C	BPR5EA-L	.031	Electronic		8B	8B	156	3.0-4.5	750	750	0.008	0.012
	2T-C	BPR5EA-L	.031	Electronic		10B⑪	10B⑪	171	3.0-4.5	850	850	0.008	0.013
	4M	BPR5EA-L	.031	Electronic		12B	12B	156	⑨	800	800	0.011	0.014
	20R	BP5EA-L⑫	.031	Electronic		8B	8B	156	⑨	750⑬	920	0.008	0.012
	2E	BP5EA⑭	031	Electronic		7B	7B	⑨	⑨	800	800	0.008	0.012

NOTE: If the information given in this chart disagrees with the information on the engine tune-up decal, use the specifications on the decal—they are current for the engine in your car.

▲ With manual transmission in Neutral and automatic transmission in Drive (D) (1973) or Neutral (1974-80).
③ Except Calif.
④ California only
⑤ Dual point—main 57°; sub 52°
⑨ Not available at the time of publication
⑩ California model Celica GT equipped with transistorized ignition
⑪ Calif.: 8B
⑫ Celica: BPR5EA-L
⑬ 870 Calif.
⑭ Canada BPR5EA
MT Manual transmission
AT Automatic transmission
TDC Top Dead Center
B Before top dead center
A After top dead center

TIMING MARK LOCATIONS

Engine Type	Location	Type of mark
3K-C and 21T-C	Crankshaft pulley	Notch and number scale
18R-C, 20R	Crankshaft pulley	Pointer and painted slot
4M	Crankshaft pulley	Slot and number scale
F, 2F	Flywheel	Ball and pointer

FAST IDLE ADJUSTMENT

Engine	Throttle Valve to bore clearance (in.)	Primary throttle angle (deg)	To adjust fast idle:
3K-C	0.040①	9②	Bend the fast idle lever
2T-C	0.032③	7	Turn the fast idle adjusting screw
18R-C	0.041	13—from closed	Turn the fast idle adjusting screw
20R	0.047	—	Turn the fast idle screw
4M	—	16—from closed④	Turn the fast idle adjusting screw
F	—	30—from closed	Bend the fast idle lever
2F	0.051	30—from closed	Bend the fast idle lever

— Not availabe
① 0.051 in 1976; 0.056 in 1977; 0.037 in 1978-79
② 20° open
③ 1976-79: 0.043
④ 1977-79: 9°

CAR SERIAL NUMBER AND ENGINE IDENTIFICATION

All models have the vehicle identification number (VIN) stamped on a plate which is attached to the left side of the instrument panel. This number consists of model and serial numbers.

The engine serial number consists of an engine series identification number, followed by a six-digit production number. The location of this serial number varies from one engine type to another. Serial numbers may be found in the following locations:
1200 cc (3KC) the serial number is stamped on the right side of the engine, below the spark plugs.
1600 cc (2T-C) the serial number is stamped on the left side of the engine, behind the oil dipstick.
2200 cc (2OR) the serial number is stamped on the left side of the engine, behind the alternator.
2600 cc (4M) the serial number is stamped on the right side of the cylinder block, below the oil filter.
Landy Cruiser (2F) the serial number is stamped on the front right side of the engine.

Model	Year	Displacement Cu. In. (cm³)	No. of Cylinders	Type	Engine Series Identification
Corolla					
1200	1976-80	71.2 (1166)	4	OHV	3K-C
1600	1976-80	96.9 (1588)	4	OHV	2T-C
Corona					
2200	1976-80	133.6 (2189)	4	OHC	20R
Mark II & Cressida					
2600	1976-80	156.4 (2563)	6	OHC	4M
Celica					
2200	1976-80	133.6 (2189)	4	OHC	20R
Hi-Lux					
2200	1976-80	133.6 (2189)	4	OHC	20R
Land Cruiser					
	1976-80	256.00 (4200)	6	OHV	2F

OHV—Overhead valve
OHC—Overhead cam

U.S. EMISSION EQUIPMENT

Positive crankcase ventilation
Throttle positioner system
Spark control system
Exhaust gas recirculation system
Air injection system
Catalytic converter system
Auxiliary system
Vacuum limiter system
Air suction system
High altitude compensation system

EMISSION CONTROL SYSTEMS

POSITIVE CRANKCASE VENTILATION SYSTEM

A PCV valve is used in the line to prevent the gases in the crankcase from being ignited in case of a backfire. The amount of blow-by gases entering the mixture is also regulated by the PCV valve, which is spring-loaded and has a variable orifice.

The valve is ether mounted on the valve cover or in the line which runs from the intake manifold to the crankcase.

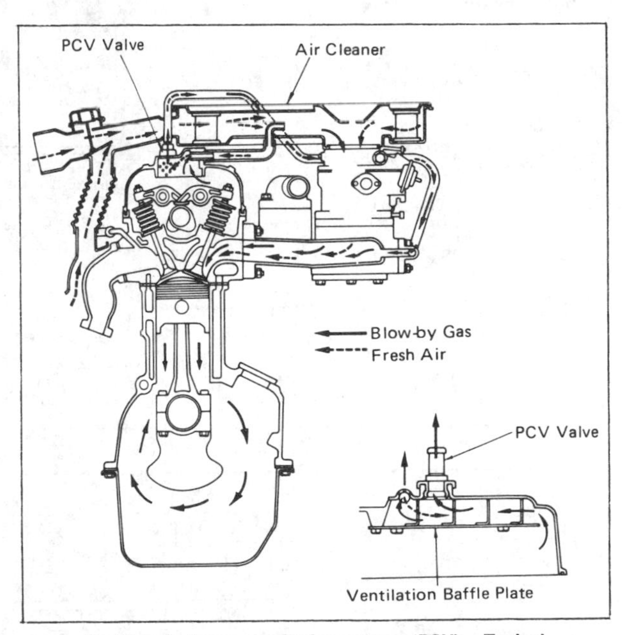

Positive crankcase ventilation system (PCV)—Typical

AIR INJECTION SYSTEM

A belt-driven air pump supplies air to an injection manifold which has nozzles in each exhaust port. Injection of air at this point causes combustion of unburned hydrocarbons in the exhaust manifold rather than allowing them to escape into the atmosphere. An antibackfire valve controls the flow of air from the pump to prevent backfiring which results from an overly rich mixture under closed throttle conditions.

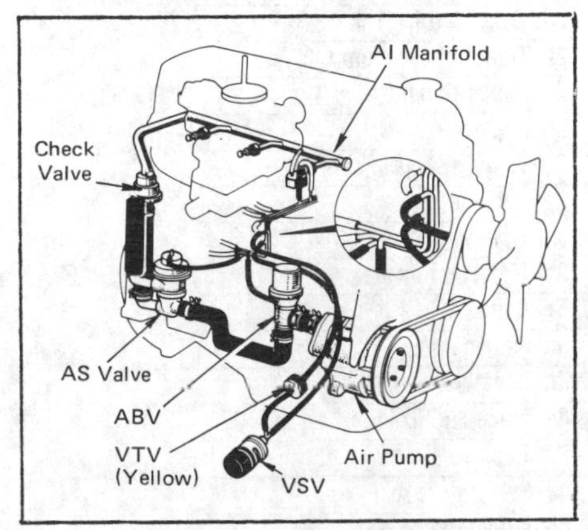

Air injection system (AI)—Typical Calif.

A check valve prevents hot exhaust gas back-flow into the pump and hoses, in case of a pump failure, or when the antibackfire valve is working.

AIR SUCTION SYSTEM

The Air Suction System, available on certain engines, brings fresh, filtered air into the exhaust ports to

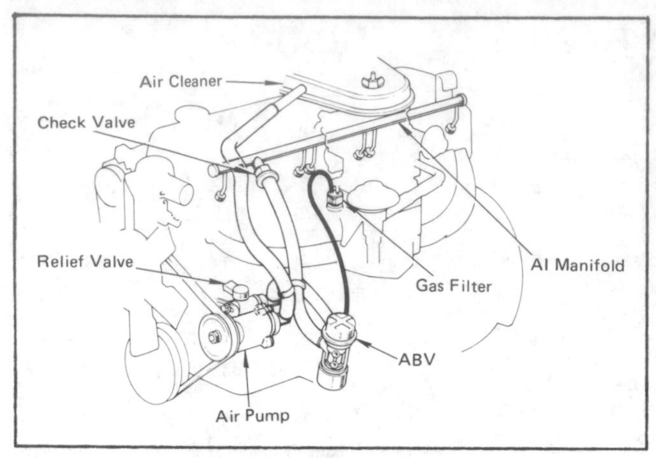

Air injection system (AI)—Typical

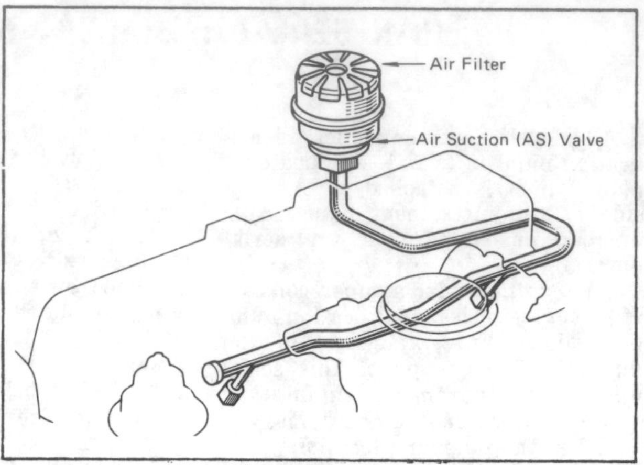

Air suction system (AS)—Typical

promote better burning of hydrocarbons. It also supplies the air necessary for the oxidizing reaction in the catalytic converter.

There are no adjustments on the system and, should it malfunction, the unit must be replaced as a whole.

To check the system, look over all lines for cracks or damage. If checks indicate no problems, start the engine and put a thin sheet of paper over the inlet port of the filter. If it is drawn to the opening, the unit is operating. If not, remove the filter and test the valve opening the same way. Replace the filter, if necessary.

FUEL EVAPORATIVE EMISSION CONTROL SYSTEM

Toyota vehicles use evaporative emission control (EEC) systems. All models use a "charcoal canister" storage system.

The charcoal canister storage system stores the fuel vapors in a canister filled with activated charcoal. All models use a vacuum switching valve to purge the system. The air filter is an integral part of the charcoal canister.

Some models equipped with the 4M six-cylinder engine have a canister-mounted purge control valve.

The purge control valve is connected to a carburetor port, which is located above the throttle control valve. When the engine is stopped or idling, there is no vacuum signal at the purge control valve so that it remains closed. When the throttle valve opens, the carburetor port is uncovered and a vacuum signal is sent to the purge control valve, which opens and allows the vapors stored in the canister to be pulled into the carburetor.

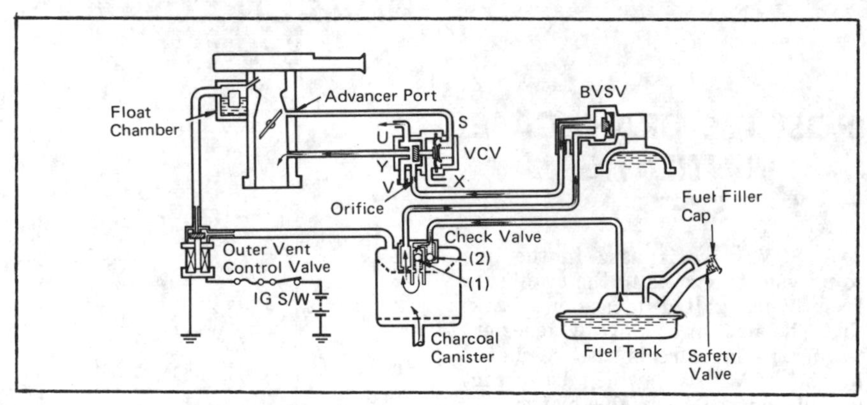

Fuel evaporative emission control system (EVAP) 4M engine—Typical

AIR INJECTION SYSTEM DIAGNOSIS CHART

Problem	Cause	Cure
1. Noisy drive belt	1a Loose belt	1a Tighten belt
	1b Seized pump	1b Replace
2. Noisy pump	2a Leaking hose	2a Trace and fix leak
	2b Loose hose	2b Tighten hose clamp
	2c Hose contacting other parts	2c Reposition hose
	2d Diverter or check valve failure	2d Replace
	2e Pump mounting loose	2e Tighten securing bolts
	2g Defective pump	2g Replace
3. No air supply	3a Loose belt	3a Tighten belt
	3b Leak in hose or at fitting	3b Trace and fix leak
	3c Defective anti-backfire valve	3c Replace
	3d Defective check valve	3d Replace
	3e Defective pump	3e Replace
4. Exhaust backfire	4a Vacuum or air leaks	4a Trace and fix leak
	4b Defective anti-backfire valve	4b Replace
	4c Sticking choke	4c Service choke
	4d Choke setting rich	4d Adjust choke

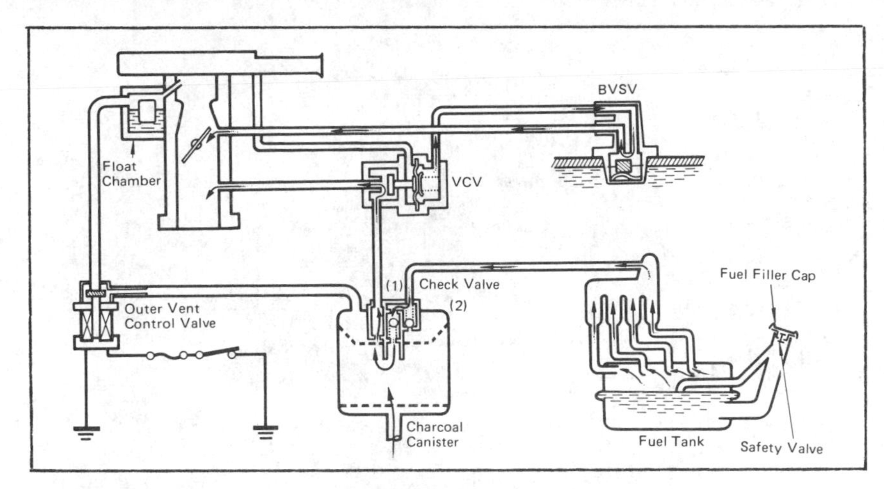

Fuel evaporative emission control system (EVAP) 2F engine

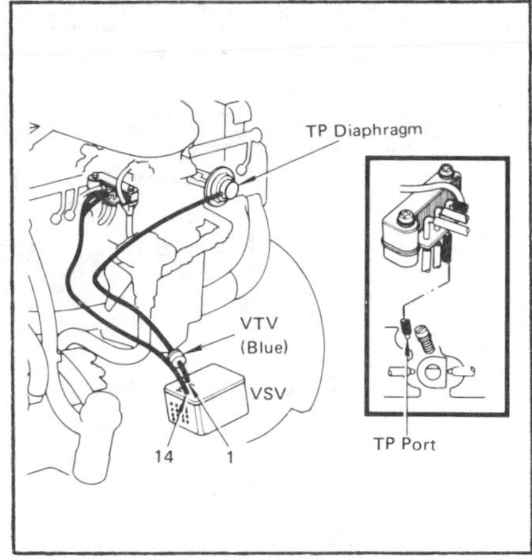

Throttle positioner system (TP)—Typical (2F engine)

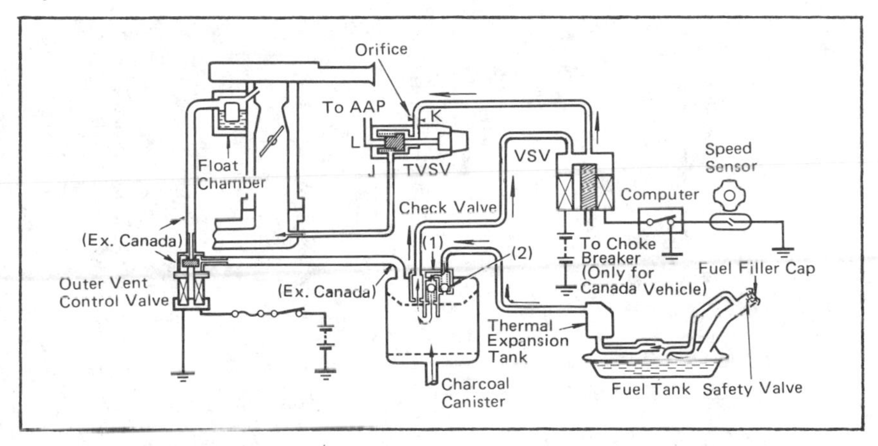

Fuel evaporative emission control system (EVAP) 20R engine—Typical

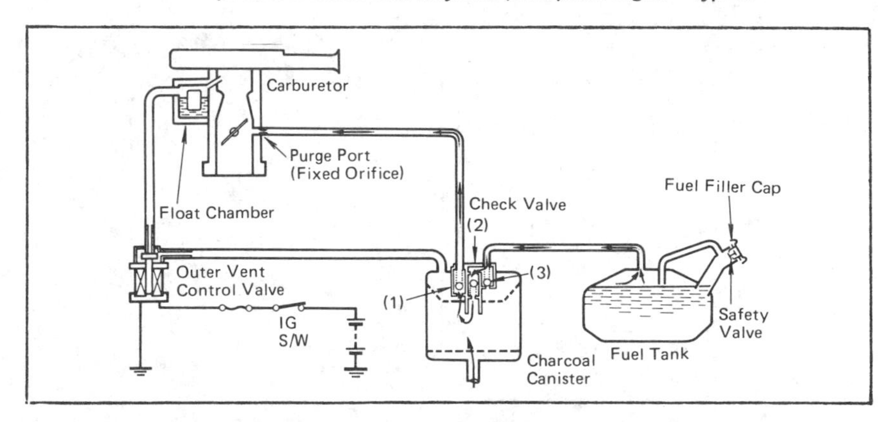

Fuel evaporation emission control system (EVAP) 2T-C engine—Typical

THROTTLE POSITIONER

On Toyotas with an engine modification system, a throttle positioner is included to reduce exhaust emissions during deceleration. The positioner prevents the throttle from closing completely. Vacuum is reduced under the throttle valve which, in turn, acts on the retard chamber of the distributor vacuum unit (if so equipped). This compensates for the loss of engine braking caused by the partially opened throttle.

Once the vehicle drops below a predetermined speed the vacuum switching valve provides vacuum to the throttle positioner diaphragm; the throttle positioner retracts allowing the throttle valve to close completely. The distributor also is returned to normal operation.

MIXTURE CONTROL SYSTEM

The mixture control valve, used on all 1978 and later 3K-C and 2T-C engines aids in combustion of unburned fuel during periods of deceleration. The mixture control valve is operated by the vacuum switching valve during periods of deceleration to admit additional fresh air into the intake manifold. The extra air allows more complete combustion of the fuel, thus reducing hydrocarbon emissions.

EXHAUST GAS RECIRCULATION

The Exhaust Gas Recirculation System (EGR) is used on all U.S. engines, except the 2T-C engine used in the Corolla. The Corolla did not adapt to the EGR system until 1977. The EGR valve is controlled by the same computer and vacuum switching valve that is used to operate other emissions control system components.

On all engines there are several conditions, determined by the computer and vacuum switching valve, which permit exhaust gas recirculation to take place.

1. Vehicle speed
2. Engine coolant temperature
3. EGR valve temperature, (F engine)

2F Thermal Reactor System and Heat Control Valve

Installed in place of the exhaust manifold on the Land Cruiser 2F engine for California, is the Thermal Reactor System. It collects the exhaust gases in a common area in order to keep their temperatures higher for a longer period of time to increase the efficiency of the exhaust gas and secondary air, restricting the release of unburned emissions.

Choke Return System

Because of the chance of seriously damaging the catalytic converter by operating the automobile with the choke out for long periods of time, the 3K-C engine is equipped with a choke return system.

Utilizing a holding coil, a holding plate and a return spring, the system

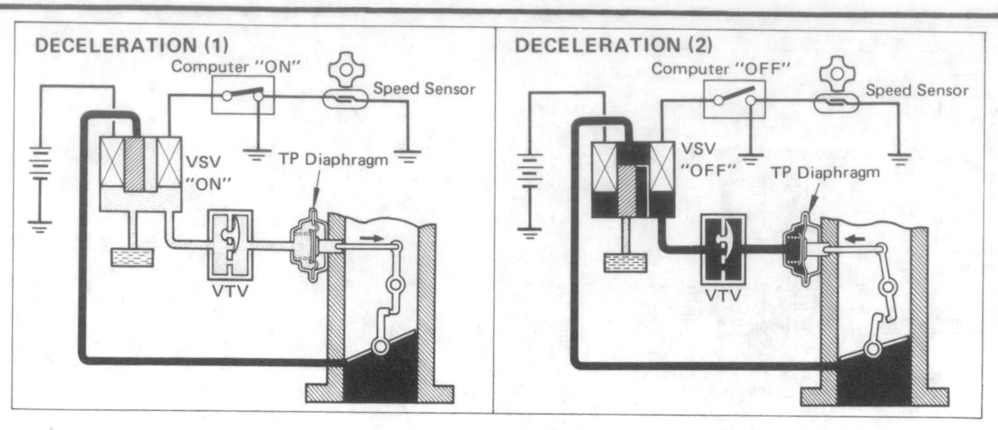

Throttle positioner operation—Typical (2F engine)

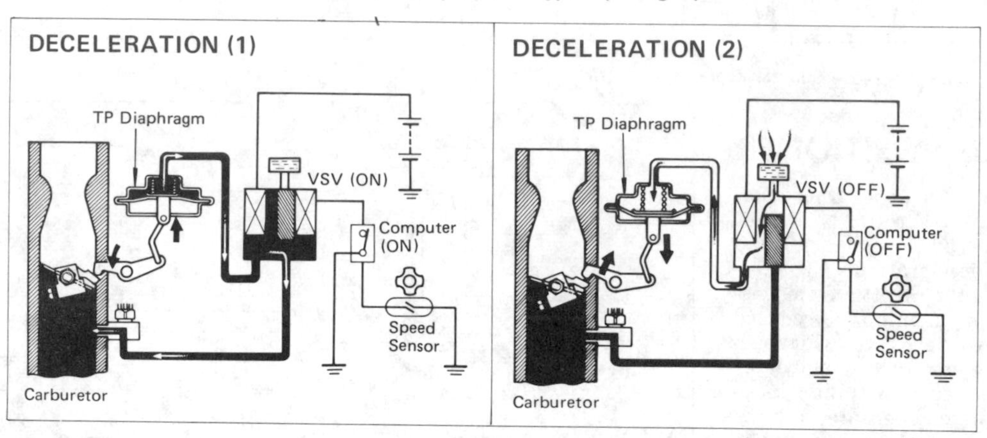

Throttle positioner operation Calif. typical 2OR engine (manual trans.)

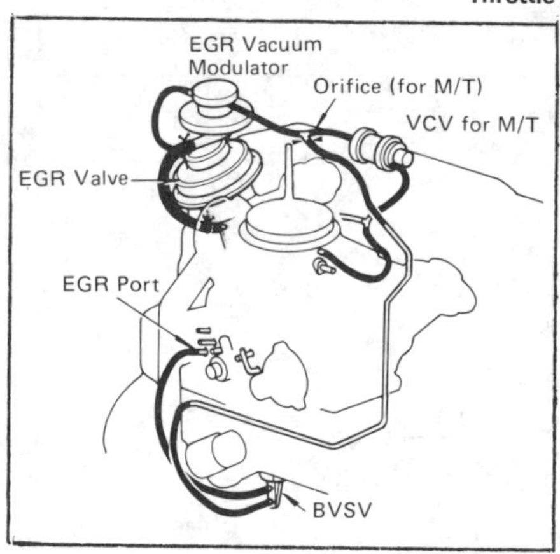

Exhaust gas recirculation system (EGR)—Typical

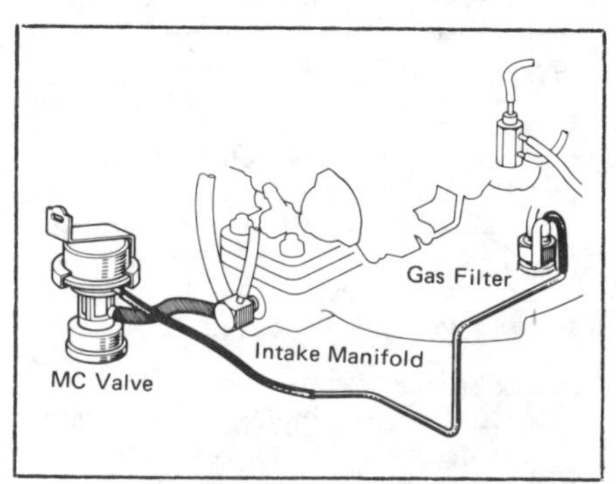

Mixture control system (MC) Calif.—Typical

generates a magnetic force when the coolant temperature is below 104°F, holding the choke plate open.

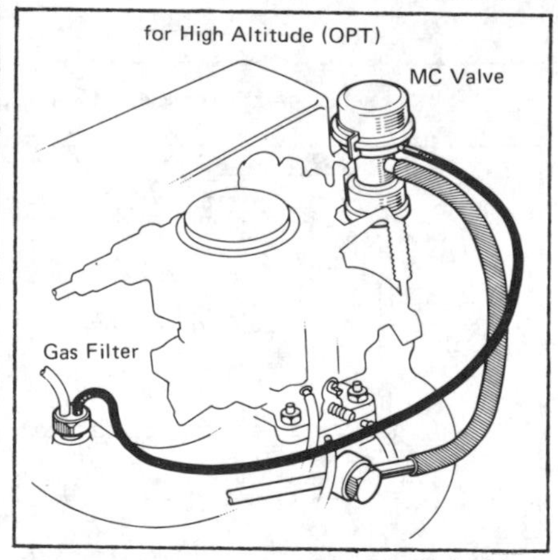

Mixture control system (MC) high alt. (opt.)—Typical

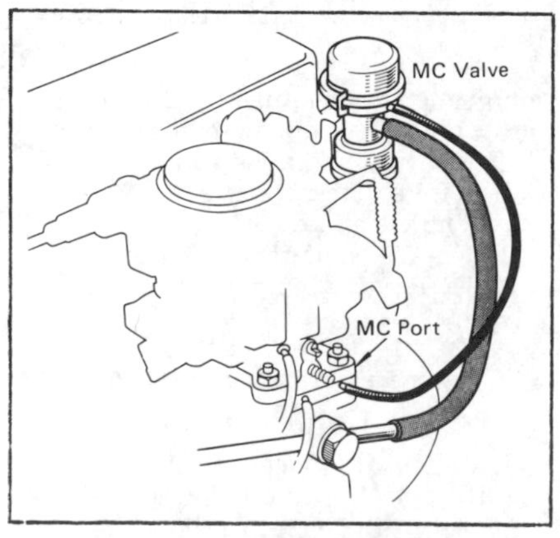

Mixture control system (MC)—Typical

However, when the coolant exceeds 104°F., the thermo switch opens, cutting off the current flow and allows the return spring to pull the choke plate open.

Most problems in this system will be electrical. Should the system malfunction, check for continuity in all circuits and replace the part not operating.

Choke Opener System

If a cold engine is driven soon after starting, the automatic choke system will close the choke plate, resulting in high levels of emissions. To combat that situation 1976-1977 California and High Altitude 2T-C engines and all 1978-80 engines are equipped with a system that forcibly holds the choke plate open.

When the coolant is below 140° F, the thermo wax in the TVSV closes the valve and prohibits any vacuum from acting on the choke diaphragm. This keeps the choke open. Above 140°F, the wax expands, opening the valve, and allows the choke plate to operate normally. Should the system malfunction, repace the TVSV.

When the engine is cold, an auxiliary enrichment circuit in the carburetor is operated to squirt extra fuel into the acceleration circuit in order to prevent the mixture from becoming too lean.

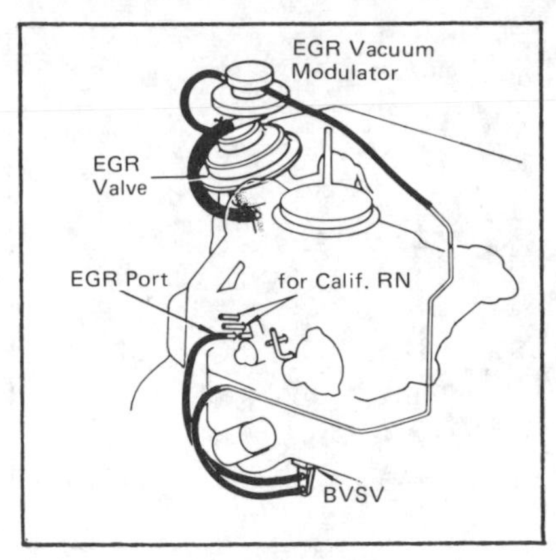

Exhaust gas recirculation system (EGR) Calif. —Typical

DUAL-DIAPHRAGM DISTRIBUTOR

Some Toyota models are equipped with a dual-diaphragm distributor unit. This distributor has a retard diaphragm, as well as a diaphragm for advance. Retarding the timing helps to reduce exhaust emissions, as well as making up for the lack of engine braking on models equipped with a throttle positioner.

SPARK CONTROL SYSTEM

Some Toyotas are equipped with the spark control system. The valve has a small orifice in it, which slows down the vacuum flow to the vacuum advance unit on the distributor. By delaying the vacuum to the distributor, a reduction in HC and CO emission is possible.

When the coolant temperature reaches a certain degree a coolant temperature operated vacuum control valve is opened, allowing the distributor to receive undelayed, ported vacuum through a separate vacuum line. Above 95°F-140°F this line is blocked, and all ported vacuum must go through the spark delay valve.

Spark Control Valve Operating Temperatures

Closed	Degrees F
1976-77 non-Cal. Corolla	95
1978-80 3K-C	95
1978-80 4M	122
1978-80 20R	129
1978-80 2T-C	86
Open	
1976-77 non-Cal.	140
1978-80 3K-C	120
1978-80 4M	147
1978-80 20R	104
1978-80 2T-C	111

ENGINE MODIFICATIONS SYSTEM

Toyota also uses an assortment of engine modifications to regulate exhaust emissions. Most of these devices fall into the category of engine vacuum controls. There are three principal components used on the engine modifications system, as well as a number of smaller parts. The three major components are: a speed sensor; a computer (speed marker); and a vacuum switching valve.

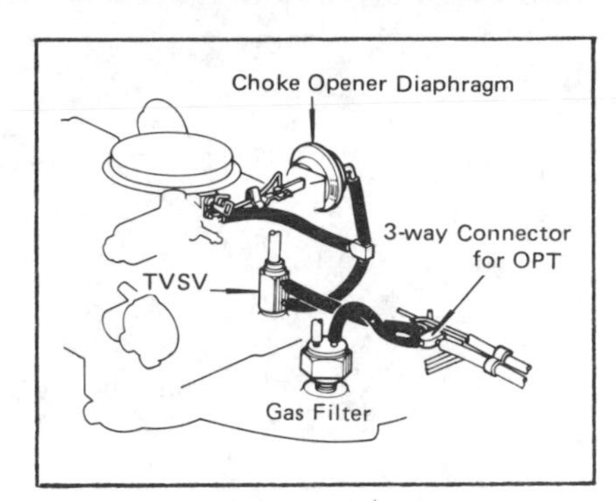

Choke opener system—Typical

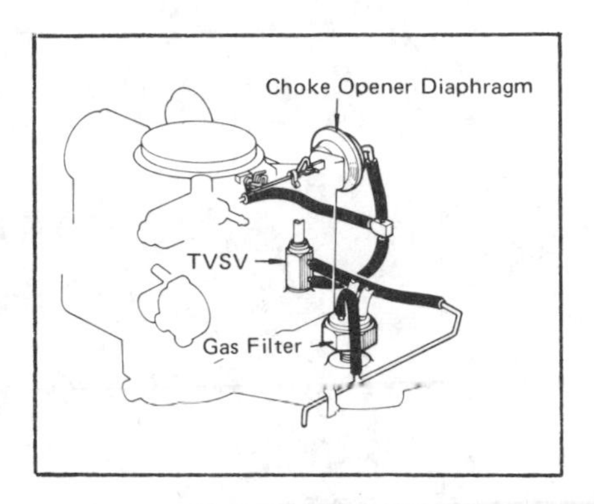

Choke opener system Calif.—Typical

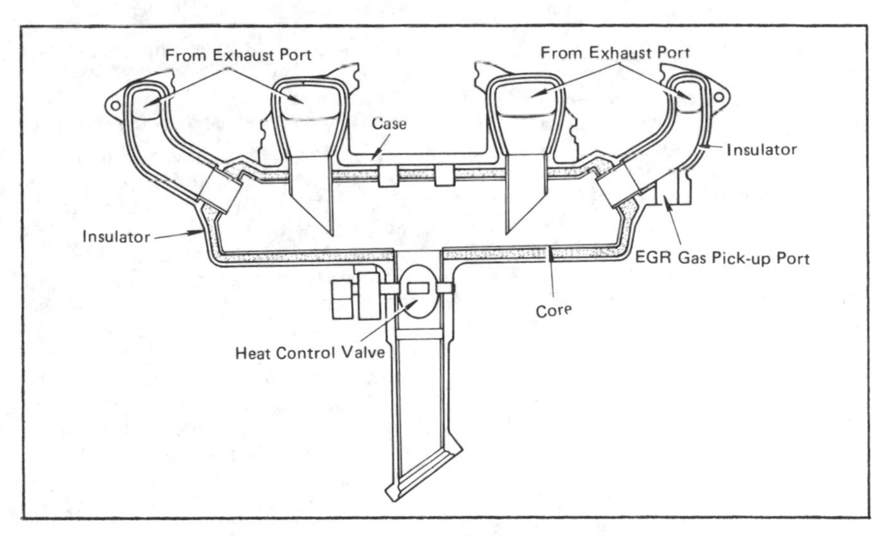

Thermal reactor system Calif. 2F engine only

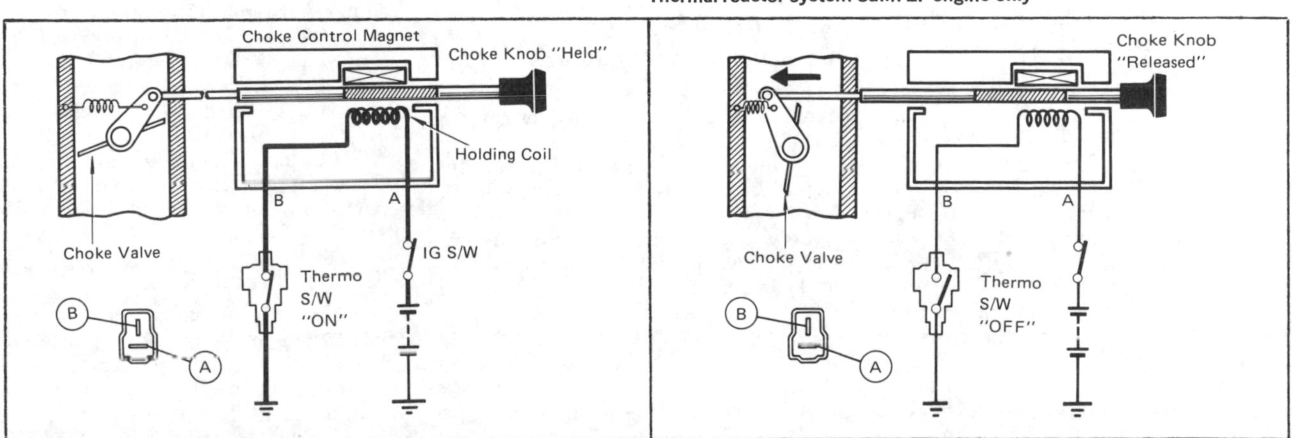

Choke return system—Cold position

Choke return system—Hot position

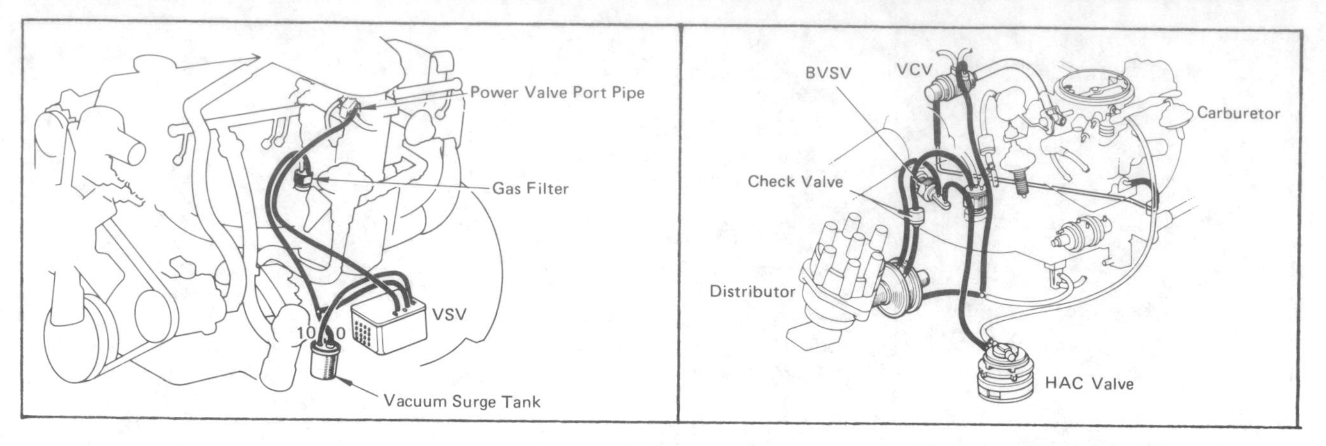

Power valve control system Calif. 2F engine—Typical

Spark control system (SC)—Typical

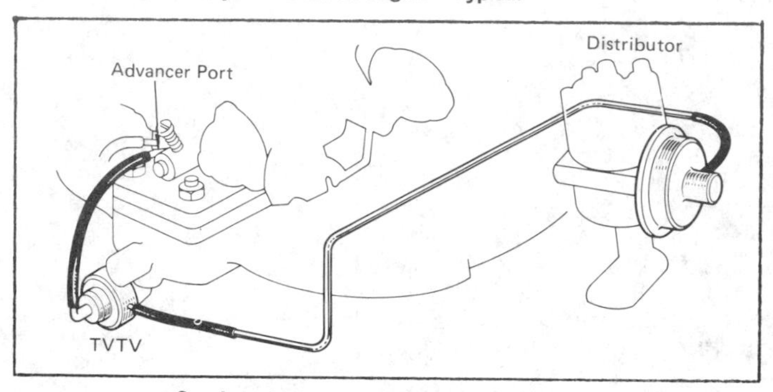

Spark control system (SC) Calif.—Typical

The vacuum switching valve and computer circuit operates most of the emission control components. Depending upon year and engine usage, the vacuum switching valve and computer may operate the purge control for the evaporative emission control system; the transmission controlled spark (TCS) or speed controlled spark (SCS); the dual-diaphragm distributor; and the throttle positioner systems.

The major difference between the transmission controlled spark and speed controlled spark systems is in the manner in which system operation is determined.

Below a predetermined speed, or any gear other than fourth, the vacuum advance unit on the distributor is rendered inoperative or, on F engines, timing is retarded. By changing the distributor advance curve in this manner, it is possible to reduce emissions of oxides of nitrogen (NOx).

NOTE: *Some engines are equipped with a thermo-sensor so that the TCS on SCS system only operates when the coolant temperature is 140°-212° F.*

Aside from determining the conditions outlined above, the vacuum switching valve computer circuit operates other devices in the emission control system.

The computer acts as a speed marker; at certain speeds it sends a signal to the vacuum switching valve which acts as a gate, opening and closing the emission control system vacuum circuits.

Power Valve Control System 2F

In order to minimize CO while ensuring good driveability, the Land Cruiser 2F engines for California are equipped with a power valve control system on the carburetor.

Dependent on coolant temperature, the power system opens or closes turning the VSV "On." It is also influenced by the amount of pressure on the accelerator pedal. Various combinations of pedal pressure and temperature will result in more or less fuel available for use. Any breakdown in the system requires replacement of the part involved.

High Altitude Compensation System

For all engines to be sold in areas over 4,000 ft. in altitude, a system has been installed to automatically lean out the fuel mixture by supplying additional air. This also results in lower emissions.

Low atmospheric pressure allows the bellows in the system to expand and close a port, allowing more air to enter from different sources.

In some engines, this also results in a timing advance to improve driveability.

All parts in this system must be replaced. The only adjustment available is in the timing.

Hot Air Intake—All Engines

In order to keep the temperature of the air drawn into the carburetor as constant as possible, all engines are equipped with a Hot Air Intake System (HAI).

CATALYTIC CONVERTERS

All Toyota vehicles sold in this country are equipped with a catalytic converter, except the 1976 Hi-Lux and Land Cruiser sold in California and the 1976 Mark IIs sold in the United States.

The catalysts are made of noble metals (platinum and palladium) which are bonded to individual pellets. These catalysts cause the HC and CO to break down into water and carbon dioxide (CO_2) without taking part in the reaction; hence, a catalyst life of 50,000 miles may be expected under normal conditions.

An air pump is used to supply air to the exhaust system to aid in the reaction. A thermosensor, inserted into the converter, shuts off the air supply if the catalyst temperature becomes excessive.

The same sensor circuit also causes a dash warning light labeled "EXH TEMP" to come on when the catalyst temperature gets too high.

NOTE: *It is normal for the light to come on temporarily if the car is being driven downhill for long periods of time (such as descending a mountain).*

The light will come on and stay on if the air injection system is malfunctioning or if the engine is misfiring.

Catalyst Precautions

1. Use only unleaded fuel.
2. Avoid prolonged idling; the engine should run no longer than 20 minutes at curb idle, nor longer than 10 minutes at fast idle.
3. Reduce the fast idle speed, by quickly depressing and releasing the

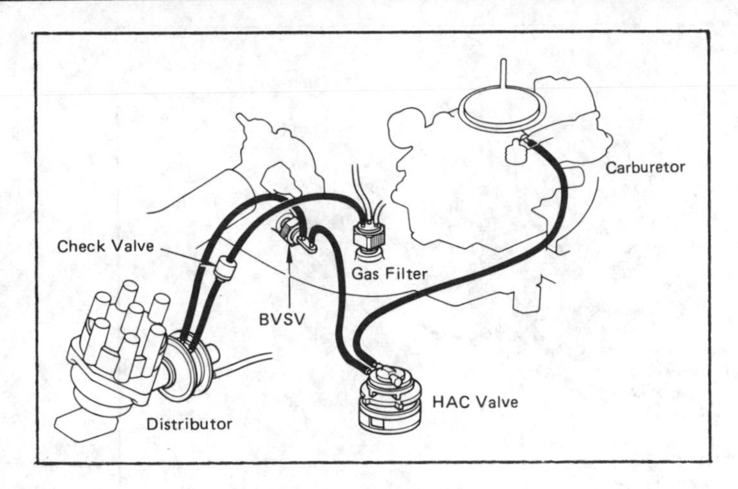

High altitude compensation system—Typical

VACUUM LIMITER SYSTEM

Some vehicles are equipped with the vacuum limiter system, which is used to reduce HC and CO emissions. This system allows fresh air to enter the air intake chamber (surge tank) on sudden deceleration.

To inspect the vacuum limiter system disconnect the air inlet hose of the vacuum limiter from the air connecter and plug the air connecter port. Start the engine. Close the inlet hose of the vacuum limiter with your finger. When the throttle valve is opened and closed, you should momentarily feel a vacuum. If no vacuum is felt check the vacuum limiter.

accelerator pedal, as soon as the coolant temperature reaches 120°F.

4. Do not disconnect any spark plug leads while the engine is running.

5. Make engine compression checks as quickly as possible.

6. Do not dispose of the catalyst in a place where anything coated with grease, gas, or oil is present; spontaneous combustion could result.

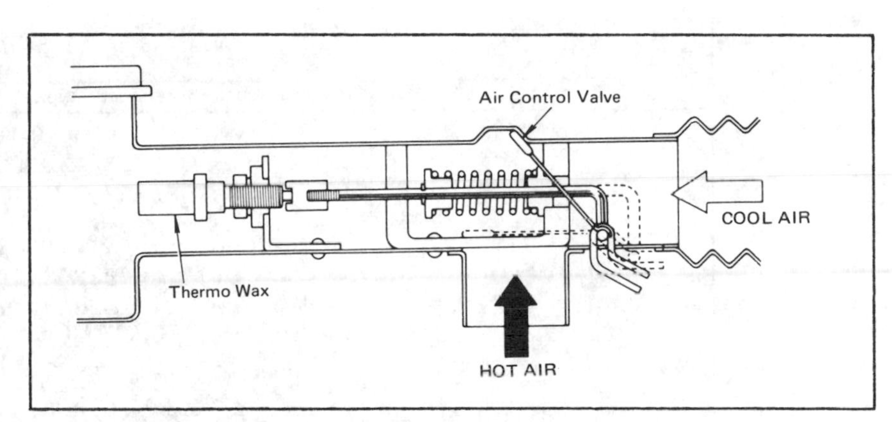

Automatic hot air intake system (HAI)—Typical

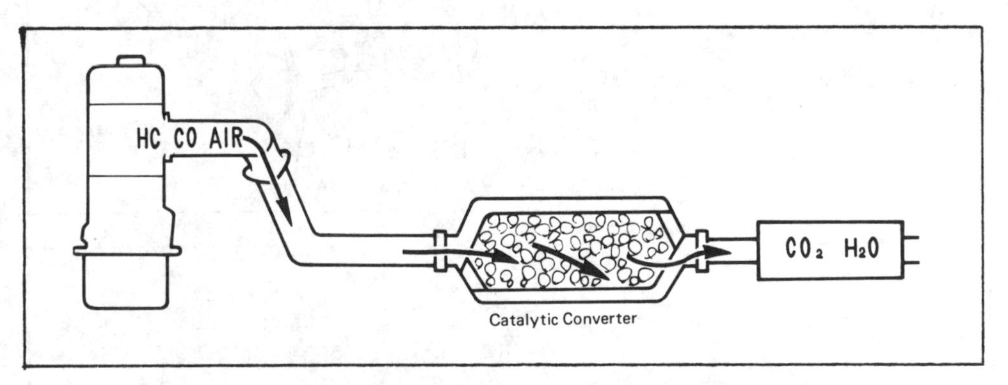

Catalaytic converter operation—Typical

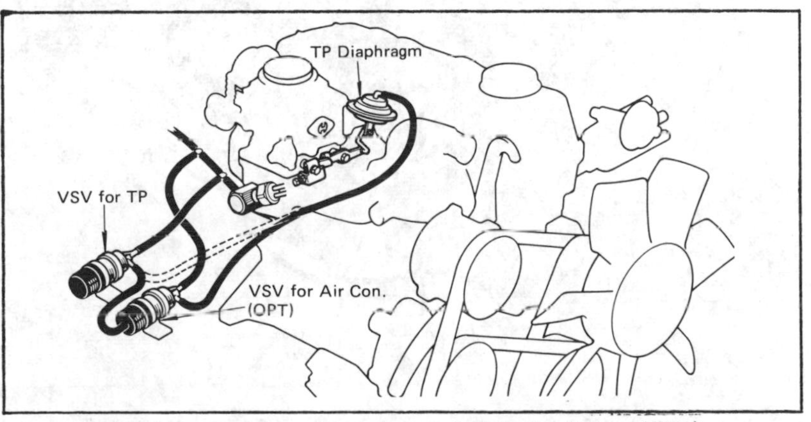

Throttle positioner system (TP) Calif. typical 20R engine (manual trans.)

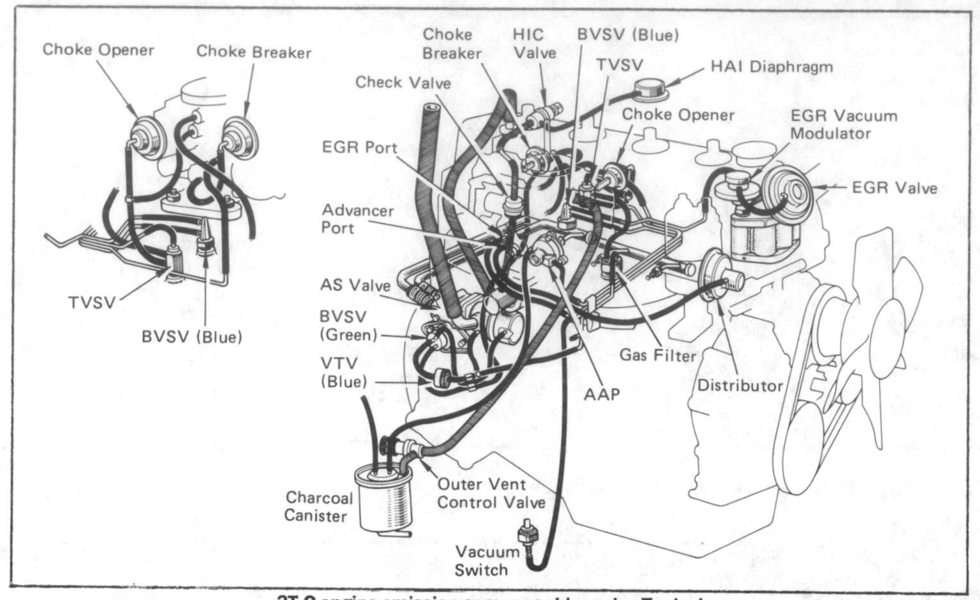

2T-C engine emission component layout—Typical

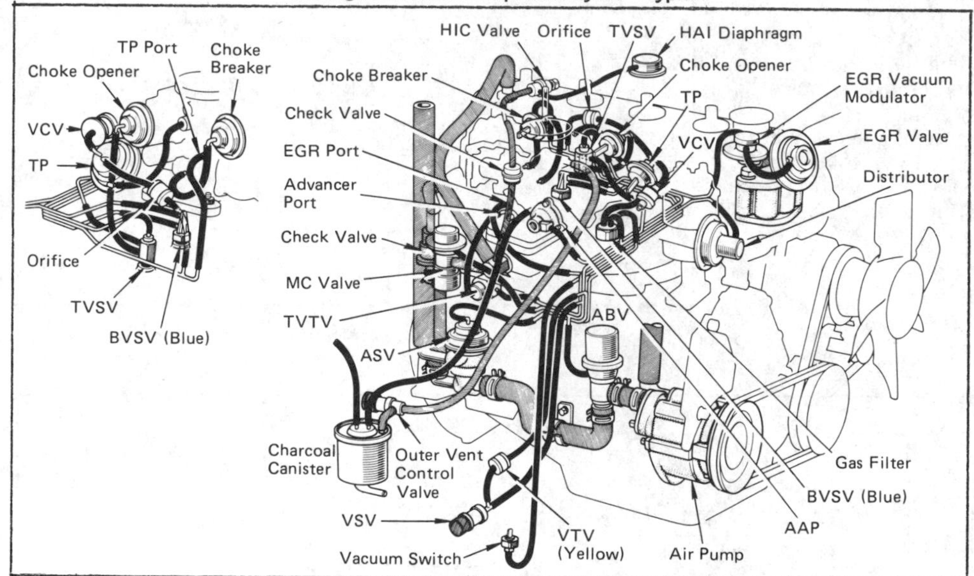

2T-C engine emission component layout Calif.—Typical

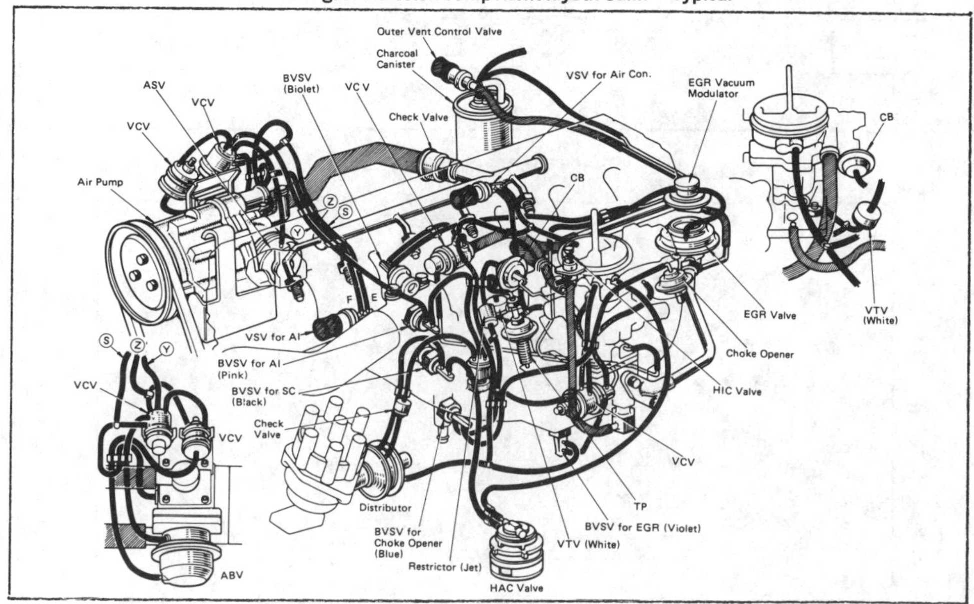

4M engine emission component layout—Typical

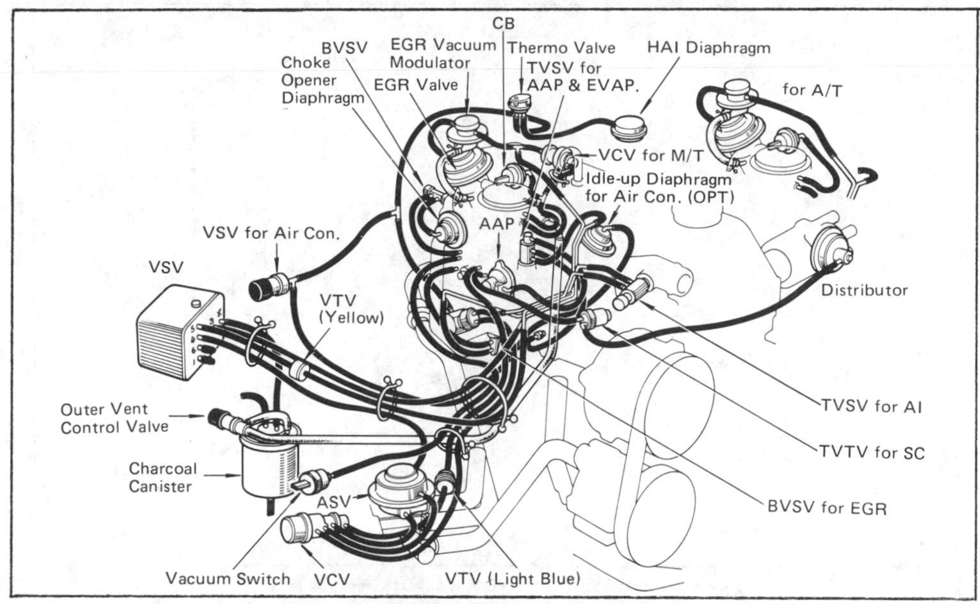

2OR engine emission component layout—Typical

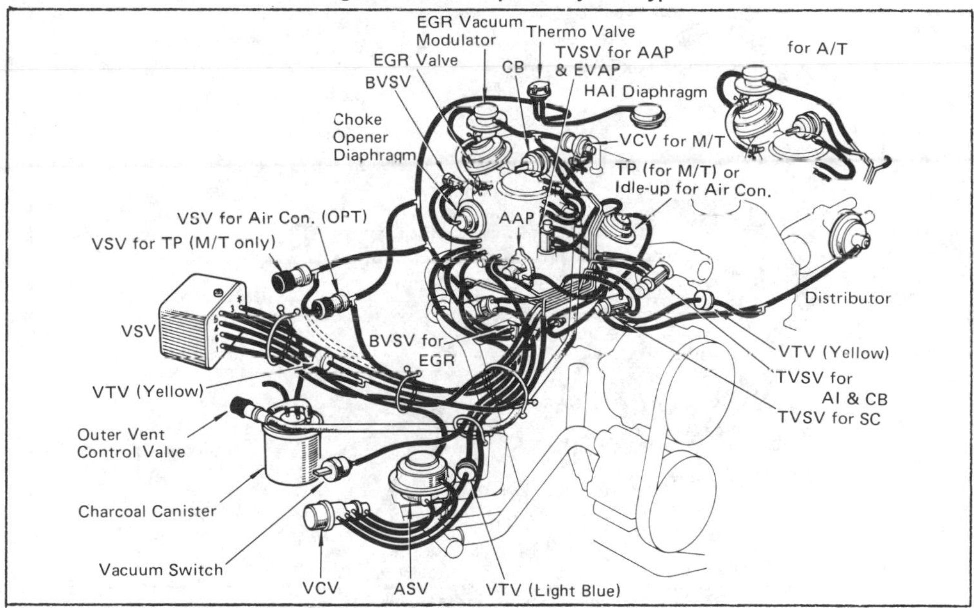

2OR engine emission component layout Calif.—Typical

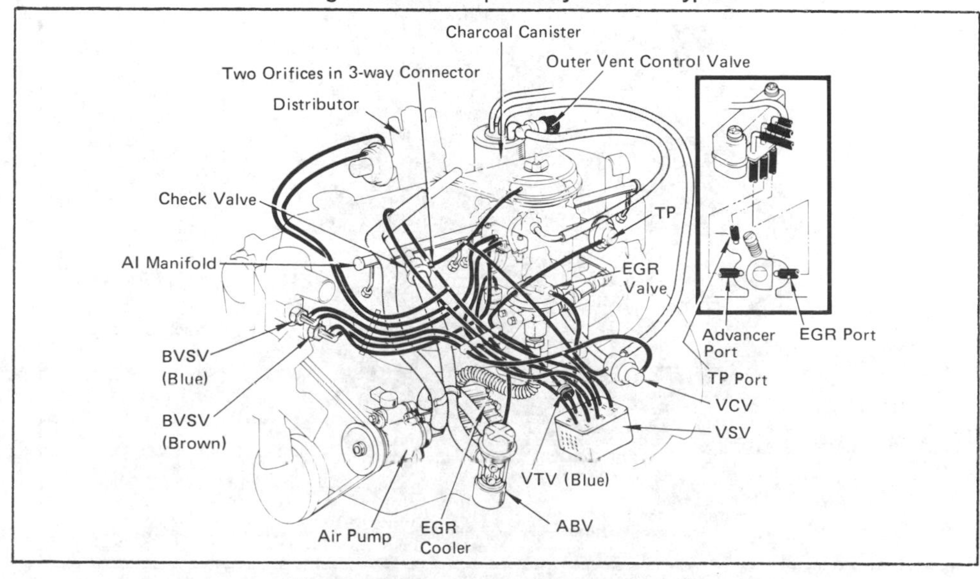

2F engine emission component layout—Typical

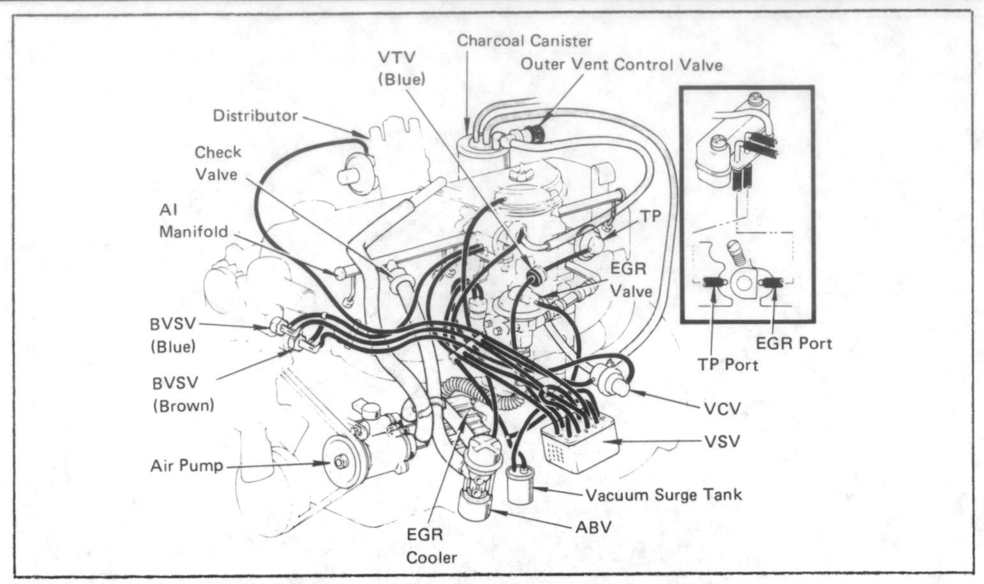

2F engine emission component layout Calif.—Typical

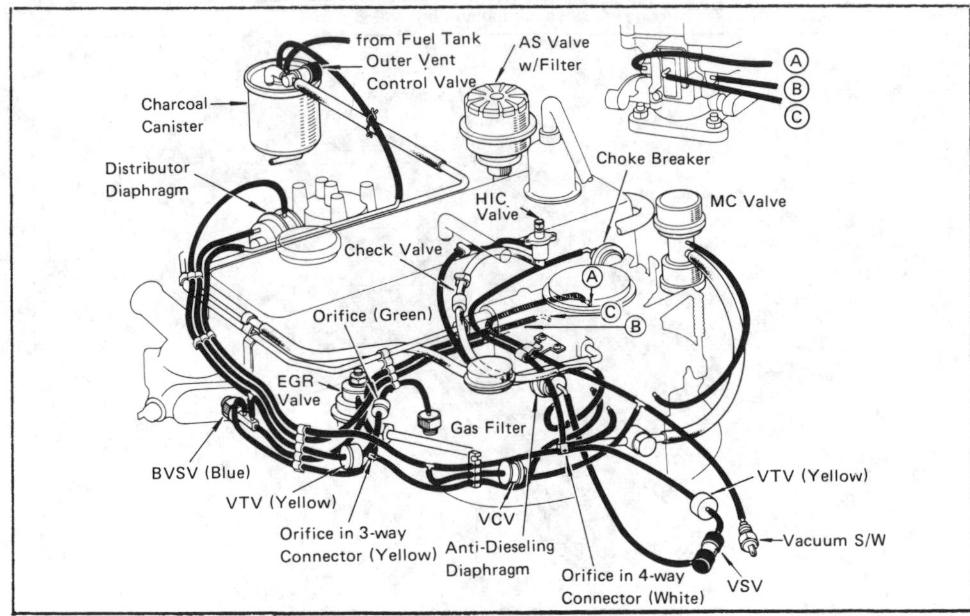

3K-C engine emission component layout—Typical

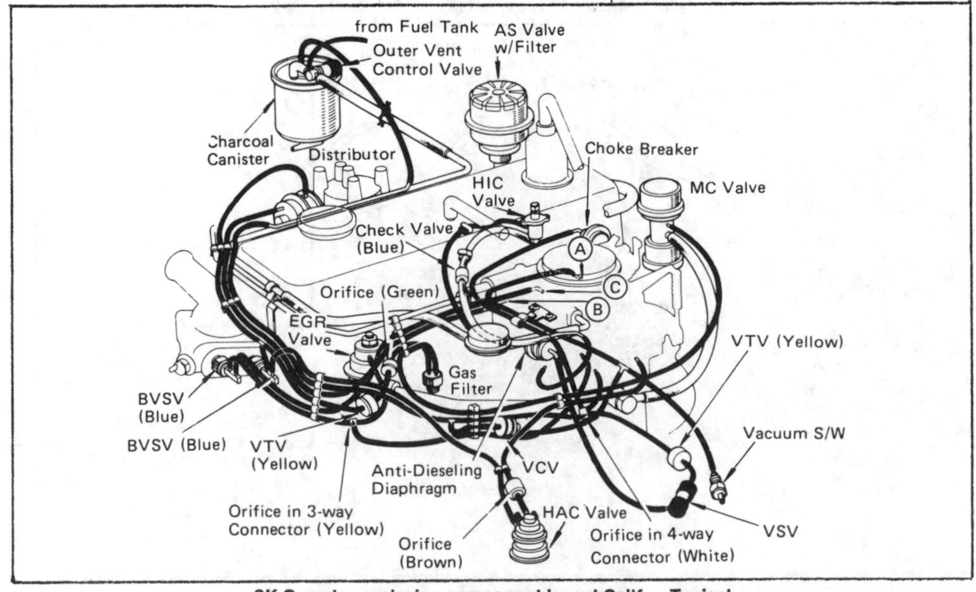

3K-C engine emission component layout Calif.—Typical

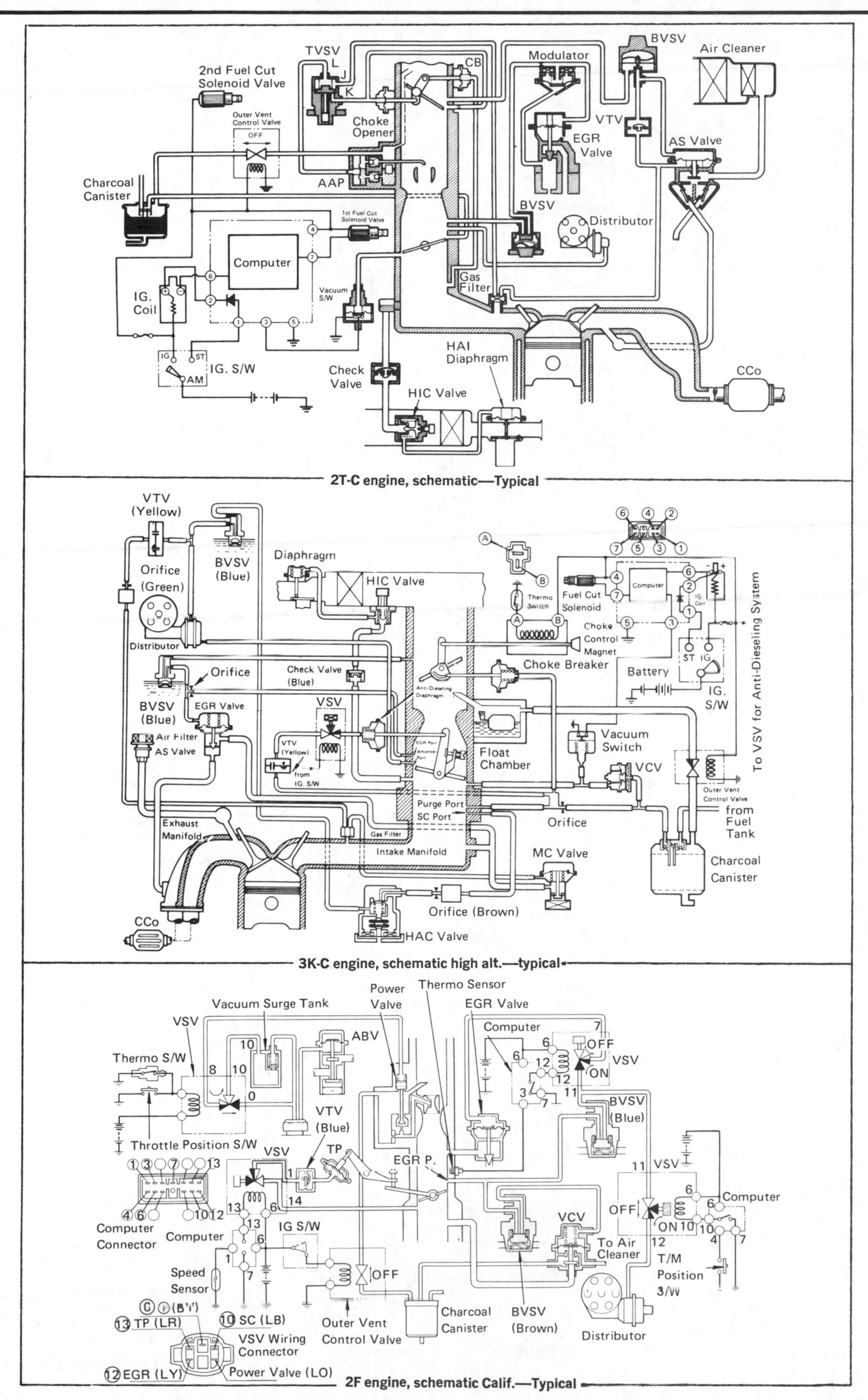

2T-C engine, schematic—Typical

3K-C engine, schematic high alt.—typical

2F engine, schematic Calif.—Typical

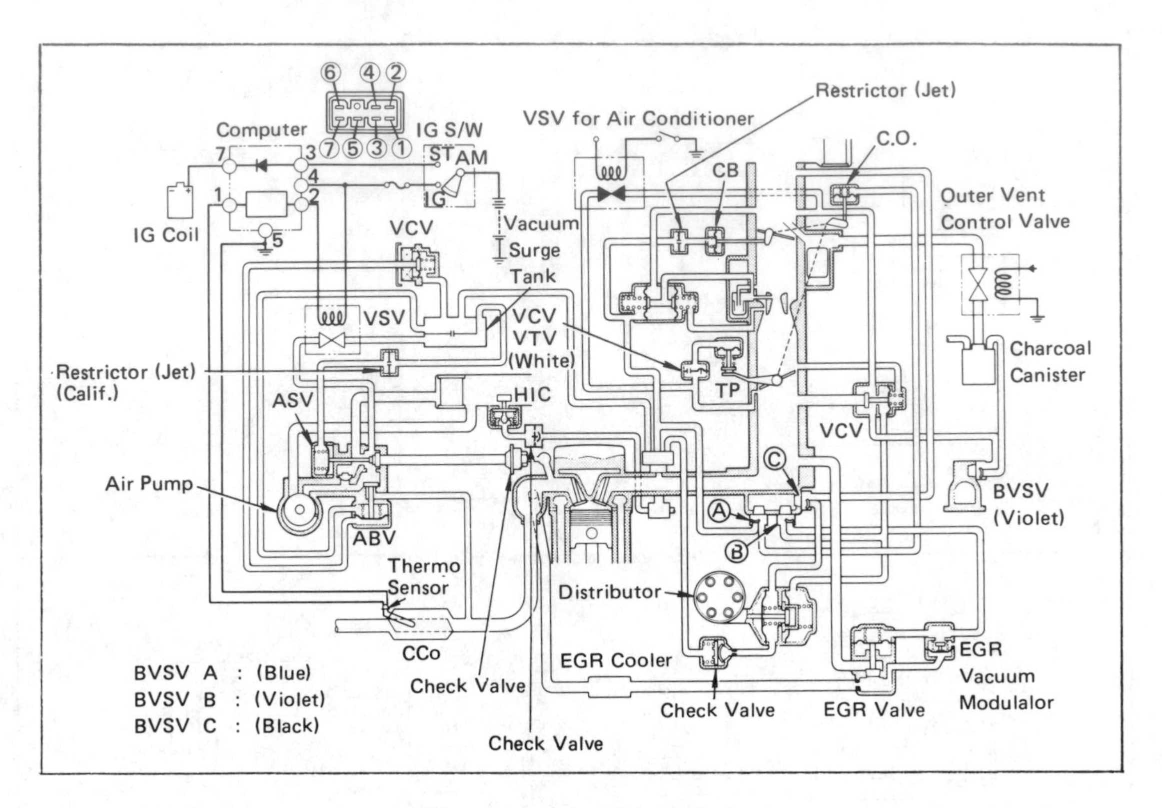

4M engine, schematic—Typical

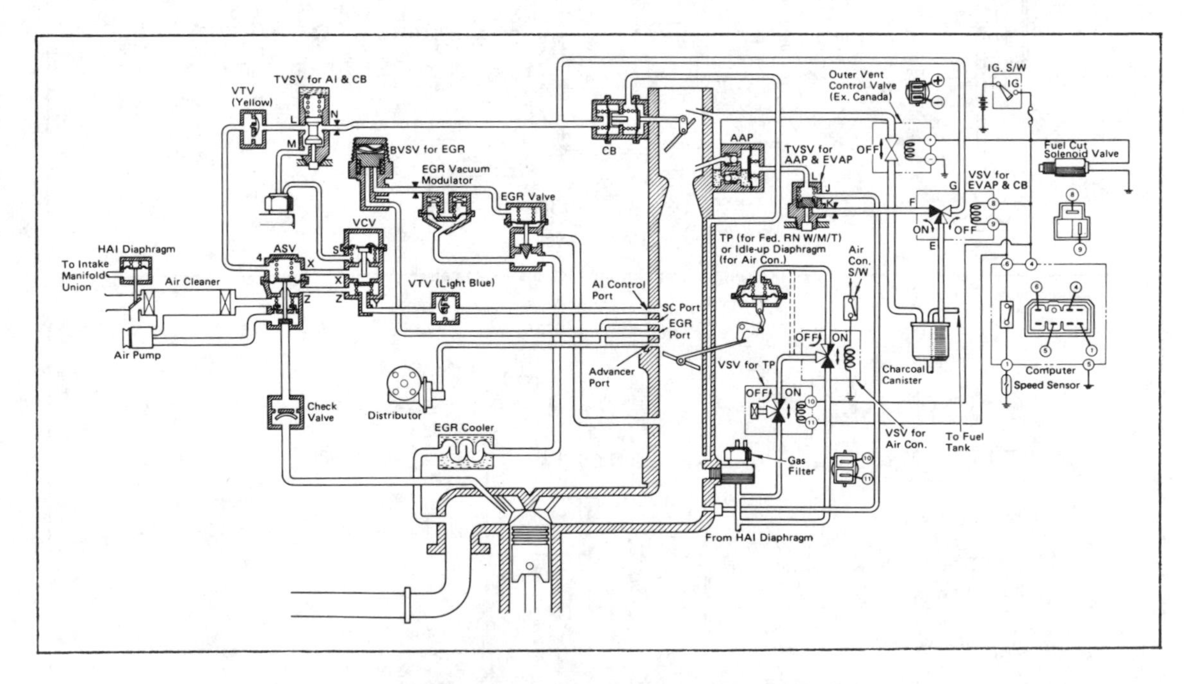

2OR engine, schematic U.S. series—Typical

TUNE-UP SPECIFICATIONS

	Code	Type	Common Designation	Spark Plugs Type	Spark Plugs Gap (in.)	Distributor Point Dwell (deg)	Distributor Point Gap (in.)	Ignition Timing (deg) MT	Ignition Timing (deg) AT	Fuel Pump Pressure (psi) @ 4000 rpm	Compression Pressure (psi)	Idle Speed (rpm) MT	Idle Speed (rpm) AT	Valve Clearance (in.) Cold In	Valve Clearance (in.) Cold Ex
1975	AJ	1	1600	W145M1 L288	.024	44-50	.016	5ATDC⑤	TDC⑤	28	85-135	875	875	.006	.006
	ED	2	1800	W145M2 N288	.024	44-50	.016	5ATDC⑤	5ATDC⑤	28	85-135	900	900	.006	.006
1976	AJ	1	1600	Bosch W145M1 Champ L288	.024	44-50	.016	5ATDC⑤	TDC⑤	28	85-135	875	925	.006	.006
	GD	2	2000	Bosch W145M2 Champ N288	.028	44-50	.016	7½BTDC ⑤	7½BTDC ⑤	28	85-135	900	950	.006	.006
1977	AJ	1	1600	Bosch M145M1 Champ N288	.028	44-50	.016	5ATDC⑤	5ATDC⑤	28	85-135	800- 950	800- 950	.006	.006
	GD	2	2000	Bosch M145M2 Champ N288	.028	44-50	.016	7½BTDC ⑤	7½BTDC ⑤	28	85-135	800- 950	850- 1000	.006	.006
1978	AJ	1	1600	Bosch W145M1 Champ L288	.028 .028	44-50	.016	5ATDC	5ATDC	28	85-135	800- 950	800- 950	.006	.006
	GE	2	2000	Bosch W145M2 Champ N288	.028 .028	44-50	.016	7½BTDC	7½BTDC	28	85-135	800- 950	900- 1000	Hydraulic	Hydraulic
1979	AJ	1	1600	Bosch W145M1 Champ L288	.028 .028	44-50	.016	5ATDC	5ATDC	28	85-135	800- 950	800- 950	.006	.006
	GE	2	2000	Bosch W145M2	.028	44-50	.016	7½BTDC ⑥	7½BTDC ⑥	28	85-135	800- 950	900- 1000	Hydraulic	Hydraulic
1980	Vanagon		—	Bosch W145M2	.028	44-50 ⑦	.016 ⑦	7½BTDC	7½BTDC	28	85-135	800- 950	900- 1000	Hydraulic	Hydraulic

① At idle, throttle valve closed (Types 1 & 2), vacuum hose(s) on
② At idle, throttle valve closed (Types 1 & 2), vacuum hose(s) off
③ At 3,500 rpm, vacuum hose(s) off
④ From March 1973, vehicles with single diaphragm distributor (one vacuum hose); adjust timing to 7½° BTDC with hose disconnected and plugged. The starting serial numbers for those type 1 vehicles using the single diaphragm distributors are # 113 2674 897 (manual trans.) and 113 2690 032 (auto. stick shift)
⑤ Carbon canister hose at air cleaner disconnected; at idle; vacuum hose(s) on
MT Manual Transmission
AT Automatic Transmission
BTDC Before Top Dead Center
ATDC After Top Dead Center
⑥ 5ATDC—California
⑦ California—Electronic Ignition

CAUTION: **When checking RPM with a tachometer on "Hall Ignition," VW electronic ignition, a special adapter must be used to avoid damage to the ignition system.**

DASHER-RABBIT-SCIROCCO

TUNE-UP SPECIFICATIONS

Year, Model	Engine Displacement cm³	Spark Plugs Type	Spark Plugs Gap (in.)	Distributor Point Dwell (deg)	Distributor Point Gap (in)	Ignition Timing (deg)	Intake Valve Opens (deg)	Com-pression Pressure (psi)	Idle Speed (rpm)	Valve Clearance (in) In ▲	Valve Clearance (in) Ex ▲
1974 Dasher	(1,471)	W175 T30 N8Y	0.024-0.028	44-50 ①	0.016	3 ATDC @ idle	4 BTDC	142-184	850-1000	0.008-0.012	0.016-0.020
1975 Dasher	(1,471)	W200 T30 N8Y	0.024-0.028	44-50	0.016	3 ATDC @ idle	4 BTDC	142-184	850-1000	0.008-0.012	0.016-0.020
1975 Scirocco, Rabbit	(1,471)	W200 T30 N8Y	0.024-0.028	44-50	0.016	3 ATDC @ idle	4 BTDC	142-184	900-1000	0.008-0.012	0.016-0.020
1976-80 Dasher	(1,588)	W215 T30 N7Y	0.024-0.028	44-50	0.016	3 ATDC @ idle	4 BTDC	142-184	850-1000	0.008-0.012	0.016-0.020
1976-77 Rabbit, Scirocco	(1,588)	W215 T30 N7Y	0.024-0.028	44-50	0.016	3 ATDC @ idle	4 BTDC	142-184	900-1000	0.008-0.012	0.016-0.020
1978-79 Rabbit	(1,457)	W175 T30 N8Y	0.024 0.028	44-50	0.016	3 ATDC @ idle	4 BTDC	142-184	850-1000	0.008-0.012	0.016-0.020
1978 Scirocco	(1,457)	W175 T30 N8Y	0.024 0.028	44-50	0.016	3 ATDC @ idle	4 BTDC	142-184	850-1000	0.008-0.012	0.016-0.020
1979-80 Scirocco	(1,588)	W175 T30 N8Y	0.024 0.028	44-50	0.016	3 ATDC @ idle	4 BTDC	142-184	850-1000	0.008-0.012	0.016-0.020
1980 Rabbit 49 states	(1,457)	W175 T30 N8Y	0.024-0.028	Electronic② Ignition		71/2BTDC③ @ idle	—	—	800-1000	—	—
1980 Rabbit California	(1,588)	W145 T30 N10Y	0.024-0.028	Electronic Ignition		3 ATDC @ idle	—	—	880-1000	—	—
1980 Scirocco California	(1,588)	W175 T30 N8Y	0.024-0.028	Electronic Ignition		3 ATDC @ idle	—	—	880-1000	—	—
1980 Dasher California	(1,588)	WR7DS N8GY	0.024-0.028	Electronic Ignition		3 ATDC @ idle	—	—	880-1000	—	—

NOTE: **The underhood specifications sticker often reflects tune-up specification changes made in production. Sticker figures must be used if they disagree with those in this chart.**

① 47°-53°—California

② "Hall Ignition"

③ Vacuum retard hose disconnected and plugged. Vacuum lines removed from idle stabilization unit and connected together Timing mark on flywheel—remove plug in housing to set timing.

— Not available

▲ NOTE: **Valve clearance need not be adjusted unless it varies more than 0.002 in. from specification.**

DASHER-RABBIT-SCIROCCO

DIESEL TUNE-UP SPECIFICATIONS

Model	Valve Clearance (cold) ① Intake (in.)	Valve Clearance (cold) ① Exhaust (in.)	Intake valve opens (deg)	Injection pump setting (deg)	Injection Nozzle Pressure (psi) New	Injection Nozzle Pressure (psi) Used	Idle speed (rpm) ③	Cranking compression pressure (psi)
1977-79 Diesel Rabbit	0.008-0.012	0.016-0.020	N.A.	Align marks	1849	1706	770-870	398 minimum
1979 Dasher Diesel	0.008-0.012	0.016-0.020	N.A.	Align marks	1849	1706	770-870	398 minimum

① Warm clearance given—Cold clearance: Intake 0.006-0.010
Exhaust 0.014-0.018

Valve clearance need not be adjusted unless it varies more than 0.002 in. from specification.

N.A. Not Available

ENGINE IDENTIFICATION CHART

Engine Code Letter	Type Vehicle	First Production Year	Last Production Year	Engine Type	Common Designation
AH (Calif.)	1	1972	1974	①	1600
AK	1	1973	1974	①	1600
AM (181)	1	1973	1974	①	1600
AJ	1	1975	In production	①	1600
CB	2	1972	1973	②	1700
CD	2	1973	1973	②	1700
AW	2	1974	1974	②	1800
ED	2	1975	1975	②	1800
GD	2	1976	1978	②	2000
GE	2	1979	In production	②	2000
U	3	1968	1973	②	1600
X	3	1972	1973	②	1600
EA	4	1972	1974	②	1700
EB (Calif.)	4	1973	1973	②	1700
EC	4	1974	1974	②	1800

① Fan driven by generator
② Fan driven by crankshaft

–DASHER, RABBIT, SCIROCCO

ENGINE IDENTIFICATION CHART

Model	Year	Code	Engine
Rabbit	1975-78	FC, FG, FN	1.5 Liter Gasoline
	1975-78	EF EE	1.6 Liter Fuel Injection
	1979-80	EH	1.5 Liter Fuel Injection
	1979-80	CK	1.5 Liter Diesel
	1980	FX	1.6 Liter Gasoline
Scirocco	1975-78	FC, FG, FN	1.5 Liter Gasoline
	1975-78	EF, EE	1.6 Liter Fuel Injection
	1979-80	EJ	1.6 Liter Fuel Injection
Dasher	1974	XW, WV, XZ, XY	1.5 Liter Gasoline
	1975	YG, YH, XS, XR	1.6 Liter Gasoline
	1976-80	YK, YH, YG	1.6 Liter Gasoline①
	1979-80	CK	1.5 Liter Diesel

① 1979-80 Fuel Injected gasoline.

CHASSIS NUMBER CHART

Model Year	Vehicle	Model No.	Chassis Number From	Chassis Number To
1975	Beetle	11	115 2000 001	115 3200 000
	Beetle Convertible	15	155 2000 001	155 3200 000
	Thing	181	185 2000 001	185 3200 000
	Van	21	215 2000 001	215 2300 000
	Bus	22	225 2000 001	225 2300 000
	Camper, Kombi	23	235 2000 001	235 2300 000
1976	Beetle	11	116 2000 001	—
	Beetle Convertible	15	156 2000 001	—
	Bus	22	226 2000 001	—
	Camper, Kombi	23	236 2000 001	—
1977	Beetle	11	117 2000 001	—
	Beetle Convertible	15	157 2000 001	—
	Bus	22	227 2000 001	—
	Camper, Kombi	23	237 2000 001	—
1978	Beetle Convertible	15	158 2000 001	—
	Bus	22	228 2000 001	—
	Camper	23	238 2000 001	—
1979	Beetle Convertible	15	159 2000 001	—
	Bus	22	229 2000 001	—
	Camper	23	239 2000 001	—

EMISSION CONTROLS

VOLKSWAGEN—TYPES 1 AND 2

CRANKCASE VENTILATION SYSTEM

All models are equipped with a crankcase ventilation system.

Type 1 and 2 crankcase vapors are recirculated from the oil breather through a rubber hose to the air cleaner. The vapors then join the air/fuel mixture and are burned in the engine. Fuel injected cars mix crankcase vapors into the air/fuel mixture to be burned in the combustion chambers. Fresh air is forced through the engine to evacuate vapors and recirculate them into the oil breather, intake air distributor, and then to be burned.

The only maintenance required on the crankcase ventilation system is a periodic check. At every tune-up, examine the hoses for clogging or deterioration. Clean or replace the hoses as required.

EVAPORATIVE EMISSION CONTROL SYSTEM

Required by law, this system prevents raw fuel vapors from entering the atmosphere. The various systems for different models are similar. They consist of an expansion chamber, activated charcoal filter, and connecting lines. Fuel vapors are vented to the charcoal filter where hydrocarbons are deposited on the element. The engine fan forces fresh air into the filter when the engine is running. The air purges the filter and the hydrocarbons are forced into the air cleaner to become part of the air/fuel mixture and burned.

Maintenance of this system consists of checking the condition of the various connecting lines and the charcoal filter at 10,000 mile intervals. The charcoal filter, which is located under the engine compartment, should be replaced at 48,000 mile intervals.

AIR INJECTION SYSTEM

Type 2 vehicles, are equipped with the air injection system, or air pump as it is sometimes called. In this system, an engine driven air pump delivers fresh air to the engine exhaust ports. The additional air is used to promote afterburning of any unburned mixture as they leave the combustion chamber. In addition, the system supplies fresh air to the in-

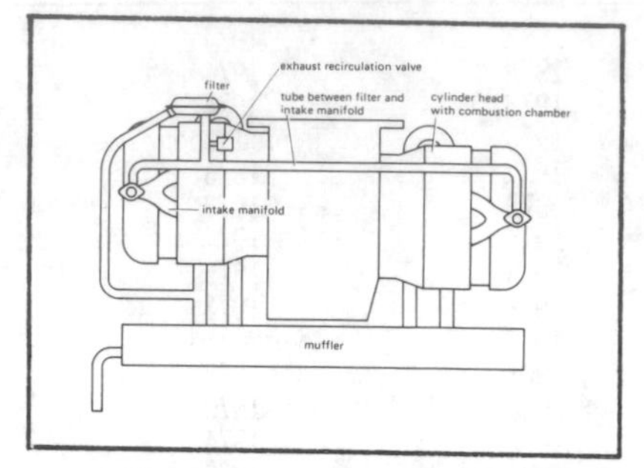

EGR system, 1975 and later Type 2

take manifold during gear changes to provide more complete combustion of the air/fuel mixture.

Check the air pump belt tension and examine the hoses for deterioration as a regular part of your tune-up procedure. The filter element adjacent to the pump should be replaced every 18,000 miles or at least every two years.

EXHAUST GAS RECIRCULATION SYSTEM

Type 1

EGR is installed on all 1976 and later models. All applications use the element type filter and single stage EGR valve. Recirculation occurs during part throttle applications as before. The system is controlled by a throttle valve switch which measures throttle position, and an intake air sensor which reacts to engine vacuum. An odometer actuated EGR reminder light (on the dashboard) is used to inform the driver that it is time to service the EGR system. The reminder light measures elapsed mileage and lights at 15,000 mile intervals. A reset button is located behind the switch.

Type 2

All 1976 and later Type 2 models utilize an EGR system. A single stage EGR valve and element type filter are used on all applications. Recirculation occurs during part throttle openings, and is controlled by throttle position, engine vacuum, and engine compartment temperature. At 15,000 mile intervals, a dash mounted EGR service reminder light is activated to warn that EGR service is now due. A reset button is located behind the switch.

CATALYTIC CONVERTER SYSTEM

All 1976 and later Type 1 and 2

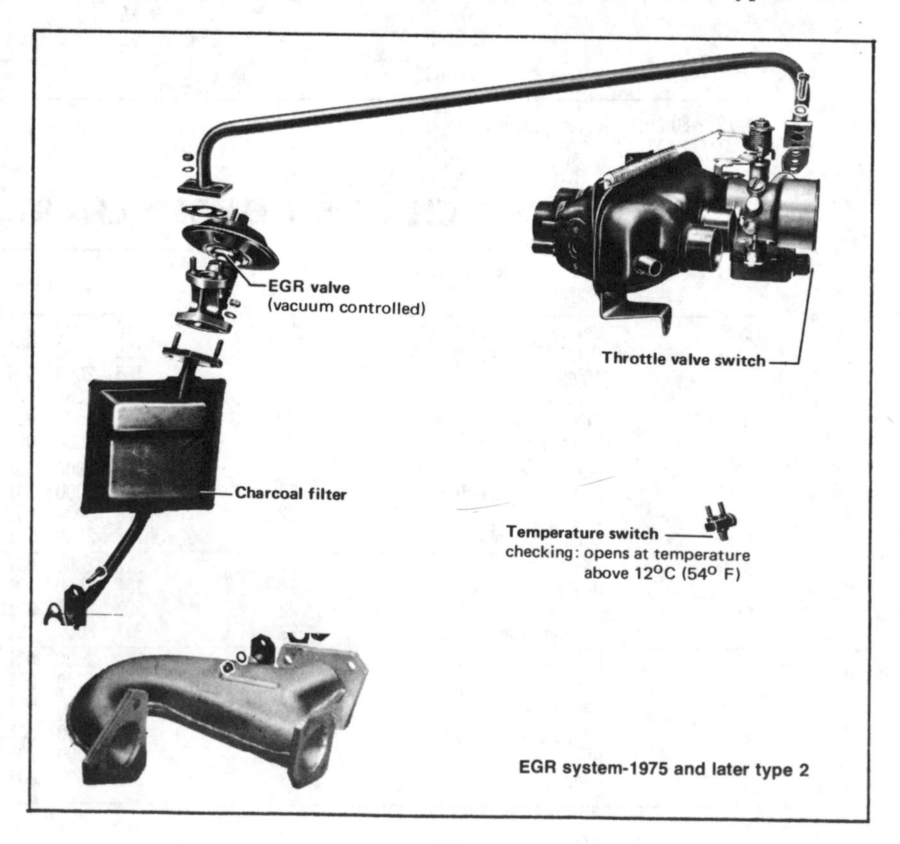

EGR system-1975 and later type 2

models sold in California are equipped with a catalytic converter. The converter is installed in the exhaust system, upstream and adjacent to the muffler.

Catalytic converters change noxious emission of hydrocarbons (HC) and carbon monoxide (CO) into harmless carbon dioxide and water vapor. The reaction takes place inside the converter at great heat using platinum and palladium metals as the catalyst. If the engine is operated on lead-free fuel, they are designed to last 50,000 miles before replacement.

DECELERATION CONTROL

All 1976 and later Type 2 models, as well as those 1976 and later Type 1 models equipped with manual transmission, are equipped with deceleration control to prevent an overly rich fuel mixture from reaching the exhaust. During deceleration, a vacuum valve (manual transmission) or electrical transmission switch (automatic transmission) opens, bypassing the closed throttle plate and allowing air to enter the combustion chambers.

EMISSION CONTROLS

DASHER - RABBIT - SCIROCCO

CRANKCASE VENTILATION

The crankcase ventilation system keeps harmful vapor byproducts of combustion from escaping into the atmosphere and prevents the building of crankcase pressure which can lead to oil leaking. Crankcase vapors are recirculated from the camshaft cover through a hose to the air cleaner. Here they are mixed with the air/fuel mixture and burned in the combustion chamber.

The only maintenance required on the crankcase ventilation system is a periodic check. At every tune up, examine the hoses for clogging or deterioration. Clean or replace the hoses as necessary.

EVAPORATIVE EMISSION CONTROL SYSTEM

This system prevents the escape of raw fuel vapors (unburned hydrocarbons or HC) into the atmosphere. The system consists of a sealed carburetor, unvented fuel tank filter cap, fuel tank expansion chamber, an activated charcoal filter canister and connector hoses. Fuel vapors which reach the filter deposit hydrocarbons on the surface of the charcoal filter element. Fresh air enters the filter when the engine is running and forces the hydrocarbons to the air cleaner where they join the air/fuel mixture and are burned.

Maintenance of the system requires checking the condition of the various connector hoses and the charcoal filter at 10,000 mile intervals. The charcoal filter should be replaced at 50,000 mile intervals.

DUAL DIAPHRAGM DISTRIBUTORS

The distributor is equipped with a vacuum retard diaphragm as well as vacuum advance.

Advance Diaphragm

1. Check the ignition timing.
2. Remove the retard hose from the distributor and plug it. Increase the engine speed. The ignition timing should advance. If it doesn't, the vacuum unit is faulty and must be replaced.

Temperature Valve

1. Remove the temperature valve and place the threaded portion in hot water.
2. Create a vacuum by sucking on the angled connection.
3. The valve must be open above approximately 130°F.

EXHAUST GAS RECIRCULATION (EGR)

1975-76 Models

1975 models have an EGR filter and a 2-stage EGR valve. The first stage is controlled by the temperature valve. The second stage is controlled by the micro-switch on the carburetor throttle valve. The switch opens the valve when the throttle valve is open between 30°-67° (manual transmission) or 23°-63° (automatic transmission).

The EGR filter was discontinued on 1976 models but the 2-stage EGR valve was retained. On Federal vehicles, only the first stage is connected; California vehicles use both stages.

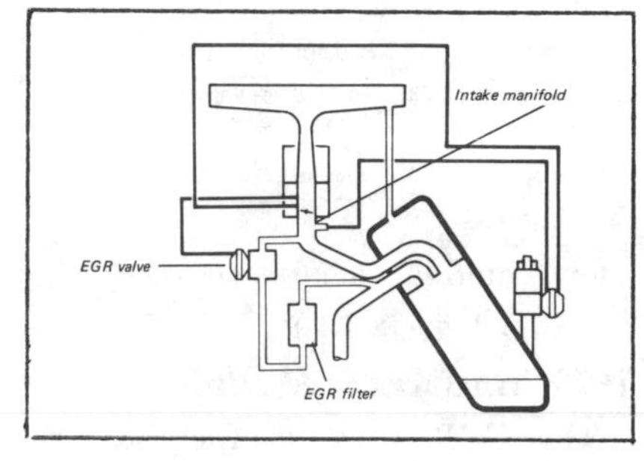

EGR system schematic

First stage EGR is controlled by engine vacuum and coolant temperature. The EGR valve is open above approximately 120°F. coolant temperature and below approximately 80° F. At idle and during full throttle acceleration (engine hot), there is no EGR since the engine vacuum is too low to open the valve.

The second stage is controlled by temperature, engine vacuum and micro-switch on the carburetor throttle valve. Vacuum is always present at the second stage and the valve is opened at about 120°F. coolant temperature. When the throttle valve opens between 25° and 67°, the micro-switch activates the 2-way valve and

Resetting the EGR elapsed mileage odometer

Resetting the catalytic converter elapsed mileage odometer

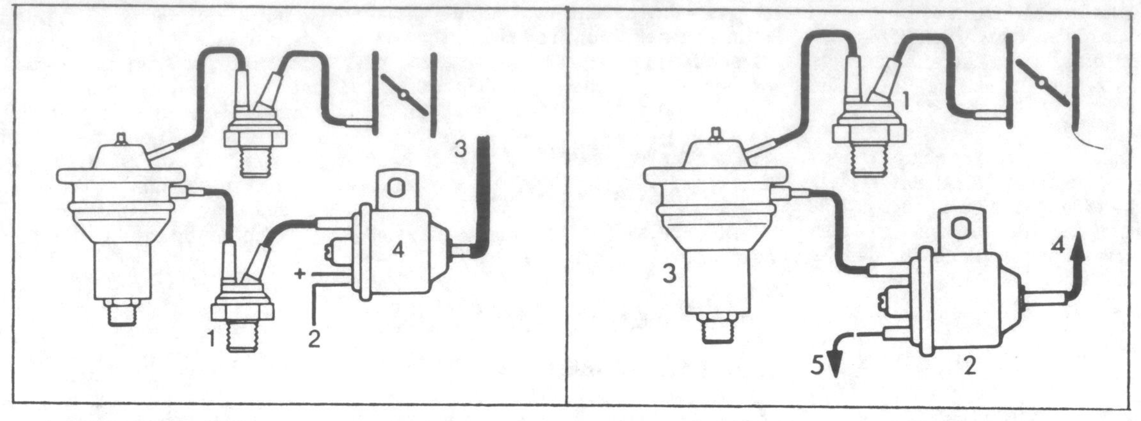

EGR operation-1975

1. Temperature valve
2. Two-way valve
3. EGR valve
4. To brake booster
5. To micro switch

EGR operation-1976 and later

1. Temperature valve for EGR 2nd stage
2. To micro switch on throttle valve
3. Vacuum hose to brake booster
4. Two way valve

allows engine vacuum to reach the second stage.

1977 and Later Models

The EGR valve on fuel injected models is controlled by a temperature valve and a vacuum amplifier. The valve is located at the front of the intake manifold.

AIR INJECTION

This system includes pump, filter, check valve and anti-backfire (gulp) valve. The required maintenance on the air pump involves visually checking the pump, control valves, hoses and lines every 10,000 miles. Clean the pump filter silencer at this interval, and replace every 20,000 miles.

Air Pump

Clean or replace the air filter or air manifolds at the required intervals.

1. Blow compressed air into the anti-back fire valve in the direction of the air flow.
2. Start the engine. Exhaust gas should flow equally from each air inlet.
3. With the engine idling, block the relief valve air outlet—only a slight pressure should be felt if the system is operating properly.

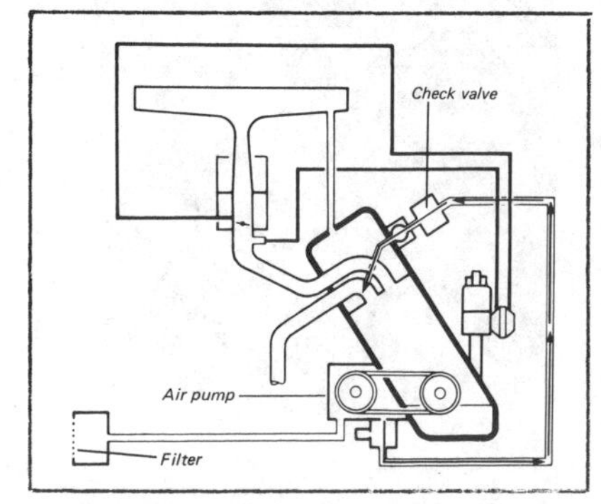

Air injection system schematic. The arrows indicate air flow

Anti-Backfire Valve

1. Disconnect the air pump filter line from the anti-backfire valve.
2. Briefly disconnect the anti-backfire valve vacuum line with the engine running. There should be a noticeable vacuum.
3. Replace the anti-backfire valve if the engine backfires.

Required maintenance involves replacing the catalyst when a malfunction occurs. Some early models may have a catalyst indicator light that will glow every 30,000 miles. The catalyst odometer can be reset by pushing the white button marked "CAT" on the mileage odometer in the engine compartment. Later models do not require the previous 30,000 mile check and can have the wire disconnected to disable the warning light.

CATALYTIC CONVERTER

All models are equipped with a catalytic converter located in the exhaust system.

Overheating of the catalytic converter is indicated by the CAT light in the speedometer flickering. This can temporarily be caused by a straining engine (trailer pulling, driving on steep grades, etc.) or high speed driving at high temperatures. Either easing the load or slowing down will stop the light from flickering.

More permanent causes include:
Misfiring (faulty or worn plugs),
Faulty ignition timing,
CO valve too high,
EGR not shutting off, or
Defective temperature sensor in converter.

Symptoms of a faulty converter include poor engine output, drop in the idle speed or continually stalling engine, rattle in the exhaust system (ceramic insert broken or loose) and high CO reading.

The converter is unbolted and removed from the car after disconnecting the temperature sensor.

Hold the converter up to a strong light and look through both ends, checking for blockages. If the converter is blocked, replace it.

Install the converter in the reverse order of removal. Reset the elapsed mileage odometer by pushing the white button marked "CAT".

CAUTION: *Do not drop or strike the converter assembly or damage to the ceramic insert will result.*

Damage and overheating of the catalytic converter, indicated by the flickering of the "CAT" warning light, can be caused by the following:

1. Engine misfire,
2. Improper ignition timing,
3. CO valve set too high,
4. Faulty air pump diverter valve,
5. Faulty temperature sensor, or
6. Engine strain caused by trailer hauling, high speed driving in hot weather.

A faulty converter is indicated by one of the following symptoms.

1. Poor engine performance.
2. The engine stalls.
3. Rattling in the exhaust system.
4. A CO reading greater than 0.4% at the tail pipe.

Resetting the CAT Warning Light

The CAT warning light in the speedometer should come on at 50,000 mile intervals to remind you to have the converter serviced.

The light can be reset by pushing the button marked CAT on the switch. The light on the speedometer should go out.

NOTE: *Vehicles equipped with the Solex 34 PICT-5 carburetor have no "CAT" warning light.*

OXYGEN SENSOR SYSTEM

An oxygen sensor picks up a signal from the exhaust and sends it to a control unit. The control unit reg-

ulates the air/fuel ratio by the frequency valve and fuel distributor.
Main components:

1. Frequency valve — regulates quantity of fuel delivered to the injectors by the fuel distributor.
2. Thermoswitch—cuts in the system whenever the coolant temperature is above 77° F.
3. Relay—supplies voltage to the control unit and frequency valve.
4. Warning light—Marked OXS and comes on in the instrument panel when oxygen sensors must be replaced. The replacement interval is 30,000 miles.
5. Control unit—receives a signal from the oxygen sensor and the thermoswitch. Regulates air/fuel ratio accordingly.
6. Oxygen sensor—sends a voltage signal to the control unit.

1979-80 SCIROCCO
1980 RABBIT

49 STATES—CARBURETOR AND ELECTRONIC IGNITION VEHICLES

Vacuum and electrical carburetor controls are used.

VW electronic ignition is called "Hall Ignition" and they give several cautions when working on the system:

1. Connect/disconnect test equipment only when the ignition is switched off.
2. If a conventional tachometer is used, an adapter must be used to prevent system damage.
3. Do not crank the engine for a compression test or other reason; until the high tension coil wire is grounded.
4. Do not use a battery booster longer than 1 minute.

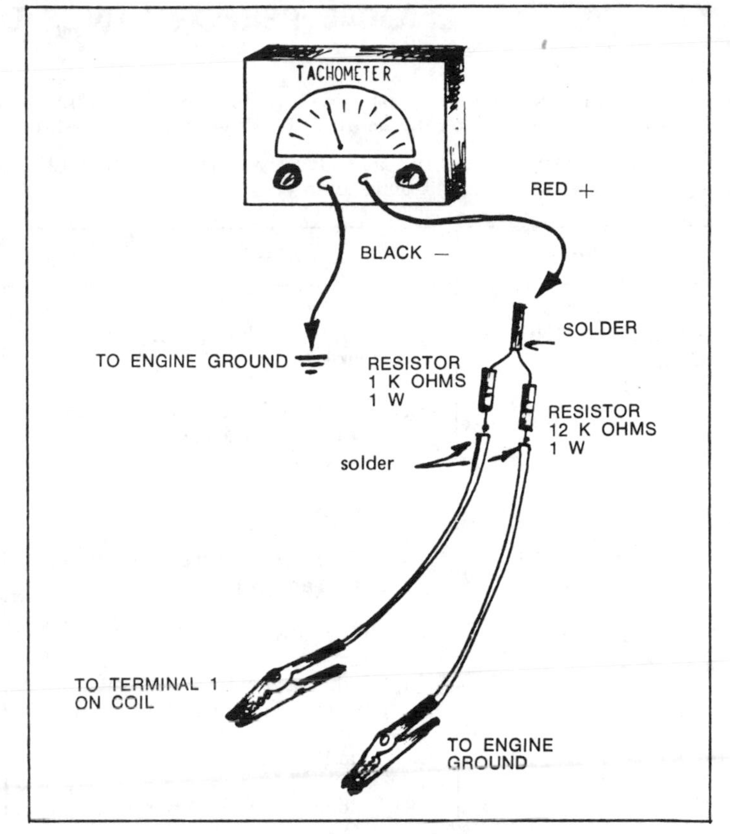

VW electronic ignition tachometer adaptor

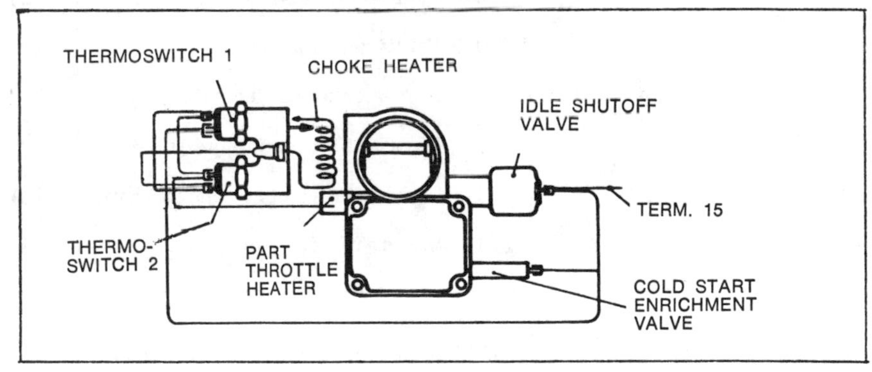

Carburetor electrical controls

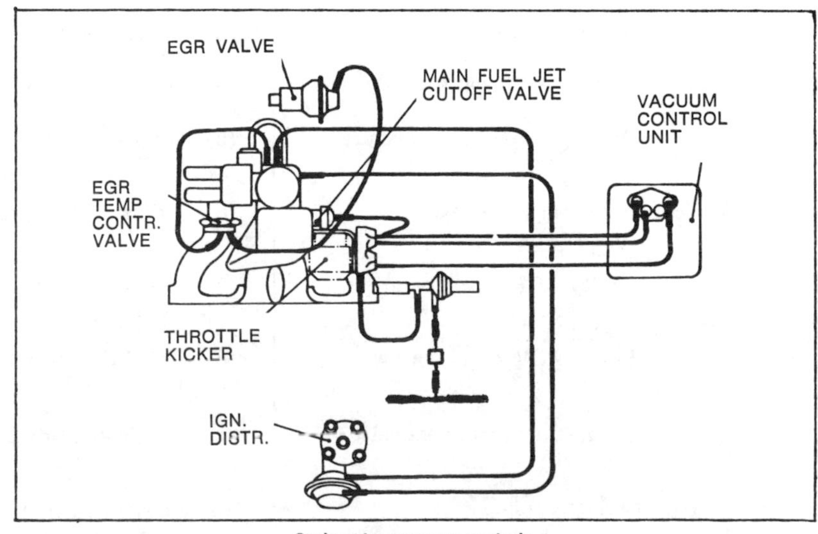

Carburetor vacuum controls

ENGINE PERFORMANCE DIAGNOSIS

The following table lists causes of service problems in descending order of probability. It is more likely a problem results from the first listed "possible cause" than the tenth, for instance.

However, visual examination often leads directly to the correct solution and all service procedures should begin with a careful look at any suspected part or assembly.

Condition	Possible Cause	Correction
HARD STARTING (ENGINE CRANKS NORMALLY)	(1) Binding linkage, choke valve or choke piston.	(1) Repair as necessary.
	(2) Restricted choke vacuum and hot air passages, where applicable.	(2) Clean passages.
	(3) Improper fuel level.	(3) Adjust float level.
	(4) Dirty, worn or faulty needle valve and seat.	(4) Repair as necessary.
	(5) Float sticking.	(5) Repair as necessary.
	(6) Exhaust manifold heat valve stuck. (6- and 8-cylinder only).	(6) Lubricate or replace.
	(7) Faulty fuel pump.	(7) Replace fuel pump.
	(8) Incorrect choke cover adjustment.	(8) Adjust choke cover.
	(9) Inadequate unloader adjustment.	(9) Adjust unloader.
	(10) Faulty ignition coil.	(10) Test and replace as necessary.
	(11) Improper spark plug gap.	(11) Adjust gap.
	(12) Incorrect initial timing.	(12) Adjust timing.
	(13) Incorrect dwell (4-cylinder only).	(13) Adjust dwell.
	(14) Incorrect valve timing.	(14) Check valve timing; repair as necessary.
ROUGH IDLE OR STALLING	(1) Incorrect curb or fast idle speed.	(1) Adjust curb or fast idle speed.
	(2) Incorrect initial timing.	(2) Adjust timing to specifications.
	(3) Incorrect dwell (4-cylinder only).	(3) Adjust dwell.
	(4) Improper idle mixture adjustment.	(4) Adjust idle mixture.
	(5) Damaged tip on idle mixture screw.	(5) Replace mixture screw.
	(6) Improper fast idle cam adjustment.	(6) Adjust fast idle cam.
	(7) Faulty EGR valve operation.	(7) Test EGR system and replace as necessary.
	(8) Faulty PCV valve air flow.	(8) Test PCV valve and replace as necessary.
	(9) Exhaust manifold heat valve inoperative.	(9) Lubricate or replace heat valve as necessary.
	(10) Choke binding.	(10) Locate and eliminate binding condition.

ENGINE PERFORMANCE DIAGNOSIS—Continued

Condition	Possible Cause	Correction
ROUGH IDLE OR STALLING (Continued)	(11) Improper choke setting.	(11) Adjust choke.
	(12) Faulty TAC unit.	(12) Repair as necessary.
	(13) Vacuum leak.	(13) Check manifold vacuum and repair as necessary.
	(14) Improper fuel level.	(14) Adjust fuel level.
	(15) Faulty distributor rotor or cap.	(15) Replace rotor or cap.
	(16) Leaking engine valves.	(16) Check cylinder leakdown rate or compression, repair as necessary.
	(17) Incorrect ignition wiring.	(17) Check wiring and correct as necessary.
	(18) Faulty coil.	(18) Test coil and replace as necessary.
	(19) Clogged air bleed or idle passages.	(19) Clean passages.
	(20) Restricted air cleaner.	(20) Clean or replace air cleaner.
	(21) Faulty choke vacuum diaphragm.	(21) Repair as necessary.
FAULTY LOW-SPEED OPERATION	(1) Clogged idle transfer slots.	(1) Clean transfer slots.
	(2) Restricted idlc air bleeds and passages.	(2) Clean air bleeds and passsages.
	(3) Restricted air cleaner.	(3) Clean or replace air cleaner.
	(4) Improper fuel level.	(4) Adjust fuel level.
	(5) Faulty spark plugs.	(5) Clean or replace spark plugs.
	(6) Dirty, corroded, or loose secondary circuit connections.	(6) Clean or tighten secondary circuit connections.
	(7) Faulty ignition cable.	(7) Replace ignition cable.
	(8) Faulty distributor cap.	(8) Replace cap.
	(9) Incorrect dwell (4-cylinder only).	(9) Adjust dwell.
FAULTY ACCELERATION	(1) Improper pump stroke.	(1) Adjust pump stroke.
	(2) Incorrect ignition timing.	(2) Adjust timing.
	(3) Inoperative pump discharge check ball or needle.	(3) Clean or replace as necessary.
	(4) Faulty elastomer valve.	(4) Replace valve.
	(5) Worn or damaged pump diaphragm or piston.	(5) Replace diaphragm or piston.
	(6) Leaking main body cover gasket.	(6) Replace gasket.
	(7) Engine cold and choke too lean.	(7) Adjust choke.
	(8) Improper metering rod adjustment (YF Model carburetor or BBD Model carburetor).	(8) Adjust metering rod.
	(9) Faulty spark plug(s).	(9) Clean or replace spark plug(s).

ENGINE PERFORMANCE DIAGNOSIS—Continued

Condition	Possible Cause	Correction
FAULTY ACCELERATION	(10) Leaking engine valves.	(10) Check cylinder leakdown rate or compression, repair as necessary.
	(11) Faulty coil.	(11) Test coil and replace as necessary.
FAULTY HIGH SPEED OPERATION	(1) Incorrect ignition timing.	(1) Adjust timing.
	(2) Excessive ignition point gap (4-cylinder only).	(2) Adjust dwell.
	(3) Defective TCS system.	(3) Test TCS system; repair as necessary.
	(4) Faulty distributor centrifugal advance.	(4) Check centrifugal advance and repair as necessary.
	(5) Faulty distributor vacuum advance.	(5) Check vacuum advance and repair as necessary.
	(6) Low fuel pump volume.	(6) Replace fuel pump.
	(7) Wrong spark plug gap; wrong plug.	(7) Adjust gap; install correct plug.
	(8) Faulty choke operation.	(8) Adjust choke.
	(9) Partially restricted exhaust manifold, exhaust pipe, muffler or tailpipe.	(9) Eliminate restriction.
	(10) Clogged vacuum passages.	(10) Clean passages.
	(11) Improper size or obstructed main jet.	(11) Clean or replace as necessary.
	(12) Restricted air cleaner.	(12) Clean or replace as necessary.
	(13) Faulty distributor rotor or cap.	(13) Replace rotor or cap.
	(14) Faulty coil.	(14) Test coil and replace as necessary.
	(15) Leaking engine valve(s).	(15) Check cylinder leakdown rate or compression, repair as necessary.
	(16) Faulty valve spring(s).	(16) Inspect and test valve spring tension and replace as necessary.
	(17) Incorrect valve timing.	(17) Check valve timing and repair as necessary.
	(18) Intake manifold restricted.	(18) Remove restriction or replace manifold.
	(19) Worn distributor shaft.	(19) Replace shaft.
MISFIRE AT ALL SPEEDS	(1) Faulty spark plug(s).	(1) Clean or replace spark plug(s).
	(2) Faulty spark plug cable(s).	(2) Replace as necessary.
	(3) Faulty distributor cap or rotor.	(3) Replace cap or rotor.
	(4) Faulty coil.	(4) Test coil and replace as necessary.
	(5) Trigger wheel too high (6- and 8-cylinder only).	(5) Set to specifications.

ENGINE PERFORMANCE DIAGNOSIS—Continued

Condition	Possible Cause	Correction
MISFIRE AT ALL SPEEDS (Continued)	(6) Incorrect dwell (4-cylinder only).	(6) Adjust dwell.
	(7) Faulty condenser (4-cylinder only).	(7) Replace condenser.
	(8) Primary circuit shorted or open intermittently.	(8) Trace primary circuit and repair as necessary.
	(9) Leaking engine valve(s).	(9) Check cylinder leakdown rate or compression, repair as necessary.
	(10) Faulty hydraulic tappet(s) (6- and 8-cylinder only).	(10) Clean or replace tappet(s).
	(11) Incorrect valve adjustment (4-cylinder only).	(11) Adjust valves.
	(12) Out-of-round or cracked tappets (4-cylinder only).	(12) Replace tappets.
	(13) Faulty valve spring(s).	(13) Inspect and test valve spring tension, repair as necessary.
	(14) Worn lobes on camshaft.	(14) Replace camshaft.
	(15) Vacuum leak.	(15) Check manifold vacuum and repair as necessary.
	(16) Improper carburetor settings.	(16) Adjust carburetor.
	(17) Fuel pump volume or pressure low.	(17) Replace fuel pump.
	(18) Blown cylinder head gasket.	(18) Replace gasket.
	(19) Intake or exhaust manifold passage(s) restricted.	(19) Pass chain through passages.
	(20) Wrong trigger wheel.	(20) Install correct wheel.
POWER NOT UP TO NORMAL	(1) Incorrect ignition timing.	(1) Adjust timing.
	(2) Faulty distributor rotor.	(2) Replace rotor.
	(3) Incorrect dwell (4-cylinder only).	(3) Adjust dwell.
	(4) Trigger wheel positioned too high or loose on shaft (6- and 8-cylinder only).	(4) Reposition or replace trigger wheel.
	(5) Incorrect spark plug gap.	(5) Adjust gap.
	(6) Faulty fuel pump.	(6) Replace fuel pump.
	(7) Incorrect valve timing.	(7) Check valve timing and repair as necessary.
	(8) Faulty coil.	(8) Test coil and replace as necessary.
	(9) Faulty ignition.	(9) Test cables and replace as necessary.
	(10) Leaking engine valves.	(10) Check cylinder leakdown rate or compression and repair as necessary.
	(11) Blown cylinder head gasket.	(11) Replace gasket.

ENGINE PERFORMANCE DIAGNOSIS—Continued

Condition	Possible Cause	Correction
POWER NOT UP TO NORMAL	(12) Leaking piston rings.	(12) Check compression and repair as necessary.
	(13) Worn distributor shaft.	(13) Replace shaft.
INTAKE BACKFIRE	(1) Improper ignition timing.	(1) Adjust timing.
	(2) Incorrect dwell (4-cylinder only).	(2) Adjust dwell.
	(3) Faulty accelerator pump discharge.	(3) Repair as necessary.
	(4) Improper choke operation.	(4) Repair as necessary.
	(5) Defective EGR CTO.	(5) Replace EGR CTO.
	(6) Defective TAC unit.	(6) Repair as necessary.
	(7) Lean fuel mixture.	(7) Check float level or manifold vacuum for vacuum leak. Remove sediment from bowl.
EXHAUST BACKFIRE	(1) Vacuum leak.	(1) Check manifold vacuum and repair as necessary.
	(2) Faulty diverter valve.	(2) Test diverter valve and replace as necessary.
	(3) Faulty choke operation.	(3) Repair as necessary.
	(4) Exhaust leak.	(4) Locate and eliminate leak.
PING OR SPARK KNOCK	(1) Incorrect ignition timing.	(1) Adjust timing.
	(2) Distributor centrifugal or vacuum advance malfunction.	(2) Check advance and repair as necessary.
	(3) Excessive combustion chamber deposits.	(3) Use combustion chamber cleaner.
	(4) Carburetor set too lean.	(4) Adjust carburetor.
	(5) Vacuum leak.	(5) Check manifold vacuum and repair as necessary.
	(6) Excessively high compression.	(6) Check compression and repair as necessary.
	(7) Fuel octane rating excessively low.	(7) Try alternate fuel source.
	(8) Heat riser stuck in heat ON position (6- and 8-cylinder only).	(8) Free-up or replace heat riser.
	(9) Sharp edges in combustion chamber.	(9) Grind smooth.
SURGING (CRUISING SPEEDS TO TOP SPEEDS)	(1) Low fuel level.	(1) Adjust fuel level.
	(2) Low fuel pump pressure or volume.	(2) Replace fuel pump.
	(3) Metering rod(s) not adjusted properly (YF Model Carburetor or BBD Model Carburetor).	(3) Adjust metering rod.
	(4) Improper PCV valve air flow.	(4) Test PCV valve and replace as necessary.

C-4 SYSTEM DIAGNOSIS

Chart 1—System Operational Check

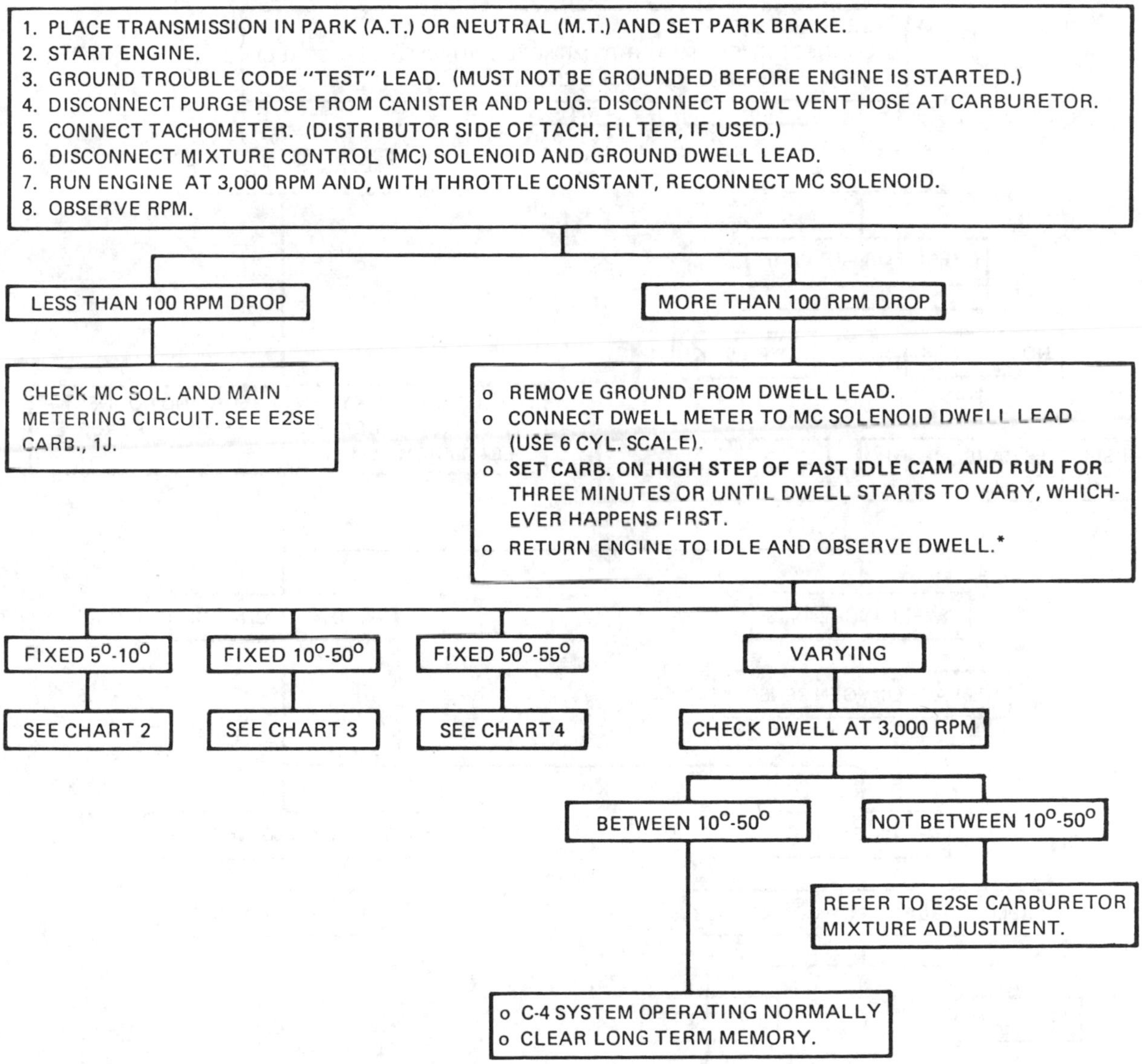

1. PLACE TRANSMISSION IN PARK (A.T.) OR NEUTRAL (M.T.) AND SET PARK BRAKE.
2. START ENGINE.
3. GROUND TROUBLE CODE "TEST" LEAD. (MUST NOT BE GROUNDED BEFORE ENGINE IS STARTED.)
4. DISCONNECT PURGE HOSE FROM CANISTER AND PLUG. DISCONNECT BOWL VENT HOSE AT CARBURETOR.
5. CONNECT TACHOMETER. (DISTRIBUTOR SIDE OF TACH. FILTER, IF USED.)
6. DISCONNECT MIXTURE CONTROL (MC) SOLENOID AND GROUND DWELL LEAD.
7. RUN ENGINE AT 3,000 RPM AND, WITH THROTTLE CONSTANT, RECONNECT MC SOLENOID.
8. OBSERVE RPM.

LESS THAN 100 RPM DROP

CHECK MC SOL. AND MAIN METERING CIRCUIT. SEE E2SE CARB., 1J.

MORE THAN 100 RPM DROP

- o REMOVE GROUND FROM DWELL LEAD.
- o CONNECT DWELL METER TO MC SOLENOID DWELL LEAD (USE 6 CYL. SCALE).
- o SET CARB. ON HIGH STEP OF FAST IDLE CAM AND RUN FOR THREE MINUTES OR UNTIL DWELL STARTS TO VARY, WHICHEVER HAPPENS FIRST.
- o RETURN ENGINE TO IDLE AND OBSERVE DWELL.*

FIXED 5°-10°

SEE CHART 2

FIXED 10°-50°

SEE CHART 3

FIXED 50°-55°

SEE CHART 4

VARYING

CHECK DWELL AT 3,000 RPM

BETWEEN 10°-50°

- o C-4 SYSTEM OPERATING NORMALLY
- o CLEAR LONG TERM MEMORY.

NOT BETWEEN 10°-50°

REFER TO E2SE CARBURETOR MIXTURE ADJUSTMENT.

*OXYGEN SENSOR TEMPERATURE MAY COOL AT IDLE CAUSING THE DWELL TO CHANGE FROM VARYING TO A FIXED INDICATION BETWEEN 10°-50°. IF THIS HAPPENS, RUN THE ENGINE AT FAST IDLE TO HEAT THE SENSOR.

C-4 SYSTEM DIAGNOSIS—Continued

Chart 2—Dwell Fixed Between 5 Degrees and 10 Degrees

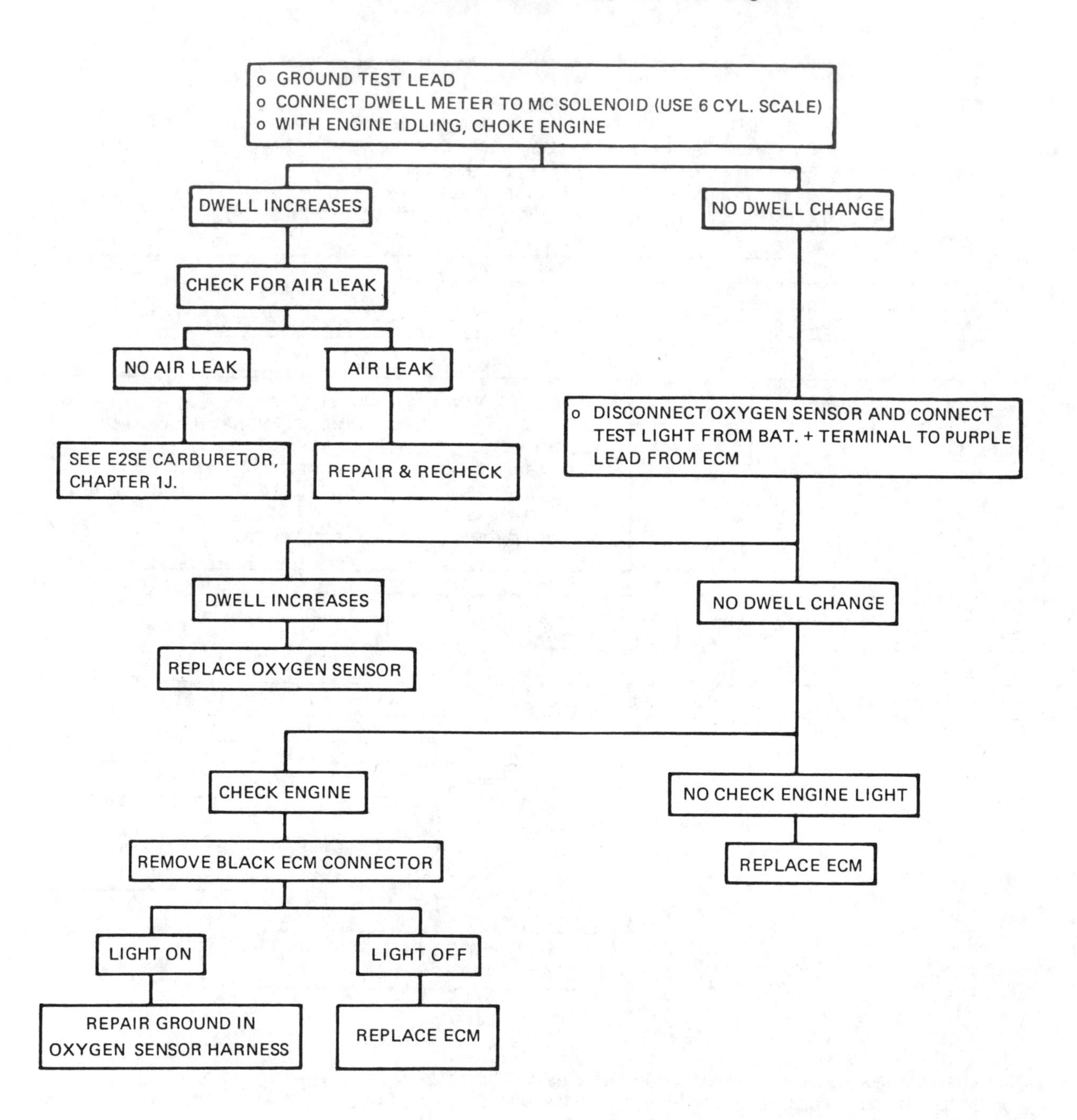

C-4 SYSTEM DIAGNOSIS—Continued

Chart 3—Dwell Fixed Between 10 Degrees and 50 Degrees

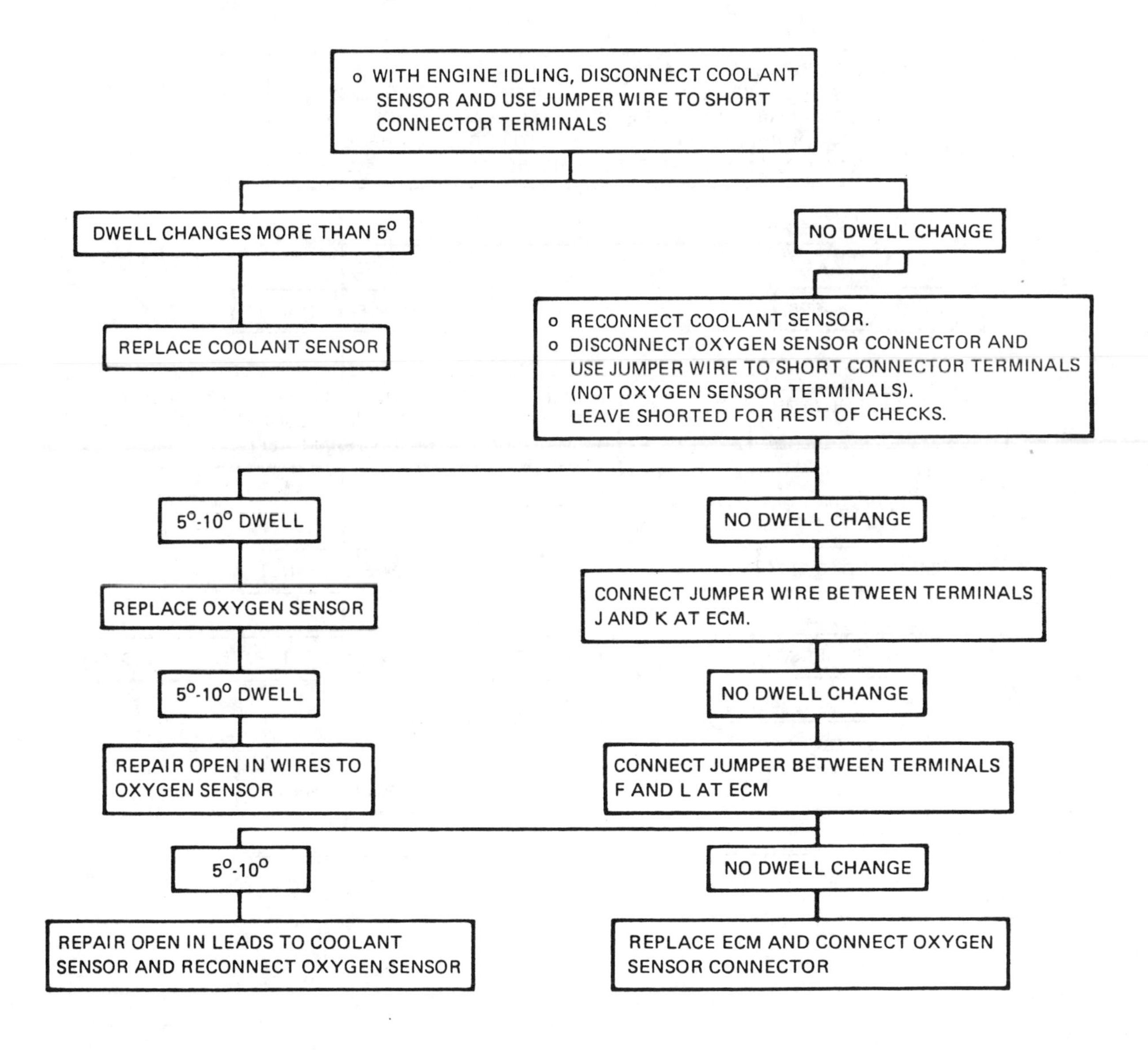

C-4 SYSTEM DIAGNOSIS—Continued

Chart 4—Dwell Fixed Between 50 Degrees and 55 Degrees

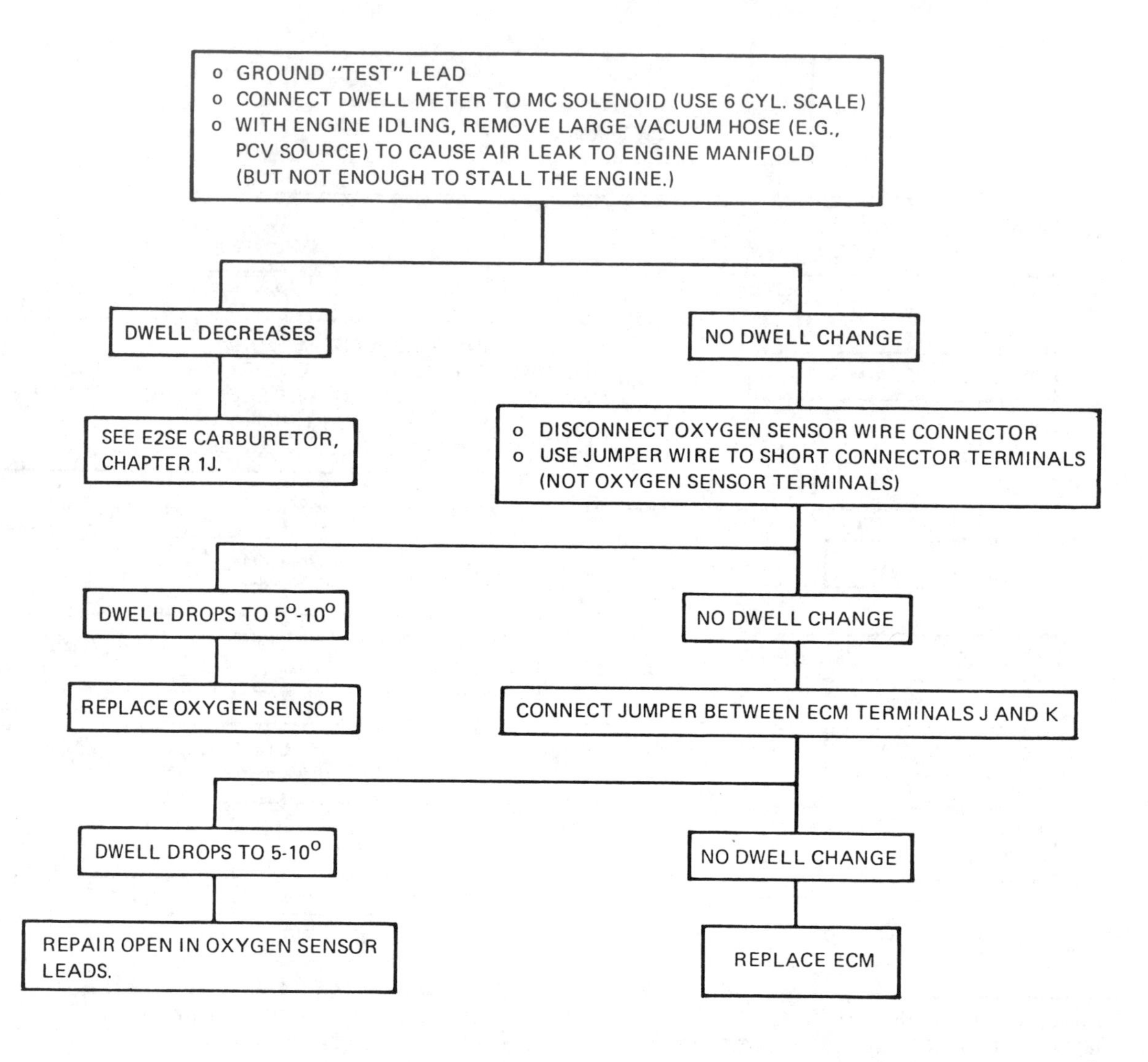

C-4 SYSTEM DIAGNOSIS—Continued

Chart 5—Self-Diagnostic Circuit Check

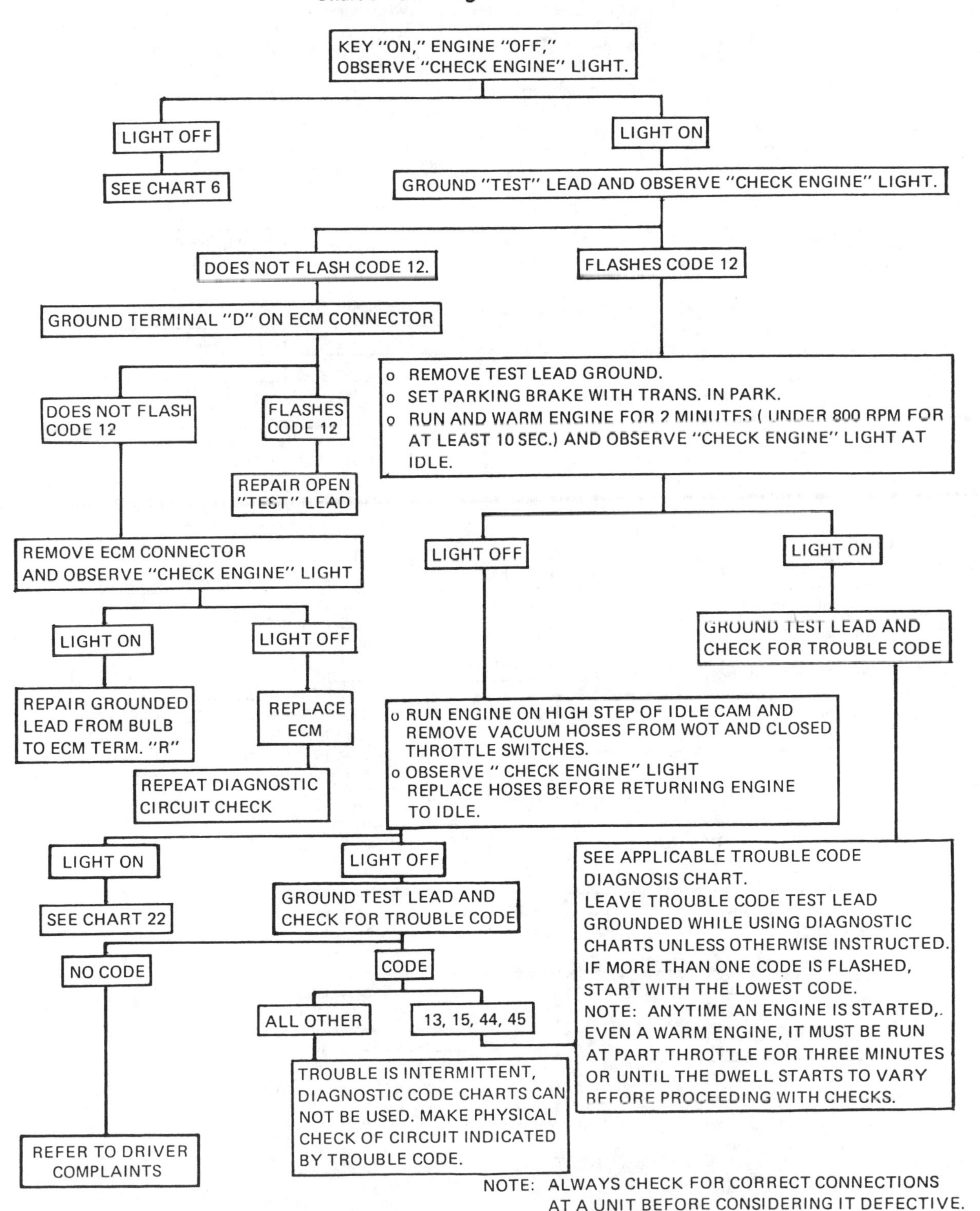

NOTE: ALWAYS CHECK FOR CORRECT CONNECTIONS AT A UNIT BEFORE CONSIDERING IT DEFECTIVE.

C-4 SYSTEM DIAGNOSIS—Continued

Chart 6—Check Engine Display Inoperative

NOTE: LIGHT WILL NOT BE ILLUMINATED IF "TEST" LEAD IS GROUNDED BEFORE THE IGNITION IS TURNED ON. THE MC SOLENOID WILL ALSO CLICK.

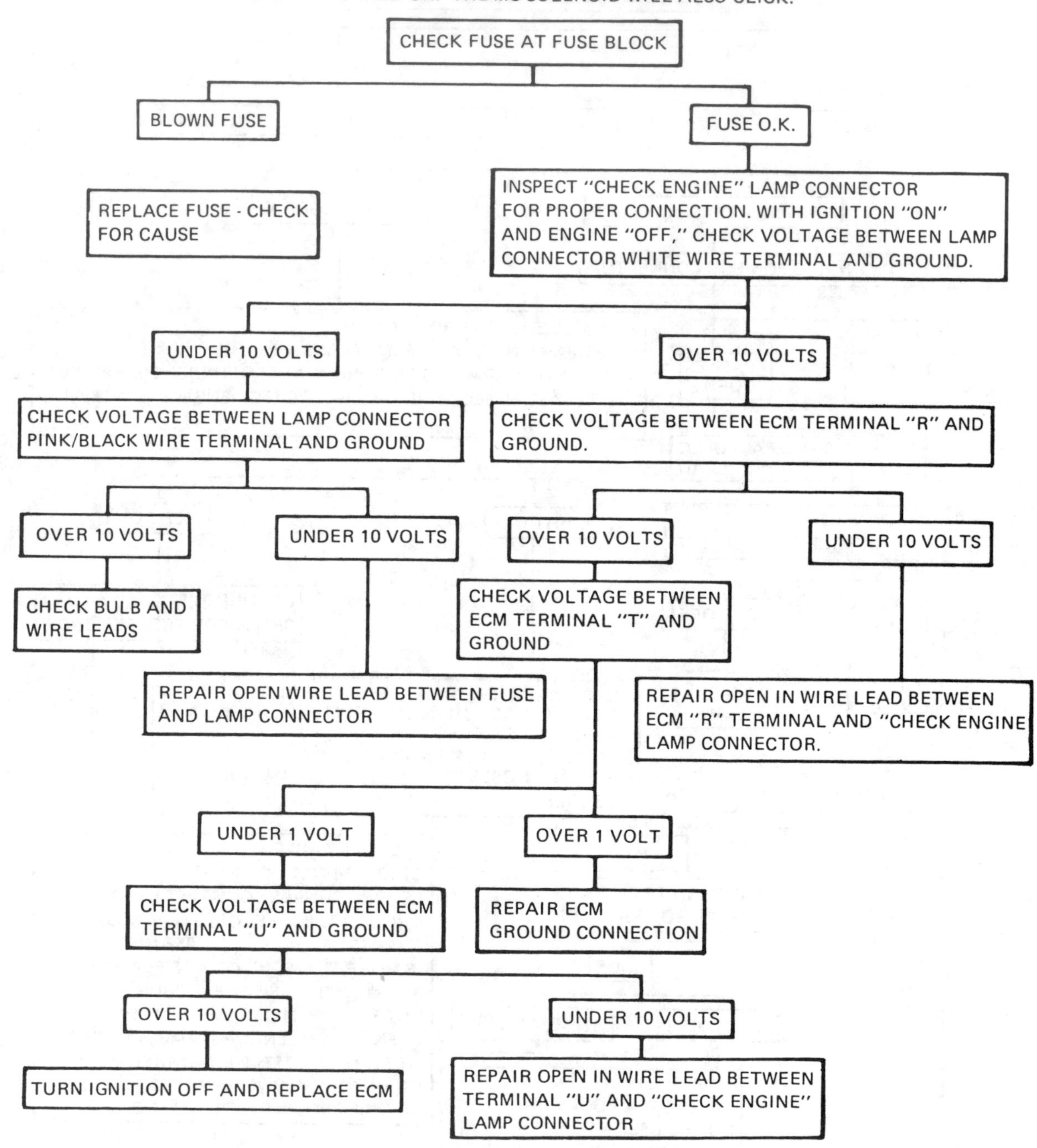

C-4 SYSTEM DIAGNOSIS—Continued

Chart 7—Adaptive Vacuum Switch Circuit Check

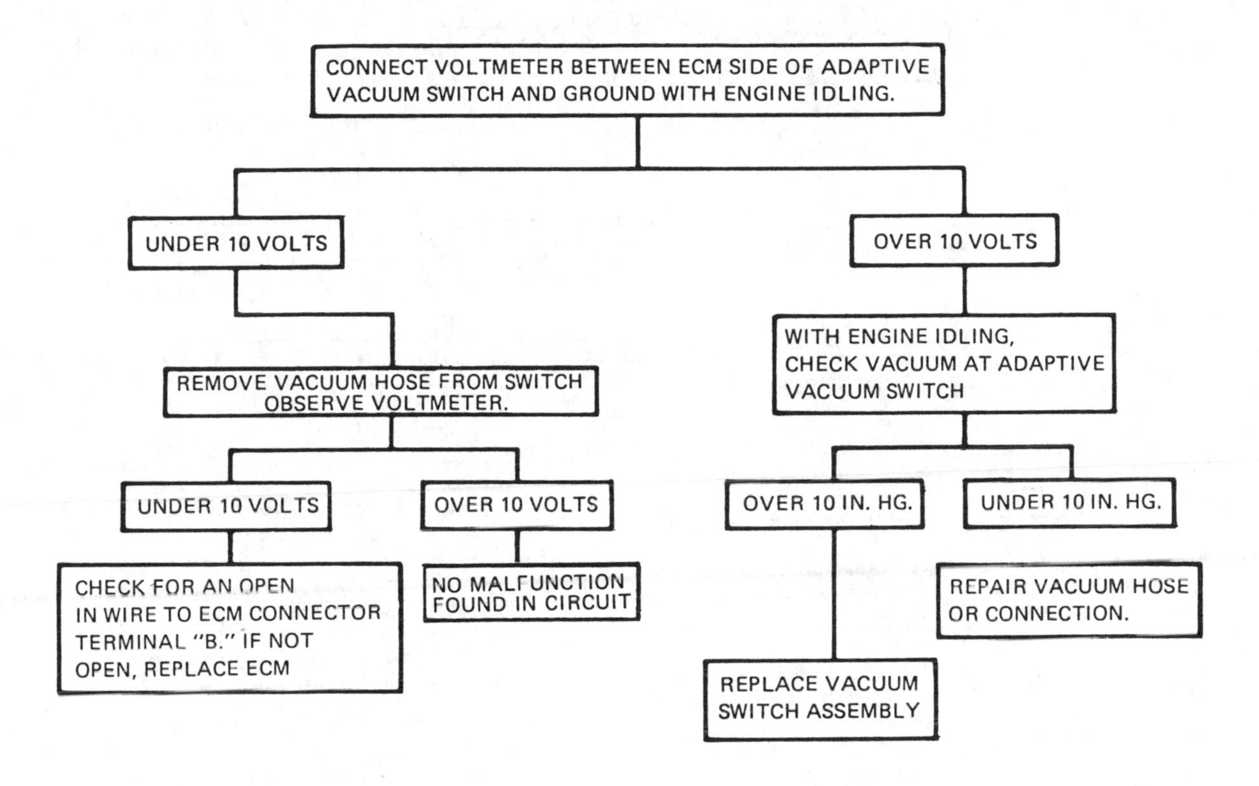

Trouble Code 13

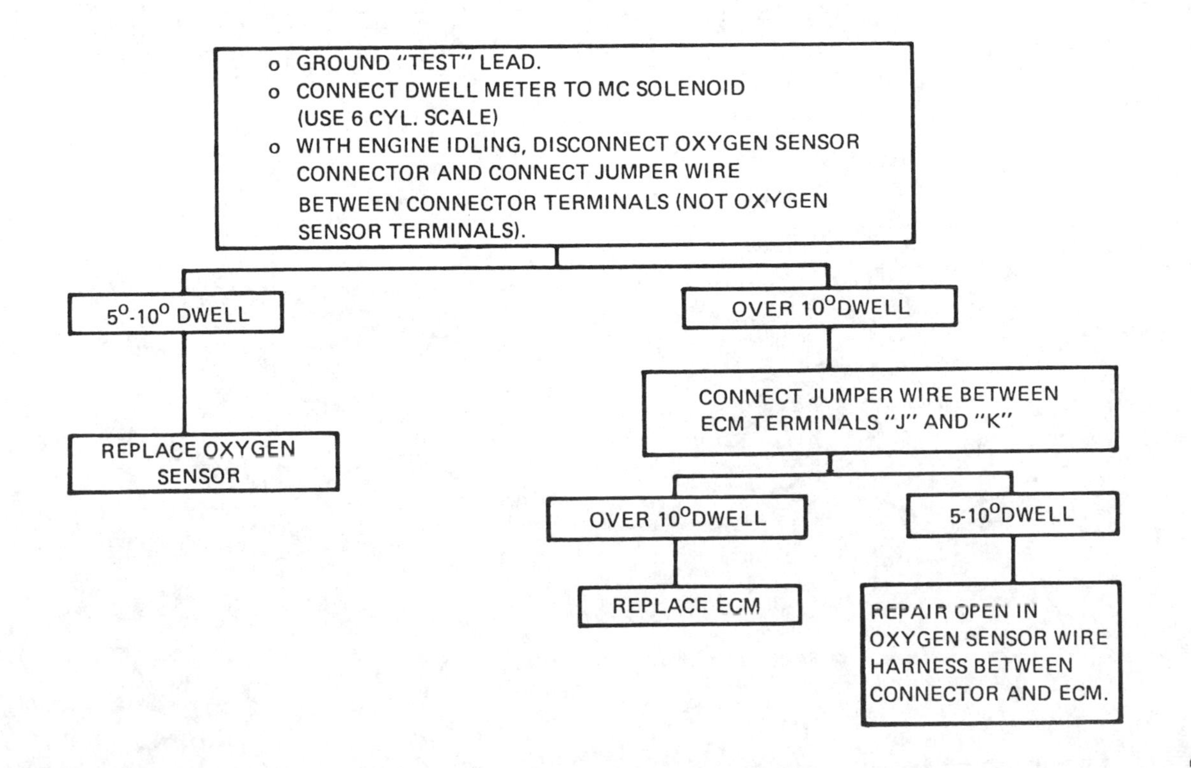

C-4 SYSTEM DIAGNOSIS—Continued

Trouble Code 12

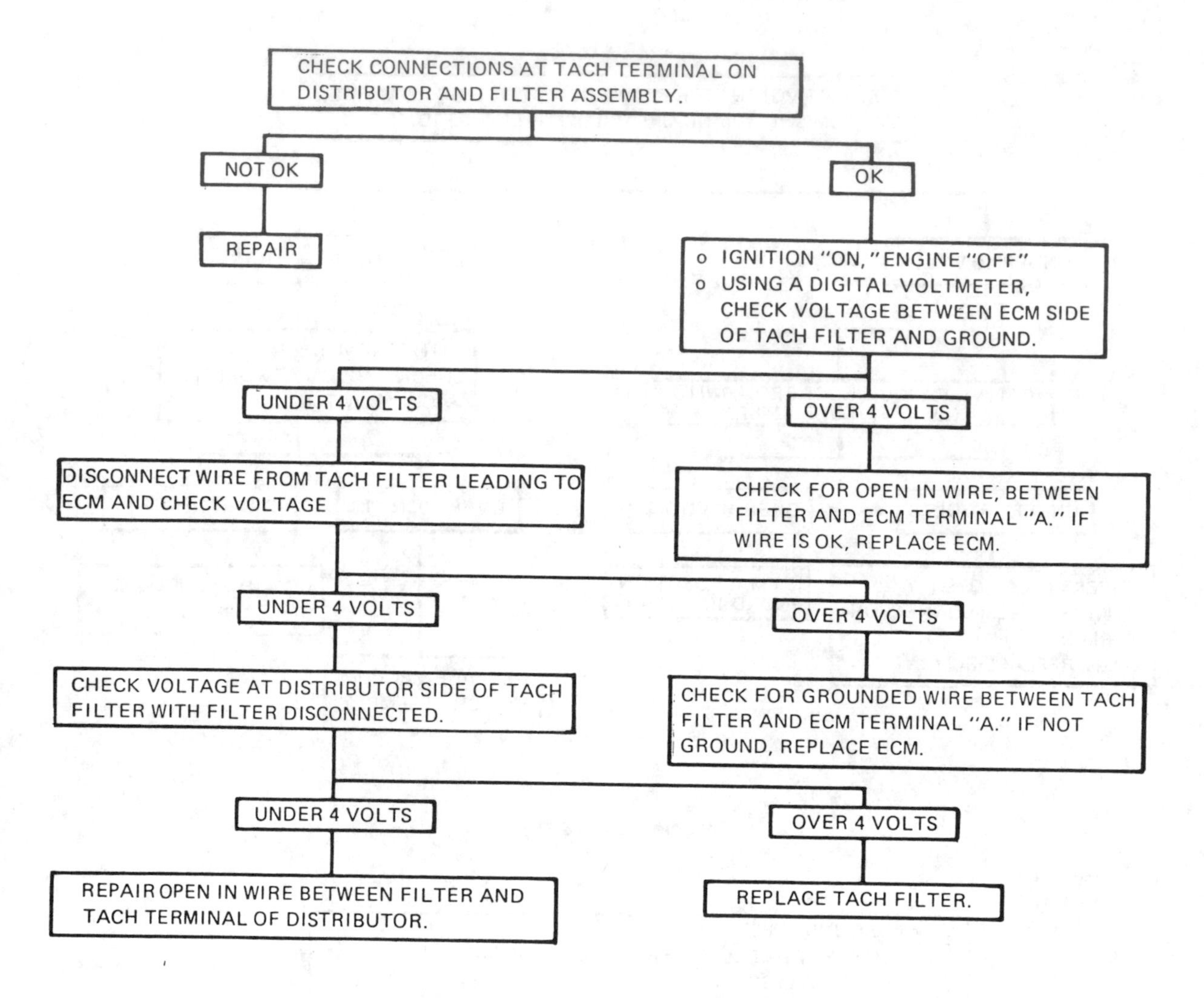

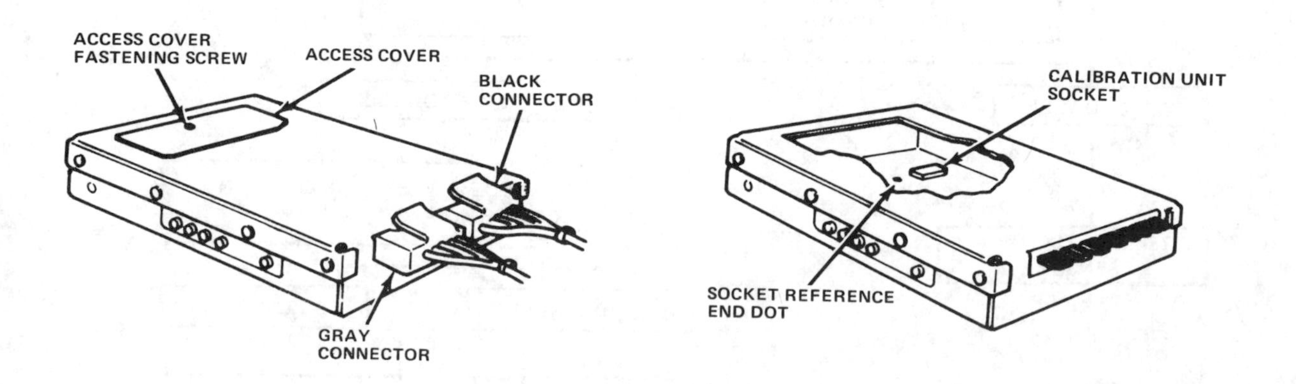

Calibration unit socket location (© American Motors)

C-4 SYSTEM DIAGNOSIS—Continued

Trouble Code 14

NOTE: IF THE ENGINE COOLANT WARNING LIGHT IS "ON," CHECK FOR AN OVERHEATING CONDITION BEFORE PERFORMING THE FOLLOWING TEST.

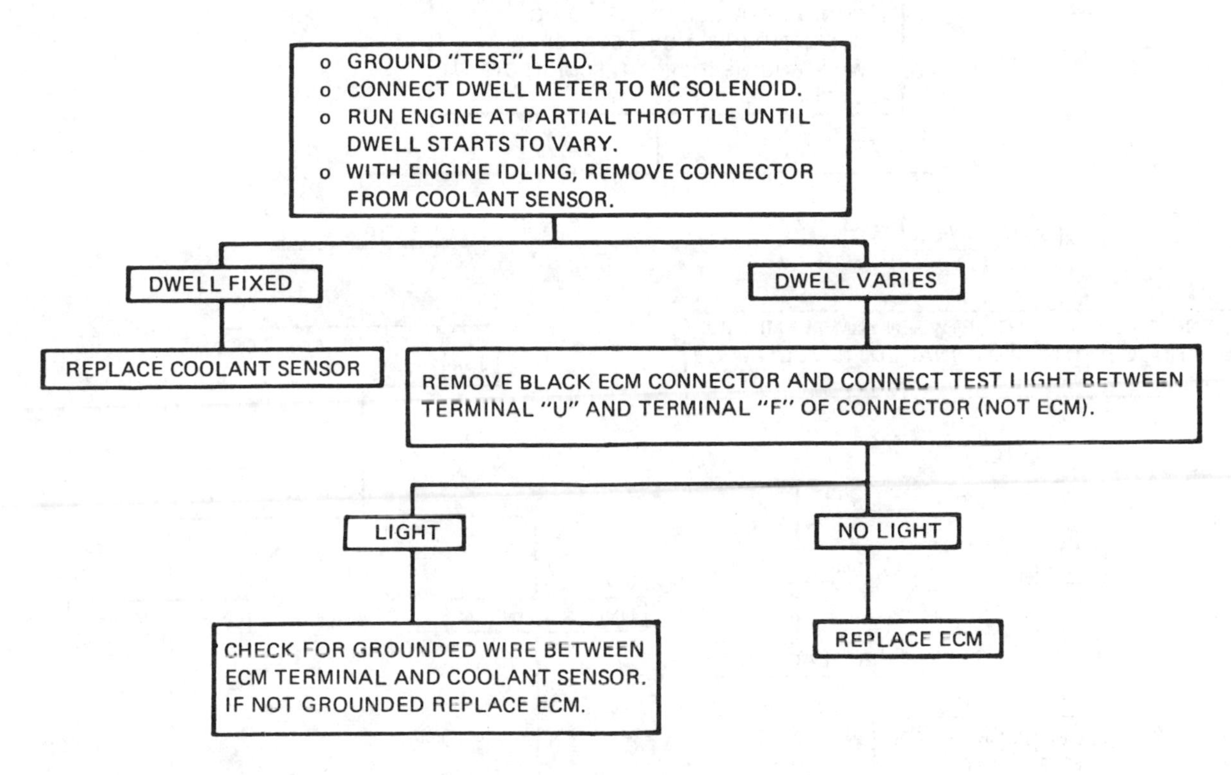

Trouble Code 15

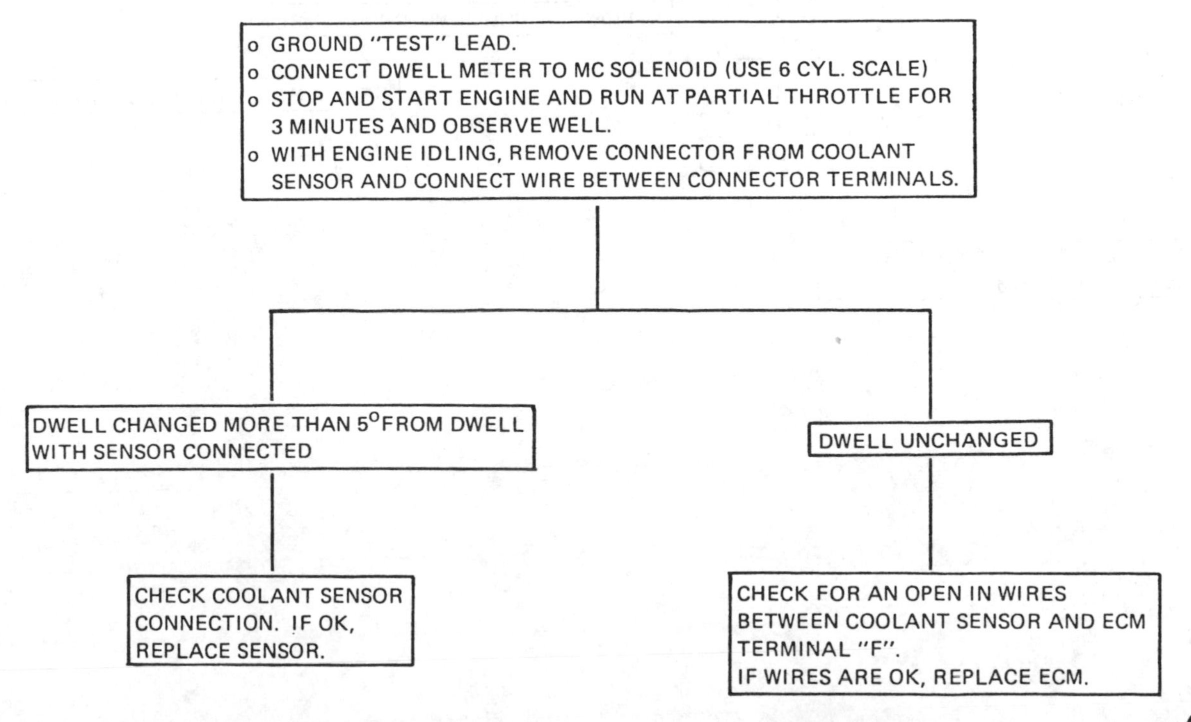

C-4 SYSTEM DIAGNOSIS—Continued

Trouble Code 21 and 22 Together

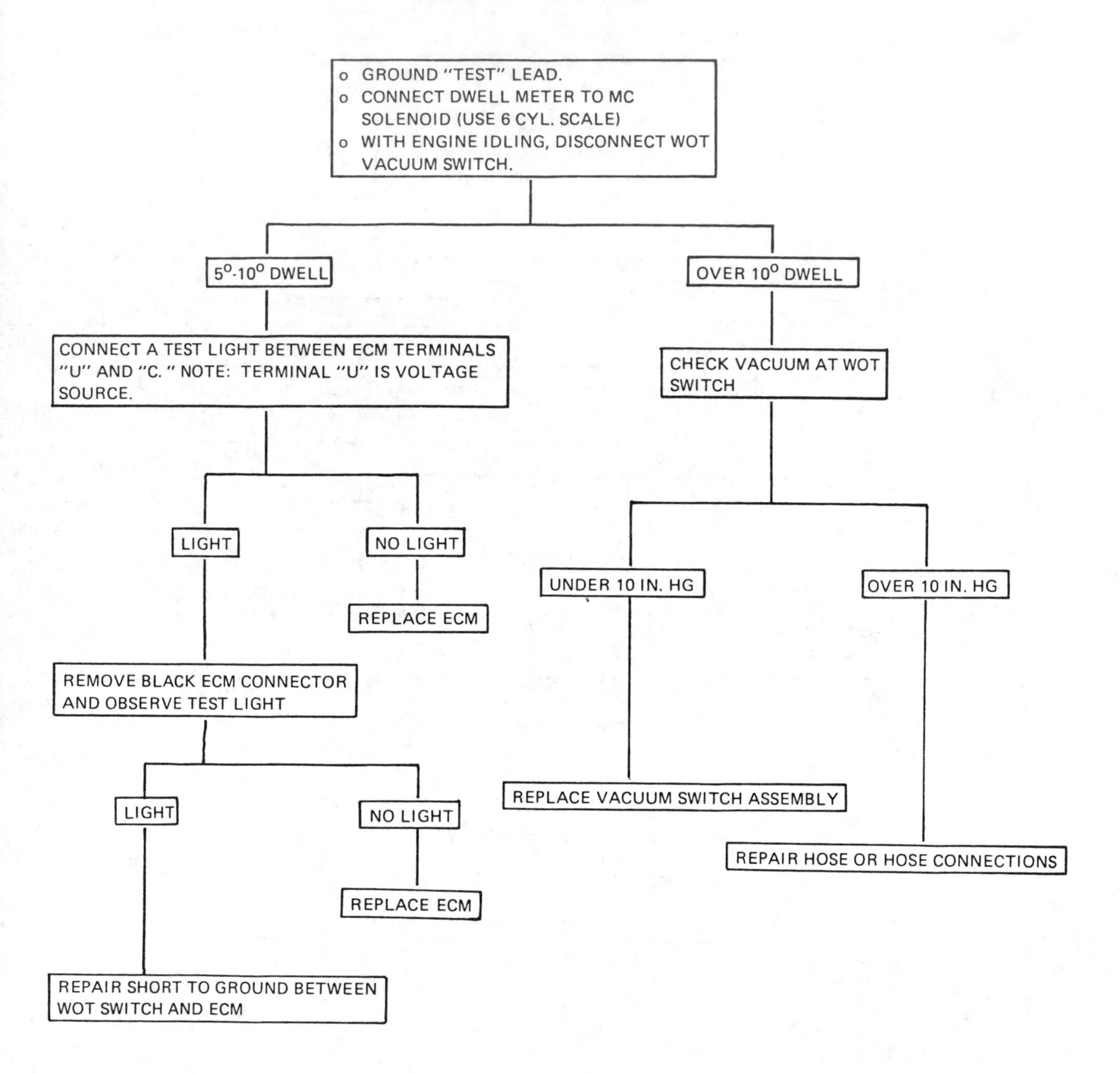

C-4 SYSTEM DIAGNOSIS—Continued

Trouble Code 22

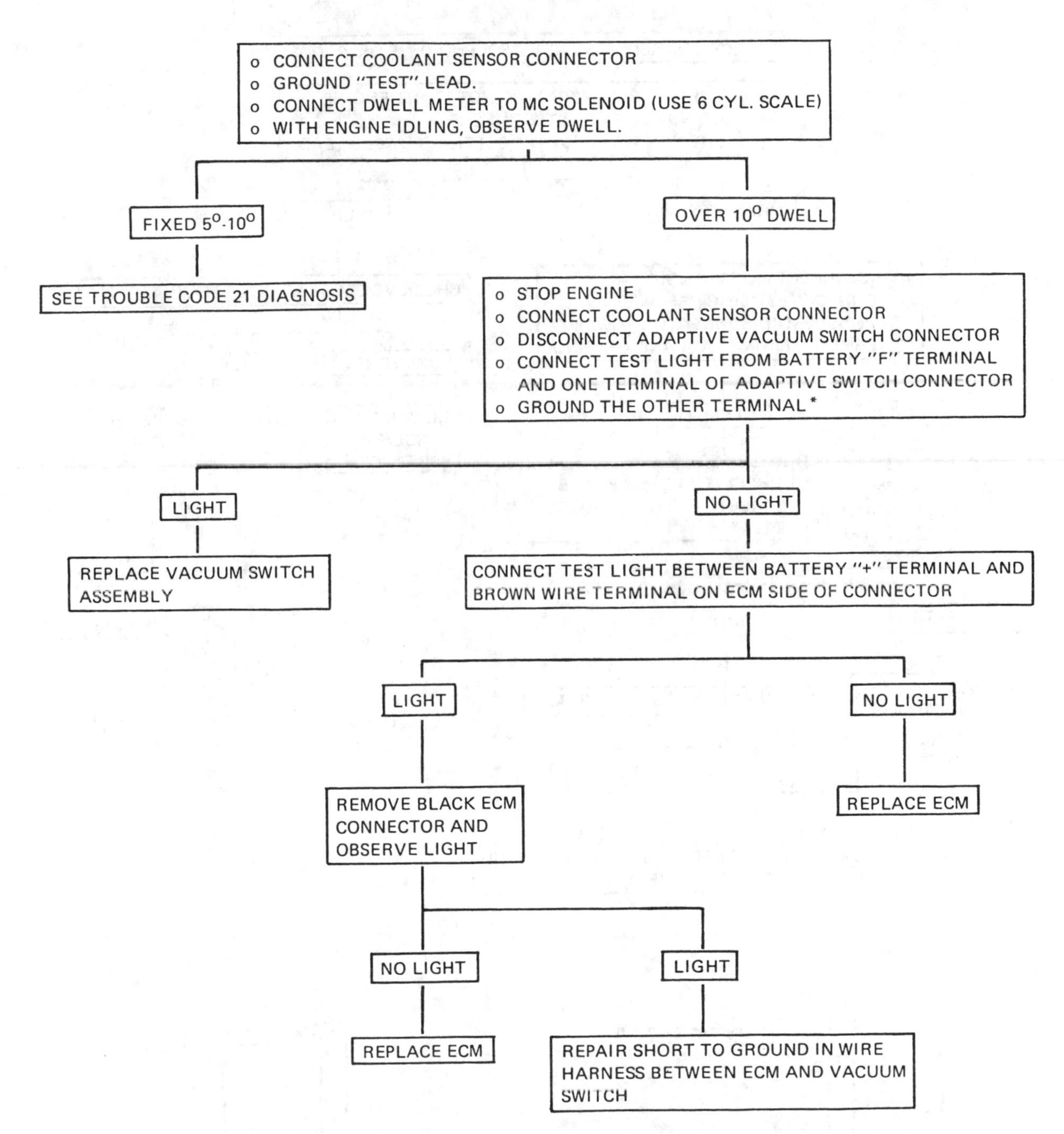

*OXYGEN SENSORS MAY COOL OFF AT IDLE AND THE DWELL CHANGE FROM VARYING TO FIXED BETWEEN 10–50°. IF THIS OCCURS, RUNNING THE ENGINE AT FAST IDLE WILL WARM IT UP AGAIN.

C-4 SYSTEM DIAGNOSIS—Continued

Trouble Code 23

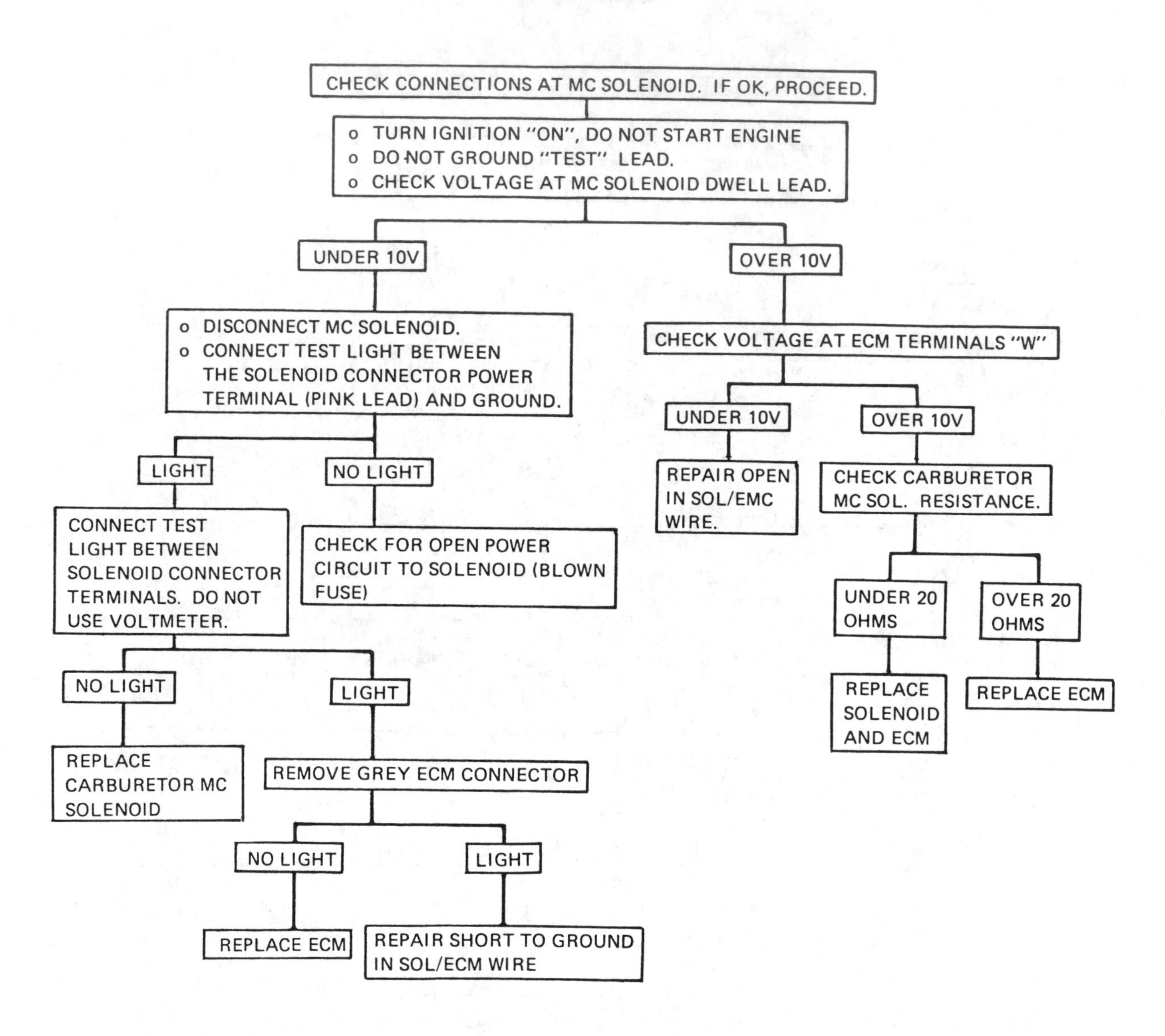

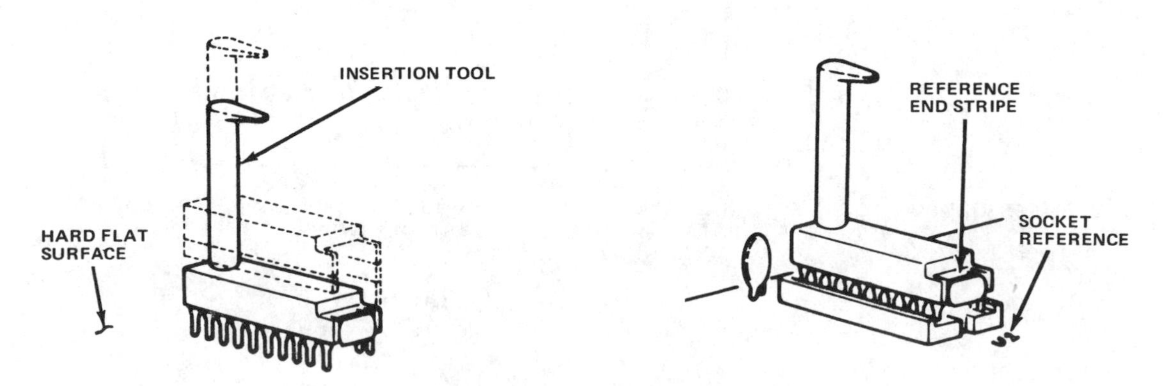

Calibration unit installation (© American Motors)

C-4 SYSTEM DIAGNOSIS—Continued

Trouble Code 44

NOTE: IF CODE 44 AND 55 ARE BOTH FLASHED,REPLACE OXYGEN SENSOR AND PERFORM SYSTEM OPERATIONAL CHECK.

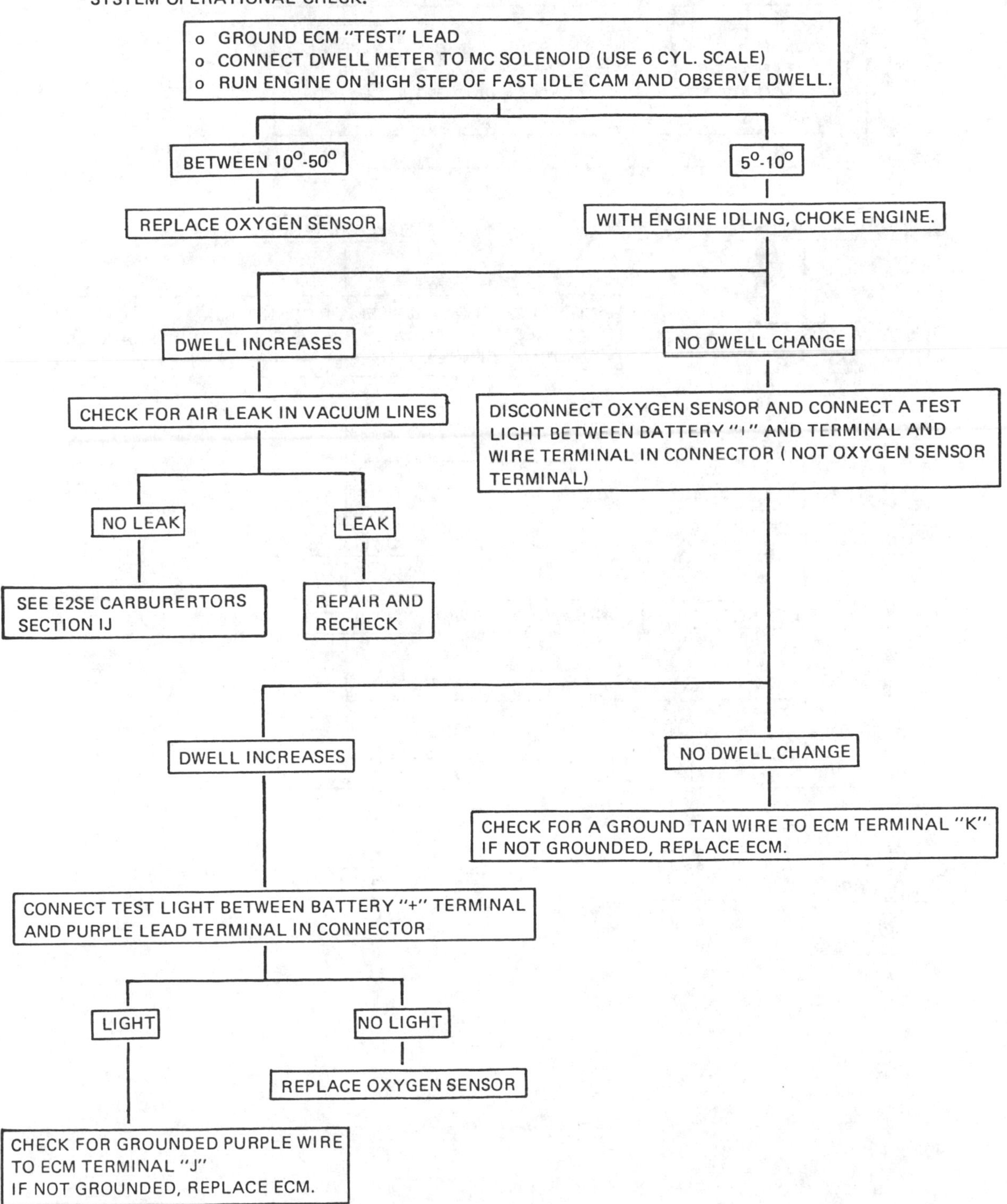

C-4 SYSTEM DIAGNOSIS—Continued

Trouble Code 45

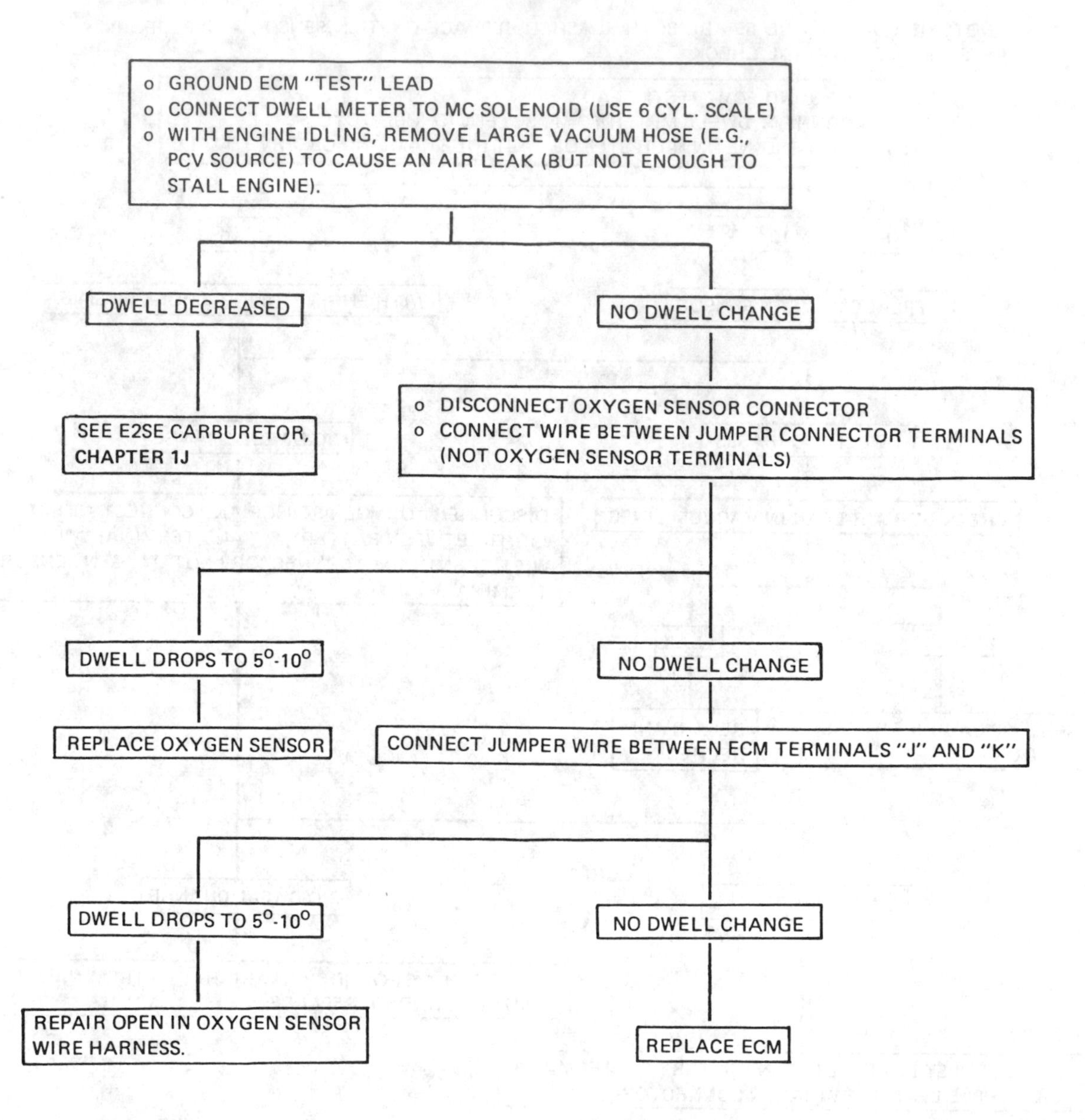

C-4 SYSTEM DIAGNOSIS—Continued

Trouble Code 51-55

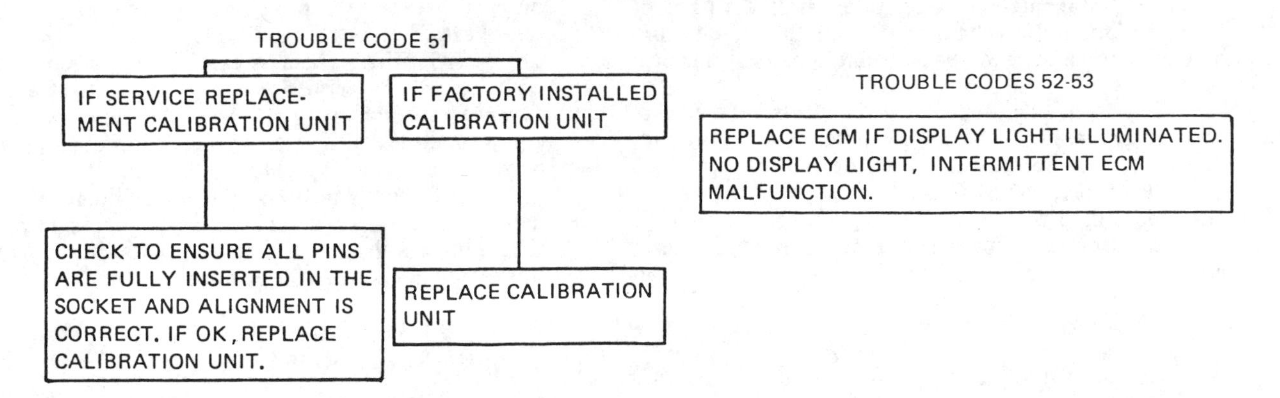

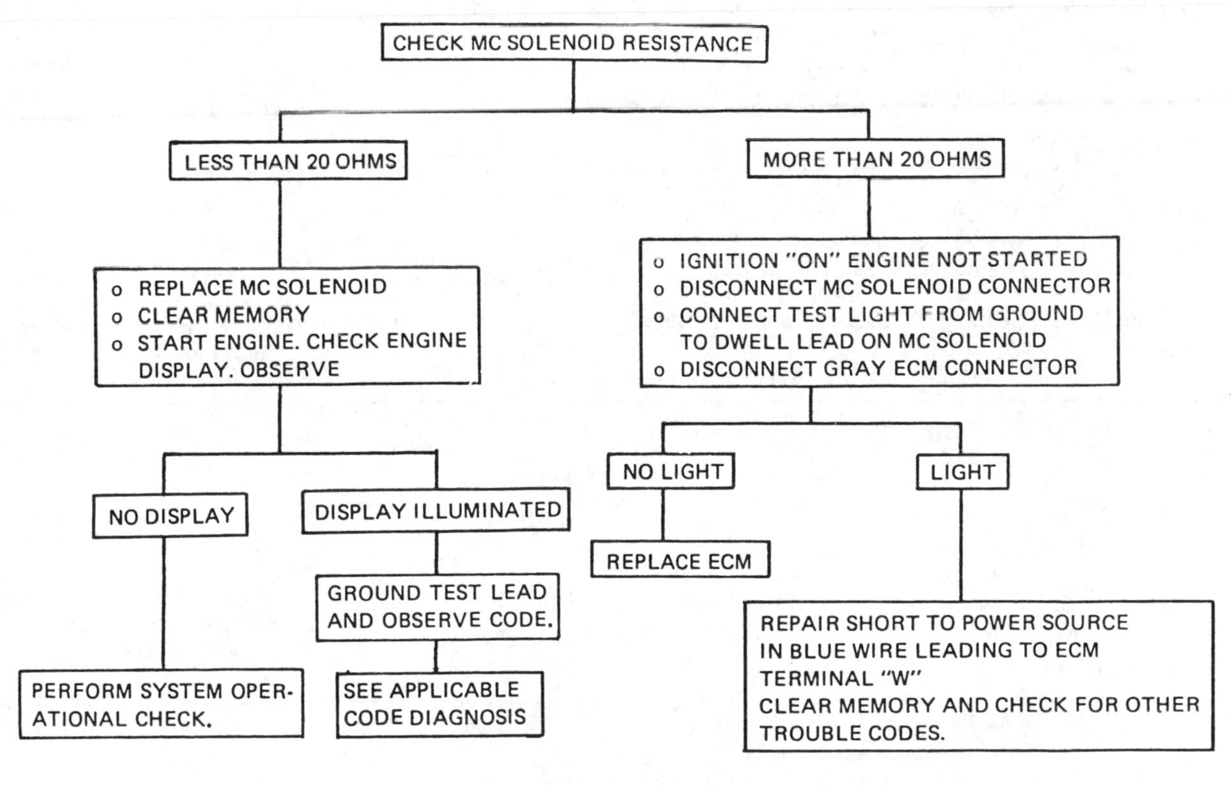

TROUBLE CODE 55

CHECK FOR AN OPEN IN SENSOR RETURN WIRE TO ECM TERMINAL "L". IF OK, REPLACE ECM.

ENGINE PERFORMANCE DIAGNOSIS

Introduction

Engine Performance Diagnosis procedures are guides that will lead to the most probable causes of engine performance complaints. They consider all of the parts of the fuel, ignition, and mechanical systems that could cause a particular complaint, and then outline repairs in a logical sequence.

Each Sympton is defined, and it is vital that the correct one be selected based on the complaints reported or found.

Review the Symptoms and their definition to be sure that only the correct terms are used.

The words used may not be what you are used to in all cases, but because these terms have been used interchangeably for so long, it was necessary to decide on the most common usage and then define them. If the definition is not understood, and the exact Symptom is not used, the Diagnostic procedure will not work.

It is important to keep two facts in mind:

1. The procedures are written to diagnose problems on cars that have "run well at one time" and that time and wear have created the condition.

2. All possible causes cannot be covered, particularly with regard to emission controls that affect vacuum advance. If doing the work prescribed does not correct the complaint, then either the wrong Symptom was used, or a more detailed analysis will have to be made.

All of the Symptoms can be caused by worn out or defective parts such as Spark Plugs, Ignition Wiring, etc. If time and/or mileage indicate that parts should be replaced, it is recommended that it be done.

Symptom	Definition	Symptom	Definition
Dieselling	Engine continues to run after the switch is turned off. It runs unevenly and may make knocking noises. The exhaust stinks.	Surges	Engine Power variation under steady throttle or cruise. Feels like the car speeds up and slows down with no change in the accelerator pedal. Can occur at any speed.
Detonation	A mild to severe ping, usually worse under acceleration. The engine makes sharp metallic knocks that change with throttle opening. Sounds like pop corn popping.	Sluggish	Engine delivers limited power under load or at high speed. Won't accelerate as fast as normal; loses too much speed going up hills; or has less top speed than normal.
Stalls	The engine quits running. It may be at idle or while driving.	Spongy	Less than the anticipated response to increased throttle opening. Little or no increase in speed when the accelerator pedal is pushed down a little to increase cruising speed. Continuing to push the pedal down will finally give an increase in speed.
Rough Idle	The engine runs unevenly at idle. If bad enough, it may make the car shake.		
Miss	Steady pulsation or jerking that follows engine speed, usually more pronounced as engine load increases. Not normally felt above 1500 RPM or 30 mph. The exhaust has a steady spitting sound at idle or low speed.	Poor Gas Mileage	Self describing.
Hesitates	Momentary lack of response as the accelerator is depressed. Can occur at all car speeds. Usually most severe when first trying to make the car move, as from a stop sign. May cause the engine to stall if severe enough.	Cuts Out	Temporary complete loss of power. The engine quits at sharp, irregular intervals. May occur repeatedly, or intermittently. Usually worse under heavy acceleration.

ENGINE PERFORMANCE DIAGNOSIS—Continued

CONDITION	POSSIBLE CAUSE	CORRECTION
Engine starts and runs but misses rough or stalls at idle speed only.	Engine vacuum leak.	Check engine vacuum connections and hoses for leaks. Repair or replace as necessary.
	Spark plug malfunction. (Fouled, cracked or incorrect gap.)	Inspect, clean, and adjust or install new plugs as necessary.
	Ignition wire malfunction	Clean, check continuity and inspect wires, replace if brittle, cracked or worn. Refer to HEI diagnosis.
	HEI malfunction	Refer to HEI diagnosis
	EGR malfunction	Refer to EGR diagnosis.
	PCV malfunction.	Refer to PCV diagnosis.
	Engine valve leakage.	
	Intake manifold or cylinder head gasket leaks.	Check intake manifold gaskets and seals for leaks. Make necessary repairs or replacements.
	Carburetor malfunction.	Refer to carburetor diagnosis.
Engine starts and runs, misses at all speeds.	Spark plug malfunction.	Inspect, clean, and adjust or replace as required.
	HEI malfunction.	Refer to HEI diagnosis
	Engine valve leakage.	Refer to Section 6C (compression check).
	EGR malfunction.	Refer to EGR diagnosis.
	EFE valve malfunction.	Make necessary repairs or replacements.
	Malfunctioning fuel pump or fuel system.	Refer to fuel pump diagnosis and/or fuel delivery system function.
	Faulty carburetion (including fuel filter).	Refer to carburetor diagnosis.
	Contaminated fuel.	Remove and install fresh fuel. Clean fuel system and carburetor as required.
Engine starts and runs but misses at high speeds.	Spark plug malfunction.	Inspect, clean and adjust or install new plugs.
	Engine valve leakage.	
	HEI malfunction.	Refer to HEI diagnosis

ENGINE PERFORMANCE DIAGNOSIS—Continued

CONDITION	POSSIBLE CAUSE	CORRECTION
Engine starts and runs but misses at high speeds.	Ignition wire malfunction.	Clean, check resistance, and inspect wires, replace if brittle, cracked or worn.
	Malfunctioning fuel pump or fuel system.	Refer to fuel pump diagnosis and/or fuel delivery system.
	Faulty carburetion (including fuel filters).	Refer to carburetor and/or fuel system diagnosis.
	EFE valve malfuction.	Refer to EFE diagnosis.
	PCV malfunction.	Refer to PCV diagnosis.
	Faulty carburetor inlet air temperature regulation.	Refer to air cleaner diagnosis.
	Exhaust system restricted.	Make necessary repairs or replacements.

HIGH ENERGY IGNITION (H.E.I.) DIAGNOSIS

CONDITION	POSSIBLE CAUSE	CORRECTION
Engine cranks but will not start	Low battery.	Charge battery and/or check generator.
	No spark at spark plugs.	Disconnect one spark plug lead from spark plug, hold 1/4″ from dry area of engine block while cranking engine. If spark occurs while cranking, check condition of spark plugs (if plugs are okay, problem is in fuel system). If no spark occurs: a. Refer to HEI diagnosis.
	a. Open circuit between "Bat" terminal and battery.	b. Repair open circuit between "Bat" terminal and battery.
	Engine timing out of adjustments.	Reset timing to specifications. For procedure to rough-set timing on an engine that won't run, see setting ignition timing.
	Internal engine problems.	Refer to engine diagnosis.
	Out of gas.	Put a supply of gas in tank and start engine.
	Malfunctioning carburetor.	Refer to carburetor diagnosis.
	Malfunctioning fuel pump.	Refer to fuel system diagnosis.
	Battery cables loose or corroded.	Tighten and/or clean cables.
Engine starts but stops when ignition switch is released to "RUN POSITION"	An "OPEN" in the ignition circuit or a defective ignition switch.	Use a 12 volt test lamp and check IGN-1 terminal of ignition switch in "RUN POSITION". a. If lamp lights, locate and repair "OPEN" in circuit to HEI "Bat" terminal. This includes pink wire from switch connector to eowl connector. Black wire with double white stripe from cowl connector to distributor connector. b. If lamp does not light, replace ignition switch.

HIGH ENERGY IGNITION (H.E.I.) DIAGNOSIS—Continued

Engine will not crank.	Loose or corroded wire or cable connections.	Inspect for and clean and/or tighten connections as necessary.
	Discharged battery.	Refer to diagnosis in battery section.
	Starter assembly.	Test and/or repair starter as necessary. See Section 12.
	Neutral start switch malfunctioning out of adjustment or poor connection.	Make certain connector terminals are clean and connector is properly installed. Connect 12 volt test lamp to purple wire with white stripe and ground. Lamp should light in start position with shift in neutral or park. If lamp lights, ignition and neutral start switch are okay. If lamp did not light on purple wire with white stripe, connect it to light green wire with double black stripe. If lamp now lights in start position in neutral or park, adjust or replace neutral start switch. If lamp did not light, see following ignition switch test.
	Loose connection at or defective ignition switch.	Inspect connector to assure clean terminals and proper connection. Using 12 volt test lamp check both purple wires at neutral start switch. If lamp does not light at either wire, repair "OPEN" circuit in purple wire with white stripe to the ignition switch and/or replace ignition switch as required.
	"OPEN" circuit in wiring to solenoid or defective solenoid.	Using 12 volt test lamp between purple wire at solenoid and ground, lamps should light in start position in neutral or park. a. If lamp does not light, locate and correct "OPEN" in circuit. b. If lamp lights replace solenoid.
	Burned out fusible link.	Using 12 volt test lamp between No. 10 red wire at cowl connector and ground, lamp should light. a. If lamp does not light, replace fusible link.

HIGH ENERGY IGNITION (H.E.I.) DIAGNOSIS—Continued

Engine runs rough, poor power and fuel economy.	Faulty spark plugs.	Inspect, clean and adjust or adjust and install new plugs.
	Incorrect timing.	Adjust timing to specifications.
	Inoperative vacuum advance. (Engines equipped with vaccum advance)	With engine running at part throttle, hold finger on vacuum advance rod and remove vacuum hose. Rod should move toward distributor. If rod does not move, replace vacuum advance unit.
	Faulty centrifugal advance on engine.	Check centrifugal advance on engine.
	HEI distributor.	Refer to HEI diagnosis.
	Faulty plug wires.	Clean, check resistance, and inspect wires for brittle, cracked or loose insulation condition. Also inspect for burned, corroded terminals. Clean or replace wire assemblies as necessary including deteriorated nipples or boots.
	Faulty carburetion.	Refer to carburetion diagnosis.
	Faulty engine parts such as valves, rings, etc.	Refer to engine diagnosis.

CARBURETOR DIAGNOSIS

The following diagnostic procedures are directed toward carburetor related problems and their effects on car performance. It is understood that other systems of the car can also cause similar problems and should be investigated in conjunction with the carburetor. In all instances, the complaint item should be verified by competent service personnel. The problem areas described are:

1. Engine cranks normally. Will not start.
2. Engine starts and stalls.
3. Engine starts hard.
4. Engine idles abnormally and/or stalls.
5. Inconsistent engine idle speeds.
6. Engine diesels (after-run) upon shut off.
7. Engine hesitates on acceleration.
8. Engine has less than normal power at low speeds.
9. Engine has less than normal power on heavy acceleration or at high speed.
10. Engine surges.
11. Fuel economy complaints.

CARBURETOR DIAGNOSIS—Continued

CONDITION	POSSIBLE CAUSE	CORRECTION
Engine Cranks Normally Will Not Start.	Improper starting procedure used.	Check with the customer to determine if proper starting procedure is used, as outlined in the Owner's Manual.
	Choke valve not operating properly.	Adjust the choke thermostatic coil to specification. Check the choke valve and/or linkage as necessary. Replace parts if defective. If caused by foreign material and gum, clean with suitable non-oil base solvent. NOTE: After any choke system work, check choke vacuum break settings and correct as necessary.
	No fuel in carburetor.	Remove fuel line at carburetor. Connect hose to fuel line and run into metal container. Remove the wire from the "bat" terminal of the distributor. Crank over engine - if there is no fuel discharge from the fuel line, test fuel pump as outlined in Section 6B. If fuel supply is okay, check the following: a. Inspect fuel inlet filter. If plugged, replace. b. If fuel filter is okay, remove air horn and check for a bind in the float mechanism or a sticking inlet needle. If okay, adjust float as specified.
	Engine flooded. To check for flooding, remove the air cleaner with the engine immediately shut off and look into the carburetor bores. Fuel will be dripping off nozzles.	Remove the air horn. Check fuel inlet needle and seat for proper seal. If a needle and seat tester is not available, apply vacuum to the needle seat with needle installed. If the needle is leaking, replace. Check float for being loaded with fuel, bent float hanger or binds in the float arm. A solid float can be checked for fuel absorption by lightly squeezing between fingers. If wetness appears on surface or float feels heavy (check with known good float), replace the float assembly. If foreign material is in fuel system, clean the system and replace fuel filters as necessary. If excessive foreign material is found, completely disassemble and clean.

CARBURETOR DIAGNOSIS—Continued

Condition	Possible Cause	Correction
Engine Starts - Will Not Keep Running	Fuel pump.	Check fuel pump pressure and volumn, replace if necessary.
	Idle speed.	Adjust idle to specifications.
	Choke heater system malfunctioning (may cause loading).	Chec vacuum supply at hot air inlet to choke housing. Should be not less than manifold vacuum minus 3" Hg. with engine running at idle. (Exc. IMV)
		Check for plugged, restricted, or broken heat tubes.
		Check routing of all hot air parts.
	Loose, broken or incorrect vacuum hose routing.	Check condition and routing of all vacuum hoses - correct as necessary.
	Engine does not have enough fast idle speed when cold.	Check for free movement of fast idle cam. Clean and/or realign as necessary.
	Choke vacuum break units are not adjusted to specification or are defective.	Adjust both vacuum break assemblies to specification. If adjusted okay, check the vacuum break units for proper operation as follows:
		To check the vacuum break units, apply a constant vacuum source of at least 10" Hg., plungers should slowly move inward and hold vacuum. If not, replace the unit.
		Always check the fast idle cam adjustment when adjusting vacuum break units.
	Choke valve sticking and/or binding.	Clean and align linkage or replace if necessary. Readjust all choke settings, see Section 6M, if part replacement or realignment is necessary.
	Insufficient fuel in carburetor.	Check fuel pump pressure and volume.
		Check for partially plugged fuel inlet filter. Replace if contaminated.
(NOTE: The EGR system diagnosis should also be performed.)		Check the float level adjustment and for binding condition. Adjust as specified.
Engine Starts Hard (Cranks Normally)	Loose, broken or incorrect vacuum hose routing.	Check condition and routing of all vacuum hoses - correct as necessary.
	Incorrect starting procedure.	Check to be sure customer is using the starting procedure outlined in Owner's Manual.
	Malfunction in accelerator pump system.	Check accelerator pump adjustment and operation.
		Check pump discharge ball for sticking or leakage.

CARBURETOR DIAGNOSIS—Continued

	Choke valve not closing.	Adjust choke thermostatic coil. Check choke valve and linkage for binds and alignment. Clean and repair or replace as necessary.
	Vacuum breaks misadjusted or malfunctioning.	Check for adjustment and function of vacuum breaks. Correct as necessary.
	Insufficient fuel in bowl.	Check fuel pump pressure and volume. Check for partially plugged fuel inlet filter. Replace if dirty. Check float mechanism. Adjust as specified.
	Flooding.	Check float and needle and seat for proper operation.
	Where used, check to see if vent valve is inoperative or misadjusted.	Check for operation and adjustment of vent valve (if used).
	Slow engine cranking speed.	Refer to starting system diagnosis.
(NOTE: The EGR system diagnosis should also be performed.)		
Engine Idles Abnormally	Incorrect idle speed.	Reset idle speed per instructions on underhood label.
	Air leaks into carburetor bores beneath throttle valves, manifold leaks, or vacuum hoses disconnected or installed improperly.	Check all vacuum hoses and restrictors leading into the manifold or carburetor base for leaks or being disconnected. Install or replace as necessary. Torque carburetor to manifold bolts to 15 ft. lbs. (L-6), 10 ft. lbs. (exc. L-6). Using a pressure oil can, spray light oil or kerosene around manifold to head surfaces and carburetor throttle body. NOTE: Do not spray at throttle shaft ends. If engine RPM changes, tighten or replace the carburetor or manifold gaskets as necessary.
	Clogged or malfunctioning PCV system.	Check PCV system. Clean and/or replace as necessary.
	Carburetor flooding. Check by using procedure outlined under "Engine Flooded".	Remove air horn and check float adjustments. Check float needle and seat for proper seal. If a needle and seat tester is not available, apply vacuum to the needle seat with needle installed. If the needle is leaking or damaged, replace. Check float for being loaded with fuel. Check for bent float hanger or binds in the float arm. A solid float can be checked for fuel. Check for bent float hanger or binds in the float arm.

CARBURETOR DIAGNOSIS—Continued

		A solid float can be checked for fuel absorption by lightly squeezing between fingers. If wetness appears on surface or float feels heavy (check with known good float), replace the float assembly.
		If foreign material is found in the carburetor, clean the fuel system and carburetor. Replace fuel filters as necessary.
	Restricted air cleaner element.	Replace as necessary.
	Idle system plugged or restricted.	Clean per section 6M.
	Incorrect idle mixture adjustment.	Readjust per specified procedure
	Defective idle stop solenoid, idle speed-up solenoid or wiring.	Check solenoid and wiring.
	Throttle blades or linkage sticking and/or binding.	Check throttle linkage and throttle blades (primary and secondary) for smooth and free operation. Correct problem areas.
(NOTE: The EGR system diagnosis should also be performed.)		
Engine Diesels (After Run – Upon Shut Off)	Loose, broken or improperly routed vacuum hoses.	Check condition and routing of all vacuum hoses. Correct as necessary.
	Incorrect idle speed.	Reset idle speed per instructions on label in engine compartment.
	Malfunction of idle stop solenoid, idle speed-up solenoid or dashpot.	Check for correct operation of idle solenoid. Check for sticky or binding solenoid.
	Excessively lean idle mixture caused by air leaks into carburetor beneath throttle valves, manifold vacuum leaks, or failed PCV system.	See corrections listed causes 2 and 3 under "Engine Idles Abnormally and/or Stalls".
	Fast idle cam not fully off.	Check fast idle cam for freedom of operation. Clean, repair, or adjust as required. Check choke heated air tubes for routing, fittings being tight or tubes plugged. Check choke linkage for bending. Clean and correct as necessary.
	Excessively lean condition caused by maladjusted carburetor idle mixture.	Adjust carburetor idle mixture as described in Section 6M.
	Ignition timing retarded.	Set to specifications.

CARBURETOR DIAGNOSIS—Continued

Engine Hesitates On Acceleration	Loose, broken or incorrect vacuum hose routing	Check condition and routing of all vacuum hoses - correct or replace.
	Accelerator pump not adjusted to specification or inoperative.	Adjust accelerator pump, or replace.
	Inoperative accelerator pump system.	Remove air horn and check pump cup. If cracked, scored or distorted, replace the pump plunger.
	NOTE: A quick check of the pump system can be made as follows: With the engine off, look into the carburetor bores and observe pump shooters while briskly opening throttle lever. A full stream of fuel should emit from each pump shooter.)	Check the pump discharge ball for proper seating and location.
	Foreign matter in pump passages.	Clean and blow out with compressed air.
	Float level too low.	Check and reset float level to specification.
	Front vacuum break diaphragm not functioning properly.	Check adjustment and operation of vacuum break diaphragm.
	Air valve malfunction.	Check operation of secondary air valve. Check spring tension adjustment.
	Power enrichment system not operating correctly.	Check for binding or stuck power piston(s) - correct as necessary.
	Inoperative air cleaner heated air control.	Check operation of thermostatic air cleaner system.
	Fuel filter dirty or plugged.	Replace filter and clean fuel system as necessary.
	Distributor vacuum or mechanical advance malfunctioning.	Check for proper operation.
	Timing not to specifications.	Adjust to specifications.
	Choke coil misadjusted (cold operation.)	Adjust to specifications
	EGR valve stuck open.	Inspect and clean EGR valve.
Engine Has Less Than Normal Power At Normal Accelerations.	Loose, broken or incorrect vacuum hose routing.	Check condition and routing of all vacuum hoses.
	Clogged or defective PCV system.	Clean or replace as necessary.
	Choke sticking.	Check complete choke system for sticking or binding.
		Clean and realign as necessary.
		Check adjustment of choke thermostatic coil.
		Check connections and operation of choke hot air system.

CARBURETOR DIAGNOSIS—Continued

		Check jets and channels for plugging; clean and blow out passages.
	Clogged or inoperative power system.	Remove air horn and check for free operation of power pistons.
	Air cleaner temperature regulation improper.	Check regulation and operation of air cleaner system.
	Transmission malfunction.	Refer to transmission diagnosis.
	Ignition system malfunction.	Check ignition timing. Reset to specification.
		Refer to H.E.I. diagnosis.
	Excessive brake drag.	Refer to Section 5.
	Exhaust system.	Check for restrictions. Correct as required.

(NOTE: An engine tune-up should be conducted in conjunction with the carburetor diagnosis. The EGR system diagnosis should also be performed.)

Less Than Normal Power On Heavy Acceleration Or At High Speed	Carburetor throttle valves not going wide open. Check by pushing accelerator pedal to floor.	Correct throttle linkage to obtain wide open throttle in carburetor.
	Secondary throttle lockout not allowing secondaries to open.	Check for binding or sticking lockout lever.
		Check for free movement of fast idle cam.
		Check choke heated air system for proper and tight connections plus flow through system.
		Check adjustment of choke thermostatic coil.
	Spark plugs fouled, incorrect gap.	Clean, regap, or replace plugs.
	Plugged air cleaner element.	Replace element.
	Air valve malfunction. (Where applicable)	Check for free operation of air valve.
		Check spring tension adjustment. Make necessary adjustments and corrections.
	Contaminated fuel inlet filter.	Replace with a new filter element.
	Insufficient fuel to carburetor.	Check fuel pump and system, run pressure and volume test.
	Vapor lock.	Eliminate cause.
	Power enrichment system not operating correctly.	Remove the air horn and check for free operation of both power piston(s), clean and correct as necessary.
	Choke closed or partially closed.	Free choke valve or linkage.

CARBURETOR DIAGNOSIS—Continued

Condition	Possible Cause	Correction
		Check for loose jets.
	Float level too low.	Check and reset float level to specification.
	Transmission malfunction.	Refer to transmission diagnosis.
	Ignition system malfunction.	Check ignition timing. Reset to specification. Refer to H.E.I. diagnosis.
	Fuel metering jets restricted.	If the fuel metering jets are restricted and excessive amount of foreign material is found in the fuel bowl, the carburetor should be completely disassembled and cleaned.
	Fuel pump.	Check fuel pump pressure and volumn, inspect lines for leaks and restrictions.
	Exhaust system.	Check for restrictions. Correct as required.

(NOTE: Complete engine tune-up should be performed in conjunction with carburetor diagnosis.)

Condition	Possible Cause	Correction
Engine Surges	Loose, broken or incorrect vacuum hose routing.	Check condition and routing of all vacuum hoses. Correct as necessary.
	PCV system clogged or malfunctioning.	Check PCV system. Clean or replace as necessary.
	Loose carburetor, EGR or intake manifold bolts and/or leaking gaskets.	Torque carburetor to manifold bolts to 15 ft. lbs. (L-6), 10 ft. lbs. (All exc. L-6). Using a pressure oil can, spray light oil or kerosene around manifold to head mounting surface and carburetor base. If engine RPM changes, tighten or replace the carburetor or manifold gaskets as necessary. Check EGR mounting bolt torque.
	Low or erratic fuel pump pressure.	Check fuel delivery and pressure.
	Contaminated fuel.	Check for contaminants in fuel. Clean system if necessary.
	Fuel filter plugged.	Check and replace as necessary.
	Float level too low.	Check and reset float level to specification.
	Malfunctioning float and/or needle and seat.	Check operation of system. Repair or replace as necessary.
	Power piston stuck or binding.	Check for free movement of power piston(s). Clean and correct as necessary.
	Fuel jets or passages plugged or restricted.	Clean and blow out with compressed air.
	Ignition system malfunction.	Check ignition timing. Correct as necessary.
	Exhaust system.	Check for restrictions. Correct as necessary.

(NOTE: EGR system diagnosis should also be performed.)

CARBURETOR DIAGNOSIS—Continued

Fuel Economy Complaints	Customer driving habits.	Run mileage test with customer driving if possible. Make sure car has 2000-3000 miles for the "break-in" period.
	Loose, broken or improperly routed vacuum hoses.	Check condition of all vacuum hose routings. Correct as necessary.
	Engine needs complete tune-up.	Check engine compression, examine spark plugs; if fouled or improperly gapped, clean and regap or replace. Check ignition wire condition and check and reset ignition timing. Replace air cleaner element if dirty. Check for restricted exhaust system and intake manifold for leakage. Check carburetor mounting bolt torque. Check vacuum and mechanical advance.
	Fuel leaks.	Check fuel tank, fuel lines and fuel pump for any fuel leakage.
	High fuel level in carburetor.	Check fuel inlet needle and seat for proper seal. Test, using suction from a vacuum source. If needle is leaking, replace.
		Check for loaded float. Reset float level to specification.
		If excessive foreign material is present in the carburetor bowl, the carburetor should be cleaned.
	Power system in carburetor not functioning properly. Power piston(s) sticking or metering rods out of jets.	Remove air horn and check for free movement of power piston(s). Clean and correct as necessary.
	Choke system.	Check choke heated air tubes for routing and/or plugging which would restrict hot air flow to choke housing. Check choke linkage for binding. Clean or repair as required. Check adjustment of thermostatic coil. Readjust to specification as required.
	Plugged air cleaner element.	Replace element.
	Exhaust system.	Check for restrictions. Correct as required.
	Low tire pressure or incorrect tire size.	Inflate tires to specifications and use correct size tires.
	Transmission malfunction.	Refer to transmission diagnosis.

THERMOSTATIC AIR CLEANER SYSTEM DIAGNOSIS

CONDITION	POSSIBLE CAUSE	CORRECTION
Hesitation, sag, and stalling during cold start.	Failure of air cleaner in cold air mode.	Inspect and correct as required all parts connected with the air cleaner system.
Fuel Economy Complaints	Failure of air cleaner in hot air mode.	Inspect and correct as required all parts connected with the air cleaner.

EXHAUST GAS RECIRCULATION SYSTEM DIAGNOSIS

CONDITION	POSSIBLE CAUSE	CORRECTION
Engine idles abnormally rough and/or stalls	EGR valve vacuum hoses misrouted.	Check EGR valve vacuum hose routing. Correct as required.
	Leaking EGR valve.	Check EGR valve for correct operation.
	Incorrect idle speed.	Set idle RPM per engine label specification. Remove EGR vacuum hose from valve and observe effect on engine RPM. If speed is affected, reset RPM to specification and reconnect hose.
	EGR valve gasket failed or loose EGR attaching bolts.	Check EGR attaching bolts for tightness. Tighten as required. If not loose, remove EGR valve and inspect gasket. Replace as required.
	EGR thermal control valve and/or EGR-TVS.	Check vacuum into valve from carburetor EGR port with engine at normal operating temperature and at curb idle speed. Then check the vacuum out of the EGR thermal control valve to EGR valve. If the two vacuum readings are not equal within + 1/2 in. Hg., then proceed to EGR vacuum control diagnosis. See Section 6C.
	Improper vacuum to EGR valve at idle.	Check vacuum from carburetor EGR port with engine at stabilized operating temperature and at curb idle speed. If vacuum is more than 1.0 in. Hg., refer to carburetor idle diagnosis.
Cars with a back pressure transducer valve.	Improper carburetor vacuum to exhaust back pressure transducer at idle.	Check vacuum from carburetor to exhaust back pressure transducer with engine at stabilized operating temperature and at curb idle speed. If vacuum is more than 1.0 in. Hg., refer to carburetor idle diagnosis.
Engine runs rough on light throttle acceleration, poor part load performance and poor fuel economy	EGR valve vacuum hose misrouted.	Check EGR valve vacuum hose routing. Correct as required.
	Failed EGR vacuum control valve.	Same as listing in "Engine Idles Rough" condition.
	EGR flow unbalanced due to deposit accumulation in EGR passages or under carburetor.	Clean EGR passages of all deposits.
	Sticky or binding EGR valve.	Remove EGR valve and inspect. Clean or replace as required.
Cars with a back pressure transducer valve (No.'s 5, 6 and 7)	Failed exhaust back pressure transducer valve.	Functional check per procedure.
	Wrong or no EGR gaskets.	Check and correct as required.
	Exhaust system restricted causing excessive back pressure.	Inspect exhaust system for restriction and replace or repair as required.

EXHAUST GAS RECIRCULATION SYSTEM DIAGNOSIS—Continued

CONDITION	POSSIBLE CAUSE	CORRECTION
Engine stalls on decelerations.	Restriction in EGR vacuum line.	Check EGR vacuum lines for kinks, bends, etc. Remove or replace hoses as required. Check EGR vacuum control valve function.
		Check EGR valve for excessive deposits causing sticky or binding operation. Clean or repair as required.
	Sticking or binding EGR valve.	Remove EGR valve and inspect clean or repair as required.
Part throttle engine detonation	Insufficient exhaust gas recirculation flow during part throttle accelerations.	Check EGR valve hose routing. Check EGR valve operation. Repair or replace as required. Check EGR thermal control valve and/or EGR–TVS as listed in "Engine Idles Rough" section. Replace valve as required. Check EGR passages and valve for excessive deposit Clean as required.
	Exhaust back pressure transducer failed.	Check function per service procedure. (Section 6C)

(NOTE: Detonation can be caused by several other engine variables. Perform ignition and carburetor related diagnosis.)

CONDITION	POSSIBLE CAUSE	CORRECTION
Engine starts but immediately stalls when cold	EGR valve hoses misrouted.	Check EGR valve hose routings.
	EGR system malfunctioning when engine is cold.	Perform check to determine if the EGR thermal control valve and/or EGR-TVS are operational. Replace as required.
Cars with a back pressure transducer valve.	Exhaust back pressure transducer failed.	Check function per service manual procedure.

(NOTE: Stalls after start can also be caused by carburetor problems. Refer to carburetor diagnosis section.)

CATALYTIC CONVERTER DIAGNOSIS

CONDITION	POSSIBLE CAUSE	CORRECTION
Exhaust system noisy	Exhaust pipe joints loose at catalytic converter.	Tighten clamps at joints.
	Catalytic converter ruptured.	Replace catalytic converter.
		Ignition system and AIR system (if used) should also be diagnosed and repairs made if necessary.

E.F.E.-E.G.R: THERMO VACUUM SWITCH DIAGNOSIS

Condition	Possible Causes	Correction
Rough idle or stall during warm-up.	1. No vacuum to EFE vacuum actuator with engine coolant temperature below 120°F. ±3°F. (49°C.±2°) for 231 California and 350 49 State. 90°F. (32°C.) 231 and 196 49 State.	1. Check vacuum source for vacuum of 8″ hg. or above. 2. Correct improper vacuum hose routing, leak in connecting system, or EFE vacuum actuator diaphragm. Replace if required.
		3. Failed EFE-EGR thermo vacuum switch. Replace.
	2. Vacuum to EGR valve below 120° ±3°F. (49°.±2°) for 231 California and 350 49 State. 90°F. (32°C.) 231 and 196 49 State.	1. Correct improper vacuum hose routing if necessary. 2. Failed EFE-EGR thermo vacuum switch. Replace.
Rough idle, lack of performance, surge after warm-up period.	1. Vacuum to EFE vacuum actuator with engine coolant temperature above 120°F. ±3°F. (49°C.±2°) for 231 California and 350 49 State. 90°F (32°C.) 231 and 196 49 State.	1. Correct improper vacuum hose routing. 2. Failed EFE-EGR thermo vacuum switch. Replace.
Improper EGR operation.	1. Vacuum to EGR valve with engine coolant temperature below 120°F. ±3°F. (49°C.±2°) for 231 California and 350 49 State. 90°F(32°C.) 231 and 196 49 state.	1. Correct improper vacuum hose routing if necessary. 2. Failed EFE-EGR thermo vacuum switch. Replace.

EARLY FUEL EVAPORATION (E.F.E.) SYSTEM DIAGNOSIS

Condition	Possible Cause	Correction
Poor Operation during warm-up such as — rough idle, stumble, etc.	1. No vacuum to EFE valve during warm-up period for cold start.	1. Check vacuum source for vacuum of 8" hg. or above. Repair improper vacuum hose routing, leak in connecting system, diaphragm, or EFE, TVS. Failed EFE, TVS. Replace.
	2. EFE valve linkage bent or binding.	1. Repair EFE valve linkage.
	3. EFE valve linkage disconnected.	1. Reconnect linkage.
	4. EFE valve shaft seized in bearing.	1. Replace EFE valve.
	5. EFE valve loose on shaft.	1. Replace EFE valve.
Poor Operation after warm-up rough idle -lack of high speed performance -surge, misses at all speeds	1. Failed EFE TVS -vacuum present at vacuum actuator.	1. Replace EFE, TVS.
	2. EFE valve asm. shaft seized in bearing.	1. Replace EFE valve.
	3. EFE valve to housing interference.	1. Repair EFE valve.
	4. Vacuum actuator linkage bent or binding.	1. Repair EFE valve linkage.
	5. EFE valve separated from shaft.	1. Repair EFE valve linkage.
Noisy EFE valve asm.	1. Linkage stop failed.	1. Repair linkage stop tab.
	2. No vacuum actuator linkage over travel.	1. Replace vacuum actuator.
	3. Valve loose on shaft.	1. Replace EFE valve.
	4. Shaft loose in bushing. or bushing loose in housing.	1. Replace EFE valve.

FUEL SYSTEM DIAGNOSIS

CONDITION	POSSIBLE CAUSE	CORRECTION
Car feels like it is running out of gas-surging occurs in mid-speed range	Plugged fuel filters.	Remove and replace filters.
	Faulty fuel pump.	Perform diagnostic tests on the fuel pump as described in Section 6B. Remove and replace fuel pump as required.
	Foreign material in fuel system or kinked fuel pipes or hoses.	Inspect pipes and hoses for kinks and bends, blow out to check for plugging. Remove and replace as required.
Engine starts but will not continue to run or will run but surges and back fires.	Faulty fuel pump.	Perform diagnostic tests on the fuel pump as described in Section 6B. Remove and replace fuel pump as required.
Engine will not start	Faulty fuel pump.	Perform diagnostic tests on the fuel pump as described in Section 6B. Remove and replace fuel pump as required.

EVAPORATION EMISSION CONTROL SYSTEM DIAGNOSIS

CONDITION	POSSIBLE CAUSE	CORRECTION
Fuel odor	Vapor leak from evap. system.	Inspect and correct as necessary fuel and evap. hoses and pipes, fuel sender sealing gasket, fuel cap.

POSITIVE CRANKCASE VENTILATION SYSTEM DIAGNOSIS

CONDITION	POSSIBLE CAUSE	CORRECTION
Rough idle	PCV valve stuck open.	Test - remove and replace PCV system as required.
Oil in air cleaner	PCV system plugged.	Test - remove and replace PCV system as required.
	Leak in closed ventilation system.	Check and correct as necessary, for leaks to atmosphere of the closed crankcase ventilation system.
	Oil return holes in cylinder head restricted.	Remove valve covers, inspect and clean as required.
	Valve cover oil baffle restricted.	Remove, inspect and repair baffle as required.

(NOTE: See P.C.V. Diagnosis Chart using CT-3 Tester, Section 6C.)

AIR INJECTOR REACTOR SYSTEM (A.I.R.) DIAGNOSIS V-8 ENGINE

(NOTE: The AIR system is not completely noiseless. Under normal conditions, noise rises in pitch as engine speed increases. To determine if excessive noise is the fault of the air injection system, disconnect the drive belt and operate the engine. If noise now does not exist, proceed with diagnosis.

CONDITION	POSSIBLE CAUSE	CORRECTION
Excessive belt noise	Loose belt	Tighten to spec.
	Seized pump	Replace pump.
Excessive pump noise. Chirping	Insufficient break-in	Run car 10-15 miles at turnpike speeds – recheck.
Excessive pump noise, chirping, rumbling, or knocking	Leak in hose	Locate source of leak using soap solution and correct.
	Loose hose	Reassemble and replace or tighten hose clamp.
	Hose touching other engine parts.	Adjust hose position.
	Diverter valve inoperative	Replace diverter valve.
	Check valve inoperative	Replace check valve.
	Pump mounting fasteners loose	Tighten mounting screws as specified.
	Pump failure	Replace pump.
No air supply (accelerate engine to 1500 rpm and observe air flow from hoses. If the flow increases as the rpm's increase, the pump is functioning normally. If not, check possible cause.	Loose drive belt	Tighten to specs.
	Leaks in supply hose	Locate leak and repair or replace as required.
	Leak at fitting(s)	Tighten or replace clamps.
	Diverter valve leaking	If air is expelled through diverter muffler with engine at idle, replace diverter valve.
	Diverter valve inoperative	Usually accompanied by backfire during deceleration. Replace diverter valve.
	Check valve inoperative	Blow through hose toward air manifold. If air passes, function is normal. If air can be sucked from manifold, replace check valve.
	Pump pressure relief plug leaking or damaged.	Replace pressure relief plug.
Centrifugal filter fan damaged or broken.	Mechanical damage	Replace centrifugal filter fan.
Poor idle or driveability.	A defective AIR pump cannot cause poor idle or driveability.	Do NOT replace AIR pump.

AIR INJECTOR REACTOR SYSTEM (A.I.R.) DIAGNOSIS V-6 ENGINE

Condition	Possible Cause	Correction
No air supply(accelerate engine to 1500 rpm and observe air flow from hoses. If the flow increases as the rpm's increase, the pump is functioning normally. If not, check possible cause.)	1. Loose drive belt. 2. Leaks in supply hose. 3. Leak at fittings. 4. Air expelled through by-pass valve. 4a. Connect a vacuum line directly from engine manifold vacuum to by-pass valve. 4b. Connect vacuum line from engine manifold vacuum source to by-pass valve through vacuum differential valve directly, by passing the differential vacuum delay and separator valve. 5. Check valve inoperative. 6. Pump failure.	1. Tighten to specifications. 2. Locate leak and repair. 3. Tighten or replace clamps. 4a. If this corrects the problem go to step b. If not, replace air by-pass valve. 4b. If this corrects the problem, check differential vacuum delay and separator valve and vacuum source line for plugging. Replace as required. If it doesn't, replace vacuum differential valve. 5. Disconnect hose and blow through hose toward check valve. If air passes, function is normal. If air can be sucked from check valve, replace check valve. 6. Replace pump.
Excessive pump noise, chirping, rumbling, knocking, loss of engine performance.	1. Leak in hose. 2. Loose hose. 3. Hose touching other engine parts. 4. Vacuum differential valve inoperative. 5. By-pass valve inoperative. 6. Pump mounting fasteners loose. 7. Pump failure. 8. Check valve inoperative. 9. Loose pump outlet elbow on rear cover of pump. Turbo engines.	1. Locate source of leak using soap solution and correct. 2. Reassemble and replace or tighten hose clamp. 3. Adjust hose position. 4. Replace vacuum differential valve. 5. Replace by-pass valve. 6. Tighten mounting screws as specified. 7. Replace pump. 8. Replace check valve. 9. Torque bolts to 16 N·m (12 lb. ft.)
Excessive belt noise.	1. Loose belt. 2. Seized pump.	1. Tighten to spec. 2. Replace pump.
Excessive pump noise. Chirping.	1. Insufficient break-in 2. Loose pump outlet elbow on rear cover of pump. Turbo engines.	1. Run vehicle 10-15 miles at interstate speeds and recheck. 2. Torque bolts to 16 N·m (12 lb. ft.)
Poor idle or driveability.	1. A defective AIR system cannot cause poor idle or driveability.	1. Do not replace AIR system.

POSITIVE CRANKCASE VENTILATION SYSTEM DIAGNOSIS—USING CT-3 TESTER

WINDOW READING	PROBABLE TROUBLE	CORRECTION
GREEN	System Satisfactory Vent valve partially plugged. Blow-by close to capacity of valve	 Check valve
YELLOW	Tester hose kinked or blocked Crankcase not sealed properly Tester "selector knob" set incorrectly Vent-valve partially plugged Slight kink in CT-2 tester hose	Reposition or clean hose Check tester plugs and other seal-off points Check setting Check vent valve Reposition tester hose
YELLOW-GREEN	Slight engine blow-by Crankcase not sealed properly Tester "selector knob" set incorrectly Vent valve partially or fully plugged	Check vent valve Check tester plugs and other seal-off points Check setting Check vent valve
RED-YELLOW	Engine blow-by exceeds valve capacity Rubber vent hose collapsed or plugged	Engine overhaul indicated Clean or replace hose
RED	Vent valve plugged Vent valve stuck at engine off position Rubber vent hose collapsed or plugged Extreme engine blow-by	Check vent valve Check vent valve Replace hose Engine requires major overhaul

CHOKE AIR MODULATOR SYSTEM DIAGNOSIS

Condition	Possible Cause	Correction
Poor driveability after warm-up (choke not releasing).	1. Choke air modulator system plugged.	1. Locate restriction and correct as necessary.
Poor driveability during warm-up (choke releasing too soon).	1. Modulator valve stuck in open position.	1. Replace valve.

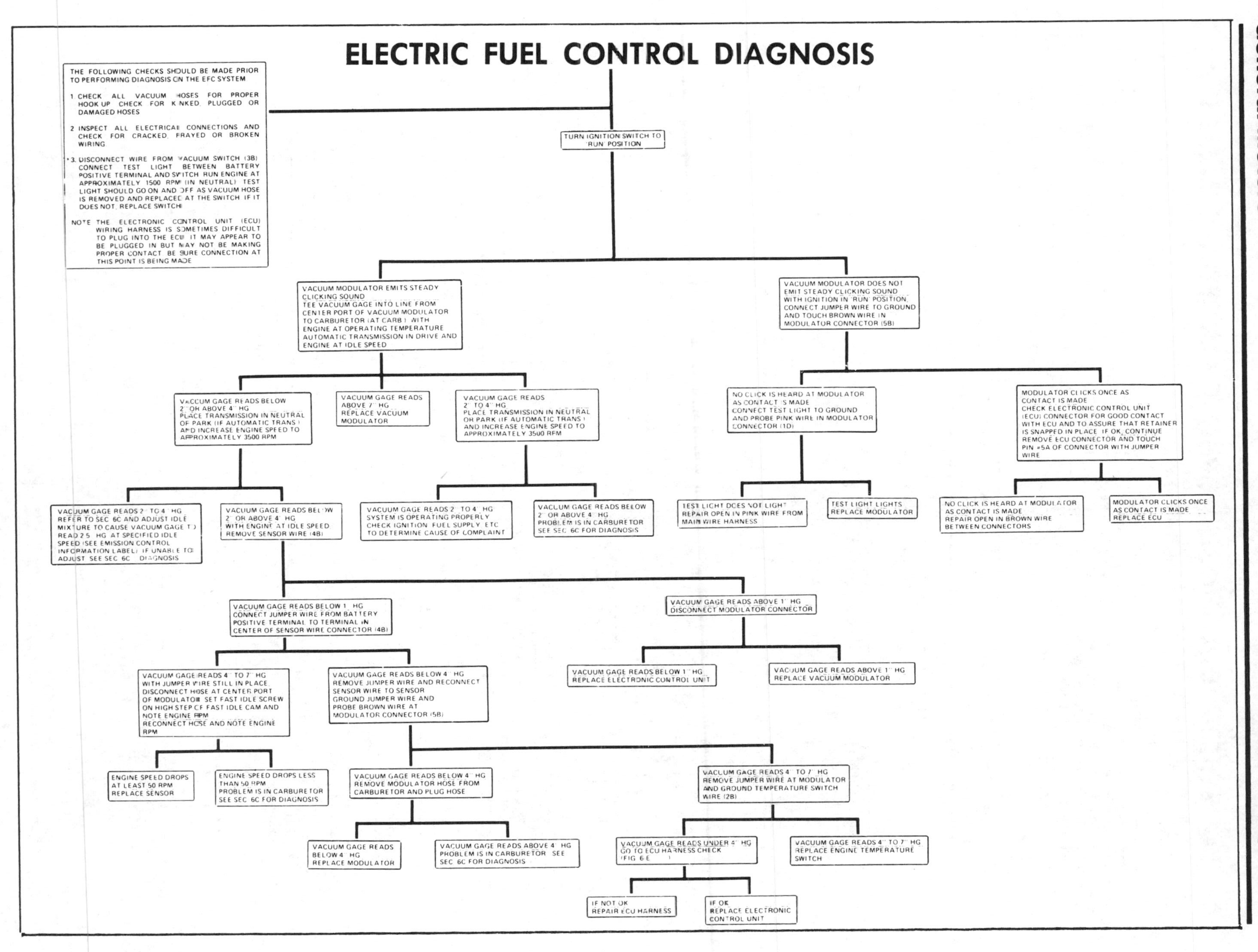
ELECTRIC FUEL CONTROL DIAGNOSIS
THE FOLLOWING CHECKS SHOULD BE MADE PRIOR TO PERFORMING DIAGNOSIS ON THE EFC SYSTEM
1 CHECK ALL VACUUM HOSES FOR PROPER HOOK UP CHECK FOR KINKED, PLUGGED OR DAMAGED HOSES
2 INSPECT ALL ELECTRICAL CONNECTIONS AND CHECK FOR CRACKED, FRAYED OR BROKEN WIRING
*3 DISCONNECT WIRE FROM VACUUM SWITCH (3B) CONNECT TEST LIGHT BETWEEN BATTERY POSITIVE TERMINAL AND SWITCH RUN ENGINE AT APPROXIMATELY 1500 RPM (IN NEUTRAL) TEST LIGHT SHOULD GO ON AND OFF AS VACUUM HOSE IS REMOVED AND REPLACED AT THE SWITCH IF IT DOES NOT, REPLACE SWITCH
NOTE THE ELECTRONIC CONTROL UNIT (ECU) WIRING HARNESS IS SOMETIMES DIFFICULT TO PLUG INTO THE ECU IT MAY APPEAR TO BE PLUGGED IN BUT MAY NOT BE MAKING PROPER CONTACT BE SURE CONNECTION AT THIS POINT IS BEING MADE
TURN IGNITION SWITCH TO 'RUN' POSITION
VACUUM MODULATOR EMITS STEADY CLICKING SOUND TEE VACUUM GAGE INTO LINE FROM CENTER PORT OF VACUUM MODULATOR TO CARBURETOR (AT CARB) WITH ENGINE AT OPERATING TEMPERATURE AUTOMATIC TRANSMISSION IN DRIVE AND ENGINE AT IDLE SPEED
VACUUM MODULATOR DOES NOT EMIT STEADY CLICKING SOUND WITH IGNITION IN 'RUN' POSITION, CONNECT JUMPER WIRE TO GROUND AND TOUCH BROWN WIRE IN MODULATOR CONNECTOR (5B)
VACUUM GAGE READS BELOW 2" OR ABOVE 4" HG PLACE TRANSMISSION IN NEUTRAL OR PARK (IF AUTOMATIC TRANS) AND INCREASE ENGINE SPEED TO APPROXIMATELY 3500 RPM
VACUUM GAGE READS ABOVE 7" HG REPLACE VACUUM MODULATOR
VACUUM GAGE READS 2" TO 4" HG PLACE TRANSMISSION IN NEUTRAL OR PARK (IF AUTOMATIC TRANS) AND INCREASE ENGINE SPEED TO APPROXIMATELY 3500 RPM
NO CLICK IS HEARD AT MODULATOR AS CONTACT IS MADE CONNECT TEST LIGHT TO GROUND AND PROBE PINK WIRE IN MODULATOR CONNECTOR (1D)
MODULATOR CLICKS ONCE AS CONTACT IS MADE CHECK ELECTRONIC CONTROL UNIT (ECU) CONNECTOR FOR GOOD CONTACT WITH ECU AND TO ASSURE THAT RETAINER IS SNAPPED IN PLACE IF OK, CONTINUE REMOVE ECU CONNECTOR AND TOUCH PIN #5A OF CONNECTOR WITH JUMPER WIRE
VACUUM GAGE READS 2" TO 4" HG REFER TO SEC 6C AND ADJUST IDLE MIXTURE TO CAUSE VACUUM GAGE TO READ 2.5" HG AT SPECIFIED IDLE SPEED (SEE EMISSION CONTROL INFORMATION LABEL) IF UNABLE TO ADJUST SEE SEC 6C DIAGNOSIS
VACUUM GAGE READS BELOW 2" OR ABOVE 4" HG WITH ENGINE AT IDLE SPEED REMOVE SENSOR WIRE (4B)
VACUUM GAGE READS 2" TO 4" HG SYSTEM IS OPERATING PROPERLY CHECK IGNITION, FUEL SUPPLY, ETC TO DETERMINE CAUSE OF COMPLAINT
VACUUM GAGE READS BELOW 2" OR ABOVE 4" HG PROBLEM IS IN CARBURETOR SEE SEC 6C FOR DIAGNOSIS
TEST LIGHT DOES NOT LIGHT REPAIR OPEN IN PINK WIRE FROM MAIN WIRE HARNESS
TEST LIGHT LIGHTS REPLACE MODULATOR
NO CLICK IS HEARD AT MODULATOR AS CONTACT IS MADE REPAIR OPEN IN BROWN WIRE BETWEEN CONNECTORS
MODULATOR CLICKS ONCE AS CONTACT IS MADE REPLACE ECU
VACUUM GAGE READS BELOW 1" HG CONNECT JUMPER WIRE FROM BATTERY POSITIVE TERMINAL TO TERMINAL IN CENTER OF SENSOR WIRE CONNECTOR (4B)
VACUUM GAGE READS ABOVE 1" HG DISCONNECT MODULATOR CONNECTOR
VACUUM GAGE READS 4" TO 7" HG WITH JUMPER WIRE STILL IN PLACE DISCONNECT HOSE AT CENTER PORT OF MODULATOR SET FAST IDLE SCREW ON HIGH STEP OF FAST IDLE CAM AND NOTE ENGINE RPM RECONNECT HOSE AND NOTE ENGINE RPM
VACUUM GAGE READS BELOW 4" HG REMOVE JUMPER WIRE AND RECONNECT SENSOR WIRE TO SENSOR GROUND JUMPER WIRE AND PROBE BROWN WIRE AT MODULATOR CONNECTOR (5B)
VACUUM GAGE READS BELOW 1" HG REPLACE ELECTRONIC CONTROL UNIT
VACUUM GAGE READS ABOVE 1" HG REPLACE VACUUM MODULATOR
ENGINE SPEED DROPS AT LEAST 50 RPM REPLACE SENSOR
ENGINE SPEED DROPS LESS THAN 50 RPM PROBLEM IS IN CARBURETOR SEE SEC 6C FOR DIAGNOSIS
VACUUM GAGE READS BELOW 4" HG REMOVE MODULATOR HOSE FROM CARBURETOR AND PLUG HOSE
VACUUM GAGE READS 4" TO 7" HG REMOVE JUMPER WIRE AT MODULATOR AND GROUND TEMPERATURE SWITCH WIRE (2B)
VACUUM GAGE READS BELOW 4" HG REPLACE MODULATOR
VACUUM GAGE READS ABOVE 4" HG PROBLEM IS IN CARBURETOR SEE SEC 6C FOR DIAGNOSIS
VACUUM GAGE READS UNDER 4" HG GO TO ECU HARNESS CHECK (FIG 6E)
VACUUM GAGE READS 4" TO 7" HG REPLACE ENGINE TEMPERATURE SWITCH
IF NOT OK REPAIR ECU HARNESS
IF OK REPLACE ELECTRONIC CONTROL UNIT

C-4 SYSTEM DIAGNOSIS TURBO CHARGED E.F.E.-E.G.R:

Each valve must be diagnosed separately

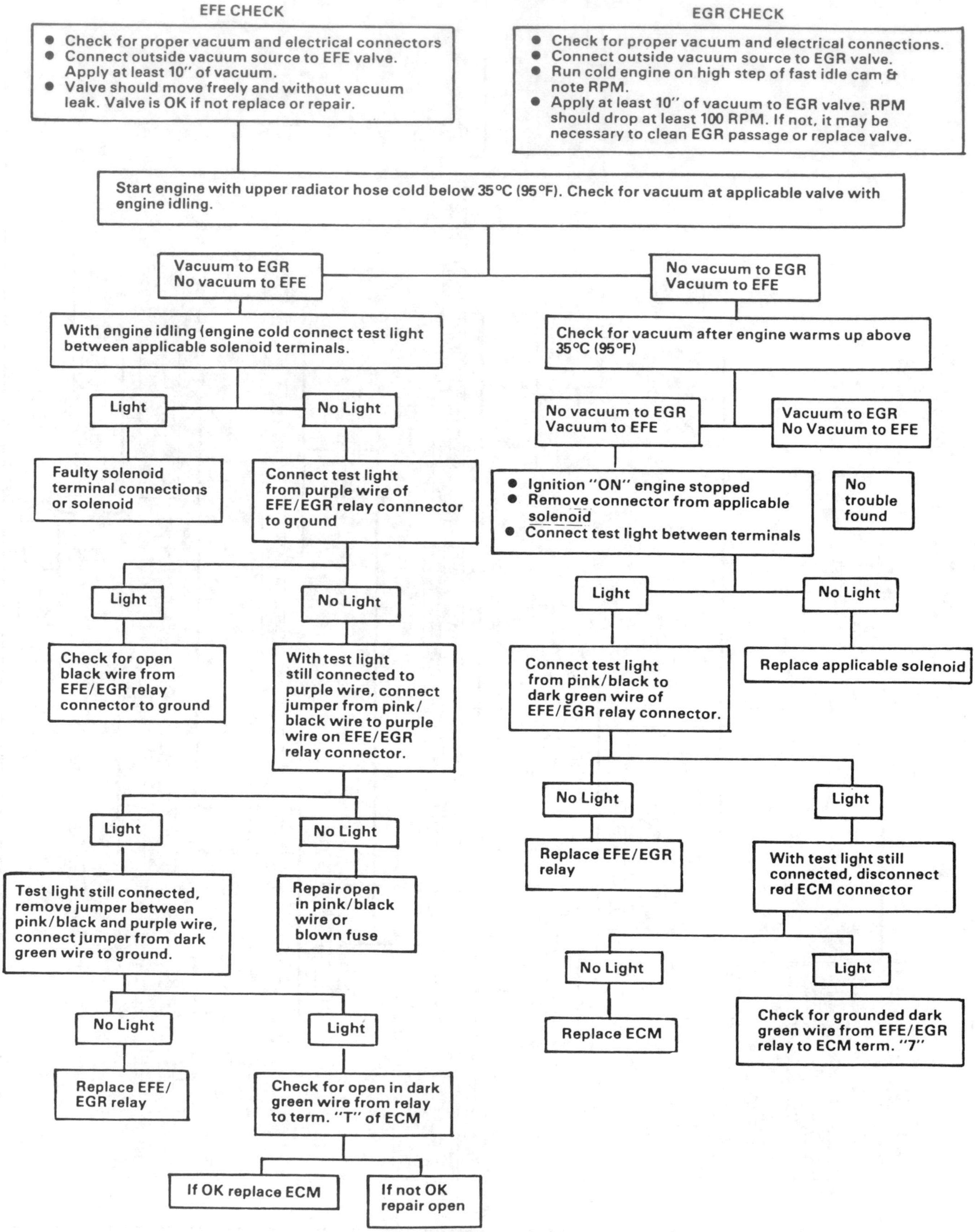

EFE CHECK

- Check for proper vacuum and electrical connectors
- Connect outside vacuum source to EFE valve. Apply at least 10" of vacuum.
- Valve should move freely and without vacuum leak. Valve is OK if not replace or repair.

EGR CHECK

- Check for proper vacuum and electrical connections.
- Connect outside vacuum source to EGR valve.
- Run cold engine on high step of fast idle cam & note RPM.
- Apply at least 10" of vacuum to EGR valve. RPM should drop at least 100 RPM. If not, it may be necessary to clean EGR passage or replace valve.

Start engine with upper radiator hose cold below 35°C (95°F). Check for vacuum at applicable valve with engine idling.

Vacuum to EGR
No vacuum to EFE

With engine idling (engine cold connect test light between applicable solenoid terminals.

Light

Faulty solenoid terminal connections or solenoid

No Light

Connect test light from purple wire of EFE/EGR relay connnector to ground

Light

Check for open black wire from EFE/EGR relay connector to ground

No Light

With test light still connected to purple wire, connect jumper from pink/black wire to purple wire on EFE/EGR relay connector.

Light

Test light still connected, remove jumper between pink/black and purple wire, connect jumper from dark green wire to ground.

No Light

Replace EFE/EGR relay

Light

Check for open in dark green wire from relay to term. "T" of ECM

If OK replace ECM

If not OK repair open

No Light

Repair open in pink/black wire or blown fuse

No vacuum to EGR
Vacuum to EFE

Check for vacuum after engine warms up above 35°C (95°F)

Vacuum to EGR
No Vacuum to EFE

No trouble found

No vacuum to EGR
Vacuum to EFE

- Ignition "ON" engine stopped
- Remove connector from applicable solenoid
- Connect test light between terminals

No Light

Replace applicable solenoid

Light

Connect test light from pink/black to dark green wire of EFE/EGR relay connector.

No Light

Replace EFE/EGR relay

Light

With test light still connected, disconnect red ECM connector

No Light

Replace ECM

Light

Check for grounded dark green wire from EFE/EGR relay to ECM term. "7"

EXHAUST GAS RECIRCULATION SYSTEM DIAGNOSIS GASOLINE ENGINES

Condition	Possible Cause	Correction
Engine idles abnormally rough and/or stalls.	EGR valve vacuum hoses misrouted.	Check EGR valve vacuum hose routing. Correct as required.
	Leaking EGR valve.	Check EGR valve for correct operation.
	EGR valve gasket failed or loose EGR attaching bolts.	Check EGR attaching bolts for tightness. Tighten as required. If not loose, remove EGR valve and inspect gasket. Replace as required.
	EGR thermal control valve and/or EGR-TVS.	Check vacuum into valve from carburetor EGR port with engine at normal operating temperature and at curb idle speed. Then check the vacuum out of the EGR thermal control valve to EGR valve. If the two vacuum readings are not equal within ± 1/2 in. Hg. (1.7 kPa), then proceed to EGR vacuum control diagnosis.
	Improper vacuum to EGR valve at idle.	Check vacuum from carburetor EGR port with engine at stabilized operating temperature and at curb idle speed. If vacuum is more than 1.0 in. Hg., refer to carburetor idle diagnosis.
Engine runs rough on light throttle acceleration and has poor part load performance.	EGR valve vacuum hose misrouted.	Check EGR valve vacuum hose routing. Correct as required.
	Check for loose valve.	Torque valve.
	Failed EGR vacuum control valve.	Same as listing in "Engine Idles Rough" condition.
	(TVS open below 130°F.)	Clean EGR passage of all deposits.
	Sticky or binding EGR valve.	Remove EGR valve and inspect. Replace as required.
	Wrong or no EGR gasket(s) and/or Spacer.	Check and correct as required. Install new gasket(s), install spacer (if used), torque attaching parts.
Engine stalls on decelerations.	Control valve blocked or air flow restricted.	Check internal control valve function per service procedure.
	Restriction in EGR vacuum line or valve vacuum signal tube.	Check EGR vacuum lines for kinks, bends, etc. Remove or replace hoses as required. Check EGR vacuum control valve function.
		Check EGR valve for excessive deposits causing sticky or binding operation. Replace valve.
	Sticking or binding EGR valve.	Remove EGR valve and replace valve.
Part throttle engine detonation.	Control valve blocked or air flow restricted.	Check internal control valve function per service procedure.
	Insufficient exhaust gas recirculation flow during part throttle accelerations.	Check EGR valve hose routing. Check EGR valve operation. Repair or replace as required. Check EGR thermal control valve and/or EGR-TVS as listed in "Engine Idles Rough" section. Replace valve as required. Check EGR passages and valve for excessive deposit. Clean as required.
(NOTICE: Non-Functioning EGR valve could contribute to part throttle detonation.)		
	Control valve blocked or flow restricted.	Check EGR per service procedure.
(NOTICE: Detonation can be caused by several other engine variables. Perform ignition and carburetor related diagnosis.)		
Engine starts but immediately stalls when cold.	EGR valve hoses misrouted.	Check EGR valve hose routings.
	EGR TVS system malfunctioning when engine is cold.	Perform check to determine if the EGR thermal control valve and/or EGR-TVS are operational. Replace as required.
(NOTICE: Stalls after start can also be caused by carburetor problems.)		

EXHAUST GAS RECIRCULATION SYSTEM DIAGNOSIS DIESEL ENGINES

CONDITION	POSSIBLE CAUSE	CORRECTION
EGR valve will not open (No noticable effect to driver	Binding or stuck EGR valve.	Replace EGR valve.
	No vacuum to EGR valve	Check EGR vacuum solenoid, EGR vacuum switch, vacuum regulator valve and vacuum pump for proper operation.
EGR valve will not close (Heavy smoke on acceleration)	Binding or stuck EGR valve (check with engine "OFF")	Replace EGR valve.
	Vacuum to EGR at W.O.T.	Check EGR vacuum solenoid (open when energized), EGR vacuum switch (N.O. below preset vacuum), and vacuum regulator valve.
EGR valve closes early (no noticable effect to driver)	Vacuum switch opening circuit above calibration (8 inches).	Replace vacuum switch.
	Vacuum regulator valve misadjusted.	Adjust vacuum regulator valve
EGR valve closes late (smoke on medium acceleration but not at W.O.T.	Vacuum switch opening circuit below calibration (8 inches)	Replace vacuum switch
	Vacuum regulator valve misadjusted.	Adjust vacuum regulator valve
Loss of engine performance and smokey exhaust	Exhaust restriction	Replace restricted part.
	Air cleaner filter element restricted	Replace air cleaner filter element.

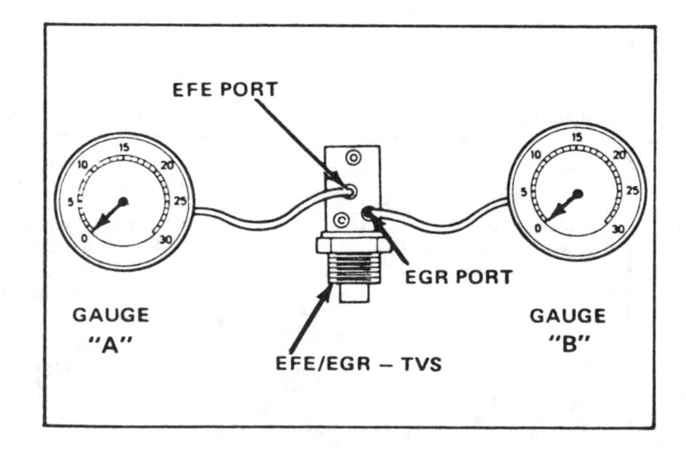

Vacuum gauge hook-up (© General Motors Corp.)

DIVERTER VALVE SYSTEM DIAGNOSIS

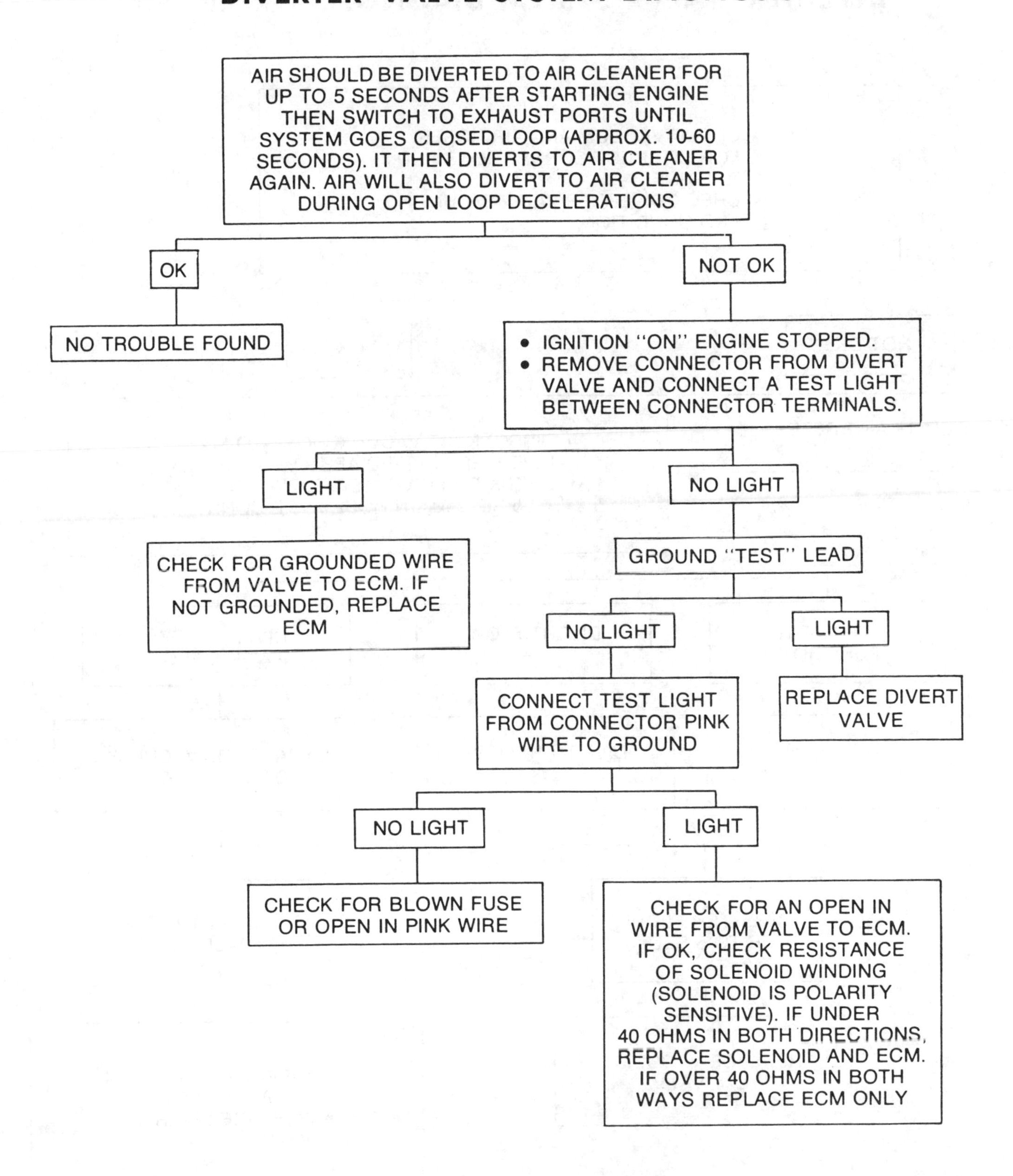

DIVERTER VALVE SYSTEM DIAGNOSIS—Continued

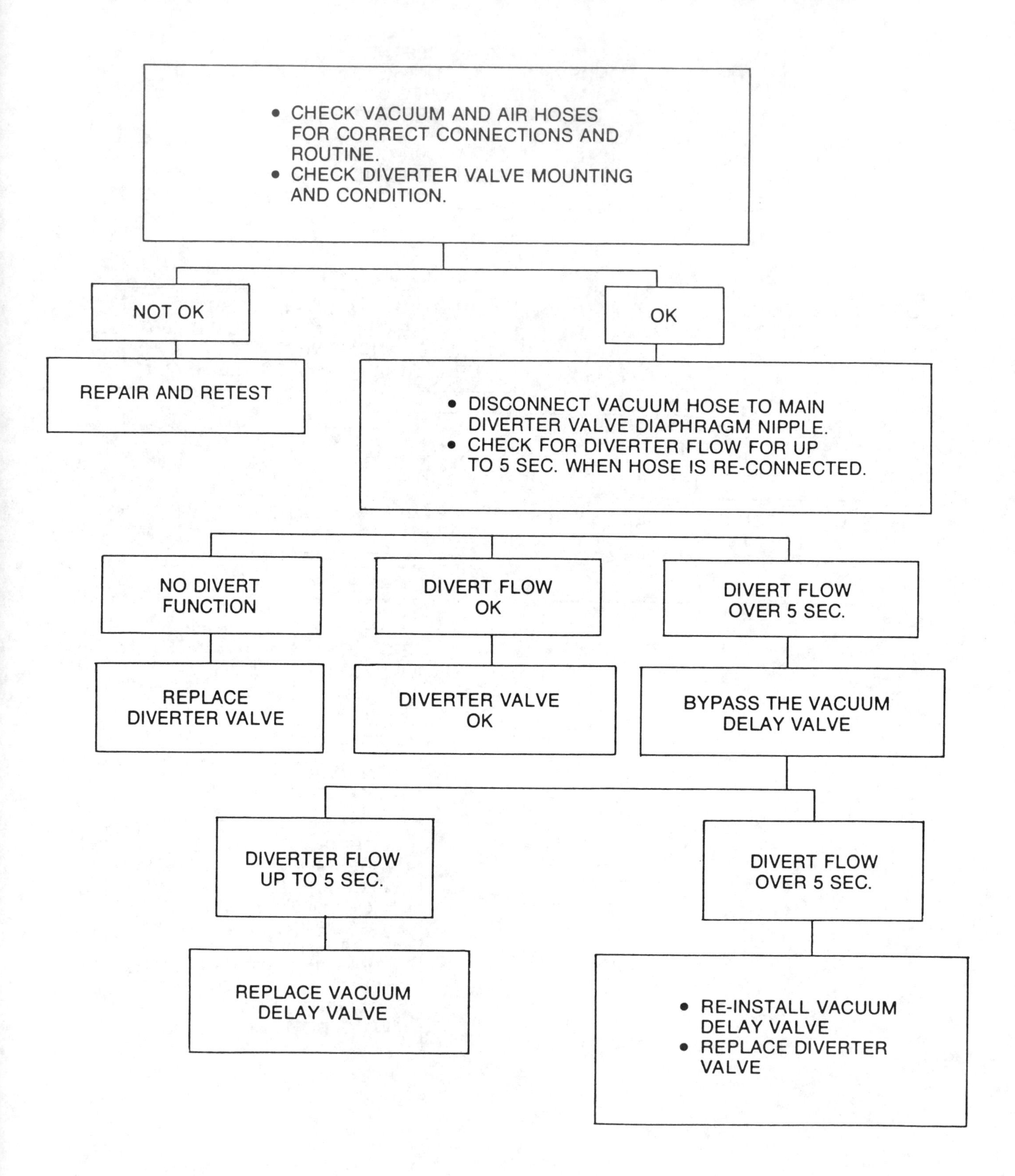

AIR MANAGEMENT SYSTEM DIAGNOSIS

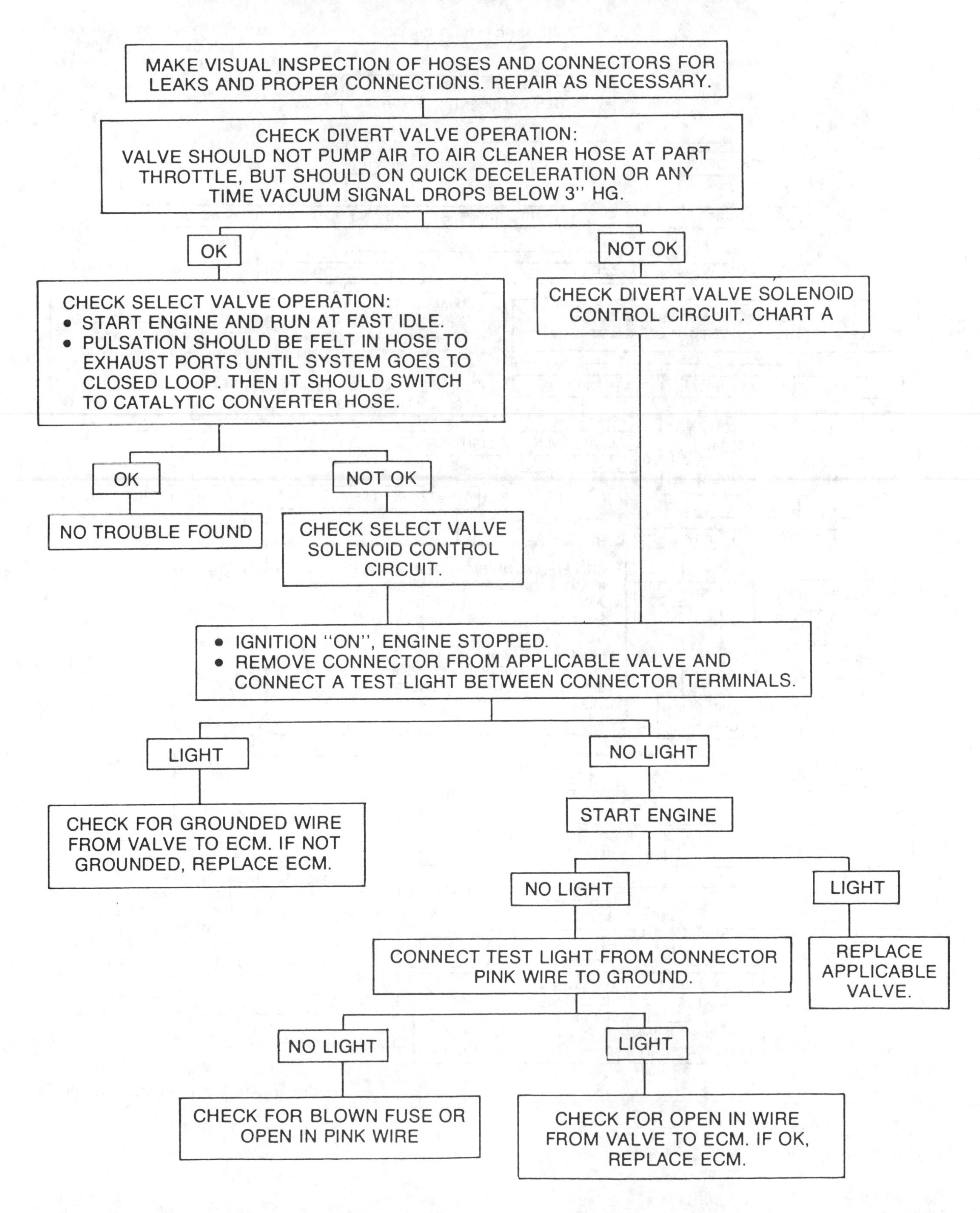

C-4 SYSTEM DIAGNOSIS

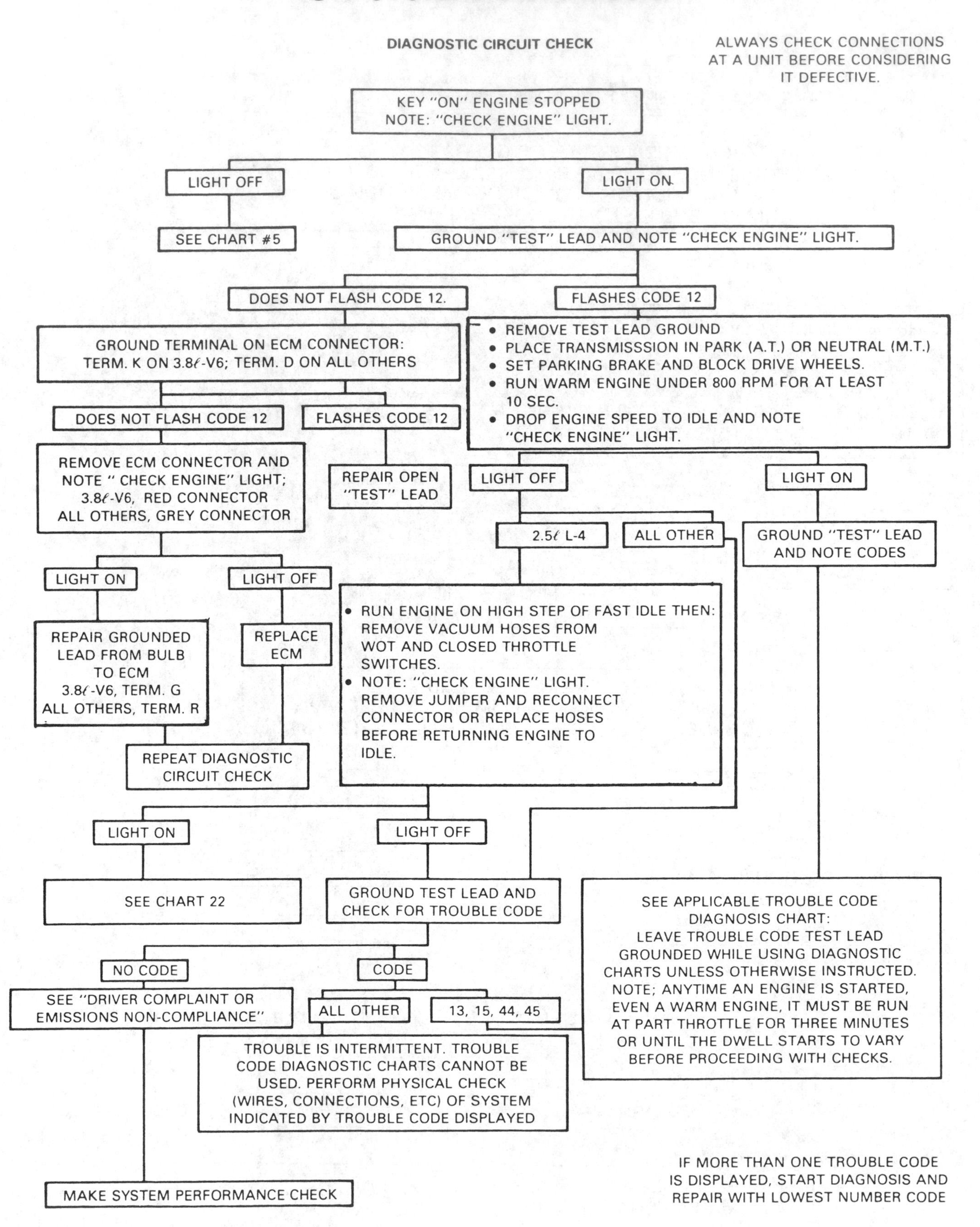

C-4 SYSTEM DIAGNOSIS—Continued

DRIVER COMPLAINT OR EMISSION NON-COMPLIANCE
ENGINE PERFORMANCE PROBLEM (ODOR, SURGE, FUEL ECONOMY. . .)

IF THE "CHECK ENGINE" LIGHT IS NOT ON, NORMAL CHECKS THAT WOULD BE PERFORMED ON CARS WITHOUT C-4 SHOULD BE DONE FIRST. IF GENERATOR OR COOLANT LIGHT IS ON WITH THE CHECK ENGINE LIGHT, THEY SHOULD BE DIAGNOSED FIRST. INSPECT FOR POOR CONNECTIONS AT COOLANT SENSOR, M/C SOLENOID, ETC., AND POOR OR LOOSE VACUUM HOSES AND CONNECTIONS. REPAIR AS NECESSARY.

ANYTIME AN ENGINE IS STARTED, EVEN A WARM ENGINE, IT MUST BE RUN AT PART THROTTLE FOR THREE (3) MINUTES OR UNTIL THE DWELL STARTS TO VARY BEFORE PROCEEDING WITH CHECKS.

- INTERMITTENT CHECK ENGINE LIGHT BUT NO TROUBLE CODE STORED. IF FOR ANY REASON ENGINE RPM DROPS BELOW 200, THE CHECK ENGINE LIGHT WILL COME ON UNTIL RPM EXCEEDS 200.

  CHECK FOR POOR CONNECTION IN CIRCUIT FROM:

  ALL BUT 3.8L V-6:

  DISTRIBUTOR TACH TERMINAL TO ECM TERMINAL "A", IGNITION "1" TO ECM, AND ECM TERMINAL "T" TO GROUND.

  CHECK TACH FILTER – SHOULD BE 14,000 TO 18,000 OHMS BETWEEN TERMINALS (WITH ONE END DISCONNECTED) AND AN OPEN CIRCUIT TO GROUND.

  3.8L V-6:

  DISTRIBUTOR MODULE TERMINAL "R" TO ECM TERMINAL "10", BAT. TO ECM TERMINAL "C" AND "R", AND ECM TERMINAL "U" TO GROUND.

  LOW BATTERY VOLTAGE (UNDER 9 VOLTS) AT ECM.

  LOSS OF LONG TERM MEMORY.

  MOMENTARILY GROUNDING DWELL LEAD WITH ENGINE IDLING SHOULD GIVE CODE 23 WHICH SHOULD BE RETAINED AFTER ENGINE IS STOPPED AND RESTARTED. IF VOLTAGE IS PRESENT AT THE LONG TERM MEMORY TERMINAL BUT THE CODE WAS NOT STORED, THE ECM IS DEFECTIVE.

  ELECTRICAL INTERFERENCE.

- COLD OPERATION COMPLAINT.

  SEE CHART 6 FOR COLD IDLE LEAN LIMIT SWITCH CHECK, IF APPLICABLE.

- ACCELERATION STUMBLE 2.5L L-4. SEE CHART 8.

- POOR MILEAGE, 5.7L V-8, SEE SECTION 6D.

- FULL THROTTLE PERFORMANCE COMPLAINT. SEE APPLICABLE CHART 4.

- ALL OTHER COMPLAINTS: MAKE SYSTEM PERFORMANCE CHECK ON WARM ENGINE. (UPPER RADIATOR HOSE HOT.)

THE SYSTEM PERFORMANCE CHECK SHOULD BE PERFORMED AFTER ANY REPAIRS TO THE C-4 SYSTEM HAVE BEEN MADE.

C-4 SYSTEM DIAGNOSIS—Continued

SYSTEM PERFORMANCE CHECK

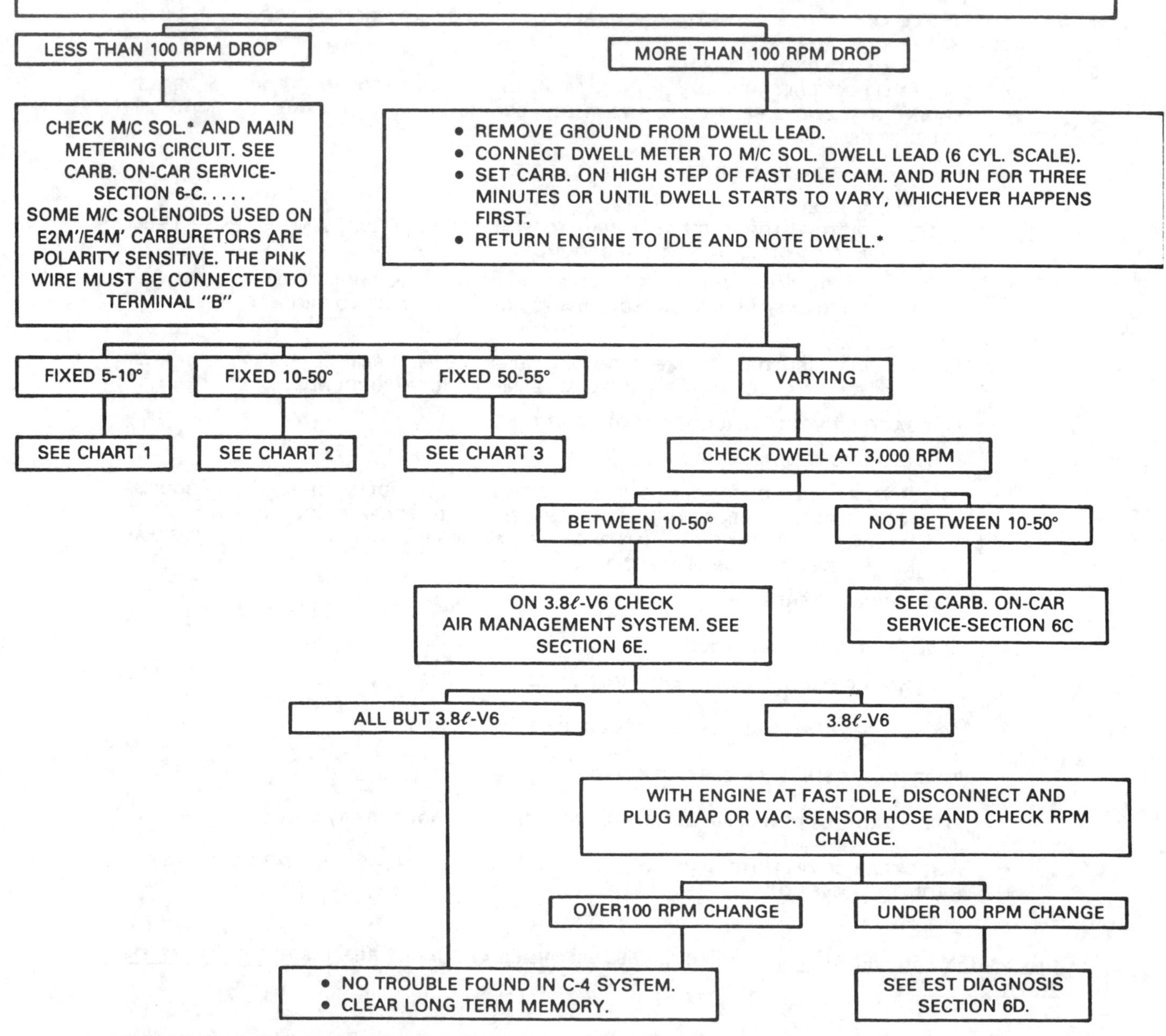

*OXYGEN SENSORS MAY COOL OFF AT IDLE AND THE DWELL CHANGE FROM VARYING TO FIXED BETWEEN 10-50°. IF THIS HAPPENS RUNNING THE ENGINE AT FAST IDLE WILL WARM IT UP AGAIN.

C-4 SYSTEM DIAGNOSIS—Continued

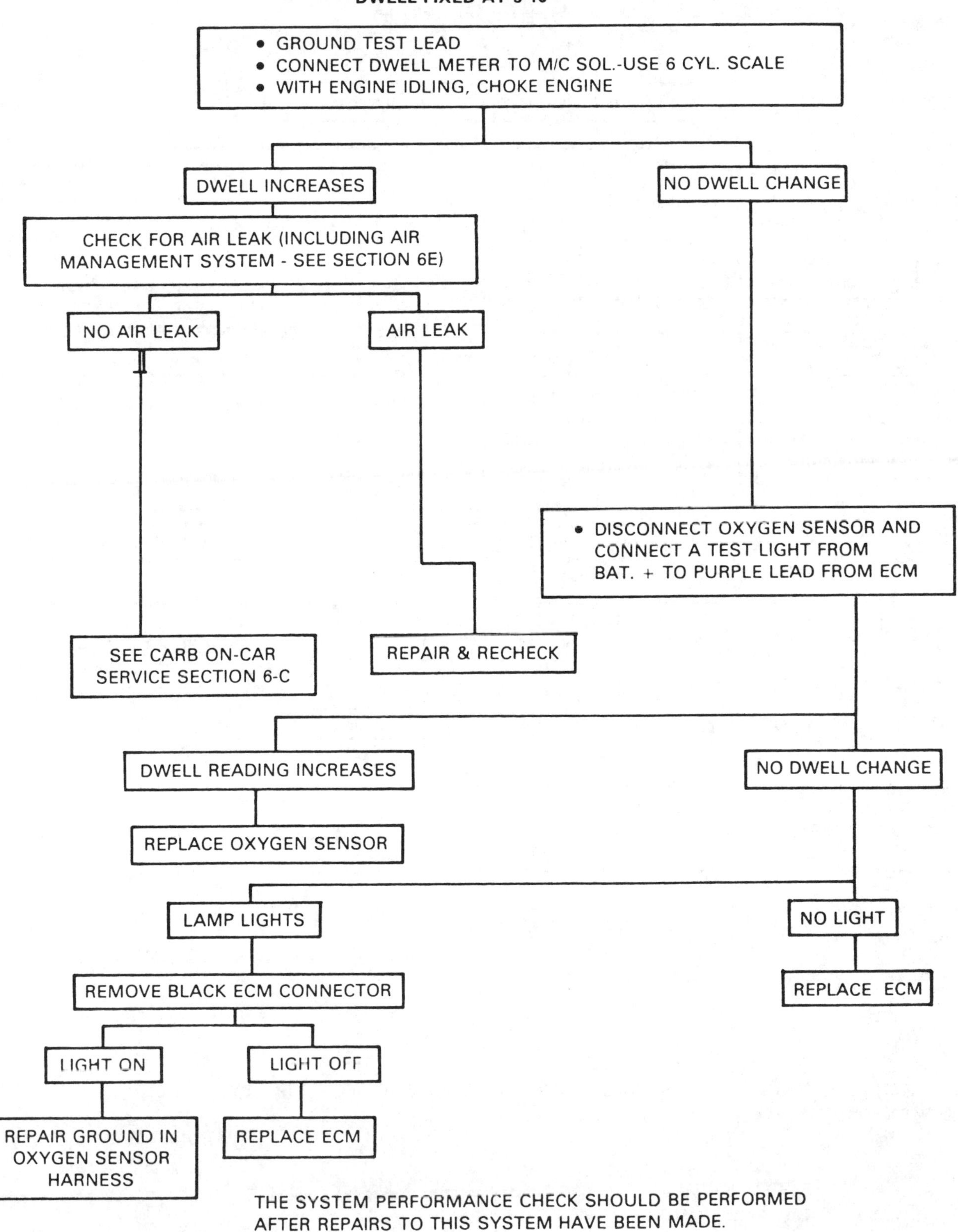

THE SYSTEM PERFORMANCE CHECK SHOULD BE PERFORMED AFTER REPAIRS TO THIS SYSTEM HAVE BEEN MADE.

C-4 SYSTEM DIAGNOSIS—Continued

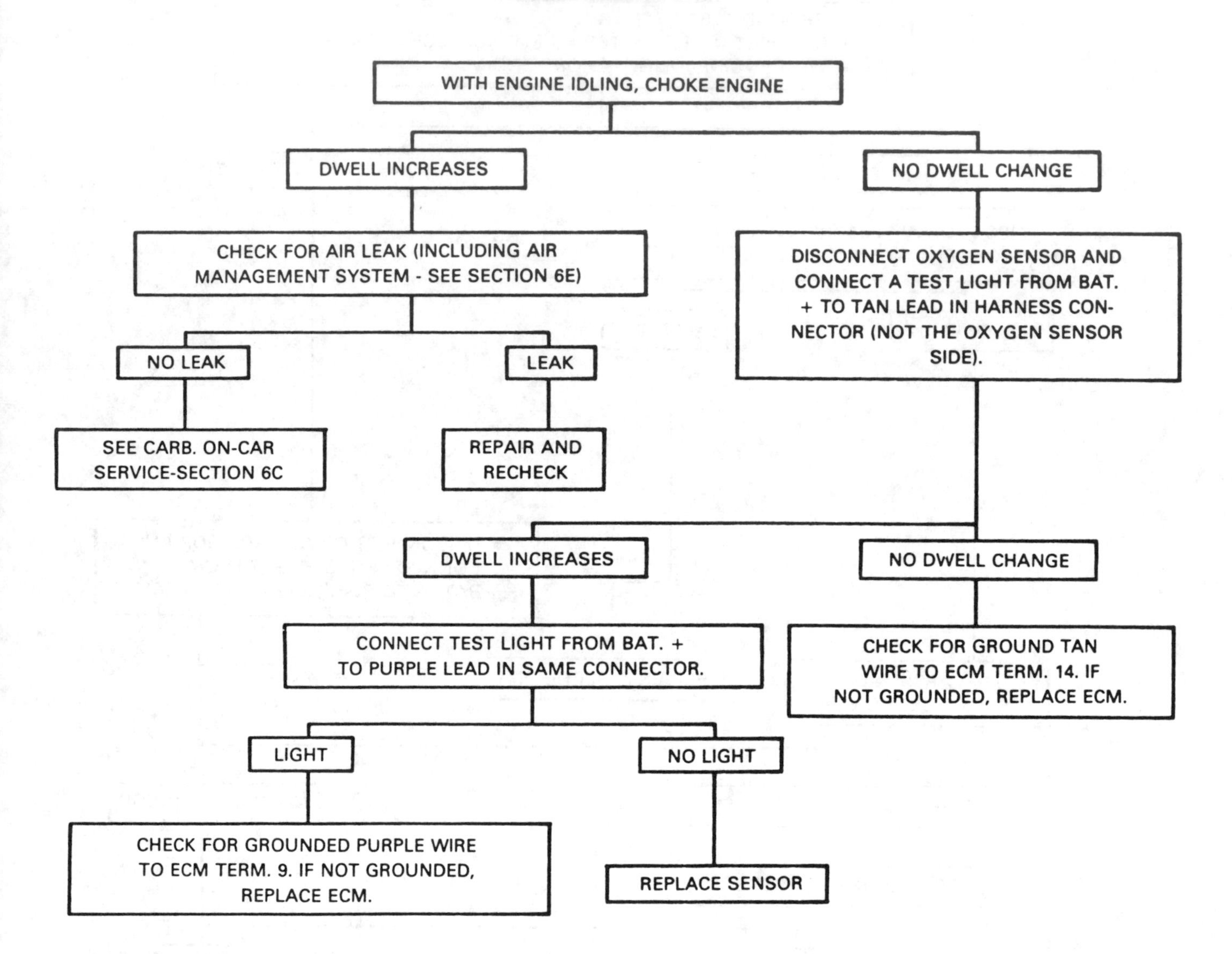

THE SYSTEM PERFORMANCE CHECK SHOULD BE PERFORMED AFTER REPAIRS TO THIS SYSTEM HAVE BEEN MADE.

C-4 SYSTEM DIAGNOSIS—Continued

CHART #2
DWELL FIXED BETWEEN 10-50°

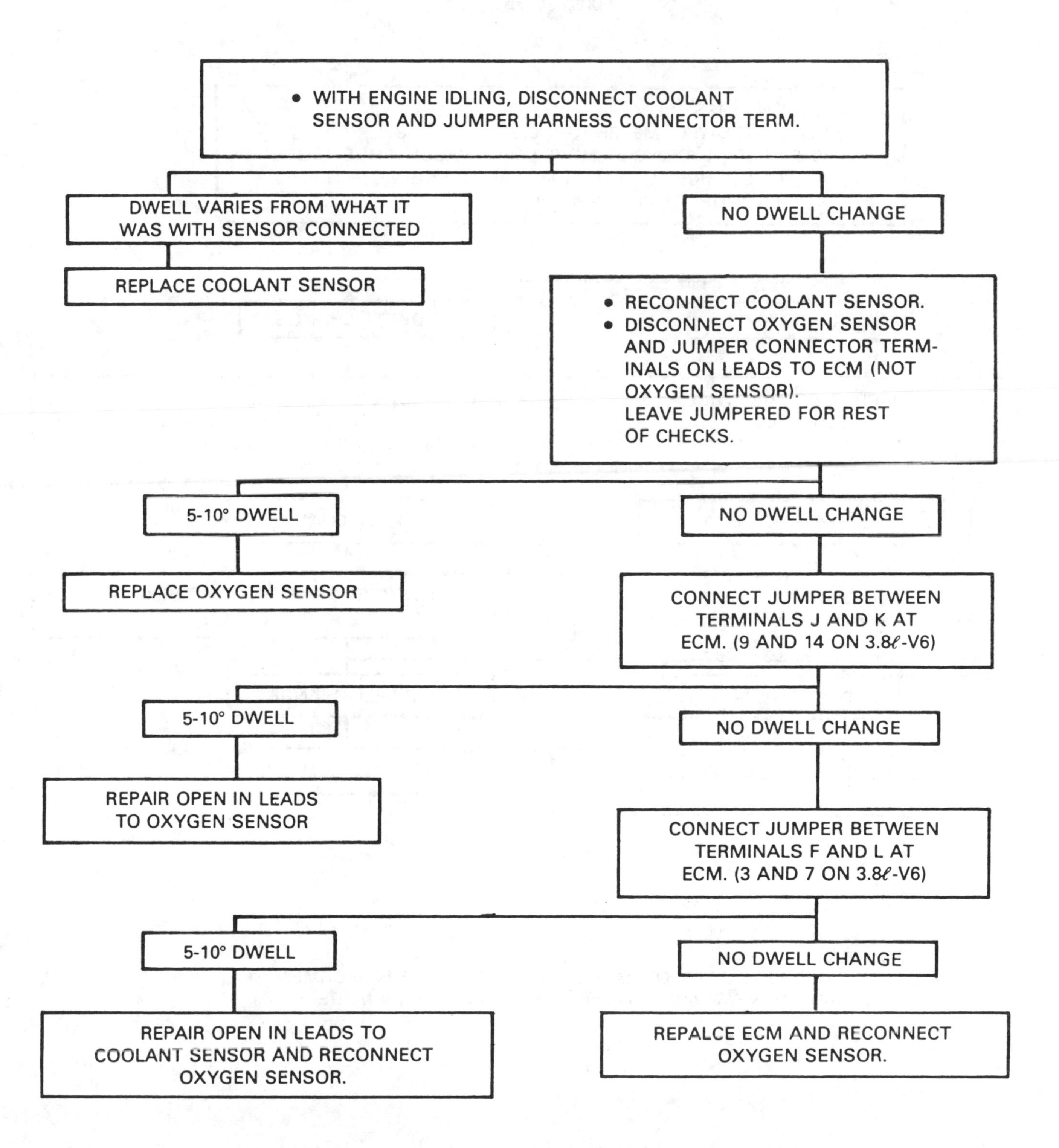

THE SYSTEM PERFORMANCE CHECK SHOULD BE PERFORMED AFTER ANY REPAIRS TO THIS SYSTEM HAVE BEEN MADE.

C-4 SYSTEM DIAGNOSIS—Continued

CHART #3
FIXED 50-55°

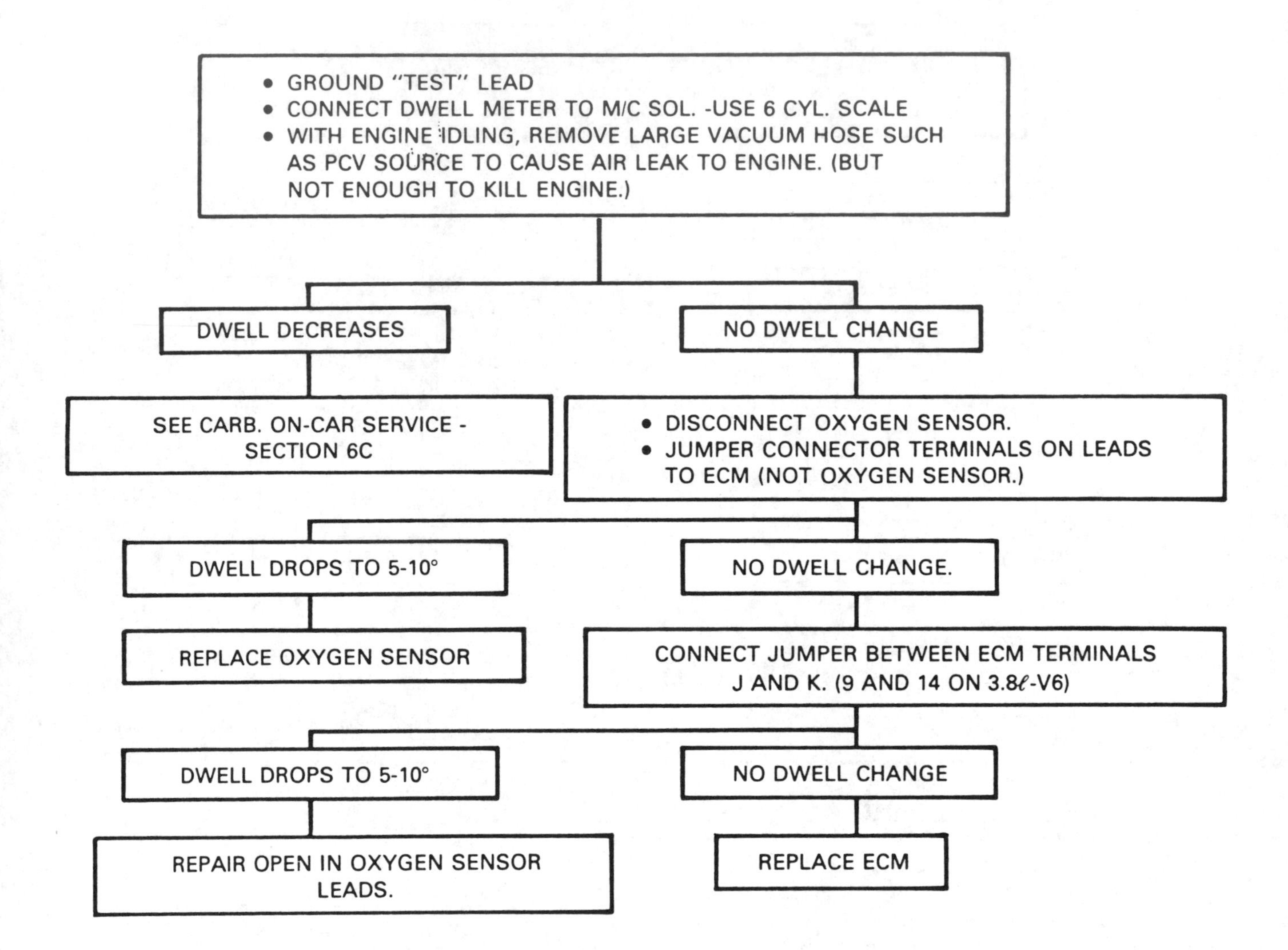

THE SYSTEM PERFORMANCE CHECK SHOULD BE PERFORMED AFTER ANY REPAIRS TO THIS SYSTEM HAVE BEEN MADE.

C-4 SYSTEM DIAGNOSIS—Continued

PRELIMINARY
CHART 4
2.5ℓ L4
W.O.T. ENRICHMENT CIRCUIT CHECK

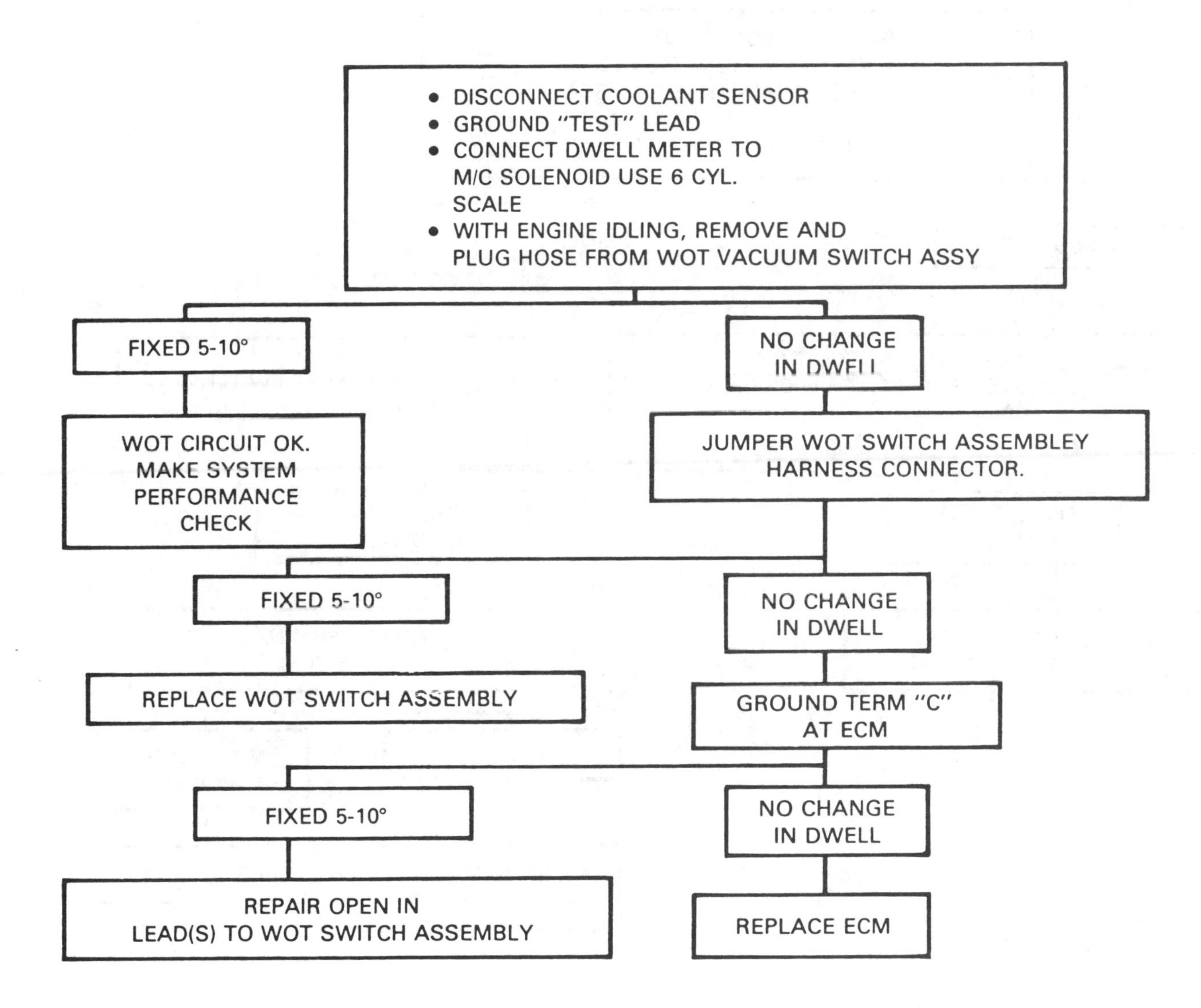

THE SYSTEM PERFORMANCE CHECK SHOULD BE PERFORMED AFTER ANY REPAIRS TO THIS SYSTEM HAVE BEEN MADE.

C-4 SYSTEM DIAGNOSIS—Continued

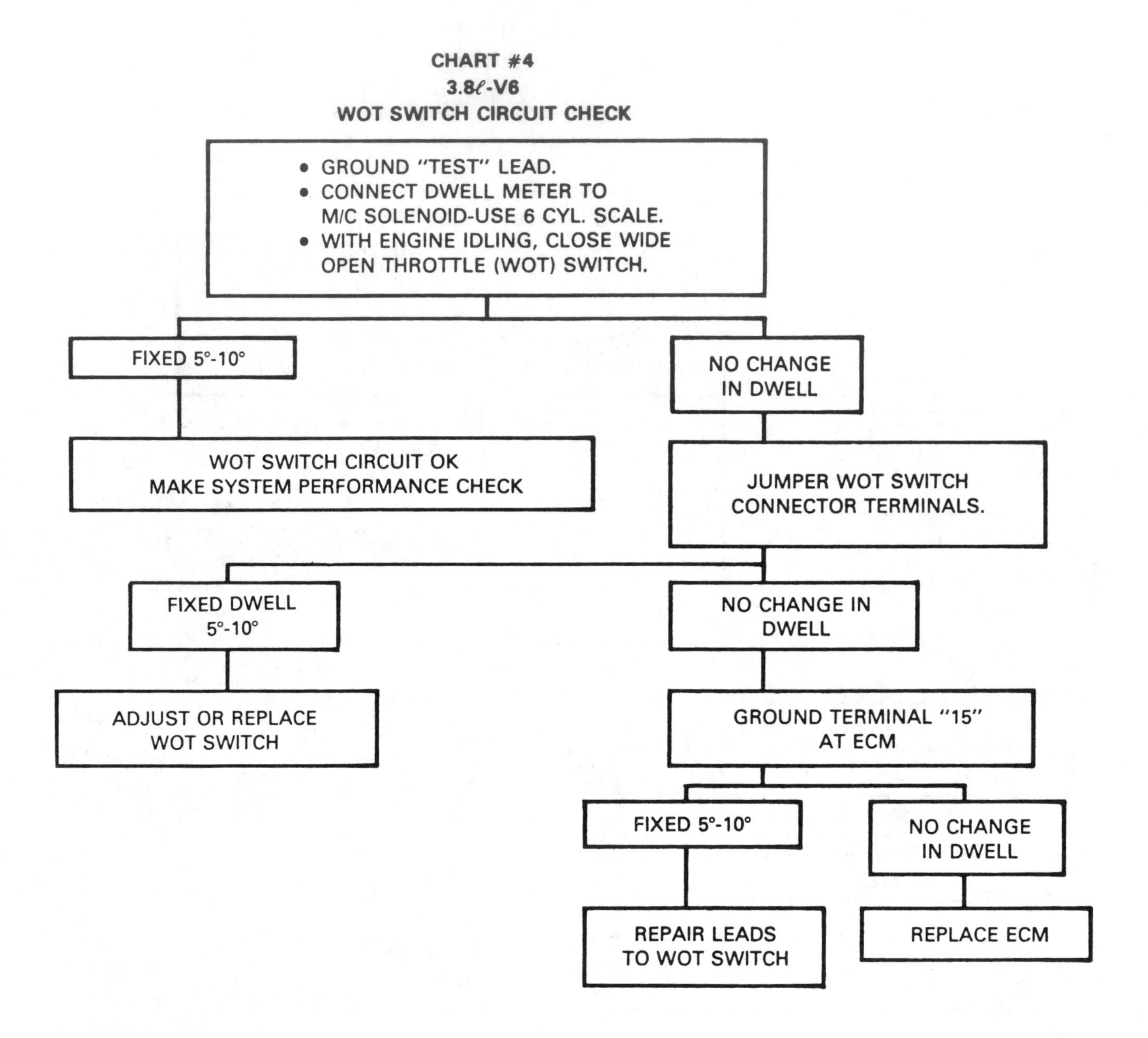

THE SYSTEM PERFORMANCE CHECK SHOULD BE PERFORMED AFTER ANY REPAIRS TO THIS SYSTEM HAVE BEEN MADE.

C-4 SYSTEM DIAGNOSIS—Continued

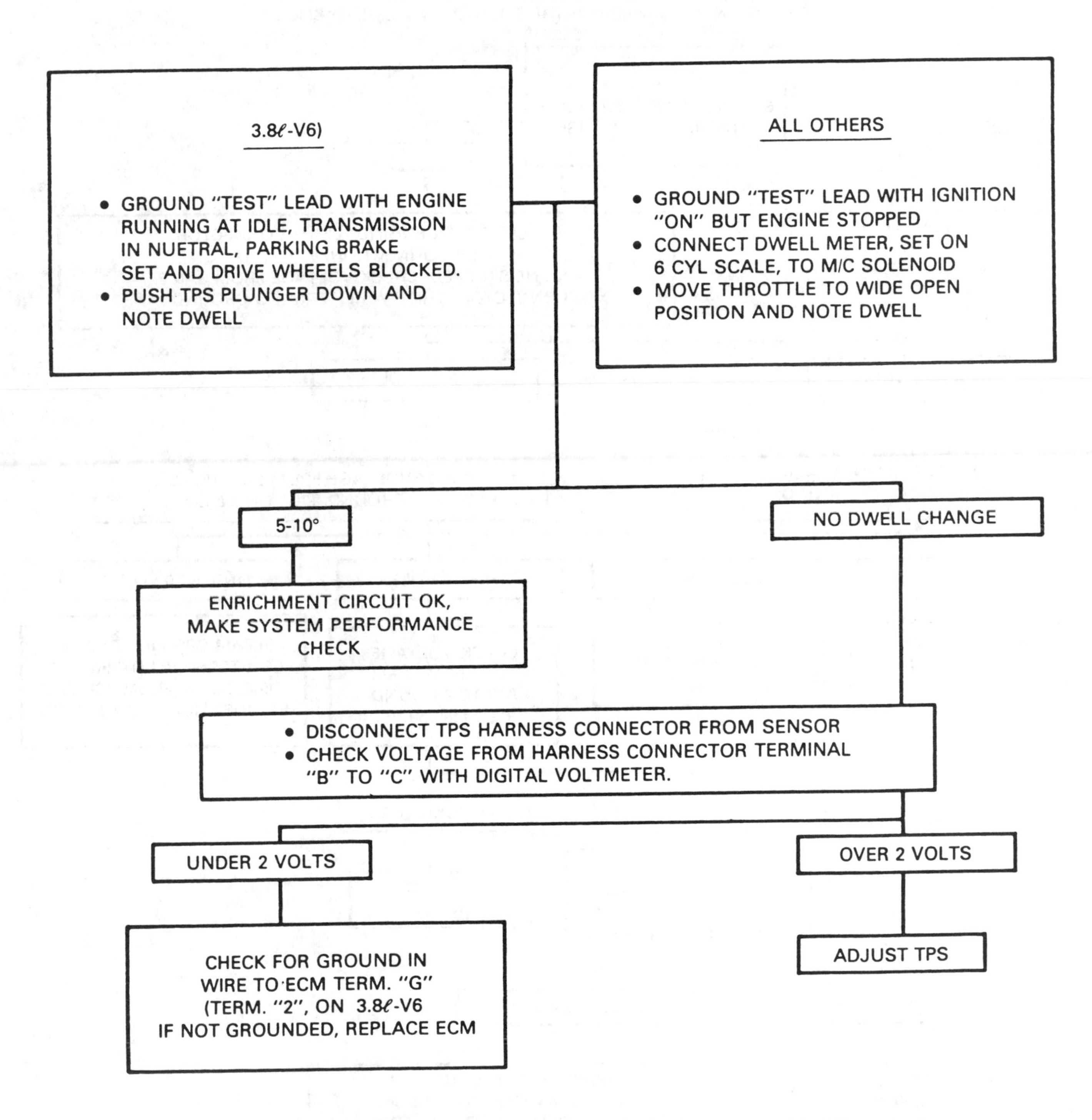

THIS SYSTEM PERFORMANCE CHECK SHOULD BE PERFORMED AFTER ANY REPAIRS TO THIS SYSTEM HAVE BEEN MADE.

C-4 SYSTEM DIAGNOSIS—Continued

CHART #5
"CHECK ENGINE" LIGHT INOPERATIVE

NOTE: LIGHT WILL NOT COME ON IF "TEST" LEAD IS GROUNDED AND THEN THE IGNITION IS TURNED ON; THE SOLENOID WILL ALSO CLICK.

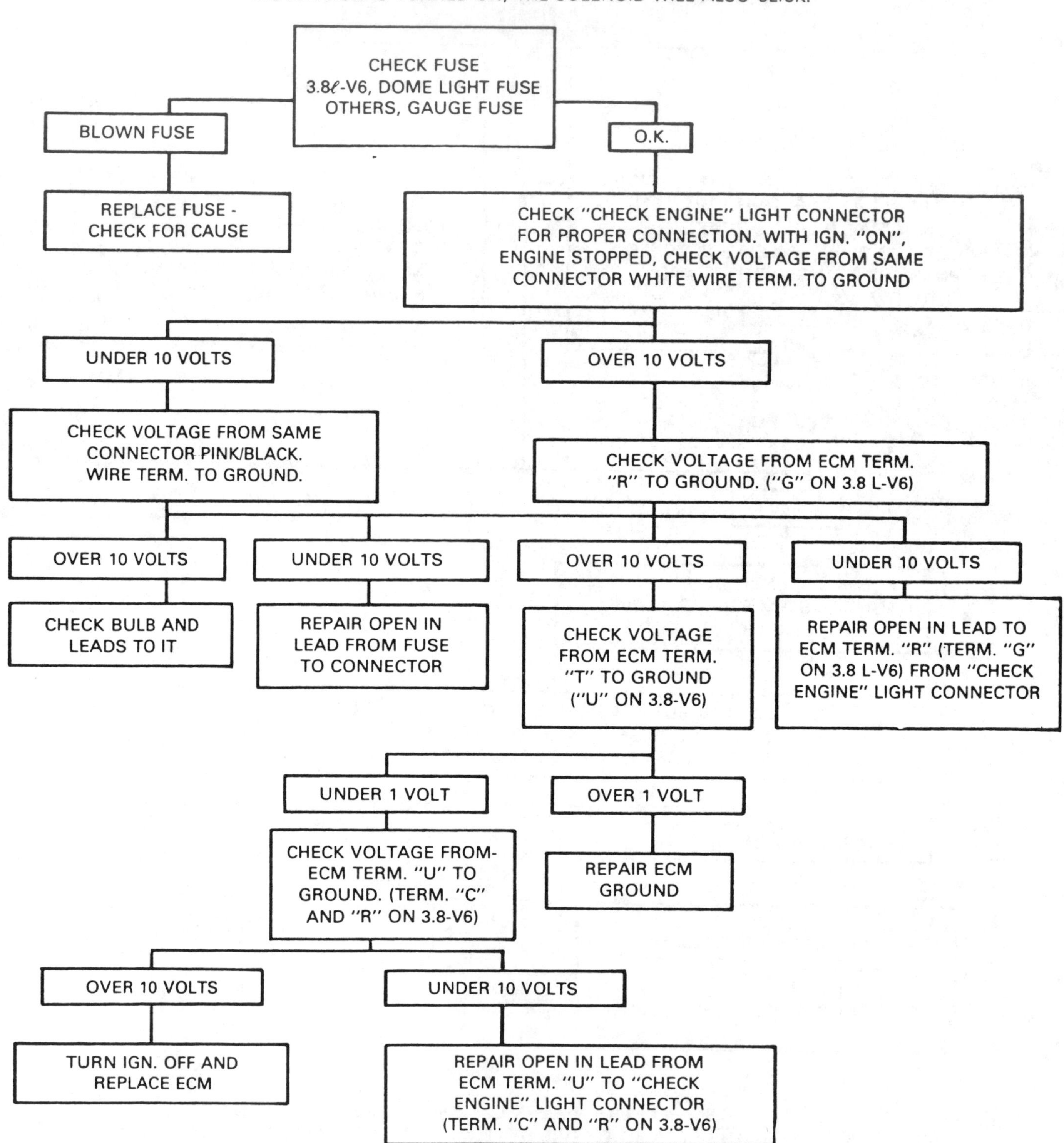

THE SYSTEM PERFORMANCE CHECK SHOULD BE PERFORMED AFTER ANY REPAIRS TO THIS SYSTEM HAVE BEEN MADE.

C-4 SYSTEM DIAGNOSIS—Continued

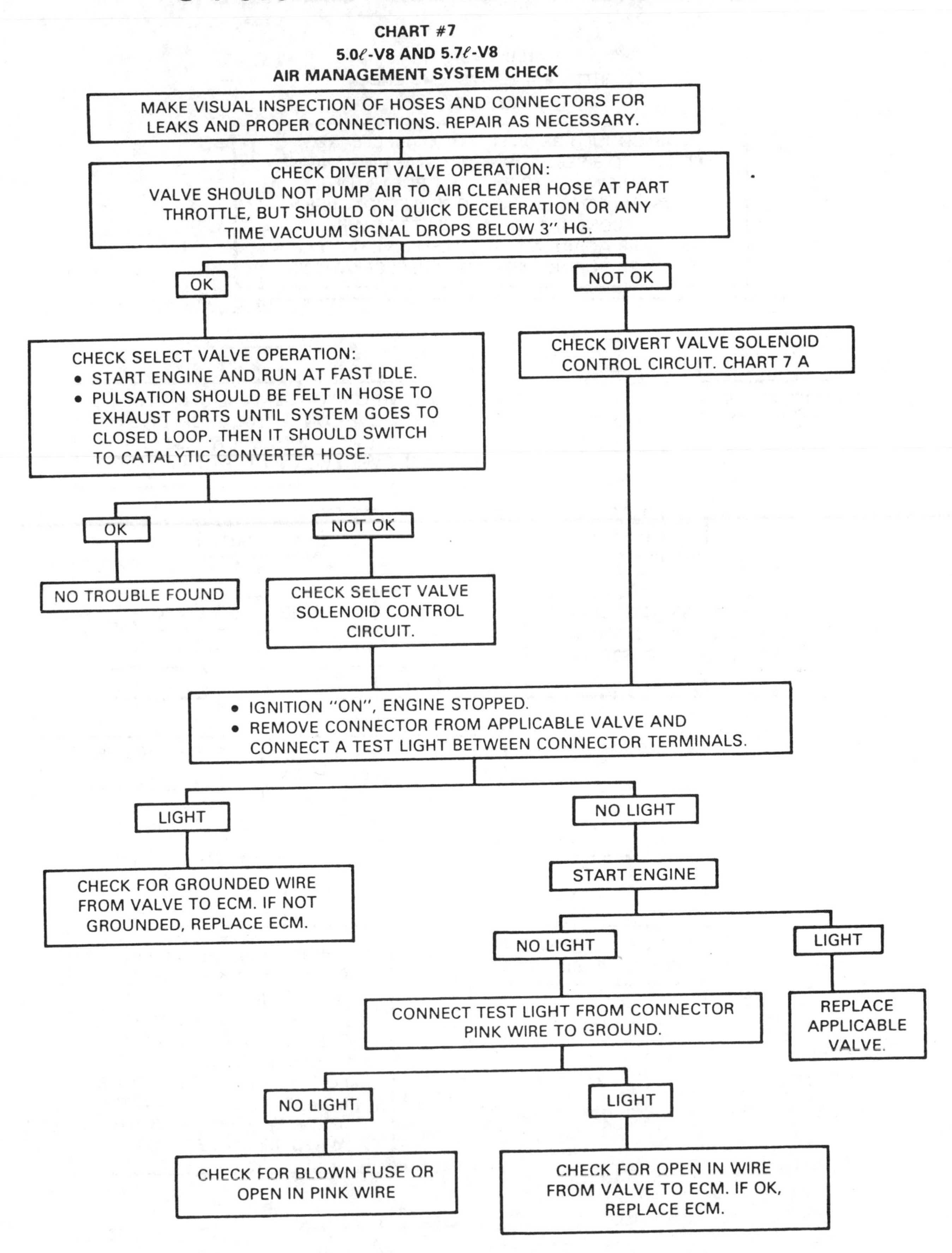

C-4 SYSTEM DIAGNOSIS—Continued

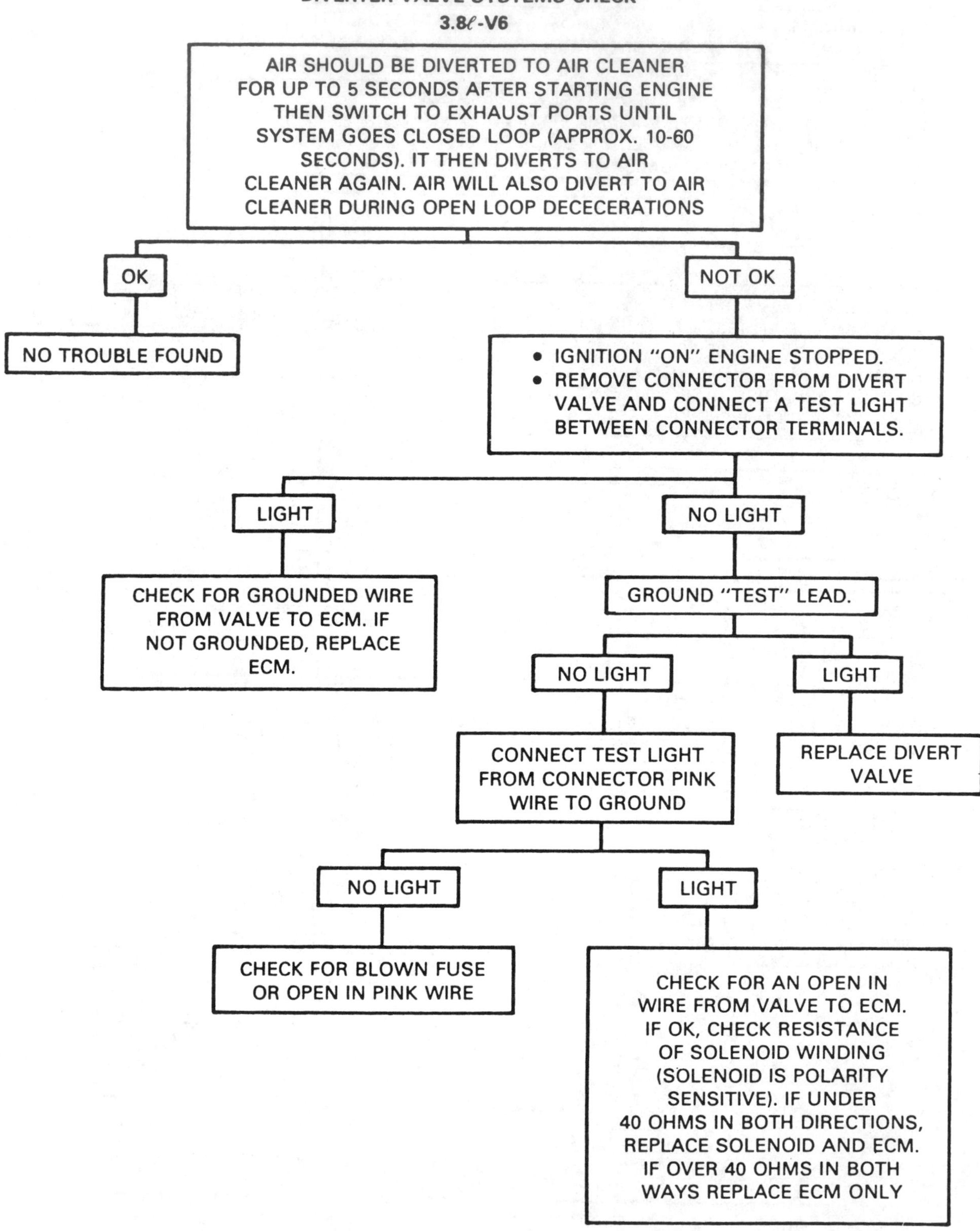

C-4 SYSTEM DIAGNOSIS—Continued

CHART #7B
DIVERTER VALVE CHECK
3.8ℓ-V6

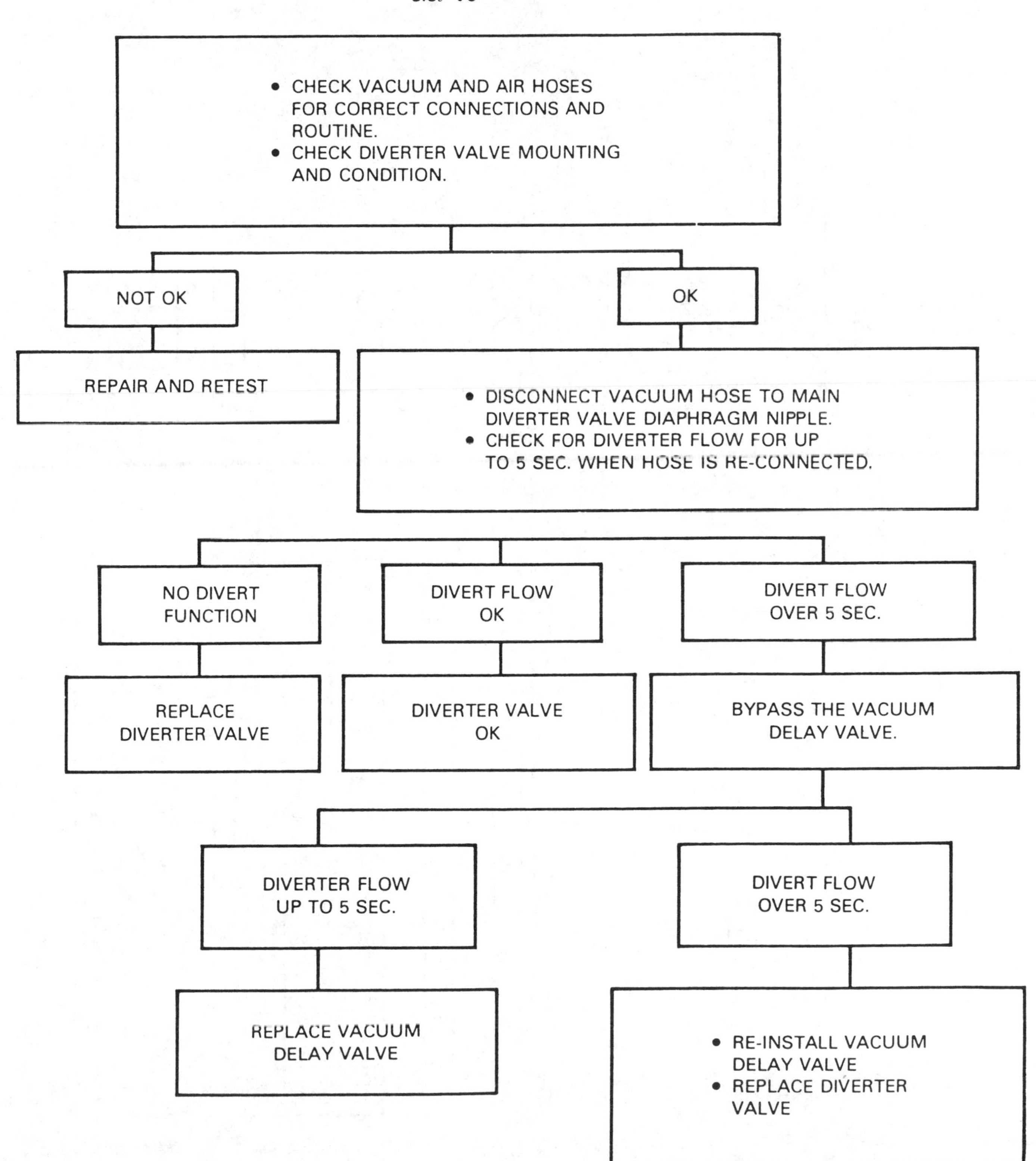

C-4 SYSTEM DIAGNOSIS—Continued

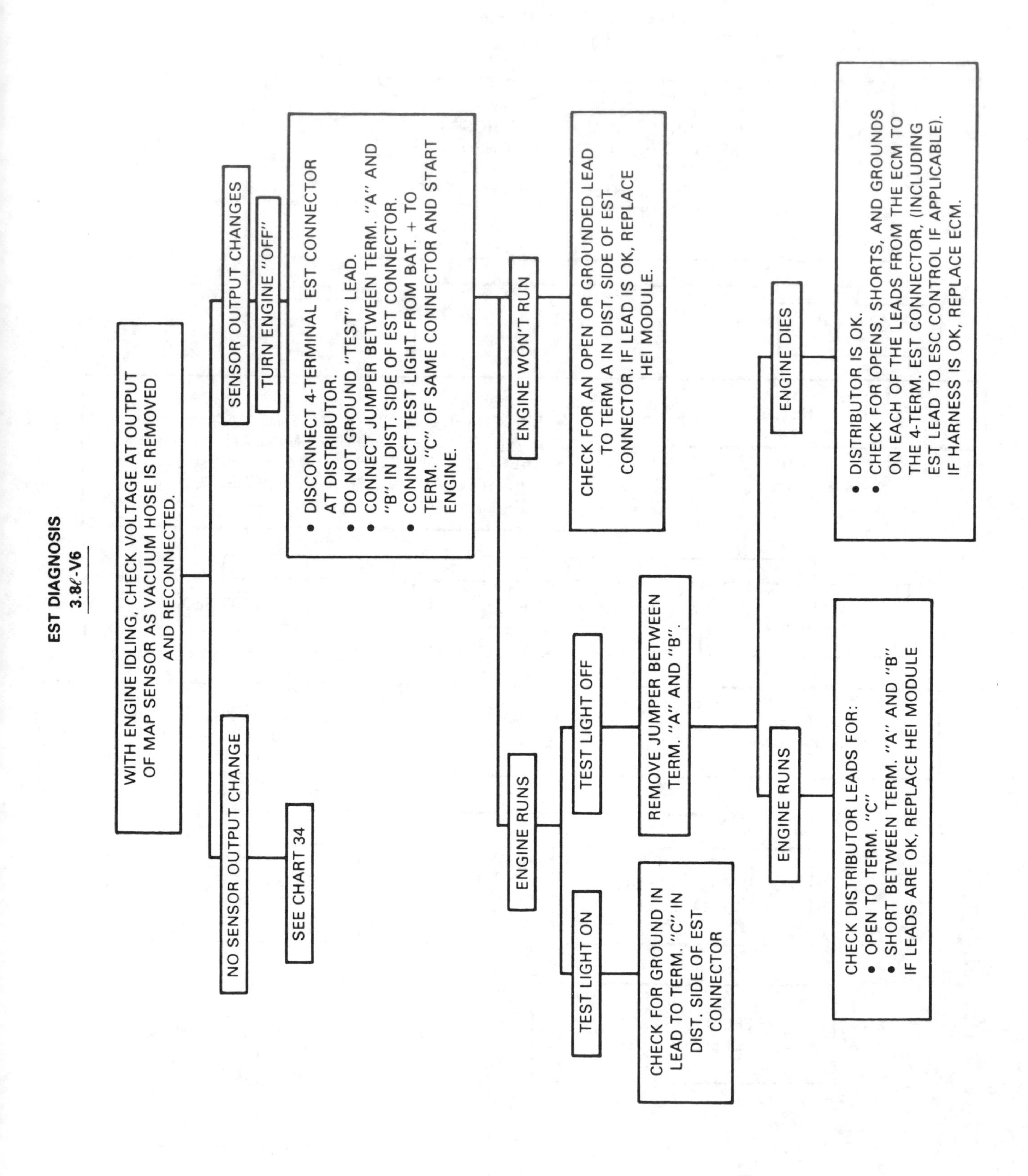

C-4 SYSTEM DIAGNOSIS—Continued

TROUBLE CODE 12
2.5ℓ-L4, 5.0ℓ-V8, 5.7ℓ-V8

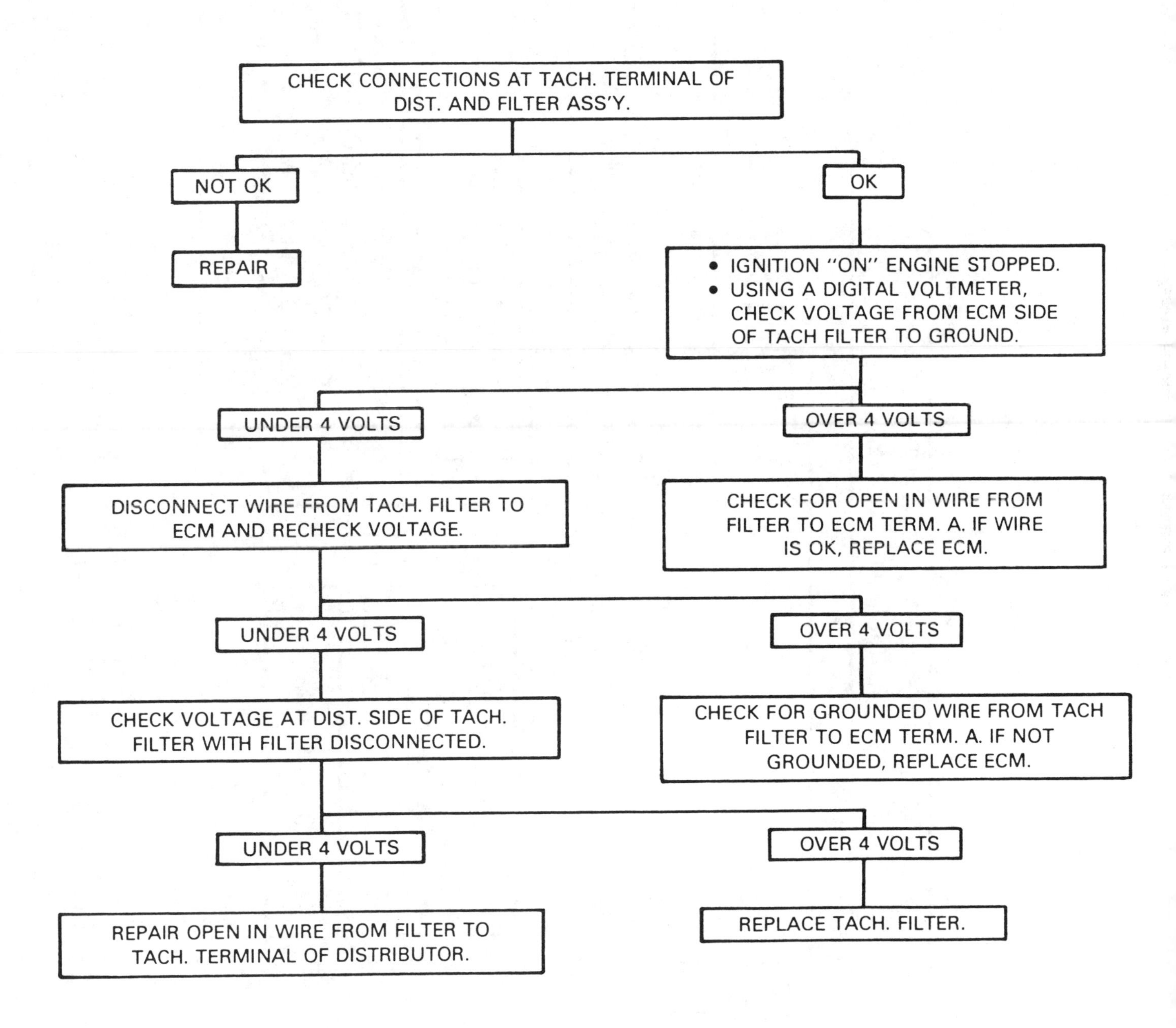

THE SYSTEM PERFORMANCE CHECK SHOULD BE PERFORMED AFTER ANY REPAIRS TO THIS SYSTEM HAVE BEEN MADE.

C-4 SYSTEM DIAGNOSIS—Continued

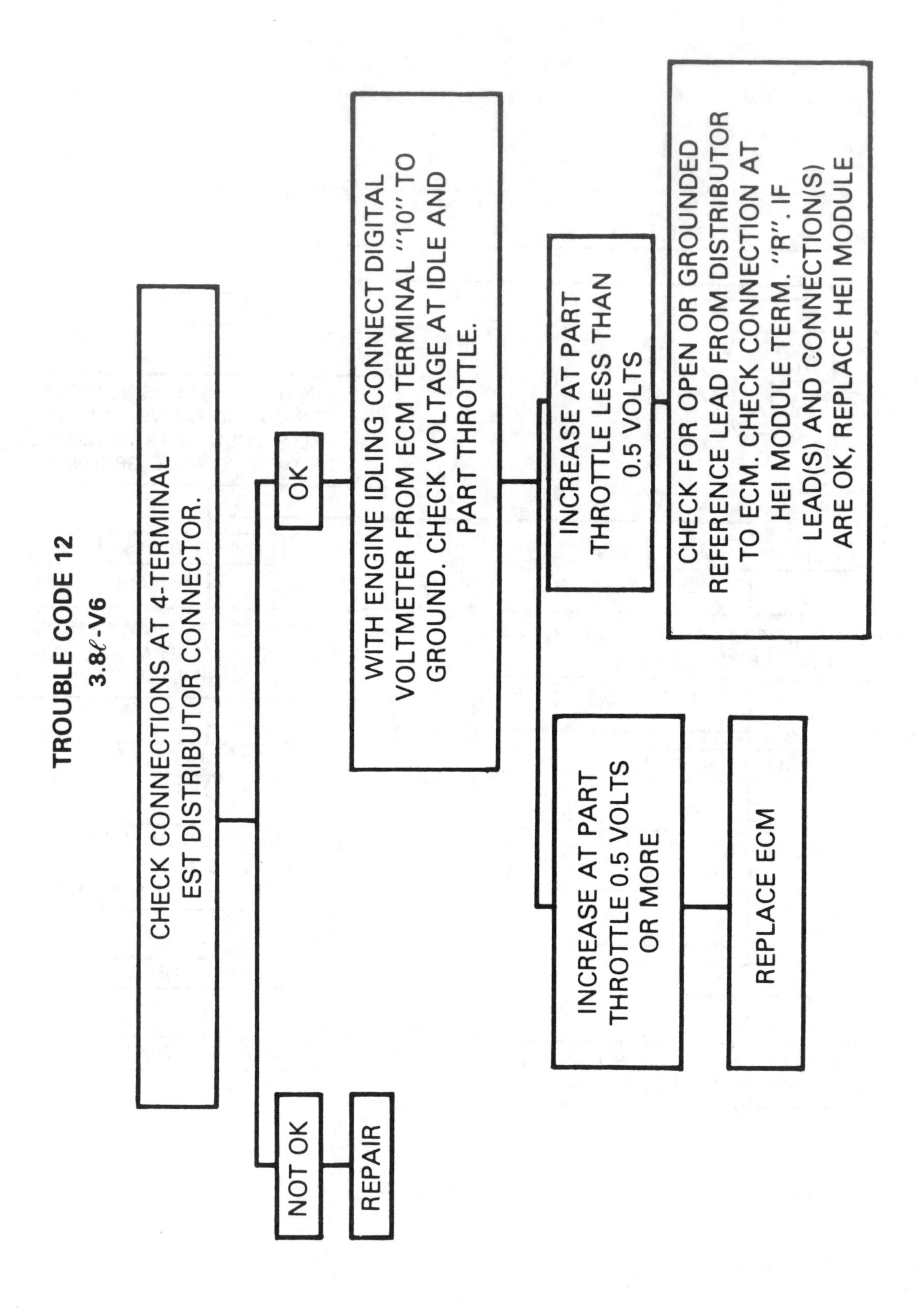

C-4 SYSTEM DIAGNOSIS—Continued

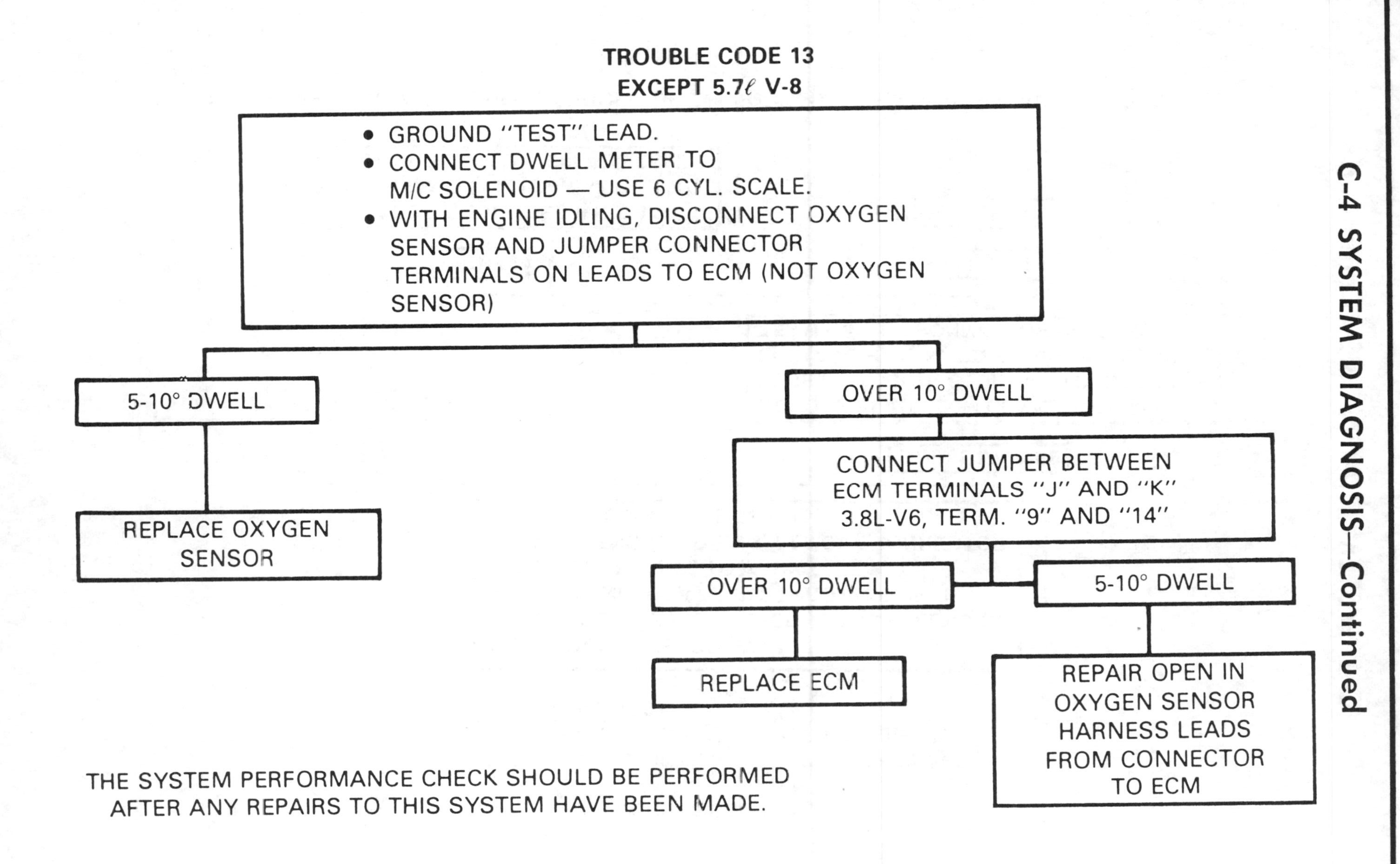

C-4 SYSTEM DIAGNOSIS—Continued

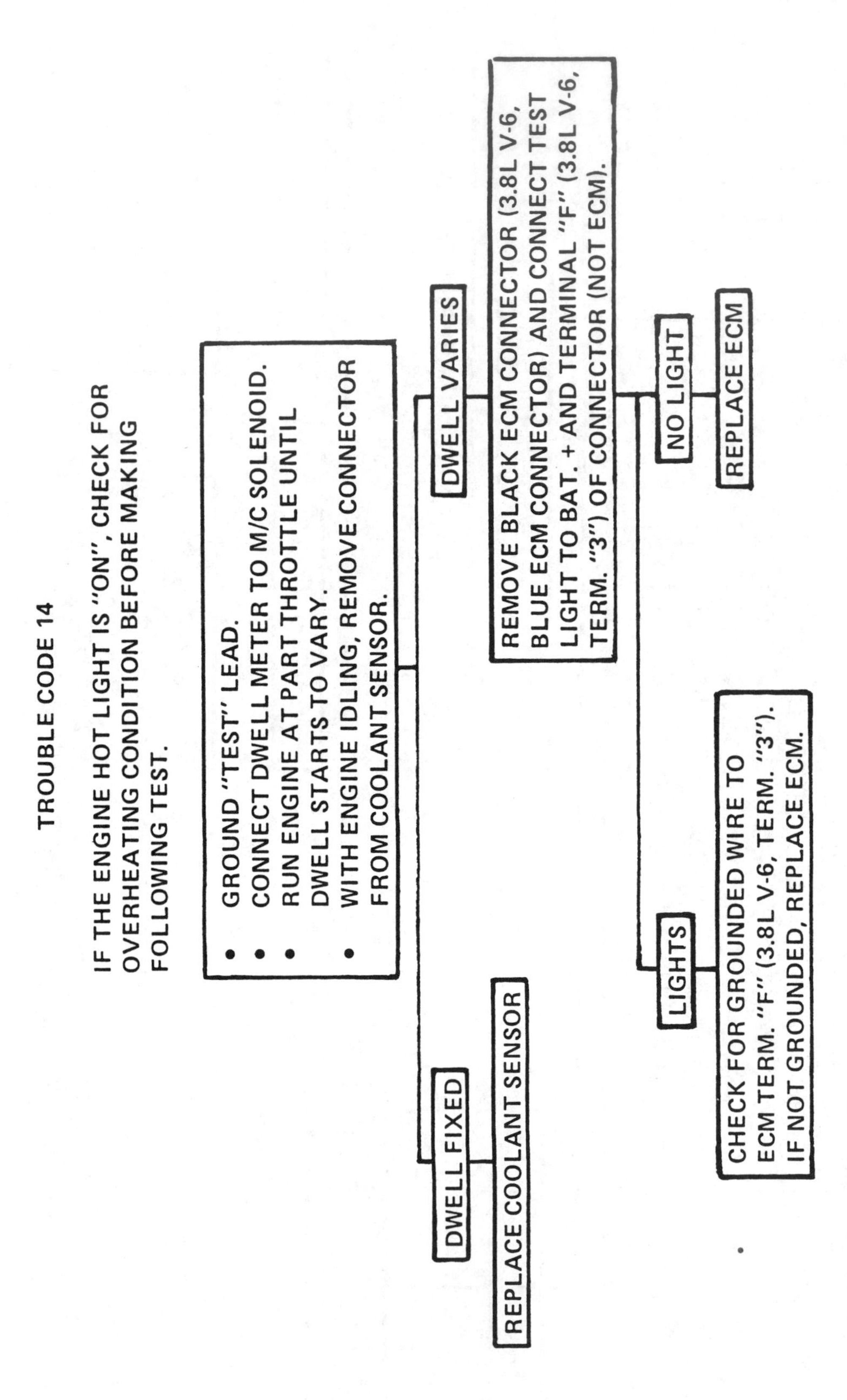

THE SYSTEM PERFORMANCE CHECK SHOULD BE PERFORMED AFTER ANY REPAIRS TO THIS SYSTEM HAVE BEEN MADE.

C-4 SYSTEM DIAGNOSIS—Continued

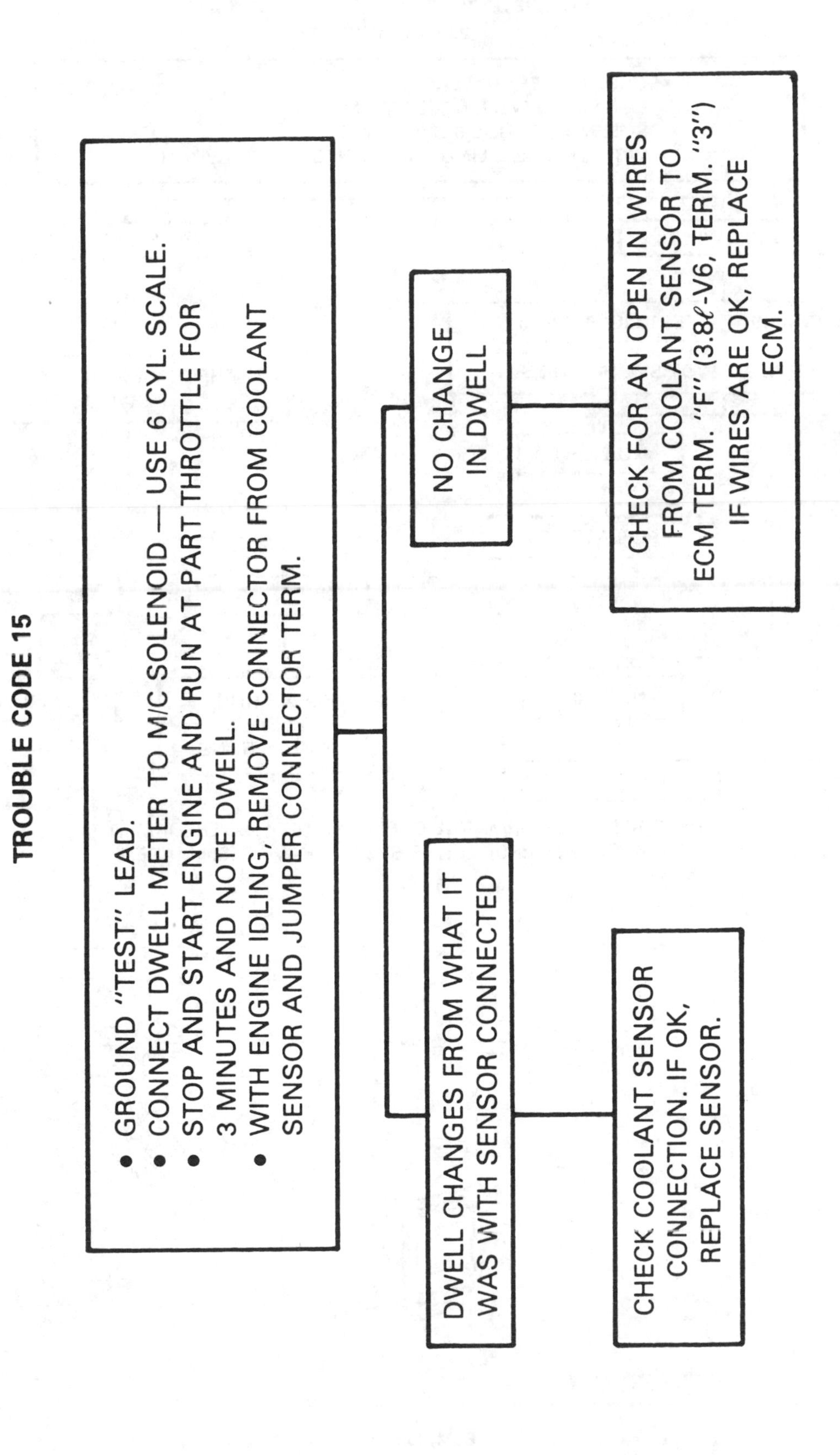

C-4 SYSTEM DIAGNOSIS—Continued

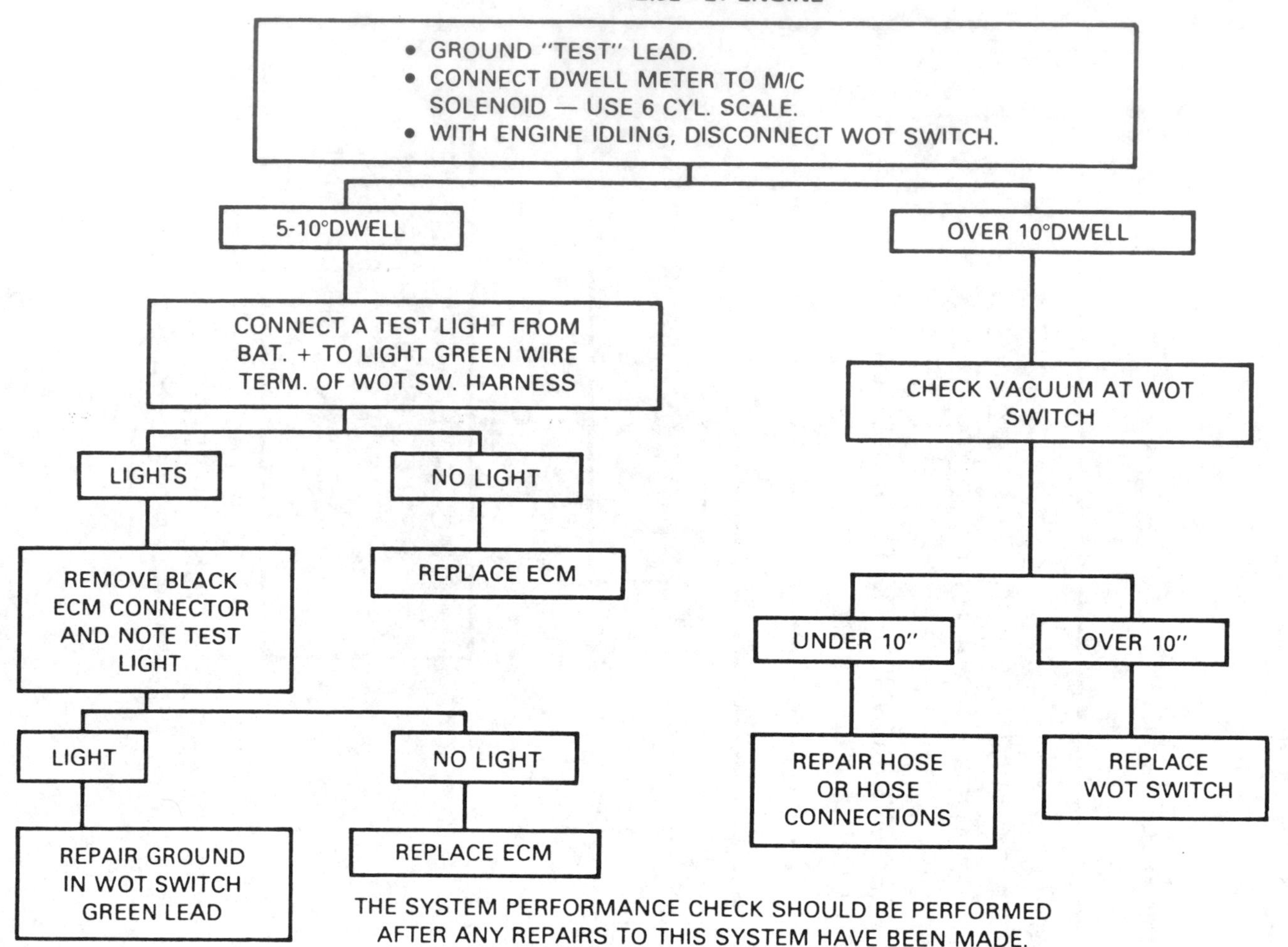

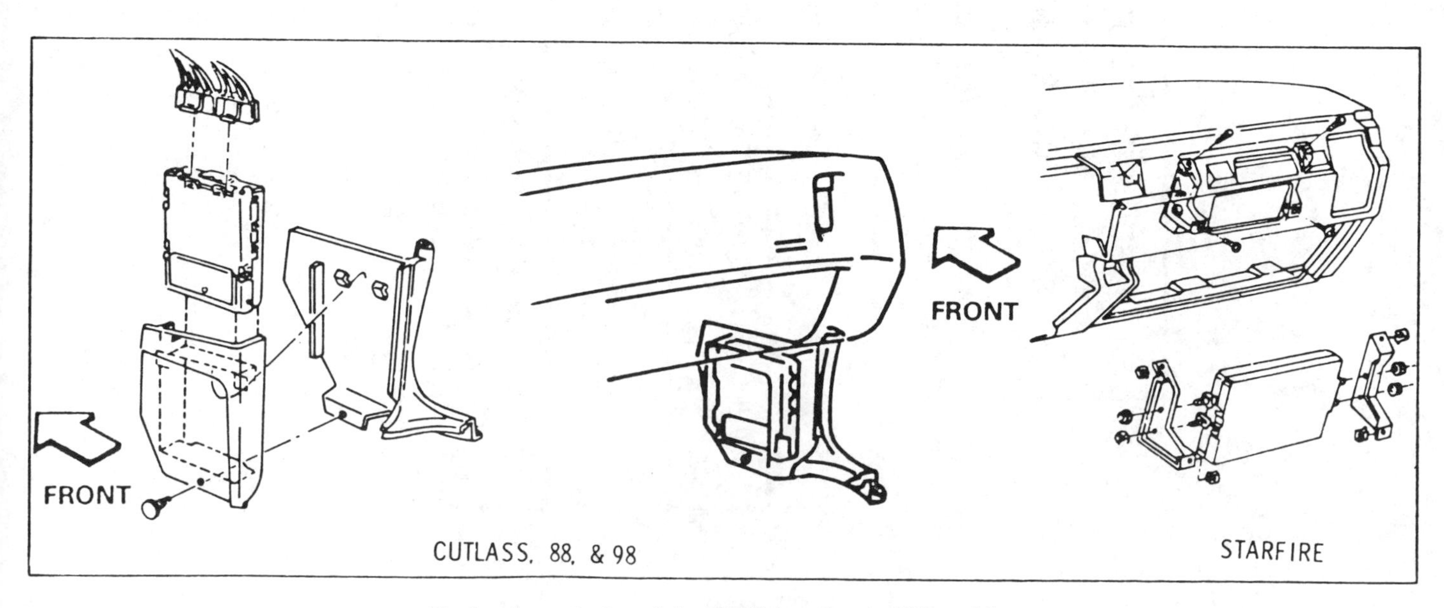

Electronic control module (ECM) locations—Oldsmobile
(© General Motors Corp.)

C-4 SYSTEM DIAGNOSIS—Continued

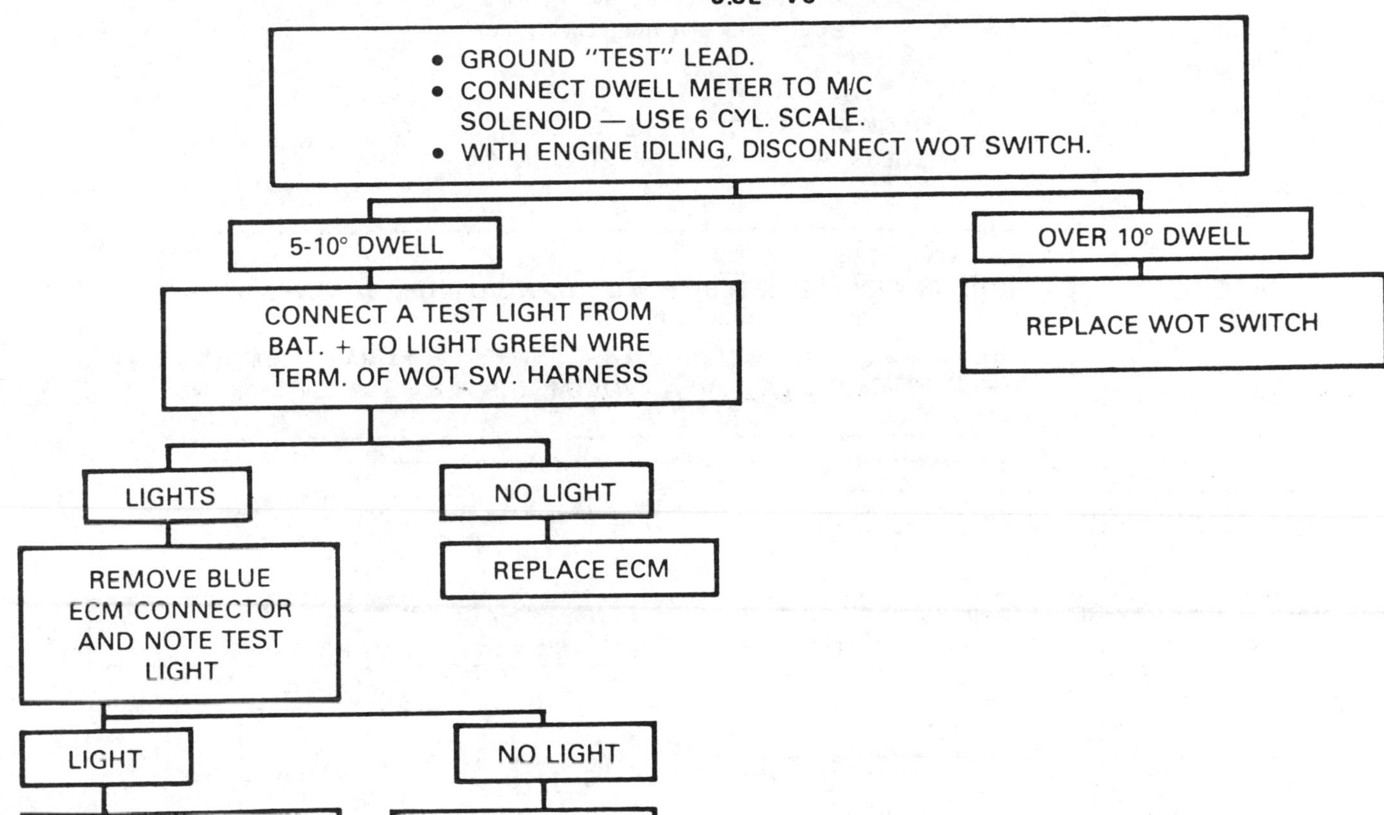

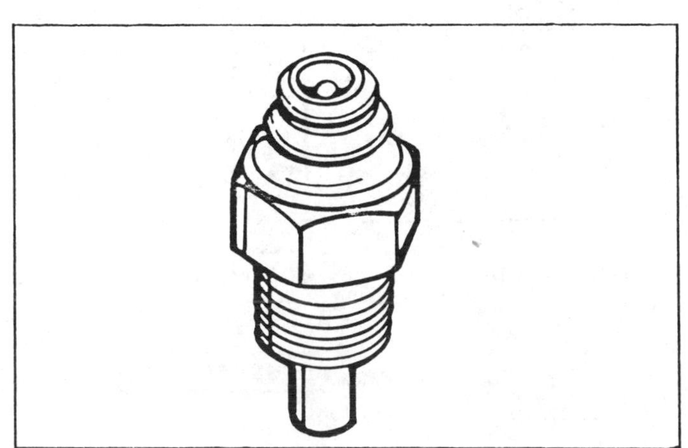

Engine temperature sensor (© General Motors Corp.)

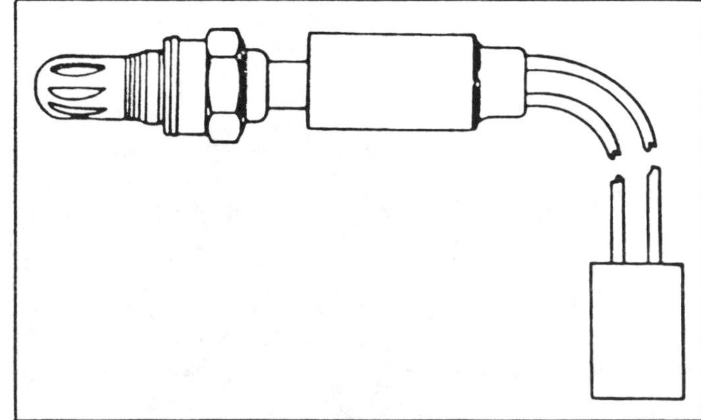

Oxygen sensor (© General Motors Corp.)

C-4 SYSTEM DIAGNOSIS—Continued

TROUBLE CODE 21
THROTTLE POSITION SENSOR (TPS)
EQUIPPED ENGINES ONLY

CHECK FOR STUCK TPS PLUNGER; REPAIR AS NECESSARY. IF OK, PROCEED BELOW:

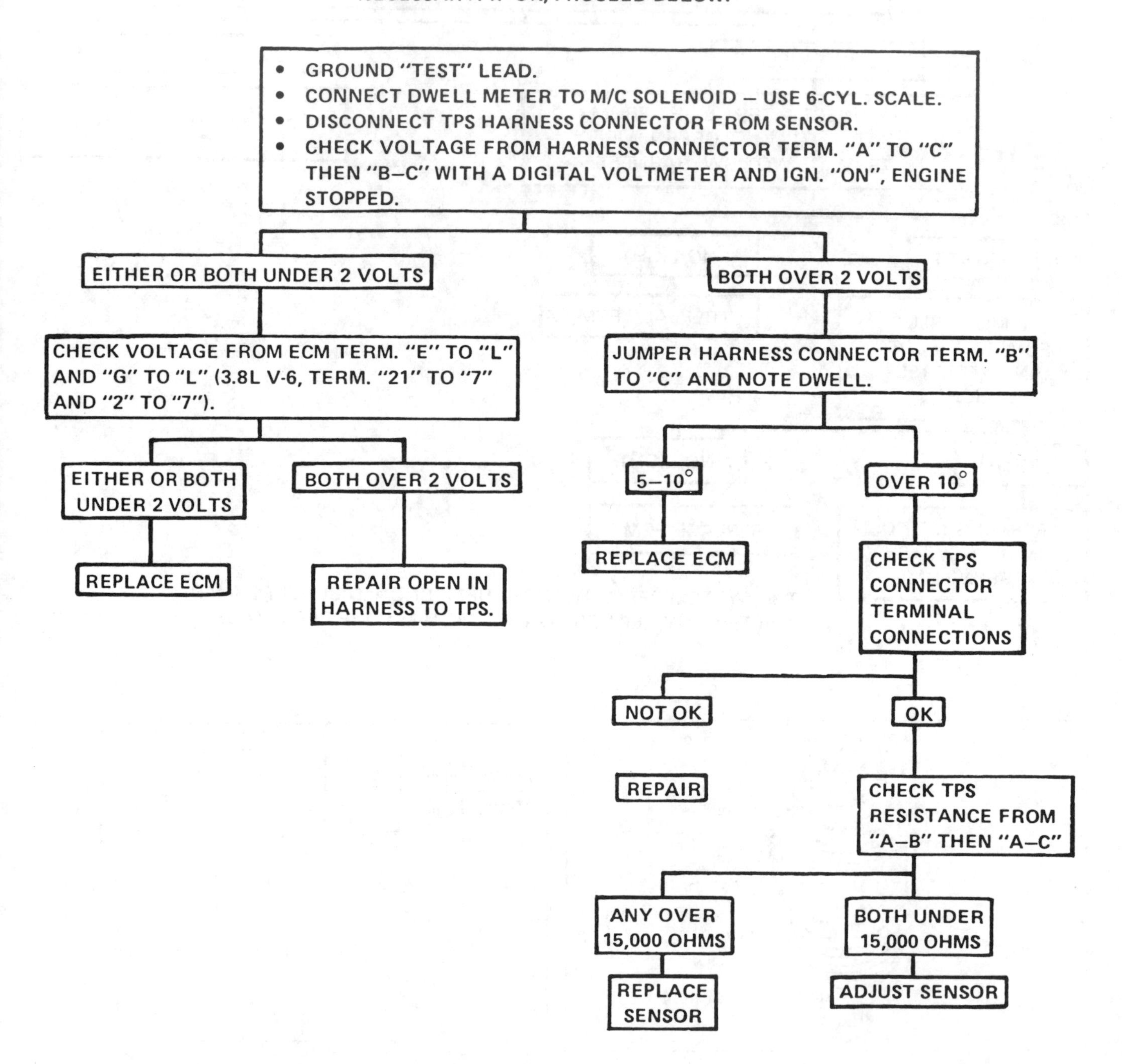

THE SYSTEM PERFORMANCE CHECK SHOULD BE PERFORMED AFTER ANY REPAIRS TO THIS SYSTEM HAVE BEEN MADE.

C-4 SYSTEM DIAGNOSIS—Continued

TROUBLE CODE 22
2.5ℓ-L4 ENGINE

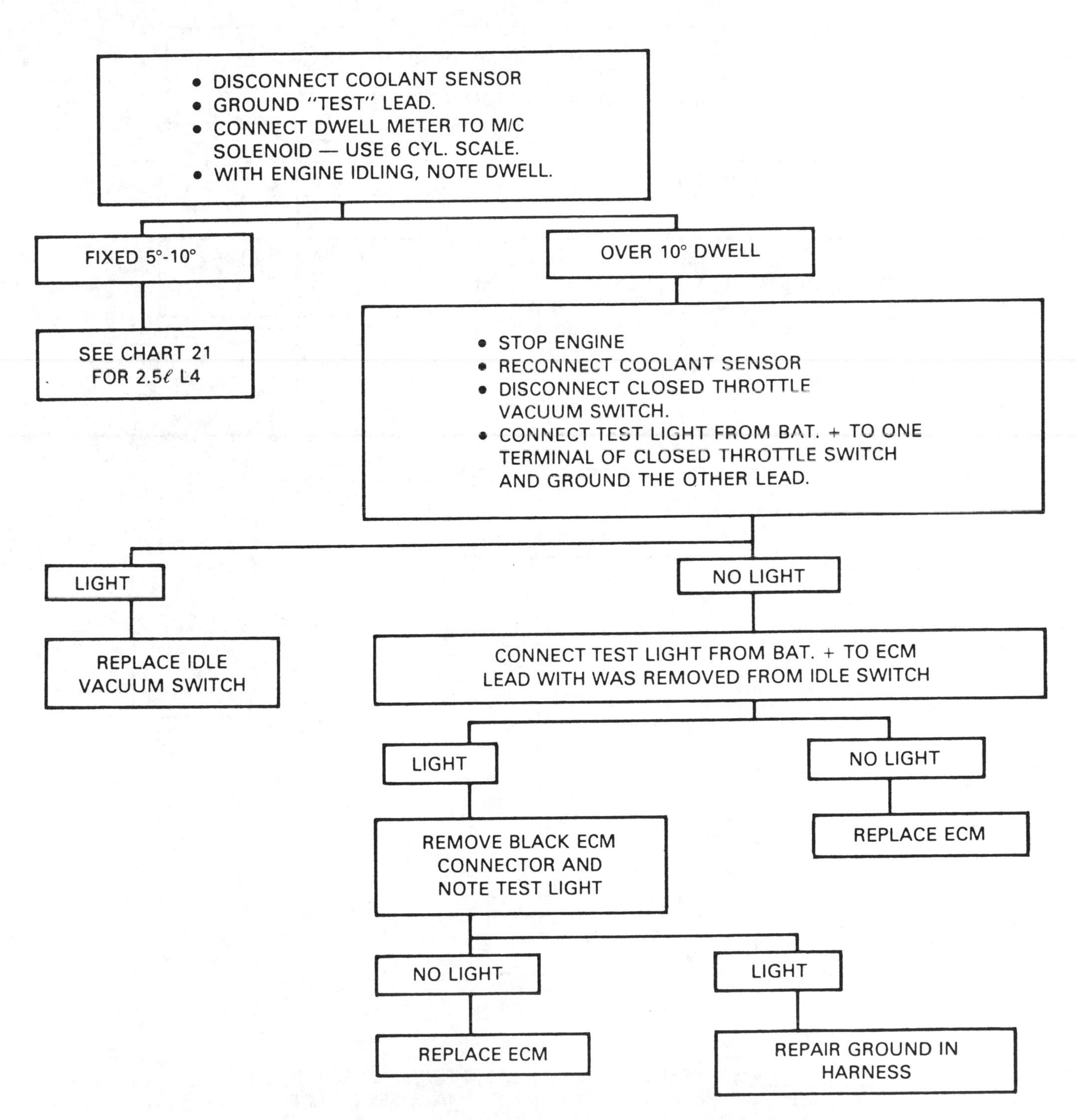

THE SYSTEM PERFORMANCE CHECK SHOULD BE PERFORMED AFTER ANY REPAIRS TO THIS SYSTEM HAVE BEEN MADE.

C-4 SYSTEM DIAGNOSIS—Continued

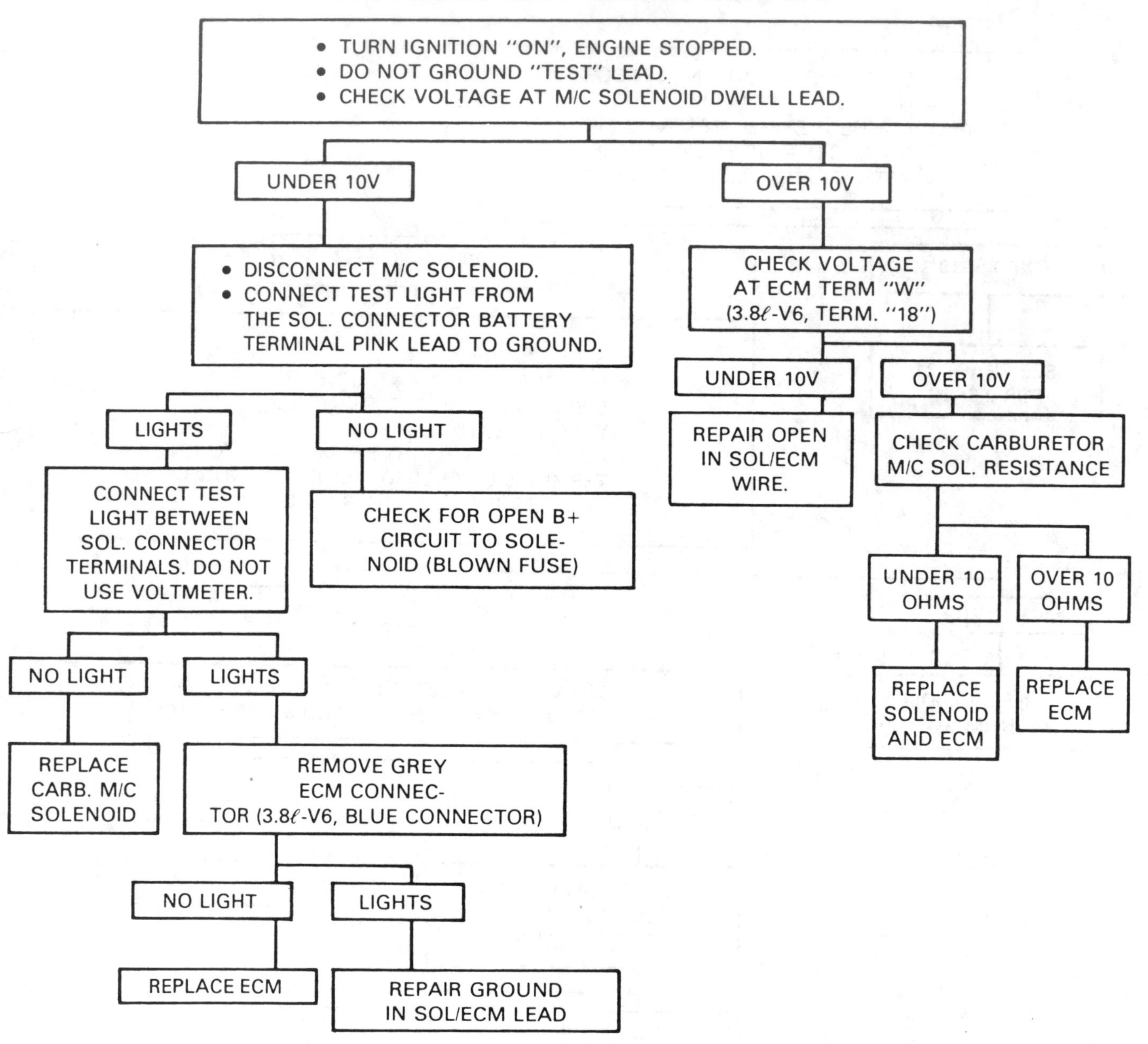

THE SYSTEM PERFORMANCE CHECK SHOULD BE PERFORMED AFTER ANY REPAIRS TO THIS SYSTEM HAVE BEEN MADE.

C-4 SYSTEM DIAGNOSIS—Continued

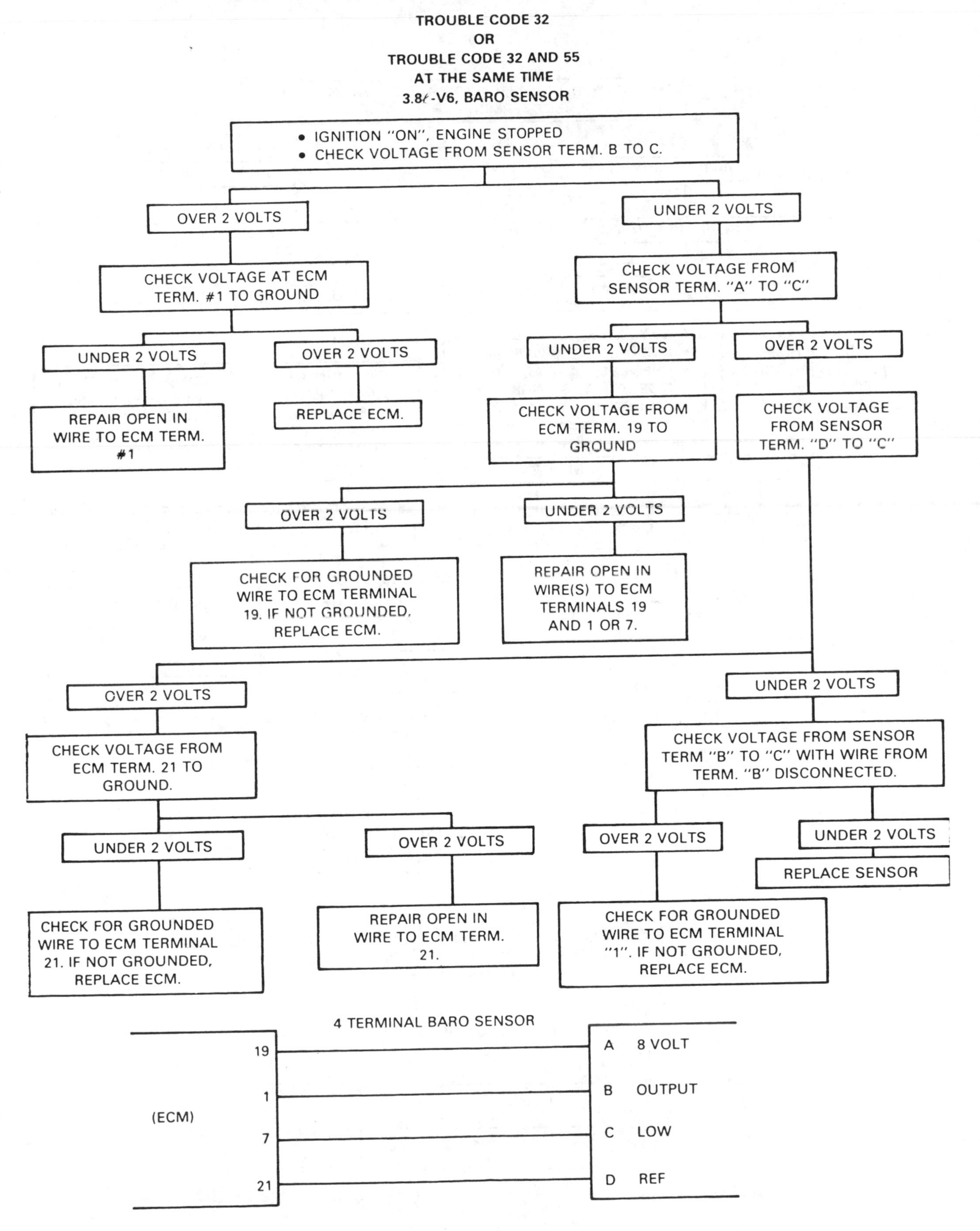

C-4 SYSTEM DIAGNOSIS—Continued

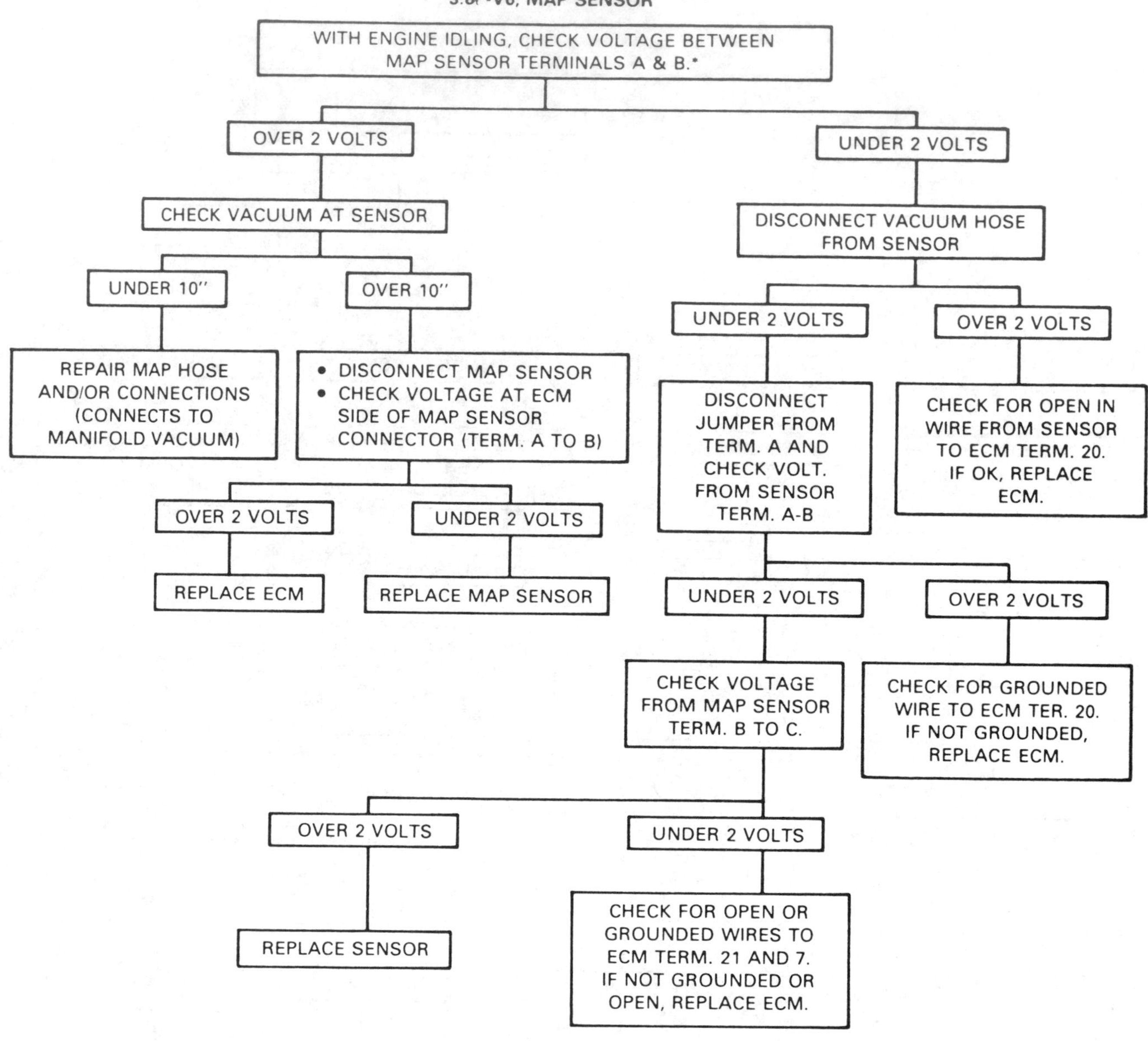

*THIS REQUIRES USE OF THREE JUMPERS BETWEEN THE CONNECTOR AND THE SENSOR TO GAIN ACCESS TO THE TERMINALS. THESE CAN BE MADE BY USING PACKARD WEATHERPAK TERMINALS 12014836 AND 12014837 OR EQUIVALENT.

3 TERMINAL MAP SENSOR

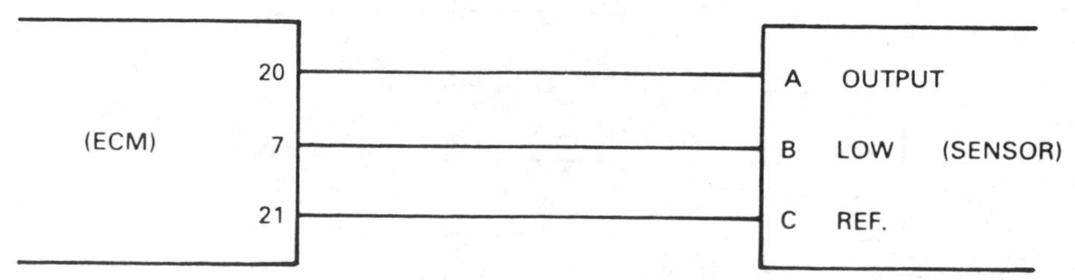

C-4 SYSTEM DIAGNOSIS—Continued

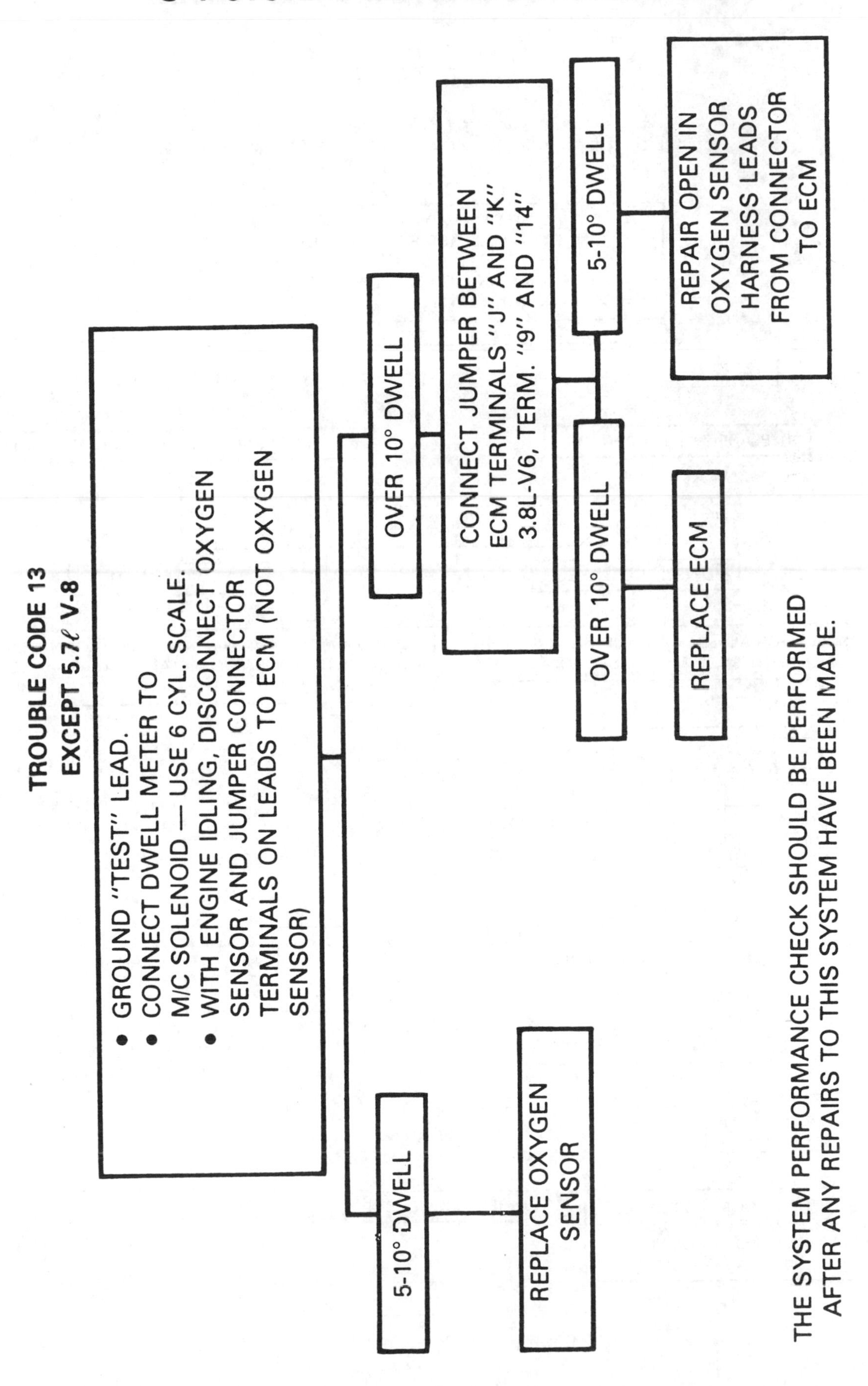

C-4 SYSTEM DIAGNOSIS—Continued

TROUBLE CODE 44

CODE 44 AND 55 AT SAME TIME
REPLACE OXYGEN SENSOR AND MAKE SYSTEM PERFORMANCE CHECK.

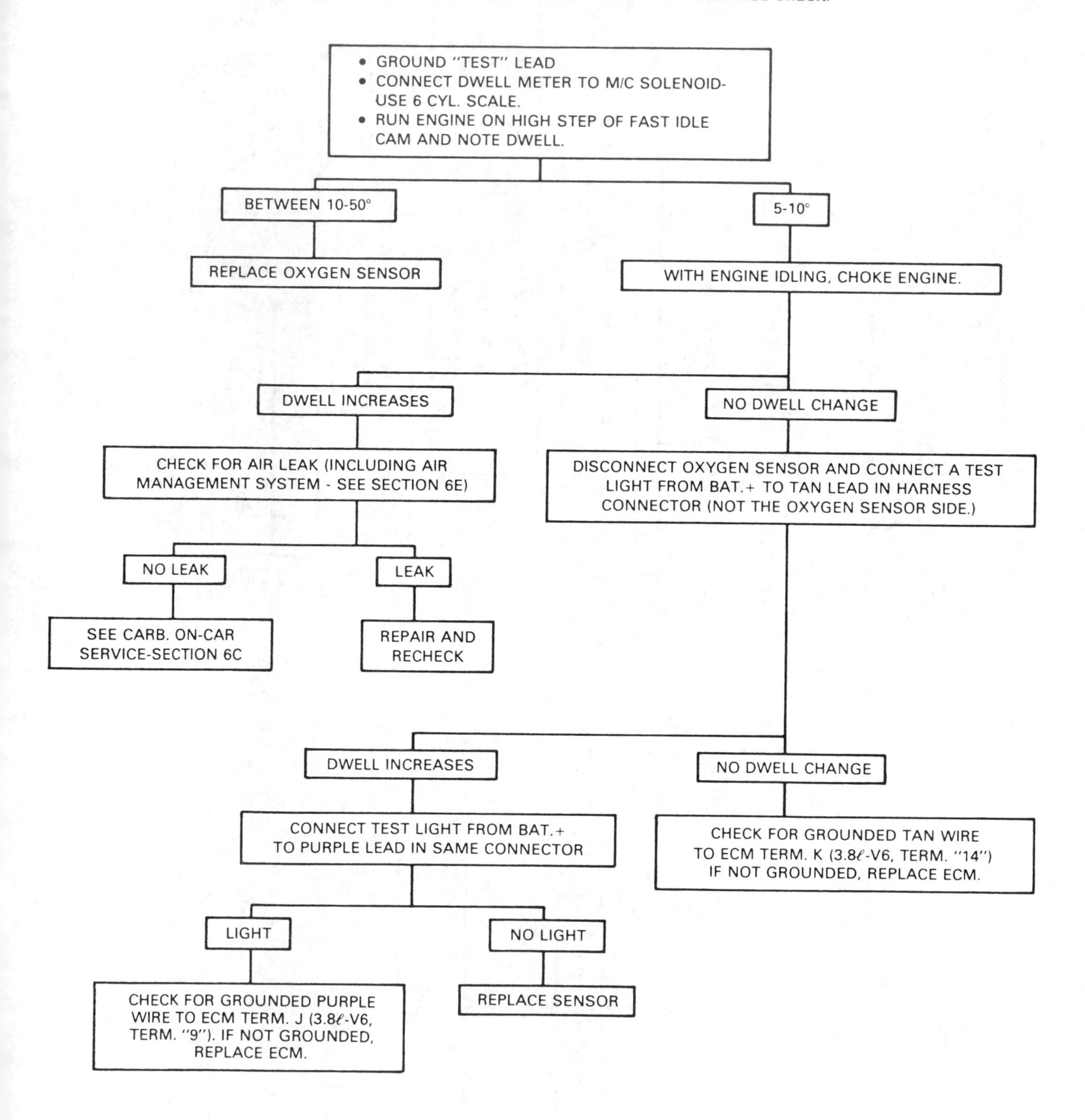

C-4 SYSTEM DIAGNOSIS—Continued

TROUBLE CODE 45

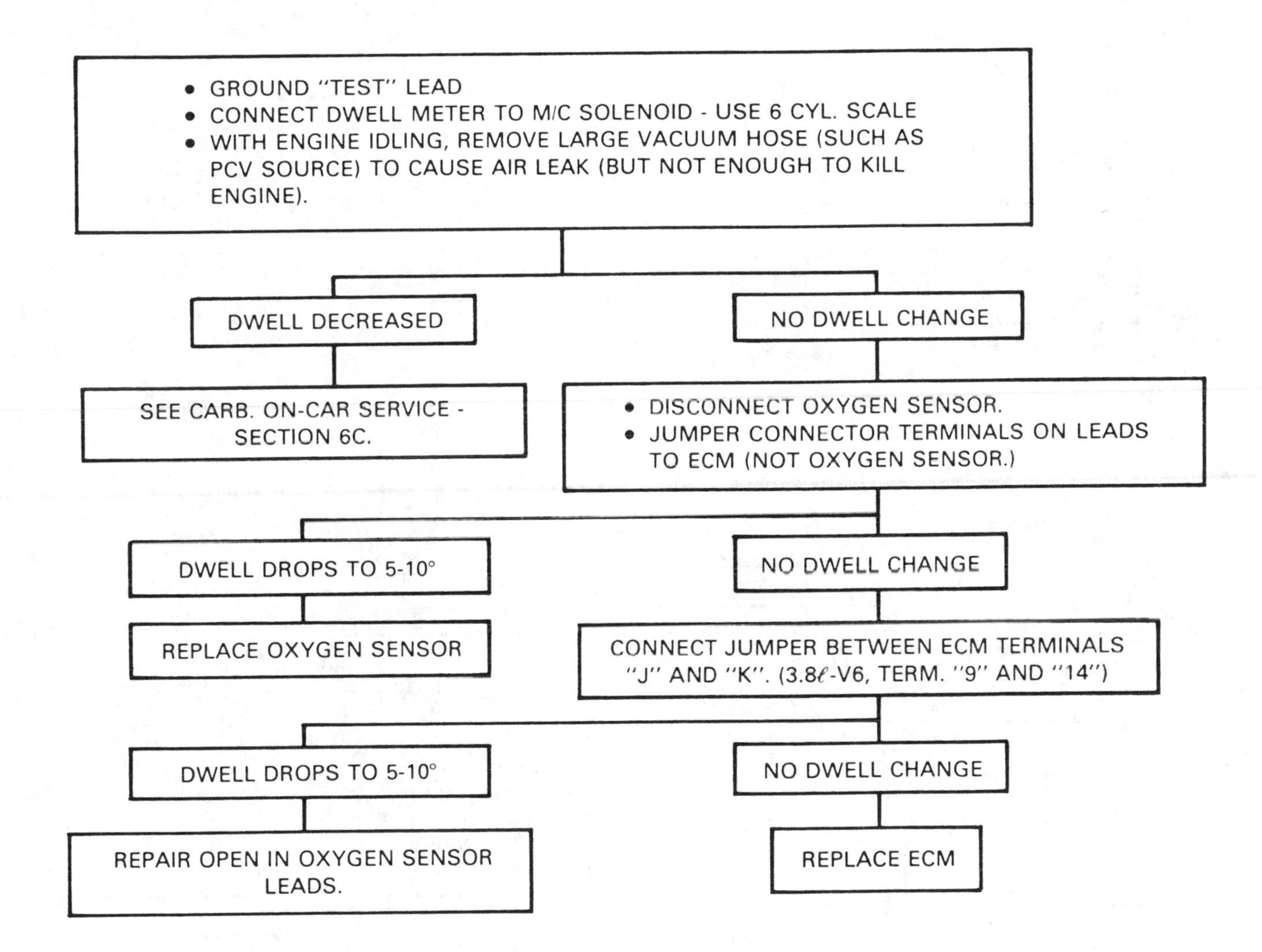

THE SYSTEM PERFORMANCE CHECK SHOULD BE PERFORMED AFTER REPAIRS TO THIS SYSTEM HAVE BEEN MADE.

C-4 SYSTEM DIAGNOSIS—Continued

TROUBLE CODES 51 THRU 54

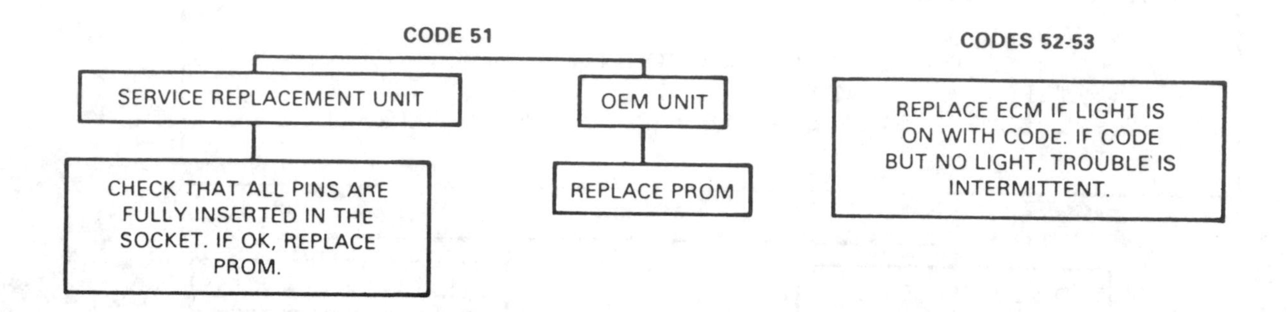

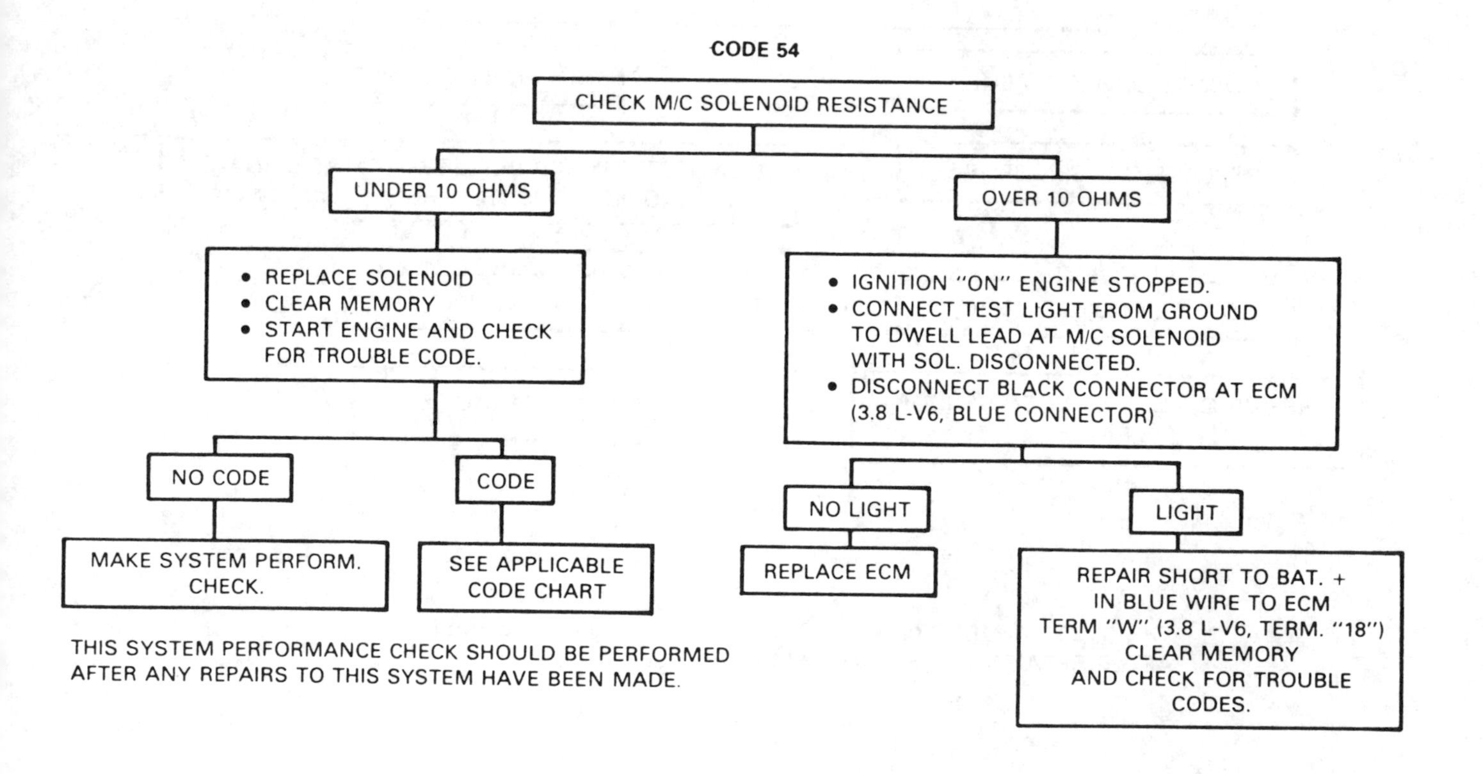

C-4 SYSTEM DIAGNOSIS—Continued

TROUBLE CODE 55

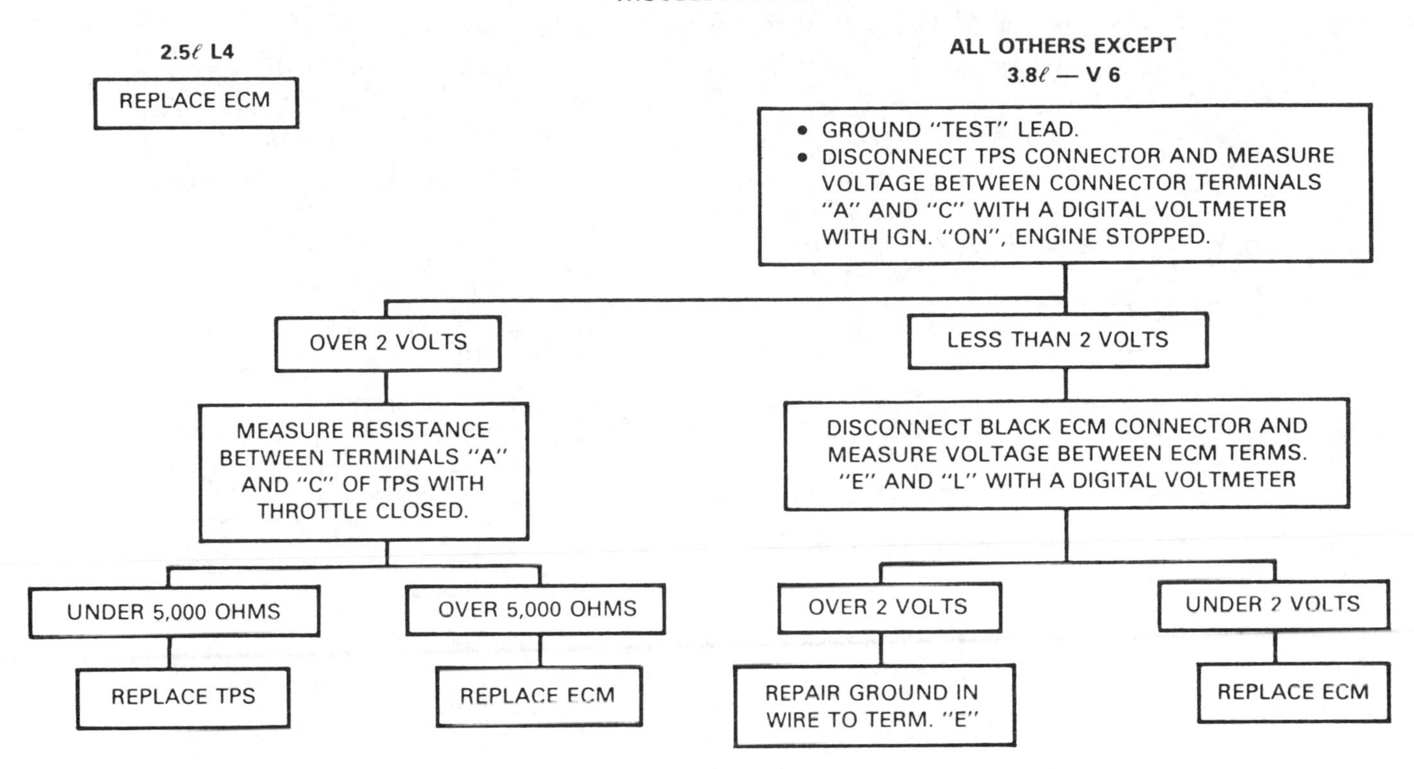

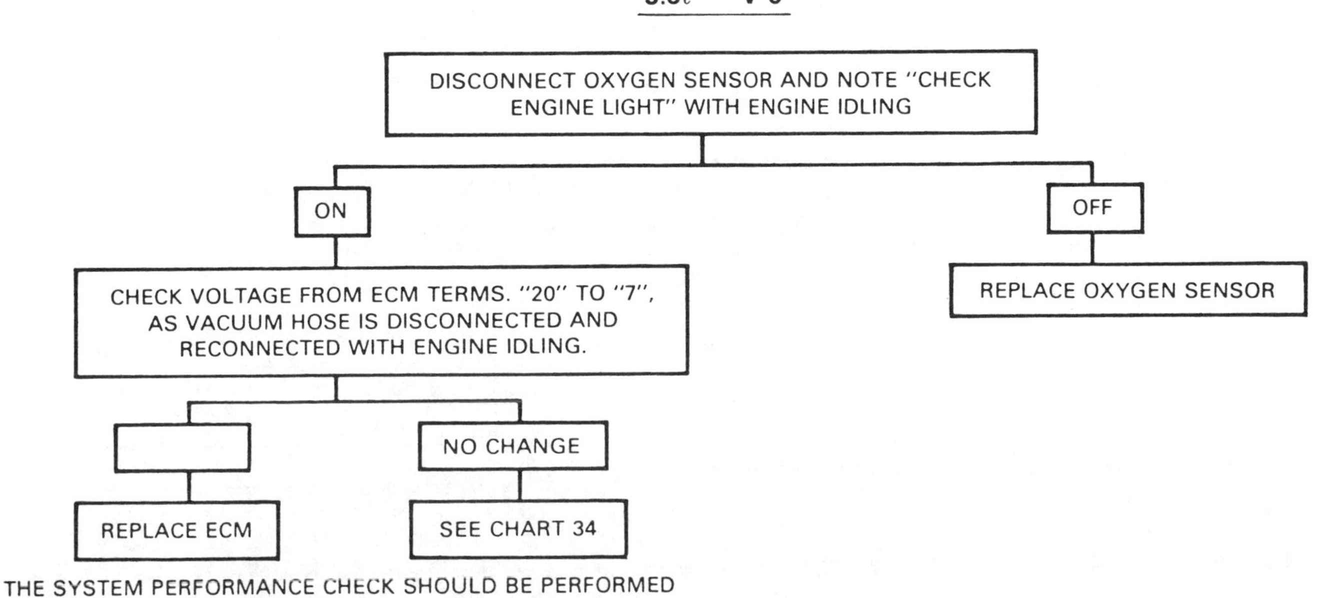

THE SYSTEM PERFORMANCE CHECK SHOULD BE PERFORMED AFTER ANY REPAIRS TO THIS SYSTEM HAVE BEEN MADE.

ENGINE PERFORMANCE DIAGNOSIS

Condition	Possible Cause	Correction
EGR VALVE STEM DOES NOT MOVE ON SYSTEM TEST.	(a) Cracked, leaking, disconnected or plugged hoses.	(a) Verify correct hose connections and leak check to confirm that all hoses are open. If defective hoses are found, replace hose harness.
	(b) Defective EGR valve, ruptured diaphragm or valve stem frozen.	(b) Disconnect hose harness from EGR valve. Connect external vacuum source, 10 inches Hg or greater, to valve diaphragm. If no valve movement occurs, replace valve. If valve opens, approximately 1/8 inch (3mm) travel, pinch off supply hose to check for diaphragm leakage. Valve should remain open 30 seconds or longer. If leakage occurs, replace valve.
EGR VALVE STEM DOES NOT MOVE ON SYSTEM TEST, OPERATES NORMALLY ON EXTERNAL VACUUM SOURCE.	(a-1) Defective CTS	(a-1) Bypass EGR solenoid and connect amplified venturi vacuum directly to EGR valve. If normal operation of EGR stem is observed, further diagnosis of the CTSEGR system is required. With engine warmed up, remove wire from center terminal of the CTS. Wait 90 seconds. If EGR valve operates, replace CTS. To check CTS with engine cold, remove both wires from the CTS. Use an ohmmeter to check the resistance of the switch. If the resistance is greater than 10 ohms, replace the switch. If the resistance is not greater than 10 ohms, check the continuity of the ground circuit.
	(a-2) Defective CCEGR valve.	(a-2) By pass valve so that amplifier is connected directly to EGR valve. If normal operation of EGR valve stem is restored, replace CCEGR valve.
	(b) Defective vacuum control unit (amplifier).	(b) **Venturi Vacuum Control System:** Remove venturi signal hose from nipple on carburetor. With engine operating at idle, apply a vacuum of one to two inches Hg to venturi signal hose. Engine speed should drop a minimum of 150 RPM and EGR valve stem should visibly move 1/8 inch (3mm) or more. If this does not occur, replace vacuum control unit.
	(c) Plugged carburetor venturi signal passage.	(c) If vacuum control unit (amplifier), operates normally in previous test, plugged vacuum tap to carburetor venturi is indicated. Use suitable carburetor solvent to remove deposits from passage and use light air pressure to verify that passage is clear.

Do not use drills or wires to clean carburetor control passages for either type of control system as calibration of precision control orifices may be altered resulting in unsatisfactory vehicle operation.

Condition	Possible Cause	Correction
ENGINE WILL NOT IDLE, DIES OUT ON RETURN TO IDLE OR IDLE IS VERY ROUGH OR SLOW.	(a) Control system defective—EGR valve open.	(a) Disconnect hose from EGR valve and plug hose—recheck idle. If satisfactory, replace vacuum control unit (amplifier).
	(b) High EGR valve leakage in closed position.	(b) If vacuum hose removal does not correct, remove EGR valve and inspect to insre poppet is seated. Clean deposits if necessary, or replace valve if found defective.
WEAK WIDE OPEN THROTTLE PERFORMANCE.	(a) Defective vacuum control unit (amplifier).	(a) Disconnect hose from EGR valve and plug the hose. Road test vehicle, if performance is restored, replace vacuum control unit.
UNDER COLD START CONDITIONS WITH ENGINE WATER TEMP. BELOW 90°F, (32.2°C) VEHICLE EXHIBITS POOR DRIVEABILITY, OR CHOKE LOADING CHARACTERISTICS, OR IDLES ROUGHLY, OR STALLS ON RETURN TO IDLE AFTER STEADY SPEED DRIVING	(a) Leaking CCEGR or EGR control valve.	(a) Leak test and if necessary replace CCEGR valve.

AIR PUMP DIAGNOSIS

Condition	Possible Cause	Correction
EXCESSIVE BELT NOISE	(a) Loose belt.	(a) Tighten belt
	(b) Seized pump.	(b) Replace pump.
EXCESSIVE PUMP NOISE. CHIRPING	(a) Insufficient break-in.	(a) Recheck for noise after 1000 miles (1600 km) of operation.
EXCESSIVE PUMP NOISE CHIRPING, RUMBLING, OR KNOCKING	(a) Leak in hose.	(a) Locate source of leak using soap solution and correct.
	(b) Loose hose.	(b) Reassemble and replace or tighten hose clamp.
	(c) Hose touching other engine parts.	(c) Adjust hose position.
	(d) Diverter valve inoperative.	(d) Replace diverter valve.
	(e) Check valve inoperative.	(e) Replace check valve.
	(f) Pump mounting fasteners loose.	(f) Tighten mounting screws as specified.
	(g) Pump failure.	(g) Replace pump.
NO AIR SUPPLY (ACCELERATE ENGINE TO 1500 RPM AND OBSERVE AIR FLOW FROM HOSES. IF THE FLOW INCREASES AS THE RPM'S INCREASE, THE PUMP IS FUNCTIONING NORMALLY. IF NOT, CHECK POSSIBLE CAUSE.	(a) Loose drive belt.	(a) Tighten to specifications.
	(b) Leaks in supply hose.	(b) Locate leak and repair or replace as required.
	(c) Leak at fitting(s).	(c) Tighten or replace clamps.
	(d) Diverter valve leaking.	(d) If air is expelled through diverter exhaust with vehicle at idle, replace diverter valve.
	(e) Diverter valve inoperative.	(e) Usually accompanied by backfire during deceleration. Replace diverter valve.
	(f) Check valve inoperative.	(f) Replace check valve.

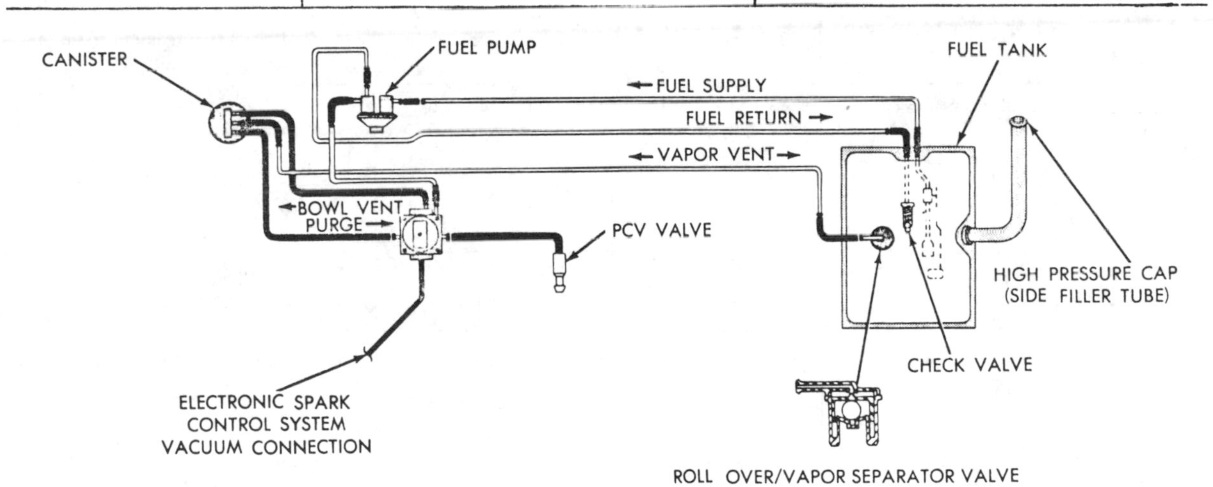

Evaporation control system (© Chrysler Corp.)

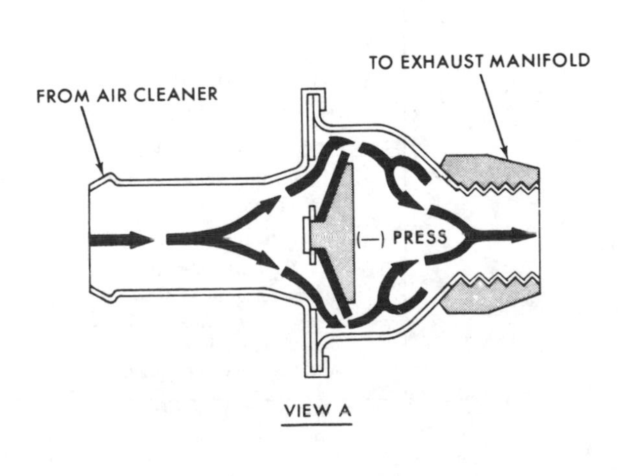

Aspirator valve (© Chrysler Corp.)

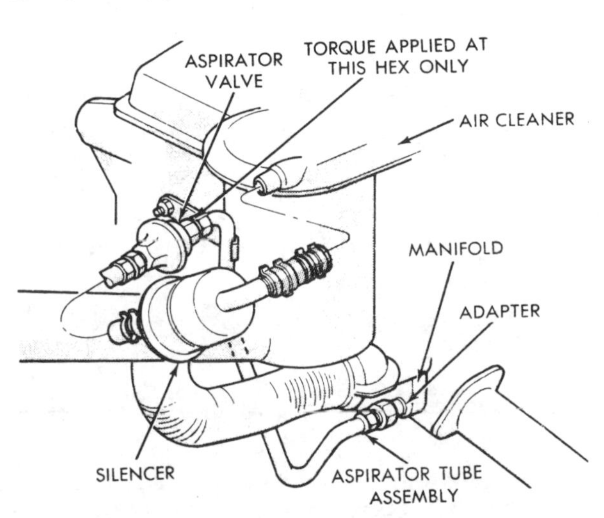

Aspirator system (© Chrysler Corp.)

THERMACTOR SYSTEM NOISE DIAGNOSIS

NOTE: *The thermactor system is not completely noiseless. Under normal conditions, noise rises in pitch as engine speed increases. To determine if excess noise is the fault of the air injection system, disconnect the drive belt and operate the engine. If the noise disappears, proceed with the following diagnosis:*

CONDITION	POSSIBLE CAUSE	RESOLUTION
1. Excessive Belt Noise	a. Loose belt	• Tighten to specification using tool T75L-9480-A or equivalent to hold belt tension and Belt Tension Gage T63L-8620-A or equivalent. CAUTION: Do not use a pry bar to move the air pump.
	b. Seized pump	• Replace pump.
	c. Loose pulley	• Replace pulley and/or pump if damaged. Torque bolts to specification (130-180 in-lbs.)
	d. Loose or broken mounting brackets or bolts	• Replace parts as required and torque bolts to specification.
2 Excessive Pump Noise (Chirps, Squeaks & Ticks)	a. Insufficient break-in or worn or damaged pump	• Check the thermactor system for wear or damage and make any necessary corrections. If less than 500 miles in service, allow for a 500 mile break-in period. Replace the pump if the noise continues after a 500 mile break-in period.
3. Excessive Mechanical Clicking Noise	a. Over torqued mounting bolt	• Torque to 25 ft-lbs.
	b. Excessive flash on the air pump adjusting arm boss.	• Remove flash from the boss.
	c. Distorted adjusting arm	• Replace adjusting arm.
	d. Adjusting arm bolt too long (bottom of the bolt boss is pushed in)	• Replace pump using a proper length bolt.
4. Excessive Thermactor System Noise (Putt-Putt, Whirling or Hissing)	a. Leak in hose	• Locate source of leak using soap solution and replace hoses as necessary.
	b. Loose, pinched or kinked hose	• Reassemble, straighten or replace hose and clamps as required.
	c. Hose touching other engine parts	• Adjust hose to prevent contact with other engine parts.
	d. By-pass valve inoperative	• Test by-pass valve and replace as necessary.
	e. Check valve inoperative	• Test check valve and replace as necessary.
	f. Pump mounting fasteners loose	• Torque fasteners to specification.
	g. Restricted or bent pump outlet fitting	• Inspect fitting and remove any flash blocking the air passage way. Replace bent fittings.
	h. Pump worn or damaged	• Replace pump.
	i. Air dumping through bypass valve (at idle only)	• On many vehicles the thermactor system has been designed to dump air at idle to prevent overheating the catalyst. This condition is normal. Determine that the noise persists at operating speeds before replacing the air pump.

DURA SPARK I SYSTEM DIAGNOSIS

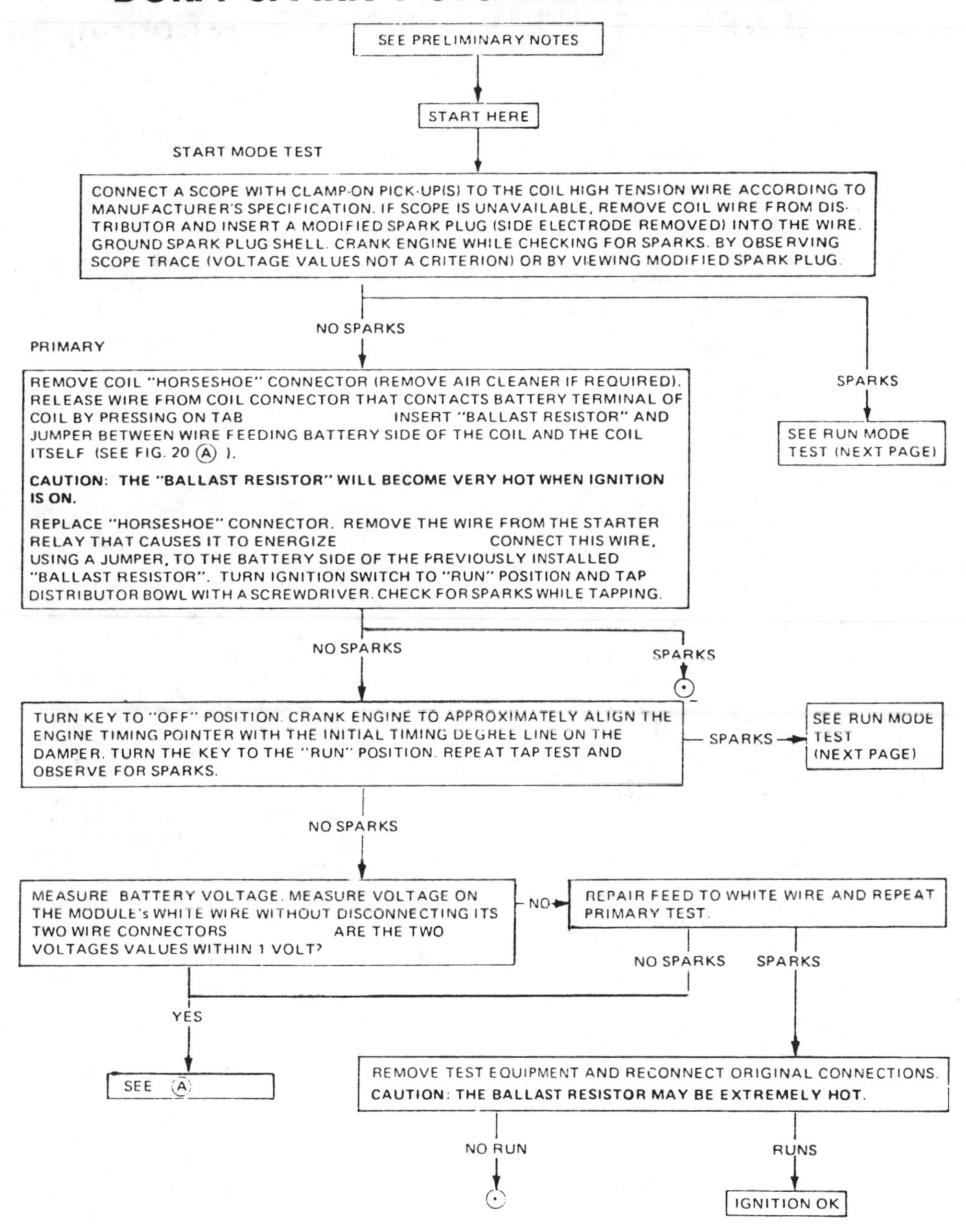

(•) SUGGESTIONS

. REPEAT PROCEDURE (RETURN TO RUN MODE TEST)
. SEE SUGGESTIONS FOR INTERMITTENT OPERATION

IMPORTANT

. WHEN REINSTALLING COIL WIRE, COAT THE INSIDE OF THE BOOT WITH SILICONE GREASE (D7AZ-19A331-A OR EQUIVALENT) USING A SMALL, CLEAN SCREWDRIVER BLADE.

(A) CONTINUE WITH DIAGNOSIS

PRELIMINARY NOTES:

- THE DURA SPARK I SYSTEM CAN BE IDENTIFIED BY THE PREFIX "D7AE" ON THE MODULE.
- EQUIPMENT NECESSARY TO CONDUCT THIS TEST IS:

ROTUNDA DIAGNOSTIC CONSOLE NO. 360036 (OR EQUIVALENT) OR A "MODIFIED SPARK PLUG" (SIDE ELECTRODE REMOVED).
BALLAST RESISTOR # B8A-12250-A 6" JUMPER WIRE FOR BALLAST RESISTOR WITH TWO D5AB-14474-AA TERMINALS (OR EQUIVALENT). A 2' TO 3' JUMPER WIRE WITH #24 A MUELLER CLIPS FOR SPARK PLUG GROUND (OR EQUIVALENT). TWO ORDINARY STRAIGHT PINS. ONE ORDINARY PAPER CLIP.
ONE KNOWN GOOD BREAKERLESS TYPE DISTRIBUTOR. ONE KNOWN GOOD D7AE MODULE.
ONE IGNITION COIL #D7AE-12029-AA (OR EQUIVALENT). ONE VOLTMETER, 0-20V RANGE.

Low voltage test, California (© Ford Motor Co.)

DURA SPARK I SYSTEM DIAGNOSIS—Continued

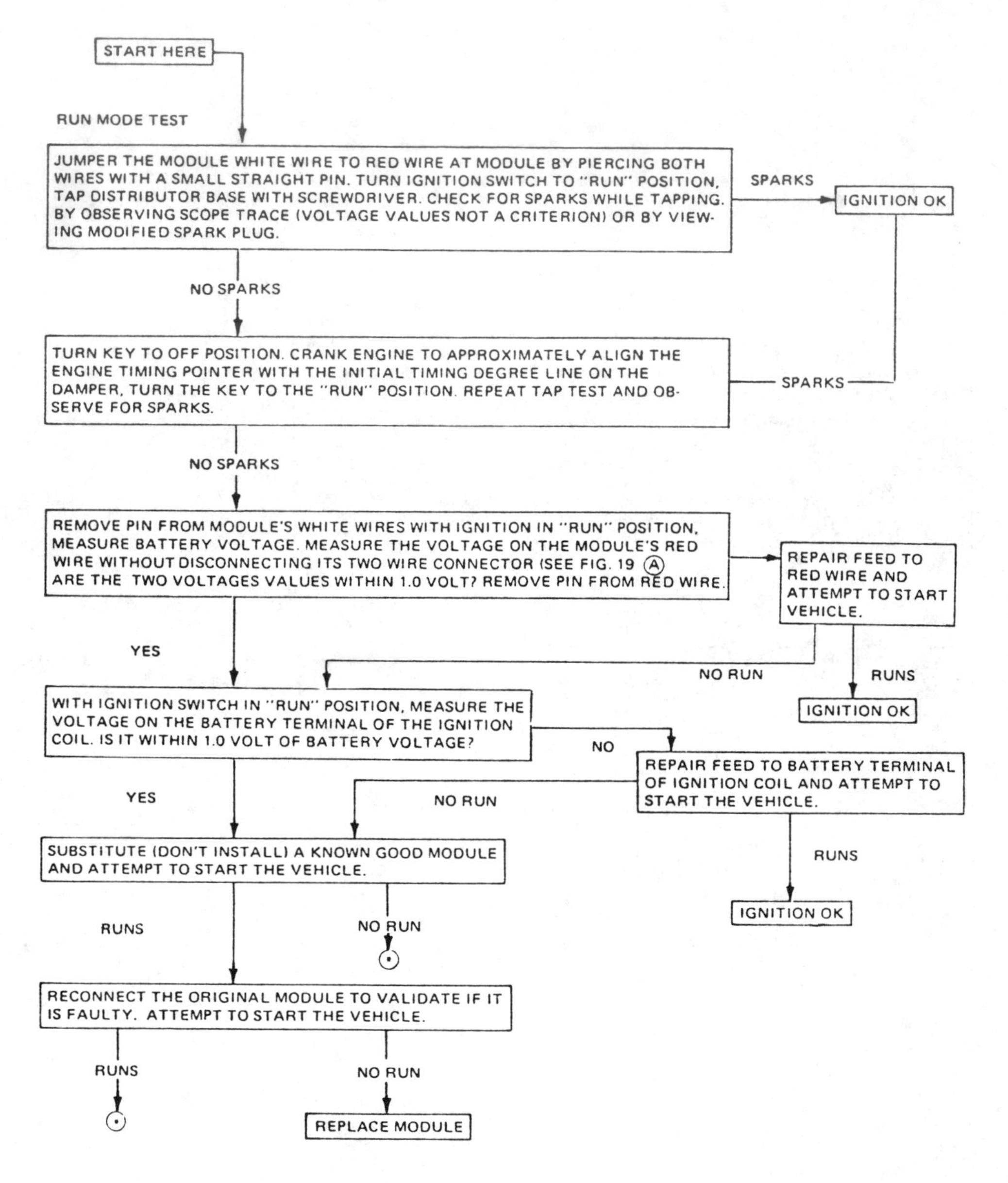

⊙ SUGGESTIONS

- REPEAT PROCEDURE (RETURN TO RUN MODE TEST).
- SEE SUGGESTIONS FOR INTERMITTENT OPERATION.

IMPORTANT

- WHEN REINSTALLING COIL WIRE, COAT THE INSIDE OF BOOT WITH SILICONE GREASE (D7AZ-19A331-A OR EQUIVALENT) USING A SMALL, CLEAN SCREWDRIVER BLADE.

Low voltage test, California (© Ford Motor Co.)

DURA SPARK I SYSTEM DIAGNOSIS—Continued

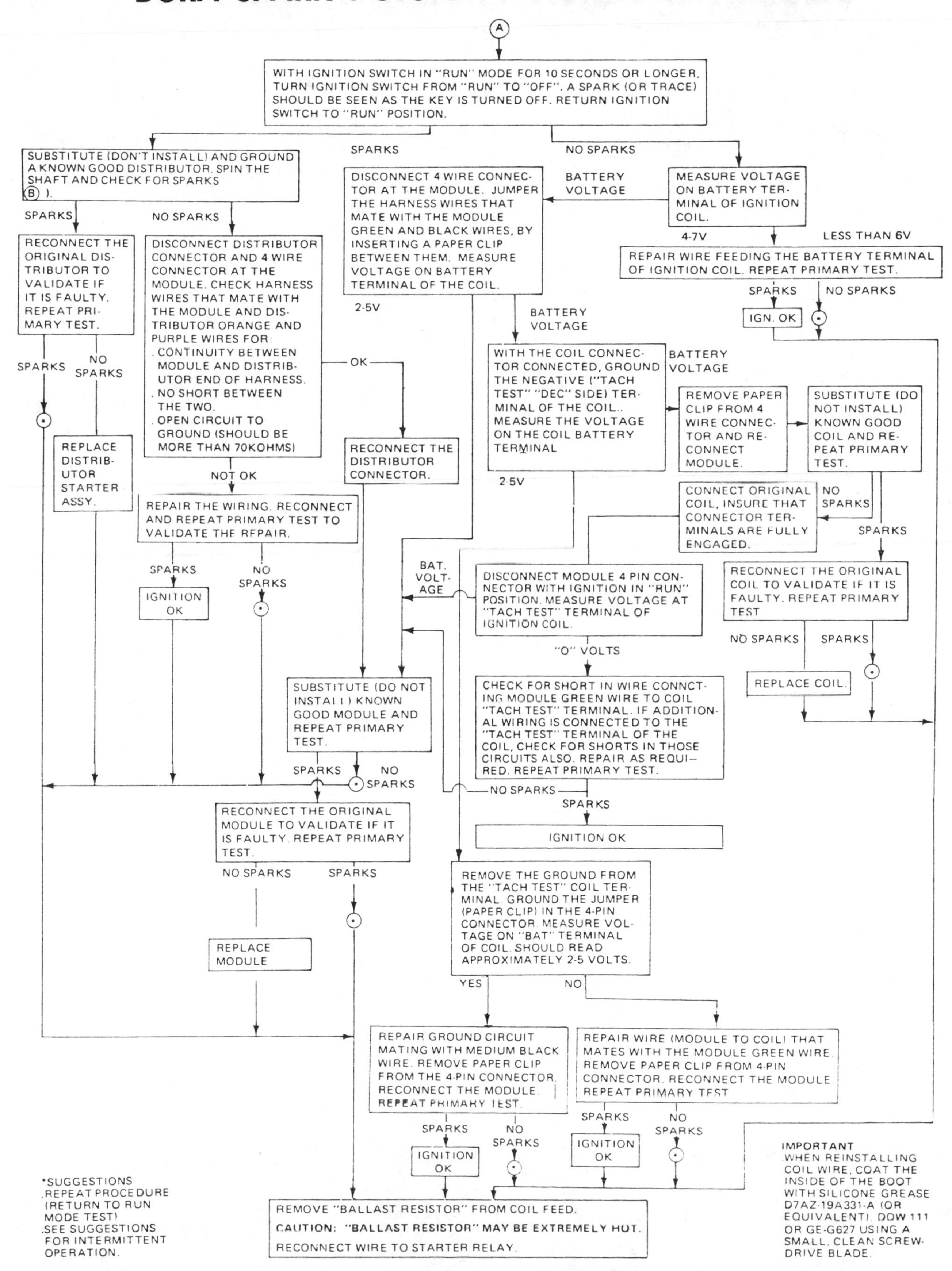

Low voltage test, California—cont'd.

DURA SPARK I SYSTEM DIAGNOSIS—Continued

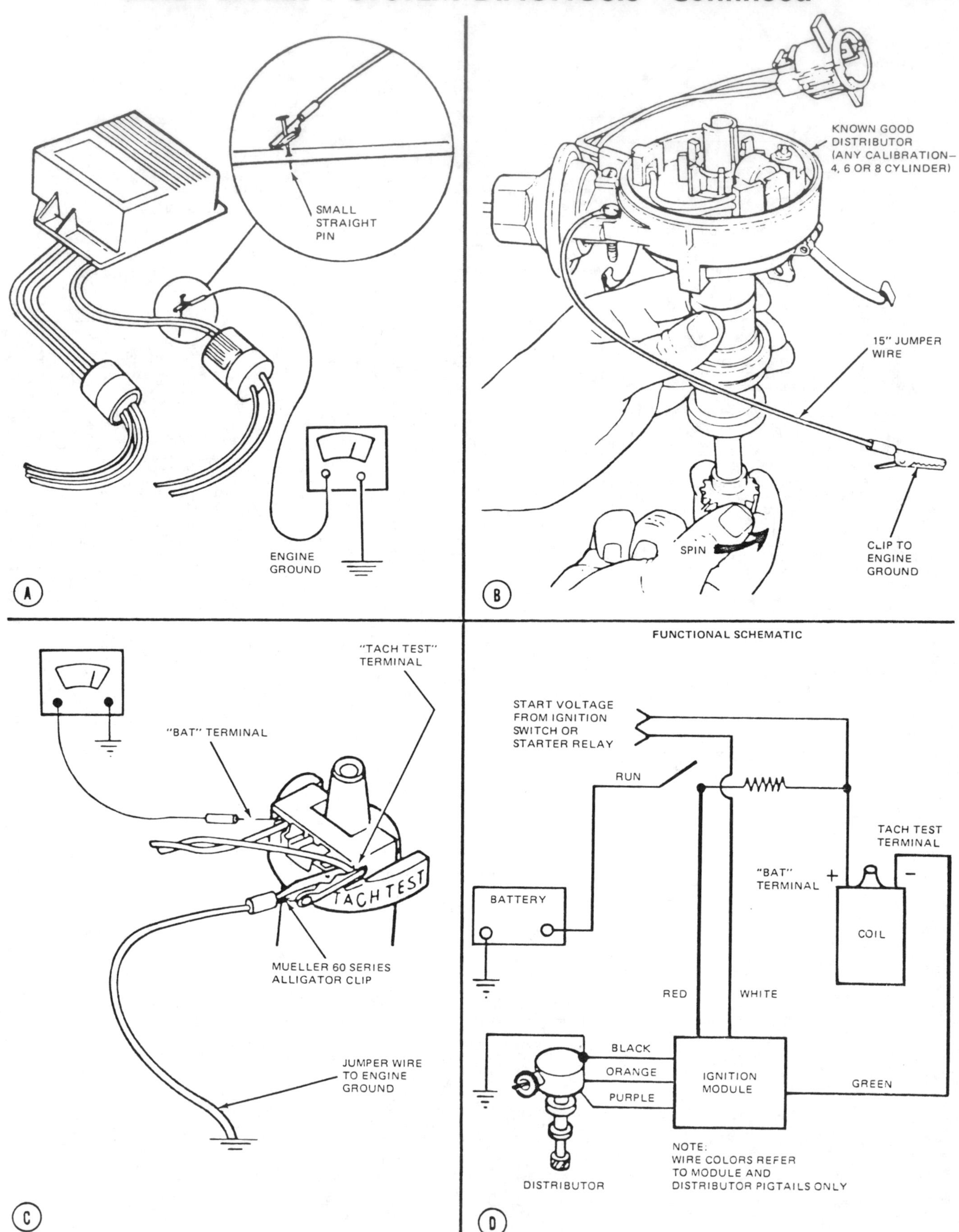

Dura Spark I test connections (© Ford Motor Co.)

DURA SPARK II SYSTEM DIAGNOSIS

PRELIMINARY NOTES

THE DURA SPARK II SYSTEM IS USED FOR ALL APPLICATIONS EXCEPT CALIFORNIA 302 (DURA SPARK I) AND EEC EQUIPPED VEHICLES.

EQUIPMENT NECESSARY TO CONDUCT THIS TEST IS:

- ROTUNDA DIAGNOSTIC CONSOLE NO. 360036 (OR EQUIVALENT) OR A "MODIFIED SPARK PLUG" (SIDE ELECTRODE REMOVED).
- 2' TO 3' JUMPER WIRE WITH #60 MUELLER ALLIGATOR CLIPS OR EQUIVALENT.
- 2' TO 3' JUMPER WIRE WITH #24A MUELLER ALLIGATOR CLIPS (FOR SPARK PLUG GROUND) OR EQUIVALENT.
- ONE ORDINARY STRAIGHT PIN.
- ONE ORDINARY PAPER CLIP.
- ONE KNOWN GOOD BREAKERLESS TYPE DISTRIBUTOR.
- ONE KNOWN GOOD IGNITION COIL #D5AE-12029-AB (OR EQUIVALENT).
- ONE VOLTMETER, 0-20V RANGE.

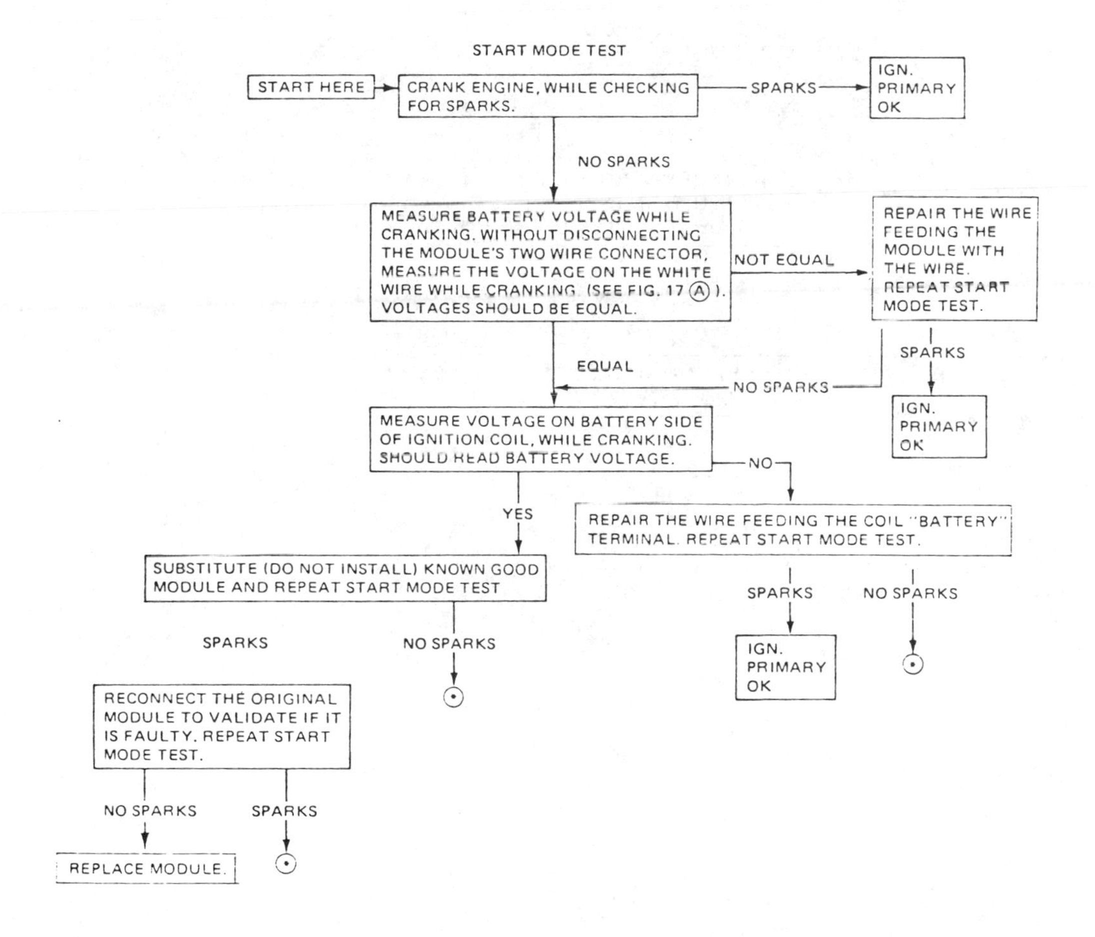

⊙ SUGGESTIONS

. REPEAT PROCEDURE (RETURN TO RUN MODE TEST).
. SEE SUGGESTIONS FOR INTERMITTENT OPERATION.

IMPORTANT

. WHEN REINSTALLING COIL WIRE, COAT THE INSIDE OF THE BOOT WITH SILICONE GREASE (D7AZ-19A331-A OR EQUIVALENT) USING A SMALL, CLEAN SCREWDRIVER BLADE.

Low voltage test (© Ford Motor Co.)

DURA SPARK II SYSTEM DIAGNOSIS—Continued

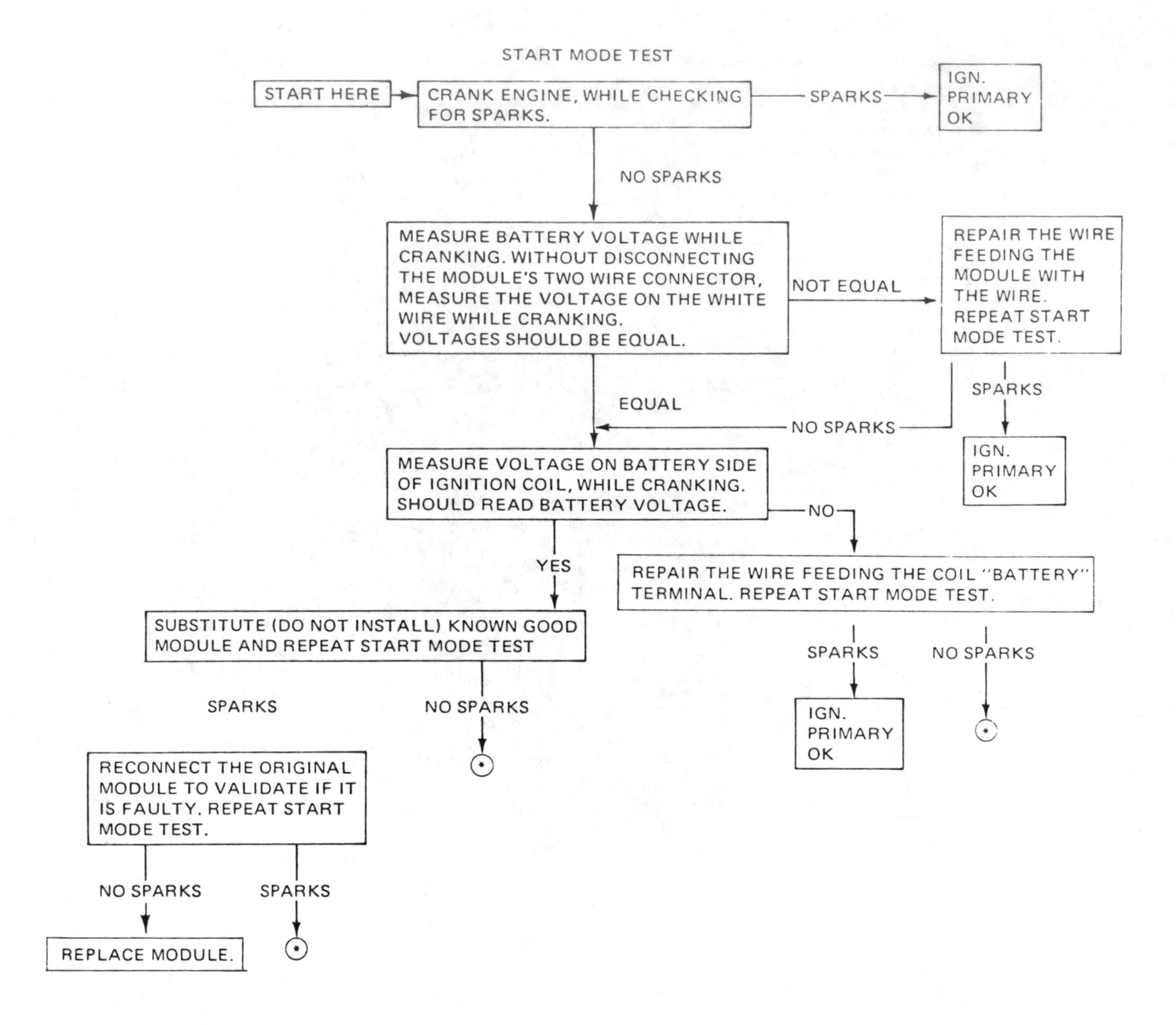

⊙ SUGGESTIONS

. REPEAT PROCEDURE (RETURN TO RUN MODE TEST).
. SEE SUGGESTIONS FOR INTERMITTENT OPERATION.

IMPORTANT

. WHEN REINSTALLING COIL WIRE, COAT THE INSIDE OF THE BOOT WITH SILICONE GREASE (D7AZ-19A331-A OR EQUIVALENT) USING A SMALL, CLEAN SCREWDRIVER BLADE.

Low voltage test—cont'd. (© Ford Motor Co.)

DURA SPARK II SYSTEM DIAGNOSIS—Continued

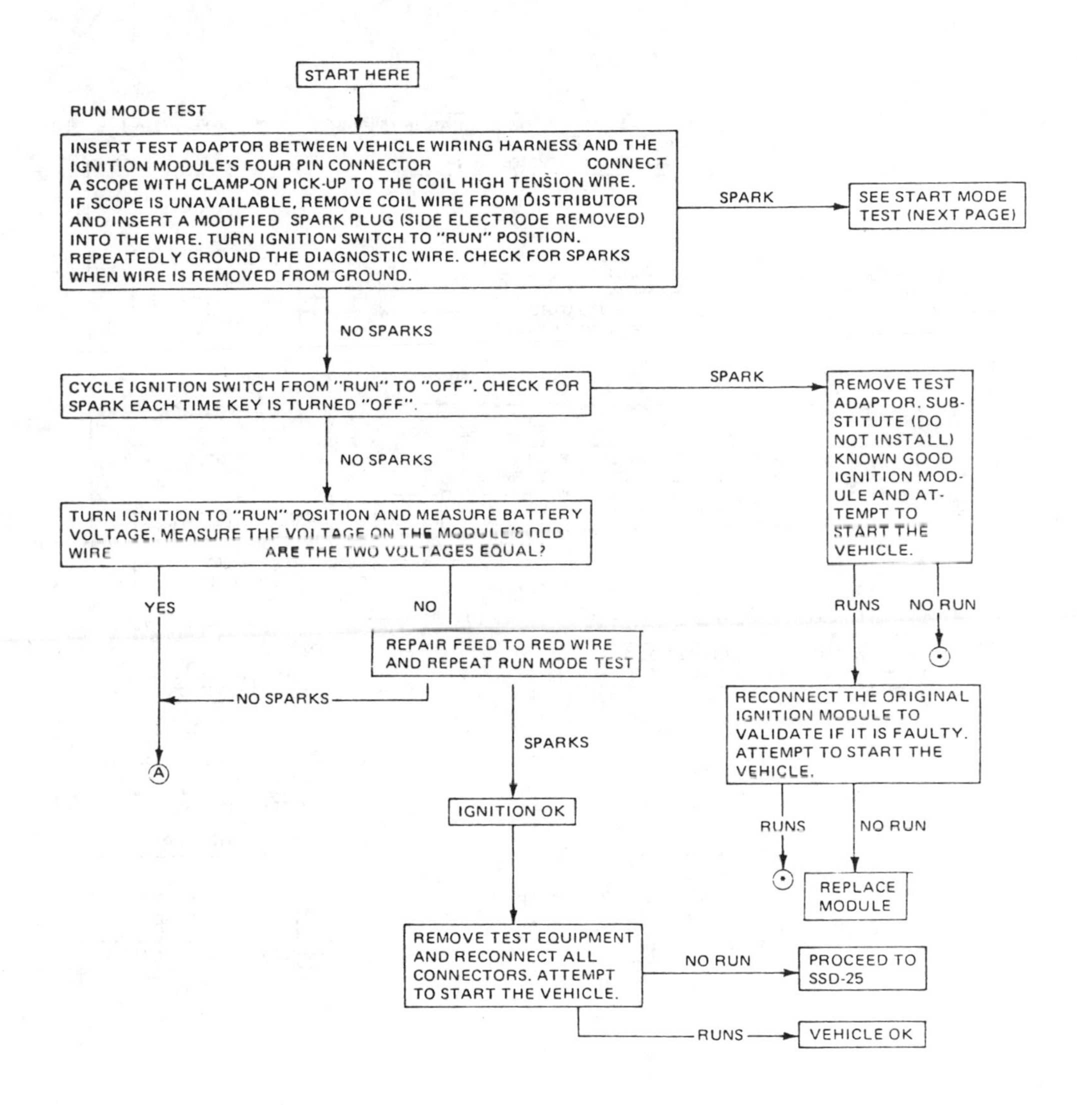

⊙ SUGGESTIONS

- REPEAT PROCEDURE (RETURN TO RUN MODE TEST)
- SEE SUGGESTIONS FOR INTERMITTENT OPERATION.

IMPORTANT

- WHEN REINSTALLING COIL WIRE, COAT THE INSIDE OF THE BOOT WITH SILICONE GREASE (D7AZ-19A331-A OR EQUIVALENT) USING A SMALL, CLEAN SCREWDRIVER BLADE.

Low voltage (Primary) E.E.C. (© Ford Motor Co.)

DURA SPARK II SYSTEM DIAGNOSIS—Continued

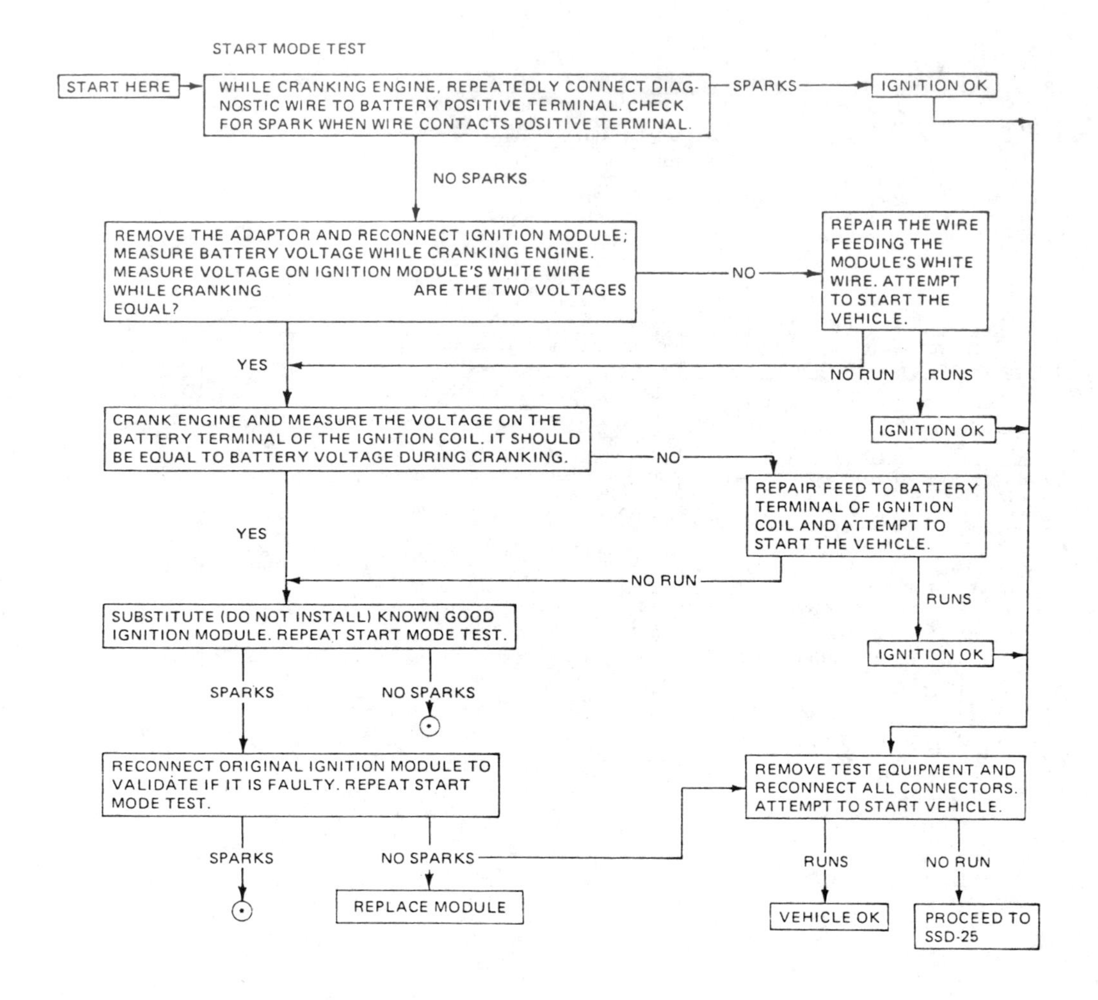

⊙ SUGGESTIONS

- REPEAT PROCEDURE (RETURN TO RUN MODE TEST)
- SEE SUGGESTIONS FOR INTERMITTENT OPERATION (FIG. 29)

IMPORTANT:

WHEN REINSTALLING COIL WIRE, COAT THE INSIDE OF THE BOOT WITH SILICONE GREASE (D7AZ-19A331-A OR EQUIVALENT) USING A SMALL, CLEAN SCREWDRIVER BLADE.

Low voltage (Primary) E.E.C.—cont'd. (© Ford Motor Co.)

DURA SPARK II SYSTEM DIAGNOSIS—Continued

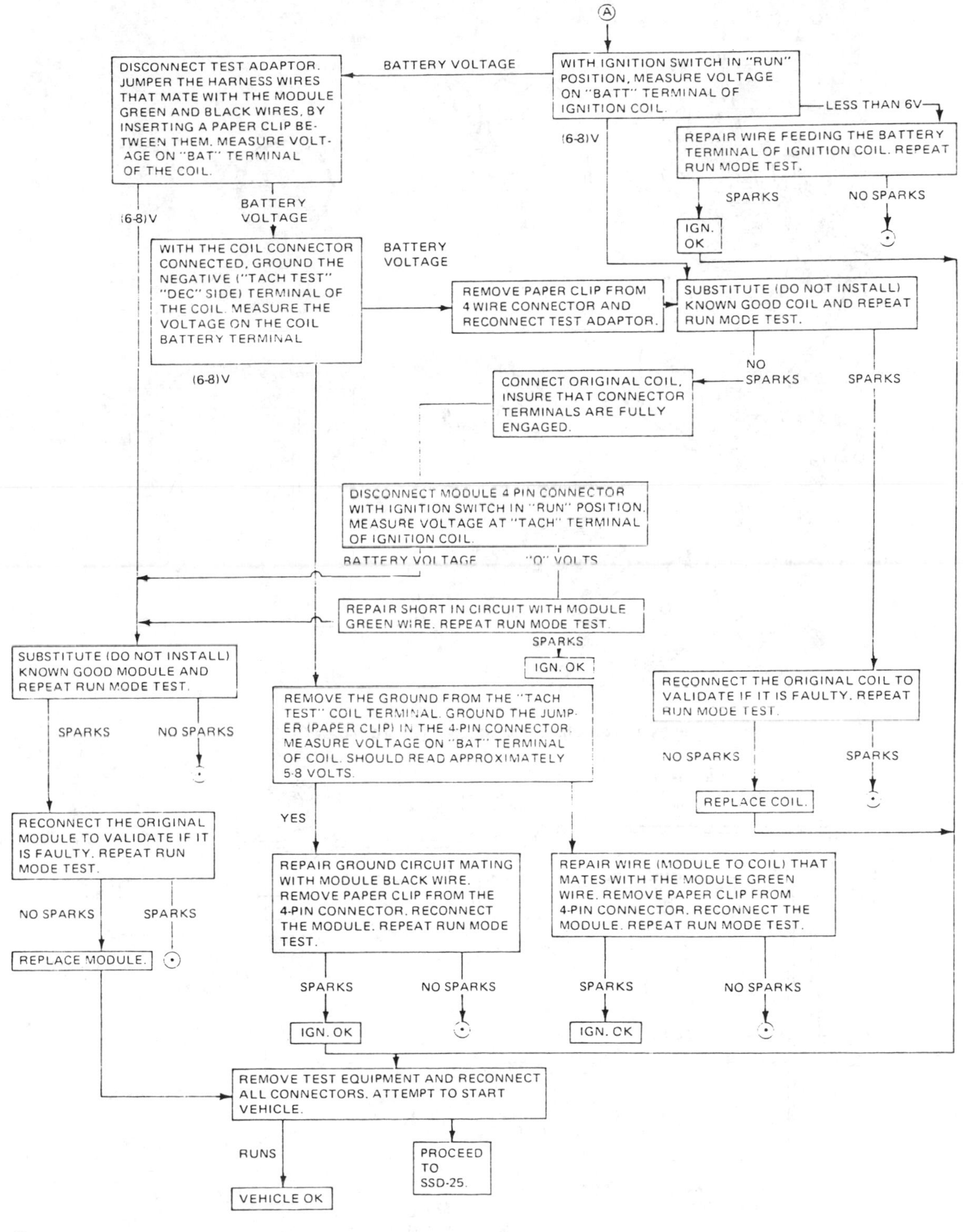

(•) SUGGESTIONS

. REPEAT PROCEDURE (RETURN TO RUN MODE TEST)
. SEE SUGGESTIONS FOR INTERMITTENT OPERATION

IMPORTANT

. WHEN REINSTALLING COIL WIRE, COAT THE INSIDE OF THE BOOT WITH SILICONE GREASE (D7AZ-19A331-A OR EQUIVALENT) USING A SMALL, CLEAN SCREWDRIVER BLADE.

Low voltage (Primary) E.E.C.—cont'd. (© Ford Motor Co.)

ELECTRONIC ENGINE CONTROL SYSTEM DIAGNOSIS

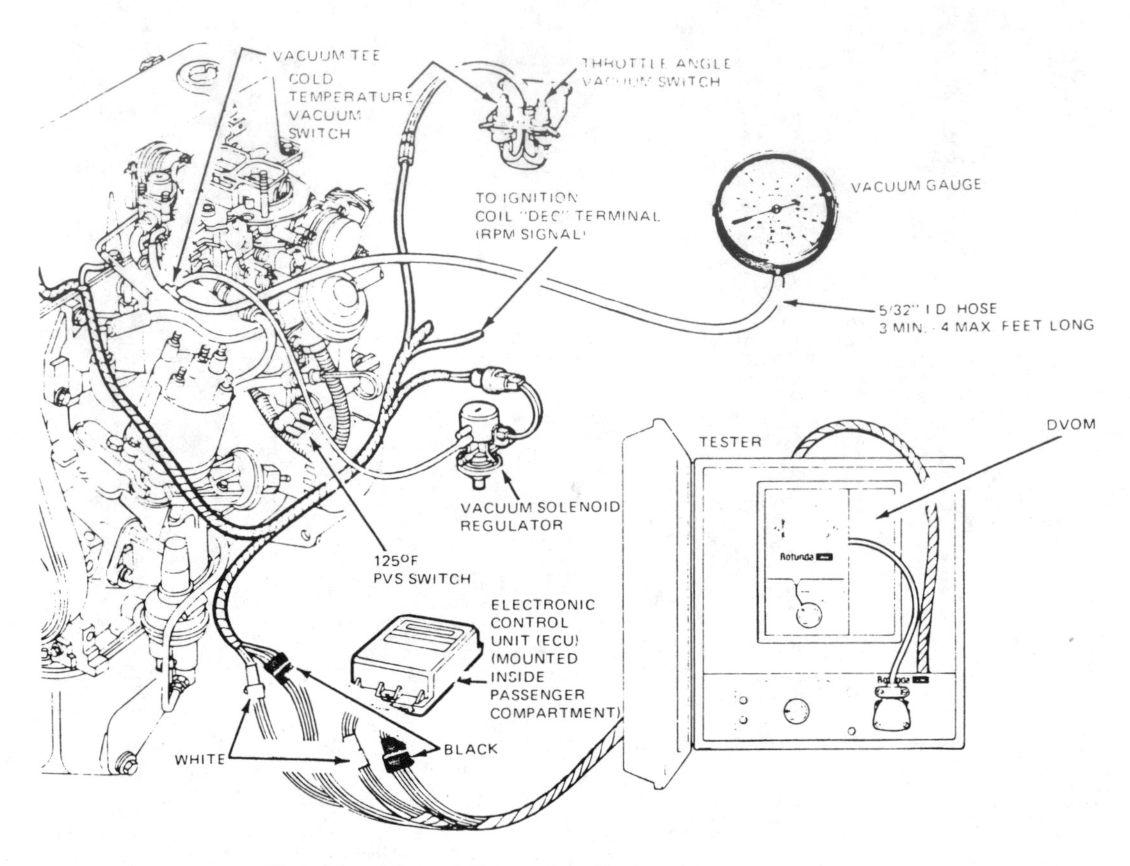

Test connections 2.3 L engine (© Ford Motor Co.)

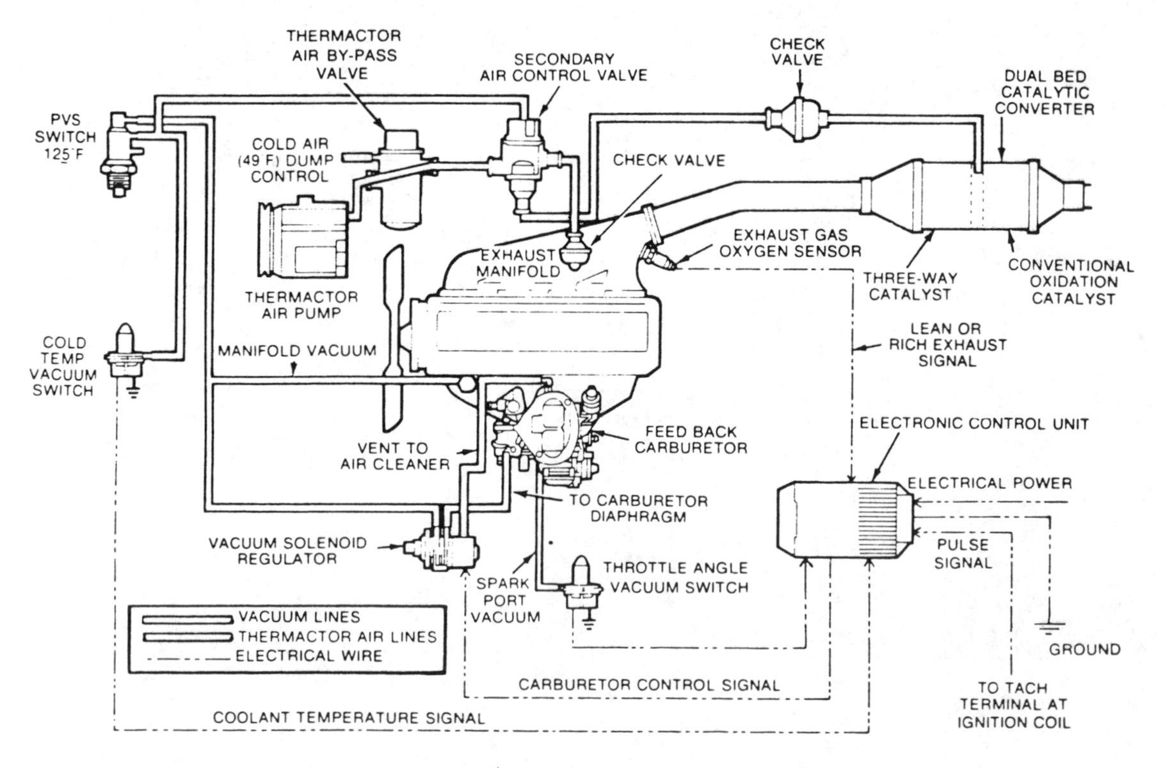

Typical system connections 2.3 L engine (© Ford Motor Co.)

ELECTRONIC ENGINE CONTROL SYSTEM DIAGNOSIS—Continued

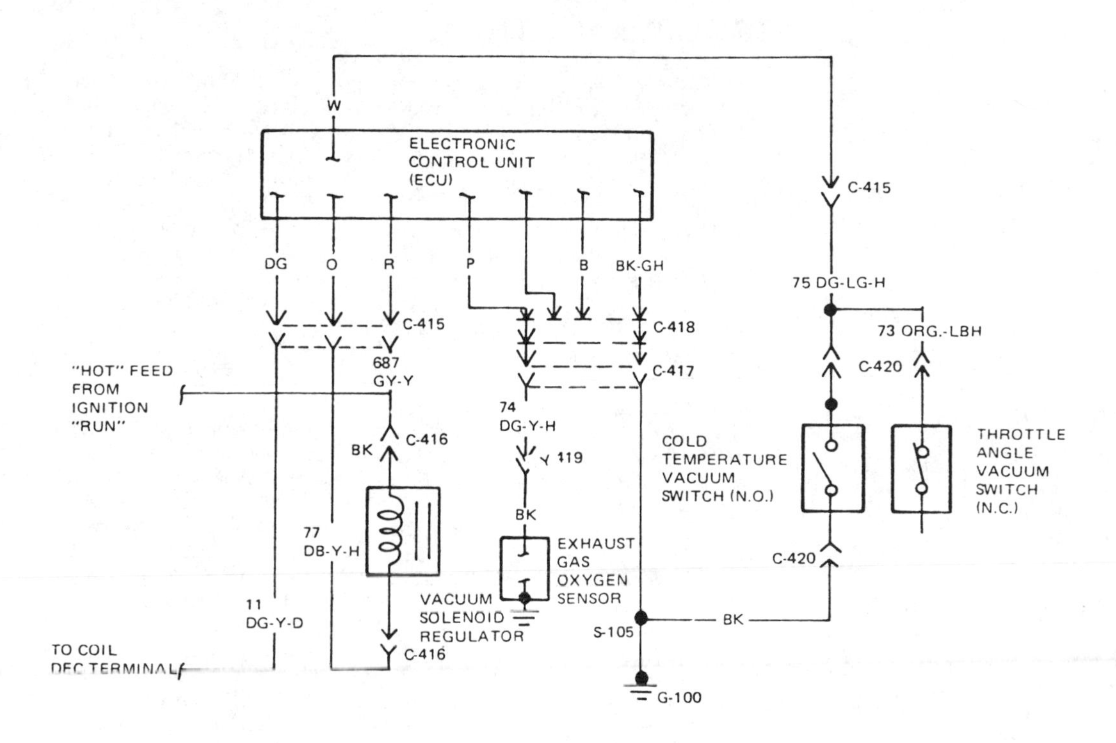

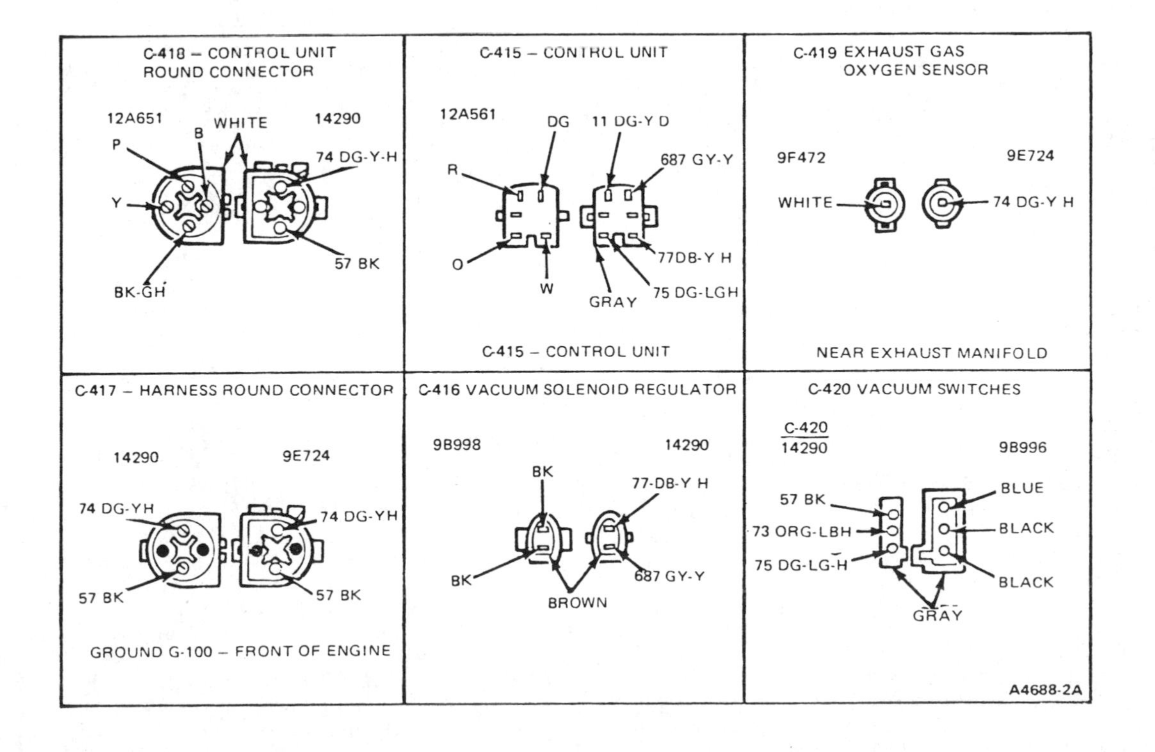

Wiring diagram 2.3 L engine (© Ford Motor Co.)

ELECTRONIC ENGINE CONTROL SYSTEM DIAGNOSIS

COMPONENT TROUBLESHOOTING

1. Reference Voltage (Vref) Not Within Limits
   a. Turn the TSS switch to position 9, ignition key in RUN position and DVOM switch to TESTER.
      • If the battery voltage is less than 10.5 (volts) check for:
      • A discharged battery
      • Open wire from the battery to the ECA power relay (circuit 175).
      • Open wire from ignition switch to the ECA power relay coil (circuit 20).
      • Open wire from the ECA to the power relay (circuit 361).
      • Inoperative ECA power relay.
      • Repair or replace the defective item-as-required and retest.*
      • If the battery voltage is greater than 10.5 (volts) proceed to step VI-D1b.
   b. Disconnect the Engine Coolant Temperature (ECT) sensor at its connector and set the TSS switch to position 1.
      • If the reference voltage is between 8.5 and 9.5 volts, replace ECT sensor and retest.*
      • If the reference voltage does not come within limits, proceed to the next step.
   c. Disconnect the Inlet Air Temperature Sensor (IAT) at its connector.
      • If the reference voltage is between 8.5 and 9.5 volts, replace IAT Sensor and retest.*
      • If the reference voltage does not come within limits, proceed to the next step.
   d. Disconnect the EGR Valve Position (EVP) Sensor at its connector.
      • If the reference voltage is between 8.5 and 9.5 (volts), replace the EGR Valve assembly and retest.*
      • If the reference voltage does not come within limits, proceed to the next step.
   e. Disconnect the Throttle Position (TP) Sensor at its connector.
      • If the reference voltage is between 8.5 and 9.5 (volts), replace TP Sensor and retest.*
      • If the reference voltage does not come within limits, proceed to the next step.
   f. Disconnect the Manifold Absolute Pressure (MAP) Sensor at its connector.
      • If the reference voltage is between 8.5 and 9.5 (volts), replace MAP Sensor and retest.*
      • If the reference voltage does not come within limits, proceed to the next step.
   g. Disconnect the Barometric Pressure (BP) Sensor at its connector.
      • If the reference voltage is between 8.5 and 9.5 (volts), replace BP Sensor and retest.*
      • If the reference voltage does not come within limits, proceed to the next step.
   h. Disconnect the blue (10 pin) harness connector from the TESTER.
      • If the reference voltage is within 8.5-9.5 (volts) check circuits 351 (A, B, C and D), 354 or 357 for a short or check circuits 359 (A thru F) for an open. Repair the circuit and restest.*
      (1) If reference voltage now within limits, install new ECA Calibration Assembly and retest.
      (2) If reference voltage does not come within limits, install new ECA Processor Assembly (Reinstall original Calibration Assembly). Retest.*

* Whenever a component has been repaired or replaced a retest is required.

   i. Reconnect all sensors to the harness and the harness to the TESTER. Complete gathering and recording data per TEST SEQUENCES.

2. Manifold Absolute Pressure (MAP) Sensor — Not Within Limit
   a. On the DIAGNOSTIC DATA CHART, compare the MAP and BP sensor readings of TEST SEQUENCE.
      • If the MAP reading is within 0.75 (volts) of the BP reading, both senors are ok. DO NOT CHANGE THEM—the out of tolerance condition is the result of variations in local air pressure.
   b. Verify that the Ignition switch is "OFF" and disconnect MAP sensor. Connect the yellow CONTINUITY TEST FIXTURE to the wiring harness. Set the TSS switch to position 13, and DVOM switch to TESTER.
      • If the reading is between 170 and 230 (ohms), proceed to the next step.
      • If the reading is less than 170 (ohms) or more than 230 (ohms), repair short or open in circuit 358 and/or 359 of wiring harness and retest.*
   c. Set the TSS switch to position 12.
      • If the reading is less than 170 (ohms) or more than 230 (ohms), repair short or open in circuit 351 and/or 359 of wiring harness and retest.*
      • If the reading is between 170 and 230 (ohms), inspect the sensor connector. If bad, repair. If good, replace MAP sensor and retest.*
      • If the voltage is still out of limits, replace ECA Processor Assembly (Reinstall the original Calibration Assembly). Retest.*
   d. Turn TESTER FUNCTION switch to "OFF" and disconnect the test equipment.
      CAUTION: *Remove CONTINUITY TEST FIXTURE.*

* Whenever a component has been repaired or replaced a retest is required.

   e. If MAP reading is not within limits during TEST SEQUENCE II,
      (1) Turn Ignition key "OFF."
      (2) Remove YELLOW CONTINUITY TEST FIXTURE from MAP connector of harness.
      (3) Using DVOM on 2000 (ohm) range, check resistance between adjacent pins of YELLOW TEST FIXTURE.
         • If resistance is less than 170 (ohms) or greater than 230 (ohms), repair or replace TEST FIXTURE (refer to TESTER INSTRUCTION MANUAL) and repeat TEST SEQUENCE.
         • If resistance is within 170 to 230 (ohms), carefully align pins of YELLOW TEXT FIXTURE with MAP connector of harness and reconnect. Repeat TEST SEQUENCE.

3. Throttle Position Sensor (TP)—Not Within Limits
   a. Verify that the ignition switch is "OFF." Set the TSS switch to position 12, and the DVOM switch to "TESTER." Disconnect the TP connector and connect the yellow CONTINUITY TEST FIXTURE to the TP

ELECTRONIC ENGINE CONTROL SYSTEM DIAGNOSIS—Continued

sensor harness connector.
- If the reading is between 170 and 230 (ohms), proceed to the next step.
- If the reading is less than 170 (ohms) or more than 230 (ohms), repair the short or open in circuit 351 and/or 359 and retest.*

b. Set the TSS switch to position 14.
- If the reading is between 170 and 230 (ohms), proceed to the next step.
- If the reading is less than 170 (ohms) or more than 230 (ohms), repair the short or open in circuit 355 and/or 359 and retest.*

c. Connect DVOM between circuit 351 and circuit 359 of the TP Sensor connector. Set DVOM switch to 200 x 1000 (ohms) range.
- If the reading is less than 3 (x 1000 ohms) or more than 5 (x 1000 ohms), replace the sensor and retest.*
- If the reading is between 3 and 5 (x 1000 ohms), proceed to the next step.

d. Connect DVOM between circuit 355 and circuit 359 of the connector. Set DVOM switch to 2000 (ohms) range and verify the throttle is in the closed position (i.e., off high cam).
- If the reading is less than 580 (ohms), or greater than 1100 (ohms), readjust by loosening the bolt and turning the TP sensor until the reading is within limits. If not able to set to this range replace the TP sensor and retest.*

NOTE: *Torque mounting screws to 8-10 in-lbs (.904-1.13 N•m).*
- If the reading is between 580 and 1100 (ohms) reconnect the sensor to the harness. Set the TSS switch to position 3, key to "RUN" position and DVOM to "TESTER." Adjust the sensor by loosening the bolt and turning the TP sensor until the VOLTMETER reads 1.82 ± 0.11 (volts). Retest.*

NOTE: *Torque mounting screws to 8-10 in-lbs (.904-1.13 N•m).*
- If the sensor cannot be adjusted to obtain the proper voltage, replace the ECA Processor Assembly (Reinstall the original Calibration Assembly). Retest.*

e. Turn DVOM switch to "OFF" and disconnect the test equipment.

CAUTION: *Remove CONTINUITY TEST FIXTURE.*

4. Barometric Pressure (BP) Sensor—Not Within Limits

a. On the BP and DIAGNOSTIC DATA CHART review the MAP reading of TEST SEQUENCE.
- If the BP reading is within 0.75 (volts) of the MAP reading both sensors are OK. DO NOT CHANGE THEM—the out of tolerance condition is the result of variations in local air pressure.
- If the BP reading differs from the MAP reading by 0.75 (volts) or more, proceed to step VI-D4b.

b. Verify that the ignition switch is "OFF" and disconnect BP sensor. Connect the yellow CONTINUITY TEST FIXTURE to wiring harness. Set the TSS switch to position 12 and DVOM to TESTER.
- If the reading is less than 170 (ohms) or more than 230 (ohms), repair short or open in circuit 351 and/or 359 of wiring harness and retest.*
- If the reading is between 170 and 230 (ohms), proceed to next step.

c. Set the TSS switch to position 15.
- If reading is less than 170 (ohms) or more than 230 (ohms), repair short or open in circuit 356 and/or 359 of wiring harness and retest.*
- If the reading is between 170 and 230 (ohms), inspect the sensor leads and connector. If bad, repair. If good, replace BP sensor and retest.*
- If the voltage is still out of limits, replace the ECA Processory Assembly (Reinstall the original Calibration Assembly). Retest.*

d. Turn DVOM switch to "OFF" and disconnect test equipment.

CAUTION: *Remove CONTINUITY TEST FIXTURE.*

5. Engine Coolant Temperature (ECT) Sensor — Not Within Limits

a. Verify the ignition switch is "OFF." Set the TSS switch to position 16, DVOM to TESTER. Disconnect the ECT connector and connect the blue CONTINUITY TEST FIXTURE to the wiring harness ECT connector.
- If the reading is between 170 and 230 (ohms), proceed to the next step.
- If the reading is less than 170 (ohms) or more than 230 (ohms), repair the short or open in circuit 354 or 359 and retest.*

b. Connect the DVOM between the two pins of the ECT sensor connector. Set DVOM switch to 200 x 1000 (ohms) scale.
- If reading is greater than 6 x 1000 (ohms), verify engine coolant temperature, per diagnostics routine No. 124. If coolant temperature is below 160°F (71°C) follow Diagnostic routine No. 124 and retest. If coolant temperature is above 160°F (71°C) replace the sensor and retest.*
- If reading is less than 1.5 (x 1000 ohms) verify engine coolant temperature per diagnostic routine No. 125. If coolant temperature is above 220°F (104°C) follow Diagnostic routine and retest.*
- If coolant temperature is below 220°F (104°C), replace the sensor and retest.*
- If reading is between 1.5 and 6.0 (x 1000 ohms), replace ECA Processor Assembly (Reinstall original Calibration Assembly). Retest.*

c. Turn TESTER FUNCTION switch to "OFF" and disconnect test equipment.

CAUTION: *Remove CONTINUITY TEST FIXTURE. Connect sensor to harness.*

6. Inlet Air Temperature (IAT)—Not Within Limits

a. Verify the Ignition switch is "OFF." Set the TSS switch to position 17. Disconnect the IAT connector and connect the blue CONTINUITY TEST FIXTURE to the wiring harness.
- If the reading is between 170 and 250 (ohms), proceed to the next step.
- If the reading is less than 170 (ohms), or more than 230 (ohms), repair the short or open in circuit 357 and/or 359 and retest.*

b. Connect the DVOM between circuit 357 and circuit 359 of the IAT sensor connector. Set DVOM switch to 200 x 1000 (ohms) scale.
- If the reading is less than 6.5 (x 1000 ohms) or greater than 45 (x 1000 ohms), check air cleaner temperature control per SSD-9, if the temperature control is ok replace the IAT sensor and retest.*

* Whenever a component has been repaired or replaced a retest is required.

ELECTRONIC ENGINE CONTROL SYSTEM DIAGNOSIS—Continued

CAUTION: *Remove continuity text Fixture. Connect sensor to harness.*

• If the reading is between 6.5 and 45 (x 1000 ohms), replace the ECA Processor Assembly (Reinstall original Calibration Assembly). Retest.*

c. Turn TESTER FUNCTION switch to "OFF" and disconnect test equipment.

d. If IAT reading is *not* within limits during TEST SEQUENCE II,

• Turn Ignition key "OFF."

• Remove BROWN CONTINUITY TEST FIXTURE from IAT connector of harness.

• Using DVOM an 200 x 1000 (ohms) range, check resistance between pins of BROWN TEST FIXTURE.

• If resistance is less than 23.3 (x 1000 ohms) or greater than 25.3 (x 1000 ohms), repair or replace TEST FIXTURE (refer to TESTER INSTRUCTION MANUAL) and repeat TEST SEQUENCE.

• If resistance is within 23.3 to 25.3 (x 1000 ohms), carefully align pins of BROWN TEST FIXTURE with IAT connector of harness and reconnect. Repeat TEST SEQUENCE.

7. Thermactor Air-Bypass Solenoid Actuator—Not Within Limits

a. To verify Throttle Position sensor detects changes in throttle position, set TSS switch to position 3, DVOM to Tester position and the ignition key to the RUN position.

b. Set throttle to the Wide Open Throttle position and observe DVOM reading.

• If reading is 5.0 (volts) or greater, TP sensor is OK. Proceed to paragraph VI-D7c.

• If reading is less than 5.0 (volts). Remove and replace TP sensor and retest.*

c. Turn Ignition Key "OFF." Set DVOM switch to the 200 OHM position and install external test leads in DVOM test jacks.

d. Check TAB solenoid coil as follows:

(1) Disconnect harness connector from TAB solenoid.

(2) Measure resistance of solenoid coil by inserting DVOM probes into solenoid electrical contacts and observing DVOM reading.

• If reading is between 50 and 90 (ohms), solenoid coil is OK. Remove DVOM leads from both coil and DVOM and re-install harness connector on solenoid. Proceed to paragraph VI-D7e.

• If reading is less than 50 (ohms) or more than 90 (ohms), install new TAB solenoid, reconnect harness and retest.*

e. Set the TSS switch to position 7, DVOM to the TESTER position and the ignition key in the RUN position.

• If reading is greater than 10.5 (volts) proceed to VI-D7f.

• If reading is less than 10.5 (volts) check circuit 361. With a DVOM measure the voltage between circuit 361 and the engine block. If reading is less than 10.5 (volts) repair open or short in circuit 361 and retest.*

• If circuit 361 reading is greater than 10.5 (volts), repair open or short in circuit 100 and retest.*

f. Temporarily install known, good ECA Processor Assembly (Reinstall original Calibration Assembly) and retest.*

• If system tests OK, install new ECA Processor (Reinstall original Calibration Assembly) and retest.*

• If problem still exists, temporarily install known good ECA Calibration Assembly on to original Processor Assembly and retest.*

(1) If system tests OK, install new Calibration Assembly and retest.*

(2) If problem still exists, install new ECA Processor and new ECA Calibration Assembly and retest.*

8. EGR Valve Position (EVP) Sensor—Not Within Limits

a. If EVP Sensor reading is not within limits during Test Sequence I, proceed to Section VI.D.8.b. If not within limits during Test Sequence II, proceed to Section VI.D.8.e.

b. Turn "OFF" engine and disconnect the EVP sensor. Connect the yellow CONTINUITY TEST FIXTURE to the wiring harness. Set the TSS switch to position "12," DVOM switch to TESTER.

• If the reading is less than 170 (ohms) or more than 230 (ohms), repair the short or open in circuit 351 and/or 359 of the harness and retest.*

• If the reading is between 170 and 230 (ohms) proceed to the next step.

c. Set the TSS switch to position 18.

• If the reading is less than 170 (ohms) or more than 230 (ohms), repair the short or open in circuit 352 and/or 359 of the harness and retest.*

• If the reading is between 170 and 230 (ohms), proceed to the next step.

d. CAUTION: Remove continuity test fixture before proceding.

e. Disconnect the EGR air hose from the AIR BYPASS VALVE (small hose on side of valve) and connect the PRESSURE GAGE to the fitting. Leave EGR hose disconnected during this step. Run the engine at about 1600 rpm in "park."

• If the pressure is greater than 1.5 PSI proceed to VI-D8f.

• If the pressure is less than 1.5 PSI check the AIR BYPASS VALVE and THERMACTOR AIR PUMP per SSD-5. Repair or replace as required and retest.*

f. Connect a known good test hose between the small port at the side of AIR BYPASS VALVE and the EGR valve assembly. Run the engine at about 1600 RPM in "park" with the TSS switch in position 8.

• If the voltage is 6.4 (volts) or less, replace the EGR valve and sensor assembly and retest.*

• If the voltage is greater than 6.4 (volts), reinstall the original hoses and proceed. Reconnect all pressure and vacuum hoses to their proper connection.

9. EGR Solenoid Valve Check

a. Turn the key to "RUN" (engine "OFF"). Set the TSS to position 19. Depress and hold the "EGR PRESSURE" test button on TESTER while reading DVOM.

(1) If the reading is less than 1.0 (volt), check for open in harness circuits 361 or 362 or short in harness circuit 362 to ground. Repair or replace as required and retest.*

• If harness circuits 361 and 362 are OK, install new solenoid valve assembly and retest.*

(2) If the reading is greater than 5.0 (volts), check for short between harness circuits 362 and 361. Repair or replace as required and retest.*

• If harness is OK, replace the solenoid valve assembly

ELECTRONIC ENGINE CONTROL SYSTEM DIAGNOSIS—Continued

and retest.*
(3) If the reading is between 1 and 5 (volts), proceed to the next step.

b. Set the TSS to position 20 and depress and hold the "EGR VENT" button on TESTER.
(1) If the reading is less than 1.0 (volt), check for open in harness circuits 360 or 361 or short in harness circuit 360 to ground. Repair or replace as required and retest.*
• If harness circuits 360 and 361 are OK, install new solenoid valve assembly and retest.*
(2) If reading is greater than 5.0 (volts), check for short between circuits 360 and 361. Repair or replace as required and retest.
• If harness is OK, replace the solenoid valve assembly and retest.*
(3) If the reading is between 1.0 and 5.0 (volts), proceed to the next step.

c. Disconnect the pressure hose from the input side of the EGR "PRESSURE" solenoid and insert pressure gage into hose. Operate engine, in "NEUTRAL," at 1100 ± 100 rpm and observe pressure gage.
• If pressure is less than 0.5 psi, check Thermactor system per SSD-5. Repair or replace Thermactor system parts as required and retest.*
• If pressure is greater than 0.5 psi, proceed to next step.

d. Disconnect the pressure hose from the output side of the EGR pressure solenoid (lower fitting) and connect a pressure gage to the solenoid. Leave pressure hose disconnected. Run the engine at 1100 ± 100 rpm with the transmission selector in "neutral" or "park."
• If the pressure is greater than 0.5 psi after 5 seconds, replace the solenoid valve assembly and retest.*
• If the pressure is less than 0.5 psi, proceed to the next step.

e. Reconnect the pressure hose to the output side of EGR pressure solenoid. Disconnect the EGR valve pressure hose at the hose tee and connect the pressure gage to the tee. Leave EGR valve hose disconnected. Run the engine at about 1100 ± 100 rpm with the transmission selector in "neutral" or "park." Depress and hold the EGR PRESSURE and EGR VENT test buttons, on the TESTER, until the pressure stabilizes. Then release the EGR PRESSURE button.
• If the maximum pressure is less than 0.5 psi or decreases more than 0.5 psi in 5 seconds, replace the EGR solenoid assembly or hoses and fittings and retest.*
• If the maximum pressure is greater than 0.5 psi and decreases less than 0.5 psi in 5 seconds, proceed to the next step.
With the engine at 1100 ± 100 rpm and EGR valve hose still disconnected, press and hold the EGR PRESSURE and VENT buttons until the pressure stabilizes. Release EGR PRESSURE button, then release the EGR VENT button. The pressure should drop IMMEDIATELY.
• If the pressure remains within 0.5 psi of the maximum reading for 5 seconds replace the EGR solenoid assembly and retest.
• If the pressure drops immediately to less than half of the maximum reading, replace the EGR valve assembly and retest.*
• If after retest, the EVP data is still out of limits, proceed to the next step.

f. Temporarily install known good ECA Calibration Assembly on to ECA Processor Assembly and retest.*
(1) If system now operates properly, install new Calibration Assembly onto original Processor Assembly and retest.*
(2) If the problem still exists, temporarily install known good ECA Processor (Reinstall original Calibration Assembly). Retest.*
• If system now operates properly, install new ECA Processor Assembly (reinstall original Calibration Assembly) and retest.*
• If problem still exists, install new ECA Processor Assembly and new ECA Calibration Assembly. Retest.*

10. Spark Advance (in degrees) Not Within Limits

NOTE: *If vehicle is equipped with Dual Mode Ignition System, verify correct dual mode operation before performing the following paragraphs. If repair or replacement of Ignition system components(s) was necessary, retest EEC sub-system (1) to verify original problem has been corrected.*

a. Run engine at idle, in "PARK," to check base engine timing per the following steps.

b. Depress and hold "NO START" button on TESTER while reading SPARK ADVANCE using Timing Light.
(1) If SPARK ADVANCE is less than 8 degrees BTDC or more than 12 degrees BTDC, check alignment of Timing Pointer and position of Crankshaft Position Pulse Ring (see Engine Shop Manual).
(2) If SPARK ADVANCE is within 8 degrees to 12 degrees BTDC, temporarily install known good ECA Calibration Assembly on original ECA Processor and Retest.*
• If SPARK ADVANCE is now in limits in Test Sequence III, install new Calibration Assembly on to original Processor Assembly and retest.*
• If SPARK ADVANCE is still not within limits of Test Sequence III, temporarily install a known good ECA Processor assembly (re-install original Calibration Assembly) and retest.*
• If SPARK ADVANCE is now in limits in Test Sequence III, install new ECA Processor Assembly (reinstall original Calibration Assembly) and retest.*
• If SPARK ADVANCE is still not within limits of Test Sequence III, install new ECA Processor and new ECA Calibration Assembly and retest.*

11. Crank Signal—Not Within Limits

a. Verify correct vehicle battery voltage.

b. Disconnect harness wire from the "S" terminal on the Starter Relay and set TESTER TSS to position "22."

c. With Ignition Key in "START" position, verify reading is 9.0 (volts) or greater.
• If reading is less than 9.0 (volts), check for an open or short in harness circuits 32 and 57. Repair as required and retest.*

CAUTION: *Reconnect harness wire to "S" terminal on Starter Relay.*

* Whenever a component has been repaired or replaced a retest is required.

ELECTRONIC ENGINE CONTROL SYSTEM DIAGNOSIS—Continued

Vehicle Condition	Test Selector Switch Position 1 Vref	 2 MAP	 3 TP	 4 BP	 5 ECT	 6 IAT	 7 TAB	 8 EVP	Spark Advance (degrees)
TEST SEQUENCE I 1. Parking brakes "on" 2. Engine, "off" & "hot", key "run" 3. All accessories "off" 4. Crank signal checked per VI.A.4.	Limits 8.5 – 9.5V ** Reading ___V	Limits table A Reading ___V	Limits 1.65 2.14V Reading ___V	Limits table A Reading ___V	Limits 1.9 – 3.7V Reading ___V	Limits 4.1 – 6.3V Reading ___V	Limits Greater Than 10.5V Reading ___V	Limits 1.09 – 1.61V Reading ___V	
TEST SEQUENCE II 1. Parking brakes "on" 2. Install test fixtures: MAP (yellow), IAT (brown) 3. Engine "hot" and "running" in "park" at "curb idle". All accessories "off". (4.) Observe MAP and IAT data. (5.) Raise RPM to 1600 – 1800. Start timing TAB when throttle returns to curb idle. (6.) Set engine to 1600 ± 50 RPM. Observe EVP and Spark Advance. (See Emission Label for Limits).		Limits (4) 3.5 – 5.5V ** Reading ___V				(4) Limits 5.2 – 7.2V ** Reading ___V	(5) See Limits Less than 1.6V changing to greater than 10.5V within 60 ± 5 sec. yes no	(6) See Limits to engine emission label Reading ___V	(6) See Limits to engine emission label Reading ___V
DATA ANALYSIS:	If out of limits** Step VI.D.1.	If out of limits* Step VI.D.2.	If out of limits* Step VI.D.3.	If out of limits* Step VI.D.4.	If out of limits* Step VI.D.5.	If out of limits* Step VI.D.6.	If out of limits* Step VI.D.7.	If out of limits* Step VI.D.8.	If out of limits* Step VI.D.10.

5.0 L (302 eng) diagnostic data chart (© Ford Motor Co.)

* Complete all data gathering per Test Sequence I and II prior to performing any Diagnostic Subroutine except as noted with "**"

** If reading is out of limits, perform indicated diagnostic Step prior to continuing Test Sequence.

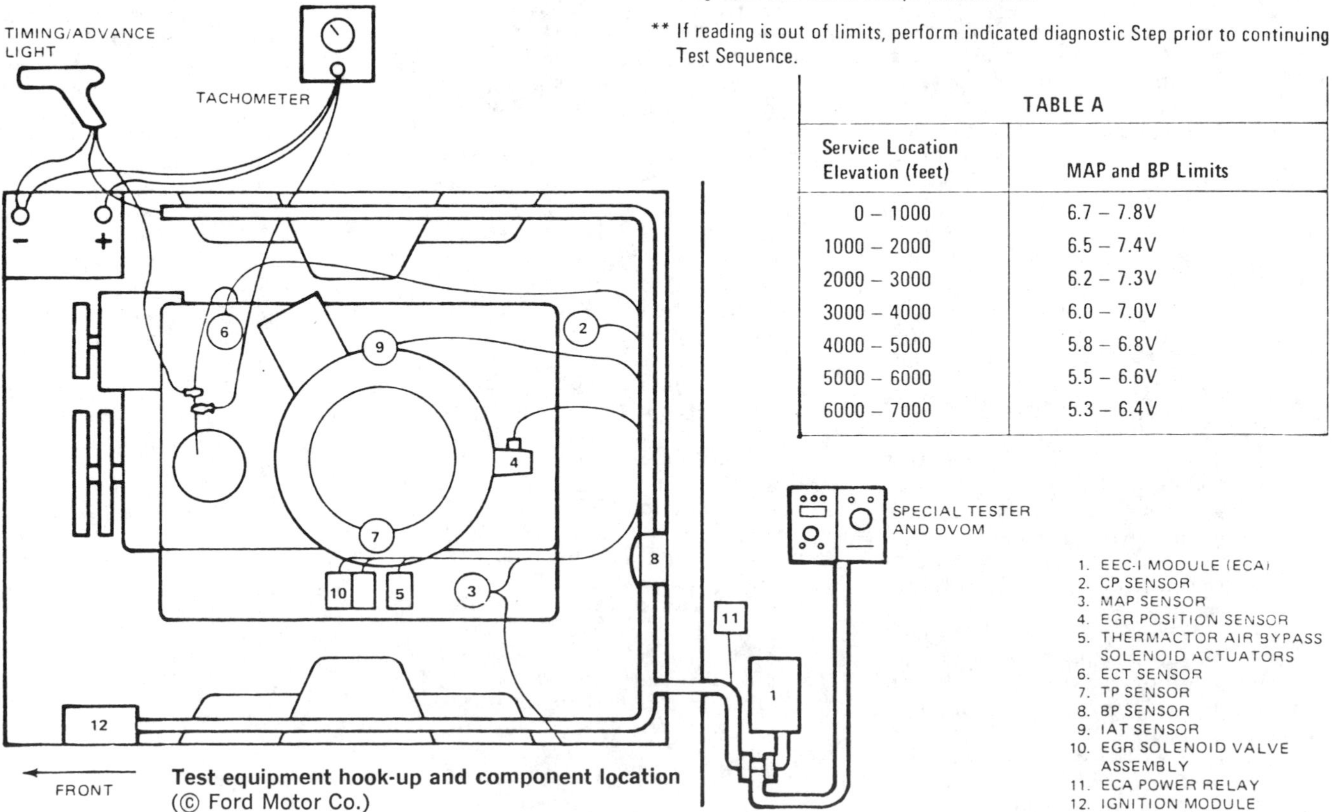

Test equipment hook-up and component location (© Ford Motor Co.)

TABLE A

Service Location Elevation (feet)	MAP and BP Limits
0 – 1000	6.7 – 7.8V
1000 – 2000	6.5 – 7.4V
2000 – 3000	6.2 – 7.3V
3000 – 4000	6.0 – 7.0V
4000 – 5000	5.8 – 6.8V
5000 – 6000	5.5 – 6.6V
6000 – 7000	5.3 – 6.4V

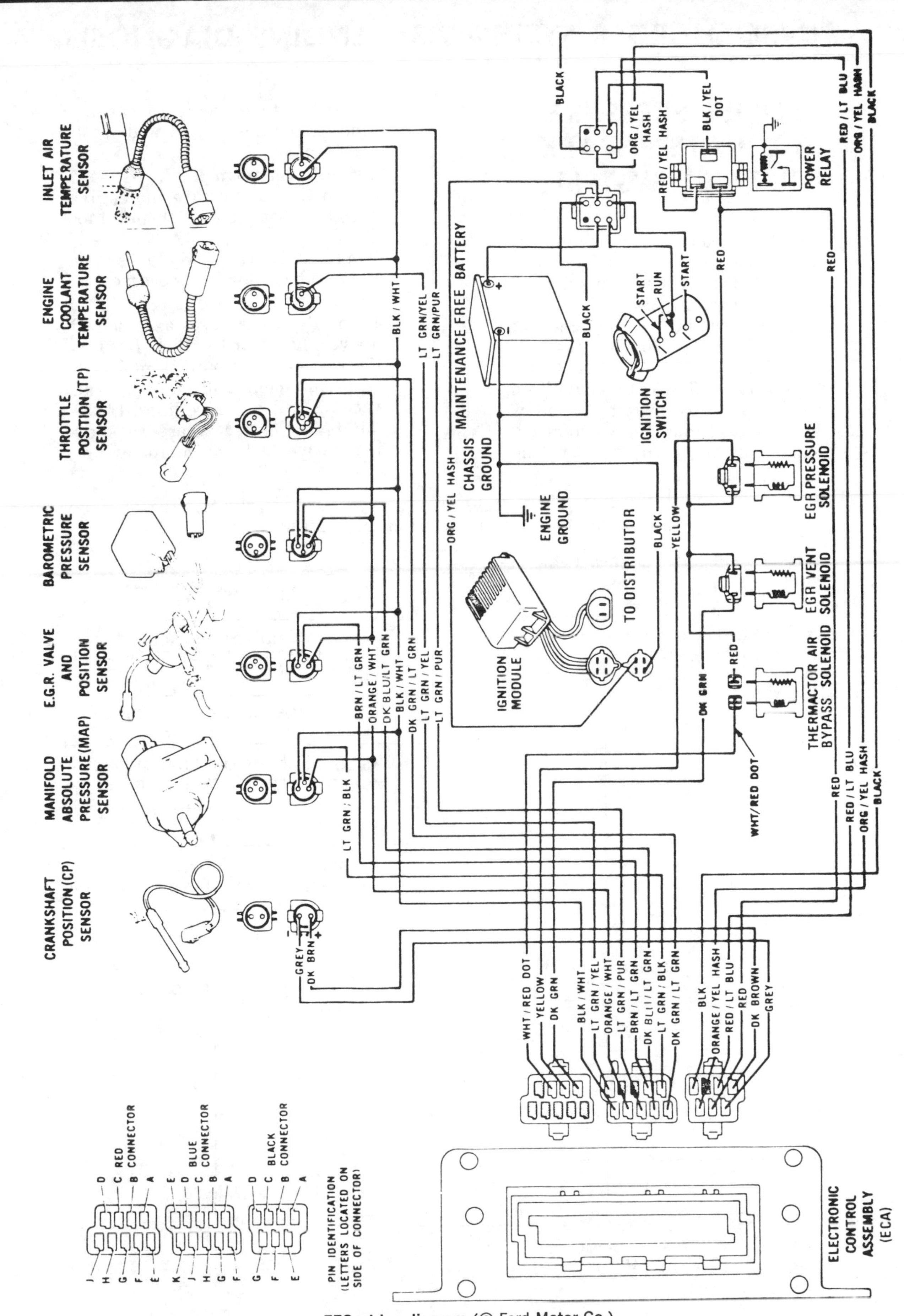

EEC wiring diagram (© Ford Motor Co.)

TURBOCHARGER SYSTEM 2.3 L ENGINE DIAGNOSIS

BOOST PRESSURE ACTIVATED SPARK RETARD SYSTEM

1. Check the basic engine timing, if not previously done.
2. Remove the pressure supply line to the ignition timing pressure switch assembly (3 switches on one bracket). Cap the line and install an external pressure source (Tool T79P-6634-A) to the ignition timing pressure switch assembly.
3. Remove and plug the vacuum advance line to the distributor and connect a tachometer to the engine. Start the engine and let it warm up.
4. Using a Snap-On Tool #GA-437 or equivalent, increase the engine speed to between 1350 and 1400 RPM.
5. Apply pressure slowly as shown in the table below and observe rpm changes:

   If the RPM decreases as specified above, the retard system is functioning satisfactorily, remove all test equipment and reconnect all vacuum and pressure lines. If the RPM does not decrease as specified above perform the following:

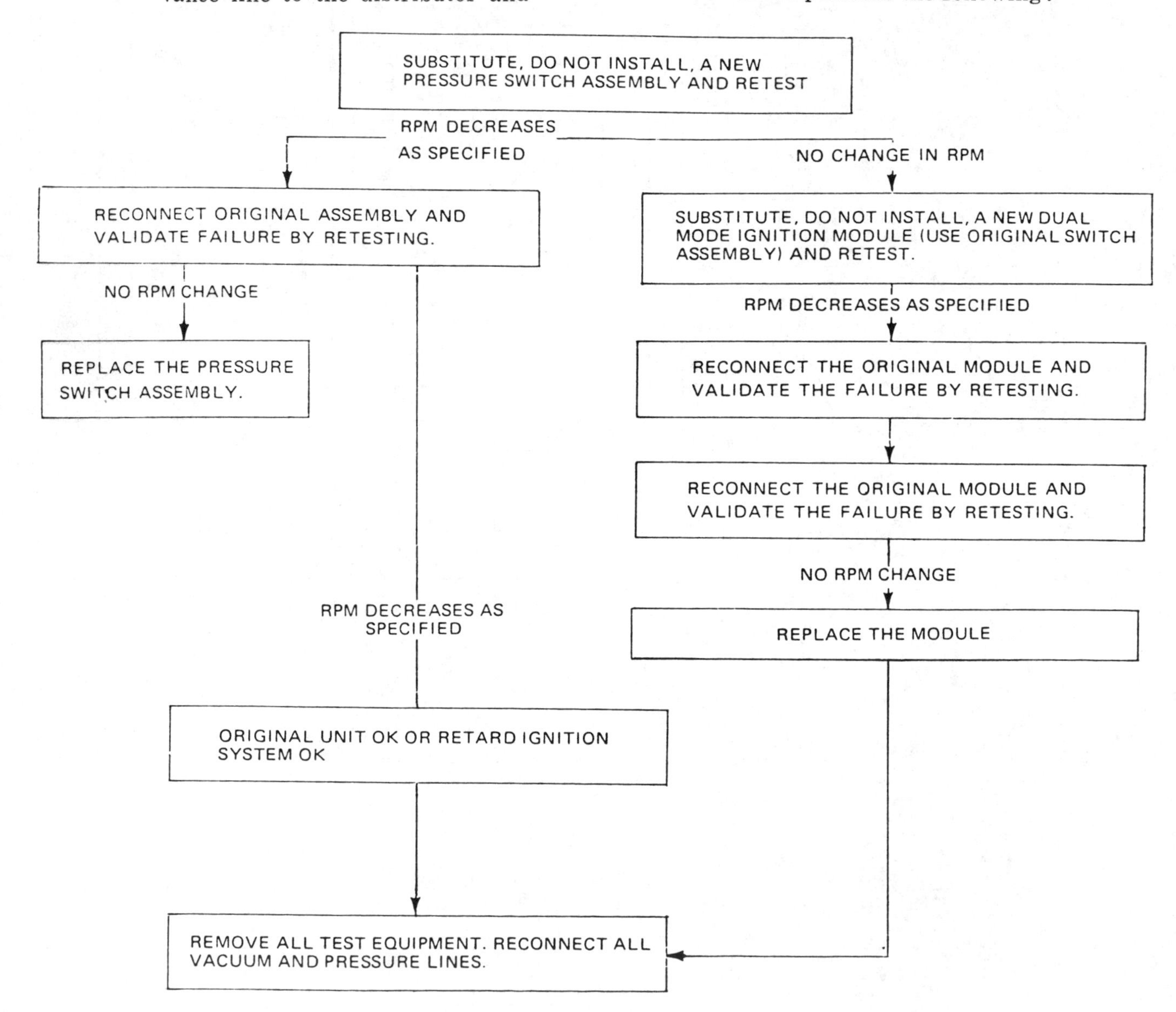

CARBURETOR FEEDBACK SYSTEM 2.3 L ENGINE DIAGNOSIS

FEEDBACK CONTROL VACUUM NOT WITHIN LIMITS

1. To check for manifold vacuum to Vacuum Solenoid Regulator, disconnect the manifold vacuum input line from the VACUUM SOLENOID REGULATOR and connect the vacuum gauge to that line. Start the engine, let idle in neutral, and measure the vacuum.
   NOTE: *Verify that the throttle is off the fast idle cam.*
   a. If the vacuum is less than 12" Hg., check for leakage in the vacuum system or blocked vacuum hose. Repair and retest.
   b. If the vacuum is greater than 12" Hg., proceed to the next step.
2. Reconnect the manifold vacuum to the VACUUM SOLENOID REGULATOR. To check for feedback control vacuum, disconnect the control vacuum line from the output port and connect the vacuum gage to this port.
   a. If the vacuum is between 1.75 and 3.25" Hg. average, check the carburetor per SSD-31 and retest.*
   b. If the vacuum is less than 1.75" Hg. average, check the vacuum output line for leakage or blockage.
      (1) If leakage or blockage exists here, repair and retest.*
      (2) If no problem is found here, replace the Vacuum Solenoid Regulator and retest.*
   c. If the vacuum is greater than 3.25" Hg. average, check the vent line per SSD-35-3 for blockage.
      (1) If an obstruction is found here, repair or replace hose
      (2) If no obstruction is found here, replace Vacuum Solenoid Regulator and retest.* and retest.*

2.3L FEEDBEACK CARBURETOR EEC SYSTEM DIAGNOSTIC DATA CHART

		TEST POINT SELECTOR SWITCH				
Test Sequence	Vehicle Condition	No. 1 Vehicle Battery Voltage	No. 2 Cold Temperature Switch and Throttle Angle Switch	EGO Sensor Indicator	Vacuum Solenoid Light	Feedback Control Vacuum (Vacuum Gauge)
I Step V-D 1	1. Front wheels blocked 2. Parking brake "ON" 3. Engine "Hot" in neutral 4. RPM = 2200 – 2800	Limits 11.5 – 16.0 (Volts)	Limits 11.5 (Volts) Min.	Limits Alternating between "Lean" and "Rich"	Limits "Blinking"	Limits Oscillating about 1.5 to 3.5" Hg average
II Step V-D 2	1. Remove the carburetor feedback vacuum control line from TEE and immediately cover the TEE opening. 2. RPM = 2200 – 2800				Limits From "Blinking" to "ON" in 11 seconds or less	Limits From Oscillating to steady at 4.5 to 5.5" in 11 seconds or less
III Step V-D 3	1. Replace carburetor feedback vacuum control line. RPM = 2200 – 2800 2. Apply 5" vacuum to Cold Temperature Switch		Limits Condition 2: 0.3 (Volts) Max.		Limits Condition 2: Blinking	Limits (Condition 2) Oscillating about 1.75 to 3.25" Hg. average
	3. While covering Carburetor idle air bleed hole, remove vacuum from Cold Temperature Switch.		Condition 3: 11.5 (Volts) Min.		Condition 3: "ON"	
	4. Uncover idle air bleed hole.		Condition 4: 11.5 (Volts) Min.		Condition 4: Blinking	
	5. Return to curb idle		Condition 5: 0.3 (Volts) Max.			
		If out of limits, Stop Sequence Proceed to step VI-A.	If out of limits* See VI-B.	If out of limits* See VI-C.	If out of limits* See VI-D.	If out of limits* See VI-E.

* Complete all data gathering per test sequence I, II, and III prior to performing any diagnostic subroutines.

CARBURETOR FEEDBACK SYSTEM 2.3 L ENGINE DIAGNOSIS

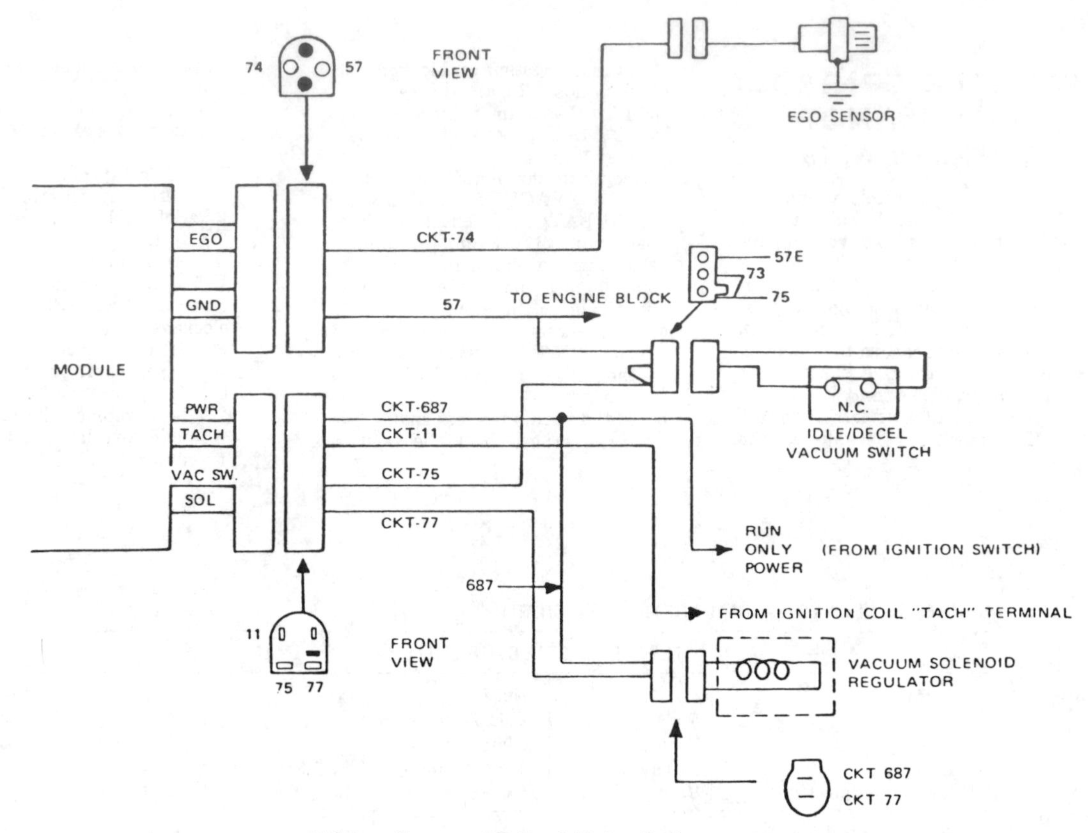

Wiring diagram (© Ford Motor Co.)

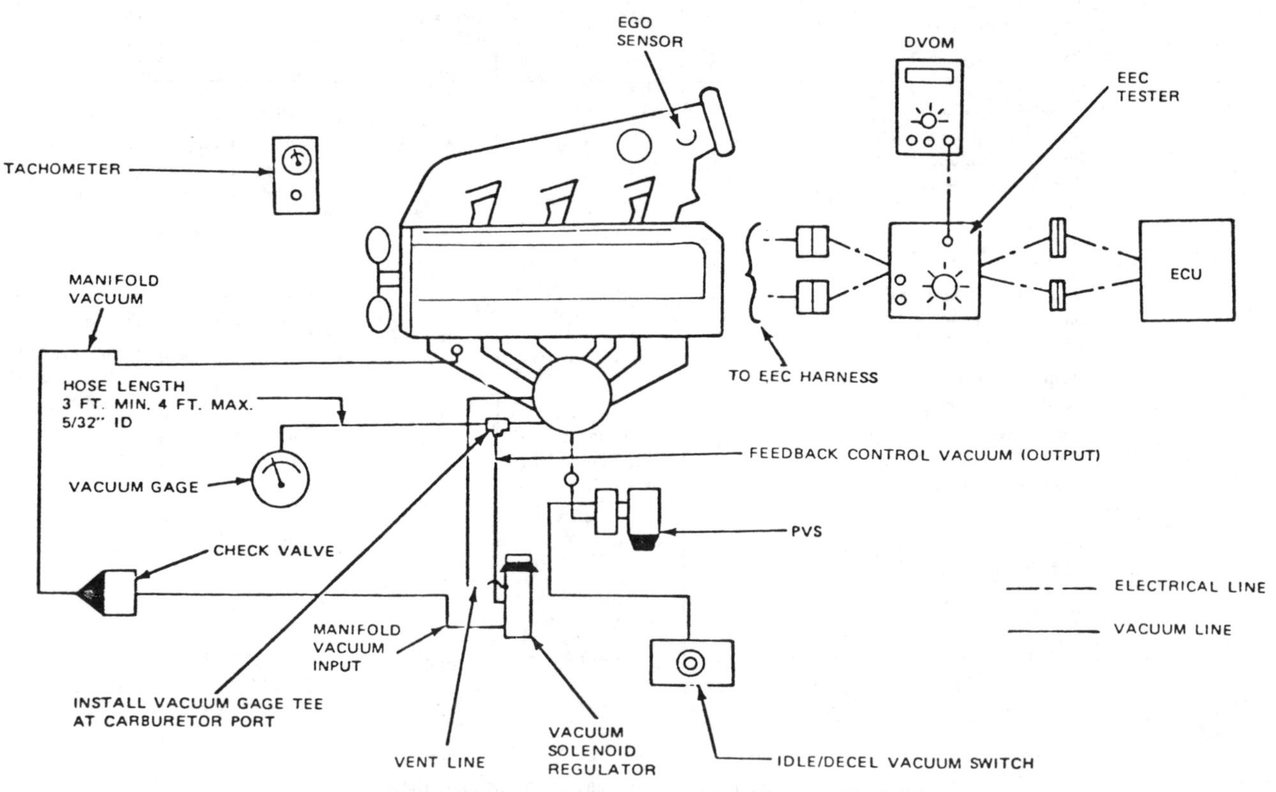

Test equipment hook-up and vacuum diagram (typical)
(© Ford Motor Co.)

E.G.R. SYSTEM OPERATION

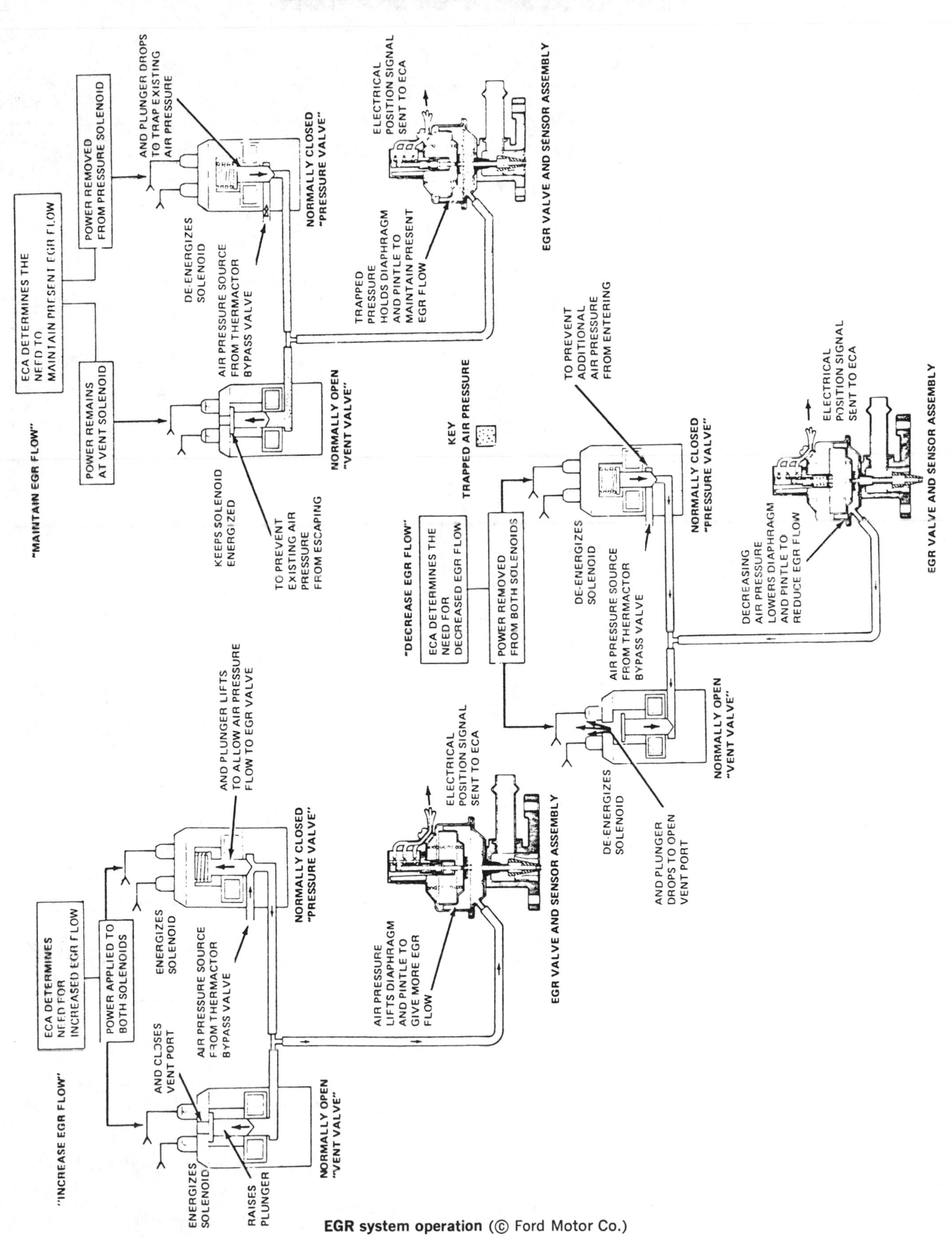

EGR system operation (© Ford Motor Co.)

E.G.R. SYSTEM OPERATION

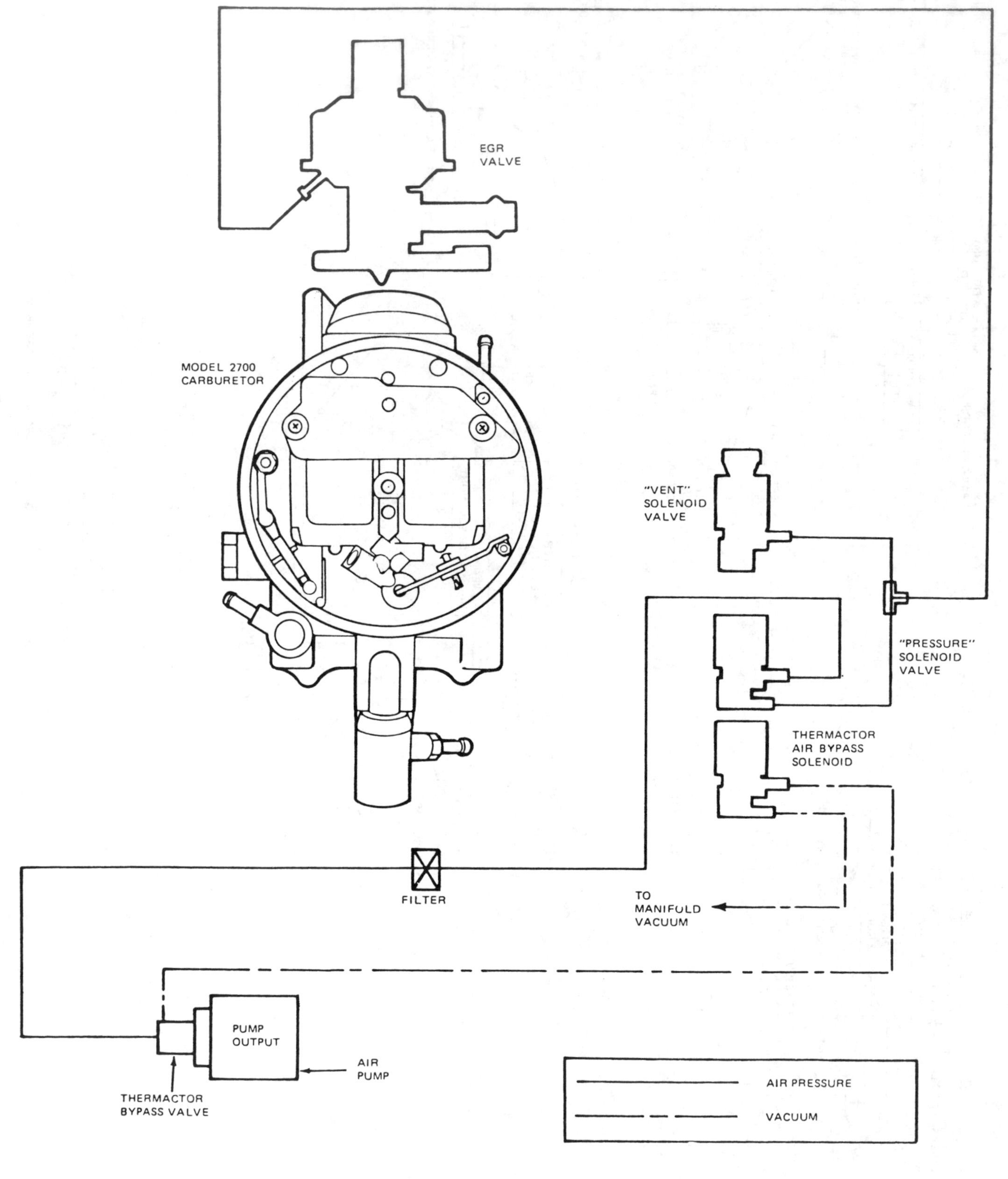

EGR system vacuum/pressure line schematic
(© Ford Motor Co.)

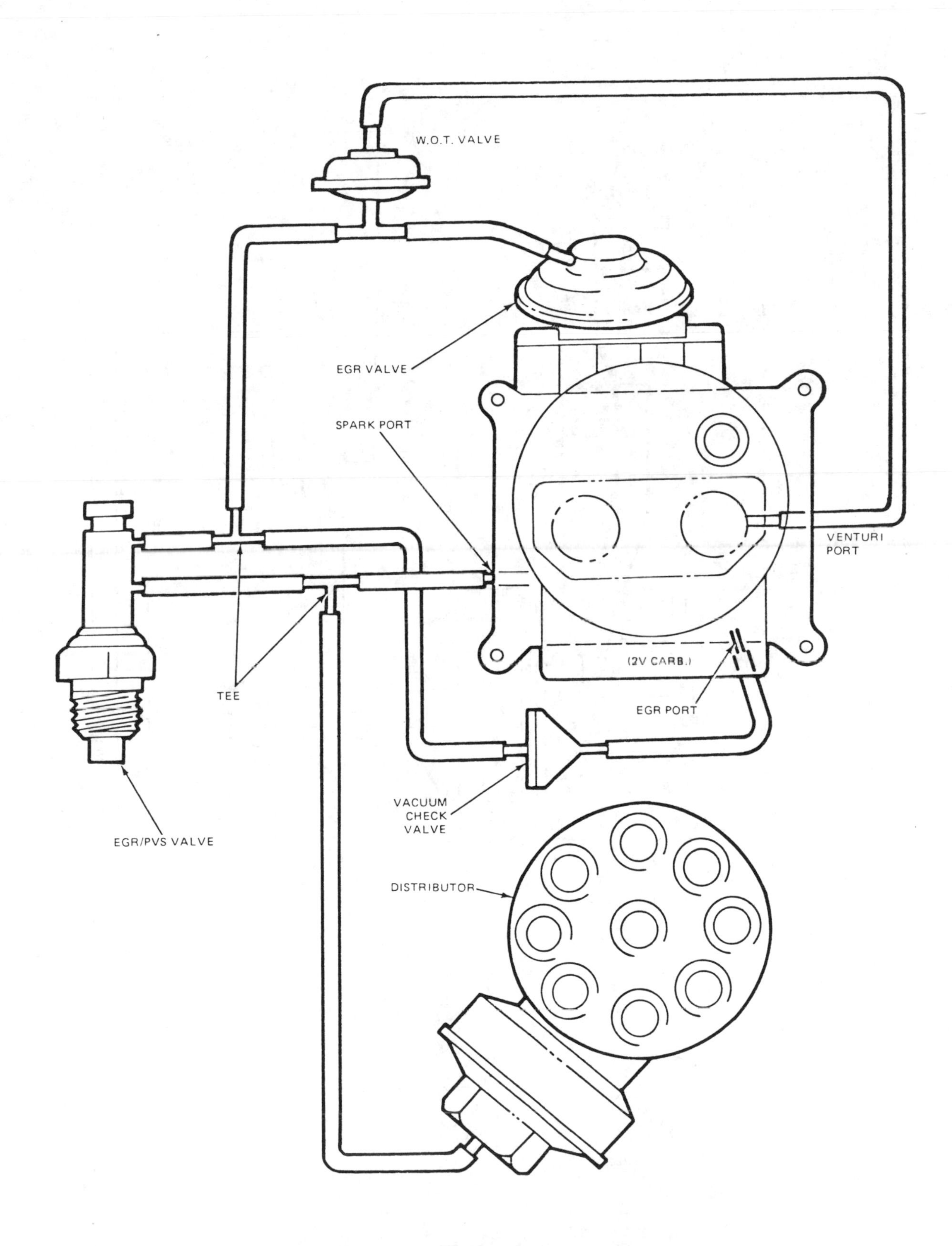

Typical modified combined vacuum (modified co vac) installation with W.O.T. valve (© Ford Motor Co.)

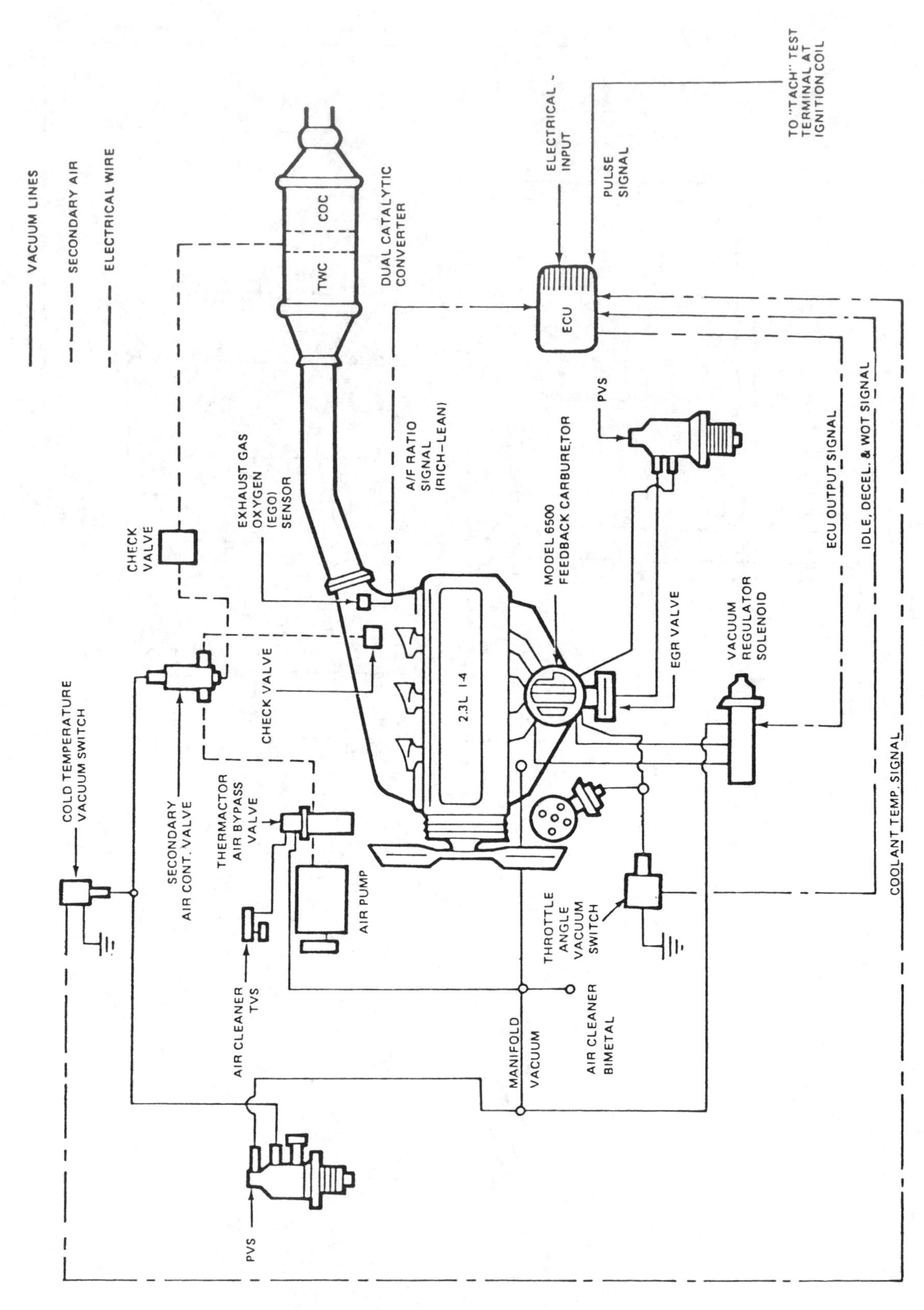

Feedback carburetor 2.3 L engine electronic engine control system schematic (© Ford Motor Co.)

EEC III ELECTRONIC ENGINE CONTROL

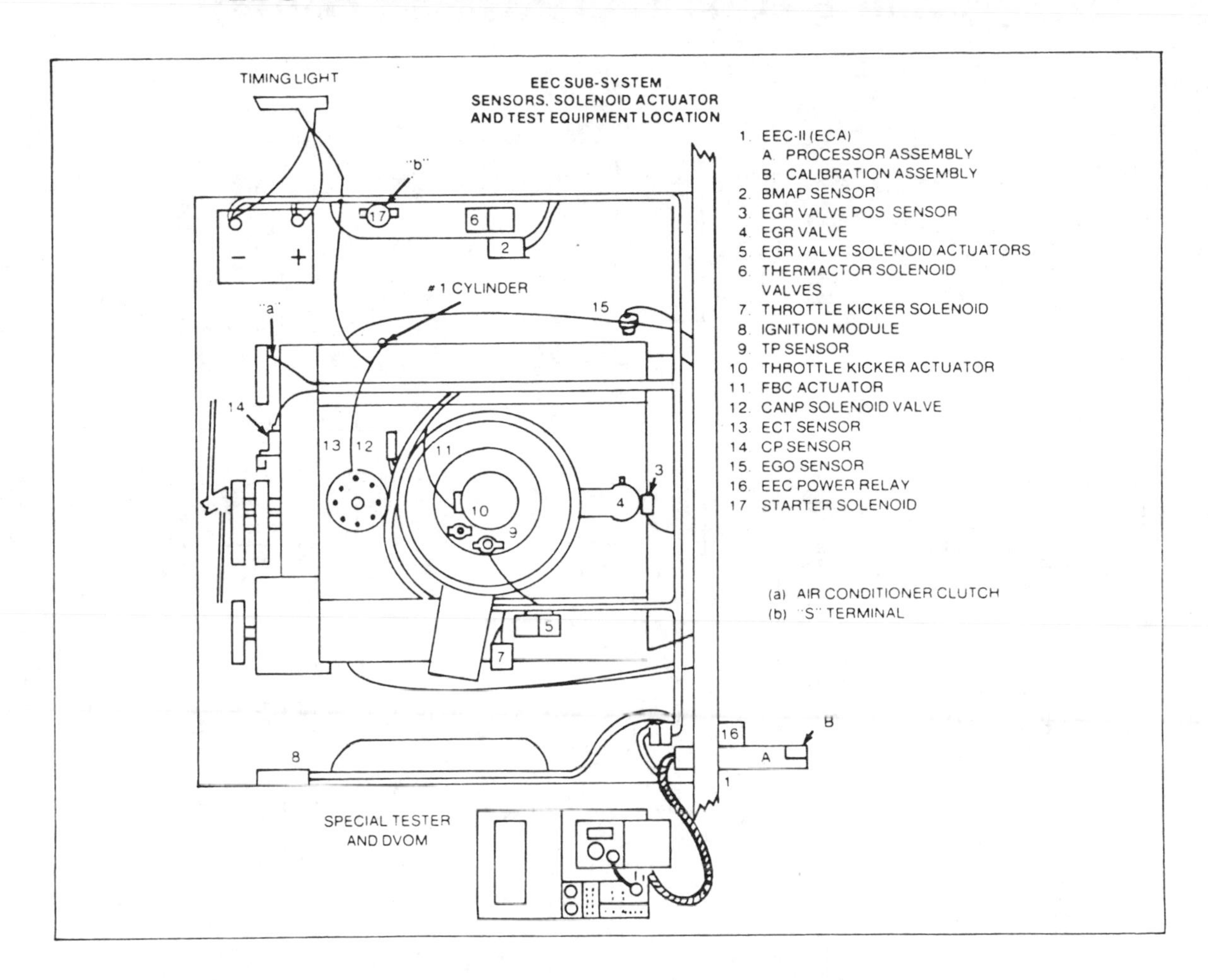

Test connections and component locations (© Ford Motor Co.)

VISUAL INSPECTION— ALL VEHICLES

1. Remove air cleaner assembly, check it for excessive contamination, and check EEC Sub-System vacuum hoses for:
   a) Proper connection to fittings.
   b) Broken, cracked or pinched hoses or fittings.

2. Check the EEC Sub-System harness connector electrical connections for:
   a) Proper connection to the Sensors and Solenoids.
   b) Loose or detached connectors.
   c) Broken or detached wires.
   d) Terminals not completely seated in the connector.
   e) Partially broken or frayed wires at connectors.
   f) Shorting between wires.
   g) Corrosion.
3. Check the sensors for evidence of physical damage.
4. Check the exhaust manifold and around the Exhaust Gas Oxygen (EGO) Sensor for leaks, while engine is operating (if engine starts).
5. Service items as required, reinstall air cleaner assembly.

EEC III ELECTRONIC ENGINE CONTROL

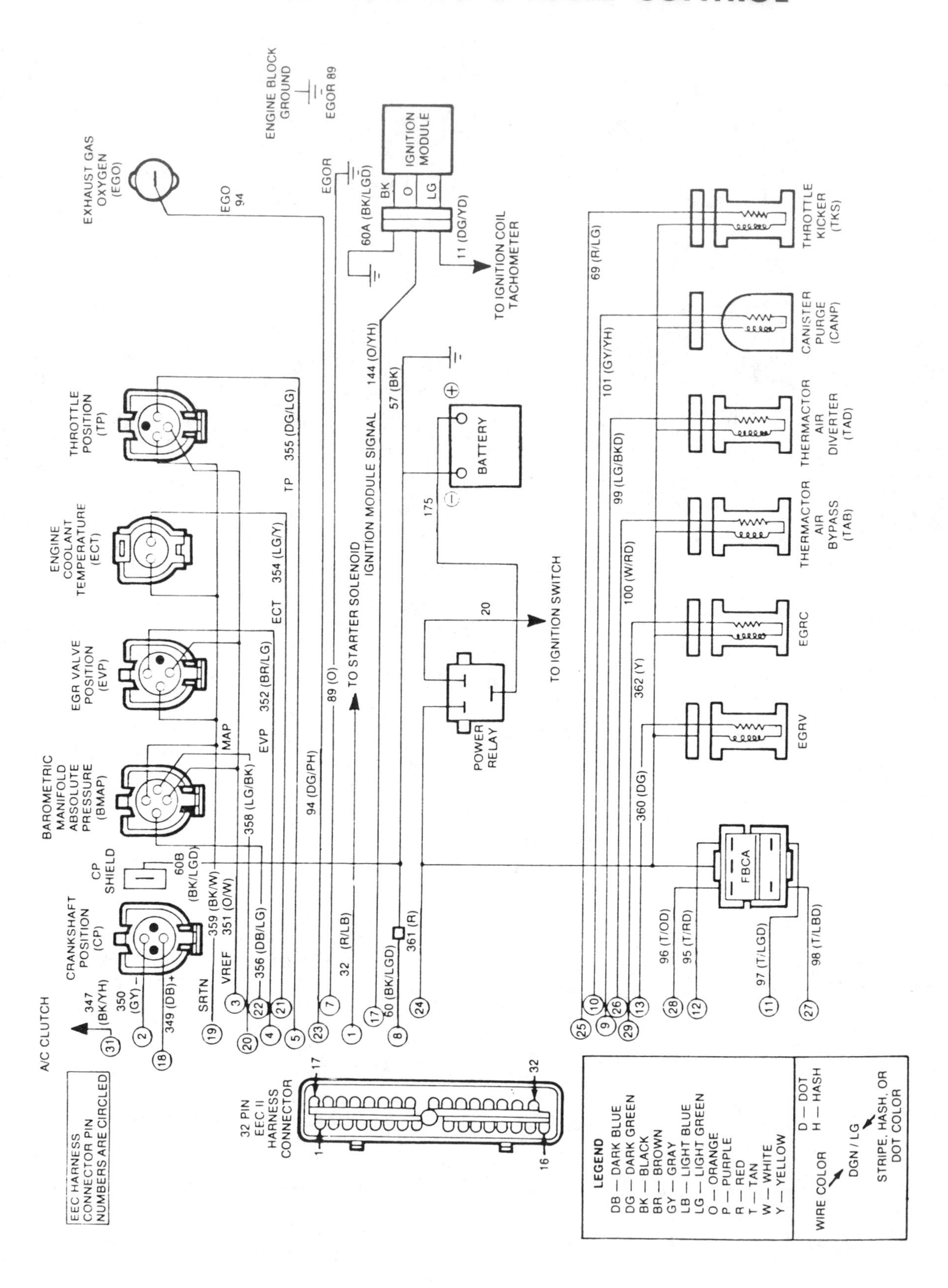

EEC III sub-system wiring diagram (© Ford Motor Co.)

EEC III FEEDBACK CARBURETOR DIAGNOSIS 2.3 L ENGINE

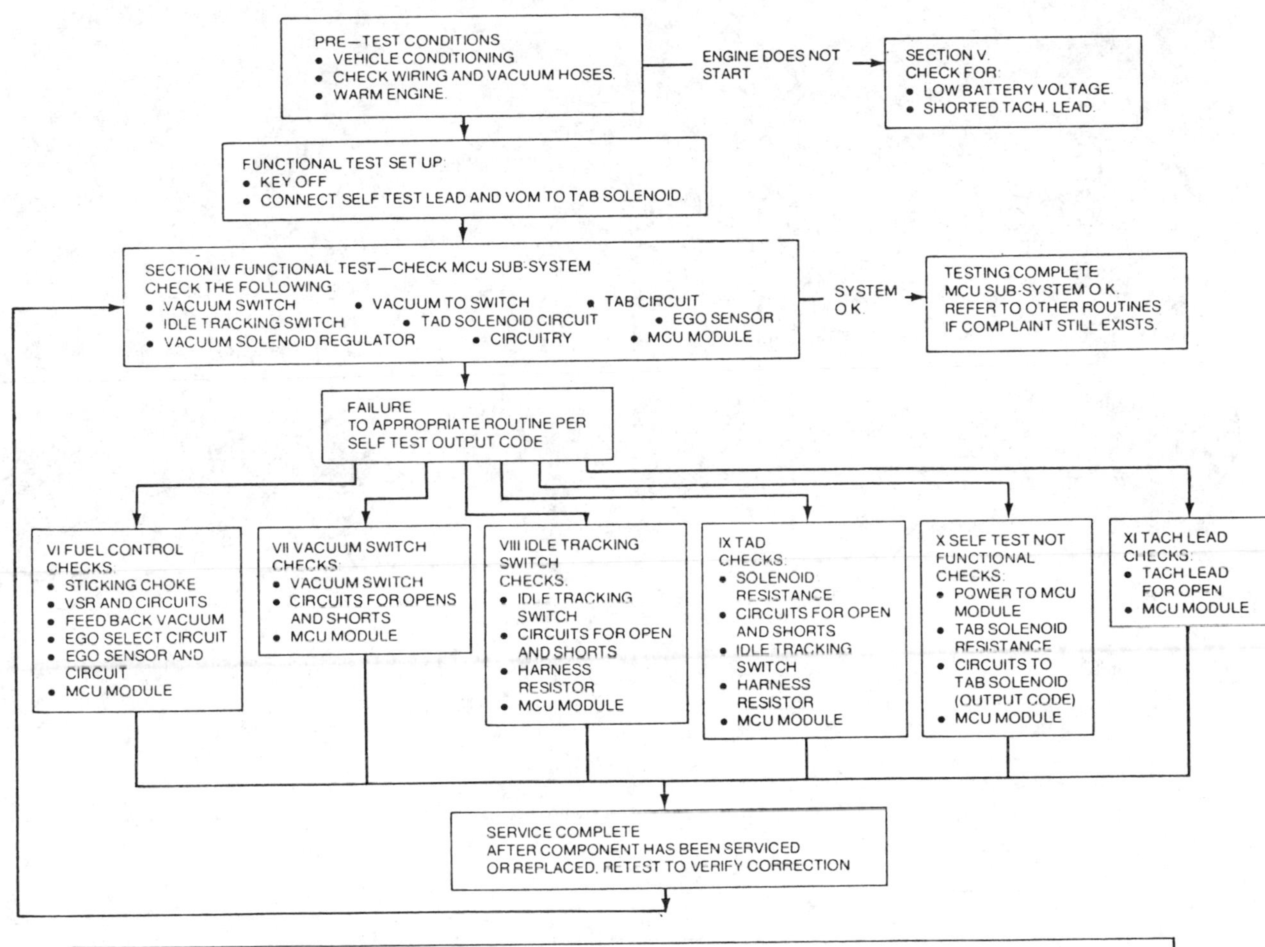

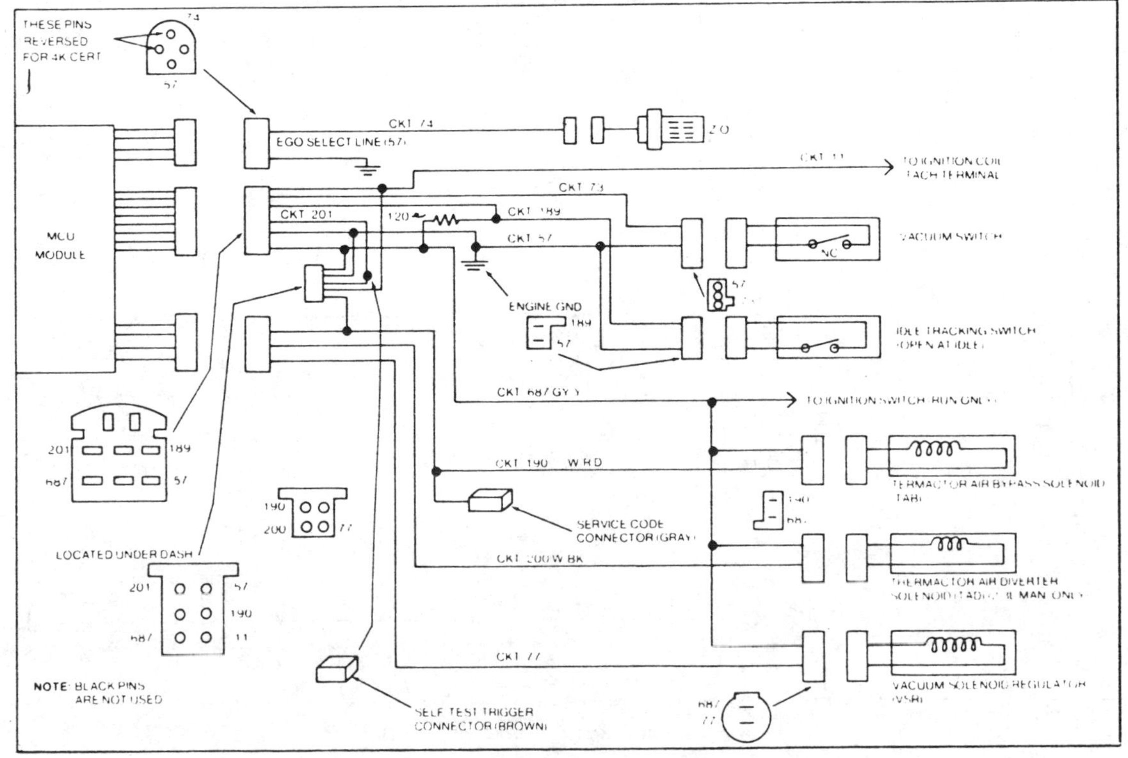

EEC III FEEDBACK CARBURETOR DIAGNOSIS 3.3 L ENGINE

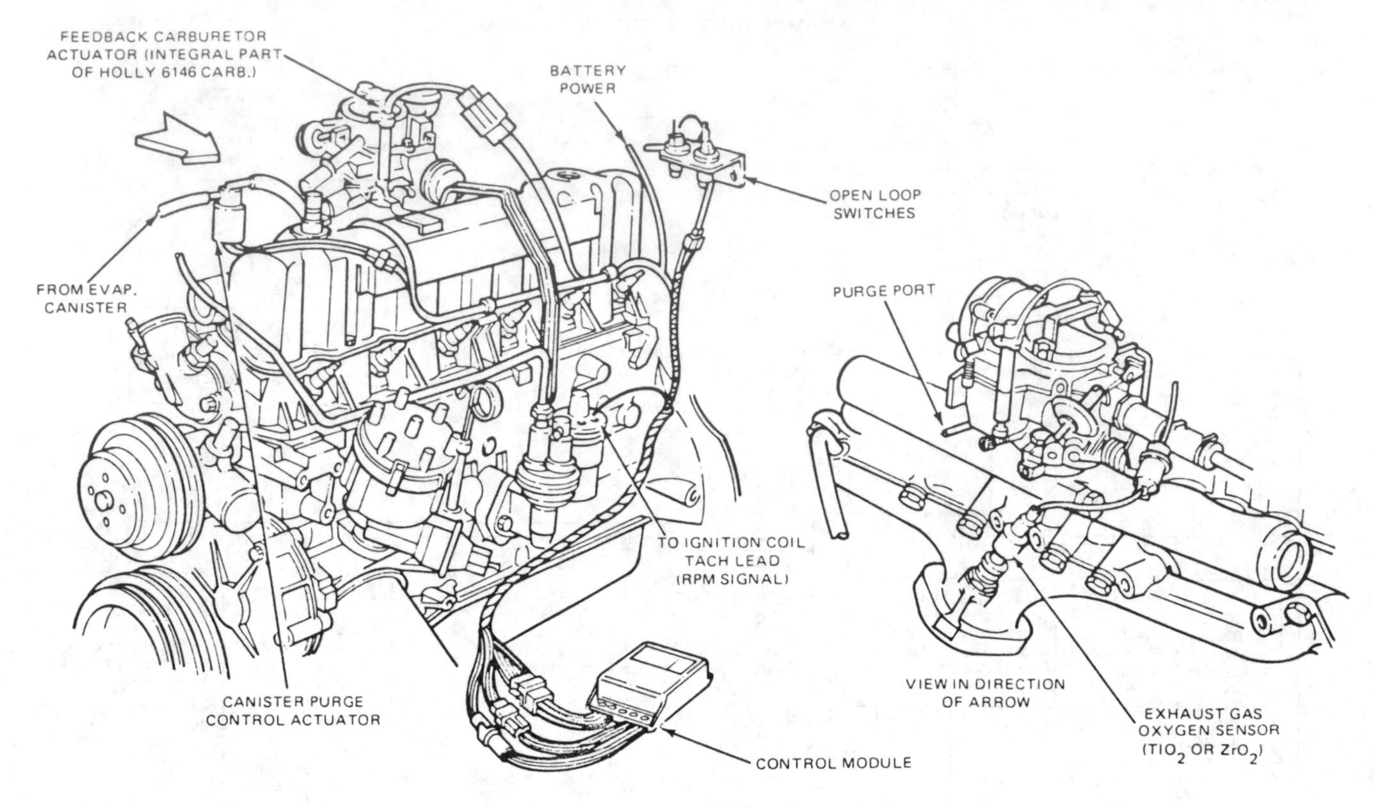

Component locations (© Ford Motor Co.)

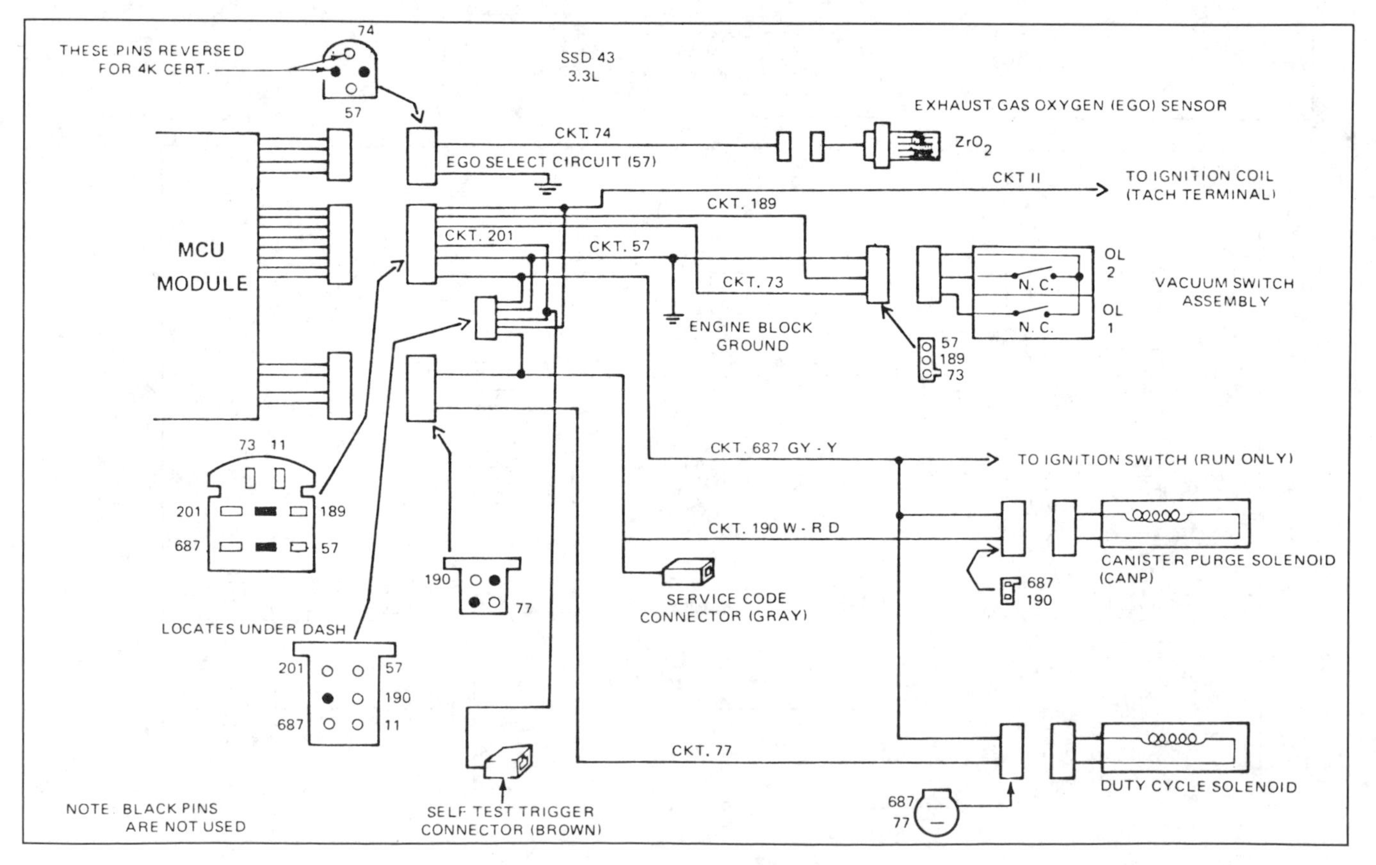

Electrical schematic (© Ford Motor Co.)

EEC III FEEDBACK CARBURETOR DIAGNOSIS 3.3 L ENGINE

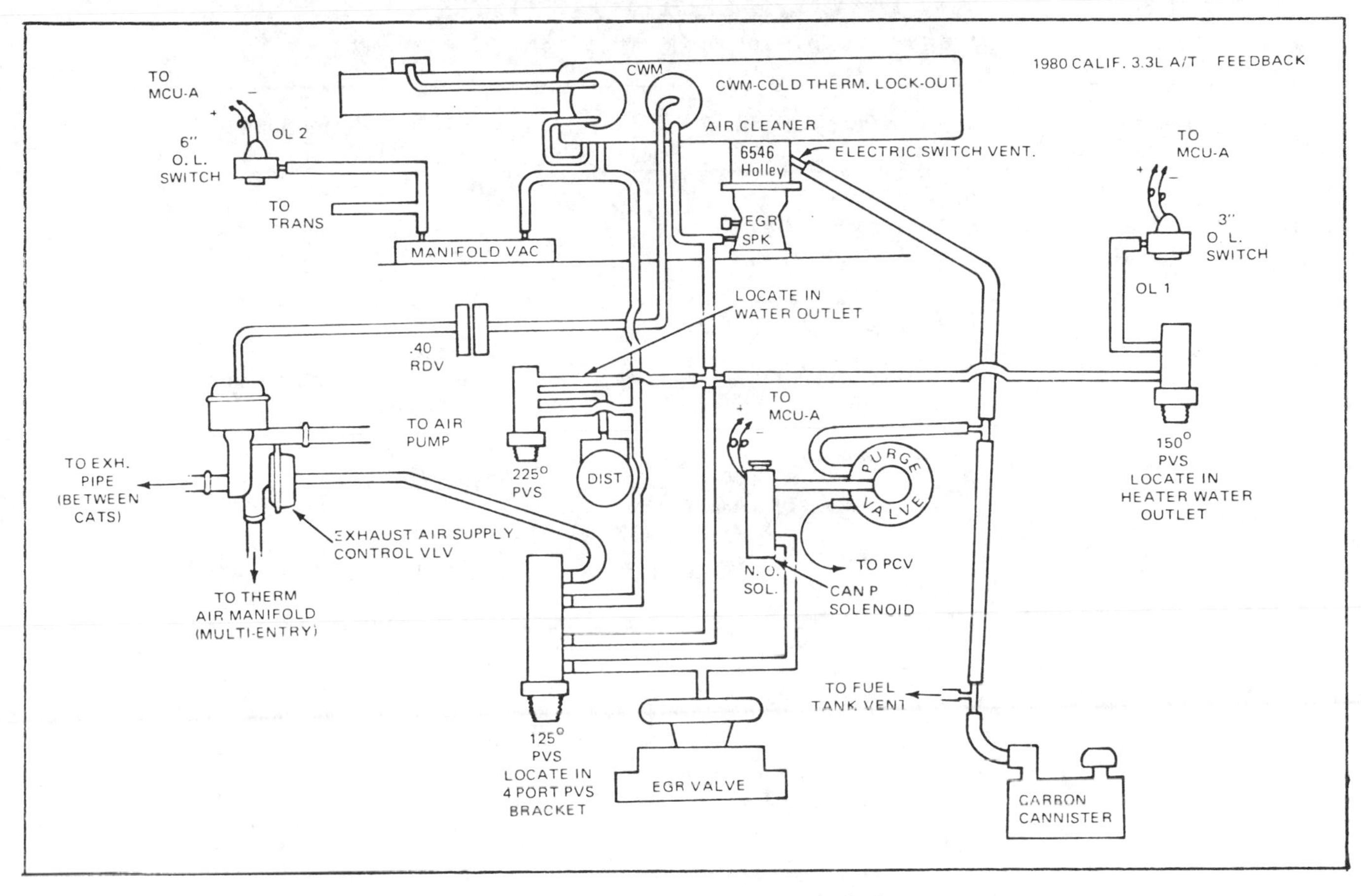

Vacuum schematic (© Ford Motor Co.)

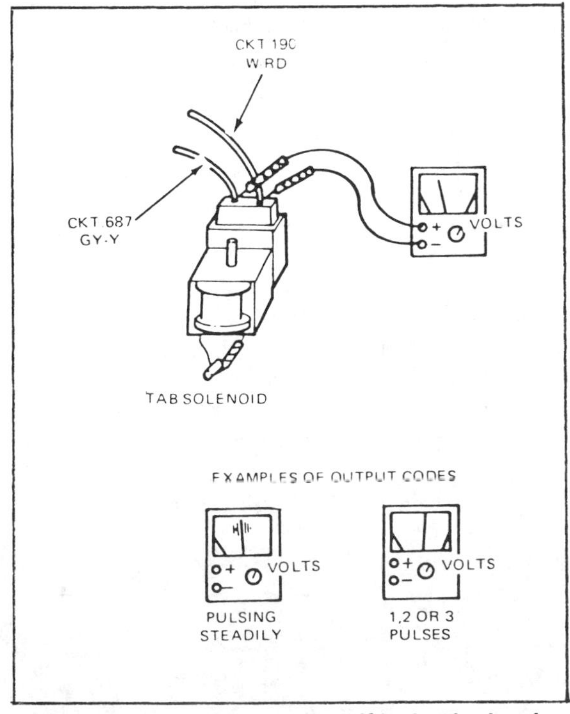

Connections of voltmeter for self-test output code (© Ford Motor Co.)

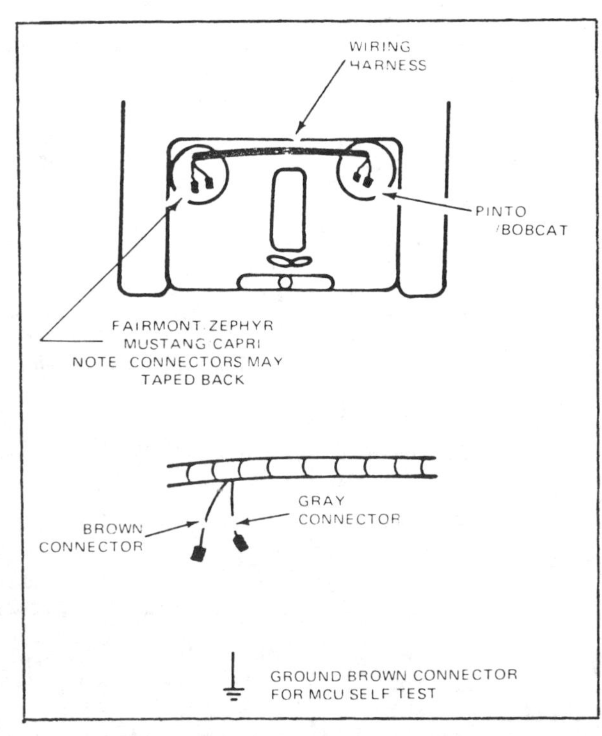

Self test pigtail locations (© Ford Motor Co.)

AMC PRESTOLITE IGNITION

TROUBLESHOOTING AMC PRESTOLITE IGNITION

The Condition	Possible Cause	Correction
Engine Fails to Start (No Spark at Plugs).	No voltage to ignition system.	Check battery, ignition switch and wiring. Repair as needed.
	Electronic ignition control ground lead open, loose or corroded.	Clean, tighten, or repair as needed.
	Primary wiring connectors not fully engaged.	Make sure connectors are clean and firmly seated.
	Coil open or shorted.	Test coil. Replace if faulty.
	Damaged trigger wheel or sensor	Replace damaged part.
	Electronic ignition control faulty.	Replace electronic ignition control.
Engine Backfires but Fails to Start.	Incorrect ignition timing.	Check timing. Adjust as needed.
	Moisture in distributor cap.	Dry cap and rotor.
	Distributor cap faulty (shorting out).	Check cap for loose terminals, cracks and dirt. Clean or replace as needed.
	Wires not in correct firing order.	Reconnect in proper firing order.
Engine Does Not Operate Smoothly and/or Engine Misfires at High Speed.	Spark plugs fouled or faulty.	Clean and regap plugs. Replace if needed.
	Spark plug cables faulty.	Check cables. Replace if needed.
	Spark advance system(s) faulty.	Check operation of advance system(s). Repair as needed.
Excessive Fuel Consumption.	Incorrect ignition timing.	Check timing. Adjust as needed.
	Spark advance system(s) faulty.	Check operation of advance system(s). Repair as needed.
Erratic Timing Advance.	Faulty vacuum advance assembly.	Check operation of advance diaphragm and replace if needed.
Basic Timing Not Affected by Vacuum. (Disconnected)	Misadjusted, weak or damaged mechanical advance springs.	Readjust or replace springs as needed.
	Worn distributor shaft bushings.	Check for worn bushings. Replace distributor.

AMC PRESTOLITE IGNITION

TROUBLESHOOTING AMC PRESTOLITE IGNITION

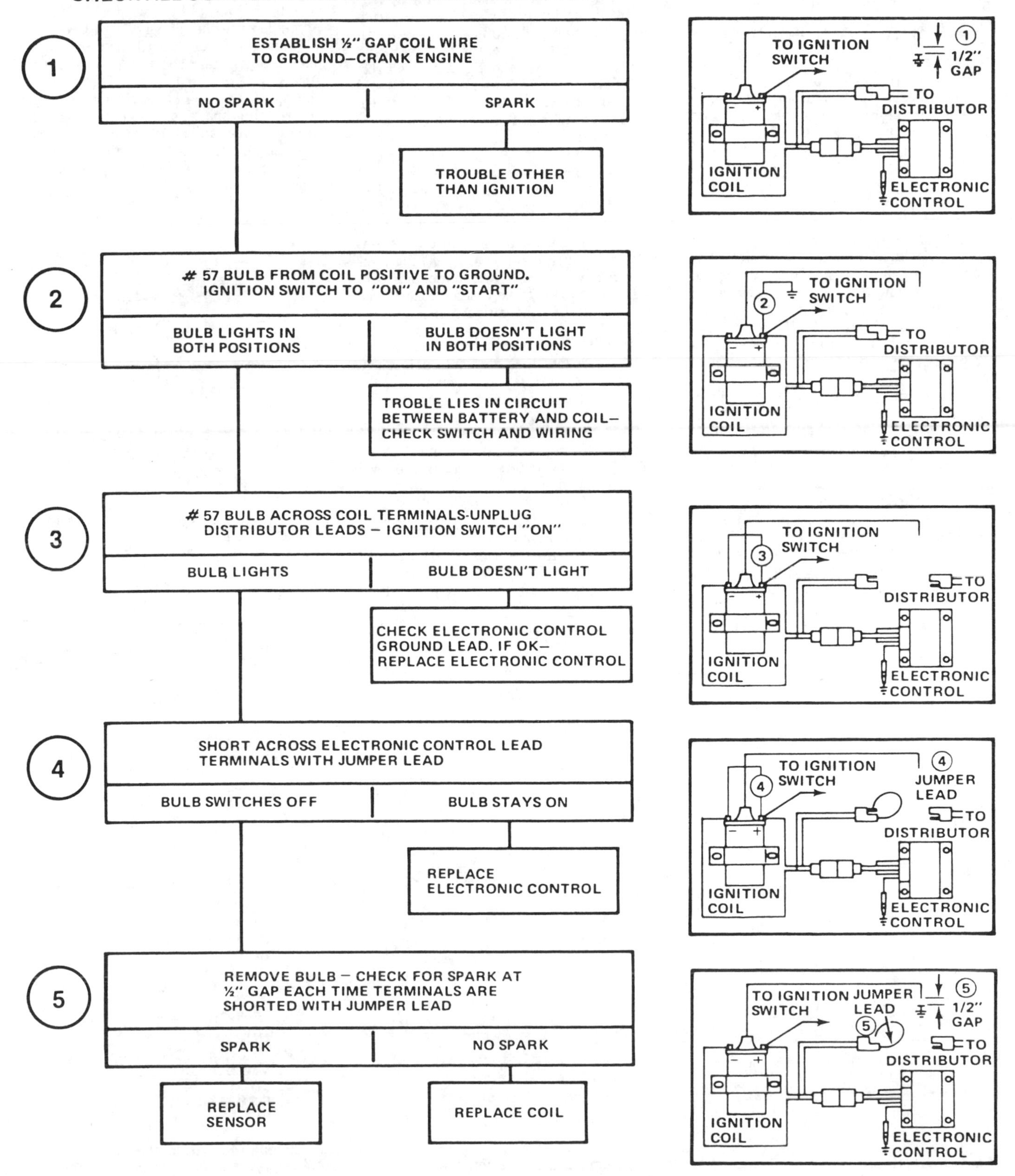

AMC PRESTOLITE IGNITION

AMC Prestolite Breakerless Inductive Discharge (BID) Ignition System

The American Motors BID Ignition System consists of five major components: an electronic ignition control unit, an ignition coil, a distributor, high tension wires, and spark plugs.

Control Unit

The electronic control unit is a solid-state, moisture-resistant module. The component parts are permanently sealed in a potting material to resist vibration and environmental conditions. All connections are waterproof. The unit has built-in current regulation, reverse polarity protection and transient voltage protection.

Because the control unit has built-in current regulation, there is no resistance wire or ballast resistor used in the primary circuit. Battery voltage is present at the ignition coil positive terminal whenever the ignition key is in the ON or START position; therefore, there is no need for an ignition system bypass during cranking. The primary (low voltage) coil current is electronically regulated by the control unit. The control unit is not repairable and must be serviced as a unit.

Ignition Coil

The ignition coil is an oil-filled, hermetically-sealed unit (standard construction). Ignition coils do not require special service other than keeping terminals and connections clean and tight. For correct polarity, the coil positive terminal should be connected to the battery ignition feed.

The function of the ignition coil in the BID ignition system is to transform battery voltage in the primary winding to a high voltage for the secondary system.

When an ignition coil is suspected of being defective, it should be checked on the car. A coil may break down after it has reached operating temperature; it is important that the coil be at operating temperature when tests are made. Perform the test following the instructions of the Test Equipment Manufacturer.

Distributor

The distributor is conventional except that a sensor and trigger wheel replace the usual contact points, condenser, and distributor cam.

The distributor uses two spark advance systems (mechanical and vacuum) to establish the spark timing setting required for various engine speed and load conditions. The two systems operate independently, yet work together to provide proper spark advance.

The mechanical (centrifugal) advance system is built internally into the distributor and consists of two flyweights which pivot on long-life, low-friction bearings and are controlled by calibrated springs which tend to hold the weights in the no-advance position. The flyweights respond to changes in engine (distributor shaft) speed, and rotate the trigger wheel with respect to the distributor shaft to advance the spark as engine speed increases and retard the spark as engine speed decreases. Mechanical advance characteristics can be adjusted by bending the hardened spring tabs to alter the spring tension.

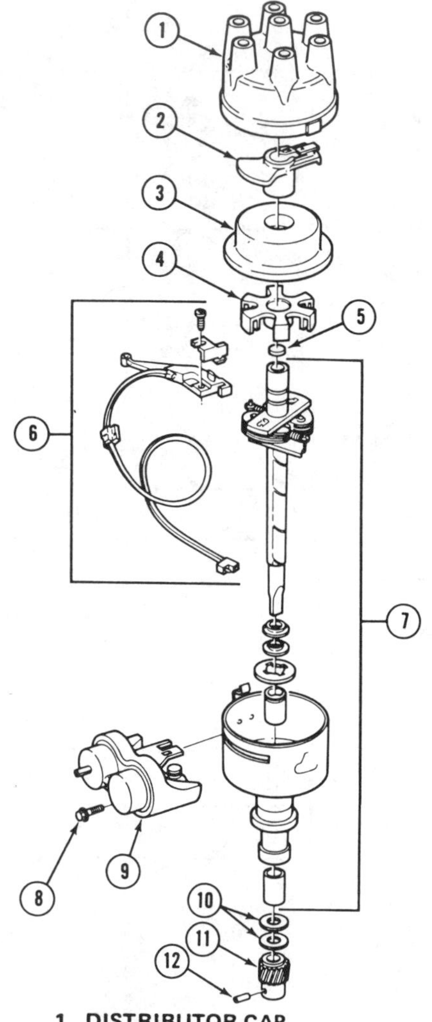

AMC Breakerless Inductive Discharge ignition system distributor, exploded view

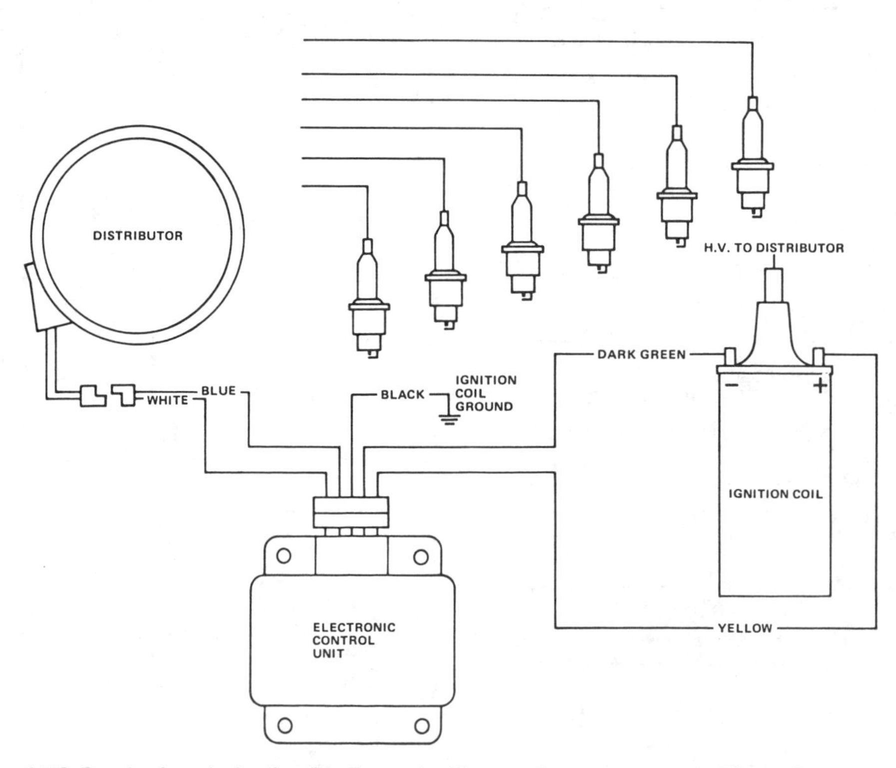

AMC Breakerless Inductive Discharge ignition system component schematic

AMC PRESTOLITE IGNITION

The vacuum advance system incorporates a vacuum diaphragm unit which moves the distributor sensor in response to the changes in carburetor throttle bore vacuum.

Sensor/Trigger Wheel

The sensor (a component of the distributor) is a small coil, wound of fine wire, which received an alternating current signal from the electronic control unit.

The sensor develops an electromagnetic field which is used to detect the presence of metal. The sensor detects the edges of the metal in the teeth of trigger wheel. When a leading edge of a trigger wheel tooth aligns with the center of the sensor coil, a signal is sent to the control unit to open the coil primary circuit. There are no wearing surfaces between the trigger wheel and sensor, dwell angle remains constant and requires no adjustment. The dwell angle is determined by the control unit and the angle between the trigger wheel teeth.

Operation

With the ignition switch in the START or RUN position, the control unit is activated. At this time, an oscillator, contained in the control unit, excites the sensor which is contained in the distributor. When the sensor is excited, it develops an electromagnetic field. As the leading edge of a tooth of the trigger wheel enters the sensor field, the tooth reduces the strength of oscillation in the sensor. As the oscillator strength is reduced to a predetermined level, the demodulator circuit switches. The demodulator switching signal controls a power transistor which is in series with the coil primary circuit. The power transistor switches the coil primary circuit off, thereby inducing the high voltage in the coil secondary winding. High voltage is then distributed to the spark plugs by the distributor cap, rotor, and ignition wires.

The following procedures can be used to check operation of the components of the BID ignition system.

Electrical components of the ignition system (sensor, coil, and electronic ignition control unit) are not repairable. If the operation test indicates that they are faulty, replace them.

The following equipment is required to make this test: ohmmeter, DC voltmeter, jumper wire (12 to 18 inches long) with clip at each end, Tester (distributor sensor substitute) J-25331, insulated pliers (grippers) for handling high tension cables.

BID System Test

1 Test battery using DC voltmeter. Voltage should be 12 to 13 volts for a fully charged battery. If necessary charge or replace battery.
2 Inspect ignition primary (low voltage) circuit for loose or damaged wiring. Inspect connectors for proper fit. Spread male connector with punch or awl and crimp female connectors to ensure proper fit. Reconnect connectors.
3 Inspect secondary (high voltage) cables for cracks and deterioration. Replace any defective wiring. Be sure ignition cables are routed correctly.
4 Disconnect high tension cable from one spark plug. (always grasp the spark plug boot and use a twisting motion when removing plug cables so as not to destroy the resistance wire termination.) Using insulated pliers, hold plug cable to create approximately ½ to a ¾-inch gap between cable terminal and engine. Crank engine and observe spark. If a spark jumps the gap, ignition system is satisfactory. If no spark occurs, reinstall spark plug cable and proceed to the next step.
5 Disconnect high tension cable from center tower terminal of distributor cap. Set up a spark gap of approximately ½ to ¾ inch by clipping end of jumper wire over the high tension cable ½ to ¾ inch away from the metal tip at distributor end of cable. Ground other end of jumper wire to engine. Crank engine and observe for spark between jumper wire clip and ignition cable terminal. If spark now occurs, distributor cap or rotor is faulty. Replace faulty part and recheck for spark at spark plug. If no spark occurs between jumper wire clip and cable terminal, check coil secondary wire with the ohmmeter for 5,000 to 10,000 ohms resistance. If coil wire checks satisfactory, proceed to the next step. If coil wire is faulty, replace wire, then proceed to the next step.
6 Disconnect the distributor primary wires (black and dark green) from the control unit connector (blue and white).
7 Visually inspect the distributor primary wire connectors for proper fit. Spread male connector with a punch or awl and slightly crimp the female

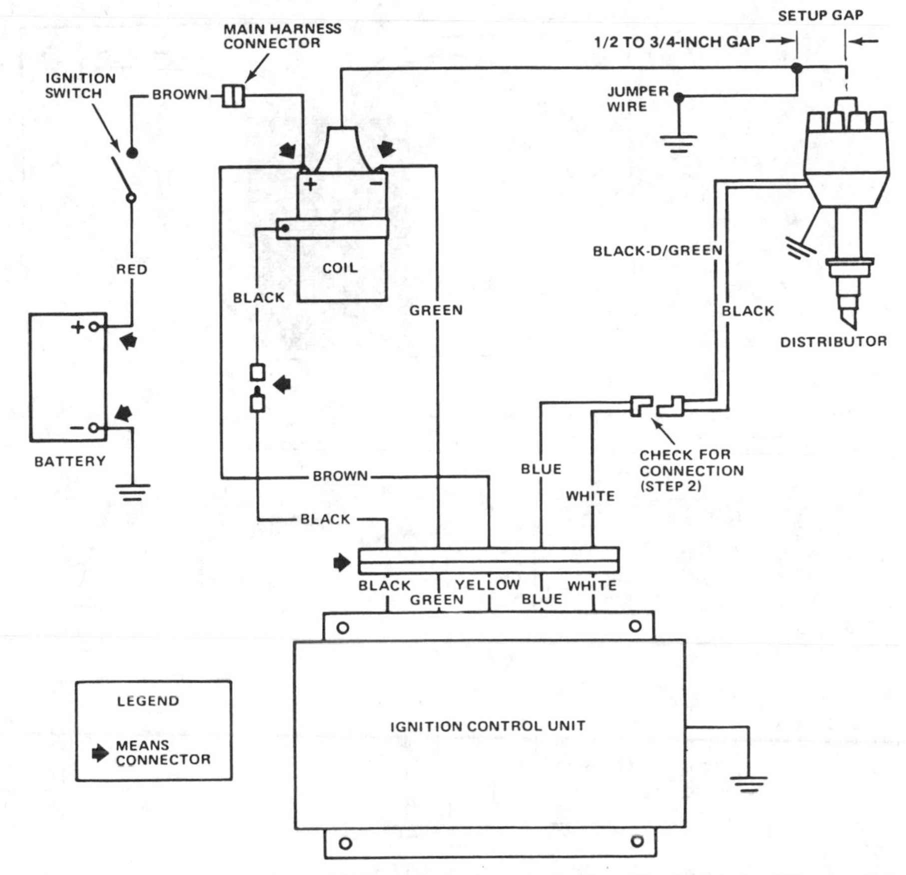

Checking the spark gap on AMC's Breakerless Inductive Discharge ignition system

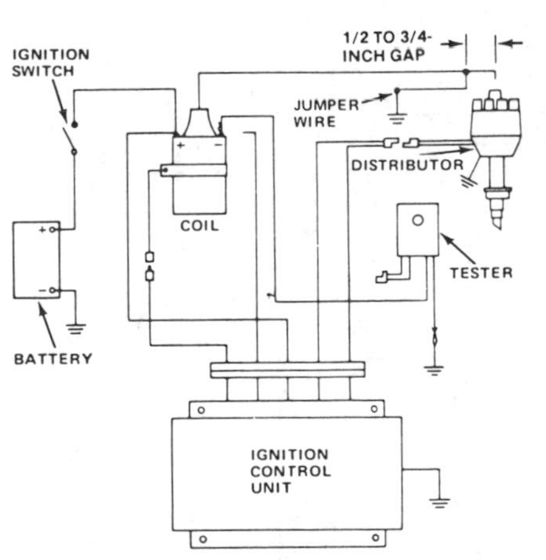

Tester connected to coil negative terminal

AMC PRESTOLITE IGNITION

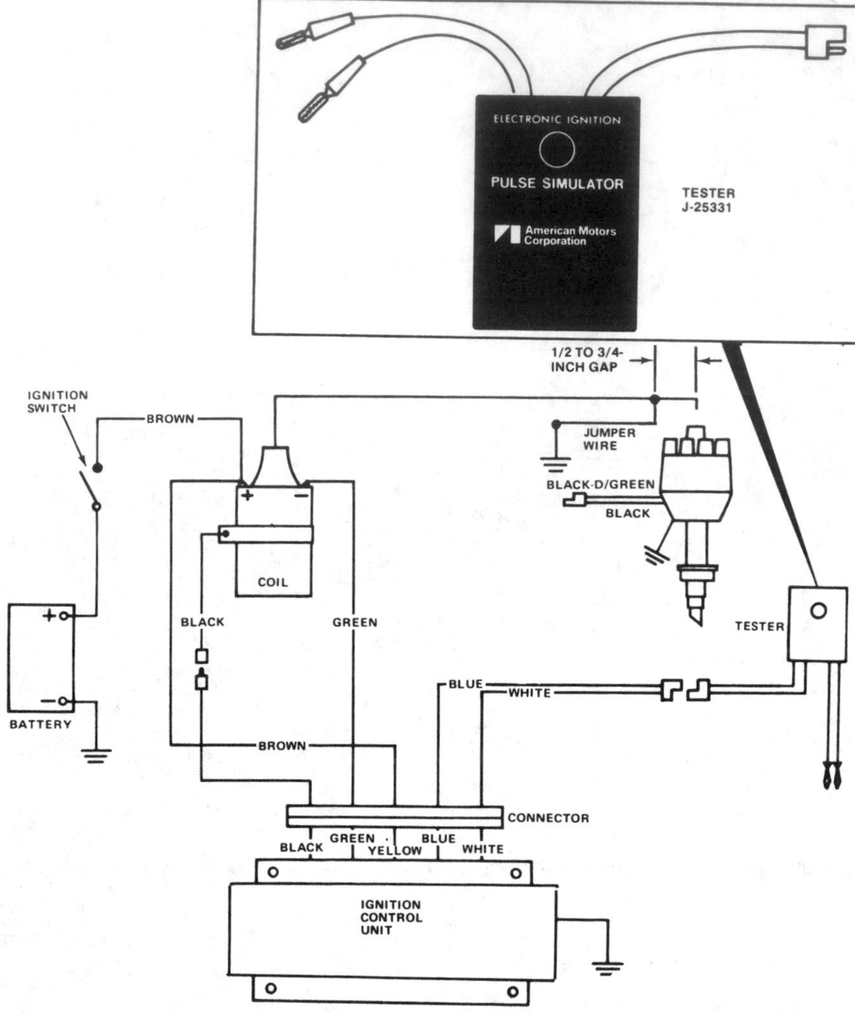

Tester connected into AMC's Breakerless Inductive Discharge ignition system

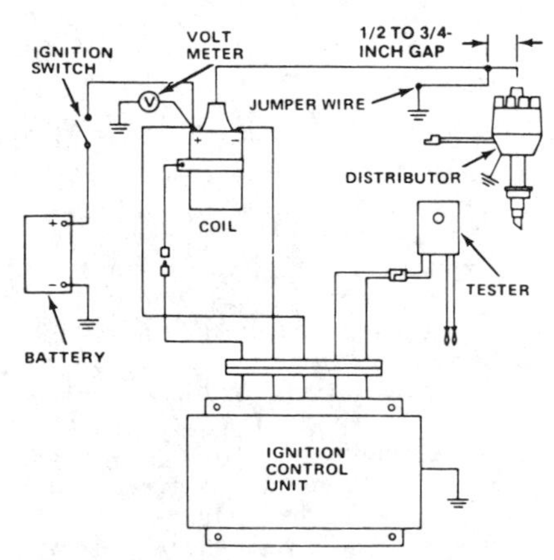

Voltmeter connected to coil positive terminal

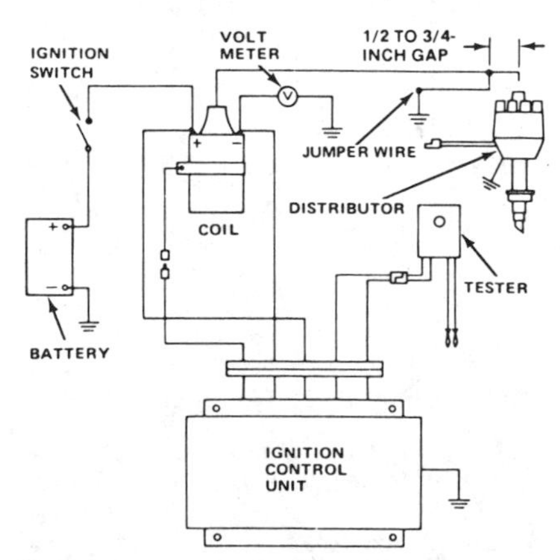

Voltmeter connected to coil negative terminal

connector to ensure proper fit.

8 Connect distributor primary wires to control unit connector and crank engine. Observe for spark between jumper wire clip and ignition cable terminal. If spark now jumps the gap, the ignition system is satisfactory. If no spark occurs between jumper wire clip and cable terminal, proceed to the next step.

9 Disconnect the distributor primary wires (black and dark green) and plug Tester J)25331 into wire harness. Turn ignition switch on. Cycle test button and observe for spark between jumper wire clip and ignition cable terminal. If spark occurs, distributor sensor unit is faulty and must be replaced. If no spark occurs, proceed to the next step.

10 Connect voltmeter between coil positive (+) terminal and ground. With ignition switch ON, voltmeter should read battery voltage. If voltage at coil positive terminal is noticeably lower than battery (through ignition switch) and the coil. Before proceeding, the resistance must be corrected. If voltage at coil positive terminal equals battery voltage, proceed to the next step.

11 Connect voltmeter between coil negative (−) terminal and ground. With ignition switch ON, voltage should read 5 to 8 volts. A reading under 5 volts or over 8 volts indicates a bad coil which must be replaced. If voltage is satisfactory, press button on tester and observe voltmeter. Voltage reading should increase to battery voltage (12 to 13 volts). Release button on tester. Voltage should drop to 5 to 8 volts. If voltage does not switch up and down, the electronic ignition control is faulty and must be replaced. If voltage switches up and down but there is no spark between jumper wire clip and ignition cable terminal, proceed to the next step.

12 Disconnect tester from control unit.

13 Turn off ignition switch. Remove wire from the negative terminal of the ignition coil.

14 Connect one clip lead from tester to negative terminal of ignition coil and the other clip lead to an engine ground.

15 Turn on ignition switch. Cycle test button.

16 Spark should jump the gap. If spark does not, test the ignition coil. The coil can be tested on any conventional coil tester or with an ohmmeter. (A coil tester is preferable as it will detect faults that an ohmmeter will not.) The coil primary resistance should be 1 to 2 ohms. Coil secondary resistance should be 8,000 to 12,000 ohms. Coil open-circuit output should exceed 20 kv. If the coil does not pass these tests, it must be replaced.

Distributor Disassembly

1 Place distributor in suitable holding device.

2 Remove rotor and dust shield.

3 Remove trigger wheel using a small gear puller. Be sure the puller jaws are gripping the inner shoulder of the trigger wheel or the trigger wheel may be damaged during removal. Use a thick flat washer or nut as a spacer. Do not press against the small center shaft.

4 Loosen sensor locking screw about three turns. The sensor locking screw has a tamper proof head design which requires a Special Driver Bit Tool J-25097. If a driver bit is not available, use a small needlenose pliers to re-

AMC PRESTOLITE IGNITION

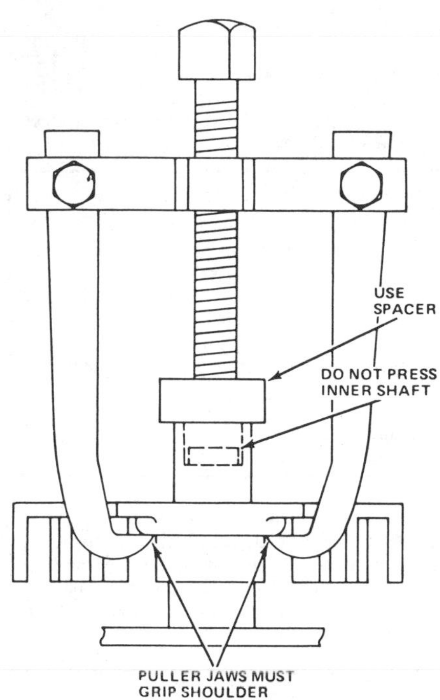

Trigger wheel removal with puller

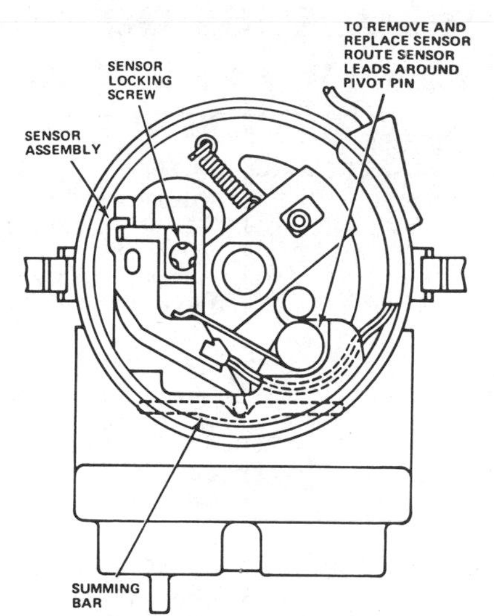

Sensor assembly R&R details

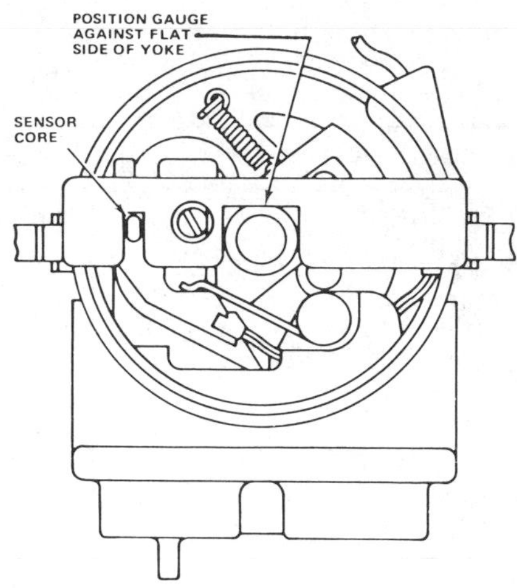

Sensor positioning details

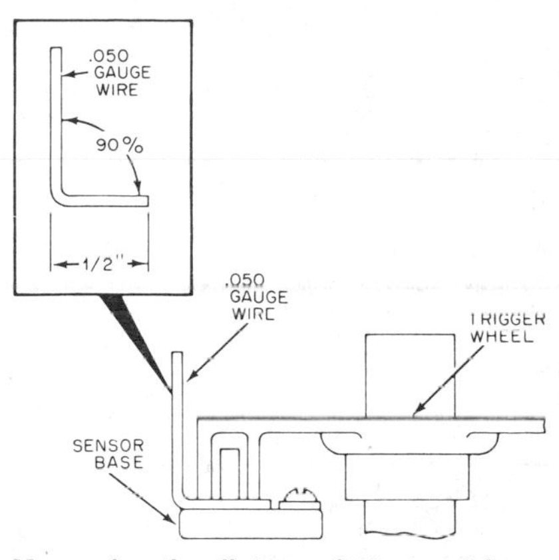

Measuring the distance between trigger wheel legs and sensor base

move screw. The service sensor has a standard slotted head screw.

Lift the sensor lead grommet out of the distributor bowl. Pull sensor leads out of the slot around sensor spring pivot pin. Lift and release sensor spring, making sure it clears the leads, then slide the sensor off bracket.

5 If the vacuum chamber is to be replaced, remove the retaining screw and slide the vacuum chamber out of the distributor. DO NOT remove the vacuum chamber unless replacement is required.
6 Clean dirt or grease off of the vacuum chamber bracket. Clean and dry sensor and bracket. The material used for sensor and vacuum chamber requires no lubrication.
7 With the vacuum chamber installed, assemble sensor, sensor guide, flat washer, and retaining screw. Install retaining screw only far enough to hold assembly together and be sure it does not project beyond the bottom of sensor.
8 If the vacuum chamber has been replaced and the original sensor is being used, substitute new screw for original special head screw to facilitate sensor positioning. Use existing flat washer.
9 Install sensor assembly on vacuum chamber bracket, making certain that the tip of the sensor is located properly in summing bar. Place sensor spring in its proper position on sensor, then route sensor leads around spring pivot pin. Install sensor lead grommet in distributor bowl, then make certain the leads are positioned so they cannot be caught by the trigger wheel.
10 Place sensor positioning gauge over yoke (be sure gauge is against flat of shaft) and move sensor sideways until the gauge can be positioned. With the gauge in place, use a small blade screwdriver to snug down retaining screw. Check sensor position by removing and installing gauge. When properly positioned, it should be possible to remove and replace gauge without any sensor side movement. Tighten the retaining screw to 5 to 10 oz.-in., then recheck the sensor position as before.
11 Remove gauge and set trigger wheel in place on yoke. Visually check to make certain the sensor core is positioned approximately in the center of trigger wheel legs and that trigger wheel legs cannot touch sensor core.
12 Support distributor shaft and press trigger wheel onto yoke. Using).050 gauge wire, bend wire gauge to the dimension shown. Use gauge to measure the distance between trigger wheel legs and the sensor base. Install trigger wheel until it just touches the gauge.
13 Add about 3 to 5 drops of SAE 20 oil to the felt wick in the top of the yoke.
14 Install dust shield and rotor. Distributor is ready for installation. Install the distributor and time the engine to specification.

TROUBLESHOOTING CHRYSLER TYPE IGNITION

CONDITION	POSSIBLE CAUSE	CORRECTION
ENGINE WILL NOT START (Fuel and Carburetion Known to be OK)	a) Dual Ballast	Check resistance of each section: Compensating resistance: .50-.60 ohms @ 70°-80°F Auxiliary Ballast: 4.75-5.75 ohms Replace if faulty. Check wire positions.
	b) Faulty Ignition Coil	Check for carbonized tower. Check primary and secondary resistances: Primary: 1.41-1.79 ohms @ 70°-80°F Secondary: 9,200-11,700 ohms @ 70°-80°F Check in coil tester.
	c) Faulty Pickup or Improper Pickup Air Gap	Check pickup coil resistance: 400-600 ohms Check pickup gap. .010 in. feeler gauge should not slip between pickup coil core and aligned reluctor blade. No evidence of pickup core striking reluctor blades should be visible. To reset gap, tighten pickup adjustment screw with a .008 in. feeler gauge held between pickup core and an aligned reluctor blade. After resetting gap, run distributor on test stand and apply vacuum advance, making sure that the pickup core does not strike the reluctor blades.
	d) Faulty Wiring	Visually inspect wiring for brittle insulation. Inspect connectors. Molded connectors should be inspected for rubber inside female terminals.
	e) Faulty Control Unit	Replace if all of the above checks are negative. Whenever the control unit or dual ballast is replaced, make sure the dual ballast wires are correctly inserted in the keyed molded connector.
ENGINE SURGES	a) Wiring	Inspect for loose connection and/or broken conductors in harness.
SEVERELY (Not Lean	b) Faulty Pickup Leads	Disconnect vacuum advance. If surging stops, replace pickup.
Carburetor	c) Ignition Coil	Check for intermittent primary.
ENGINE MISSES (Carburetion OK)	a) Spark Plugs b) Secondary Cable c) Ignition Coil d) Wiring e) Faulty Pickup Lead f) Control Unit	Check plugs. Clean and regap if necessary. Check cables with an ohmmeter, or observe secondary circuit performance with an oscilloscope. Check for cabonized tower. Check in coil tester. Check for loose or dirty connections. Disconnect vacuum advance. If miss stops, replace pickup. Replace if the above checks are negative.

CHRYSLER TYPE IGNITION

Chrysler Electronic Ignition

Testing Ignition

ALL CARS

To properly test the Electronic Ignition System, special testors should be used. But in the event they are not available, the system may be tested using a voltmeter with a 20,000 ohm/volt rating and an ohmmeter which uses a 1½ volt battery for its operation. Both meters should be in calibration. When Ignition System problems are suspected, the following procedure should be followed:

1 Visually inspect all secondary cables at the coil, distributor and spark plugs for cracks and tightness.
2 To check wiring harness and connections, check primary wire at the ignition coil and ballast resistor for tightness. If the above checks do not determine the problem, the following steps will determine if a component is faulty.
3 Check and note battery voltage reading using voltmeter. Battery voltage should be at least 12 volts.
4 Remove the multi-wiring connector from the control unit.

CHILTON CAUTION: *Whenever removing or installing the wiring harness connector to the control unit, the ignition switch must be in the "Off" position.*

5 Turn the ignition switch "On".
6 Connect the negative lead of a voltmeter to a good ground.
7 Connect the positive lead of the voltmeter to the wiring harness connector cavity #1. Available voltage at cavity #1 should be within 1 volt of battery voltage with all accessories off. If there is more than a 1 volt difference, the circuit must be checked between the battery and the connector.
8 Connect the positive lead of the voltmeter to the wiring harness connector cavity #2. Available voltage at cavity

CHRYSLER TYPE IGNITION

#2 should be within 1 volt of battery voltage with all accessories off. If there is more than a 1 volt difference, the circuit must be checked back to the battery.

9 Connect the positive lead of the voltmeter to the wiring harness connector cavity #3. Available voltage at cavity #3 should be within 1 volt of battery voltage with all accessories off. If there is more than a 1 volt difference, the circuit that must be checked back to the battery.

10 Turn ignition switch "Off".

11 To check distributor pickup coil connect an ohmmeter to wiring harness connector cavity #4 and #5. The ohmmeter resistance should be between 150 and 900 ohms.

If the readings are higher or lower than specified, disconnect the dual lead connector coming from the distributor. Using the ohmmeter, check the resistance at the dual lead connector. If the reading is not between the prementioned resistance values, replace the pickup coil assembly in the distributor.

12 Connect one ohmmeter lead to a good ground and the other lead to either connector of the distributor. Ohmmeter should show an open circuit (infinity). If the ohmmeter does show a reading less than infinity the pick up coil in the distributor must be replaced.

13 To check electronic control unit ground circuit connect one ohmmeter lead to a good ground and the other lead to the control unit connector pin #5. The ohmmeter should show continuity between the ground and the connector pin. If continuity does not exist, tighten the bolts holding the control unit to the fire wall. Then recheck. If continuity does still not exist, control unit must be replaced.

14 Reconnect wiring harness at control unit and distributor.

NOTE: Whenever removing or installing the wiring harness connector to the control unit, the ignition switch must be in the "Off" position.

15 Check air gap between reluctor tooth and pick up coil. To set the gap refer to Air Gap Adjustment.

16 Check ignition secondary; remove the high voltage cable from the center tower of the distributor. Hold the cable approximately 3/16 inch from engine. Crank engine. If arcing does not occur, replace the control unit.

17 Crank the engine again. If arcing still does not occur, replace the ignition coil.

18 If a problem does not show up when making the voltage checks, coil resistance checks, or ground continuity checks it is likely the control unit or coil is faulty. It is unlikely that both units would fail simultaneously. However, before replacing the control unit

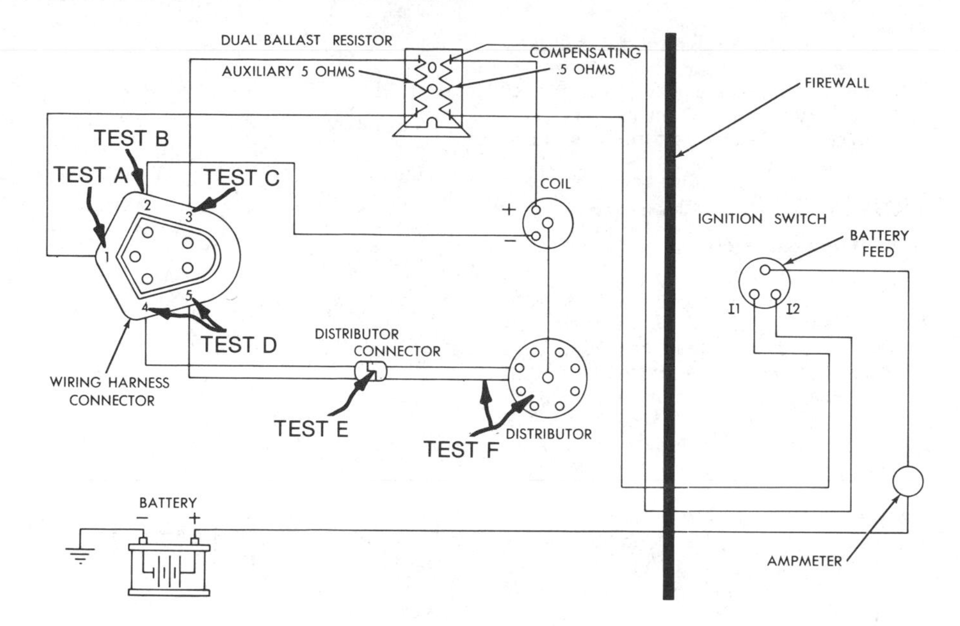

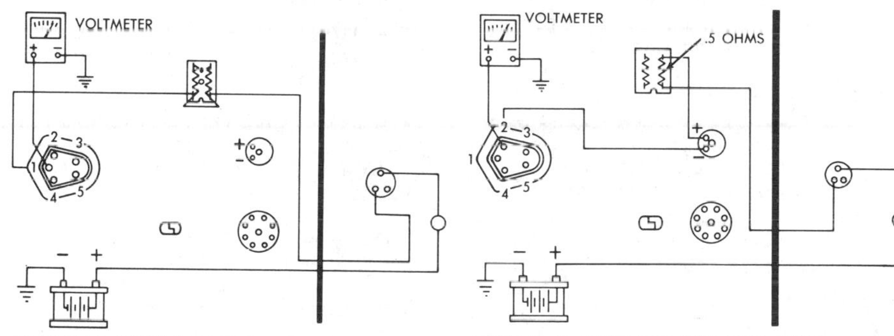

TESTING CAVITY NUMBER ONE

TESTING CAVITY NUMBER TWO

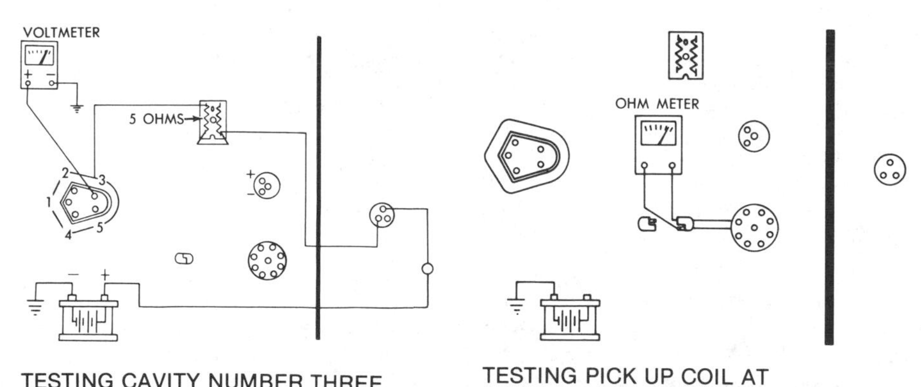

TESTING CAVITY NUMBER THREE

TESTING PICK UP COIL AT DISTRIBUTOR LEAD CONNECTOR

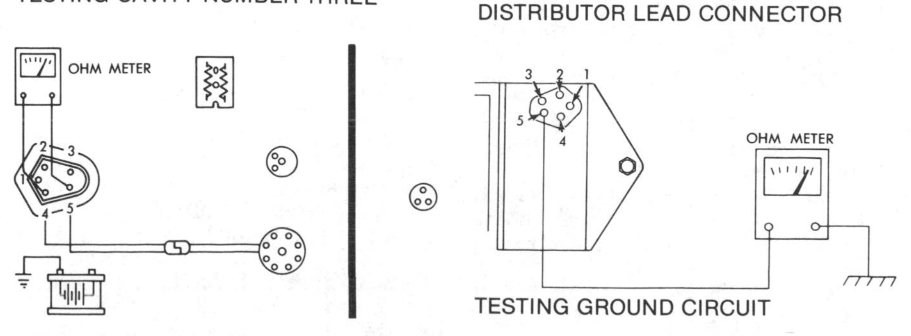

TESTING GROUND CIRCUIT

TESTING PICK UP COIL AT WIRING HARNESS CONNECTOR, CAVITIES FOUR AND FIVE

CHRYSLER TYPE IGNITION

make sure no foreign matter is lodged in or blocking the female terminal cavities in the harness connector. If clear, try replacing control unit or coil to see which one restores secondary ignition voltage.

Servicing Procedures

Pick-Up Coil R&R

1. Remove the distributor.
2. Remove the two screws and lockwashers attaching the vacuum control unit to the distributor housing. Disconnect the arm and remove the vacuum unit.
3. Remove the reluctor by pulling it off with your fingers, or use two small screwdrivers to pry it off. Be careful not to distort or damage the teeth on the reluctor.
4. Remove the two screws and lockwashers attaching the lower plate to the housing and lift out the lower plate, upper plate, and pick-up coil as an assembly.
5. Remove the upper plate and pick-up coil assembly from the lower plate by depressing the retaining clip and moving it away from the mounting stud.
6. Remove the upper plate and pick-up coil assembly. The pick-up coil is not removable from the upper plate, and is serviced as an assembly. On early models, the coil was removable from the plate.
7. To install the pick-up coil assembly, put a little distributor cam lube on the upper plate pivot pin and lower plate support pins.
8. Position the upper plate pivot pin through the smallest hole in the lower plate.
9. Install the retaining clip. The upper plate must ride on the three support pins on the lower plate.
10. Install the lower plate, upper plate, and pickup coil assembly into the distributor and install screws.
11. Attach the vacuum advance arm to the pick-up plate, then install the vacuum unit attaching screws and washers.
12. Position the reluctor keeper pin in place on the reluctor sleeve, then slide the reluctor down the sleeve and press firmly into place.

Air Gap Adjustment

1. Align one reluctor tooth with the pick-up coil tooth.
2. Loosen the pick-up coil hold-down screw.
3. Insert a 0.008 in. nonmagnetic feeler gauge between the reluctor tooth and the pick-up coil tooth.
4. Adjust the air gap so that contact is made between the reluctor tooth, the feeler gauge, and the pick-up coil tooth.
5. Tighten the hold-down screw.
6. Remove the feeler gauge.

NOTE: No force should be required in removing the feeler gauge.

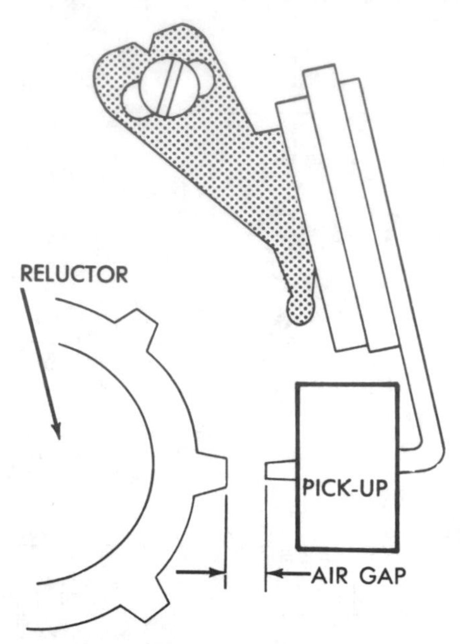

Air gap adjustment

7. A 0.010 in. feeler gauge should not fit into the air gap. Do not force the feeler gauge.

CHILTON CAUTION: ***A 0.010 in. feeler gauge can be forced into the air gap. DO NOT FORCE THE FEELER GAUGE INTO THE AIR GAP.***

8. Apply vacuum to the vacuum unit and rotate the governor shaft. The pick-up pole should not hit the reluctor teeth. The gap is not properly adjusted if any hitting occurs. If hitting occurs on only one side of the reluctor, the distributor shaft is probably bent, and the governor and shaft assembly should be replaced.

Shaft and Bushing Wear Test

1. Remove distributor and rotor.
2. Clamp distributor is a vise equipped with soft jaws and apply only enough pressure to restrict any movement of the distributor during the test.
3. Attach a dial indicator to distributor housing so indicator plunger arm rests against reluctor.
4. Wiggle the shaft and read the total movement of the dial indicator plunger. If the movement exceeds .006 in. replace the housing or shaft.

Distributor Overhaul

1. Remove distributor rotor.
2. Remove the two screws and lockwashers attaching the vacuum control unit to distributor housing, disconnect the vacuum control arm from upper plate, and remove control.
3. Remove reluctor by prying up from the bottom of the reluctor with two screwdrivers. Be careful not to distort or damage the teeth on the reluctor.
4. Remove two screws and lockwashers attaching the lower plate to the housing and lift out the lower plate, upper plate, and pick-up coil as an assembly. Distributor cap clamp springs are held in place by peened metal around the openings and should not be removed.
5. If the side play exceeds .006 inch in "Shaft and Bushing Wear Test", replace distributor housing assembly or shaft and governor assembly as follows: Remove distributor drive gear retaining pin and slide gear off end of shaft.

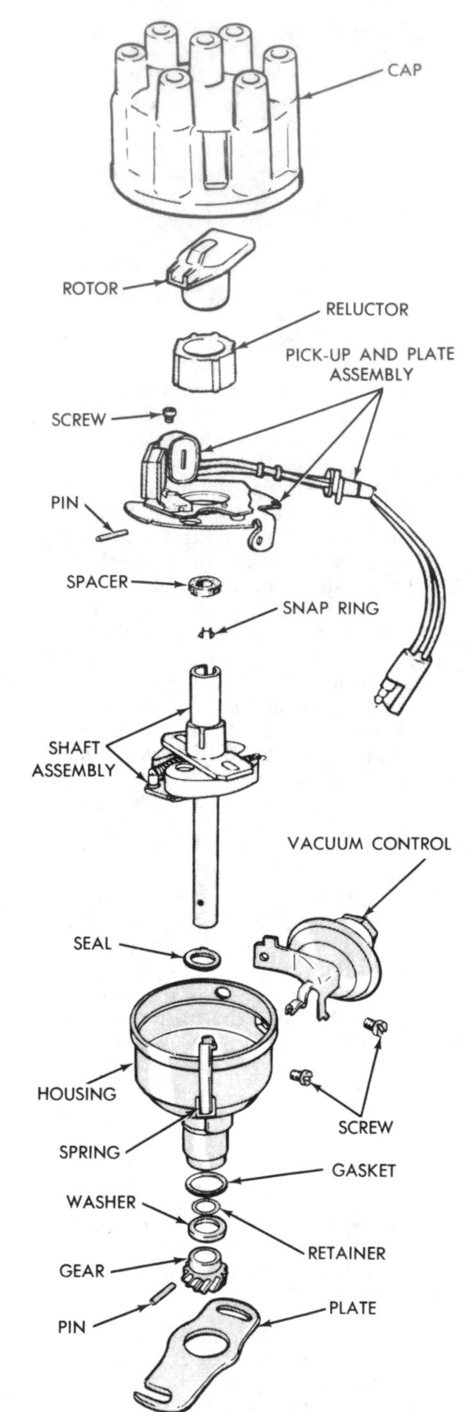

Chrysler electronic distributor

CHRYSLER TYPE IGNITION

CHILTON CAUTION: ***Support hub of gear in a manner that pin can be driven out of gear and shaft without damaging gear teeth.***

Use a file to clean burrs, from around pin hole in the shaft and remove the lower thrust washer. Push shaft up and remove shaft through top of distributor body.

6 If gear is worn or damaged, replace as follows: Install lower thrust washer and old gear on lower end of shaft and temporarily install rollpin. Scribe a line on the end of the shaft from center to edge, so line is centered between two gear teeth as shown in. **Do not Scribe completely across the shaft.** Remove rollpin and gear. Use a fine file to clean burrs from around pin hole. Install new gear with thrust washer in place. Drill hole in gear and shaft approximately 90 degrees from old hole in shaft and with scribed line centered between the two gear teeth as shown. Before drilling through shaft and gear, place a .007 feeler gauge between gear and thrust washer and after again observing that the centerline between two of the gear teeth is in line with centerline of rotor electrode drill a .124–.129 inch hole and install the rollpin.

CHILTON CAUTION: ***Support hub of gear when installing roll-pin so that gear teeth will not be damaged.***

7 Test operation of governor weights and inspect weight springs for distortion.
8 Lubricate governor weights.
9 Inspect all bearing surfaces and pivot pins for roughness, binding or excessive looseness.
10 Lubricate and install upper thrust washer (or washers) on the shaft and slide the shaft into the distributor body.
11 Install lower plate, upper plate and pick-up coil assembly and install attaching screws.
12 Slide shaft into distributor body, then align scribe marks and install gear and rollpin.
13 Attach vacuum advance unit arm to the pick-up plate.
14 Install vacuum unit attaching screws and washers.
15 Position reluctor keeper pin into place on reluctor sleeve.
16 Slide reluctor down reluctor sleeve and press firmly into place.
17 Lubricate the felt pad in top of reluctor sleeve with 1 drop of light engine oil and install the rotor.

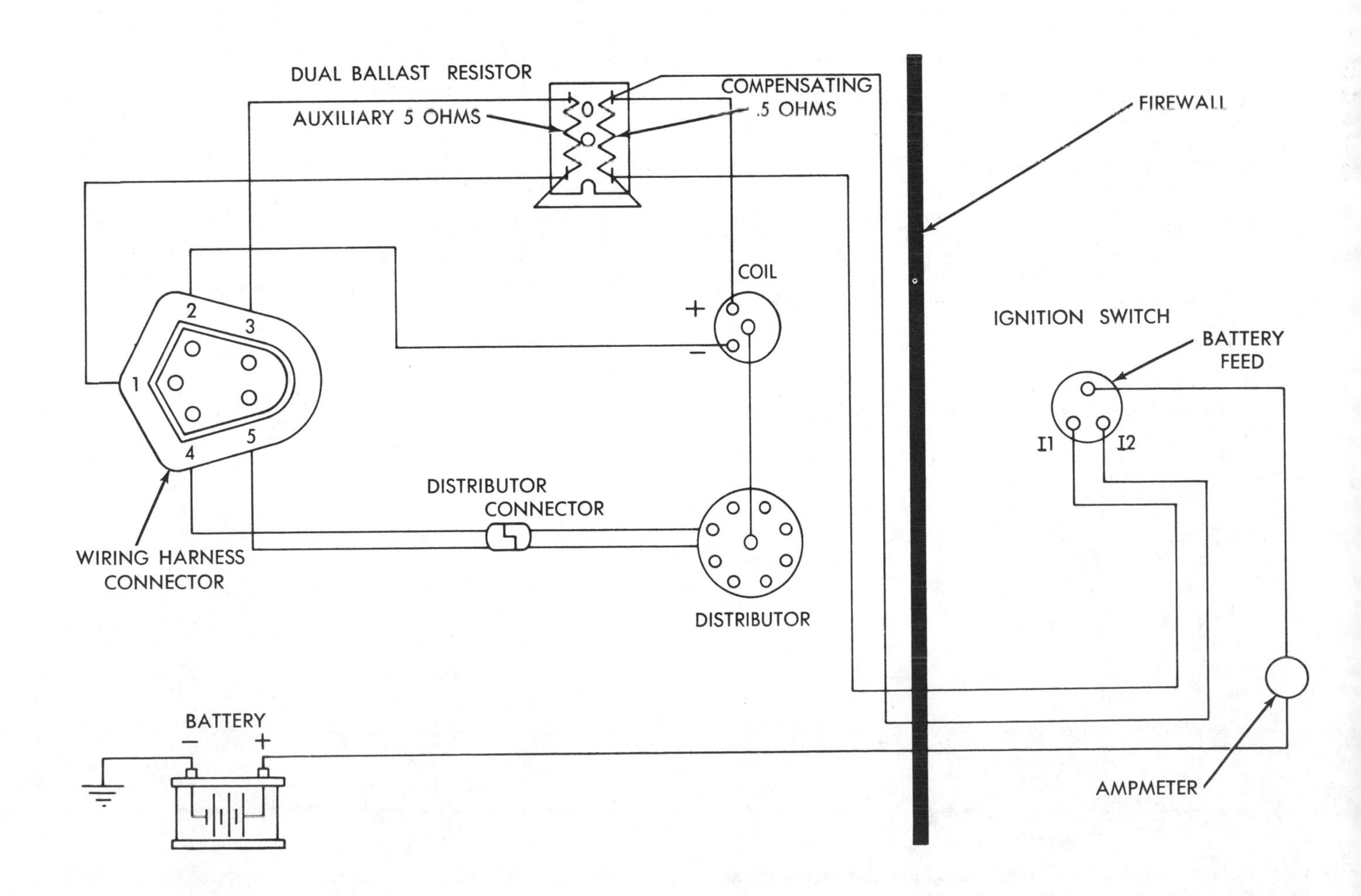

FORD ELECTRONIC IGNITION

Ford Motorcraft Dura-Spark 1 & 2

Also Used By AMC

OPERATION

Dura Spark I features special vehicle primary wiring, a new design coil, an all-new electronic control module, a new distributor cap and adapter, new high energy secondary wiring and special wide-gap spark plugs. In operation, it senses current flow through the coil, adjusting "dwell" for maximum spark intensity. Thus, coil "on" time is adjusted for best efficiency related to engine RPM. If the module senses that the ignition switch is ON and the distributor is not turning, it will automatically turn the coil current OFF after about one second. To reestablish the module cycle, turn the key to START or to OFF and then to ON again.

Dura Spark II is basically a solid state ignition system with two major differences: the ballast resistor value has been changed from 1.35 to 1.10 ohms to boost coil current and energy output, and the system now uses the new Dura Spark I rotor, distributor cap and adapter, ignition secondary wires and wide-gap spark plugs to take advantage of the higher energy produced.

Both systems use the same distributor and both control modules have the same exterior appearance.

It is important to note that the Dura Spark II amplifier module and coil are "on" when the ignition switch is "on". Because of this, the ignition system will generate a spark when the key is turned "off". This feature may be used as a diagnostic tool, to verify continuity of circuit, coil and ignition switch. Certain other service actions such as removing the distributor cap with the ignition switch "on" may also cause the system to fire. For this reason, the ignition switch should remain "off" during any underhood operations unless the intent is to start the vehicle engine, or to perform a specific test that requires the ignition switch to be "on".

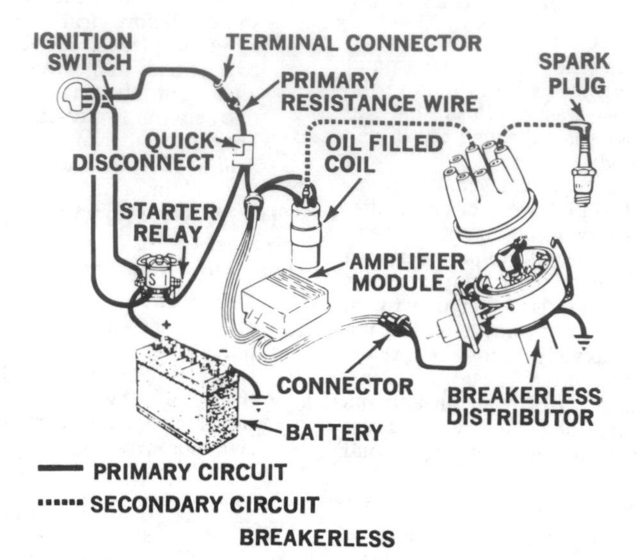

Components and connectors, Ford Motorcraft breakerless ignition system

This procedure will prevent the possibility of inadvertent engine rotation caused by triggering of the ignition system during servicing.

Dura Spark I systems automatically shut down when the system senses no distributor rotation within one second.

Primary Circuit Testing

A breakdown or energy loss in the primary circuit can be caused by: defective primary wiring; loose or corroded connections; inoperative or defective magnetic pick-up coil assembly; or defective amplifier module.

A complete test of the primary circuit consists of checking the circuits in the ignition coil, the magnetic pick-up coil assembly and the amplifier module. Wiring harness checks should be included as a part of basic component circuit tests.

Always inspect connectors for dirt, corrosion or poor fit before assuming you have spotted a possible problem.

Troubleshooting

Make sure the battery is fully charged before beginning tests. Perform a spark intensity test. If no spark is observed, make sure that the high tension coil wire is good. Disconnect the 3-way and 4-way connectors at the electronic module.

The first trouble isolation test will be conducted on the harness terminals, with the electronic module disconnected from the circuit. The pin numbers shown in the schematic correspond to those shown in the trouble isolation test table.

Make the following tests with a sensitive volt-ohmmeter. These tests will direct you to the proper follow-up test to determine the actual problem.

If the circuit checks good at all these test points, connect a known good electronic module in place of the vehicle module and again perform the spark intensity test. If the substitution corrects the malfunction again reconnect the vehicle module and perform the spark intensity test. If the malfunction still exists, the problem is in the module and it must be replaced. If the problem is gone, it may be in the wiring connectors.

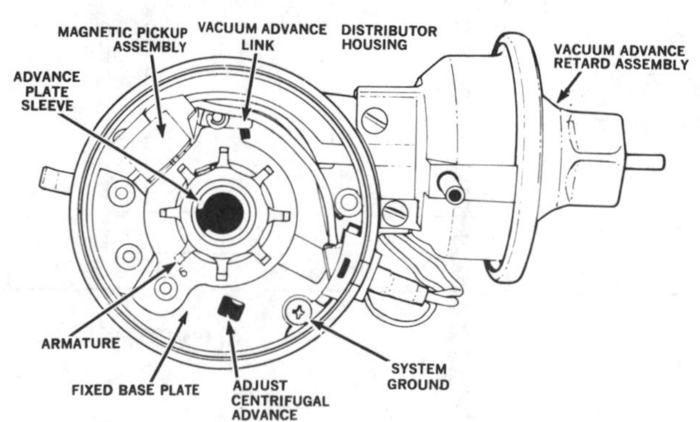

Ford Motorcraft solid state breakerless ignition system V8 engine distributor top view and components

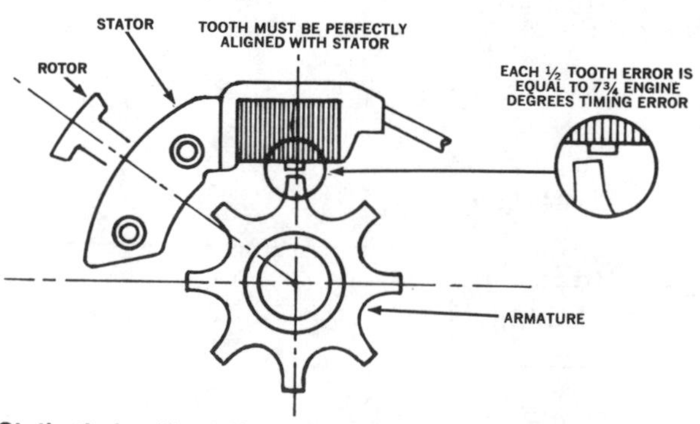

Static timing Ford Motorcraft breakerless ignition system

FORD ELECTRONIC IGNITION

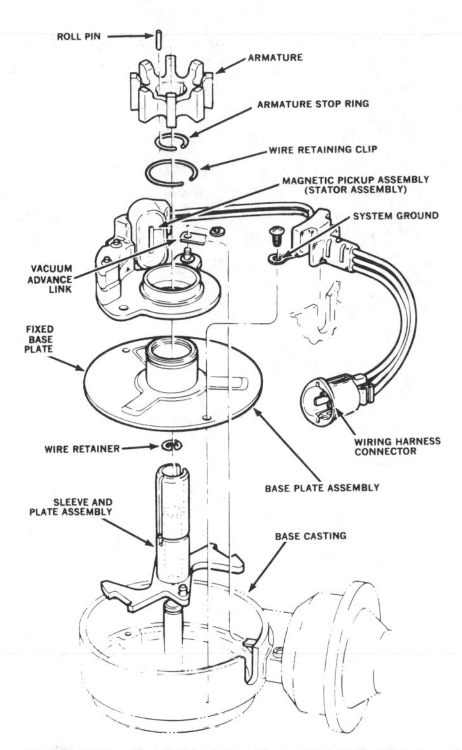

The Ford Motorcraft solid state breakerless ignition system V8 engine distributor assembly

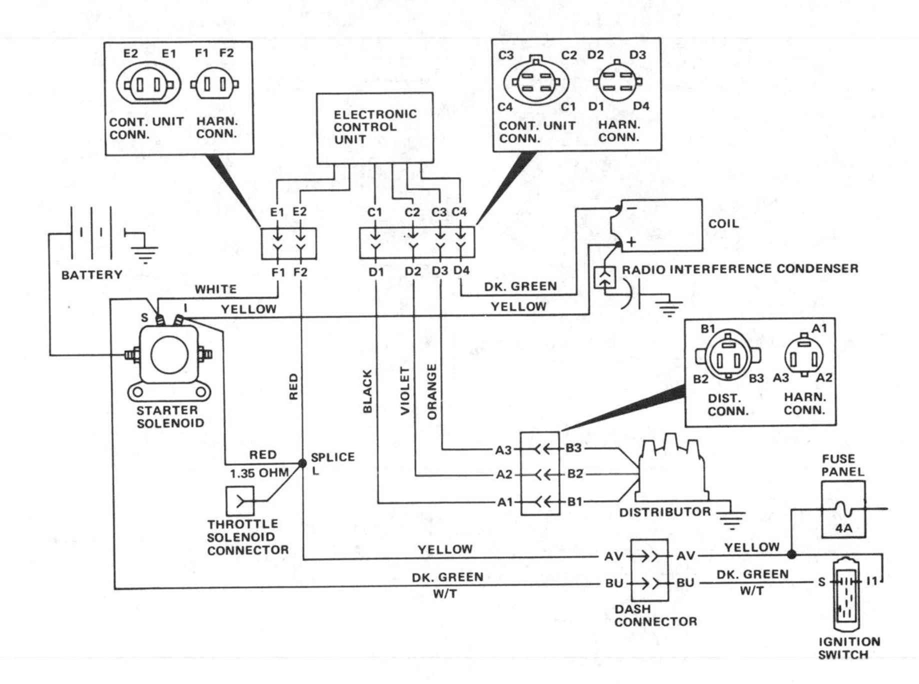

Dura Spark Ignition system wiring schematic

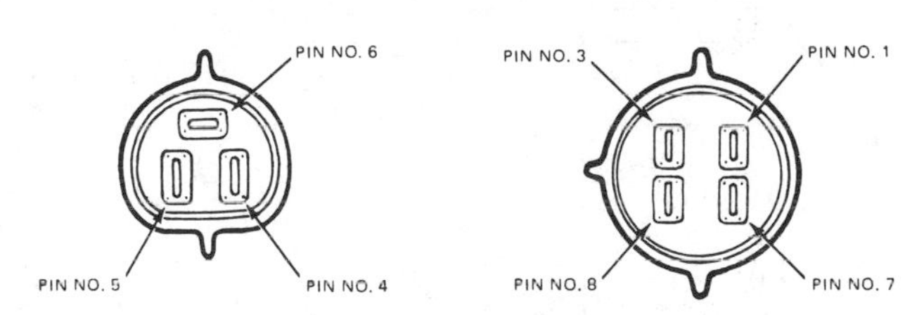

Electronic module connectors-harness side, 1974-76

If the substitute module does not correct the problem, reconnect the original module and make repairs elsewhere in the system.

Module Bias Test

Measure the voltage at Pin 3 to engine ground with the ignition key on. If the voltage observed is less than battery voltage, repair the voltage feed wiring to the module for running conditions (re-wire).

Battery Source Test

1. Connect the voltmeter leads from the battery terminal at the coil to engine ground, without disconnecting the coil from the circuit.
2. Install a jumper wire from the DEC terminal of the coil to a good engine ground.
3. Turn the lights and all accessories off.
4. Turn the ignition switch on.
5. If the voltmeter reading is between 4.9 and 7.9 volts, the primary circuit from the battery is satisfactory.
6. If the voltmeter reading is less than 4.9 volts, check the following:
   a. The primary wiring for worn insulation, broken strands, and loose or corroded terminals.
   b. The resistance wiring for defects.
7. If the voltmeter reading is greater than 7.9 volts, the resistance wire should be replaced after verifying a defect.

Cranking Test

Measure the voltage at Pin 1 to engine ground with the engine cranking. If the

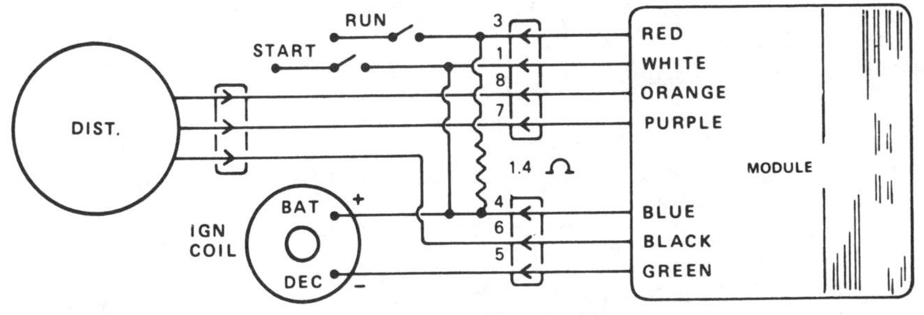

1974 primary circuit-Ford breakerless ignition system

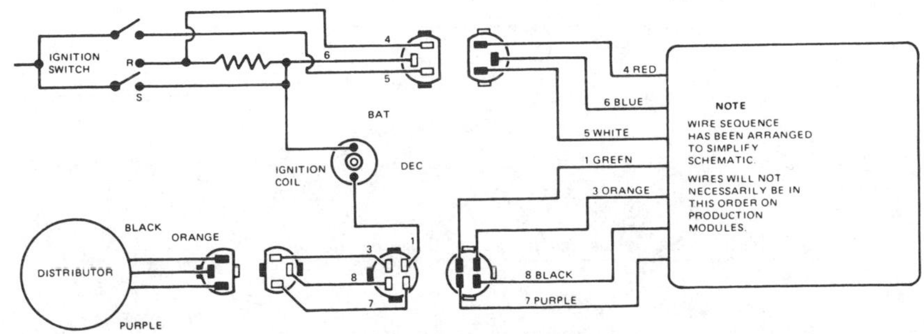

1975 primary circuit-Ford breakerless ignition system

FORD ELECTRONIC IGNITION

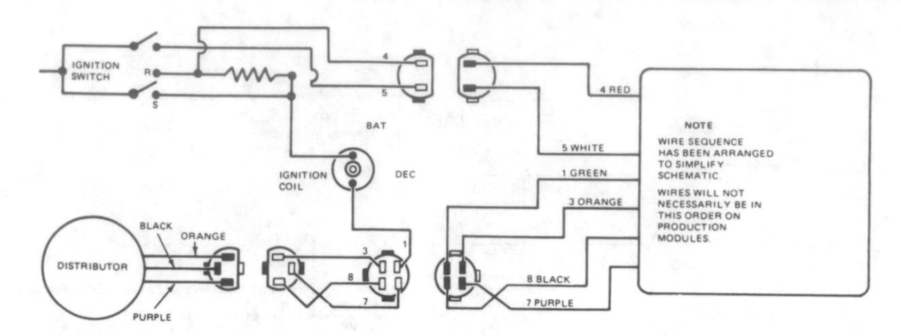

1976 primary circuit-Ford breakerless ignition system

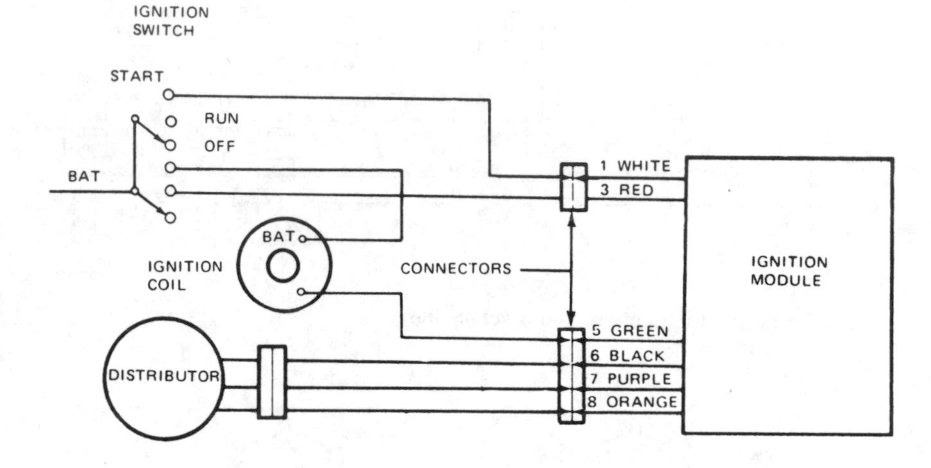

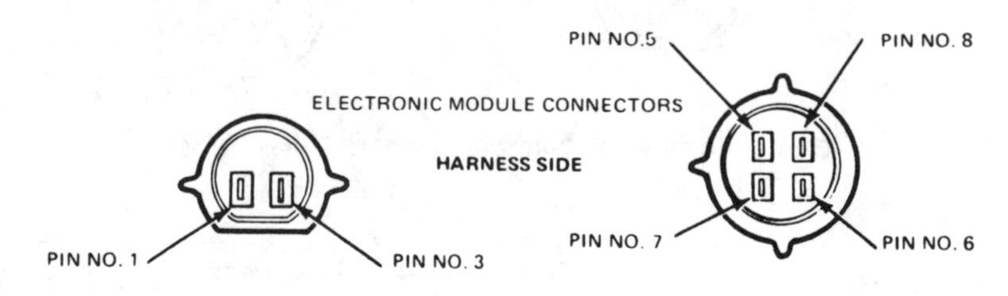

Dura Spark I breakerless ignition primary circuit

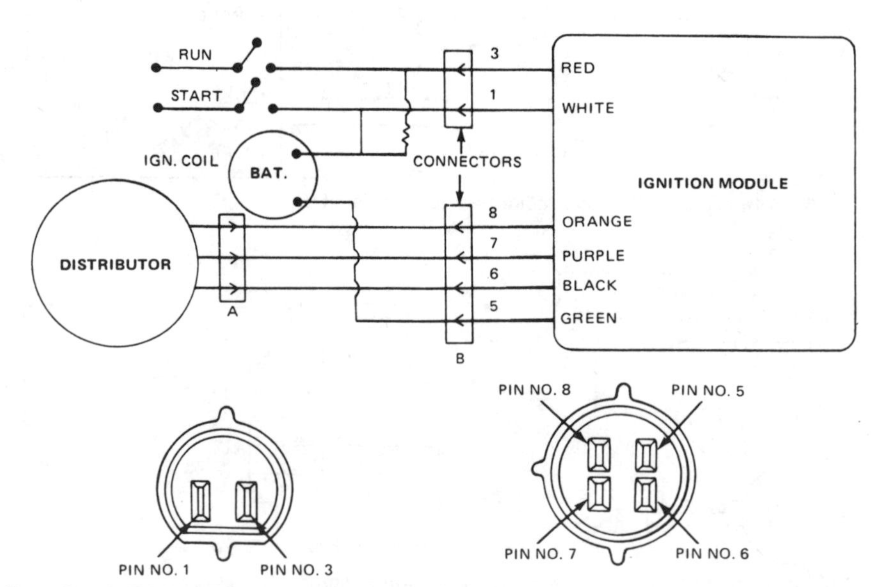

Dura Spark II breakerless ignition primary circuit

voltage observed is not 8 to 12 volts, repair the voltage feed to the module for starting conditions (white wire).

Starting Circuit Test

If the reading is not between 8 and 12 volts, the ignition by-pass circuit is open or grounded from either the starter solenoid or the ignition switch to Pin 5. Check the primary connections at the coil.

Distributor Hardware Test

1 Disconnect the 3-wire weatherproof connector at the distributor pigtail.
2 Connect a DC voltmeter on a 2.5 volt scale to the two parallel blades. With the engine cranking, the meter needle should oscillate.
3 Remove the distributor cap and check for visual damage or misassembly.
   a. Sintered iron armature (6 or 8-toothed wheel) must be tight on the sleeve, and the roll pin aligning the armature must be in position.
   b. Sintered iron stator must not be broken.
   c. Armature must rotate when the engine is cranked.
4 If the hardware is OK, but the meter doesn't oscillate, replace the magnetic pick-up assembly.

Magnetic Pick-Up Tests

1 Resistance of pick-up coil measured between two parallel pins in the distributor connector must be 400–800 ohms.
2 Resistance between the third blade (ground) and the distributor bowl must be zero ohms.
3 Resistance between either parallel blade and engine ground must be greater than 70,000 ohms.
4 If any test fails, the distributor stator assembly is defective and must be replaced.
5 If the above readings are not the same as measured in the original test, check for a defective harness. If the readings are the same, proceed.
6 If these tests check OK, the signal generator portion of the distributor is working properly.

Ignition Coil Test

The breakerless ignition coil must be diagnosed separately from the rest of the ignition system.

1 Primary resistance must be 1.0–2.0 ohms, measured from the BAT to the DEC terminals.
2 Secondary resistance must be 7,000–13,000 ohms, measured from the BAT or DEC terminal to the center tower of the coil.
3 If resistance tests are OK, but the coil is still suspected, test the coil on a coil tester by following the test equipment manufacturer's instructions for a standard coil. If the reading differs from the original test, check for a defective harness.

FORD ELECTRONIC IGNITION

DIAGNOSIS CHARTS

DURA-SPARK 1 & 2 BREAKERLESS IGNITION SYSTEMS

	Test Voltage Between	Should Be	If Not, Conduct
KEY ON	Pin #3 and Engine Ground	Battery Voltage ± 0.1 volts	Module Bias Test
	Pin #5 and Engine Ground	Battery Voltage ± 0.1 volts	Battery Source Test
CRANKING	Pin #1 and Engine Ground	8 to 12 volts	Cranking Test
	Jumper #5 to #6—Read Coil "Bat" Term. & Engine Ground**	More than 6 volts	Starting Circuit Test
	Pin #7 and Pin #8	½ volt minimum wiggle	Distributor Hardware Test

	Test Resistance Between	Should Be	If Not, Conduct**
KEY OFF	Pin #7 and Pin #8 Pin #6 and Engine Ground Pin #7 and Engine Ground Pin #8 and Engine Ground	400 to 800 ohms 0 ohms More than 70,000 ohms More than 70,000 ohms	Magnetic Pick-up (Stator) Test
	Pin #3 and Coil Tower	7,000 to 13,000 ohms	Coil Test
	Pin #5 and Coil "Bat" Term.	1.0 to 2.0 ohms Dura Spark II 0.5 to 1.5 ohms Dura Spark I	
	Pin #5 and Engine Ground	More than 4 ohms	Short Test
	Pin #3 and Coil "Bat" Term. (Except Dura Dura Spark I)	0.7 to 1.7 ohms Dura Spark II	Resistance Wire Test

**Test duration shall be less than 30 seconds (Dura Spark I)

1974–76 FORD BREAKERLESS IGNITION SYSTEMS

	Test Voltage Between	Should Be	If Not, Conduct
KEY ON	Pin #3 and Engine Ground	Battery Voltage	Module Bias Test
	Pin #5 and Engine Ground	Battery Voltage	Battery Source Test
CRANKING	Pin #1 and Engine Ground	8 to 12 volts	Cranking Test
	Pin #5 and Engine Ground	8 to 12 volts	Starting Circuit Test
	Pin #7 and Pin #8	½ volt A.C or D.C. volt wiggle	Distributor Hardware Test

	Test Resistence Between	Should Be	If Not, Conduct
KEY ON	Pin #7 and Pin #8 Pin #6 and Engine Ground Pin #7 and Engine Ground Pin #8 and Engine Ground	400 to 800 ohms 0 ohms more than 70,000 ohms more than 70,000 ohms	Magnetic Pick-up (Stator) Test
	Pin #3 and Coil Tower Pin #5 and Pin #4	7000 to 13000 ohms 1.0 to 2.0 ohms	Coil Test
	Pin #5 and Engine Ground	more than 10.0 ohms	Short Test
	Pin #3 and Pin #4	1.0 to 2.0 ohms	Resistance Wire

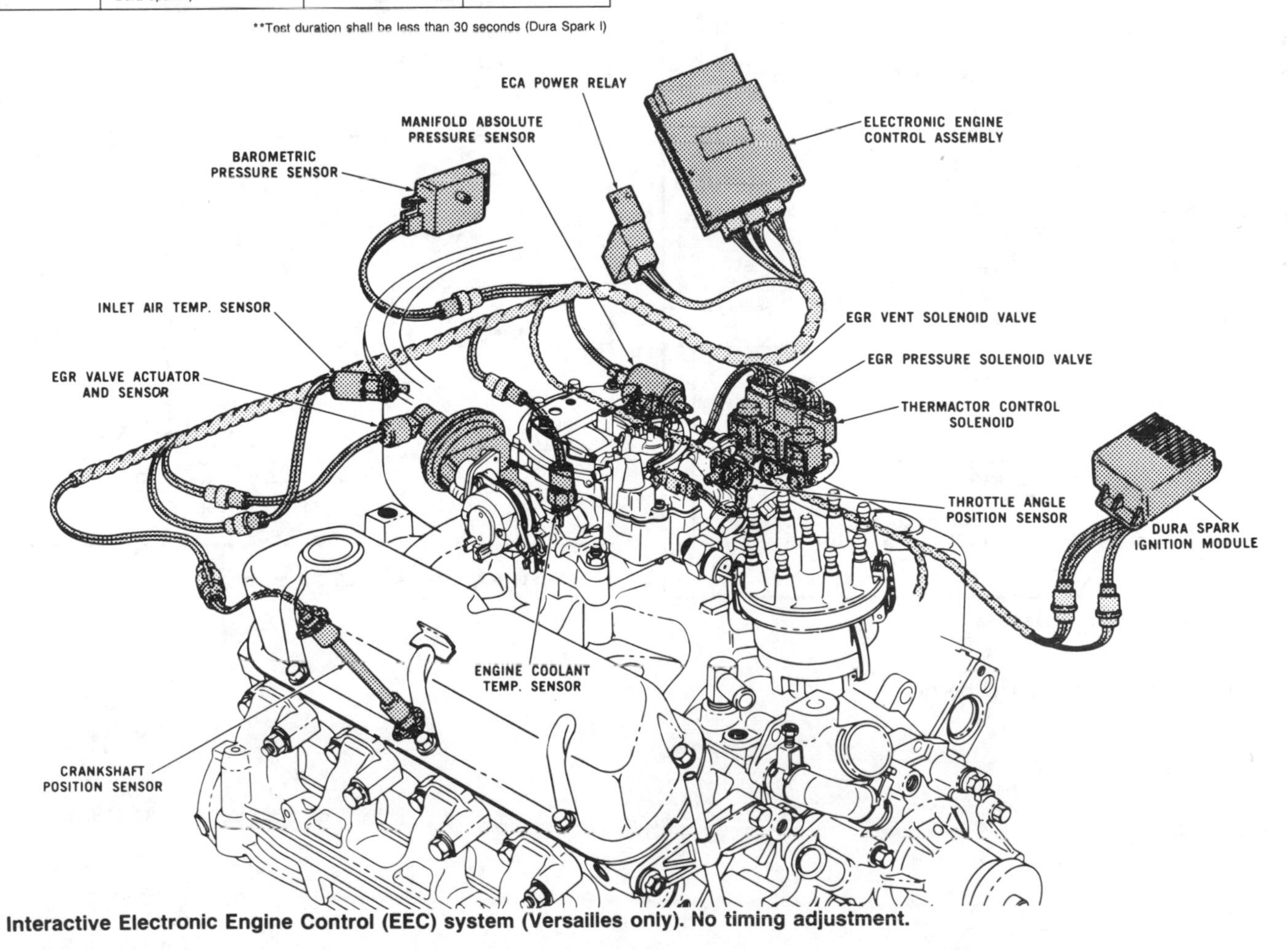

Interactive Electronic Engine Control (EEC) system (Versailles only). No timing adjustment.

FORD ELECTRONIC IGNITION

Short Test

If the resistance from Pin 5 to ground is less than 10 ohms, check for a short to ground at the DEC terminal of the ignition coil or in the connection wiring to that terminal.

Resistance Wire Test

Replace the resistance wire if it is out of specifications

Adjustments

The air gap between the armature and magnetic pick-up coil in the distributor is not adjustable, nor are there any adjustments for the amplifier module. Inoperative components are simply replaced. Any attempt to connect components outside the vehicle may result in component failure.

Magnetic Pick-up Assembly R&R

1 Remove the distributor cap and rotor and disconnect the distributor harness plug.
2 Using a small gear puller or two screwdrivers, lift or pry the armature from the advance plate sleeve. Remove the roll pin.
3 Remove the large wire retaining clip from the base plate annular groove.
4 Remove the snap-ring which secures the vacuum advance link to the pick-up assembly.
5 Remove the magnetic pick-up assembly ground screw and lift the assembly from the distributor.
6 Lift the vacuum advance arm off the post on the pick-up assembly. Move it out against the distributor housing.
7 Place the new pick-up assembly in position over the fixed base plate and slide the wiring in position through the slot in the side of the distributor housing.
8 Install the fine wire snap-ring securing the pick-up assembly to the fixed base plate.
9 Position the vacuum advance arm over the post on the pick-up assembly and install the snap-ring.
10 Install the grounding screw through the tab on the wiring harness and into the fixed base plate.
11 Install the armature on the advance plate sleeve making sure that the roll pin is engaged in the matching slots.
12 Install the distributor rotor cap.
13 Connect the distributor wiring plug to the vehicle harness.

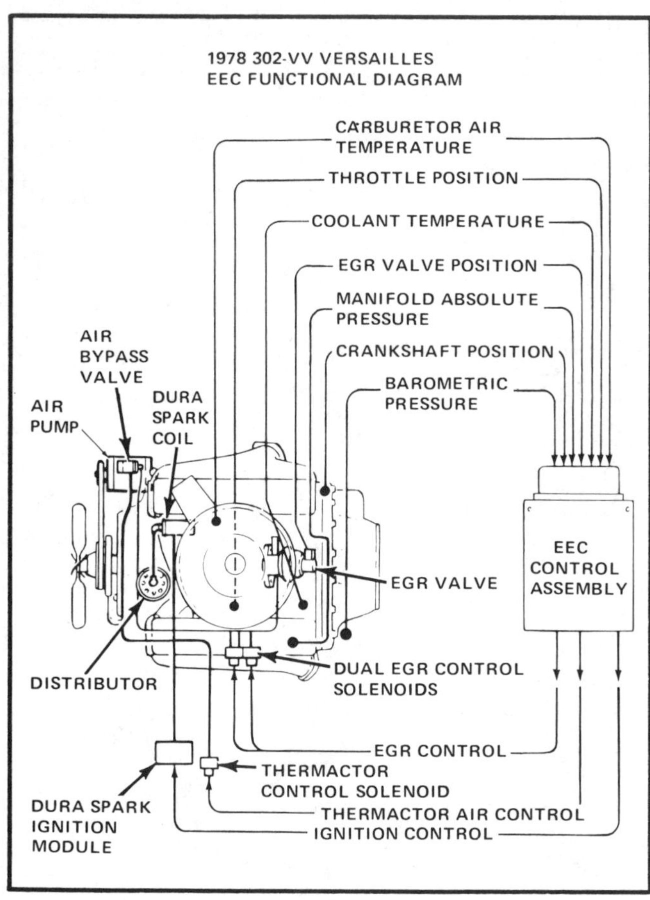

Diagram of the EEC System

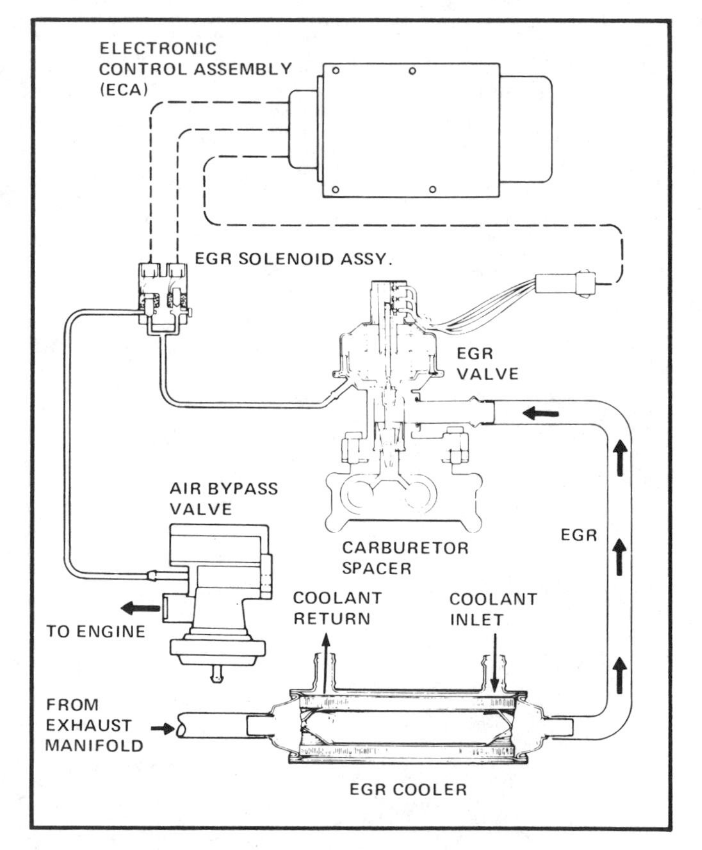

EGR System for use with EEC System

NOTE: See Computer Control Section

GM DELCO-REMY HIGH ENERGY (H.E.I.) IGNITION

TROUBLESHOOTING GM DELCO-REMY HIGH ENERGY IGNITION (H.E.I.)

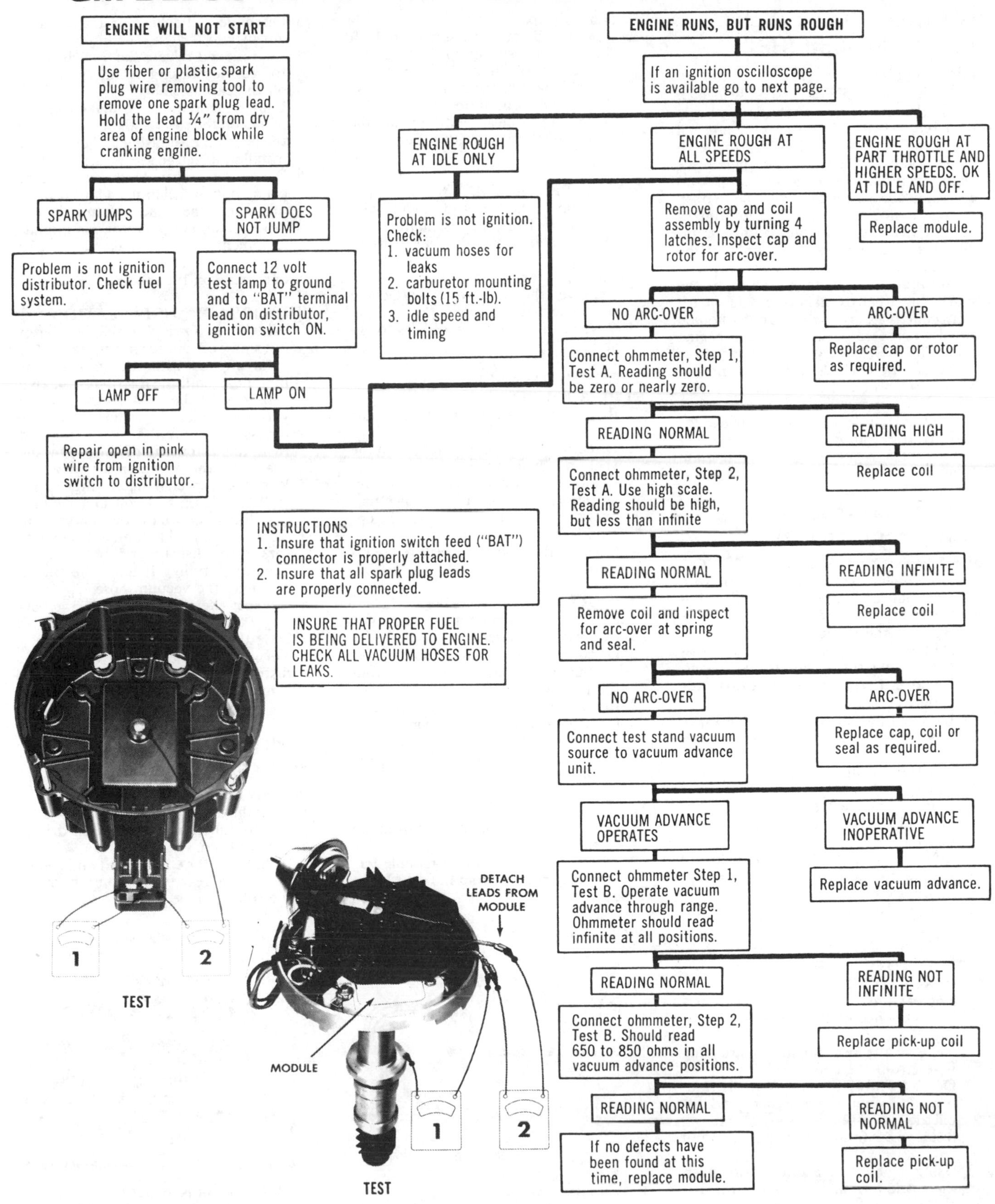

GM DELCO-REMY HIGH ENERGY IGNITION (H.E.I.)

GM Delco-Remy High Energy Ignition (HEI)

OPERATION

The magnetic pick-up assembly located inside the distributor contains a permanent magnet, a pole piece with internal teeth, and a pick-up coil. When the teeth of the rotating timer core and pole piece align, an induced voltage in the pick-up coil signals the electronic module to open the coil primary circuit. As the primary current decreases, a high voltage is induced in the secondary windings of the ignition coil, directing a spark through the rotor and high voltage leads to fire the spark plugs. The dwell period is automatically controlled by the electronic module and is increased with increasing engine rpm. The HEI System features a longer spark duration which is instrumental in firing lean and EGR diluted fuel/air mixtures. The condenser (capacitor) located within the HEI distributor is provided for noise (static) suppression purposes only and is not a regularly replaced ignition system component.

Major Repair Procedures (distributor in engine)

IGNITION COIL REPLACEMENT

1 Disconnect the feed and module wire terminal connectors from the distributor cap.
2 Remove the ignition set retainer.
3 Remove the four coil cover-to-distributor cap screws and the coil cover.
4 Remove the four coil-to-distributor cap screws.
5 Using a blunt drift, press the coil wire spade terminals up out of distributor cap.
6 Lift the coil up out of the distributor cap.
7 Remove and clean the coil spring, rubber seal washer and coil cavity of the distributor cap.
8 Coat the rubber seal with a dielectric lubricant furnished in the replacement ignition coil package.
9 Reverse the above procedures to install.

DISTRIBUTOR CAP REPLACEMENT

1 Remove the feed and module wire terminal connectors from the distributor cap.
2 Remove the retainer and spark plug wires from the cap.
3 Depress and release the four distributor cap-to-housing retainers and lift off the cap assembly.
4 Remove the four coil cover screws and cover.
5 Using a finger or a blunt drift, push the spade terminals up out of the distributor cap.
6 Remove all four coil screws and lift the coil, coil spring and rubber seal washer out of the cap coil cavity.
7 Using a new distributor cap, reverse the above procedures to assemble being sure to clean and lubricate the rubber seal washer with dielectric lubricant.

ROTOR REPLACEMENT

1 Disconnect the feed and module wire connectors from the distributor.
2 Depress and release the four distributor cap to housing retainers and lift off the cap assembly.
3 Remove the two rotor attaching screws and rotor.
4 Reverse the above procedure to install.

VACUUM ADVANCE REPLACEMENT

1 Remove the distributor cap and rotor as previously described.
2 Disconnect the vacuum hose from the vacuum advance unit.
3 Remove the two vacuum advance retaining screws, pull the advance unit outward, rotate and disengage the operating rod from its tang.
4 Reverse the above procedure to install.

MODULE REPLACEMENT

1 Remove the distributor cap and rotor as previously described.
2 Disconnect the harness connector and pick-up coil spade connectors from the module.
3 Remove the two screws and module from the distributor housing.
4 Coat the bottom of the new module with dielectric lubricant. Reverse the above procedure to install.

DISTRIBUTOR REMOVAL

1 Disconnect the ground cable from the battery.
2 Disconnect the feed and module terminal connectors from the distributor cap.
3 Disconnect the hose at the vacuum advance.
4 Depress and release the four distributor cap-to-housing retainers and lift off the cap assembly.
5 Using crayon or chalk, make locating marks on the rotor and module and on the distributor housing and engine for installation purposes.
6 Loosen and remove the distributor clamp bolt and clamp, and lift distributor out of the engine. Noting the relative position of the rotor and module alignment marks, make a second mark on the rotor to align it with the one mark on the module.

DISTRIBUTOR INSTALLATION

1 With a new O-ring on the distributor housing and the second mark on the rotor aligned with the mark on the module, install the distributor, taking care to align the mark on the housing with the one on the engine. It may be necessary to lift the distributor and turn the rotor slightly to align the gears and the oil pump driveshaft.
2 With the respective marks aligned, install the clamp and bolt finger-tight.
3 Install and secure the distributor cap.
4 Connect the feed and module connectors to the distributor cap.
5 Connect a timing light to the engine and plug the vacuum hose.
6 Connect the ground cable to the battery.
7 Start the engine and set the timing.
8 Turn the engine off and tighten the distributor clamp bolt. Disconnect the timing light and unplug and connect the hose to the vacuum advance.

Service Procedures (distributor removed)

DRIVEN GEAR REPLACEMENT

1 With the distributor removed, use a 1/8 in. pin punch and tap out the driven gear roll pin.
2 Hold the rotor end of shaft and rotate the driven gear to shear any burrs in the roll pin hole.

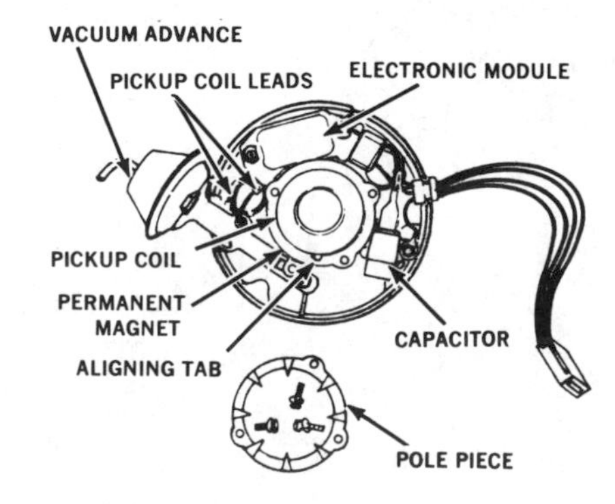

1: pole piece removal

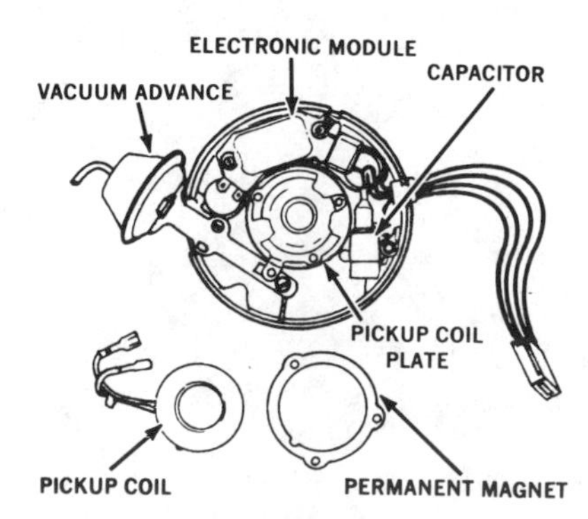

2: coil & magnet removal

GM DELCO-REMY HIGH ENERGY (H.E.I.) IGNITION

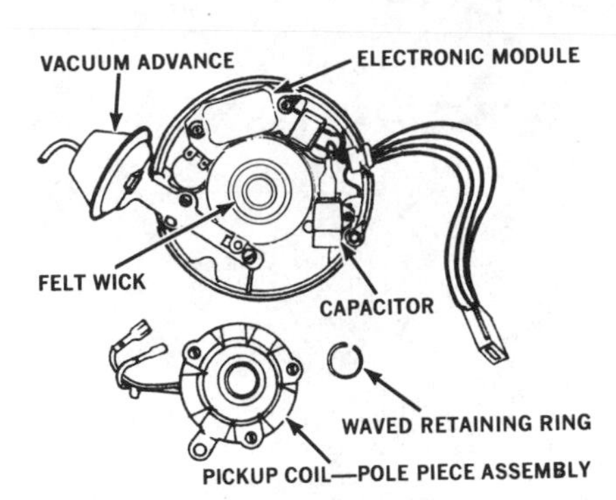

3: pick-up coil removal

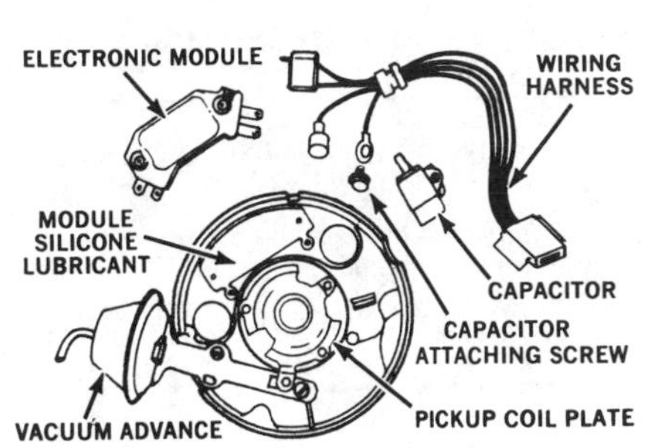

4: module & harness removal

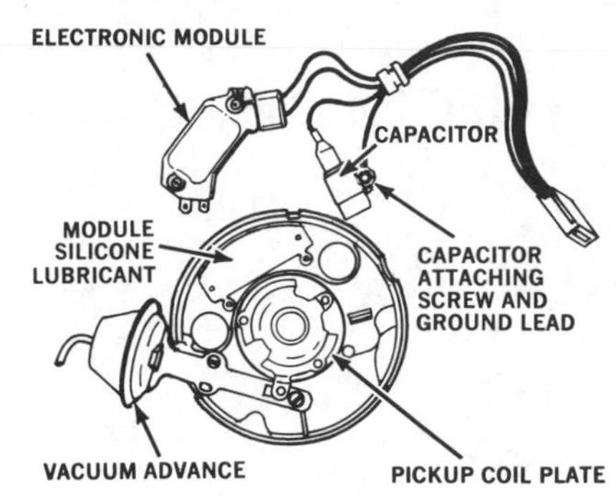

5: module & harness installation

3 Remove the driven gear from the shaft.
4 Reverse the above procedure to install.

MAINSHAFT REPLACEMENT

1 With the driven gear and rotor removed, gently pull the mainshaft out of the housing.
2 Remove the advance springs, weights and slide the weight base plate off the mainshaft.
3 Reverse the above procedure to install.

POLE PIECE, MAGNET OR PICK-UP COIL REPLACEMENT

1 With the mainshaft out of its housing, remove the three retaining screws, pole piece and magnet and/or pick-up coil.
2 Reverse the removal procedure to install making sure that the pole piece teeth do not contact the timer core teeth by installing and rotating the mainshaft. Loosen the three screws and realign the pole piece as necessary.

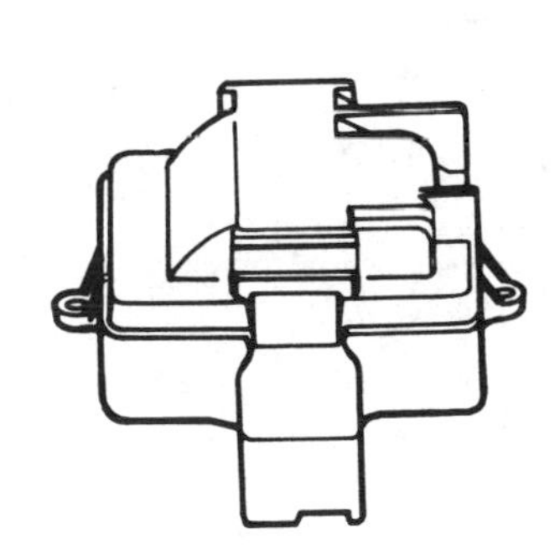

Coil Cover

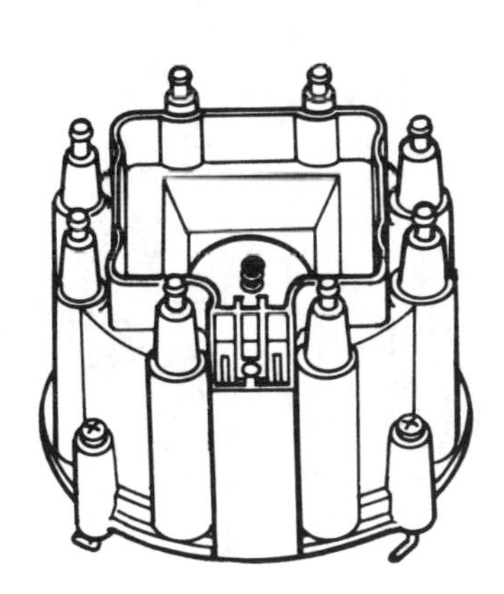

Cap

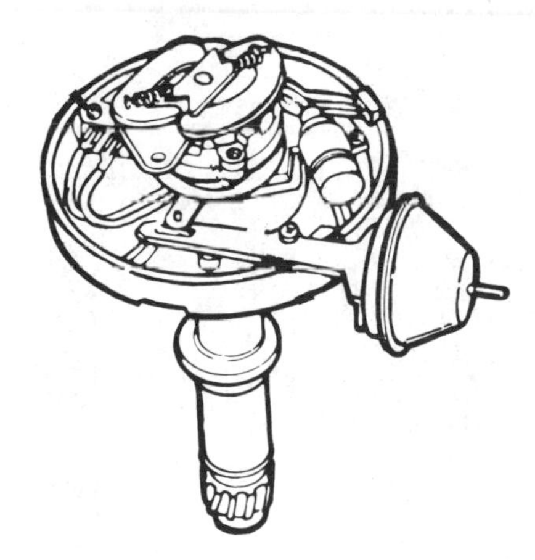

Distributor

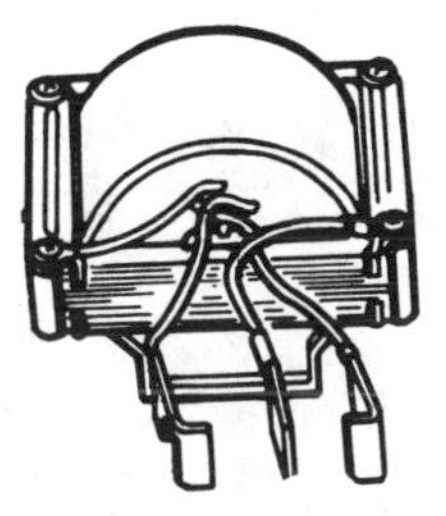

Coil

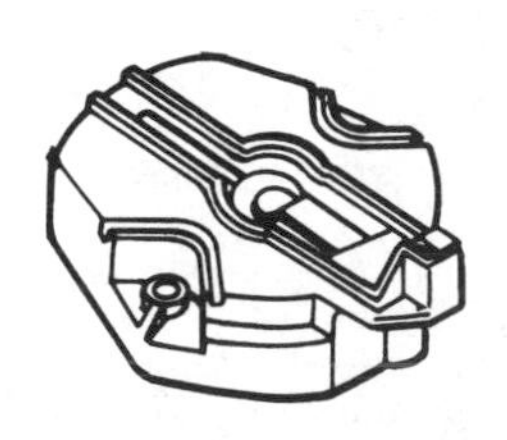

Rotor

TROUBLESHOOTING GM DELCO-REMY MAGNETIC PULSE IGNITION

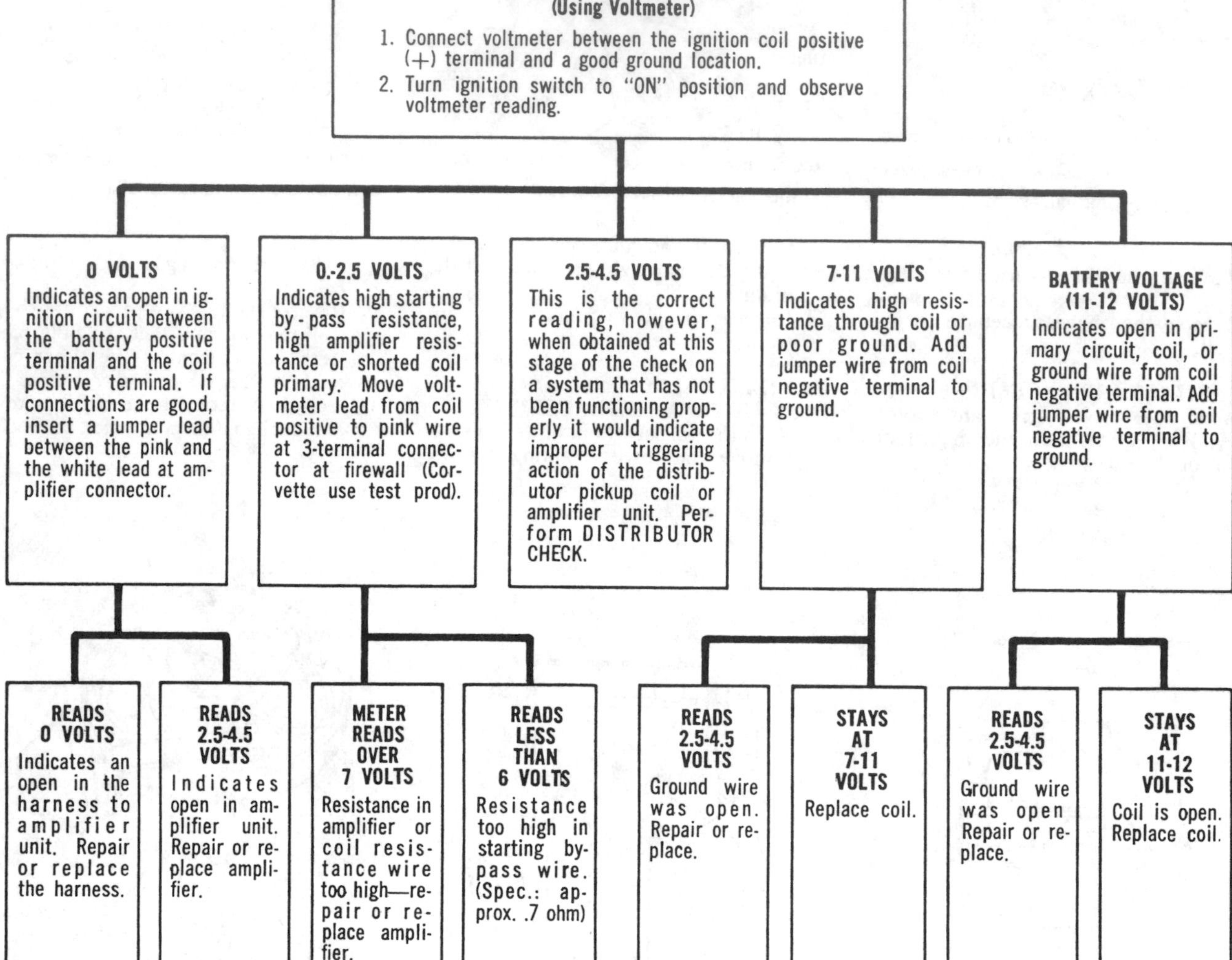

GM DELCO-REMY MAGNETIC PULSE IGNITION

GM Delco-Remy Magnetic Pulse Ignition System

OPERATION

The ignition primary circuit is connected from the battery, through the ignition switch, through the ignition pulse amplifier assembly, through the primary side of the ignition coil, and back to the amplifier housing where it is grounded externally. The secondary circuit is the same as in conventional ignition systems: the secondary side of the coil, the coil wire to the distributor, the rotor, the spark plug wires and the spark plugs.

The magnetic pulse distributor is also connected to the ignition pulse amplifier. As the distributor shaft rotates, the distributor rotating pole piece turns inside the stationary pole piece. As the rotating pole piece turns inside the stationary pole piece, the eight teeth on the rotating pole piece align with the eight teeth on the stationary pole piece eight times during each distributor revolution (two crankshaft revolutions since the distributor runs at one-half crankshaft speed). As the rotating pole piece teeth move close to, and align with, the teeth on the stationary pole piece, the magnetic rotating pole piece induces voltage into the magnetic pole piece through the stationary pole piece. This voltage pulse is sent to the ignition pulse amplifier from the magnetic pole piece. When the pulse enters the amplifier, it signals the ignition pulse amplifier to interrupt the ignition primary circuit. This causes the primary circuit to collapse and begins the induction of the magnetic lines of force from the primary side of the coil into the secondary side of the coil. This induction provides the required voltage to fire the spark plugs.

The advantages of this system are that the transistors in the ignition pulse amplifier can make and break the primary ignition circuit much faster than conventional ignition points. Higher primary voltage also can be utilized since this system can be made to handle higher voltage without adverse effects, whereas ignition breaker points cannot. The shorter switching time of this system allows longer coil primary circuit saturation time and longer induction time when the primary circuit collapses. This increased

GM DELCO-REMY MAGNETIC PULSE IGNITION

time allows the primary circuit to build up more current and the secondary circuit to discharge more current.

Troubleshooting

CHILTON CAUTIONS:

1 *Don't use 18 volts or 24 volts for emergency starting.*
2 *Never crank engine with coil high-tension lead or more than three spark plug leads disconnected.*
3 *Don't short circuit between coil positive terminal and ground.*
4 *On any repair that necessitates replacement of control unit or ignition resistor, perform complete charging system check before releasing the unit. Basic cause of trouble may be high or uncontrolled charging rate.*

ENGINE SURGE OR INTERMITTENT MISS

Since there are so many possible causes for this problem, all other possible defects must be ruled out before the specialized components of the electronic ignition system are judged defective.

As a general rule, a miss or surge that is caused by an ignition problem will be much more pronounced than a similar problem that is caused by carburetion. Also, carburetion is usually affected by temperature more than the ignition system is. A carburetor or intake manifold vacuum leak is often compensated for by the choke when the engine is cold. When the engine warms up and the choke is released, the engine surge will show up.

If the ignition system is found to be the source of the problem, first check all connections in the system to make sure that they are *clean and tight.* Check the coil and spark plug high-tension wires with an ohmmeter to be sure they have the correct resistance. Check the inside and outside of the distributor cap and the tower on the ignition coil for cracks which would allow the high voltage intended for the spark plugs to short to ground.

If none of the above checks uncovers a defective component, the distributor pickup coil leads may be reversed in the connector, or the pick-up coil itself may have an intermittent open.

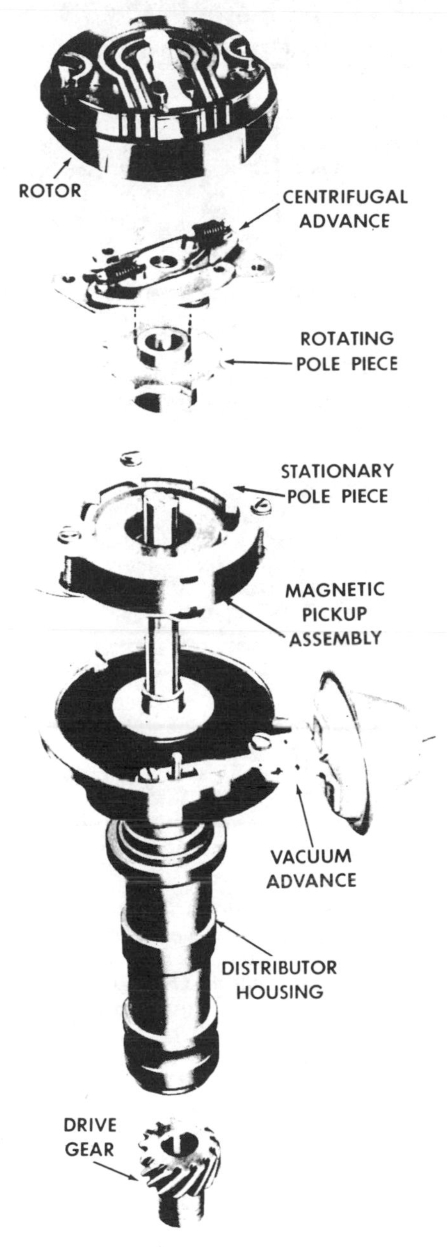

GM Delco-Remy Magnetic Pulse distributor and components

ENGINE WILL NOT START OR IS HARD TO START

1 Disconnect a spark plug wire from one spark plug and hold the wire ¼ in. from a good ground with a pair of insulated pliers.
2 Crank the engine and observe whether a spark jumps from the plug wire to ground.
3 *If spark occurs,* the problem is not in the ignition system.
4 *If spark does not occur,* reconnect the spark plug wire that was disconnected and connect a tachometer between the positive (+) coil primary terminal and the pink wire in the 3-wire connector to the ignition pulse amplifier.
5 Crank the engine over and observe the tachometer.
6 *If the tachometer needle deflects* while cranking the engine, perform "Ignition Distributor Test" to locate the problem.
7 *If the tachometer needle does not deflect* while cranking the engine, perform "Circuit Resistance Test" to pinpoint the problem.

IGNITION DISTRIBUTOR CHECK

1 Disconnect the distributor leads from the engine wiring harness.
2 Connect the two leads of an ohmmeter to the distributor leads at the connector.
3 Rotate the magnetic pick-up assembly in the distributor through full vacuum advance travel and read the ohmmeter. If the reading is not within a range of 500–700 ohms, replace the magnetic pick-up assembly.
4 If the reading is within the 500–700 ohms range, disconnect one ohmmeter lead from the distributor connector and connect it to a good ground. If the reading is less than infinity (needle moves to end of scale), replace the magnetic pick-up assembly.
5 If the reading is infinite, and there was no spark when the spark plug wire was disconnected from the plug, the amplifier is defective.

TROUBLESHOOTING GM DELCO-REMY UNIT IGNITION

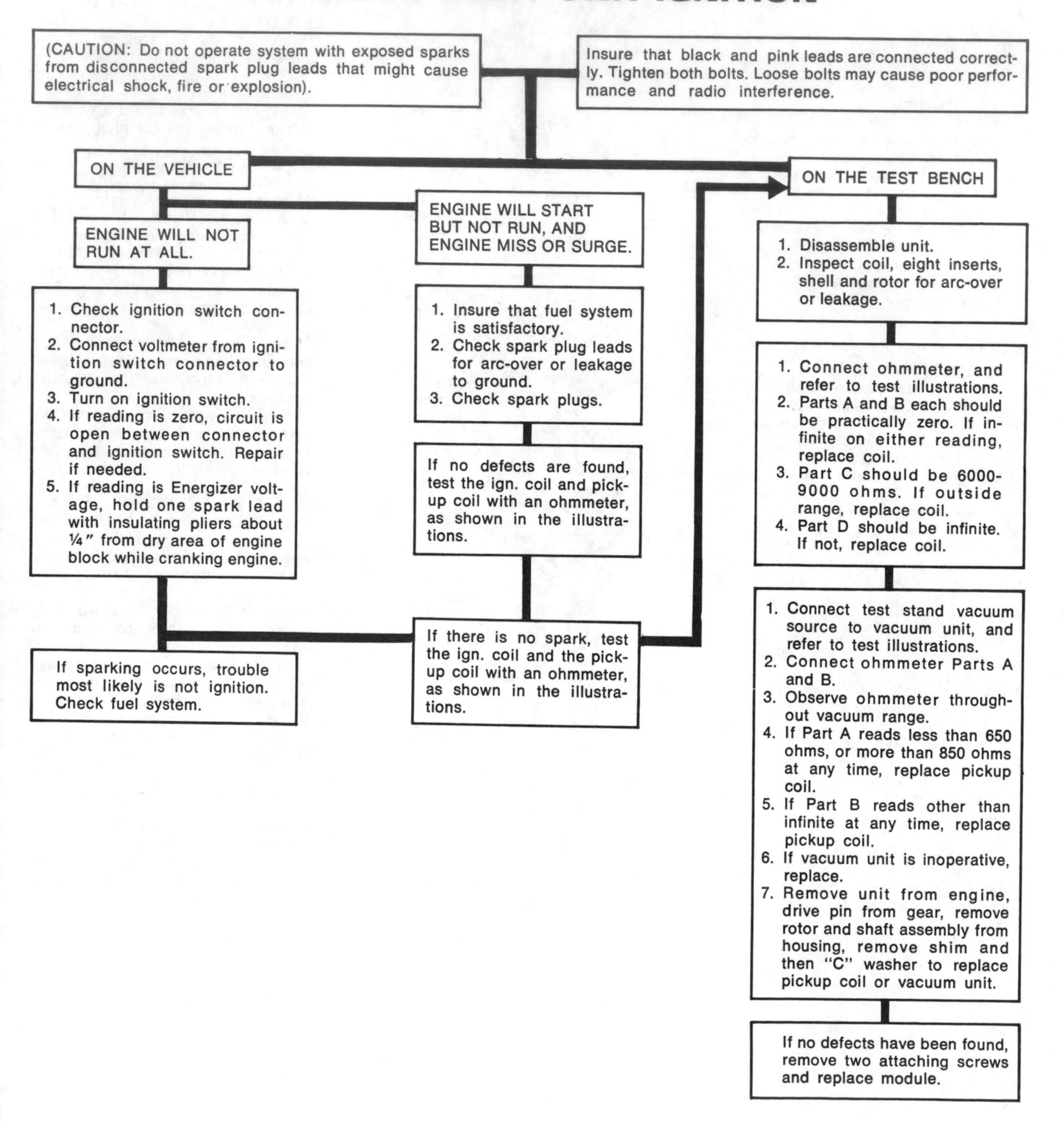

GM DELCO-REMY UNIT IGNITION

GM Delco-Remy Unit Ignition System

OPERATION

This ignition system was available as optional equipment on 1974 Pontiac cars equipped with V8 engines. The term "unit", or unitized, is used to describe this hardware because the coil and distributor are both contained in one unit.

This system is similar to other pointless ignition systems in that it uses a magnetic pulse distributor. The big difference, however, is that all the ignition components, except the spark plugs and wires, are built into the distributor. The distributor cap is a hollow shell without any electrodes. A harness assembly, with the wires and electrodes permanently molded in, sits on top of the cap so that the electrodes stick down through holes in the cap. The coil sits on top of the harness assembly, and the whole sandwich of coil, harness assembly, and cap is held on top of the distributor with two long bolts.

The control box, called an electronic module, is bolted to the distributor underneath the bowl. The module is greatly reduced in size from previous electronic controls. There is no ballast or other primary resistor in the wiring to the unit. All that is required to hook up the system is a single wire from the ignition switch.

The magnetic pulse part of the distributor has a magnetic pickup assembly located over the shaft. The assembly consists of a permanent magnet, a pole piece with internal teeth, and a pickup coil. When the teeth of the timer core, rotating inside the pole piece, line up with the teeth of the pole piece, an induced voltage in the pickup coil signals the all-electronic module to open the ignition coil primary circuit. The primary current decreases and a high voltage is induced in the ignition coil secondary winding, which is directed through the rotor and high voltage leads to fire the spark plugs.

The magnetic pickup assembly is mounted over the main bearing on the distributor housing, and is made to rotate by the vacuum control unit, providing vacuum advance. The timer core is made to rotate about the shaft by conventional advance weights, which gives centrifugal advance.

NOTE: If you want to crank the engine without having the ignition fire, disconnect the primary lead so the unit does not receive any current. Cranking the engine with all the spark plug wires disconnected and the ignition firing is not recommended, as it may damage the electronic parts.

Troubleshooting

The first thing to look for in case of trouble is loose connections or broken wires. If the wiring looks OK, make ohmmeter checks of the pickup coil and the ignition coil, as shown in our illustrations. If the pickup coil and ignition coil check out OK, the only other thing to do is replace the electronic module with a new, good unit. There is no field test for the electronic module, other than a trial replacement.

GM DELCO-REMY UNIT IGNITION

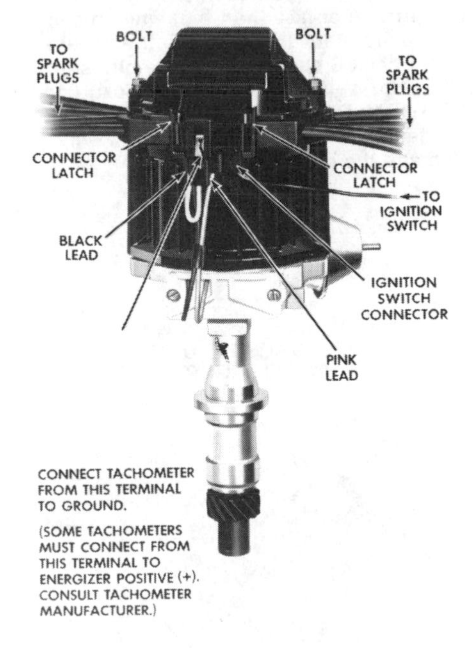

An exposed terminal on the ignition is used for a test tachometer connection. The two wires that come out of the electronic module are plugged into the coil. The wire from the ignition switch is a separate connector that also plugs into the coil.

COMPONENTS & TESTS

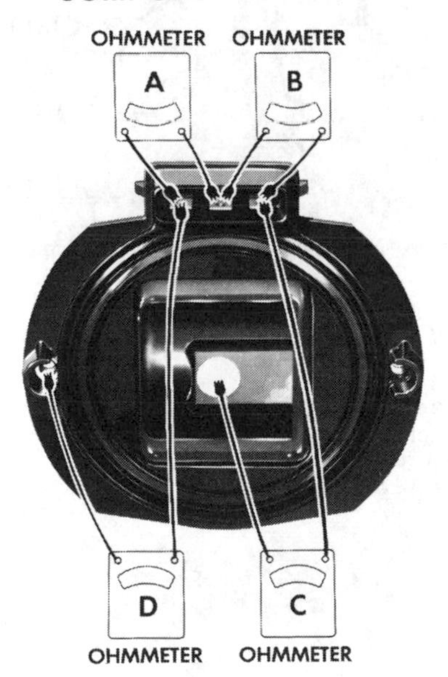

The four ohmmeter checks of the coil are as follows. Use the high ohmmeter scale.

Check Letter	Good Reading	Bad Reading
A	Zero ohms	Infinite ohms
B	Zero ohms	Infinite ohms
C	6000-9000 ohms	Below 6000 or over 9000 ohms
D	Infiinite ohms	Anything less than infinite

GM DELCO-REMY UNIT IGNITION

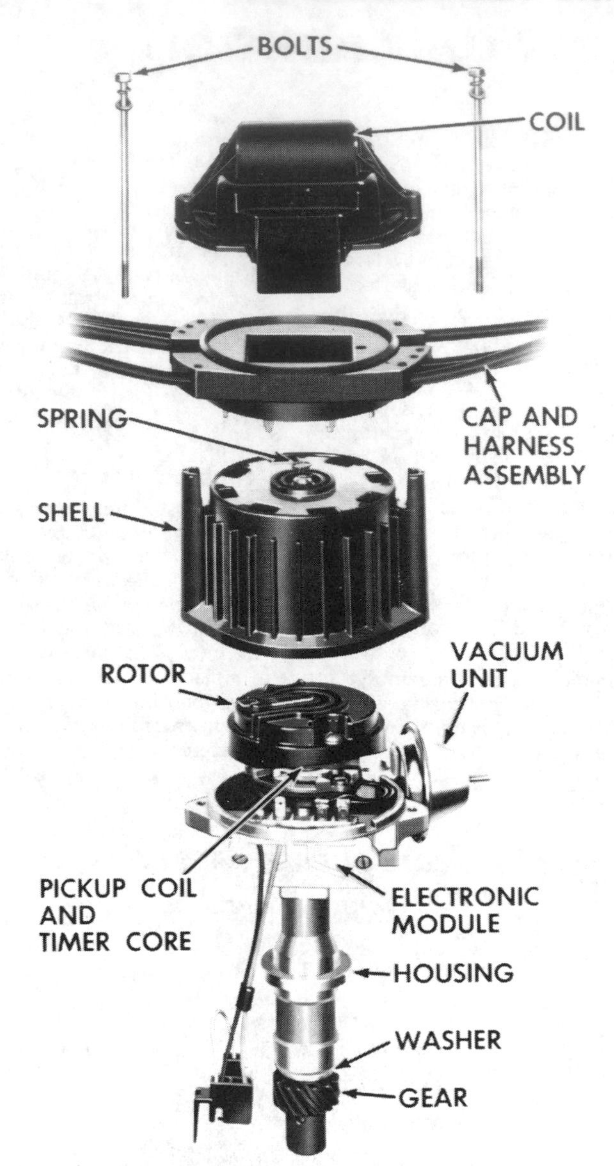

Two long bolts hold the cap, harness, and coil on top of the distributor. The pickup coil leads must be unplugged before removing the electronic module.

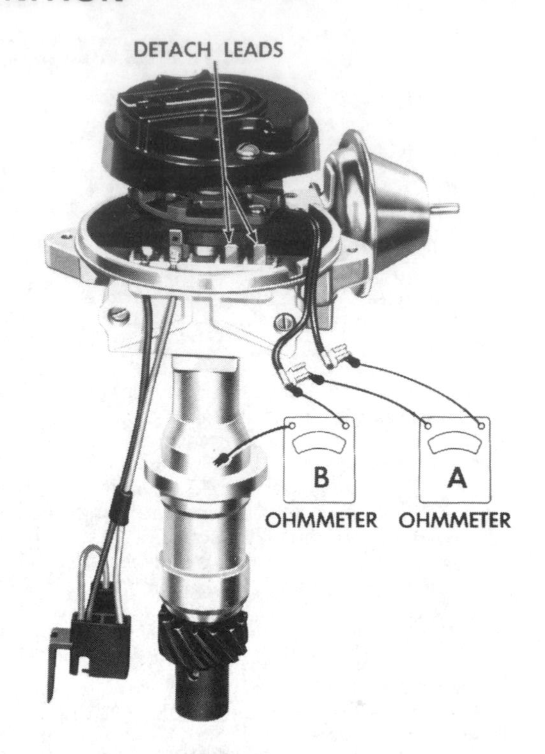

Ohmmeter checks of the pickup coil should be made while moving the coil back and forth through the entire vacuum advance range with a vacuum pump. Check A should be 650-850 ohms. If it reads more or less than that while moving the vacuum advance, replace the pickup coil. Check B should be infinite ohms at all times. If not, replace the pickup coil. CAUTION: Ohmmeter checks are made on dead circuits only. Live circuits will damage the ohmmeter.

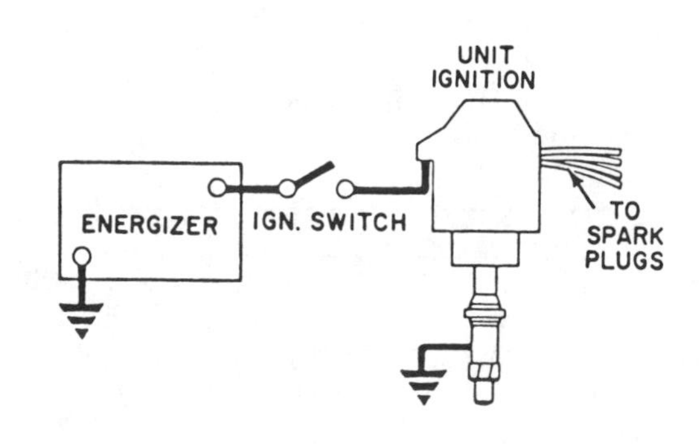

This basic wiring diagram emphasizes that the feed for the system is direct from the ignition switch, without any resistor.

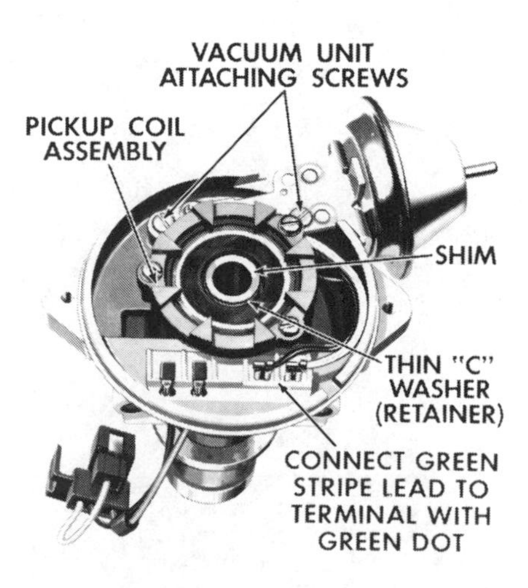

Distributor body with shaft removed. Note that the pickup coil lead with the green stripe connects to the terminal with the green dot.

TROUBLESHOOTING GM DELCO-REMY CRANKSHAFT SENSOR IGNITION

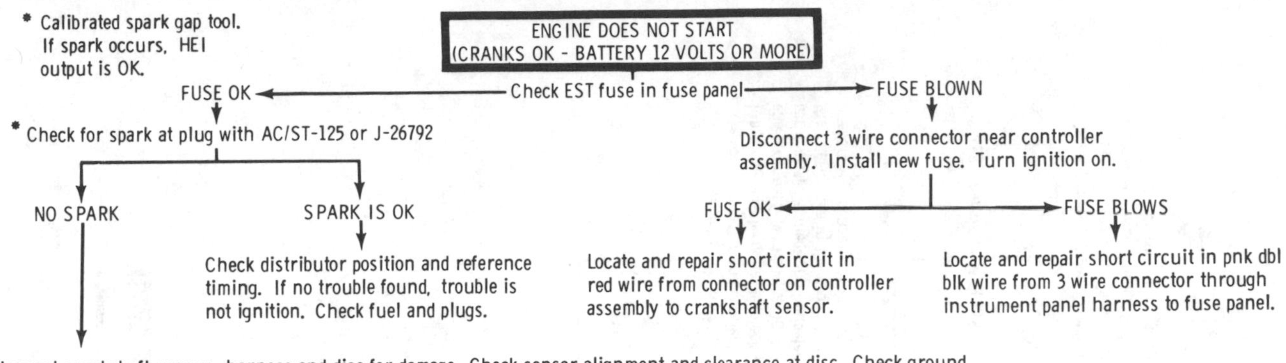

Inspect crankshaft sensor, harness and disc for damage. Check sensor alignment and clearance at disc. Check ground screw (black wire) in distributor. If connection is OK, turn ignition key to "RUN" and check for battery voltage at locations listed below. Look at circuit diagram for wire connections when more information is needed.

1. Ignition wire (blk pnk str) at connector on distributor - 12 volts or more, OK. Less than 12 volts, check ignition wire from distributor to ignition switch for loose connections or open circuit, also check ignition switch.
2. Terminal J (2 wires, pnk and red) in connector at controller assembly - 12 volts or more, OK. Less than 12 volts, check pink wire for loose connection or open circuit from connector at controller to 3 wire connector near controller then pnk/dbl blk str wire through instrument panel harness to fuse panel. (See circuit diagram).
3. Tan wire in 2 wire connector near distributor. Do not disconnect. Voltage should be .5 to 2 volts while cranking.

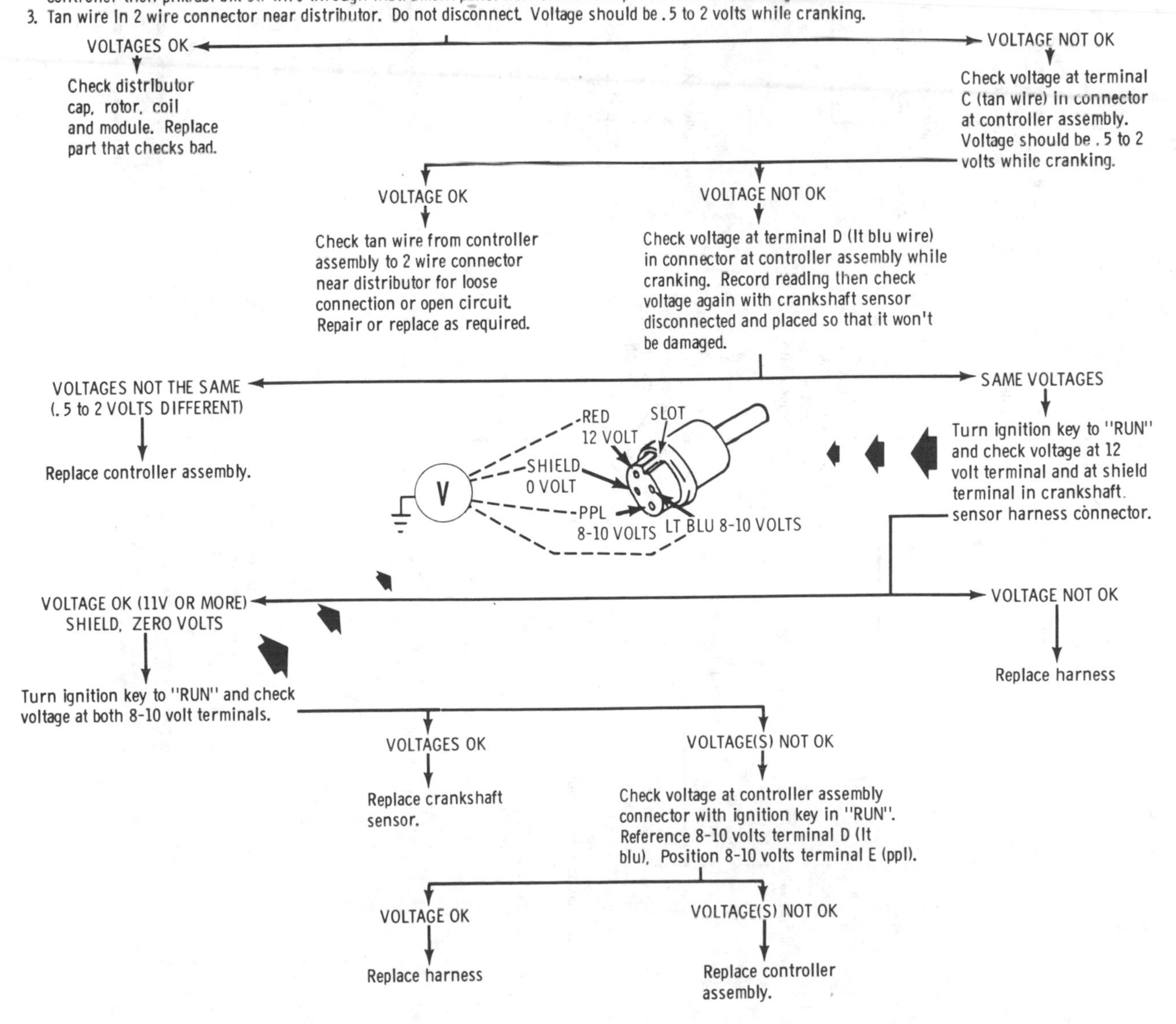

TROUBLESHOOTING
GM DELCO-REMY CRANKSHAFT SENSOR IGNITION

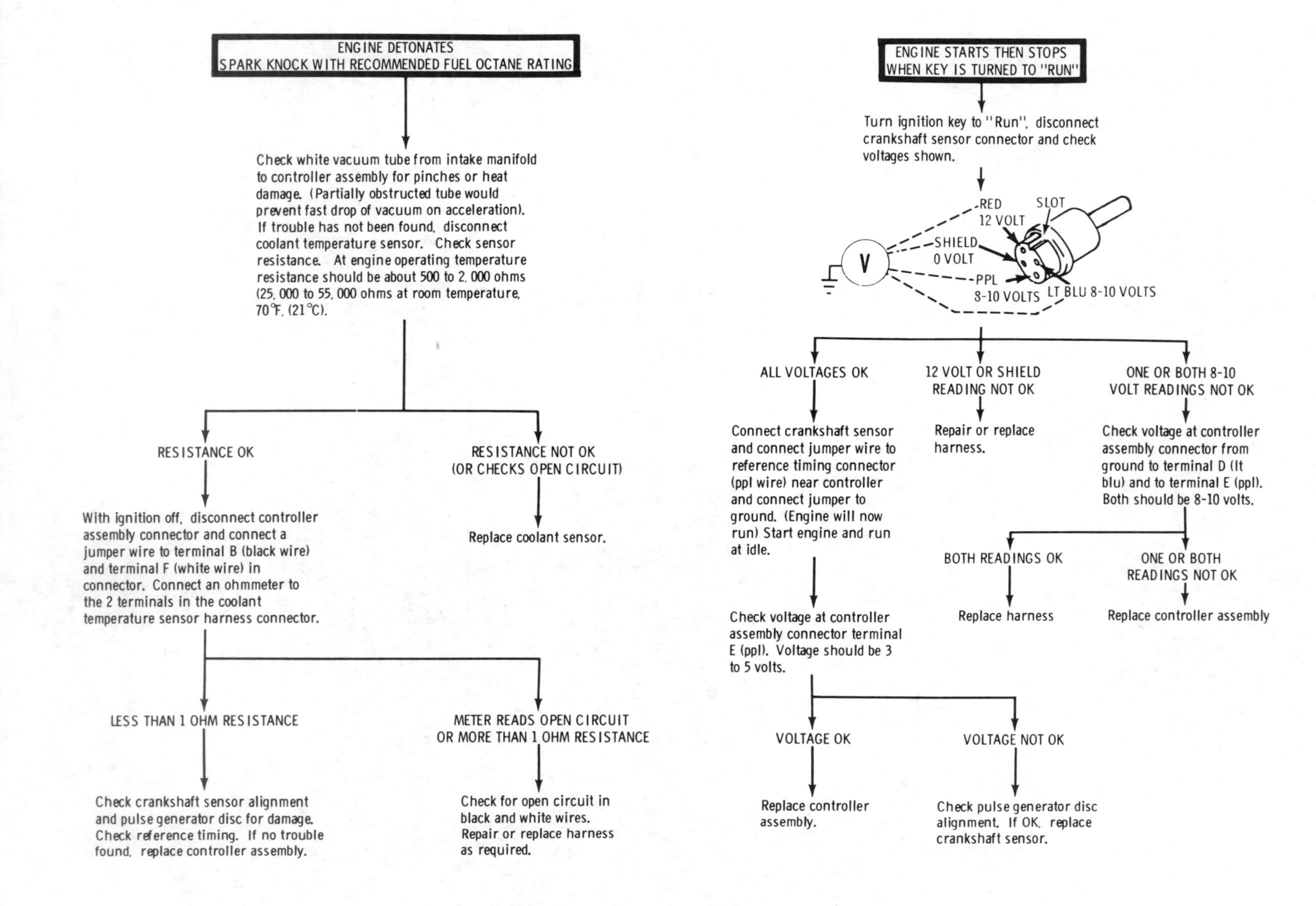

TROUBLESHOOTING
GM DELCO-REMY CRANKSHAFT SENSOR IGNITION

HARD STARTING, ROUGH ENGINE, POOR PERFORMANCE
(BATTERY FULLY CHARGED)

1. Check fuel system, choke, spark plugs and cables.
2. Make sure harness connections to distributor, coolant temperature sensor and controller assembly are good.
3. Inspect crankshaft sensor for alignment and clearance between sensor and pulse generator disc, (.045" to .055" top and bottom). Make sure harness and connector are good. Check disc for proper installation. Button on disc should be visible in hole in pulley.
4. Check all vacuum hoses for pinches, or disconnects. Check white vacuum tube from intake manifold to controller.
5. With engine at idle, transmission in park and parking brakes applied, connect voltmeter to ground and touch probe to ignition wire (blk-pnk str) in connector on distributor. Voltage should be 12 volts or more. If less, check for loose connection between distributor connector and ignition switch.
6. Connect voltmeter to ground and touch probe to terminal J in controller assembly connector (two wires, pink and red). Voltage should be 12 volts or more. If less, check for loose connection through instrument panel extension harness connector to fuse panel. (Refer to circuit diagram).
7. Remove distributor cap, check rotor and cap for signs of arcing. Check ground wire (screw) in distributor. Check module with J-24642. Check distributor position and reference timing.
   IF REFERENCE TIMING CANNOT BE SET to 20 degrees or if engine will not run at fast idle or timing light gives double flash causing timing mark to change position, replace crankshaft sensor.
8. If trouble has not been found, go to step 9.

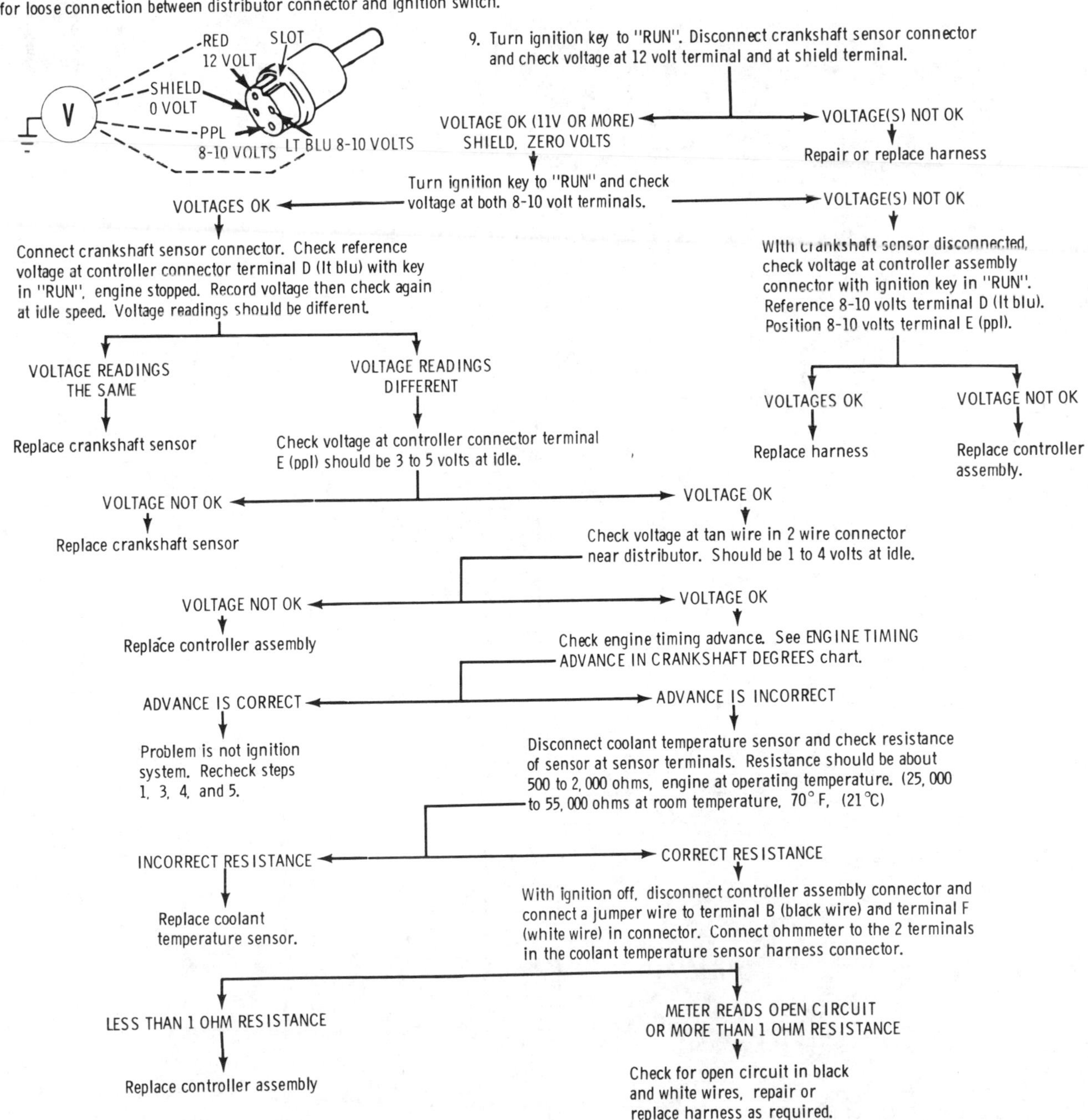

TROUBLESHOOTING
GM DELCO-REMY CRANKSHAFT SENSOR IGNITION

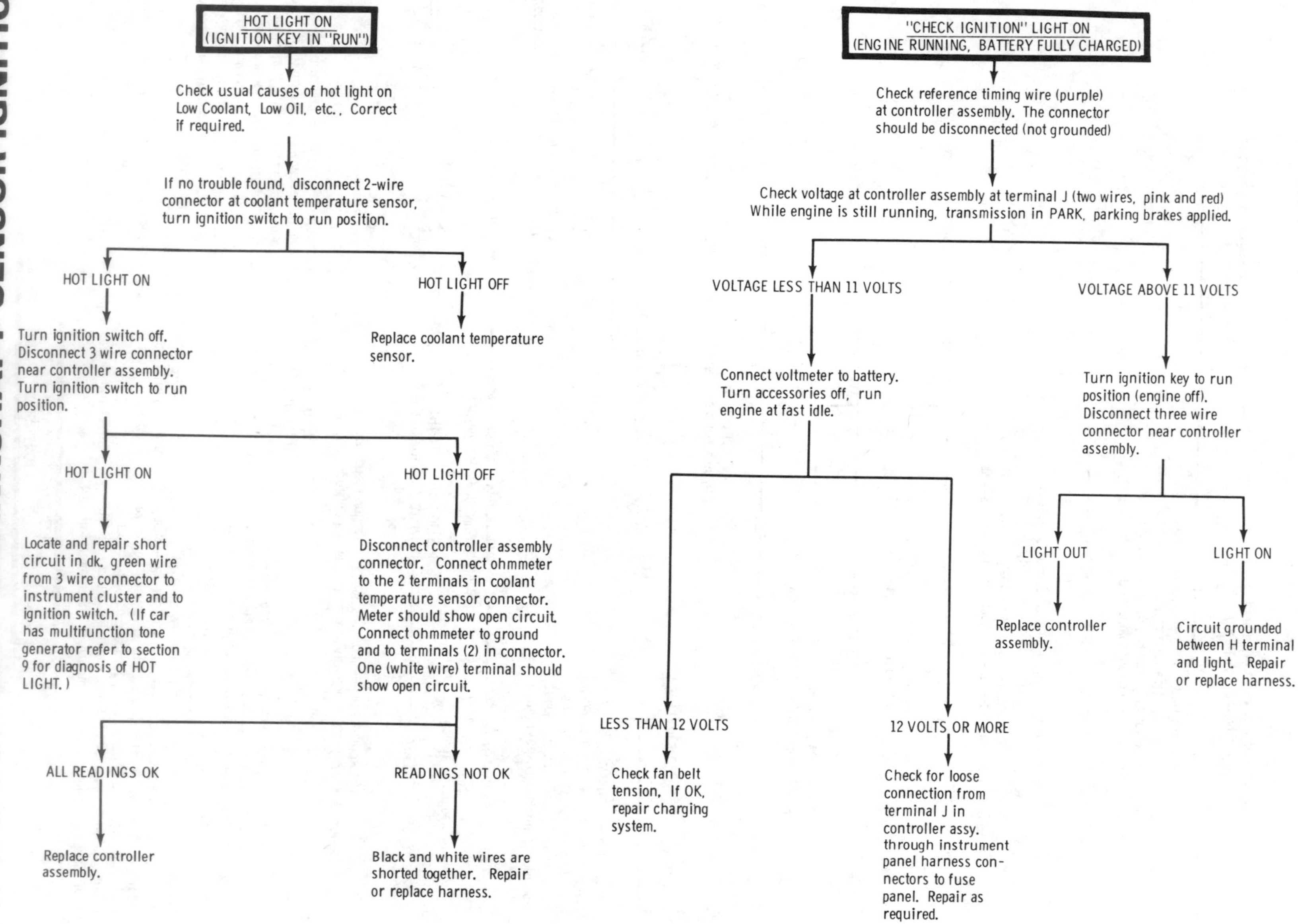

Hot Light and Check Ignition Light diagnosis

GM DELCO-REMY CRANKSHAFT SENSOR IGNITION

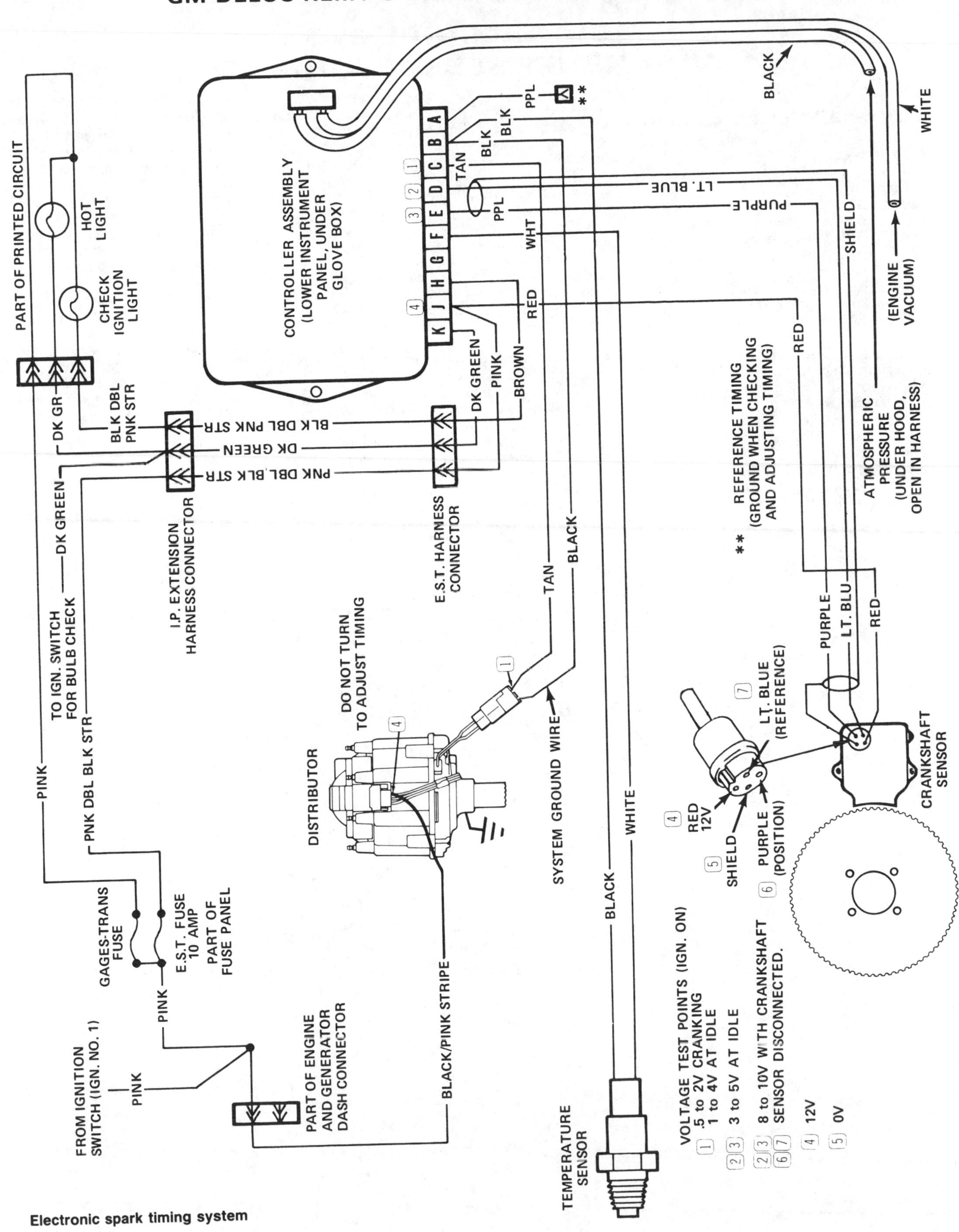

Electronic spark timing system

GM DELCO-REMY CRANKSHFT SENSOR IGNITION

GM Delco-Remy Crankshaft Sensor Ignition

GM's crankshaft sensor type ignition is on the following cars:
Cadillac
Oldsmobile

OPERATION

This ignition system does not use a standard HEI distributor assembly. Absent from the distributor are a vacuum advance unit, mechanical advance weights pick up coil and pole piece. The cap, coil and module are the same, the rotor is different and timing is not adjusted by turning the distributor. Instead, a crankshaft sensor, engine coolant sensor, a controller and electrical harness are used to control timing.

The engine coolant temperature sensor is different from the on-off switches used in other cars and is part of the ignition system Resistance in the sensor changes with changes in coolant temperature.

The controller assembly is an electronic unit, under the glove box, that recieves signals from the crankshaft sensor (crankshaft position and RPM), engine coolant temperature sensor, engine vacuum and atmospheric pressure. The controller assembly decides the most efficient advance based on the sensor signals and sends the signal to the distributor module to fire the spark plugs.

The electrical harness connecting these units together and to the car harness contains two vacuum tubes, both connected to the controller assembly; the white one is connected to manifold vacuum and the black one (atmospheric pressure) is not connected. The open end of the black one is in the engine compartment so that it will not be open to inside car pressure.

There are three different controller assemblies and two different harnesses. Altitude cars have a different controller assembly and harness. California cars have a different controller assembly, but the harness is the same as all other cars except Altitude. All except Altitude and California cars have a different controller assembly; harness is the same as California cars.

The Altitude controller assembly and harness has a black wire and connector at terminal G. If the car is driven to lower altitude areas (below 4000 feet) and detonation is a problem, the connector can be disconnected. This will retard timing about 4 degrees.

A "Check Ignition" light is located in the instrument panel cluster and will light, under the following conditions:

1 Ignition switch in the start position (bulb check).
2 If electrical system voltage is low and there is a heavy electrical load such as operation of power door lock, power windows, power seat, cigar lighter, rear window defogger, etc. The "check ignition" light will go off as soon as the electrical load is removed if the system voltage returns to normal.
3 When checking the reference timing and the controller circuit is grounded.
4 If there should be a controller failure so the spark timing would not advance.

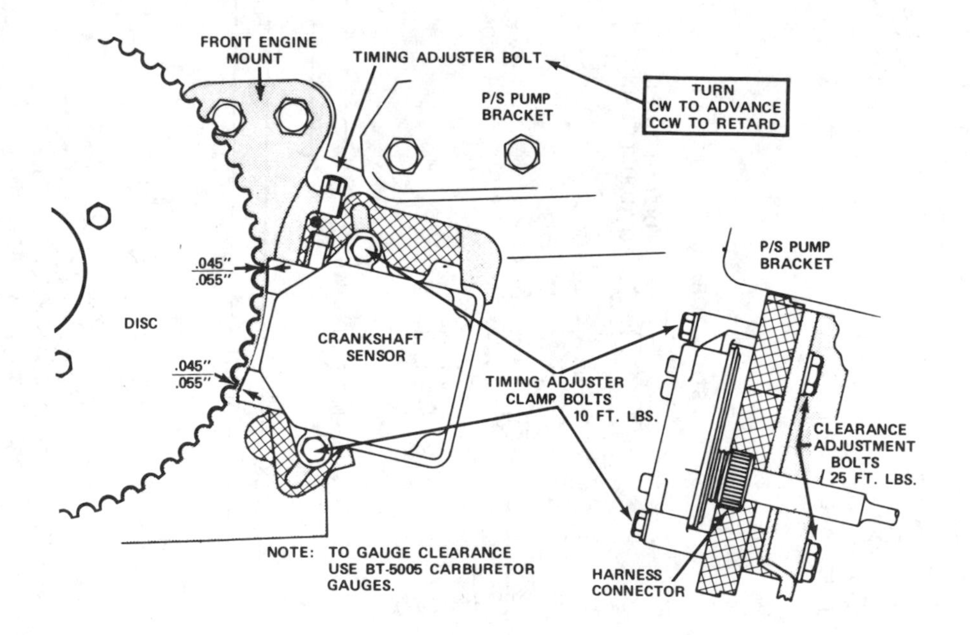

TURBOCHARGER SECTION

TURBOCHARGER TROUBLESHOOTING

PROBLEM	POSSIBLE CAUSE	CORRECTION
ENGINE DETONATION	Electronic Spark Control	Refer to ESC Diagnostic Procedure.
	EGR	Refer to Back Pressure EGR.
	Carburetor or Turbocharger	Correct air inlet restrictions. Air cleaner duct. Thermac door operation. Air cleaner dirty. Eliminate actuator overboost. Mechanical linkage jammed or blocked. Hose from compressor housing to actuator assembly or return hose from actuator to carburetor tee damaged or loose. Wastegate not operating-Refer to Wastegate-Boost Pressure Test Procedure. Service Carburetor Power System. Refer to P.E.V.R. Test Procedure. Refer to M4ME Quadrajet. Inspect turbocharger-Refer to Turbocharger Internal Inspection Procedure.
	Other Causes.	Refer to naturally-aspirated engine diagnosis.
ENGINE LACKS POWER	Air Inlet Restriction	Air cleaner duct. Thermac door operation. Air cleaner element dirty.
	Exhaust System Restriction	Repair exhaust pipes, if damaged. Repair or replace catalytic converter, if damaged. Check for correct muffler.
	Transmission	Check for correct shifting. Refer to transmission diagnosis.
	Electronic Spark Control	Refer to ESC Diagnostic Procedure.
	EFE	Refer to EFE.
	EGR	Refer to Back Pressure EGR.
	Carburetion	Refer to M4ME Quadrajet.
	Turbocharger	Check for exhaust leaks or restrictions. Refer to P.E.V.R. Test Procedure. Inspect for collapsed or kinked plenum coolant hoses. Check wastegate operation.-Refer to Wastegate-Boost Pressure Test Procedure. Refer to Turbocharger Internal Inspection Procedure.
	Other Causes	Refer to naturally-aspirated engine diagnosis.
ENGINE SURGES	Electronic Spark Control	Refer to ESC Diagnostic Procedure.
	Carburetion	Refer to P.E.V.R. Test Procedure. Refer to M4ME Quadrajet.
	EGR	Refer to Back Pressure EGR.
	Turbocharger	Inspect Turbocharger for loose bolts on compressor side of assembly, tighten.
	Other causes	Refer to naturally-aspirated engine diagnosis.
EXCESSIVE OIL CONSUMPTION OR BLUE EXHAUST SMOKE	External Turbocharger Oil Leaks	Inspect turbocharger oil inlet for proper connection. Inspect turbocharger oil drain hose for leaks or restriction.
	PCV	Refer to PCV.
	Other causes	Refer to naturally-aspirated engine diagnosis.
	Turbocharger	Refer to Turbocharger Internal Inspection Procedure.

TURBOCHARGER TROUBLESHOOTING

PROBLEM	POSSIBLE CAUSE	CORRECTION
BLACK EXHAUST SMOKE	Carburetion	Refer to P.E.V.R. Test Procedure. Refer to M4ME Quadrajet.
	Other Causes	Refer to naturally-aspirated engine diagnosis.
ENGINE NOISE EXCESSIVE	EFE	Refer to EFE.
	Exhaust System	Inspect for incorrect or loose mountings.
	AIR System	Refer to AIR System.
	Other Causes	Refer to naturally-aspirated engine diagnosis.
	Turbocharger	Check for exhaust leaks. Inspect for restriction of turbocharger oil supply. Refer to Turbocharger Internal Inspection Procedure

General Instructions

Before starting any turbocharger unit repair procedure several general cautions should be considered:

- Clear area around turbocharger with non-caustic solution before removal of assembly.
- When removing turbocharger assembly, take special care not to bend, nick, or in any way damage compressor or turbine wheel blades. Any damage may result in rotating assembly imbalance, failure of center housing rotating assembly (CHRA), and failure of compressor and/or turbine housings.
- Before disconnecting center housing rotating assembly from either compressor housing or turbine housing, scribe the components in order that they may be re-assembled in the same relative position.
- If silastic sealer, or equivalent, is found at any point in turbocharger disassembly (such as between center housing rotating assembly backplate and compressor housing) the area should be cleaned and sealed with an equivalent sealer during reassembly.

Test Procedures

Wastegate/Boost Pressure Test Procedure

- Visually inspect wastegate-actuator mechanical linkage for damage.
- Check hose from compressor housing to actuator assembly and return tubing from actuator to PCV tee.
- Attach hand operated vacuum/pressure pump J-23738, in series with compound gage J-28474 to actuator assembly, replacing compressor housing to actuator assembly hose.
- Apply pressure to actuator assembly. At approximately 9 psi (8.5 to 9.5 psi) the actuator rod end should move .015 inch, actuating the wastegate linkage. If not, replace the actuator assembly and check that opening calibration pressure is 9 psi. Crimp threads on actuator rod to maintain correct calibration.
- Remove test equipment and reconnect compressor housing to actuator assembly hose.
- An alternative method of checking wastegate operation is to perform a road test which measures boost pressure.

Power Enrichment Vacuum Regulator Test Procedure

- Visually check the PEVR and attaching hoses for deterioration, cracking or other damage.
- Tee one hose from manometer J-23951 between the yellow-striped input hose and the input port. Connect the other manometer hose directly to the output port of the PEVR.
- Start the engine and let it idle. There should be no more than a 14″ H^2O difference. If there is, replace the PEVR.
- If the PEVR passes the preceding test and is still considered to be a possible problem source, remove the PEVR from the intake manifold.
- Plug the intake manifold and connect the input and output hoses to the PEVR.
- Tee compound gage J-28474 into the output hose of the PEVR.
- Start the engine and let it idle. The compound gage reading from the output port should be 7.0″ to 9.0″ Hg.
- Apply 3 psi to the manifold signal port of the PEVR. The vacuum reading from the output port should be 1.4″ to 2.6″ Hg. If there is difficulty in measuring this low level of vacuum output, an additional requirement can be used. Apply a minimum of 5 psi to the manifold signal port of the PEVR. There should be no vacuum output from the PEVR.
- If the PEVR does not meet requirements 7 and 8, replace the PEVR.
- Check the journal bearings for *radial* clearance as follows:
  a. Attach a dial indicator, with a two inch long, ¾ to one inch offset extension rod to the center housing such that the indicator plunger extends through the oil outlet port and contacts the shaft of the rotating assembly. If required, a dial indicator mounting adapter can be used.
  b. Manually apply pressure equally and at the same time to both the compressor and turbine wheels, as required, to move the shaft away from the dial indicator plunger as far as it will go.
  c. Set the dial indicator to zero.
  d. Manually apply pressure equally and at the same time to both the compressor and turbine wheels to move the shaft toward the dial indicator plunger as far as it will go. Note the maximum value on the indicator dial.

Turbocharger Internal Inspection Procedure

- Remove turbocharger exhaust outlet pipe from the elbow assembly. Using a mirror, observe movement of wastegate while manually operating actuator linkage. Replace elbow assembly if wastegate fails to open or close.
- Remove turbocharger assembly from engine following removal procedure for Center Housing Rotating Assembly, omitting the last step, which involves separation of the CHRA from the turbine housing.
- Check for loose backplate to CHRA bolts and missing gasket or "O" ring.
- Gently spin compressor wheel. If rotating assembly binds, replace CHRA.
- Remove oil drain from CHRA. Check CHRA for sludging in oil drain area. Clean, if minor. Replace CHRA if severely sludged or coked.
- Inspect compressor wheel area for oil leakage from CHRA. If leakage is present, replace CHRA.
- If compressor wheel is damaged, or severely coked, replace CHRA.
- If CHRA is being replaced, pre-lubricate with clean engine oil and proceed to step 5. If CHRA is **not** being replaced, proceed to step 6.
- Inspect compressor housing (still attached to engine) and turbine housing. Replace either housing if gouged, nicked, or distorted.

- If CHRA is not being replaced, remove turbine housing from CHRA and check journal bearing radial clearance and thrust bearing axial clearance.

NOTE: Make sure that the dial indicator reading noted is the maximum reading obtainable, which can be verified by rolling the wheels slightly in both directions while applying pressure.

e. Manually apply pressure equally and at the same time to the compressor and turbine wheels, as required, to move the shaft away from the dial indicator plunger as far as it will go. Note that the indicator pointer returns exactly to zero.

f. Repeat steps b. through e., as required, to make sure that the maximum clearance between the center housing bores and the shaft bearing diameters, as indicated by the maximum shaft travel, has been obtained.

g. If the maximum bearing radial clearance is less than 0.003 inch or greater than 0.006 inch, replace CHRA and inspect housings as indicated in step 5.

NOTE: Continued operation of a turbocharger having improper bearing radial clearance will result in severe damage to the compressor wheel and housing or to the turbine wheel and housing.

- Check for thrust bearing *axial* clearance as follows:

a. Mount a dial indicator at the turbine end of the turbocharger such that the dial indicator tip rests on the end of the turbine wheel.

b. Manually move the compressor wheel and turbine wheel assembly alternately toward and away from the dial indicator plunger. Note the travel of the shaft in each direction, as shown on the dial indicator.

c. Repeat step b. as required, to make sure that the maximum clearance between the thrust bearing components has been obtained.

d. If the maximum thrust bearing axial clearance is less than 0.001 inch or greater than 0.003 inch, replace CHRA and inspect housings as indicated in step 5.

NOTE: Continued operation of a turbocharger having an improper amount of thrust bearing axial clearance will result in severe damage to the compressor wheel and housing or to the turbine wheel and housing.

- Install oil drain on CHRA.
- Install turbocharger assembly to engine following the procedure for the Center Housing Rotating Assembly.

NOTE: Before connecting turbocharger exhaust outlet pipe to elbow assembly, gently spin the turbine wheel to be certain that the rotating assembly (turbine wheel, connecting shaft, and compressor wheel) does not bind.

Road Test

- Tee compound gage J-28474 into tubing between compressor housing and boost gage switches with sufficient length of hose to place gage in passenger compartment.

CHILTON CAUTION: ***Determine that hose and compound gage are in proper operating condition to avoid possible leakage of air-fuel mixture into pasenger compartment during road test.***

- Conditions and speed limits permitting, perform a zero to 40 to 50 mph wide open throttle acceleration. Boost pressure as measured by the compound gage during road testing should reach 9-10 psi. If not, replace actuator assembly and check for proper calibration. Actuator rod end should move .015″ at approximately 9 psi.

Wastegate Actuator

- Disconnect hoses, remove the clip attaching wastegate linkage to actuator rod, and remove mounting bolts.

Turbocharger wastegate actuator

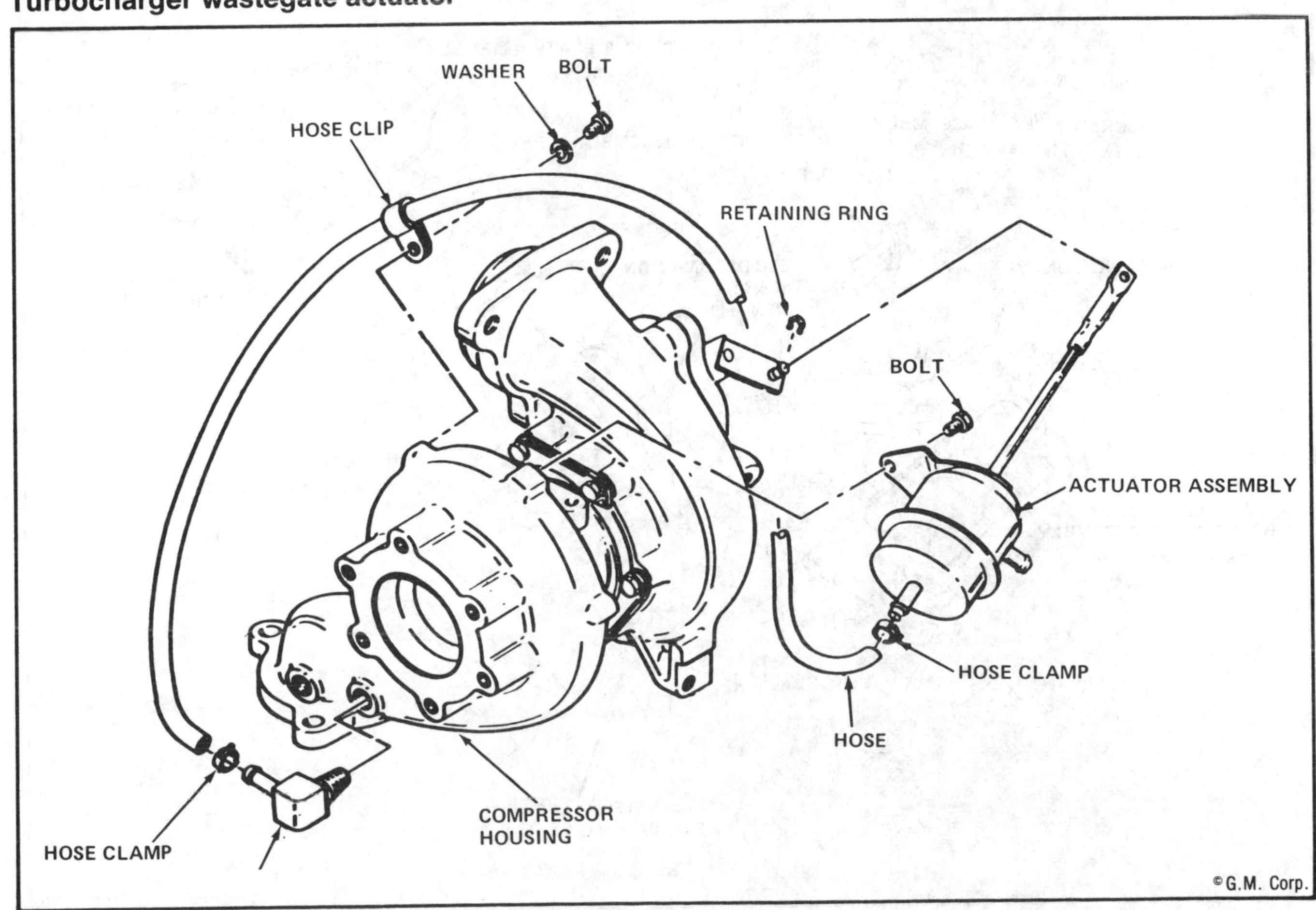

Turbocharger elbow assembly

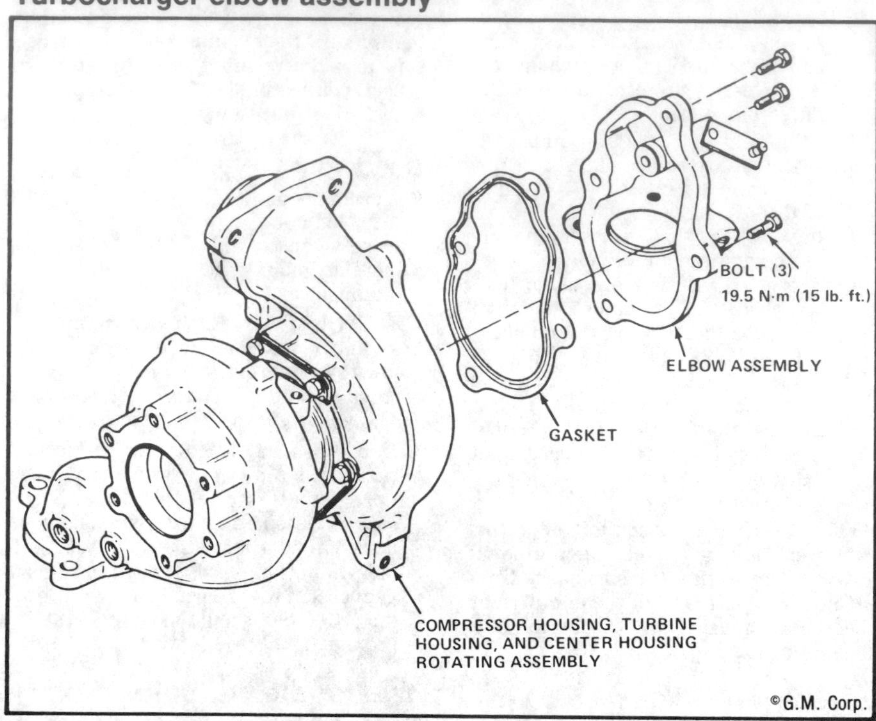

Electronic Spark Control Detonation Sensor

- Squeeze sides of metal connector crosswire to wire from controller and gently pull straight up to remove connector. Do *not* pull up on the wire.
- Unbolt and remove the sensor.
- When installing sensor, torque to 14 ft./lbs. Do not use an impact tool, apply a side load to the sensor, or attempt to repair the tapped hole in intake manifold.

Elbow Assembly

- Loosen turbocharger exhaust outlet pipe at the catalytic converter, and disconnect it from the elbow assembly.
- Remove the clip attaching wastegate linkage to actuator rod, and remove the bolts which mount the elbow to the turbine housing.

Turbocharger Electronic Spark Control (ESC) detonation sensor

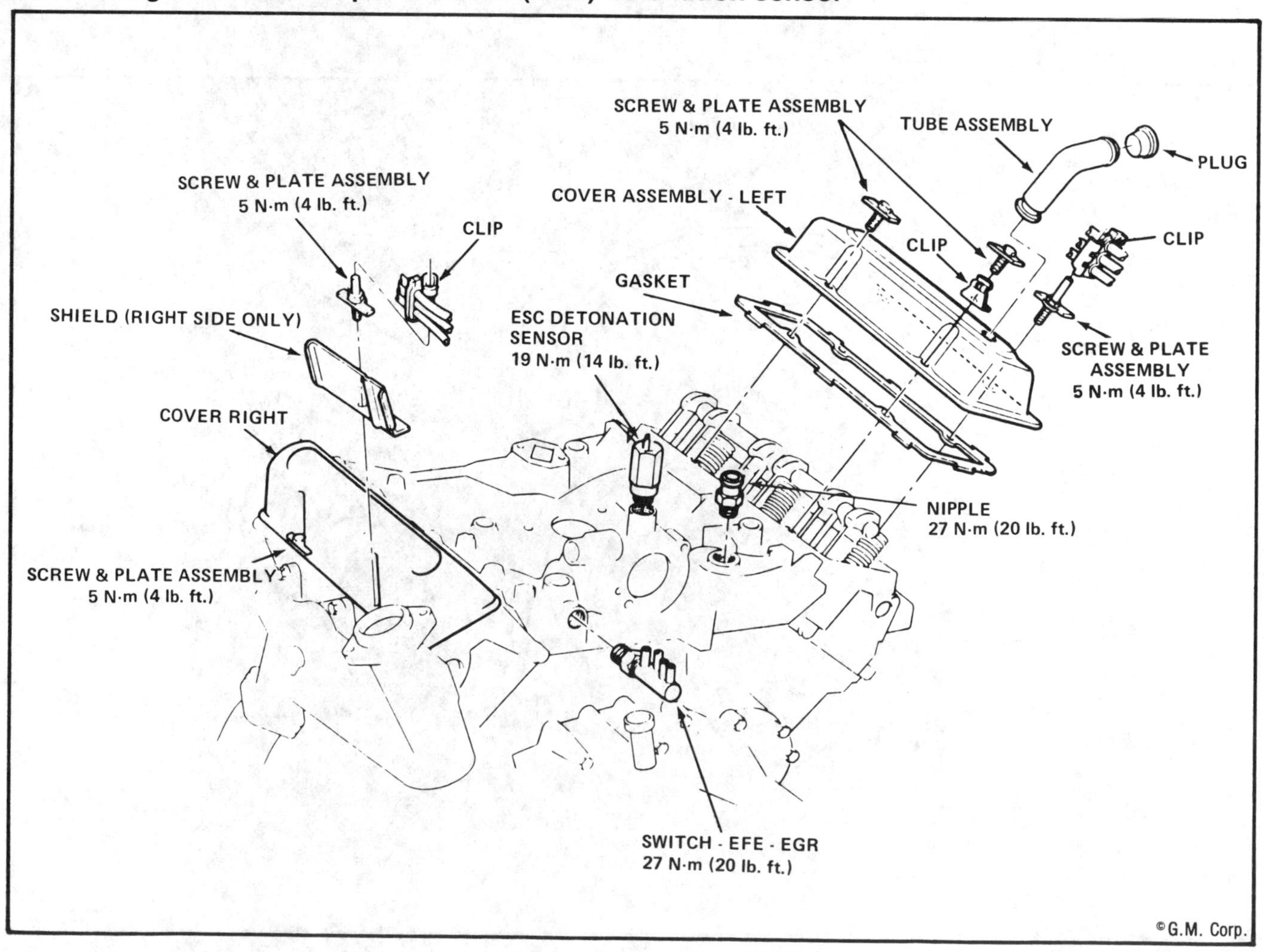

Turbocharger components

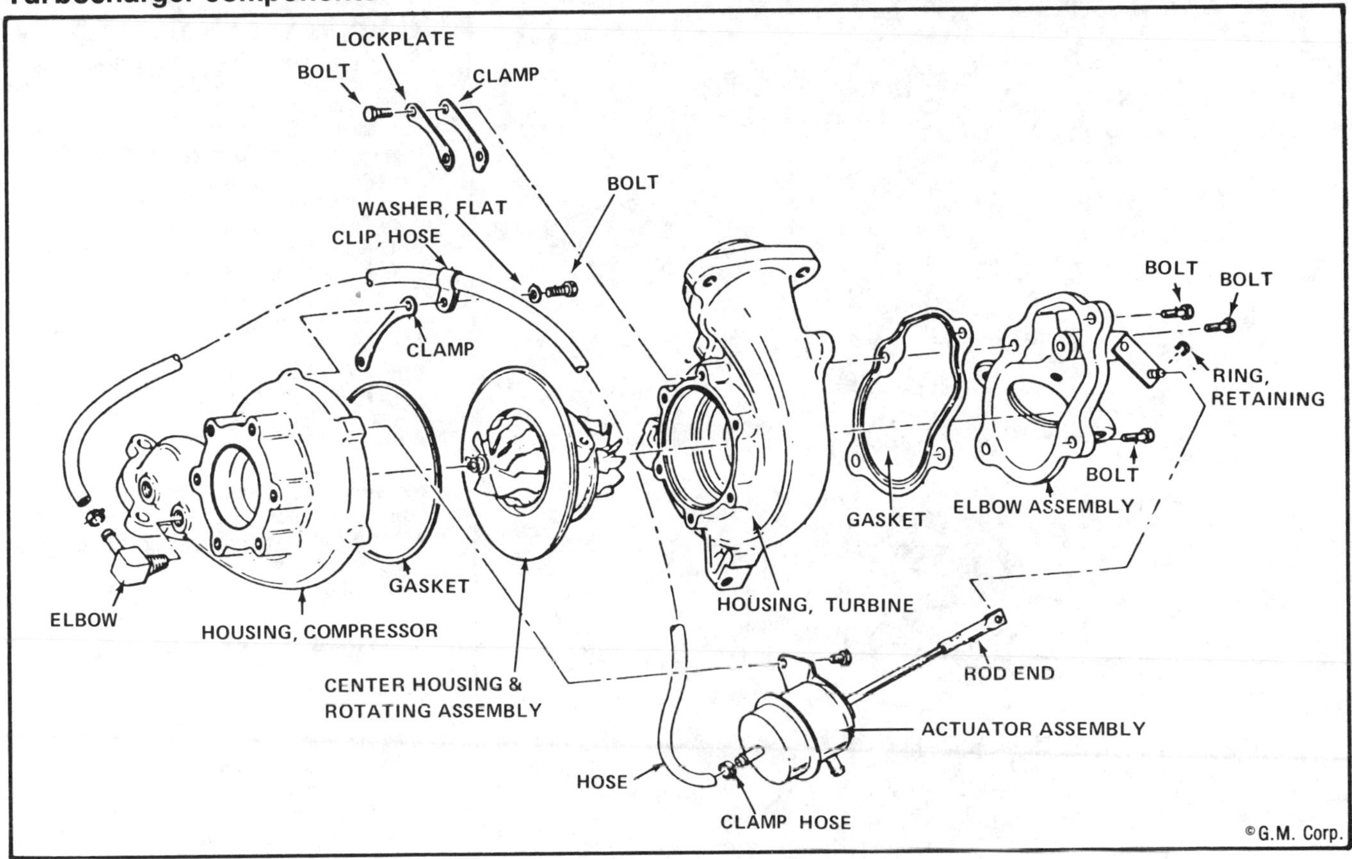

Turbine Housing and Elbow Assembly/Center Housing Rotating Assembly

- Disconnect turbocharger exhaust outlet pipe from the catalytic converter and the elbow assembly.
- Disconnect turbocharger exhaust inlet pipe from the turbine housing and the exhaust manifold.
- Remove bolts attaching turbine housing to bracket on intake manifold.
- Disconnect oil feed pipe from center housing rotating assembly, and remove oil drain hose from oil drain pipe.
- Remove clip attaching wastegate linkage to actuator rod.
- Remove bolts and clamps attaching CHRA backplate to compressor housing.
- Remove bolts, clamps, etc. attaching turbine housing to CHRA.
- Installation is the reverse of removal.

Turbocharger and plenum assembly

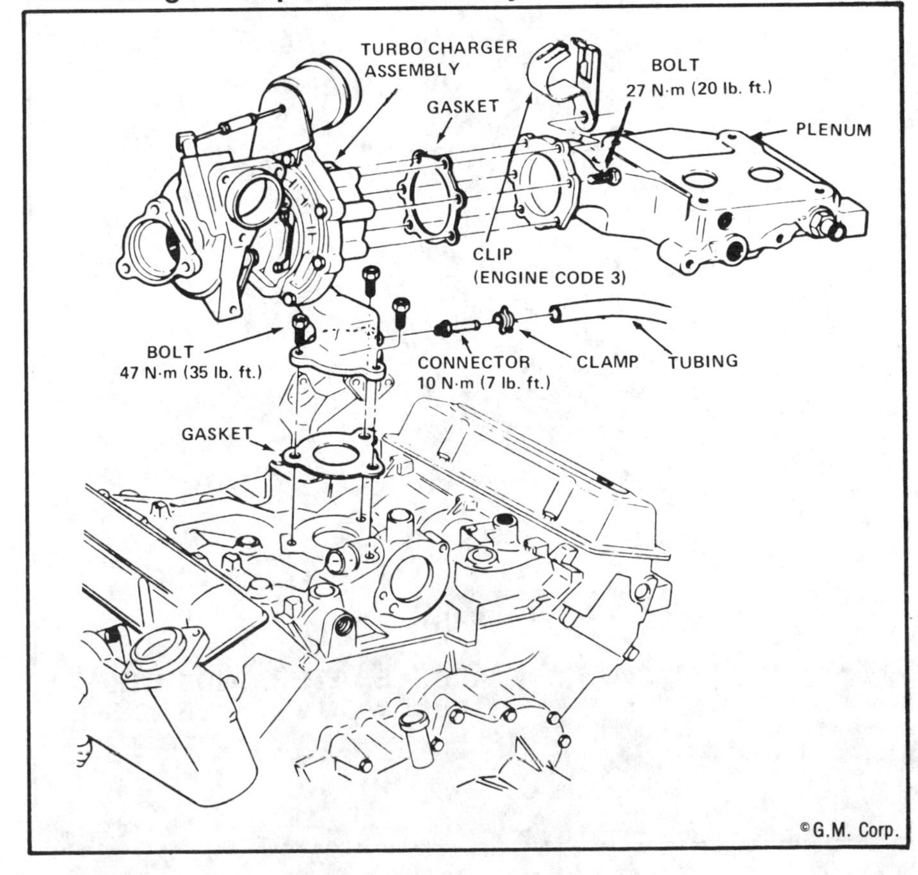

Carburetor-to-plenum mounting

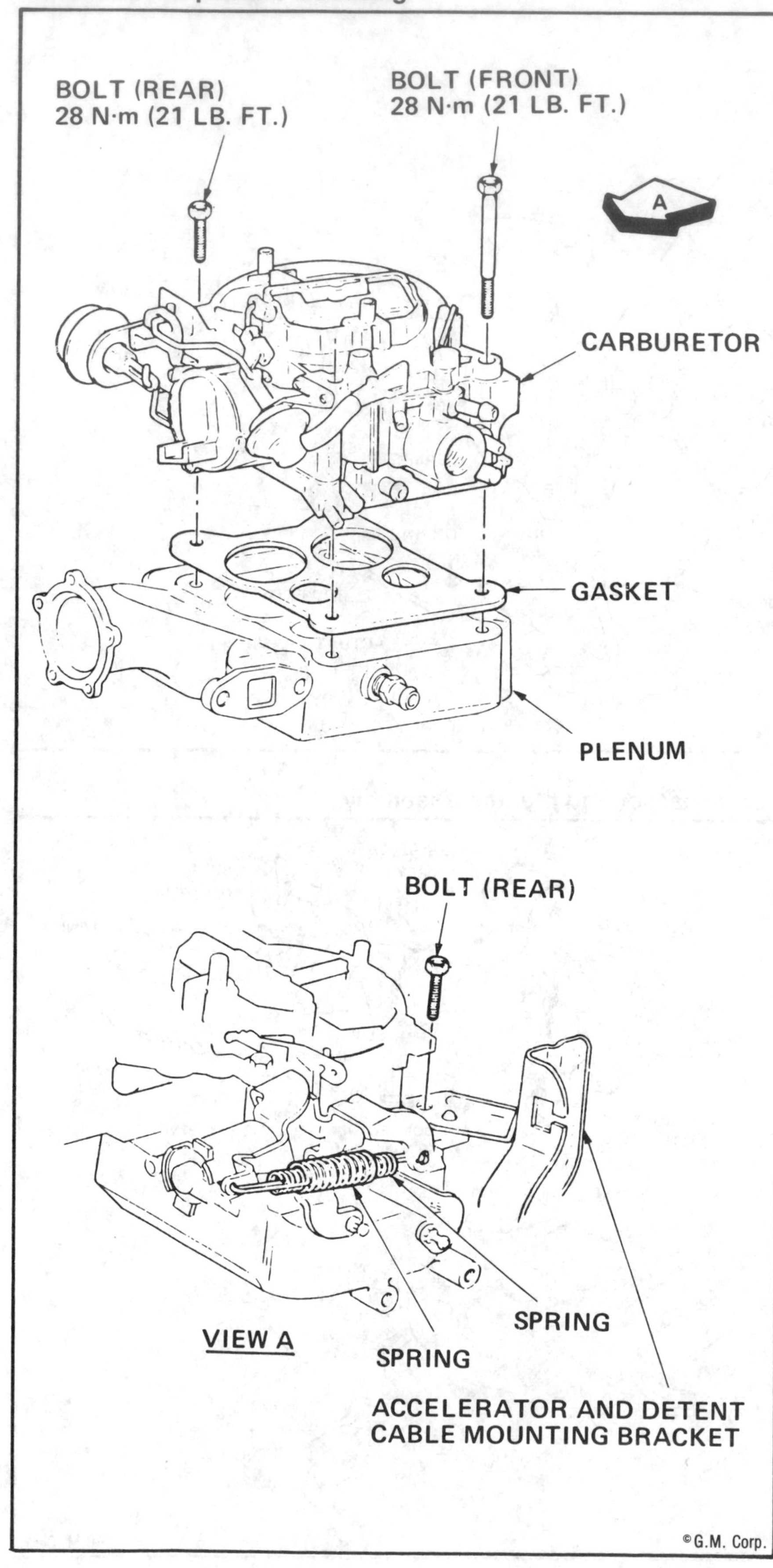

Compressor Housing

- Disconnect turbocharger exhaust outlet pipe from the catalytic converter and the elbow assembly.
- Disconnect turbocharger exhaust inlet pipe from the turbine housing and the exhaust manifold.
- Remove bolts attaching turbine housing to bracket on intake manifold.
- Disconnect oil feed pipe from center housing rotating assembly, and remove oil drain hose from oil drain pipe.
- Remove clip attaching wastegate linkage to actuator rod.
- Remove bolts and clamps attaching CHRA backplate to compressor housing.
- Remove bolts attaching compressor housing to plenum.
- Disconnect boost gauge hose from housing connector.
- Remove bolts attaching compressor housing to intake manifold.
- Installation is the reverse of removal.

Turbocharger and Actuator Assembly

- Disconnect exhaust inlet and outlet pipes at the turbocharger.
- Disconnect oil feed pipe from center housing rotating assembly.
- Remove air intake elbow.
- Disconnect accelerator, cruise and detent linkages at carburetor. Disconnect linkage bracket from plenum.
- Remove bolts attaching plenum to side bracket.
- Disconnect carburetor fuel and vacuum lines.
- Drain the cooling system and disconnect coolant hoses from plenum.
- Disconnect plenum from bracket by removing the attaching bolt at the intake manifold.
- Remove bolts attaching turbine housing to bracket on intake manifold.
- Remove bolts attaching EGR valve manifold to plenum, and loosen the bolts attaching the EGR valve manifold to the intake manifold.
- Remove AIR bypass-to-pipe-to-check valve hose.
- Remove bolts attaching compressor housing to intake manifold.
- Remove turbocharger and actuator, still attached to carburetor and plenum, from the engine. Separate components as necessary.
- Installation is the reverse of removal.

Plenum

- Use "Turbocharger and Actuator Assembly" procedure for plenum removal and replacement.

EGR Valve Manifold

- Disconnect vacuum line and unbolt from EGR manifold, plenum and intake manifold.

Turbocharger Center Housing Rotating Assembly (CHRA)

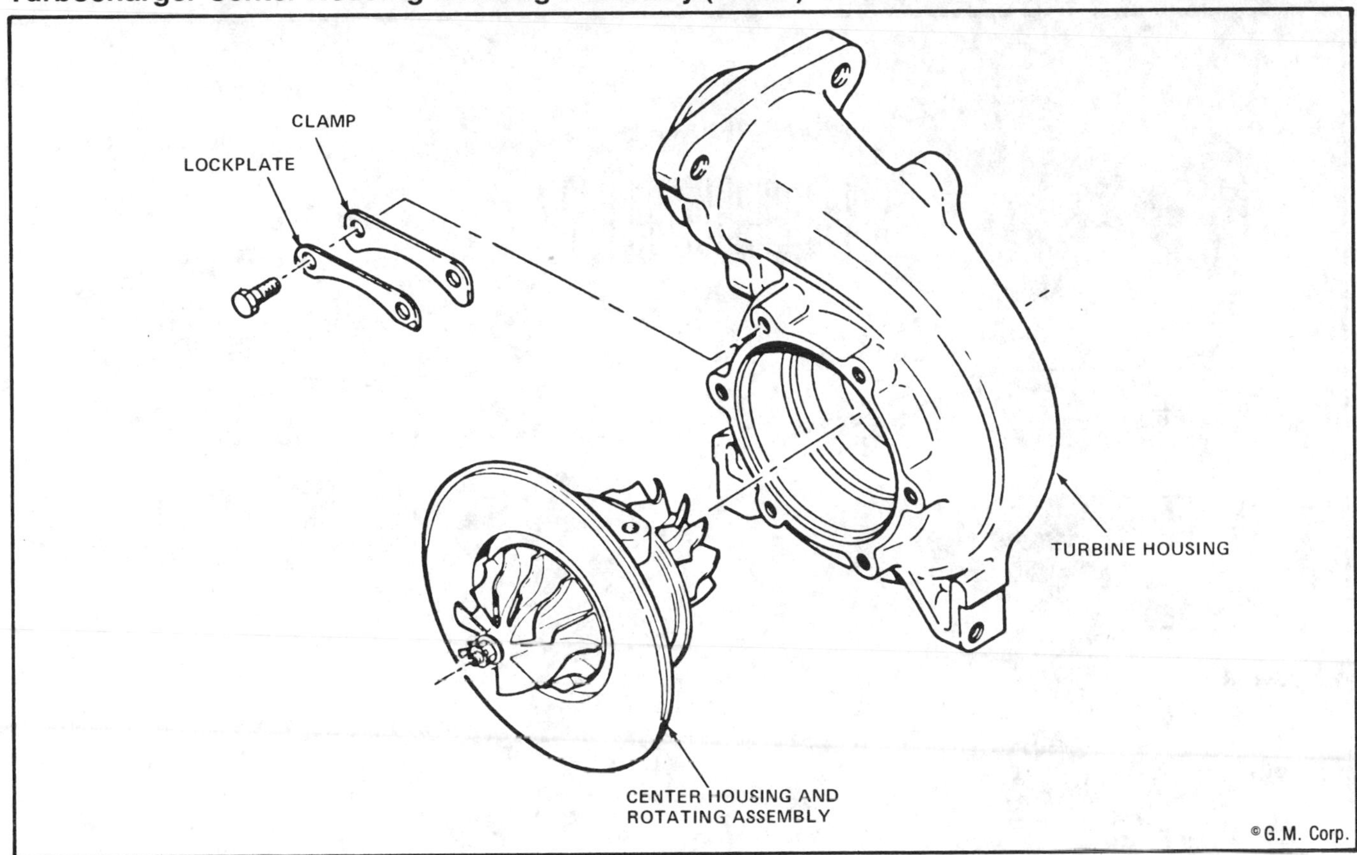

EGR valve manifold mounting

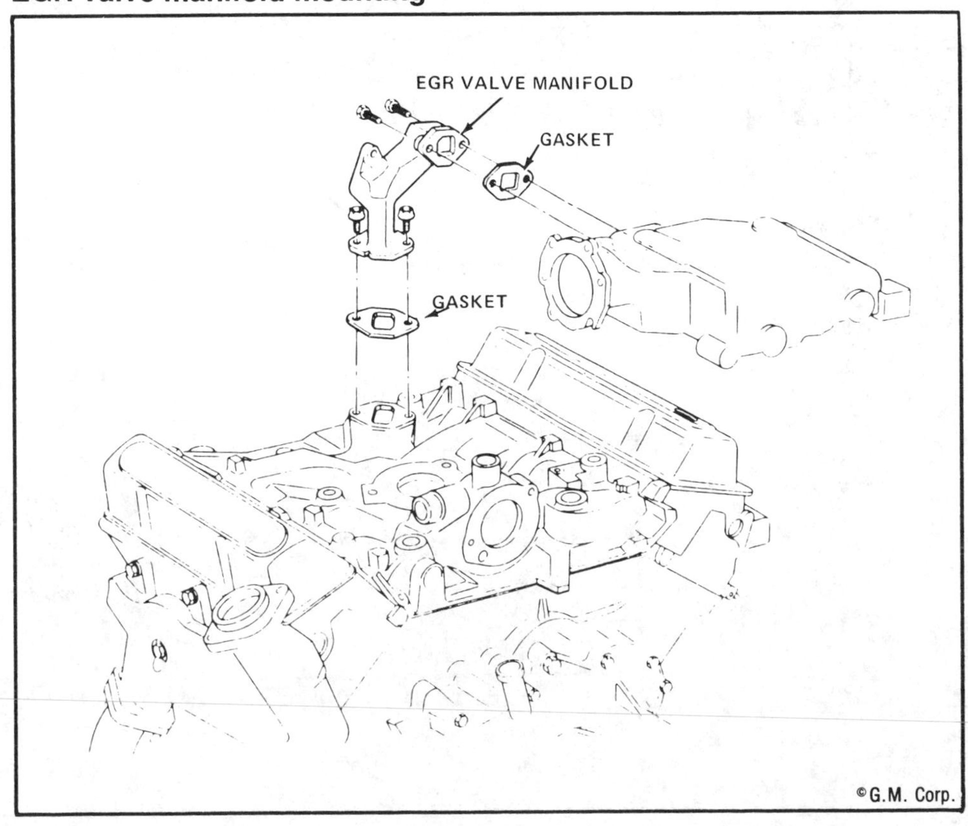

Electronic Spark Control (ESC) system used with turbocharged engine

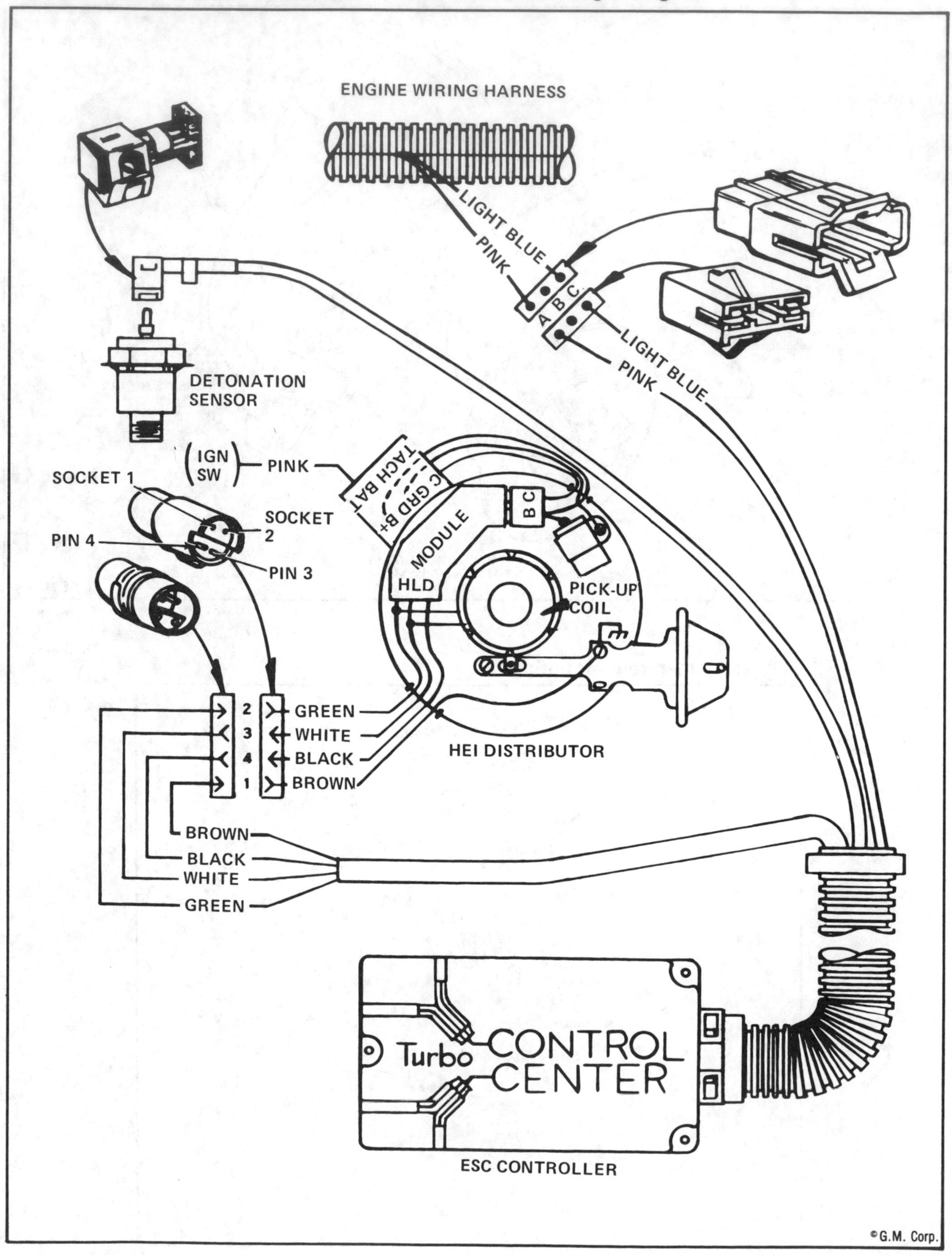

Popular Replacement Spark Plugs for Late Model American Cars

Following are spark plug manufacturers recommendations for 1978-79 unmodified engines in normal service, using the vehicle manufacturers settings for ignition timing and carburetion. Check the Tune-Up Specifications chart for correct OEM plug gap.

AUTOMOBILE MAKE/MODEL	ENGINE TYPE/CYL. CID	CARB. BBLS.	AC	AUTOLITE
AMERICAN MOTORS CORP.				
1978-79 AMX, Concord, Spirit, Gremlin	L4-121	2	—	912
1978-79 All Models	L6-232	1	—	915[1]
1979 All Models	L6-258	2	—	915
1978 All Models	L6-232, 258	1	—	915
1978 All Models	L6-258	2	—	915[3]
1978-79 All Models	V8-304	2	R44XLS[11]	55
1978-79 Concord, Hornet, Matador	V8-360	2 & 4	R44XLS[11]	55
BUICK				
1979 Skyhawk, Century/Regal	V6-196	2	R45TSX	666
1978 Skyhawk, Century/Regal	V6-196	2	R46TSX	667
1979 Century/Regal, LeSabre, Skyhawk, Skylark	V6-231	2	R45TSX	666
1979 Century/Regal, LeSabre, Skyhawk, Skylark, Skyhawk (Calif. & Hi-Alt.)	V6-231	2	R46TSX	667
1979 Century/Regal, LeSabre, Riviera	V6-231 Turbo	2 & 4	R44TS	665
1978 Century/Regal, LeSabre	V6-231 Turbo	2 & 4	R44TSX	665
1978-79 Century/Regal, LeSabre, Skylark	V8-301	2	R46TSX	667
1979 Century/Regal	V8-301	4	R45TSX	667
1978 Century/Regal, LeSabre, LeSabre (Calif. & Hi-Alt.), Skylark	V8-305	2 & 4	R45TS	26
1978-79 Century/Regal, LeSabre, Skylark	V8-350L	4	R45TS	26
1978-79 Electra, LeSabre, Riviera	V8-350R	4	R46SZ	847
1979 Century/Regal, Electra, LeSabre	V8-350X	4	R45TSX	666
1978 Century/Regal, Electra, LeSabre, Riviera	V8-350X	4	R45TSX	667
1978-79 Electra, LeSabre, Riviera	V8-403	4	R46SZ	847
CADILLAC				
1978-79 Seville, Eldorado	V8-350	FI	R47SX	847
1978-79 Cadillac	V8-425	4	R45NSX	646
1979 Cadillac	V8-425	FI	R45NSX	567
1978 Cadillac	V8-425	FI	R45NSX	646
CHEVROLET				
1979 Chevette	L4-98	2	R42TS	23
1978 Chevette	L4-98	1	R43TS	24
1979 Monza	L4-151	2	R43TSX	664
1978 Monza	L4-151	2	R43TSX	664
1979 Monza	V6-196	2	R45TSX	666
1978 Monza	V6-196	2	R46TSX	667
1979 Malibu, Monte Carlo	V6-200	2	R45TS	26
1979 Malibu, Monte Carlo, Monza	V6-231	2	R45TSX	666
1978 Malibu, Monte Carlo, Monza	V6-231	2	R46TSX	667
1978 Caprice, Impala, Nova, Camaro, Chevette	L6-250	1	R46TS	27
1979 Camaro	L6-250	1	R46TS	27
1979 Malibu, Monte Carlo	V8-267	2	R45TS	26
1978-79 Caprice, Impala, Camaro, Malibu, Monte Carlo, Monza, Nova	V8-305	2 & 4	R45TS	26
1978-79 Caprice, Impala, Camaro	V8-350L, V8-350 Z28	4	R45TS	26
1978 Caprice, Impala	V8-350X	4	R45TS	—
1978-79 Caprice, Impala, Corvette, Malibu, Nova, Camaro, Chevelle, Monte Carlo	V8-350 Reg. & Hi Perf.	4	R45TS	26
CHRYSLER				
1978-79 Chrysler, Cordoba, LeBaron	L6-225	1 & 2	R45TS	27
1978-79 Chrysler, Cordoba, LeBaron	V8-318, 360	2 & 4	R44XLS	65
1978 Chrysler, Cordoba,, LeBaron	V8-400, 440	4	R44S	85
DODGE				
1978-79 Omni	L4-104.7	2	R44XLS	65
1978-79 Aspen, Diplomat, Magnum XE, St. Regis, Monaco	L6-225	1 & 2	R45TS	27
1978-79 Aspen, Charger, Diplomat, Magnum, Magnum XE, St. Regis, Monaco, Royal Monaco	V8-318, 360	2 & 4	R44XLS	65
1978 Charger, Magnum, Monaco, Royal Monaco	V8-400	4	R44S	85
1978 Monaco	V8-440 Hi-Perf	4	R44S	84
FORD				
1978-79 Fiesta	L4-98	2	R43LTSX	884
1978-79 Fairmont, Mustang, Pinto	L4-140	2	R43LTS	765
1979 Mustang II	L4-140Turbo	2	R43LTS	764
1978-79 Mustang II, Pinto	V6-171	2	R43LTS	765
1978-79 Fairmont	L6-200	1	R85TS	746
1978-79 Granada	L6-250	1	R85TS	746
1979 Fairmont, Granada, LTD II, Mustang II, Thunderbird, Ford	V8-302	2	R45TSX	726
1979 LTD II, Thunderbird	V8-302 Calif.	2	R45TSX	3606
1978 Fairmont, Ford, Granada, LTD II, Maverick, Mustang II, Thunderbird, Ford	V8-302	2	R45TSX	26

[1] Except California
[2] Cadillac 425 FI Autolite spark plug gapped at .080 in.
[3] With automatic and 2.53:1 axle use 55 gapped at .035 in.
[4] Camaro L6-250 Autolite spark plug gapped at .045 in.
[5] Horizon L4-105 Autolite spark plug gapped at .040 in.
[6] V6-231 (turbo) Bosch, Motorcraft, NGK spark plug gapped at .060 in.
[7] California use N14LY
[8] L4-140 engine Ford and Mercury spark plug gapped at .035 in.

Popular Replacement Spark Plugs for Late Model American Cars

Whenever the letter H, J, L, P, R or X follows a V8-350 engine, it refers to the 5th digit of the VIN, which can be found on the dash panel. Check the Tune-Up Specifications chart for the individual car/engine to be sure of the correct plug gap.

BOSCH (NEW #)	CHAMPION	MOPAR	MOTORCRAFT	NGK	NIPPON-DENSO	PRESTOLITE RESISTOR	GAP
W9H	N8L	—	AG14	—	—	—	.035
W9H	N13L	—	AG44	—	—	—	.035
W9H	N13L[7]	—	AG44	—	—	—	.035
W9H	N13L	—	AG44	—	—	—	.035
W9H	N13L	—	AG44	—	—	—	.035
W9D	N12Y	—	AG42	BP5ES	W16EX-U	14GR42	.035
W9D	N12Y	—	AG42	BP5ES	W16EX-U	14GR42	.035
HR10BY	RBL15Y6	—	ARF52-6	BPR4FS-15	T16PR15	14RF52A	.060
HR10BY	RBL17Y6	—	ARF62-6	BPR4FS-15	T16PR15	14RF52A	.060
HR9BX	RBL15Y6	—	ARF52-6	BPR4FS-15	T16PR15	14RF52A	.060
HR9BX	RBL17Y6	—	ARF62-6	BPR4FS-15	T16PR15	14RF52A	.060
HR9BX	RBL11Y6	—	ARF42-6	BPR5FS-15	—	14RF42	.040[6]
HR9BX	RBL11Y6	—	ARF42-6	BPR5FS-15	—	14RF42A	.060
HR10BY	RBL17Y6	—	ARF62-6	BPR4FS-15	T16PR15	14RF52A	.060
HR10BY	RBL15Y6	—	ARF52-6	BPR4FS-15	T16PR15	14RF52A	.060
HR10BX	RBL15Y4	—	ARF52	BPR4FS-11	T16PR11	14RF52A	.045
HR10BX	RBL15Y4	—	ARF52	BPR4FS-11	T16PR11	14RF52A	.045
WR10FY	RJ18Y6	—	AR82-6	BPR4S-15	W14P	14R52A	.060
HR10BY	RBL17Y6	—	ARF52-6	BPR4FS-15	T16PR15	14RF52A	.060
HR10BY	RBL17Y6	—	ARF62-6	BPR4FS-15	T16PR15	14RF52A	.060
WR10FY	RJ18Y6	—	AR82-6	BPR4S-15	W14P	14R52A	.060
WR10FY	RJ18Y6	—	AR82-6	BPR4S-15	—	14R52A	.060
WR9DY	RN14Y6	—	AGR52-6	BPR5ES-15	W16EX-U	14GR52A	.060
WR9DY	RN14Y6	—	AGR52-6	BPR5ES-15	W16EX-U	14GR52A	.060[2]
WR9DY	RN14Y6	—	AGR52-6	BPR5ES-15	W16EX-U	14GR52A	.060
HR8B	RBL11Y	—	ARF22	BPR6FS	T20PR	14RF23	.035
HR8B	RBL11Y	—	ARF32	BPR5FS	T20PR	14RF32	.035
HR10BY	RBL15Y6	—	ARF32-6	BPR5FS-15	T20PR15	14RF32A	.060
HR10BY	RBL15Y6	—	ARF42-6	BPR5FS-15	T20PR15	14RF32A	.060
HR10BY	RBL15Y6	—	ARF52-6	BPR4FS-15	T16PR15	14RF52A	.060
HR10BY	RBL17Y6	—	ARF62-6	BPR4FS-15	T16PR15	14RF52A	.060
HR10BX	RBL17Y	—	ARF52	BPR4FS-11	T16PR11	14RF52A	.045
HR9BX	RBL15Y6	—	ARF52-6	BPR4FS-15	T16PR15	14RF52A	.060
HR9BX	RBL17Y6	—	ARF62-6	BPR4FS-15	T16PR15	14RF52A	.060
HR10B	RBL17Y	—	ARF62	BPR4FS	T16PR	14RF52	.035
HR10B	RBL17Y	—	ARF62	BPR4FS	T16PR	14RF52	.035
HR10BX	RBL15Y4	—	ARF52	BPR4FS-11	T16PR11	14RF52A	.045
HR10BX	RBL15Y4	—	ARF52	BPR4FS-11	T16PR11	14RF52A	.045
HR10BX	RBL15Y4	—	ARF52	BPR4FS-11	T16PR11	14RF52A	.045
HR10BX	RBL15Y4		—	—	T16PR11	—	.060
HR10BX	RBL15Y4		ARF52	BPR4FS-11		14RF52A	.045
HR10B	RBL16Y	P-560PR	ARF62	BPR4FS	—	14RF52	.035
WR9D	RN12Y	P-65PR	AGR42	BPR5ES	W16EX-U	14GR42	.035
WR10F	OJ13Y[12]	P-35PX	AR42	BPR5S	W14P	14R42	.035
WR8D	RN12Y	P-65PR	AGR42	BPR5ES	W16EX-U	14GR42	.035
HR10B	RBL16Y	P-560PR	ARF62	BPR4FS	—	14RF52	.035
WR9D	RN12Y	P-65PR	AGR42	BPR5ES	W16EX-U	14GR42	.035
WR10F	OJ13Y[12]	P-35PX	AR42	BPR5S	W17P	14R42	.035
—	OJ11Y[13]	P-34PX	AR32	BPR5S	W17P	14R32	.035
HR9DX	RZN12Y5	—	AWRF32-5	BPR6EFS-13	—	14GR32	.050
HR9D	RZN12Y	—	AWSF42	BPR5EFS	—	14GRF52	.034[8]
HR9D	RZN12Y	—	AWSF32	—	—	14GRF32	.034[8]
HR9D	RZN12Y	—	AWSF42	BPR5EFS	—	14GRF52	.034
DR10B	RF14Y4	—	BSF82	APR5FS-15	MA14P	18RF82	.050
DR10B	RF14Y4	—	BSF82	APR5FS-15	MA14P	18RF82	.050
HR10BX	RBL17Y6	—	ASF52	BPR4FS-15	T16PR15	14RF52A	.050
HR10BY	RBL17Y6	—	ASF52-6	BPR4FS-15	T16PR15	14RF52A	.060
HR10BX	RBL17Y6	—	ARF52	BPR4FS-15	T16PR15	14RF52A	.050

[9] V6-171 engine Mercury spark plug gapped at .035 in.

[10] Not used

[11] 1978 only

[12] Use RJ13Y if OJ13Y is unavailable

[13] Use RJ11Y if OJ11Y is unavailable

Popular Replacement Spark Plugs for Late Model American Cars

Following are spark plug manufacturers recommendations for 1978-79 unmodified engines in normal service, using the vehicle manufacturers settings for ignition timing and carburetion. Check the Tune-Up Specifications chart for correct OEM plug gap.

AUTOMOBILE MAKE/MODEL	ENGINE TYPE/CYL. CID	CARB. BBLS.	AC	AUTOLITE
1978 Fairmont, Ford, Granada, LTD II, Maverick, Mustang, Thunderbird, Ford	V8-302 Calif.	2	R45TSX	666
1978-79 LTD II, Thunderbird, Ford	V8-351M	2	R45TSX	726
1978-79 LTD II, Thunderbird, Granada, Ford	V8-351W	2	R45TSX	726
1978 LTD II, Thunderbird, Ford	V8-400	2	R45TSX	726
1978 LTD II	V8-400 Calif.	2	—	26
1978 LTD (Police), Ford, Ford (Police)	V8-460	4	R45TSX	26
LINCOLN-CONTINENTAL				
1979 Versailles	V8-302	2	R45TSX	726
1978 Versailles	V8-302	2	R45TSX	26
1979 Versailles	V8-302 Calif.	2	R45TSX	666
1978 Versailles	V8-302 Calif.	2	R45TSX	666
1979 Continental, Mark V	V8-400	4	R45TSX	726
1978 Continental, Mark V	V8-400	4	R45TSX	726
1978 Continental, Mark V	V8-460	4	R45TSX	26
MERCURY				
1978-79 Bobcat, Capri, Zephyr	L4-140	2	R43LTS	765
1979 Capri	L4-140 Turbo	2	R43LTS	764
1978-79 Bobcat, Capri	V6-171	2	R43LTS	765
1978-79 Zephyr	L6-200	1	R85TS	746
1978-79 Monarch	L6-250	1	R85TS	746
1979 Capri, Cougar, Marquis (Exc Cal.), Monarch, XR-7, Zephyr, Mercury	V8-302	2	R45TSX	726
1978 Cougar, XR-7, Monarch, Mercury	V8-302	2	R45TSX	26
1979 Marquis	V8-302 Calif.	2	R45TSX	3606
1978 Cougar, XR-7, Mercury, Zephyr, Monarch	V8-302 Calif.	2	R45TSX	666
1978 Cougar, XR-7, Marquis, Monterey, Mercury	V8-351M	2	R45TSX	726
1979 Cougar, Marquis, XR-7	V8-351	2	R45TSX	726
1978 Cougar, XR-7, Mercury, Monarch	V8-351W	2	R45TSX	26
1978 Cougar, Mercury	V8-400	2	R45TSX	726
1978 Mercury	V8-460	4	R45TSX	26
OLDSMOBILE				
1979 Starfire	L4-151	2	R43TSX	664
1978 Starfire	L4-151	2	R43TSX	664
1979 Starfire	L4-151 Calif.	2	—	664
1979 Cutlass, Omega, Starfire, Delta 88	V6-231	2	R45TSX	666
1978 All Models Exc. Toronado	V6-231	2	R46TSX	667
1979 All Exc. Oldsmobile, Omega	V8-260	2	R46SZ	847
1978-79 Oldsmobile, Omega,	V8-260	2	R46SZ	847
1979 Delta 88	V8-301	2	R46TSX	667
1978-79 Cutlass, Omega, Starfire	V8-305	2 & 4	R45TS	26
1978-79 Cutlass, Omega, Delta 88 & 98	V8-350L	4	R45TS	26
1978-79 Delta 88 & 98, Omega, Toronado, Cutlass	V8-350R	4	R46SZ	847
1978 Delta 88 & 98	V8-350X	4	—	—
1978-79 Delta 88, 98, Toronado	V8-403	4	R46SZ	847
PLYMOUTH				
1978-79 Horizon	L4-105	1	R44XLS	65
1978-79 Fury, Volare, Duster	L6-225	1 & 2	R45TS	27
1978-79 Fury, Grand Fury, Volare, Duster	V8-318	2 & 4	R44XLS	65
1978-79 Fury, Grand Fury, Volare	V8-360	2 & 4	R44XLS	65
1978 Fury, Grand Fury	V8-400	4	R44S	85
1978 Fury	V8-440	4	R44S	84
PONTIAC				
1979 LeMans, Grand AM, Phoenix, Sunbird	L4-151	2	R43TSX	664
1978 LeMans, Grand AM, Phoenix, Sunbird	L4-151	2	R43TSX	66
1979 LeMans, Grand AM, Phoenix	L4-151 Calif.	2	R43TSX	6
1979 Bonneville, Catalina, Firebird, Trans AM, Grand Prix, LeMans, Grand Am, Phoenix, Sunbird	V6-231	2	R45TSX	[illegible]
1978 Bonneville, Catalina, Firebird, Trans AM, Grand Prix, LeMans, Grand Am, Phoenix, Sunbird, Ventura	V6-231	2	R46TSX	[illegible]
1978-79 Bonneville, Catalina, Firebird, Trans AM, Grand Prix, LeMans, Grand AM, Phoenix, Ventura	V8-301	2	R46TSX	[illegible]
1978-79 Bonneville, Catalina, Grand Prix, LeMans, Grand AM	V8-301	4	R45TSX	[illegible]
1978-79 Firebird, Trans AM, Grand Prix, Phoenix, Sunbird, LeMans, Grand AM	V8-305	2 & 4	R45TS	[illegible]
1978-79 Firebird, Trans AM, LeMans, Grand AM, Phoenix, Ventura, Grand Prix	V8-350L	4	R45TS	[illegible]
1978-79 Bonneville, Catalina, Firebird, Grand Prix, LeMans, Phoenix, Ventura	V8-350R	4	R46SZ	[illegible]
1979 Bonneville, Catalina	V8-350X	4	R45TSX	[illegible]
1978 Bonneville, Catalina	V8-350X	4	R46TSX	[illegible]
1978-79 Bonneville, Catalina, Firebird, Trans AM, Grand Prix	V8-400	4	R45TSX	[illegible]
1978-79 Bonneville, Catalina, Firebird, Trans AM, Grand Prix, LeMans	V8-403	4	RH6SZ	[illegible]

[1] Except California
[2] Cadillac 425 FI Autolite spark plug gapped at .080 in.
[3] With automatic and 2.53:1 axle use 55 gapped at .035 in.
[4] Camaro L6-250 Autolite spark plug gapped at .045 in.
[5] Horizon L4-105 Autolite spark plug gapped at .040 in.
[6] V6-231 (turbo) Bosch, Motorcraft, NGK spark plug gapped at .060 in.
[7] California use N14LY
[8] L4-140 engine Ford and [illegible] gapped at .035 in.

Popular Replacement Spark Plugs for Late Model American Cars

Whenever the letter H, J, L, P, R or X follows a V8-350 engine, it refers to the 5th digit of the VIN, which can be found on the dash panel. Check the Tune-Up Specifications chart for the individual car/engine to be sure of the correct plug gap.

BOSCH (NEW #)	CHAMPION	MOPAR	MOTORCRAFT	NGK	NIPPON-DENSO	PRESTOLITE RESISTOR	GAP
HR10BY	RBL17Y6	—	ARF52-6	BPR4FS-15	T16PR15	14RF52A	.060
HR10BX	RBL17Y6	—	ASF52	BPR4FS-15	T16PR15	14RF52A	.050
HR10BX	RBL17Y6	—	ARF52	BPR4FS-15	T16PR15	14RF52A	.050
HR10BX	RBL17Y6	—	ASF52	BPR4FS-15	T16PR15	14RF52A	.050
—	—	—	ASF52	—	—	14RF52A	.060
HR10BX	RBL17Y6	—	ARF52	BPR4FS-15	T16PR15	14RF52A	.050
HR10BX	RBL17Y6	—	ASF52	BPR4FS-15	T16PR15	14RF52A	050
HR10BX	RBL17Y6	—	ARF52	BPR4FS-15	T16PR15	14RF52A	.050
HR10BY	RBL17Y6	—	ASF52-6	BPR4FS-15	T16PR15	14RF52A	.060
HR10BY	RBL17Y6	—	ASF52-6	BPR4FS-15	T16PR15	14RF52A	.060
HR10BX	RBL17Y6	—	ASF52	BPR4FS-15	T16PR15	14RF52A	050
HR10BX	RBL17Y6	—	ASF52	BPR4FS-15	T16PR15	14RF52A	.050
HR10BY	RBL17Y6	—	ARF52	BPR4FS-15	T16PR15	14RF52A	050
HR9D	RZN12Y	—	AWSF42	BPR5EFS	—	14GRF52	.034[8]
HR9D	RZN12Y	—	AWSF42	BPR6EFS	—	14GRF32	.034[8]
HR9D	RZN12Y	—	AWSF42	BPR5EFS	W16EX-U	14GRF52	.034[9]
DR10B	RF14YF	—	BSF82	APR5FS-15	MA14P	18RF82	.050
DR10B	RF14Y4	—	BSF82	APR5FS-15	MA14P	18RF82	.050
HR10BX	RBL17Y6	—	ASF52	BPR4FS-15	T16PR15	14RF52A	.050
HR10BX	RBL17Y6	—	ARF52	BPR4FS-15	T16PR15	14RF52A	.050
HR10BY	RBL17Y6	—	ARF52-6	BPR4FS-15	T16PR15	14RF52A	.060
HR10BY	RBL17Y6	—	ARF52-6	BPR4FS-15	T16PR15	14RF52A	.060
HR10BX	RBL17Y6	—	ASF52	BPR4FS-15	T16PR15	14RF52A	.050
HR10BX	RBL17Y6	—	ASF52	BPR4FS-15	T16PR15	14RF52A	.050
HR10BX	RBL17Y6	—	ARF52	BPR4FS-15	T16PR15	14RF52A	.050
HR10BX	RBL17Y6	—	ASF52	BPR4FS-15	T16PR15	14RF52A	.050
HR10BX	RBL17Y6	—	ARF52	BPR5FS-15	T16PR15	14RF52A	.050
HR10BY	RBL15Y6	—	ARF32-6	BPR5FS-15	T20PR15	14RF32A	.060
HR10BY	RBL15Y6	—	ARF42-6	BPR5FS-15	T20PR15	14RF32A	.060
—	RBL15Y6	—	—	—	T20PR15	—	.060
HR9BX	RBL15Y6		ARF52-6	BPR4FS-15	T16PR15	14RF52A	.060
HR9BX	RBL17Y6	—	ARF62-6	BPR4FS-15	T16PR15	14RF52A	.060
WR10FY	RJ18Y6	—	AR82-6	BPR4S-15	W14R	14R52A	.060
WR10FY	RBL18Y6	—	AR82-6	BPR4S-15	W14R	14R52A	.060
HR10BY	RBL17Y6	—	—	BPR4FS-15	T16PR15	14RF52A	.060
HR10BX	RBL15Y4	—	ARF52	BPR4FS-11	T16PR11	14RF52A	.045
HR10BX	RBL15Y4	—	ARF52	BPR4FS-11	T16PR11	14RF52A	.045
WR10FY	RBL18Y6	—	AR82-6	BPR4S-15	W14P	14R52A	.060
—	—	—	ARF62-6	BPR4FS-15	—	—	.060
WR10FY	RJ18Y6	—	AR82-6	BPR4S-15	W14P	14R52A	.060
WR8D	RN12Y	P-65PR	AGR42	BPR5ES	W16EX-U	14GR42	.035[5]
HR10B	RBL16Y	P-560PR	ARF62	BPR4FS	—	14RF52	.035
WR9D	RN12Y	P-65PR	AGR42	BPR5ES	W16EX-U	14GR42	.035
WR9D	RN12Y	P-65PR	AGR42	BPR5ES	W16EX-U	14GR42	.035
[illegible]10F	OJ13Y[12]	P-35PX	AR42	BPR5S	W14P	14R42	.035
[illegible]	OJ11Y[13]	P-34PX	AR32	BPR5S	W17P	14R32	.035
[illegible]	RBL15Y6	—	ARF32-6	BPR5FS-15	T20PR15	14RF32A	.060
[illegible]	RBL15Y6	—	ARF42-6	BPR5FS-15	T20PR15	14RF32A	.060
[illegible]	RBL15Y6	—	ARF42-6	—	T20PR15	—	.060
[illegible]	RBL15Y6	—	ARF52-6	BPR5FS-15	T16PR15	14RF52A	.060
[illegible]	RBL17Y6	—	ARF62-6	BPR5FS-15	T16PR15	14RF52A	.060
[illegible]	[illegible]L17Y6	—	ARF62-6	BPR4FS-15	T16PR15	14RF52A	.060
[illegible]	[illegible]5Y6	—	ARF52-6	BPR4FS-15	T16PR15	14RF52A	.060
[illegible]	[illegible]	—	ARF52	BPR4FS-11	T16PR11	14RF52A	.045
[illegible]	[illegible]	—	ARF52	BPR4FS-11	T16PR11	14RF52A	.045
[illegible]	[illegible]	—	ARF82-6	BPR4S-15	W14P	14R52A	.060
[illegible]	[illegible]	—	ARF52-6	BPR4FS-15	T16PR15	14RF52A	.060
[illegible]	[illegible]	—	ARF62-6	BPR4FS-15	T16PR15	14RF52A	.060
[illegible]	[illegible]	—	ARF52-6	BPR4FS-15	T16PR15	14RF52A	.060
[illegible]	[illegible]	—	AR82-6	BPR4S-15	W14P	14R52A	.060

[illegible] in.

[12] Use RJ13Y if OJ13Y is unavailable
[13] Use RJ11Y if OJ11Y is unavailable